HANDBOOK OF CONSTRUCTIVE MATHEMATICS

Constructive mathematics – mathematics in which 'there exists' always means 'we can construct' – is enjoying a renaissance. Fifty years on from Bishop's groundbreaking account of constructive analysis, constructive mathematics has spread out to touch almost all areas of mathematics and to have profound influence in theoretical computer science. This handbook gives the most complete overview of modern constructive mathematics, with contributions from leading specialists surveying the subject's myriad aspects. Major themes include: constructive algebra and geometry, constructive analysis, constructive topology, constructive logic and foundations of mathematics, and computational aspects of constructive mathematics. A series of introductory chapters provides graduate students and other newcomers to the subject with foundations for the surveys that follow. Edited by four of the most eminent experts in the field, this is an indispensable reference for constructive mathematicians and a fascinating vista of modern constructivism for the increasing number of researchers interested in constructive approaches.

Encyclopedia of Mathematics and Its Applications

This series is devoted to significant topics or themes that have wide application in mathematics or mathematical science and for which a detailed development of the abstract theory is less important than a thorough and concrete exploration of the implications and applications.

Books in the **Encyclopedia of Mathematics and Its Applications** cover their subjects comprehensively. Less important results may be summarized as exercises at the ends of chapters. For technicalities, readers can be referred to the bibliography, which is expected to be comprehensive. As a result, volumes are encyclopedic references or manageable guides to major subjects.

All the titles listed below can be obtained from good booksellers or from Cambridge University Press. For a complete series listing visit
www.cambridge.org/mathematics

135 V. Berthé and M. Rigo (eds.) *Combinatorics, Automata and Number Theory*
136 A. Kristály, V. D. Radulescu and C. Varga *Variational Principles in Mathematical Physics, Geometry, and Economics*
137 J. Berstel and C. Reutenauer *Noncommutative Rational Series with Applications*
138 B. Courcelle and J. Engelfriet *Graph Structure and Monadic Second-Order Logic*
139 M. Fiedler *Matrices and Graphs in Geometry*
140 N. Vakil *Real Analysis through Modern Infinitesimals*
141 R. B. Paris *Hadamard Expansions and Hyperasymptotic Evaluation*
142 Y. Crama and P. L. Hammer *Boolean Functions*
143 A. Arapostathis, V. S. Borkar and M. K. Ghosh *Ergodic Control of Diffusion Processes*
144 N. Caspard, B. Leclerc and B. Monjardet *Finite Ordered Sets*
145 D. Z. Arov and H. Dym *Bitangential Direct and Inverse Problems for Systems of Integral and Differential Equations*
146 G. Dassios *Ellipsoidal Harmonics*
147 L. W. Beineke and R. J. Wilson (eds.) with O. R. Oellermann *Topics in Structural Graph Theory*
148 L. Berlyand, A. G. Kolpakov and A. Novikov *Introduction to the Network Approximation Method for Materials Modeling*
149 M. Baake and U. Grimm *Aperiodic Order I: A Mathematical Invitation*
150 J. Borwein et al. *Lattice Sums Then and Now*
151 R. Schneider *Convex Bodies: The Brunn–Minkowski Theory (Second Edition)*
152 G. Da Prato and J. Zabczyk *Stochastic Equations in Infinite Dimensions (Second Edition)*
153 D. Hofmann, G. J. Seal and W. Tholen (eds.) *Monoidal Topology*
154 M. Cabrera García and Á. Rodríguez Palacios *Non-Associative Normed Algebras I: The Vidav–Palmer and Gelfand–Naimark Theorems*
155 C. F. Dunkl and Y. Xu *Orthogonal Polynomials of Several Variables (Second Edition)*
156 L. W. Beineke and R. J. Wilson (eds.) with B. Toft *Topics in Chromatic Graph Theory*
157 T. Mora *Solving Polynomial Equation Systems III: Algebraic Solving*
158 T. Mora *Solving Polynomial Equation Systems IV: Buchberger Theory and Beyond*
159 V. Berthé and M. Rigo (eds.) *Combinatorics, Words and Symbolic Dynamics*
160 B. Rubin *Introduction to Radon Transforms: With Elements of Fractional Calculus and Harmonic Analysis*
161 M. Ghergu and S. D. Taliaferro *Isolated Singularities in Partial Differential Inequalities*
162 G. Molica Bisci, V. D. Radulescu and R. Servadei *Variational Methods for Nonlocal Fractional Problems*
163 S. Wagon *The Banach–Tarski Paradox (Second Edition)*
164 K. Broughan *Equivalents of the Riemann Hypothesis I: Arithmetic Equivalents*
165 K. Broughan *Equivalents of the Riemann Hypothesis II: Analytic Equivalents*
166 M. Baake and U. Grimm (eds.) *Aperiodic Order II: Crystallography and Almost Periodicity*
167 M. Cabrera García and Á. Rodríguez Palacios *Non-Associative Normed Algebras II: Representation Theory and the Zel'manov Approach*
168 A. Yu. Khrennikov, S. V. Kozyrev and W. A. Zúñiga-Galindo *Ultrametric Pseudodifferential Equations and Applications*
169 S. R. Finch *Mathematical Constants II*
170 J. Krajícek *Proof Complexity*
171 D. Bulacu, S. Caenepeel, F. Panaite and F. Van *Oystaeyen Quasi-Hopf Algebras*
172 P. McMullen *Geometric Regular Polytopes*
173 M. Aguiar and S. Mahajan *Bimonoids for Hyperplane Arrangements*
174 M. Barski and J. Zabczyk *Mathematics of the Bond Market: A Lévy Processes Approach*
175 T. R. Bielecki, J. Jakubowski and M. Niewęglowski *Structured Dependence between Stochastic Processes*
176 A. A. Borovkov, V. V. Ulyanov and Mikhail Zhitlukhin *Asymptotic Analysis of Random Walks: Light-Tailed Distributions*
177 Y.-K. Chan *Foundations of Constructive Probability Theory*
178 L. W. Beineke, M. C. Golumbic and R. J. Wilson (eds.) *Topics in Algorithmic Graph Theory*
179 H.-L. Gau and P. Y. Wu *Numerical Ranges of Hilbert Space Operators*
180 P. A. Martin *Time-Domain Scattering*
181 M. D. de la Iglesia *Orthogonal Polynomials in the Spectral Analysis of Markov Processes*
182 A. E. Brouwer and H. Van Maldeghem *Strongly Regular Graphs*
183 D. Z. Arov and O. J. Staffans *Linear State/Signal Systems*
184 A. A. Borovkov *Compound Renewal Processes*
185 D. Bridges, H. Ishihara, M. Rathjen and H. Schwichtenberg (eds.) *Handbook of Constructive Mathematics*

ENCYCLOPEDIA OF MATHEMATICS AND ITS APPLICATIONS

Handbook of Constructive Mathematics

Edited by

DOUGLAS BRIDGES
University of Canterbury

HAJIME ISHIHARA
Japan Advanced Institute of Science and Technology

MICHAEL RATHJEN
University of Leeds

HELMUT SCHWICHTENBERG
Ludwig Maximilian University of Munich

CAMBRIDGE
UNIVERSITY PRESS

CAMBRIDGE
UNIVERSITY PRESS

Shaftesbury Road, Cambridge CB2 8EA, United Kingdom

One Liberty Plaza, 20th Floor, New York, NY 10006, USA

477 Williamstown Road, Port Melbourne, VIC 3207, Australia

314–321, 3rd Floor, Plot 3, Splendor Forum, Jasola District Centre, New Delhi - 110025, India

103 Penang Road, #05-06/07, Visioncrest Commercial, Singapore 238467

Cambridge University Press is part of Cambridge University Press & Assessment, a department of the University of Cambridge.

We share the University's mission to contribute to society through the pursuit of education, learning and research at the highest international levels of excellence.

www.cambridge.org
Information on this title: www.cambridge.org/9781316510865
DOI: 10.1017/9781009039888

First published 2023

A catalogue record for this publication is available from the British Library.

ISBN 978-1-316-51086-5 Hardback

Contents

Part VI Aspects of Computation

Contributors

Josef Berger *Department of Mathematics, University of Munich*
Ulrich Berger *Department of Computer Science, Swansea University*
Mark Bickford *Department of Computer Science, Cornell University*
Douglas Bridges *School of Mathematics & Statistics, University of Canterbury*
Yuen-Kwok Chan *Mortgage Analytics, Citigroup (retired)*
Francesco Ciraulo *Department of Mathematics, University of Padua*
Thierry Coquand *Computer Science Department, University of Gothenburg*
Laura Crosilla *Department of Philosophy, IFIKK, University of Oslo*
Robin S. Havea *Tonga Campus, University of the South Pacific*
Matthew Hendtlass *School of Mathematics & Statistics, University of Canterbury*
Hajime Ishihara *School of Information Science, Japan Advanced Institute of Science and Technology*
Gerhard Jäger *Institute of Computer Science, University of Bern*
Tatsuji Kawai *Japan Advanced Institute of Science and Technology, Asahidai*
Davorin Lešnik *Faculty of Mathematics and Physics, University of Ljubljana*
Henri Lombardi *Department of Mathematics, University of Franche-Comté*
Robert S. Lubarsky *Department of Mathematical Sciences, Florida Atlantic University*
Maria Emilia Maietti *Department of Mathematics, University of Padua*
Mark Mandelkern *Department of Mathematics, New Mexico State University*
Kenji Miyamoto *Mathematics Institute, Ludwig Maximilian University of Munich*
Takako Nemoto *Department of Architectural Design, Hiroshima Institute of Technology*
Iosif Petrakis *Mathematics Institute, Ludwig Maximilian University of Munich*
Michael Rathjen *Department of Pure Mathematics, University of Leeds*
Fred Richman *Department of Mathematics, Florida Atlantic University*
Giovanni Sambin *Department of Mathematics, University of Padua*

Helmut Schwichtenberg *Mathematics Institute, Ludwig Maximilian University of Munich*
Monika Seisenberger *Department of Computer Science, Swansea University*
Gregor Svindland *Institute of Actuarial and Financial Mathematics, Leibniz University Hanover*
Ihsen Yengui *Department of Mathematics, University of Sfax*

Preface

Constructive mathematics, in which 'there exists' is interpreted strictly as 'we can find/construct/compute', can be traced back at least to Kronecker and was first taken up systematically by Brouwer [6] and his 'intuitionist' followers. For various reasons, Brouwer's intuitionistic mathematics (INT), other than its underlying intuitionistic logic, garnered relatively little interest outside parts of Europe. In the Soviet Union in the late 1940s, A. A. Markov began a research programme on recursive constructive mathematics (RUSS), in which 'constructive' was interpreted as 'applying recursion theory and intuitionistic logic to analysis'. Markov's programme, too, failed to convince mathematicians, other than logicians, that it had much significance for the working mathematician.

The tipping point for constructive mathematics was the publication, in 1967, of Errett Bishop's groundbreaking monograph *Foundations of Constructive Analysis* [3], in which, confounding the predictions of Hilbert and the majority of active research mathematicians, he presented a fully algorithmic development of deep analysis, including functional analysis and measure theory. Moreover, he did so in the natural style of an analyst, resorting to neither the non-classical principles of Brouwer nor Markov's framework of recursion theory. The key to his development was the use of intuitionistic logic and an informal set theory (one formalisation of which is described in Chapter 2), the former capturing the Brouwer–Heyting–Kolmogorov (BHK) interpretation of the logical connectives and quantifiers; this meant that his work read like normal analysis rather than mathematical logic.

In a certain sense, intuitionistic logic, which is discussed in Chapter 1, is weaker than classical logic: with the former one cannot prove, for example, the law of excluded middle, De Morgan's law, or even the seemingly trivial limited principle of omniscience, which states that for every binary sequence, either all the terms are 0 or else there exists a term equal to 1. However, as the Curry–Howard isomorphism shows, we can ensure constructivity in mathematics by using intuitionistic logic. Moreover, we can extract programs from intuitionistic-logic-based proofs (see Part IV).

In the 50-plus years since the appearance of his book, there has been considerable progress in the continuing development of Bishop's analysis [4, 8, 10]; but the constructive banner has also been raised by algebraists [11, 13], topologists [5, 14], researchers into formal set- and type-theoretic foundations for Bishop-style mathematics (BISH) ([1, 2, 12], Chapter 2), and computer scientists working on program extraction from proofs in BISH [7, 15]. Following the initiative of Veldman [17] and Ishihara [9], there is now also a substantial body of research in constructive reverse mathematics, in which theorems and principles of classical, intuitionistic, and (constructive) recursive mathematics are classified constructively by those principles that are necessary and sufficient additions to BISH in order to derive them (see Chapters 23 and 24).

There is another aspect of constructive mathematics that is increasingly regarded as a *sine qua non*: predicativity. This means ensuring that we avoid self-referential, or *impredicative*, definitions such as

$$A \equiv \{n \in \mathbf{N} : \forall S \subset \mathbf{N}\, \phi(S, n)\},$$

in which the criterion for membership of n in the definiendum A involves universal quantification over all subsets of $\mathbf{N}$, including A itself. In the past three decades there has been increasing research activity in *formal topology* [14], with its emphasis on predicativity and point-free constructive mathematics. Formal-topological methods are being applied far more widely than the word topology would suggest, with a considerable body of research into point-free methods in analysis (see Chapters 15–17). More recently, the late Fields Medallist Vladimir Voyevodsky introduced homotopy type theory, an approach to constructive mathematics that has attracted a great deal of attention. However, given the length of our Handbook, we refer our readers to the comprehensive treatise *Homotopy Type Theory: Univalent Foundations of Mathematics* [16] for more information on Voyevodsky's approach.

The aim of this Handbook is two-fold:

- to provide an accessible introduction to constructive mathematics – its foundations (Parts I and V), its practice within mathematics itself (Parts II–IV), and its significance for computation (Part VI);
- to demonstrate how far mathematics can be developed with the requirements of constructivity and predicativity.

We hope that our compilation will encourage mathematicians of all persuasions to appreciate the power, subtlety, and growing reach of the constructive mathematical enterprise.

Acknowledgements

The editors wish to thank the following.

Iosif Petrakis, for initiating the handbook project and assisting with its development.

The anonymous referees of our chapters.

Tom Harris and David Tranah, from Cambridge University Press, for their patient guidance over the preparation and production of the Handbook.

The Hausdorff Research Institute of Mathematics, Bonn, for the Trimester *Types, Sets and Constructions* (May–August, 2018).

The European Union, for the projects *Computing with Infinite Data* (2017–2021) and *Correctnesss by Construction* (2014–2017).

The Japan Society for the Promotion of Science, Core-to-Core Program (A. Advanced Research Networks), for the project *Mathematical Logic and its Applications* (2015–2020).

The John Templeton Foundation, for the project *A New Dawn of Intuitionism: Mathematical and Philosophical Advances* (2017–2020).

References

[1] Aczel, P., and Rathjen, M. 2001. *Notes on Constructive Mathematics*. Technical report 40. Royal Swedish Academy of Sciences.

[2] Alps, R. A., and Bridges, D. S. *Morse Set Theory as a Foundation for Constructive Mathematics*. Monograph in preparation.

[3] Bishop, E. 1967. *Foundations of Constructive Analysis*. New York: McGraw-Hill.

[4] Bridges, D. S., and Vîţă, L. S. 2006. *Techniques of Constructive Analysis*. Springer.

[5] Bridges, D. S., and Vîţă, L. S. 2011. *Apartness and Uniformity: A Constructive Development*. CiE series Theory and Applications of Computability. Springer-Verlag.

[6] Brouwer, L. E. J. 1907. Over de Grondslagen der Wiskunde. PhD Thesis, University of Amsterdam.

[7] Constable, R. L., et al. 1986. *Implementing Mathematics with the NUPRL Proof Development System*. Englewood Cliffs, New Jersey: Prentice-Hall.

[8] Ishihara, H. 2001. Locating subsets of a Hilbert space. *Proc. Amer. Math. Soc.*, **129**(5), 1385–2390.

[9] Ishihara, H. 2005. Constructive reverse mathematics: compactness properties. Pages 245–267 of: *From Sets and Types to Topology and Analysis*. Oxford Logic Guides, vol. 48. Oxford: Clarendon Press.

[10] Ishihara, H., and Vîţă, L. S. 2003. Locating subsets of a normed space. *Proc. Amer. Math. Soc.*, **131**(10), 3231–3239.

[11] Lombardi, H., and Quitté, C. 2011. *Algèbre Commutative, Méthodes constructives (Modules projectifs de type fini)*. Montrouge, France: Calvage et Mounet. English translation, with additions and corrections: *Commutative Algebra, Constructive Methods (Finitely Generated Projective Modules)*, Springer-Verlag, 2015.

[12] Martin-Löf, P. 1998. An intuitionistic theory of types. Pages 127–172 of: *Twenty-five Years of Constructive Type Theory*. Oxford Logic Guides, vol. 36. Oxford: Clarendon Press.

[13] Mines, R., Richman F., and Ruitenburg, W. 1988. *A Course in Constructive Algebra*. Universitext. Heidelberg: Springer-Verlag.

[14] Sambin, G. 2023. *Positive Topology and the Basic Picture. New Mathematics Emerging from Dynamic Constructivism*. Oxford Logic Guides. Oxford: Oxford University Press. (In the press.)

[15] Schwichtenberg, H. 2009. Program extraction in constructive analysis. Pages 255–275 of: *Logicism, Intuitionism, and Formalism – What has Become of them?* Synthese Library, vol. 341. Berlin: Springer-Verlag.

[16] Univalent Foundations Program. 2013. *Homotopy Type Theory: Univalent Foundations of Mathematics*. Princeton, New Jersey: Institute for Advanced Study. Available at `https://homotopytypetheory.org/book`.

[17] Veldman, W. 2001. Brouwer's Fan Theorem as an axiom and as a contrast to Kleene's Alternative. *Arch. Math. Logic*, **5–6**, 621–693.

PART I

INTRODUCTORY

1

An Introduction to Intuitionistic Logic

Michael Rathjen

1.1 Introduction

The constructive existence embodied in intuitionistic logic is very desirable in mathematics as it supports the computational view of mathematics. For a classically trained mathematician, though, it is often not that easy to switch to a mode of reasoning that maintains constructivity. For instance, arguing by making case distinctions is an almost automatic habit in classical mathematics, but one that is liable to introduce illegitimate employments of the *law of excluded middle*, LEM. Several examples will be discussed in Section 1.2.

The main aim of this chapter is to present an informal and intuitive approach to intuitionistic[1] logic – also known as constructive logic – for the working constructive mathematician.[2] The guiding idea is that this will be furnished via the *Brouwer–Heyting–Kolmogorov interpretation* (henceforth the BHK-interpretation) of the logical connectives and quantifiers. This is presented in Section 1.3.

Sometimes, however, uncertainties as to the constructive validity of an argument might still arise as the BHK-interpretation is based on an unexplained notion of constructive function. Moreover, it can also be cumbersome to ascertain constructivity of a mode of reasoning by means of the BHK-interpretation, or, venturing in the other direction, to demonstrate that an argument doesn't hold under an intuitionistic lens. In such situations, a more formal approach may be called for. For this reason, and for other equally important purposes, this chapter also features two formal proof systems in Sections 1.4 and 1.5: Gentzen's *natural deductions* and a *Hilbert-style calculus* for intuitionistic predicate logic. In the natural deduction style of reasoning there are no axioms, only rules of inference. This lack of axioms is compensated for by having permission to introduce any formula as a hypothesis at any time. To be able to get rid of such formulas at later stages in the proof, there are rules that allow

[1] No pun intended.

[2] There are several excellent introductions to intuitionistic logic; for example, [26, Chapter 2]. This one is for the reader's convenience, namely to have one to hand in the same volume.

one to discharge hypotheses. A Hilbert-style proof calculus, on the other hand, has a number of axioms but few inference rules. The latter are, moreover, of a local nature in that they do not involve the sometimes burdensome regime of tracking open and discharged hypotheses over the entire proof. Both formalizations have important roles to play. Natural deductions beautifully exhibit the connection between intuitionistic logic and computations known as the *Curry–Howard correspondence* (or *isomorphism*) and the *formulae-as-types interpretation*, whereas a Hilbert-style calculus is very useful in demonstrating that crucial concepts (e.g., realizability) are preserved under intuitionistic logic. The Curry–Howard correspondence will be briefly discussed in the final section, Section 1.7.

The penultimate section, Section 1.6 is devoted to Kleene's *1945-realizability* of intuitionistic number theory (or Heyting arithmetic), **HA**. The concept and technique of realizability is another nice illustration of the fact that intuitionistic proofs encapsulate numerical information, and it reveals how it is extractable from them. Furthermore, realizability bears out the fact that one option of instantiating the unexplained notion of constructive function of the BHK-interpretation consists in equating it with the notion of partial computable (= partial recursive = partial Turing machine computable) function.

1.2 Constructive Existence

Constructive mathematics is both old and new for the reason that, with few exceptions, mathematicians thought constructively until the 1870s, that is, before the set-theoretic shift initiated by Dedekind, Cantor, and others, while a substantial development of modern mathematics from a constructive base (largely thought to be impossible) had to await the work of Errett Bishop in the second half of the twentieth century.

The meaning of 'existence' in mathematics in the first phase was essentially what we equate with constructive existence nowadays. In general, the requirement for the latter is the demand that E be respected:

> (E) *The correctness of an existential claim* $(\exists x \in A)\varphi(x)$ *is to be guaranteed by warrants from which both an object* $x \in A$ *and a further warrant for* $\varphi(x)$ *are constructible.*

Or as Bishop ([2, p. 2]) put it:

> *When a man proves a positive integer to exist, he should show how to find it. If God has mathematics of his own that needs to be done, let him do it himself.*

The year 1888 saw Hilbert's proof of the basis theorem (Gordan's problem of invariants). Hilbert demonstrated the existence of a finite basis via a proof by contradiction. It is telling that he had to convince Cayley and Gordan that he had really proved the theorem, since they, like other mathematicians, expected a solution along the lines of E that exhibited a finite basis.[3]

The later part of the nineteenth century and the first part of the twentieth century was a period of great advances in mathematics, but also one of uncertainty and opposing views. A central role in the discussions about mathematical existence was played by Zermelo's proof that the reals can be well-ordered, presented at the International Congress of Mathematicians in 1905. While many mathematicians were apt to dismiss the paradoxes as peripheral to mathematics, contradictions of a somewhat philosophical nature, Zermelo's result concerned a core object of mathematics: $\mathbb{R}$. His proof notoriously used the axiom of choice, AC. While Zermelo argued that AC was self-evident, it was also criticized as an excessively non-constructive principle by some of the most distinguished analysts of the day.[4] Zermelo's proof furnishes absolutely no idea as to how a well-ordering of $\mathbb{R}$ can be defined (let alone be constructed). At the time it was natural to single out AC as the sole villain that engenders undefinable mathematical entities. With the advent and tools of modern mathematical logic, however, it emerged that the venerable logical principle (or law) of excluded middle, $\phi \vee \neg\phi$, suffices to produce such examples. For example, one can produce existential statements of the form $\exists x \subseteq \mathbb{R}^2 \varphi(x)$ that are provable in pure logic with the aid of excluded middle, however, **ZFC** (Zermelo–Fraenkel set theory with the axiom of choice), even when augmented by the generalized continuum hypothesis, GCH, cannot prove that there is a definable such set.[5] In a similar vein, there are number-theoretic statements such that first-order number theory **PA** (Peano arithmetic) proves $\exists x\, \theta(x)$ but for no term t does **PA** prove $\theta(t)$.

Brouwer made his famous criticism of the law of excluded middle, LEM, in his 1907 dissertation [4] and his 1908 article 'De onbetrouwbaarheid der logische principes'[6] [5]. He was not the first person, though, to raise doubts about its validity. The German mathematician Paul du Bois-Reymond in his book [9] *Die allgemeine Functionentheorie* from 1882 clearly separated actual infinities from potential infinities and argued that the logic governing potential but non-actual infinite sets would not countenance LEM.

Brouwer called his mathematics *intuitionistic mathematics*. The formal logic that drops LEM and related principles such as the double negation shift $\neg\neg A \to A$ is

[3] Hence Gordan's famous remark, 'this is not mathematics, this is theology', although later Gordan came to appreciate 'theology' in mathematics.

[4] At the end of a note sent to the *Mathematische Annalen* in December 1905, Borel writes about the axiom of choice: *'It seems to me that the objection against it is also valid for every reasoning where one assumes an arbitrary choice made an uncountable number of times, for such reasoning does not belong in mathematics.'* ([3, pp. 1251–1252]; translation by H. Jervell, cf. [16, p. 96]).

[5] This means that for no set-theoretic formulas $\psi(x)$ does one have $\mathbf{ZFC} + \mathsf{GCH} \vdash \exists! x[x \subseteq \mathbb{R}^2 \wedge \varphi(x) \wedge \psi(x)]$. The latter follows from a result of Feferman [11] obtained by forcing in 1963.

[6] The Unrealiability of the Logical Principles.

called *intuitionistic logic* and sometimes *constructive logic* or *Heyting's predicate calculus*. The first name is well ingrained, but Brouwer did not develop intuitionistic logic. The first explicit formulation of the laws of intuitionistic logic is due to the Russian logician Kolmogorov [18]. Kolmogorov accepted Brouwer's critique of LEM when applied to infinite domains. He then took Hilbert's formalization of classical logic [13] as the starting point for his investigation, deselecting those axioms that have validity only in the domain of the finitary. With the exception of the axiom $A \to (\neg A \to B)$ (which is not valid in minimal logic), Kolmogorov arrived at a complete formalization of intuitionistic logic. The main achievement of his paper, though, was to prove that classical logic is translatable into intuitionistic logic, thereby largely anticipating the independent discoveries of translations by Gentzen and Gödel in 1933. The full formalization of intuitionistic logic was obtained in 1930 by Heyting [12], who was unaware of Kolmogorov's work.

Here is an example of a non-constructive existence proof that one finds in almost every book and article concerned with constructive issues.[7]

Proposition 1.1 *There exist irrational numbers $\alpha, \beta \in \mathbb{R}$ such that α^β is rational.*

Proof We know that $\sqrt{2}$ is irrational, and ${\sqrt{2}}^{\sqrt{2}}$ is either rational or irrational. If it is rational, let $\alpha := \beta := \sqrt{2}$. If not, put $\alpha := {\sqrt{2}}^{\sqrt{2}}$ and $\beta := \sqrt{2}$. Thus in either case a solution exists. □

This proof provides two pairs of candidates for solving the equation $x^y = z$ with x and y irrational and z rational, without giving a means of determining which. From a non-trivial result of Gelfand and Schneider, it is known that ${\sqrt{2}}^{\sqrt{2}}$ is transcendental, and thus the second pair provides an explicit answer.

Similarly, classical proofs of disjunctions can be unsatisfactory. H. Friedman pointed out that classically either $e - \pi$ or $e + \pi$ is a irrational number since assuming that both $e - \pi$ and $e + \pi$ are rational entails the contradiction that e is rational. But to this day we don't know which of these numbers is irrational.

Another example is the standard proof of the Bolzano–Weierstraß Theorem.

Example 1.2 If S is an infinite subset of the closed interval $[a, b]$, then $[a, b]$ contains at least one point of accumulation of S.

Proof We construct an infinite nested sequence of intervals $[a_i, b_i]$ as follows.
Put $a_0 = a$, $b_0 = b$. For each i, consider two cases:

(i) if $\left[a_i, \frac{1}{2}(a_i + b_i)\right]$ contains infinitely many points of S, put $a_{i+1} = a_i$, $b_{i+1} = \frac{1}{2}(a_i + b_i)$;

[7] Dummett [10] writes that this example is due to Peter Rososinski and Roger Hindley.

(ii) if $\left[a_i, \frac{1}{2}(a_i + b_i)\right]$ contains only finitely many points of S, put $a_{i+1} = \frac{1}{2}(a_i + b_i)$, $b_{i+1} = b_i$.

With the help of LEM, it is plain that each interval $[a_i, b_i]$ contains infinitely many points of S. This being a sequence of nested intervals, $(a_i)_{i\in\mathbb{N}}$ converges to a point every neighbourhood of which contains infinitely many points of S. □

The foregoing proof specifies a 'method' which, in general, a constructivist cannot carry out.

1.2.1 Counterexamples from Analysis

Certain basic principles of classical mathematics, which are taboo for the constructive mathematician, were called *principles of omniscience* by Bishop. They can be stated in terms of binary sequences, where a binary sequence is a function $\alpha : \mathbb{N} \to \{0, 1\}$. Below, the quantifier $\forall\alpha$ is supposed to range over all binary sequences and the variables n, m range over natural numbers. Let $\alpha_n := \alpha(n)$.

Definition 1.3 *Limited Principle of Omniscience* (LPO):

$$\forall\alpha\, [\exists n\, \alpha_n = 1 \ \lor \ \forall n\, \alpha_n = 0].$$

Weak Limited Principle of Omniscience (WLPO):

$$\forall\alpha\, [\forall n\, \alpha_n = 0 \ \lor \ \neg\forall n\, \alpha_n = 0].$$

Lesser Limited Principle of Omniscience (LLPO):

$$\forall\alpha\, \big(\forall n, m[\alpha_n = \alpha_m = 1 \to n = m] \to [\forall n\, \alpha_{2n} = 0 \ \lor \ \forall n\, \alpha_{2n+1} = 0]\big).$$

Theorem 1.4 *The following implications hold constructively:*

$$\mathsf{LPO} \Rightarrow \mathsf{WLPO} \Rightarrow \mathsf{LLPO}. \tag{1.1}$$

Proof The first implication is obvious. For the second, assume $\forall n, m[\alpha_n = \alpha_m = 1 \to n = m]$. Applying WLPO to $\beta(n) := \alpha_{2n}$, we have $\forall n\, \beta_n = 0$ or $\neg\forall n\, \beta_n = 0$. Clearly, the first case yields $\forall n\, \alpha_{2n} = 0$. So assume $\neg\forall n\, \beta_n = 0$. From $\alpha_{2k+1} = 1$ one obtains $\beta_n = 0$ for all n, contradicting the latter assumption. Hence $\alpha_{2k+1} \neq 1$, whence $\alpha_{2k+1} = 0$ for all k since $\forall k\, [\alpha_{2k+1} = 0 \lor \alpha_{2k+1} = 0]$. □

Classically one has the principle

$$\forall x, y \in \mathbb{R}\, [x = y \lor x \neq y].$$

This principle entails WLPO and is thus not acceptable constructively. Many well-known theorems of classical analysis only require LPO or just WLPO. The story with LLPO, though, is much more subtle.[8] One of the best-known consequences of LLPO is $\forall x, y \in \mathbb{R}\,[x \leq y \vee y \leq x]$.

At this point it is worth mentioning that LPO is still much weaker than LEM. Indeed, it is interesting to study semi-intuitionistic systems with LPO. Particularly noteworthy seems to be the fact that adding LPO to constructive Zermelo–Fraenkel set theory, **CZF**, does not change the proof-theoretic strength whereas adding LEM to **CZF** yields classical **ZF** (for details see [21, Section 2.7] in this volume).

One way to refute all of these principles is via a recursive reading of the BHK-interpretation.

1.3 The Brouwer–Heyting–Kolmogorov Interpretation

The difference between the classical and the intuitionistic understanding of the logical connectives and quantifiers is partiularly well illuminated by the BHK-interpretation, to which we turn next.

In a first approach, a mathematical assertion could be construed as a meaningful statement describing a state of affairs, which traditionally is something that is either true or false. In the case of mathematical statements involving quantifiers ranging over infinite domains, however, by adopting such a view one is compelled to postulate an objective transcendent realm of mathematical objects which determines their meaning and truth value. Most schools of constructive mathematics reject such an account as unconvincing. Kolmogorov observed that the laws of the constructive propositional calculus become evident upon conceiving propositional variables as ranging over problems or tasks. The constructivist's restatement of the meaning of the logical connectives is known as the *BHK-interpretation*. It is couched in terms of an informal notion of proof. It is instructive to view such proofs as pieces of *evidence* sometimes referred to as *proof objects*.

Definition 1.5

(i) p proves $\varphi \wedge \psi$ iff p is pair $\langle a, b \rangle$, where a is proof for φ and b is proof for ψ.
(ii) p proves $\varphi \vee \psi$ iff p is pair $\langle n, q \rangle$, where $n = 0$ and q proves φ, or $n = 1$ and q proves ψ.
(iii) p proves $\varphi \to \psi$ iff p is a function (or rule) which transforms any proof s of φ into a proof $p(s)$ of ψ.
(iv) p proves $\neg\varphi$ iff p proves $\varphi \to \bot$.
(v) p proves $(\exists x \in A)\varphi(x)$ iff p is a pair $\langle a, q \rangle$ where a is a member of the set A and q is a proof of $\varphi(a)$.

[8] LPO and LLPO can be separated at the level of full intuitionistic Zermelo–Fraenkel. For this and more references see [6, Section 9].

(vi) p proves $(\forall x \in A)\varphi(x)$ iff p is a function (rule) such that for each member a of the set A, $p(a)$ is a proof of $\varphi(a)$.
(vii) p proves $\bot$ is impossible, so there is no proof of $\bot$.

Many objections can be raised against the above definition. The explanations offered for implication and universal quantification are notoriously imprecise because the notion of function (or rule) is left unexplained. Another problem is that the notions of set and set membership are in need of clarification. But in practice these rules suffice to codify the arguments which mathematicians want to call constructive. Note also that the above interpretation (except for $\bot$) does not address the case of atomic formulas.

Definition 1.6 We say that a formula φ is *valid under the BHK-interpretation*, if a construction (or proof object) p can be exhibited that is a proof of φ in the sense of the BHK-interpretation.

Example 1.7 Here are some examples of the BHK-interpretation. We sometimes use λ-notation for functions.

(i) The identity map, $\lambda x.x$, is a proof of any proposition of the form $\varphi \to \varphi$ since $(\lambda x.x)(p) = p$.
(ii) A proof of $\varphi \wedge \psi \to \psi \wedge \varphi$ is provided by the function $f(\langle a, b\rangle) = \langle b, a\rangle$.
(iii) Perhaps a bit wondrous, but any function is a proof of $\bot \to \varphi$ as $\bot$ has no proof.
(iv) $(*)$ $(\varphi \to \psi) \to [(\psi \to \theta) \to (\varphi \to \theta)]$ is valid under the BHK-interpretation. Note that the latter entails as a special case the law of contraposition,

$$(\varphi \to \psi) \to (\neg\psi \to \neg\varphi)$$

as $\neg\vartheta$ is $\vartheta \to \bot$. To find a BHK-proof of $(*)$, assume that f proves $\varphi \to \psi$, g proves $\psi \to \theta$, and p proves φ. Then $f(p)$ proves ψ, and hence $g(f(p))$ proves θ. Consequently, $\lambda x.g(f(x))$ proves $\varphi \to \theta$, and therefore $\lambda g.\lambda x.g(f(x))$ proves $(\psi \to \theta) \to (\varphi \to \theta)$. Thus, $\lambda f.\lambda g.\lambda x.g(f(x))$ is a proof of $(*)$.
(v) The law of excluded middle is not valid under any reasonable reading of the BHK-interpretation. Given a sentence θ, we might not be able to find a proof of θ nor a proof of $\neg\theta$. Wondrously, the double negation of that principle is valid under the BHK-interpretation. This may be seen as follows. Suppose g proves $\neg(\psi \vee \neg\psi)$. One easily constructs functions $\mathfrak{f}_0$ and $\mathfrak{f}_1$ such that $\mathfrak{f}_0$ transforms a proof of ψ into a proof of $\psi \vee \neg\psi$ and $\mathfrak{f}_1$ transforms a proof of $\neg\psi$ into a proof of $\psi \vee \neg\psi$, respectively. Thus, $\lambda a.g(\mathfrak{f}_0(a))$ is a proof of $\neg\psi$ while $\lambda b.g(\mathfrak{f}_1(b))$ is a proof of $\neg\psi \to \bot$. Consequently, $g(\mathfrak{f}_1(\lambda a.g(\mathfrak{f}_0(a))))$ is a proof of $\bot$. As a result, $\lambda g.g(\mathfrak{f}_1(\lambda a.g(\mathfrak{f}_0(a))))$ proves $\neg\neg(\psi \vee \neg\psi)$ for any formula ψ.

1.4 Natural Deductions

Constructive mathematics, just as classical mathematics, is mostly carried out informally by humans. Going back to the BHK-interpretation provides a good tool for testing whether a piece of mathematical reasoning holds under the constructive lens. Still, it may be desirable and convenient to have a set of formal logical rules available, should questions about the constructive validity of a proof be raised. Even with the BHK-interpretation at one's disposal, doubts can arise, due to BHK being based on an unexplained notion of function.

This section presents two formal systems of axioms and rules for intuitionistic logic, the natural deduction calculus invented by Gentzen and the intuitionistic Hilbert-style calculus.

Definition 1.8 In the following it is assumed that we are given a language $\mathcal{L}$ of predicate logic (also called first-order logic) with equality $=$. The logical primitives are $\wedge, \vee, \rightarrow, \perp, \forall, \exists$, where $\perp$ stands for absurdity and the negation $\neg\psi$ of a formula ψ is defined by $\psi \rightarrow \perp$. Such a language is further specified by its constant, function and relation symbols, together with their arities in the latter two cases. It is convenient to use different symbols for free $a, b, c, a_0, a_1, a_2, \ldots$ and *bound* $x, y, z, x_0, x_1, x_2, x_3, \ldots$ variables.[9]

Terms are generated from constants and free variables via function symbols. Bound variables aren't terms. We use the convention that metavariables $s, t, s_0, s_1, \ldots$ range over terms.

Formulas are then mostly defined as usual, the exception being the quantifiers. It is convenient to use notations such as $\phi(), \psi(), \theta(), \ldots$ as metavariables ranging over finite strings made up of symbols from $\mathcal{L}$ and a place-holder symbol $\star$, where $\star$ is assumed not to belong to $\mathcal{L}$. They were called *nominal forms* by Schütte [23]. The purpose of these nominal forms is to describe substitutions succinctly. If $\mathfrak{s}$ is any string of symbols, $\phi(\mathfrak{s})$ is obtained from $\phi()$ by replacing every occurrence of $\star$ by $\mathfrak{s}$.

The formation rule for formulas commencing with a quantifier is the following. If $\phi(a)$ is a formula with free variable a and x is a bound variable that does not occur in $\phi(a)$, then $\forall x\phi(x)$ and $\exists x\phi(x)$ are formulas.

Note that in a formula, a bound variable x can only occur within the scope of a quantifier $\forall x$ or $\exists x$.

We say that the variable a is *fully indicated* in $\phi(a)$ if a does not occur in $\phi()$.

A *closed formula* is one without free variables.

9 Using different symbols for free and bound variables is not absolutely essential but it is extremely useful and simplifies arguments a great deal. Terms can be freely substituted for both kinds of variables since variables occurring in them are always free and thus cannot be captured by quantifiers.

Definition 1.9 *Natural deductions* are pictorially presented as trees labelled with formulas. We want to give a formal definition of *deduction* as well as its *open assumptions*, the set of which will be denoted by $o(\mathcal{D})$. We use $\mathcal{D}, \mathcal{D}_1, \mathcal{D}_2, \ldots$ to range over deductions and write

$$\begin{array}{c}\mathcal{D}\\ \psi\end{array}$$

to convey that ψ is the conclusion of $\mathcal{D}$.

Deductions are defined inductively as follows.

Basis: For any formula ψ, the single-node tree $\bullet_\psi$ with label ψ is a deduction whose sole open assumption is ψ; that is, $o(\bullet_\psi) = \{\psi\}$.

Inductive step: Let $\mathcal{D}, \mathcal{D}_1, \mathcal{D}_2, \mathcal{D}_3$ be deductions. Then a deduction may be constructed from these by any of the rules below. Some of these rules are subject to restrictions to be specified afterwards.

For $\perp$ we have the following *intuitionistic absurdity rule*.

$$\frac{\begin{array}{c}\mathcal{D}\\ \perp\end{array}}{\psi}\ \perp_i$$

For the other logical constants the rules can be nicely grouped into introduction and elimination rules.

Introduction rules (I-rules) | Elimination rules (E-rules)

$$\frac{\begin{array}{c}\mathcal{D}_1\\ \varphi\end{array} \qquad \begin{array}{c}\mathcal{D}_2\\ \psi\end{array}}{\varphi \wedge \psi}\ \wedge\mathrm{I} \qquad\qquad \frac{\begin{array}{c}\mathcal{D}\\ \varphi \wedge \psi\end{array}}{\varphi}\ \wedge\mathrm{E}_r \qquad \frac{\begin{array}{c}\mathcal{D}\\ \varphi \wedge \psi\end{array}}{\psi}\ \wedge\mathrm{E}_l$$

$$\frac{\begin{array}{c}[\varphi]\\ \mathcal{D}\\ \psi\end{array}}{\varphi \to \psi}\ \to\mathrm{I} \qquad\qquad \frac{\begin{array}{c}\mathcal{D}_1\\ \varphi \to \psi\end{array} \qquad \begin{array}{c}\mathcal{D}_2\\ \varphi\end{array}}{\psi}\ \to\mathrm{E}$$

$$\frac{\begin{array}{c}\mathcal{D}\\ \varphi\end{array}}{\varphi \vee \psi}\ \vee\mathrm{I}_r \qquad \frac{\begin{array}{c}\mathcal{D}\\ \psi\end{array}}{\varphi \vee \psi}\ \vee\mathrm{I}_l \qquad\qquad \frac{\begin{array}{c}\mathcal{D}_1\\ \varphi \vee \psi\end{array} \qquad \begin{array}{c}[\varphi]\\ \mathcal{D}_2\\ \theta\end{array} \qquad \begin{array}{c}[\psi]\\ \mathcal{D}_3\\ \theta\end{array}}{\theta}\ \vee\mathrm{E}$$

$$\frac{\begin{array}{c}\mathcal{D}\\ \varphi(a)\end{array}}{\forall x\,\varphi(x)}\,\forall\mathrm{I} \qquad \frac{\begin{array}{c}\mathcal{D}\\ \forall x\,\varphi(x)\end{array}}{\varphi(t)}\,\forall\mathrm{E}$$

$$\frac{\begin{array}{c}\mathcal{D}\\ \varphi(t)\end{array}}{\exists x\,\varphi(x)}\,\exists\mathrm{I} \qquad \frac{\begin{array}{c}\mathcal{D}_1\\ \exists x\,\varphi(x)\end{array} \qquad \begin{array}{c}[\varphi(a)]\\ \mathcal{D}_2\\ \theta\end{array}}{\theta}\,\exists\mathrm{E}$$

Next come the rules for equality.

$$\frac{\begin{array}{c}\mathcal{D}\\ t = t \to \psi\end{array}}{\psi}\,\mathrm{Eq}_{\mathit{refl}} \qquad \frac{\begin{array}{c}\mathcal{D}_1\\ \varphi(t)\end{array} \qquad \begin{array}{c}\mathcal{D}_2\\ t = s\end{array}}{\varphi(s)}\,\mathrm{Eq}_{\mathit{repl}}$$

The *open* assumptions of the above deductions are declared as follows.

(i) The deduction with last rule $\perp_i$ has the same open assumptions as its immediate subdeduction $\mathcal{D}$.

(ii) In the deduction whose last inference rule is $\to$I, the open assumptions are those of $\mathcal{D}$ without φ. Here φ is a cancelled assumption of the deduction. This is indicated by putting φ in square brackets on top of the deduction.

In the deduction whose last inference rule is $\vee$E, its set of open assumptions is $o(\mathcal{D}_1) \cup o(\mathcal{D}_2) \setminus \{\varphi\} \cup o(\mathcal{D}_3) \setminus \{\psi\}$.

The set of open assumptions of the deduction whose last inference rule is $\exists$E is $o(\mathcal{D}_1) \cup o(\mathcal{D}_2) \setminus \{\varphi\}$.

If the last inference rule of a deduction is different from $\to$I, $\vee$E, and $\exists$E, then the open assumptions are those of the immediate subdeductions combined.

A formula in a deduction $\mathcal{D}$ that is not an open assumption of $\mathcal{D}$ but is open in a subdeduction of $\mathcal{D}$ will be called a a *cancelled* or *discharged assumption* of $\mathcal{D}$.

The inference rules $\forall$I and $\exists$E are subject to the following eigenvariable conditions.

(iii) In the deduction whose last inference is $\forall$I, the variable a is an eigenvariable; that is, a is fully indicated in $\varphi(a)$ and a must not occur in any of the open assumptions of $\mathcal{D}$.

In the deduction whose last inference is $\exists$E, a is an eigenvariable; that is, a is fully indicated in $\varphi(a)$ and a must not occur in any of the open assumptions of $\mathcal{D}_2$ other than $\varphi(a)$.

If ϑ is among the open assumptions of a deduction $\mathcal{D}$ with conclusion ψ, the conclusion ψ is said to *depend on* ϑ in $\mathcal{D}$. A deduction without open assumptions is said to be *closed*. A formula θ is *deducible* if there is a closed deduction with conclusion θ. We shall convey this by writing $\vdash \theta$.

Example 1.10 Our first example is a natural deduction of the law of contraposition.

$$\dfrac{\dfrac{\dfrac{\dfrac{\neg\psi \qquad \dfrac{\varphi\to\psi \qquad \varphi}{\psi}{\to}\mathrm{E}}{\bot}{\to}\mathrm{E}}{\neg\varphi}{\to}\mathrm{I}}{\neg\psi\to\neg\varphi}{\to}\mathrm{I}}{(\varphi\to\psi)\to(\neg\psi\to\neg\varphi)}{\to}\mathrm{I}$$

The second example is a deduction of the double negation of the law of excluded middle.

$$\dfrac{\dfrac{\neg(\psi\vee\neg\psi) \qquad \dfrac{\dfrac{\dfrac{\neg(\psi\vee\neg\psi) \qquad \dfrac{\psi}{\psi\vee\neg\psi}\vee\mathrm{I}}{\bot}{\to}\mathrm{E}}{\neg\psi}{\to}\mathrm{I}}{\psi\vee\neg\psi}\vee\mathrm{I}}{\bot}{\to}\mathrm{E}}{\neg\neg(\psi\vee\neg\psi)}{\to}\mathrm{I}$$

The third example features an application of the intuitionistic absurdity rule $\bot_i$.

$$\dfrac{\dfrac{\dfrac{\dfrac{\psi\wedge\neg\psi}{\psi}\wedge\mathrm{E}_r \qquad \dfrac{\psi\wedge\neg\psi}{\neg\psi}\wedge\mathrm{E}_l}{\bot}{\to}\mathrm{E}}{\theta}\bot_i}{\psi\wedge\neg\psi\to\theta}{\to}\mathrm{I}$$

Lemma 1.11 *Here is a list of intuitionistic laws that (of course) have natural deductions and are useful to know.*

(i) $\neg\neg(\psi\vee\neg\psi)$
(ii) $\varphi\to\neg\neg\varphi$
(iii) $\neg\neg\neg\varphi\leftrightarrow\neg\varphi$
(iv) $(\neg\neg\psi\to\neg\neg\varphi)\leftrightarrow\neg\neg(\psi\to\varphi)\leftrightarrow(\psi\to\neg\neg\varphi)$
(v) $(\psi\to\varphi)\to(\neg\varphi\to\neg\psi)$
(vi) $\neg\neg(\psi\to\varphi)\to(\psi\to\neg\neg\varphi)$.
(vii) $\neg\neg(\psi\wedge\varphi)\leftrightarrow(\neg\neg\varphi\wedge\neg\neg\psi)$.
(viii) $\neg\neg\forall x\varphi(x)\to\forall x\neg\neg\varphi(x)$
(ix) $\neg\exists x\varphi(x)\leftrightarrow\forall x\neg\varphi(x)$
(x) $\neg\forall x\neg\varphi(x)\leftrightarrow\neg\neg\exists x\varphi(x)$.

(xi) $(\psi \vee \neg\psi) \rightarrow ([\psi \rightarrow \exists x\varphi(x)] \rightarrow \exists x[\psi \rightarrow \varphi(x)])$ *where* x *is not to occur in* ψ.

Definition 1.12 Thus far, we have only considered deductions in pure intuitionistic predicate logic with equality. Given a theory T, that is, a collection of closed formulas (i.e., without free variables) in a first-order language $\mathcal{L}$ with equality, we say that a formula θ is *intuitionistically deducible in* T if there is a deduction $\mathcal{D}$ with conclusion θ whose open assumptions are in T. We shall convey this by writing $T \vdash \theta$.

1.5 A Hilbert-Style System for Intuitionistic Logic

For certain metamathematical purposes, such as showing that a structure satisfies the laws of intuitionistic logic, it is more convenient to work with a system based on axioms and a few rules, where the rules just act locally on the conclusions of derivations and do not involve sequences of formulae nor cancellation of open assumptions elsewhere in the derivation. Such codifications of logic are known by the generic name of Hilbert-type systems. Hilbert introduced such a system for classical logic in [13]. In 1925, Kolmogorov [18] used this calculus to show that classical logic can be interpreted in intuitionistic logic.

Definition 1.13 We introduce a Hilbert-style system for intuitionistic predicate logic with equality.

Axioms

(A1) $\varphi \rightarrow (\psi \rightarrow \varphi)$
(A2) $(\varphi \rightarrow (\psi \rightarrow \chi)) \rightarrow ((\varphi \rightarrow \psi) \rightarrow (\varphi \rightarrow \chi))$
(A3) $\varphi \rightarrow (\psi \rightarrow (\varphi \wedge \psi))$
(A4) $(\varphi \wedge \psi) \rightarrow \varphi$
(A5) $(\varphi \wedge \psi) \rightarrow \psi$
(A6) $\varphi \rightarrow (\varphi \vee \psi)$
(A7) $\psi \rightarrow (\varphi \vee \psi)$
(A8) $(\varphi \vee \psi) \rightarrow ((\varphi \rightarrow \chi) \rightarrow ((\psi \rightarrow \chi) \rightarrow \chi))$
(A9) $(\varphi \rightarrow \psi) \rightarrow ((\varphi \rightarrow \neg\psi) \rightarrow \neg\varphi)$
(A10) $\varphi \rightarrow (\neg\varphi \rightarrow \psi)$
(A11) $\forall x\, \varphi(x) \rightarrow \varphi(t)$
(A12) $\varphi(t) \rightarrow \exists x\, \varphi(x)$
(Eq1) $t = t$
(Eq2) $s = t \rightarrow (\varphi(s) \rightarrow \varphi(t))$

Inference Rules $\vdash \varphi$ conveys that φ is deducible.

(Ax) All axioms are deducible.
(MP) If $\vdash \varphi$ and $\vdash \varphi \to \psi$, then $\vdash \psi$.
($\forall$I) If $\vdash \psi \to \varphi(a)$, then $\vdash \psi \to \forall x\, \varphi(x)$.
($\exists$I) If $\vdash \varphi(a) \to \psi$, then $\vdash \exists x\, \varphi(x) \to \psi$.

Here (MP) stands for 'modus ponens'. In ($\forall$I) and ($\exists$I), a is an eigenvariable, which means that a is fully indicated in $\varphi(a)$ and must not occur in ψ.

1.6 Realizability

The BHK-interpretation provides an intuitive and readily applicable account of constructive reasoning that relieves one from having to resort to formal rules. It is based, though, notably in the cases of implication and universal quantification, on an unexamined notion of function or constructive function. There is also Bishop's program to give numerical meaning to large swaths of mathematics. In many ways, the technique of *realizability*, developed by Kleene in 1945 [17], is a systematic method for illuminating the role of computable functions in constructive reasoning as well as making the numerical import of constructively proved theorems explicit.

Below we will treat the example of intuitionistic first-order number theory, known as Heyting arithmetic, $\mathbf{HA}$. $\mathbf{HA}$ is just the intuitionistic version of classical first-order number theory, known as Peano arithmetic, $\mathbf{PA}$.

The *language* of these theories has a constant 0 (for zero) and function symbols $\mathsf{S}, +, \cdot$ for the successor ($n \mapsto n+1$), addition and multiplication function, respectively.[10] All natural numbers n have a name $\bar{n}$ in this language, where zero barred is the same as the constant 0 and $\overline{n+1}$ is $\mathsf{S}(\bar{n})$.

The *axioms* of the two theories are the same. They comprise the usual elementary school laws for computing with $\mathsf{S}, +, \cdot$ and the induction scheme

$$\phi(0) \ \wedge\ \forall x\, [\phi(x) \to \phi(x+1)] \to \forall x \phi(x)$$

for all formulas $\phi(x)$ of the language, where we used the more familiar $x+1$ for $\mathsf{S}(x)$.

Kleene's 1945 realizability for a closed formula is defined by induction on the complexity of the formula. It employs partial computable functions as realizers for implications and for formulas starting with a universal quantifier $\forall$. As such functions can be represented via Turing machine programs (or other forms of description), which in turn can be encoded by a single natural number (their Gödel number), realizers for these can be taken to be naturals. The other interesting cases are disjunctive and existential formulas. For a disjunction a realizer has to pick a

[10] One can also add symbols for all primitive recursive functions and incorporate their defining equations as axioms.

realizer for one of the disjuncts whereas in the case of $\exists x \phi(x)$ a realizer has to choose a witness n and also supply a realizer for $\phi(\bar{n})$. Also in these cases, realizers can be engineered to be single naturals via a binary coding function.

Definition 1.14 Let $\langle\,,\,\rangle$ be a binary coding function[11] with inverses π_1, π_2, that is, $\pi_1(\langle m, n\rangle) = m$ and $\pi_2(\langle m, n\rangle) = n$.

Let $e \in \mathbb{N}$ be the program code of a Turing machine M_e. We then write $\{e\}(k) \simeq m$ to convey that when M_e is run on input k, it reaches a halting state and outputs m. Below $\{e\}(k) \Vdash \phi$ will be an abbreviation for

$$\exists m\, [\{e\}(k) \simeq m \;\wedge\; m \Vdash \phi].$$

A closed term t of the language of $\mathbf{HA}$, using the obvious interpretation of the function symbols and the constant 0, can be evaluated to a natural, which we denote by $t^{\mathbb{N}}$.

With the foregoing coding machinery in place, the clauses for realizability are spelled out as follows (where '⇒', of course, stands for 'implies').

$$\begin{aligned}
e \Vdash s = t &\quad \text{iff} \quad s^{\mathbb{N}} = t^{\mathbb{N}} \\
e \Vdash \phi \wedge \psi &\quad \text{iff} \quad \pi_1(e) \Vdash \phi \text{ and } \pi_2(e) \Vdash \psi \\
e \Vdash \phi \vee \psi &\quad \text{iff} \quad [\pi_1(e) = 0 \Rightarrow \pi_2(e) \Vdash \phi] \text{ and } [\pi_1(e) \neq 0 \Rightarrow \pi_2(e) \Vdash \psi] \\
e \Vdash \phi \to \psi &\quad \text{iff} \quad \forall d \in \mathbb{N}\, [d \Vdash \psi \Rightarrow \{e\}(n) \Vdash \psi] \\
e \Vdash \forall x \psi(x) &\quad \text{iff} \quad \forall n \in \mathbb{N}\, \{e\}(n) \Vdash \psi(\bar{n}) \\
e \Vdash \exists x \psi(x) &\quad \text{iff} \quad \pi_1(e) \Vdash \psi(\bar{n}) \text{ with } n := \pi_2(e)
\end{aligned}$$

There is no clause pertaining to $\perp$ in this list since in the context of arithmetic $\perp$ can be identified with $0 = \mathsf{S}(0)$.

Theorem 1.15 (Kleene 1945) *Let ψ be a closed formula. If $\mathbf{HA} \vdash \psi$, then one can effectively construct a realizer e such that $e \Vdash \psi$.*

Proof The proof proceeds by induction on the deduction of ψ. It is convenient to use the Hilbert calculus for intuitionistic logic. For details see [17]. □

There are more realizable formulas than there are theorems of $\mathbf{HA}$. Considerable interest is attached to the so-called *Church's thesis*, which is the schema

$$\mathsf{CT} \qquad \forall x\, \exists y\, \varphi(x, y) \to \exists u\, \forall x\, \varphi(x, \{u\}(x)),$$

declaring that every total relation expressible in the language contains the graph of a computable function,[12] and also to Markov's schema

$$\mathsf{M} \qquad \forall x\, [\psi(x) \vee \neg\psi(x)] \;\wedge\; \neg\neg\exists x \psi(x) \to \exists x \psi(x).$$

[11] For example, $\langle m, n\rangle = \frac{1}{2}((n+m)^2 + 3m + n)$.
[12] Whence a form of choice together with 'every function is computable'.

Realizability shows that both CT and M are realizable, confirming the coherence of a Markovian world, also known as 'Russian constructivism'.

Realizability has been extended to analysis, formal theories of second-order arithmetic, higher-type arithmetic and set theories.[13] Moreover, there are many different and subtle forms of realizability. Venturing a comparison with classical set theory, one could say that this technique is as important for intuitionistic systems as forcing is for classical set theory.[14]

1.7 The Curry–Howard Correspondence

We conclude this article with a few remarks about what came to be called the Curry–Howard correspondence (or, more ambitiously, 'isomorphism') and its relationship with the topics broached in this article. It evolved from observations relating certain standard models of computation – such as combinatory logic and the simply typed lambda calculus – to intuitionistic proof systems. Curry in 1934 [7] and 1958 [8], noted that there is an analogy between the laws governing the combinators k and s of combinatory logic and the respective axioms (A1) and (A2) of the implicational fragment of the intuitionistic Hilbert-style system of Definition 1.13.[15] In 1968, Howard [15], drew a connection between intuitionistic natural deductions and the simply typed lambda calculus, whereby a proof is interpreted as a λ-term and the formula it proves emerges as the type of that term. At a more abstract level, that is, leaving the pecularities of proof systems aside, one could say that a proof of a formula is a program whose type is revealed by that formula. The essence of the Curry–Howard correspondence is thus that the seemingly unrelated areas of proof systems and models of computation are manifestations of the same underlying mathematical structures.

There is an obvious relationship between this famous correspondence and the BHK-interpretation. The latter provides an operational semantics for intuitionistic logic, rendering proofs as functions, but does not delineate the class of required functions. If one takes functions definable by λ-terms for this class, one arrives at Howard's correspondence between natural deductions and such functions. If, on the other hand, one equates this class with certain partial recursive functions, then the CH-correspondence, specialized to deductions in Heyting arithmetic, appears to tell a similar story to Kleene's 1945 realizability.

13 See [21, Proposition 2.8] this volume, and [1, 19, 20, 24, 25, 27] for more background.

14 However, the technique of forcing is not wedded to classical theories; it is also important in investigations of intuitionistic theories. Recall that Cohen used ideas from intuitionism.

15 See [14, 22] for more background and also [21, Definition 2.59] in this volume for k and s.

References

[1] Beeson, M. J. 1985. *Foundations of Constructive Mathematics*. Ergebnisse der Mathematik und ihrer Grenzgebiete, vol. 6. Berlin: Springer-Verlag.

[2] Bishop, E. 1967. *Foundations of Constructive Analysis*. New York: McGraw-Hill.

[3] Borel, É. 1972. *Œuvres de Émil Borel*. Paris: Centre National de la Recherche Scientifique.

[4] Brouwer, L. E. J. 1907. Over de grondslagen der wiskunde. Ph.D. thesis, University of Amsterdam.

[5] Brouwer, L. E. J. 1908. De onbetrouwbaarheid der logische principes. *Tijdschr. Wijsbegeerte*, 152–158.

[6] Chen, R.-M., and Rathjen, M. 2012. Lifschitz realizability for intuitionistic Zermelo–Fraenkel set theory. *Arch. Math. Logic*, **51**, 789–818.

[7] Curry, H. B. 1934. Functionality in combinatory logic. *Proc. Natl Acad. Sci. USA*, **20**, 584–590.

[8] Curry, H. B., and Feys, R. 1958. *Combinatory Logic*. Amsterdam: North-Holland.

[9] du Bois-Reymond, P. 1882. *Die allgemeine Functionentheorie*. Tübingen: Laupp.

[10] Dummett, M. 2000. *Elements of Intuitionism*. Oxford: Clarendon Press.

[11] Feferman, S. 1965. Some applications of the notions of forcing and generic sets. *Fund. Math.*, **56**, 325–345.

[12] Heyting, A. 1930. *Die formalen Regeln der intuitionistischen Logik*. Sitzungsber. Preuss. Akad. Wissensch. Physikalisch-mathematische Klasse, 42–56.

[13] Hilbert, D. 1922. Die logischen Grundlagen der Mathematik. *Math. Annalen*, **88**, 151–165.

[14] Hindley, J. R., and Seldin, J. P. 2008. *Lambda-Calculus and Combinators, an Introduction*. Cambridge: Cambridge University Press.

[15] Howard, W. A. 1980. The formulae-as-types notion of construction. Pages 479–490 of: Hindley, J. R., and Seldin, J. P. (eds), *To H. B. Curry, Essays on Combinatory Logic, Lambda Calculus and Formalism*. London: Academic Press.

[16] Jervell, H. 1996. From the axiom of choice to choice sequences. *Nordic J. Philos. Logic*, **1**, 95–98.

[17] Kleene, S. C. 1945. On the interpretation of intuitionistic number theory. *J. Symbol. Logic*, **10**, 109–124.

[18] Kolmogorov, A. 1925. O principe tertium non datur. *Matematiceskij Sbornik*, **32**, 646–667.

[19] McCarty, D.C. 1986. Realizability and recursive set theory. *Ann. Pure Appl. Logic*, **32**, 153–183.

[20] Rathjen, M. 2006. Realizability for constructive Zermelo–Fraenkel set theory. Pages 228–314 of: Väänänen, J., and Stoltenberg-Hansen, V. (eds.), *Logic*

Colloquium '03. Lecture Notes in Logic, vol. 24. Wellesley, MA: A. K. Peters.

[21] Rathjen, M. 2023. Introduction to constructive set theory: an appetizer. Chapter 2 of: Bridges, D., Ishihara, H., Rathjen, M., and Schwichtenberg, H. (eds.), *Handbook of Constructive Mathematics*. Cambridge: Cambridge University Press.

[22] Sørenson, M. H., and Urzyczyn, P. 2006. *Lectures on the Curry–Howard isomorphism*. Studies in Logic and the Foundations of Mathematics, vol. 149. Elsevier Science.

[23] Schütte, K. 1977. *Proof Theory*. Berlin: Springer.

[24] Troelstra, A. S. 1973. *Metamathematical Investigations of Intuitionistic Arithmetic and Analysis*. Lecture Notes in Mathematics, vol. 344. Berlin: Springer.

[25] Troelstra, A. S. 1998. Realizability. Pages 407–473 of: Buss, S. (ed.), *Handbook of Proof Theory*. Studies in Logic and the Foundations of Mathematics, vol. 137. Amsterdam: North-Holland.

[26] Troelstra, A. S., and van Dalen, D. 1988. *Constructivism in Mathematics, Volumes I, II*. Amsterdam: North Holland.

[27] van Oosten, J. 2008. *Realizability: An Introduction to its Categorical Side*. Studies in Logic and Foundations of Mathematics, vol. 152. Amsterdam: Elsevier.

2

An Introduction to Constructive Set Theory: An Appetizer

Michael Rathjen

2.1 Introduction

The primary purpose of this contribution is to provide an axiomatic framework and reference system for the development of constructive mathematics, in the same vein as Zermelo–Fraenkel set theory **ZF** furnishes such a framework for classical ('ordinary') mathematics. Peter Aczel [1] has called this theory *Constructive Zermelo–Fraenkel Set Theory*, **CZF**. Of course, the underlying logic of **CZF** is intuitionistic logic.

2.1.1 Overview

Here is a brief summary of the contents of this chapter. Section 2.2 presents the axiomatic framework. Section 2.3 demonstrates that many set-theoretic constructions can be carried out in **CZF**, and that important mathematical structures, including the natural numbers, the Cauchy and Dedekind real numbers, as well as function spaces, can be developed in **CZF**. This is followed by Section 2.4, dealing with genuinely set-theoretic topics, including the definition of functions via transfinite recursion and inductively defined classes.

The last four sections address more-advanced topics. Section 2.5 is concerned with large sets in **CZF**. Although **CZF** accomodates inductively defined classes, it is sometimes lacking axiomatic strength to show that certain important inductive definitions, such as the W-types and universes of Martin-Löf type theory, give rise to sets. To achieve this, new axioms about the existence of larger sets are required. This is a familiar topic in modern set theory. Remarkedly, it turns out that large set axioms, classically corresponding to regular and inaccessible cardinals, engender only modest additional proof-theoretic strength when considered in the context of **CZF**. In Section 2.6, the ambiguous role of the axiom of choice in constructive theories is discussed. It explores which restricted forms of choice can be added to **CZF** without jeopardizing its essential constructive nature. Along the way, Section 2.6 presents rarely seen choice principles that are in some sense maximal for the constructive context. Section 2.7 broaches the interesting topic of semi-intuitionistic set theories. It investigates what happens if one adds a little bit of classical reasoning to **CZF**, namely, in the guise of what Bishop called the *limited principle of omniscience*, LPO. Unexpectedly, **CZF** + LPO is still conservative over **CZF** for statements of the complexity of the twin prime conjecture. The final section, Section 2.8, provides some insight into how wondrous non-classical worlds, such as a Brouwerian world, in which all endofunctions of $\mathbb{R}$ are continuous, or a Markovian world, in which all functions from $\mathbb{N}$ to $\mathbb{N}$ are computable, can be defined and be explored within **CZF** via the technique of *realizability*.

2.1.2 Background: the 'Intellectual' Landscape

If we prove a mathematical theorem, claiming that our proof is actually constructive, we should be able to declare, at least in principle, which axiom system our proof is based on.[1] Axiomatic approaches to intuitionistic mathematics and other forms of constructive mathematics are not new. Whereas intuitionistic logic was formalized even before[2] Heyting's 1930 paper [34], the first formalizations of intuitionistic mathematics are due to Heyting [35, 36], also in 1930. A major turning point in constructive mathematics was Bishop's publication of *Foundations of Constructive Analysis* [11] in 1967, in which Bishop presented an informal constructive development of analysis. What was remarkable about his work was that it went strikingly further in the development of mathematics than anything that had been done hitherto by intuitionists and constructivists. Surprisingly, large chunks of modern functional analysis were shown to be amenable to constructive treatment. What was also novel about Bishop's work was that it broke with the intuitionistic past in that it could be read as a straight piece of classical mathematics as well, albeit one with a keen interest in exhibiting and preserving computational information.[3]

While Bishop [11] works with informal notions of set, family of sets and function, he did not provide an axiomatic approach in his book, although he was clearly interested in the foundations, formalization, and axiomatization of constructive mathematics (as attested by his writings [12, 13, 14]). The early 1970s saw the development of several foundational frameworks for constructive mathematics that were to no small extent galvanized by Bishop's work, notably the following.

[1] One can imagine that some of the founding fathers of constructivism would have balked at this request, insisting that the right frame of mind and its emanations are the shibboleths and guarantors of constructiveness.

[2] Kolmogorov's formalization [40] in 1925, though excluding ex falso quodlibet, was followed by Glivenko's [29, 30] in 1928 and 1929. Kolmogorov proceeded by assaying the axioms and rules of classical logic in Hilbert–Ackermann's [37] as to their constructive validity.

[3] Between classical mathematics, regarded as largely proof irrelevant, and Martin-Löf's type theory, as the one making all witnessing information explicit, Bishop's mathematics seems to pursue a middle path, on the whole staying closer to classical mathematics.

(i) S. Feferman, Explicit Mathematics, $\mathbf{T}_0$ [23].
(ii) J. Myhill [48] and Friedman [26], Constructive Set Theory.
(iii) P. Martin-Löf, Intuitionistic Type Theory, **MLTT** [44].

In this chapter I will follow an approach to constructivism based on set theory as commenced by Myhill [48] with his system **CST** and also pursued by Friedman [26]. All three approaches, though, are exciting in their own right. Explicit mathematics and Martin-Löf's type theory are of great importance to the philosophy of mathematics and the theory of computation.[4] They provide careful analyses of the ontology of mathematics and the nature of computation and construction. However, there is really no idea of singling one of the theories out as the gold standard.[5] Moreover, all three theories have been studied by proof theorists and a lot of insight has been obtained as to how these theories are related to each other (see [33, 52, 62]). The upshot of that work is that they are in actuality closely related. There is one feature, however, that distinguishes **CST** from the other two. Every mathematician knows the basics of presenting mathematics in set theory. $\mathbf{T}_0$ and **MLTT**, though, have a complicated syntax that needs to be learned and mastered, and my impression is that they are only known to relatively small bands of logicians and computer theoreticians. Myhill wanted to single out the principles that undergird Bishop's conceptions of what sets and functions are, adding that he wanted 'these principles to be such as to make the process of formalization completely trivial, as it is in the classical case' ([48, p. 347]).

Of particular interest to this article is the embedding of set theory into Martin-Löf type theory found and propounded by Aczel [1, 2, 3].[6] In the process of validating the axioms of Myhill's **CST** in **MLTT** under this interpretation, he realized that the ontological distinctions between numbers, functions, and sets that Myhill had made in **CST** weren't strictly necessary nor particularly helpful. As a consequence, he based his Constructive Zermelo–Fraenkel Set Theory, **CZF**, on the same language as the standard classical set theory **ZF**. Moreover, the interpretation in type theory validated some new principles, notably Subset Collection, thereby giving rise to axioms of **CZF** that **CST** lacks.

A very apt description of the approach adopted for this chapter is given in the prefaces of [5, 6]. As I don't know how to better that, let's just quote it:

> *It is distinctive in that it uses the standard first order language of classical axiomatic set theory and makes no explicit use of specifically constructive ideas. Of course its logic is intuitionistic, but there is no special notion of*

[4] For a presentation of mathematics in explicit mathematics see, for example, [7].
[5] Feferman used to emphasize this point.
[6] It bears stressing that this embedding or interpretation of set theory in **MLTT**, rather than being a mere technical feat, is in Aczel's view the correct way of assigning meaning to set-theoretic statements; for more details see [2] and also [16, Section 1.3.3].

construction or constructive object. There are just the sets, as in classical set theory. This means that mathematics in constructive set theory can look very much like ordinary classical mathematics. The advantage of this is that the ideas, conventions and practice of the set theoretical presentation of ordinary mathematics can be used also in the set theoretical development of constructive mathematics, provided that a suitable discipline is adhered to. In the first place only the methods of logical reasoning available in intuitionistic logic should be used. In addition only the set theoretical axioms allowed in constructive set theory can be used. With some practice it is not difficult for the constructive mathematician to adhere to this discipline.

Constructivism and set theory are sometimes depicted as antipodes, putting forward such reasons as the extensional treatment of sets and the way functions are formalized just as sets of ordered pairs as points of divergence.[7] These misgivings are misguided. The above quote really means what it says. The notion of function is to be treated in the same way as in classical set theory. That the same concept behaves very differently in constructive set theory is purely a consequence of the underlying intuitionistic logic. Quite dramatically the difference between the Cantor set $2^{\mathbb{N}}$ and the powerset of $\mathbb{N}$ can be demonstrated by the fact that even full intuitionistic Zermelo–Fraenkel set theory possesses realizability models that satisfy the uniformity principle

$$\mathrm{UP}_1 \qquad \forall x \subseteq \mathbb{N}\, \exists f \in 2^{\mathbb{N}}\, \varphi(x,f) \to \exists f \in 2^{\mathbb{N}}\, \forall x \subseteq \mathbb{N}\, \varphi(x,f)$$

for all formulae $\varphi(x,f)$.

2.2 The Axiomatic Framework

Bishop [14, p. 60] writes:

Another important foundational problem is to find a formal system that will efficiently express predictive mathematics. I think we should keep the formalism as primitive as possible, starting with a minimal system and enlarging it only if the enlargement serves a genuine mathematical need. In this way the formalism and the mathematics will hopefully interact to the advantage of both.[8]

The formalism here will be the set theory **CZF** whose language just has the primitive relation symbols $\in$ and $=$ for membership and equality, respectively,

[7] For instance, Martin-Löf ([45, p. 23]) writes that: '*The reason that B^A can be constructed as a set is that we take the notion of function as primitive, instead of defining a function as a set of ordered pairs or a binary relation satisfying the usual existence and uniqueness conditions, which would make it a category (like $\mathcal{P}(A)$) instead of a set*'.

[8] For more on Bishop's views on open-ended formalization see Crosilla's chapter [17] in this volume.

and only intuitionistic reasoning will be allowed.[9] As mathematics is gradually developed in this framework, it will be pointed out which axioms are required and some subtheories that are sufficient onto the task of developing various larger connected chunks of mathematics and set theory will be singled out, although the number of these subsystems will be kept small lest the patience of the reader gets strained too much.

For the formalization of most of elementary mathematics a small part of **CZF** suffices, which in actuality is no stronger than intuitionistic number theory, also known as Heyting arithmetic, **HA** (see [59]).

Still, the idea is to start out by exhibiting the axioms of **CZF** all at once.

2.2.1 The Axioms of CZF

The axioms of Extensionality, Pairing, and Union are exactly as in classical set theory. The Infinity axiom comes in a slightly beefed-up form, the reason being that, although the usual version suffices in the presence of elementhood induction, without the latter we need to strengthen it in order to have access to ordinary induction over the naturals. There are several reasons for interest in a set theory without elementhood induction. One is that it yields a set theory no stronger than **HA**; another is that it can serve as a basis for a constructive set theory with the antifoundation axiom(see [6, 53, 54]).

The Separation axiom is also altered. It is solely allowed for *bounded* formulas, also called Δ_0-*formulas*. In these formulas, quantifiers are permissible only when they are bounded. This means that they have to be of the form $\forall x \in a$ and $\exists x \in a$, respectively. More formally, these restricted quantifiers in formulas $\forall x \in a\, \varphi(x)$ and $\exists x \in a\, \varphi(x)$ are to be read as $\forall x[x \in a \to \varphi(x)]$ and $\exists x[x \in a \wedge \varphi(x)]$, respectively.

An axiom native to **CZF** is the Subset Collection axiom. It is the most unfamiliar axiom. One can readily see that it is a consequence of the Powerset axiom. It will later be shown that it is equivalent to a statement roughly saying the following: *For all sets A, B there exists a 'sufficiently large' set of multi-valued functions from A to B*. It will then also become clear that it is an extension of Myhill's Exponentiation axiom. The latter asserts that, given two sets A and B, the collection of all functions from B to A (notated by A^B) is a set too.

So here are the axioms.

Extensionality

$$\forall x \forall y\, [\forall u\, (u \in x \leftrightarrow u \in y) \to x = y]$$

Pairing

$$\forall x \forall y\, \exists z\, \forall u [u \in z \leftrightarrow (u \in x \vee u \in y)]$$

[9] An introduction to intuitionistic logic intended for the working constructivist can be found in this volume [63] and many other books, for example [68, Chapter 2] .

Union

$$\forall x \exists y \forall u [u \in y \leftrightarrow \exists z \in x \, u \in z]$$

Bounded Separation

$$\exists x \forall u [u \in x \leftrightarrow (u \in a \wedge \theta(u))]$$

where $\theta(u)$ is Δ_0 and x is not free in $\theta(u)$.

Strong Infinity

$$\exists a [Ind(a) \wedge \forall b [Ind(b) \rightarrow \forall x \in a (x \in b)]]$$

where we use the following abbreviations.

(i) $Succ(x, y)$ for $\forall z [z \in y \ \leftrightarrow \ z \in x \vee z = x]$,
(ii) $Ind(a)$ for $(\exists y \in a)(\forall z \in y)\bot \wedge (\forall x \in a)(\exists y \in a) Succ(x, y)$. Here $\bot$ stands for falsum or 'absurdity'.[10]

Subset Collection

$$\begin{aligned} &\exists c \, \forall u \left[\forall x{\in}a \, \exists y{\in}b \, \psi(x, y, u) \ \rightarrow \right. \\ &\qquad \left. \exists d{\in}c \, \left(\forall x{\in}a \, \exists y{\in}d \ \psi(x, y, u) \ \wedge \ \forall y{\in}d \, \exists x{\in}a \ \psi(x, y, u)\right)\right] \end{aligned}$$

for all formulas $\psi(x, y, u)$.

Strong Collection

$$\begin{aligned} &(\forall x \in a) \, \exists y \ \varphi(x, y) \ \rightarrow \\ &\exists b \, [\, (\forall x \in a) \, (\exists y \in b) \ \varphi(x, y) \ \wedge \ (\forall y \in b) \, (\exists x \in a) \ \varphi(x, y) \,] \end{aligned}$$

for all formulas $\varphi(x, y)$.

Set Induction

$$\forall x [\forall u \in x \, \psi(u) \rightarrow \psi(x)] \rightarrow \forall x \psi(x)$$

for all formulas $\psi(u)$.

A few preliminary comments are in order. The reasons for strengthening the Inifinity axiom have already been adumbrated. Set Induction is a positive way of rendering the classical Foundation axiom and it forestalls the undesirable forms of excluded middle implied by the Foundation axiom (due to Myhill; see [47, 48] or [6, Proposition 10.4.1]). The usual axiomatization of set theory via the Replacement axiom incurs disadvantages in an intuitionistic context as a premise $\forall x \, \exists! y \varphi(x, y)$ is often difficult to ascertain because classically legitimate moves that circumvent this problem (such as 'Scott's trick') are not available. But why is Strong Collection better than Collection? In a set theory with full Separation the former is deducible from the latter. However, with just Bounded Separation at one's disposal, one would

[10] In intuitionistic logic, $\bot$ is used to express negation $\neg\phi$ as $\phi \rightarrow \bot$. Crucially, one has ex falso quodlibet, that is, $\bot \rightarrow \psi$ for every formula ψ.

not get the part $(\forall y \in b)\,(\exists x \in a)\ \varphi(x, y)$, and thus b would be liable to contain a lot of 'garbage', that is, elements y unrelated to any $x \in a$ via $\varphi(x, y)$, making Collection rather unworkable for many mathematical purposes.

2.3 Elementary Mathematics in CZF

As is often the case, a little bit goes a long way. For elementary mathematics just a few of the axioms of **CZF** suffice. This is a valid reason to single out an axiom system that will be called *Elementary Constructive Set Theory*, **ECST**. It is like **CZF** except for the following changes.

(i) It drops the Set Induction and Subset Collection Schemes.
(ii) It uses the Replacement Scheme instead of the Collection Scheme, where the former is the scheme

$$\forall x \in a \exists ! y \phi(x, y) \ \rightarrow \ \exists b \forall y\, [y \in b \ \leftrightarrow \ \exists x \in a\, \phi(x, y)]$$

for all formulae $\phi(x, y)$, where b is not free in $\phi(x, y)$.

In particular the omission of Set Induction has quite a dramatic effect on proof-theoretic strength, well known to proof theorists. Indeed, if $\mathbf{CZF}\!\restriction$ denotes the theory **CZF** without Set Induction, then this engenders a theory not stronger than intuitionistic number theory.

Theorem 2.1 ([59]) $\mathbf{CZF}\!\restriction$ *is conservative over Heyting arithmetic* **HA**, *that is, the theories prove the same arithmetical theorems.*[11] *Here we assume the natural interpretation of the language of arithmetic into set theory with ω playing the role of the set of natural numbers.*

Another, but related system of set theory, was shown to be of the strength of **HA** by Friedman [26].

2.3.1 Operations on Sets and Classes

First steps will be taken to develop some of the standard apparatus for representing mathematical ideas in **CZF**.

In carrying out mathematics in this system, it will be convenient to exploit the use of **class notation** and terminology, just as in classical set theory. Given a formula $\phi(x)$ in the language of set theory we use the expression $\{x \mid \phi(x)\}$ for the collection of all sets x satisfying $\phi(x)$. Of course, there may not exist a set of the form $\{x \mid \phi(x)\}$ but there is nothing wrong with thinking about such a

[11] In [59], conservativity is claimed just for Π^0_2-formulas. However, with Goodman-style realizability [31] it can be extended to all arithmetic formulas.

collection. The formula $\phi(x)$ may have additional free variables other than x that are considered parameters upon which the class depends. As is customary in set theory, collections $\{x \mid \phi(x)\}$ will be referred to as *classes* and they will be treated extensionally, that is, two classes A and B that happen to have the same elements will be viewed as equal, which we notate by $A = B$. It is also convenient to talk about elementhood in classes by allowing expressions of the form $y \in A$, where A is a class term. Obviously, one can augment the official language of set theory by permitting atomic class formulas such as $A = B$ and $y \in A$ without really going beyond that language since $y \in \{x \mid \phi(x)\}$ and $\{x \mid \phi(x)\} = \{y \mid \psi(y)\}$ can be translated back into the official language by rewriting them as $\phi(y)$ and $\forall z\, [\phi(z) \leftrightarrow \psi(z)]$, respectively (with z not in $\phi(x)$ and $\psi(y)$).

Sets can also be viewed as classes in that each set b is identified with the class $\{x \mid x \in b\}$. The Extensionality axiom then guarantees that equality of sets amounts to the same as their equality as classes.

As usual, A is a subclass of B, written $A \subseteq B$, if $\forall x \in A\; x \in B$.

Now, several basic and familar class constructions can be performed without the need for any non-logical axioms. Letting A, B, C be classes and letting $a, a_1, \ldots, a_n$ be sets we obtain the following classes.

(i) $\{a_1, \ldots, a_n\} = \{x \mid x = a_1 \vee \cdots \vee x = a_n\}$. When $n = 0$ this is the empty class $\emptyset$.
(ii) $\bigcup A = \{x \mid \exists y \in A\; x \in y\}$.
(iii) $A \cup B = \{x \mid x \in A \vee x \in B\}$.
(iv) $a^+ = a \cup \{a\}$.
(v) $\mathcal{P}(A) = \{x \mid x \subseteq A\}$.
(vi) $V = \{x \mid x = x\}$.

The **Union** axiom asserts that the class $\bigcup A$ is a set for each set A. So, using the **Pairing** axiom one can infer that the class $A \cup B$ is a set whenever A, B are sets and hence that $\{a_1, \ldots, a_n\}$ is a set whenever $a_1, \ldots, a_n$ are sets for $n > 0$.

If A is a class and $\theta(x, y)$ is a formula in the language of set theory, then we may form a **family of classes** $(B_a)_{a \in A}$ over A, where for each $a \in A$

$$B_a = \{y \mid \theta(a, y)\}.$$

If $(B_a)_{a \in A}$ is a family of classes then we may form the classes

$$\bigcup_{a \in A} B_a = \{y \mid \exists a \in A\; y \in B_a\},$$

$$\bigcap_{a \in A} B_a = \{y \mid \forall a \in A\; y \in B_a\}.$$

2.3.2 Reasoning Intuitionistically, for Example, Russell's Paradox

In **CZF** one is bound to reason intuitionistically. This might take a bit of practice. Arguing by case distinctions is a typical mode of reasoning one ought to wean oneself off as it often entails a tacit use of excluded middle. A point in case is Russell's paradox, RP. In almost all book and article treatments of RP the proof employs classical logic, and in the literature one even finds statements to the effect that RP requires classical logic. So it is perhaps instructive to see why such opinions are misguided.

Russell's paradox asserts that the general comprehension principle leads to a contradiction. In the language of classes, this can be rephrased as saying that not every class is a set. There is also a positive rendering of RP.

Lemma 2.2 (**ECST**) *For every set A there is a set A_R such that $A_R \notin A$.*

Proof Let $A_R = \{x \in A \mid x \notin x\}$. Note that A_R is a set by Bounded Separation. From $A_R \in A_R$ we get that $A_R \notin A_R$ by the very definition of A_R. Now, in intuitionistic logic $\neg\psi$ is treated as $\psi \rightarrow \perp$, where $\perp$ is absurdity. As a result of the above,

$$A_R \in A_R \rightarrow (A_R \in A_R \rightarrow \perp)$$

which yields $A_R \in A_R \rightarrow \perp$, that is, $A_R \notin A_R$. As $A_R \in A \rightarrow A_R \in A_R$ holds too, one infers $A_R \notin A$. □

Russell's paradox follows from Lemma 2.2 by assuming that the universal class $A := \{x \mid x = x\}$ is a set.

2.3.3 Class Relations and Functions

Relations and functions are salient concepts in mathematics. To be able to express them in set theory one needs a pairing function which allows one to code a pair of sets as a single set in such a way that the pair can be retrieved from its code. There are many ways of doing this in set theory, notably the von Neumann and the Wiener pairing functions. As long as it works with intuitionistic logic as well, it is pretty immaterial which of these gadgets one uses. So, let the ordered pair of sets a, b be the set $\langle a, b \rangle$ defined by

$$\langle a, b \rangle := \{\{a\}, \{a, b\}\};$$

$\langle a, b \rangle$ is a set by a couple of applications of the Pairing axiom while its uniqueness is guaranteed by the Extensionality axiom. Now, the crucial property of $\langle a, b \rangle$ is spelled out as follows ([5, Proposition 3.1], [6, Proposition 4.1.1]).

Proposition 2.3 (**ECST**) *If $\langle a, b \rangle = \langle c, d \rangle$ then $a = c$ and $b = d$.*

Proof The usual classical argument divides the proof up into cases as to whether or not $a = b$. This method is not available in intuitionistic logic.

Instead we can argue as follows. Assume that $\langle a, b\rangle = \langle c, d\rangle$.

As $\{a\}$ is an element of the left-hand side it is also an element of the right-hand side and so either $\{a\} = \{c\}$ or $\{a\} = \{c, d\}$. In either case $a = c$.

As $\{a, b\}$ is an element of the left-hand side it is also an element of the right-hand side and so either $\{a, b\} = \{c\}$ or $\{a, b\} = \{c, d\}$. In either case $b = c$ or $b = d$. If $b = c$ then $a = c = b$ so that the two sets in $\langle a, b\rangle$ are equal and hence $\{c\} = \{c, d\}$ giving $c = d$ and hence $b = d$. So in either case $b = d$. □

If R is a class of ordered pairs then we use aRb for $\langle a, b\rangle \in R$. If A, B are classes and $R \subseteq A \times B$ such that

$$\forall x \in A \exists! y \in B\ xRy$$

then we use the standard notation

$$R : A \to B,$$

and for each $a \in A$ we write $R(a)$ for the unique $b \in B$ such that aRb. If $R : A \to B$ we will say that R is a **class function** or **map**.

Note that $\mathsf{dom}(R)$ and $\mathsf{ran}(R)$ are the classes $\{x \mid \exists y\ xRy\}$ and $\{y \mid \exists x\ xRy\}$, respectively.

Lemma 2.4 (**ECST**) *If A is a set and $F : A \to B$ then F is a set.*

Proof Since $\forall x \in A\, \exists! y\ (\langle x, y\rangle \in F)$ it follows that there is a function f with $\mathsf{dom}(f) = A$ and $\forall x \in A\ (\langle x, f(x)\rangle \in F)$. Hence $F = f$, so F is a set. □

2.3.4 Some Consequences of Replacement

With the help of Replacement it can be shown that the universe of sets is closed under several important mathematical constructions.

Lemma 2.5 (**ECST**) *Let A be a set and $(B_a)_{a\in A}$ be a family of sets over A. Then, $\bigcup_{a\in A} B_a$ is a set and if A is inhabited, $\bigcap_{a\in A} B_a$ is a set also.*

Proof Replacement entails that $\{B_a \mid a \in A\}$ is a set, and hence, by Union, $\bigcup_{a\in A} B_a$ is a set. Now suppose that A is inhabited. Let $a_0 \in A$. By Lemma 2.4, there is a function f with domain A such that $\forall a \in A\, f(a) = B_a$. Then

$$\bigcap_{a\in A} B_a = \{u \in a_0 \mid \forall x \in A\, u \in f(x)\},$$

so it is a set by Bounded Separation. □

2.3.5 Cartesian Products and Sums of Classes

For classes A, B let $A \times B$ be the class given by

$$A \times B = \{z \mid \exists a \in A \exists b \in B \; z = \langle a, b\rangle\}.$$

For r a natural number greater than 0, the r-fold **Cartesian product** of a class A, A^r, is defined by $A^1 = A$ and $A^{k+1} = A^k \times A$.

If $F : A \times B \to C$ is a class function we will write $F(a, b)$ rather than $F(\langle a, b\rangle)$ for $\langle a, b\rangle \in A \times B$. Similarly, if $G : A^r \to B$ is a class function defined on the r-fold Cartesian product of a class A, we will write $F(a_1, \ldots, a_r)$ for $F(\langle a_1, \ldots, a_r\rangle)$ whenever $\langle a_1, \ldots, a_r\rangle \in A^r$.

Proposition 2.6 (**ECST**) *If A, B are sets then so is the class $A \times B$.*

Proof Let A, B be sets. Then, as

$$\{a\} \times B = \{\langle a, b\rangle \mid b \in B\}$$

is a set, by Replacement, so is

$$\{\{a\} \times B \mid a \in A\},$$

by another application of Replacement. Thus

$$A \times B = \bigcup_{a \in A} (\{a\} \times B)$$

is a set by Union. □

Definition 2.7 *Let I be a class and $(A_i)_{i\in I}$ be a family of classes over I. The* **disjoint union** *or* **dependent sum** *of $(A_i)_{i\in I}$ is the class*

$$\sum_{i\in I} A_i = \{\langle i, a\rangle \mid a \in A_i \;\wedge\; i \in I\}.$$

Note that the cartesian product $A \times B$ is a special case of disjoint union as $A \times B = \sum_{i\in A} B_i$, where $B_i = B$ for all $i \in A$.

Proposition 2.8 (**ECST**) *If I is a set and $(A_i)_{i\in I}$ is a family of sets over I, then $\sum_{i\in I} A_i$ is a set.*

Proof We know that $\{i\} \times A_i$ is a set for every $i \in I$. As

$$\sum_{i\in I} A_i = \bigcup_{i\in I} \{i\} \times A_i$$

it follows by Replacement combined with Union that $\sum_{i\in I} A_i$ is a set. □

2.3.6 Quotients

Quotient constructions are a familiar sight in mathematics and there is nothing intrinsically unconstructive about them.[12]

Let A be a class and let R be a subclass of $A \times \mathrm{A}$. R is said to be an **equivalence relation on** A if the following hold for all $a, b, c \in A$:

(i) aRa (R is **reflexive**),
(ii) if aRb then bRa (R is **symmetric**),
(iii) if aRb and bRc then aRc (R is **transitive**).

Then for each $a \in A$ we may form its **equivalence class**

$$[a]_R = \{x \in A \mid xRa\}.$$

Lemma 2.9 (ECST) *If A and R are sets, where $R \subseteq A \times A$, then for each $a \in A$, $[a]_R$ is a set and, moreover, the* **quotient of** A **with respect to** R,

$$A/R = \{[a]_R \mid a \in A\},$$

is a set.

Proof This is an immediate consequence of Bounded Separation and Replacement. □

2.3.7 The Naturals in ECST

The Strong Infinity axiom furnishes a set that plays the role of the natural numbers in set theory. Recall that this axiom asserts that $\exists a\, \theta(a)$ where

$$\theta(a) \equiv [Ind(a) \wedge \forall y(Ind(y) \rightarrow a \subseteq y)].$$

Here

$$Ind(a) \equiv [\emptyset \in a \wedge \forall x \in a\ (x^+ \in a)],$$

where $\emptyset$ denotes the set without elements and $x^+ = x \cup \{x\}$.

So $\theta(a)$ expresses that a is the smallest set containing $\emptyset$ that is closed under the operation $x \mapsto x^+$.

Lemma 2.10 (ECST) *If $\theta(a)$ and $\theta(b)$ then $a = b$.*

Henceforth ω will denote the unique set a satisfying $\theta(a)$.

Proof Immediate by the definition of $Ind(a)$ and Extensionality. □

The leastness of ω entails the following induction principle.

12 Unless one wants to use Zermelo's axiom of choice whereby a representative can be picked from each equivalence class. See Proposition 2.46.

Proposition 2.11 (ECST) *For any bounded formula $\varphi(x)$,*

$$\varphi(\emptyset) \wedge (\forall x \in \omega)[\varphi(x) \to \varphi(x^+)] \to (\forall x \in \omega)\varphi(x).$$

Proof As $\varphi(x)$ is a bounded formula, the class $b = \{x \in \omega \mid \phi(x)\}$ is a set by Bounded Separation. From $\varphi(\emptyset) \wedge (\forall x \in \omega)[\varphi(x) \to \varphi(x^+)]$ one gets $Ind(b)$ so that $\omega \subseteq b$, and consequently $(\forall x \in \omega)\varphi(x)$. □

Here is a straightforward application of Proposition 2.11.

Proposition 2.12 (ECST) $(\forall x \in \omega)[x = \emptyset \vee (\exists y \in x)(x = y^+)]$.

A more structural approach to the naturals is provided by the **Dedekind–Peano Axioms**.

Definition 2.13 $\mathcal{A} = (A, a_0, F)$ is a DP-*structure* if A is a set and

(DP1) $a_0 \in A$,
(DP2) $F : A \to A$.

$\mathcal{A}$ is a *Dedekind–Peano model* (DP-model) if also

(DP3) $a_0 \neq F(x)$ for $x \in A$,
(DP4) F is injective; that is, $F(x_1) = F(x_2) \Rightarrow x_1 = x_2$ for $x_1, x_2 \in A$,
(DP5) if Y is a subset of A such that $a_0 \in Y$ and $F(x) \in Y$ for all $x \in Y$ then $x \in Y$ for all $x \in A$.

Note: the assertions $(\mathbf{DP1}) - (\mathbf{DP5})$ are the *Dedekind–Peano axioms* for a structure $\mathcal{A} = (A, a_0, F)$.

Definition 2.14 Let $\mathcal{A} = (A, a_0, F)$ and $\mathcal{A}' = (A', a_0', F')$ be DP-structures and let $\pi : A \to A'$. Then π is a DP-*map* $\mathcal{A} \to \mathcal{A}'$ if $\pi a_0 = a_0'$ and $\pi(F(x)) = F'(\pi x)$ for all $x \in A$. It is an *isomorphism* if π is a bijection.

The next result states that any DP-model can serve as the incarnation of the naturals.

Proposition 2.15 (ECST) *Let $\mathcal{A}$ be a* DP-*model and let $f : \mathcal{A} \to \mathcal{A}'$ be a* DP-*map, where $\mathcal{A}'$ is a* DP-*structure.*

(i) The DP-*map f is unique.*
(ii) The DP-*map f is an isomorphism if $\mathcal{A}'$ is also a* DP-*model.*

The Strong Infinity axiom provides us with the set ω. In order to show that ω gives rise to a DP-model the following result is useful.

Lemma 2.16 (ECST) *For all $x \in \omega$,*

(i) $(\forall y \in x)\, y \subseteq x$,

(ii) $x \notin x$, *and*

(iii) $x \subseteq \omega$.

Proof In each case this can be proved by the induction principle of Proposition 2.11. The details are left as an exercise. □

Proposition 2.17 (ECST) *The structure* $\mathcal{N}_\omega = (\omega, 0, \mathsf{S})$ *with* $0 := \emptyset$ *and* $\mathsf{S}(n) = n^+$ *for* $n \in \omega$ *is a* DP*-model.*

Proof Surely $\mathcal{N}_\omega$ satisfies (DP1)–(DP3). Also, (DP5) is a consequence of Proposition 2.11. So it remains to prove (DP4). Let $x, y \in \omega$ such that $x^+ = y^+$. Since $x \in x^+$, $x \in y^+$ so that either $x \in y$ or $x = y$. Likewise, either $y \in x$ or $y = x$. If $x \in y$ and $y \in x$ then, by part (i) of Lemma 2.16, $x \in x$ contradicting part (ii) of the Lemma, leaving $x = y$ as the only remaining possibility. □

2.3.8 Exponentiation and Subset Collection

A crucial construction in mathematics is the formation of function sets, that is if A, B are sets one forms the collection of all functions from A to B. There is no problem in talking about function spaces as classes when working in **ECST**. However, in general, if we want to treat this class as a set we need to augment **ECST** by Myhill's **Exponentiation** axiom. An important application of the latter will be made in showing that the class of constructive Cauchy reals actually form a set. For other notions of reals, as for example the constructive Dedekind reals, even the Exponentiation axiom is too weak (see [42, 43]) but with the aid of Subset Collection they can be shown to form a set, too.

The Subset Collection axiom of **CZF** is a rather complicated and unfamiliar principle. It implies Exponentiation. It can be rendered more memorable, though, by viewing it as a statement, dubbed *Fullness*, ascertaining the existence of large sets of multi-valued functions between two given sets.

Definition 2.18 For sets A, B let ${}^A B$ be the class of all functions with domain A and with range contained in B. Let $\mathsf{mv}({}^A B)$ be the class of all sets $R \subseteq A \times B$ satisfying $\forall u \in A\, \exists v \in B\, \langle u, v\rangle \in R$. A set C is said to be **full** in $\mathsf{mv}({}^A B)$ if $C \subseteq \mathsf{mv}({}^A B)$ and

$$\forall R \in \mathsf{mv}({}^A B)\, \exists S \in C\; S \subseteq R.$$

The collection $\mathsf{mv}({}^A B)$ will be called the class of **multi-valued functions** (or **multi-functions**) from the set A to the set B.

The additional axiom we consider is as follows.

Fullness For all sets A, B there exists a set C such that C is full in $\mathsf{mv}({}^A B)$.

Theorem 2.19 ([6, Theorem 5.1.2])
(i) $(\mathbf{ECST})$ *Subset Collection implies Fullness.*
(ii) $(\mathbf{ECST} +$ *Strong Collection*$)$ *Fullness implies Subset Collection.*
(iii) $(\mathbf{ECST})$ *Fullness implies Exponentiation.*

Proof (i) Suppose A, B are sets. Let $\phi(x, y, u)$ be the formula $y \in u \wedge \exists z \in B\,(y = \langle x, z\rangle)$. Using the relevant instance of Subset Collection and noticing that for all $R \in \mathsf{mv}({}^A B)$ we have

$$\forall x \in A\,\exists y \in A \times B\,\phi(x, y, R),$$

there exists a set C such that $\forall R \in \mathsf{mv}({}^A B)\,\exists S \in C\; S \subseteq R$.

For (ii), let A, B be sets. Pick a set C which is full in $\mathsf{mv}({}^A B)$. Assume $\forall x \in A \exists y \in B \phi(x, y, u)$. Define $\psi(x, w, u) \;:=\; \exists y \in B\,\big[w = \langle x, y\rangle \wedge \phi(x, y, u)\big]$. Then $\forall x \in A \exists w$
$\psi(x, w, u)$. Thus, by Strong Collection, there exists $v \subseteq A \times B$ such that

$$\forall x \in A\,\exists y \in B\,\big[\langle x, y\rangle \in v \wedge \phi(x, y, u)\big] \;\wedge\; \forall x \in A\,\forall y \in B\,\big[\langle x, y\rangle \in v \rightarrow \phi(x, y, u)\big].$$

As C is full, we find $w \in C$ with $w \subseteq v$. Consequently, we can infer that $\forall x \in A \exists y \in \mathsf{ran}(w)\phi(x, y, u)$ and $\forall y \in \mathsf{ran}(w) \exists x \in A\,\phi(x, y, u)$, where $\mathsf{ran}(w) := \{v \mid \exists z\,\langle z, v\rangle \in w\}$. Whence $D := \{\mathsf{ran}(w) : w \in C\}$ witnesses the truth of the instance of Subset Collection pertaining to ϕ. (iii) Let C be full in $\mathsf{mv}({}^A B)$. If now $f \in {}^A B$, then $\exists R \in C\; R \subseteq f$. But then $R = f$. Therefore ${}^A B = \{f \in C : f \text{ is a function}\}$. □

At this point, the reader might wonder whether, instead of the rather fickle Subset Collection axiom, one should just adopt an axiom asserting that $\mathsf{mv}({}^A B)$ is a set for all sets A, B. This, however, would lead to an impredicative theory much stronger than classical Zermelo set theory. The reason for this is the following.

Lemma 2.20 $\mathbf{ECST} + \forall A \forall B$ '$\mathsf{mv}({}^A B)$ *is a set' proves the Powerset axiom.*

Proof See [6, Proposition 5.1.6]. □

An important infinitary operation in set theory is the dependent product or function spaces construction.

Definition 2.21 Let I be a set and $(A_i)_{i \in I}$ be a family of classes over I. The **dependent product** of $(A_i)_{i \in I}$ is the class

$$\prod_{i \in I} A_i = \left\{ f \mid f : I \rightarrow \bigcup_{i \in I} A_i \;\wedge\; (\forall i \in I) f(i) \in A_i \right\}.$$

Proposition 2.22 $(\mathbf{ECST} + \text{Exponentiation})$ *If I is a set and $(A_i)_{i \in I}$ is a family of sets over I, then $\prod_{i \in I} A_i$ is a set.*

Proof We know that $\bigcup_{i\in I} A_i$ is a set by Lemma 2.5, and hence, by Exponentiation, $\{f \mid f : I \to \bigcup_{i\in I} A_i\}$ is a set. Thus, Bounded Separation ensures that $\prod_{i\in I} A_i$ is a set. □

2.3.9 The Cauchy Real Numbers, $\mathbb{R}_c$

A **rational number** is a pair of integers m/n, with $n \neq 0$. Two rational numbers m_1/n_1 and m_2/n_2 are equal if $m_1n_2 = m_2n_1$. The familiar operations and relations on the rational numbers remain unchanged in a constructive context.

Analysis begins with the real numbers. A **Cauchy real number** is presented by rational approximations, and may be identified with a sequence $x = (x_n)$ of rational numbers that is **regular**, in the sense that

$$|x_m - x_n| \leq \frac{1}{m} + \frac{1}{n}$$

for all positive integers m and n. Any Cauchy sequence of rational numbers can be considered to be a real number, but by requiring that the sequence $x = (x_n)$ be regular, we avoid having to provide an auxiliary sequence that tells how large n must be for x_n to approximate x to a specific accuracy.

Two Cauchy real numbers (x_n) and (y_n) are **equal**, notated $(x_n) \approx (y_n)$, if

$$|x_m - y_m| \leq \frac{2}{m}$$

for each positive integer m. Because x and y are regular, this is equivalent to demanding that $|x_n - y_n|$ be arbitrarily small for sufficiently large values of n. It therefore follows that equality on the reals is an **equivalence relation**.

We denote the collection of Cauchy real numbers by $\mathbb{R}^c$. Exponentiation rather than Subset Collection suffices for the following.

Theorem 2.23 (**ECST** + Exponentiation) *$\mathbb{R}^c$ is a set.*

Proof The Cauchy reals are a subclass of ${}^{\mathbb{N}}\mathbb{Q}$ which can be carved out of the latter set by Bounded Separation, whence a set. See also [5, Theorem 3.20]. □

2.3.10 The Constructive Dedekind Reals

Another important account of the reals was introduced by Dedekind. The constructive rendering of the notion of Dedekind real is more delicate than in a classical setting.

Definition 2.24 Let $S \subseteq \mathbb{Q}$. S is called a **(constructive) left cut** or **(constructive) Dedekind real** if the following conditions are satisfied:

(i) $\exists r(r \in S) \wedge \exists r'(r' \notin S)$ (*boundedness*),
(ii) $\forall r \in S\, \exists r' \in S\, (r < r')$ (*openness*),
(iii) $\forall rs \in \mathbb{Q}\, [r < s \rightarrow r \in S \vee s \notin S]$ (*locatedness*).

For $X \subseteq \mathbb{Q}$ define

$$X^{<} := \{u \in \mathbb{Q} \mid \exists r \in X\ u < r\}.$$

If S is a left cut it follows from openness and locatedness that $S = S^{<}$.

We use $\mathbb{R}_d$ to denote the class of Dedekind reals.

Lemma 2.25 *Let $r = (r_n)_{n\in\mathbb{N}}$ and $r' = (r'_n)_{n\in\mathbb{N}}$ be regular sequences of rationals. Define*

$$X_r := \left\{ s \in \mathbb{Q} \mid \exists n\, s < \max_{1\le m\le n} \left(r_m - \frac{1}{m} \right) \right\}.$$

We then have

(i) *X_r is a Dedekind real.*
(ii) *$X_r = X_{r'}$ if and only if $(r_n)_{n\in\mathbb{N}} \approx (r'_n)_{n\in\mathbb{N}}$.*
(iii) *$\mathbb{R}_c$ is a subfield of $\mathbb{R}_d$ via the mapping $(r_n)_{n\in\mathbb{N}}/\approx\ \mapsto\ X_r$.*

Proof See [5, Proposition 3.25]. □

With the help of Subset Collection it can be shown that Dedekind reals form a set.

Theorem 2.26 (**CZF**) *$\mathbb{R}_d$ is a set.*

Proof This is from [5, Theorem 3.24] or [6, Theorem 7.3.1].

Employing Subset Collection, let D be a set full in $\mathsf{mv}({}^{\mathbb{Q}}2)$, where $2 = \{0, 1\}$. For a Dedekind real X, let $X_0 := X$ and $X_1 := \mathbb{Q}\backslash X$. Put

$$R_X := \{\langle q, i\rangle \mid q \in X_i\}.$$

Then $R_X \in \mathsf{mv}({}^{\mathbb{Q}}2)$ on account of X's locatedness. So there there exists $S \in D$ such that $S \subseteq R_X$. However, on inspection this entails that $S = R_X$. As a result, $X \in \{\{q \mid \langle q, 0\rangle \in B\} \mid B \in D\}$, yielding that the Dedekind reals form a set. □

Indeed, Exponentiation is not enough to establish the latter result as was shown by Lubarsky and Rathjen. Let $\mathbf{CZF}_{Exp}$ be the version of **CZF** with Subset Collection replaced by Exponentiation.

Theorem 2.27 $\mathbf{CZF}_{Exp}$ *does not prove that $\mathbb{R}_d$ is a set.*

Proof See [42, 43]. □

On the other hand, in the presence of countable choice, AC_ω (see Definition 2.47), the two versions of reals are isomorphic.

Theorem 2.28 (**CZF** + AC_ω) $\mathbb{R}_c \cong \mathbb{R}_d$.

Proof See [5, Proposition 3.21]. □

2.4 The Development of Set Theory in CZF

Next, we revisit some of the standard constructions and concepts of classical set theory. Central notions include transitive closure, definition by transfinite recursion, ordinals, and inductively defined classes.

2.4.1 Transfinite Recursion

The technique of defining functions and classes by transfinite recursion pervades classical set theory. It carries over to **CZF** with little change. Axiomatically, this requires the Set Induction scheme, and thus is not available in **ECST**.

Theorem 2.29 (**CZF**) (Definition by $\in$-Recursion) *Let* $\vec{x} = x_1, \ldots, x_n$. *If* G *is a total* $(n+2)$*-ary class function, that is,*

$$\forall \vec{x} y z \exists ! u\, G(\vec{x}, y, z) = u,$$

then there is a total $(n+1)$*-ary class function* H *such that*

$$\forall \vec{x} y [H(\vec{x}, y) = G(\vec{x}, y, (H(\vec{x}, z) | z \in y))],$$

where $(H(\vec{x}, z)|z \in y) := \{\langle z, H(\vec{x}, z)\rangle : z \in y\}$.

In particular, for every set A, H *restricted to* A^{n+2} *is a set function.*

Proof The proof is similar to the classical case. For details see [6, Proposition 9.3.3]. □

A nice application of Theorem 2.29 uses the notion of transitivity.

Definition 2.30 A set A is said to be **transitive** if elements of elements of A are elements of A, in symbols: $\forall x \in A \forall y \in x\; y \in A$.

Given a set B, a set C is said to be the **transitive closure of** B if $B \subseteq C$, C is transitive, and whenever X is transitive set with $B \subseteq X$, then $C \subseteq X$.

Clearly, the transitive closure of a set, if it exists, is unique. If it exists, we denote the transitive closure of a set a by $\mathsf{TC}(a)$.

Lemma 2.31 (**CZF**) *For every set* a, $\mathsf{TC}(a)$ *exists. Moreover,*

$$\mathsf{TC}(a) = a \cup \bigcup \{\mathsf{TC}(z) \mid z \in a\}. \tag{2.1}$$

Proof This is a consequence of Theorem 2.29. □

There is a slight generalization of Theorem 2.29, whereby instead of recurring to the elements of a set, the function recurs to the elements of its transitive closure.

Proposition 2.32 (CZF) (Definition by TC-Recursion.) *If G is a total $(n+2)$-ary class function, that is,*

$$\forall \vec{x} yz \exists! u\, G(\vec{x}, y, z) = u,$$

then there is a total $(n+1)$-ary class function F such that

$$\forall \vec{x} y [F(\vec{x}, y) = G(\vec{x}, y, (F(\vec{x}, z) | z \in \mathsf{TC}(y)))],$$

where $(F(\vec{x}, z) | z \in y) := \{\langle z, F(\vec{x}, z)\rangle : z \in y\}$.

Proof See, for example, [5, Proposition 4.1] or [6, Proposition 9.3.2]. □

It has been stressed [6, Section 10.4] that the foregoing results use Set Induction in an essential way. Strong Collection also figures in their proofs, though the more restrictive Replacement also suffices.

2.4.2 Ordinals

Ordinals are an important concept in classical set theory. They provide a scale along which transfinite processes can be iterated. This can be done almost without change when transferring to the constructive context. A difference, though, occurs with regards to the linearity (or order trichotomy) of ordinals and the trichotomy classification of ordinals into 0, successor, and limit. This is no longer possible. Indeed, both trichotomies yield unwanted inconstructivities in that they are linked to forms of excluded middle and both imply Powerset on the basis of **CZF** (see, e.g., [6, Section 10.4]).

Definition 2.33 An **ordinal** α is a transitive set of transitive sets, that is, α and every element of α will be transitive.

Observe that an element of an ordinal is an ordinal as well.

In future, variables $\alpha, \beta, \gamma, \delta, \ldots$ will be assumed to range over ordinals. ON will denote the class of ordinals.

Every set can be assigned an ordinal rank.

Definition 2.34 The *rank* function, rank, is obtained as follows. For a set a, define

$$\mathrm{rank}(a) := \bigcup \{\mathrm{rank}(u) + 1 : \ u \in a\}.$$

This definition is justified by the previous Proposition 2.29, letting

$$G(y, z) := \bigcup \{u + 1 \mid \exists v\, \langle v, u\rangle \in z\},$$
$$F(y) := G(y, (F(\vec{x}, z) | z \in y))$$

since then $F(a) = \bigcup \{F(u) + 1 \mid u \in a\}$.

Note that the values of the rank function are always ordinals.

Ordinal arithmetic, namely, addition, multiplication, and exponentiation of ordinals, can be defined by transfinite recursion on ordinals, invoking Theorem 2.29, albeit case distinctions, as in the trichotomy classification of ordinals into 0, successor and limit, have to be circumvented (for details see [6, Exercise 9.7.8]).

2.4.3 Inductive Definitions of Classes

A very interesting aspect of **CZF** is that classes can be defined inductively, regardless of the complexity of the formula that defines the class. Here, in addition to Set Induction, the Strong Collection axiom comes into its own.

Definition 2.35

(i) We define an **inductive definition** to be any class of ordered pairs.
(ii) If Φ is an inductive definition and $(X, a) \in \Phi$ then we prefer to write

$$X/a \in \Phi$$

and call X/a an **(inference) step** of Φ, with set X of **premisses** and **conclusion** a.
(iii) We associate with an inductive definition Φ the operator Γ_Φ on classes that assigns to each class Y the class $\Gamma_\Phi(Y)$ of all conclusions a of inference steps X/a of Φ, with a set X of premisses that is a subset of Y; in other words (or rather symbols),

$$\Gamma_\Phi(Y) = \{a \mid \exists X\,[X \subseteq Y \wedge X/a \in \Phi]\}.$$

We define a class Y to be **Φ-closed** if $\Gamma_\Phi(Y) \subseteq Y$.
(iv) The class **inductively defined by** Φ, if it exists, is the smallest Φ-closed class and will be denoted by $I(\Phi)$.

Example 2.36

(i) $H(A)$ is the smallest class X such that for each set a that is an image of a set in A

$$a \subseteq X \Rightarrow a \in X.$$

Note that $H(A) = I(\Phi)$, where Φ is the class of all pairs (a, a) such that a is an image of a set in A.
(ii) If R is a subclass of $A \times A$ such that $R_a = \{x \mid xRa\}$ is a set for each $a \in A$ then $Wf(A, R)$ is the smallest subclass X of A such that

$$\forall a \in A\,[R_a \subseteq X \Rightarrow a \in X].$$

Note that $Wf(A, R) = I(\Phi)$, where Φ is the class of all pairs (R_a, a) such that $a \in A$.

(iii) If B_a is a set for each $a \in A$ then $W_{a \in A} B_a$ is the smallest class X such that

$$a \in A \;\&\; f : B_a \to X \;\Rightarrow\; (a, f) \in X.$$

Note that $W_{x \in A} B_a = I(\Phi)$ where Φ is the class of all pairs $(\mathrm{ran}(f), (a, f))$ such that $a \in A$ and $f : B_a \to V$.

2.4.4 Class Inductive Definition Theorem

Theorem 2.37 (CZF) (Class Inductive Definition Theorem) *For any inductive definition Φ there is a smallest Φ-closed class $I(\Phi)$.*

Proof This was proved in [3]. Intuitively, the proof involves the iteration of the class operator $\Gamma := \Gamma_\Phi$ until it closes up at its least fixed point, which turns out to be the required class $I(\Phi)$. As an inductive definition needs not be finitary; that is, it can have steps with infinitely many premisses, we will need transfinite iterations of Γ in general. But is this possible with proper classes?

Note that Γ is monotone; that is, for classes Y_1, Y_2

$$Y_1 \subseteq Y_2 \Rightarrow \Gamma(Y_1) \subseteq \Gamma(Y_2).$$

Call a class J of ordered pairs an **iteration class for** Φ if for each ordinal α,

$$J^\alpha = \Gamma(J^{\in\alpha}),$$

where $J^\alpha = \{x \mid (\alpha, x) \in J\}$ and $J^{\in\alpha} = \bigcup_{\beta\in\alpha} J^\beta$.

Auxilliary Lemma *Every inductive definition has an iteration class.*

Proof Call a set G of ordered pairs **good** if

$$(*) \qquad (\alpha, y) \in G \Rightarrow y \in \Gamma(G^{\in\alpha}),$$

where

$$G^{\in\alpha} = \{y' \mid \exists\beta \in \alpha\; (\beta, y') \in G\}.$$

Let $J = \bigcup\{G \mid G \text{ is good}\}$. We must show that for each α

$$J^\alpha = \Gamma(J^{\in\alpha}).$$

First, let $y \in J^\alpha$. Then $(\alpha, y) \in G$ for some good set G and hence by $(*)$, above, $y \in \Gamma(G^{\in\alpha})$. As $G^{\in\alpha} \subseteq J^{\in\alpha}$ it follows that $y \in \Gamma(J^{\in\alpha})$. Thus $J^\alpha \subseteq \Gamma(J^{\in\alpha})$. $\square$

For the converse inclusion let $y \in \Gamma(J^{\in\alpha})$. Then $Y/y \in \Phi$ for some set $Y \subseteq J^{\in\alpha}$. It follows that $\forall y' \in Y \exists \beta \in \alpha\ y' \in J^{\beta}$ so that

$$\forall y' \in Y \exists G\ [\ G \text{ is good and } y' \in G^{\in\alpha}].$$

By **Strong Collection**, there is a set Z of good sets such that

$$\forall y' \in Y\ \exists G \in Z\ y' \in G^{\in\alpha}.$$

Let $G = \{(\alpha, y)\} \cup \bigcup Z$. Then $\bigcup Z$ is good and, as $Y/y \in \Phi$ and $Y \subseteq G^{\in\alpha}$, G is good. As $(\alpha, y) \in G$ we get that $y \in J^{\alpha}$. Thus $\Gamma(J^{\in\alpha}) \subseteq J^{\alpha}$, finishing the proof of the auxilliary lemma. □

Proof of the theorem It only remains to show that

$$J^{\infty} = \bigcup_{\alpha \in V} J^{\alpha}$$

is the smallest Φ-closed class. To show that J^{∞} is Φ-closed let $Y/y \in \Phi$ for some set $Y \subseteq J^{\infty}$. Then $\forall y' \in Y \exists \beta\ y' \in J^{\beta}$. So, by Collection, there is an ordinal α such that

$$\forall y' \in Y \exists \beta \in \alpha\ y' \in J^{\beta}; \qquad \text{i.e., } Y \subseteq J^{\in\alpha}.$$

Hence $y \in \Gamma(J^{\in\alpha}) = J^{a} \subseteq J^{\infty}$. Thus J^{∞} is Φ-closed.

Now let I be a Φ-closed class. We show that $J^{\infty} \subseteq I$. It suffices to show that $J^{\alpha} \subseteq I$ for all α. We do this by Set Induction on α. So we may assume, as induction hypothesis, that $J^{\beta} \subseteq I$ for all $\beta \in \alpha$. It follows that $J^{\in\alpha} \subseteq I$ and hence

$$J^{\alpha} = \Gamma(J^{\in\alpha}) \subseteq \Gamma(I) \subseteq I,$$

the inclusions holding because Γ is monotone and I is Φ-closed. □

Theorem 2.37 guarantees that, for instance, the inductively defined classes of Definition 2.36, namely $H(A)$, $Wf(A, R)$, and $W_{x \in A} B_a$, exists in **CZF** (as definable classes). A further application of Theorem 2.37 will be made in Section 2.8 on realizability for set theory, where it is used to define the realizability structure of Definition 2.60 in the absence of Powerset.

It is also possible to show the existence of largest fixed points in **CZF**, although a little help from choice seems to be required, but Set Induction is not needed. The form of choice, RDC, dubbed relativized dependent choice, will be introduced in Definition 2.47. Recall that **CZF**$\restriction$ signifies **CZF** without Set Induction.

Theorem 2.38 (**CZF**$\restriction$ + RDC) *For every inductive definition Φ, there is a largest fixed point $I^*(\Phi)$. Indeed,*

$$I^*(\Phi) \ = \ \bigcup\{x \mid x \textit{ set and } x \subseteq \Gamma_{\Phi}(x)\}.$$

Proof See [6, Theorem 1.1.5]. □

Largest fixed points are especially useful if one wants to define an internal model of set theory with the Antifoundation axiom (see [4, 8, 53, 54]). The latter axiom enables one to model a plethora of circular phenomena in semantics, modal logics and computer science. A version of **CZF** without Set Induction, but with the Antifoundation axiom added, yields a very nice theory in which most of the mathematical results of [4, 8] have constructive proofs as was shown in [53, 54].

2.4.5 Bounded Inductive Definitions

Least fixed points of inductive definitions, more often than not, yield proper classes. However, there are special inductive definitions for which it is possible to show that their least fixed point indeed gives rise to a set.

Definition 2.39 A **set** B is a *bound* for an inductive definition Φ if, whenever $X/a \in \Phi$, then X is an image of a set $b \in B$; that is, there is a function from b onto X.

So Φ is said to be **bounded** if Φ has a bound and

$$\{y \mid X/y \in \Phi\}$$

is a set for all sets X.

Recall that the operator Γ_Φ associated with Φ is defined by

$$\Gamma_\Phi(Y) = \{y \mid \exists X\,(X/y \in \Phi\ \wedge\ X \subseteq Y)\}.$$

Theorem 2.40 (**CZF**) *If Φ is bounded, then $\Gamma_\Phi(Y)$ is a set for each set Y.*

Proof Let B be a bound for Φ. If $Y/y \in \Phi$ then for some $b \in B$ there is a surjective $f : b \to Y$. So if X is a set then

$$\Gamma_\Phi(X) = \bigcup_{f \in C} \{y \mid \mathrm{ran}(f)/y \in \Phi\},$$

where $C = \bigcup_{b \in B} {}^bX$. By Exponentiation and Replacement combined with Union, C is a set. As Φ is bounded, $\{y \mid ran(f)/y \in \Phi\}$ is always a set, so that, by Union–Replacement, $\Gamma_\Phi(X)$ is a set. □

It is, however, not possible to show that the inductive definitions of Example 2.36 are bounded. This will require the existence of sets bigger than those available on the basis of **CZF** alone. One needs a constructive analogue of the notion of regular cardinal. This will be addressed in the next section via the axiom REA.

The proof-theoretic strength of **CZF** + REA is considerably higher than that of **CZF** (see [33, 52]), however, Aczel's interpretation in type theory still validates REA (see [3]) on the strength of the W-type of Martin-Löf type theory.

2.5 Large Sets in CZF

The first large set axiom proposed in the context of **CZF** was the **Regular Extension axiom**, REA [3]. It can be used to show that the inductive definitions of Example 2.36 are bounded, and thus give rise to inductively defined sets as shown in Lemma 2.40.

Definition 2.41 A set C is said to be **regular** if it is transitive, inhabited (i.e., $\exists u\; u \in C$) and for any $u \in C$ and $R \in \mathsf{mv}({}^{u}C)$ there exists a set $v \in C$ such that

$$\forall x \in u\, \exists y \in v\; \langle x, y\rangle \in R \;\wedge\; \forall y \in v\, \exists x \in u\; \langle x, y\rangle \in R.$$

We write $\mathsf{Reg}(C)$ to express that C is regular.

REA is the principle

$$\forall x\, \exists y\; (x \subseteq y \;\wedge\; \mathsf{Reg}(y)).$$

With the help of REA it is possible to show that the inductively defined classes of Example 2.36 are sets.

Proposition 2.42 (**CZF** + REA) *Let A be a set and let $(B_a)_{a \in A}$ be a family of sets over A. Then*

$$W_{a \in A} B_a$$

is a set.

Proof Recall that $\mathrm{W}_{a \in A} B_a$ is the smallest class X such that

$$a \in A \;\&\; f : B_a \to X \;\Rightarrow\; (a, f) \in X.$$

Note that $\mathrm{W}_{x \in A} B_a = I(\Phi)$, where Φ is the class of all pairs $(\mathrm{ran}(f), (a, f))$ such that $a \in A$ and $f : B_a \to V$.

Now take a regular set C such that $A \in C$ and $(B_a)_{a \in A} \in C$. Note that C satisfies

$$a \in A \;\&\; f : B_a \to C \;\Rightarrow\; (a, f) \in C.$$

As a result, Φ defines the same inductive class as the modification Φ' consisting of all pairs $(\mathrm{ran}(f), (a, f))$ such that $a \in A$ and $f : B_a \to C$. Thus C is a bound for the inductive definition Φ' and Φ' is also bounded. So it follows from Theorem 2.40 that $I(\Phi) = I(\Phi')$ is a set. Similarly, the other inductive definitions of Example 2.36 can be seen to be bounded in the presence of REA. □

2.5.1 Other Notions of Large Set

The theory of large cardinals has been a central theme of modern classical set theory. In an intuitionistic context, rather than talking about large cardinals, one speaks about large sets. The guiding idea is that a large cardinal notion should give rise to a large set notion by discerning the properties of V_κ when a cardinal κ falls under this notion. In this vein, several large set notions have been studied in the context of intuitionistic **ZF**, including analogues of inaccessible, Mahlo, and weakly compact cardinals. Here we shall just briefly look at two notions of inaccessibility.

Definition 2.43

(i) A set I is said to be **weakly inaccessible** if I is a regular set and the structure $(I, \in_I)$ satisfies all the axioms of $\mathbf{CZF}\!\restriction$, where $\in_I$ is the restriction of the elementhood relation to $I \times I$.

(ii) A set I will be called **inaccessible** if I is weakly inaccessible and for all $x \in I$ there exists a regular set $y \in I$ such that $x \in y$.

If the background universe satisfies Set Induction, then every weakly inaccessible set I will automatically satisfy the latter axiom. So in the definition of weak inaccessibilty, instead of requiring the axioms of $\mathbf{CZF}\!\restriction$ to hold we could just have demanded this for all axioms of $\mathbf{CZF}$. As it turns out, however, it is also very interesting to study these notions in the context of $\mathbf{CZF}\!\restriction$, yielding rather surprising results.

A first observation is that the concept of weak inaccessibility has a nice 'algebraic' characterization, that is to say, one that doesn't hark back to the syntax of $\mathbf{CZF}\!\restriction$.

2.5.2 An 'Algebraic' Characterization of 'Inaccessibility'

Proposition 2.44 ($\mathbf{CZF}\!\restriction$) *A set I is weakly inaccessible* **iff** *I is a regular set such that the following are satisfied:*

(i) $\omega \in I$,
(ii) $\forall a \in I \bigcup a \in I$,
(iii) $\forall a \in I\, [a \textit{ inhabited } \Rightarrow \bigcap a \in I]$,
(iv) $\forall A, B \in I\, \exists C \in I \;\; C \textit{ is full in } \mathsf{mv}({}^{A}B)$.

Proof [5, Proposition 10.26]. □

In a classical context, the existence of weakly inaccessible sets is a strong statement that entails the existence of strongly inaccessible cardinals, thus surpassing **ZF** (see [18, Lemma 2.8]). This contrasts starkly with the following result due to Crosilla and Rathjen.

Theorem 2.45 *The theory*

$$\mathbf{CZF}\!\restriction + \forall x\, \exists I\, [x \in I \wedge I \textit{ is weakly inaccessible}]$$

is of the same strength as the theory $\mathbf{ATR}_0$ *of reverse mathematics. Its proof-theoretic ordinal is the Feferman–Schütte ordinal* Γ_0.

As a result, this theory is proof-theoretically weaker than $\mathbf{CZF}$.

Proof This result and more details can be found in [18]. For the theory of arithmetical transfinite recursion $\mathbf{ATR}_0$ see [67, I11]. □

One can also study the theory $\mathbf{CZF}+\mathsf{INACC}$, where INACC stands for $\forall x\, \exists I\, [x \in I \wedge I$ is inaccessible$]$. $\mathbf{CZF} + \mathsf{INACC}$ is interpretable in Martin-Löf type theory with Palmgren's superuniverse (cf. [50, 65]). As for strength, $\mathbf{CZF} + \mathsf{INACC}$ still amounts to a tiny fragment of classical set theory $\mathbf{ZF}$ and there are aeons between it and $\mathbf{ZF}$ with inaccessible cardinals. For more details on these topics see [62].

2.6 Axioms of Choice in Constructive Set Theory

Certain forms of the axiom of choice, AC, are routinely and often tacitly used in Bishop-style mathematics as well as Brouwer's intuitionistic mathematics. Myhill included countable and dependent choice in his set theory [48]. One also encounters views to the effect that contructivists not only take familiar choice principles to be true, but also believe them to be a consequence of their constructive reading of the quantifiers (as, for instance, in Martin-Löf type theory). Bishop remarked that:

> *This axiom is unique in its ability to trouble the conscience of the classical mathematician, but in fact it is not a real source of the unconstructivities of classical mathematics. A choice function exists in constructive mathematics, because a choice is implied by the very meaning of existence.* [11, p. 9]

However, the axiom of choice tends to induce a certain apprehensiveness in the constructivist's mind, too, as it is known to lead to undesirable forms of excluded middle or omniscience principles. Bishop was well aware of this. How, then, can the impression that the nature of AC is given to erratic changeableness be explained? It has to do with what constitutes a mathematical object. According to some tenets, the starting point of mathematics can be found in certain symbolic or mental objects that are viewed as representations of mathematical objects. The mathematical objects arise, as it were, from the intensional 'soup' of symbolic representations by making various identifications, namely by imposing an equivalence relation on a collection of representations. An example is the case of Cauchy sequences of rationals that give rise to the real numbers. Assuming that $\forall x \in A\, \exists y \in B\, R(x, y)$ holds, the intensional version of AC asserts that there is a map $f : A \to B$ such that

$\forall x \in A\, R(x, f(x))$. In Bishop's mathematics, every set A comes equipped with its own identity relation $=_A$, an equivalence relation germane to the set A. The extensional version of AC for A would require that f additionally satisfied for all $x, u \in A$ that $f(x) =_B f(u)$ ought to follow from $x =_A u$, rendering f a proper extensional function with regard to the relations $=_A$ and $=_B$. Whereas intensional AC is reconcilable with the constructive stance, it is extensional AC that leads to unwanted instances of excluded middle.

Turning to the context of set theory, a function f has to be extensional in the sense that if both sets x, y are in the domain of f and x and y have the same elements, then $f(x)$ and $f(y)$ have the same elements, owing to the axiom of Extensionality. As a result, one can expect that adding the general axiom of choice to $\mathbf{CZF}$ would 'destroy' $\mathbf{CZF}$. This was observed by Diaconescu [21] and Goodman and Myhill [32].

Proposition 2.46

(i) *For bounded* ϕ, $\mathbf{ECST} + \mathsf{AC}$ *proves* $\phi \vee \neg\phi$.

(ii) $\mathbf{CZF} + \mathbf{AC} \vdash$ Powerset.

Proof Point (i) follows from [21] and [32]; however, we will use a slightly different proof from [56, Proposition 3.2] couched in terms of equivalence relations and classes.

Let ϕ be an arbitrary bounded formula. Define an equivalence relation $\sim_\phi$ on 2 by

$$\begin{aligned} a \sim_\phi b &:\Leftrightarrow a = b \vee \phi \\ [a]_{\sim_\phi} &:= \{b \in 2 : a \sim_\phi b\} \\ 2/_{\sim_\phi} &:= \{[0]_{\sim_\phi}, [1]_{\sim_\phi}\}. \end{aligned}$$

Note that $[0]_{\sim_\phi}$ and $[1]_{\sim_\phi}$ are sets by full Separation and thus $2/_{\sim_\phi}$ is a set, too. One easily verifies that $\sim_\phi$ is an equivalence relation.

We have

$$\forall z \in 2/_{\sim_\phi}\, \exists k \in 2\, (k \in z).$$

Using AC, there is a choice function f defined on $2/_{\sim_\phi}$ such that

$$\forall z \in 2/_{\sim_\phi}\, [f(z) \in 2 \wedge f(z) \in z],$$

in particular, $f([0]_{\sim_\phi}) \in [0]_{\sim_\phi}$ and $f([1]_{\sim_\phi}) \in [1]_{\sim_\phi}$. Next, we are going to exploit the important fact

$$\forall n, m \in 2\, (n = m \vee n \neq m). \tag{2.2}$$

As $\forall z \in 2/_{\sim_\phi}\, [f(z) \in 2]$, we obtain

$$f([0]_{\sim_\phi}) = f([1]_{\sim_\phi}) \vee f([0]_{\sim_\phi}) \neq f([1]_{\sim_\phi})$$

by (2.2). If $f([0]_{\sim_\phi}) = f([1]_{\sim_\phi})$, then $0 \sim_\phi 1$ and hence ϕ holds. So assume $f([0]_{\sim_\phi}) \neq f([1]_{\sim_\phi})$. As ϕ would imply $[0]_{\sim_\phi} = [1]_{\sim_\phi}$ (this requires Extensionality) and thus $f([0]_{\sim_\phi}) = f([1]_{\sim_\phi})$, we must have $\neg\phi$. Consequently, $\phi \vee \neg\phi$.

Point (ii) follows from (i) because with the help of excluded middle for bounded formulas, the powerset of a set A is in 1-1 correspondence with the function space ${}^{A}2$. □

If one wants to add a restricted form AC^r of AC to **CZF** without changing its fundamental nature, one can take guidance from answers to the following questions.

(i) Does the canonical interpretation of **CZF** in Martin-Löf type theory validate AC^r?
(ii) Is there an inner 'model' of **CZF** + AC^r definable within **CZF**?
(iii) Does **CZF** + AC^r remain conservative over **CZF** with respect to arithmetic theorems?

A positive answer to the first question would perhaps yield the most foundational reason to adopt AC^r whereas a positive answer to (iii) yields an instrumentalist reason for using it, very much in a Hilbertian spirit in that AC^r posits the existence of ideal objects, that is, choice functions, to further the deduction of concrete statements.

In the remainder of this section, various AC^r will be considered. In point of fact, for all them the foregoing questions will be answered positively.

Definition 2.47 The weakest constructive choice principle we shall consider is the **Axiom of Countable Choice**, AC_ω, namely, whenever F is a function with domain ω such that $\forall i \in \omega\, \exists y \in F(i)$, then there exists a function f with domain ω such that $\forall i \in \omega\, f(i) \in F(i)$.

Let xRy stand for $\langle x, y\rangle \in R$. A mathematically very useful axiom to have in set theory is the **Dependent Choices Axiom**, DC, that is, for all sets a and (set) relations $R \subseteq a \times a$, whenever

$$(\forall x \in a)\,(\exists y \in a)\, xRy$$

and $b_0 \in a$, then there exists a function $f : \omega \to a$ such that $f(0) = b_0$ and

$$(\forall n \in \omega)\, f(n) R f(n+1).$$

Even more useful in constructive set theory is the *Relativized Dependent Choices Axiom*, RDC, also included by Myhill [48]. It asserts that, for arbitrary formulae ϕ and ψ, whenever

$$\forall x\big[\phi(x) \to \exists y\big(\phi(y) \wedge \psi(x, y)\big)\big]$$

and $\phi(b_0)$, then there exists a function f with domain ω such that $f(0) = b_0$ and

$$(\forall n \in \omega)\big[\phi(f(n)) \wedge \psi(f(n), f(n+1))\big].$$

Here are some well-known relationships.

Proposition 2.48 (**ECST**)

(i) DC *implies* AC_ω.
(ii) RDC *implies* DC.

Proof (i) If z is an ordered pair $\langle x, y\rangle$ let $1^{st}(z)$ denote x and let $2^{nd}(z)$ denote y.
Suppose F is a function with domain ω such that $\forall i \in \omega\, \exists x \in F(i)$. Let $A = \{\langle i, u\rangle \mid i \in \omega \wedge u \in F(i)\}$. A is a set by Union, Cartesian Product, and restricted Separation. We then have

$$\forall x \in A\, \exists y \in A\; xRy,$$

where $R = \{\langle x, y\rangle \in A \times A \mid 1^{st}(y) = 1^{st}(x) + 1\}$. Pick $x_0 \in F(0)$ and let $a_0 = \langle 0, x_0\rangle$. Using DC there exists a function $g : \omega \to A$ satisfying $g(0) = a_0$ and

$$\forall i \in \omega\, \big[g(i) \in A \wedge 1^{st}(g(i+1)) = 1^{st}(g(i)) + 1\big].$$

Letting f be defined on ω by $f(i) = 2^{nd}(g(i))$ one gets $\forall i \in \omega\; f(i) \in F(i)$.
Point (ii) is obvious. □

AC_ω does not imply DC, not even on the basis of **ZF**.

Proposition 2.49 **ZF** + AC_ω *does not prove* DC.

Proof This was shown by Jensen [38]. □

An interesting consequence of RDC, not implied by DC, is the following.

Proposition 2.50 ($\mathbf{CZF}^-$ + RDC) *Suppose* $\forall x \exists y \phi(x, y)$. *Then for every set* d *there exists a transitive set* A *such that* $d \in A$ *and*

$$\forall x \in A\, \exists y \in A\, \phi(x, y).$$

Moreover, for every set d *there exists a transitive set* A *and a function* $f : \omega \to A$ *such that* $f(0) = d$ *and* $\forall n \in \omega\, \phi\big(f(n), f(n+1)\big)$.

Proof The assumption yields that $\forall x \in b\, \exists y \phi(x, y)$ holds for every set b. Thus, by Collection and the existence of the transitive closure of a set, we get

$$\forall b\, \exists c\, [\theta(b, c) \wedge Tran(c)],$$

where $Tran(c)$ means that c is transitive and $\theta(b, c)$ is the formula

$\forall x \in b\, \exists y \in c\, \phi(x, y)$. Let B be a transitive set containing d. Employing RDC, there exists a function g with domain ω such that $g(0) = B$ and

$$\forall n \in \omega\, \theta(g(n), g(n+1)).$$

Obviously $A = \bigcup_{n\in\omega} g(n)$ satisfies our requirements.

The existence of the function f follows from the latter since RDC entails DC. □

2.6.1 The Presentation Axiom

The *Presentation axiom*, PAx, is an example of a choice principle which is validated upon interpretation in type theory. In category theory, it is also known as the *existence of enough projective sets*, EPsets (cf. [15]).

A set P is *projective* if, for any P-indexed family $(X_a)_{a\in P}$ of inhabited sets X_a, there exists a function f with domain P such that, for all $a \in P$, $f(a) \in X_a$.

PAx (or EPsets) is the statement that every set is the surjective image of a projective set.

Alternatively, projective sets have also been called *bases*, and we shall follow that usage henceforth. In this terminology, AC_ω expresses that ω is a base whereas AC amounts to saying that every set is a base.

Proposition 2.51 (**CZF**) PAx *implies* DC.

Proof See [1] or [15, Theorem 6.2]. □

The preceding implication cannot be reversed, not even on the basis of **ZF**.

Theorem 2.52 **ZF** + DC *does not prove* PAx.

Proof See [56, Proposition 5.2]. □

2.6.2 'Maximal' Choice Principles

Through the interpretation of constructive set theory in type theory, Aczel [2] found several new forms of AC which have not been considered anywhere else in the literature. These choice principles are also in a certain sense 'maximal'. It was shown in [64] that they furnish a characterization of all set-theoretic principles validated in type theory.

Lemma 2.53 (**CZF**) *There exists a smallest* $\Pi\Sigma\mathsf{I}$*-closed class, that is, a smallest class* Y *such that the following hold:*

(i) $n \in \mathsf{Y}$ *for all* $n \in \omega$;
(ii) $\omega \in \mathsf{Y}$;

(iii) $\prod_{x\in A} B_x \in \mathsf{Y}$ *and* $\sum_{x\in A} B_x \in \mathsf{Y}$ *whenever* $A \in \mathsf{Y}$ *and* $B_x \in \mathsf{Y}$ *for all* $x \in A$*;*
(iv) *if* $A \in \mathsf{Y}$ *and* $a, b \in A$*, then* $\mathsf{I}(a,b) \in \mathsf{Y}$*, where* $\mathsf{I}(a,b) = \{z \in 1 \mid a = b\}$.

Proof This is mainly a consequence of the class inductive definition Theorem 2.37. For more details see [3] and [64, Lemma 1.2]. □

Similarly one has the following.

Lemma 2.54 (**CZF** + REA) *There exists a smallest* $\Pi\Sigma\mathsf{IW}$*-closed class, that is, a smallest class* Y *such that* Y *is* $\Pi\Sigma\mathsf{I}$*-closed and also contains* $W_{x\in A}B_a$ *whenever* $A \in \mathsf{Y}$ *and* $B_x \in \mathsf{Y}$ *for all* $x \in A$.

Definition 2.55 The $\Pi\Sigma\mathsf{I}$-generated sets ($\Pi\Sigma\mathsf{IW}$-generated sets) are the sets in the smallest $\Pi\Sigma\mathsf{I}$-closed class ($\Pi\Sigma\mathsf{IW}$-closed class).

So $\Pi\Sigma\mathsf{I}-\mathsf{AC}$ ($\Pi\Sigma\mathsf{IW}-\mathsf{AC}$) is the statement that every $\Pi\Sigma\mathsf{I}$-generated ($\Pi\Sigma\mathsf{IW}$-generated) set is a base.

Note that $\Pi\Sigma\mathsf{I}-\mathsf{AC}$ has as its immediate consequence that all the sets in the finite-type hierarchy above ω (which Gödel used in his functional interpretation) are bases.

Theorem 2.56

(i) *The axioms* $\Pi\Sigma\mathsf{I}-\mathsf{AC}$ *and* $\Pi\Sigma\mathsf{IW}-\mathsf{AC}$ *are both validated by the standard interpretation in Martin-Löf type theory.*
(ii) **CZF** + RDC + $\Pi\Sigma\mathsf{I}-\mathsf{AC}$ + PAx *is conservative over* **CZF** *for arithmetical statements.*
(iii) **CZF** + REA + RDC + $\Pi\Sigma\mathsf{IW}-\mathsf{AC}$ + PAx *is conservative over* **CZF** + REA *for arithmetical statements.*

Proof Point (i) is shown in [2]. For (i) and (ii) see [62]. □

It is noteworthy that the class Y gives rise to an interpretation of type theory in **CZF**. As one has AC in type theory, there is this nice harmony between AC in type theory and $\Pi\Sigma\mathsf{I}-\mathsf{AC}$ in set theory with the latter stating AC exactly for those sets that can be generated by type-theoretic operations.

There are even stronger forms of choice validated by the interpretation in type theory if one works in the extension of **CZF** with the regular extension axiom, a context in which one considers the axiom $\Pi\Sigma\mathsf{IW}-\mathsf{AC}$. The latter asserts that all sets in the wider class of $\Pi\Sigma\mathsf{IW}$-generated sets are bases. Here W refers to the operation defined in Example 2.36(iii), which is the set-theoretic rendering of the W-type of Martin-Löf type theory. For details see [3, 5, 6, 57, 62, 64].

2.7 CZF and the Limited Principle of Omniscience

The law of excluded middle, LEM, was designated a principle of omniscience by Bishop. **CZF** + LEM equates to classical Zermelo–Fraenkel set theory. And even if one adds the law of excluded middle to **CZF** just for bounded formulas one gets a veritable explosion in strength, arriving at a theory stronger than Zermelo set theory (see [60]). There is, however, another classical principle discussed by Bishop, which one might consider adding to **CZF**.

> *This principle* [LEM], *which we shall call the principle of omniscience, lies at the root of most of the unconstructivities of classical mathematics. This is already true of the principle of omniscience in its simplest form: if* $\{n_k\}$ *is a sequence of integers, then either* $n_k = 0$ *for some* k *or* $n_k \neq 0$ *for all* k*. We shall call this the* limited principle of omniscience.
>
> *Theorem after theorem of classical mathematics depends in an essential way on the limited principle of omniscience, and is therefore not constructively valid.*
>
> *Some instances of this are the theorem that a continuous real-valued function on a closed bounded interval attains its maximum, the fixedpoint theorem for a continuous map of a closed cell into itself, the ergodic theorem, and the Hahn–Banach theorem.* [11, p. 9]

Surprisingly, the limited principle of omniscience, LPO, has no effect on the proof-theoretic strength of **CZF**.[13]

Theorem 2.57 **CZF** *and* **CZF** + LPO + RDC *prove the same* Π^0_2*-theorems of arithmetic.*

Proof This was shown in [61]. □

The syntactic class Π^0_2 includes many famous theorems and conjectures such as Fermat's last theorem (actually Wiles' theorem), the twin prime conjecture, and the Riemann hypothesis.

If one wants to engineer a theory of low proof-theoretic strength but with great expressive power, that is, in which it is easy to formalize 'ordinary' mathematics, **CZF** + LPO + RDC seems to be the ideal candidate as one can accomodate most of the classical proofs in it with little change. Nick Weaver has proposed a semi-intuitionistic theory **CM** of third-order arithmetic for axiomatizing what he calls mathematical conceptualism.

> *The philosophical approach we adopt, mathematical conceptualism, is a refinement of the predicativist philosophy of Poincaré and Russell. The basic idea is that we accept as legitimate only those structures that can be*

[13] Friedman [27] contains similar results.

> *constructed, but we allow constructions of transfinite length. What makes this "conceptual" is that we are concerned not only with those constructions that we can actually physically carry out, but more broadly with all those that are conceivable (perhaps supposing our universe had different properties than it does).* [70]

In [70] it is shown that it is unexpectedly easy to formalize a great deal of modern functional analysis in $\mathbf{CM}$. The interesting connection between $\mathbf{CZF}+\mathsf{LPO}+\mathsf{RDC}$ and $\mathbf{CM}$ is the following.

Theorem 2.58 $\mathbf{CM}$ *can be interpreted in* $\mathbf{CZF}+\mathsf{LPO}+\mathsf{RDC}$.

Proof See [51]. □

A fundamental difference between $\mathbf{CM}$ and $\mathbf{CZF}+\mathsf{LPO}+\mathsf{RDC}$, though, concerns the (formally) uncountable. In the conceptualist's framework $\mathbf{CM}$, uncountable structures (e.g., the collection of all reals) can only be partially realized as potential unfinished structures whereas in $\mathbf{CZF}$ the structure of the reals and function spaces form sets. This leads to the intriguing thought that Hilbert's program (originally conceived to give a finististic justification of the infinite) can be successfully carried out when transferred to the countable/uncountable hiatus. In the framework of $\mathbf{CZF}+\mathsf{LPO}+\mathsf{RDC}$, several uncountable structures exist in their entirety (i.e., as sets), while at the same time the latter theory can be reduced to intuitionistic Kripke–Platek set theory, which is a proof-theoretically weak theory compatible with the axiom of countability, namely that every set is countable. Hence the uncountable sets of $\mathbf{CZF}$ amount to an evocative idealization in Hilbert's sense that can be dispensed with in proofs of concrete Π^0_2 statements. Of course, a crucial tool to achieve this reduction, which Hilbert, in all likehood, didn't presage, is the employment of (semi)-intuitionistic systems; but I am sure he would have appreciated it.

2.8 Models of CZF and Axiomatic Freedom

Constructivism in mathematics, rather than being a single movement, seems to split into various 'schools', characterized by differing philosophical approaches. Bishop was suspicious of some of these philosophical rationalizations and adverse to using principles in constructive mathematics that engender contradictions in classical mathematics. However, these non-classical philosophies can be very interesting from a neutral and even classical point of view when one adopts an axiomatic approach. As it turns out, they give rise to structures that can model phenomena which are impossible to account for in a classical setting, precisely owing to their internal logic being intuitionistic. David McCarty, I believe, coined the expression *axiomatic freedom* in [46] for this phenomenon.

One of the great discoveries made possible by modern logic is the multitude of models 'out there'. Turning to **CZF**, this means that internally in **CZF** one can show, for example, the existence of models that validate a Brouwerian world or a Markovian world and a plethora of other worlds. There are many different techniques for finding models of **CZF**, for instance, realizability, forcing, sheaf, and topological models; however, this section will just focus on realizability models.

Realizability semantics for intuitionistic theories were first proposed by Kleene in 1945 [39]. Inspired by Kreisel and Troelstra's [41] definition of realizability for higher-order Heyting arithmetic, realizability was first applied to systems of set theory by Myhill [49] and Friedman [28]. More recently, realizability models of set theory were investigated by Beeson [9, 10] (for non-extensional set theories) and McCarty [46] (directly for extensional set theories). Reference [46] is concerned with realizability for intuitionistic Zermelo–Fraenkel set theory, **IZF**, and employs transfinite iterations of the Powerset operation through all the ordinals in defining the realizability (class) structure $\mathrm{V}(\mathcal{A})$ over any applicative structure $\mathcal{A}$. Moreover, in addition to the Powerset axiom the approach in [46] also avails itself of unfettered separation axioms. At first blush, this seems to render the approach unworkable for **CZF** with its lack of Powerset. However, these obstacles can be overcome.

Combinatory algebras are the brainchild of Moses Schönfinkel [66] who presented his ideas in Göttingen in 1920. The quest for an optimization of this setup, singling out a minimal set of axioms, engendered much work and writings from 1929 onwards, notably by H. B. Curry [19, 20], under the heading of *combinatory logic*. Curiously, a very natural generalization of Schönfinkel's structures, where the application operation is not required to be always defined, was axiomatically characterized only in 1975 by Solomon Feferman in the shape of the most basic axioms of his theory T_0 of explicit mathematics [23] and in [24, p. 70]. Feferman called these structures *applicative structures*.

Definition 2.59 In order to introduce the notion of a **partial combinatory algebra**, we shall start with a formal theory, $\mathbf{PCA}^*$. $\mathbf{PCA}^*$ is formulated in the logic of partial terms, wherein term and formula formation proceed as in ordinary predicate logic with equality, except there is one more rule: if t is a term, then $t\downarrow$ is an atomic formula. Intutively, it is read as 't is defined', or 't denotes'. In this logic variables and constants denote, whereas complex terms may fail to denote. For this reason, the equality axioms and quantifier rules have to be modified in the obvious way (see [10, VI1] or [68, Section 2.2] for details). To state the axioms of $\mathbf{PCA}^*$, we introduce the very helpful abbreviation $t \simeq s$ for $(t\downarrow \vee\, s\downarrow \rightarrow s = t)$.

$\mathbf{PCA}^*$ has five constants $\mathsf{k}, \mathsf{s}, \mathsf{p}, \mathsf{p}_0, \mathsf{p}_1$ and a binary function symbol $\cdot$ which we write in infix notation $s \cdot t$. The symbol $\cdot$ will however mostly be omitted, so we write st for $s \cdot t$. Moreover, we adopt a the convention of 'association to the left'. Hence exy means $(e \cdot x) \cdot y$.

With these conventions in place, the (non-logical) axioms of $\mathbf{PCA}^*$ can be stated:

(i) $x \downarrow$ for every variable x; $\mathsf{k} \downarrow \wedge \mathsf{s} \downarrow \wedge \mathsf{p} \downarrow \wedge \mathsf{p}_0 \downarrow \wedge \mathsf{p}_1 \downarrow$.
(ii) $\mathsf{k}xy \simeq x$.
(iii) $\mathsf{s}xyz \simeq xz(yz)$ and $\mathsf{s}xy \downarrow$.
(iv) $\mathsf{p}xy \downarrow \wedge \mathsf{p}_0(\mathsf{p}xy) \simeq x \wedge \mathsf{p}_1(\mathsf{p}xy) \simeq y$.

A partial combinatory algebra (pca) is structure that models the axioms of $\mathbf{PCA}^*$.

The combinators k and s are due to Schönfinkel [66] while the axiomatic treatment, although formulated just in the total case, is due to Curry [20]. For more information on partial combinatory algebras see [10, 23, 25, 69].

Definition 2.60 For a PCA $\mathcal{A}$, let $|\mathcal{A}|$ be its domain. We define classes

$$\mathrm{V}(\mathcal{A})_\alpha = \bigcup_{\beta \in \alpha} \mathcal{P}(|\mathcal{A}| \times \mathrm{V}(\mathcal{A})_\beta), \tag{2.3}$$

$$\mathrm{V}(\mathcal{A}) = \bigcup_{\alpha} \mathrm{V}(\mathcal{A})_\alpha. \tag{2.4}$$

As the power set operation is not available in $\mathbf{CZF}$ it is not clear whether the universe $\mathrm{V}(\mathcal{A})$ can be formalized in $\mathbf{CZF}$. However, this type of definition is an example of defining classes inductively, as accounted for by Theorem 2.37 (see [58, Lemma 3.4]).

We now proceed to define a notion of extensional realizability over $\mathrm{V}(\mathcal{A})$, that is, $e \Vdash_{\mathcal{A}} \phi$ for $e \in |\mathcal{A}|$ and sentences ϕ with parameters in $\mathrm{V}(\mathcal{A})$. For $e \in |\mathcal{A}|$ we shall write $(e)_0$ and $(e)_1$ rather than $\mathsf{p}_0 e$ and $\mathsf{p}_1 e$, respectively. An expression $ef \Vdash_{\mathcal{A}} \phi$ is to be read as $\exists u \in |\mathcal{A}| \, (ef \simeq u \wedge u \Vdash_{\mathcal{A}} \phi)$.

Definition 2.61 Bounded quantifiers will be treated as quantifiers in their own right, that is, bounded and unbounded quantifiers are treated as syntactically different kinds of quantifiers. Let $a, b \in \mathrm{V}(\mathcal{A})$ and $e \in |\mathcal{A}|$.

$$\begin{aligned}
e \Vdash_{\mathcal{A}} a \in b \ &\text{ iff } \ \exists c \left[\langle (e)_0, c\rangle \in b \wedge (e)_1 \Vdash_{\mathcal{A}} a = c\right] \\
e \Vdash_{\mathcal{A}} a = b \ &\text{ iff } \ \forall f, d \left[\left(\langle f, d\rangle \in a \rightarrow (e)_0 f \Vdash_{\mathcal{A}} d \in b\right)\right. \\
&\qquad\qquad \left.\wedge \ \left(\langle f, d\rangle \in b \rightarrow (e)_1 f \Vdash_{\mathcal{A}} d \in a\right)\right] \\
e \Vdash_{\mathcal{A}} \phi \wedge \psi \ &\text{ iff } \ (e)_0 \Vdash_{\mathcal{A}} \phi \wedge (e)_1 \Vdash_{\mathcal{A}} \psi \\
e \Vdash_{\mathcal{A}} \phi \vee \psi \ &\text{ iff } \ \left[(e)_0 = \mathbf{0} \wedge (e)_1 \Vdash_{\mathcal{A}} \phi\right] \vee \left[(e)_0 = \mathbf{1} \wedge (e)_1 \Vdash_{\mathcal{A}} \psi\right]
\end{aligned}$$

$$\begin{aligned}
e \Vdash_{\mathcal{A}} \neg\phi \quad & \text{iff } \forall f \in |\mathcal{A}|\ \neg f \Vdash_{\mathcal{A}} \phi \\
e \Vdash_{\mathcal{A}} \phi \to \psi \ & \text{iff } \forall f \in |\mathcal{A}|\ \big[f \Vdash_{\mathcal{A}} \phi\ \to\ ef \Vdash_{\mathcal{A}} \psi\big] \\
e \Vdash_{\mathcal{A}} \forall x \in a\, \phi(x) \ & \text{iff } \forall \langle f, c\rangle \in a\ ef \Vdash_{\mathcal{A}} \phi(c) \\
e \Vdash_{\mathcal{A}} \exists x \in a\, \phi(x) \ & \text{iff } \exists c \big(\langle (e)_0, c\rangle \in a\ \wedge\ (e)_1 \Vdash_{\mathcal{A}} \phi(c)\big) \\
e \Vdash_{\mathcal{A}} \forall x \phi(x) \quad & \text{iff } \forall c \in \mathrm{V}(\mathcal{A})\ e \Vdash_{\mathcal{A}} \phi(c) \\
e \Vdash_{\mathcal{A}} \exists x \phi(x) \quad & \text{iff } \exists c \in \mathrm{V}(\mathcal{A})\ e \Vdash_{\mathcal{A}} \phi(x)
\end{aligned}$$

Note that this notion of realizability is strikingly different from Kleene's 1945 realizability in the case of unbounded quantifiers (see [39] and in this volume [63, Section 1.6]). Realizing a universal statement means that the realizer is realizing all instances generically.

Theorem 2.62

(i) *For every theorem θ of* **CZF**, *there exists a closed application term t such that*

$$\mathbf{CZF} \vdash\ \forall \mathcal{A}\, [\mathcal{A}\, pca\ \to t^{\mathcal{A}} \downarrow\ \wedge t^{\mathcal{A}} \Vdash_{\mathcal{A}} \theta].$$

(ii) *For every theorem θ of* $\mathbf{CZF} + \mathsf{REA}$, *there exists a closed term t of* $\mathbf{PCA}^*$ *such that*

$$\mathbf{CZF} + \mathsf{REA} \vdash\ \forall \mathcal{A}\, [\mathcal{A}\, pca\ \to t^{\mathcal{A}} \downarrow\ \wedge t^{\mathcal{A}} \Vdash_{\mathcal{A}} \theta].$$

(iii) *If T stands for any of the theories* **CZF** *or* $\mathbf{CZF} + \mathsf{REA}$ *and Φ stands for any combination of the choice axioms* AC_ω, DC, PAx, *and* RDC, *then, for every theorem θ of $T + \Phi$, there exists a closed term t of* $\mathbf{PCA}^*$ *such that*

$$T + \Phi \vdash\ \forall \mathcal{A}\, [\mathcal{A}\, pcs\ \to t^{\mathcal{A}} \downarrow\ \wedge t^{\mathcal{A}} \Vdash_{\mathcal{A}} \theta].$$

Proof Point (i) and (ii) follow from [58] while (iii) is shown in [22]. With **IZF** in **CZF**'s stead, (i) is due to McCarty [46]. □

Theorem 2.62 exhibits those principles of the background universe that always transfer to the realizability model regardless of which pca one takes. Moreover, the realizing term t is independent of the pca (of course, its interpretation depends on a given pca). Other principles which are not valid in the background may hold in $\mathrm{V}(\mathcal{A})$, owing to combinatorial and computational properties germane to the pca. Here we will just look at Kleene's first, $\mathcal{K}_1$, and second algebra, $\mathcal{K}_2$. In $\mathcal{K}_1$, the underlying set is $\mathbb{N}$ and the application is a Turing machine application, that is, $e \cdot n$ means $\{e\}(n)$. For $\mathcal{K}_2$, the underlying set is Baire space $\mathbb{N}^{\mathbb{N}}$ and the application, $f \bullet g$, is continuous function application defined as follows.

Definition 2.63 *For $f, g \in \mathbb{N}^{\mathbb{N}}$ let*

$$f(g) = x \text{ stand for } \exists y \left[f(\bar{g}(y)) = x + 1 \wedge \forall y' < y\, f(\bar{g}(y')) = 0\right],$$
$$f \bullet g = h \text{ stand for } \forall x[\lambda n.f(\langle x\rangle * \langle n\rangle)(g) = h(x)],$$

where $\langle\rangle = 0$, $\bar{f}(0) = \langle\rangle$, and $\bar{f}(k+1) = \langle f(0), \ldots, f(n)\rangle$ with $\langle\ldots\rangle$ being some injective primitive recursive tuple coding, and $\langle\rangle * n = \langle n\rangle$, $\langle s_0, \ldots s_k\rangle * n = \langle s_0, \ldots, s_k, n\rangle$.

Note that $\bullet$ is merely a partial operation on $\mathbb{N}^{\mathbb{N}}$.

Let's write $\mathrm{V}(\mathcal{A}) \models \phi$ if there exists $e \in \mathrm{V}(\mathcal{A})$ such that $e \Vdash_{\mathcal{A}} \phi$. Kleene's pcas give rise to the following 'worlds'.

Theorem 2.64

(i) $V(\mathcal{K}_1) \models$ *Markov's world.*

(ii) $V(\mathcal{K}_1) \models$ *Brouwer's world.*

Proof For (i) see McCarty's [46] and [58]. For (ii) see [55]. □

Admittedly, Theorem 2.64 is stated in a rather sloppy manner. By Markov's world, I mean that in this world all functions from $\mathbb{N}$ to $\mathbb{N}$ are computable (dubbed *Church's thesis* in this context) and also that Markov's principle obtains, providing it holds in the background set theory. Brouwer's world refers to a mathematical world in which Brouwer's continuity principles hold, entailing that all functions from $\mathbb{R}$ to $\mathbb{R}$ are continuous. Moreover, it also means that Brouwerian principles such as bar induction and the fan theorem hold if they are valid in the background set theory. For more details see [68] and [55].

Further very interesting realizability models $\mathrm{V}(\mathcal{A})$ are obtainable from total pcas $\mathcal{A}$, especially models of the λ-calculus such as the graph model and Scott's D_∞.

References

[1] Aczel, P. 1978. The type theoretic interpretation of constructive set theory. Pages 55–66 of: MacIntyre, A., Pacholski, L., and Paris, J. (eds.), *Logic Colloquium '77*. Amsterdam: North-Holland.

[2] Aczel, P. 1982. The type theoretic interpretation of constructive set theory: Choice Principles. Pages 1–40 of: Troelstra, A. S., and van Dalen, D. (eds.), *The L. E. J. Brouwer Centenary*. Amsterdam: North-Holland.

[3] Aczel, P. 1986. The type theoretic interpretation of constructive set theory: Inductive definitions. Pages 17–49 of: R. B. Marcus (ed.), *Logic, Methodology and Philosophy of Science VII*. Amsterdam: North-Holland.

[4] Aczel, P. 1988. *Non-well-founded sets*. CSLI Lecture Notes, vol. 14. Center for the Study of Language and Information.

[5] Aczel, P., and Rathjen, M. 2001. *Notes on Constructive Set Theory*. Technical Report 40. Institut Mittag-Leffler, The Royal Swedish Academy of Sciences, Stockholm.

[6] Aczel, P., and Rathjen, M. 2010. *Constructive Set Theory*. Book draft. Available at http://www1.maths.leeds.ac.uk/~rathjen/book.pdf.

[7] Jäger, G. 2023. Identity, equality, and extensionality in explicit mathematics. In: Bridges, D., Ishihara, H., Rathjen, M., and Schwichtenberg, H. (eds.), *Handbook of Constructive Mathematics*. Cambridge: Cambridge University Press.

[8] Barwise, J., and Moss, L. 1996. *Vicious Circles*. CSLI Lecture Notes, vol. 60. Stanford: CSLI Publications.

[9] Beeson, M. J. 1979. Continuity in intuitionistic set theories. Pages 1–52 of: *Studies in Logic and the Foundations of Mathematics*, vol. 97. Amsterdam: Elsevier.

[10] Beeson, M. J. 1985. *Foundations of Constructive Mathematics*. Ergebnisse der Mathematik und ihrer Grenzgebiete (3) [Results in Mathematics and Related Areas (3)], vol. 6. Berlin: Springer-Verlag.

[11] Bishop, E. 1967. *Foundations of Constructive Analysis*. New York: McGraw-Hill.

[12] Bishop, E. 1968?a. A general language. `https://u.math.biu.ac.il/~katzmik/bishop18aa.pdf`

[13] Bishop, E. 1968?b. How to compile mathematics in Algol. `https://u.math.biu.ac.il/~katzmik/bishop18bb.pdf`

[14] Bishop, E. 1970. Mathematics as a numerical language. Pages 53–71 of: Kino, A., Myhill, J., and Vesley, R. E. (eds.), *Intuitionism and Proof Theory*. Amsterdam: North-Holland.

[15] Blass, A. 1979. Injectivity, projectivity, and the axiom of choice. *Trans. AMS*, **255**, 31–59.

[16] Crosilla, L. 2014. Constructive and intuitionistic ZF. In: *Stanford Encyclopedia of Philosophy*. `http://plato.stanford.edu/entries/set-theory-constructive/`. Substantive revision of the entry with the same title first published in 2009.

[17] Crosilla, L. 2023. Bishop's mathematics: a philosophical perspective. Chapter 3 of: Bridges, D., Ishihara, H., Rathjen, M., and Schwichtenberg, H. (eds.), *Handbook of Constructive Mathematics*. Cambridge: Cambridge University Press.

[18] Crosilla, L., and Rathjen, M. 2002. Inaccessible set axioms may have little consistency strength. *Ann. Pure Appl. Logic*, **115**, 33–70.

[19] Curry, H. B. 1929. An analysis of logical substitution. *Am. J. Math.*, **51**, 509–536, 789–834.

[20] Curry, H. B. 1930. Grundlagen der kombinatorischen Logik. *Am. J. Math.*, **51**, 363–384.

[21] Diaconescu, R. 1975. Axiom of choice and complementation. *Proc. Amer. Math. Soc.*, **51**, 176–178.

[22] Dihoum, E., and Rathjen, M. 2019. Preservation of choice principles under realizability. *Logic J. IGPL*, **27**(5), 746–765.

[23] Feferman, S. 1975. A language and axioms for explicit mathematics. Pages 87–139 of: Crossley, J. (ed.), *Algebra and Logic*. Lecture Notes in Mathematics, vol. 450. Berlin: Springer.

[24] Feferman, S. 1978. Recursion theory and set theory: a marriage of convenience. Pages 55–98 of: *Generalized Recursion Theory II*. Studies in Logic and Foundation of Mathematics. Amsterdam: North-Holland.

[25] Feferman, S. 1979. Constructive theories of functions and classes. In: *Logic Colloquium '78*. Amsterdam: North-Holland.

[26] Friedman, H. 1977. Set theoretic foundations for constructive analysis. *Ann. Math.*, **105**, 1–28.

[27] Friedman, H. 1980. A strong conservative extension of Peano arithmetic. Pages 113–122 of: Barwise, J., Keisler, H. J., and Kunen, K. (eds.), *The Kleene Symposium*. Amsterdam: North-Holland.

[28] Friedman, H. 1973. Some applications of Kleene's methods for intuitionistic systems. Pages 113–170 of: *Cambridge Summer School in Mathematical Logic*. Berlin: Springer.

[29] Glivenko, V. I. 1928. Sur la logique de M. Brouwer. *Acad. Roy. Belg. Bull. Cl. Sci.*, **14**, 225–228.

[30] Glivenko, V. I. 1929. Sur quelques points de la logique de M. Brouwer. *Acad. Roy. Belg. Bull. Cl. Sci.*, **15**, 183–188.

[31] Goodman, N. 1978. Relativized realizability in intuitionistic arithmetic of all finite types. *J. Symbol. Logic*, **43**, 23–44.

[32] Goodman, N., and Myhill, J. 1978. Choice implies excluded middle. *Zeitschr. math. Logik Grundlagen Math.*, **24**, 461.

[33] Griffor, E., and Rathjen, M. 1994. The strength of some Martin-Löf type theories. *Arch. Math. Logic*, **33**(5), 347–385.

[34] Heyting, A. 1930a. Die ormalen Regeln der intuitionistischen Logik. *Sitzungsber. Preuss. Akad. Wissensch. Phys.-math. Klasse*, 42–56.

[35] Heyting, A. 1930b. Die formalen Regeln der intuitionistischen Mathematik II. *Sitzungsber. Preuss. Akad. Wissensch. Physikalisch-mathematische Klasse*, 57–71.

[36] Heyting, A. 1930c. Die formalen Regeln der intuitionistischen Mathematik III. *Sitzungsber. Preuss. Akad. Wissensch. Physikalisch-mathematische Klasse*, 158–169.

[37] Hilbert, D, and Ackermann, W. 1928. *Grundzüge der theoretischen Logik*, 1st ed. Berlin: Springer.

[38] Jensen, R.B. 1966. Independence of the axiom of dependent choices from the countable axiom of choice (abstract). *J. Symbol. Logic*, **31**, 294.

[39] Kleene, S.C. 1945. On the interpretation of intuitionistic number theory. *J. Symbol. Logic*, **10**, 109–124.

[40] Kolmogorov, A. 1925. O principe tertium non datur. *Matematičeskij Sbornik*, **32**, 646–667.

[41] Kreisel, G., and Troelstra, A. S. 1970. Formal systems for some branches of intuitionistic analysis. *Ann. Math. Logic*, **1**, 229–387.

[42] Lubarsky, R., and Rathjen, M. 2007. On the constructive Dedekind reals. Pages 349–362 of: Artemov, S., and Nerode, A. (eds.), *Logical Foundations of Computer Science*. Lecture Notes in Computer Science, vol. 4514. Berlin: Springer.

[43] Lubarsky, R., and Rathjen, M. 2008. On the constructive Dedekind reals. *Logic and Analysis*, **1**, 131–152.

[44] Martin-Löf, P. 1975. An intuitionistic theory of types: predicative part. In: Rose, H. E., and Shepherdson, J. C. (eds.), *Logic Colloquium 1973*. Amsterdam: North-Holland.

[45] Martin-Löf, P. 1984. *Intuitionistic Type Theory*. Naples: Bibliopolis.

[46] McCarty, D. C. 1985. Realizability and Recursive Mathematics. Ph.D. thesis, University of Edinburgh.

[47] Myhill, J. 1973a. Some properties of intuitionistic Zermelo–Fraenkel set theory. Pages 206–231 of: *Cambridge Summer School in Mathematical Logic*. Lecture Notes in Mathematics, vol. 337. Berlin: Springer.

[48] Myhill, J. 1975. Constructive set theory. *J. Symbol. Logic*, **40**, 347–382.

[49] Myhill, J. 1973b. Some properties of intuitionistic Zermelo–Frankel set theory. Pages 206–231 of: *Cambridge Summer School in Mathematical Logic*. Lecture Notes in Mathematics, vol. 337. Berlin: Springer.

[50] Palmgren, E. 1998. On universes in type theory. Pages 191–204 of: Sambin, G., and Smith, J. (eds), *Twenty-Five Years of Constructive Type Theory (Venice, 1995)*. Oxford Logic Guides. Oxford: Oxford University Press.

[51] Rathjen, M. 2021. Hilbert's program and (semi) intuitionism. *Slides from Workshop on Gödel's incompleteness theorems, Wuhan.*

[52] Rathjen, M. 1993. The strength of some Martin-Löf type theories. Preprint, Department of Mathematics, Ohio State University, available at http://www1.maths.leeds.ac.uk/~rathjen/typeOHIO.pdf.

[53] Rathjen, M. 2003. The anti-foundation axiom in constructive set theories. Pages 87–108 of: Mints, G., and Muskens, R. (eds.), *Games, Logic, and Constructive Sets*. Stanford: CSLI Publications.

[54] Rathjen, M. 2004. Predicativity, circularity, and anti-foundation. Pages 191–219 of: Link, G. (ed.), *One Hundred Years of Russell's Paradox*. de Gruyter Series in Logic and its Applications, vol. 6. Berlin: de Gruyter.

[55] Rathjen, M. 2005. Constructive set theory and Brouwerian principles. *J. Univ. Comp. Sci.*, **11**, 2008–2033.

[56] Rathjen, M. 2006a. Choice principles in constructive and classical set theories. In: Chatzidakis, Z., Koepke, P., and Pohlers, W. (eds.), *Logic Colloquium '02*. Lecture Notes in Logic, vol. 27. Wellesley, MA: A. K. Peters.

[57] Rathjen, M. 2006b. The formulae-as-classes interpretation of constructive set theory. Pages 279–322 of: Schwichtenberg, H., and Spies, K. (eds), *Proof Technology and Computation*. NATO Science Series III: Computer and System Sciences, vol. 200. Amsterdam: IOS Press.

[58] Rathjen, M. 2006c. Realizability for constructive Zermelo–Fraenkel set theory. Pages 228–314 of: Väänänen, J., and Stoltenberg-Hansen, V. (eds.), *Logic Colloquium '03*. Lecture Notes in Logic, vol. 24. Wellesley, MA: A. K. Peters.

[59] Rathjen, M. 2008. The natural numbers in constructive set theory. *Math. Logic Quart.*, **54**, 83–97.

[60] Rathjen, M. 2012. *Constructive Zermelo–Fraenkel Set Theory, Power Set, and the Calculus of Constructions*. Pages 313–349 of Logic, Epistemology, and the Unity of Science, vol. 27. Dordrecht: Springer.

[61] Rathjen, M. 2014. Constructive Zermelo–Fraenkel set theory and the limited principle of omniscience. *Ann. Pure Appl. Logic*, **165**, 563–572.

[62] Rathjen, M. 2017. Proof theory of constructive systems: Inductive types and univalence. Pages 385–419 of: Jäger, G., and Sieg, W. (eds.), *Feferman on Foundations*. Outstanding Contributions to Logic, vol. 13. Berlin: Springer.

[63] Rathjen, M. 2023. An introduction to intuitionistic logic. In: Bridges, D., Ishihara, H., Rathjen, M., and Schwichtenberg, H. (eds.), *Handbook of Constructive Mathematics*. Cambridge: Cambridge University Press.

[64] Rathjen, M., and Tupailo, S. 2006. Characterizing the interpretation of set theory in Martin-Löf type theory. *Ann. Pure Appl. Logic*, **141**(3), 442–471.

[65] Rathjen, M., Griffor, E., and Palmgren, E. 1998. Inaccessibility in constructive set theory and type theory. *Ann. Pure Appl. Logic*, **94**, 181–200.

[66] Schönfinkel, M. 1924. Über die Bausteine der mathematischen Logik. *Math. Annalen*, **92**, 305–316.

[67] Simpson, S. 2009. *Subsystems of Second Order Arithmetic*, 2nd ed Cambridge: Cambridge University Press.

[68] Troelstra, A. S., and van Dalen, D. 1988. *Constructivism in Mathematics, Volumes I, II*. Amsterdam: North-Holland.

[69] van Oosten, J. 2008. *Realizability: An Introduction to its Categorical Side*. Studies in Logic and Foundations of Mathematics, vol. 152. Amsterdam: Elsevier.

[70] Weaver, N. 2009. Axiomatizing mathematical conceptualism in third order arithmetic. ArXiv:0905.1675v1, 31 pages.

3

Bishop's Mathematics: A Philosophical Perspective

Laura Crosilla

3.1 Introduction

The past 50 years have seen the flourishing of constructive approaches to mathematics and the growth of a variety of research groups working on constructive mathematics. This has given rise to a rich literature witnessing the depth and breadth of constructive mathematics, of which this volume is further proof. A principal drive for these developments has been the stimulus derived from the natural alliance between constructive mathematics and computer-aided computation, at first only an envisaged possibility and, more recently, a fact. The spark that started the present abundance of constructive mathematics was the publication in 1967 of Errett Bishop's *Foundations of Constructive Analysis* [9]. Bishop's book also prompted pivotal work in logic, with the emergence of new foundational frameworks, such as intuitionistic set and type theories, which, in turn, fostered fundamental work in computer-assisted theorem proving.[1]

Notwithstanding the extensive progress in constructive mathematics since 1967, there has been no corresponding advance in its philosophy. With very few exceptions, Bishop's mathematics has at most received a quick mention, but no thorough consideration.[2] My overarching aim in this chapter is to enliven the philosophical debate about constructive mathematics Bishop-style.[3] There are a number of reasons to stimulate the philosophical reflection on constructive mathematics. First of

[1] See, for example, [14, 15, 22, 60]. See also [1, 3, 40, 43, 44, 59, 62, 63, 75] and [2, 25, 27, 65]. Note that Constable [64, p. 83] credits directly Bishop's influence for the work that lead to the design of 'a large computing system that would execute constructive proofs'.

[2] For example, Stewart Shapiro mentions Bishop very briefly in the well-known textbook [70] (e.g. at pages 184, in a footnote, 187, and 189) and in [71, 72]. The only exceptions I am aware of are [7, 45] and an exchange on the possibility of developing constructively the mathematical foundations of quantum physics [6, 8, 16, 47, 48, 49, 68, 69]. The lack of progress on the philosophy of Bishop's mathematics is quite surprising, since there has been instead sustained analysis of other variants of constructivism inspired especially by the work of Brouwer, Gentzen, Dummett, Prawitz, and Martin-Löf.

[3] In the following, I write 'constructive mathematics' to denote the form of mathematics initiated by Bishop, and 'intuitionistic mathematics' to refer to other forms of mathematics that use intuitionistic logic.

all, the significant advances of this form of mathematics in recent times and its natural bond with computation make constructive mathematics much more prominent within today's mathematical landscape than at its birth. An analysis of constructive mathematics is therefore important and pressing for a philosopher of mathematics who aims genuinely to engage with contemporary mathematics.[4] Second, Bishop's mathematics, as will be further argued, requires altogether different philosophical considerations compared with better-known approaches to intuitionistic mathematics, such as Brouwer's. Third, a sympathetic analysis of Bishop's philosophical remarks presents us with intriguing foundational ideas that deserve to be better understood and further developed.

The chapter is organised as follows. I begin by examining key elements of Bishop's philosophical remarks, especially focussing on Bishop's assessment of Brouwer, as it illuminates some of the most important aspects of Bishop's own foundational reflection. I then briefly sketch the most salient features of what I would like to call 'traditional' arguments for intuitionistic mathematics and consider whether these arguments could also support today's constructive mathematics. I argue that this is not the case, as traditional arguments are in tension with both Bishop's remarks and the constructive mathematical practice. This observation raises fundamental questions for the philosophy of constructive mathematics, and indicates that a thoroughly new approach is required. I conclude with the suggestion to look anew at Bishop's own remarks for inspiration.

3.2 Bishop on Brouwer

In this section, I review some prominent themes from Bishop's reflection on mathematics, especially focussing on the relation between Bishop and his main predecessor, Brouwer.[5] My aim is to understand Bishop's thought rather than defend his views. Bishop often combined severe criticism of Brouwer with recognition of his achievements. This complex relationship with the founder of intuitionism is aptly portrayed by Gabriel Stolzenberg in his review of [9], where Bishop's 'constructive framework' is presented as 'intimately related to Brouwer's intuitionism – though with important differences' [73].

An important and often emphasised difference between Bishop's and Brouwer's mathematics concerns the greater extent of the new form of mathematics. According to Bishop [9, p. ix], Brouwer and other constructivists were more successful in their criticism of classical mathematics than in replacing it with a better alternative. Soon after the publication of the fundamental [9], Bishop's work was celebrated for 'going substantially further mathematically' [41, p. 170]. For this reason, mathematicians

[4] A similar point is made in [33], in agreement with the spirit of the 'philosophy of the mathematical practice' (see, e.g., [24, 42, 58]).

[5] I especially draw from [9, 10, 11, 12].

sympathetic to Bishop's project took it to refute the most prominent mathematical objection to Brouwerian intuitionism, famously emphasised by Hilbert and other critics of intuitionism, for which relinquishing the principle of excluded middle is tantamount to relinquishing the science of mathematics altogether.[6] Bishop himself writes that 'Hilbert's implied belief that there are a significant number of interesting theorems whose statements (standing alone) are constructive but whose proofs are not constructive (or cannot easily be made constructive) has not been justified. In fact we do not know of even one such theorem' [9, p. 354].

The greater extent of Bishop's mathematics is only one of the points of difference with Brouwer. Others relate to specific aspects of Brouwer's own approach, such as his treatment of the continuum and his philosophical ideas. It is plausible that when Bishop distanced himself from Brouwer and his followers, he aimed at separating aspects of the Brouwerian revolution he agreed with from others he found problematic. In so doing, he probably also hoped to attract to his new mathematics classical mathematicians who did care about constructivity but were sceptical of Brouwer's own approach. In the preface to his book, Bishop [9, p. ix] mentions previous attempts to constructivise mathematics, 'the most sustained' of which was made by L. E. J. Brouwer. He then notes that

> [t]he movement Brouwer founded has long been dead, killed partly by compromises of Brouwer's disciples with the viewpoint of idealism, partly by extraneous peculiarities of Brouwer's system which made it vague and even ridiculous to practising mathematicians, but chiefly by the failure of Brouwer and his followers to convince the mathematical public that abandonment of the idealistic viewpoint would not sterilize or cripple the development of mathematics. ([9], p. ix)

In this passage, Bishop criticises Brouwer not only for 'extraneous peculiarities' of his mathematics, but also for his perceived inability to communicate with and involve classical mathematicians.[7] In the following sections, I consider each point in turn.

3.3 Brouwer's Mathematics

In the 'Constructivist Manifesto' that opens [9], Bishop criticises Brouwer's mathematics especially for its use of free choice sequences in analysis. According to

[6] See [52, p. 476]. As noted by Myhill in [61], also constructivists such as Heyting thought that the "mutilation" of mathematics was an inevitable consequence of their standpoint [51, p. 74]. See [61, 73] for examples of favourable reception of Bishop's book. See Beeson's introduction to [13] for a more general discussion of the overall reception of Bishop's book.

[7] Note that in the quote above, Bishop claims that intuitionistic mathematics has long been dead. A similar image makes its way in Douglas Bridges's foreword to [34], where he writes that Bishop 'single-handedly showed that deep mathematics could be developed constructively, and thereby pulled the subject back from the edge of the grave'.

Bishop, free choice sequences make the continuum 'not well enough defined' and the resulting mathematics 'so bizarre it becomes unpalatable to mathematicians' [9, p. 6]. As further mentioned in Section 3.4, a significant characteristic of Bishop's mathematics is its full *compatibility* with classical mathematics: every proof in Bishop's mathematics of a statement is also a classical proof of it.[8] In fact, one of the characteristics of Bishop's approach emphasised from the start is that it is not only compatible with classical mathematics, but its 'spirit' and 'execution' are 'much more like everyday modern mathematics than anything previously done in a systematic constructive way' [41, p. 171]. According to Bishop, Brouwer's treatment of the continuum, with the introduction of free choice sequences, marks instead a more drastic departure from the standard classical approach, and this, alone, makes Brouwer's mathematics less appealing, or even 'unpalatable' to mathematicians.[9]

Bishop also seems to think that Brouwer's free choice sequences represent an undesirable interference of 'metaphysical speculation' in mathematics, as they are dictated by Brouwer's philosophical view of the continuum rather than by the needs of the mathematical practice. Bishop further claims that Brouwer's mathematics (in general) shows 'a preoccupation with the philosophical aspects of constructivism at the expense of concrete mathematical activity'[10] [9, p. 6]. On the contrary, Bishop's book aimed to develop a large portion of abstract analysis within a constructive framework 'with an absolute minimum of philosophical prejudice concerning the nature of constructive mathematics' [9, p. ix].[11]

In subsequent texts, Bishop is more specific on what makes Brouwer's theory of the continuum problematic from his own point of view. For example, Bishop [10, p. 53] deplores the *lack of numerical interpretation* of free choice sequences: 'Brouwer's intuitionism at first glance contains elements that are extremely dubious; free choice sequences and allied concepts admit no ready numerical interpretation'. Similarly, in notes posthumously published as [12], Bishop writes that there seem to be at least two motivations for Brouwer's introduction of free choice sequences. First, 'it appears that Brouwer was troubled by a certain aura of the discrete clinging to the constructive real number system **R**'.[12] Second,

[8] See [21], especially Chapter 6.

[9] Feferman [41, p. 171] goes on writing: 'Indeed, a (philosophically unprepared) mathematician could pick up Bishop 1967 and read it as a straight piece of classical Cantorian mathematics. What would be puzzling to him is the more involved choice of certain notions and proofs, unless he also saw in what sense these were dictated by constructive requirements'. A similar point is made in Myhill's review [61, p. 744].

[10] Bishop [9, p. 6] writes: 'Brouwer became involved in metaphysical speculation by his desire to improve the theory of the continuum. A bugaboo of both Brouwer and the logicians has been compulsive speculation about the nature of the continuum. In the case of the logicians this leads to contortions in which various formal systems, all detached from reality, are interpreted within one another in the hope that the nature of the continuum will somehow emerge. In Brouwer's case there seems to have been a nagging suspicion that unless he personally intervened to prevent it the continuum would turn out to be discrete'.

[11] I discuss this point further in Section 3.6.

[12] See also the quote from [9, p. 6], in footnote [10].

> Brouwer had hopes of proving that every function from $\mathbf{R} \to \mathbf{R}$ is continuous, using arguments involving free choice sequences. [...] My objection to this is, that by introducing such a theorem as "all $f : \mathbf{R} \to \mathbf{R}$ are continuous" in the guise of axioms, we have lost contact with numerical meaning. Paradoxically this terrible price buys little or nothing of real mathematical value. The entire theory of free choice sequences seems to me to be made of very tenuous mathematical substance. [12, p. 26]

Summarising, Bishop seems to think that Brouwer's introduction of free choice sequences marks too drastic a departure from ordinary mathematics and is not sufficiently well motivated *mathematically* as it is not needed in practice. More importantly, free choice sequences cannot be directly explained in terms of *finitely performable computations with the integers*, therefore lacking clear *numerical meaning*. This is seen as a very substantial defect by Bishop, for whom the possibility of endowing mathematics with numerical meaning is a principal motive for developing his constructive mathematics.

One may wonder whether notwithstanding Bishop's disparaging remarks on free choice sequences, it would be beneficial for mathematics as a whole not to neglect forms of mathematics such as Brouwer's that countenance more abstract notions of construction.[13] As noted by Kreisel [56, pp. 146-147], since Bishop focusses on lawlike sequences, he does not offer a general explanation of why Brouwer's principles of continuity 'does not really affect mathematical practice'.[14] As further mentioned in Section 3.4, today's constructive mathematicians pledge to take a more encompassing approach compared with Bishop, as they hope to analyse from their general point of view a number of mathematical approaches, among which are the classical one, the Brouwerian one, and the recursion-theoretic approach of the Russian school of constructivism.[15]

Kreisel [56] further argues for the fruitfulness and the *naturalness* from a *mathematical* perspective of focusing on *abstract* (rather than more explicit) notions, and in particular on the most general notion of construction.[16] The interplay between

[13] I would like to thank a referee for asking this question and for drawing my attention to the relevant passage of [56]. A related point is made by Veldman in [77, p. 61]: 'The principles proposed by Brouwer, even if one does not want to subscribe to the way he defends them, deserve to be discussed as possible starting points for our common mathematical discourse'.

[14] It should be noted that Bishop [10, pp. 67–68] does consider the question of how one could accommodate Brouwer's theory of free choice sequences within a formal system. This, however, does not satisfy Kreisel who writes [56, pp. 146–147]: 'It is one thing to point out (correctly), as Bishop does, that Brouwer's assertion concerning the continuity of (extensional) functions does not really affect mathematical practice [...], if we simply take our functions as given *together* with a modulus of continuity. But it is a separate matter to explain this step by showing that any definition satisfying some abstract condition is bound to provide the additional information; in other words by analysing (when possible) the most general notion of construction, not merely definitions in some formal system such as Gödel's T'.

[15] See [21] for a discussion of the most well-known varieties of constructive mathematics. Note that very recent work suggests that, contrary to what Bishop thought, there is the potential for practical results ensuing a Brouwerian approach to mathematics (see, e.g., [5]).

[16] Note that the word 'abstract' is used here in its mathematical rather than philosophical sense. Kreisel [56, p

more abstract and more explicit notions and their respective roles within different approaches to mathematics is a very significant point that ought to be central to the philosophy of mathematics, and especially to the philosophy of constructivism. Discussion of this point clearly exceeds the aims of this chapter. I wish, however, to point out the fact that Kreisel's comments clearly indicate that one of the most significant differences between a Brouwerian approach and Bishop's own, is the crucial foundational role the domain of the natural numbers has for Bishop. The role of the natural numbers for Bishop is the main focus of Section 3.6.

3.4 Persuasion and Dialogue

We saw Bishop's criticism of Brouwer's mathematics, but also of his inability to convince the mathematical public of the viability of his intuitionistic project. Bishop [9, p. 6] writes that Brouwer's programme 'failed to gain support' as Brouwer was

> an indifferent expositor and an inflexible advocate, contending against the great prestige of Hilbert and the undeniable fact that idealistic mathematics produced the most general results with the least effort.

Bishop was hopeful that constructive mathematics would eventually prevail over classical mathematics, to such an extent that in his preface [9, p. x], he writes that his ultimate goal is 'to hasten the inevitable day when constructive mathematics will be the accepted norm'. But he was aware that for this transformation to occur he needed to persuade his fellow mathematicians that the constructive program could succeed. This seems to motivate his criticism of Brouwer's 'inflexible' attitude towards classical mathematics.

Bishop was keen to reach out to the mathematical community, so much so that soon after the publication of [9], he embarked on a series of lectures on constructive mathematics across the USA. Although his lectures attracted large audiences, he had limited success in persuading classical mathematicians to join constructive mathematics.[17] In fact, Bishop thought that he had not been understood.

Notwithstanding his clear desire to communicate with classical mathematicians, some of Bishop's statements on classical mathematics may well have made the

122] claims that Bishop's 1967 book in some respects witnesses this attitude, too, as Bishop does not pin down his notion of algorithm to a specific notion like, for example, that of recursive function, working instead with a primitive notion of constructive function. The significance of Bishop's very abstract approach to the notion of computation is also emphasised, for example, in [17].

[17] See [12, p. 1] where Bishop states that his general impression was that in those lectures he failed to communicate a real feeling for the philosophical issues involved. Nerode [64] recalls that after his tour of the eastern universities in the USA, Bishop told Nerode that he felt the trip might have been counterproductive, as the audiences did not take his work seriously. Bishop also thought that the difficulties experienced during the lecture tour contributed to the deterioration of his health, resulting in a heart attack [64, p. 80]. See also Beeson's foreword to [13].

communication problem with the classical mathematician worse.[18] For example, Bishop [9, p. ix] claims that his book is a 'piece of constructivist propaganda' and goes on to write:

> Our program is simple: to give numerical meaning to as much as possible of classical abstract analysis. Our motivation is the well-known scandal, exposed by Brouwer (and others) in great detail, that classical mathematics is deficient in numerical meaning.

More forceful are the remarks in [12], such as the claim that there is a 'philosophical deficit of major proportions' in contemporary mathematics, and that the latter manifests the 'debasement of meaning'.

A more conciliatory attitude characterises [11]. Here Bishop imagines an ideal dialogue between Hilbert and Brouwer in which the two mathematicians amicably discuss and compare their divergent foundational views.[19] Bishop [11, p. 510] claims that '[p]erhaps Brouwer should not have denounced the mathematics that Hilbert wished to do as meaningless'. In that text, Bishop strongly advocates a key role for constructive mathematics as *enhancing* or *deepening* the classical practice. The idea is that *within constructive mathematics* one can express distinctions in meaning which are not available to the classical mathematician, such as, for example, the distinction between statements that have a computational interpretation from those that lack one. Furthermore, constructive mathematics can be taken to be the basis over which one expresses and analyses a classically valid theorem T by means of implications of the form, for example, LPO $\rightarrow$ T*, with T* a constructive substitute of the classical theorem T.[20] Incidentally, Bishop [11, p. 512] claims that implications such as LPO $\rightarrow$ T* are 'ugly' and that we should try to obtain 'an implication which is natural and reflects the nature of the problem', namely one that is related to the structure of a particular theorem in some special way. If we do that, then working within a constructive context allows us to clearly single out any non-constructive assumption and identify and bring to the fore important aspects invisible from a classical perspective, especially the computational content of mathematics.

We see here the emergence of an idea that has been profitably refined in recent years giving rise to the *constructive reverse mathematics programme*. Constructive mathematics here is the ***core*** of a number of *varieties* of mathematics, among

[18] For example, [9] is portrayed as 'pure ideology' in [57] (p. 228) and the introduction to that book is termed 'embarrassing' (p. 239).

[19] Note that in this very text, Bishop [11, pp. 513–514] expresses, without argumentation, a very harsh opinion of non-standard analysis: 'It is difficult to believe that debasement of meaning could be carried so far'.

[20] Bishop calls 'Limited Principle of Omniscience' (LPO) the following statement: if $\{a_n\}$ is a binary sequence, then either there exists n such that $a_n = 1$, or else $a_n = 0$ for each n.

which are **classical**, **Brouwerian**, and **Russian** constructive mathematics.[21] Indeed, each of the latter three forms of mathematics may be developed on the basis of some suitable extension of Bishop's mathematics by characteristic principles. For example, the principle of excluded middle and the axiom of choice can be added to Bishop's constructive mathematics, giving rise to a context for developing classical mathematics, while adding the principle of continuous choice to Bishop's mathematics, and the fan theorem allows one to develop a Brouwerian form of mathematics.[22] The constructive mathematician claims that due to its privileged position, constructive mathematics allows us to study from a 'neutral' perspective relations between mathematical notions belonging to these varieties, as well as comparing these varieties with each other.

3.5 Formalisation

As we saw on page 63, Bishop [9] also criticised Brouwer's disciples. His concern in that respect was especially Heyting's formalisation of intuitionistic logic and subsequent work in mathematical logic on intuitionistic formal systems. In this respect, Bishop's anti-formalist attitude appears particularly close to Brouwer's views. It is here that we find some of Bishop's strongest words of appreciation for Brouwer. Brouwer is often praised for his realisation of the defects of classical mathematics, especially its lack of numerical content, and his opposition to formalism. Bishop [9, p. 6] credits to Brouwer the 'disengagement of mathematics from logic':

> Brouwer fought the advance of formalism and undertook the disengagement of mathematics from logic. He wanted to strengthen mathematics by associating to every theorem and every proof a pragmatically meaningful interpretation.

As to the criticism of Brouwer's disciples, Bishop (ibid.) writes that Brouwer's precepts were

> formalized, giving rise to so-called intuitionistic number theory, and [...] the formal system so obtained turned out not to be of any constructive value. In fairness to Brouwer it should be said that he did not associate himself with these efforts to formalize reality [...].

Bishop's views on formalisation changed in some respects after the completion of *Foundations of Constructive Analysis*, as witnessed, for example, by [10]. Bishop did not give up criticising formalism and the dry study of formal systems as opposed to contentful mathematical practice (see especially [12]). However, by 1970 he seemed to have reached the conclusion that formalisation could be employed to

[21] See [21] for a comparison of these varieties of mathematics and see [35, 53, 54, 55, 76] for the constructive reverse mathematics programme.

[22] It should be noted that the constructive reverse mathematics programme is often developed informally, without fixing specific formal systems. This makes the above claims not completely precise and raises important questions. See the discussion in [35, p. 100], and especially footnote 1.

the benefit of mathematics. For example, in [10], he employs Gödel's Dialectica interpretation to clarify the numerical content of mathematical statements. One of Bishop's main concerns in that text is the constructive interpretation of conditional statements. Bishop [10, p. 53] begins by giving a clear characterisation of constructive mathematics which makes more explicit ideas already hinted at in [9].[23] Constructive mathematics is here termed *predictive* since it

> describes or predicts the result of certain finitely performable, albeit hypothetical, computations within the set of integers.

This interpretation of *constructive* in terms of *finitely performable operations with the integers* is at the heart of Bishop's approach to constructive mathematics and his insistence on the *numerical content* of mathematical statements. After discussing some characteristic examples of mathematical problems, Bishop writes [10, p. 54]:

> The most urgent task of the constructivist is to give predictive embodiment to the ideas and techniques of classical mathematics. Classical mathematics is not totally divorced from reality. On the contrary, most of it has a strongly constructive cast.

A key step in the task of giving predictive embodiment to classical mathematics, is to endow conditional statements with suitable numerical meaning and it is here that Bishop employs Gödel's Dialectica interpretation. Bishop expresses his dissatisfaction with the standard Brouwer–Heyting–Kolmogorov (BHK) interpretation of implication as well as with a variant he proposed in [9]. He therefore suggests to use Gödel's Dialectica interpretation to offer a more satisfactory computational interpretation of conditional statements.[24] While Bishop's proposal in [10] deserves more careful analysis, I cannot go into more detail in the present context. The main point I wish to highlight is that [10] witnesses an apparent change of attitude, as formal systems are now taken to offer the means to tackle these urgent tasks and Bishop envisages the possibility of a fruitful cooperation between formalisation and mathematics. He writes [10, p. 60]:

> Another important foundational problem is to find a formal system that will efficiently express predictive mathematics. I think we should keep the formalism as primitive as possible, starting with a minimal system and enlarging it only if the enlargement serves a genuine mathematical need. In this way the formalism and the mathematics will hopefully interact to the advantage of both.

It is possible that the difficulties Bishop encountered in conveying his ideas to mainstream mathematicians and the more favourable reception of his mathematics

[23] See, for example, [9, p. viii].

[24] Bridges [18] reports a conversation with Bishop that suggests that Bishop's dissatisfaction with material implication was a major motive for his "conversion" to constructive mathematics.

among logicians had an impact on his apparent change in attitude.[25] It seems also likely that in the meantime Bishop had become more aware of the potential of applying constructive mathematics to computer programming. This was already prefigured in Appendix B of [9].[26] The remarkable point is that while in his 1967 book, formalisation was mainly seen as an artificial obstacle, distracting us from mathematics' genuine content, by 1970, Bishop appears more interested in formalisation, as long as it engages with questions of meaning. Even after 1967, formalization for the sake of formalization is strongly criticised. Now, however, Bishop thinks that when properly employed, formal systems *can* be a useful tool for clarifying issues of meaning and fostering possible applications to computers.

3.6 Philosophy

We have seen Bishop's concerns in [9] for Brouwer's introduction of free choice sequences and for his inflexible attitude. Furthermore, in his 1967 book, Bishop complained that Brouwer was preoccupied with philosophical aspects of constructivism at the expense of concrete mathematical activity and criticised his 'metaphysical speculation' over the nature of the continuum. This suggests a rather bleak view of philosophy and its relation with mathematics. Philosophy, however, has a more prominent and positive role in subsequent texts by Bishop. For example, in [11] Bishop clearly sees a role for philosophical thought in mathematics.[27] The article starts with the following very powerful statement.

> There is a crisis in contemporary mathematics and anybody who has not noticed it is being willfully blind. *The crisis is due to our neglect of philosophical issues.*[28]

Bishop [11, p. 507] complains that university courses in the foundations of mathematics focus on formal systems and their analysis 'at the expense of philosophical substance'. He writes that we need to change emphasis from proving theorems to knowing what they mean, 'from the mechanics of the assembly line which keeps grinding out the theorems, to an examination of what is being produced'. Philosophical reflection ought to contribute to this shift of focus and clarify the *meaning* of mathematical statements. Bishop writes that '[t]here is only one basic criterion to justify the philosophy of mathematics, and that is, does it contribute to making mathematics more meaningful'. [11, p. 508]

[25] See [64].

[26] There has been a recent discussion among constructive mathematicians on two unpublished manuscripts by Bishop, 'A general language' and 'How to compile mathematics into Algol'. These texts also witness Bishop's interest for formalisation as a tool for the application of constructive mathematics to computer programming. See, for example, http://www.cs.bham.ac.uk/∼mhe/Bishop/

[27] See also [10, p. 57], where Bishop claims that there must be a *philosophical* explanation of the empirical fact that intuitionistic implication admits a numerical interpretation.

[28] Italics in the original text.

To explain why he takes issues of meaning as central to mathematics and philosophy, Bishop [11, p. 507] asks the question 'What do we mean by an integer?' He considers three possible answers:

(i) an integer that we can actually compute, for example, 3,
(ii) one that we can compute in principle only, for example, 9^{9^9},
(iii) one that is not computable by known techniques, even in principle, for example, the integer that is defined to be 1 if φ is true and 0 otherwise, where φ is an open problem such as the Riemann hypothesis.

A constructive approach to mathematics, so argues Bishop, is necessary if we want to bring to light important distinctions such as that between (i) and (ii) on the one side and (iii) on the other. Bishop [11, p. 507] adds:

> To my mind, it is a major defect of our profession that we refuse to distinguish, in a systematic way, between integers that are computable in principle and those that are not. We even refuse to do mathematics in such a way so as to permit one to make the distinction.[29]

Philosophy therefore can help clarify the computational meaning of mathematical statements and distinguish different statements depending on their meaning. In fact, Bishop's philosophy of mathematics rests on two crucial assumptions: the foundational role of the natural numbers within mathematics and the constructive interpretation of the logical constants. As to the *natural numbers*, Bishop [9, p. 2] asserts that 'the primary concern of mathematics is number, and this means the positive integers'. Bishop also mentions Kant, Kronecker, and echoes Brouwer, when he claims that

> the development of the theory of the positive integers from the primitive concept of the unit, the concept of adjoining a unit, and the process of mathematical induction carries complete conviction.

In later texts [10, 11, 12], it becomes even clearer that Bishop's insistence on the meaning (or lack of it) of mathematical statements is very much related to the availability (or not) of an interpretation of each statement in terms of some finitely performable operation with the natural numbers.[30] This fundamental role of the natural numbers within mathematics reminds us not only of Kronecker, but also of predicativism, especially as developed by Hermann Weyl [78, 79].[31]

With regard to the *constructive interpretation of the logical constants*, the philosophical import of this choice becomes apparent especially when Bishop considers

[29] One may wonder whether we should also pay attention to the distinction between (i) and (ii). Bishop [12, pp. 9–10] briefly discusses this question, noting the difficulties involved in demarcating (i) and (ii). See also [37, 45, 81].

[30] See Section 3.5.

[31] Bishop mentions Weyl, for example, in [9, p. 10].

the interpretation of statements that quantify over infinite domains. Bishop often stresses the fact that we are *finite* beings, and claims that, for this reason, we should only be concerned with forms of mathematics that a finite being can carry out, at least *in principle*. Bishop's qualification 'in principle' is important, as it clarifies that his aim is not to ban infinite domains, rather to give prominence to the infinite domain of the natural numbers.

The thought that we should be concerned only with those forms of mathematics that a finite being can, in principle, carry out, brings Bishop to question the meaningfulness of *classical* quantification over infinite domains, which he implicitly assimilates to the doings of an *infinite mind*. Once more, this is already hinted at in [9, p. 2], where we read:

> We are not interested in properties of the positive integers that have no descriptive meaning for finite man. When a man proves a positive integer to exist, he should show how to find it. If God has mathematics of his own that needs to be done, let him do it himself.

These ideas are developed in more detail in [12], where Bishop explicitly frames the distinction between *classical* and *constructive* mathematics in terms of the opposition between agents with finite and infinite powers. For example, at page 12, he writes that while constructive mathematics describes mathematical operations that can be carried out by finite beings, 'classical mathematics concerns itself with operations that can be carried out by God'. Subsequently he considers the question of what powers should God (or a being with 'non-finite powers') have. A minimum requirement, according to Bishop, is a form of limited omniscience, that enables such an agent to search through a sequence of integers to determine whether they are all equal to 0 or not. In other terms, a minimum requirement is the principle LPO.[32]

To summarise, for Bishop, a classical interpretation of the truth of a universal statement whose quantifiers range over an infinite domain involves an *infinite search* through the domain to check each individual element. An aspect I find particularly fascinating is that this interpretation of classical quantification bears surprising similarities with how it is often framed by both predicativists and intuitionists. In this way the debate over classical versus constructive mathematics is brought back to the traditional theme of the opposition between finite and infinite domains, which was central to the thought of intuitionists and predicativists alike. For example, a predicativist would consider quantification (i.e., classical quantification) over an infinite domain justified only if some constraints are met (e.g., if a step-by-step specification of the domain is available). For an intuitionist, quantification over infinite domains has to be intuitionistic rather than classical.[33]

[32] See footnote 20 for the statement LPO.

[33] See, for example, [78, p. 23] and [39, p. 41] for a similar interpretation of classical quantification. See also

I will be returning to the role of Bishop's fundamental assumptions regarding the natural numbers and logic at the end of this chapter. Here I wish to get back to Bishop's views on philosophy. After Bishop's unfavourable comments on philosophy in [9], it is surprising to read Bishop's claims that mathematics is experiencing a crisis which is due to our *neglect of philosophical issues* and that philosophy can help clarify fundamental distinctions in meaning [11]. There is clearly a change in emphasis between the earlier and the later texts, and it is natural to ask if this signals also a deep change in Bishop's views on philosophy. I am inclined to think that there is no direct disagreement between [9] and subsequent texts. My impression is that Bishop may have thought he was focussing on different points. On the one hand, as already mentioned in Section 3.1, Bishop's most prominent criticism of Brouwer's philosophy is the charge of 'metaphysical speculation'. Though Bishop's remarks are not only sharp but also very brief, and therefore difficult to interpret, it is plausible that Bishop took certain philosophical questions, for example, whether there are mathematical entities and whether they are mind-dependent or not, as largely irrelevant to the mathematical practice, or 'superfluous'.[34] His criticism of Brouwer can therefore be explained by supposing that he thought Brouwer's mathematics was deeply bound up with Brouwer's views on these matters, while Bishop's own mathematics did not share these characteristics. On the other hand, Bishop's more positive comments on philosophy in [11] relate to its possible role in clarifying the *meaning* of mathematical statements, by distinguishing classical and constructive interpretations of the logical constants and highlighting the key foundational role of the natural numbers. It is possible that Bishop thought that his views on this matter did not require him to take a stance on the nature of mathematical entities, for example with regard to their existence and their mind-dependence (or independence). Issues of meaning, however, have for Bishop deep mathematical consequences, as they determine whether a piece of mathematics has computational content or not. To gain a computationally meaningful mathematics, Bishop thinks, we need to abandon non-constructive methods of proofs and reform mathematics constructively. These are the philosophical questions that deserve to be pursued and it is in pursuing them that philosophy can contribute to a fruitful development of mathematics.

3.7 Traditional Philosophical Arguments for Intuitionistic Logic

We have seen Bishop's thoughts on Brouwer, his criticism and, simultaneously, the appreciation for his predecessor's achievements. The philosophical literature presents us with a vast and important chapter in the philosophy of mathematics

[32] for a discussion of the key role of this interpretation of quantification within the predicativist literature. See [31] for a discussion of the relation between logic and infinite domains.

[34] See especially [12, pp. 10–11].

on arguments for intuitionistic logic and their critique. A central element of this debate is a number of arguments or argument schemas which are usually taken to be the most common defences of intuitionistic mathematics. Their key elements are inspired especially by the thought of Brouwer, Heyting, and Dummett.[35] Let us call them *traditional philosophical arguments for intuitionistic logic*. A natural question to ask is whether Bishop's views are compatible with traditional philosophical arguments for intuitionistic logic. In fact, I am interested in a more general question, as I would like to understand whether today's constructive mathematicians could employ traditional arguments for intuitionistic logic to support a shift from classical to intuitionistic mathematics, or if entirely different considerations are required. I take Bishop's views, as described above, as my starting point.

For our purposes, it is helpful to single out the most general characteristics of traditional philosophical arguments for intuitionistic logic. A typical feature of traditional arguments for intuitionistic logic is that they move from philosophical considerations and reach the conclusion that the general use of classical logic in mathematics is *illegitimate*. As a consequence, these arguments reject classical mathematics and propose its replacement with intuitionistic mathematics. The philosophical considerations may concern, for example, the nature of the mathematical entities, the nature of our mathematical activity, or important features of our mathematical language. Indeed, these traditional arguments are sometimes taken to entail not only that classical logic is illegitimate, but also that it is meaningless, incoherent, or even unintelligible.[36]

For example, one traditional 'Brouwerian' argument for intuitionistic logic starts from a view of mathematics as an essentially languageless activity of the mind and may also see mathematical entities as mental constructions.[37] This brings to the forefront the notion of *proof* of a mathematical statement (or construction), because to ascertain the truth of a mathematical statement, the mathematician needs to perform a certain mental construction by producing a proof of it. A purported proof of the principle of excluded middle is interpreted as a construction which either proves or reduces to absurdity any mathematical statement, the availability of which is highly implausible. Therefore the argument is seen to entail the rejection of the principle of excluded middle (and similar essentially classical laws). More precisely, the Brouwerian mathematician may accept the validity of the principle of excluded middle for finitary statements within a thoroughly finitary context, but objects to its assumption in infinitary contexts.[38]

[35] See, for example, [26].

[36] See, for example, [46].

[37] See, for example, [23, p. 141], [50, p. 53] and Dummett's discussion of traditional intuitionism in [36, 38].

[38] For example, Brouwer [23, p. 141] writes that 'every construction of a bounded finite character in a finite mathematical system can be attempted only in a finite number of ways, and each attempt can either be carried through to completion, or be continued until further progress is impossible. It follows that every assertion of

Another kind of traditional argument for intuitionistic logic is inspired by Dummett and proceeds from semantic considerations to a rejection of classical logic in favour of intuitionistic logic. A key element of this kind of argument is a view of language, and therefore meaning, as communicable and observable, with the related thought that use exhaustively determines meaning. This brings once more the focus on *proofs* as instruments of verification of mathematical statements. Classical logic is seen as embodying a verification-transcendent notion of truth, and for this reason rejected, while intuitionistic logic is seen as fully satisfying the requirement of meaning as communicable and observable.

These arguments' focus on proofs and constructions is clearly in agreement with the perspective of both Bishop and today's constructive mathematicians. However, in light of Bishop's criticism of Brouwer, it is important to see whether these traditional arguments would overall be acceptable to a constructive mathematician. In Sections 3.8 and 3.9, I consider three possible complaints that may be raised against traditional arguments for intuitionistic logic and argue that the third one highlights a conflict between these arguments and the very practice of constructive mathematics.

3.8 Philosophical Objections

A prominent reason for constructivism's lack of popularity among philosophers today is the fact that traditional arguments for intuitionistic logic are typically bound up with forms of *anti-realism* which are rather unpopular today. For example, the first kind of argument starts from a view of mathematics as free activity of the mind (and possibly of mathematical entities as mental constructions), and therefore is committed from the start to the mind-dependence of mathematical proofs (and possibly also of mathematical entities). The second kind of argument also gives rise to a form of anti-realism, as it focusses, once more, on proofs as instruments of verification and rejects a verification-transcendent notion of truth. Given that within today's philosophy of mathematics these forms of anti-realism are widely considered either unattractive or untenable, traditional arguments for intuitionistic logic do not enjoy widespread support among contemporary philosophers.

One may wonder whether the constructive mathematician would deem these arguments unfit for the same reason. Looking again at Bishop's criticism of Brouwer's philosophy, we see that it focusses especially on those parts of philosophy that Bishop seemed to consider 'superfluous' to the mathematical practice. Bishop, on the contrary, pledged to develop his mathematics with an absolute minimum of

possibility of a construction of a bounded finite character in a finite mathematical system can be judged. So, in this exceptional case, application of the principle of the excluded third is permissible'. Here 'judged' means 'either proved or reduced to absurdity'. Brouwer then goes on to use the example of 'fleeing properties' to argue that the principle of excluded middle is not permissible for 'infinite systems' such as the natural numbers.

philosophical prejudice concerning the nature of constructive mathematics. Therefore, it is plausible that Bishop would have found traditional arguments for intuitionistic logic unpalatable in view of their alliance with anti-realism. I think that the same is probably true of many constructive mathematicians today who work in the tradition initiated by Bishop. It is, however, important to stress that Bishop and, plausibly, a constructive mathematician more generally, would object to traditional arguments for very different reasons compared with contemporary philosophers. While many contemporary philosophers find these arguments' anti-realism problematic, the constructive mathematician would not want to commit to a specific view on the nature of mathematics (mathematical entities, mathematical discourse) and for *that* reason would probably find the alliance with anti-realism unattractive.[39]

Philosophers of mathematics sometimes raise a different kind of objection against intuitionism: that it is a paradigmatic example of *philosophy-first*. For example, in Chapter 2 of [71], Shapiro discusses the relation between philosophy and mathematics and presents intuitionism, both in the Brouwerian and the Dummettian traditions, as paradigmatic examples of philosophy-first: 'the view that philosophical considerations should set the stage for and determine the proper practice of mathematics' [71, p. 21]. Traditional arguments for intuitionistic logic would seem to exemplify philosophy-first since they move from philosophical considerations, for example, specific thesis in the philosophy of mind or in the philosophy of language, and conclude with the rejection of classical mathematics. In the following, I review the key ideas of philosophy-first and argue that, although prima facie appealing, a rejection of traditional arguments for intuitionistic logic on the ground that they exemplify philosophy-first is problematic.

Shapiro claims that a philosophy-first approach to mathematics was once common, as exemplified, for example, by Plato's thought. Shapiro and other contemporary philosophers find philosophy-first approaches to mathematics problematic because purely philosophical considerations are taken to determine and fix the way mathematics is done. Many find this even more problematic when the philosophical conclusions, as in the case of intuitionism, impose a *revision* of standard mathematical practice. Shapiro [71, p. 30] writes: 'Many contemporary philosophers, including me, believe that scientists and mathematicians usually know what they are doing, and that what they are doing is worthwhile'. This has brought some philosophers to lean towards the opposite to philosophy-first, the thesis that philosophy is irrelevant to mathematics. Shapiro himself proposes a form of anti-revisionism, but does not go all the way to support what he calls a 'philosophy-last-if-at-all' approach. Furthermore, he objects to the exclusive use of philosophical considerations

[39] It is natural to ask whether the constructive mathematician's hope to maintain a neutral stance on crucial metaphysical issues can be sustained. While I cannot discuss this issue in this note, in my conclusion I suggest further work that could help clarify this point.

to restrict one's practice in general, thus even if, on their basis, one were to reject intuitionistic logic in favour of the more standard classical logic.

We have seen that Bishop, especially in [11, 12], did see a role for philosophical considerations in mathematics. He thought that disagreement over meaning has to be settled prior to disagreement over specific assumptions and techniques. He may therefore have found no fault with philosophy-first, as long as the philosophical considerations were prompted by issues of meaning, rather than based on what he considered 'speculation' regarding the nature of mathematics. What about today's constructive mathematicians? Would they find this objection to traditional arguments for intuitionistic logic compelling?

It is natural to expect that constructive mathematicians would be sympathetic to the thought that it should be the mathematician rather than the philosopher to decide which principles and techniques to employ in mathematics.[40] However, notwithstanding the appeal of this thought, I think that talk of philosophy-first may oversimplify the complex interaction between philosophy and mathematics. One may note, for example, that the debate on philosophy-first often artificially opposes mathematicians and philosophers, while historically many major mathematicians were also major philosophers (or philosophers of mathematics).[41] There is, of course, an obvious reply to this worry. One may observe that even if the same person, say Brouwer, engaged simultaneously in mathematical and philosophical inquiry, we may carefully distinguish between philosophical and mathematical components of his thought. For example, one may claim that Brouwer pursued philosophical rather than mathematical thoughts when he introduced his notion of free choice sequences. We have seen that Bishop probably thought of Brouwer along similar lines.

I find this attempt to rescue the philosophy-first objection to traditional arguments unconvincing, since more needs to be said on how to draw the line between mathematical and philosophical thinking. I tend to think that if we look more carefully at mathematicians' reasons for choosing a certain methodology or introducing some new concepts, it is likely that a number of different factors will have a role, some mathematical, some philosophical and some, furthermore, sociological in character. Typical discussions on mathematical methodology are a mixed bag, a blend of different issues that are difficult to categorise as exclusively mathematical or exclusively philosophical. This suggests that either the very notion of philosophy-first is hopelessly imprecise, or that one should offer a very careful formulation of it and, consequently, of this objection. One possible strategy would be to try and formulate

40 This does not mean that they would also support a form of anti-revisionism which sanctions classical mathematics, as they would rather claim that there are good mathematical reasons to revise the standard classical practice.

41 This point is acknowledged in [71, p. 31]. I would like to thank a referee for suggesting to further develop this point.

the latter in terms of reasons that are *predominantly* philosophical or mathematical in character, rather than exclusively so. It is unclear to me if this could be done in a satisfactory way, but perhaps, if it can be done, it would suffice to express the worry that traditional arguments for intuitionistic logic are examples of philosophy-first.

Let us suppose, for the sake of argument, that we can meaningfully talk of philosophy-first and that traditional arguments for intuitionistic logic do constitute examples of it. Does this imply that the constructive mathematician should reject those arguments on this ground? I think the constructive mathematician should clearly say 'no'. Whatever the reasons for intuitionistic mathematics, the key question the mathematician will ask is whether the resulting mathematics is interesting and fruitful. Banning philosophy-first arguments a priori could then result in obstructing the development of fruitful and interesting mathematics, making mathematical progress more difficult. Brouwer's intuitionism is an emblematic example. One may claim that Brouwer's reasons for intuitionistic mathematics were predominantly philosophical in character, and find this unsatisfactory in some respect. However, it is clear that, this notwithstanding, the arguments Brouwer adduced for intuitionism had important and useful mathematical consequences, as they gave rise to the discovery of intuitionistic logic and opened up a whole new realm of mathematics. For these reasons, I think the constructive mathematician should not object to traditional arguments for intuitionistic logic on the sole basis that they are examples of philosophy-first.

On reflection, it seems that the philosophical discussion on philosophy-first highlights a different point. We have seen that those who object to traditional arguments for intuitionistic logic because they consider them paradigmatic examples of philosophy-first often express concerns for the revisionary spirit of these arguments. This suggests that they are concerned not only with the motives supporting the premises of these arguments, but also with these arguments' consequences, namely the fact that they demand a thorough change of the mathematical practice. In the next section, I focus, although from a different perspective, on crucial consequences of traditional arguments for intuitionistic logic and argue that they suggest the need for new arguments for constructive mathematics.

3.9 Too Strong

I believe that a better reason for objecting to traditional arguments for intuitionistic logic is that they are *too strong*. These arguments, as we have seen, entail the outright *rejection* of classical logic and, consequently, of classical mathematics. They are often taken to imply that classical mathematics is incoherent, and are sometimes also read as entailing the thorough unintelligibility of classical mathematics. For example, in his famous article 'The philosophical basis of intuitionist logic'

[36, p. 215], Dummett asks 'what plausible rationale can there be for repudiating, within mathematical reasoning, the canons of classical logic in favour of intuitionistic logic?'. Dummett clarifies that he is not concerned with 'justifications of intuitionistic mathematics from an eclectic point of view, that is, one which admits intuitionistic mathematics as a legitimate and interesting form of mathematics alongside classical mathematics'. Dummett's concern is rather the standpoint of the intuitionists themselves, who took classical mathematics to employ forms of reasoning which are *invalid on any legitimate way of construing mathematical statements*. Similarly to the view that Dummett examines in his article, also the traditional arguments for intuitionistic logic we considered above are usually taken to completely reject classical mathematics as illegitimate.

According to Bishop, classical mathematics is *deficient in meaning* so much so that he hopes that constructive mathematics will eventually replace it. Bishop does not, however, maintain that classical mathematics is outright illegitimate. The constructive mathematician may stress, like Bishop, that there are good, indeed, better reasons to work constructively, compared with working classically, and that classical mathematics as a whole cannot be given constructive meaning, as not every classical theorem can be given a computational interpretation. However, there are parts of classical mathematics that do have computational content, and we can make some sense of the rest, for example, in terms of conditional statements that prefix a suitable classical statement to a constructively meaningful one.[42] Furthermore, at least initially, classical mathematics is seen as a guide that helps the constructive mathematician develop new mathematics. For example, Bishop writes [9, p. x]:

> We are not contending that idealistic mathematics is worthless from the constructive point of view. This would be as silly as contending that unrigorous mathematics is worthless from the classical point of view. Every theorem proved with idealistic methods presents a challenge: to find a constructive version, and to give it a constructive proof.

For these reasons, I believe, Bishop should consider traditional arguments for intuitionistic logic not viable, as they imply the outright rejection of classical mathematics, if not its unintelligibility. I would think that many constructive mathematicians (Bishop-style) would also see things in essentially this way. One reason is that there is a tension between these arguments' conclusions and the contemporary constructive practice. First of all, claims of utter unintelligibility of classical mathematics are implausible given the above-mentioned use by the constructive mathematician of classical proofs as an initial guide. Second, constructive mathematicians believe that much of classical mathematics does not possess the same clear constructive meaning as a piece of constructive mathematics, and find classical

42 See Section 3.4.

mathematics unappealing for that reason. But they would certainly claim that they understand a classical theorem as clearly as any classical mathematician. In fact, as mentioned in Section 3.4, they would argue that they can offer a more precise analysis of a classical theorem, separating its constructive core from (possibly) an essentially classical component, such as, for example, LPO. More importantly, the constructive reverse mathematics programme, which was mentioned in Section 3.4, also requires a more tolerant approach to classical mathematics. One of its stated aims is to clarify the relation between concepts and theorems belonging to a number of mathematical practices, among which the classical one. A crucial claim of the constructive reverse mathematics programme is that constructive mathematics offers a 'neutral' perspective, on the basis of which to analyse classical mathematics.[43] Classical mathematics therefore is not to be rejected and, I would suggest, also not completely devoid of interest from the perspective of the constructive mathematician. Constructive mathematics is taken to be highly preferable, among other reasons for its computational content and because it offers an ideal ground from which to carry out a fine comparison between mathematical notions, theorems and proofs in different contexts. However, the constructive mathematician cannot on these sole grounds outlaw classical mathematics.

3.10 Concluding Remarks

If I am right to think that the constructive mathematician cannot accept traditional arguments for intuitionistic logic because their consequences are too strong, this raises a pressing question for the philosopher of mathematics: are there other arguments for constructive mathematics that can play a similar role as traditional arguments for intuitionistic mathematics? As a first step towards answering this question, we may consider what are the reasons for doing mathematics constructively. We have seen that Bishop's aim was to develop a 'meaningful' form of mathematics, one that 'predicts the results of certain finitely performable, albeit hypothetical, computations with the set of integers' [10, p. 53]. For Bishop, this meant that working constructively also deepens mathematics by making available important distinctions that a classical mathematician does not perceive. Furthermore, the constructive approach makes it possible to develop a computational form of mathematics which has systematic application to real-life computers. In fact, for today's constructive mathematician, the principal reason for working constructively is the *direct computational content* of constructive mathematics. The other reasons Bishop mentions are also important. Thirty-odd years ago, Fred Richman [66, 67] further developed some of Bishop's remarks, arguing that constructive mathematics has the advantage of being more *general* than classical mathematics. Since

[43] This is a strong claim. See, for example, [74] for criticism.

constructive mathematics avoids the use of the excluded middle (and cognate principles such as LPO) but does not introduce principles that diverge from classical mathematics (contrary to Brouwerian mathematics), all of its theorems are also classically true. Therefore, constructive mathematics is more general than classical mathematics. Richman's notion of generality may be clarified by his comparison with algebra [66, p. 126]:

> Because intuitionistic mathematics is the weaker theory, its theorems have more models, so they are more general: for example, a theorem that holds for all groups is more general than one holding only for abelian groups.

Similarly to Bishop, Richman also stresses the importance of the possibility of distinguishing between mathematical concepts which are routinely identified in classical contexts.[44] I would like to call this feature of constructive mathematics *refinement*, as it allows for finer distinctions compared with classical mathematics. To summarise, the analysis of Bishop and Richman suggest that there are at least three main reasons for working constructively: (i) the possibility of giving direct *computational meaning* to mathematical statements, in particular one that can be readily applied to computers, (ii) the greater *generality* (in the sense above) of the resulting theorems, and (iii) *refinement*, namely the availability of significant distinctions that are unavailable within a classical context.

Can these reasons support a new argument for constructive mathematics? Billinge [7] seems to think so, but argues that it would be a mathematical rather than a philosophical argument. Billinge considers generality and refinement, and thinks of them primarily as *mathematical* rather than philosophical motives for constructive mathematics.[45] She also distinguishes between a *liberal* and a *radical* constructivist.[46] The first 'believes that constructive mathematics is preferable to classical mathematics, but that classical mathematics is at least coherent'. The second 'takes it that classical mathematics is absolutely illegitimate and cannot be rendered coherent under any interpretation' [7, p. 177]. She argues that Bishop did his mathematics in a constructive manner for explicit philosophical reasons and that he was a liberal rather than a radical constructivist. But she also argues that Bishop's philosophical comments cannot be fleshed out into an adequate philosophy of constructive mathematics. Billinge [7, p. 188] claims that the basic premises of Bishop's position are controversial, in particular so are Bishop's main assumptions – that all mathematical statements should have numerical content and that existence claims should be interpreted constructively. Her main complaint is that Bishop does not

[44] See the example on page 76.

[45] Surprisingly, in her concluding discussion in [7], Billinge does not mention the computational content as a key reason for doing mathematics constructively, although she discusses it in relation to the special status of the natural numbers within Bishop's philosophy.

[46] See also [47, p. 222].

give good enough grounds for accepting these controversial assumptions, and that, as a consequence, Bishop's philosophical remarks cannot be taken to fully support liberal constructivism.

Billinge [7, p. 192] thinks that generality and refinement, as spelled out by Richman, are key *mathematical* advantages of working constructively and provide 'the most promising argument for liberal constructivism at the moment'. However, such an argument would not be a *philosophical* argument. In fact, Billinge [7, pp. 190–191] claims that she cannot see 'how one could give purely philosophical arguments for the superiority of constructive mathematics without overplaying one's hand and concluding that constructive mathematics is the only acceptable way of doing mathematics'. Succinctly: 'any adequate *philosophical* defence of constructive mathematics will justify *radical* constructivism'. [7, p. 192].

I agree with Billinge that overall Bishop's texts suggest a liberal rather than a radical constructivist position and also that Bishop does not offer a full philosophical defence of his claims. I think that this should not be surprising, as Bishop was a mathematician whose main focus was the concrete mathematical activity and whose philosophical views are briefly presented in remarks which appear primarily in introductions to technical work or in lecture notes. Furthermore, I also agree with Billinge that generality and refinement are important motives for constructive mathematics and that they are primarily motivated by mathematical needs, rather than explicit philosophical considerations. I am not persuaded, however, that it is utterly implausible that these reasons could play a key role in a philosophical argument for liberal constructivism. I cannot argue for this here due to space constraints. I will rather focus on a different point that is more relevant in the present context.

As already mentioned, for many contemporary constructive mathematicians the direct computational content of constructive mathematics is the primary motive for developing this form of mathematics. In view of Billinge's discussion, one may wonder whether this should count as primarily philosophical, mathematical or, perhaps, neither, for example, as an external practically motivated reason for doing mathematics constructively? I can think of at least two different ways of looking at the computational content of mathematical statements. First, there is the fact that if a mathematical statement can be given computational meaning then this can be employed in computer applications. In this sense, arguing for constructive mathematics on the basis of the availability of direct computational meaning seems to rely on pragmatic considerations, external to the mathematical practice. Second, the focus on the computational content may instead be determined by a preference for algorithmic proofs, independently from the possibility of applications.[47] Some

[47] See also [17, 20] for a discussion of the algorithmic nature of constructive mathematics. See [29] for a similar distinction.

mathematicians have a preference for proofs that are more algorithmic and explicit, proofs that construct their witnesses step-by-step. We have seen Bishop's focus on the natural numbers and on finite operations over the natural numbers. For Bishop the natural numbers have a key foundational role in mathematics, that he assimilates to the role they played for Kronecker. It is in this sense that the computational content can be taken as independent of practical considerations and not merely a mathematical but also as a philosophical reason for doing mathematics constructively.

Billinge takes the computational content (in this second sense) to be one of Bishop's main controversial assumptions, and argues that not only does Bishop inadequately support this assumption but, in fact, that it cannot be given adequate support in a way that coheres with Bishop's overall views. My impression is that Billinge reaches this conclusion because she takes Bishop to suggest an ontological reduction of mathematical entities to the natural numbers, that is, the thought that every mathematical entity can ultimately be reduced to some combination of natural numbers. She also suggests that for Bishop we have direct epistemic access to the natural numbers in a way that is analogous to our access to the physical world via sense perception. She then argues that this can only be supported if we take mathematical entities to be mental constructions, which would contradict Bishop's desire to remain neutral on metaphysical issues.

I do not think Bishop proposes such a reductive strategy, exactly because he argues against taking a specific stance on the nature of mathematical entities. This is also evidenced by the fact that Bishop frequently stresses the role in constructive mathematics of finitely performable *operations* with the natural numbers. This is suggestive that Bishop's focus is not the natural number themselves, but the possibility of dynamically developing mathematics via finite operations with the natural numbers, whatever the natural numbers may be. I take this to support the view that Bishop is not arguing, as Billinge [7, p. 188] claims, that every mathematical entity should be reduced to the natural numbers, that is, that the rational numbers, the real or the complex numbers are *really* just sets of natural numbers. I rather think that Bishop's remarks are better read in epistemological terms, leaving unanswered the question of the real nature of mathematical entities (including the natural numbers).

In fact, even if Bishop's own remarks are not to be read as I suggest, I believe that if we were to expand and build on his philosophical remarks more generally, the best strategy would be to focus on the epistemological claim that the natural numbers have a fundamental role in our understanding of mathematics in general. I take Bridges [19] to give a somewhat similar interpretation of contemporary constructivism, though without the commitment to the primality of the natural numbers that characterises Bishop's approach. Bridges [19] distinguishes between an ontological and an epistemological form of constructivism. He associates the first one with Brouwer and sees it as motivated by the belief that mathematical objects are mental creations. The second one focusses on methodological issues and takes them to motivate the shift to intuitionistic logic. Bridges takes today's constructivists of the Bishop school to be epistemological constructivists, rather than ontological constructivists.

I am tempted to think that there is the possibility of giving philosophical substance to a Bishop-inspired form of epistemological constructivism, i.e. one which focuses on the methodology of mathematics and reaches constructivism on the basis of a blend of mathematical and philosophical considerations. If understood in this way, Bishop's discussion of the distinction between finite and infinite domains gains a new prominence. A natural way to expand Bishop's remarks would be to look at the predicativist tradition, and especially Weyl's thought.[48] I am inclined to think that in this way Bishop's philosophical remarks may be enriched to give a new argument for constructive mathematics that takes the natural numbers as fundamental without rejecting *tout court* classical mathematics.[49]

Acknowledgements

I would like to thank a referee for insightful comments that helped sharpen some of my claims and for drawing my attention to [56, 57]. My thanks to Douglas Bridges for comments on a draft of this chapter. I would also like to thank the organisers of the Trimester *Types, Sets and Constructions* at the Hausdorff Research Institute for Mathematics (2018), Douglas Bridges, Michael Rathjen, Peter Schuster, and Helmut Schwichtenberg, for their invitation to speak at the conference 'Constructive Mathematics' and the audience of that meeting for helpful feedback. Finally, my thanks to the organisers and the audience of the conference 'Constructive Mathematics: Foundations and Practice' held in Niŝ (2013) for discussions that initiated some of the thoughts in the second part of this article. The research leading to this chapter has received funding from the European Union's Horizon 2020 research and innovation programme under the Marie Sklodowska-Curie grant agreement No. 838445.

References

[1] Aczel, P. 1978. The type theoretic interpretation of constructive set theory. Pages 55–66 of: MacIntyre, A., Pacholski, L., and Paris, J. (eds), *Logic Colloquium '77*. Amsterdam, New York: North–Holland.

[48] See [28, 30] for discussion and references.

[49] This will be developed in future work. Note that there is a substantial difference between Bishop and Weyl, as Bishop's constructivism has a dynamic component, informed by the advent of the computer, that does nor figure in Weyl.

[2] AGDA. 2020. *AGDA Wiki.* Available at http://wiki.portal.chalmers.se/agda/pmwiki.php.

[3] Beeson, M. 1985. *Foundations of Constructive Mathematics.* Berlin: Springer Verlag.

[4] Benacerraf, P., and Putnam, H. 1983. *Philosophy of Mathematics: Selected Readings.* Cambridge: Cambridge University Press.

[5] Bickford, M., Cohen, L., Constable, R. L., and Rahli, V. 2018. Computability beyond Church–Turing via choice sequences. Pages 245–254 of: *Proceedings of the 33rd Annual ACM/IEEE Symposium on Logic in Computer Science.* LICS '18. New York: Association for Computing Machinery.

[6] Billinge, H. 1998. *A defence of constructive mathematics.* Ph.D. thesis, Department of Philosophy, University of Leeds.

[7] Billinge, H. 2003. Did Bishop have a philosophy of mathematics? *Philos. Math.*, **11**(2), 176–194.

[8] Billinge, H. 1997. A constructive formulation of Gleason's theorem. *J. Philos. Logic*, **26**(6), 661–670.

[9] Bishop, E. 1967. *Foundations of Constructive Analysis.* New York: McGraw-Hill.

[10] Bishop, E. 1970. Mathematics as a numerical language. Pages 53–71 of: Kino, A., Myhill, J., and Vesley, R. E. (eds.), *Intuitionism and Proof Theory.* Amsterdam: North-Holland

[11] Bishop, E. 1975. The crisis in contemporary mathematics. *Historia Math.*, **2**, 507–517.

[12] Bishop, E. 1985. Schizophrenia in contemporary mathematics. Pages 1–32 of: Rosenblatt, M. (ed.), *Errett Bishop: Reflections on Him and His Research.* Contemporary Mathematics, vol. 39. Providence, RI: American Mathematical Society.

[13] Bishop, E. 2012. *Foundations of Constructive Analysis.* New York: Ishi Press International.

[14] Bishop, E., and Bridges, D. S. 1985. *Constructive Analysis.* Berlin, Heidelberg: Springer.

[15] Bishop, E., and Cheng, H. 1972. *Constructive Measure Theory.* Memoirs of the American Mathematical Society, no. 116. Providence, RI: American Mathematical Society.

[16] Bridges, D. S. 1995. Constructive mathematics and unbounded operators – A reply to Hellman. *J. Philos. Logic*, **24**, 549–561.

[17] Bridges, D. S. 1999. Constructive Mathematics: a Foundation for Computable Analysis. *Theor. Comput. Sci.*, **219**, 95–109.

[18] Bridges, D. S. 2005. Errett Bishop. In: Crosilla, L., and Schuster, P. (eds.), *From Sets and Types to Topology and Analysis: Towards Practicable Foundations for Constructive Mathematics.* Oxford Logic Guides, vol. 48. Oxford: Oxford University Press.

[19] Bridges, D. S. 2009. Constructive mathematics. In: Zalta, E. N. (ed.), *The Stanford Encyclopedia of Philosophy*. The Metaphysics Research Lab Center for the Study of Language and Information Stanford University.

[20] Bridges, D. S., and Reeves, S. 1999. Constructive mathematics, in theory and programming practice. *Phil. Math.*, **7**, 65–104.

[21] Bridges, D. S., and Richman, F. 1987. *Varieties of Constructive Mathematics*. Cambridge: Cambridge University Press.

[22] Bridges, D. S., and Vîţă, L. S. 2006. *Techniques of Constructive Analysis*. Universitext. Berlin: Springer–Verlag.

[23] Brouwer, L. E. J. 1952. Historical background, principles and methods of intuitionism. *S. Afr. J. Sci.*, **49**, 139–146.

[24] Carter, J. 2019. Philosophy of mathematical practice – motivations, themes and prospects. *Philos. Math.*, **27**(1), 1–32.

[25] Constable, R. L., Allen, S. F., Bromley, H. M., et al. 1986. *Implementing Mathematics with the Nuprl Proof Development System*. Englewood Cliffs, NJ: Prentice-Hall.

[26] Cook, R. 2005. Intuitionism Reconsidered. Pages 387–411 of: Shapiro, S. (ed.), *Oxford Handbook of Philosophy of Mathematics and Logic*. Oxford: Oxford University Press.

[27] Coquand, T., and Huet, G. 1986 (May). *The calculus of constructions*. Technial report RR-0530. INRIA.

[28] Crosilla, L. 2015. Error and predicativity. Pages 13–22 of: Beckmann, A., Mitrana, V., and Soskova, M. (eds.), *Evolving Computability*. Lecture Notes in Computer Science, vol. 9136. Cham: Springer International Publishing.

[29] Crosilla, L. 2016. Matematica costruttiva. *APhEx: portale italiano di filosofia analitica*, **9136**. APhEx / Rivista con periodicità semestrale / ISSN 2036-9972.

[30] Crosilla, L. 2017. Predicativity and Feferman. Pages 423–447 of: Jäger, G., and Sieg, W. (eds.), *Feferman on Foundations: Logic, Mathematics, Philosophy*. Cham: Springer International Publishing.

[31] Crosilla, L. 2019. The entanglement of logic and set theory, constructively. *Inquiry*, **65**(6), 638–659.

[32] Crosilla, L. 2020. *Predicativity and indefinite extensibility*. Unpublished manuscript.

[33] Crosilla, L. 2022. Predicativity and constructive mathematics. In: Oliveri, G., Ternullo, C., and Boscolo, S. (eds.), *Objects, Structures, and Logics*. Boston Studies in the Philosophy and History of Science, vol. 339. Cham: Springer International Publishing.

[34] Crosilla, L., and Schuster, P. (eds.). 2005. *From Sets and Types to Topology and Analysis: Towards Practicable Foundations for Constructive Mathematics*. Oxford Logic Guides, vol. 48. Oxford: Oxford University Press.

[35] Diener, H. 2020. *Constructive Reverse Mathematics*. arXiv preprint arXiv:1804.05495.

[36] Dummett, M. 1975. The philosophical basis of intuitionistic logic. Pages 5–40 of: Rose, H. E., and Shepherdson, J. C. (eds.), *Logic Colloquium '73*, vol. 80. Studies in Logic and the Foundations of Mathematics. Amsterdam: North-Holland.

[37] Dummett, M. 1975. Wang's paradox. *Synthese*, **30**(3–4), 201–32.

[38] Dummett, M. 1977. *Elements of Intuitionism*. Oxford: Oxford University Press.

[39] Dummett, M. 2000. *Elements of Intuitionism*, 2nd ed. Oxford: Oxford University Press.

[40] Feferman, S. 1975. A language and axioms for explicit mathematics. Pages 87–139 of: Crossley, J. (ed.), *Algebra and Logic*. Lecture Notes in Mathematics, vol. 450. Berlin: Springer.

[41] Feferman, S. 1979. Constructive theories of functions and classes. In: *Logic Colloquium '78*. Amsterdam: North-Holland.

[42] Ferreiros, J. 2016. *Mathematical Knowledge and the Interplay of Practices*. Princeton, NJ: Princeton University Press.

[43] Friedman, H. 1973. The consistency of classical set theory relative to a set theory with intuitionistic logic. *J. Symbol. Logic*, **38**, 315–319.

[44] Friedman, H. 1977. Set-theoretical foundations for constructive analysis. *Ann. Math.*, **105**, 1–28.

[45] Goodman, N. D. 1981. Reflections on Bishop's philosophy of mathematics. Pages 135–145 of: Richman, F. (ed.), *Constructive Mathematics*. Lecture Notes in Mathematics, vol. 873. Berlin, Heidelberg: Springer.

[46] Hellman, G. 1989. Never say "Never"!: On the communication problem between intuitionism and classicism. *Philos. Topics*, **17**(2), 47–67.

[47] Hellman, G. 1993a. Constructive mathematics and quantum mechanics: Unbounded operators and the spectral theorem. *J. Philos. Logic*, **22**, 221–248.

[48] Hellman, G. 1993b. Gleason's theorem is not constructively provable. *J. Philos. Logic*, **22**(2), 193–203.

[49] Hellman, G. 1998. Mathematical constructivism in spacetime. *Brit, J. Philos. Sci.*, **49**(3), 425–450.

[50] Heyting, A. 1931. Die Intuitionistische Grundlegung der Mathematik. *Erkenntnis*, **2**(1), 106–115. Reprinted in [4], pages 52–61, with the title *The Intuitionist Foundations of Mathematics*. (Page references are to the reprinting).

[51] Heyting, A. 1956. *Intuitionism an Introduction*. Amsterdam: North-Holland.

[52] Hilbert, D. 1927. Die Grundlagen der Mathematik. *Abhandlungen aus dem Seminar der Hamburgischen Universität*, **6**, 65–85. Reprinted and translated as *The Foundations of Mathematics*, in van Heijenoort, 1967, pp. 464–479.

[53] Ishihara, H. 2005. Constructive reverse mathematics: compactness properties. In: Crosilla, L., and Schuster, P. (eds.), *From Sets and Types to Topology and Analysis: Towards Practicable Foundations for Constructive Mathematics*. Oxford Logic Guides, vol. 48. Oxford: Oxford University Press.

[54] Ishihara, H. 2006. Reverse mathematics in Bishop's constructive mathematics. *Philos. Scient.*, **6**, 43–59.

[55] Ishihara, H. 2022. An introduction to constructive reverse mathematics. In: Bridges, D., Ishihara, H., Rathjen, M., and Schwichtenberg, H. (eds.), *Handbook of Constructive Mathematics*. Cambridge: Cambridge University Press.

[56] Kreisel, G. 1970. Church's thesis: a kind of reducibility axiom for constructive mathematics. Pages 121–150 of: Kino, A., Myhill, J., and Vesley, R. E. (eds.), Studies in Logic and the Foundations of Mathematics, vol. 60. Amsterdam: Elsevier.

[57] Kreisel, G., and MacIntyre, A. 1982. Constructive logic versus algebraization I. Pages 217–260 of: Troelstra, A. S., and van Dalen, D. (eds.), *The L. E. J. Brouwer Centenary Symposium*. Studies in Logic and the Foundations of Mathematics, vol. 110. Amsterdam: Elsevier.

[58] Mancosu, P. 2008. *The Philosophy of Mathematical Practice*. Oxford: Oxford University Press.

[59] Martin-Löf, P. 1975. An intuitionistic theory of types: predicative part. In: Rose, H. E., and Shepherdson, J. C. (eds.), *Logic Colloquium 1973*. Amsterdam: North–Holland.

[60] Mines, R., Richman, F., and Ruitenburg, W. 1988. *A Course in Constructive Algebra*. Universitext. New York: Springer.

[61] Myhill, J. 1972. Review: Errett Bishop, Foundations of Constructive Analysis. *J. Symbol. Logic*, **37**(4), 744–747.

[62] Myhill, J. 1973. Some properties of intuitionistic Zermelo–Fraenkel set theory. Pages 206–231 of: *Cambridge Summer School in Mathematical Logic*. Lecture Notes in Mathematics, vol. 337. Berlin: Springer.

[63] Myhill, J. 1975. Constructive set theory. *J. Symbol. Logic*, **40**, 347–382.

[64] Nerode, A., Metakides, G., and Constable, R. 1985. Remembrance of Errett Bishop. Pages 79–84 of: Rosenblatt, M. (ed.), *Errett Bishop: Reflections on Him and His Research*. Contemporary Mathematics, vol. 39. American Mathematical Society.

[65] Nordström, B., Petersson, K., and Smith, J. M. 1990. *Programming in Martin-Löf's Type Theory: an Introduction*. Oxford: Clarendon Press.

[66] Richman, F. 1990. Intuitionism as generalization. *Philos. Math.*, **5**, 124–128.

[67] Richman, F. 1996. Interview with a constructive mathematician. *Modern Logic*, **6**, 247–271.

[68] Richman, F. 2000. Gleason's theorem has a constructive proof. *J. Philos. Logic*, **29**, 425–431.

[69] Richman, F., and Bridges, D. 1999. A constructive proof of Gleason's theorem. *J. Funct. Anal.*, **162**, 287–312.

[70] Shapiro, S. 2000. *Thinking About Mathematics: The Philosophy of Mathematics*. Oxford: Oxford University Press.

[71] Shapiro, S. 1997. *Philosophy of Mathematics: Structure and Ontology*. Oxford: Oxford University Press.

[72] Shapiro, S. 2014. *Varieties of Logic*. Oxford: Oxford University Press.
[73] Stolzenberg, G. 1970. Review: Errett Bishop, Foundations of Constructive Analysis. *Bull. Amer. Math. Soc.*, **76**(2), 301–323.
[74] Sundholm, G. 2014. *Constructive Recursive Functions, Church's Thesis, and Brouwer's Theory of the Creating Subject: Afterthoughts on a Parisian Joint Session*, pages 1–35. Dordrecht: Springer Netherlands.
[75] The Univalent Foundations Program. 2013. *Homotopy Type Theory: Univalent Foundations of Mathematics*. Institute for Advanced Study.
[76] Veldman, W. 2014. Brouwer's fan theorem as an axiom and as a contrast to Kleene's alternative. *Arch. Math. Logic*, **53**, 621–693.
[77] Veldman, W. 2021. Intuitionism: an inspiration? *Jahresber. Dtsch. Math. Ver.*, **123**, 221–284.
[78] Weyl, H. 1918. *Das Kontinuum. Kritische Untersuchungen über die Grundlagen der Analysis*. Lepizig: Veit. Translated in [80].
[79] Weyl, H. 1921. Über die neue Grundlagenkrise der Mathematik. *Math. Zeitschr.*, **10**, 39–79.
[80] Weyl, H. 1994. *The Continuum: a Critical Examination if the Foundations of Analysis*, translated by S. Pollard and T. Bole. New York: Dover.
[81] Wright, C. 1982. Strict finitism. *Synthese*, **51**(2), 203–282.

PART II

ALGEBRA AND GEOMETRY

4

Algebra in Bishop's style: A Course in Constructive Algebra

Henri Lombardi

4.1 Introduction

The reception of the book *A Course in Constructive Algebra* by Mines, Richman, and Ruitenburg [7] (cited as [CCA] in the sequel) in France is even more confidential than that of Bishop's book [2]. I have hardly ever met a French mathematician who has but heard of the existence of the book.

The Computer Algebra community could be expected to be a little more up-to-date since all theorems in [CCA] have a computational content, and could, at least in principle, be implemented in the usual Computer Algebra softwares.

Some years ago I submitted an article of constructive algebra to the section 'Computer Algebra' of the *Journal of Algebra*, a section whose recommendations to the authors explicitly indicate the interest of the journal for constructive mathematics. I was surprised when the referee asked me to explain the precise meaning of 'or' in constructive mathematics, because he was confused and did not understand some arguments. The article was finally rejected in this section of the *Journal of Algebra*, apparently because of the impossibility of finding a competent referee.

Nevertheless I have recently discovered the following article by Sebastian Posur, *A constructive approach to Freyd categories*.

`https://arxiv.org/abs/1712.03492`

Here is an excerpt from Section 2, 'Constructive category theory'. This article seems to me to be a salutary and expected turning point.

> To present our algorithmic approach to Freyd categories, we chose the language of constructive mathematics (see, e.g., [MRR88]). We did that for the following reasons: the language of constructive mathematics
>
> (i) reveals the algorithmic content of the theory of Freyd categories,

(ii) is perfectly suited for describing generic algorithms, i.e., constructions not depending on particular choices of data structures,
(iii) allows us to express our algorithmic ideas without choosing some particular model of computation (like Turing machines),
(iv) encompasses classical mathematics, i.e., all results stated in constructive mathematics are also valid classically,
(v) does not differ very much from the classical language in our particular setup.

In constructive mathematics the notions of data types and algorithms (or operations) are taken as primitives and every property must have an algorithmic interpretation. For example given an additive category $\mathbf{A}$ we interpret the property

$$\mathbf{A} \text{ has kernels}$$

as follows: we have algorithms that compute for given

- $A, B \in \mathrm{Obj}_{\mathbf{A}}$, $\alpha \in \mathrm{Hom}_{\mathbf{A}}(A, B)$, an object $\ker(\alpha) \in \mathrm{Obj}_{\mathbf{A}}$ and a morphism
$$\mathrm{KernelEmbedding}(\alpha) \in \mathrm{Hom}_{\mathbf{A}}(\ker(\alpha), A)$$
for which $\mathrm{KernelEmbedding}(\alpha) \cdot \alpha = 0$,
- $A, B, T \in \mathrm{Obj}_{\mathbf{A}}$, $\alpha \in \mathrm{Hom}_{\mathbf{A}}(A, B)$, $\tau \in \mathrm{Hom}_{\mathbf{A}}(T, A)$ such that $\tau \cdot \alpha = 0$ a morphism $u \in \mathrm{Hom}_{\mathbf{A}}(T, \ker(\alpha))$ such that
$$u \cdot \mathrm{KernelEmbedding}(\alpha) = \tau,$$
where u is uniquely determined (up to $=$) by this property.

Another important example is given by *decidable equality*, where we interpret the property that for all objects $A, B \in \mathbf{A}$, we have

$$\forall \alpha, \beta \in \mathrm{Hom}_{\mathbf{A}}(A, B) : (\alpha = \beta) \vee (\alpha \neq \beta)$$

as follows: we are given an algorithm that decides or disproves equality of a given pair of morphisms. . .

On the other hand, we allow ourselves to work classically whenever we interpret Freyd categories in terms of finitely presented functors. The reason for this is pragmatic: we want to demonstrate the usefulness of having Freyd categories computationally available, and we believe that this can be done by interpreting Freyd categories in terms of other categories that classical mathematicians care about.

4.2 Revisiting Bishop's Set Theory

The authors of [CCA] introduce a philosophy of mathematics that differs slightly from that of [2]. This point of view is probably expressed more directly in the papers [11, 12] and in the book [4].

First of all, as in Bishop, the point of view is not that of formalized mathematics, but of mathematics open to unpredictable developments, and for which the only criterion of truth is the conviction given by a proof.

The mathematical universe is thus not preexisting; it is, on the contrary. a properly human construction for the use of the human community.

A novelty is the following. The general point of view is to consider that all mathematics, classical as well as constructive, deal with the same ideal objects. The unique difference is in the tools used for the investigation of this universe. Constructive mathematics are more general than classical mathematics since they use neither Law of Excluded Middle (LEM) nor Choice. In exactly the same way, the theory of groups is more general than the theory of abelian groups, since commutativity is not assumed.

Let us quote a passage.

> Our notion of what constitutes a **set** is a rather liberal one.
>
> **[I.]2.1 Definition.** A set S is defined when we describe how to construct its members from objects that have been, or could have been, constructed prior to S, and describe what it means for two members of S to be equal.
>
> Following Bishop we regard the **equality relation** on a set as conventional: something to be determined when the set is defined, subject only to the requirement that it be an equivalence relation.
>
> ....
>
> A unary relation P on S defines a **subset** $A = \{x \in S : P(x)\}$ of S : an element of A is an element of S that satisfies P, and two elements of A are equal if and only if they are equal as elements of S. If A and B are subsets of S, and if every element of A is an element of B, then we say that A is **contained** in B, and write $A \subseteq B$. Two subsets A and B of a set S are **equal** if $A \subseteq B$ and $B \subseteq A$; this is clearly an equivalence relation on subsets of S. We have described how to construct a subset of S, and what it means for two subsets of S to be equal. Thus we have defined the set of all subsets, or the **power set**, of S.

This is rather surprising for a follower of Bishop. The authors of [CCA] think that the notion of 'a unary relation defined on a given set' is so clear that we may consider a well-defined set of all these unary relations. In other words, we know how to construct these unary relations, in a similar way as for example we know how to construct a nonnegative integer, or a real number, or a real function. But this seems problematic since nobody thinks that it is possible to have a universal language for mathematics allowing us to codify these relations. In particular, if the set Ω of subsets of the singleton $\{0\}$ exists, this means that truth values form a set rather than a class. This seems to say that we know a priori all the truth values that may appear in the future development of mathematics.

In fact, it seems that each time a 'set of all subsets of ... ' is used in the book, this happens in a context where only a well-defined set of subsets (in the usual, Bishop, meaning) is necessary. So the set of *all* subsets is not really needed. Or sometimes the quantification over this *set* is not needed.[1]

For example, let us see the following theorem, whose proof is incredibly simple and elegant.[2]

> The decomposition in Theorem [V.]2.3 is essentially unique over an arbitrary commutative ring.
>
> **[V.]2.4 Theorem.** *Let R be a commutative ring, $m \leq n$ positive integers, and $I_1 \supseteq I_2 \supseteq \cdots \supseteq I_m$ and $J_1 \supseteq J_2 \supseteq \cdots \supseteq J_n$ ideals of R. Suppose M is an R-module that is isomorphic to $\Sigma_{i=1}^{m} R/I_i$ and to $\Sigma_{j=1}^{n} R/J_j$. Then*
>
> (a) $J_1 = J_2 = \cdots = J_{n-m} = R.$
> (b) $I_i = J_{n-m+i}$ *for* $i = 1, \ldots, m.$

Here there is no hypothesis on the ideals I_i and J_j. If you would want to formalize completely the discourse, you need the quantification over all ideals of R, but you don't really need this complete formalization. Similarly, we do not need to quantify over the class of all commutative rings when we write: "Let R be a commutative ring". See [9] for a formal system using class quantification.

Note, however, the following passage, which deals with the category of sets, and where the set Ω of all subsets of $\{0\}$ plays a crucial role. Note also that the nice Theorem I.4.1 seems to be mainly aesthetic, without more concrete applications, within the framework of the theory of the categories.

[1] The most important exception is in the definition of well-founded sets and ordinals (see later in the present section).

[2] This theorem is not found in classical textbooks. Bourbaki, N. 2003. *Elements of Mathematics, Algebra II*, Chapters 4–7. Berlin, Heidelberg: Springer (*Algebra*, Chapter VII, Paragraph 4, Section 1)], perhaps the best text for this problem, gives the theorem only for the case $m = n$, $I_1 \neq R$, and $J_1 \neq R$. And the proof is less beautiful than in [CCA].

[...] The categorical property corresponding to a function f being one-to-one is that if g and h are maps from any set C to A, and $fg = fh$, then $g = h$; that is, f is **left cancellable**. It is routine to show that f is one-to-one if and only if it is left cancellable.

A map f from A to B is onto if for each b in B there exists a in A such that $f(a) = b$. The corresponding categorical property is that f be **right cancellable**, that is, if g and h are maps from B to any set C, and $gf = hf$, then $g = h$. The proof that a function f is right cancellable if and only if it is onto is less routine than the proof of the corresponding result for left cancellable maps.

[I.]4.1 Theorem. *A function is right cancellable in the category of sets if and only if it is onto.*

Proof Suppose $f : A \to B$ is onto and $gf = hf$. If $b \in B$, then there exists a in A such that $f(a) = b$. Thus $g(b) = g(f(a)) = h(f(a)) = h(b)$, so $g = h$. Conversely suppose $f : A \to B$ is right cancellable, and let Ω be the set of all subsets of $\{0\}$. Define $g : B \to \Omega$ by $g(b) = \{0\}$ for all b, and define $h : B \to \Omega$ by

$$h(b) = \{x \in \{0\} : b = f(a) \text{ for some } a\}.$$

Thus $h(b)$ is the subset of $\{0\}$ such that $0 \in h(b)$ if and only if there exists a such that $b = f(a)$. Clearly $gf = hf$ is the map that takes every element of A to the subset $\{0\}$. So $g = h$, whence $0 \in h(b)$, which means that $b = f(a)$ for some a. □

In fact, an original feature of [CCA] is the consideration of a notion of category as a fully fledged mathematical object and not as a simple 'manière de parler'.

We deal with two sorts of collections of mathematical objects: sets and categories.

. . . .

Given two groups, or sets, on the other hand, it is generally incorrect to ask if they are equal; the proper question is whether or not they are *isomorphic*, or, more generally, what are the homomorphisms between them.

A **category,** like a set, is a collection of objects. An equality relation on a set constructs, given any two objects a and b in the set, a *proposition* '$a = b$'. To specify a category $\mathcal{C}$, we must show how to construct, given any two objects A and B in $\mathcal{C}$, a *set* $\mathcal{C}(A, B)$.

A primary interest of categories is to generalize the notion of a family of objects (indexed by a set). For the category of sets, Bishop [2] considers only families of subsets of a given set. But in usual mathematical practice, and particularly in algebra, we sometimes need a more general notion, which corresponds to the notion of dependent types in the constructive theory of types.

> Using the notion of a functor, we can extend our definition of a family of elements of a set to a family of objects in a category $\mathcal{C}$. Let I be a set. A **family** A **of objects of** $\mathcal{C}$ **indexed by** I is a functor from I, viewed as a category, to the category $\mathcal{C}$. We often denote such a family by $\{A_i\}_{i \in I}$. If $i = j$, then the map from A_i to A_j is denoted by A^i_j, and is an isomorphism.

With these tools, it is possible to construct important objects in today's mathematics, as

- limits and colimits (e.g., products and coproducts) in some categories,
- some algebraic structures freely generated by general sets (not necessarily discrete),
- many operations on ordinals (see the definition of ordinals in [CCA]).

For example, one proves that a module freely generated by a set S is flat; but it is not necessarily projective [CCA, Exercise IV.4.9]. The classical theorem saying that every module is a quotient of a free module remains valid; the effective consequence is not that the module is a quotient of a projective module, but rather a quotient of a flat module. Thus, by forcing the sets to be discrete (with the aid of LEM), classical mathematics oversimplifies the notion of a free module and leads to conclusions impossible to satisfy algorithmically.

A natural notion of ordinal[3] is also introduced in chapter I of [CCA], and it is used in classification problems of abelian groups (in chapter XI of [CCA]).

Note that the definition below of a well-founded set uses the quantification over all subsets of W.

> Let W be a set with a relation $a < b$. A subset S of W is said to be **hereditary** if $w \in S$ whenever $w' \in S$ for each $w' < w$. The set W (or the relation $a < b$) is **well founded** if each hereditary subset of W equals W. A discrete partially ordered set is well founded if the relation $a < b$ (that is, $a \leq b$ and $a \neq b$) on it is well-founded. An **ordinal**, or a **well-ordered set**, is a discrete, linearly ordered, well-founded set.

[3] This notion is different from the ones given by Brouwer or Martin-Löf. See also [5].

Well-founded sets provide the environment for arguments by induction.

....

If λ and μ are ordinals, then an **injection** of λ into μ is a function ρ from λ to μ such that if $a < b$ then $\rho a < \rho b$, and if $c < \rho b$, then there is $a \in \lambda$ such that $\rho a = c$. We shall show that there is at most one injection from λ to μ.

[I.]6.5 Theorem. *If λ and μ are ordinals, and ρ and σ are injections of λ into μ, then $\rho = \sigma$.*

....

If there is an injection from the ordinal λ to the ordinal μ we write $\lambda \leq \mu$. Clearly compositions of injections are injections, so this relation is transitive. By [Theorem] 6.5 it follows that if $\lambda \leq \mu$ and $\mu \leq \lambda$, then λ and μ are isomorphic, that is, there is an invertible order preserving function from λ to μ. It is natural to say that two isomorphic ordinals are **equal**.

We are here in a framework close to the constructive theory of dependent types, where all types are created via inductive definitions.

4.3 The Corpus of Classical Abstract Algebra Treated in the Book

Basic classical algebra is fairly widely covered by the various chapters of [CCA]. This can be seen from the table of contents of the book.

In the following sections we comment on some significant examples of classical theorems to which the constructive reformulation brings a new light and precise additional information.

We also give some examples of theorems which are trivial in classical mathematics and yet very important from the algorithmic point of view.

4.4 Principal Ideal Domains and Finitely Generated Modules on these Rings

In classical mathematics, a principal ideal domain (PID) is an integral ring in which all ideals are principal. From a constructive point of view, even the two-element field does not satisfy this definition: consider an ideal generated by a binary sequence; finding a generator of this ideal is the same thing as deciding if the sequence is identically zero, which amounts to LPO.

An algorithmically relevant definition, classically equivalent to the classical one, is that of a discrete Bézout integral ring that satisfies a precisely formulated Noetherian condition.

A **GCD-monoid** is a cancellation [commutative] monoid in which each pair of elements has a greatest common divisor. A **GCD-domain** is a discrete domain whose nonzero elements form a GCD-monoid.

....

A **principal ideal** of a commutative monoid M is a subset I of M such that $I = Ma = \{ma : m \in M\}$ for some a in M. We say that M satisfies the **divisor chain condition** if for each ascending chain $I_1 \subseteq I_2 \subseteq I_3 \subseteq \cdots$ of principal ideals, there is n such that $I_n = I_{n+1}$.

A discrete domain is said to satisfy the divisor chain condition if its monoid of nonzero elements does.

[IV.]2.7 Definition. A **Bézout domain** is a discrete domain such that for each pair of elements a, b there is a pair s, t such that $sa + tb$ divides a and b. A **principal ideal domain** is a Bézout domain which satisfies the divisor chain condition.

The classical structure theorem says that a finitely generated module on a PID is a direct sum of a finite-rank free submodule and of the torsion submodule, itself equal to a direct sum of modules $R/(a_i)$ with the non-zero a_i put in an order where each a_i divides the next one.

The purest algorithmic form of this theorem is the theorem of reduction of a matrix into a Smith normal form.

A matrix $A = (a_{ij})$ is in **Smith normal form** if it is diagonal and $a_{ii} | a_{i+1,i+1}$ for each i.

[V.]1.2 Theorem. *Each matrix over a principal ideal domain is equivalent to a matrix in Smith normal form.*

[V.]1.4 Theorem. *Two $m \times n$ matrices in Smith normal form over a GCD-domain are equivalent if and only if corresponding elements are associates.*

The structure theorem for finitely presented modules follows directly from Theorem V.1.2.

[V.]2.3 Theorem (Structure theorem). *Let M be a finitely presented module over a principal ideal domain R. Then there exist principal ideals $I_1 \supseteq I_2 \supseteq \cdots \supseteq I_n$ such that M is isomorphic to the direct sum $R/I_1 \oplus R/I_2 \oplus \cdots \oplus R/I_n$.*

Since the ring is discrete by definition, we can separate the sum into two pieces: the beginning, for indices from 1 to k, say, is the torsion submodule, with $I_k = (a_k) \neq 0$; and the second piece, for $j > k$ with $a_j = 0$, is a free module of rank $n - k$. On the other hand, in order to know which I_js ($j \leq k$) are equal to R (and thus could be removed without damage), we need to have a test of invertibility for elements of R, which in this case is equivalent to having a divisibility test between two elements.

In classical mathematics, Theorem V.2.3 is stated for finitely generated modules. From a classical point of view the finitely generated modules over a PID are finitely presented, while from a constructive point of view it is clearly impossible to have an algorithm to achieve this implication, even in the simple case of the $\mathbf{Z}$-module $\mathbf{Z}/I$, where I is countably generated (e.g., generated by a binary sequence).

The way in which Bourbaki (*Algebra*, Chapter VII) treats these theorems deserves to be compared. The structure theorem is given before the Smith reduction theorem for matrices. And the proof, which uses LEM, fails to produce an algorithm to make the theorem explicit.

4.5 Factorization Problems

Theorem IV.4.7(i) in [CCA] is usually shown for unique factorization domains, but the underlying Noetherian condition is in fact useless.

> **[IV.]4.7 Theorem.** *Let R be a discrete domain.*
> *(i) If R is a GCD-domain, then so is $R[X]$.*

The reader is invited to appreciate the elegance of the proof in [CCA].

The classical theorem of factorization of an element into a product of prime factors in a GCD monoid satisfying the divisor chain condition is inaccessible from an algorithmic point of view. It is replaced in constructive mathematics by a slightly more subtle theorem. This new theorem can generally be used instead of the classical one when needed to obtain concrete results.

> **[IV.]1.8 Theorem (Quasi-factorization).** *Let $x_1, \ldots, x_k$ be elements of a GCD-monoid M satisfying the divisor chain condition. Then there is a family P of pairwise relatively prime elements of M such that each x_i is an associate of a product of elements of P.*

Let M be a cancellation monoid. An element $a \in M$ is said to be **bounded by** n if whenever $a = a_0 \cdots a_n$ with $a_i \in M$, then a_i is a unit for some i. An element of M is **bounded** if it is bounded by n for some $n \in \mathbf{N}$; the monoid M is **bounded** if each of its elements is bounded.
A discrete domain is **bounded** if its nonzero elements form a bounded monoid.
A GCD-domain satisfying the divisor chain condition is called a **quasi-UFD**.

The quasi-UFDs (quasi-unique factorization domains) and the bounded GCD-domains are two constructive versions (that are not constructively equivalent) of the classical notion of a UFD. In fact, we find in [CCA] three other constructive versions of this classical notion.

[IV.]2.1 Definition. A discrete domain R is called a **unique factorization domain**, or **UFD**, if each nonzero element r in R is either a unit or has an essentially unique factorization into irreducible elements, that is, if $r = p_1 \cdots p_m$ and $r = q_1 \cdots q_n$ are two factorizations of r into irreducible elements, then $m = n$ and we can reindex so that $p_i \sim q_i$ for each i. We say that R is **factorial** if $R[X]$ is a UFD.
Call a discrete field k **fully factorial** if any finite-dimensional extension of k is factorial.

The five constructive versions are, in classical mathematics, equivalent to the classical notion, but they introduce algorithmically relevant distinctions, totally invisible in classical mathematics, due to the use of LEM, which annihilates these relevant distinctions. In Theorem IV.4.7 the points (ii) (attached to the point (i)) and (vi) (i.e., (i) and (v)) are two distinct, inequivalent versions of the same classical theorem about UFDs.

[IV.]4.7 Theorem. *Let R be a discrete domain.*

(i) If R is a GCD-domain, then so is $R[X]$.
(ii) If R is bounded, then so is $R[X]$.
(iii) If R has recognizable units, then so does $R[X]$.
(iv) If R has decidable divisibility, then so does $R[X]$.
(v) If R satisfies the divisor chain condition, then so does $R[X]$.
(vi) If R is a quasi-UFD, then so is $R[X]$.

Concerning factorization problems for polynomials over a discrete field, the algorithmic situation is not correctly described by classical mathematics. For example, factorization of polynomials in $k[X]$, where k is a discrete field, is not a trivial thing, contrary to what is stated in classical mathematics.

Chapter VII of [CCA] explores the factorization problems in polynomial rings in great detail.

The basic constructive theorem on this subject is given in Chapter VI of [CCA]. If the characteristic of a field or a ring is not known in advance, but can be revealed during a construction, some precautions are necessary in the statements, as in point (i) of Theorem VI.6.3. Note that if we discover a prime number p equal to zero in a ring k, it is necessarily unique (unless the ring is trivial).

In Theorem VI.6.3 of [CCA], if k is a discrete field, then we simply drop the alternative 'k has a nonzero nonunit'. But it happens in [CCA] that the theorem is used in the precise form given here, for example, in Chapter IX about the structure of finite-dimensional algebras.

[VI.]6.3 Theorem. *Let k be a discrete commutative ring with recognizable units, and S a finite set of monic polynomials in $k[X]$. Then either k has a nonzero nonunit or we can construct a finite set T of monic polynomials in $k[X]$ such that*

(i) Each element of T is of the form $f(X^q)$ where f is separable, and $q = 1$ or q is a power of a prime that is zero in k.
(ii) Distinct elements of T are strongly relatively prime.
(iii) Every polynomial in S is a product of polynomials in T.

When k is a discrete field, we thus obtain, starting from a given family of univariate polynomials, a family of separable strongly relatively prime monic polynomials which gives a more precise version of the quasi-factorization theorem, Theorem IV.1.8 (which deals with quasi-UFDs).

4.6 Noetherian Rings, Primary Decompositions and the Principal Ideal Theorem

An R-module is said to be **strongly discrete** if finitely generated submodules are detachable.[4] It is said to be **coherent**[5] if any finitely generated submodule is finitely

[4] In [CCA], the terminology is 'module with detachable submodules'; it was later replaced by 'strongly discrete module'. See, for example, [13].

[5] Bourbaki (*Algebra*, Chapter X, or *Commutative Algebra*, Chapter I) calls *pseudo coherent module* what [CCA] calls coherent module (as in quasi, all texts in English literature), and *coherent module* what [CCA] calls finitely presented coherent module. This is to be linked to 'Faisceaux Algébriques Cohérents' by J.-P. Serre. Note also that the Stacks Project (Collective work, `http://stacks.math.columbia.edu`) uses Bourbaki's definition for coherent modules.

presented. The notion of a strongly discrete coherent ring is fundamental from the algorithmic point of view in commutative algebra, in particular for the following reason: on a strongly discrete coherent ring, linear systems are perfectly understood and mastered.[6]

In standard textbooks in classical mathematics, this notion is usually hidden behind that of a *Noetherian* ring, and rarely put forward. In classical mathematics every Noetherian ring R is coherent because every submodule of R^n is finitely generated, and every finitely generated module is coherent for the same reason. Furthermore, we have the Hilbert basis theorem, which states that *if R is Noetherian, then every finitely presented R-algebra is also a Noetherian ring,* whereas the same statement does not hold if one replaces 'Noetherian' with 'coherent' (see [20]).

From an algorithmic point of view, however, it seems impossible to find a satisfying constructive formulation of Noetherianity which implies coherence, and coherence is often the most important property from an algorithmic point of view. Consequently, from a constructive point of view, coherence must be added when we use the notion of a Noetherian ring or module.

The definition adopted for a **Noetherian module** in [CCA] is: a module in which any ascending chain of finitely generated submodules admits two equal consecutive terms. It is a constructively acceptable definition, equivalent in classical mathematics to the usual definition.

The classical theorem stating that over a Noetherian ring every finitely generated A-module is Noetherian is often advantageously replaced by the following constructive theorems.

Over a coherent ring (resp. strongly discrete coherent) every finitely presented A-module is coherent (resp. strongly discrete coherent).
Over a Noetherian coherent ring every finitely presented A-module is Noetherian coherent.

Two important classical results about Noetherian rings have constructive proofs within the framework given by [CCA].

[VIII.]2.7 Theorem (Artin–Rees). *Let I be a finitely generated ideal of a coherent commutative Noetherian ring R. Let N be a finitely generated submodule of a finitely presented R-module M. Then there is k such that for all $n \geq k$ we have*

$$I^{n-k}(I^k M \cap N) = I^n M \cap N.$$

[6] In the article of Posur cited in Section 4.1, these rings are called 'computable'.

[VIII.]2.8 Theorem (Krull intersection theorem). *Let M be a finitely presented module over a coherent commutative Noetherian ring R, and let I be a finitely generated ideal of R. Let $A = \bigcap_n I^n M$. Then $a \in Ia$ for each $a \in A$, so $IA = A$.*

4.6.1 Hilbert-Basis Theorem

Which are the coherent rings R such that the polynomial rings $R[X_1, \ldots, X_n]$ are also coherent? From a constructive point of view, we know two classes of rings sharing this property: coherent Noetherian rings and Prüfer domains (see [21, Chapter 4]).

The Hilbert-basis theorem for the definition of Noetherianity given in [CCA] is Theorem VIII.1.5 below. Proofs go back to 1974 [10, 16]; see also [14, 15] for polynomial rings over a discrete field. These proofs are very clearly laid out in [CCA].

[VIII.]1.5 Theorem (Hilbert basis theorem). *If R is a coherent Noetherian ring, then so is $R[X]$. If, in addition, R has detachable left ideals,[7] then so does $R[X]$.*

There is an analogous theorem in Computer Algebra (see [1, Theorem 4.2.8]) saying that for a coherent Noetherian strongly discrete ring R, there is a 'Gröbner basis algorithm' computing the leading ideal of a finitely generated ideal in $R[X_1, \ldots, X_n]$ for a given monomial order. In fact, this Computer Algebra theorem and Theorem VIII.1.5 in [CCA] are essentially the same result. One is easily deduced from the other.

Nevertheless, we note that algorithms for these theorems are quite different from each other. Moreover, the authors of [1] in 1994 seem to ignore that the problem was solved essentially in 1974, and algorithms in [1] are not certified constructively (in fact, from the proof, no bound can be estimated for the number of steps as depending on the data.)

4.6.2 Primary Decomposition Theorem

An adequate constructive theory of primary decompositions is given in [CCA]. This is based on the work of Seidenberg [17, 18]. In [CCA] this work is made more simple and synthetic.

[7] i.e., is left strongly discrete

Let R be a commutative ring. An ideal Q of R is said to be **primary** if $xy \in Q$ implies $x \in Q$ or $y^n \in Q$ for some n. One sees that $\sqrt{Q}$ is a prime ideal P.

A variant with respect to the usual terminology is given in [CCA] with no importance in the case of Noetherian rings for classical mathematics: ideals are all finitely generated. A **primary decomposition** of an ideal I in a commutative ring is a finite family of finitely generated primary ideals $Q_1, \ldots, Q_n$ such that the $\sqrt{Q_i}$ are finitely generated and $I = \bigcap_i Q_i$. In this case the ideal I is said to be **decomposable**. In classical mathematics, every ideal of a Noetherian ring has a primary decomposition.

In a constructive framewok, which convenient hypotheses do we have to add for a coherent Noetherian strongly discrete ring in order to get primary decompositions? A possible answer is the following one, given in [CCA].

> A **Lasker–Noether** ring is a coherent Noetherian ring with detachable ideals such that the radical of each finitely generated ideal is the intersection of a finite number of finitely generated prime ideals.

This definition is constructively acceptable and applies to usual examples like $\mathbf{Z}$, $\mathbf{Q}[X]$, and $k[X]$ when k is an algebraically closed discrete field: they are clearly constructively Lasker–Noether for this definition. Many other usual examples are also available, as explained below.

In fact, when k is a discrete field, $k[X]$ is easily seen to be Lasker–Noether if and only if k is a factorial field. This equivalence has no meaning in classical mathematics since all fields are factorial. Nevertheless it should be possible to state an analogous result for mechanical computations using Turing machines.

The first properties of Lasker–Noether rings are summarized in three theorems.

> **[VIII.]8.1 Theorem.** *Let S be a multiplicative submonoid of a Lasker–Noether ring R such that $I \cap S$ is either empty or nonempty for each finitely generated ideal I of R. Then $S^{-1}R$ is a Lasker–Noether ring.*

If $S = R \setminus P$ for a prime ideal P, condition '$I \cap S$ is either empty or nonempty' means that 'I either is contained in P or is not'. Since I is finitely generated, the test exists if and only if P is detachable. So, Theorem VIII.8.1 implies that for each detachable prime ideal, and so for each finitely generated prime ideal, the localization R_P is Lasker–Noether.

[VIII.]8.2 Theorem. *Let R be a Lasker–Noether ring, and let I be a finitely generated ideal of R. Then R/I is a Lasker–Noether ring.*

[VIII.]8.5 Theorem (Primary decomposition theorem). *Let R be a Lasker–Noether ring. Then each finitely generated ideal of R has a primary decomposition.*

4.6.3 Principal Ideal Theorem

A more elaborate property of Lasker–Noether rings is the famous principal ideal theorem of Krull and the fact that finitely generated proper prime ideals have a well-defined height.

[VIII.]10.4 Theorem (Generalized principal ideal theorem). *Let R be a Lasker–Noether ring. Let $I = (a_1, \ldots, a_n)$. Then every minimal prime ideal over I has height at most n.*

[VIII.]10.5 Theorem. *Let P be a finitely generated proper prime ideal of a Lasker–Noether ring R. Then there is m such that P has height m, and P is a minimal prime over some ideal generated by m elements.*

4.6.4 Fully Lasker–Noether Rings

Finally, it is important to give a constructive answer to the following: which convenient hypotheses do we have to add for a Lasker–Noether ring R in order to get that $R[X_1, \ldots, X_n]$ is also Lasker–Noether?

Call R a **fully Lasker–Noether** ring if it is a Lasker–Noether ring and if for each finitely generated prime ideal P of R, the field of quotients of R/P is fully factorial. Note that the ring of integers $\mathbf{Z}$ is a fully Lasker–Noether ring, as is any fully factorial field.

The following three theorems (with the previous theorems about Lasker–Noether rings) show that in this context (i.e., with this constructively acceptable definition equivalent to the definition of a Noetherian ring in classical mathematics), a very large number of classical theorems concerning Noetherian rings now have a constructive proof and a clear meaning. It sounds like a 'miracle' of the same kind as Bishop's book.

[VIII.]9.1 Theorem. *Let I be a finitely generated ideal of a fully Lasker–Noether ring R. Then R/I is a fully Lasker–Noether ring.*

[VIII.]9.2 Theorem. *If P is a detachable prime ideal of a fully Lasker–Noether ring R, then R_P is a fully Lasker–Noether ring.*

[VIII.]9.6 Theorem. *If R is a fully Lasker–Noether ring, then so is $R[X]$.*

Note. The paper by Perdry [8] defines a notion of Noetherianity which is constructively stronger than the one in [CCA]. The usual examples of Noetherian rings are Noetherian in this meaning. With this notion, the definition of a Lasker–Noether ring becomes more natural: it is a Noetherian coherent strongly discrete ring in which we have a primality test for finitely generated ideals. The paper gives a nice theory of fully Lasker–Noether rings in this context.

Note. The computation of primary decompositions in polynomial rings over discrete fields or over $\mathbf{Z}$ is an active area of research in Computer Algebra. The seminal paper of Seidenberg [18] is sometimes cited, but not the book [CCA].

4.7 Wedderburn Structure Theorem for Finite-Dimensional k-Algebras

We deal here with unitary associative k-algebras, which are finite-dimensional k-vector spaces on a discrete field k. In other words, these algebras are isomorphic to a finitely generated subalgebra of an algebra of matrices $E_k(k^n)$ (the algebra of k-endomorphisms of the vector space k^n). We shorten the terminology by speaking of 'k-algebra of finite dimension'.

If A is a not necessarily commutative ring, its **Jacobson radical** is the set I of elements x such that $1 + xA \subseteq A^\times$. It is a (two-sided) ideal and the Jacobson radical of the quotient A/I is zero.

When A is a k-algebra of finite dimension, this radical can also be defined as the **nilpotent radical**: $\mathrm{rad}(A)$ is the set of elements x such that the left ideal xA is nilpotent, that is, there exists an integer n such that every product $xa_1 \cdots xa_n$ is zero.

Let A be a k-algebra of finite dimension. We can construct a basis of the centre of A as well as the minimal polynomial over k of an arbitrary element of A. We can also construct a basis of the left ideal and another of the two-sided ideal generated by a finite part of A. But it may be difficult to construct a basis of the radical, and we cannot generally state that the radical is finite-dimensional (over k).

Nevertheless, we know how to construct objects whose counterparts are trivial in classical mathematics (if we do not try to construct them). For example, as an alternative to the construction of the radical, we have the following theorem.

[IX.]3.3 Theorem. *Let A be a finite-dimensional k-algebra and L a finite-dimensional (left) ideal of A. Then either $L \cap \operatorname{rad} A \neq 0$ or $A = L \oplus N$ for some (left) ideal N.*

A module M is **reducible** if it has a non-trivial submodule – otherwise it is **irreducible** (or **simple**).

A k-algebra is said to be **simple** if each two-sided ideal is trivial. When the algebra is discrete (as in the present context) the definition amounts to saying that if an element is non-zero, the (two-sided) ideal it generates contains 1.

The first part of Wedderburn's structure theorem says that every finite-dimensional k-algebra with zero radical is a product of simple algebras. Here is the constructive reformulation given in [CCA]. A field k is called separably factorial when separable polynomials in $k[X]$ have a prime decomposition.

We now characterize separably factorial fields in terms of decomposing algebras into products of simple algebras. This is the first part of Wedderburn's theorem.

[IX.]4.3 Theorem. *A discrete field k is separably factorial if and only if every finite-dimensional k-algebra with zero radical is a product of simple algebras.*

A clarification concerning the ability to construct a basis of the radical is given in the following corollary.

[IX.]4.5 Corollary. *A discrete field k is fully factorial if and only if every finite-dimensional algebra A over k has a finite-dimensional nilpotent ideal I such that A/I is a product of simple k-algebras.*

The second part of Wedderburn's structure theorem for semi-simple algebras says that a finite-dimensional simple algebra is isomorphic to a full ring of matrices over a division algebra.

The constructive version of this theorem given in [CCA] elucidates in a surprising way the computational content of this classical theorem.

[IX.]5.1 Theorem. *Let A be a finite-dimensional k-algebra, and L a nontrivial left ideal of A. Then either*

(i) A has a nonzero radical
(ii) A is a product of finite-dimensional k-algebras
(iii) A is isomorphic to a full matrix ring over some k-algebra of dimension less than A.

. . . .

The fundamental problem is to be able to recognize whether a given finite-dimensional algebra is a division algebra or not, in the sense of being able either to assert that it is a division algebra or to construct a nontrivial left ideal. If we could do that, then Theorem IX.5.1 would imply that every finite-dimensional k-algebra has a finite-dimensional radical, and modulo its radical it is a product of full matrix rings over division algebras. This condition is equivalent to being able to recognize whether an arbitrary finite-dimensional representation of a finite-dimensional k-algebra is reducible.

[IX.]5.2 Theorem. *The following conditions on a discrete field k are equivalent.*

(i) Each finite-dimensional k-algebra is either a division algebra or has a nontrivial left ideal.
(ii) Each finite-dimensional left module M over a finite dimensional k-algebra A is either reducible or irreducible.
(iii) Each finite-dimensional k-algebra A has a finite-dimensional radical, and $A/\operatorname{rad} A$ is a product of full matrix rings over division algebras.

And we remain a little disappointed with these questions at the end of Chapter IX in [CCA].

For what fields k do the conditions of Theorem 5.2 hold? Finite fields and algebraically closed fields provide trivial examples. The field of algebraic real numbers admits only three finite-dimensional division algebras, and a constructive proof of this statement shows that this field satisfies the conditions of Theorem 5.2.

[IX.]5.3 Theorem. *Let k be a discrete subfield of $\mathbf{R}$ that is algebraically closed in $\mathbf{R}$, and $H = k(i, j)$ the quaternion algebra over k. If A is a finite-dimensional algebra over k, then either A has a zero-divisor, or A is isomorphic to k, to $k(i)$, or to H.*

Does the field $\mathbf{Q}$ of rational numbers satisfy the conditions of Theorem 5.2?

Certainly we are not going to produce a Brouwerian counterexample when $k = \mathbf{Q}$. Probably a close analysis of the classical theory of division algebras over $\mathbf{Q}$, in analogy with Theorem 5.3, will yield a proof.

4.8 Dedekind Domains

Although it is commonly felt that algebraic number theory is essentially constructive in its classical form, even those authors who pay particular attention to the constructive aspects of the theory employ highly nonconstructive techniques which nullify their efforts. In [3], for example, it is assumed that every polynomial can be factored into a product of irreducible polynomials (every field is factorial) and that given a non-empty subset of the positive integers you can find its least element.

The constructive theory of Dedekind domains in [CCA] allows us to give an explicit version of the classical statements of number theory and algebraic geometry concerning local fields, for example in the book of J.-P. Serre [19]. This theory also gives the appropriate hypotheses to account for the classical results concerning Dedekind domains, as found, for example, in Bourbaki.

This requires giving sufficiently precise and binding definitions, beginning with those in the theory of (rank-one) valuations.

For example, let us see the definitions concerning Dedekind domains.

[XIII.]1.1 Definition. A nonempty discrete set S of nontrivial discrete valuations on a Heyting field k is a **Dedekind set** if

(i) For each $x \in k$ there is a finite subset T of S so that $|x|_p \leq 1$ for each $p \in S \setminus T$.
(ii) If q and q' are distinct valuations of S, and $\varepsilon > 0$, then there exists $x \in k$ with $|x|_p \leq 1$ for each $p \in S$, such that $|x - 1|_q < \varepsilon$ and $|x|_{q'} < \varepsilon$. Hence distinct valuations are inequivalent.

Let S be a Dedekind set of valuations on a Heyting field k. If $p \in S$, then, because p is nonarchimedean, the set $R(p) = \{x \in k : |x|_p \leq 1\}$ is a ring, which is local as p is discrete. We call $R(p)$ the **local ring at** p. The elements of the ring $\bigcap_{p \in S} R(p)$ are called the **integers at** S. A ring is a **Dedekind domain** if it is the ring of integers at a Dedekind set of valuations on a Heyting field.

If the strong point is to give a constructive account of most of the classical theorems, a weak point is that, for example, a PID is a Dedekind domain only in the case where we have algorithms of factorization of principal ideals into prime ideals.

We can compare this for example with the exposition in [6], where a definition is given that is constructively weaker but closer to the usual classical definition (see Definition XII.7.7 and Theorem XII.7.9). In [6], Dedekind domains have quasi-factorization of finite sets of finitely generated ideals, and the total factorization Dedekind domains correspond to the discrete Dedekind domains of [CCA].

Acknowledgements

I am indebted to Thierry Coquand and Stefan Neuwirth for many relevant comments and suggestions.

References

[1] Adams, W. W., and Loustaunau, P. 1994. *An Introduction to Gröbner Bases.* Providence, RI: American Mathematical Society.

[2] Bishop, E. 1967. *Foundations of Constructive Analysis.* New York: McGraw-Hill.

[3] Borevich, Z. I., and Shafarevich, I. R. 1966. *Number Theory.* Pure and Applied Mathematics, 20. New York: Academic Press. Translated from the Russian by Newcomb Greenleaf.

[4] Bridges, D., and Richman, F. 1987. *Varieties of Constructive Mathematics.* London Mathematical Society Lecture Note Series, 97. Cambridge: Cambridge University Press.

[5] Coquand, T., Lombardi, H., and Neuwirth, S. 2023. A constructive theory of ordinals. In: Benini, M., Beyersdorff, O., Rathjen, M., and Schuster, P. (eds.), *Mathematics for Computation – M4C.* Singapore: World Scientific. (In the press.)

[6] Lombardi, H., and Quitté, C. 2015. *Commutative algebra: Constructive Methods. Finite Projective Modules.* Algebra and Applications, 20. Dordrecht: Springer.

[7] Mines, R., Richman, F., and Ruitenburg, W. 1988. *A Course in Constructive Algebra.* Universitext. New York: Springer.

[8] Perdry, H. 2004. Strongly Noetherian rings and constructive ideal theory. *J. Symbolic Comput.*, **37**(4), 511–535.

[9] Petrakis, I. 2018. Dependent sums and dependent products in Bishop's set theory. Technical report, Hausdorff Research Institute for Mathematics. www.hcm.uni-bonn.de/fileadmin/him/Preprints/Types18.pdf.

[10] Richman, F. 1974. Constructive aspects of Noetherian rings. *Proc. Amer. Math. Soc.*, **44**, 436–441.

[11] Richman, F. 1994. Confessions of a formalist, Platonist intuitionist. http://math.fau.edu/Richman/html/Confess.htm.

[12] Richman, F. 1996. Interview with a constructive mathematician. *Modern Logic*, **6**(3), 247–271.

[13] Richman, F. 1998. The regular element property. *Proc. Amer. Math. Soc.*, **126**(7), 2123–2129.

[14] Seidenberg, A. 1971. On the length of a Hilbert ascending chain. *Proc. Amer. Math. Soc.*, **29**, 443–450.

[15] Seidenberg, A. 1973. Constructive proof of Hilbert's theorem on ascending chains. *Trans. Amer. Math. Soc.*, **174**, 305–312.

[16] Seidenberg, A. 1975. What is Noetherian? *Rend. Semin. Mat. Fis. Milano*, **44**, 55–61.

[17] Seidenberg, A. 1978. Constructions in a polynomial ring over the ring of integers. *Amer. J. Math.*, **100**, 685–706.

[18] Seidenberg, A. 1984. On the Lasker–Noether decomposition theorem. *Amer. J. Math.*, **106**, 611–638.

[19] Serre, J.-P. 1968. *Corps Locaux.* Deuxième édition, Publications de l'Université de Nancago, VIII. Paris: Hermann.

[20] Soublin, J.-P. 1970. Anneaux et modules cohérents. *J. Algebra*, **15**, 455–472.

[21] Yengui, I. 2015. *Constructive Commutative Algebra: Projective Modules Over Polynomial Rings and Dynamical Gröbner Bases.* Lecture Notes in Mathematics, 2138. Cham: Springer.

5

Constructive Algebra: The Quillen–Suslin Theorem

Ihsen Yengui

5.1 Introduction

Constructive algebra can be seen as an abstract version of computer algebra. In computer algebra, on the one hand, one attempts to construct efficient algorithms for solving concrete problems given in an algebraic formulation, where a problem is understood to be concrete if its hypotheses and conclusion have computational content. Constructive algebra, on the other hand, can be understood as a "pre-processing" step for computer algebra that leads to general algorithms, even if they are sometimes not efficient. In constructive algebra, one tries to give general algorithms for solving "virtually any" theorem of abstract algebra.

Therefore, a first task in constructive algebra is to define the computational content hidden in hypotheses that are formulated in a very abstract way. For example, what is a good constructive definition (gcd) of a local ring (i.e., a ring with a unique maximal ideal), a valuation ring (i.e., a ring in which all elements are comparable under division), an arithmetical ring (i.e., a ring which is locally a valuation ring), a ring of Krull dimension $\leq n$ (i.e., a ring in which every chain $\mathfrak{p}_0 \subset \mathfrak{p}_1 \subset \cdots \subset \mathfrak{p}_k$ of prime ideals has length $k \leq n$), and so on? A good constructive definition must be equivalent to the usual definition within classical mathematics; it must have computational content; and it must be fulfilled by "usual" objects that satisfy the definition.

As a typical example, let us consider the classical theorem "any polynomial P in $\mathbf{K}[X]$ is a product of irreducible polynomials ($\mathbf{K}$ a field)." This leads to an interesting problem: it seems like no general algorithm produces the irreducible factors. What, then, is the constructive content of this theorem? A possible answer is as follows: When performing computations with P, proceed as if its decomposition into irreducible polynomials were known (at the beginning, proceed as if P were irreducible). When something strange happens (e.g., when the greatest common divisor (gcd) of P and another polynomial Q is a strict divisor of P), use this fact to improve the decomposition of P. This trick was invented in Computer Algebra as the D5-philosophy (see [8, 9]), and later taken up in the form of the dynamical proof method in algebra [5]. It indeed enables one to carry out computations inside the algebraic closure $\widetilde{\mathbf{K}}$ of $\mathbf{K}$ even if it is not possible to effectively construct $\widetilde{\mathbf{K}}$, for in general this would require transfinite methods such as Zorn's Lemma. The foregoing has been referred to as "dynamical evaluation" of the algebraic closure.

From a logical point of view, the "dynamical evaluation" gives a constructive substitute for two highly nonconstructive tools of abstract algebra: the Law of Excluded Middle and Zorn's Lemma. For instance, these tools are required in order to "construct" the complete prime factorization of an ideal in a Dedekind domain (i.e., in a Noetherian domain which is locally a valuation domain), while the dynamical method reveals the computational content of this "construction." We refer to [5] for more details on the dynamical proof method in algebra, including a wealth of examples.

Following this "dynamical" philosophy, the main goal of these notes is to find the constructive content hidden in abstract proofs (namely, those of Quillen and Suslin) of the now-so-called Quillen–Suslin theorem. The problem of freeness of projective modules over polynomial rings originally raised by Serre [40] is approached constructively. Serre remarked that it was not known whether there exist finitely generated projective modules over $\mathbf{A} = \mathbf{K}[X_1, \ldots, X_n]$, where $\mathbf{K}$ is a field, which are not free. This remark turned into the so-called "Serre's conjecture" or "Serre's problem," stating that indeed there were no such modules. Proved independently by Quillen [36] and Suslin [43], it became known subsequently as the Quillen–Suslin theorem. Quillen and Suslin's proofs had a big effect on the subsequent development of the study of projective modules. The books of Lam [17, 18] are nice expositions of Serre's conjecture. It was known by Serre [41] well before Serre's conjecture was settled that finitely generated projective modules over $\mathbf{A}$ are *stably free*, that is, every finitely generated projective $\mathbf{A}$-module P is isomorphic to the kernel of an $\mathbf{A}$-epimorphism T from $\mathbf{A}^m$ onto $\mathbf{A}^s$. In that situation the matrix T is unimodular; that is, the maximal minors of T generate the unit ideal in $\mathbf{A}$. The fact that P is free is nothing but the case that the matrix T can be completed (we say that it is *completable*) to an invertible matrix by adding a suitable number of new rows. For $s = 1$, we speak of a *unimodular row* $(b_1, \ldots, b_m) \in \mathbf{A}^{1\times m}$ (${}^{\mathrm{t}}(b_1, \ldots, b_m)$ is called a *unimodular vector*), that is, such that $\langle b_1, \ldots, b_m \rangle = \mathbf{A}$.

The Quillen–Suslin theorem finds natural applications either in signal processing (see, e.g., the work of Park [32, 33, 34, 35]) or in mathematical systems theory and control theory (see, e.g., the work of Fabiańska and Quadrat [11]). There are several papers ([4, 11, 13, 19, 20, 22]) in the literature proposing algorithms for the Quillen–Suslin (see, for example, the `MAPLE` package `QuillenSuslin` [12]). All the proposed algorithms rely on the fact that for a discrete field $\mathbf{K}$, the ring

$\mathbf{K}[X_1, \ldots, X_k]$ is Noetherian and has an effective Nullstellensatz. As a matter of fact, roughly speaking, in order to eliminate one variable, say X_k, via comaximal resultants $r_1, \ldots, r_m$ (i.e., such that $1 \in \langle r_1, \ldots, r_m \rangle$, see the previously mentioned lemma of Suslin), they compute a maximal ideal $\mathfrak{M}$ of $\mathbf{K}[X_1, \ldots, X_{k-1}]$ containing the ideal $\langle r_1, \ldots, r_j \rangle$ generated by the current list $[r_1, \ldots, r_j]$ of resultants, and then, enter a "local loop" by localizing $\mathbf{K}[X_1, \ldots, X_{k-1}]$ at $\mathfrak{M}$ in order to find a new resultant $r_{j+1} \notin \langle r_1, \ldots, r_j \rangle$. The fact that $\mathbf{K}[X_1, \ldots, X_{k-1}]$ is Noetherian ensures the termination of this search for comaximal resultants. In Section 5.3.5, we will give a method avoiding this heavy use of maximal ideals and

It is worth pointing out that dynamical method was also used successfully in order to find a constructive substitute to the very elegant theorems such as [21] (a wide generalization of the Quillen–Suslin theorem) and [42] stating (when combined) that for any arithmetical ring $\mathbf{R}$, all finitely generated projective $\mathbf{R}[X_1, \ldots, X_n]$-modules are extended from $\mathbf{R}$. As a matter of fact, a key simple trick coined in [10] to decipher constructively this far-reaching result, is that for any ring $\mathbf{R}$ with Krull dimension $\leq d$, the ring $\mathbf{R}\langle X \rangle$ (the localization of $\mathbf{R}[X]$ at monic polynomials) "dynamically behaves like the ring $\mathbf{R}(X)$ (the localization of $\mathbf{R}[X]$ at primitive polynomials) or a localization of a polynomial ring of type $(S^{-1}\mathbf{R})[X]$ with S a multiplicative subset of $\mathbf{R}$ and the Krull dimension of $S^{-1}\mathbf{R}$ is $\leq d - 1$."

The present chapter is written in the framework of Bishop-style constructive mathematics (see [1, 2, 26, 27, 30, 48]).

5.2 Quillen's Proof of Serre's Problem

5.2.1 Finitely Generated Projective Modules

Definition 5.1 Let $\mathbf{R}$ be a ring.

(1) Let M be an $\mathbf{R}$-module and $k \in \mathbb{N}$. We say that M is *free of rank k* if it is isomorphic to $\mathbf{R}^k$.

(2) Let N be a submodule of an $\mathbf{R}$-module M. We say that N is a *direct summand* in M if there exists a submodule L of M such that $M = L \oplus N$. The interesting situation is when N if a finite-rank free module (see Proposition 5.4).

Definition 5.2 Let P be a finitely generated module over a ring $\mathbf{R}$. We say that P is a *projective* $\mathbf{R}$-module if any surjective $\mathbf{R}$-module homomorphism $\alpha\colon M \to P$ has a right inverse $\beta\colon P \to M$; or equivalently, if it is isomorphic to a direct summand in $\mathbf{R}^n$ for some n.

Example 5.3

(i) Every free module is projective.

(ii) Suppose that m and n are coprime natural numbers. Then as abelian groups (and also as $(\mathbb{Z}/mn\mathbb{Z})$-modules), we have $\mathbb{Z}/mn\mathbb{Z} \cong \mathbb{Z}/n\mathbb{Z} \oplus \mathbb{Z}/m\mathbb{Z}$. Thus, $\mathbb{Z}/m\mathbb{Z}$ is a projective $(\mathbb{Z}/mn\mathbb{Z})$-module which is not free as it contains fewer than mn elements.

(iii) A finitely generated ideal I of an integral domain $\mathbf{R}$ is projective if and only if it is invertible. Integral domains in which every ideal is invertible are known as Dedekind domains, and they are important in number theory. For example, the ring of integers in any algebraic number field is a Dedekind domain. So, by considering a Dedekind domain which is not a principal ideal domain (PID), one can find an example of a projective module (an invertible ideal) which is not free (not principal).

(iv) Let e be a nontrivial idempotent of a ring $\mathbf{R}$, that is, $e^2 = e$, $e \neq 0$, and $e \neq 1$ (for example, one can consider the generic case $\mathbf{R} = \mathbf{K}[u]/\langle u^2 - u\rangle = \mathbf{K}[\bar{u}]$ and $e = \bar{u}$ where $\mathbf{K}$ is a field). As $\mathbf{R} \cong e\,\mathbf{R} \oplus (1-e)\mathbf{R}$, $e\,\mathbf{R}$ is a projective $\mathbf{R}$ module. Note that $e\,\mathbf{R}$ is not free since $(1-e)(e\,\mathbf{R}) = 0$ while $(1-e)\mathbf{R} \neq 0$.

Definition and Proposition 5.4 (Finitely presented modules)

(1) Let $\mathbf{R}$ be a ring. A *finitely presented* $\mathbf{R}$-module is an $\mathbf{R}$-module given by a finite number of generators and relations. Thus, it is a finitely generated $\mathbf{R}$-module having a finitely generated relations module. Equivalently, it is an $\mathbf{R}$-module isomorphic to the cokernel of a linear application

$$\gamma : \mathbf{R}^m \to \mathbf{R}^q.$$

The matrix $G \in \mathbf{R}^{q\times m}$ of γ has as columns a generating set of the relations module between the generators g_i which are the images of the canonical basis by the epimorphism $\pi : \mathbf{R}^q \to M$. The matrix G is called a presentation matrix of the module M for the generating system $(g_1, \ldots, g_q)$. We have:

- $[g_1 \cdots g_q]\, G = 0$, and
- each relation between the g_i is a linear combination of the columns of G, that is, if $[g_1 \cdots g_q]\, C = 0$ with $C \in \mathbf{R}^{q\times 1}$ then there exists $C' \in \mathbf{R}^{m\times 1}$ such that $C = G\,C'$.

For example, a free module of rank k (i.e., isomorphic to $\mathbf{R}^k$) is finitely presented. Its presentation matrix is a column matrix formed by k zeros. More generally, if P is a finitely generated projective module then, as it is isomorphic to the image of an idempotent matrix $F \in \mathbf{R}^{n\times n}$ for some $n \in N^*$ (see Remark 5.6) and $\mathbf{R}^n = \mathrm{Im}(F) \oplus \mathrm{Im}(\mathrm{I}_n - F)$, we get $P \cong \mathrm{Coker}(\mathrm{I}_n - F)$ and thus P is finitely presented.

(2) The definition above can be rephrased as follows: An $\mathbf{R}$-module M is finitely presented if there is an epimorphism $\pi : \mathbf{R}^q \to M$ for some $q \in \mathbb{N}^*$ (and thus, $\mathbf{R}^q/\mathrm{Ker}(\pi) \cong M$) whose kernel $\mathrm{Ker}(\pi)$ is finitely generated. The module M is specified using finitely many generators (the images of the q generators of $\mathbf{R}^q$) and finitely many relations (the generators of $\mathrm{Ker}(\pi)$).

For example, for $a \in \mathbf{R}$, the ideal $a\,\mathbf{R}$ is finitely presented if and only if the annihilator $\mathrm{Ann}(a) := \{b \in \mathbf{R} \mid b\,a = 0\}$ of a is finitely generated. The epimorphism π corresponds to the multiplication by a and its kernel is $\mathrm{Ann}(a)$. More generally, a finitely generated ideal $\langle a_1, \ldots, a_n\rangle$ of $\mathbf{R}$ is finitely presented if and only if the syzygy module

$$\mathrm{Syz}(a_1, \ldots, a_n) := \{(b_1, \ldots, b_n) \in \mathbf{R}^n \mid b_1 a_1 + \cdots + b_n a_n = 0\}$$

is finitely generated.

(3) Two matrices $G \in \mathbf{R}^{q\times m}$ and $H \in \mathbf{R}^{r\times m}$ present the same module, that is, their cokernels are isomorphic, if and only if the following two matrices are equivalent:

$$\begin{pmatrix} G & 0_{q,r} & 0_{q,q} & 0_{q,n} \\ 0_{r,m} & \mathrm{I}_r & 0_{r,q} & 0_{r,n} \end{pmatrix},$$
$$\begin{pmatrix} 0_{q,m} & 0_{q,r} & \mathrm{I}_q & 0_{q,n} \\ 0_{r,m} & 0_{r,r} & 0_{r,q} & H \end{pmatrix}.$$

Definition 5.5 (1) A ring $\mathbf{R}$ is said to be *discrete* if there is an algorithm deciding if $x = 0$ or $x \neq 0$ for an arbitrary element of $\mathbf{R}$.

(2) A ring is said to be *strongly discrete* if it is equipped with a membership test for finitely generated ideals.

Remark 5.6

(i) **Projective modules via idempotent matrices** by Rosenberg [38] There is another approach to finitely generated projective modules which is more concrete and therefore more convenient for our constructive approach. If P is a finitely generated projective $\mathbf{R}$-module, we may assume (replacing P by an isomorphic module) that $P \oplus Q = \mathbf{R}^n$ for some n, and we consider the idempotent matrix M of the $\mathbf{R}$-module homomorphism p from $\mathbf{R}^n$ to itself which is the identity on P and 0 on Q written in the standard basis. So, P can be seen (up to isomorphism) as the image of an idempotent matrix M. Conversely, different idempotent matrices can give rise to the same isomorphism class of projective modules. As a matter of fact, if M and N are idempotent matrices over a ring $\mathbf{R}$ (of possibly different sizes m and n, respectively), the corresponding finitely generated projective modules are isomorphic if and only if it is possible to enlarge the sizes of M and N (by adding zeros in the lower right-hand corner) so that they have the same size $s \times s$ and conjugate under the group $\mathrm{GL}_s(\mathbf{R})$.

In more detail ([38, Lemma 1.2.1] or [26, Lemma V.2.10]), if the isomorphism from $\mathrm{Im}(M)$ to $\mathrm{Im}(N)$ is coded by U and its inverse is coded by U', we obtain a matrix

$$A = \begin{pmatrix} \mathrm{I}_m - F & -U' \\ U & \mathrm{I}_n - G \end{pmatrix} = \begin{pmatrix} \mathrm{I}_m & 0 \\ U & \mathrm{I}_n \end{pmatrix} \begin{pmatrix} \mathrm{I}_m & -U' \\ 0 & \mathrm{I}_n \end{pmatrix} \begin{pmatrix} \mathrm{I}_m & 0 \\ U & \mathrm{I}_n \end{pmatrix}$$

in $\mathrm{GL}_{n+m}(\mathbf{R})$ with $\begin{pmatrix} 0_m & 0 \\ 0 & G \end{pmatrix} = A \begin{pmatrix} F & 0 \\ 0 & 0_n \end{pmatrix} A^{-1}$ (with usual block matrix notation).

The matrix $\begin{pmatrix} 0_m & 0 \\ 0 & G \end{pmatrix}$ is obviously conjugate to $\begin{pmatrix} G & 0 \\ 0 & 0_m \end{pmatrix}$ by a permutation matrix.

We will embed $\mathrm{M}_n(\mathbf{R})$ in $\mathrm{M}_{n+1}(\mathbf{R})$ by $M \mapsto \begin{pmatrix} M & 0 \\ 0 & 0 \end{pmatrix}$, $\mathrm{GL}_n(\mathbf{R})$ in $\mathrm{GL}_{n+1}(\mathbf{R})$ by the group homomorphism $M \mapsto \begin{pmatrix} M & 0 \\ 0 & 1 \end{pmatrix}$, so that we can define by $\mathrm{M}(\mathbf{R})$ (resp., $\mathrm{GL}(\mathbf{R})$) as the infinite union of the $\mathrm{M}_n(\mathbf{R})$ (resp., $\mathrm{GL}_n(\mathbf{R})$). Denoting by $\mathrm{Idem}(\mathbf{R})$ the set of idempotent matrices in $\mathrm{M}(\mathbf{R})$, $\mathrm{Proj}\,\mathbf{R}$ may be identified with the set of conjugaison orbits of $\mathrm{GL}(\mathbf{R})$ on $\mathrm{Idem}(\mathbf{R})$. The monoid operation is induced by $(M, N) \mapsto \begin{pmatrix} M & 0 \\ 0 & N \end{pmatrix}$ and $\mathrm{K}_0(\mathbf{R})$ is the Groethendieck group of this monid. Denoting by $M = (m_{i,j})_{i,j \in I}$ and $N = (n_{k,\ell})_{k,\ell \in J}$, the Kronecker product $M \otimes N := (r_{(i,k),(j,\ell)})_{(i,k),(j,\ell) \in I \times J}$, where $r_{(i,k),(j,\ell)} = m_{i,j} n_{k,\ell}$, corresponds to the tensor product $\mathrm{Im}\, M \otimes \mathrm{Im}\, N$.

(ii) **Projective modules via Fitting ideals** by Lombardi and Quitté [26, 27] The theory of Fitting ideals of finitely presented modules is an extremely efficient computing machinery from a theoretical constructive point of view. Recall that if G is a presentation matrix of a module T given by q generators related by m relations, the Fitting ideals of T are the ideals

$$\mathcal{F}_n(T) := \mathcal{D}_{q-n}(G),$$

where for any integer k, $\mathcal{D}_k(G)$ denotes the determinantal ideal of G of order k, that is, the ideal generated by all the minors of G of size k, with the convention that for $k \leq 0$, $\mathcal{D}_k(G) = \langle 1 \rangle$, and for $k > \min(m, n)$, $\mathcal{D}_k(G) = \langle 0 \rangle$. It is worth pointing out that the Fitting ideals of a finitely presented module T do not depend on its presentation matrix G and that one has

$$\langle 0 \rangle = \mathcal{F}_{-1}(T) \subseteq \mathcal{F}_0(T) \subseteq \mathcal{F}_q(T) = \langle 1 \rangle .$$

Projectivity can be tested via the Fitting ideals as follows: *a finitely presented* $\mathbf{R}$*-module is projective if and only if its Fitting ideals are principal generated by idempotent elements* (see Theorems V.6.1 and V.8.14 in [26]).

Recall that a ring $\mathbf{R}$ is *local* if it satisfies:

$$\forall x \in \mathbf{R}, \quad x \in \mathbf{R}^\times \vee 1 - x \in \mathbf{R}^\times .$$

Theorem 5.7 *If* $\mathbf{R}$ *is a local ring, then every finitely generated projective* $\mathbf{R}$*-module is free. In particular,* $\mathrm{K}_0(\mathbf{R}) \cong \mathbb{Z}$ *(since* $\mathrm{Proj}\,\mathbf{R} \cong \mathbb{N}$*) with the generator the isomorphism class of a free module of rank* 1 *(*$\cong \mathbf{R}$*).*

Proof Let $F = (f_{i,j})_{1 \le i,j \le m}$ be an idempotent matrix with coefficients in a local ring $\mathbf{R}$. Let us prove that F is conjugate to a standard projection matrix

$$\mathrm{I}_{r,m} := \begin{pmatrix} \mathrm{I}_r & 0_{r,m-r} \\ 0_{m-r,r} & 0_{m-r,m-r} \end{pmatrix}.$$

Two cases may arise.

- If $f_{1,1}$ is invertible, then one can find $G \in \mathrm{GL}_m(\mathbf{R})$ such that

$$GFG^{-1} = \begin{pmatrix} 1 & 0_{1,m-1} \\ 0_{m-1,1} & F_1 \end{pmatrix},$$

  where F_1 is an idempotent matrix of size $(m-1) \times (m-1)$, and an induction on m applies.
- If $1 - f_{1,1}$ is invertible, then one can find $H \in \mathrm{GL}_m(\mathbf{R})$ such that

$$HFH^{-1} = \begin{pmatrix} 0 & 0_{1,m-1} \\ 0_{m-1,1} & F_2 \end{pmatrix},$$

  where F_2 is an idempotent matrix of size $(m-1) \times (m-1)$, and again an induction on m applies. □

The following theorem gives a local characterization of projective modules.

Theorem 5.8 *An* $\mathbf{R}$*-module* P *is projective if and only if there exist comaximal elements* $s_1, \ldots, s_k \in \mathbf{R}$ *(i.e., satisfying* $\langle s_1, \ldots, s_k \rangle = \mathbf{R}$*) such that for each* $1 \le i \le k$, $P_{s_i} := P \otimes \mathbf{R}[\frac{1}{s_i}]$ *is a free* $\mathbf{R}[\frac{1}{s_i}]$*-module.*

Definition 5.9 (Extended modules) A module M over $\mathbf{R}[X_1, \ldots, X_n] = \mathbf{R}[\underline{X}]$ is said to be *extended from* $\mathbf{R}$ (or simply, *extended*) if it is isomorphic to a module $N \otimes_{\mathbf{R}} \mathbf{R}[\underline{X}]$ for some $\mathbf{R}$-module N. Necessarily

$$N \simeq \mathbf{R} \otimes_{\mathbf{R}[\underline{X}]} M \text{ through } \rho : \mathbf{R}[\underline{X}] \to \mathbf{R},\ f \mapsto f(0),$$

that is, $N \simeq M/(X_1M + \cdots + X_nM)$. In particular, if M is finitely presented, denoting by $M^0 = M[0, \ldots, 0]$ the $\mathbf{R}$-module obtained by replacing the X_i by 0 in a relation matrix of M, then M is extended if and only if

$$M \simeq M^0 \otimes_{\mathbf{R}} \mathbf{R}[\underline{X}],$$

or equivalently, if the matrices M and M^0 are equivalent (once properly enlarged, see Definition and Proposition 5.4) using invertible matrices with entries in $\mathbf{R}[\underline{X}]$.

If M is given as the image of an idempotent matrix $F = F(X_1, \ldots, X_n)$, then M is extended if and only if F is conjugate to $F(0, \ldots, 0)$.

Definition 5.10 (Finitely generated projective modules of constant rank)

(i) **Classical approach by** Lam [18] The rank of a nonzero free module $\mathbf{R}^m$ is defined by $\mathrm{rk}_{\mathbf{R}}(\mathbf{R}^m) = m$. If P is a finitely generated projective module, as it is locally free (i.e., $P_{\mathfrak{p}} := P \otimes_{\mathbf{R}} R_{\mathfrak{p}}$ is a free $R_{\mathfrak{p}}$-module for any $\mathfrak{p} \in \mathrm{Spec}(\mathbf{R})$, where $\mathrm{Spec}(\mathbf{R})$ denotes the set of prime ideals of $\mathbf{R}$), we define the (rank) map $\mathrm{rk}(P) : \mathrm{Spec}(\mathbf{R}) \to \mathbb{N}$ by $\mathrm{rk}(P)(\mathfrak{p}) = \mathrm{rk}_{R_{\mathfrak{p}}}(P_{\mathfrak{p}})$. The map $\mathrm{rk}(P)$ is locally constant. Especially, if $\mathrm{Spec}(\mathbf{R})$ is connected, that is, if $\mathbf{R}$ is not a direct product of nontrivial rings (or equivalently, if $\mathbf{R}$ has no nontrivial idempotents), then $\mathrm{rk}(P)$ is constant.

(ii) **Constructive approach** by Lombardi and Quitté [26, 27] Roughly speaking, if $\varphi\colon P \to P$ is an endomorphism of a finitely generated projective $\mathbf{R}$-module P, then supposing that $P \oplus Q$ is isomorphic to a free module, then the determinant of $\varphi_1 := \varphi \oplus \mathrm{Id}_Q$ depends only on φ; it is called the determinant of φ. Now, let us consider the $\mathbf{R}[X]$-module $P[X] := P \otimes_{\mathbf{R}} \mathbf{R}[X]$. The polynomial $R_P(X) := \det(X\mathrm{Id}_P)$ is called the rank polynomial of the module P. If P is free of rank k, then clearly $R_P(X) = X^k$. Moreover, $R_{P\oplus Q}(X) = R_P(X)R_Q(X)$, $R_P(X)R_P(Y) = R_P(XY)$, and $R_P(1) = 1$, in such a way the coefficients of $R_P(X)$ form a fundamental system of orthogonal idempotents ($\sum e_i = 1$ and $e_ie_j = 0$ for $i \neq j$).

Now, this terminology being established, a finitely generated projective $\mathbf{R}$-module P is said to have rank equal to h if $R_P(X) = X^h$. If we do not specify h, we say that P has a constant rank.

For any finitely generated projective $\mathbf{R}$-module P, denoting by $R_P(X) = \sum_{h=0}^{n} r_h X^h$ (as said above, the r_hs form a fundamental system of orthogonal idempotents), we have $P = \bigoplus_{h=0}^{n} r_h P$ as $\mathbf{R}$-modules, and each module $r_h P$ is a constant rank projective $\mathbf{R}/\langle 1 - r_h\rangle$-module of rank h (recall that $\mathbf{R}/\langle 1 - r_h\rangle \cong \mathbf{R}[\frac{1}{r_h}]$).

(iii) **Projective modules of rank one** To any ring $\mathbf{R}$, we can associate its *Picard group* $\mathrm{Pic}\,\mathbf{R}$, that is, the group of projective $\mathbf{R}$-modules of rank one equipped with tensor product as group operation. The inverse of P is its dual $P^\star$. If

$P \simeq \operatorname{Im} M$ then $P^\star \simeq \operatorname{Im} {}^{\mathrm{t}}M$. In particular, if M is a rank-1 idempotent matrix, then $M \otimes^{\mathrm{t}} M$ is an idempotent matrix whose image is a rank-1 free module.

If that $\mathbf{R}$ is an integral domain, $\operatorname{Pic} \mathbf{R}$ is isomorphic to the *class group* of $\mathbf{R}$, the group of invertible fractional ideals in the field of fractions of $\mathbf{R}$, modulo the principal ideals. So, this generalizes to an arbitrary ring the class group introduced originally by Kummer.

Example 5.11 Let e be a nontrivial idempotent of a ring $\mathbf{R}$, that is, $e^2 = e$, $e \neq 0$, and $e \neq 1$, and consider the projective $\mathbf{R}$-module $P = e\mathbf{R}$. We have $R_P(X) = 1 - e + e\,X$. The module P does not have a constant rank: it is of rank 1 over the ring $\mathbf{R}[\frac{1}{e}] \cong \mathbf{R}/\langle 1 - e\rangle$ and of rank 0 over the ring $\mathbf{R}[\frac{1}{1-e}] \cong \mathbf{R}/\langle e\rangle$.

5.2.2 Finitely Generated Stably Free Modules

Definition 5.12 An $\mathbf{R}$-module P is said to be *finitely generated stably free* (of rank $n - m$) if $P \oplus \mathbf{R}^m \cong \mathbf{R}^n$ for some m, n. This amounts to saying that P is isomorphic to the kernel of an epimorhism $f : \mathbf{R}^n \to \mathbf{R}^m$. If M is the $m \times n$ matrix associated with f, then M is right invertible, that is, there exists an $n \times m$ matrix N such that $MN = \mathrm{I}_m$. Conversely, the kernel of any right invertible matrix defines a stably free module. So, the study of finitely generated stably free $\mathbf{R}$-modules becomes equivalent to the study of right invertible rectangular matrices over $\mathbf{R}$.

Example 5.13

(i) Every free module is stably free.

(ii) Every stably free module is projective. The converse does not hold. To see this, it suffices to consider a nonprincipal ideal in a Dedekind domain (e.g., the ideal $\langle 3, 2 + \sqrt{-5}\rangle$ in the Dedekind domain $\mathbb{Z}[\sqrt{-5}]$). It is a rank-1 projective module (as it is an invertible ideal) but not a stably free module since as will be seen in Theorem 5.19, stably free modules of rank one are free.

Note that for a ring $\mathbf{R}$, the fact that every projective $\mathbf{R}$-module is stably free is equivalent to the fact that $\mathrm{K}_0(\mathbf{R}) = \mathbb{Z}$.

(iii) Let $\mathbf{R} = \mathbb{R}[X_1, X_2, X_3]/\langle X_1^2 + X_2^2 + X_3^2 - 1\rangle = \mathbb{R}[x_1, x_2, x_3]$ be the affine coordinate ring of the real 2-sphere, and consider the syzygy module

$$T = \operatorname{Syz}(x_1, x_2, x_3) := \{(y_1, y_2, y_3) \in \mathbf{R}^3 \mid x_1y_1 + x_2y_2 + x_3y_3 = 0\}.$$

As (x_1, x_2, x_3) is unimodular (see Definition 5.15 below), we have $T \oplus \mathbf{R} \cong \mathbf{R}^3$, and thus T is a rank-2 stably free module. It is well known that T is not free for topological reasons but this remarkable fact is still lacking a simple algebraic proof.

The following gives a criterion for the freeness of finitely generated stably free modules in matrix terms.

Proposition 5.14 *For any right invertible $r \times n$ matrix M with entries in a ring* $\mathbf{R}$*, the (stably free) solution space of M is free if and only if M can be completed to an invertible matrix by adding a suitable number of new rows.*

Definition 5.15 Let $b_1, \ldots, b_n \in \mathbf{R}$. Recall that a row $(b_1, \ldots, b_n)$ is said to be *unimodular* (or that ${}^t(b_1, \ldots, b_n)$ is a unimodular vector) if the row matrix $(b_1, \ldots, b_n)$ is right invertible, that is, if $\langle b_1, \ldots, b_n \rangle = \mathbf{R}$. The set of such unimodular rows will be denoted by $\mathrm{Um}_n(\mathbf{R})$ (in order to lighten the notation, we use the same notation for unimodular vectors).
If a unimodular row over $\mathbf{R}$ can be completed to an invertible matrix (i.e., can be written as the first row of an invertible matrix with entries in $\mathbf{R}$), we say that it is *completable* over $\mathbf{R}$. For example, every unimodular row (a, b) of length 2 is completable. As a matter of fact, writing $ac + bd = 1$, the matrix $\begin{pmatrix} a & b \\ -d & c \end{pmatrix}$ has determinant 1.

The following gives a criterion for the freeness of all finitely generated stably free modules over a ring $\mathbf{R}$ in terms of unimodular rows. It is a consequence of Proposition 5.14.

Proposition 5.16 *For any ring* $\mathbf{R}$ *and integer $d \geq 0$, the following are equivalent.*

(i) *Any finitely generated stably free module of rank $> d$ is free.*
(ii) *Any unimodular row over* $\mathbf{R}$ *of length $q \geq d + 2$ is completable.*
(iii) *For any unimodular row v over* $\mathbf{R}$ *of length $q \geq d + 2$, there exists $G \in \mathrm{GL}_q(\mathbf{R})$ such that $v\,G = (1, 0, \ldots, 0)$.*
(iv) *For $q \geq d + 2$, $\mathrm{GL}_q(\mathbf{R})$ acts transitively on $\mathrm{Um}_q(\mathbf{R})$.*
(v) *For any unimodular row v over* $\mathbf{R}$ *of length $q \geq d + 2$, we have $\mathbf{R}^q \cong \mathbf{R}\,v \oplus \mathbf{R}^{q-1}$.*

In fact, when studying finitely generated stably free modules, one has only to care about stably free modules of rank ≥ 2, since as will be seen in Theorem 5.19, stably free modules of rank 1 are free.

Notation 5.17 Let $\mathbf{R}$ be a ring and $A \in \mathbf{R}^{n \times m}$ an $n \times m$ matrix with entries in $\mathbf{R}$. Denote by $A_1, \ldots, A_m$ the columns of A, so that we can write $A = [A_1, \ldots, A_m]$. If $I = (i_1, \ldots, i_r)$ is a sequence of natural numbers with $1 \leq i_1 < \cdots < i_r \leq m$, we denote by A_I the matrix $[A_{i_1}, \ldots, A_{i_r}]$.

Binet–Cauchy Formula 5.18 Let $\mathbf{R}$ be a ring and consider two matrices $M \in \mathbf{R}^{s \times r}$ and $N \in \mathbf{R}^{r \times s}$, $r \leq s$. Then

$$\det(MN) = \sum_I \det(M_I)\,\det(N_I),$$

where I runs through all sequences of natural numbers $(i_1, \ldots, i_r)$ with $1 \leq i_1 < \cdots < i_r \leq s$.

Theorem 5.19 *For any ring* $\mathbf{R}$*, any stably free* $\mathbf{R}$*-module of rank* 1 *is free* ($\cong \mathbf{R}$).

Proof Let P be a stably free $\mathbf{R}$-module of rank 1 (i.e., $P \oplus \mathbf{R}^{n-1} \cong \mathbf{R}^n$ for some $n \geq 2$) represented as the solution space of a right invertible $(n-1) \times n$ matrix M. That is, $P = \operatorname{Ker} M$ and $\exists\, N \in \mathbf{R}^{n\times(n-1)}$ such that $MN = \mathrm{I}_{n-1}$. Proving that P is free is nothing else than proving that M can be completed to an invertible matrix (see Proposition 5.14). This clearly amounts to proving that the maximal minors $b_1, \ldots, b_n$ are comaximal, namely, $1 \in \langle b_1, \ldots, b_n \rangle$. As a matter of fact, if $a_1 b_1 + \cdots + a_n b_n = 1$ then M can be completed to a matrix of determinant 1 by adding a last row $[a_1, \ldots, a_n]$ with appropriate signs. Thus, our task is reduced to prove that $1 \in \langle b_1, \ldots, b_n \rangle$.

Classical approach Let $\mathfrak{m}$ be a maximal ideal of $\mathbf{R}$. Then, modulo $\mathfrak{m}$, we have $\bar{M}\bar{N} = \mathrm{I}_{n-1}$. Since $\bar{M}$ is right invertible, it has rank $n-1$ and can be completed by linear algebra to an invertible matrix $M_{\mathfrak{m}} \in \mathrm{GL}_n(\mathbf{R}/\mathfrak{m})$. Thus, det $M_{\mathfrak{m}} \neq \bar{0}$ and a fortiori $\langle b_1, \ldots, b_n \rangle \nsubseteq \mathfrak{m}$.

Constructive approach Reasoning modulo $\langle b_1, \ldots, b_n \rangle$, the fact that $\bar{M}\bar{N} = \mathrm{I}_{n-1}$ together with the Binet–Cauchy Formula 5.18 give that $\bar{1} = \bar{0}$. Thus, $1 \in \langle b_1, \ldots, b_n \rangle$. □

Definition and Proposition 5.20 (Finite free resolution) Let $\mathbf{R}$ be a ring.

(1) A *complex* $\mathcal{F}$ of $\mathbf{R}$-modules is a sequence of modules F_i and maps $\varphi_i : F_i \to F_{i-1}$ such that $\varphi_i \circ \varphi_{i+1} = 0$ for all i. The module $H_i := \operatorname{Ker}(\varphi_i)/\operatorname{Im}(\varphi_{i+1})$ is called the *homology* of this complex at F_i.

If $H_i = 0$ for all i, we say the complex $\mathcal{F}$ is *exact*. For example, if U is a submodule of a module M then the complex $0 \to U \xrightarrow{\mathrm{i}} M \xrightarrow{\pi} M/U \longrightarrow 0$ is exact, where i is inclusion and π is the canonical projection.

For $a \in \mathbf{R}$, the homology of the complex $0 \longrightarrow \mathbf{R} \xrightarrow{\varphi_a} \mathbf{R}$ (called Koszul complex of length 1), where $\varphi_a(x) = a\,x$, is the annihilator $\operatorname{Ann}(a) := \{b \in \mathbf{R} \mid b\,a = 0\}$ of a.

A *finite free resolution* of length n of a module M is a complex

$$0 \longrightarrow \mathbf{R}^{r_n} \xrightarrow{\varphi_n} \cdots \xrightarrow{\varphi_2} \mathbf{R}^{r_1} \xrightarrow{\varphi_1} \mathbf{R}^{r_0} \longrightarrow 0$$

which is exact except at $\mathbf{R}^{r_0}$ and such that $M = \operatorname{Coker}(\varphi_1)$ and $r_i \in \mathbb{N}^*$.

(2) **Hilbert Syzygy Theorem** *If* $\mathbf{K}$ *is a field then every finitely generated module over* $\mathbf{K}[X_1, \ldots, X_k]$ *has a finite free resolution.*

Note that such finite free resolution over a field can be computed effectively using Gröbner bases and following Schreyer's algorithm [39]. In [14],

Schreyer's algorithm has been extended to Bézout domains of Krull dimension ≤ 1 ($\mathbb{Z}$ for example) and coherent Bézout rings with zero-divisors and of Krull dimension ≤ 0 ($\mathbb{Z}/N\mathbb{Z}$, for example).

(3) **Serre's Theorem** *Any projective module with a finite free resolution is stably free.*

To see this, if

$$0 \longrightarrow \mathbf{R}^{r_n} \xrightarrow{\varphi_n} \cdots \xrightarrow{\varphi_2} \mathbf{R}^{r_1} \xrightarrow{\varphi_1} \mathbf{R}^{r_0} \xrightarrow{\varphi_0} P \longrightarrow 0$$

is a free resolution of the projective module P, then

$$P \text{ projective} \Rightarrow \mathrm{Ker}(\varphi_0) \text{ projective} \Rightarrow \mathrm{Im}(\varphi_1) = \mathrm{Ker}(\varphi_0) \text{ projective} \Rightarrow \cdots$$

$$P \oplus \mathbf{R}^{r_1} \oplus \mathbf{R}^{r_3} \oplus \cdots \cong \mathbf{R}^{r_0} \oplus \mathbf{R}^{r_2} \oplus \cdots \quad \text{and} \quad \mathrm{rk}(P) = \sum_{i=0}^{n} (-1)^i r_i.$$

(4) Combining (2) and (3) we get the following.

If $\mathbf{K}$ *is a field then every finitely-generated projective module over* $\mathbf{K}[X_1, \ldots, X_k]$ *is stably free.*

Readers interested in new constructive techniques in finite free resolutions can refer to the nice papers [6] and [7] in which the authors greatly simplify the main proofs given in the book *Finite Free Resolutions* [31] without any use of minimal prime ideals. It is well known that the existence of minimal prime ideals is equivalent to the axiom of choice.

5.2.3 Concrete Local–Global Principle

We give here general explanations about how to decipher constructively classical proofs in commutative algebra using local-global principles. This section comes essentially from [25].

From Local to Quasi-global

The classical reasoning by localization works as follows. When the ring is local, a property P is satisfied by virtue of a quite concrete proof. When the ring is not local, the same property remains true (from a classical nonconstructive point of view) as it suffices to check it locally.

When carefully examining the first proof, some computations come into view. These computations are feasible thanks to the following principle:

$$\forall x \in \mathbf{R}, \quad x \in \mathbf{R}^{\times} \vee x \in \mathrm{Rad}(\mathbf{R}).$$

This principle is in fact applied to elements coming from the proof itself. In case of a nonnecessarily local ring, we repeat the same proof, replacing at each disjunction

"x is a unit or x is in the radical" in the passage of the proof we are considering, by the consideration of two rings $\mathbf{T}_x := \mathbf{T}[\frac{1}{x}]$ and $\mathbf{T}_{1+x\mathbf{T}}$ (the localization of $\mathbf{T}$ at the monoid $1 + x\mathbf{T}$), where $\mathbf{T}$ is the "current" localization of the ring $\mathbf{R}$ we start with. When the initial proof is completely unrolled, we obtain a finite number (since the proof is finite) of localizations $\mathbf{R}_{S_i}$, for each of them the property is true. Moreover, the corresponding Zariski open subsets U_{S_i} cover $\mathrm{Spec}(\mathbf{R})$, implying that the property P is true for $\mathbf{A}$, and this time in an entirely explicit way.

It is worth pointing out that, in order to roll out the method described above, one needs Lemma 5.27 which guarantees that an element remains in the radical once it is forced into being in.

Definition 5.21 (Constructive definition of the radical) Recall that a ring $\mathbf{R}$ is said to be *discrete* if there is an algorithm deciding if $x = 0$ or $x \neq 0$ for an arbitrary element of $\mathbf{R}$.

Constructively, the *radical* $\mathrm{Rad}(\mathbf{R})$ of a ring $\mathbf{R}$ is the set of all $x \in \mathbf{R}$ such that $1 + x\mathbf{R} \subseteq \mathbf{R}^\times$, where $\mathbf{R}^\times$ is the group of units of $\mathbf{R}$. A ring $\mathbf{R}$ is *local* if it satisfies

$$\forall x \in \mathbf{R}, \quad x \in \mathbf{R}^\times \ \vee \ 1 + x \in \mathbf{R}^\times. \tag{5.1}$$

It is *residually discrete local* if it satisfies

$$\forall x \in \mathbf{R}, \quad x \in \mathbf{R}^\times \ \vee \ x \in \mathrm{Rad}(\mathbf{R}). \tag{5.2}$$

From a classical point of view, we have $(5.1) \Leftrightarrow (5.2)$, but the constructive meaning of (5.2) is stronger than that of (5.1). Constructively a *discrete field* is defined as a ring in which each element is zero or invertible, with an explicit test for the "or." A *Heyting field* (or a field) is defined as a local ring whose Jacobson radical is 0. So $\mathbf{R}$ is residually discrete local exactly when it is local and the residue field $\mathbf{R}/\mathrm{Rad}(\mathbf{R})$ is a discrete field.

Definition 5.22 (Monoids and saturations)

(i) We say that S is a *multiplicative subset* (or a *monoid*) of a ring $\mathbf{R}$ if

$$\begin{cases} 1 \in S \\ \forall\, s, t \in S, \quad st \in S. \end{cases}$$

For example, for $a \in \mathbf{R}$, $a^{\mathbb{N}} := \{a^n;\ n \in \mathbb{N}\}$ is a monoid of $\mathbf{R}$.

(ii) A monoid S of a ring $\mathbf{R}$ is said to be *saturated* if we have the implication

$$\forall\, s, t \in \mathbf{R}, \quad (st \in S \ \Rightarrow \ s \in S).$$

(iii) The localization of $\mathbf{R}$ at S will be denoted by $S^{-1}\mathbf{R}$ or $\mathbf{R}_S$. If S is generated by $s \in \mathbf{R}$, we denote $\mathbf{R}_S$ by $\mathbf{R}_s$ or $\mathbf{R}[1/s]$. Note here that $\mathbf{R}_s$ is isomorphic to the ring $\mathbf{R}[T]/(sT - 1)$. Saturating a monoid S (that is, replacing S by its saturation $\bar{S} := \{s \in \mathbf{R},\ \ \exists\, t \in \mathbf{R} \mid \ st \in S\}$) does not change the localization $\mathbf{R}_S$. Two monoids are said to be *equivalent* if they have the same saturation.

We keep the same notation for the localization of an $\mathbf{R}$-module.

Definition 5.23 (Comaximal elements and monoids) Let $\mathbf{R}$ be a ring.

(1) Let $s_1, \ldots, s_k \in \mathbf{R}$. We say that the elements $s_1, \ldots, s_k$ are *comaximal* if $\langle s_1, \ldots, s_k \rangle = \mathbf{R}$.
(2) Let $S, S_1, \ldots, S_n$ be monoids of $\mathbf{R}$.

(i) We say that the monoids $S_1, \ldots, S_n$ are *comaximal* if any ideal of $\mathbf{R}$ meeting all the S_i must contain 1. In other words, if we have

$$\forall s_1 \in S_1, \ldots,\ \forall s_n \in S_n,\ \ \exists\, a_1, \ldots, a_n \in \mathbf{R} \ \Bigg|\ \sum_{i=1}^{n} a_i s_i = 1,$$

that is, $s_1, \ldots, s_k$ are comaximal elements in $\mathbf{R}$.
For example, if $u_1, \ldots, u_m$ are comaximal elements in $\mathbf{R}$, then the monoids $u_1^{\mathbb{N}}, \ldots, u_m^{\mathbb{N}}$ are comaximal.

(ii) We say that the monoids $S_1, \ldots, S_n$ cover the monoid S if S is contained in the S_i and any ideal of $\mathbf{R}$ meeting all the S_i must meet S. In other words, if we have

$$\forall s_1 \in S_1 \ \cdots\ \forall s_n \in S_n\ \ \exists a_1, \ldots, a_n \in \mathbf{R} \ \Bigg|\ \sum_{i=1}^{n} a_i s_i \in S.$$

We remark that comaximal monoids remain comaximal when we replace the ring by a bigger one or the multiplicative subsets by smaller ones.

In classical algebra (with the axiom of the prime ideal) this amounts to saying, in the first case, that the Zariski open subsets U_{S_i} cover $\mathrm{Spec}(\mathbf{R})$ and, in the second case, that the Zariski open subsets U_{S_i} cover the open subset U_S. From a constructive point of view, $\mathrm{Spec}(\mathbf{R})$ is a topological space via its open subsets U_S but whose points are often hardly accessible.

We have the following immediate result.

Lemma 5.24 (Associativity and transitivity of coverings)

(1) *(Associativity) If monoids $S_1, \ldots, S_n$ of a ring $\mathbf{R}$ cover a monoid S and each S_ℓ is covered by some monoids $S_{\ell,1}, \ldots, S_{\ell,m_\ell}$, then the $S_{\ell,j}$s cover S.*
(2) *(Transitivity) Let S be a monoid of a ring $\mathbf{R}$ and let $S_1, \ldots, S_n$ monoids of the ring $\mathbf{R}_S$. For $\ell = 1, \ldots, n$, let V_ℓ be the monoid of $\mathbf{R}$ formed by the denominators of the elements of S_ℓ. Then the monoids $V_1, \ldots, V_n$ cover S.*

Definition and Notation 5.25 Let I and U two subsets of a ring $\mathbf{R}$. We denote by $\mathcal{M}(U)$ the monoid generated by U, $\mathcal{I}_{\mathbf{R}}(I)$ or $\mathcal{I}(I)$ the ideal generated by I, and $\mathcal{S}(I;U)$ the monoid $\mathcal{M}(U)+\mathcal{I}(I)$. If $I=\{a_1,\ldots,a_k\}$ and $U=\{u_1,\ldots,u_\ell\}$, we denote $\mathcal{M}(U)$, $\mathcal{I}(I)$, and $\mathcal{S}(I;U)$ by $\mathcal{M}(u_1,\ldots,u_\ell)$, $\mathcal{I}(a_1,\ldots,a_k)$, and $\mathcal{S}(a_1,\ldots,a_k;u_1,\ldots,u_\ell)$, respectively.

Remark 5.26

(1) It is clear that if u is equal to a product $u_1\cdots u_\ell$, then the monoids $\mathcal{S}(a_1,\ldots,a_k;u_1,\ldots,u_\ell)$ and $\mathcal{S}(a_1,\ldots,a_k;u)$ are equivalent.
(2) When we localize at $S=\mathcal{S}(I;U)$, the elements of U are forced into being invertible and those of I end up on the radical of $\mathbf{R}_S$.

According to Henri Lombardi, the "good category" would be that whose objects are couples $(\mathbf{R},I)$, where $\mathbf{R}$ is a commutative ring and I is an ideal contained in the radical of $\mathbf{R}$. Arrows from $(\mathbf{R},I)$ onto $(\mathbf{R}',I')$ are ring homomorphisms $f:\mathbf{R}\to\mathbf{R}'$ such that $f(I)\subseteq I'$. Thus, one can retrieve usual rings by taking $I=0$ and local rings (equipped with the notion of local homomorphism) by taking I equal to the maximal ideal. In order to "localize" an object $(\mathbf{A},I)$ in this category, we use a monoid U and an ideal J in such a way we form the new object $(\mathbf{R}_{\mathcal{S}(J_1;U)},J_1\mathbf{R}_{\mathcal{S}(J_1;U)})$, where $J_1=I+J$.

The following lemma will play a crucial role when we want to reread constructively with an arbitrary ring a proof given in the local case.

Lemma 5.27 (A trick by Lombardi [23, 24]) *Let U and I be two subsets of a ring $\mathbf{R}$ and consider $a\in\mathbf{R}$. Then the monoids $\mathcal{S}(I;U,a)$ and $\mathcal{S}(I,a;U)$ cover the monoid $\mathcal{S}(I;U)$.*

Proof For $x\in\mathcal{S}(I;U,a)$ and $y\in\mathcal{S}(I,a;U)$, we have to find a linear combination of the form $x_1x+y_1y\in\mathcal{S}(I;U)$ $(x_1,y_1\in\mathbf{R})$. Write $x=u_1a^k+j_1$, $y=(u_2+j_2)-(az)$ with $u_1,u_2\in\mathcal{M}(U)$, $j_1,j_2\in\mathcal{I}(I)$, $z\in\mathbf{R}$. The classical identity $c^k-d^k=(c-d)\times\cdots$ gives a $y_2\in\mathbf{A}$ such that $y_2y=(u_2+j_2)^k-(az)^k=(u_2^k+j_3)-(az)^k$. Just write $z^kx+u_1y_2y=u_1u_2^k+u_1j_3+j_1z^k=u_4+j_4$. □

It is worth pointing out that, in the lemma above, we have

$$a\in\left(\mathbf{R}_{\mathcal{S}(I;U,a)}\right)^\times \quad\text{and}\quad a\in\operatorname{Rad}\left(\mathbf{R}_{\mathcal{S}(I,a;U)}\right).$$

Having this lemma in hands, we can state the following general deciphering principle allowing to automatically get a quasi-global version of a theorem from its local version.

General local–Global Principle 5.28 (Lombardi [24]) When rereading an explicit proof given when $\mathbf{R}$ is local, with an arbitrary ring $\mathbf{R}$, start with $\mathbf{R}=\mathbf{R}_{\mathcal{S}(0;1)}$.

Then, at each disjunction (for an element a produced when computing in the local case)

$$a \in \mathbf{R}^\times \ \vee \ a \in \mathrm{Rad}(\mathbf{R}),$$

replace the "current" ring $\mathbf{R}_{\mathcal{S}(I;U)}$ by both $\mathbf{R}_{\mathcal{S}(I;U,a)}$ and $\mathbf{R}_{\mathcal{S}(I,a;U)}$ in which the computations can be pursued. At the end of this rereading, one obtains a finite family of rings $\mathbf{R}_{\mathcal{S}(I_j;U_j)}$ with comaximal monoids $\mathcal{S}(I_j;U_j)$ and finite sets $I_j,\ U_j$.

The following examples are frequent and ensue immediately from Lemmas 5.24 and 5.27, except the first one which is an easy exercise.

Examples 5.29 Let $\mathbf{R}$ be a ring, U and I subsets of $\mathbf{R}$, and $S = \mathcal{S}(I;U)$.

(1) Let $s_1, \ldots, s_n \in \mathbf{R}$ be comaximal elements. Then the monoids $S_i = \mathcal{M}(s_i) = s_i^{\mathbb{N}}$ are comaximal.
More generally, if $t_1, \ldots, t_n \in \mathbf{R}$ are comaximal elements in $\mathbf{R}_S$, then the monoids $\mathcal{S}(I;U,t_i)$ cover the monoid S.

(2) Let $s_1, \ldots, s_n \in \mathbf{R}$. The monoids $S_1 = \mathcal{S}(0;s_1)$, $S_2 = \mathcal{S}(s_1;s_2)$, $S_3 = \mathcal{S}(s_1,s_2;s_3)$, …, $S_n = \mathcal{S}(s_1,\ldots,s_{n-1};s_n)$ and $S_{n+1} = \mathcal{S}(s_1,\ldots,s_n;1)$ are comaximal.
More generally, the monoids

$$\begin{aligned} &V_1 = \mathcal{S}(I;U,s_1), V_2 = \mathcal{S}(I,s_1;U,s_2), V_3 = \mathcal{S}(I,s_1,s_2;U,s_3),\ldots,\\ &V_n = \mathcal{S}(I,s_1,\ldots,s_{n-1};U,s_n), V_{n+1} = \mathcal{S}(I,s_1,\ldots,s_n;U)\end{aligned}$$

cover the monoid S.

(3) If $S, S_1, \ldots, S_n \subseteq \mathbf{R}$ are comaximal monoids and if $b = \frac{a}{s} \in \mathbf{R}_S$, then the monoids

$$\mathcal{S}(I;U,a),\ \mathcal{S}(I,a;U), S_1, \ldots,\ S_n$$

are comaximal.

From Quasi-global to Global

Different variant versions of the abstract local–global principle in commutative algebra can be reread constructively: the localization at each prime ideal is replaced by the localization at a finite family of comaximal monoids. In other words, in these "concrete" versions, we affirm that some properties pass from the quasi-global to the global. As an illustration, we cite the following results which often permit us to finish our constructive rereading.

Concrete local–Global Principle 5.30 Let $S_1, \ldots, S_n$ be comaximal monoids in a ring $\mathbf{R}$ and let $a, b \in \mathbf{R}$. Then we have the following equivalences.

(1) Concrete gluing of equalities:

$$a = b \text{ in } \mathbf{R} \iff \forall i \in \{1,\ldots,n\},\ a/1 = b/1 \text{ in } \mathbf{R}_{S_i}.$$

(2) Concrete gluing of nonzero-divisors:

$$a \text{ is not a zero-divisor in } \mathbf{R} \quad \Longleftrightarrow$$
$$\forall i \in \{1, \ldots, n\}, \quad a/1 \text{ is not a zero-divisor in } \mathbf{R}_{S_i}.$$

(3) Concrete gluing of units:

$$a \text{ is a unit in } \mathbf{R} \quad \Longleftrightarrow$$
$$\forall i \in \{1, \ldots, n\}, \quad a/1 \text{ is a unit in } \mathbf{R}_{S_i}.$$

(4) Concrete gluing of solutions of linear systems: Let B be a matrix $\in \mathbf{R}^{m \times p}$ and C a column vector $\in \mathbf{R}^{m \times 1}$.

$$\text{The linear system } BX = C \text{ has a solution in } \mathbf{R}^{p \times 1} \quad \Longleftrightarrow$$
$$\forall i \in \{1, \ldots, n\}, \text{ the linear system } BX = C \text{ has a solution in } \mathbf{R}_{S_i}^{p \times 1}.$$

(5) Concrete gluing of direct summands: Let M be a finitely generated submodule of a finitely presented module N.

$$M \text{ is a direct summand of } N \quad \Longleftrightarrow$$
$$\forall i \in \{1, \ldots, n\}, M_{S_i} \text{ is a direct summand of } N_{S_i}.$$

Concrete local–Global Principle 5.31 (Concrete gluing of module finiteness properties) Let $s_1, \ldots s_n$ be comaximal elements of a ring $\mathbf{R}$, and let M be an $\mathbf{R}$-module. Then we have the following equivalences.

(1) M is finitely generated if and only if each of the M_{s_i} is a finitely generated $\mathbf{R}_{s_i}$-module.
(2) M is finitely presented if and only if each of the M_{s_i} is a finitely presented $\mathbf{R}_{s_i}$-module.
(3) M is a finitely generated projective module if and only if each of the M_{s_i} is a finitely generated projective $\mathbf{R}_{s_i}$-module.
(4) M is projective of rank k if and only if each of the M_{s_i} is a projective $\mathbf{R}_{s_i}$-module of rank k.

One can rarely find such principles in classical literature. In Quillen's style, the corresponding general principle is in general stated using localizations at all prime ideals, but the proof often brings in a crucial lemma which has exactly the same signification as the corresponding concrete local–global principle. For example, we can state the concrete local–global Principle 5.31 "à la Quillen" under the following form.

Lemma 5.32 (Propagation lemma for some module finiteness properties) *Let M be an $\mathbf{R}$-module. The following subsets I_k of $\mathbf{R}$ are ideals.*

(1) $I_1 = \{s \in \mathbf{R} \colon M_s$ *is a finitely generated* $\mathbf{R}_s$*-module*$\}$.
(2) $I_2 = \{s \in \mathbf{R} \colon M_s$ *is a finitely presented* $\mathbf{R}_s$*-module*$\}$.
(3) $I_4 = \{s \in \mathbf{R} \colon M_s$ *is a finitely generated projective* $\mathbf{R}_s$*-module*$\}$.
(4) $I_5 = \{s \in \mathbf{R} \colon M_s$ *is a rank-k projective* $\mathbf{R}_s$*-module*$\}$.

5.2.4 The patchings of Quillen and Vaserstein

Here we give a detailed constructive proof by Lombardi and Quitté of the Quillen patching. This comes essentially from [16]. The localization at maximal ideals is replaced by localization at comaximal monoids. In [28], the constructive Quillen patching (Concrete local–global Principle 4) is given with only a sketch of proof.

Lemma 5.33 *Let S be a multiplicative subset of a ring $\mathbf{R}$ and consider three matrices A_1, A_2, A_3 with entries in $\mathbf{R}[X]$ such that the product A_1A_2 is defined and has the same size as A_3. If $A_1A_2 = A_3$ in $\mathbf{R}_S[X]$ and $A_1(0)A_2(0) = A_3(0)$ in $\mathbf{R}$, then there exists $s \in S$ such that $A_1(sX)A_2(sX) = A_3(sX)$ in $\mathbf{R}[X]$.*

Proof All the coefficients of the matrix $A_1A_2 - A_3$ are multiple of X and become zero after localization at S. Thus, there exists $s \in S$ annihilating all of them. Write $A_1A_2 - A_3 = B(X) = XB_1 + X^2B_2 + \cdots + X^kB_k$. We have $sB_1 = sB_2 = \cdots = sB_k = 0$ and, thus, $sB_1 = s^2B_2 = \cdots = s^kB_k = 0$, that is, $B(sX) = A_1(sX)A_2(sX) - A_3(sX) = 0$. □

Lemma 5.34 *Let S be a multiplicative subset of a ring $\mathbf{R}$ and consider a matrix $C(X) \in \mathrm{GL}_r(\mathbf{R}_S[X])$. Then there exists $s \in S$ and $U(X,Y) \in \mathrm{GL}_r(\mathbf{R}[X,Y])$ such that $U(X,0) = \mathrm{I}_r$, and, over $\mathbf{R}_S[X,Y]$, $U(X,Y) = C(X+sY)C(X)^{-1}$.*

Proof Set $E(X,Y) = C(X+Y)C(X)^{-1}$ and denote $F(X,Y)$ the inverse of $E(X,Y)$. We have $E(X,0) = \mathrm{I}_r$ and, thus, $E(X,Y) = \mathrm{I}_r + E_1(X)Y + \cdots + E_k(X)Y^k$. For some $s_1 \in S$, the $s_1^jE_j$s can be written without denominators and, thus, we obtain a matrix $E'(X,Y) \in \mathbf{R}[X,Y]^{r\times r}$ such that $E'(X,0) = \mathrm{I}_r$, and, over $\mathbf{R}_S[X,Y]$, $E'(X,Y) = E(X,s_1Y)$. We do the same with F (we can choose the same s_1). Hence we obtain $E'(X,Y)F'(X,Y) = \mathrm{I}_r$ in $\mathbf{R}_S[X,Y]^{r\times r}$ and $E'(X,0)F'(X,0) = \mathrm{I}_r$. Applying Lemma 5.33 in which we replace X by Y and $\mathbf{R}$ by $\mathbf{R}[X]$, we obtain $s_2 \in S$ such that $E'(X,s_2Y)F'(X,s_2Y) = \mathrm{I}_r$. Taking $U = E'(X,s_2Y)$ and $s = s_1s_2$, we obtain the desired result. □

Lemma 5.35 *Let S be a multiplicative subset of a ring $\mathbf{R}$ and $M \in \mathbf{R}[X]^{p\times q}$. If $M(X)$ and $M(0)$ are equivalent over $\mathbf{R}_S[X]$ then there exists $s \in S$ such that $M(X+sY)$ and $M(X)$ are equivalent over $\mathbf{R}[X,Y]$.*

Proof Writing $M(X) = C(X)M(0)D(X)$ with $C(X) \in \mathrm{GL}_q(\mathbf{R}_S[X])$ and $D(X) \in \mathbf{GL}_p(\mathbf{R}_S[X])$, we get

$$M(X+Y) = C(X+Y)C(X)^{-1}M(X)D(X)^{-1}D(X+Y).$$

Applying Lemma 5.34, we find $s_1 \in S, U(X,Y) \in \mathrm{GL}_q(\mathbf{R}[X,Y])$ and $V(X,Y) \in \mathrm{GL}_p(\mathbf{R}[X,Y])$ such that $U(X,0) = \mathrm{I}_q$, $V(X,0) = \mathrm{I}_p$, and, over $\mathbf{R}_S[X,Y]$, $U(X,Y) = C(X+s_1Y)C(X)^{-1}$ and $V(X,Y) = D(X)^{-1}D(X+s_1Y)$. It follows that $M(X) = U(X,0)M(X)V(X,0)$, and over $\mathbf{R}_S[X,Y]$, $M(X+s_1Y) = U(X,Y)M(X)V(X,Y)$.

Applying Lemma 5.33 (as in Lemma 5.34), we get $s_2 \in S$ such that $M(X + s_1s_2Y) = U(X, s_2Y)M(X)V(X, s_2Y)$. The desired result is obtained by taking $s = s_1s_2$. □

Theorem 5.36 (Vaserstein) *Let M be a matrix in $\mathbf{R}[X]$ and consider $S_1, \dots, S_n$ comaximal multiplicative subsets of $\mathbf{R}$. Then $M(X)$ and $M(0)$ are equivalent over $\mathbf{R}[X]$ if and only if, for each $1 \le i \le n$, they are equivalent over $\mathbf{R}_{S_i}[X]$.*

Proof It is easy to see that the set of $s \in \mathbf{R}$ such that $M(X + sY)$ is equivalent to $M(X)$ is an ideal of $\mathbf{R}$. Applying Lemma 5.35, this ideal meets S_i for each $1 \le i \le n$, and, thus, contains 1. This means that $M(X + Y)$ is equivalent to $M(X)$. To finish, just take $X = 0$. □

Theorem 5.37 (Quillen patching) *Let P be a finitely presented module over $\mathbf{R}[X]$ and consider $S_1, \dots, S_n$ comaximal multiplicative subsets of $\mathbf{R}$. Then P is extended from $\mathbf{R}$ if and only if for each $1 \le i \le n$, P_{S_i} is extended from $\mathbf{R}_{S_i}$.*

Proof This is a corollary of the previous theorem since, by Definition and Proposition 5.4, the isomorphism between $P(X)$ and $P(0)$ is nothing but the equivalence of two matrices $A(X)$ and $A(0)$ constructed from a relation matrix $M \in \mathbf{R}^{q\times m}$ of $P \simeq \mathrm{Coker}\, M$:

$$A(X) = \begin{pmatrix} M(X) & 0_{q,q} & 0_{q,q} & 0_{q,m} \\ 0_{q,m} & \mathrm{I}_q & 0_{q,q} & 0_{q,m} \end{pmatrix}.$$

□

5.2.5 Horrocks' Theorem

The local Horrocks' theorem is the following result.

Theorem 5.38 (Local Horrocks extension theorem) *If $\mathbf{R}$ is a residually discrete local ring and P is a finitely generated projective module over $\mathbf{R}[X]$ which is free over $\mathbf{R}\langle X\rangle$, then it is free over $\mathbf{R}[X]$ (i.e., extended from $\mathbf{R}$).*

Note that it is straightforward to see that the hypothesis $M \otimes_{\mathbf{R}[X]} \mathbf{R}\langle X\rangle$ is a free $\mathbf{R}\langle X\rangle$-module is equivalent to the fact that M_f is a free $\mathbf{R}[X]_f$-module for some monic polynomial $f \in \mathbf{R}[X]$. The detailed proof given by [16] is elementary and constructive, except Lemma 3.13 whose proof is abstract since it uses maximal ideals. In fact this lemma asserts if P is a projective module over $\mathbf{R}[X]$ which becomes free of rank k over $\mathbf{R}\langle X\rangle$, then its kth Fitting ideal equals $\langle 1\rangle$. This result has the following elementary constructive proof. If $P\oplus Q \simeq \mathbf{R}[X]^m$ then $P\oplus Q_1 = P\oplus(Q\oplus\mathbf{R}[X]^k)$ becomes isomorphic to $\mathbf{R}\langle X\rangle^{m+k}$ over $\mathbf{R}\langle X\rangle$ with Q_1 isomorphic to $\mathbf{R}\langle X\rangle^m$ over $\mathbf{R}\langle X\rangle$. So, we may assume $P \simeq \mathrm{Im}F$, where $G = \mathrm{I}_n - F \in \mathbf{R}[X]^{n\times n}$ is an idempotent matrix, conjugate to a standard projection matrix of rank $n-k$ over $\mathbf{R}\langle X\rangle$. We deduce that $\det(\mathrm{I}_n + TG) = (1+T)^{n-k}$ over $\mathbf{R}\langle X\rangle$.

Since $\mathbf{R}[X]$ is a subring of $\mathbf{R}\langle X\rangle$ this remains true over $\mathbf{R}[X]$. So the sum of all $(n-k)$ principal minors of G is equal to 1 (i.e., the coefficient of T^{n-k} in $\det(\mathrm{I}_n+TG)$). Hence we conclude by noticing that G is a relation matrix for P. For more details see, for example, [26, 27].

A global version is obtained from a constructive proof of the local one by the Quillen patching theorem 5.37 and applying the general local–global principle 5.28.

Theorem 5.39 (Global Horrocks extension theorem) *Let S be the multiplicative set of monic polynomials in $\mathbf{R}[X]$, where $\mathbf{R}$ is a ring. If P is a finitely generated projective module over $\mathbf{R}[X]$ such that P_S is extended from $\mathbf{R}$, then P is extended from $\mathbf{R}$.*

Proof Apply the general local–global Principle 5.28 and conclude with the concrete Quillen patching theorem 5.37. □

5.2.6 Quillen Induction Theorem

Let $\mathbf{R}$ be a ring. We denote by S the multiplicative subset of $\mathbf{R}[X]$ formed by monic polynomials. Let

$$\mathbf{R}\langle X\rangle := S^{-1}\mathbf{R}[X].$$

The interest in the properties of $\mathbf{R}\langle X\rangle$ branched in many directions and is attested by the abundance of articles on $\mathbf{R}\langle X\rangle$ appearing in the literature (see [15] for a comprehensive list of papers dealing with the ring $\mathbf{R}\langle X\rangle$). The ring $\mathbf{R}\langle X\rangle$ played an important role in Quillen's solution to Serre's problem [36] and its succeeding generalizations to non-Noetherian rings [3, 21, 29], as can be seen in these notes.

Classical Quillen induction is the following one.

Theorem 5.40 (Quillen induction) *Suppose that a class of rings $\mathcal{P}$ satisfies the following properties.*

(i) *If $\mathbf{R}\in\mathcal{P}$ then $\mathbf{R}\langle X\rangle\in\mathcal{P}$.*
(ii) *If $\mathbf{R}\in\mathcal{P}$ then $\mathbf{R}_\mathfrak{m}\in\mathcal{P}$ for any maximal ideal $\mathfrak{m}$ of $\mathbf{R}$.*
(iii) *If $\mathbf{R}\in\mathcal{P}$ and $\mathbf{R}$ is local, and if M is a finitely generated projective $\mathbf{R}[X]$-module, then M is extended from $\mathbf{R}$ (that is, free).*

Then, for each $\mathbf{R}\in\mathcal{P}$, if M is a finitely generated projective $\mathbf{R}[X_1,\ldots,X_n]$-module, then M is extended from $\mathbf{R}$.

Quillen induction needs maximal ideals; it works in classical mathematics but it cannot be fully constructive. One has to replace Quillen patching with maximal ideals by the constructive form (Theorem 5.37) with comaximal multiplicative subsets. In contrast, the "inductive step" in the proof is elementary and is based only on hypotheses (i) and (iii′) below (induct on n and use the global Horrocks extension theorem 5.39).

Theorem 5.41 (Concrete induction à la Quillen) *Suppose that a class of rings $\mathcal{P}$ satisfies the following properties.*

(i) *If* $\mathbf{R} \in \mathcal{P}$ *then* $\mathbf{R}\langle X \rangle \in \mathcal{P}$.
(ii′) *If* $\mathbf{R} \in \mathcal{P}$ *then* $\mathbf{R}_a \in \mathcal{P}$ *for any* $a \in \mathbf{R}$.
(iii′) *If* $\mathbf{R} \in \mathcal{P}$ *and* M *is a finitely generated projective* $\mathbf{R}[X]$*-module, then* M *is extended from* $\mathbf{R}$.

Then, for each $\mathbf{R} \in \mathcal{P}$, *if* M *is a finitely generated projective* $\mathbf{R}[X_1, \ldots, X_n]$*-module, then* M *is extended from* $\mathbf{R}$.

In the case of Serre's problem, $\mathbf{R}$ is a discrete field. So (i) and (iii′) are well known. We remark that (iii′) is also given by the global Horrocks extension theorem 5.39. So Quillen's proof is deciphered in a fully constructive way. Moreover, since a zero-dimensional reduced local ring is a discrete field, we obtain the following well-known generalization (see [3]).

Theorem 5.42 (Quillen–Suslin, non-Noetherian version)

(1) *If* $\mathbf{R}$ *is a zero-dimensional reduced ring then any finitely generated projective module* P *over* $\mathbf{R}[X_1, \ldots, X_n]$ *is extended from* $\mathbf{R}$ *(i.e., isomorphic to a direct sum of modules* $e_i\mathbf{R}[\underline{X}]$ *where the* e_i*s are idempotent elements of* $\mathbf{R}$*).*
(2) *As a particular case, any finitely generated projective module of constant rank over* $\mathbf{R}[X_1, \ldots, X_n]$ *is free.*
(3) *More generally the results work for any zero-dimensional ring.*

Proof The first point can be obtained from the local case by the constructive Quillen patching theorem 5.37. It can also be viewed as a concrete application of the general local–global principle 5.28.

Let us denote by $\mathbf{R}_{\text{red}}$ the reduced ring associated to a ring $\mathbf{R}$. Recall that $\mathrm{K}_0(\mathbf{R})$ is the set isomorphism classes of finitely generated projective $\mathbf{R}$-modules.

The third point follows from the fact that the canonical map $M \mapsto M_{\text{red}}$, $\mathrm{K}_0(\mathbf{R}) \to \mathrm{K}_0(\mathbf{R}_{\text{red}})$ is a bijection. Moreover $\mathbf{R}_{\text{red}}[X_1, \ldots, X_n] = \mathbf{R}[X_1, \ldots, X_n]_{\text{red}}$. □

5.3 Suslin's Proof of Serre's Problem

5.3.1 Making Use of the Maximal Ideals Constructive

The purpose of this subsection is to decipher constructively a lemma of [44] which played a central role in his second solution of Serre's problem. This lemma says that for a commutative ring $\mathbf{R}$, if $\langle v_1(X), \ldots, v_n(X) \rangle = \mathbf{R}[X]$, where v_1 is monic and

$n \geq 3$, then there exist $\gamma_1, \ldots, \gamma_\ell \in \mathrm{E}_{n-1}(\mathbf{R}[X])$ (the subgroup of $\mathrm{SL}_{n-1}(\mathbf{R}[X])$ generated by elementary matrices) such that

$$\left\langle \mathrm{Res}(v_1, e_1.\gamma_1^{\mathrm{t}}(v_2, \ldots, v_n)), \ldots, \mathrm{Res}(v_1, e_1.\gamma_\ell^{\mathrm{t}}(v_2, \ldots, v_n)) \right\rangle = \mathbf{R}.$$

By the constructive proof we give, Suslin's proof of Serre's problem becomes fully constructive. As a matter of fact, the lemma cited above is the only nonconstructive step in Suslin's elementary proof of Serre's problem. Moreover, the new method with which we treat this academic example may be a model for mimicking constructively abstract proofs in which one works modulo each maximal ideal to prove that a given ideal contains 1. The concrete local–global principle developed in Subsection 5.2.3 cannot be used here since the proof we want to decipher constructively, instead of passing to the localizations at each maximal ideal, passes to the residue fields modulo each maximal ideal.

5.3.2 A Reminder about the Resultant

In this subsection, we content ourselves with a brief outline of *resultant*: This is an important idea in constructive algebra whose development owes considerably to famous pioneers such as Bézout, Cayley, Euler, Herman, Hurwitz, Kronecker, Macaulay, Noether, and Sylvester, among others.

This subsection will be focused on the few properties of the resultant that we need in our constructive view toward projective modules over polynomial rings.

Definition 5.43 Let $\mathbf{R}$ be a ring, $f = a_0 X^\ell + a_1 X^{\ell-1} + \cdots + a_\ell \in \mathbf{R}[X]$, $a_0 \neq 0,\ a_i \in \mathbf{R}$, and $g = b_0 X^m + b_1 X^{m-1} + \cdots + b_m \in \mathbf{R}[X],\ b_0 \neq 0,\ b_i \in \mathbf{R}$.

The resultant of f and g, denoted by $\mathrm{Res}_X(f, g)$, or simply $\mathrm{Res}(f, g)$ if there is no risk of ambiguity, is the determinant of the following $(m + \ell) \times (m + \ell)$ matrix (called the Sylvester matrix of f and g with respect to X):

$$\mathrm{Syl}(f,\, g,\, X) = \underbrace{\begin{pmatrix} a_0 & & & \\ a_1 & a_0 & & \\ a_2 & a_1 & \ddots & \\ \vdots & & \ddots & a_0 \\ & \vdots & & a_1 \\ a_\ell & & & \\ & a_\ell & & \vdots \\ & & \ddots & \\ & & & a_\ell \end{pmatrix}}_{m \text{ columns}} \underbrace{\begin{pmatrix} b_0 & & & \\ b_1 & b_0 & & \\ b_2 & b_1 & \ddots & \\ \vdots & & \ddots & b_0 \\ & \vdots & & b_1 \\ b_m & & & \\ & b_m & & \vdots \\ & & \ddots & \\ & & & b_m \end{pmatrix}}_{\ell \text{ columns}}.$$

The resultant is an efficient tool for eliminating variables, as can be seen in the following proposition. We apply this proposition in the particular case where $\mathbf{R}[X] = \mathbf{K}[X_1, \ldots, X_n]$, $\mathbf{K}$ a field, and $\mathrm{Res}_{X_n}(f, g)$ is in the first elimination ideal $\langle f, g\rangle \cap \mathbf{K}[X_1, \ldots, X_{n-1}]$.

Proposition 5.44 *Let* $\mathbf{R}$ *be a ring. Then, for any* $f,\ g \in \mathbf{R}[X]$, *there exist* $h_1,\ h_2 \in \mathbf{R}[X]$ *such that*

$$h_1 f + h_2 g = \mathrm{Res}_X(f, g) \in \mathbf{R},$$

with $\deg(h_1) \leq m - 1$ *and* $\deg(h_2) \leq \ell - 1$.

Proof First notice that

$$(X^{\ell+m-1}, \ldots, X, 1)\ \mathrm{Syl}(f,\ g,\ X) = (X^{m-1} f, \ldots, f, X^{\ell-1} g, \ldots, g).$$

Thus, by Cramer's rule, considering 1 as the $(\ell + m - 1)$th unknown of the linear system whose matrix is $\mathrm{Syl}(f, g,\ X)$, $\mathrm{Res}_X(f, g)$ is the determinant of the Sylvester matrix of f and g in which the last row is replaced by $(X^{m-1} f, \ldots, f, X^{\ell-1} g, \ldots, g)$. □

Corollary 5.45 *Let* $\mathbf{K}$ *be a discrete field and* $f,\ g \in \mathbf{K}[X] \setminus \{0\}$. *Then*

(i) $1 \in \langle f, g\rangle \Leftrightarrow \gcd(f, g)$ *is constant* $\Leftrightarrow \mathrm{Res}(f, g) \neq 0$,
(ii) f *and* g *have a common factor* $\Leftrightarrow \gcd(f, g)$ *is nonconstant* $\Leftrightarrow \mathrm{Res}(f, g) = 0$.

Because, in this chapter, we are concerned with the general setting of multivariate polynomials over a ring, we are tempted to say that Corollary 5.45 remains valid for any ring $\mathbf{R}$, where the condition "$\mathrm{Res}(f, g) \neq 0$" is replaced by "$\mathrm{Res}(f, g) \in \mathbf{R}^{\times}$". Of course the implication "$\mathrm{Res}(f, g) \in \mathbf{R}^{\times} \Rightarrow 1 \in \langle f, g\rangle$" is always true by Proposition 5.44. Unfortunately, the converse does not hold, as will be shown by the following example. This is essentially due to the fact that if I is an ideal of a ring $\mathbf{R}$, then modulo I, we do not have $\overline{\mathrm{Res}(f, g)} = \mathrm{Res}(\bar{f}, \bar{g})$ for any $f,\ g \in \mathbf{R}[X]$. For the purpose of generalizing Corollary 5.45 to rings, we have to suppose that either f or g is monic.

Proposition 5.46 *Let* $\mathbf{R}$ *be a ring and* $f, g \in \mathbf{R}[X] \setminus \{0\}$ *with* f *monic. Then*

$$1 \in \langle f, g\rangle \text{ in } \mathbf{R}[X] \quad \Longleftrightarrow \quad \mathrm{Res}(f, g) \in \mathbf{R}^{\times}.$$

Proof **A classical nonconstructive proof** We have only to prove the implication "$\Rightarrow$", the implication "$\Leftarrow$" being immediate by virtue of Proposition 5.44. For this, let $\mathfrak{m}$ be a maximal ideal of $\mathbf{R}$. As f is monic, we have $\overline{\mathrm{Res}(f, g)} = \mathrm{Res}(\bar{f}, \bar{g})$ modulo $\mathfrak{m}$. Moreover, since $\mathbf{R}/\mathfrak{m}$ is a field, then using Corollary 5.45, we infer that

$\overline{\mathrm{Res}(f,g)} \neq \bar{0}$, that is, $\mathrm{Res}(f,g) \notin \mathfrak{m}$. Since this is true for any maximal ideal of $\mathbf{R}$, then necessarily $\mathrm{Res}(f,g) \in \mathbf{R}^{\times}$.

A constructive proof Let $h_1, h_2 \in \mathbf{R}[X]$ such that $h_1 f + h_2 g = 1$. Since f is monic, we have $\mathrm{Res}(f, h_2 g) = \mathrm{Res}(f, h_2)\,\mathrm{Res}(f,g)$ and $\mathrm{Res}(f, h_2 g) = \mathrm{Res}(f, h_1 f + h_2 g) = \mathrm{Res}(f, 1) = 1$. □

5.3.3 A Lemma of Suslin

Recall that for any ring $\mathbf{B}$ and $n \geq 1$, an $n \times n$ elementary matrix $E_{i,j}(a)$ over $\mathbf{B}$, where $i \neq j$ and $a \in \mathbf{B}$, is the matrix with ones on the diagonal, a on position (i,j) and zeros elsewhere; that is, $E_{i,j}(a)$ is the matrix corresponding to the elementary operation $L_i \to L_i + aL_j$. $\mathrm{E}_n(\mathbf{B})$ will denote the subgroup of $\mathrm{SL}_n(\mathbf{B})$ generated by elementary matrices.

Theorem 5.47 (Suslin [44])
Let $\mathbf{A}$ be a commutative ring. If $\langle v_1(X), \ldots, v_n(X)\rangle = \mathbf{A}[X]$, where v_1 is monic and $n \geq 2$, then there exist $\gamma_1, \ldots, \gamma_\ell \in \mathrm{E}_{n-1}(\mathbf{A}[X])$ such that, denoting by w_i the first coordinate of $\gamma_i^{\,t}(v_2, \ldots, v_n)$, we have

$$\langle \mathrm{Res}(v_1, w_1), \ldots, \mathrm{Res}(v_1, w_\ell)\rangle = \mathbf{A}.$$

Proof For $n = 2$, we have $\mathrm{Res}(f,g) \in \mathbf{A}^{\times}$ by Proposition 5.46.

Suppose $n \geq 3$. We can, without loss of generality, suppose that all the v_is, for $i \geq 2$, have degrees $< d = \deg v_1$. For the sake of simplicity, we write v_i instead of $\overline{v_i}$. We will use the notation $e_1.x$, where x is a column vector, to denote the first coordinate of x.

Suslin's original proof This consists in solving the problem modulo an arbitrary maximal ideal $\mathfrak{M}$ using a unique matrix $\gamma^{\mathfrak{M}} \in \mathrm{E}_{n-1}(\mathbf{A}/\mathfrak{M})[X]$ which transforms ${}^{t}(v_2, \ldots, v_n)$ into ${}^{t}(g, 0 \ldots, 0)$, where g is the gcd of $v_2, \ldots, v_n$ in $(\mathbf{A}/\mathfrak{M})[X]$. This matrix is given by a classical algorithm using elementary operations on ${}^{t}(v_2, \ldots, v_n)$. One starts by choosing a minimum degree component, say v_2, then the v_is, $3 \leq i \leq n$, are replaced by their remainders modulo v_2. By iterations, we obtain a column whose all components are zero except the first one. The matrix $\gamma^{\mathfrak{M}}$ lifts as a matrix $\gamma_{\mathfrak{M}} \in \mathrm{E}_{n-1}(\mathbf{A}[X])$. It follows that the first component $w_{\mathfrak{M}}$ of $\gamma_{\mathfrak{M}}\,{}^{t}(v_2, \ldots, v_n)$ is equal to the gcd of $v_2, \ldots, v_n$ in $(\mathbf{A}/\mathfrak{M})[X]$. Thus, $\mathrm{Res}(v_1, w_{\mathfrak{M}}) \notin \mathfrak{M}$.

Constructive rereading of Suslin's proof (Yengui [46])
Let $u_1(X), \ldots, u_n(X) \in \mathbf{A}[X]$ such that $v_1u_1 + \cdots + v_nu_n = 1$. Set $w = v_3u_3 + \cdots + v_nu_n$ and $V = {}^{t}(v_2, \ldots, v_n)$. We suppose that v_1 has degree d and, for $2 \leq i \leq n$, the formal degree of v_i is $d_i < d$. This means that v_i has no coefficient

of degree $> d_i$ but one does not guarantee that $\deg v_i = d_i$ (it is not necessary to have a zero test inside $\mathbf{A}$).

We proceed by induction on $\min_{2\leq i\leq n}\{d_i\}$. To simplify, we always suppose that $d_2 = \min_{2\leq i\leq n}\{d_i\}$.

For $d_2 = -1$, $v_2 = 0$, and by one elementary operation, we put w in the second coordinate. We have $\mathrm{Res}(v_1, w) = \mathrm{Res}(v_1, v_1u_1 + w) = \mathrm{Res}(v_1, 1) = 1$ and we are done.

Now, suppose that we can find the desired elementary matrices for $d_2 = m - 1$ and let show that we can do the job for $d_2 = m$.

Let a be the coefficient of degree m of v_2, and consider the ring $\mathbf{B} = \mathbf{A}/\langle a\rangle$. In $\mathbf{B}$, all the induction hypotheses are satisfied without changing the v_i nor the u_i. Thus, we can obtain $\Gamma_1, \ldots, \Gamma_k \in \mathrm{E}_{n-1}(\mathbf{B}[X])$ such that

$$\langle \mathrm{Res}(v_1, e_1.\Gamma_1 V), \ldots, \mathrm{Res}(v_1, e_1.\Gamma_k V)\rangle = \mathbf{B}.$$

It follows that, denoting by $\Upsilon_1, \ldots, \Upsilon_k$ the matrices in $\mathrm{E}_{n-1}(\mathbf{A}[X])$ lifting respectively $\Gamma_1, \ldots, \Gamma_k$, we have

$$\langle \mathrm{Res}(v_1, e_1.\Upsilon_1 V), \ldots, \mathrm{Res}(v_1, e_1.\Upsilon_k V), a\rangle = \mathbf{A}.$$

Let $b \in \mathbf{A}$ such that

$$ab \equiv 1 \bmod \langle \mathrm{Res}(v_1, e_1.\Upsilon_1 V), \ldots, \mathrm{Res}(v_1, e_1.\Upsilon_k V)\rangle = J$$

and consider the ring $\mathbf{C} = \mathbf{A}/J$. Note that in $\mathbf{C}$, we have $ab = 1$.

By an elementary operation, we replace v_3 by its remainder modulo v_2, say v_3', and then we exchange v_2 and $-v_3'$. The new column V' obtained has as first coordinate a polynomial with formal degree $m - 1$. The induction hypothesis applies and we obtain $\Delta_1, \ldots, \Delta_r \in \mathrm{E}_{n-1}(\mathbf{C}[X])$ such that

$$\big\langle \mathrm{Res}(v_1, e_1.\Delta_1 V'), \ldots, \mathrm{Res}(v_1, e_1.\Delta_r V')\big\rangle = \mathbf{C}.$$

Since V' is the image of V by a matrix in $\mathrm{E}_{n-1}(\mathbf{C}[X])$, we obtain matrices $\Lambda_1, \ldots, \Lambda_r \in \mathrm{E}_{n-1}(\mathbf{C}[X])$ such that

$$\langle \mathrm{Res}(v_1, e_1.\Lambda_1 V), \ldots, \mathrm{Res}(v_1, e_1.\Lambda_r V)\rangle = \mathbf{C}.$$

The matrices Λ_j lift in $\mathrm{E}_{n-1}(\mathbf{A}[X])$ as, say $\Psi_1, \ldots, \Psi_r$.

Finally, we obtain

$$\langle \mathrm{Res}(v_1, e_1.\Psi_1 V), \ldots, \mathrm{Res}(v_1, e_1.\Psi_r V)\rangle + J = \mathbf{A},$$

the desired conclusion. □

Example 5.48 Take $\mathbf{A} = \mathbb{Z}$ and

$$V = {}^{\mathrm{t}}(v_1, v_2, v_3) = {}^{\mathrm{t}}(x^2 + 2x + 2,\ 3,\ 2x^2 + 11x - 3) \in \mathrm{Um}_3(\mathbb{Z}[x]),$$

(taking $u_1 = -2x + 2$, $u_2 = -3x^2 + x - 1$, $u_3 = x$, we have $u_1 v_1 + u_2 v_2 + u_3 v_3 = 1$). Following the algorithm given in the proof of Theorem 5.47 and keeping the same notation, one has to perform a Euclidean division of v_3 by v_1, so that ${}^{\mathrm{t}}(v_1, v_2, v_3) \overset{E_{3,1}(-2)}{\longrightarrow} {}^{\mathrm{t}}(v_1, v_2, \tilde{v}_3 = 7x - 7)$, and then passes to the ring $(\mathbb{Z}/3\mathbb{Z})[x]$. This yields $\ell = 2$, $\gamma_1 = \begin{pmatrix} 1 & 1 \\ 0 & 1 \end{pmatrix}$, $\gamma_2 = \mathrm{I}_2 = \begin{pmatrix} 1 & 0 \\ 0 & 1 \end{pmatrix}$, and finally

$$\left\langle \mathrm{Res}\left(v_1, e_1.\gamma_1 {}^{\mathrm{t}}(v_2, v_3)\right),\ \mathrm{Res}\left(v_1, e_1.\gamma_2 {}^{\mathrm{t}}(v_2, v_3)\right) \right\rangle = \langle 170,\ 9 \rangle = \mathbb{Z}.$$

This example will be pursued in Section 5.3.5, where, as a result of the computations above, we will obtain a free basis for the syzygy module

$$\mathrm{Syz}(v_1, v_2, v_3) := \left\{ (w_1, w_2, w_3) \in \mathbb{Z}[x]^3 \mid w_1 v_1 + w_2 v_2 + w_3 v_3 = 0 \right\}.$$

5.3.4 A More General Strategy (by "Backtracking")

As already mentioned, contrary to the local–global principles explained in Section 5.2.3, we do not reread a proof in which one localizes at a generic prime ideal $\mathfrak{P}$ or at a generic maximal ideal $\mathfrak{M}$ but a proof in which one passes modulo a generic maximal ideal $\mathfrak{M}$ in order to prove that an ideal $\mathfrak{a}$ of a ring $\mathbf{A}$ contains 1. The classical proof is very often by contradiction: For a generic maximal ideal $\mathfrak{M}$, if $\mathfrak{a} \subseteq \mathfrak{M}$ then $1 \in \mathfrak{M}$. But, in fact, this reasoning hides a concrete fact: $1 = 0$ in the residue field $\mathbf{A}/\mathfrak{M}$ (see [37]). Consequently, this reasoning by contradiction can be converted dynamically into a constructive proof as follows. One has to do the necessary computations as if $\mathbf{A}/\mathfrak{a}$ was a field. Every time one needs to know if an element x_i is null or a unit modulo $\mathfrak{a}$, one has just to force it into being null by adding it to $\mathfrak{a}$. Suppose, for example, that we have established that $1 \in \mathfrak{a} + \langle x_1, x_2, x_3 \rangle$ (this corresponds in the classical proof to the fact that $x_1, x_2, x_3 \in \mathfrak{M} \Rightarrow 1 \in \mathfrak{M}$). This means that x_3 is a unit modulo $\mathfrak{a} + \langle x_1, x_2 \rangle$ and, thus, one has to follow the classical proof in case $x_1, x_2 \in \mathfrak{M}$ and x_3 is a unit modulo $\mathfrak{M}$. It is worth pointing out that there is no need of $\mathfrak{M}$ since one has already computed an inverse of x_3 modulo $\mathfrak{a} + \langle x_1, x_2 \rangle$.

For the purpose of illustrating this strategy, let us consider an example of a binary tree corresponding to the computations produced by a local–globalrereading.

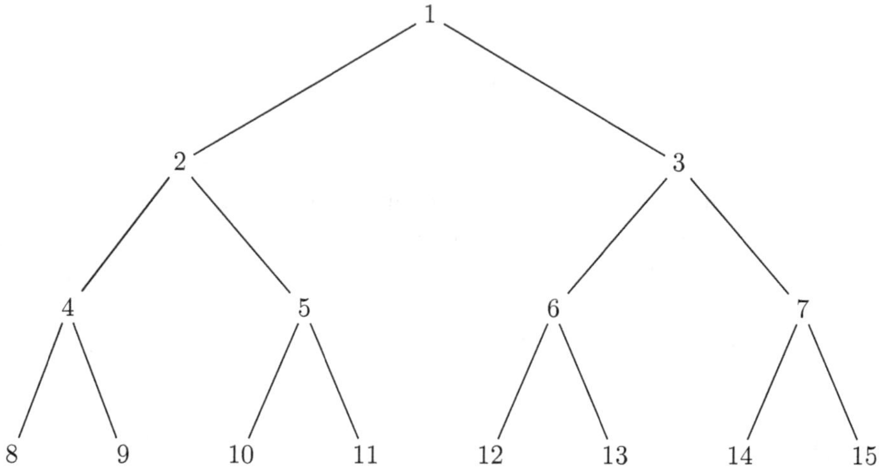

In the tree, the disjunctions correspond to a test:

$$x \in \mathbf{A}_i^\times \quad \vee \quad 1 - x \ \in \mathbf{A}_i^\times,$$

and each node corresponds to a localization $\mathbf{A}_i$ of the initial ring $\mathbf{A}$. In order to glue the local solutions (at the terminal nodes, that is, at the leaves), one has to go back from the leaves to the root in a "parallel" way. Now imagine that these disjunctions correspond to a test:

$$x \in \mathbf{A}_i^\times \quad \vee \quad x = 0 \text{ in } \mathbf{A}_i,$$

and each node i corresponds to a quotient $\mathbf{A}_i$ of the initial ring $\mathbf{A}$. Following the classical proof which proves that an ideal $\mathfrak{a}$ of $\mathbf{A}$ contains 1, one has to start with the leaf which is completely on the right (leaf 15); that is, to follow the path $1 \to 3 \to 7 \to 15$ by considering the successive corresponding quotients $\mathbf{A} = \mathbf{A}/\langle 0\rangle$, $\mathbf{A}/\langle a_1\rangle$, $\mathbf{A}/\langle a_1, a_3\rangle$, and $\mathbf{A}/\langle a_1, a_3, a_7\rangle$. Using just the information at leaf 15 where the considered ring is $\mathbf{A}/\langle a_1, a_3, a_7\rangle$ (this information corresponds in the classical proof to the fact that: $a_1, a_3, a_7 \in \mathfrak{M} \Rightarrow 1 \in \mathfrak{M}$), one obtains an element $b_{15} \in \mathbf{A}$ such that $1 \in \langle a_1, a_3, a_7, b_{15}\rangle$, or equivalently, a_7 is a unit modulo $\langle a_1, a_3, b_{15}\rangle$. Now, we go back to node 7 but with a new quotient $\mathbf{A}/\langle a_1, a_3, b_{15}\rangle$ (note that at the first passage through 7 the considered quotient ring was $\mathbf{A}/\langle a_1, a_3\rangle$) and we can follow the branch $7 \to 14$ (this corresponds in the classical proof to the fact that $a_1, a_3 \in \mathfrak{M}$ and a_7 is a unit modulo $\mathfrak{M} \Rightarrow 1 \in \mathfrak{M}$). This will produce an element b_{14} such that $1 \in \langle a_1, a_3, b_{14}, b_{15}\rangle$, or equivalently, a_3 is a unit modulo $\langle a_1, b_{14}, b_{15}\rangle$. Thus, we can go back to node 3 through the branch $14 \to 7 \to 3$, and so on. In the end, the entire path followed is

$$1 \to 3 \to 7 \to 15 \to 7 \to 14 \to 7 \to 3 \to 6 \to 13 \to 6 \to 12 \to 6 \to 3 \to 1 \to 2 \to 5 \to 11 \to 5 \to 10 \to 5 \to 2 \to 4 \to 9 \to 4 \to 8 \to 4 \to 2 \to 1.$$

Finally, at the root of the tree (node 1), we get that $1 \in \langle b_8, \ldots, b_{15} \rangle$ in the ring $\mathbf{A}/\langle 0 \rangle = \mathbf{A}$. It is worth pointing out that, as can be seen above, another major difference between a "local–global tree" and the tree produced by our method is that the quotient ring changes at each new passage through the considered node. For example, in the first passage through 7, the ring was $\mathbf{A}/\langle a_1, a_3 \rangle$, in the second passage it becomes $\mathbf{A}/\langle a_1, a_3, b_{15} \rangle$, and in the last one the ring is $\mathbf{A}/\langle a_1, a_3, b_{14}, b_{15} \rangle$.

We can sum up this new method as follows.

Elimination of maximal ideals by backtracking 5.49 (Yengui [46]) When rereading dynamically the original proof, follow systematically the branch $x_i \in \mathfrak{M}$ any time you find a disjunction $x_i \in \mathfrak{M} \;\vee\; x_i \notin \mathfrak{M}$ in the proof until getting $1 = 0$ in the quotient. That is, in the corresponding leaf of the tree, you get $1 \in \langle x_1, \ldots, x_k \rangle$ for some $x_1, \ldots, x_k \in \mathbf{A}$. This means that at the node $\langle x_1, \ldots, x_{k-1} \rangle \subseteq \mathfrak{M}$, you know a concrete $a \in \mathbf{A}$ such that $1 - a x_k \in \langle x_1, \ldots, x_{k-1} \rangle$. So you can follow the proof.

If the proof given for a generic maximal ideal is sufficiently "uniform," you know a bound for the depth of the (infinite branching) tree. For example, in Suslin's lemma, the depth is $\deg(v_1)$. So your "finite branching dynamical evaluation" is finite: You get an algorithm.

5.3.5 Suslin's Algorithm

For any ring $\mathbf{B}$, when we say that a matrix $N \in \mathrm{M}_n(\mathbf{B})$ $(n \geq 3)$ is in $\mathrm{SL}_2(\mathbf{B})$, we mean that it is of the form

$$\begin{pmatrix} N' & 0 & \ldots & 0 \\ 0 & 1 & & \\ \vdots & & \ddots & \\ 0 & & & 1 \end{pmatrix}$$

with $N' \in \mathrm{SL}_2(\mathbf{B})$.

Lemma 5.50 (Translation by the resultant; [44], Lemma 2.1)
Let $\mathbf{R}$ be a ring, f_1, $f_2 \in \mathbf{R}[X]$, $b, d \in \mathbf{R}$, and $r = \mathrm{Res}(f_1, f_2) \in \mathbf{R}$. Then there exists $B \in \mathrm{SL}_2(\mathbf{R}[X])$ such that

$$B \begin{pmatrix} f_1(b) \\ f_2(b) \end{pmatrix} = \begin{pmatrix} f_1(b + rd) \\ f_2(b + rd) \end{pmatrix}.$$

Proof Take $g_1, g_2 \in \mathbf{R}[X]$ such that $f_1\, g_1 + f_2\, g_2 = r$, denote by s_1, s_2, t_1, t_2 the polynomials in $\mathbf{R}[X, Y, Z]$ such that

$$\begin{aligned} f_1(X + YZ) &= f_1(X) + Y s_1(X, Y, Z), \\ f_2(X + YZ) &= f_2(X) + Y s_2(X, Y, Z), \\ g_1(X + YZ) &= g_1(X) + Y t_1(X, Y, Z), \\ g_2(X + YZ) &= g_2(X) + Y t_2(X, Y, Z), \end{aligned}$$

and set

$$\begin{aligned} B_{1,1} &= 1 + s_1(b, r, d)\, g_1(b) + t_2(b, r, d)\, f_2(b), \\ B_{1,2} &= s_1(b, r, d)\, g_2(b) - t_2(b, r, d)\, f_1(b), \\ B_{2,1} &= s_2(b, r, d)\, g_1(b) - t_1(b, r, d)\, f_2(b), \\ B_{2,2} &= 1 + s_2(b, r, d)\, g_2(b) + t_1(b, r, d)\, f_1(b). \end{aligned}$$

Then, one can take

$$B = \begin{pmatrix} B_{1,1} & B_{1,2} \\ B_{2,1} & B_{2,2} \end{pmatrix}.$$

□

The following algorithm is due to Suslin [44].

Algorithm 5.51 (An algorithm for eliminating variables from unimodular polynomial vectors with coefficients in a ring, general case) Input A column $\mathcal{V} = \mathcal{V}(X) = {}^{\mathrm{t}}(v_1(X), \ldots, v_n(X)) \in \mathrm{Um}_n(\mathbf{A}[X])$ such that v_1 is monic.

Output A matrix $\mathcal{B} \in \mathrm{SL}_n(\mathbf{A}[X])$ such that $\mathcal{B}\,\mathcal{V} = \mathcal{V}(0)$.

Step 1 Find $\gamma_0, \ldots, \gamma_s \in \mathrm{E}_{n-1}(\mathbf{A}[X])$ such that, denoting $w_i = e_1.\gamma_i\, {}^{\mathrm{t}}(v_2, \ldots, v_n)$ and $r_i = \mathrm{Res}(v_1, w_i)$, we can find $\alpha_0, \ldots, \alpha_s \in \mathbf{A}$ such that $\alpha_0 r_0 + \cdots + \alpha_s r_s = 1$ (here we use the algorithm given in the proof of Theorem 5.47).

For $0 \le i \le s$, compute $f_i, g_i \in \mathbf{A}[X]$ such that $f_i v_1 + g_i w_i = r_i$ (use Proposition 5.44).

Step 2 Set

$$\begin{aligned} b_{s+1} &:= 0, \\ b_s &:= \alpha_s\, r_s X, \\ b_{s-1} &:= b_s + \alpha_{s-1} r_{s-1} X, \\ &\ \vdots \\ b_0 &:= b_1 + \alpha_0 r_0 X = X \end{aligned}$$

(the last equality follows from the fact that $X = \sum_{i=0}^{s} \alpha_i r_i X$).

Step 3 For $1 \le i \le s+1$, find $\mathcal{B}_i \in \mathrm{SL}_n(\mathbf{A}[X])$ such that $\mathcal{B}_i \mathcal{V}(b_{i-1}) = \mathcal{V}(b_i)$. In more detail, for $3 \le j \le n$, set $F_{i,j} := \frac{v_j(b_{i-1}) - v_j(b_i)}{b_{i-1} - b_i} = \frac{v_j(b_{i-1}) - v_j(b_i)}{\alpha_i r_i X} \in \mathbf{A}[X]$, so that one obtains

$$\begin{aligned}v_j(b_{i-1}) - v_j(b_i) = \alpha_i r_i X F_{i,j} &= \alpha_i X F_{i,j}\, f_i(b_{i-1}) v_1(b_{i-1}) \\ &\quad + \alpha_i X F_{i,j}\, g_i(b_{i-1}) w_i(b_{i-1}) \\ &= \sigma_{i,j}\, v_1(b_{i-1}) + \tau_{i,j}\, w_i(b_{i-1}),\end{aligned}$$

with

$$\sigma_{i,j} := \alpha_i X F_{i,j}\, f_i(b_{i-1}), \quad \tau_{i,j} := \alpha_i X F_{i,j}\, g_i(b_{i-1}) \in \mathbf{A}[X].$$

Let $\Gamma_i \in \mathrm{E}_n(\mathbf{A}[X])$ be the matrix corresponding to the elementary operations:

$$L_j \to L_j - \sigma_{i,j} L_1 - \tau_{i,j} L_2, \quad 3 \le j \le n;$$

that is,

$$\Gamma_i := \prod_{j=3}^{n} E_{j,1}(-\sigma_{i,j})\, E_{j,2}(-\tau_{i,j}).$$

Set

$$B_{i,2} := \Gamma_i\, \gamma_i \in \mathrm{E}_n(\mathbf{A}[X]),$$

so that we have

$$B_{i,2} \mathcal{V}(b_{i-1}) = \begin{pmatrix} v_1(b_{i-1}) \\ w_i(b_{i-1}) \\ v_3(b_i) \\ \vdots \\ v_n(b_i) \end{pmatrix}.$$

Following Lemma 5.50, set

$$\begin{aligned}s_{i,1}(X,Y,Z) &:= \frac{v_1(X+YZ) - v_1(X)}{Y} \in \mathbf{A}[X,Y,Z], \\ s_{i,2}(X,Y,Z) &:= \frac{w_i(X+YZ) - w_i(X)}{Y} \in \mathbf{A}[X,Y,Z], \\ t_{i,1}(X,Y,Z) &:= \frac{f_i(X+YZ) - f_i(X)}{Y} \in \mathbf{A}[X,Y,Z], \\ t_{i,2}(X,Y,Z) &:= \frac{g_i(X+YZ) - g_i(X)}{Y} \in \mathbf{A}[X,Y,Z], \in \mathbf{A}[X],\end{aligned}$$

$$\begin{aligned}C_{i,1,1} &:= 1 + s_{i,1}(b_{i-1}, r_i, -\alpha_i X)\, f_i(b_{i-1}) + t_{i,2}(b_{i-1}, r_i, -\alpha_i X)\, w_i(b_{i-1}) \\ C_{i,1,2} &= s_{i,1}(b_{i-1}, r_i, -\alpha_i X)\, g_i(b_{i-1}) - t_{i,2}(b_{i-1}, r_i, -\alpha_i X)\, v_1(b_{i-1}) \in \mathbf{A}[X], \\ C_{i,2,1} &= s_{i,2}(b_{i-1}, r_i, -\alpha_i X)\, f_i(b_{i-1}) - t_{i,1}(b_{i-1}, r_i, -\alpha_i X)\, w_i(b_{i-1}) \in \mathbf{A}[X], \\ C_{i,2,2} &= 1 + s_{i,2}(b_{i-1}, r_i, -\alpha_i X)\, g_i(b_{i-1}) + t_{i,1}(b_{i-1}, r_i, -\alpha_i X)\, v_1(b_{i-1}) \in \mathbf{A}[X],\end{aligned}$$

$$C_i := \begin{pmatrix} C_{i,1,1} & C_{i,1,2} \\ C_{i,2,1} & C_{i,2,2} \end{pmatrix} \in \mathrm{SL}_2(\mathbf{A}[X]).$$

Note that

$$C_i \begin{pmatrix} v_1(b_{i-1}) \\ w_i(b_{i-1}) \end{pmatrix} = \begin{pmatrix} v_1(b_i) \\ w_i(b_i) \end{pmatrix}.$$

Set

$$B_{i,1} := \gamma_i^{-1} \begin{pmatrix} C_i & 0 \\ 0 & I_{n-2} \end{pmatrix},$$

and

$$\mathcal{B}_i := B_{i,1}\, B_{i,2} \in \mathrm{SL}_n(\mathbf{A}[X]),$$

so that $\mathcal{B}_i \mathcal{V}(b_{i-1}) = \mathcal{V}(b_i)$.

Step 4 $\mathcal{B} := \mathcal{B}_{s+1} \cdots \mathcal{B}_1$.

Example 5.52 Take $\mathbf{A} = \mathbb{Z}$ and $V = {}^{\mathrm{t}}(x^2 + 2x + 2,\ 3,\ 2x^2 + 11x - 3) \in \mathrm{Um}_3(\mathbb{Z}[x])$.

A generating set for $\mathrm{Syz}(v_1, v_2, v_3)$ can be obtained by computing a dynamical Gröbner basis (see [45, 47, 48]) for the ideal $\langle v_1, v_2, v_3 \rangle$. A dynamical computation gives

$$\begin{aligned} &\mathrm{Syz}(v_1, v_2, v_3) \\ &= \left\langle \begin{pmatrix} 3 \\ -X^2 - 2X - 2 \\ 0 \end{pmatrix}, \begin{pmatrix} 0 \\ -2X^2 - 11X + 3 \\ 3 \end{pmatrix}, \begin{pmatrix} -2X^3 - 11X^2 - 18X \\ 7X^3 + 14X^2 + 14X \\ X^3 + 2X^2 + 2X \end{pmatrix}, \right. \\ &\qquad \left. \begin{pmatrix} -21 - 6X \\ 14 + 21X \\ 3X \end{pmatrix}, \begin{pmatrix} -4X^3 - 36X^2 - 71X + 21 \\ 14X^3 + 77X^2 - 21X \\ 2X^3 + 11X^2 - 3X + 14 \end{pmatrix} \right\rangle. \end{aligned}$$

But of course this is not a minimal generating set for $\mathrm{Syz}(v_1, v_2, v_3)$ as it is a rank-2 free $\mathbb{Z}[x]$-module (by the Lequain–Simis–Vasconcelos Theorem, [48]). Following Algorithm 5.51 and doing the computations by hand (assisted by the computer algebra system `MAPLE`), we get a matrix $G \in \mathrm{SL}_3(\mathbb{Z}[x])$ such that

$$G\,V = \begin{pmatrix} 1 \\ 0 \\ 0 \end{pmatrix}.$$

$> V :=$ `matrix`$(3, 1, [x^2 + 2 * x + 2, 3, 2 * x^2 + 11 * x - 3])$;
$> G :=$ `matrix`$([[2+29142*x^2+340*x+4788*x^3, -25686*x^2-2394*x^3-272*x-1, -6192*x^2-2394*x^3-44*x], [-3-43713*x^2-510*x-7182*x^3, 38529*x^2+3591*x^3+408*x+2, 9288*x^2+3591*x^3+66*x], [12+204092*x^2+2975*x+33516*x^3, -179851*x^2-16758*x^3-2429*x-7, -43393*x^2-16758*x^3-434*x+1]])$
$> \det(G)$;
1
$> F :=$ `expandvector`(`multiply`(G, V));
$F :=$ `matrix`$([[1], [0], [0]])$.

Thus,

$$\left(\begin{pmatrix} -3 - 43713x^2 - 510x - 7182x^3 \\ 38529x^2 + 3591x^3 + 408x + 2 \\ 9288x^2 + 3591x^3 + 66x \end{pmatrix}, \begin{pmatrix} 12 + 204092x^2 + 2975x + 33516x^3 \\ -179851x^2 - 16758x^3 - 2429x - 7 \\ -43393x^2 - 16758x^3 - 434x + 1 \end{pmatrix}\right)$$

is a free basis for $\mathrm{Syz}(v_1, v_2, v_3)$.

$>$ `inverse`(G);
`matrix`$([[x^2+2*x+2, 5586*x^3+14465*x^2+146*x+1, 1197*x^3+3096*x^2+22*x], [3, 2, 0], [2*x^2+11*x-3, 11172*x^3+68032*x^2+999*x+2, 2394*x^3+14571*x^2+170*x+1]])$.

The matrix G^{-1} is a completion of V into an invertible matrix as V is the first column of G^{-1}.

5.3.6 Suslin's Solution to Serre's Problem

Theorem 5.53 (Unimodular completion theorem) *Let* $\mathbf{K}$ *be a field,* $\mathbf{R} = \mathbf{K}[X_1, \ldots, X_r]$ *and consider a unimodular vector*

$$f = {}^{\mathrm{t}}(f_1(X_1, \ldots, X_r), \ldots, f_n(X_1, \ldots, X_r)),$$

in $\mathbf{R}^{n\times 1}$*. Then there exists a matrix* $H \in \mathrm{SL}_n(\mathbf{R})$ *such that* $Hf = {}^{\mathrm{t}}(1, 0, \ldots, 0)$.
In other words, f *is the first column of a matrix in* $\mathrm{SL}_n(\mathbf{R})$.

Proof If $n = 1$ or 2, the result is straightforward. If $n > 2$ and $r = 1$, the result comes from the fact that $\mathbf{R}$ is a principal ideal domain (PID). It is explicitly given by a Smith reduction of the column matrix f. For $r \geq 2$, we make an induction on r. If the field $\mathbf{K}$ has enough elements (for example, if it is infinite), we can make a linear change of variables so that one of the f_i becomes monic. Otherwise, we make a change of variables "à la Nagata": $Y_r = X_r$, and for $1 \leq j < r$, $Y_j = X_j + X_r^{d^j}$, with a sufficiently large integer d. It suffices now to use Algorithm 5.51. □

Theorem 5.54 (Suslin's solution to Serre's problem) *Let* $\mathbf{K}$ *be a field,* $\mathbf{R} = \mathbf{K}[X_1, \ldots, X_r]$ *and let* M *be a finitely generated projective* $\mathbf{R}$*-module. Then* M *is free.*

Proof By virtue of Definition and Proposition 5.20, we know that M is stably free, that is, we have an isomorphism

$$\varphi : \mathbf{R}^k \oplus M \longrightarrow \mathbf{R}^{\ell+k}$$

for some integers k and ℓ. If $k = 0$, there is nothing to prove. Suppose that $k > 0$. The vector $f = \varphi((e_{k,1}, 0_M))$ (where $e_{k,1}$ is the first vector in the canonical basis of $\mathbf{R}^k$) is unimodular. To see this, just consider the linear form λ over $\mathbf{R}^{\ell+k}$ mapping y ($y \in \mathbf{R}^{\ell+k}$) to the first coordinate of $\varphi^{-1}(y)$. We have $\lambda(y_1, \ldots, y_{k+\ell}) = u_1 y_1 + \cdots + u_{k+\ell} y_{k+\ell}$ and $\lambda(f) = 1$.

Consider f as a column vector. Taking the composition of φ with the isomorphism given in Theorem 5.53, we obtain an isomorphism ψ mapping $(e_{k,1}, 0_M)$ to $e_{k+\ell,1}$. By passing modulo $\mathbf{A}(e_{k,1}, 0_M)$ and modulo $\mathbf{A}e_{k+\ell,1}$, we get an isomorphism

$$\theta : \mathbf{R}^{k-1} \oplus M \longrightarrow \mathbf{R}^{\ell+k-1}. \qquad \square$$

References

[1] Bishop, E. 1967. *Foundations of Constructive Analysis*. New York: McGraw-Hill.

[2] Bishop, E., and Bridges, D. 1985. *Constructive Analysis*. Grundlehren der mathematischen Wissenschaften, 279. Berlin: Springer.

[3] Brewer, J., and Costa, D. 1987. Projective modules over some non-Noetherian polynomial rings. *J. Pure Appl. Algebra*, **13**, 157–163.

[4] Caniglia, L., Cortiñas, G., Danón, S., et al. 1993. Algorithmic aspects of Suslin's proof of Serre's conjecture. *Comput. Complexity*, **3**, 31–55.

[5] Coste, M., Lombardi, H., and Roy, M.-F. 2001. Dynamical method in algebra: effective Nullstellensätze. *Ann. Pure Appl. Logic*, **111**, 203–256.

[6] Coquand, T., Lombardi, H., Quitté, C., and Tête, C. 2019. Résolutions libres finies. Méthodes constructives. Preprint. (see `https://arxiv.org/abs/1811.01873`)

[7] Coquand, T., and Quitté, C. 2011. Constructive finite free resolutions. *Manu. Math.*, **137**, 331–345.

[8] Della Dora, J., Dicrescenzo, C., and Duval, D. 1985. About a new method for computing in algebraic number fields. In: Caviness, B. F. (ed.), EUROCAL '85. *Lecture Notes in Comp. Sci.*, **204**, 289–290.

[9] Duval, D., and Reynaud, J.-C. 1994. Sketches and computation (Part II). Dynamic evaluation and applications. *Math. Struct. Comp. Sci.*, **4**, 239–271.

[10] Ellouz, A., Lombardi, H., and Yengui, I. 2008. A constructive comparison of the rings $\mathbf{R}(X)$ and $\mathbf{R}\langle X\rangle$ and application to the Lequain–Simis induction theorem. *J. Algebra*, **320**, 521–533.

[11] Fabiańska, A., and Quadrat, A. 2007. Applications of the Quillen–Suslin theorem to the multidimensional systems theory. Pages 23–106 in: *INRIA Report 6126 (2007), published in Gröbner Bases in Control Theory and Signal Processing*, Park, H., and Regensburger, G. (eds.), Radon Series on Computation and Applied Mathematics 3, Berlin: De Gruyter.

[12] Fabiańska, A. 2009. *A Maple QuillenSuslin package*: www.math.rwth-aachen.de/QuillenSuslin/

[13] Fitchas, N., Galligo, A. 1990. Nullstellensatz effectif et conjecture de Serre (Théorème de Quillen–Suslin) pour le calcul formel. *Math. Nachr.*, **149**, 231–253.

[14] Gamanda, M., Lombardi, H., Neuwirth, S., and Yengui, I. 2020. The syzygy theorem for Bézout rings. *Math. Comp.*, **89**, 941–964.

[15] Glaz, S. 2001. Finite conductor rings with zero divisors. In: Chapman, S. T. (ed.), *Non-Noetherian Commutative Ring Theory*. Dordrecht: Kluwer Academic Publishers. (*Math. Appl., Dordr.*, **520**, 251–269 (2000).)

[16] Kunz, E. 1991. *Introduction to Commutative Algebra and Algebraic Geometry*. Birkh'auser.

[17] Lam, T. Y. 1978. *Serre's Conjecture*. Lecture Notes in Mathematics, no. 635. Berlin, New York: Springer-Verlag.

[18] Lam, T. Y. 2006. *Serre's Problem on Projective Modules*. Springer Monographs in Mathematics. Berlin, Heidelberg: Springer.

[19] Laubenbacher, R. C., and Woodburn, C. J. 1997. An algorithm for the Quillen–Suslin theorem for monoid rings. *J. Pure Appl. Algebra*, **117–118**, 395–429.

[20] Laubenbacher, R. C., and Woodburn, C. J. 2000. A new algorithm for the Quillen–Suslin theorem. *Beitr'age Algebra Geom.*, **41**, 23–31.

[21] Lequain, Y., and Simis, A. 1980. Projective modules over $R[X_1, \ldots, X_n]$, R a Prüfer domain. *J. Pure Appl. Algebra*, **18**(2), 165–171.

[22] Logar, A., and Sturmfels, B. 1992. Algorithms for the Quillen–Suslin theorem. *J. Algebra*, **145**(1), 231–239.

[23] Lombardi, H. 1997. *Le contenu constructif d'un principe local–global avec une application à la structure d'un module projectif de type fini*. Publications Mathématiques de Besançon. Théorie des nombres.

[24] Lombardi, H. 2001. *Platitude, localisation et anneaux de Prüfer, une approche constructive*. Publications Mathématiques de Besançon. Théorie des nombres.

[25] Lombardi, H., and Quitté, C. 2002. Constructions cachées en algèbre abstraite (2) Le principe local–global. Pages 461–476 in: *Commutative Ring Theory and Applications*, Fontana, M., Kabbaj, S.-E., and Wiegand, S. (eds.), Lecture Notes in Pure and Applied Mathematics, vol. 131. New York: Marcel Dekker.

[26] Lombardi, H., and Quitté, C. 2011. *Algèbre Commutative. Méthodes Constructives. Modules Projectifs de Type Fini. Cours et Exercices*. Calvage et Mounet.

[27] Lombardi, H., and Quitté, C. 2015. *Commutative Algebra: Constructive Methods. Finite Projective Modules*. Dordrecht: Springer.

[28] Lombardi, H., Quitté, C., and Yengui, I. 2008. Hidden constructions in abstract algebra (6) The theorem of Maroscia, Brewer and Costa. *J. Pure Appl. Algebra*, **212**, 1575–1582.

[29] Maroscia, P. 1977. Modules projectifs sur certains anneaux de polynomes. *C.R.A.S. Paris*, **285** série A, 183–185.

[30] Mines, R., Richman, F., and Ruitenburg, W. 1988. *A Course in Constructive Algebra*. Universitext. Berlin: Springer-Verlag.

[31] Northcott, D. G. 1976. *Finite Free Resolutions*. Cambridge: Cambridge University Press.

[32] Park, H. 1995. A computational theory of Laurent polynomial rings and multidimensional FIR systems. Ph.D. thesis, University of Berkeley.

[33] Park, H. 1999. A realization algorithm for $\mathrm{SL}_2(\mathbf{R}[X_1, \ldots, X_m])$ over the euclidean domain. *SIAM J. Matrix Anal. Appl.*, **21**, 178–184.

[34] Park, H. 2004. Symbolic computations and signal processing. *J. Symb. Comp.*, **37**, 209–226.

[35] Park, H. 2006. *Generalizations and variations of Quillen–Suslin theorem and their applications. Workshop Gröbner Bases in Control Theory and Signal Processing. Special semester on Gröbner bases and related methods*, University of Linz (Austria).

[36] Quillen, D. 1976. Projective modules over polynomial rings. *Invent. Math.*, **36**, 167–171.

[37] Richman, F. 1988. Nontrivial use of trivial rings. *Proc. Amer. Math. Soc.*, **103**, 1012–1014.

[38] Rosenberg, J. 1996. *Algebraic K-Theory and its Applications*. Graduate Texts in Mathematics. New York: Springer-Verlag.

[39] Schreyer, F. O. 1980. Die Berechnung von Syzygien mit dem verallgemeinerten Weierstrass'schen Divisionssatz. Diploma thesis, University of Hamburg.

[40] Serre, J. P. 1955. Faisceaux algébriques cohérents. *Ann. Math.*, **61**, 191–278.

[41] Serre, J. P. 1958. *Modules projectifs et espaces fibrés à fibre vectorielle*. Sém. Dubreil-Pisot, no. 23, Paris.

[42] Simis, A., and Vasconcelos, W. 1971. Projective modules over $\mathbf{R}[X]$, $\mathbf{R}$ a valuation ring are free. *Not. Amer. Math. Soc.*, **18**(5), 944.

[43] Suslin, A. A. 1976. Projective modules over a polynomial ring are free. *Soviet Math. Dokl.*, **17**, 1160–1164.

[44] Suslin, A. A. 1977. On the structure of the special linear group over polynomial rings. *Math. USSR-Izv.*, **11**, 221–238.

[45] Yengui, I. 2006. Dynamical Gröbner bases. *J. Algebra*, **301**, 447–458.

[46] Yengui, I. 2008. Making the use of maximal ideals constructive. *Theor. Comp. Sci.*, **392**, 174–178.

[47] Yengui, I. 2011. Corrigendum to "Dynamical Gröbner bases" [J. Algebra 301 (2) (2006) 447–458] and to "Dynamical Gröbner bases over Dedekind rings" [J. Algebra 324 (1) (2010) 12–24]. *J. Algebra*, **339**, 370–375.

[48] Yengui, I. 2015. *Constructive Commutative Algebra: Projective Modules over Polynomial Rings and Dynamical Gröbner Bases.* Lecture Notes in Mathematics, no. 2138. Berlin: Springer.

6

Constructive Algebra and Point-Free Topology

Thierry Coquand

6.1 Introduction

Topological ideas play an important rôle in algebra, bringing geometrical intuitions and powerful methods from algebraic topology, such as the use of sheaf-theoretical notions, as in Serre's classical paper on coherent sheaves [62]. The goal of this chapter is to survey some recent constructive interpretations of these methods.

One *constructive* issue in using these topological ideas in algebra is that the various spaces one considers, such as Zariski spectrum, space of valuations, etc., may fail to have enough points.[1] For instance, Tierney [69] describes a suitable topos, where there is a non-trivial ring without any point in its Zariski spectrum. Since intuitionistic logic is valid in any topos, this means that there is no hope of proving constructively the existence of a point in the Zariski spectrum of a non-trivial ring. In classical algebra, these topological spaces are often used by introducing a generic point, which is shown to exist using classical methods, such as Zorn's Lemma. It is then not clear how to interpret such arguments constructively.[2]

There is, however, one general method to 'force' a space to have a (generic) point, which works in a constructive setting, presented, for example, in [18, 38], and which is the following. While a space may fail to have a point constructively, it *always* gets a generic point when working in the *sheaf model* over this space. In several cases, it is possible to analyse proof theoretically the presentation of the space and show that we can 'eliminate' the use of this generic point. This is actually similar to the technique of elimination of choice sequences in intuitionistic mathematics

[1] To some extent, there are similarities with the worries that there might not be, in a constructive setting, 'enough' points in Cantor space or the real lines, worries that motivated Brouwer's 'second act of intuitionism' [10], and the introduction of the notion of choice sequences. The methods we present in this survey also have some analogy with this development.

[2] Typical examples of such arguments are provided by several proofs in Nothcott's book [55], which simplifies the treatment of Buchsbaum and Eisenbud [11]. The method, which we describe in the present chapter, provides a way to interpret these non-constructive arguments, and to obtain an effective and elementary presentation of the theory of finite free resolutions [24].

and presented in [70]. This method works as well for the generalized notion of space, as a topos given by a site, introduced by Grothendieck. For instance, the algebraic closure of a field can be represented as a *point* of a suitable sheaf topos. As stressed in Joyal [39], this is also reminiscent of Hilbert's notion of *introduction* and *elimination* of ideal elements.[3]

This approach comes from two complementary lines of research: on one side, the idea of using point-free topology [26, 38, 60], for representing topological spaces in constructive mathematics,[4] and on the other side, the *dynamical* method in algebra [30, 43, 44]. The dynamical method originated first in computer algebra in [31], where it was used to explain how to do computations in an algebraic closure of an arbitrary computable field. As stressed in [43], this is quite paradoxical, since it is well known that such an object *cannot* exist in general in constructive mathematics (see [9, 52]).

This chapter is organized as follows. We first study some examples corresponding to this analysis of *topological* spaces, such as the Zariski spectrum or the space of minimal or maximal primes, and then we present an example corresponding to the analysis of more general *Grothendieck sites*. We end by listing some open questions and research directions.

6.2 Zariski Spectrum

6.2.1 Point-Free Representation

The Zariski spectrum of a commutative ring R is, in classical mathematics, the topological space $\mathrm{Sp}(R)$. The *points* of this space are prime ideals. The basic open are the subsets $D(a) = \{I \in \mathrm{Sp}(R) \mid a \notin I\}$. We have by definition

$$D(1) = \mathrm{Sp}(R) \quad D(0) = \emptyset \quad D(ab) = D(a) \cap D(b) \quad D(a+b) \subseteq D(a) \cup D(b).$$

Let us write $D(b_1, \ldots, b_m)$ for $D(b_1) \cup \cdots \cup D(b_m)$. It is also a classical result (Krull's theorem) that a is nilpotent[5] if, and only if, $D(a)$ is empty. A corollary of this result is that $D(a) \subseteq D(b_1, \ldots, b_m)$ if, and only if, a belongs to the radical of the ideal $\langle b_1, \ldots, b_m \rangle$.

[3] Yet another connection is with the notion of 'descent' in algebraic geometry, see [13, 63].

[4] In several ways, this line of research was already suggested in the work of Lorenzen [20, 46, 47]. In particular, the paper [48] suggests a point-free analysis of Cantor–Bendixson. Also, one important tool in analysing the presentation of a space is the notion of *entailment relation*, explained in [12, 44], which already appears in Lorenzen's paper on cut elimination [46]. (The concept of entailment relation is now being used for an abstract development of proof theory, see [57, 58, 61].) One can also mention some remarks in [70], suggesting the use of elimination of choice sequences for a constructive reading of some classical proofs, which is strongly reminiscent of the present method based on point-free topology.

[5] This means that we have $a^n = 0$ for some n.

Krull's theorem relies on Zorn's Lemma.[6] In constructive mathematics, as we have seen, a non-trivial ring may fail to have prime ideals, and it thus seems impossible to use Zariski spectrum in this setting.

As we wrote in Section 6.1, the solution of this problem has some similarity to Brouwer's analysis of the notion of choice sequences, explained in [14], for example. We consider $\mathrm{Sp}(R)$ as a *point-free* space, defined simply by the lattice of its compact open subsets. We see then $D(a)$ as a *pure symbol*, generating a distributive lattice given by the following relations:

$$D(1) = 1 \qquad D(0) = 0 \qquad D(ab) = D(a) \wedge D(b) \qquad D(a+b) \leqslant D(a) \vee D(b).$$

This lattice $\mathrm{Sp}(R)$ is thus presented by generators and relations, and not as a collection of subsets of a given set of points. This definition originates in the work of Joyal [39]. We check that we can realize this lattice as the lattice of the *radicals* of finitely generated ideals.[7]

A corollary is that we have $D(a) \leqslant D(b_1, \dots, b_m)$ if, and only if, a belongs to the radical of the ideal $\langle b_1, \dots, b_m \rangle$. In particular, $D(a) = 0$ if, and only if, a is nilpotent. This argument was obtained by pure universal algebra, without ever having to build any prime ideals!

It is then possible to develop notions connected classically to the Zariski spectrum in a constructive setting. A typical example is provided by the notion of *Krull dimension* of a ring, which we examine next.

6.2.2 An Example: Krull Dimension

Classically, the Krull dimension n of a ring is the maximal length of strict chains of prime ideal $I_0 \subsetneqq I_1 \subsetneqq \cdots \subsetneqq I_n$. So a ring is of dimension 0 if we cannot have $I \subsetneqq J$ for two prime ideals I and J, that is, if any prime ideal is a maximal ideal.

A field, or a Boolean algebra, is a ring of dimension 0.

This notion can be analysed in a point-free way. It is actually simpler to define first the Krull dimension of a distributive lattice L. Such a definition goes back to works of Boileau and Joyal, and Espanol [8, 35], but it was realized later [23] that it can be seen as a simple case of Menger's dimension of topological space. Define the *boundary* $B(a)$ of an element a. It is the ideal generated by a and the ideal of elements x such that $x \wedge a = 0$. We say then[8] that L is of dimension < 0 if $1 = 0$ in L, and L is of dimension $< n + 1$ if each $L/B(a)$ is of dimension $< n$.

[6] Stricly speaking, it relies on a weaker form of the axiom of choice, the Boolean prime ideal theorem, but it is usually proved using Zorn's Lemma.

[7] In general, finitely generated ideals may not be closed by intersection, while if $D(a_1, \dots, a_n)$ is the *radical* of the ideal $\langle a_1, \dots, a_n \rangle$ we always have $D(a_1, \dots, a_n) \cap D(b_1, \dots, b_m) = D(a_1 b_1, \dots, a_n b_m)$.

[8] In a constructive setting, the dimension is not given as a natural number, but as a downward set of natural number.

The Krull dimension of a ring R is then defined to be $< n$ if, and only if, $\mathrm{Sp}(R)$ is of dimension $< n$.

Since we have $D(ab) = D(a) \wedge D(b)$, any element of $\mathrm{Sp}(R)$ is of the form $D(a_1, \dots, a_n)$. A natural question is if we can make n as small as possible. It can be checked that $D(a, b) = D(a + b, ab)$. In particular, if $ab = 0$ we have $D(a, b) = D(a + b)$. Using this remark, one can show that if R is a Boolean algebra, any element of $\mathrm{Sp}(R)$ can be written on the form $D(a)$. This can be generalized by the following version of Kronecker's Theorem [15].

Theorem 6.1 *If R is of dimension $< n$ any element of* $\mathrm{Sp}(R)$ *can be written* $D(a_1, \dots, a_m)$ *with* $m \leqslant n$.

The point here is that this has a simple effective proof, following the fact that all notions involved are defined in an elementary and effective way.

Reformulating basic notions of algebra in this point-free setting may reveal connections that were hidden in a classical setting. For instance, the following result, which also has an elementary proof [22, 44], contains *both* Forster–Swan and Serre splitting off theorem, while the classical version of these results look quite different. If F is a matrix over R we write $\Delta_n(F)$ the ideal generated by the $n \times n$ minors of F and we say that a vector is *unimodular* if the ideal generated by its elements contains 1.

Theorem 6.2 *If R is of dimension $< n$ and F a rectangular matrix such that* $\Delta_n(F) = 1$ *then some linear combination of the column of F is unimodular.*

Serre's splitting off theorem is the special case when the matrix is a square idempotent matrix.

6.3 Minimal and Maximal Primes

This 'phenomenological' approach to prime ideals extends to the notion of *maximal* and *minimal* prime ideals. We restrict ourselves to explaining the case of minimal prime ideals, and only mention a spectacular application of the analysis of maximal ideals: the computational interpretation by Yengui [72] of a lemma of Suslin used in his proof of Serre's problem (which states that any idempotent matrix over a polynomial ring is similar to a canonical projection matrix). The existence of a maximal ideal is the only non-effective element in Suslin's proof.[9]

The lattice $\mathrm{Sp}(R)$ can be seen as a point-free presentation of the Zariski spectrum. This presentation is *finitary* and this corresponds to the fact that this space is coherent, see [38, 66]. For the space of *minimal* prime ideals, we need a presentation

[9] It is even possible to realize this constructive version as a functional program and run it on small examples [42].

of a general point-free space with finite conjunctions and *infinitary* disjunctions, as in [26, 60]. To find the presentation of this space, we give first a *classical* characterization of minimal prime ideals. We define a *multiplicative monoid* of a ring to be a subset closed by multiplication and containing 1. This monoid is *proper* if it does not contain 0.

Lemma 6.3 *A subset of a ring is a minimal prime ideal if, and only if, its complement is a maximal proper multiplicative monoid. Furthermore, such a maximal proper multiplicative monoid D is exactly a mutiplicative monoid such that if $a \notin D$ then there exists b in D such that ab is nilpotent.*

We refer to Northcott [55] for the proof.

We can then use as a presentation of the space of the minimal prime ideals the following theory:

$$D(0) = 0 \qquad D(1) = 1 \qquad D(a) \wedge D(b) \leqslant D(ab) \qquad 1 = D(a) \vee \textstyle\bigvee_{b \in N(a)} D(b),$$

where $N(a)$ is the ideal of elements b such that ab is nilpotent.

One can then show in an elementary way, as in [19], the following result.

Theorem 6.4 *We have $D(a) \leqslant D(b_1) \vee \cdots \vee D(b_m)$ if, and only if, for all x, xa is nilpotent as soon as all $xb_1, \ldots, xb_m$ are nilpotent.*

Corollary 6.5 *We have $D(a) = 0$ if, and only if, a is nilpotent.*

This corresponds to the classical fact that the intersection of all minimal prime ideals is the set of nilpotent elements.

A further example where minimal prime ideals are used, and can be eliminated in this way, is Swan–Traverso's characterization of seminormal rings: see [17, 44]. Here again, it is actually possible to implement the constructive proof and run it on small examples, as in [2].

We explain here in detail a simpler example: a theorem of Vasconcellos, presented in [55], the proof of which uses a generic minimal prime ideal. For this example, this analysis produces an elementary argument, which does not mention minimal prime ideals.

If E is a module over R we define $Ann_R(E)$ to be the ideal of elements a in R such that $aE = 0$. We say that an ideal I of R is *regular* if $Ann_R(I) = 0$.

Theorem 6.6 *Let E be a module over R which admits a finite free resolution*

$$0 \to F_m \to F_{m-1} \to \cdots \to F_0 \to E \to 0$$

Each F_i is of the form R^{p_i} and we define [10] $\mathsf{Char}_R(E)$ *to be $p_0 - p_1 + p_2 - \cdots$. Then*

[10] It is shown in [55] that this number $\mathsf{Char}_R(E)$ is the same for any given finite free resolution of E.

- *if* $\mathsf{Char}_R(E) = 0$ *then* $Ann_R(E)$ *is regular,*
- *if* $\mathsf{Char}_R(E) > 0$ *then* $Ann_R(E) = 0$,
- *if* $\mathsf{Char}_R(E) < 0$ *then the ring* R *is trivial.*

This corresponds to Theorem 12 of Chapter 4 of [55], which is proved using minimal prime ideals.

We will analyse a special case of this theorem, which then suggests a direct proof of the general case, that can be found in [24].

We assume that we have an exact sequence $0 \to R \to R^2 \to E \to 0$ and we analyse a non-effective proof that $Ann_R(E) = 0$.

The fact that we have an exact sequence $0 \to R \to R^2 \to E \to 0$ can be unfolded as follows: E is generated by two elements e_1, e_2 and we have a_1, a_2 *regular* such that $b_1e_1 + b_2e_2 = 0$ if, and only if, (b_1, b_2) is a multiple of (a_1, a_2).

We will make use of the following lemma, which has a direct proof.

Lemma 6.7 *Let* $a_1, \ldots, a_n$ *be a regular sequence. Then* $a_1^l, \ldots, a_n^l$ *is also regular for all* l.

Let x be an element of $Ann_R(E)$. We need to show that we have $x = 0$. The classical reasoning in [55] proceeds as follows. We assume $x \neq 0$ and we consider a minimal prime ideal M over $(0 : x) = \{a \in R \mid ax = 0\}$. Using Lemma 6.3, we know that, if b is in M, then there exists a not in M and n such that ab^n is in $(0 : x)$.

If a_1 is not in M, then a_1 is invertible in R_M, the localization of R at the prime ideal M. Since $xe_2 = 0$ we have $(0, x) = r(a_1, a_2)$ for some r, and so $ra_1 = 0$ and $x = ra_2$. This implies $r = 0$ since a_1 is invertible and so $x = 1x = 0$ in R_M. We thus have 1 in $(0 : x)_M$ and so 1 in M, a contradiction. So a_1 is in M, and similarly a_2 is in M.

Using Lemma 6.3, this implies a_1^n and a_2^n in $(0 : x)_M$ for some n. Since a_1^n, a_2^n is regular by Lemma 6.7, this implies $1x = 0$ in R_M, and we have 1 in $(0 : x)_M$ and so 1 in M and a contradiction.

This reasoning was using a minimal prime ideal over $(0 : x)$ in a generic way and Lemma 6.3. We can follow it and give the following derivation of a contradiction in the theory T_M of minimal prime ideal over $(0 : x)$, where we add the axiom $\neg D(a)$ for $ax = 0$. This reasoning is now *constructive* and we can furthermore *eliminate* the use of the theory T_M.

We first prove $\neg D(a_1)$. We have $xe_2 = 0$ and hence $(0, x) = r(a_1, a_2)$ for some r. This means that we have $0 = ra_1$ and $x = ra_2$. We get then $xa_1 = 0$ which implies $\neg D(a_1)$. Similarly we have $\neg D(a_2)$. Using the axioms

$$1 = D(a_1) \vee \vee_{b \in N(a_1)} D(b) \qquad 1 = D(a_2) \vee \vee_{b \in N(a_2)} D(b)$$

of T_M, this means that we can find $b_1,\ b_2$ such that $D(b_1),\ D(b_2)$ and $a_1b_1,\ a_2b_2$ are nilpotent mod. $(0 : x)$. If $b = b_1b_2$ we have $D(b)$ and $a_1b,\ a_2b$ nilpotent mod. $(0 : x)$. Using Lemma 6.7, we get b nilpotent mod. $(0 : x)$, which contradicts $D(b)$ in the theory T_M.

Let us now eliminate the reference to this theory T_M. The fact that we can prove $\neg D(a_1)$ in T_M means that we can show $a_1^n x = 0$ for some n. Indeed, since we have $xe_2 = 0$ we get $0 = ra_1$ and $x = ra_2$ for some r, which implies that we have $x = 0$ in $R[1/a_1]$, that is $a_1^n x = 0$ for some n. Similarly, $xa_2^n = 0$ for some n, and then $x = 0$ by Lemma 6.7.

So the core of the argument is the following global–local principle for regular ideals, which does not mention any minimal prime ideal (and is used systematically in [24]).

Lemma 6.8 *Let $a_1, \ldots, a_n$ be a regular sequence. If $x = 0$ in the localizations $R[1/a_1], \ldots, R[1/a_n]$ then $x = 0$ in R. If I is an ideal of R which becomes regular in the localizations $R[1/a_1], \ldots, R[1/a_n]$ then I is regular.*

Proof The first statement follows from Lemma 6.7, since $x = 0$ in $R[1/a_i]$ if, and only if, we can find l such that $xa_i^l = 0$. For the second statement, assumes $xI = 0$. We then have $x = 0$ in $R[1/a_1], \ldots, R[1/a_n]$ and hence $x = 0$ by the first statement. □

The proof of the *general* case of Theorem 6.6 follows from this lemma: we look at the first column of the matrix corresponding to the map $F_m \to F_{m-1}$. This column is regular, and we prove Theorem 6.6 by induction by localizing over each element of this column and applying Lemma 6.8.

Here is one simple application.

Corollary 6.9 *If each principal ideal of R has a finite free resolution then R is an integral domain.*

Proof Indeed, by Theorem 6.6, each element is either 0 or is regular. □

Following this method, the paper [24] presents an effective proof of the following non-Noetherian version of the classical result of Auslander and Buchsbaum that any regular Noetherian local ring is a unique factorization domain: if each finitely generated ideal of a ring R has a finite free resolution, then R is a greatest common divisor (gcd) domain. For further developments, see [29, 68].

6.4 Forcing over a Site

In all previous examples, we can interpret what is going on as follows. We have a space X, described in a point-free way, and we 'force' the existence of a point by working inside $Sh(X)$, the collection of *sheaves* over X. So we move from the usual framework of usual sets to the frame of sheaves over the space X. We can then 'descend' what is going on in $Sh(X)$ back to the usual frame of sets.[11]

Grothendieck has generalized the notion of topological space to the notion of topos over a *site*. We can also use this notion in a constructive setting. A *point* for this notion of topos now becomes an *algebraic structure*. We can then adapt the previous method to the case where X is a such a generalized space given by a topos. A prime example of this situation is to 'force' the existence of a separable algebraic closure of a given field, by using the fact that such an algebraic closure can be seen as a point of a suitable topos. As before, by moving from the framework of sets to the framework of sheaves over the given site, it is 'as if' we had access to this algebraic structure. As in the case of topological spaces, the main problem is how to 'descend' from this framework of sheaves to the framework of sets. For the case of separable algebraic closure of a field, this problem is actually similar to the notion of Galois descent (see, for example, [63, Chapter X]), going back to the work of Châtelet [13].

For constructive mathematics, this method was first suggested by Joyal in two short papers [8, 39]). The method in [39] can be described as an elegant purely algebraic presentation of quantifier elimination. What we present is a variation, which does not proceed via quantifier elimination. This variation can be directly connected with the notion of *dynamical* algebra as presented in [30]. This notion was first introduced in computer algebra [31], for computing inside the algebraic closure of an *arbitrary* computable field.

6.4.1 Algebraic Closure in Constructive Mathematics

First, we recall the problem in constructive mathematics for building the algebraic closure of a given field. The first step in building such a closure is to add a root of a given monic polynomial P over a field F. This is simple if P is irreducible: $F[X]/\langle P\rangle$ is the desired extension of F containing a root of P. But if P is not irreducible, we need to consider an irreducible factor Q of P, and we add a root by working with $F[X]/\langle Q\rangle$.

The problem is that, in general, for a computable field, we *cannot* decide if a given polynomial is irreducible or not and cannot compute in general an irreducible factor of a given polynomial. This observation goes back at least to the paper by van der Waerden [71] (formulated in term of a Brouwerian counter-example, since this was done before the introduction of the notion of recursive functions and computable structures). One possible formulation of this result is the following.

[11] See [3] for a suggestive analogy with the notion of change of frames in physics.

Theorem 6.10 *In the intuitionistic theory of field, we cannot show* $\exists_x\ (x^2+1=0) \vee \forall_x\ (x^2+1 \neq 0)$.

Proof The theory of fields is the theory of rings, together with the axioms $1 \neq 0$ and $x = 0 \vee \exists_y\ (xy = 1)$. Consider the following Kripke counter-model. At time 0, we have $F_0 = \mathbb{Q}$, and at time 1, we have $F_1 = \mathbb{Q}[i]$ with $i^2+1=0$. Then we do not have $\exists_x\ (x^2+1=0)$ at time 0, and we do not have $\forall_x\ (x^2+1 \neq 0)$ at time 0 either, since x^2+1 has a root at time 1. □

This means that we have problem in constructive mathematics in adding a root of a given polynomial P (even a simple polynomial such as X^2+1) since we cannot decide in general if this polynomial is irreducible or not.

In [9, 52], a solution of this problem is given when the field F is *countable*. But, in the general case, there are actually results, explained in [52, Chapter VI, Section 3 Exercise 1], , that, in intuitionistic mathematics, a field may fail to have an algebraic closure.

6.4.2 Dynamical Method

Given this impossibility result, it is quite surprising that a technique has been developped, originally in computer algebra, showing how to compute in an algebraic closure of an arbitrary computable field!

This technique was introduced in [31], following a suggestion of Daniel Lazard. It replaces the 'computation', which is impossible in general, as explained above, of an irreducible factor of a polynomial, by computations of the gcd of two polynomials, which are always possible. This method might be interesting even in the case where we can decide irreducibility (e.g., over algebraic extensions of $\mathbb{Q}$), since deciding irreducibility might be computationally difficult compared with computations of the gcd of two polynomials.

The main idea is best explained with examples. Assume we want to add to a field F a root of X^2-3X+2 without deciding irreducibility. We work in the formal extension $F[a]$, $a^2-3a+2=0$, proceeding 'lazily'. If we are required to compute an inverse, for example, the inverse of $a+1$, we compute the gcd of $X+1$ and X^2-3X+2, producing the equality $X^2-3X+2=(X+1)(X-4)+6$ and this gives us that the inverse of $a+1$ is $(4-a)/6$. If we want to compute the inverse of $a-1$, we compute the gcd of $X-1$ and X^2-3X+2, which produces the equality $X^2-3X+2=(X-1)(X-2)$. We *discover* in this way the factorization $X^2-3X+2=(X-1)(X-2)$, without the need of a factorization algorithm. This factorization was simply discovered by the attempt to compute the inverse of $a-1$. We then open two branches: one with $a-1=0$ and one with $a-2=0$, and continue the computations.

In this way, we can proceed *as if* we were working in a field, and computing only the gcd of polynomials, but we may have to open some branches: the computation is 'dynamic'. This method is presented in depth in [30, 44, 53]. In [43], this method is used to build constructively the real algebraic closure of an ordered field, and in [41], to build the algebraic closure of a valued field.

6.4.3 Topos Theoretic Formulation of the Dynamical Method

In [8, 39], Joyal suggested instead the following approach to solve the problem of algebraic closure in an effective way: the algebraic closure may not exist in the framework of sets, but it *always* exist in a suitable *sheaf extension*. The argument suggested in [39] is an algebraic version of quantifier elimination, but this can also be explained in a way which stresses the analogy with the dynamical method described above.

For the Zariski spectrum, the point-free description of the space was a *propositional* theory. For the algebraic closure, it will be a *geometric* theory: the geometric theory of the algebraic closure of a given field F. The language is the language of the theory of rings, with a constant for each element of F. The axioms are the axioms of rings with the diagram of F and the following axioms:

(a) $x = 0 \vee \exists_y\ (xy = 1)$,
(b) $\exists_x\ (x^n + a_1 x^{n-1} + \cdots + a_0 = 0)$,
(c) $\vee_P P(x) = 0$ where P varies over monic polynomials.

Note that the last axiom is infinitary.

There is *always* a sheaf extension in which these axioms are satisfied. This is the *classifying topos* of this theory, as explained in [8]. This topos might be degenerate, however. What happens in the present case is that we have a *direct* description of a site which defines this classifying topos. From this direct description follows in particular the consistency of the theory.[12]

We now present this direct description of the site and, to simplify the discussion, we will limit ourselves to the case where the base field F is of characteristic 0.

A *triangular* algebra over F is an F-algebra obtained by a sequence of monic separable extensions. We can then prove this, as in [49, 50].

Lemma 6.11 *If R is a triangular F-algebra then for any element a of R both $R/\langle a\rangle$ and $R[1/a]$ are product of triangular F-algebras. We also have $R = R/\langle a\rangle \times R[1/a]$ for all a in R.*

[12] This is similar to the case of the Zariski spectrum of a ring, where we have a direct description of the Zariski lattice in term of radical of finitely generated ideals. As in this previous analysis, this also can be presented as a cut-elimination result, see [12, 18, 44].

We then consider the following site $\mathcal{S}_F$. The base category is the category of triangular F-algebras. A basic *covering* of R is given by decomposing R as a product by $R = R_1 \times \cdots \times R_n$ or by adding a formal root of a monic separable polynomial $R \to R[X]/\langle P\rangle$. We obtain, in an elementary and constructive way the following result; see [50].

Theorem 6.12 *The topos defines by this site $\mathcal{S}_F$ is the classifying topos of the theory of algebraic closure of F. The presheaf $L(R) = Hom(F[X], R)$ is a sheaf and is separably closed in the internal logic of this topos.*

One can think of a triangular algebra as an *approximation* of the algebraic closure of F. Intuitively, we do not consider the algebraic closure given 'actually', but we proceed by adding roots of polynomials as needed, and, at any point, we only have added finitely many roots. This is strongly reminiscent of the description by Edwards, in [33], of Kronecker's versus Dedekind's approach of the theory of algebraic curves:

> *The necessity of using an algebraically closed ground field introduced – and has perpetuated for 110 years – a fundamentally* transcendental *construction at the foundation of the theory of* algebraic curves. *Kronecker's approach, which calls for adjoining new constants algebraically as they are needed, is much more consonant with the nature of the subject.*

It is possible to interpret computationally in this way any argument which makes use of an algebraic closure of a field F, by working in the sheaf topos over $\mathcal{S}_F$. An example is a variation of Abhyankar's proof of the Newton–Puiseux Theorem in [1]. We first prove constructively the following result.

Theorem 6.13 *If L is separably closed, then $\cup_n L((x^{1/n}))$ is separably closed.*

By reading such a proof in the topos over the site $\mathcal{S}_{\mathbb{Q}}$, we get an effective way to compute Puiseux series. The interpretation of $L[[X]]$ is given by the exponential $L^{\mathbb{N}}$ in the sheaf model. Since we have $L^{\mathbb{N}}(R) = R^{\mathbb{N}}$, this also gives a purely logical a priori explanation of the fact that we do not need to keep adding new algebraic numbers when computing a Puiseux expansion: the *existence* of an element of $L[[X]]$ provides a *finite* triangular extension of $\mathcal{S}$ which contains all the coefficients of this series.

For instance, we can solve in this way a polynomial equation over $\mathbb{Q}$ such as $y^4 - 3y^2 + xy + x^2 = 0$ finding y as a formal series in some $x^{1/n}$. Since this interpretation is *effective*, we can 'run' the proof, as in [49, 50], and actually find a triangular algebra $\mathbb{Q}[a, b]$ with $a^2 = 13/36$ and $b^2 = 3$ over which we can write y as a power series in x. This illustrates the following point: in this approach, the algebraic closure L is given only *potentially*, but *finite approximations* of L become *actual* for solving specific questions.

Yet another remark about this constructive analysis of sheaf models is that it represents a combination of the 'computational' aspect of constructive mathematics and the 'epistemological' aspect present in sheaf models, where a basic open set represents a 'stage' of knowledge. For instance, a computational problem, such as the problem of finding an inverse of $a - 2$ in $\mathbb{Q}[a]$, $a^2 - 3a + 2 = 0$, provides the knowledge of a factorization $X^2 - 3X + 2 = (X - 2)(X - 1)$, knowledge which itself may simplify further computations. There is thus a feedback between 'computational' and 'epistemological' aspects of constructive mathematics. This is reminiscent of some remarks in [37] about the combination of forcing and realizability.

This method of introducing and eliminating the algebraic closure of a field can also been used for a constructive reading of the theory of simple central algebras over a field. See [21], where we give a dynamical reading of Wedderburn's Theorem representing central simple algebras as matrix algebras over a division algebra.[13] We can prove constructively in this way that any central simple algebra is split over an algebraically closed field. We deduce from this that the dimension of a central simple algebra over any field F is a *square*, by 'descending' the fact that its dimension is the one of a matrix algebra in the sheaf topos over $\mathcal{S}_F$.

6.5 Concluding Remarks

This chapter presents some applications of point-free topology and sheaf models to the constructive analysis of several concepts and proofs in algebra. Note that this is *different* from, but complementary to, the use of sheaf models suggested in [8, 54]. What is observed there is that, if we prove intuitionistically a theorem, this theorem will hold in any sheaf models. By looking externally at this proof, we may then get a new result, or a new proof of a classical result.[14] Here, instead, sheaf models are used in order to get a *computational interpretation* of *classical proofs*.

This technique can also be used in abstract functional analysis, where spaces are now *compact separated*. For instance, the work [28] analyses a representation theorem, and [27] gives a constructive proof of the Peter–Weyl Theorem. These works rely on the paper [16], which presents an analysis of some general representation results due to Stone [64, 65]. This in turn uses a constructive reading of some results presented in [38] and a fundamental result from Krivine [40], which implies

[13] This result can also be represented using negative translation, as has been done in formalisation of algebra in type theory, see [4, 36].

[14] The example given in [8] is the Weierstrass preparation theorem, which if it is proved intuitionistically in *one* variable, gives the multivariable version by this method of interpretation in sheaf models (see [59]). Another recent example in [7] is a simple proof of Grothendieck's generic freeness lemma.

a suitable cut-elimination result. The paper [12] presents a constructive reading of the Hahn–Banach extension theorem.

A general remark about this approach is that it avoids non-canonical enumerations, which are used in the algebraic closure of a countable field [9], or in representation theorems for separable spaces[15] in [5, 6]. The computations associated to our arguments thus feel more natural, not relying on an arbitrary enumeration. As we have seen, in most cases, it is actually possible to carry the out by hand or with the help of a computer algebra system, for small examples, as in [2, 29, 42, 50]. The notion of dynamical computations, connected to the idea of lazy evaluation, is also very interesting algorithmically (see [31]), and Yengui's book [73] presents several algorithms inspired by this technique of dynamical algebra.

We end by presenting some specific open problems, and a possible future research direction.

The first problem is about the theory algebraic curves, and in particular the proof of the Riemann–Roch Theorem. This is covered in [32, 34], but relying on an irreducibility algorithm. Is it possible instead to follow a dynamical approach, without such irreducibility algorithm? If so, can we have a constructive treatment which avoids, as in the classical approach, an actual computation of an integral basis?

The second problem is about valuation domains. A remarkable consequence of the paper [56] is that if V is a valuation domain, then $V[X_1, \ldots, X_n]$ is coherent. While this result has a constructive proof in [45], this proof was found directly without relying on the paper of Raynaud and Gruson. The problem is to understand the computational content of this highly non-effective argument and its connection with the algorithm presented in [45].

The third problem is similar. Merkurjev's Theorem, presented in [51], provides a complete description by generators and relations of the 2-torsion part of the Brauer group of a field. While the argument in [51] is effective, the first version of this proof was non-constructive and relied on a paper of Suslin [67], itself based on arguments of Quillen using the highly non-effective homotopy theory of simplical sets. What is the computational content of this non-effective proof?

Finally, one can hope that the constructive approach to sheaf models we have presented here can be generalized to higher toposes, providing in particular an effective treatment of cohomology groups which avoids the use of injective resolutions; for preliminary results in this direction, see the paper [25].

References

[1] Abhyankar, S. S. 1990. *Algebraic Geometry for Scientists and Engineers*. vol. 35. Providence, RI: American Mathematical Society.

[15] Compare, for instance, the statements and proofs of the Gelfand representation theorem in [28] and in [5, 6].

[2] Barhoumi, S. 2009. Seminormality and polynomial rings. *J. Algebra*, **322**, 1974–1978.

[3] Bell, J. L. 1986. From absolute to local mathematics. *Synthese*, **69**, 409–426.

[4] Bernard, S., Cohen, C., Mahboubi, A., and Strub, P.-Y. 2021. Unsolvability of the quintic formlized in dependent type theory. Pages 8.1–8.18 of: Cohen, L., and Kaliszyk, C. (eds.), *ITP 2021 – 12th International Conference on Interactive Theorem Proving, June 29–July 1, 2021, Rome, Italy (Virtual Conference), vol. 193*. Wadern: Schloss Dagstuhl – Leibniz-Zentrum.

[5] Bishop, E. 1967. *Foundations of Constructive Analysis*. New York: McGraw-Hill, New York.

[6] Bishop, E., and Bridges, D. 1985. *Constructive Analysis*. Berlin, Heidelberg: Springer.

[7] Blechschmidt, I. 2017 Using the internal language of toposes in algebraic geometry. Doctoral dissertation, University of Augsburg.

[8] Boileau, A., and Joyal, A. 1981. La logique des topos. *J. Symbol. Logic*, **46**, 6–16.

[9] Bridges, D., and Richman, F. 1987. *Varieties of Constructive Mathematics*. London Mathematical Society Lecture Note Series, vol. 97. Cambridge: Cambridge University Press, Cambridge.

[10] Brouwer, L.E. J. 1952. Historical background, principles and methods of intuitionism. *S. Afr. J. Sci.*, 139–146.

[11] Buchsbaum, D. A., and Eisenbud, D. 1974. Some structure theorems for finite free resolutions. *Adv. Math.*, **12**(1), 84–139.

[12] Cederquist, J., and Coquand, T. 2000. Entailment relations and distributive lattices. Pages 127–139 of: Buss, S. R., Hájek, P., and Pudlák, P. (eds.), *Logic Colloquium '98. Proceedings of the Annual European Summer Meeting of the Association for Symbolic Logic, Prague, Czech Republic, August 9–15, 1998. Lect. Notes Logic*, vol. 13.

[13] Châtelet, F. 1944. Variations sur un thème de H. Poincaré. *Ann. Sci. Éc. Norm. Supér. (3)*, **61**, 249–300.

[14] Coquand T. 2004. About Brouwer's fan theorem. *Revue Int. Philos.*, **230**, 483–489.

[15] Coquand, T. 2004. Sur un théorème de Kronecker concernant les variétés algébriques. *C. R. Math. Acad. Sci. Paris*, **338**(4), 291–294.

[16] Coquand, T. 2005. About Stone's notion of spectrum. *J. Pure Appl. Algebra*, **197**(1–3), 141–158.

[17] Coquand, T. 2006. On seminormality. *J. Algebra*, **305**(1), 577–584.

[18] Coquand, T. 2009. Space of valuations. *Ann. Pure Appl. Logic*, **157**, 97–109.

[19] Coquand, T., and Lombardi, H. 2006. A logical approach to abstract algebra. *Math. Struct. Comput. Sci.*, **16**(5), 885–900.

[20] Coquand, T., Lombardi, H., and Neuwirth, S. 2019. Lattice-ordered groups generated by an ordered group and regular systems of ideals. *Rocky Mt. J. Math.*, **49**(5), 1449–1489.

[21] Coquand T., Lombardi H., and Neuwirtz, S. 2021. Constructive basic theory of central simple algebras. https://doi.org/10.48550/arxiv.2102.12775.

[22] Coquand T., Lombardi H., and Quitté, C. 2004. Generating non noetherian modules constructively. *Manuscripta Math.*, **115**, 513–520.

[23] Coquand T., Lombardi H., and Roy, M.-F. 2005. An elementary characterisation of Krull dimension. Pages 239–244 of: Crosilla, L., and Schuster, P. (eds.), *From Sets and Types to Topology and Analysis*. Oxford Logic Guides, vol. 48. Oxford: Oxford University Press.

[24] Coquand T., and Quitté C. 2012. Constructive finite free resolutions. *Manuscr. Math.*, **137**(3–4), 331–345.

[25] Coquand T., Ruch F., and Sattler, C. 2020. Constructive sheaf models of type theory. Preprint, https://arxiv.org/abs/1912.10407.

[26] Coquand T., Sambin G., Smith, J., and Valentini, S. 2003. Inductively generated formal topologies. *Ann. Pure Appl. Logic*, **124**, 71–106.

[27] Coquand T., and Spitters, B. 2005. A constructive proof of the Peter–Weyl theorem. *Math. Log. Q.*, **51**(4), 351–359.

[28] Coquand, T., and Spitters B. 2005. Formal topology and constructive mathematics: the Gelfand and Stone–Yosida representation theorems. *J. UCS*, **11**(12), 1932–1944.

[29] Coquand, T., and Tête, C. 2018. An elementary proof of Wiebe's Theorem. *J. Algebra*, **499**, 103–110.

[30] Coste, M., Lombardi, H., and Roy, M.-F. 2001. Dynamical method in algebra: effective Nullstellensätze. *Ann. Pure Appl. Logic*, **111**(3), 203–256.

[31] Della Dora, J., Dicrescenzo, C., and Duval, D. 1985. About a new method for computing in algebraic number fields. In *European Conference on Computer Algebra (2)*, pp. 289–290.

[32] Edwards, M. H. 1990. *Divisor Theory*. Boston, MA: Birkhäuser, Boston, MA.

[33] Edwards, H. M. 1992. Mathematical ideas, ideals, and ideology. *Math. Intelligencer*, **14**(2), 6–19.

[34] Edwards, M. H. 2005. *Essays in Constructive Mathematics*. New York: Springer.

[35] Espanol, L. 1983. Le spectre d'un anneau dans l'algèbre constructive et applications à la dimension. *Cahiers Topologie Géom. Différentielle*, **24**(2), 133–144.

[36] Gonthier, G. 2011. Point-free, set-free concrete linear algebra. In *Interactive Theorem Proving*. Pages 103–118 of: *Proceedings of Second International Conference, ITP 2011, Berg en Dal, The Netherlands, August 22–25, 2011*.

[37] Goodman, N. D. 1978. Relativized realizability in intuitionistic arithmetic of all finite types. *J. Symbol. Logic*, **43**, 23–44.

[38] Johnstone, P. T. 1982. *Stone Spaces*. Cambridge Studies in Advanced Mathematics, no. 3. Cambridge: Cambridge University Press.

[39] Joyal, A. 1976. Les théoremes de Chevalley–Tarski et remarques sur l'algèbre constructive. *Cah. Topol. Géom. Différ. Catég.*, **16**, 256–258.

[40] Krivine, J. L. 1964. Anneaux preordonnes. *J. Anal. Math.*, **12**, 307–326.

[41] Kuhlmann, F.-V., Lombardi, H., and Perdry, H. 2003. Dynamic computations inside the algebraic closure of a valued field. Pages 133–156 of: Kuhlmann, S., Kuhlmann, F.-V., and Marshall, M. (eds.), *Valuation Theory and its Applications, vol. II. Proceedings of the International Conference and Workshop, University of Saskatchewan, Saskatoon, Canada, July 28–August 11, 1999*. Providence, RI; American Mathematical Society.

[42] Leino, A. 2011. An implementation of Quillen–Suslin Theorem. Master's thesis, Chalmers University.

[43] Lombardi, H. 1998. Relecture constructive de la théorie d'Artin–Schreier. *Ann. Pure Appl. Logic*, **91**(1), 59–92.

[44] Lombardi, H., and Quitté, C. 2012. *Algèbre Commutative. Méthodes Constructives. Modules Projectifs de Type Fini.* Paris: Calvage & Mounet.

[45] Lombardi, H., Quitté, C., and Yengui, I. 2014. Un algorithme pour le calcul des syzygies sur $V[X]$ dans le cas où V est un domaine de valuation. *Commun. Algebra*, **42**(9), 3768–3781.

[46] Lorenzen, P. 1951. Algebraische und logistische Untersuchungen über freie Verbände. *J. Symbol. Logic*, **16**(2), 81–106.

[47] Lorenzen, P. 1953. Die Erweiterung halbgeordneter Gruppen zu Verbandsgruppen. *Math. Z.*, **58**(1), 15–24.

[48] Lorenzen, P. 1959. Logical reflection and formalism. *J. Symbol. Log.*, **23**, 241–249.

[49] Mannaa, B., and Coquand, T. 2013. Dynamic Newton–Puiseux theorem. *J. Log. Anal.*, **5**. `http://logicandanalysis.org/index.php/jla/article/view/152/86`

[50] Mannaa, B., and Coquand, T. 2014. A sheaf model of the algebraic closure. Pages 18–32 of Olva, P. (ed.), *Proceedings Fifth International Workshop on Classical Logic and Computation, CL&C 2014, Vienna, Austria, July 13, 2014*. *EPTCS*, vol. 164.

[51] Merkurjev, A. 2006. On the norm residue homomorphism of degree two. In *Proceedings of the Pages 103–107 of: Petersburg Mathematical Society, vol. XII*. Translated from Russian.

[52] Mines, R., Richman, F., and Ruitenburg, W. 1988. *A Course in Constructive Algebra*. Universitext. New York: Springer.

[53] Mora, T. 2003. *Solving Polynomial Equation Systems. I: The Kronecker–Duval Philosophy*, vol. 88. Cambridge: Cambridge University Press.

[54] Mulvey, C. 1974. Intuitionistic algebra and representations of rings. Pages 3–57 of: *Recent Advances in the Representation Theory of Rings and C^*-algebras by Continuous Sections. A Seminar held at Tulane University, New Orleans, LA, USA, March 28–April 5, 1973*. Providence, RI: American Mathematical Society (AMS).

[55] Northcott, D. G. 1976. *Finite Free Resolutions*, vol. 71. Cambridge: Cambridge University Press.

[56] Raynaud, M., and Gruson, L. 1971. Critères de platitude et de projectivité. Techniques de "platification" d'un module. (Criterial of flatness and projectivity. Techniques of "flatification" of a module.) *Invent. Math.*, **13**, 1–89.

[57] Rinaldi, D., and Schuster, P. 2016. A universal Krull–Lindenbaum theorem. *J. Pure Appl. Algebra*, **220**, 3207–3232.

[58] Rinaldi, D., Schuster, P., and Wessel, D. 2018. Eliminating disjunctions by disjunction elimination. *Indag. Math. (N.S.)*, **29**(1), 226–259. Virtual Special Issue – L.E.J. Brouwer, fifty years later. Communicated first in *Bull. Symb. Logic* **23** (2017), 181–200.

[59] Rousseau, C. 1978. Topos theory and complex analysis. *J. Pure Appl. Algebra*, **10**, 299–313.

[60] Sambin, G. 1987. Intuitionistic formal spaces – a first communication. Pages 187–204 of: Skordev, D. (ed.), *Mathematical Logic and its Applications, Proc. Adv. Internat. Summer School Conf., Druzhba, Bulgaria, 1986*. New York: Plenum.

[61] Schuster, P., and Wessel, D. 2020. Resolving finite indeterminacy: A definitive constructive universal prime ideal theorem. Pages 820–830 of: *LICS '20: 35th Annual ACM/IEEE Symposium on Logic in Computer Science, Saarbrücken, Germany, July 8–11, 2020*. ACM.

[62] Serre, J.-P. 1955. Faisceaux algébriques cohérents. *Ann. Math. (2)*, **61**, 197–278, 1955.

[63] Serre, J.-P. 1962. Corps Locaux. Publications de l'Institut de Mathématique de l'Université de Nancago. 8; Actualités Scientifiques et Industrielles 1296. Paris: Hermann & Cie. 243 pages.

[64] Stone, M. H. 1940. A general theory of spectra. I. *Proc. Natl Acad. Sci. USA*, **26**, 280–283.

[65] Stone, M. H. 1941. A general theory of spectra. II. *Proc. Natl Acad. Sci. USA*, **27**, 83–87.

[66] Stone, M. H. 1938. Topological representations of distributive lattices and Brouwerian logics. *Časopis pro pěstování matematiky a fysiky*, **067**(1), 1–25.

[67] Suslin, A. A. 1982. The quaternion homomorphism for the function field on a conic. *Sov. Math., Dokl.*, **26**, 72–77.

[68] Tête, C. 2014. Profondeur, dimension et résolutions en algèbre commutative : quelques aspects effectifs. Ph.D. thesis, University of Poitier.

[69] Tierney, M. 1976. On the spectrum of a ringed topos. *Algebra, Topol., Category Theory; Collect. Pap. Honor S. Eilenberg*, 189–210.

[70] Troelstra, A. S. 1977. *Choice Sequences. A Chapter of Intuitionistic Mathematics*. Oxford: Oxford University Press, Oxford.

[71] van der Waerden, B. L. 1930. Eine Bemerkung über die Unzerlegbarkeit von Polynomen. *Math. Ann.*, **102**, 738–739.

[72] Yengui, I. 2008. Making the use of maximal ideals constructive. *Theor. Comput. Sci.*, **392**(1–3), 174–178.

[73] Yengui, I. 2015. *Constructive Commutative Algebra. Projective Modules over Polynomial Rings and Dynamical Gröbner Bases*, vol. 2138. Cham: Springer.

7

Constructive Projective Geometry

Mark Mandelkern

7.1 Introduction

Of the great theories of classical mathematics, projective geometry, with its powerful concepts of symmetry and duality, has been exceptional in continuing to intrigue investigators. The "Constructivist Manifesto" of Errett Bishop (1928–1983) (see [4, Chapter 1] and [7, Chapter 1]) and the challenge put forth by Bishop

> Every theorem proved with [nonconstructive] methods presents a challenge: to find a constructive version, and to give it a constructive proof. [4, p. x], [7, p. 3]

motivate a large portion of current constructive work. One way to answer this challenge is to discover the hidden constructive content of classical projective geometry. Here we briefly outline, with few details, recent constructive work concerning the real projective plane, and projective extensions of affine planes. Special note is taken of a number of interesting open problems that remain; these show that constructive projective geometry is still a theory very much in need of further effort.

There has been a considerable amount of work in the constructivization of geometry, on various topics, in different directions, and from diverse standpoints. Intuitionistic axioms for projective geometry were introduced by Heyting [17], with further work by van Dalen [36]. Work in constructive geometry by Beeson [1, 2] uses Markov's Principle, a nonconstructive principle which is accepted in recursive function theory, but not in Bishop-type strict constructivism, which is followed here. Lombard and Vesley [22] construct an axiom system for intuitionistic plane geometry, and study it with the aid of recursive function theory. The work of von Plato [38, 39, 40] in constructive geometry, proceeding from the viewpoint of formal logic, is related to type theory, computer implementation, and combinatorial analysis. The work of Pambuccian [28, 29, 30], also proceeding within formal logic, covers a wide range of topics concerning axioms for constructive geometry.

Bishop-type constructive mathematics proceeds from a viewpoint well-nigh opposite that of either formal logic or recursive function theory. For further details concerning this distinction, see [3, 4, 5, 6, 7, 8]. For stricly constructive work on the coordinatization of a plane, see [23]; on the extension of an affine plane to a projective plane, see [18, 24, 25, 35]; on the structure of a real projective plane, see [17, 26, 27].

7.2 Real Projective Plane

Arend Heyting (1898–1980), in his doctoral dissertation [17], began the constructivization of projective geometry. Heyting's work involves both synthetic and analytic theories. Axioms for projective space are adopted; since a plane is then embedded in a space of higher dimension, it is possible to include a proof of Desargues's Theorem. For the coordinatization of projective space, axioms of order and continuity are assumed. The theory of linear equations is included, and results in analytic geometry are obtained. Later, Heyting discussed the role of axiomatics in constructive mathematics as follows.

> At first sight it may appear that the axiomatic method cannot be used in intuitionistic mathematics, because there are only considered mathematical objects which have been constructed, so that it makes no sense to derive consequences from hypotheses which are not yet realized. Yet the inspection of the methods which are actually used in intuitionistic mathematics shows us that they are for an important part axiomatic in nature, though the significance of the axiomatic method is perhaps somewhat different from that which it has in classical mathematics. [18]

Recent work [26], briefly outlined in the following, constructivized the synthetic theory of the real projective plane as far as harmonic conjugates, projectivities, the axis of homology, conics, Pascal's Theorem, and polarity. The axioms adopted are only for a plane. The basis for the constructivization is the extensive literature concerning the classical theory, including works of Veblen and Young [37, 43], Coxeter [12], Lehmer [21], Cremona [13], and Pickert [32]. An entertaining history of the classical theory is found in Lehmer's last chapter.

7.2.1 Axioms

For nearly 200 years a sporadic and sometimes bitter debate has continued, concerning the value of synthetic versus analytic methods. In his Erlangen program of 1872, Felix Klein sought to mediate the dispute.

> The distinction between modern synthetic and modern analytic geometry must no longer be regarded as essential, inasmuch as both subject-matter and methods of reasoning have gradually taken a similar form in both. We chose therefore as common designation of them both the term *projective geometry*. Although the synthetic method has more to do with space-perception and thereby imparts a rare charm to its first simple developments, the realm of space-perception is nevertheless not closed to the analytic method, and the formulae of analytic geometry can be looked upon as a precise and perspicuous statement of geometrical relations. On the other hand, the advantage to original research of a well formulated analysis should not be underestimated, an advantage due to its moving, so to speak, in advance of the thought. But it should always be insisted that a mathematical subject is not to be considered exhausted until it has become intuitively evident, and the progress made by the aid of analysis is only a first, though a very important, step. [20]

In the synthetic work summarized here, axioms are formulated which can be traced to an analytic model based on constructive properties of the real numbers, the resulting axiom system is used to construct a synthetic projective plane $\mathbb{P}$, and the consistency of the synthetic axiom system is then verified by means of the analytic model. In this sense, the construction of the synthetic plane $\mathbb{P}$ takes into account Bishop's thesis.

> All mathematics should have numerical meaning.
> [4, p. ix] [7, p. 3]

Axiom Group C

The constructivization of [26], resulting in the synthetic projective plane $\mathbb{P}$, uses only axioms for a plane. There exist non-Desarguesian projective planes; see [42]. This means that Desargues's Theorem must be taken as an axiom; it is required to establish the essential properties of harmonic conjugates, and further portions of the theory. Other special features of the axiom system are also required, to obtain constructive versions of the most important classical results. The consistency of the axiom system is verified by means of the analytic model discussed in Section 7.2.7; the properties of this model have guided the choice of axioms.

The constructive axiom group C, adopted for the projective plane $\mathbb{P}$ in [26, Section 2], has seven initial axioms. The first four are those usually seen for a classical projective plane; for example, two points determine a line, and two lines intersect at a point. The last three axioms, which have special constructive significance, are discussed here.

For the construction of the projective plane $\mathbb{P}$, there is given a family $\mathscr{P}$ of points and a family $\mathscr{L}$ of lines, along with equality and inequality relations for each family. The *inequality relations* assumed for the families $\mathscr{P}$ and $\mathscr{L}$, both denoted $\neq$, are *tight apartness relations*; thus, for any elements x, y, z, the following conditions are satisfied.

(i) $\neg(x \neq x)$.
(ii) If $x \neq y$, then $y \neq x$.
(iii) If $x \neq y$, then either $z \neq x$ or $z \neq y$.
(iv) If $\neg(x \neq y)$, then $x = y$.

The notion of an apartness relation was introduced by L. E. J. Brouwer (1881–1966) [10], and developed further by Heyting [19]. Property (iii) is known as *cotransitivity*, and (iv) is known as *tightness*. The implication "$\neg(x = y)$ implies $x \neq y$" is invalid in virtually all constructive theories; the inequality, representing distinctness, is the stronger of the two conditions. For example, with real numbers considered constructively, $x \neq 0$ means that there exists an integer n such that $|x| > 1/n$, whereas $x = 0$ means merely that it is contradictory that such an integer exists. For more details concerning the constructive properties of the real numbers, see [4, 7, 9]; for a comprehensive treatment of constructive inequality relations, see [8, Section 1.2].

A given *incidence relation,* written $P \in l$, links the two families; we say that *the point P lies on the line l*, or that *the line l passes through the point P*. A line is not viewed as a set of points; the families $\mathscr{P}$ and $\mathscr{L}$ used for the construction are independent; this contributes to the duality of the resulting system. The set $\bar{l}$ of points that lie on a line l is a *range of points*, while the set Q^* of lines that pass through a point Q is a *pencil of lines*. The outside relation $P \notin l$ is obtained by a definition.

Definition 7.1 *Outside relation.* For any point P on the projective plane $\mathbb{P}$, and any line l, it is said that *P lies outside l* (and *l avoids P*), and written $P \notin l$, if $P \neq Q$ for all points Q that lie on l [26, Definition 2.3].

This condition for the relation $P \notin l$, when viewed classically, is simply the negation of the condition $P \in l$, when written as the tautology "there exists $Q \in l$

such that $P = Q$." Constructively, however, the condition acquires a strong, positive significance, derived from the character of the condition $P \neq Q$.

Several axioms connect these relations.

Axiom C5 *For any lines l and m on the projective plane $\mathbb{P}$, if there exists a point P such that $P \in l$, and $P \notin m$, then $l \neq m$.*

The implication "If $\neg(P \in l)$, then $P \notin l$" is nonconstructive. However, we have the following.

Axiom C6 *For any point P on the projective plane $\mathbb{P}$, and any line l, if $\neg(P \notin l)$, then $P \in l$.*

Axiom C6 would be immediate in a classical setting, where $P \notin l$ means $\neg(P \in l)$; applying the law of excluded middle, a double negation results in an affirmative statement. For a constructive treatment, where the condition $P \notin l$ is not defined by negation, but rather by the affirmative definition above, Axiom C6 must be assumed; it is analogous to the tightness property of the inequality relations that are assumed for points and lines.

For the metric plane $\mathbb{R}^2$, the condition of Axiom C6 follows from the analogous constructive property of the real numbers: "For any real number α, if $\neg(\alpha \neq 0)$, then $\alpha = 0$," interpreting the outside relation in terms of distance. For the analytic model $\mathbb{P}^2(\mathbb{R})$, which motivates the axiom system, Axiom C6 is verified using this constructive property of the real numbers.

The following axiom has an exceptional role; it is indispensable for virtually all constructive proofs involving the projective plane $\mathbb{P}$. The point of intersection of distinct lines l and m is denoted $l \cdot m$.

Axiom C7 *If l and m are distinct lines on the projective plane $\mathbb{P}$, and P is a point such that $P \neq l \cdot m$, then either $P \notin l$ or $P \notin m$.*

This axiom is a strongly worded, yet classically equivalent, constructive form of a classical axiom: "distinct lines have a *unique* common point," which means only that if the points P and Q both lie on both lines, then $P = Q$. Axiom C7, a (classical) contrapositive of the classical axiom, is significantly stronger, since the condition $P \notin l$ is an affirmative condition.

Heyting and van Dalen have used an apparently weaker version of Axiom C7; it is Heyting's Axiom VI [17], and van Dalen's Lemma 3(f), obtained using his axiom Ax5 [36]. This weaker version states: "If l and m are distinct lines, P is a point such that $P \neq l \cdot m$, and $P \in l$, then $P \notin m$." However, the two versions are equivalent, as the reader can easily show.

Axiom C4 states that at least three distinct points lie on any given line; this is the usual classical axiom. Then, for the study of projectivities, Axiom E is added, increasing the required number of points to six. Recently, a constructive proof in [27], of an essential result concerning harmonic conjugates, required at least eight points on a line; thus we have the following.

Problem 7.2 Determine the minimum number of points on a line that are required for the various constructive proofs concerning the projective plane $\mathbb{P}$. Examine the propositions involved for the exceptional small finite planes.

The axioms and definitions of constructive projective geometry can be given a variety of different arrangements; see [17, 36, 38]. For example, in [36] the outside relation $P \notin l$ is taken as a primitive notion, and the condition of Axiom C6 above becomes the definition of the incidence relation $P \in l$.

The axiom system could be extended; thus we have the following.

Problem 7.3 Extend the constructive axiom group C to projective space, and derive constructive versions of the main classical theorems.

Desargues's Theorem

Desargues's Theorem is assumed as an axiom; the converse is then proved as a consequence.

Two triangles are *distinct* if corresponding vertices are distinct and corresponding sides are distinct; it is then easily shown that the three lines joining corresponding vertices are distinct, and the three points of intersection of corresponding sides are distinct. Distinct triangles are said to be *perspective from the center* O if the lines joining corresponding vertices are concurrent at the point O, and O lies outside each of the six sides. Distinct triangles are said to be *perspective from the axis* l if the points of intersection of corresponding sides are collinear on the line l, and l avoids each of the six vertices.

Axiom D (Desargues's Theorem) *If two triangles are perspective from a center, then they are perspective from an axis.*

The proof of the converse is included below, as an example of constructive methods in geometry; note that it uses Axiom C7 six times.

Theorem 7.4 Converse of Desargues's Theorem. *If two triangles are perspective from an axis, then they are perspective from a center* [26, Theorem 3.2].

Proof We are given distinct triangles PQR and $P'Q'R'$, with collinear points of intersection of corresponding sides, $A = QR \cdot Q'R'$, $B = PR \cdot P'R'$, $C =$

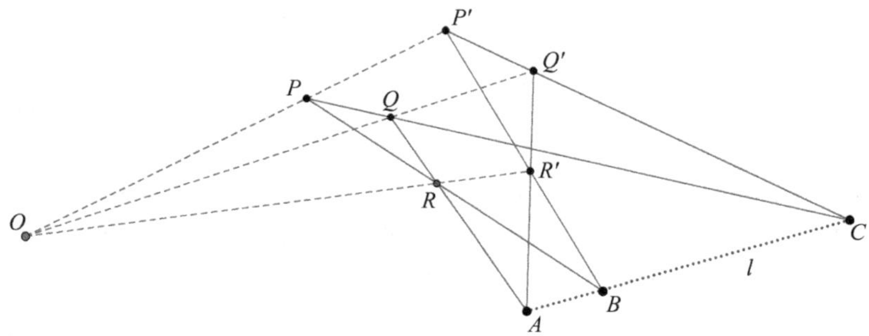

Figure 7.1 Converse of Desargues's Theorem.

$PQ \cdot P'Q'$, and with all six vertices lying outside the axis $l = AB$ (Fig. 7.1). Set $O = PP' \cdot QQ'$.

The points A, Q, Q' are distinct, and the points B, P, P' are distinct. Since $Q \neq A = QR \cdot Q'R'$, it follows from Axiom C7 that $Q \notin Q'R' = AQ'$; thus the points A, Q, Q' are noncollinear, and similarly for B, P, P'. Since $P \notin AB$, we have $AB \neq BP$. Since $A \neq B = AB \cdot BP$, it follows that $A \notin BP$, so $AQ \neq BP$. By symmetry, $AQ' \neq BP'$. This shows that the triangles AQQ', BPP' are distinct.

The lines AB, PQ, $P'Q'$, joining corresponding vertices of the triangles AQQ', BPP', are concurrent at C. Since $Q \notin AB$, we have $AB \neq AQ$. From $C \neq A = AB \cdot AQ$, it follows that $C \notin AQ$; by symmetry, $C \notin AQ'$. Since $Q' \neq C = PQ \cdot P'Q'$, it follows that $Q' \notin PQ$; thus $QQ' \neq PQ$, that is, $CQ \neq QQ'$. From $C \neq Q = CQ \cdot QQ'$, we have $C \notin QQ'$. Thus C lies outside each side of triangle AQQ', and similarly for triangle BPP'. Thus the triangles AQQ', BPP' are perspective from the center C.

It follows from Axiom D that the triangles AQQ', BPP' are perspective from the axis $(AQ \cdot BP)(AQ' \cdot BP') = RR'$, the axis avoids all six vertices, and $O \in RR'$. Thus the lines PP', QQ', RR', joining corresponding vertices of the given triangles, are concurrent at O. Since $Q \notin RR'$, we have $Q \neq O$. From $O \neq Q = QQ' \cdot PQ$, it follows that $O \notin PQ$. By symmetry, O lies outside each side of the given triangles.

Hence the triangles PQR and $P'Q'R'$ are perspective from the center O. □

Duality

Given any statement, the *dual statement* is obtained by interchanging the words "point" and "line." For example, we have the following.

Dual of Axiom C5 *For any points P and Q on the projective plane $\mathbb{P}$, if there exists a line l such that $P \in l$, and $Q \notin l$, then $P \neq Q$.*

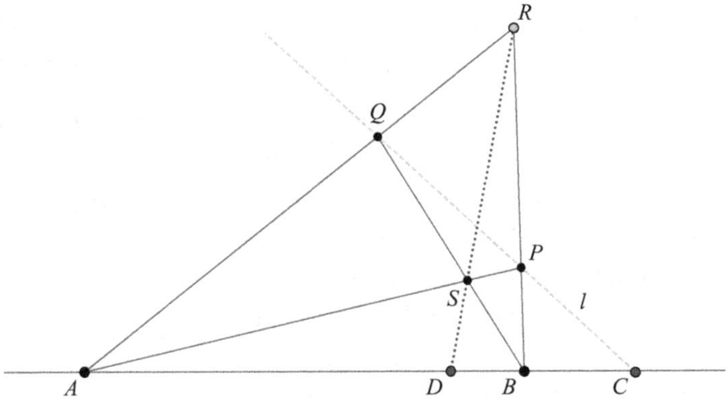

Figure 7.2 Construction of harmonic conjugates.

Dual of Axiom C7 *Let A and B be distinct points on the projective plane $\mathbb{P}$. If l is a line such that $l \neq AB$, then either $A \notin l$ or $B \notin l$.*

Clearly, Axiom C6 is self-dual. *Duality* in a given system is the principle that the dual of any true statement is also true. Duality of the construction of the plane $\mathbb{P}$, and of the axiom system, is verified as follows:

Theorem 7.5 *The definition of the projective plane $\mathbb{P}$ is self-dual. The dual of each axiom in axiom group C is valid on $\mathbb{P}$* [26, Theorem 2.10].

The dual of the definition of the outside relation $P \notin l$ is also verified:

Theorem 7.6 *Let P be any point on the projective plane $\mathbb{P}$, and let l be any line on $\mathbb{P}$. Then $P \notin l$ if and only if $l \neq m$ for any line m that passes through P* [26, Theorem 2.11].

7.2.2 Harmonic Conjugates

In the theory of the projective plane $\mathbb{P}$, harmonic conjugates have an essential role, with applications to projectivities, involutions, and polarity. In Fig. 7.2, the quadrangle $PQRS$, which is often used classically, appears to determine the harmonic conjugate D of the point C, with respect to the base points A and B. However, this is only valid when C is distinct from each base point. Thus we must use a constructive definition that applies to every point on the base line AB.

Definition 7.7 Let A and B be distinct points on the projective plane $\mathbb{P}$. For any point C on the line AB, select a line l through C, distinct from AB, and select a point R lying outside each of the lines AB and l. Set $P = BR \cdot l$, $Q = AR \cdot l$, and

$S = AP \cdot BQ$. The point $D = AB \cdot RS$ is called the *harmonic conjugate of* C *with respect to the points* A, B; we write $D = h(A, B; C)$ [26, Definition 4.1].

Since the construction of a harmonic conjugate requires the selection of auxiliary elements, it must be verified, by an invariance theorem, that the result is independent of the choice of these auxiliary elements. The proof given in [26] for the invariance theorem is incorrect; apart from the error, the proof there is excessively complicated, and objectionable on several counts. A correct proof appears in a later paper [27]. This theorem is important for the further development of the theory; for example, the construction of a polar, in Section 7.2.6, uses harmonic conjugates and relies explicitly on this invariance theorem to establish the uniqueness of the polar construction.

Theorem 7.8 (Invariance Theorem) *Let* C *be any point on the line* AB*, and let auxiliary element selections* (l, R) *and* (l', R') *be used to construct harmonic conjugates* D *and* D' *of the point* C*. Then* $D = D'$*; the harmonic conjugate construction is independent of the choice of auxiliary elements* [27, Theorem 3.2].

In the special case of a point distinct from both base points, constructive harmonic conjugates can be related to the traditional quadrangle configuration, due to Philippe de La Hire (1640–1718) [14].

Corollary 7.9 *Let* A, B, C, D *be collinear points, with* $A \neq B$*, and* C *distinct from each of the points* A *and* B*. Then* $D = h(A, B; C)$ *if and only if there exists a quadrangle with vertices outside the line* AB*, of which two opposite sides intersect at* A*, two other opposite sides intersect at* B*, while the remaining two sides meet the base line* AB *at* C *and* D [26, Theorem 3.2].

7.2.3 Projectivities

The elementary mappings of a projective plane are bijections relating a pencil of lines with a range of points; these are called *sections.* Certain combinations of sections result in *projections,* which map a range of points onto another range, projecting from a center, or map a pencil of lines onto another pencil, projecting from an axis. These sections and projections are the *perspectivities* of the plane (see Fig. 7.3).

The product (composition) of two perspectivities need not be a perspectivity. For the projective plane $\mathbb{P}$, a finite product of perspectivities is called a *projectivity;* this is the definition used by Jean-Victor Poncelet (1788–1867) [33]. Subsequently, Karl Georg Christian von Staudt (1798–1867) [41] defined a *projectivity* as a mapping of a range or a pencil that preserves harmonic conjugates. Classically, the two notions of perspectivity are equivalent. Constructively, we have the following.

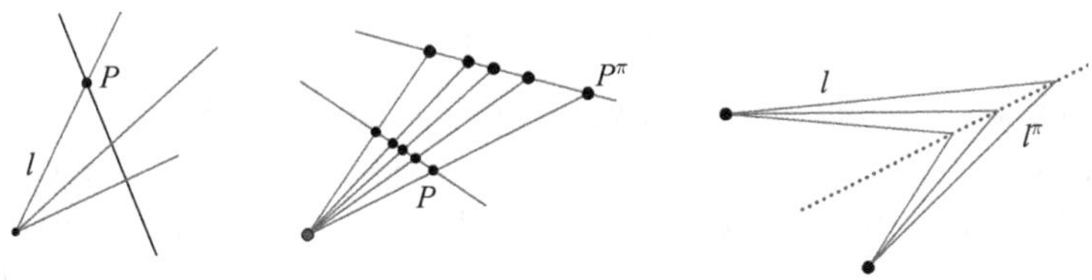

Figure 7.3 Perspectives.

Theorem 7.10 *A projectivity of the projective plane $\mathbb{P}$ preserves harmonic conjugates. Thus every Poncelet projectivity is a von Staudt projectivity* [26, Theorem 5.3].

However, the constructive content of the converse is not known; thus we have the following.

Problem 7.11 On the projective plane $\mathbb{P}$, show that every von Staudt projectivity is a Poncelet projectivity, or construct a counterexample.

It is necessary to establish the existence of projectivities.

Theorem 7.12 *For any three distinct points P, Q, R in a range $\overline{l}$, and any three distinct points P', Q', R' in a range $\overline{m}$, there exists a projectivity $\pi : \overline{l} \to \overline{m}$ such that the points P,Q,R map into the points P', Q', R', in the order given* [26, Theorem 5.6].

Classically, the projectivity produced by this theorem is the product of at most three perspectivities. However, the constructive proof in [26] requires six perspectivities; thus we have the following.

Problem 7.13 Determine the minimum number of perspectivities required for a constructive proof of the above theorem.

If a projectivity π, of a range or pencil onto itself, is of order 2 (i.e., π^2 is the identity ι), it is called an *involution*; this term was first used by Girard Desargues (1591–1661) [15]. Desargues introduced 70 new geometric terms; they were considered highly unusual, and met with sharp criticism and ridicule by his contemporaries. Of these 70 terms, *involution* is the only one to have survived. One example of an involution is the harmonic conjugate relation.

Theorem 7.14 *Let A and B be distinct points in a range $\overline{l}$, and let υ be the mapping of harmonic conjugacy with respect to the base points A, B; that is, set $X^\upsilon = h(A, B; X)$, for all points X in the range $\overline{l}$. Then υ is an involution* [26, Theorem 7.2].

7.2.4 Fundamental Theorem

The fundamental theorem of projective geometry, due to von Staudt [41], is required for many results, including Pascal's Theorem. Classically, the fundamental theorem is derived from axioms of order and continuity. For the projective plane $\mathbb{P}$, since no axioms of order and continuity have been adopted, the crucial component of the fundamental theorem must be derived directly from an axiom.

Axiom T *If a projectivity π of a range or pencil onto itself has three distinct fixed elements, then it is the identity ι.*

Classically, Axiom T is often given the following equivalent form: *Let π be a projectivity from a range onto itself, with $\pi \neq \iota$, and distinct fixed points M and N. If Q is a point of the range distinct from both M and N, then $Q^\pi \neq Q$.* Constructively, this appears to be a stronger statement, since the implication "$\neg(Q^\pi = Q)$ implies $Q^\pi \neq Q$" is constructively invalid; thus we have the following.

Problem 7.15 Give a proof of the apparently stronger alternative statement for Axiom T, or show that it is constructively stronger.

To prove that the alternative statement is constructively stronger would require a Brouwerian counterexample. To determine the specific nonconstructivities in a classical theory, and thereby to indicate feasible directions for constructive work, Brouwerian counterexamples are used, in conjunction with nonconstructive omniscience principles. A *Brouwerian counterexample* is a proof that a given statement implies an omniscience principle. In turn, an *omniscience principle* would imply solutions or significant information for a large number of well-known unsolved problems. This method was introduced by Brouwer [11] to demonstrate that use of the law of excluded middle inhibits mathematics from attaining its full significance. A statement is considered *constructively invalid* if it implies an omniscience principle. For a typical Brouwerian counterexample, see [26, Example 1.1].

The omniscience principles can be expressed in terms of real numbers; the following are most often utilized.

Limited principle of omniscience (LPO) *For any real number α, either $\alpha = 0$ or $\alpha \neq 0$.*

Weak limited principle of omniscience (WLPO) *For any real number α, either $\alpha = 0$ or $\neg(\alpha = 0)$.*

Lesser limited principle of omniscience (LLPO) *For any real number* α*, either* $\alpha \leq 0$ *or* $\alpha \geq 0$.

Markov's principle *For any real number* α*, if* $\neg(\alpha = 0)$*, then* $\alpha \neq 0$.

For work according to Bishop-type strict constructivism, as followed here, these principles, consequences of the law of excluded middle, are used only to demonstrate the nonconstructive nature of certain classical statements, and are not accepted for developing a constructive theory. Markov's Principle, however, is accepted for work in recursive function theory.

Theorem 7.16 (Fundamental Theorem) *Given any three distinct points* P, Q, R *in a range* $\overline{l}$*, and any three distinct points* P', Q', R' *in a range* $\overline{m}$*, there exists a unique projectivity* $\pi : \overline{l} \to \overline{m}$ *such that the points* P, Q, R *map into the points* P', Q', R'*, in the order given* [26, Theorem 6.1].

Proof The existence of the required projectivity is provided by Theorem 7.12 in Section 7.2.3. Uniqueness, however, requires Axiom T. □

Classically, the fundamental theorem is derived from axioms of order and continuity; thus we have the following.

Problem 7.17 Introduce constructive axioms of order and continuity for the projective plane $\mathbb{P}$; then derive Axiom T and the fundamental theorem.

It follows from the fundamental theorem that any projectivity between distinct ranges or pencils that has a fixed element is a perspectivity [26, Corollary 6.2]. A projectivity π such that $x^{\pi} \neq x$, for all elements x, is called *nonperspective.*

The concept of projectivity can be extended to the entire plane. A *collineation* of the projective plane $\mathbb{P}$ is a bijection of the family $\mathscr{P}$ of points, onto itself, that preserves collinearity and noncollinearity. A collineation σ induces an analogous bijection σ' of the family $\mathscr{L}$ of lines. A collineation is *projective* if it induces a projectivity on each range and each pencil of the plane.

The following theorem is a constructivization of one of the main results in the classical theory.

Theorem 7.18 *A projective collineation with four distinct fixed points, any three of which are noncollinear, is the identity* [26, Proposition 6.7].

Proof Let the collineation σ have the fixed points P, Q, R, S as specified; thus the three distinct lines PQ, PR, PS are fixed. The mapping σ' induces a projectivity on the pencil P^*; by the fundamental theorem, this projectivity is the identity. Thus

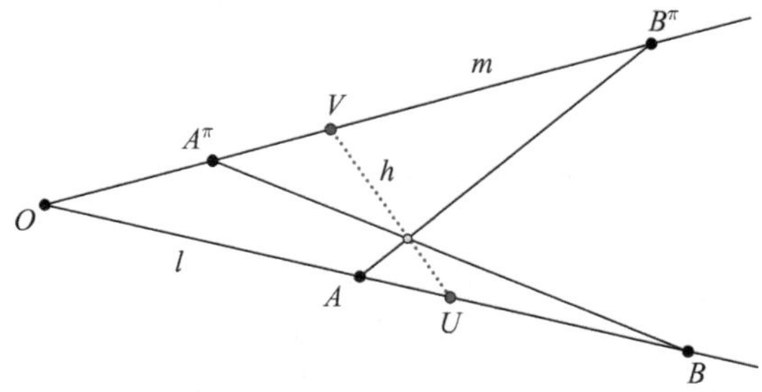

Figure 7.4 Axis of homology.

every line through P is fixed under σ'; similarly, the same is true for the other three points.

Now let X be any point on the plane. By three successive applications of cotransitivity for points, we may assume that X is distinct from each of the points P, Q, R. Since $PQ \neq PR$, using cotransitivity for lines we may assume that $XP \neq PQ$. Since $Q \neq P = XP \cdot PQ$, it follows from Axiom C7 that $Q \notin XP$, and thus $XP \neq XQ$. Since $X = XP \cdot XQ$, and the lines XP and XQ are fixed under σ', it follows that $\sigma X = X$. □

Problem 7.19 Theorem 7.18 ensures the uniqueness of a projective collineation that maps four distinct points, any three of which are noncollinear, into four distinct specified points, any three of which are also noncollinear. Establish the existence of such a collineation for the projective plane $\mathbb{P}$.

The classical theory of the axis of homology has also been constructivized.

Definition 7.20 Let $\pi : \bar{l} \to \overline{m}$ be a nonperspective projectivity between distinct ranges on the projective plane $\mathbb{P}$. Set $O = l \cdot m$, $V = O^\pi$, and $U = O^{\pi^{-1}}$; then the line $h = UV$ is called the *axis of homology for* π. [26, Definition 6.4]

Theorem 7.21 is the main result concerning the axis of homology; the proof requires the fundamental theorem.

Theorem 7.21 *Let $\pi : \bar{l} \to \overline{m}$ be a nonperspective projectivity between distinct ranges on the projective plane $\mathbb{P}$. If A and B are distinct points on l, each distinct from the common point O, then the point $AB^\pi \cdot BA^\pi$ lies on the axis of homology h; see Fig. 7.4* [26, Theorem 6.5].

7.2.5 Conics

The conic sections have a long history; they were discovered by Menaechmus (c. 340 BCE) and studied by the Greek geometers to the time of Pappus of Alexandria

(c. 320 CE). The motivation for Menaechmus's discovery was a geometrical problem, put forth by the oracle on the island of Delos, the solution of which would have provided a remedy for the Athenian plague of 430 BCE. Unfortunately, Menaechmus's solution was too late; see [12, p. 79] for details.

In the seventeenth century, an intense new interest in the conics arose in connection with projective geometry. On a projective plane there is no distinction between the hyperbola, parabola, and ellipse; these arise only in the affine plane that results after a line is removed from the projective plane. The missing line is then considered to be "at infinity." Which of the three forms of affine conic that results depends on whether the line at infinity meets the conic at two, one, or no points.

Construction of a Conic

On the projective plane $\mathbb{P}$, conics are defined by means of projectivities, using the method of Jakob Steiner (1796–1863) [34]. Alternatively, in classical works, conics are often defined by means of polarities, using the method of von Staudt [41]; see Problem 7.38 at the end of Section 7.2.6.

Definition 7.22 (Steiner) Let $\pi : U^* \to V^*$ be a nonperspective projectivity between distinct pencils of lines on the projective plane $\mathbb{P}$. The *conic* $\kappa = \kappa(\pi; U, V)$ *defined by* π is the locus of points $\{l \cdot l^\pi : l \in U^*\}$. For any point X on $\mathbb{P}$, we will say that X *lies outside* κ, written $X \notin \kappa$, if $X \neq Y$ for all points Y on κ [26, Definition 8.1]. See Fig. 7.5.

Problem 7.23 This definition, with the assumption that the given projectivity is nonperspective, produces what is usually called a *non-singular* conic. Singular conics await constructive investigation.

Properties of Conics

The basic properties of conics are constructively valid on the plane $\mathbb{P}$; for example, *any three distinct points on a conic are noncollinear* [26, Proposition 8.2(b)]. Theorem 7.24 establishes an essential property of a conic. Although classically trivial, it requires constructivization; it is an analogue of the tightness property for inequalities, and can be viewed as an extension of Axiom C6: "If $\neg(P \notin l)$, then $P \in l$."

Theorem 7.24 *Let* $\kappa = \kappa(\pi; U, V)$ *be a conic on the projective plane* $\mathbb{P}$*. For any point* X *on* $\mathbb{P}$*, if* $\neg(X \notin \kappa)$*, then* $X \in \kappa$ [26, Proposition 8.2(d)].

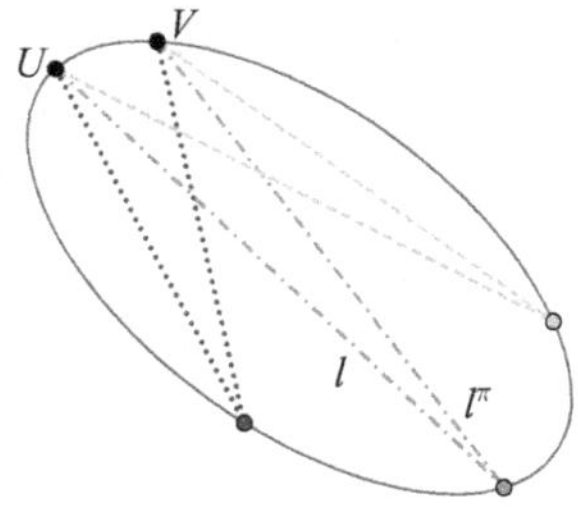

Figure 7.5 Construction of a conic.

Proof Let X be a point on the plane such that $\neg(X \notin \kappa)$. By cotransitivity and symmetry, we may assume that $X \neq U$. Set $z = UX$; then $Z = z \cdot z^\pi$ is a point of κ. Suppose that $X \neq Z$.

We now show that $X \neq Y$ for any point Y of κ. Either $Y \neq X$ or $Y \neq U$. We need to consider only the second case; set $y = UY$, it follows that $Y = y \cdot y^\pi$. Either $Y \neq X$ or $Y \neq Z$; again, it suffices to consider the second case. Since $Y \neq Z = z \cdot z^\pi$, it follows from Axiom C7 that either $Y \notin z$ or $Y \notin z^\pi$. In the first subcase, $y \neq z$. In the second subcase, $y^\pi \neq z^\pi$, and since π is a bijection we again have $y \neq z$. Since $X \neq U = y \cdot z$, it follows that $X \notin y$, and thus $X \neq Y$.

The preceding paragraph shows that $X \notin \kappa$, contradicting the hypothesis. It follows that $X = Z$, and hence $X \in \kappa$. □

Using Theorem 7.24 and other preliminary results, many well-known classical results are obtained constructively; for example, we have the following basic result.

Theorem 7.25 *There exists a unique conic containing any given five distinct points, each three of which are noncollinear* [26, Proposition 8.3].

Pascal's Theorem

Perhaps the most widely known classical result concerning conics is the following; it is due to Blaise Pascal (1623–1662) [31], and now has a constructive proof.

Theorem 7.26 Pascal's Theorem. *Let a simple hexagon $ABCDEF$ be inscribed in a conic κ. Then the three points of intersection of the pairs of opposite sides are distinct and collinear; see Fig. 7.6* [26, Theorem 9.2].

According to legend, Pascal gave in addition some 400 corollaries. Only one has been constructivized; it recalls a traditional construction method for drawing a conic "point by point" on paper; see [43, p. 68] and Fig. 7.7.

Corollary 7.27 *Let A, B, C, D, E be five distinct points of a conic κ. If l is a line through E that avoids each of the other four points, and l passes through a distinct sixth point F of κ, then*

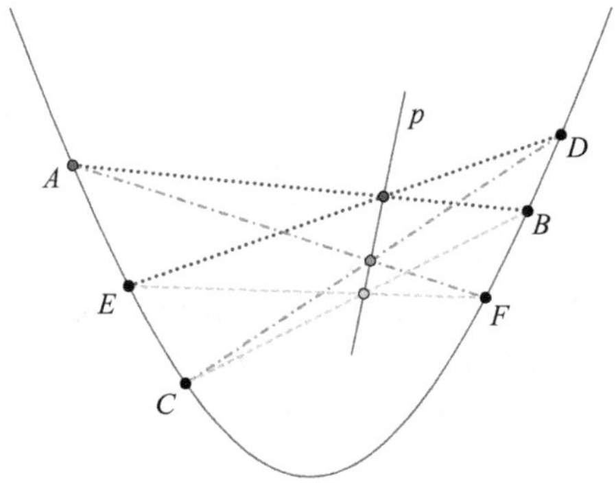

Figure 7.6 Pascal's Theorem.

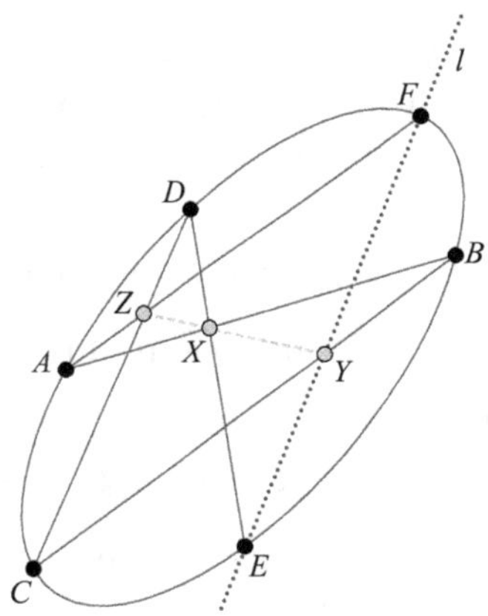

Figure 7.7 Drawing a conic.

$$F = l \cdot A(CD \cdot (AB \cdot DE)(BC \cdot l))$$

[26, Corollary 9.3].

Proof The *Pascal line* p of the hexagon $ABCDEF$ passes through the three distinct points $X = AB \cdot DE$, $Y = BC \cdot EF$, and $Z = CD \cdot AF$. Since $A \notin CD$, we have $A \neq Z$, and it follows that $AF = AZ$. Since $B \notin CD$, we have $BC \neq CD$, so by cotransitivity for lines either $p \neq BC$ or $p \neq CD$. In the first case, since $C \notin EF$, we have $C \neq Y = BC \cdot p$, and it follows from Axiom C7 that $C \notin p$. Thus in both cases we have $CD \neq p$, and $Z = CD \cdot p$. Finally,

$$\begin{aligned} F &= EF \cdot AF = l \cdot AZ = l \cdot A(CD \cdot p) \\ &= l \cdot A(CD \cdot XY) = l \cdot A(CD \cdot (AB \cdot DE)(BC \cdot l)). \end{aligned}$$ □

7.2.6 Polarity

The role of symmetry in projective geometry reaches a peak of elegance in the theory of polarity, introduced by von Staudt in 1847 [41].

A *correlation* is a mapping of the points of the projective plane to the lines, together with a mapping of the lines to the points, that preserves collinearity and concurrence. The correlation is *involutory* if it is of order 2, and is then called a *polarity.* A conic determines a polarity; each point of the plane has a corresponding polar, and each line has a corresponding pole.

Tangents and Secants

The construction of poles and polars determined by a conic is dependent upon the existence of tangents and secants. A line t that passes through a point P on a conic κ is said to be *tangent to* κ *at* P if P is the unique point of κ that lies on t. A line that passes through two distinct points of a conic κ is a *secant of* κ. For the construction of poles and polars, it has been necessary to adopt an additional axiom.

Axiom P The tangents at any three distinct points of a conic are nonconcurrent.

Problem 7.28 Determine whether this axiom can be derived from the others.

The tangents and secants to a conic are related by means of projectivities; the tangent at a point on a conic is the projective image of any secant through the point.

Theorem 7.29 *Let κ be a conic on the projective plane $\mathbb{P}$, let P be a point on κ, and let t be a line passing through P. The line t is tangent to κ at P if and only if for any point Q of κ with $Q \neq P$, if s is the secant QP, and π is the nonperspective projectivity such that $\kappa = \kappa(\pi; Q, P)$, then $t = s^{\pi}$* [26, Proposition 10.2(b)].

This theorem ensures the existence and uniqueness of tangents. To establish the existence of secants, it is first shown that a line through a point on a conic, if not the tangent, is a secant.

Lemma 7.30 *Let κ be a conic on the projective plane $\mathbb{P}$, let P be a point on κ, and let t be the tangent to κ at P. If l is a line passing through P, and $l \neq t$, then l passes through a second point R of κ, distinct from P; thus l is a secant of κ* [26, Lemma 10.9].

With this lemma, Theorem 7.31 will provide the secants required for the construction of polars. The need for this theorem contrasts with complex geometry, where every line meets every conic.

Theorem 7.31 *Let κ be a conic on the projective plane $\mathbb{P}$. Through any given point P of the plane, at least two distinct secants of κ can be constructed* [26, Theorem 10.10(a)].

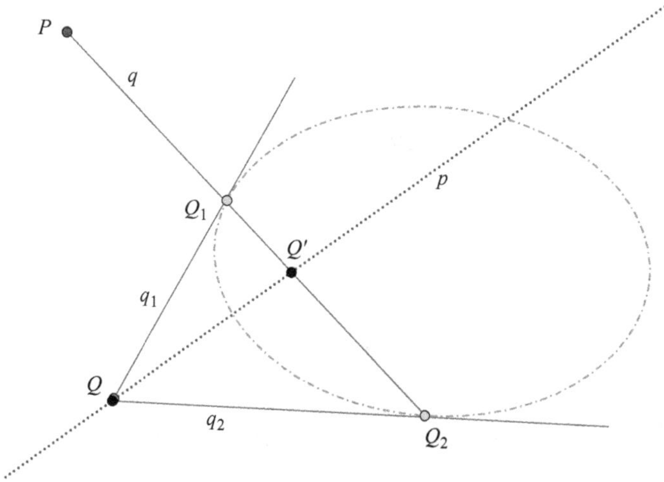

Figure 7.8 Construction of a polar.

Proof Select distinct points A, B, C on κ, with tangents a, b, c. By Axiom P, these tangents are nonconcurrent; thus the points $E = a \cdot b$ and $F = b \cdot c$ are distinct. Either $P \neq E$ or $P \neq F$; it suffices to consider the first case. By Axiom C7, either $P \notin a$ or $P \notin b$. It suffices to consider the first subcase; thus $P \neq A$ and $PA \neq a$. It follows from the lemma that PA is a secant.

Denote the second point of PA that lies on κ by R, and choose distinct points A', B', C' on κ, each distinct from both A and R. With these three points, construct a secant through P using the above method; we may assume that it is PA'. Since $A' \notin AR = PA$, it follows that $PA' \neq PA$. □

Construction of Polars and Poles

The traditional classical method for defining a polar is based on a quadrangle inscribed in the conic, and must consider separately points on or outside the conic. Constructively, this method is not available; polars on the constructive projective plane $\mathbb{P}$ are defined by means of harmonic conjugates.

The discussion of harmonic conjugates in Section 7.2.2 included an invariance theorem to show that the result of the construction is independent of the selection of auxiliary elements. Now, the definition of the polar of a point must be shown to be independent of the choice of an auxiliary secant; the proof requires the invariance theorem for harmonic conjugates (see Fig. 7.8).

Theorem 7.32 Construction of a polar. *Let κ be a conic on the projective plane $\mathbb{P}$, and let P be any point on the plane. Through the point P, construct a secant q of κ. Denote the intersections of q with κ by Q_1 and Q_2, and let the tangents at these*

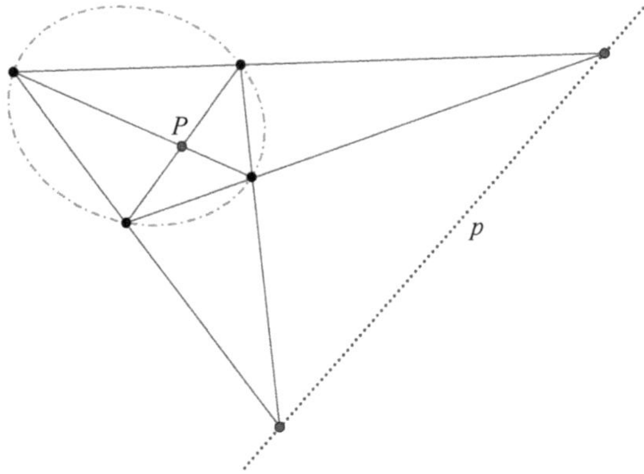

Figure 7.9 Quadrangle inscribed in a conic.

points be denoted q_1 *and* q_2. *Set* $Q = q_1 \cdot q_2$. *Set* $Q' = h(Q_1, Q_2; P)$, *the harmonic conjugate of* P *with respect to the base points* Q_1, Q_2. *Then the line* $p = QQ'$ *is independent of the choice of the secant* q [26, Theorem 11.1].

Definition 7.33 Let κ be a conic on the projective plane $\mathbb{P}$, and let P be any point on the plane. The line $p = QQ'$ in Theorem 7.32 is called the *polar of* P *with respect to* κ [26, Definition 11.2].

Note that if P lies on κ, then the polar of P produced by the construction is the tangent to κ at P. When P is a point outside κ, Corollary 7.34 will relate the construction of Theorem 7.32 to the classical construction of a polar that is based on quadrangles. The three *diagonal points* of a quadrangle are the intersection points of the three pairs of opposite sides; see Fig. 7.9. For this corollary we will require:

Fano's Axiom *The diagonal points of any quadrangle are noncollinear.*

Gino Fano (1871–1952) studied finite projective planes [16], some of which do not satisfy Fano's Axiom.

Corollary 7.34 *Let* κ *be a conic on the projective plane* $\mathbb{P}$, *and let* P *be any point outside* κ. *Inscribe a quadrangle in* κ *with* P *as one diagonal point. Then the line* p *joining the other two diagonal points is the polar of* P [26, Corollary 11.4].

Since any three distinct points on a conic are noncollinear [26, Proposition 8.2(b)], if a point P is a diagonal point of a quadrangle inscribed in a conic κ, it follows that $\neg(P \in \kappa)$. However, it does not immediately follow that P lies outside κ; thus we have the following.

Problem 7.35 If κ is a conic on the projective plane $\mathbb{P}$, and P is a diagonal point of a quadrangle inscribed in κ, show that $P \notin \kappa$.

Definition 7.36 Let κ be a conic on the projective plane $\mathbb{P}$, and let l be any line on $\mathbb{P}$. A construction analogous to that of the above theorem results in a point L, called the *pole of l with respect to κ* [26, Definition 11.5].

The following theorem shows that any conic on the plane $\mathbb{P}$ determines a polarity.

Theorem 7.37 *Let κ be a conic on the projective plane $\mathbb{P}$. If the line p is the polar of the point P, then the point P is the pole of the line p, and conversely* [26, Theorem 11.6(a)].

The definition of conic in Section 7.2.5 used the Steiner method [34], based on a projectivity. Later, von Staudt [41] defined a conic by means of a polarity: a point lies on the conic if its polar passes through the point. Classically, the two definitions produce the same conics; thus we have the following.

Problem 7.38 Construct correlations and polarities based on the axioms for the projective plane $\mathbb{P}$, develop the theory of conics constructively using the von Staudt definition, and prove that von Staudt conics are equivalent to the Steiner conics constructed in Section 7.2.5 above.

7.2.7 Consistency of the Axiom System

The consistency of the axiom system for the synthetic projective plane $\mathbb{P}$ is established by an analytic model. A projective plane $\mathbb{P}^2(\mathbb{R})$ is built from subspaces of the linear space $\mathbb{R}^3$, using only constructive properties of the real numbers. The axioms adopted for the synthetic plane $\mathbb{P}$ have been chosen to reflect the properties of the analytic plane $\mathbb{P}^2(\mathbb{R})$, taking note of Bishop's thesis: "All mathematics should have numerical meaning" [4, p. ix] [7, p. 3].

The model is built following a well-known classical method, adding constructive refinements to the definitions and proofs. The analytic plane $\mathbb{P}^2(\mathbb{R})$ consists of a family $\mathscr{P}_2$ of points and a family $\mathscr{L}_2$ of lines; a *point* P in $\mathscr{P}_2$ is a subspace of dimension 1 of the linear space $\mathbb{R}^3$, while a *line* λ in $\mathscr{L}_2$ is a subspace of dimension 2. The inequality relations, the incidence relation, and the outside relation are defined by means of vector operations. All the essential axioms adopted for the synthetic plane $\mathbb{P}$, and all the required properties, such as cotransitivity, tightness, and duality, are verified.

Theorem 7.39 *Axiom Group C, Fano's Axiom, and Axioms D, E, T, are valid on the analytic projective plane $\mathbb{P}^2(\mathbb{R})$* [26, Theorem 14.2].

Axiom C3 specifies the existence of a common point for any two distinct lines. On the plane $\mathbb{P}^2(\mathbb{R})$, this property is dependent on the restriction to distinct lines, as the Brouwerian counterexample below will show. By duality, the two statements of the example are equivalent. The proof of the first statement is easier to visualize, and can be described informally as follows: On $\mathbb{R}^2$, thought of as a portion of $\mathbb{P}^2(\mathbb{R})$, consider two points which are extremely near or at the origin, with P on the x-axis, and Q on the y-axis. If P is very slightly off the origin, and Q is at the origin, then the x-axis is the required line. In the opposite situation, the y-axis would be required. In any conceivable constructive routine, such a large change in the output, resulting from a minuscule variation of the input, would reveal a severe discontinuity, and a strong indication that the statement in question is constructively invalid.

Example 7.40 On the analytic projective plane $\mathbb{P}^2(\mathbb{R})$, the following statements are constructively invalid.

(i) *Given any points P and Q, there exists a line that passes through both points.*

(ii) *Given any lines λ and μ, there exists a point that lies on both lines* [26, Example 14.1].

Proof By duality, it will suffice to consider the second statement. When the nonzero vector $t = (t_1, t_2, t_3)$ spans the point T in $\mathbb{P}^2(\mathbb{R})$, we write $T = \langle t \rangle = \langle t_1, t_2, t_3 \rangle$. When the vectors u, v span the line λ, and $w = u \times v$, we write $\lambda = [w] = [w_1, w_2, w_3]$. The incidence relation $T \in \lambda$ is defined by the inner product, $t \cdot w = 0$.

Let α be any real number, and set $\alpha^+ = \max\{\alpha, 0\}$, and $\alpha^- = \max\{-\alpha, 0\}$. Define lines $\lambda = [\alpha^+, 0, 1]$ and $\mu = [0, \alpha^-, 1]$. By hypothesis, we have a point $T = \langle t \rangle = \langle t_1, t_2, t_3 \rangle$ that lies on both lines. Thus $\alpha^+ t_1 + t_3 = 0$, and $\alpha^- t_2 + t_3 = 0$. If $t_3 \neq 0$, then we have both $\alpha^+ \neq 0$ and $\alpha^- \neq 0$, an absurdity; thus $t_3 = 0$. This leaves two cases. If $t_1 \neq 0$, then $\alpha^+ = 0$, so $\alpha \leq 0$, while if $t_2 \neq 0$, then $\alpha^- = 0$, so $\alpha \geq 0$. Hence LLPO results. □

Problem 7.41 Develop the theory of conics for the analytic plane $\mathbb{P}^2(\mathbb{R})$; compare the results with those for the synthetic plane $\mathbb{P}$. On the plane $\mathbb{P}^2(\mathbb{R})$, determine the constructive validity of Axiom P of Section 7.2.6.

Problem 7.42 For the analytic projective plane $\mathbb{P}^2(\mathbb{R})$, apply constructive methods to the study of harmonic conjugates, cross ratios, and other topics of classical projective geometry.

Problem 7.43 Although the analytic model $\mathbb{P}^2(\mathbb{R})$ establishes the consistency of the axiom system used for the synthetic plane $\mathbb{P}$, it remains for us to prove the independence of that axiom system, or to reduce it to an independent system.

7.3 Projective Extensions

The notion of infinity has mystified finite humans for millennia. On the analytic projective plane $\mathbb{P}^2(\mathbb{R})$, where points and lines are merely lines and planes through the origin in $\mathbb{R}^3$, it is no surprise to notice that any two distinct lines meet at a unique point. However, to envision two parallel lines on $\mathbb{R}^2$ meeting at infinity requires some imagination. Johannes Kepler (1571–1630) invented the term "focus" in regard to ellipses, and stated that a parabola also has two foci, with one at infinity. This idea was extended by Poncelet, leading to the concepts of a line at infinity, and a projective plane.

In the classical theory, a projective extension of an affine plane is a fairly simple matter: Each pencil of parallel lines determines a point at infinity, at which the lines meet, and these points form the line at infinity. A projective plane results, and the required projective axioms are satisfied. The extension of the metric plane $\mathbb{R}^2$ to a projective plane can be described heuristically, with lamps and shadows; see [12, Section 1.3].

There have been at least three constructive attempts to extend an affine plane to a projective plane. An extension by A. Heyting [18] uses elements called "projective points" and "projective lines." The extension constructed in [25] uses elements called "prime pencils" and "virtual lines," resulting in a projective plane with different properties. The analytic projective plane $\mathbb{P}^2(\mathbb{R})$ discussed in Section 7.2.7, constructed using subspaces of $\mathbb{R}^3$, can be viewed as an extension of the metric plane $\mathbb{R}^2$; it also has distinctive properties.

The differences between these several extensions involve the crucial axiom concerning the existence of a point common to two lines, and the cotransitivity property. The statement that any two distinct lines have a common point is called the *Common Point Property (CPP),* while the *Strong Common Point Property (SCPP)* is the same statement without the restriction to distinct lines. The analytic extension $\mathbb{P}^2(\mathbb{R})$ of $\mathbb{R}^2$ satisfies both the CPP and cotransitivity, but *not* the SCPP. Neither synthetic extension satisfies both cotransitivity and the CPP. The Heyting extension satisfies cotransitivity, but the essential axiom CPP has not been verified. On the virtual line extension, the CPP is satisfied, and even the SCPP; however, cotransitivity is constructively invalid, and this is now seen as a serious limitation. The analytic extension $\mathbb{P}^2(\mathbb{R})$ could be taken as a standard; one might demand that the basic properties of $\mathbb{P}^2(\mathbb{R})$ hold in any acceptable synthetic extension, and then neither of the two synthetic extensions discussed here would suffice.

Problem 7.44 Construct a synthetic projective extension of an affine plane which has the usual properties of a projective plane, including both the common point property and cotransitivity.

7.3.1 Heyting Extension

In [18], Heyting adopts axioms for both affine and projective geometry. Then, from a plane affine geometry $(\mathscr{P}, \mathscr{L})$, Heyting constructs an extension (Π, Λ), consisting of *projective points* of the form

$$\mathfrak{P}(l, m) = \{n \in \mathscr{L} : n \cap l = l \cap m \text{ or } n \cap m = l \cap m\},$$

where $l, m \in \mathscr{L}$ with $l \neq m$, and *projective lines* of the form

$$\lambda(\mathfrak{A}, \mathfrak{B}) = \{\mathfrak{Q} \in \Pi : \mathfrak{Q} \cap \mathfrak{A} = \mathfrak{A} \cap \mathfrak{B} \text{ or } \mathfrak{Q} \cap \mathfrak{B} = \mathfrak{A} \cap \mathfrak{B}\},$$

where $\mathfrak{A}, \mathfrak{B} \in \Pi$ with $\mathfrak{A} \neq \mathfrak{B}$.

Under the Heyting definition of projective point, if the original two lines l and m intersect, then $\mathfrak{P}(l, m)$ is the pencil of all lines passing through the point of intersection, while if the lines are parallel, then $\mathfrak{P}(l, m)$ is the pencil of all lines parallel to the original two. In these cases, the definition determines either a finite point of the extension, or a point on the line at infinity. More significant is the fact that even when the status of the two original lines is not known constructively, still a projective point is (potentially) determined. Heyting comments on the need for this provision as follows:

> . . . serious difficulties . . . are caused by the fact that not only points at infinity must be adjoined to the affine plane, but also points for which it is unknown whether they are at infinity or not. [18, p. 161]

A projective line is determined by two distinct projective points. The definition is based on the lines common to the two projective points; namely, the lines common to two pencils of lines. In the simplest case, when the two projective points are finite, these are the pencils of lines through distinct points in the original affine plane, and there is a single common line, connecting these finite points, of which the projective line is an extension. In the case of two distinct pencils of parallel lines; the pencils have no common line, each determines a point at infinity, and the resulting projective line is the line at infinity. Again, even when the status of the original projective points is not known constructively, still a projective line is determined. The distinctive, and perhaps limiting, features of the Heyting extension are the requirements that the construction of a projective point depends on a given pair of distinct finite lines, and the construction of a projective line depends on a pair of distinct projective points previously constructed.

Nearly all the axioms for a projective plane are then verified, although the most essential axiom, which states that two distinct lines have a common point, escapes proof. The axiom considered in [18] is the weaker version, designated above as the common point property, CPP, involving distinct lines. In [24], Heyting's axioms for affine geometry are verified for $\mathbb{R}^2$, and a Brouwerian counterexample is given for the Heyting extension, showing that the stronger form of the axiom, SCPP, involving arbitrary lines, is constructively invalid, with the following attempted justification.

> This counterexample concerns the full common point axiom, rather than the limited Axiom P3 as stated in [18], where only distinct lines are considered. An investigation into the full axiom is necessary for a constructive study based upon numerical meaning, as proposed by Bishop. Questions of distinctness are at the core of constructive problems; any attempted projective extension of the real plane is certain to contain innumerable pairs of lines which may or may not be distinct. [24, p. 113]

However, taking note of the analytic model $\mathbb{P}^2(\mathbb{R})$, for which CPP is verified, but SCPP is constructively invalid, CPP now appears as a reasonable goal for an extension. Thus we have:

Problem 7.45 Complete the study of the projective extension of [18]; verify Heyting's Axiom P3 (CPP), or construct a Brouwerian counterexample.

7.3.2 Virtual Line Extension

Any attempt to build a constructive projective extension of an affine plane encounters difficulties due to the indeterminate nature of arbitrary pencils of lines. Classically, a pencil of lines is either the family of lines passing through a given point, or a family of parallel lines. An example of a family of lines is easily formed from two lines which might be distinct, intersecting or parallel, or might be identical. To obtain the SCPP in a constructive projective extension, the corresponding pencil must include both these lines, so that it will determine a point of the extension common to both extended lines, whether distinct or not. Thus the definition of pencil must not depend upon a pair of lines previously known to be distinct.

In the projective extension of [25], the definition of pencil is further generalized; rather than depending upon specific finite lines, it involves the intrinsic properties of a family of lines. Included are pencils of unknown type, with nonspecific properties, and pencils for which no lines are known to have been previously constructed.

The definition of line in the extension is independent of the definition of point; it will depend directly upon a class of generalized lines in the finite plane, called virtual lines.

Pencils

The virtual line extension of [25] is based on an incidence plane $\mathscr{G} = (\mathscr{P}, \mathscr{L})$, consisting of a family $\mathscr{P}$ of points and a family $\mathscr{L}$ of lines, with constructive axioms, definitions, conventions, and results as delineated in [23].

Definition 7.46 Let $\mathscr{G} = (\mathscr{P}, \mathscr{L})$ be an incidence plane.

(i) For any point $Q \in \mathscr{P}$, define

$$Q^* = \{l \in \mathscr{L} : Q \in l\}.$$

(ii) For any line $l \in \mathscr{L}$, define

$$l^* = \{m \in \mathscr{L} : m \parallel l\}.$$

(iii) A family of lines of the form Q^*, or of the form l^*, is called a *regular pencil*.

(iv) A family of lines α is called a *pencil* if it contains no fewer than two lines, and satisfies the following condition: *If l and m are distinct lines in α with $l, m \in \rho$, where ρ is a regular pencil, then $\alpha \subset \rho$.*

(v) A pencil of the form Q^* is called a *point pencil*.

(vi) A pencil α with the property that $l \parallel m$, for any lines l and m in α, is called a *parallel pencil* [25, Definition 2.1].

In the extension, the point pencil Q^*, consisting of all lines through Q, will represent the original finite point Q. The pencil l^*, consisting of all lines parallel to the line l, will result in an infinite point. However, the extension also admits parallel pencils which need not arise from given lines, but which nevertheless result in points at infinity.

Virtual Lines

A problem that arises in the construction of a projective extension is the difficulty in determining the nature of an arbitrary line in the extension, by means of an object in the original plane. If a line λ on the extended plane contains a finite point, then the set λ_0, of all finite points on λ, is a line in the original plane. However, if λ is the line at infinity, then λ_0 is void. Since constructively it is in general not known which is the case, we adopt the following.

Definition 7.47 A set p of points in $\mathscr{P}$ is said to be a *virtual line* if it satisfies the following condition: *If p is inhabited, then p is a line* [25, Definition 3.1].

It is necessary to ensure the existence of pencils derived from virtual lines.

Theorem 7.48 *Given any virtual lines p and q, one can construct a pencil $\varphi(p, q)$ that contains each of the virtual lines p and q, if it is a line* [25, Theorem 3.4].

The notion of virtual line also helps in resolving a problem that arises in connection with pencils of lines. The family of lines common to two distinct pencils may consist of a single line (as in the case of two point pencils, or a point pencil and a regular parallel pencil), or it may be void (as in the case of two regular parallel pencils); constructively, it is in general unknown which alternative holds. The following definition provides a tool for dealing with this situation.

Definition 7.49 For any distinct pencils α and β, define

$$\alpha \sqcap \beta = \{Q \in \mathscr{P} : Q \in l \in \alpha \cap \beta \text{ for some line } l \in \mathscr{L}\}.$$

The set of points $\alpha \sqcap \beta$ is called the *core* of the pair of pencils α, β [25, Definition 3.2].

The core, as a set of finite points (which might be void), is a constructive substitute for a possible line that is common to two pencils.

Theorem 7.50 *For any distinct pencils α and β, the core $\alpha \sqcap \beta$ is a virtual line* [25, Lemma 3.3].

Definition

Points of the extension, called *e-points*, are defined using a selected class of pencils of lines, called *prime pencils;* the prime pencil α determines the e-point $\overline{\alpha}$. Lines in the extension are not formed from previously constructed e-points; they are direct extensions of objects in the original plane. Lines of the extension, called *e-lines*, are defined using a selected class of virtual lines, called *prime virtual lines;* the prime virtual line p in the finite plane determines the e-line λ_p in the extended plane.

The projective plane $\mathscr{G}^* = (\mathscr{P}^*, \mathscr{L}^*)$, where $\mathscr{P}^*$ is the family of e-points, and $\mathscr{L}^*$ is the family of e-lines, is the *virtual line projective extension of the incidence plane* $\mathscr{G} = (\mathscr{P}, \mathscr{L})$. The axioms of projective geometry are verified for the extension. The following theorems are the main results; the proof outlines will exhibit the symmetry of the construction, and the utility of adopting independent definitions for e-points and e-lines.

Theorem 7.51 *On the projective extension $\mathscr{G}^*$ of the plane $\mathscr{G}$, there exists a unique e-line passing through any two distinct e-points* [25, Theorem 5.3].

Proof outline The given e-points $\overline{\alpha}$ and $\overline{\beta}$ originate from pencils α and β on the finite plane; the core $p = \alpha \sqcap \beta$ of these pencils is a virtual line on the finite plane. This virtual line p determines the e-line λ_p in the extension, which passes through both e-points $\overline{\alpha}$ and $\overline{\beta}$.

Theorem 7.52 *On the projective extension $\mathscr{G}^*$ of the plane $\mathscr{G}$, any two e-lines have an e-point in common. If the e-lines are distinct, then the common e-point is unique* [25, Theorem 5.5].

Proof outline The given e-lines λ_p and λ_q originate from virtual lines p and q on the finite plane; these virtual lines determine a pencil of lines $\gamma = \varphi(p, q)$ on the finite plane. This pencil γ determines the e-point $\overline{\gamma}$ in the extension, which lies on both e-lines λ_p and λ_q.

Several definitions in [25] involve negativistic concepts, including Definition 2.1 for pencil, and Definition 3.1 for distinct virtual lines. Can this be avoided? Generally in constructive mathematics one tries to avoid negativistic concepts, but perhaps some are unavoidable in constructive geometry; thus we have the following.

Problem 7.53 Modify the virtual line extension so as to avoid negativistic concepts as far as possible.

The Cotransitivity Problem

There is what might be called an irregularity of the extension plane $\mathscr{G}^*$, the constructive invalidity of cotransitivity; this is revealed by a Brouwerian counterexample.

Example 7.54 On the virtual line projective extension of the plane $\mathbb{R}^2$, the cotransitivity property for e-points is constructively invalid [25, p. 705].

Proof Given any real number c, construct the virtual line

$$p = \{(t, 0) : t \in \mathbb{R} \text{ and } c = 0\} \cup \{(0, t) : t \in \mathbb{R} \text{ and } c \neq 0\}$$

and consider the e-point $\overline{\gamma}$ determined by the pencil $\gamma = \varphi(p, p)$.

Let the x-axis be denoted by l_0; the pencil l_0^* of horizontal lines then determines the e-point $\overline{l_0^*}$. Similarly, we have the y-axis m_0, the pencil m_0^* of vertical lines, and the e-point $\overline{m_0^*}$. By hypothesis, $\overline{\gamma} \neq \overline{l_0^*}$ or $\overline{\gamma} \neq \overline{m_0^*}$. In the first case, suppose that $c = 0$. Then p is the x-axis and $\overline{\gamma} = \overline{l_0^*}$, a contradiction; thus we have $\neg(c = 0)$. In the second case, we find that $c = 0$. Hence WLPO results. □

Problem 7.55 Modify the virtual line extension, so that CPP and cotransitivity are both valid. It is then likely that SCPP will be constructively invalid; in that case, provide a Brouwerian counterexample.

7.3.3 Analytic Extension

The analytic projective plane $\mathbb{P}^2(\mathbb{R})$ of Section 7.2.7, formed from subspaces of $\mathbb{R}^3$, is based on constructive properties of the real numbers; it can be viewed as an extension of the affine plane $\mathbb{R}^2$.

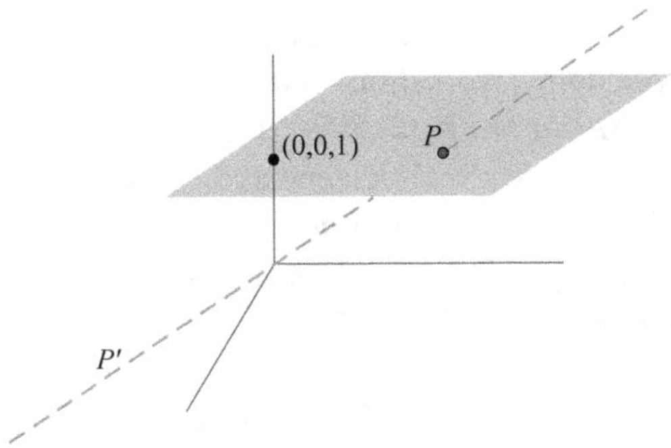

Figure 7.10 Analytic projective extension of the real plane.

The plane $z = 1$ in $\mathbb{R}^3$ is viewed as a copy of $\mathbb{R}^2$. A point P on the plane $z = 1$ corresponds to the point P' of the extension $\mathbb{P}^2(\mathbb{R})$ that, as a line through the origin in $\mathbb{R}^3$, contains P (see Fig. 7.10). Every horizontal line through the origin of $\mathbb{R}^3$ is an infinite point of the extension $\mathbb{P}^2(\mathbb{R})$. A line l on the plane $z = 1$ corresponds to the line l' of $\mathbb{P}^2(\mathbb{R})$ that, as a plane through the origin in $\mathbb{R}^3$, contains l. The line of intersection of this plane with the xy-plane is the point at infinity on l'. In this way, $\mathbb{P}^2(\mathbb{R})$ is seen as a projective extension of $\mathbb{R}^2$, with the xy-plane as the line at infinity.

The plane $\mathbb{P}^2(\mathbb{R})$ satisfies both the common point property and cotransitivity. However, as a projective extension of the specific plane $\mathbb{R}^2$, it does not provide an extension of an arbitrary affine plane; thus we have the following.

Problem 7.56 Construct a synthetic projective extension of an arbitrary affine plane, which has both the common point property and the cotransitivity property.

References

[1] Beeson, M. 2010. Constructive geometry. Pages 19–84 of: Arai, T., et al. (eds.), *Proceedings of the 10th Asian Logic Conference*. Singapore: World Scientific.

[2] Beeson, M. 2016. Constructive geometry and the parallel postulate. *Bull. Symb. Logic*, **22**, 1–104.

[3] Bishop, E. 1973. Schizophrenia in contemporary mathematics. *AMS Colloquium Lectures, Missoula, Montana, 1973*. Reprint: *Contemp. Math.*, **39**, 1–32 (1985).

[4] Bishop, E. 1967. *Foundations of Constructive Analysis*. New York: McGraw-Hill.

[5] Bishop, E. 1975. The crisis in contemporary mathematics. In: Proceedings of the American Academy Workshop on the Evolution of Modern Mathematics, Boston, 1974. *Historia Math.*, **2**, 507–517.

[6] Bishop, E. 1965. Review; S. C. Kleene and R. E. Vesley, The Foundations of Intuitionistic Mathematics. *Bull. Amer. Math. Soc.*, **71**(6), 850–852.

[7] Bishop, E., and Bridges, D. 1985. *Constructive Analysis*. Berlin: Springer.

[8] Bridges, D., and Richman, F. 1987. *Varieties of Constructive Mathematics*. Cambridge: Cambridge University Press.

[9] Bridges, D., and Vîţă, L. 2007. *Techniques of Constructive Analysis*. Universitext. New York: Springer.

[10] Brouwer, L. E. J. 1974. Intuitionistische Zerlegung mathematischer Grundbegriffe, 1924. Pages 275–280 of: *L. E. J. Brouwer Collected Works, Vol. 1*. Amsterdam, New York: North-Holland/American Elsevier.

[11] Brouwer, L. E. J. 1908. De onbetrouwbaarheid der logische principes. *Tijdschrift voor Wijsbegeerte*, **2**, 152–158. Translation: The unreliability of the logical principles, A. Heyting (ed.), *L. E. J. Brouwer Collected Works, Vol. 1*, pp. 107–111. Amsterdam, New York: North-Holland/American Elsevier (1974).

[12] Coxeter, H. S. M. 1955. *The Real Projective Plane*, 2nd ed. Cambridge: Cambridge University Press. Reprint: Lowe and Brydone, London, 1960.

[13] Cremona, L. 1873. *Elementi di Geometria Projettiva*. GB Paravia e comp., Torino. Translation by C. Leudesdorf: *Elements of Projective Geometry*. Oxford: Clarendon Press (1985). Reprint: *Elements of Projective Geometry*. Forgotten Books, Hong Kong (2012).

[14] de La Hire, P. 1685. *Sectiones Conicae in IX Libros Distributae. Cum Appendice De Sectionibus Conicis Omnium Generum Eadem et Universali Methodo*. Seph. Michallet.

[15] Desargues, G. 1864. *Oeuvres de Desargues: réunies et analysées*. Leiber. Reprint: *Oeuvres de Desargues, vol. I, vol. II*. Cambridge: Cambridge University Press (2011).

[16] Fano, G. 1892. Sui postulati fondamentali della geometria proiettiva. *Giornale Matemat.*, **30**, 106–132.

[17] Heyting, A. 1928. Zur intuitionistischen Axiomatik der projektiven Geometrie. *Math. Ann.*, **98**(1), 491–538.

[18] Heyting, A. 1959. Axioms for intuitionistic plane affine geometry. Pages 160–173 of: *Studies in Logic and the Foundations of Mathematics*, vol. 27. Elsevier.

[19] Heyting, A. 1966. *Intuitionism: an Introduction*. Amsterdam: North-Holland.

[20] Klein, F. 1893. Vergleichende Betrachtungen über neuere geometrische Forschungen. *Math. Ann.*, **43**(1), 63–100. Originally published by A. Deichert, Erlangen, 1872. Translation by M. W. Haskell: A comparative review of researches in geometry. *Bull. New York Math. Soc.*, **2**, 215–249 (1893).

[21] Lehmer, D. 1917. *An Elementary Course in Synthetic Projective Geometry*. Boston, MA: Ginn.

[22] Lombard, M., and Vesley, R. 1998. A common axiom set for classical and intuitionistic plane geometry. *Ann. Pure Appl. Logic*, **95**, 229–255.

[23] Mandelkern, M. 2007. Constructive coördinatization of Desarguesian planes. *Beiträge zur Algebra und Geometrie*, **48**(2), 547–589.

[24] Mandelkern, M. 2013. The common point problem in constructive projective geometry. *Indag. Math. (N.S.)*, **24**(1), 111–114.

[25] Mandelkern, M. 2014. Constructive projective extension of an incidence plane. *Trans. Amer. Math. Soc.*, **366**(2), 691–706.

[26] Mandelkern, M. 2016. A constructive real projective plane. *J. Geometry*, **107**(1), 19–60.

[27] Mandelkern, M. 2019. Constructive harmonic conjugates. *Beiträge zur Algebra und Geometrie*, **60**(2), 391–398.

[28] Pambuccian, V. 1998. Zur konstruktiven Geometrie euklidischer Ebenen. *Abhandl. math. Seminar Universität Hamburg*, **68**, 7–16.

[29] Pambuccian, V. 2001. Constructive axiomatization of plane hyperbolic geometry. *Math. Logic Q.*, **47**(4), 475–488.

[30] Pambuccian, V. 2003. Constructive axiomatization of non-elliptic metric planes. *Bull. Pol. Acad. Sci. Math.*, **51**, 49–57.

[31] Pascal, B. 1639. *Essai pour les Coniques*. `https://fr.wikisource.org/wiki/Essay_pour_les_coniques`. Texte établi par Léon Brunschvicg et Pierre Boutroux, Hachette, 1923 (2e éd.) (pp. 252–260). Translation: `http://euclid.trentu.ca/math/sb/3820H/Fall-2016/Essay_on_Conics_Pascal.pdf`

[32] Pickert, G. 1975. *Projektive Ebenen, 2. Aufl.* Berlin, New York: Springer-Verlag.

[33] Poncelet, J. 1822. *Traité des propriétés projectives des figures*. Paris: Gauthier-Villars.

[34] Steiner, J. 1832. *Systematische entwicklung der Abhängigkeit geometrischer Gestalten von einander, mit Berücksichtigung der Arbeiten alter und neuer Geometer über Porismen, Projections-Methoden, Geometrie der Lage, Transversalen, Dualität, und Reciprocität.* Berlin: G. Fincke.

[35] van Dalen, D. 1963. Extension problems in intuitionistic plane projective geometry I, II. *Indag. Math.*, **25**, 349–383.

[36] van Dalen, D. 1996. 'Outside' as a primitive notion in constructive projective geometry. *Geom. Dedicata*, **60**, 107–111.

[37] Veblen, O., and Young, J. W. 1910. *Projective Geometry, Vol. 1*. Boston, MA: Ginn.

[38] von Plato, J. 1995. The axioms of constructive geometry. *Ann. Pure Appl. Logic*, **76**, 169–200.

[39] von Plato, J. 1998. A constructive theory of ordered affine geometry. *Indag. Math. (N.S.)*, **9**, 549–562.

[40] von Plato, J. 2010. Combinatorial analysis of proofs in projective and affine geometry. *Ann. Pure Appl. Logic*, **162**, 144–161.

[41] Von Staudt, K. G. C. 1847. *Geometrie der Lage*. Insingen: Bauer und Raspe.

[42] Weibel, C. 2007. Survey of non-Desarguesian planes. *Not. Amer. Math. Soc.*, **54**(10), 1294–1303.

[43] Young, J. W. 1930. *Projective Geometry*. Open Court, Chicago: Mathematical Association of America.

PART III

ANALYSIS

8

Elements of Constructive Analysis

Hajime Ishihara

8.1 Introduction

During more than 50 years since the publication of Bishop's monograph *Foundation of Constructive Analysis* [3] in 1967, which changed the landscape of constructive mathematics, several books on Bishop-style constructive mathematics have been published. Bishop and Bridges' *Constructive Analysis* (1985) [4] is based on Bishop's monograph, but contains much new material and is essentially a new book. Bridges and Richman's *Varieties of Constructive Mathematics* (1987) [6] is a concise introduction to the spirit and practice of constructive mathematics and the varieties. Bridges and Vîţă's *Techniques of Constructive Analysis* (2006) [7] covers much new material from activity on constructive analysis in the intervening years.

This chapter summarises elementary constructive analysis and is intended to help the reader understand the other chapters in the book. It is not intended to be a comprehensive introduction, but to be a key that opens the door to constructive analysis. The reader should consult the aforementioned references for a broader and more-detailed appreciation of the subject.

In Section 8.2, we give a construction of the set $\mathbb{R}$ of real numbers. Then we deal with the elementary order-theoretic and algebraic properties of real numbers, and we look at the Cauchy completeness of $\mathbb{R}$ and its consequences. In Section 8.3, we generalise the notions and consequences in the reals to metric spaces. We examine the notion of a totally bounded space, yielding many constructive consequences, in many of which the constructively fundamental property of locatedness of a subset plays an important role. We also look at locally compact spaces which generalise the euclidean spaces $\mathbb{R}^n$. In Section 8.4, we introduce the fundamental notions in normed linear spaces, such as Banach spaces, quotient spaces, linear mappings and their continuity, and the notion of a normable linear mapping. We show that the geometric properties of normed linear spaces and inner product spaces bring us some interesting consequences. The material of this chapter could be formalised in the constructive Zermelo–Fraenkel set theory **CZF** [1, 2] together with the axioms of countable and dependent choice.

8.2 Real Numbers

Throughout the chapter, we use the following (constructive) set theoretical notions: a set S is *inhabited* if there exists an element of S; a set S is *finitely enumerable* if there exist an n and a surjection $f : \{0, \ldots, n-1\} \to S$; sets S and T *overlap* if $S \cap T$ is inhabited, and we then write $S \between T$. Note that if a set S is inhabited, then S is non-empty, but the converse does not hold constructively.

Most structures in constructive analysis are given with setoids: a *setoid* is a set S equipped with an equivalence relation $=_S$ on S for the equality on S.

In this section, we start with constructing the set $\mathbb{R}$ of real numbers. Then we deal with the elementary order-theoretic and algebraic properties of real numbers, and we discuss the Cauchy completeness of $\mathbb{R}$ and its consequences.

8.2.1 Cauchy Reals

The set $\mathbb{Z}$ of *integers* is the set $\mathbb{N} \times \mathbb{N}$ with the equality

$$(n, m) =_{\mathbb{Z}} (n', m') \Leftrightarrow n + m' = n' + m.$$

The arithmetical relations and operations are defined on $\mathbb{Z}$ in a straightforward way; natural numbers are embedded into $\mathbb{Z}$ by the mapping $n \mapsto (n, 0)$. The set $\mathbb{Q}$ of *rationals* is the set $\mathbb{Z} \times \mathbb{N}$ with the equality

$$(a, m) =_{\mathbb{Q}} (b, n) \Leftrightarrow a \cdot (n+1) =_{\mathbb{Z}} b \cdot (m+1).$$

The arithmetical relations and operations are defined on $\mathbb{Q}$ in a straightforward way; integers are embedded into $\mathbb{Q}$ by the mapping $a \mapsto (a, 0)$.

Definition 8.1 A *real number* is a sequence $(p_n)_{n\in\mathbb{N}}$ of rationals such that

$$\forall m, n \left(|p_m - p_n| < 2^{-m} + 2^{-n} \right).$$

We shall write $\mathbb{R}$ for the set of real numbers as usual. Note that rationals are embedded into $\mathbb{R}$ by the mapping $p \mapsto p^* = (p, p, \ldots)$.

Definition 8.2 The *ordering relation* $<$ between real numbers $x = (p_n)_{n\in\mathbb{N}}$ and $y = (q_n)_{n\in\mathbb{N}}$ is defined by

$$x < y \Leftrightarrow \exists n \left(2^{-n+2} < q_n - p_n \right).$$

Proposition 8.3 *Let $x, y, z \in \mathbb{R}$. Then*

(i) $\neg(x < y \wedge y < x)$,
(ii) $x < y \Rightarrow x < z \vee z < y$.

Proof (i) Trivial.

(ii) Let $x = (p_n)_{n\in\mathrm{N}}$, $y = (q_n)_{n\in\mathrm{N}}$ and $z = (r_n)_{n\in\mathrm{N}}$, and suppose that $x < y$. Then there exists n such that $2^{-n+2} < q_n - p_n$. Let $N = n + 3$. Then either $(p_n + q_n)/2 < r_N$ or $r_N \leq (p_n + q_n)/2$; in the former case, we have

$$2^{-N+2} < 2^{-n+1} - (2^{-(n+3)} + 2^{-n}) < \frac{q_n - p_n}{2} - (p_N - p_n)$$
$$= \frac{p_n + q_n}{2} - p_N < r_N - p_N,$$

and hence $x < z$; in the latter case, we have

$$2^{-N+2} < -(2^{-(n+3)} + 2^{-n}) + 2^{-n+1} < (q_N - q_n) + \frac{q_n - p_n}{2}$$
$$= q_N - \frac{p_n + q_n}{2} \leq q_N - r_N,$$

and so $z < y$. □

Definition 8.4 We define the *apartness* $\#$, the *equality* $=_\mathbb{R}$, and the ordering relation $\leq$ between real numbers x and y by

(i) $x \# y \Leftrightarrow (x < y \vee y < x)$,
(ii) $x =_\mathbb{R} y \Leftrightarrow \neg(x \# y)$,
(iii) $x \leq y \Leftrightarrow \neg(y < x)$.

Lemma 8.5 *Let $x, y, z \in \mathbb{R}$. Then*

(i) $x \# y \Leftrightarrow y \# x$,
(ii) $x \# y \Rightarrow x \# z \vee z \# y$.

Proof Straightforward by Proposition 8.3. □

Proposition 8.6 *Let $x, y, z \in \mathbb{R}$. Then*

(i) $x =_\mathbb{R} x$,
(ii) $x =_\mathbb{R} y \Rightarrow y =_\mathbb{R} x$,
(iii) $x =_\mathbb{R} y \wedge y =_\mathbb{R} z \Rightarrow x =_\mathbb{R} z$.

Proof Straightforward by Lemma 8.5. □

In the following, we omit the subscript from the equality $=_\mathbb{R}$ between reals.

Definition 8.7 The *canonical bound* K_x of a real number $x = (p_n)_{n\in\mathbb{N}}$ is the least natural number k such that $|p_0| + 2 < 2^k$; clearly, $|p_n| < 2^{K_x}$ for all n.

The following proposition shows how to define the arithmetical operations on $\mathbb{R}$ in terms of operations on sequences of rational numbers.

Proposition 8.8 *For real numbers $x = (p_n)_{n\in\mathbb{N}}$ and $y = (q_n)_{n\in\mathbb{N}}$, define*

(i) $x + y = (p_{n+1} + q_{n+1})_{n\in\mathbb{N}}$;
(ii) $-x = (-p_n)_{n\in\mathbb{N}}$;
(iii) $xy = (p_{n+k+1}q_{n+k+1})_{n\in\mathbb{N}}$, *where* $k = \max\{K_x, K_y\}$;
(iv) $\max\{x, y\} = (\max\{p_n, q_n\})_{n\in\mathbb{N}}$.

Then $x + y$, $-x$, xy and $\max\{x, y\}$ are real numbers. Moreover, $(xy) \mapsto x + y$, $x \mapsto -x$, $(x, y) \mapsto xy$, $(x, y) \mapsto \max\{x, y\}$, are mappings: that is, $x+y = x'+y'$ whenever $x = x'$ and $y = y'$, and so on.

Proof Straightforward; see [3, Chapter 2, Proposition 2], [4, Chapter 2, (2.4) and (2.5)], [6, Chapter 1, (5.1)], and [10, Chapter 5] for details. □

Proposition 8.9 *For a real number $x = (p_n)_{n\in\mathbb{N}}$ with $x \# 0$, choose N such that $|p_n| \geq 2^{-N}$ for all $n \geq N$, and define*

$$x^{-1} = \left(\frac{1}{p_{2N+\max\{n,N\}}}\right)_{n\in\mathbb{N}}.$$

Then x^{-1} is the unique real number z such that and $xz = 1$. Moreover, $x \mapsto x^{-1}$ is a mapping on $\{x \in \mathbb{R} \mid x \# 0\}$.

Proof See [3, Chapter 2, Proposition 6] and [4, Chapter 2, (2.13)]. □

Proposition 8.10 *Let $x_0, \ldots, x_{n-1}$ be real numbers. If $0 < \sum_{k<n} x_k$, then $0 < x_k$ for some $k < n$.*

Proof We proceed by induction on n. It is trivial for $n = 0$. If $0 < \sum_{k<n+1} x_k$, then either $0 < \sum_{k<n} x_k$ or $\sum_{k<n} x_k < \sum_{k<n+1} x_k$, by Proposition 8.3 (ii); in the former case, $0 < x_k$ for some $k < n$, by the induction hypothesis; in the latter case, we have $0 < x_n$. □

8.2.2 Completeness

Definition 8.11 A sequence $(x_n)_{n\in\mathbb{N}}$ of real numbers

- is a *Cauchy sequence* if for each $\epsilon > 0$ there exists N such that

$$\forall m, n \geq N(|x_m - x_n| \leq \epsilon);$$

- *converges to a limit* $x \in \mathbb{R}$ if for each $\epsilon > 0$ there exists N such that

$$\forall n \geq N(|x_n - x| \leq \epsilon),$$

and we then write either $x_n \to x$ as $n \to \infty$ or $x = \lim_{n\to\infty} x_n$.

Theorem 8.12 *A sequence of real numbers converges if and only if it is a Cauchy sequence.*

Proof See [3, Chapter 2, Theorem 2], [4, Chapter 2, (3.3)], and [6, Chapter 2, (5.1)]. □

Definition 8.13 Let S be an inhabited subset of $\mathbb{R}$. A real number z is an *upper bound* of S if $s \leq z$ for all $s \in S$; a *supremum* of S if it is an upper bound of S, and for each k there exists $s \in S$ such that $z < s + 2^{-k}$, and we then write $z = \sup S$. A *lower bound* and an *infimum* of S are defined similarly.

An application of the completeness of the real numbers, we have the following *constructive supremum principle*.

Theorem 8.14 *Let S be an inhabited subset of $\mathbb{R}$ with an upper bound. If for all $a, b \in \mathbb{R}$ with $a < b$, either there exists $s \in S$ with $a < s$ or else $s < b$ for all $s \in S$, then a supremum of S exists.*

Proof Let $s_0 \in S$, and let t_0 be an upper bound of S with $s_0 < t_0$. Define sequences $(s_n)_{n\in\mathbb{N}}$ and $(t_n)_{n\in\mathbb{N}}$ of real numbers by

$$s_{n+1} = \begin{cases} \frac{2s_n+t_n}{3} & \text{if } \exists s \in S\left[\frac{2s_n+t_n}{3} < s\right] \\ s_n & \text{if } \forall s \in S\left[s < \frac{s_n+2t_n}{3}\right] \end{cases},$$

and

$$t_{n+1} = \begin{cases} t_n & \text{if } \exists s \in S\left[\frac{2s_n+t_n}{3} < s\right] \\ \frac{s_n+2t_n}{3} & \text{if } \forall s \in S\left[s < \frac{s_n+2t_n}{3}\right] \end{cases}.$$

Note that for each n, $s_n < t_n$; there exists $s \in S$ with $s_n \leq s$; and $s \leq t_n$ for all $s \in S$. Then $(s_n)_{n\in\mathbb{N}}$ and $(t_n)_{n\in\mathbb{N}}$ converge to the same limit which is a supremum of S. □

Definition 8.15 A set S of real numbers is *totally bounded* if for each $\epsilon > 0$ there exists a finitely enumerable subset $\{s_0, \ldots, s_{n-1}\}$ of S such that

$$\forall x \in S \exists i < n(|x - s_i| \leq \epsilon).$$

Proposition 8.16 *An inhabited totally bounded set S of real numbers has a supremum and an infimum.*

Proof Note that S has an upper bound: there exists a finitely enumerable subset $\{t_0, \ldots, t_{m-1}\}$ of S such that $\forall x \in S \exists j < m(x - t_j \leq 1)$, and hence $\max\{t_j \mid j < m\} + 1$ is an upper bound of S. Consider $a, b \in \mathbb{R}$ with $a < b$. Then there exists a finitely enumerable subset $\{s_0, \ldots, s_{n-1}\}$ of S such that

$$\forall x \in S \exists i < n \left(|x - s_i| \leq \frac{b-a}{2} \right).$$

Either $a < \max\{s_i \mid i < n\}$ or $\max\{s_i \mid i < n\} < (a+b)/2$; in the first case, there exists $s \in S$ such that $a < s$; in the second case, for each $s \in S$ there exists i such that $|s - s_i| \leq (b-a)/2$, and hence

$$s < s_i + |s - s_i| < \frac{a+b}{2} + \frac{b-a}{2} = b.$$

Similarly, S has an infimum. □

Notes

In the literature, it is common to use the symbol $\neq$ instead of $\#$ for an inequality relation. However, we make explicit here the difference between $x \# y$ and $\neg(x = y)$, that is, $\neg\neg(x \# y)$. *Principles of omniscience* are principles which are derivable in classical logic, but underivable in intuitionistic logic and hence constructively unacceptable; most of them are instances of classical principles, such as the *law of excluded middle* LEM, *De Morgan's law* DML, and *double negation elimination* DNE. An example which shows that a theorem implies an omniscience principle is called a *Brouwerian counterexample to* the theorem. Note that it does not show that the theorem is *false*, but shows that the theorem is constructively underivable. For example, the statement

$$\forall x, y \in \mathbb{R}[x \# y \vee x = y]$$

implies (is even equivalent to) an omniscience principle which is an instance of the LEM and is called the *limited principle of omniscience* (LPO):

> for each sequence $(a_n)_{n \in \mathbb{N}}$ of natural numbers, either there exists $n \in \mathbb{N}$ with $a_n \neq 0$ or else $a_n = 0$ for all $n \in \mathbb{N}$;

the statement

$$\forall x, y \in \mathbb{R}[x \leq y \vee y \leq x]$$

is equivalent to an omniscience principle which is an instance of DML and is called the *lesser limited principle of omniscience* (LLPO):

> for each sequence $(a_n)_{n \in \mathbb{N}}$ of natural numbers with at most one nonzero term, either $a_{2n} = 0$ for all $n \in \mathbb{N}$ or else $a_{2n+1} = 0$ for all $n \in \mathbb{N}$;

the statement

$$\forall x, y \in \mathbb{R}[\neg(x = y) \Rightarrow x \# y]$$

is equivalent to an omniscience principle which is an instance of DNE and is called *Markov's principle* (MP):

- for each sequence $(a_n)_{n \in \mathbb{N}}$ of natural numbers, if it is impossible that $a_n = 0$ for all $n \in \mathbb{N}$, then there exists $n \in \mathbb{N}$ with $a_n \neq 0$;

see [9, 3.2] for omniscience principles, and [9, Proposition 3] for other counterexamples on the relations between real numbers.

A real number z is a *least upper bound* of an inhabited subset S of $\mathbb{R}$ if z is an upper bound and $z \leq y$ for each upper bound y of S. Classically, the notions of least upper bound and supremum coincide, and every subset with an upper bound has a least upper bound. However, the statement 'Every inhabited subset of $\mathbb{R}$ with a least upper bound has a supremum' implies DNE: for if $\neg\neg\varphi$, then the inhabited subset $S = \{0\} \cup \{1 \mid \varphi\}$ of $\mathbb{R}$ has a least upper bound 1, but it has a supremum if and only if φ holds; see also [9, Propositions 4, 5, and 6].

A proof of the Cauchy completeness without the axiom of countable choice is given in [10, 5.4.2]. A treatment of Dedekind reals and the difference between Cauchy reals and Dedekind reals can be found in [10, 5.5]. A construction of the real numbers using a form of interval arithmetic is given in [7, 2.1].

Calculus on the reals, including series, continuous functions, differentiation and integration, is developed in [3, Chapter 2] and [4, Chapter 2]; see also [10, Chapter 6].

8.3 Metric Spaces

In this section, we generalise the notions and consequences in the reals to metric spaces. In particular, we look at locally compact spaces, which generalise the euclidean spaces $\mathbb{R}^n$.

8.3.1 Fundamental Definitions

Definition 8.17 A *metric space* is a set X equipped with a *metric* $d : X \times X \to \mathbb{R}$ such that

(i) $d(x, y) = 0 \Leftrightarrow x = y$,
(ii) $d(x, y) = d(y, x)$,
(iii) $d(x, y) \leq d(x, z) + d(z, y)$,

for all $x, y, z \in X$. For $x, y \in X$ and $r > 0$, we write $x \# y$ for $0 < d(x, y)$, and $B_X(x, r)$ to denote the *open ball* with *centre* x and *radius* r, that is,

$$B_X(x, r) = \{y \in X \mid d(x, y) < r\}.$$

Definition 8.18 Let S be a subset of a metric space X. Then the *interior* S° of S is defined by

$$S^\circ = \{x \in S \mid \exists r > 0(B_X(x, r) \subseteq S)\};$$

the *closure* $\overline{S}$ of S is defined by

$$\overline{S} = \{x \in X \mid \forall r > 0(B_X(x, r) \between S)\}.$$

A subset S of a metric space X is *open* if $S = S^\circ$; *closed* if $S = \overline{S}$; *dense* in X if $\overline{S} = X$. A metric space is *separable* if there exists a countable dense subset.

Definition 8.19 A subset S of a metric space X is *located* if

$$d(x, S) = \inf\{d(x, s) \mid s \in S\}$$

exists for all $x \in X$.

Definition 8.20 A *metric subspace* of a metric space X with a metric d is a subset S of X taken with the restriction of d to $S \times S$.

8.3.2 Completeness

Definition 8.21 A sequence $(x_n)_{n \in \mathbb{N}}$ in a metric space X

- is a *Cauchy sequence* if for each $\epsilon > 0$ there exists N such that

$$\forall m, n \geq N(d(x_m, x_n) \leq \epsilon);$$

- *converges to a limit* $x \in X$ if for each $\epsilon > 0$ there exists N such that

$$\forall n \geq N(d(x_n, x) \leq \epsilon).$$

In the latter event, we write either $x_n \to x$ as $n \to \infty$ or $x = \lim_{n \to \infty} x_n$. A metric space is *complete* if every Cauchy sequence converges; a subset of a metric space is *complete* if it is complete as a metric subspace.

Definition 8.22 A sequence $x = (x_n)_{n \in \mathbb{N}}$ of a metric space X is *regular* if

$$\forall m, n \left(d(x_m, x_n) \leq 2^{-m} + 2^{-n}\right).$$

Proposition 8.23 *Let $\tilde{X}$ be the set of regular sequences of a metric space X, and define*

$$\tilde{d}(x, y) = \lim_{n \to \infty} d(x_n, y_n)$$

for each $x = (x_n)_{n \in \mathrm{N}}$ and $y = (y_n)_{n \in \mathrm{N}}$ in $\tilde{X}$. Let $\iota : X \to \tilde{X}$ be the inclusion map, *where $\iota(x)$ is the sequence $(x_n)_{n \in \mathrm{N}}$ with each $x_n = x$. Then $\tilde{X}$ with the equality relation*

$$x =_{\tilde{X}} y \Leftrightarrow \tilde{d}(x, y) = 0$$

is a complete metric space with the metric $\tilde{d}$, called the completion *of X. Moreover, $\tilde{d}(\iota(x), \iota(y)) = d(x, y)$ for all $x, y \in X$, and ι embeds X as a dense subset of $\tilde{X}$.*

Proof See [3, Chapter 4, Theorem 1 and Theorem 2] and [4, Chapter 4, (3.2) and (3.4)]. □

We have the following constructive version of the Baire theorem.

Theorem 8.24 *If $(U_n)_{n \in \mathrm{N}}$ is a sequence of open dense subsets of a complete metric space X, then $\bigcap_{n=0}^{\infty} U_n$ is dense in X.*

Proof See [3, Chapter 4, Theorem 4], [4, Chapter 4, (3.9)], and [6, Chapter 2, (1.3)]. □

Corollary 8.25 *If $(x_n)_{n \in \mathrm{N}}$ is a sequence of real numbers, then there exists $a \in \mathbb{R}$ such that $x_n \# a$ for each n.*

Proof Let $U_n = \{x \in \mathbb{R} \mid x \# x_n\}$ for each n. Then U_n is open and dense. Therefore there exists $a \in \bigcap_{n=0}^{\infty} U_n$. □

8.3.3 Total Boundedness and Compactness

Definition 8.26 A metric space X is *bounded* if there exists $M > 0$ such that $d(x, y) \leq M$ for all $x, y \in X$; a subset S of an inhabited metric space X is *bounded* if there exist $M > 0$ and $x \in X$ such that $d(x, s) \leq M$ for all $s \in S$.

Definition 8.27 An *ϵ-approximation* to a metric space X is a subset S of X such that

$$\forall x \in X \exists s \in S (d(x, s) \leq \epsilon).$$

A metric space X is *totally bounded* if for each $\epsilon > 0$ there exists a finitely enumerable ϵ-approximation to X, that is, there exists a finitely enumerable subset $\{x_0, \ldots, x_{n-1}\}$ of X such that

$$\forall x \in X \exists i < n (d(x, x_i) \leq \epsilon).$$

Note that every totally bounded metric space is separable and bounded. A metric space is *compact* if it is totally bounded and complete. A subset of a metric space is *totally bounded* (respectively, *compact*) if it is totally bounded (respectively, compact) as a metric subspace.

Proposition 8.28 *Every located subset S of a totally bounded metric space X is totally bounded.*

Proof Given an $\epsilon > 0$, let $\{x_0, \dots, x_{n-1}\}$ be an $\epsilon/3$-approximation to X, and choose $s_0, \dots, s_{n-1} \in S$ such that $d(x_i, s_i) < d(x_i, S) + \epsilon/3$ for each $i < n$. Then for each $y \in S$ there exists $i < n$ such that $d(x_i, y) < \epsilon/3$, and hence

$$\begin{aligned} d(s_i, y) &\leq d(s_i, x_i) + d(x_i, y) < d(x_i, S) + \frac{\epsilon}{3} + d(x_i, y) \\ &\leq d(x_i, y) + \frac{\epsilon}{3} + d(x_i, y) \leq \epsilon. \end{aligned}$$

Therefore $\{s_0, \dots, s_{n-1}\}$ is an ϵ-approximation to S. □

Definition 8.29 A mapping f between metric spaces X and Y is

- *uniformly continuous* if for each $\epsilon > 0$ there exists $\delta > 0$ such that

$$\forall x, y \in X[d(x, y) < \delta \Rightarrow d(f(x), f(y)) \leq \epsilon];$$

- (pointwise) *continuous* if for each $x \in X$ and each $\epsilon > 0$ there exists $\delta > 0$ such that

$$\forall y \in X[d(x, y) < \delta \Rightarrow d(f(x), f(y)) \leq \epsilon].$$

Note that, trivially, every uniformly continuous mapping is continuous.

Proposition 8.30 *Let D be a dense subset of a metric space X, and let $f : D \to Y$ be a uniformly continuous mapping of D into a complete metric space Y. Then there exists a uniformly continuous extension $g : X \to Y$ of f.*

Proof Since D is dense in X, for each $x \in X$ there exists a sequence $(x_n)_{n \in \mathbb{N}}$ in D converging to x. Since $(x_n)_{n \in \mathbb{N}}$ is a Cauchy sequence and f is uniformly continuous on D, the sequence $(f(x_n))_{n \in \mathbb{N}}$ is a Cauchy sequence in Y, whose limit we denote by $g(x)$. Let $(x_n)_{n \in \mathbb{N}}$ and $(x'_n)_{n \in \mathbb{N}}$ be sequences in D converging respectively to x and x' in X. Then for each $\epsilon > 0$ there exists $\delta > 0$ such that

$$d(x_n, x'_n) < \delta \Rightarrow d(f(x_n), f(x'_n)) \leq \epsilon$$

for each n. If $d(x, x') < \delta$, then $d(x_n, x'_n) < \delta$, and hence $d(f(x_n), f(x'_n)) \leq \epsilon$, for sufficiently large n; whence $d(g(x), g(x')) \leq \epsilon$. Therefore if $x = x'$, then $g(x) = g(x')$, so $g : X \to Y$ is uniformly continuous. □

Proposition 8.31 *If f is a uniformly continuous mapping of a totally bounded metric space X into a metric space, then*

$$f(X) = \{f(x) \mid x \in X\}$$

is totally bounded.

Proof For each $\epsilon > 0$ there exists $\delta > 0$ such that

$$\forall x, y \in X[d(x, y) < \delta \Rightarrow d(f(x), f(y)) \leq \epsilon],$$

and there exists a $\delta/2$-approximation $\{x_0, \ldots, x_{n-1}\}$ to X. Then for each $y \in X$ there exists $i < n$ such that $d(x_i, y) \leq \delta/2 < \delta$, and hence $d(f(x_i), f(y)) \leq \epsilon$. Therefore $\{f(x_0), \ldots, f(x_{n-1})\}$ is an ϵ-approximation to $f(X)$. □

Corollary 8.32 *If f is a uniformly continuous mapping of a totally bounded metric space X into $\mathbb{R}$, then the* supremum *of f,*

$$\sup f = \sup\{f(x) \mid x \in X\},$$

and the infimum *of f,*

$$\inf f = \inf\{f(x) \mid x \in X\},$$

exist.

Proof Since $\{f(x) \mid x \in X\}$ is a totally bounded subset of $\mathbb{R}$, by Proposition 8.31, the supremum and the infimum of f exist, by Proposition 8.16. □

Proposition 8.33 *A totally bounded subset S of a metric space X is located.*

Proof Let $x \in X$. Then the mapping $y \mapsto d(x, y)$ is a uniformly continuous mapping from S into $\mathbb{R}$, and hence $\inf\{d(x, y) \mid y \in S\}$ exists, by Corollary 8.32. □

8.3.4 Locally Compact Spaces

Definition 8.34 An inhabited metric space X is

- *locally totally bounded* if each bounded subset of X is contained in a totally bounded subset;
- *locally compact* if it is locally totally bounded and complete.

Note that a metric space X is locally compact if and only if each bounded subset of X is contained in a compact subset. A subset of a metric space X is *locally totally bounded* (respectively, *locally compact*) if it is locally totally bounded (respectively, locally compact) as a metric subspace.

Proposition 8.35 *A locally totally bounded subset S of a metric space X is located.*

Proof Let $x \in X$, and choose $s_0 \in S$. Then

$$B = \{s \in S \mid d(x, s) \leq d(x, s_0)\}$$

is a bounded subset of S, and hence is contained in a totally bounded subset K of S. Since K is located, by Proposition 8.33, it suffices to show that $d(x, K) \leq d(x, s)$ for all $s \in S$. If, for some $s \in S$, $d(x, s) < d(x, K)$, then either $d(x, s) < d(x, s_0)$ or $d(x, s_0) < d(x, K)$; in the former case, since $s \in B \subseteq K$, we have $d(x, s) < d(x, K) \leq d(x, s)$, a contradiction; in the latter case, since $s_0 \in B \subseteq K$, we have $d(x, s_0) < d(x, K) \leq d(x, s_0)$, a contradiction. Therefore $d(x, K) \leq d(x, s)$ for all $s \in S$, and so $d(x, S) = d(x, K)$. □

Lemma 8.36 *Let S be a located subset of a metric space X, and let T be a totally bounded subset of X. Then there exists a totally bounded subset K of X such that $S \cap T \subseteq K \subseteq S$.*

Proof See [6, Chapter 2, Lemma 4.10]. □

Proposition 8.37 *A located subset S of a locally totally bounded space X is locally totally bounded.*

Proof For each bounded subset B of S, since B is bounded in X, there exists a totally bounded subset T of X containing B, and there exists a totally bounded subset K of X such that $B \subseteq S \cap T \subseteq K \subseteq S$, by Lemma 8.36. Therefore S is locally totally bounded. □

Notes

The notion of a located set is peculiar to intuitionistic and constructive analysis, since every set in a metric space is automatically located in classical mathematics; the notion for compact metric spaces is introduced in Brouwer [8].

A detailed analysis of the Baire theorem from a constructive point of view is given in [6, Chapter 2.2]; applications of the theorem are discussed in [7, 6.6]

Bishop originally defined a mapping between metric spaces X and Y to be continuous if it is uniformly continuous on each compact subset of X, and later, if it is uniformly continuous near each compact image. The theory based on this notion is developed in [3, 4]. The advantage and the disadvantage of the Bishop continuity have been discussed; see, for example, [7, Chapter 2, Notes].

A Brouwerian counterexample to the statement

Every subset of a totally bounded metric space is totally bounded

can be found in [9, Proposition 9].

The one-point compactification of a locally compact space and the Tietze extension theorem are given in [3, Chapter 4, Theorem 9] and [4, Chapter 4, (6.8)]. Spaces of functions, including the Stone–Weierstraß theorem, are dealt in [4, Chapter 4.5].

8.4 Normed Linear Spaces

In this section, we look at the fundamental notions in the theory of normed linear spaces. In particular, we reveal some interesting consequences of the geometric properties of normed linear spaces.

8.4.1 Normed and Banach Spaces

For the remainder of this chapter, 'linear space' will mean 'real linear space'.

Definition 8.38 A *normed linear space* is a linear space E equipped with a *norm* $\|\cdot\| : E \to \mathbb{R}$ such that

(i) $\|x\| = 0 \Leftrightarrow x = 0$,
(ii) $\|ax\| = |a|\|x\|$,
(iii) $\|x + y\| \leq \|x\| + \|y\|$,

for all $x, y \in E$ and $a \in \mathbb{R}$. Note that a normed linear space E is a metric space with the metric

$$d(x, y) = \|x - y\|.$$

A *Banach space* is a normed linear space which is complete with respect to this metric. We write $B_E(r)$ to denote the ball $B_E(0, r)$ relative to this metric, and B_E to denote the *open unit ball* $B_E(1)$.

Proposition 8.39 *The completion $\tilde{E}$ of a normed linear space E as a metric space is a Banach space with the norm*

$$\|x\| = \lim_{n\to\infty} \|x_n\|$$

for each $x = (x_n)_{n\in\mathbb{N}} \in \tilde{E}$. The inclusion map $\iota : E \to \tilde{E}$ preserves norms and realises E as a dense subspace of $\tilde{E}$.

Proof See [3, Chapter 9, Proposition 4] and [4, Chapter 7, (1.12)]. □

Definition 8.40 A *subspace* M of a linear space E is a linear subset of E.

Proposition 8.41 *Let Y be a located subspace of a normed linear space E, and define*

$$\|x\|_{E/Y} = d(x, Y).$$

Then E with the equality relation

$$x =_{E/Y} y \Leftrightarrow \|x - y\|_{E/Y} = 0$$

is a normed linear space with the norm $\|\cdot\|_{E/Y}$, called the quotient space *of E by Y, and written E/Y. Moreover, if E is a Banach space, then E/Y is a Banach space.*

Proof It is straightforward to see that E/Y is a normed linear space with a norm $\|\cdot\|_{E/Y}$. Suppose that E is a Banach space, and $(x_n)_{n\in\mathbb{N}}$ is a Cauchy sequence in E/Y. Then there exists a subsequence $(x_{n_k})_{k\in\mathbb{N}}$ of $(x_n)_{n\in\mathbb{N}}$ such that $\|x_{n_k} - x_{n_{k+1}}\|_{E/Y} < 2^{-(k+1)}$, and we can then inductively choose a sequence $(y_k)_{k\in\mathbb{N}}$ of Y such that $\|(x_{n_k} - y_k) - (x_{n_{k+1}} - y_{k+1})\| < 2^{-k}$. Since E is complete, the Cauchy sequence $(x_{n_k} - y_k)_{k\in\mathbb{N}}$ converges to a limit $z \in E$, and we have

$$\|x_{n_k} - z\|_{E/Y} \le \|(x_{n_k} - y_k) - z\| \to 0$$

as $k \to \infty$. Therefore, since $(x_n)_{n\in\mathbb{N}}$ is a Cauchy sequence and has a convergent subsequence, $(x_n)_{n\in\mathbb{N}}$ must converge. □

8.4.2 Bounded and Normable Linear Mappings

Definition 8.42 A mapping T between linear spaces E and F is *linear* if

(i) $T(ax) = aTx$,
(ii) $T(x + y) = Tx + Ty$

for all $x, y \in E$ and $a \in \mathbb{R}$. A *linear functional* f on a linear space E is a linear mapping from E into $\mathbb{R}$. The *kernel* of a linear mapping T between linear spaces E and F is defined by

$$\ker(T) = \{x \in E \mid Tx = 0\}.$$

Definition 8.43 A linear mapping T between normed linear spaces E and F is

- *non-zero* if there exists $x \in E$ such that $Tx \# 0$;
- *bounded* if $T(B_E) = \{Tx \mid x \in B_E\}$ is bounded;
- *compact* if $T(B_E)$ is totally bounded;
- *open* if $T(B_E)$ is open.

Proposition 8.44 *Let T be a linear mapping between normed linear spaces E and F. Then the following are equivalent.*

(i) *T is continuous,*
(ii) *T is uniformly continuous,*
(iii) *T is bounded.*

Proof Suppose that T is continuous. Then there exists $\delta > 0$ such that

$$\forall x \in E[\|x\| < \delta \Rightarrow \|Tx\| \leq 1].$$

If $x \in B_E$, then $\|\delta x\| < \delta$, and hence $\|\delta Tx\| \leq 1$; whence $\|Tx\| \leq 1/\delta$. Therefore (i) implies (iii).

Suppose that T is bounded. Then there exists $M > 0$ such that $\|Tx\| \leq M$ for all $x \in B_E$. For each $\epsilon > 0$, if $\|x - y\| < \epsilon/M$, then $(M/\epsilon)(x - y) \in B_E$, and hence $\|(M/\epsilon)T(x - y)\| \leq M$; whence $\|Tx - Ty\| \leq \epsilon$. Therefore (iii) implies (ii).

It is clear that (ii) implies (i). □

Definition 8.45 A linear mapping T between normed linear spaces E and F is *normable* if

$$\|T\| = \sup\{\|Tx\| \mid x \in B_E\}$$

exists. Note that every compact mapping is normable.

Proposition 8.46 *A non-zero bounded linear functional f on a normed linear space E is normable if and only if its kernel* $\ker(f)$ *is located.*

Proof See [3, Chapter 9, Proposition 8], [4, Chapter 7, (1.10)], and [7, Proposition 2.3.6]. □

8.4.3 Uniformly Convex Spaces

Definition 8.47 A subset C of a linear space E is *convex* if

$$\lambda x + (1 - \lambda)y \in C$$

for all $x, y \in C$ and $\lambda \in [0, 1]$.

Definition 8.48 A normed linear space E is *uniformly convex* if for each $\epsilon > 0$ there exists $\delta > 0$ such that

$$\|x - y\| > r\epsilon \Rightarrow \left\|\frac{x + y}{2}\right\| \leq r(1 - \delta)$$

for all $x, y \in B_E(r)$ and $r > 0$.

Lemma 8.49 *Let C be a convex subset of a uniformly convex normed linear space E, and let x be an element of E such that*

$$d(x, C) = \inf\{\|x - z\| \mid z \in C\}$$

exists. Then there exists strongly at most one *element y of C such that $d(x, C) = \|x - y\|$, in the sense that for each $y, z \in C$, if $y \# z$, then $d(x, C) < \|x - y\|$ or $d(x, C) < \|x - z\|$.*

Proof Note that, since

$$0 < \|y - z\| \leq \|x - y\| + \|x - z\| \leq 2\max\{\|x - y\|, \|x - z\|\},$$

we have $0 < r = \max\{\|x - y\|, \|x - z\|\}$. Choose $\epsilon > 0$ such that $r\epsilon < \|y - z\|$. Then there exists $\delta > 0$ such that

$$d(x, C) \leq \left\| x - \frac{y + z}{2} \right\| \leq \left\| \frac{(x - y) + (x - z)}{2} \right\| \leq r(1 - \delta) < r,$$

and hence either $d(x, C) < \|x - y\|$ or $d(x, C) < \|x - z\|$. □

The following proposition shows that closest points in a closed, convex subset in a uniformly convex Banach space exist.

Proposition 8.50 *Let C be a closed, convex subset of a uniformly convex Banach space E, and let x be an element of E such that*

$$d(x, C) = \inf\{\|x - z\| \mid z \in C\}$$

exists. Then there exists a unique element y of C such that $d(x, C) = \|x - y\|$.

Proof Let $d = d(x, C)$, and let $(z_n)_{n \in \mathbb{N}}$ be a sequence in C such that

$$\|x - z_n\| \leq d + 2^{-n}.$$

Then for each $\epsilon > 0$, either $0 < d$ or $d < \epsilon/4$. In the former case, for $\epsilon' = \epsilon/(d+1)$, there exists $\delta > 0$ such that if $m \geq n$ and $(d + 2^{-n})\epsilon' < \|z_m - z_n\|$, then

$$d \leq \left\| x - \frac{z_m + z_n}{2} \right\| \leq (d + 2^{-n})(1 - \delta).$$

Hence for all sufficiently large n with $(d + 2^{-n})(1 - \delta) < d$, we have

$$\|z_m - z_n\| \leq (d + 2^{-n})\epsilon' \leq (d + 1)\epsilon' = \epsilon$$

for $m \geq n$. In the case $d < \epsilon/4$, for all sufficiently large n with $2^{-n} < \epsilon/4$, we have

$$\|z_m - z_n\| \leq \|x - z_m\| + \|x - z_n\| < \epsilon$$

for $m \geq n$. Therefore $(z_n)_{n \in \mathbb{N}}$ is a Cauchy sequence and so converges to a limit $y \in C$. Clearly $d(x, C) = \|x - y\|$. □

Proposition 8.51 *Let f be a non-zero normable linear functional on a uniformly convex Banach space E. Then there exists $x \in E$ such that $f(x) = \|f\|$ and $\|x\| = 1$.*

Proof See [4, Chapter 7, (3.23)] and [7, Proposition 2.3.7]. □

8.4.4 Hilbert Spaces

Definition 8.52 An *inner product space* is a linear space E equipped with an *inner product* $\langle \cdot, \cdot \rangle : E \times E \to \mathbb{R}$ such that

(i) $\langle x, x \rangle \geq 0$, and $\langle x, x \rangle = 0 \Leftrightarrow x = 0$,
(ii) $\langle x, y \rangle = \langle y, x \rangle$,
(iii) $\langle ax, y \rangle = a\langle x, y \rangle$,
(iv) $\langle x + y, z \rangle = \langle x, z \rangle + \langle y, z \rangle$

for all $x, y, z \in E$ and $a \in \mathbb{R}$. Note that an inner product space E is a normed linear space with the norm

$$\|x\| = \langle x, x \rangle^{1/2}.$$

A *Hilbert space* is an inner product space which is a Banach space.

Proposition 8.53 *Let E be an inner product space. Then for each $x, y \in E$, the Cauchy–Schwarz inequality*

$$|\langle x, y \rangle| \leq \|x\| \|y\|$$

and the parallelogram identity

$$\|x + y\|^2 + \|x - y\|^2 = 2\|x\|^2 + 2\|y\|^2$$

hold.

Proof See [3, Chapter 9, Theorem 5] and [4, Chapter 7, (8.4)]. □

Note that if E is an inner product space, then, by the parallelogram identity, we have

$$\left\| \frac{x + y}{2} \right\|^2 = \frac{\|x\|^2}{2} + \frac{\|y\|^2}{2} - \frac{\|x - y\|^2}{4} \leq r^2 - \frac{\|x - y\|^2}{4}$$

for each $x, y \in B_E(r)$ and $r > 0$, and so E is uniformly convex.

We have the following constructive version of the Riesz theorem for normable linear functionals.

Theorem 8.54 *Let f be a bounded linear functional on a Hilbert space H. Then f is normable if and only if there exists $x_0 \in H$ such that $f(x) = \langle x, x_0 \rangle$ for each $x \in H$.*

Proof See [4, Chapter 8, (2.3)] and [7, Theorem 4.3.6]. □

The following proposition characterises the closest point of a convex subset of an inner product space; the classical proof of [5, Theorem 5.2] works constructively.

Proposition 8.55 *Let C be a convex subset of an inner product space E, and let $x \in E$ and $y \in C$. Then $\|x - y\| \le \|x - z\|$ for all $z \in C$ if and only if $0 \le \langle x - y, y - z\rangle$ for all $z \in C$.*

Proof Suppose that $\|x - y\| \le \|x - z\|$ for all $z \in C$. Then, for λ with $0 < \lambda < 1$ and $z \in C$, we have

$$\begin{aligned}\|x - y\|^2 &\le \|x - (\lambda y + (1 - \lambda)z)\|^2 = \|x - y - \lambda(z - y)\|^2 \\ &\le \|x - y\|^2 - 2\lambda\langle x - y, z - y\rangle + \lambda^2\|z - y\|^2,\end{aligned}$$

Therefore $-\lambda^2\|z-y\|^2 \le 2\lambda\langle x-y, y-z\rangle$, and so $(-\lambda/2)\|z-y\|^2 \le \langle x-y, y-z\rangle$. Thus, letting $\lambda \to 0$, we have $0 \le \langle x - y, y - z\rangle$.

Conversely, suppose that $0 \le \langle x - y, y - z\rangle$ for all $z \in C$. Then, for each $z \in C$, we have

$$\begin{aligned}\|x - z\|^2 &= \|x - y + y - z\|^2 \\ &= \|x - y\|^2 + 2\langle x - y, y - z\rangle + \|y - z\|^2 \ge \|x - y\|^2.\end{aligned}$$ □

Since Hilbert spaces have very good geometric properties, we have the following simple, direct proofs of the separation theorem and the continuous extension theorem.

Theorem 8.56 *Let A and B be convex subsets of a Hilbert space H such that*

$$d = d(0, B - A) = \inf\{\|y - x\| \mid x \in A, y \in B\}$$

exists. Then there exists a normable linear functional f on H such that $\|f\| = d$ and

$$f(x) + d^2 \le f(y)$$

for each $x \in A$ and $y \in B$.

Proof Let $z \in \overline{B - A}$ be such that $\|z\| = d$, and let $f(u) = \langle u, z\rangle$. Then, since $0 \le \langle -z, z - (y - x)\rangle$ for each $x \in A$ and $y \in B$, we have $f(x) + d^2 \le f(y)$ for each $x \in A$ and $y \in B$. □

Theorem 8.57 *Let M be a subspace of a Hilbert space H, and let f be a normable linear functional on M. Then there exists a normable linear functional g on H such that $\|g\| = \|f\|$ and $g(x) = f(x)$ for each $x \in M$.*

Proof Let $\overline{M}$ be the closure of M. Then there exists a normable extension $\overline{f}$ of f on $\overline{M}$. Since $\overline{M}$ is a Hilbert space, there exists $x_0 \in \overline{M}$ such that

$$\overline{f}(x) = \langle x, x_0\rangle$$

for each $x \in \overline{M}$. Let $g(x) = \langle x, x_0\rangle$ for each $x \in H$. Then it is straightforward to show that $g(x) = f(x)$ for each $x \in M$ and $\|g\| = \|f\|$. □

Notes

Brouwerian counterexamples to the statements

Every bounded linear functional is normable

and

The set of normable linear functionals is linear

can be found in [9, Propositions 16 and 17].

Finite-dimensional spaces are discussed in [4, Chapter 7.2] and [7, Chapter 4]. Projections onto closed and located subspaces of a Hilbert space are dealt in [3, Chapter 9, Theorem 6], [4, Chapter 7, (8.7)], and [7, Theorem 4.3.1]; see also [9, Theorem 6].

In [3, Chapter 9], [4, Chapter 7], and [7, Chapter 5], the reader will find advanced material on normed linear spaces, such as: L_p spaces and the Radon–Nikodym theorem; the L_∞ space; the separation theorem, the Hahn–Banach extension theorem; dual spaces; extreme points and the Krein–Milman theorem; and the spectral theorem in Hilbert space. Some of these topics are dealt with in Chapter 9 in this Handbook.

References

[1] Aczel, P., and Rathjen, M. 2000/2001. *Notes on constructive set theory*. Technical report 40. Institut Mittag-Leffler.

[2] Aczel, P., and Rathjen, M. August, 2010. *CST Book draft*. http://www1.maths.leeds.ac.uk/~rathjen/book.pdf.

[3] Bishop, E. 1967. *Foundations of Constructive Analysis*. New York, Toronto, London: McGraw-Hill Book Co.

[4] Bishop, E., and Bridges, D. 1985. *Constructive Analysis*. Grundlehren der Mathematischen Wissenschaften [Fundamental Principles of Mathematical Sciences], vol. 279. Berlin: Springer-Verlag.

[5] Brezis, H. 2011. *Functional Analysis, Sobolev Spaces and Partial Differential Equations*. Universitext. New York: Springer.

[6] Bridges, D., and Richman, F. 1987. *Varieties of Constructive Mathematics*. London Mathematical Society Lecture Note Series, vol. 97. Cambridge: Cambridge University Press.

[7] Bridges, D. S., and Vîţă, L. S. 2006. *Techniques of Constructive Analysis*. Universitext. New York: Springer.

[8] Brouwer, L. E. J. 1926. Intuitionistische Einführung des Dimensionsbegriffes. *Nederl. Akad. Wetensch. Proc.*, **29**, 855–863.

[9] Ishihara, H. 2018. Constructive functional analysis: an introduction. Pages 109–165 of: Mainzer, K., Schuster, P., and Schwichtenberg, H. (eds.), *Proof and Computation: Digitization in Mathematics, Computer Science and Philosophy*. Singapore: World Scientific.

[10] Troelstra, A. S., and van Dalen, D. 1988. *Constructivism in mathematics. Vol. I.* Studies in Logic and the Foundations of Mathematics, vol. 121. Amsterdam: North-Holland.

9
Constructive Functional Analysis

Hajime Ishihara

9.1 Introduction

In his 1967 monograph *Foundations of Constructive Analysis* [3], Errett Bishop clearly demonstrated that the constructive program can succeed, by actually developing *mathematics*, as distinct from logic, constructively. His book covered a very wide range of analysis, from calculus to commutative Banach algebras. His framework, which we call *neutral constructivism*, is without ontological or idealistic objects and principles, in contrast with Brouwer's intuitionism and Markov's constructive recursive mathematics. Several books on constructive analysis have been published following Bishop: Bridges, *Constructive Functional Analysis* (1979) [10]; Bishop and Bridges, *Constructive Analysis* (1985) [4]; and Bridges and Vîţă, *Techniques of Constructive Analysis* (2006) [11]. The second of these is based on Bishop's monograph but contains much new material; the third covers results obtained in the years since 1985.

Considerable activity in constructive mathematics has taken place over the past 30 or so years. This chapter is intended to present a general theory of constructive functional analysis with some new results and improved proofs. The reader should consult [3] and [4] for specific theories, such as measure, integration, limit operations in measure theory, locally compact abelian groups and commutative Banach algebras; see also corresponding chapters in this Handbook.

In Section 9.2, we introduce fundamental notions, notations and results in the theory of metric and normed linear spaces. In Section 9.3, we discuss the (constructive) existence of Minkowski functionals, and apply it to locating the kernel of a linear map. Then we review the uniform boundedness theorem, before proving new versions of the open mapping and closed graph theorems. In Section 9.4, we move to the Hahn–Banach theorem including a proof of an approximate version of the separation theorem (for two *unbounded* convex subsets) using the Baire theorem and a proof of an approximate version of the (one-dimensional) dominated extension theorem. We also deal with the weak topology in Hilbert spaces and show that a convex set which is totally bounded with respect to the weak topology is located. In Section 9.5, we discuss adjoints of operators on a Hilbert space, the Hellinger–Toeplitz theorem, the closed range theorem, and compact operators.

The material of this chapter could be formalised in the constructive Zermelo–Fraenkel set theory **CZF** [1, 2] together with the axioms of countable and dependent choice.

9.2 Preliminaries

Throughout the chapter, we use the following (constructive) set theoretical notions: a set A is *inhabited* if there exists an element of A; a set A is *finitely enumerable* if there exist an n and a surjection $f : \{0, \ldots, n-1\} \to A$; sets A and B *overlap* if $A \cap B$ is inhabited, and we then write $A \between B$. Note that if a set A is inhabited, then it is non-empty, but the converse does not hold constructively.

In this section, we assume basic facts about metric and normed linear spaces, as found in Chapter 8 in this Handbook.

9.2.1 Metric Spaces

Let X be a metric space with a metric d. Then, for $x, y \in X$ and $r > 0$, we write $x \# y$ for $0 < d(x, y)$, and $B_X(x, r)$ to denote the *open ball* with *centre* x and *radius* r, that is,

$$B_X(x, r) = \{y \in X \mid d(x, y) < r\}.$$

Definition 9.1 Let A be a subset of a metric space X. The *complement* $\sim A$ of A is given by

$$\sim A = \{x \in X \mid \forall a \in A(x \# a)\};$$

the *metric complement* $-A$ of A is the set of all points $x \in X$ that are *bounded away* from A, that is,

$$-A = \{x \in X \mid \forall a \in A(d(x, a) \geq r) \text{ for some } r > 0\}.$$

Note that $-A$ is an open subset of X.

Recall that a subset A of a metric space X is (metrically) *located* if

$$d(x, A) = \inf\{d(x, y) \mid y \in A\}$$

exists for all $x \in X$.

Definition 9.2 A subset A of a metric space X is *topologically located* if A is inhabited and for each $x \in X$ and each $r > 0$, either $B_X(x, r) \between A$ or $x \in -A$.

Note that for each topologically located subset A of a metric space X and each $x \in X$, if $x \notin -A$, then $x \in \overline{A}$.

Lemma 9.3 *Let A be a subset of a metric space.*

(i) *If A is located, then it is topologically located.*
(ii) *A is topologically located if and only if $\overline{A}$ is topologically located.*

Proof Straightforward; see [36, Propositions 7.3.3 and 7.3.5]. □

Recall that we have the following constructive version of the Baire theorem.

Theorem 9.4 *If $(U_n)_{n \in \mathbf{N}}$ is a sequence of open dense subsets of a complete metric space X, then $\bigcap_{n=0}^{\infty} U_n$ is dense in X.*

Proof See [3, Chapter 4, Theorem 4], [4, Chapter 4, (3.9)], and [7, Chapter 2, (1.3)]. □

Definition 9.5 A mapping f between metric spaces X and Y is

- *sequentially continuous* if for each sequence $(x_n)_{n \in \mathbf{N}}$ in X converging to a limit $x \in X$,
$$f(x_n) \to f(x) \text{ as } n \to \infty;$$
- *strongly extensional* if for each $x, y \in X$,
$$f(x) \# f(y) \Rightarrow x \# y.$$

Trivially, every continuous mapping is sequentially continuous.

Lemma 9.6 *Every sequentially continuous mapping f of a metric space X into a metric space Y is strongly extensional.*

Proof Consider x and y in X with $f(x) \# f(y)$, and construct an increasing binary sequence $(\lambda_n)_{n \in \mathbf{N}}$ such that

$$\lambda_n = 0 \Rightarrow d(x, y) < 1/(n+1),$$
$$\lambda_n = 1 \Rightarrow d(x, y) > 0.$$

Define a sequence $(z_n)_{n \in \mathbf{N}}$ in X as follows: if $\lambda_n = 0$, set $z_n = x$; if $\lambda_n = 1$, set $z_n = y$. Then $(z_n)_{n \in \mathbf{N}}$ converges to y. Since f is sequentially continuous, the sequence $(f(z_n))_{n \in \mathbf{N}}$ in Y converges to the limit $f(y)$. Choose a natural number N such that $d(f(z_N), f(y)) < d(f(x), f(y))$. If $\lambda_N = 0$, then $z_N = x$; whence $d(f(x), f(y)) = d(f(z_N), f(y)) < d(f(x), f(y))$, a contradiction. Therefore $\lambda_N = 1$, and so $x \# y$. □

Lemma 9.7 *Let f be a strongly extensional mapping of a complete metric space X into a metric space Y, and let $(x_n)_{n\in\mathbb{N}}$ be a sequence in X converging to a limit x. Then for all positive numbers s, t with $s < t$, either $d(f(x_n), f(x)) > s$ for infinitely many n or $d(f(x_n), f(x)) < t$ for all sufficiently large n.*

Proof See [11, Lemma 3.2.2], [19, Lemma 2], and [24, Lemma 5]. □

9.2.2 Normed Linear Spaces

For the remainder of this chapter, 'linear space' will mean 'real linear space'.

Let E be a normed linear space. Then we write $B_E(r)$ to denote the ball $B_E(0, r)$ relative to the metric associated with the norm on E, and B_E to denote the *open unit ball* $B_E(1)$. For subsets S and S' of a linear space and $a \in \mathbb{R}$, we write $S + S'$ and aS for the subsets

$$S + S' = \{x + y \mid x \in S, y \in S'\}$$

and

$$aS = \{ax \mid x \in S\},$$

respectively.

Definition 9.8 A subset C of a linear space is *convex* if $\lambda x + (1 - \lambda)y \in C$ for all $x, y \in C$ and $\lambda \in [0, 1]$; the *convex hull* $\mathrm{co}A$ of a subset A of a linear space is the set of all *convex combinations* of elements of A, that is, sums

$$\lambda_0 x_0 + \cdots + \lambda_{n-1} x_{n-1},$$

where $x_i \in A$ and $\lambda_i \geq 0$ for all $i < n$, and $\sum_{i<n} \lambda_i = 1$. A subset C of a linear space E is *absorbing* if for each x in E there exists $r > 0$ with $x \in rC$.

Note that if C and D are convex subsets of a linear space E and $r > 0$, then $C + D$ and rC are convex subsets of E. If C is absorbing, then $0 \in C$; if C is convex and absorbing, then $\lambda C \subseteq C$ for all $\lambda \in [0, 1]$, and hence $sC \subseteq tC$ if $0 \leq s < t$.

Lemma 9.9 *Let C be a convex, absorbing subset of a Banach space E. Then $0 \notin \overline{-C}$, where $-C$ is the metric complement of C.*

Proof Suppose that $0 \in \overline{-C}$. We prove that $-nC$ is dense in E for all n. To that end, fix n, let $y \in E$, and let $\epsilon > 0$. Then there exist $r > 0$, $x \in -C$ and $s > 0$ such that $-y \in rC$, $\|x\| < (n + r)^{-1}\epsilon$, and $\|x - w\| \geq s$ for all $w \in C$. Set $y' = y + (n + r)x$. Then $\|y - y'\| = (n + r)\|x\| < \epsilon$. Moreover, for each $z \in nC$, we have

$$(n + r)^{-1}(z - y) \in (n + r)^{-1}(nC + rC) = C,$$

and therefore

$$\|y' - z\| = (n+r)\|x - (n+r)^{-1}(z-y)\| \geq (n+r)s > 0.$$

Hence $y' \in -nC$. It follows that for each n, $-nC$ is dense in E; clearly, it is open in E. Applying Theorem 9.4, construct a point x in $\bigcap_{n=1}^{\infty} -nC$. Choose m such that $x \in mC$. Then, as $x \in -mC$, we have a contradiction. Thus $0 \notin \overline{-C}$. □

Definition 9.10 A function p from a linear space E into $\mathbb{R}$ is

- *convex* if

$$p(\lambda x + (1-\lambda)y) \leq \lambda p(x) + (1-\lambda)p(y)$$

for all $x, y \in E$ and $\lambda \in [0,1]$;

- *sublinear* if

$$p(ax) = ap(x) \quad \text{and} \quad p(x+y) \leq p(x) + p(y)$$

for all $x, y \in E$ and $a \in \mathbb{R}$ with $a \geq 0$.

Lemma 9.11 *Let E be a linear space, and let $p : E \to \mathbb{R}$ be a convex function. Then for each $u, v \in E$ and each $t, s \in \mathbb{R}$ with $0 < t \leq s$,*

$$\begin{aligned}\frac{p(u-sv)-p(u)}{-s} &\leq \frac{p(u-tv)-p(u)}{-t}\\ &\leq \frac{p(u+tv)-p(u)}{t} \leq \frac{p(u+sv)-p(u)}{s}.\end{aligned}$$

Proof See [24, Lemma 55]. □

Lemma 9.12 *Let E be a normed linear space, and let $p : E \to \mathbb{R}$ be a convex function. If p is continuous at $x \in E$, then there exist $r > 0$ and $L > 0$ such that $|p(u) - p(v)| \leq L\|u - v\|$ for all $u, v \in B_E(x, r)$.*

Proof See [24, Lemma 56]. □

Notes

There is another kind of complement: the *logical complement* $\neg A$ of a subset A of a set X is given by

$$\neg A = \{x \in X \mid x \notin A\} = \{x \in X \mid \forall a \in A \neg(x = a)\}.$$

For a subset A of a metric space, clearly, $-A \subseteq \sim A \subseteq \neg A$.

A notion of a topologically located set was given by Freudenthal [14], and the notion here was introduced by Troelstra in his dissertation; see also [35]. A Brouwerian counterexample (due to Grayson) to the statement

Every topologically located subset of a metric space is located

can be found in [36, Section 7.3.4].

9.3 Completeness

Many important theorems in analysis rely on completeness. The fundamental tool here is the Baire theorem for complete metric spaces, which leads to Lemma 9.9 and the uniform boundedness theorem below. Constructively, Lemma 9.7 for strongly extensional mappings on complete metric spaces plays a major role – for example, in proving the open mapping and closed graph theorems.

In this section, we start by looking into the constructive existence of Minkowski functionals and its application to locating the kernel of a linear map. Then we briefly review the uniform boundedness theorem, before proving new versions of the open mapping and closed graph theorems.

9.3.1 Minkowski Functionals

Definition 9.13 The *Minkowski functional* μ of a convex absorbing subset C of a linear space E is defined by

$$\mu(x) = \inf\{r > 0 \mid x \in rC\}$$

for each $x \in E$.

Note that if it exists, then the Minkowski functional μ of a convex absorbing subset C of a linear space E is sublinear.

Lemma 9.14 *Let C be a convex, absorbing subset of a linear space E. Then C has a Minkowski functional if and only if for each x in E and all $s, t \in \mathbb{R}$ with $0 < s < t$, either $x \notin sC$ or $x \in tC$.*

Proof Suppose that $\mu(x) = \inf\{r > 0 \mid x \in rC\}$ exists for all $x \in E$. Given $x \in E$ and $s, t \in \mathbb{R}$ with $0 < s < t$, we have either $s < \mu(x)$ or $\mu(x) < t$. In the first case, if $x \in sC$ then $\mu(x) \leq s < \mu(x)$, a contradiction, so $x \notin sC$. In the second case, clearly $x \in tC$.

Conversely, let $x \in E$ and $A = \{r > 0 \mid x \in rC\}$. Suppose that for all $s, t \in \mathbb{R}$ with $0 < s < t$, either $x \notin sC$ or $x \in tC$. For all real numbers s and t with $s < t$ we have either $s < 0$ or $0 < t$. In the first case, s is a lower bound of A. In the case $0 < t$, choose s' such that $\max\{0, s\} < s' < t$. Then either $x \notin s'C$ or $x \in tC$; in the former case, s', and hence s, is a lower bound of A; in the latter case, we have $t \in A$. It follows that $\mu(x) = \inf A$ exists, by [4, Chapter 2, (4.3)]. □

Lemma 9.15 *Let C be a topologically located subset of a normed linear space E. Then tC is topologically located for all $t > 0$.*

Proof For each $x \in E$ and each $\epsilon > 0$, either there exists $y \in C$ with $\|y-t^{-1}x\| < t^{-1}\epsilon$, in which case $ty \in tC$ and $\|ty - x\| < \epsilon$; or else there exists $r > 0$ such that $\|t^{-1}x - z\| \geq r$ for all $z \in C$. In the latter case, $\|x - tz\| \geq tr > 0$ for all $z \in C$, so $x \in -tC$. □

Lemma 9.16 *Let C be a convex, absorbing subset of a normed linear space E such that C° is inhabited. Then $0 \in C^\circ$.*

Proof Let $z \in C$ and let δ be a positive number such that $B_E(z, \delta) \subseteq C$. There exists $r > 0$ such that $-z \in rC$; so for each $x \in B_E(\delta)$, $z + x \in B_E(z, \delta)$ and therefore

$$x = (z + x) - z \in C + rC = (1 + r)C.$$

Hence $B_E(\delta/(1 + r)) \subseteq C$. □

Proposition 9.17 *Let C be a topologically located, convex, absorbing subset of a normed linear space E such that C° is inhabited. Then for each $x \in E$ and all $s, t \in \mathbb{R}$ with $0 < s < t$, either $x \in -sC$ or $x \in tC$; in particular, C has a Minkowski functional.*

Proof In view of Lemma 9.16, there exists $\delta >$ such that $B_E(\delta) \subseteq C$. Given $s, t \in \mathbb{R}$ with $0 < s < t$, set $\epsilon = \delta(t - s)$. By Lemma 9.15, for each x in E, either $B_E(x, \epsilon) \between sC$ or $x \in -sC$. In the latter case, $x \notin sC$. In the former case, there exists $y \in sC$ such that $\|x - y\| < \epsilon$ and therefore

$$x = y + (x - y) \in sC + \frac{\epsilon}{\delta}C \subseteq tC.$$

To complete the proof, we refer to Lemma 9.14. □

Lemma 9.18 *Let C be a convex, absorbing subset of a normed linear space E such that C° is inhabited and has a Minkowski functional μ. Then μ is uniformly continuous on E.*

Proof In view of Lemma 9.16, there exists $\delta > 0$ such that $B_E(\delta) \subseteq C$. Given $\epsilon > 0$ and $x, y \in E$ with $\|x - y\| < \delta\epsilon$, we have $x - y \in \epsilon C$, so $\mu(x - y) \leq \epsilon$ and therefore, by sublinearity,

$$\mu(x) = \mu(y + (x - y)) \leq \mu(y) + \mu(x - y) = \mu(y) + \epsilon.$$

Likewise, $\mu(y) - \mu(x) \leq \epsilon$, so $|\mu(x) - \mu(y)| \leq \epsilon$. □

Lemma 9.19 *Let C be a topologically located, closed, convex and absorbing subset of a Banach space E. Then for each $x \in E$, either $x \# 0$ or $x \in C$.*

Proof Consider x in E, and construct an increasing binary sequence $(\lambda_n)_{n\in\mathbb{N}}$ such that

$$\lambda_n = 0 \Rightarrow \|x\| < 1/(n+1)^2,$$
$$\lambda_n = 1 \Rightarrow \|x\| > 0.$$

We may assume that $\lambda_0 = 0$. Define a sequence $(y_n)_{n\in\mathbb{N}}$ in E as follows: if $\lambda_n = 0$, set $y_n = 0$; if $\lambda_{n+1} = 1 - \lambda_n$, set $y_k = (n+1)x$ for all $k \geq n+1$. Then $(y_n)_{n\in\mathbb{N}}$ is a Cauchy sequence in E; in fact, $\|y_m - y_n\| \leq 1/(n+1)$ whenever $m \geq n$. Since E is complete, $(y_n)_{n\in\mathbb{N}}$ converges to a limit $y \in E$. Choose N such that $y \in NC$. Then either $\lambda_n = 1$ for some $n \leq N$ or $\lambda_n = 0$ for all $n \leq N$. In the former case, we have $x \# 0$. In the latter case, assume that $x \in -C$. If $\lambda_{n+1} = 1 - \lambda_n$ for some $n \geq N$, then $y = (n+1)x \in NC$, and hence $x \in (N/(n+1))C \subseteq C$, a contradiction. Therefore $\lambda_n = 0$ for all n, and so $x = 0$, a contradiction. Thus $x \notin -C$, and, since C is topologically located and closed, we have $x \in C$. □

Proposition 9.20 *Let C be a topologically located, closed, convex and absorbing subset of a Banach space E. Then for each $x \in E$ and all $s, t \in \mathbb{R}$ with $0 < s < t$, either $x \in -sC$ or $x \in tC$; in particular, C has a Minkowski functional.*

Proof Consider $x \in E$ and $s, t \in \mathbb{R}$ with $0 < s < t$, and construct an increasing binary sequence $(\lambda_n)_{n\in\mathbb{N}}$ such that

$$\lambda_n = 0 \Rightarrow B_E(x, 1/(n+1)^2) \between sC,$$
$$\lambda_n = 1 \Rightarrow x \in -sC.$$

We may assume that $\lambda_0 = 0$. Define sequences $(z_n)_{n\in\mathbb{N}}$ in E and $(r_n)_{n\in\mathbb{N}}$ in $\mathbb{R}$ as follows: if $\lambda_n = 0$, set $z_n = 0$ and $r_n = 0$; if $\lambda_{n+1} = 1 - \lambda_n$, choose $y \in sC$ such that $\|x - y\| < 1/(n+1)^2$, and set $z_k = (n+1)(x - y)$ and $r_k = 1/(n+1)$ for all $k \geq n+1$. Then $(z_n)_{n\in\mathbb{N}}$ and $(r_n)_{n\in\mathbb{N}}$ are Cauchy sequences in E and in $\mathbb{R}$, respectively: in fact, $\|z_m - z_n\| \leq 1/(n+1)$ and $|r_m - r_n| \leq 1/(n+1)$ whenever $m \geq n$. Since E and $\mathbb{R}$ are complete, these sequences converge to limits $z \in E$ and $r \in \mathbb{R}$, respectively. We show that $x - rz \in sC$. To this end, assume that $x - rz \in -sC$. If $\lambda_{n+1} = 1 - \lambda_n$, then $z = (n+1)(x - y)$ for some $y \in sC$ and $r = 1/(n+1)$, and hence $x - rz = y \in sC$, a contradiction. Therefore $\lambda_n = 0$ for all n, and so $x - rz = x \in sC$. This contradiction ensures that $x - rz \notin -sC$, and, since sC is topologically located and closed, we have $x - rz \in sC$. Either $rz/(t-s) \# 0$ or $rz/(t-s) \in C$, by Lemma 9.19. In the first case, since $0 < r$, we have $r_n > 0$, and therefore $\lambda_n = 1$ for some n, so $x \in -sC$. In the second case we have

$$x = x - rz + rz \in sC + (t-s)C \subseteq tC.$$

□

9.3.2 Locating the Kernel of a Linear Map

We now apply the results in Section 9.3.1 to locating the kernel of a linear map. The following lemma bridges the existence of a Minkowski functional and the locatedness of the kernel.

Lemma 9.21 *Let T be a linear mapping of a normed linear space E onto a linear space F. Then* $\ker(T)$ *is located in E if and only if $T(B_E)$ has a Minkowski functional.*

Proof See [7, Proposition 5.2] and [24, Proposition 27]. □

Recall from Chapter 8 that a linear mapping T between normed linear spaces E and F is *open* if $T(B_E)$ is an open subset of F.

Proposition 9.22 *Let T be an open linear mapping of a normed linear space E onto a normed linear space F such that $T(B_E)$ is topologically located in F. Then* $\ker(T)$ *is located in E.*

Proof Apply Proposition 9.17 and Lemma 9.21. □

Lemma 9.23 *Let G be a closed subspace of the product space $E \times F$ of Banach spaces E and F such that the set*

$$S = \{y \in F \mid \exists x \in B_E((x, y) \in G)\}$$

is topologically located and absorbing, and let $y \in F$. For each $\epsilon > 0$ and each $t > 1$, if $y' \in S$ and $\|y - y'\| < (t-1)\epsilon/2$, then either $B_F(\epsilon) \between -S$ or $y \in tS$.

Proof Note that $\overline{S}$ is a topologically located, closed, convex and absorbing subset of F. Let $y \in F$, and consider $\epsilon > 0$ and $t > 1$. Suppose that $\|y - y'\| < (t-1)\epsilon/2$ for some $y' \in S$, and let $(\delta_n)_{n \in \mathbb{N}}$ be a sequence of positive numbers given by $\delta_0 = 1$ and $\delta_{n+1} = (t-1)/2^{n+1}$. We construct, inductively, an increasing binary sequence $(\lambda_n)_{n \in \mathbb{N}}$, sequences $(y_n)_{n \in \mathbb{N}}$ and $(y'_n)_{n \in \mathbb{N}}$ in F and a sequence $(x_n)_{n \in \mathbb{N}}$ in B_E such that for each n,

(i) if $\lambda_n = 0$, then $\|y_{n+1}\| < \delta_{n+1}\epsilon/(n+1)$;
(ii) if $\lambda_{n+1} = 1 - \lambda_n$, then $\delta_{n+1}^{-1} y_{n+1} \in -S$;
(iii) $(x_n, y'_n) \in G$ and $\|y'_{n+1}\| < 2\epsilon$.

For $n = 0$, set $\lambda_0 = 0$, $y_0 = y$ and $y'_0 = y'$, and choose $x_0 \in B_E$ such that $(x_0, y'_0) \in G$. Assume that we have constructed $\lambda_0, \dots, \lambda_n, y_0, \dots, y_n, y'_0, \dots, y'_n$ and $x_0, \dots, x_n$, and set $y_{n+1} = y_n - \delta_n y'_n$. If $\lambda_n = 1$, set $\lambda_{n+1} = 1$ and $y'_{n+1} = x_{n+1} = 0$. If $\lambda_n = 0$, then either

$$B_F\left(\delta_{n+1}^{-1} y_{n+1}, \frac{\epsilon}{2(n+2)}\right) \between S \quad \text{or} \quad \delta_{n+1}^{-1} y_{n+1} \in -S.$$

In the first case, set $\lambda_{n+1} = 0$, and choose $y'_{n+1} \in S$ such that

$$\|\delta_{n+1}^{-1} y_{n+1} - y'_{n+1}\| < \frac{\epsilon}{2(n+2)}$$

and choose $x_{n+1} \in E$ such that $(x_{n+1}, y'_{n+1}) \in G$. In the second case, set $\lambda_{n+1} = 1$ and $y'_{n+1} = x_{n+1} = 0$.

Note that $(x_n, y'_n) \in G$ for all n and that

$$\|y_1\| = \|y_0 - \delta_0 y'_0\| = \|y - y'\| < \frac{(t-1)\epsilon}{2} = \delta_1 \epsilon.$$

We have the following.

(i) If $\lambda_{n+1} = 0$, then

$$\begin{aligned}\|y_{n+2}\| &= \|y_{n+1} - \delta_{n+1} y'_{n+1}\| = \delta_{n+1} \|\delta_{n+1}^{-1} y_{n+1} - y'_{n+1}\| \\ &< \frac{\delta_{n+1}\epsilon}{2(n+2)} = \frac{\delta_{n+2}\epsilon}{n+2}.\end{aligned}$$

(ii) If $\lambda_{n+1} = 1 - \lambda_n$, then $\delta_{n+1}^{-1} y_{n+1} \in -S$.

(iii) If $\lambda_{n+1} = 0$, then $\lambda_n = 0$ and so

$$\|y'_{n+1}\| \leq \|\delta_{n+1}^{-1} y_{n+1}\| + \|\delta_{n+1}^{-1} y_{n+1} - y'_{n+1}\| < \frac{\epsilon}{n+1} + \frac{\epsilon}{2(n+2)} < 2\epsilon.$$

If $\lambda_{n+1} = 1$, then $\|y'_{n+1}\| = 0 < 2\epsilon$.

This completes our inductive construction.

Define a sequence $(z_n)_{n\in\mathbb{N}}$ in F as follows. Set $z_0 = 0$; if $\lambda_{n+1} = 0$, set $z_{n+1} = 0$; if $\lambda_{n+1} = 1 - \lambda_n$, set $z_k = \delta_{n+1}^{-1} y_{n+1}$ for all $k \geq n+1$. Then $(z_n)_{n\in\mathbb{N}}$ is a Cauchy sequence in F: in fact, $\|z_m - z_n\| \leq \epsilon/(n+1)$ whenever $m \geq n$. Since F is complete, $(z_n)_{n\in\mathbb{N}}$ converges to a limit $z \in F$. Either $z \# 0$ or $z \in \overline{S}$, by Lemma 9.19. In the first case, there exists n such that $\lambda_{n+1} = 1 - \lambda_n$, and hence $z = \delta_{n+1}^{-1} y_{n+1} \in -S$ and $\|z\| < \epsilon/(n+1) \leq \epsilon$. In the second case, if $\lambda_{n+1} = 1 - \lambda_n$, then $z \in -S$, a contradiction; hence $\lambda_n = 0$ for all n. Therefore the series $\sum_{n=0}^{\infty} \delta_n y'_n$ converges to y: in fact,

$$\sum_{n=0}^{\infty} \delta_n y'_n = \sum_{n=0}^{\infty} (y_n - y_{n+1}) = \lim_{n\to\infty} (y_0 - y_{n+1}) = y_0 = y.$$

Note that, since E is complete, the series $\sum_{n=0}^{\infty} \delta_n x_n$ converges to some $x \in E$, and

$$\|x\| = \left\|\sum_{n=0}^{\infty} \delta_n x_n\right\| \leq \|x_0\| + \left\|\sum_{n=1}^{\infty} \delta_n x_n\right\| < 1 + \sum_{n=1}^{\infty} \delta_n = 1 + (t-1) = t.$$

Since $(x_n, y'_n) \in G$ for all n and G is a closed subspace, we have $(x, y) \in G$.

Therefore, since $t^{-1}x \in B_E$ and $(t^{-1}x, t^{-1}y) = t^{-1}(x,y) \in t^{-1}G = G$, we have $t^{-1}y \in S$, and so $y \in tS$. □

Lemma 9.24 *Let T be a sequentially continuous linear mapping of a Banach space E onto a Banach space F such that $T(B_E)$ is topologically located. Then for each $t > 1$, $\overline{T(B_E)} \subseteq tT(B_E)$.*

Proof Let G be the graph of T, that is, $G = \{(x, Tx) \mid x \in E\}$, and define S as in the previous lemma. Then, since $S = T(B_E)$ and T is sequentially continuous, G is a closed subspace of the product space $E \times F$ such that S is topologically located and absorbing. Given $y \in \overline{T(B_E)}$ and $t > 1$, in order to prove that $y \in tT(B_E)$, we construct an increasing binary sequence $(\lambda_n)_{n\in\mathbb{N}}$ such that

$$\lambda_n = 0 \Rightarrow B_F(1/(n+1)) \between -T(B_E),$$
$$\lambda_n = 1 \Rightarrow y \in tT(B_E).$$

We may assume that $\lambda_0 = 0$. Define a sequence $(z_n)_{n\in\mathbb{N}}$ in F as follows: if $\lambda_n = 0$, choose $z_n \in -T(B_E)$ such that $\|z_n\| < 1/(n+1)$; if $\lambda_{n+1} = 1 - \lambda_n$, set $z_k = z_n$ for all $k \geq n+1$. Then $(z_n)_{n\in\mathbb{N}}$ is a Cauchy sequence in F: in fact, $\|z_m - z_n\| \leq 2/(n+1)$ whenever $m \geq n$. Since F is complete, $(z_n)_{n\in\mathbb{N}}$ converges to a limit $z \in F$. Either $z \# 0$ or $z \in \overline{T(B_E)}$, by Lemmas 9.3 and 9.19. In the latter case, if $\lambda_{n+1} = 1 - \lambda_n$ for some n, then $z = z_n \in -T(B_E)$, a contradiction; hence $\lambda_n = 0$ for all n. Therefore $0 \in \overline{-T(B_E)}$, a contradiction to Lemma 9.9. Thus we must have $z \# 0$, so $\lambda_n = 1$ for some n and therefore $y \in tT(B_E)$. □

Theorem 9.25 *Let T be a sequentially continuous linear mapping of a Banach space E onto a Banach space F such that $T(B_E)$ is topologically located. Then for each $y \in F$ and all $s, t \in \mathbb{R}$ with $0 < s < t$, either $y \in -sT(B_E)$ or $y \in tT(B_E)$. In particular,* $\ker(T)$ *is located.*

Proof Consider $y \in F$ and $s, t \in \mathbb{R}$ with $0 < s < t$, and let $t' = (s+t)/2$. Then we see from Lemma 9.3 and Proposition 9.20 that either $y \in -s\overline{T(B_E)} = -sT(B_E)$ or else $y \in t'\overline{T(B_E)}$. In the latter case, by Lemma 9.24, we have $y \in (t/t')t'T(B_E) = tT(B_E)$. Since $T(B_E)$ is convex and absorbing, it follows from by Lemma 9.14 that it has a Minkowski functional. Hence $\ker(T)$ is located, by Lemma 9.21. □

9.3.3 The Uniform Boundedness Theorem

We have the following constructive version of the uniform boundedness theorem.

Theorem 9.26 *Let $(T_m)_{m\in\mathbb{N}}$ be a sequence of bounded linear mappings from a Banach space E into a normed linear space F. If $(x_m)_{m\in\mathbb{N}}$ is a sequence in B_E*

such that $\{T_m x_m \mid m \in \mathbb{N}\}$ *is unbounded, then there exists* $x \in E$ *such that the set* $\{T_m x \mid m \in \mathbb{N}\}$ *is unbounded.*

Proof Let $(x_m)_{m\in\mathbb{N}}$ be a sequence in B_E such that $\{T_m x_m \mid m \in \mathbb{N}\}$ is unbounded, and, given n, let

$$U_n = \{x \in E \mid n < \|T_m x\| \text{ for some } m\}.$$

Then, trivially, U_n is open. Given $y \in E$ and $\epsilon > 0$, choose m such that $(2n+1)/\epsilon < \|T_m x_m\|$. Then either $n < \|T_m y\|$ or $\|T_m y\| < n+1$; in the former case, we have $y \in U_n$; in the latter, we have $\|\epsilon x_m\| < \epsilon$ and

$$\|T_m(y + \epsilon x_m)\| \geq \epsilon\|T_m x_m\| - \|T_m y\| > (2n+1) - (n+1) > n.$$

Therefore U_n is dense in E. Thus there exists $x \in \bigcap_{n=0}^{\infty} U_n$, by Theorem 9.4, and so $\{T_m x \mid m \in \mathbb{N}\}$ is unbounded. □

9.3.4 The Open Mapping Theorem

The following lemma shows that *every* linear mapping of a Banach space into a normed linear space is well behaved.

Lemma 9.27 *Let* T *be a linear mapping of a Banach space* E *into a normed linear space* F. *Then* T *is strongly extensional.*

Proof Suppose that $0 < \|Tx\|$, and construct an increasing binary sequence $(\lambda_n)_{n\in\mathbb{N}}$ such that

$$\lambda_n = 0 \Rightarrow \|x\| < 1/(n+1)^2,$$
$$\lambda_n = 1 \Rightarrow \|x\| > 0.$$

We may assume that $\lambda_0 = 0$. Define a sequence $(y_n)_{n\in\mathbb{N}}$ in E as follows: if $\lambda_n = 0$, set $y_n = 0$; if $\lambda_{n+1} = 1 - \lambda_n$, set $y_k = (n+1)x$ for all $k \geq n+1$. Then $(y_n)_{n\in\mathbb{N}}$ is a Cauchy sequence in E: in fact, $\|y_m - y_n\| \leq 1/(n+1)$ whenever $m \geq n$. Since E is complete, $(y_n)_{n\in\mathbb{N}}$ converges to a limit $y \in E$. Choose N such that $\|Ty\| < N\|Tx\|$. Then either $\lambda_n = 0$ for all $n \leq N$, or else $\lambda_n = 1$ for some $n \leq N$. In the first case, if $n \geq N$ and $\lambda_{n+1} = 1 - \lambda_n$, then $y = (n+1)x$ and $N\|Tx\| \leq n\|Tx\| = \|Ty\| < N\|Tx\|$, a contradiction; whence $\lambda_n = 0$ for all $n \geq N$, so $x = 0$ and therefore $\|Tx\| = 0$, again a contradiction. Thus we must have $\lambda_n = 1$ for some $n \leq N$, and therefore $\|x\| > 0$. □

Definition 9.28 A linear mapping T between normed linear spaces is *sequentially open* if, whenever $Tx_n \to 0$ as $n \to \infty$, there exists a sequence $(y_n)_{n\in\mathbb{N}}$ in $\ker(T)$ such that $x_n + y_n \to 0$ as $n \to \infty$.

We have the following new version of the open mapping theorem.

Theorem 9.29 *Let T be a sequentially continuous linear mapping of a Banach space E onto a Banach space F such that $T(B_E)$ is topologically located. Then T is sequentially open.*

Proof By Theorem 9.25, $\ker(T)$ is located; so the quotient space $E/\ker(T)$ is a Banach space with respect to the quotient norm defined by

$$\|x\|_{E/\ker(T)} = \inf\{\|x - y\| \mid y \in \ker(T)\}.$$

Moreover, setting $Sx = Tx$ for each $x \in E/\ker(T)$, we obtain a well-defined, one–one, linear mapping of $E/\ker(T)$ onto F. Let $(Tx_n)_{n\in\mathbf{N}}$ be a sequence in F converging to 0. The inverse $S^{-1} : F \to E/\ker(T)$ is strongly extensional, by Lemma 9.27; so by Lemma 9.7, for each $\epsilon > 0$ we have either $\|x_n\|_{E/\ker(T)} > \epsilon/2$ for infinitely many n or else $\|x_n\|_{E/\ker(T)} < \epsilon$ for all sufficiently large n. In the first case, for infinitely many n we have $\|(4/\epsilon)x_n\|_{E/\ker(T)} > 2$, so, as S is one–one, $(4/\epsilon)Tx_n = (4/\epsilon)Sx_n \notin 2T(B_{E/\ker(T)}) = 2T(B_E)$. Since Theorem 9.25 tells us that for each n, either $(4/\epsilon)Tx_n \in -T(B_E)$ or $(4/\epsilon)Tx_n \in 2T(B_E)$, we must have $(4/\epsilon)Tx_n \in -T(B_E)$ for infinitely many n. Therefore $0 \in \overline{-T(B_E)}$, contradicting Lemma 9.9. Thus we must have $\|x_n\|_{E/\ker(T)} < \epsilon$ for all sufficiently large n. Since $\epsilon > 0$ is arbitrary, $(x_n)_{n\in\mathbf{N}}$ converges to 0 in $E/\ker(T)$, and so there exists a sequence $(y_n)_{n\in\mathbf{N}}$ in $\ker(T)$ such that $x_n + y_n \to 0$ as $n \to \infty$ in E. □

As a corollary, we have the following version of the Banach inverse mapping theorem.

Corollary 9.30 *Let T be a one–one, sequentially continuous linear mapping of a Banach space E onto a Banach space F such that $T(B_E)$ is topologically located. Then T^{-1} is sequentially continuous.*

9.3.5 The Closed Graph Theorem

In this subsection, we prove a lemma and present a new version of the closed graph theorem.

Lemma 9.31 *Let T be a linear mapping of a Banach space E into a Banach space F such that the graph $G = \{(x, Tx) \mid x \in E\}$ of T is closed in the product space $E \times F$, and $T^{-1}(B_F)$ is topologically located. Then $0 \notin \overline{\sim T^{-1}(B_F)}$.*

Proof It suffices to show that $0 \notin \overline{\sim 2T^{-1}(B_F)}$. Let

$$H = \{(Tx, x) \mid x \in E\} \text{ and}$$
$$S = \{x \in E \mid \exists y \in B_F((y, x) \in H)\}.$$

Since the graph of T is a closed subspace of $E \times F$, H is a closed subspace of the Banach space $F \times E$. Moreover, $S = \overline{T^{-1}(B_F)}$ and so is topologically located, convex, and absorbing. Suppose that $0 \in \overline{\sim 2T^{-1}(B_F)}$, and consider $\epsilon > 0$. Pick $x \in \sim 2T^{-1}(B_F)$ such that $\|x\| < \epsilon$. Either $\|x - x'\| < \epsilon/2 = (2-1)\epsilon/2$ for some $x' \in T^{-1}(B_F)$ or else $x \in -T^{-1}(B_F)$. In the first case, by Lemma 9.23, either $B_E(\epsilon) \between -T^{-1}(B_F)$ or $x \in 2T^{-1}(B_F)$; but the latter is ruled out, so the former must be the case. Therefore $\|z\| < \epsilon$ for some $z \in -T^{-1}(B_F)$. Thus, since $\epsilon > 0$ is arbitrary, we have $0 \in \overline{-T^{-1}(B_F)}$, a contradiction to Lemma 9.9. Hence $0 \notin \overline{\sim 2T^{-1}(B_F)}$. □

Theorem 9.32 *Let T be a linear mapping of a Banach space E into a Banach space F such that the graph $G = \{(x, Tx) \mid x \in E\}$ of T is closed in the product space $E \times F$, and $T^{-1}(B_F)$ is topologically located. Then T is sequentially continuous.*

Proof Note that T is strongly extensional by Lemma 9.27. Let $(x_n)_{n\in\mathbb{N}}$ be a sequence in E converging to 0. Then for each $\epsilon > 0$, by Lemma 9.7, either $\|Tx_n\| > \epsilon/2$ for infinitely many n or else $\|Tx_n\| < \epsilon$ for all sufficiently large n. In the first case, $\|(4/\epsilon)Tx_n\| > 2$ and hence $(4/\epsilon)x_n \notin 2T^{-1}(B_F)$, for infinitely many n. Now for each n, either $1 < \|(4/\epsilon)Tx_n\|$ and therefore $(4/\epsilon)x_n \in \sim T^{-1}(B_F)$, or else $\|(4/\epsilon)Tx_n\| < 2$ and therefore $(4/\epsilon)x_n \in 2T^{-1}(B_F)$. It follows that $(4/\epsilon)x_n \in \sim T^{-1}(B_F)$ for infinitely many n, and so $0 \in \overline{\sim T^{-1}(B_F)}$, a contradiction to Lemma 9.31. Thus we must have $\|Tx_n\| < \epsilon$ for all sufficiently large n. Since $\epsilon > 0$ is arbitrary, $(Tx_n)_{n\in\mathbb{N}}$ converges to 0 in F. □

Notes

The material of Subsections 9.3.1 and 9.3.2 is drawn from [20]. Here we use the notion of a topologically located set, instead of the notion of a located set, and the results there are generalised. Theorem 9.25 is new. A Brouwerian counterexample to the existence of Minkowski functionals can be found in [20] and [24, Proposition 26]. Locating the kernel of a linear map onto a finite-dimensional space is dealt with in [8] and [20].

The constructive version of the uniform boundedness theorem (Theorem 9.26) is given as a problem in [3, Chapter 9, Problem 6] and [4, Chapter 7, Problem 20]. Other versions, including one due to Royden [32], for normable linear mappings are given in [11, Theorems 6.2.11 and 6.2.12]; see also [11, Chapter 6, Exercises 12].

A Brouwerian counterexample to the standard classical uniform boundedness theorem (as found in [33, Sections 2.3–2.9]) can be found in [24, Lemma 12 and Proposition 30]. The counterexample make use of a boundedness principle, BD-N, introduced in [21], which holds in intuitionistic, constructive recursive and classical mathematics, and is equivalent to various theorems in analysis. Lietz and

Streicher [28] have shown that BD-N is underivable in a natural formal system for constructive analysis.

Versions of the open mapping theorem are given in [9] and [11, Theorem 6.6.4]; sequential versions of the open mapping, Banach inverse and closed graph theorems and their generalisation to F-spaces are given in [11, Theorem 6.6.11], [22], and [27].

Theorem 9.29, Corollary 9.30 and Theorem 9.32 are new.

9.4 Convexity

The Hahn–Banach theorem is a central tool in functional analysis from both theoretical and applied points of view and has several forms: the *continuous extension theorem*, the *separation theorem*, and the *dominated extension theorem*.

The standard proof of, say, the dominated extension theorem consists of two parts. The first of these involves the supremum of a set of real numbers, which does not always exist constructively; the second part makes use of a highly nonconstructive transfinite induction such as Zorn's lemma or Hausdorff's maximality theorem. See, for example, [33, Section 3.2].

In 1967, Bishop proved an approximate form [3, Chapter 9, Theorem 3] of the separation theorem for bounded convex subsets of a separable normed linear space, obtaining the continuous extension theorem [3, Chapter 9, Theorem 4] as a corollary. In this section, we give a proof of an approximate version of the separation theorem for two *unbounded* convex subsets, using the Baire theorem. This leads to an approximate version of the (one-dimensional) dominated extension theorem. We also deal with the weak topology in Hilbert spaces and show that a convex set which is totally bounded with respect to the weak topology is located. As a consequence, the closure of a convex set with respect to the weak topology is equal to the closure with respect to the norm.

9.4.1 The Hahn–Banach Theorem

In this subsection, we give yet another proof of an approximate version of the separation theorem by using the Baire category theorem, and a proof of an approximate version of the (one-dimensional) dominated extension theorem.

Definition 9.33 A linear functional f on a linear space E is a *subderivative* of a convex function $p : E \to \mathbb{R}$ at $x \in E$ if

$$f(y - x) \leq p(y) - p(x)$$

for all $y \in E$; a convex function $p : E \to \mathbb{R}$ is *subdifferentiable* at $x \in E$ if it has a subderivative at x.

Lemma 9.34 *Let E be a normed linear space, and let $p : E \to \mathbb{R}$ be a convex function. If p is continuous at $x \in E$, then there exist $r > 0$ and $L > 0$ such that*

$$|f(y)| \leq L\|y\|$$

for each subderivative f of p at $z \in B_E(x, r)$ and all $y \in E$.

Proof See [25, Lemma 58]. □

Definition 9.35 Let E be a linear space. A convex function $p : E \to \mathbb{R}$ is *Gâteaux differentiable at $x \in E$* with the *derivative $f : E \to \mathbb{R}$* if for each $y \in E$ and $\epsilon > 0$ there exists $\delta > 0$ such that

$$\forall t \in \mathbb{R}(0 < |t| < \delta \Rightarrow |p(x + ty) - p(x) - tf(y)| < \epsilon|t|).$$

Lemma 9.36 *Let E be a linear space, and let $p : E \to \mathbb{R}$ be a convex function. If p is Gâteaux differentiable at $x \in E$, then the derivative f is linear, and is a subderivative of p at x.*

Proof See [25, Lemma 59]. □

Proposition 9.37 *Let C be a located convex subset of a normed linear space E, and let $x \in E$ be such that $d = d(x, C) > 0$. Then the following are equivalent.*

(i) *The convex function $d(\cdot, C)$ is subdifferentiable at x.*
(ii) *There exists a normable linear functional f on E with $\|f\| = 1$ such that $f(x) + d \leq f(y)$ for all $y \in C$.*

Proof See [25, Proposition 60] □

The following theorem is a variant of Mazur's theorem [29] (see also [13, V.9.8]).

Proposition 9.38 *Let E be a separable Banach space, and let $p : E \to \mathbb{R}$ be a continuous convex function. Then p is Gâteaux differentiable at each point of a dense subset of E.*

Proof Given $y \in E$ and $\epsilon > 0$, let

$$U_{y,\epsilon} = \{x \in E \mid \exists t \in \mathbb{Q}^+(p(x + ty) + p(x - ty) - 2p(x) < \epsilon t)\},$$

where $\mathbb{Q}^+ = \{t \in \mathbb{Q} \mid t > 0\}$. Since p is continuous, $U_{y,\epsilon}$ is open; we show that it is dense in E. Given $x \in E$ and $\delta > 0$, applying Lemma 9.12, we can find $r > 0$ and $L > 0$ such that

$$\forall u, v \in B_E(x, r)(|p(u) - p(v)| \leq L\|u - v\|). \tag{*}$$

Choose n such that $2L\|y\| < n\epsilon$, and then choose $s \in \mathbb{Q}^+$ such that

$$(n + 1)s\|y\| < \min\{r, \delta\}.$$

For each $k < n$, let

$$a_k = p(x + (k+1)sy) - p(x + ksy).$$

Since $x + ksy \in B_E(x, r)$ for all $k \leq n+1$, we have

$$\begin{aligned}\sum_{k<n}(a_{k+1} - a_k) &= a_n - a_0 \\ &= p(x + (n+1)sy) - p(x + nsy) - (p(x + sy) - p(x)) \\ &\leq 2Ls\|y\| < n\epsilon s.\end{aligned}$$

Hence $a_{k+1} - a_k < \epsilon s$ for some $k < n$, by [4, Chapter 2, (2.16)]. For this k, let $z = x + (k+1)sy$. Then

$$p(z + sy) + p(z - sy) - 2p(z) = a_{k+1} - a_k < \epsilon s,$$

so $z \in U_{y,\epsilon}$. Also, $\|x - z\| = (k+1)s\|y\| < \delta$. Since $x \in E$ and $\delta > 0$ are arbitrary, it follows that $U_{y,\epsilon}$ is dense in E.

Now let $(y_n)_{n\in\mathbb{N}}$ be a dense sequence of E, and let $\epsilon_m = 2^{-m}$ for each m. By Theorem 9.4, the subset

$$U = \bigcap_{n,m} U_{y_n,\epsilon_m}$$

is also dense in E. Let $x \in U$. Compute $r > 0$ and $L > 0$ such that (*) holds. Given $y \in E$ and $\epsilon > 0$, choose n and m such that $\epsilon_m + 2L\|y - y_n\| < \epsilon$. Then there exists $s \in \mathbb{Q}^+$ such that $p(x + sy_n) + p(x - sy_n) - 2p(x) < \epsilon_m s$, or

$$\frac{p(x + sy_n) - p(x)}{s} - \frac{p(x - sy_n) - p(x)}{-s} < \epsilon_m.$$

Choose $\delta > 0$ such that $\delta < s$ and $\delta \max\{\|y\|, \|y_n\|\} < r$. Then for each t with $0 < t < \delta$, since

$$x + ty, x + ty_n, x - ty, x - ty_n \in B_E(x, r),$$

we have

$$\begin{aligned}&\frac{p(x + ty) - p(x)}{t} - \frac{p(x - ty) - p(x)}{-t} \\ &\quad\leq \frac{p(x + ty_n) - p(x) + tL\|y - y_n\|}{t} - \frac{p(x - ty_n) - p(x) + tL\|y - y_n\|}{-t} \\ &\quad= \frac{p(x + ty_n) - p(x)}{t} - \frac{p(x - ty_n) - p(x)}{-t} + 2L\|y - y_n\| \\ &\quad\leq \frac{p(x + sy_n) - p(x)}{s} - \frac{p(x - sy_n) - p(x)}{-s} + 2L\|y - y_n\| \\ &\quad< \epsilon_m + 2L\|y - y_n\| < \epsilon,\end{aligned}$$

by Lemma 9.11. Therefore the limit

$$f(y) = \lim_{t \to 0} \frac{p(x + ty) - p(x)}{t},$$

exists, from which we readily see that p is Gâteaux differentiable at X. □

By virtue of the previous proposition, we have the following approximate version of the separation theorem for two *unbounded* convex subsets.

Theorem 9.39 *Let A and B be convex subsets of a separable Banach space E such that the* algebraic difference

$$B - A = \{y - x \mid x \in A, y \in B\}$$

is located and $d = d(0, B - A) > 0$. Then for each $\epsilon > 0$ there exists a normable linear functional f on E with $\|f\| = 1$ such that

$$f(x) + d \leq f(y) + \epsilon$$

for all $x \in A$ and $y \in B$.

Proof Let $z \in E$ be such that $\|z\| < \epsilon$ and $d(\cdot, B - A)$ is Gâteaux differentiable at z (such a z exists by Proposition 9.38); then $d(\cdot, B - A)$ is subdifferentiable at z, by Lemma 9.36. By Proposition 9.37, there exists a normable linear functional f on E with $\|f\| = 1$, such that for all $x \in A$ and $y \in B$, $f(z) + d \leq f(y - x)$ and therefore

$$-\epsilon + d \leq f(z) + d \leq f(y) - f(x).$$

□

Corollary 9.40 *Let A and B be convex subsets of a separable normed linear space E such that the algebraic difference*

$$B - A = \{y - x \mid x \in A, y \in B\}$$

is located and $d = d(0, B - A) > 0$. Then for each $\epsilon > 0$ there exists a normable linear functional f on E with $\|f\| = 1$ such that

$$f(x) + d \leq f(y) + \epsilon$$

for all $x \in A$ and $y \in B$.

Proof Noting that A and B are convex subsets of the Banach space completion $\tilde{E}$ of E such that $B - A$ is located in $\tilde{E}$ and $d = d(0, B - A) > 0$, apply Theorem 9.39. □

By separating two unbounded convex subsets, we can directly prove the following continuous extension theorem.

Theorem 9.41 *Let M be a subspace of a separable normed linear space E, and let f be a non-zero linear functional on M such that the kernel $\ker(f)$ is located in E. Then for each $\epsilon > 0$ there exists a normable linear functional g on E with $\|g\| \leq \|f\| + \epsilon$, such that $g(x) = f(x)$ for all $x \in M$.*

Proof See [25, Theorem 64]. □

Proposition 9.42 *Let E be a linear space, let $p : E \to \mathbb{R}$ be a sublinear function, and let $x \in E$. Then the following are equivalent.*

(i) *p is subdifferentiable at x.*
(ii) *There exists a linear functional f on E with $f(x) = p(x)$ such that $f(y) \leq p(y)$ for all $y \in E$.*

Proof See [25, Proposition 65]. □

The following theorem is a constructive (approximate) version of the one-dimensional dominated extension theorem.

Theorem 9.43 *Let E be a separable Banach space, let $p : E \to \mathbb{R}$ be a continuous sublinear function, and let $x \in E$. Then for each $\epsilon > 0$ there exists a bounded linear functional f on E with $p(x) \leq f(x) + \epsilon$ such that $f(y) \leq p(y)$ for all $y \in E$.*

Proof By Lemmas 9.12 and 9.34, since p is continuous at x, there exist $r > 0$ and $L > 0$ such that (*) holds and $|f(y)| \leq L\|y\|$ for all $y \in E$ and each subderivative f of p at $z \in B_E(x, r)$. Choose $\delta > 0$ such that $\delta < r$ and $2L\delta < \epsilon$; then choose $z \in B_E(x, \delta)$ such that p is Gâteaux differentiable, and hence subdifferentiable, at z (such a z exists by Proposition 9.38). Let f be a subderivative of p at z. Then, as in the proof of Proposition 9.42, we have $f(z) = p(z)$ and $f(y) \leq p(y)$ for all $y \in E$. Since $z \in B_E(x, r)$, we have

$$\begin{aligned} p(x) &\leq p(z) + L\|x - z\| = f(z) + L\|x - z\| \\ &\leq f(x) + 2L\|x - z\| < f(x) + 2L\delta < f(x) + \epsilon. \end{aligned}$$ □

Corollary 9.44 *Let E be a separable normed linear space E, let $p : E \to \mathbb{R}$ be a continuous sublinear function, and let $x \in E$. Then for each $\epsilon > 0$ there exists a bounded linear functional f on E with $p(x) \leq f(x) + \epsilon$ such that $f(y) \leq p(y)$ for all $y \in E$.*

Proof Note that the extension $\tilde{p}$ of p to the Banach space completion $\tilde{E}$ of E is a continuous sublinear function, and apply Theorem 9.43; see the proof of [24, Corollary 67] for details. □

9.4.2 Weak Topology in Hilbert Spaces

In what follows, and in keeping with our discussion in Chapter 8, we deal only with *real* inner product and Hilbert spaces: those in which the ground field is $\mathbb{R}$. We begin by considering weak-topological properties of convex sets in a Hilbert space.

Definition 9.45 A subset A of an inner product space E is *weakly totally bounded* if, for each finitely enumerable subset $\{y_0, \dots, y_{m-1}\}$ of E and each $\epsilon > 0$, there exists a finitely enumerable subset $\{a_0, \dots, a_{n-1}\}$ of A such that for each $x \in A$

$$\exists i < n \forall j < m(|\langle x - a_i, y_j \rangle| < \epsilon).$$

Note that every totally bounded subset of an inner product space is weakly totally bounded.

Lemma 9.46 *If A is a weakly totally bounded subset of an inner product space E, then* $\sup\{\langle x, y\rangle \mid x \in A\}$ *exists for all $y \in E$.*

Proof Note that for each $y \in E$, the subset $\{\langle x, y\rangle \mid x \in A\}$ of $\mathbb{R}$ is totally bounded, and apply [4, Chapter 2, (4.4)]. □

Lemma 9.47 *Let C be a weakly totally bounded, convex subset of an inner product space E, and let $x \in E$. Then for each $y \in C$, each ϵ with $0 < \epsilon < 1$ and each $M \geq 1$, one of the following holds:*

(i) $\|x - y\| < \|x - z\| + \epsilon$ *for all* $z \in C$;
(ii) $M < \|x - z'\|^2$ *for some* $z' \in C$;
(iii) $\|x - y'\|^2 < \|x - y\|^2 - \epsilon^4/(64M)$ *for some* $y' \in C$.

Proof If $\|x - y\| < \epsilon$, then $\|x - y\| < \|x - z\| + \epsilon$ for all $z \in C$; and if $M < \|x - y\|^2$, then there is nothing to prove. Hence we may assume that $\epsilon/2 < \|x - y\|$ and $\|x - y\|^2 < 2M$. Since $\sup\{\langle z, x - y\rangle : z \in C\}$ exists, either (a) $\langle z - y, x - y\rangle < \epsilon^2/2$ for all $z \in C$ or else (b) $\epsilon^2/4 < \langle z_0 - y, x - y\rangle$ for some $z_0 \in C$. In case (a), for each $z \in C$, since

$$\begin{aligned}\|x - y\|^2 &= \langle x - y, x - y\rangle = \langle x - z + z - y, x - y\rangle \\ &= \langle x - z, x - y\rangle + \langle z - y, x - y\rangle < \|x - y\|\|x - z\| + \frac{\epsilon^2}{2},\end{aligned}$$

we have

$$\|x - y\| < \|x - z\| + \frac{\epsilon^2}{2\|x - y\|} < \|x - z\| + \epsilon.$$

In case (b), we consider two subcases: if $M < \|x - z_0\|^2$, then there is nothing to prove; if $\|x - z_0\| < 2M$, then, setting $\lambda = \epsilon^2/(32M)$ and $y' = y + \lambda(z_0 - y)$, we have $0 < \lambda < 1$ and $y' \in C$, so

$$\begin{aligned}\|x - y'\|^2 &= \|x - y\|^2 - 2\lambda\langle x - y, z_0 - y\rangle + \lambda^2\|z_0 - y\|^2 \\ &< \|x - y\|^2 - \frac{\lambda\epsilon^2}{2} + \lambda^2(2\|x - z_0\|^2 + 2\|x - y\|^2) \\ &< \|x - y\|^2 - \frac{\lambda\epsilon^2}{2} + \lambda^2(4M + 4M) \\ &= \|x - y\|^2 - \frac{\epsilon^4}{64M} + \frac{\epsilon^4}{128M} \\ &= \|x - y\|^2 - \frac{\epsilon^4}{64M}.\end{aligned}$$

□

Theorem 9.48 *Let C be an inhabited, weakly totally bounded, bounded, convex subset of an inner product space E. Then C is located.*

Proof Let $x \in E$. It will suffice to show that for each ϵ with $0 < \epsilon < 1$ there exists $y \in C$ such that $\|x - y\| < \|x - z\| + \epsilon$ for all $z \in C$: in fact, given $s, t \in \mathbb{R}$ with $s < t$, set $\epsilon = \min\{t - s, 1\}/2$, and if $\|x - y\| < \|x - z\| + \epsilon$ for all $z \in C$, then either $(t + s)/2 < \|x - y\|$ or $\|x - y\| < t$; in the former case, we have $s < \|x - z\|$ for all $z \in C$; whence $\inf\{\|x - z\| \mid z \in C\}$ exists, by [4, Chapter 2, (4.3)]. Let $0 < \epsilon < 1$. Fixing $y_0 \in C$, choose $M \geq 1$ such that $\|x - z\|^2 \leq M$ for all $z \in C$, we construct an increasing binary sequence $(\lambda_n)_{n \in \mathbb{N}}$ with $\lambda_0 = 0$ and a sequence $(y_n)_{n \in \mathbb{N}}$ in C such that

$$\begin{aligned}\lambda_{n+1} = 0 &\Rightarrow \|x - y_{n+1}\| < \|x - y_n\| - \frac{\epsilon^4}{64M}, \\ \lambda_{n+1} = 1 &\Rightarrow \|x - y_{n+1}\| < \|x - z\| + \epsilon \text{ for all } z \in C.\end{aligned}$$

Suppose that we have constructed $\lambda_0, \ldots, \lambda_n$ and $y_0, \ldots, y_n$. If $\lambda_n = 1$, set $\lambda_{n+1} = 1$ and $y_{n+1} = y_n$. If $\lambda_n = 0$, we consider the alternatives in the conclusion of Lemma 9.47 as follows. If $\|x - y_n\| < \|x - z\| + \epsilon$ for all $z \in C$, we set $\lambda_{n+1} = 1$ and $y_{n+1} = y_n$; if $\|x - y'\| < \|x - y_n\| - \epsilon^4/(64M)$ for some $y' \in C$, we set $\lambda_{n+1} = 0$ and $y_{n+1} = y'$; the third alternative, $M < \|x - z'\|^2$ for some $z' \in C$, is ruled out by our choice of M. This completes the inductive construction.

Now choose N such that $\|x - y_0\| < N(\epsilon^4/(64M))$. If $\lambda_N = 0$, then

$$0 \leq \|x - y_N\| < \|x - y_0\| - \frac{N\epsilon^4}{64M} < 0,$$

a contradiction. Therefore $\lambda_N = 1$. □

Theorem 9.49 *Let C be an inhabited, weakly totally bounded, convex subset C of a Hilbert space H. Then C is located.*

Proof Fixing $y_0 \in C$, consider $x \in H$ and $\epsilon \in \mathbb{R}$ with $0 < \epsilon < 1$. Again, it will suffice to find $y \in C$ such that $\|x-y\| < \|x-z\|+\epsilon$ for all $z \in C$. If $\|x-y_0\| < \epsilon$, then $\|x-y_0\| < \|x-z\|+\epsilon$ for all $z \in C$; hence we may assume that $0 < \|x-y_0\|$. Construct a ternary sequence $(\lambda_n)_{n\in\mathbb{N}}$ with $\lambda_0 = 0$ and a sequence $(y_n)_{n\in\mathbb{N}}$ in C such that for each n, $\|x - y_{n+1}\| \le \|x - y_n\|$ and

$$\lambda_{n+1} = 0 \Rightarrow n+1 < \|x - z'\|^2 \text{ for some } z' \in C,$$
$$\lambda_{n+1} = 1 \Rightarrow \|x - y_{n+1}\| < \|x - z\| + \epsilon \text{ for all } z \in C,$$
$$\lambda_{n+1} = 2 \Rightarrow \|x - y_{n+1}\|^2 < \|x - y_n\|^2 - \frac{\epsilon^4}{64(n+1)}.$$

Suppose that we have constructed $\lambda_0, \dots, \lambda_n$ and $y_0, \dots, y_n$. If $\lambda_n = 1$, set $\lambda_{n+1} = 1$ and $y_{n+1} = y_n$. If $\lambda_n = 0$, we again consider the alternatives in the conclusion of Lemma 9.47. If $\|x - y_n\| < \|x - z\| + \epsilon$ for all $z \in C$, we set $\lambda_{n+1} = 1$ and $y_{n+1} = y_n$; if $(n+1) < \|x - z'\|^2$ for some $z' \in C$, we set $\lambda_{n+1} = 0$ and $y_{n+1} = y_n$; if $\|x - y'\|^2 < \|x - y_n\|^2 - \epsilon^4/64(n+1)$ for some $y' \in C$, we set $\lambda_{n+1} = 2$ and $y_{n+1} = y'$. This completes the inductive construction.

Let $(N_k)_{k\in\mathbb{N}}$ be a strictly increasing sequence of positive integers such that

$$\|x - y_0\|^2 < \sum_{n=N_k+1}^{N_{k+1}} \frac{\epsilon^4}{64(n+1)}.$$

Then for each k there exists n_k with $N_k < n_k \le N_{k+1}$ such that $\lambda_{n_k} = 0$ or $\lambda_{n_k} = 1$: in fact, if $\lambda_n = 2$ for all n with $N_k < n \le N_{k+1}$, then

$$0 \le \|x - y_{N_{k+1}}\|^2 < \|x - y_{N_k}\|^2 - \sum_{n=N_k+1}^{N_{k+1}} \frac{\epsilon^4}{64(n+1)}$$
$$\le \|x - y_0\|^2 - \sum_{n=N_k+1}^{N_{k+1}} \frac{\epsilon^4}{64(n+1)} < 0,$$

a contradiction. Define a sequence $(u_k)_{k\in\mathbb{N}}$ as follows: if $\lambda_{n_k} = 0$, choose $z' \in C$ with $(n_k + 1) < \|x - z'\|^2$ and set $u_k = x - z'$; if $\lambda_{n_k} = 1$, set

$$u_k = \frac{n_k + 1}{\|x - y_0\|}(x - y_0).$$

Then,

$$\|u_k\|^2 \ge \min\left\{(n_k+1)^2, n_k+1\right\} = n_k + 1 > N_k \ge k \to \infty \text{ as } k \to \infty.$$

Applying Theorem 9.26 to the bounded linear functionals $w \mapsto \langle u_n, w\rangle$ on H, we can find $w_0 \in H$ such that the sequence $(|\langle u_k, w_0\rangle|)_{k\in\mathbb{N}}$ is unbounded. Since C is weakly totally bounded, there exists a positive integer M such that $|\langle x-z, w_0\rangle| \leq M$ for all $z \in C$. Choose K such that $M < |\langle u_K, w_0\rangle|$. If $\lambda_{n_K} = 0$, then there exists $z' \in C$ such that $u_K = x - z'$, and hence

$$M < |\langle u_K, w_0\rangle| = |\langle x - z', w_0\rangle| \leq M,$$

a contradiction. Therefore $\lambda_{n_K} = 1$, and $\|x - y_{n_K}\| < \|x - z\| + \epsilon$ for all $z \in C$. □

Definition 9.50 A sequence $(x_n)_{n\in\mathbb{N}}$ in an inner product space E *converges weakly to a limit* $x \in E$ if for each finitely enumerable subset $\{y_0, \ldots, y_{m-1}\}$ of E and each $\epsilon > 0$, there exists N such that

$$\forall n \geq N \forall j < m(|\langle x_n - x, y_j\rangle| < \epsilon).$$

We then write $x_n \rightharpoonup x$ as $n \to \infty$.

Lemma 9.51 *Let $(x_n)_{n\in\mathbb{N}}$ be a sequence in an inner product space E converging weakly to a limit $x \in E$. Then $\mathrm{co}\{x_n \mid n \in \mathbb{N}\}$ is weakly totally bounded.*

Proof Let $(x_n)_{n\in\mathbb{N}}$ be a sequence in an inner product space E converging weakly to $x \in E$. Consider a finitely enumerable subset $\{y_0, \ldots, y_{m-1}\}$ of E and $\epsilon > 0$. Then there exists N such that $|\langle x_n - x, y_j\rangle| < \epsilon/4$ for all $n \geq N$ and $j < m$. Since $C = \mathrm{co}\{x_0, \ldots, x_N\}$ is totally bounded, it is weakly totally bounded. Therefore there exists a finitely enumerable subset $\{a_0, \ldots, a_{l-1}\}$ of C, hence of $\mathrm{co}\{x_n \mid n \in \mathbb{N}\}$, such that for each $z \in C$

$$\exists k < l \forall j < m(|\langle z - a_k, y_j\rangle| < \frac{\epsilon}{2}.$$

For each $z = \sum_{i<M} \lambda_i x_i \in \mathrm{co}\{x_n \mid n \in \mathbb{N}\}$, where $\lambda_i \geq 0$ for all $i < M$ and $\sum_{i<M} \lambda_i = 1$, since

$$z' = \sum_{i<N} \lambda_i x_i + \sum_{N\leq i<M} \lambda_i x_N \in C,$$

we can find $k < l$ such that $|\langle z' - a_k, y_j\rangle| < \epsilon/2$ for all $j < m$; for such k and each $j < m$, we have

$$\begin{aligned}
|\langle z - a_k, y_j\rangle| &\leq |\langle z' - a_k, y_j\rangle| + |\langle z - z', y_j\rangle| \\
&< \frac{\epsilon}{2} + \left|\left\langle \sum_{N\leq i<M} \lambda_i(x_i - x_N), y_i \right\rangle\right| \\
&\leq \frac{\epsilon}{2} + \sum_{N\leq i<M} \lambda_i|\langle x_i - x_N, y_i\rangle| \\
&\leq \frac{\epsilon}{2} + \sum_{N\leq i<M} \lambda_i(|\langle x_i - x, y_i\rangle| + |\langle x - x_N, y_i\rangle|)
\end{aligned}$$

$$\leq \frac{\epsilon}{2} + \sum_{N \leq i < M} \lambda_i \left(\frac{\epsilon}{4} + \frac{\epsilon}{4} \right)$$
$$\leq \frac{\epsilon}{2} + \sum_{N \leq i < M} \lambda_i \frac{\epsilon}{2} \leq \frac{\epsilon}{2} + \frac{\epsilon}{2} = \epsilon.$$
□

Theorem 9.52 *If $(x_n)_{n \in \mathbb{N}}$ is a sequence in a Hilbert space H converging weakly to $x \in H$, then there exists a sequence $(y_k)_{k \in \mathbb{N}}$ in H such that*

(i) *each y_k is a convex combination of finitely many x_n;*
(ii) *$(y_k)_{k \in \mathbb{N}}$ converges to x.*

Proof Let $(x_n)_{n \in \mathbb{N}}$ be a sequence in a Hilbert space H converging weakly to $x \in H$. Then the subset $C = \text{co}\{x_n \mid n \in \mathbb{N}\}$ of H is weakly totally bounded, by Lemma 9.51, and hence, by Theorem 9.49, located. It suffices to show that $d(x, C) = 0$. Assume that $d(x, C) > 0$. Then there exists $y \in H$ such that

$$\langle x_n, y \rangle + d(x, C)^2 \leq \langle x, y \rangle$$

for all n, by [24, Theorems 7 and 15]. This contradicts the fact that $\langle x_n, y \rangle \to \langle x, y \rangle$ as $n \to \infty$. Therefore $d(x, C) = 0$. □

Corollary 9.53 *If C is a convex subset of a Hilbert space H, then the weak closure of C is equal to its closure $\overline{C}$.*

Proof Immediate from Theorem 9.52. □

Notes

The material of Subsection 9.4.1 is drawn from [25]. A recursive counterexample to the Hahn–Banach theorem is given in [30], and a Brouwerian counterexample can be found in [24, Proposition 28]; see also [17]. In Bishop's proof [3, Chapter 9, Theorem 3], the boundedness of convex sets is crucial; see also [4, Chapter 7, (4.3)], and [11, Theorem 5.2.9]. A constructive proof of the continuous extension theorem along with a standard classical proof is given in [7, Chapter 2, (5.9), and (5.10)]. The constructive proof of an exact version of the Hahn–Banach theorem, making use of geometric properties of the space, is given in [16]. A treatment of the Hahn–Banach theorem in formal topology [34] is given in [12].

The material of Subsection 9.4.2 is drawn from [23], but the above proofs are simpler; note also [31]. A subset A of a Hilbert space H is commonly defined to be weakly totally bounded if the set $\{\langle x, y \rangle \mid x \in A\}$ is a totally bounded subset of $\mathbb{R}$ for all $y \in H$ in [6]; our Definition 9.45 is more natural from a topological point of view. A partial converse of Theorem 9.48,

If C is an inhabited, bounded, located, convex subset of an inner product space E, then $\sup\{\langle x, y\rangle \mid x \in C\}$ *exists for all $y \in E$,*

is also given in [23] and [31]; a Brouwerian example showing that boundedness is necessary in the converse is given in [31]. The full converse,

If C is an inhabited, bounded, located, convex subset of an inner product space E, then C is weakly totally bounded,

is an open problem. The results in [23] are generalised to uniformly smooth spaces in [26] and are also presented in [11, Section 6.2].

A constructive treatment of compact convex sets, including the Banach–Alaoglu theorem and the Krein–Milman theorem, is given in [3, Chapter 9.6], [4, Chapter 7.7], and [11, Chapter 5.4].

9.5 Duality in Hilbert Spaces

Since, constructively, we cannot prove that all bounded linear functionals on a Banach space are normable or that the set of normable normable linear functionals on a Banach space is linear, dealing with duality in Banach spaces is more complicated than in the classical setting. However, Hilbert spaces have good geometric properties, and the set of normable linear functionals on a Hilbert space does form a linear space.

In the remainder of this chapter, H will be a (real) Hilbert space.

9.5.1 Adjoints

In this subsection, we discuss the basic properties of adjoints of operators on a Hilbert space, and we prove the Hellinger–Toeplitz and the closed range theorems.

Definition 9.54 An *operator* T on H is a linear mapping of H into itself. An operator S on H is an *adjoint* of an operator T on H if $\langle Tx, y\rangle = \langle x, Sy\rangle$ for all $x, y \in H$.

If the operator T has an adjoint, then that adjoint is unique and is denoted by T^*; then $(T^*)^* = T$.

Lemma 9.55 *An operator T on H has an adjoint if and only if*

$$\sup\{|\langle Tx, y\rangle| \mid x \in B_H\}$$

exists for all $y \in H$.

Proof For each $y \in H$, define the linear mapping u_y on H by $u_y(x) = \langle Tx, y \rangle$. Then $\sup\{|\langle Tx, y \rangle| \mid x \in B_H\}$ exists if and only if u_y is normable, which, by the Riesz representation theorem ([4, Chapter 8, (2.3)] and [11, Theorem 4.3.6]), happens if and only if there exists a unique $z \in H$ such that $u_y(x) = \langle x, z \rangle$ for all $x \in H$. Thus $\sup\{|\langle Tx, y \rangle| \mid x \in B_H\}$ exists for all $y \in H$ if and only if there is a (clearly linear) mapping $S : H \to H$ such that $\langle Tx, y \rangle = u_y(x) = \langle x, Sy \rangle$ for all $x, y \in H$. □

Lemma 9.56 *If T is an operator on H with an adjoint, and $x \in H$, then $Tx = 0$ if and only if $T^*Tx = 0$.*

Proof If $T^*Tx = 0$, then $\|Tx\|^2 = \langle Tx, Tx \rangle = \langle x, T^*Tx \rangle = 0$, so $Tx = 0$. The converse is trivial. □

Definition 9.57 An operator T on H is

- *weakly compact* if $T(B_H)$ is a weakly totally bounded subset of H, and
- *compact* if $T(B_H)$ is a totally bounded subset of H.

Lemma 9.58 *Let $y_0, \ldots, y_{n-1}$ be elements of H. Then the mapping u of H into $\mathbb{R}^n$ given by*

$$u(x) = (\langle x, y_0 \rangle, \ldots, \langle x, y_{n-1} \rangle)$$

is compact.

Proof See [18, Proposition 1] or [24, Proposition 35]. □

Proposition 9.59 *An operator T on H has an adjoint if and only if T is weakly compact.*

Proof Suppose that T has an adjoint T^*. Then for each $y_0, \ldots, y_{m-1} \in H$, the mapping $u : H \to \mathbb{R}^m$ given by

$$u(x) = (\langle x, T^*y_0 \rangle, \ldots, \langle x, T^*y_{n-1} \rangle) = (\langle Tx, y_0 \rangle, \ldots, \langle Tx, y_{n-1} \rangle)$$

is compact, by Lemma 9.58 – that is, $u(B_H)$ is totally bounded. It follows from Definition 9.45 that $T(B_H)$ is weakly totally bounded.

Conversely, if T is weakly compact, then we see from Lemmas 9.46 and 9.55 that T^* exists. □

Theorem 9.60 *Let T and S be weakly compact operators on H, and let $a \in \mathbb{R}$. Then*

(i) *aT is weakly compact;*
(ii) *$T + S$ is weakly compact;*
(iii) *T^* is weakly compact;*
(iv) *TS is weakly compact.*

Proof Straightforward, in view of Proposition 9.59. □

Definition 9.61 Let A be a subset of an inner product space E. Then the *orthogonal complement* $A^{\perp}$ of A is defined by

$$A^{\perp} = \{x \in E \mid \langle x, a \rangle = 0 \text{ for all } a \in A\}$$

and is a closed subspace of E.

Lemma 9.62 *Let T be a weakly compact operator on H. Then* $\ker(T) = \operatorname{ran}(T^*)^{\perp}$. *Moreover,* $\operatorname{ran}(T^*)$ *is located if and only if* $\ker(T) + \operatorname{ran}(T^*)$ *is dense in H; in which case,* $\ker(T)$ *is located.*

Proof See [11, Lemma 6.5.2]. □

We have the following constructive version of the *Hellinger–Toeplitz theorem.*

Theorem 9.63 *A weakly compact operator T on H is sequentially continuous.*

Proof Let $(x_n)_{n\in\mathbb{N}}$ be a sequence in H converging to 0, and choose a strictly increasing sequence $(N_k)_{k\in\mathbb{N}}$ of natural numbers such that $\|x_n\| < 1/(k+1)^2$ for all $n \geq N_k$. We may assume that $\|x_n\| < 1$ for all n, and hence that $N_0 = 0$. Let

$$m_n = \max\{k \mid N_k \leq n\} \quad (n \in \mathbb{N}).$$

Since, by Lemma 9.27, T is strongly extensional, we see from Lemma 9.7 that for each $\epsilon > 0$, either $\|Tx_n\| > \epsilon/2$ for infinitely many n or else $\|Tx_n\| < \epsilon$ for all sufficiently large n. In the first case, since $(m_n)_{n\in\mathbb{N}}$ is increasing and $m_{N_k} = k$ for all k, the sequence $(\|(m_n+1)Tx_n\|)_{n\in\mathbb{N}}$ is unbounded; so applying Theorem 9.26 to the bounded linear functionals

$$y \mapsto \langle (m_n+1)Tx_n, y \rangle,$$

we can find $y \in H$ such that $(|\langle (m_n+1)Tx_n, y \rangle|)_{n\in\mathbb{N}}$ is unbounded. However, for each k and each $n \geq N_k$, since $N_{m_n} \leq n$ and $k \leq m_n$, we have

$$\|(m_n+1)x_n\| < \frac{m_n+1}{(m_n+1)^2} = \frac{1}{m_n+1} \leq \frac{1}{k+1}$$

and so

$$|\langle (m_n+1)Tx_n, y \rangle| = |\langle (m_n+1)x_n, T^*y \rangle| \leq \|(m_n+1)x_n\| \|T^*y\| \to 0$$

as $n \to \infty$. This contradiction ensures that we must have $\|Tx_n\| < \epsilon$ for all sufficiently large n. □

Corollary 9.64 *Let T be an operator on H. If T^* exists, then T and T^* are sequentially continuous.*

Proof If T^* exists, then by Proposition 9.59 and Theorem 9.63, T is weakly compact therefore sequentially continuous. Moreover, in that case T^* has adjoint T and is therefore sequentially continuous. □

We now turn towards the closed range theorem.

Lemma 9.65 *A sequentially continuous linear mapping T of a normed linear space E into a normed linear space F maps Cauchy sequences to Cauchy sequences.*

Proof See [11, Proposition 6.5.12]. □

Lemma 9.66 *Let T be a weakly compact operator on H with closed range. Then*

$$S = \{(T^*Tx, Tx) \mid x \in H\}$$

is a one–one, sequentially continuous linear mapping of $\mathrm{ran}(T^*T)$ *onto* $\mathrm{ran}(T)$.

Proof It follows from Lemma 9.56 that S is a well-defined, one–one mapping of $\mathrm{ran}(T^*T)$ into $\mathrm{ran}(T)$; it is clearly linear and onto $\mathrm{ran}(T)$. For the sequential continuity of S, see the proof of [11, Lemma 6.5.15]. □

Lemma 9.67 *Let T be a weakly compact operator on H. Then* $\mathrm{ran}(T^*T)$ *is dense in* $\mathrm{ran}(T^*)$.

Proof See [6, Lemma 2] and [11, Lemma 6.5.3]. □

Theorem 9.68 *Let T be a weakly compact operator on H. Then the following are equivalent:*

(i) $\mathrm{ran}(T)$ *is closed,*
(ii) $\mathrm{ran}(T^*)$ *is closed;*

in which case, $\ker(T)$, $\ker(T^*)$, $\mathrm{ran}(T)$, *and* $\mathrm{ran}(T^*)$ *are located.*

Proof Suppose that $\mathrm{ran}(T)$ is closed. Since, by Lemma 9.67, $\mathrm{ran}(T^*T)$ is dense in $\mathrm{ran}(T^*)$, to show that $\mathrm{ran}(T^*)$ is closed it suffices to show that $\mathrm{ran}(T^*T)$ is complete. Let S be as in Lemma 9.66, and let $(T^*Tx_n)_{n\in\mathbb{N}}$ be a Cauchy sequence in $\mathrm{ran}(T^*T)$. By Lemmas 9.65 and 9.66, S maps Cauchy sequences to Cauchy sequences. Hence $(Tx_n)_{n\in\mathbb{N}}$ is a Cauchy sequence in the closed, and therefore complete, space $\mathrm{ran}(T)$, and so it converges to some $Tx \in \mathrm{ran}(T)$. Since T^* is sequentially continuous, by Theorem 9.63, $(T^*Tx_n)_{n\in\mathbb{N}}$ converges to T^*Tx. Therefore $\mathrm{ran}(T^*) = \mathrm{ran}(T^*T)$ is complete.

Thus for each $x \in H$ there exists $y \in H$ such that $T^*x = T^*Ty$, and therefore $x = Ty + (x - Ty)$, where $Ty \in \mathrm{ran}(T)$ and $x - Ty \in \ker(T^*)$. It follows from Lemma 9.62 that $\mathrm{ran}(T)$ and $\ker(T^*)$ are located, □

9.5.2 Compact Operators

We end by looking at the basic properties of compact operators on our Hilbert space H.

Theorem 9.69 *If T is a compact operator on H, then T^* is compact.*

Proof Since $T(B_H)$ is totally bounded, it is weakly totally bounded, so T^* exists, by Proposition 9.59. Given $\epsilon > 0$, since T is compact, we can find a finitely enumerable subset $\{x_0, \ldots, x_{n-1}\}$ of B_H such that, for each $x \in B_H$,

$$\exists i < n(\|Tx - Tx_i\| < \epsilon/4).$$

By Theorem 9.60, T^* is weakly compact; hence there exists a finitely enumerable subset $\{y_0, \ldots, y_{m-1}\}$ of B_H such that for each $x \in B_H$,

$$\exists j < m \forall i < n(|\langle T^*(x - y_j), x_i\rangle| < \epsilon/2).$$

For each $x \in B_H$, choose $j < m$ such that $|\langle T^*(x - y_j), x_i\rangle| < \epsilon/2$ for all $i < n$. Then for each $z \in B_H$ there exists $i < n$ such that $\|Tx - Tx_i\| < \epsilon/4$, and therefore

$$\begin{aligned} |\langle T^*(x - y_j), z\rangle| &= |\langle x - y_j, Tz\rangle| \leq |\langle x - y_j, Tx_i\rangle| + |\langle x - y_j, Tz - Tx_i\rangle| \\ &\leq |\langle T^*(x - y_j), x_i\rangle| + \|x - y_j\|\|Tz - Tx_i\| \\ &< \epsilon/2 + 2\epsilon/4 = \epsilon. \end{aligned}$$

It follows that

$$\|T^*(x - y_j)\| = \sup\{|\langle T^*(x - y_j), z\rangle| \mid z \in B_H\} \leq \epsilon.$$

Thus $T^*(B_H)$ is totally bounded, and therefore T^* is compact. □

Lemma 9.70 *A sequentially continuous linear mapping T of a normed linear space E into a normed linear space F maps totally bounded sets to bounded sets.*

Proof For each $k \in \mathbf{N}$, set $\epsilon_k = 2^{-(k+2)}/(k+1)$. Given a totally bounded subset $K \subseteq E$ and $k \in \mathbf{N}$, choose n_k and a finitely enumerable subset $\{x_0^k, \ldots, x_{n_k-1}^k\}$ of K such that for each $x \in K$ there exists $i < n_k$ with $\|x - x_i^k\| < \epsilon_k$, and define $M_k = \max\{\|Tx_i^k\| \mid i < n_k\}$. To show that $(M_k)_{k\in\mathbf{N}}$ is bounded, construct a binary sequence $(\lambda_k)_{k\in\mathbf{N}}$ such that if $\lambda_k = 0$, then

$$\forall i < n_k \forall j < n_{k+1} \left(\|x_i^k - x_j^{k+1}\| > \epsilon_k \vee \|Tx_j^{k+1}\| < M_k + 2^{-k}\right),$$

and if $\lambda_k = 1$, then

$$\exists i < n_k \exists j < n_{k+1} \left(\|x_i^k - x_j^{k+1}\| < 2\epsilon_k \wedge \|Tx_j^{k+1}\| > M_k + 2^{-(k+1)}\right).$$

Then construct a sequence $(y_n)_{n\in\mathbb{N}}$ in E as follows: if $\lambda_k = 0$, set $y_k = 0$; if $\lambda_k = 1$, choose i, j with $i < n_k$ and $j < n_{k+1}$ such that $\|x_i^k - x_j^{k+1}\| < 2\epsilon_k$ and $\|Tx_j^{k+1}\| > M_k + 2^{-(k+1)}$, and set $y_k = 2^{k+1}(x_i^k - x_j^{k+1})$. Since

$$\begin{aligned}\|Ty_k\| &= 2^{k+1}\|Tx_i^k - Tx_j^{k+1}\| \geq 2^{k+1}(\|Tx_j^{k+1}\| - \|T_i^k\|)\\ &\geq 2^{k+1}(\|Tx_j^{k+1}\| - M_k) > 2^{k+1}2^{-(k+1)} = 1\end{aligned}$$

and

$$\|y_k\| = 2^{k+1}\|x_i^k - x_j^{k+1}\| < 2^{k+2}\epsilon_k = \frac{1}{k+1},$$

we have $\|y_k\| < 1/(k+1)$ for all k, so $(y_k)_{k\in\mathbb{N}}$ converges to 0. By sequential continuity, $(Ty_k)_{k\in\mathbb{N}}$ converges to 0, and therefore there exists N such that $\|Ty_k\| < 1$ for all $k \geq N$. For such k and each $j < n_{k+1}$, pick $i < n_k$ such that $\|x_i^k - x_j^{k+1}\| < \epsilon_k$; then, since $\lambda_k = 0$, we have $\|Tx_j^{k+1}\| < M_k + 2^{-k}$; thus $M_{k+1} < M_k + 2^{-k}$. It follows that

$$M_{N+n} < M_N + (2^{-N} + 2^{-N+1} + \cdots + 2^{-n}) < M_N + 1$$

for all n. Hence $\|Tx_i^k\| < M_N + 1$ for all i, k. Since $\{x_i^k \mid i < n_k, k \geq N\}$ is dense in K, $T(K)$ is bounded by $M_N + 1$. □

Lemma 9.71 *A weakly compact operator T on H maps totally bounded sets to totally bounded sets.*

Proof Given a totally bounded subset K of H, we see that by Theorem 9.60, T^*T is weakly compact, and it is sequentially continuous, by Theorem 9.63. Hence $T^*T(K)$ is bounded, by Lemma 9.70. Let $M > 0$ be a bound for $T^*T(K)$, and let $\epsilon > 0$. Choose n and a finitely enumerable subset $\{x_0, \ldots, x_{n-1}\}$ of K such that for each $x \in K$ there exists $i < n$ with $\|x - x_i\| < \epsilon^2/(2M)$. Then for such $x \in K$ and $i < n$ we have

$$\begin{aligned}\|Tx - Tx_i\|^2 &= \langle Tx - Tx_i, Tx - Tx_i\rangle = \langle T^*Tx - T^*Tx_i, x - x_i\rangle\\ &\leq \|T^*Tx - T^*Tx_i\|\|x - x_i\| \leq (\|T^*Tx\| + \|T^*Tx_i\|)\|x - x_i\|\\ &< (M + M)\frac{\epsilon^2}{2M} = \epsilon^2.\end{aligned}$$

□

Theorem 9.72 *Let T and S be respectively a weakly compact operator and a compact operator on H. Then TS and ST are compact.*

Proof Since $S(B_H)$ is totally bounded, so is $TS(B_H)$, by Lemma 9.71. Thus TS is compact. Since T^* is weakly compact, by Theorem 9.60, and S^* is compact, by Theorem 9.69, T^*S^* is compact; so $ST = (T^*S^*)^*$ is compact, by Theorem 9.69. □

Theorem 9.73 *Let T and S be compact operators on H, and let $a \in \mathbb{R}$. Then*

(i) *aT is compact;*
(ii) *$T + S$ is compact.*

Proof It is trivial to prove (i). For (ii), first note that S^* and T^* exist and are compact, by Theorem 9.69. Given $\epsilon > 0$, we can find finitely enumerable subsets $\{x_0, \ldots, x_{n-1}\}$ and $\{x'_0, \ldots, x'_{m-1}\}$ of B_H such that, for each $x \in B_H$,

$$\exists i < n \left(\|S^*x - S^*x_i\| < \frac{\epsilon}{6} \right) \quad \text{and} \quad \exists j < m \left(\|T^*x - T^*x'_j\| < \frac{\epsilon}{6} \right).$$

Since the mapping $u : H \to \mathbb{R}^{nm}$ given by

$$u(x) = \left(\langle x, S^*x_0 + T^*x_0 \rangle, \ldots, \langle x, S^*x_{n-1} + T^*x'_{m-1} \rangle \right)$$

is compact, by Lemma 9.58, there exist $y_0, \ldots, y_{l-1} \in B_H$ such that for each $x \in B_H$ there exists $k < l$ such that $|\langle x - y_k, S^*x_i + T^*x'_j \rangle| < \epsilon/3$ for all $i < n$ and $j < m$. Consider such x and k. Given $z \in B_H$, choose $i < n$ and $j < m$ such that $\|S^*z - S^*x_i\| < \epsilon/6$ and $\|T^*z - T^*x'_j\| < \epsilon/6$, and hence

$$\begin{aligned}
|\langle (S+T)(x-y_k), z \rangle| &= |\langle x - y_k, (S^* + T^*)z \rangle| \\
&\le |\langle x - y_k, S^*x_i + T^*x'_j \rangle| \\
&\quad + |\langle x - y_k, (S^*z + T^*z) - (S^*x_i + T^*x'_j) \rangle| \\
&< \frac{\epsilon}{3} + |\langle x - y_k, S^*z - S^*x_i \rangle| + |\langle x - y_k, T^*z - T^*x'_j \rangle| \\
&\le \frac{\epsilon}{3} + \|x - y_k\| \|S^*z - S^*x_i\| + \|x - y_k\| \, \|T^*z - T^*x'_j\| \\
&\le \frac{\epsilon}{3} + 2\frac{\epsilon}{6} + 2\frac{\epsilon}{6} = \epsilon.
\end{aligned}$$

It follows that

$$\|(S+T)(x-y_k)\| = \sup\{ |\langle (S+T)(x-y_k), z \rangle| \mid z \in B_H \} \le \epsilon.$$

Since x and ϵ are arbitrary, $(S+T)(B_H)$ is totally bounded, and therefore $S + T$ is compact. □

Notes

A Brouwerian counterexample to the statement

Every bounded operator has an adjoint

can be found in [18, Brouwerian Example 3] and [24, Proposition 32]. Although an operator T on a Hilbert space H is normally defined to be weakly compact if the set $\{\langle Tx, y\rangle \mid x \in B_H\}$ is a totally bounded subset of $\mathbb{R}$ for all $y \in H$ in [24, Definition 37], our Definition 9.57 is natural from a topological point of view.

A version of the Hellinger–Toeplitz theorem is given in [5, Theorem 4]; the version here is given in [11, Theorem 6.3.6]. The closed range theorem is given in [6, Theorem 1]; a gap in the proof there is filled by [11, Lemma 6.5.10, Lemma 6.5.11, and Proposition 6.5.12].

The basic properties of compact operators on a Hilbert space in Subsection 9.4.2 are drawn from [18], but here we do not require that the operator be bounded, and the proofs are new (see also [11, Proposition 4.3.7 and Corollary 4.3.8]). A Brouwerian counterexample to the statement

The composition TS of a compact operator T and a bounded operator S is compact

can be found in [18, Brouwerian Example 4] and [24, Proposition 34]. A generalisation to compact linear mappings of a normed linear space is given in [15].

References

[1] Aczel, P., and Rathjen, M. 2000/2001. *Notes on constructive set theory*. Technical report 40. Institut Mittag-Leffler.

[2] Aczel, P., and Rathjen, M. August, 2010. *CST Book draft*. https://www1.maths.leeds.ac.uk/~rathjen/book.pdf.

[3] Bishop, E. 1967. *Foundations of Constructive Analysis*. New York, Toronto, London: McGraw-Hill.

[4] Bishop, E., and Bridges, D. 1985. *Constructive Analysis*. Grundlehren der Mathematischen Wissenschaften [Fundamental Principles of Mathematical Sciences], vol. 279. Berlin: Springer-Verlag.

[5] Bridges, D., and Ishihara, H. 1992. Locating the range of an operator on a Hilbert space. *Bull. London Math. Soc.*, **24**(6), 599–605.

[6] Bridges, D., and Ishihara, H. 2000. Constructive closed range and open mapping theorems. *Indag. Math. (N.S.)*, **11**(4), 509–516.

[7] Bridges, D., and Richman, F. 1987. *Varieties of Constructive Mathematics*. London Mathematical Society Lecture Note Series, vol. 97. Cambridge: Cambridge University Press.

[8] Bridges, D., Calder, A., Julian, W., Mines, R., and Richman, F. 1981. Bounded linear mappings of finite rank. *J. Func. Anal.*, **43**(2), 143–148.

[9] Bridges, D., Julian, W., and Mines, R. 1989. A constructive treatment of open and unopen mapping theorems. *Z. Math. Logik Grundlag. Math.*, **35**(1), 29–43.

[10] Bridges, D. S. 1979. *Constructive Functional Analysis*. Research Notes in Mathematics, vol. 28. Boston, MA, London: Pitman (Advanced Publishing Program).

[11] Bridges, D. S., and Vîţă, L. S. 2006. *Techniques of Constructive Analysis*. Universitext. New York: Springer.

[12] Cederquist, J., Coquand, T., and Negri, S. 1998. The Hahn–Banach theorem in type theory. Pages 57–72 of: *Twenty-five Years of Constructive Type Theory (Venice, 1995)*. Oxford Logic Guides, vol. 36. New York: Oxford University Press.

[13] Dunford, N., and Schwartz, J. T. 1988. *Linear Operators. Part I*. Wiley Classics Library. New York: John Wiley & Sons. General theory, with the assistance of William G. Bade and Robert G. Bartle, Reprint of the 1958 original, a Wiley–Interscience Publication.

[14] Freudenthal, H. 1937. Zum intuitionistischen Raumbegriff. *Compositio Math.*, **4**, 82–111.

[15] Ishihara, H. 1989a. Constructive compact linear mappings. *Bull. London Math. Soc.*, **21**(6), 577–584.

[16] Ishihara, H. 1989b. On the constructive Hahn–Banach theorem. *Bull. London Math. Soc.*, **21**(1), 79–81.

[17] Ishihara, H. 1990. An omniscience principle, the König lemma and the Hahn–Banach theorem. *Z. Math. Logik Grundlag. Math.*, **36**(3), 237–240.

[18] Ishihara, H. 1991a. Constructive compact operators on a Hilbert space. *Ann. Pure Appl. Logic*, **52**(1–2), 31–37. International Symposium on Mathematical Logic and its Applications (Nagoya, 1988).

[19] Ishihara, H. 1991b. Continuity and nondiscontinuity in constructive mathematics. *J. Symbolic Logic*, **56**(4), 1349–1354.

[20] Ishihara, H. 1992a. Constructive existence of Minkowski functionals. *Proc. Amer. Math. Soc.*, **116**(1), 79–84.

[21] Ishihara, H. 1992b. Continuity properties in constructive mathematics. *J. Symbol. Logic*, **57**(2), 557–565.

[22] Ishihara, H. 1994. A constructive version of Banach's inverse mapping theorem. *New Zealand J. Math.*, **23**(1), 71–75.

[23] Ishihara, H. 2001. Locating subsets of a Hilbert space. *Proc. Amer. Math. Soc.*, **129**(5), 1385–1390.

[24] Ishihara, H. 2018. Constructive functional analysis: an introduction. Pages 109–165 of: Mainzer, K., Schuster, P., and Schwichtenberg, H. (eds.), *Proof and Computation: Digitization in Mathematics, Computer Science and Philosophy*. Singapore: World Scientific.

[25] Ishihara, H. 2020. The constructive Hahn–Banach theorem, revisited. Pages 638–663 of: Rezuş, A. (ed.), *Contemporary Logic and Computing*. Landscapes in Logic, vol. 1. Rickmansworth: College Publications.

[26] Ishihara, H., and Vîţă, L. 2003. Locating subsets of a normed space. *Proc. Amer. Math. Soc.*, **131**(10), 3231–3239.

[27] Ishihara, H., and Vîţă, L. 2006. A constructive Banach inverse mapping theorem in F-spaces. *New Zealand J. Math.*, **35**(2), 183–187.

[28] Lietz, P., and Streicher, T. 2012. Realizability models refuting Ishihara's boundedness principle. *Ann. Pure Appl. Logic*, **163**(12), 1803–1807.

[29] Mazur, S. 1933. Über konvexe Mengen in linearen normierten Räumen. *Studia Mathematica*, **4**(1), 70–84.

[30] Metakides, G., Nerode, A., and Shore, R. A. 1985. Recursive limits on the Hahn–Banach theorem. Pages 85–91 of: *Errett Bishop: Reflections on Him and his Research (San Diego, California, 1983)*. Contemp. Math., vol. 39. Providence, RI: American Mathematical Society.

[31] Richman, F. 2001. Adjoints and the image of the ball. *Proc. Amer. Math. Soc.*, **129**(4), 1189–1193.

[32] Royden, H. L. 1985. Aspects of constructive analysis. Pages 57–64 of: *Errett Bishop: Reflections on Him and his Research (San Diego, California, 1983)*. Contemp. Math., vol. 39. Providence, RI: American Mathematical Society.

[33] Rudin, W. 1991. *Functional Analysis*, 2nd ed. International Series in Pure and Applied Mathematics. New York: McGraw-Hill.

[34] Sambin, Giovanni. 2003. Some points in formal topology. *Theoret. Comput. Sci.*, **305**(1-3), 347–408. Topology in computer science (Schloß Dagstuhl, 2000).

[35] Troelstra, A. S. 1969. *Principles of Intuitionism*. Lectures presented at the Summer Conference on Intuitionism and Proof Theory (1968) at SUNY at Buffalo, N.Y. Lecture Notes in Mathematics, vol. 95. Berlin, New York: Springer-Verlag.

[36] Troelstra, A. S., and van Dalen, D. 1988. *Constructivism in Mathematics, Vol. II*. Studies in Logic and the Foundations of Mathematics, vol. 123. Amsterdam: North-Holland.

10

Constructive Banach Algebra Theory

Robin S. Havea and Douglas Bridges

10.1 Introduction

In his 1967 monograph [1], Errett Bishop gave the first development of his constructive theory of commutative Banach algebras. This was superseded by the final chapter of [2], which replaced the classical approach via maximal ideals by one based on finite linear combinations of elements of the Banach algebra. Bishop also proved the Gelfand Representation Theorem for certain commutative Banach algebras of normable operators on a separable complex Hilbert space [2, page 385, Corollary (8.28)].

More recently, Takamura [15] examined C*-algebras: Banach algebras with an involution whose defining properties mirror those of the adjoint mapping on bounded operators on a Hilbert space. In Bishop's constructive framework, within which our current chapter is written, he developed both the fundamentals of the abstract theory and the main theorems on states, culminating in the Gelfand–Naimark–Segal construction on cyclic representations.

In this chapter we discuss some relatively elementary results in Banach algebra theory without requiring either commutativity or the existence of an involution. In order to recover constructively as much as possible of the classical spectral mapping theorem, we need some preliminary material on monoids, metric space topology, and complex analysis. The rest of the chapter is devoted mostly to states, the state space, positivity, and Hermitian elements, and is motivated by classical counterpart material in [3, 14].

For additional research on Bishop-style constructive Banach algebra theory we refer the reader to [6, 11], the first of which contains results closely related to several in Section 10.5 below. Relevant sources of information on the classical approach to our topics can be found in [3, 10, 14].

10.2 Preliminaries

We first recall that an *inequality (relation)* on a set X is a binary relation $\neq_X$ on X such that

$$\forall_{x,y \in X}(x \neq_X y \Rightarrow \neg(x = y) \wedge y \neq_X x).$$

The inequality is

- *tight* if

$$\forall_{x,y,z \in X}(x \neq_X y \Rightarrow x \neq_X z \vee z \neq_X y);$$

- *discrete* if

$$\forall_{x,y \in X}(x = y \vee x \neq_X y).$$

If $Y \subset X$, then the set

$$X \sim Y \equiv \{x \in X : \forall_{y \in Y}(x \neq_X y)\}$$

is called the *complement* of Y in X. From now on, when the context is clear, the inequality on a set X may be denoted simply by $\neq$.

The standard inequality on a metric space[1] (X, ρ) is the *metric inequality* defined by

$$x \neq_X y \text{ if and only if } \rho(x, y) > 0.$$

This inequality is tight. It is an exercise to prove that, for the metric space $\mathbb{R}$, the statements

$$\forall_{x,y \in \mathbb{R}}(x = y \vee x \neq_{\mathbb{R}} y)$$

and

$$\forall_{x,y \in \mathbb{R}}(\neg(x = y) \Rightarrow x \neq_{\mathbb{R}} y)$$

entail, respectively, the omniscience principles LPO (the limited principle of omniscience) and Markov's principle.

By a *monoid* we mean a set S equipped with an associative binary multiplication operation $(x, y) \rightsquigarrow xy$, an identity element e, and an inequality $\neq_S$. An element a of S is *right* (respectively, *left*) *invertible* if there exists $b \in S$, a *right* (respectively, *left*) *inverse* of a, such that $ab = e$ (respectively, $ba = e$). If a has both a right and a left inverse, then those two are equal and are called the (two-sided) *inverse* of a, which is said to be *invertible*. We denote by $\mathsf{inv}\ S$ the set of invertible elements of S, and say that $a \in S$ is *non-invertible* if $a \notin \mathsf{inv}\ S$. We impose upon $\neq_S$ the additional requirement that multiplication in S be *strongly extensional*: for all

[1] We assume familiarity with the fundamentals of metric and normed space theory, as found in [1, 2, 7]. The reader should also consult Chapters 8 and 9 of this volume.

$a, b, x \in S$, if $xa \neq_S xb$ or $ax \neq_S bx$, then $a \neq b$. It follows that for all $a \in \mathsf{inv}\ S$ and $b \in S$,

$$a^{-1}b \neq_S e \Rightarrow a^{-1}b \neq_S a^{-1}a \Rightarrow b \neq_S a \Rightarrow a \neq_S b.$$

Also, if $ab^{-1} = e$, then $a = b$. Of particular interest to us is the case where the inequality has the property

$$\forall_{x \in S}(x \neq_S e \vee x \in \mathsf{inv}\ S).$$

We then say that the inequality and the monoid itself are *quasi-discrete*.

We now present a number of results that will help in our discussion of the Spectral Mapping Theorem.

Lemma 10.1 *Let S be a quasi-discrete monoid S and let a be a non-invertible element of S. Then for each $x \in S$, either $ax \neq_S e$ or $e \in {\sim}Sa$, and either $xa \neq_S e$ or $e \in {\sim}aS$.*

Proof Fixing $x \in S$, we have either $ax \neq e$ or $ax \in \mathsf{inv}\ S$. In the latter case, $a(x(ax)^{-1}) = e$, so a is right invertible. Moreover, for each $s \in S$, either $e \neq_S sa$ or $sa \in \mathsf{inv}\ S$. The second alternative implies that $((sa)^{-1}s)a = e$, so a, being both right and left invertible, belongs to $\mathsf{inv}\ S$, a contradiction. It follows that $e \neq_S sa$ for each $s \in S$, and therefore that $e \in {\sim}Sa$. The second conclusion of the lemma is proved similarly. □

Proposition 10.2 *Let S be a quasi-discrete monoid and let a be a non-invertible element of S. Then $aS \cap Sa \subset {\sim}\mathsf{inv}\ S$.*

Proof Consider $x \in aS \cap Sa$ and $y \in \mathsf{inv}\ S$. There exist $b, c \in S$ such that $ab = x = ca$. By Lemma 10.1, either $y^{-1}x = (y^{-1}c)a \neq_S e$ and therefore $y \neq_S x$, or else $e \in {\sim}aS$. In the latter case, $e \neq_S a(by^{-1}) = xy^{-1}$ and therefore $y \neq_S x$. □

Corollary 10.3 *Let S be a quasi-discrete monoid and let a be a non-invertible element of S. Then $a \in {\sim}\mathsf{inv}\ S$.*

Proof Apply Proposition 10.2, noting that $a \in aS \cap Sa$. □

Now let (X, ρ) be a metric space. We write $B_X(x, r)$ (respectively, $\overline{B}_X(x, r)$) for the open (respectively, closed) ball in X with centre a and radius r; as usual, we may drop the suffix X when there is no ambiguity in doing so.

Given a subset S of X, we adopt the following notational conventions:

- $\rho(x, S) < r$: there exists $s \in S$ with $\rho(x, s) < r$;
- $\rho(x, S) \leq r$: for each $\varepsilon > 0$ there exists $s \in S$ with $\rho(x, s) < r + \varepsilon$;
- $\rho(x, S) > 0$: there exists $r > 0$ such that $\rho(x, s) \geq r$ for each $s \in S$.

Note that these notations do not require the locatedness of S. We define the *metric complement* of S (in X) to be

$$X - S \equiv \{x \in X : \rho(x, S) > 0\}.$$

When the ambient space X is clear, we denote this metric complement simply by $-S$. Clearly, $-S \subset {\sim}S$. Note that if S is a located subset of a metric space X, then $S \cup -S$ is dense in X, since for each $\varepsilon > 0$ either $\rho(x, S) > 0$ or $\rho(x, S) < \varepsilon$.

Part (a) of the following result is known as *Bishop's lemma.*

Lemma 10.4 *Let S be a complete, located subset of a metric space X. Then*

(a) for each $x \in X$ there exists $s \in S$ such that if $x \neq_X s$, then $\rho(x, S) > 0$;
(b) $S = {\sim}(-S)$ and ${\sim}S = -S$;
(c) $\partial S = \partial(-S)$, where ∂S denotes the boundary of S in X.

Proof For the proof of (a), see [2, page 92, Lemma (3.8)]. Clearly, $S \subset {\sim}(-S)$. On the other hand, for each $x \in {\sim}(-S)$, using (a), we can find $s \in S$ such that if $\rho(x, s) > 0$, then $x \in -S$, a contradiction; hence $\rho(x, s) = 0$, $x \in \overline{S}$ and hence, as S is complete and therefore closed, $x \in S$. Thus $S = {\sim}(-S)$. Next, it readily follows from (a) that ${\sim}S \subset -S$; whence ${\sim}S = -S$. Finally, we have

$$\partial S = \overline{S} \cap \overline{{\sim}S} = S \cap \overline{(-S)} = {\sim}(-S) \cap \overline{(-S)} = \partial(-S). \qquad \square$$

We say that a subset K of S

- is *well contained* in S, written $K \subset\subset S$, if there exists $r > 0$ such that if $\rho(x, K) \leq r$, then $x \in S$;
- *approximates S internally to within ε* if $K \subset\subset S$ and $\rho(x, \partial S) < \varepsilon$ for each $x \in S - K$.

Also, if T is a class of subsets of S, we say that S is

- *approximated internally by sets of type T* if for each $\varepsilon > 0$, S is approximated internally by a set of type T;
- *coherent* if $S = -{\sim}S$.

If X is a normed space, then ρ denotes the usual metric derived from the norm. We then denote by $[x, y]$ the line segment $\{tx + (1 - t)y : 0 \leq t \leq 1\}$ joining x and y.

The proofs of the next four propositions are essentially those found in [9].

Proposition 10.5 *Let X be a finite-dimensional Banach space, let S be a subset of X, and let L be a located subset of ${\sim}S$ such that $S^\circ = X - L$. Then S is approximated internally by located subsets.*

Proposition 10.6 *Let X be a finite-dimensional Banach space and let S be a subset of X that is approximated internally by located subsets. Then $\sim S^\circ$ is located and S° is coherent.*

Proposition 10.7 *Let S be a subset of a Banach space X such that $S \cup \sim S$ is dense. Then S has the* boundary crossing property: *for each $x \in S$, each $y \in \sim S$, and each $\varepsilon > 0$, there exists $z \in \partial S$ such that $\rho(z, [x, y]) < \varepsilon$.*

Proposition 10.8 *Let S be a subset of a Banach space X. Then S and $\sim S$ are located if and only if ∂S is located and $S \cup \sim S$ is dense in X.*

Proposition 10.9 *Located subsets of a Banach space have the boundary crossing property.*

Proof Since S is located, $S \cup -S$ is dense in X, as therefore is $S \cup \sim S$, so Proposition 10.7 applies. □

Proposition 10.10 *Let S be a coherent subset of a Banach space X such that ∂S and $\sim S$ are located. Then S is located.*

Proof Since $\sim S$ is located, $\sim S \cup -\sim S$ is dense in X. But $-\sim S = S^\circ \subset S$, so $S \cup \sim S$ is dense. Hence, by Proposition 10.8, S is located. □

Let $K \subset \mathbb{C}$ be compact – that is, totally bounded and complete.[2] For each uniformly continuous $f : K \to \mathbb{C}$ we write

$$m(f, K) = \inf_{z \in K} |f(z)| .$$

A *border* of K is a compact subset B of K such that $\overline{B}_{\mathbb{C}}(z, \rho(z, B)) \subset K$ for each $z \in K$. The fundamental result about borders in complex analysis is as follows.

Proposition 10.11 *Let K be a compact subset of $\mathbb{C}$ with border B, and let $f : K \in \mathbb{C}$ be a differentiable function such that $m(f, B) > m(f, K) = 0$. Then there exists $z \in K$ with $f(z) = 0$* [2, page 156, Lemma (5.8)].

Proposition 10.12 *Let U be an open subset of $\mathbb{C}$ such that $\sim U$ and ∂U are compact. Then ∂U is a border for $\sim U$.*

Proof Fix $z_0 \in \sim U$, let $r = \rho(z_0, \partial U)$. Given $z \in \overline{B}(z_0, r)$, for each $t \in [0, 1]$ let $z_t = tz_0 + (1 - t)z$. Suppose that $z \in U$. Then there exists $\delta > 0$ such that $B(z, 3\delta) \subset U$ and therefore $\rho(z, \partial U) \geq 3\delta$. By Proposition 10.9, there exist $\zeta \in \partial U$ and $t \in [0, 1]$ such that $|\zeta - z_t| < \delta$. We have

[2] In this chapter, all (locally) totally bounded sets to be inhabited *by definition*. The the same applies to (locally) compact sets.

$$3\delta \leq \rho(z, \partial U) \leq |\zeta - z| \leq |\zeta - z_t| + |z - z_t| < \delta + |z - z_t|,$$

so $2\delta < |z - z_t|$ and therefore

$$\begin{aligned}|z_0 - \zeta| &\leq |z_0 - z_t| + |\zeta - z_t| \\ &= |z - z_0| - |z - z_t| + |\zeta - z_t| \\ &< r - 2\delta + \delta = r - \delta < r \leq |z_0 - \zeta|,\end{aligned}$$

which is absurd. We conclude that $z \notin U$ and hence, since U is open, that $z \in {\sim}U$. Thus $\overline{B}(z_0, \partial U) \subset {\sim}U$. Also, ${\sim}U$ and ∂U are compact and $\partial U \subset \overline{{\sim}U} = {\sim}U$. □

Proposition 10.13 *Let p be a polynomial function on $\mathbb{C}$ and let K be a compact subset of $\mathbb{C}$ on which p is non-vanishing (that is, $p(z) \neq_{\mathbb{C}} 0$ for each $z \in K$). Then $m(p, K) > 0$.*

Proof This is a special case of [2, page 160, Proposition (5.14)]. □

10.3 The Spectral Mapping Theorem

Let $\mathbb{K}$ stand for either $\mathbb{R}$ or $\mathbb{C}$, and let $\mathfrak{A}$ be an (associative) algebra[3] over $\mathbb{K}$ with multiplicative identity e and additive identity 0. An inequality relation $\neq_{\mathfrak{A}}$ on $\mathfrak{A}$ is said to be *compatible with the algebra structure* on $\mathfrak{A}$ if $e \neq_{\mathfrak{A}} 0$ and for all $x, y \in X$ and all $t \in \mathbb{K}$,

$$\begin{aligned}x \neq_{\mathfrak{A}} y &\Leftrightarrow x - y \neq_{\mathfrak{A}} 0, \\ x + y \neq_{\mathfrak{A}} 0 &\Rightarrow x \neq_{\mathfrak{A}} 0 \vee y \neq_{\mathfrak{A}} 0, \\ tx \neq_{\mathfrak{A}} 0 &\Leftrightarrow t \neq_{\mathbb{K}} 0 \wedge x \neq_{\mathfrak{A}} 0, \text{ and} \\ xy \neq_{\mathfrak{A}} 0 &\Leftrightarrow x \neq_{\mathfrak{A}} 0 \wedge y \neq_{\mathfrak{A}} 0.\end{aligned}$$

In that case, $\mathfrak{A}$ is a monoid with respect to its multiplication operation, and if $a, b \in X$ and either $xa \neq_{\mathfrak{A}} xb$ or $ax \neq_{\mathfrak{A}} bx$, then $x \neq_{\mathfrak{A}} 0$ and $a \neq_{\mathfrak{A}} b$. By a *norm* on such an algebra we mean a mapping $x \rightsquigarrow \|x\|$ of $\mathfrak{A}$ into the non-negative real line such that $\|e\| = 1$ and for all $x, y \in X$ and $t \in \mathbb{K}$,

$$\begin{aligned}\|x\| > 0 &\Leftrightarrow x \neq_{\mathfrak{A}} 0, \\ \|tx\| &= |t| \, \|x\|, \\ \|x + y\| &\leq \|x\| + \|y\|, \text{ and} \\ \|xy\| &\leq \|x\| \, \|y\|.\end{aligned}$$

If, regarded as a linear space with respect to addition and multiplication-by-scalars, $\mathfrak{A}$ is complete – that is, a Banach space – with respect to such a norm, then we call $\mathfrak{A}$ a *Banach algebra* over $\mathbb{K}$. It readily follows from the last two lines of the foregoing

[3] For an introduction to algebras see [13, Chapter 7].

display that the mapping $(x, y) \rightsquigarrow x+y$ is uniformly continuous on $\mathfrak{A} \times \mathfrak{A}$ and that the mapping $(x, y) \rightsquigarrow xy$ is both pointwise continuous on $\mathfrak{A} \times \mathfrak{A}$ and uniformly continuous on norm-bounded subsets of $\mathfrak{A} \times \mathfrak{A}$. Also, for each $x \in \mathsf{inv}\,\mathfrak{A}$ we have $1 = \|e\| = \left\|xx^{-1}\right\| \leq \|x\| \left\|x^{-1}\right\|$, so $x \neq_{\mathfrak{A}} 0$, $x^{-1} \neq_{\mathfrak{A}} 0$, and $\left\|x^{-1}\right\| \geq 1/\|x\|$.

Although almost everything in the remainder of this chapter deals with Banach algebras in the abstract, we shall occasionally refer to these examples of concrete Banach algebras.

- The algebra $C(X)$ of uniformly continuous, complex-valued functions on a compact metric space, taken with the *supremum norm*

$$\|f\| \equiv \sup\{|f(x) : x \in X|\}.$$

- Closed algebras of bounded operators on a Hilbert space H that include the identity operator, the multiplication in this case being composition of operators, and the norm the *operator norm*

$$\|T\| \equiv \sup\{\|Tx\| : \|x\| \leq 1\},$$

  which we require to exist for each T in the algebra.[4]

The standard classical proofs of the next two propositions are constructively acceptable [14, pages 175–176].

Proposition 10.14 *Let a be an element of the Banach algebra $\mathfrak{A}$ such that $\|a\| < 1$. Then the series $\sum_{n=0}^{\infty} a^n$ (where $a^0 = e$) converges to a limit y in $\mathfrak{A}$ such that $(e-a)y = e = y(e-a)$.*

Proposition 10.15 *Let a be an element of the Banach algebra $\mathfrak{A}$ such that $\|e - a\| < 1$. Then $a \in \mathsf{inv}\,\mathfrak{A}$.*

Proposition 10.16 *The inequality on a Banach algebra $\mathfrak{A}$ is quasi-discrete.*

Proof For each $x \in \mathfrak{A}$ either $\|x - e\| > 0$ or $\|x - e\| < 1$. In view of Proposition 10.15, we see that either $x \neq e$ or $x \in \mathsf{inv}\,\mathfrak{A}$. □

Proposition 10.17 $\mathsf{inv}\,\mathfrak{A}$ *is an open subset of* $\mathbb{C}$.

Proof Given $a \in \mathsf{inv}\,\mathfrak{A}$ and $x \in \mathfrak{A}$ such that $\|x - a\| < \left\|a^{-1}\right\|$, we have

$$\left\|a^{-1}x - e\right\| = \left\|a^{-1}x - a^{-1}a\right\| = \left\|a^{-1}(x-a)\right\| \leq \left\|a^{-1}\right\| \|x - a\| < 1,$$

so, by Proposition 10.15, $a^{-1}x$ has an inverse b. Then $ba^{-1}x = e$, and ba^{-1} is a left inverse of x. Similarly, xa^{-1} has an inverse c, and $a^{-1}c$ is a right inverse of x. Since $\mathfrak{A}$ is a monoid, it follows that $a^{-1}c = ba^{-1}$ are equal and the inverse of x. □

[4] If H is infinite dimensional, then the existence of $\|T\|$ for every $T \in B(H)$ implies LPO: see [15, page 281].

From now on, $\mathfrak{A}$ will denote a complex Banach algebra (one with $\mathbb{K} = \mathbb{C}$). For each $a \in \mathfrak{A}$ we define its *resolvent*

$$R_{\mathfrak{A}}(a) \equiv \{\lambda \in \mathbb{C} : a - \lambda e \in \mathsf{inv}\, \mathfrak{A}\}$$

and its *spectrum*

$$\sigma_{\mathfrak{A}}(a) \equiv \mathbb{C} \sim R(a).$$

Once again, we shall drop the suffix $\mathfrak{A}$ from these definienda when the underlying Banach algebra is clear.

Proposition 10.18 *Let a an element of a Banach algebra $\mathfrak{A}$ over $\mathbb{C}$. Then*

(i) $a - \lambda e \in \mathsf{inv}\, \mathfrak{A}$ *for each* $\lambda \in \mathbb{C}$ *with* $|\lambda| > \|a\|$, *and*
(ii) $\sigma(a) \subset \overline{B}_{\mathbb{C}}(0, \|a\|)$.

Proof If $|\lambda| > \|a\|$, then $\left\|\lambda^{-1}a\right\| < 1$, so, by Proposition 10.14, $e - \lambda^{-1}a$ has an inverse y. Then $a - \lambda e$ has inverse $-\lambda^{-1}y$. This proves (i), from which (ii) easily follows. □

Proposition 10.19 *For each $a \in \mathfrak{A}$, $R(a)$ is open and $\sigma(a)$ is closed.*

Proof Consider $\lambda \in \mathbb{C}$ such that $a - \lambda e \in \mathsf{inv}\, \mathfrak{A}$. By Proposition 10.17 there exists $\delta > 0$ such that if $x \in \mathfrak{A}$ and $\|x - (a - \lambda e)\| < \delta$, then $x \in \mathsf{inv}\, \mathfrak{A}$. If $\mu \in \mathbb{C}$ and $|\mu - \lambda| < \delta$, then $\|(a - \lambda e) - (a - \mu e)\| = |\mu - \lambda| \, \|e\| < \delta$, so $a - \mu e \in \mathsf{inv}\, \mathfrak{A}$ and therefore $\mu \in R(a)$. Thus $B(\lambda, \delta) \subset R(a)$. Moreover, given a sequence $(\lambda_n)_{n \geq 1}$ in $\sigma(a)$ converging to a limit λ_∞ in $\mathbb{C}$, we see that if $\lambda_\infty \in B(\lambda, \delta)$, then $\lambda_n \in B(\lambda, \delta)$, and therefore $\lambda_n \in {\sim}\sigma(a)$, for all sufficiently large n; this contradiction ensures that $|\lambda - \lambda_\infty| \geq \delta$ and therefore $\lambda_\infty \neq \lambda$. Thus for each $\lambda \in R(a)$, $\lambda \in R(a)^\circ$ and for each sequence in $\sigma(a)$ that converges to a limit λ_∞ in $\mathbb{C}$, $\lambda_\infty \neq \lambda$. The desired conclusions now follow. □

Classically, the spectrum is always compact. Constructively, we have the following proposition.

Proposition 10.20 *The following are equivalent conditions on an element a of $\mathfrak{A}$.*

(i) $\sigma(a)$ *is compact and* $R(a) = -\sigma(a)$.
(ii) $R(a)$ *is approximated internally by located sets.*

If either, and hence each, of these conditions holds, then $R(a)$ is located if and only if $\partial\sigma(a)$ is located, in which case $\partial\sigma(a) = \partial R(a)$.

Proof If (i) holds, then, by Proposition 10.19, $R(a)^\circ = R(a) = -\sigma(a)$ where $\sigma(a)$, being compact, is located. Hence, by Proposition 10.5, (ii) holds. Conversely, we see from Proposition 10.6 that if (ii) holds, then $\sigma(a)$ is located and

$R(a) = -{\sim}R(a) = -\sigma(a)$; since $\sigma(a)$ is closed in $\mathbb{C}$ and therefore complete, it is compact. Thus (ii) implies (i).

If (i) holds, then, by Lemma 10.4, $R(a) = {\sim}\sigma(a)$, so

$$\partial R(a) = \overline{R(a)} \cap \overline{{\sim}R(a)} = \overline{{\sim}\sigma(a)} \cap \overline{\sigma(a)} = \partial\sigma(a).$$

Moreover, $R(a) \cup {\sim}R(a) = {\sim}\sigma(a) \cup \sigma(a)$, which, $\sigma(a)$ being located, is dense in $\mathbb{C}$. It follows from Proposition 10.8, that $R(a)$ is located if and only if $\partial\sigma(a)$ is located. □

Proposition 10.21 *Let a be an element of a Banach algebra such that $\sigma(a)$ and $\partial\sigma(a)$ are compact subsets of $\mathbb{C}$ and $R(a) = -\sigma(a)$. Then $\partial\sigma(a)$ is a border for $\sigma(a)$.*

Proof We have $R(a)$ open, $\sigma(a) = {\sim}R(a)$, and, by Proposition 10.20, $\partial R(a) = \partial\sigma(a)$. Taking $U = R(a)$ in Proposition 10.12 yields the desired result. □

Given an element a of $\mathfrak{A}$ and a polynomial $p(z) = \sum_{n=0}^{N} c_n z^n$ over $\mathbb{C}$, we obtain an associated element of $\mathfrak{A}$ by setting $p(a) \equiv \sum_{n=0}^{N} c_n a^n$. Our next aim is to recover constructively as much as possible of the *classical Spectral Mapping Theorem*: if a is an element of a Banach algebra $\mathfrak{A}$ and p is a complex polynomial of positive degree, then $p(\sigma(a)) = \sigma(p(a))$.

A typical classical proof of this theorem, as found in [14, Proposition 3.2.10], rests on the decomposition of $p(z) - \lambda$, where $\lambda \in \mathbb{C}$, into linear factors, taken with the observation that $p(a) - \lambda e$ is non-invertible if and only if at least one of its corresponding linear factors is non-invertible. In the constructive setting, this argument fails because we may be unable to determine which of the linear factors is non-invertible. However, we can use the classical proof as a guide to proving the *constructive Spectral Mapping Theorem.*

Theorem 10.22 *Let a be an element of a Banach algebra $\mathfrak{A}$, and let $p(z) = \sum_{n=0}^{N} c_n z^n$ be a complex polynomial with $c_N \neq 0$. Then $p(\sigma(a)) \subset \overline{\sigma(p(a))}$. If also $\sigma(a)$ and $\partial\sigma(a)$ are compact, and $R(a) = -\sigma(a)$, then $\sigma(p(a)) = \overline{p(\sigma(a))}$.*

Proof Given $\lambda \in \sigma(a)$, we have

$$p(a) - p(\lambda)e = \sum_{n=1}^{N} c_n(a^n - \lambda^n e).$$

If $N = 0$, then $p(a) - p(\lambda)e = (c_0 - \lambda)e$, which is non-invertible if and only if $c_0 = \lambda = p(\lambda)$; so in this case, $p(\sigma(a)) = \sigma(p(a)$. Thus we may assume from now on that $N \geq 1$. For $1 \leq k \leq N$ we then have

$$\begin{aligned}(a^k - \lambda^k e) &= (a^{k-1} + \lambda a^{k-2} + \cdots + \lambda^{k-2} a + \lambda^{k-1} e)(a - \lambda e) \\ &= (a - \lambda e)(a^{k-1} + \lambda a^{k-2} + \cdots + \lambda^{k-2} a + \lambda^{k-1} e),\end{aligned}$$

so

$$p(a) - p(\lambda)e \in (a - \lambda e)\mathfrak{A} \cap \mathfrak{A}(a - \lambda e).$$

Since $a - \lambda e \notin \mathsf{inv}\,\mathfrak{A}$, it follows from Propositions 10.16 and 10.2 that $p(a) - p(\lambda)e \in {\sim}\mathsf{inv}\,\mathfrak{A}$ and hence that $p(\lambda) \in \sigma(p(a))$.

Now suppose that $\sigma(a)$ and $\partial\sigma(a)$ are compact and that $R(a) = -\sigma(a)$. Since p is uniformly continuous on $\sigma(a)$, both $p(\sigma(a))$ and $p(\partial\sigma(a))$ are totally bounded and hence located in $\mathbb{C}$. Given $\zeta \in \sigma(p(a))$, assume that $\rho(\zeta, p(\sigma(a))) > 0$. Then $\rho(\zeta, p(\partial\sigma(a))) > 0$, so $p(z) - \zeta \neq 0$ for each $z \in p(\partial\sigma(a))$. By Proposition 10.13, $m(p - \zeta, \partial\sigma(a)) > 0$. By the Fundamental Theorem of Algebra [2, page 156, Theorem (5.10)], there exist complex numbers $\zeta_1, \ldots, \zeta_N$ such that

$$p(z) - \zeta = c_N (z - \zeta_1) \cdots (z - \zeta_N) \quad (z \in \mathbb{C}).$$

If

$$\inf_{1 \leq k \leq N} \rho(\zeta_k, \sigma(a)) > 0,$$

then for each k, $\zeta_k \in -\sigma(a) = R(a)$, and so $a - \zeta_k e$ is invertible in $\mathfrak{A}$; hence $p(z) - \zeta$ has inverse $(a - \zeta_N)^{-1} \cdots (a - \zeta_1)^{-1}$ in $\mathfrak{A}$, and therefore $\zeta \notin \sigma(p(a))$, a contradiction. It follows that

$$\inf_{1 \leq k \leq N} \rho(\zeta_k, \sigma(a)) = 0. \tag{10.1}$$

Let

$$\alpha > \max_{1 \leq k \leq N} \sup\{|z - \zeta_k| : z \in \sigma(a)\},$$

the suprema existing since $\sigma(a)$ is compact. In view of (10.1), for each $\varepsilon > 0$ there exist $k \leq N$ and $z \in \sigma(a)$ such that $|z - \zeta_k| < c_N^{-1} \alpha^{1-N} \varepsilon$; then

$$|p(z) - \zeta| \leq c_N \alpha^{N-1} |z - \zeta_k| < \varepsilon.$$

Since $\varepsilon > 0$ is arbitrary,

$$m(p - \zeta, \sigma(a)) = 0 < m(p - \zeta, \partial\sigma(a)).$$

By Propositions 10.21 and 10.11, there exists $\lambda \in \sigma(a)$ such that $p(\lambda) = \zeta$; whence $\rho(\zeta, p(\sigma(a)) = 0$. This contradiction ensures that $\rho(\zeta, p(\sigma(a))) \leq 0$; whence $\rho(\zeta, p(\sigma(a))) = 0$ and therefore $\zeta \in \overline{p(\sigma(a))}$. From this and the first part of the proof, we obtain $p(\sigma(a)) \subset \sigma(p(a)) \subset \overline{p(\sigma(a))}$. Since $\sigma(p(a))$ is closed, it follows that $\sigma(p(a)) = \overline{p(\sigma(a))}$. □

Corollary 10.23 *Let a be an element of a Banach algebra $\mathfrak{A}$, and let $p(z) = \sum_{n=0}^{N} c_n z^n$ be a complex polynomial. Then $|p(\lambda)| \leq \|p(a)\|$ for each $\lambda \in \sigma(a)$.*

Proof Fix $\lambda \in \sigma(a)$, and suppose that $|p(\lambda)| > \|p(a)\|$. If $p(\lambda) \in \sigma(p(a))$, then by Proposition 10.18, $|p(\lambda)| \leq \|p(a)\|$, a contradiction; so $p(\lambda) \notin \sigma(p(a)$. In view of Theorem 10.22, we must have $\neg(c_N \neq 0)$; whence $c_N = 0$ and therefore p has degree at most $N-1$. Using this argument at most N times, we see that $c_k = 0$ for each k. Thus $|p(\lambda)| = 0 = \|p(a)\|$, a contradiction from which it follows that $|p(\lambda)| \ngtr \|p(a)\|$ and therefore $|p(\lambda)| \leq \|p(a)\|$. □

Consider a Banach algebra containing an element a whose resolvent set is the exterior of the closed unit ball D of $\mathbb{C}$ and whose spectrum is that ball. We have $\sigma(a)$ compact, $R(a) = -\sigma(a)$, and $\partial\sigma(a)$, the unit circle in $\mathbb{C}$, located. Let $\zeta \in [-\frac{1}{2}, \frac{1}{2}]$, and define

$$p(z) = (z + 1 - \zeta)(z - 1 - \zeta) \qquad (z \in \mathbb{C}).$$

Suppose that $p(a)$ is invertible. Then both $a - (-1+\zeta)e$ and $a - (1+\zeta)e$ are invertible, so both $-1+\zeta$ and $1+\zeta$ belong to $R(a)$; this implies that both $\zeta < 0$ and $\zeta > 0$, which is absurd. Hence $p(a)$ is not invertible, and therefore $0 \notin R(p(a))$. Since $R(p(a))$ is open, it follows that $0 \in -R(p(a)) = \sigma(p(a))$.

Now suppose that $0 \in p(\sigma(a))$. Then there exists $z \in \sigma(a)$ such that $p(z) = 0$. Either $\operatorname{Re} z > -1/2$ or else $\operatorname{Re} z < 1/2$. In the first case, if $z = -1+\zeta$, then $\operatorname{Re} z = -1 + \zeta \leq -1/2$, a contradiction, so $z \neq -1+\zeta$; since $z \neq 1+\zeta$ implies that $p(z) \neq 0$, we must have $1 + \zeta = z \in D$ and therefore $\zeta \leq 0$. In the case $\operatorname{Re} z < 1/2$, a similar argument shows that $\zeta \geq 0$. We now see that proposition

> If a is an element of a Banach algebra such that $\sigma(a)$ is compact, $R(a) = -\sigma(a)$, and $\partial\sigma(a)$ is located, and if p is a quadratic polynomial over $\mathbb{C}$, then $\sigma(p(a)) = p(\sigma(a))$

implies that for each $x \in \mathbb{R}$ either $x \geq 0$ or $x \leq 0$. This last statement is equivalent to the omniscience principle LLPO (the the lesser limited principle of omniscience).[5]

Note that in this example we can confirm Theorem 10.22 by showing directly that $0 \in \overline{p(\sigma(a))}$. For if $0 < \varepsilon < 1$, then either $\zeta \neq 0$ or $|\zeta| < \varepsilon/4$. If $\zeta < 0$, then $1 + \zeta \in D = \sigma(a)$ and $p(1+\zeta) = 0$; if $\zeta > 0$, then $-1+\zeta \in \sigma(a)$ and $p(-1+\zeta) = 0$. If $|\zeta| < \varepsilon/4$, then $1 - |\zeta| \in \sigma(a)$ and

$$|p(1 - |\zeta|)| = (2 - |\zeta| - \zeta)2\,|\zeta| \leq 4\,|\zeta| < \varepsilon.$$

Thus for each $\varepsilon > 0$ there exists $z \in \sigma(a)$ such that $|p(z)| < \varepsilon$.

[5] For LLPO, see Chapter 8 in this volume.

10.4 Approximating the State Space

At this stage we need some background in the theory of duality in normed (linear) spaces.[6] Accordingly, consider such a space $(X, \|\ \|)$, and let B_X denote its closed unit ball $\overline{B}(0,1)$. A bounded linear functional f on X is *normable*, or *normed*, if its *norm*,

$$\|f\| \equiv \sup\{|f(x)| : x \in B_X\},$$

exists; in that case, $\|f\|$ is the smallest non-negative number c such that $|f(x)| \leq c\,\|x\|$ for all $x \in X$. If X is finite-dimensional, then every linear functional on X is not only bounded but normable.

Even if f is not known to be normable, it is convenient to adopt the following notational conventions:

- $\|f\| \leq c$ means that $|f(x)| \leq c\,\|x\|$ for all $x \in X$;
- $\|f\| < c$ means that there exists $r < c$ such that $\|f\| \leq r$;
- $\|f\| > 0$ means that there exists $x \in X$ with $|f(x)| > 0$.

The *dual* of the normed space X is the set X^* of all bounded linear functionals on X. Using our notational convention, we define the *unit ball* of X^* to be

$$X_1^* \equiv \{f \in X^* : \|f\| \leq 1\}.$$

If X is infinite-dimensional, we cannot prove constructively that every member of X^* is normable. Thus we have to approach the topology on X^* more carefully. We do this best by regarding X^* as a locally convex linear space relative to the family of seminorms $(p_x)_{x\in X}$, where $p_x(f) = |f(x)|$ for each $f \in X^*$ [7, Section 5.4]. Since that would lead us too far astray, we take the less general path of restricting ourselves to separable normed spaces. *So from now on, 'normed space' will mean 'separable normed space', and we will assume that our Banach algebra $\mathfrak{A}$ is separable.*

Let $(a_n)_{n\geq 1}$ be a dense sequence in the unit ball B_X of the normed space X. The corresponding *weak* norm*[7] and *weak* metric* on X^* are defined, for $f, g \in X^*$, by

$$\|f\|_w^X \equiv \sum_{n=1}^{\infty} 2^{-n}\,|f(a_n)| \quad \text{and} \quad \rho_w^X(f,g) \equiv \|f-g\|_w\,.$$

Weak* norms defined using different dense sequences in B_X induce equivalent metrics on X^*, and X_1^* is weak* compact. For each $x \in X$, the mapping $f \rightsquigarrow f(x)$ is weak* uniformly continuous on X_1^*. Unless there is risk of confusion, we may denote the weak* norm and metric by $\|\ \|_w$ and ρ_w.

[6] See also [1, Chapter 9], [2, Chapter 7], [7, Chapters 5–6], and Chapter 9 of this volume.
[7] Bishop calls this the *double norm*.

Returning to our Banach algebra $\mathfrak{A}$, we define the *state space* of $\mathfrak{A}$ to be

$$V_{\mathfrak{A}} \equiv \{f \in \mathfrak{A}^* : f(e) = 1 = \|f\|\}$$

and call its elements *states*. A bounded linear functional u on $\mathfrak{A}$ is called a *character* if $u(e) = 1$ and u is multiplicative: that is, $u(xy) = u(x)u(y)$ for all $x, y \in \mathfrak{A}$. If $c > 0$ is a bound for u, then for each $x \in \mathfrak{A}$ and each positive integer n,

$$\begin{aligned} |u(x)| &= |u(x)^n|^{1/n} = |u(x^n)|^{1/n} \leq (c\,\|x^n\|)^{1/n} \\ &\leq c^{1/n}(\|x\|^n)^{1/n} = c^{1/n}\,\|x\| \to \|x\| \text{ as } n \to \infty \end{aligned}$$

and therefore $|u(x)| \leq \|x\|$. Hence $\|u\| = 1 = u(e)$, and the *character space* $\Sigma_{\mathfrak{A}}$, consisting of all the characters of $\mathfrak{A}$, is a subset of $V_{\mathfrak{A}}$. We note here that, as is easily proved, $u(x) \in \sigma(x)$ for all $x \in \mathfrak{A}$ and $u \in \Sigma_{\mathfrak{A}}$.

Classically, $V_{\mathfrak{A}}$ is weak* compact. Constructively, it is best handled using approximations: for $t \geq 0$ we define the *t-approximation* to $V_{\mathfrak{A}}$ by

$$V_{\mathfrak{A}}^t \equiv \{f \in \mathfrak{A}^* : \|f\| \leq 1, |1 - f(e)| \leq t\}.$$

Then $V_{\mathfrak{A}} \subset V_{\mathfrak{A}}^t$ for each $t \geq 0$, $V_{\mathfrak{A}} = \bigcap_{t>0} V_{\mathfrak{A}}^t$, and $V_{\mathfrak{A}}$ is convex.

Proposition 10.24 *For each $t > 0$, $V_{\mathfrak{A}}^t$ is a weak* compact subset of $\mathfrak{A}_1^*$.*

Proof Since the mapping $f \rightsquigarrow |1 - f(x)|$ is weak* uniformly continuous on $\mathfrak{A}_1^*$, we see from [7, Corollary 2.2.14] that for all but countably many $t > 0$, $V_{\mathfrak{A}}^t$ is either empty or weak* compact. Moreover, by [7, Proposition 5.3.1], for each t with $0 < t < 1$ there exists a normable linear functional f on $\mathfrak{A}$ such that $1 = \|f\| \geq f(e) > 1 - t$; then $0 \leq 1 - f(e) < t$, so $f \in V_{\mathfrak{A}}^t$. It readily follows that for each $t > 0$, $V_{\mathfrak{A}}^t$ is inhabited and therefore weak* compact. □

Every weak* norm on the dual of a Banach algebra $\mathfrak{A}$ has an associated *Hausdorff metric* on the set of weak* compact subsets of $\mathfrak{A}_1^*$, defined by

$$\rho_w^{\mathfrak{A}}(S, T) \equiv \max\left\{\sup_{a \in A} \rho_w^{\mathfrak{A}}(a, B), \sup_{b \in B} \rho_w^{\mathfrak{A}}(b, A)\right\}$$

[2, page 94]. We say that the state space $V_{\mathfrak{A}}$ of $\mathfrak{A}$ is *firm* if it is compact and $\rho_w^{\mathfrak{A}}\left(V_{\mathfrak{A}}^t, V_{\mathfrak{A}}\right) \to 0$ as $t \to 0$. This property is independent of the sequence with respect to which $\| \|_w$ is defined. We normally denote the Hausdorff metric simply by ρ_w.

For the next lemma we note that finite-dimensional subspaces[8] of a normed space are locally compact and therefore located, and that a located subspace of a finite-dimensional normed space is itself finite-dimensional [7, Propositions 4.1.6 and 2.2.18, and Corollary 4.1.14].

[8] In the context of a linear space, *subspace* means linear subspace.

Lemma 10.25 *Let Y be a finite-dimensional subspace of a normed space X, and y_0 a unit vector in Y. Let $0 < t < 1/2$, and let f be a linear functional on Y such that $\|f\| \leq 1$ and $|1 - f(y_0)| < t$. Then there exists a normable linear functional ϕ on X such that $\|\phi\| = 1$, $|1 - \phi(y_0)| \leqslant 3t$, and $|f(y) - \phi(y)| \leqslant 2t\|y\|$ for each $y \in Y$.*

Proof Since Y is finite-dimensional, f is normable; moreover, $\|y_0\| = 1$, so $\|f\| \geq |f(y_0)| > 1 - t$. By [2, Chapter 7, Proposition (1.10)], $\ker(f)$ is located in Y; whence $\ker(f)$ is finite-dimensional and therefore located in X. By the Hahn–Banach Theorem [7, Theorem 5.3.3], there exists an extension of f to a normable element $f^\sharp$ of X^* such that $\|f\| \leqslant \left\|f^\sharp\right\| < 1 + t$. Note that

$$-\frac{t}{1-t} < 1 - \left\|f^\sharp\right\|^{-1} < \frac{t}{1+t} < \frac{t}{1-t} < 2t.$$

Let $\phi = \left\|f^\sharp\right\|^{-1} f^\sharp$. Then $\phi \in X^*$, $\|\phi\| = 1$, and for each $y \in Y$,

$$\begin{aligned}|f(y) - \phi(y)| &= \left|f(y) - \left\|f^\sharp\right\|^{-1} f^\sharp(y)\right| = \left|1 - \left\|f^\sharp\right\|^{-1}\right| |f(y)| \\ &\leqslant \frac{t}{1-t} \|f\| \, \|y\| \leqslant 2t \, \|y\| .\end{aligned}$$

Moreover,

$$|1 - \phi(y_0)| \leqslant |1 - f(y_0)| + |f(y_0) - \phi(y_0)| \leqslant 3t.$$

□

Lemma 10.26 *Let $x_1 \neq 0, \ldots, x_n$ be elements of a normed space X, and let $\varepsilon > 0$. Then there exist disjoint P, Q such that $P \cup Q = \{1, \ldots, n\}$, $\{x_k : k \in P\}$ is a basis for a finite-dimensional subspace Y of X, and $\rho(x_k, Y) < \varepsilon$ for each $k \in Q$.*

Proof Let $P_1 = \{x_1\}$, let Y_1 be the one-dimensional space $\mathbb{C}e$ of complex multiples of x_1, and let $Q_1 = \varnothing$. Suppose that for some $k < n$ we have constructed disjoint subsets P_k, Q_k of $\{1, \ldots, k\}$, such that $\{x_j : j \in P_k\}$ is a basis for a finite-dimensional subspace Y_k of X and $\rho(x_j, Y_k) < \varepsilon$ for each $j \in Q_k$. Either $\rho(x_{k+1}, Y_k) > 0$, in which case we set $P_{k+1} = P_k \cup \{k+1\}$ and $Q_{k+1} = Q_k$; or else $\rho(x_{k+1}, Y_k) < \varepsilon$, when we set $P_{k+1} = P_k$ and $Q_{k+1} = Q_k \cup \{k+1\}$. In either case, $P_{k+1} \cap Q_{k+1} = \varnothing$, $P_{k+1} \cup Q_{k+1} = \{1, \ldots k+1\}$, P_{k+1} is a basis for a finite-dimensional subspace of X, and $\rho(x_j, Y_{k+1}) < \varepsilon$ for each $j \in Q_{k+1}$. We obtain the desired sets by setting $P = P_n$ and $Q = Q_n$. □

A subalgebra of our Banach algebra $\mathfrak{A}$ is called a *Banach subalgebra* of $\mathfrak{A}$ if it is separable, contains the identity of $\mathfrak{A}$, and is complete with respect to the norm on $\mathfrak{A}$. If $\mathfrak{B}$ is a Banach subalgebra of $\mathfrak{A}$, then for each $f \in V_{\mathfrak{A}}^t$ (respectively, $V_{\mathfrak{A}}$), the restriction of f to $\mathfrak{B}$ belongs to $V_{\mathfrak{B}}^t$ (respectively, $V_{\mathfrak{B}}$). In the classical theory, where Hahn–Banach extensions preserve norms, every element of $V_{\mathfrak{B}}^t$ (respectively, $V_{\mathfrak{B}}$) is such a restriction. The following proposition and corollary are constructive substitutes for that classical converse.

Proposition 10.27 *Let $\mathfrak{A}$ have firm state space, let $\mathfrak{B}$ be a Banach subalgebra of $\mathfrak{A}$, and let $b_1, \ldots, b_n$ be elements of $\mathfrak{B}$. For each $\varepsilon > 0$ there exists $\tau > 0$ such that if $0 < t \leq \tau$ and $f \in V_{\mathfrak{B}}^t$, then there exists $g \in V_{\mathfrak{A}}$ with $|f(b_k) - g(b_k)| < \varepsilon$ for each $k \leq n$.*

Proof We may assume that each $\|b_k\| \leq 1$. Fix a weak* norm $\| \ \|_w^{\mathfrak{A}}$ on $\mathfrak{A}$ and let $\varepsilon > 0$. Since the function $\phi \rightsquigarrow \phi(x)$ is weak* uniformly continuous on $\mathfrak{A}_1^*$ for each $x \in \mathfrak{A}$, there exists $\delta > 0$ such that if $\phi, \psi \in \mathfrak{A}_1^*$, and $\|\phi - \psi\|_w^{\mathfrak{A}} < \delta$, then $|\phi(b_k) - \psi(b_k)| < \varepsilon/4$ for each $k \leq n$. Since $V_{\mathfrak{A}}$ is firm, there exists τ such that

$$0 < \tau < \min\left\{\frac{1}{2}, \left(1 + \frac{\varepsilon}{4}\right)^{-1} \frac{\varepsilon}{8}\right\}$$

and $\rho_w(V_{\mathfrak{A}}^s, V_{\mathfrak{A}}) < \delta$ whenever $0 < s \leq 3\tau$. If each $\|b_k\| < \varepsilon/2$, then for any $t > 0$, $f \in V_{\mathfrak{B}}^t$, and $g \in V_{\mathfrak{A}}$ we have $|f(b_k) - g(b_k)| < \varepsilon$ for each k; we may therefore assume that $b_k \neq 0$ for some k. In view of Lemma 10.26, we may further assume that there exists $N \leq n$ such that $\{b_1, \ldots, b_N\}$ is a basis for a finite-dimensional subspace Y of $\mathfrak{B}$ and $\rho(b_k, Y) < \varepsilon/4$ for $N < k \leq n$. For $1 \leq k \leq N$ set $b_k' = b_k$, and for $N < k \leq n$ choose $b_k' \in Y$ with $\|b_k - b_k'\| < \varepsilon/4$; note that, in either case, $\|b_k'\| < 1 + \varepsilon/4$. Let $0 < t \leq \tau$ and $f \in V_{\mathfrak{B}}^t$. Using Lemma 10.25, construct a normable linear functional ϕ on $\mathfrak{A}$ such that $\|\phi\| = 1$, $|1 - \phi(e)| \leq 3t$, and $|f(y) - \phi(y)| \leq 2t\|y\|$ for each $y \in Y$. Then $\phi \in V_{\mathfrak{A}}^{3t}$ and $3t \leq 3\tau$, so $\rho_w(\phi, V_{\mathfrak{A}}) \leq \rho_w(V_{\mathfrak{A}}^{3t}, V_{\mathfrak{A}}) < \delta$; whence there exists $g \in V_{\mathfrak{A}}$ such that $\|\phi - g\|_w < \delta$ and therefore $|\phi(b_k) - g(b_k)| < \varepsilon/4$ for each $k \leq n$. For such k we have

$$\begin{aligned}
|f(b_k) - g(b_k)| &\leq |f(b_k - b_k')| + |f(b_k') - \phi(b_k')| \\
&\quad + |\phi(b_k') - \phi(b_k)| + |\phi(b_k) - g(b_k)| \\
&< \|b_k - b_k'\| + 2t\,\|b_k'\| + \|b_k' - b_k\| + \frac{\varepsilon}{4} \\
&< \frac{\varepsilon}{4} + 2\tau\left(1 + \frac{\varepsilon}{4}\right) + \frac{\varepsilon}{4} + \frac{\varepsilon}{4} < \varepsilon
\end{aligned}$$

and the proof is complete. □

Corollary 10.28 *Let $\mathfrak{A}$ have firm state space, let $\mathfrak{B}$ be a Banach subalgebra of $\mathfrak{A}$, and let $b_1, \ldots, b_n$ be elements of $\mathfrak{B}$. Then for each $\varepsilon > 0$ and each $f \in V_{\mathfrak{B}}$ there exists $g \in V_{\mathfrak{A}}$ with $|f(b_k) - g(b_k)| < \varepsilon$ for each $k \leq n$.*

Lemma 10.29 *Let I be an inhabited set, let $(K_i)_{i\in I}$ be a family of totally bounded subsets of a metric space X, and let $K = \bigcap_{i\in I} K_i$. Suppose that for each $\varepsilon > 0$ there exists $i \in I$ such that $\rho(x, K) < \varepsilon$ for each $x \in K_i$. Then K is totally bounded. If also each K_i is complete, then K is compact.*

Proof Given $\varepsilon > 0$, choose $i \in I$ as in the hypotheses. Let $\{x_1, \dots, x_n\}$ be a finite ε-approximation to K_i, and for each k choose $y_k \in K$ such that $\|x_k - y_k\| < \varepsilon$. Let $y \in K \subset K_i$. Then there exists k such that $\|y - x_k\| < \varepsilon$ and therefore

$$\|y - y_k\| \leqslant \|y - x_k\| + \|x_k - y_k\| < \varepsilon + \varepsilon = 2\varepsilon.$$

Thus $\{y_1, \dots, y_n\}$ is a 2ε-approximation to K. Since $\varepsilon > 0$ is arbitrary, K is totally bounded. If also each K_i is complete, then K is an intersection of complete sets and so is complete; whence it is compact. □

Theorem 10.30 *If $\mathfrak{A}$ has firm state space, then the state space of every Banach subalgebra of $\mathfrak{A}$ is firm.*

Proof Let $\mathfrak{B}$ be a Banach subalgebra of $\mathfrak{A}$, let $(b_n)_{n\geq 1}$ be a dense sequence in the unit ball of $\mathfrak{B}$, and let $\| \ \|_w^{\mathfrak{B}}$ be the corresponding double norm on $\mathfrak{B}^*$. Given $\varepsilon > 0$, choose N such that $\sum_{N+1}^{\infty} 2^{-n+1} < \varepsilon/2$. By Proposition 10.27, we can find $\tau > 0$ such that if $0 < t \leq \tau$ and $f \in V_{\mathfrak{B}}^t$, then there exists $g \in V_{\mathfrak{A}}$ such that $|f(b_n) - g(b_n)| < \varepsilon/2$ for $1 \leq n \leq N$. For such f and g we have

$$\begin{aligned}\|f - g\|_w^{\mathfrak{B}} &= \sum_{n=1}^{N} 2^{-n} |f(b_n) - g(b_n)| + \sum_{n=N+1}^{\infty} 2^{-n} |f(b_n) - g(b_n)| \\ &< \sum_{n=1}^{N} 2^{-n} \frac{\varepsilon}{2} + \sum_{n=N+1}^{\infty} 2^{-n} (|f(b_n)| + |g(b_n)|) \\ &< \frac{\varepsilon}{2} + \sum_{n=N+1}^{\infty} 2^{-n+1} < \varepsilon,\end{aligned}$$

so, since the restriction of g to $\mathfrak{B}$ belongs to $V_{\mathfrak{B}}$, $\rho_w^{\mathfrak{B}}(f, V_{\mathfrak{B}}) < \varepsilon$. Since $\varepsilon > 0$ is arbitrary and, by Proposition 10.24, $V_{\mathfrak{B}}^t$ is weak* compact for each $t > 0$, Lemma 10.29 tells us that $V_{\mathfrak{B}}$ is weak* compact and therefore weak* located. Moreover, since the mapping $\phi \rightsquigarrow \rho_w^{\mathfrak{B}}(\phi, V_{\mathfrak{B}})$ is uniformly continuous on $V_{\mathfrak{B}}^t$, $\sup\{\rho_w^{\mathfrak{B}}(\phi, V_{\mathfrak{B}}) : \phi \in V_{\mathfrak{B}}^t\}$ exists and has limit 0 as $t \to 0$. On the other hand, since $V_{\mathfrak{B}} \subset V_{\mathfrak{B}}^t$, $\sup\{\rho_w^{\mathfrak{B}}(\phi, V_{\mathfrak{B}}^t) : \phi \in V_{\mathfrak{B}}\} = 0$. Hence $\rho_w^{\mathfrak{B}}(V_{\mathfrak{B}}^t, V_{\mathfrak{B}}) \to 0$ as $t \to 0$. □

We define the *numerical range*[9] of an element x of our Banach algebra $\mathfrak{A}$ to be the set

[9] Another notion of numerical range arises in the context of a Hilbert space H. The *spatial numerical range* of an element T of $B(H)$ is

$$W(T) \equiv \{\langle Tx, x\rangle : x \in H, \|x\| = 1\}$$

$$V_{\mathfrak{A}}(x) \equiv \{f(x) : f \in V_{\mathfrak{A}}\}.$$

It is simple to prove classically that $\sigma(x) \subset V_{\mathfrak{A}}(x)$ [14, Proposition 4.3.3]; but the classical proof falls short in our constructive context, where we have the following.

Proposition 10.31 *Let $\mathfrak{A}$ have firm state space, and $x \in \mathfrak{A}$. Then $\sigma(x) \subset \overline{V_{\mathfrak{A}}(x)}$.*

Proof Let $\lambda \in \sigma(x)$. If $x \in \mathbb{C}e$, then since $(\zeta - \lambda)e$ is invertible for each complex $\zeta \neq \lambda$, we must have $x = \lambda e$; so $f(te) \equiv t$ defines a state of $\mathfrak{A}$ with $f(x) = \lambda$. Now consider the case where $\rho(x, \mathbb{C}e) > 0$ and therefore $x \neq \lambda e$. By Corollary 10.23, $|s\lambda + t| \leq \|sx + te\|$ for all $s, t \in \mathbb{C}$, so $f_0(sx + te) \equiv s\lambda + t$ defines a linear functional on the two-dimensional normed space Y spanned by $\{x, e\}$, such that $f_0(x) = \lambda$, $f_0(e) = 1 = \|e\|$, and $\ker f_0 = \mathbb{C}(x - \lambda e)$. Since $x \neq \lambda e$, $\ker f_0$ is a one-dimensional space and is therefore located in $\mathfrak{A}$. Since $V_{\mathfrak{A}}$ is firm, for each $\varepsilon > 0$ there exists $\delta > 0$ with $|\lambda|\,\delta < \varepsilon/2$, such that if $g \in V_{\mathfrak{A}}^{\delta}$, then $|g(x) - f(x)| < \varepsilon/2$ for some $f \in V_{\mathfrak{A}}$. By the Hahn–Banach theorem [2, Chapter 7, Theorem (4.6)], there exists a normable linear functional f_1 on $\mathfrak{A}$ such that $f_1(y) = f_0(y)$ for all $y \in Y$ and $\|f_1\| < 1 + \delta$. Let $g = (1 + \delta)^{-1} f_1$. Then $g(x) = (1 + \delta)^{-1}\lambda$, $g \in \mathfrak{A}^*$, $\|g\| = (1 + \delta)^{-1} < 1$, and

$$0 \leq 1 - g(e) = \frac{\delta}{1 + \delta} < \delta.$$

Hence $g \in V_{\mathfrak{A}}^{\delta}$ and there exists $f \in V_{\mathfrak{A}}$ such that $|g(x) - f(x)| < \varepsilon/2$. We have

$$\begin{aligned} |f(x) - \lambda| &\leq |g(x) - f(x)| + |g(x) - \lambda| \\ &< \frac{\varepsilon}{2} + \left|(1 + \delta)^{-1}\lambda - \lambda\right| \\ &< \frac{\varepsilon}{2} + \frac{|\lambda|\,\delta}{1 + \delta} < \varepsilon. \end{aligned}$$

We have now proved that if either $x \in \mathbb{C}e$ or $\rho(x, \mathbb{C}e) > 0$, then for each $\varepsilon > 0$ there exists $f \in V_{\mathfrak{A}}$ with $|f(x) - \lambda| < \varepsilon$. It follows that if $\rho(\lambda, V_{\mathfrak{A}}) > 0$, then we have both $\rho(x, \mathbb{C}e) = 0$ and $x \notin \mathbb{C}e$, which is absurd, as $\mathbb{C}e$, being finite-dimensional, is closed. Hence, in fact, $\rho(\lambda, V_{\mathfrak{A}}(x)) = 0$ and therefore $\lambda \in \overline{V_{\mathfrak{A}}(x)}$. □

In preparation for Section 10.5, we deal with some results on commutative Banach algebras, beginning with two fundamental results of Bishop.

Proposition 10.32 *If $\mathfrak{A}$ is commutative, then there exist weak* compact subsets $\Sigma_1 \supset \Sigma_2 \supset \cdots$ of $\mathfrak{A}_1^*$ such that $\Sigma_{\mathfrak{A}} = \bigcap_{n \geq 1} \Sigma_n$* [2, Chapter 9, Propositions (1.3) and (2.7)].

(see [10]). The classical *Toeplitz–Hausdorff theorem* states that $W(T)$ is convex. In [4] we proved that if $T \in B(H)$ has an adjoint, then $\overline{W(T)}$ is convex, and that even if H is two-dimensional, the convexity of $W(T)$ for every self-adjoint operator implies a variant of LLPO.

Bishop calls the character space $\Sigma_{\mathfrak{A}}$ *firm* if it is weak* compact and the sets Σ_n in Proposition 10.32 can be constructed such that $\rho_w(\Sigma_n, \Sigma) \to 0$ as $n \to \infty$.

Theorem 10.33 *Let $\mathfrak{A}$ be commutative, and let the sets Σ_n be as in* Proposition 10.32. *For each $x \in \mathfrak{A}$ and each positive integer n let*

$$\|x\|_{\Sigma_n} \equiv \sup\{|u(x)| : u \in \Sigma_n\}.$$

Then the sequences $\left(\|x^n\|^{1/n}\right)_{n\geq 1}$ and $\left(\|x\|_{\Sigma_n}\right)_{n\geq 1}$ are equiconvergent: *that is, for each term a_m of one sequence and each $\varepsilon > 0$, there exists a positive integer N such that $b_n \leq a_m + \varepsilon$ whenever b_n is a term of the other sequence with $n \geq N$* [2, Chapter 9, Proposition (2.9)].

By a *generating set* for the Banach algebra $\mathfrak{A}$ we mean a set $G \subset \mathfrak{A}$ such that the set of complex polynomials[10] in elements of G is dense in $\mathfrak{A}$; we then say that $\mathfrak{A}$ is *generated by* G. If $a \in \mathfrak{A}$ and $\{a\}$ is a generating set for $\mathfrak{A}$, we say that $\mathfrak{A}$ is *singly generated* by a; $\mathfrak{A}$ is then separable and commutative. In [8] the authors introduce the notion of topological independence of a finitely enumerable subset of a Banach algebra and prove the following result.[11]

Proposition 10.34 *The character space of a separable, commutative Banach algebra generated by a finite set of topologically independent vectors is firm* [8, Proposition 5].

Corollary 10.35 *The character space of a singly generated, separable, commutative Banach algebra is firm (and hence inhabited).*

Proposition 10.36 *If $\mathfrak{A}$ is singly generated by $a \in \mathfrak{A}$, then the* spectral radius *of a,*

$$\|a\|_{\Sigma_{\mathfrak{A}}} \equiv \sup\{u(a) : u \in \Sigma_{\mathfrak{A}}\}$$

exists and equals $\lim_{n\to\infty} \|a^n\|^{1/n}$.

Proof By Corollary 10.35, $\Sigma_{\mathfrak{A}}$ is firm. Let $(\Sigma_n)_{n\geq 1}$ be a sequence of weak* compact subsets of $\mathfrak{A}_1^*$ such that $\Sigma_1 \supset \Sigma_2 \supset \cdots$, $\Sigma_{\mathfrak{A}} = \bigcap_{n\geq 1}\Sigma_n$, and $\rho_w(\Sigma_n, \Sigma_{\mathfrak{A}}) \to 0$ as $n \to \infty$. Then $\|a\|_{\Sigma_n}$ and, since $\Sigma_{\mathfrak{A}}$ is weak* compact,

$$\|a\|_{\Sigma_{\mathfrak{A}}} \equiv \sup\{|u(a)| : u \in \Sigma_{\mathfrak{A}}\}$$

exist. Given $\varepsilon > 0$, choose $\delta > 0$ such that $|f(a) - g(a)| < \varepsilon$ whenever $f, g \in \mathfrak{A}_1^*$ and $\|f - g\|_w < \delta$. Then choose N such that $\sup\{\rho_w(u, \Sigma_{\mathfrak{A}}) : u \in \Sigma_n\} < \delta$ for

[10] For example, if $p(z_1, z_2) = \sum_{j=0}^{m}\sum_{k=0}^{n} c_{jk} z_1^j z_2^k$, where z_1, z_2, and c_{jk} belong to $\mathbb{C}$, is a polynomial in two variables, and a_1, a_2 are elements of $\mathfrak{A}$, then $p(a_1, a_2) \equiv \sum_{j=0}^{m}\sum_{k=0}^{n} c_{jk} a_1^j a_2^k$ is the corresponding polynomial in a_1 and a_2.

[11] Proposition 10.34 leads to constructive proofs of local Nullstellensätze for algebras of continuous functions and formal power series. For details, see [8].

all $n \geq N$. For such n and each $u \in \Sigma_n$, there exists $v \in \Sigma_{\mathfrak{A}}$ with $\|u - v\|_w < \delta$; whence $|u(a) - v(a)| < \varepsilon$ and therefore $|u(a)| < |v(a)| + \varepsilon \leq \|a\|_{\Sigma_{\mathfrak{A}}} + \varepsilon$. Since $\Sigma_{\mathfrak{A}} \subset \Sigma_n$, it follows that

$$\|a\|_{\Sigma_{\mathfrak{A}}} \leq \|a\|_{\Sigma_n} \leq \|a\|_{\Sigma_{\mathfrak{A}}} + \varepsilon \quad (n \geq N).$$

Since $\varepsilon > 0$ is arbitrary, $\left(\|a\|_{\Sigma_n}\right)_{n \geq 1}$ converges to $\|a\|_{\Sigma_n}$.

With ε, N as before, we see from Theorem 10.33 that for each n there exists k such that $\|a\|_{\Sigma_{\mathfrak{A}}} \leq \|a\|_{\Sigma_k} \leq \|a^n\|^{1/n} + \varepsilon$. On the other hand, by the same theorem, there exists $N_1 \geq N$ such that $\|a^n\|^{1/n} \leq \|a\|_{\Sigma_N} + \varepsilon \leq \|a\|_{\Sigma_{\mathfrak{A}}} + 2\varepsilon$ for all $n \geq N_1$. Thus $\|a^n\|^{1/n} - 2\varepsilon \leq \|a\|_{\Sigma_{\mathfrak{A}}} \leq \|a^n\|^{1/n} + \varepsilon$ for all $n \geq N_1$. Hence, $\varepsilon > 0$ being arbitrary, $\lim_{n\to\infty} \|a^n\|^{1/n}$ exists and equals $\|a\|_{\Sigma_{\mathfrak{A}}}$. □

10.5 Hermitian and Positive Elements

An element x of $\mathfrak{A}$ is said to be

- *Hermitian* if for each $\varepsilon > 0$ there exists $t > 0$ such that $|\operatorname{Im} f(x)| < \varepsilon$ whenever $f \in V_{\mathfrak{A}}^t$;
- *positive* if for each $\varepsilon > 0$ there exists $t > 0$ such that $\operatorname{Re} f(x) \geq -\varepsilon$ and $|\operatorname{Im} f(x)| < \varepsilon$ whenever $f \in V_{\mathfrak{A}}^t$.

We denote the set of Hermitian (respectively, positive) elements of $\mathfrak{A}$ by $\mathsf{Her}(\mathfrak{A})$ (respectively, $\mathsf{Pos}(\mathfrak{A})$). If $x \in \mathsf{Pos}(\mathfrak{A})$, we write $x \geq 0$, and if $x - y \geq 0$, we write either $x \geq y$ or $y \leq x$.

Clearly, $\mathsf{Pos}(\mathfrak{A}) \subset \mathsf{Her}(\mathfrak{A})$. It is an exercise to prove:

- if $x, y \in \mathsf{Her}(\mathfrak{A})$ and $c \in \mathbb{R}$, then $x + cy \in \mathsf{Her}(\mathfrak{A})$;
- if $x, y \in \mathsf{Pos}(\mathfrak{A})$ and $c > 0$, then $x + cy \in \mathsf{Pos}(\mathfrak{A})$; and
- $x \geq 0$ if and only if $0 \geq -x$.

The relation $\geq$ is transitive: if $x, y, z \in \mathfrak{A}$, $x \geq y$, and $y \geq z$, then $x - y \geq 0$ and $y - z \geq 0$, so $x - z = (x - y) + (y - z) \geq 0$.

The identity e of $\mathfrak{A}$ is positive since for each $t \in (0, 1)$ and each $f \in V_{\mathfrak{A}}^t$,

$$\operatorname{Re} f(e) = 1 - \operatorname{Re}(1 - f(e)) \geq 1 - |1 - f(e)| \geq 1 - t > 0 \tag{10.2}$$

and

$$|\operatorname{Im} f(e)| = |\operatorname{Im}(1 - f(e))| \leq |1 - f(e)| \leq t.$$

If $\mathfrak{B}$ is a Banach subalgebra of $\mathfrak{A}$, then $\mathsf{Her}(\mathfrak{B}) \subset \mathsf{Her}(\mathfrak{A})$ and $\mathsf{Pos}(\mathfrak{B}) \subset \mathsf{Pos}(\mathfrak{A})$ (recall that the restriction to $\mathfrak{B}$ of an element of $V_{\mathfrak{A}}^t$ belongs to $V_{\mathfrak{B}}^t$).

Proposition 10.37 *If $\mathfrak{A}$ has firm state space, then an element x of $\mathfrak{A}$ is*

(i) *Hermitian if and only if $f(x) \in \mathbb{R}$ for each $f \in V_{\mathfrak{A}}$;*
(ii) *positive if and only if $f(x) \geq 0$ for each $f \in V_{\mathfrak{A}}$.*

Proof We prove only (ii). Let $x \in \mathfrak{A}$ and suppose first that x is positive. For each $\varepsilon > 0$ there exists $\delta > 0$ such that if $f, g \in \mathfrak{A}_1^*$ and $\|f - g\|_w < \delta$, then $|f(x) - g(x)| < \varepsilon/2$. By the firmness of $V_{\mathfrak{A}}$ and the positivity of x, there exists $t > 0$ such that

- $\rho_w(V_{\mathfrak{A}}^t, V_{\mathfrak{A}}) < \delta$ and
- $\operatorname{Re} g(x) \geq -\varepsilon/2$ and $|\operatorname{Im} g(x)| < \varepsilon/2$ for all $g \in V_{\mathfrak{A}}^t$.

For each $f \in V_{\mathfrak{A}}$ pick $g \in V_{\mathfrak{A}}^t$ such that $\|f - g\|_w < \delta$; then

$$\operatorname{Re} f(x) = \operatorname{Re} g(x) - \operatorname{Re}(g(x) - f(x)) > -\frac{\varepsilon}{2} - |f(x) - g(x)| > -\varepsilon$$

and

$$|\operatorname{Im} f(x)| \leq |f(x) - g(x)| + |\operatorname{Im} g(x)| < \varepsilon.$$

Since $\varepsilon > 0$ is arbitrary, we see that $\operatorname{Im} f(x) = 0$ and $f(x) = \operatorname{Re} f(x) \geq 0$.

Conversely, suppose that $f(x) \geq 0$ for each $f \in V_{\mathfrak{A}}$. This time, given $\varepsilon > 0$, choose $\delta > 0$ such that $|f(x) - g(x)| < \varepsilon$ for all $f, g \in \mathfrak{A}_1^*$ with $\|f - g\|_w < \delta$; then choose $t > 0$ such that $\rho_w(V_{\mathfrak{A}}^t, V_{\mathfrak{A}}) < \delta$. For each $g \in V_{\mathfrak{A}}^t$ there exists $f \in V_{\mathfrak{A}}$ such that $\|f - g\|_w < \delta$. Then $f(x) \geq 0$, so

$$\operatorname{Re} g(x) = f(x) - \operatorname{Re}(f(x) - g(x)) \geq -|f(x) - g(x)| > -\varepsilon$$

and

$$|\operatorname{Im} g(x)| \leq |\operatorname{Im}(g(x) - f(x))| \leq |f(x) - g(x)| < \varepsilon.$$

Thus, $\varepsilon > 0$ being arbitrary, x is positive. □

Corollary 10.38 *Let $\mathfrak{A}$ have firm state space, and let $\mathfrak{B}$ be a Banach subalgebra of $\mathfrak{A}$. Then $\mathsf{Her}(\mathfrak{B}) = \mathfrak{B} \cap \mathsf{Her}(\mathfrak{A})$ and $\mathsf{Pos}(\mathfrak{B}) = \mathfrak{B} \cap \mathsf{Pos}(\mathfrak{A})$.*

Proof First note that, by Theorem 10.30, $V_{\mathfrak{B}}$ is firm. Let $x \in \mathfrak{B} \cap \mathsf{Her}(\mathfrak{A})$. If $f \in V_{\mathfrak{B}}$ and $\varepsilon > 0$, then, by Corollary 10.28, there exists $g \in V_{\mathfrak{A}}$ such that $|f(x) - g(x)| < \varepsilon$; then, by Proposition 10.37, $g(x) \in \mathbb{R}$, so $\rho(f(x), \mathbb{R}) < \varepsilon$. Since $\varepsilon > 0$ is arbitrary, it follows that $\rho(f(x), \mathbb{R}) = 0$, and therefore $f(x) \in \mathbb{R}$, for each $f \in V_{\mathfrak{B}}$. Thus, by Proposition 10.37, $x \in \mathsf{Her}(\mathfrak{B})$. We now see that $\mathfrak{B} \cap \mathsf{Her}(\mathfrak{A}) \subset \mathsf{Her}(\mathfrak{B}) \subset \mathfrak{B} \cap \mathsf{Her}(\mathfrak{A})$ and therefore $\mathsf{Her}(\mathfrak{B}) = \mathfrak{B} \cap \mathsf{Her}(\mathfrak{A})$. The remaining conclusion of the proposition is derived similarly. □

Corollary 10.39 *Let $\mathfrak{A}$ have firm state space. If $x \in \mathfrak{A}$ is Hermitian (respectively, positive), then every element of $\sigma_A(x)$ is real (respectively, nonnegative).*

Proof This follows from Propositions 10.31 and 10.37. □

Proposition 10.40 *If $\mathfrak{A}$ has firm state space, then* $-\|x\|\,e \leq x \leq \|x\|\,e$ *for each* $x \in \mathfrak{A}$.

Proof For each $f \in V_{\mathfrak{A}}$ we have $\|f\| = 1$ and therefore $f(\|x\|\,e \pm x) = \|x\| \pm f(x) \geq 0$. From this and Proposition 10.37 we obtain $\|x\|\,e \pm x \geq 0$ and hence the desired conclusion. □

An element f of $\mathfrak{A}^*$ is called a *positive linear functional* if $f(x) \geq 0$ for each $x \in \mathsf{Pos}(\mathfrak{A})$; in that case we write $f \geq 0$.

Proposition 10.41 *Every state of $\mathfrak{A}$ is a positive linear functional.*

Proof Let $f \in V_{\mathfrak{A}}$ and $x \in \mathsf{Pos}(\mathfrak{A})$. For each $\varepsilon > 0$ there exists $t > 0$ such that $\operatorname{Re} g(x) \geq -\varepsilon$ and $|\operatorname{Im} g(x)| < \varepsilon$ whenever $g \in V_{\mathfrak{A}}^t$. In particular, $\operatorname{Re} f(x) \geq -\varepsilon$ and $|\operatorname{Im} f(x)| < \varepsilon$. Since $\varepsilon > 0$ is arbitrary, we conclude that $f(x) = \operatorname{Re} f(x) \geq 0$. □

Proposition 10.42 *Let $\mathfrak{A}$ have firm state space. If f is a positive linear functional on $\mathfrak{A}$, then f is normable and $\|f\| = f(e)$.*

Proof Note that $f(e) \geq 0$. By Proposition 10.40, for each $x \in \mathfrak{A}$ we have $\|x\|\,e \pm x \geq 0$; whence $\|x\|\,f(e) \pm f(x) = f(\|x\|\,e \pm x) \geq 0$ and therefore $|f(x)| \leq \|x\|\,f(e)$. Moreover, $|f(e)| = \|e\|\,f(e)$, so $\|f\|$ exists and equals $f(e)$. □

Since we cannot be sure that the state space is inhabited, in dealing with the numerical range of an element x of $\mathfrak{A}$ it makes sense to consider approximations to $V_{\mathfrak{A}}$. Accordingly, for each $t > 0$ we define

$$V_{\mathfrak{A}}^t(x) \equiv \{f(x) : x \in V_{\mathfrak{A}}^t\},$$

which, in view of the weak* uniform continuity of the mapping $f \rightsquigarrow f(x)$ on the weak* compact set $V_{\mathfrak{A}}^t$, is totally bounded in $\mathbb{C}$. Working with the set $V_{\mathfrak{A}}^t$ and guided by classical arguments in [3, pages 51–57], we obtained the following.

Proposition 10.43 *If a is a Hermitian element of a separable Banach algebra $\mathfrak{A}$, then $\|\exp(\pm i\zeta x)\| = 1$ for all $\zeta \in \mathbb{R}$* [5, Proposition 19.10].

Here, naturally, $\exp(x) \equiv \sum_{n=0}^{\infty} x^n/n!$. This leads to a proof of *Sinclair's theorem* [3, page 57]:

Theorem 10.44 *If x is a Hermitian element of a separable Banach algebra $\mathfrak{A}$, then $\|x^n\|^{1/n} = \|x\|$ for each positive integer n.*

Proposition 10.45 *Let $\mathfrak{A}$ have firm state space. If $x \in \mathfrak{A}$ and $V_{\mathfrak{A}}(x) = \{0\}$, then $x = 0$.*

Proof Let $\mathfrak{B}$ be the (separable) Banach subalgebra of $\mathfrak{A}$ generated by $\{x\}$. By Corollary 10.28, if $f \in V_{\mathfrak{B}}$, then for each $\varepsilon > 0$ there exists $g \in V_{\mathfrak{A}}$ such that $|f(x) - g(x)| < \varepsilon$; but $g(x) \in V_{\mathfrak{A}}(x) = \{0\}$, so $|f(x)| < \varepsilon$. It follows that $f(x) = 0$ for each $f \in V_{\mathfrak{B}}$; in particular, $u(x) = 0$ for each $u \in \Sigma_{\mathfrak{B}}$, and therefore $\|x\|_{\Sigma_{\mathfrak{B}}} = 0$. Since $V_{\mathfrak{B}}$ is firm by Theorem 10.30, we see from Proposition 10.37 that $x \in \mathsf{Her}(\mathfrak{B})$. Therefore, by Theorem 10.44 and Proposition 10.36, $\|x\| = \lim_{n\to\infty} \|x^n\|^{1/n} = 0$. □

Proposition 10.46 *Let $\mathfrak{A}$ have firm state space. If $x \in \mathfrak{A}$, $c \geq 0$, and $0 \leq x \leq ce$, then $\|x\| \leq c$.*

Proof Let $\mathfrak{B}$ be the Banach subalgebra of $\mathfrak{A}$ generated by $\{x\}$. By Proposition 10.41, $0 \leq u(x) \leq u(ce) = c$ for each $u \in \Sigma_{\mathfrak{B}}$. On the other hand, by Corollary 10.38, $x \in \mathfrak{B} \cap \mathsf{Pos}(\mathfrak{A}) \subset \mathsf{Her}(\mathfrak{B})$, so, by Theorem 10.44, $\|x^n\|^{1/n} = \|x\|$ for each positive integer n. It follows from Proposition 10.36 that $\sup\{|u(x)| : u \in \Sigma_{\mathfrak{B}}\}$ exists and equals $\|x\|$. Hence $\|x\| \leq c$. □

Corollary 10.47 *Let $\mathfrak{A}$ have firm state space. If $x \in \mathfrak{A}$ and $0 \leq x \leq 0$, then $x = 0$.*

Corollary 10.48 *Let $\mathfrak{A}$ have firm state space. If $x, y \in \mathfrak{A}$ and $0 \leq x \leq y$, then $\|x\| \leq \|y\|$.*

Proof By Proposition 10.40, $0 \leq x \leq y \leq \|y\| e$. The result follows from Proposition 10.46. □

Corollary 10.49 *Let $\mathfrak{A}$ have firm state space. If $x, y \in \mathsf{Her}(\mathfrak{A})$ and $x = iy$, then $x = y = 0$.*

Proof For each $\varepsilon > 0$ there exists $t > 0$ such that $|\mathrm{Im}\, f(x)| < \varepsilon$ and $|\mathrm{Im}\, f(y)| < \varepsilon$ whenever $f \in V_{\mathfrak{A}}^t$. For such f we have $f(x) = if(y)$, so $\mathrm{Re}\, f(x) = -\mathrm{Im}\, f(y)$. Hence $\mathrm{Re}\, f(x) > -\varepsilon$ and $|\mathrm{Im}\, f(x)| < \varepsilon$; moreover, $\mathrm{Re}\, f(-x) = \mathrm{Im}\, f(y) > -\varepsilon$ and $|\mathrm{Im}\, f(-x)| < \varepsilon$. Since $\varepsilon > 0$ is arbitrary, we see that $x \geq 0$ and $-x \geq 0$. It follows that $0 \leq x \leq 0$, so, by Corollary 10.47, $x = 0$ and therefore $y = -ix = 0$. □

We say that an element x of $\mathfrak{A}$ is *Hermitian decomposable* if there exist $x_1, x_2 \in \mathsf{Her}(\mathfrak{A})$ such that $x = x_1 + ix_2$; in that case, the latter equation is called a *Hermitian decomposition* of x.

Proposition 10.50 *Let $\mathfrak{A}$ have firm state space. Then the Hermitian decomposition, if it exists, of a given element of $\mathfrak{A}$ is unique.*

Proof If $x = x_1 + ix_2 = x_1' + ix_2'$, where $x_1, x_2, x_1', x_2' \in \mathsf{Her}(\mathfrak{A})$, then $x_1 - x_1'$ and $x_2' - x_2$ are Hermitian and $x_1 - x_1' = i(x_2' - x_2)$. It follows from Corollary 10.49 that $x_1 - x_1' = 0 = x_2' - x_2$. □

Proposition 10.51 *Let $\mathfrak{A}$ have firm state space, and let $x \in \mathfrak{A}$ have Hermitian decomposition $x = x_1 + ix_2$. Then $\|x_k\| \leq \|x\|$ for each k.*

Proof Let $\mathfrak{B}$ be the Banach algebra generated by $\{x_1\}$. By Corollary 10.28, for each $u \in \Sigma_{\mathfrak{B}}$ and each $\varepsilon > 0$ there exists $g \in V_{\mathfrak{A}}$ such that $|u(x_1) - g(x_1)| < \varepsilon$. Then $u(x_k), g(x_k) \in \mathbb{R}$ and

$$\begin{aligned} u(x_1)^2 - g(x_1)^2 &= (|u(x_1)| + |g(x_1)|)\,(|u(x_1)| - |g(x_1)|) \\ &\leq 2\,\|x_1\|\,|u(x_1) - g(x_1)| \leq 2\,\|x_1\|\,\varepsilon. \end{aligned}$$

Hence

$$|g(x)|^2 = g(x_1)^2 + g(x_2)^2 \geq g(x_1)^2 \geq |u(x_1)|^2 - 2\,\|x_1\|\,\varepsilon$$

and therefore

$$|u(x_1)|^2 \leq |g(x)|^2 + 2\,\|x_1\|\,\varepsilon \leq \|x\|^2 + 2\,\|x_1\|\,\varepsilon.$$

Since $u \in \Sigma_{\mathfrak{B}}$ and $\varepsilon > 0$ are arbitrary, we have $|u(x_1)| \leq \|x\|$ for each $u \in \Sigma_{\mathfrak{B}}$; whence, by Proposition 10.36 and Theorem 10.44, $\|x_1\| = \|x_1\|_{\Sigma_{\mathfrak{B}}} \leq \|x\|$. □

We call the Banach algebra $\mathfrak{A}$ *quasi-stellar* if every element x of $\mathfrak{A}$ has a Hermitian decomposition. If X is a compact metric space, then $C(X)$ is quasi-stellar, the Hermitian decomposition of an element f of $C(X)$ being[12] $f = \frac{1}{2}(f + \overline{f}) + i\left(\frac{i}{2}(\overline{f} - f)\right)$. If H is a non-trivial Hilbert space, then a separable, norm-closed, self-adjoint[13] algebra of normable elements of $B(H)$ that contains the identity is quasi-stellar; a constructive proof of this can be gleaned from [14, Section 4.2]. Under certain conditions, C*-algebras are quasi-stellar [15, Section 18.3 and Proposition 18.21].

We say that $\mathfrak{A}$

- is *orderly* if $\mathsf{Pos}(\mathfrak{A})$ is closed under multiplication: that is, $xy \geq 0$ for all $x, y \in \mathsf{Pos}(\mathfrak{A})$;
- *has positive Hermitian squares* if $x^2 \geq 0$ for each $x \in \mathsf{Her}(\mathfrak{A})$.

If $\mathfrak{A}$ is orderly, then every positive integral power of a positive element of $\mathfrak{A}$ is positive; but that does not guarantee that even the square of a Hermitian element is positive.

12 When ζ is either a complex number or a complex-valued function, we denote its complex conjugate by $\overline{\zeta}$,

13 Self-adjoint means that every element of the subalgebra of $B(H)$ has an adjoint that belongs to the subalgebra. Note that if H is infinite-dimensional, the existence of the adjoint for every element of $B(H)$ implies LPO [7, page 101].

The Banach algebra $C(X)$, where X is a compact metric space, is orderly. The same holds classically for any commutative C*-algebra of operators on a Hilbert space [14, Theorems 4.2.2 and 4.3.4]. A C*-algebra has positive Hermitian squares [15, Theorems 18.11 and 18.21].

Proposition 10.52 *If $\mathfrak{A}$ has firm state space and is orderly (respectively, has positive Hermitian squares), then every Banach subalgebra of $\mathfrak{A}$ is orderly (respectively, has positive Hermitian squares).*

Proof We prove only the 'orderly' part of the proposition. Let $\mathfrak{B}$ be a Banach subalgebra of $\mathfrak{A}$, and $x, y \in \mathsf{Pos}(\mathfrak{B})$. Then $x, y \in \mathsf{Pos}(\mathfrak{A})$, so $xy \in \mathsf{Pos}(\mathfrak{A})$. By Corollary 10.28, for each $f \in V_{\mathfrak{B}}$ and each $\varepsilon > 0$ there exists $g \in V_{\mathfrak{A}}$ with $|f(xy) - g(xy)| < \varepsilon$. By Proposition 10.41, $g(xy) \geq 0$. Thus

$$\operatorname{Re} f(xy) \geq \operatorname{Re} g(xy) - |f(xy) - g(xy)| > -\varepsilon$$

and

$$|\operatorname{Im} f(xy)| \leq |\operatorname{Im} g(xy)| + |f(xy) - g(xy)| < \varepsilon.$$

Hence, $\varepsilon > 0$ being arbitrary, $f(xy) = \operatorname{Re} f(xy) \geq 0$. Since, by Theorem 10.30, $V_{\mathfrak{B}}$ is firm, it follows from Proposition 10.37 that $xy \in \mathsf{Pos}(\mathfrak{B}) \subset \mathsf{Pos}(\mathfrak{A})$. □

Lemma 10.53 *If $\mathfrak{A}$ has positive Hermitian squares, then $f(x)^2 \leq f(x^2)$ whenever $f \geq 0$, $f(e) \leq 1$, and $x \in \mathsf{Her}(\mathfrak{A})$.*

Proof Let $f \geq 0$, $f(e) \leq 1$, and $x \in \mathsf{Her}(\mathfrak{A})$. For each $t \in \mathbb{R}$, $tx + e \in \mathsf{Her}(\mathfrak{A})$ and therefore $(tx + e)^2 \geq 0$; whence

$$0 \leq f\left((tx + e)^2\right) = f\left(t^2x^2 + 2tx + e\right) = t^2 f\left(x^2\right) + 2tf(x) + f(e).$$

Thus the quadratic form in t on the right-hand side is positive semidefinite. Since $0 \leq f(e) \leq 1$ and $f(x^2) \geq 0$, it follows that $f(x)^2 \leq f(x^2)f(e) \leq f(x^2)$. □

Let C be a convex subset of a normed space X. We say that an element x_0 of C is

- a *classical extreme point of* C if $x = y = x_0$ for all $x, y \in C$ with $x_0 = \frac{1}{2}(x+y)$;
- an *extreme point of* C if for each $\varepsilon > 0$ there exists $\delta > 0$ such that $\|x - y\| < \varepsilon$ for all $x, y \in C$ with $\left\|\frac{1}{2}(x + y) - x_0\right\| < \delta$.

Extreme points are classical extreme points; the converse holds classically if C is compact.

In order to discuss the extreme points of V_A, we introduce an auxiliary subset of $\mathfrak{A}^*$:

$$C_{\mathfrak{A}}^0 \equiv \{f \in \mathfrak{A}^* : f \geq 0, f(e) \leq 1\}.$$

In view of Proposition 10.41, we see that $V_{\mathfrak{A}} \subset C^0_{\mathfrak{A}}$.

Proposition 10.54 *Let $\mathfrak{A}$ have firm state space. Then every classical extreme point of $V_{\mathfrak{A}}$ is a classical extreme point of $C^0_{\mathfrak{A}}$.*

Proof Let f_0 be a classical extreme point of $V_{\mathfrak{A}}$, and f, g elements of $C^0_{\mathfrak{A}}$ with $f_0 = \frac{1}{2}(f + g)$. Then

$$(1 - f(e)) + (1 - g(e)) = 2(1 - f_0(e)) = 0,$$

so, as $1 - f(e) \geq 0$ and $1 - g(e) \geq 0$, we have $f(e) = 1 = g(e)$. Since f and g are positive linear functionals, it follows from Proposition 10.42 that they belong to $V_{\mathfrak{A}}$, and hence that $f = g = f_0$. □

Lemma 10.55 *If x is a Hermitian element of $\mathfrak{A}$, then there exists $c > 0$ such that $0 \leq cx + \frac{1}{2}e \leq e$.*

Proof Choose $c > 0$ such that $\|cx\| < 1/4$. Given $\varepsilon > 0$, find $t \in (0, \frac{1}{2})$ such that $|\operatorname{Im} f(e)| < \varepsilon/2$ and $|\operatorname{Im} f(x)| < 3\varepsilon/4c$ for all $f \in V^t_{\mathfrak{A}}$. For all such f we have

$$\left|\operatorname{Im} f(\tfrac{1}{2}e \pm cx)\right| = \left|\operatorname{Im} f(\tfrac{1}{2}e) \pm c \operatorname{Im} f(x)\right| \leq \tfrac{1}{2} |\operatorname{Im} f(e)| + c\,|\operatorname{Im} f(x)| < \varepsilon,$$

$\operatorname{Re} f(e) > 1 - t > 1/2$ (see Eqn. (10.2)), and

$$\operatorname{Re} f(\tfrac{1}{2}e \pm cx) = \operatorname{Re} \tfrac{1}{2} f(e) \pm \operatorname{Re} f(cx) > \tfrac{1}{4} - |f(cx)| \geq \tfrac{1}{4} - \|cx\| > 0.$$

Since $\varepsilon > 0$ is arbitrary, we conclude that $cx + \frac{1}{2}e \geq 0$ and $0 \leq \frac{1}{2}e - cx = e - (cx + \frac{1}{2}e)$, the latter yielding $cx + \frac{1}{2}e \leq e$. □

Proposition 10.56 *If $\mathfrak{A}$ is orderly and f is a classical extreme point of $C^0_{\mathfrak{A}}$, then either $f = 0$, or else $f(e) = 1$ and $f(xy) = f(x)f(y)$ for all $x, y \in \mathsf{Her}(\mathfrak{A})$.* [14]

Proof For each $y \in \mathfrak{A}$ define $g_y \in \mathfrak{A}^*$ by

$$g_y(x) = f(xy) - f(x)f(y).$$

Consider first the case $y = e$. Since $f \in C^0_{\mathfrak{A}}$ and $e \geq 0$, we have $0 \leq f(e) \leq 1$. Hence, for each $x \in \mathsf{Pos}(\mathfrak{A})$,

$$\begin{aligned}(f + g_e)(x) &= (2 - f(e))f(x) \geq 0, \text{ and} \\ (f - g_e)(x) &= f(e)f(x) \geq 0.\end{aligned}$$

Thus the linear functionals $f \pm g_e$ are positive. Moreover,

$$(f + g_e)(e) = f(e)(1 - f(e)) + f(e) \leq 1 - f(e) + f(e) = 1$$

[14] The inspiration for Propositions 10.56 and 10.58 came from [12, pages 38–39].

and

$$(f - g_e)(e) = f(e)^2 \leq 1.$$

Hence $f \pm g_e \in C^0_{\mathfrak{A}}$. Since $f = \frac{1}{2}(f + g_e) + \frac{1}{2}(f - g_e)$, we see from the definition of f as a classical extreme point that $f + g_e = f = f - g_e$; whence $g_e = 0$ and therefore $f(xe) = f(x)f(e)$ for all $x \in \mathfrak{A}$. In particular, $f(e) = f(e)^2$, so either $f(e) = 0$ or $f(e) = 1$. We may now assume that $f(e) = 1$, from which we see that $g_y(e) = 0$, and hence $(f \pm g_y)(e) = f(e) = 1$, for each $y \in \mathfrak{A}$.

Next consider the case $0 \leq y \leq e$, when $0 \leq f(y) \leq f(e) \leq 1$. If $x \in \mathsf{Pos}(\mathfrak{A})$, then since $\mathfrak{A}$ is orderly, xy and $x(e - y)$ are positive,

$$(f + g_y)(x) = f(x)(1 - f(y)) + f(xy) \geq 0,$$

and

$$(f - g_y)(x) = f(x(e - y)) + f(x)f(y) \geq 0.$$

Recalling that $g_y(e) = 0$, we now see that $f \pm g_y \in C^0_{\mathfrak{A}}$. Arguing as in the first paragraph, we find that $g_y = 0$ and therefore $f(xy) = f(x)f(y)$.

Now let $y \in \mathsf{Her}(\mathfrak{A})$ and, using Lemma 10.55, find $c > 0$ such that $0 \leq cy + \frac{1}{2}e \leq e$. For each $x \in \mathsf{Pos}(\mathfrak{A})$, we see from the previous paragraph that

$$\begin{aligned} cf(xy) + \tfrac{1}{2}f(x) &= f(x(cy + \tfrac{1}{2}e)) = f(x)f(cy + \tfrac{1}{2}e) \\ &= f(x)(cf(y) + \tfrac{1}{2}) = cf(x)f(y) + \tfrac{1}{2}f(x) \end{aligned}$$

and therefore $f(xy) = f(x)f(y)$. Finally, if both x and y belong to $\mathsf{Her}(\mathfrak{A})$, find $c' > 0$ such that $0 \leq c'x + \frac{1}{2}e \leq e$ and use a similar argument to prove that $f(xy) = f(x)f(y)$. □

Corollary 10.57 *Let $\mathfrak{A}$ be orderly, quasi-stellar, and have firm state space. Then every classical extreme point of $V_{\mathfrak{A}}$ is a character of $\mathfrak{A}$.*

Proof Let f be a classical extreme point of $V_{\mathfrak{A}}$. By Proposition 10.54, f is a classical extreme point of $C^0_{\mathfrak{A}}$. Since $f(e) = 1$, we see from Proposition 10.56 that $f(xy) = f(x)f(y)$ for all $x, y \in \mathsf{Her}(\mathfrak{A})$. Let $x, y \in \mathfrak{A}$ have respective Hermitian decompositions $x_1 + ix_2,\ y_1 + iy_2$. Then

$$\begin{aligned} f(xy) &= f((x_1 + ix_2)(y_1 + iy_2)) \\ &= f(x_1y_1 + ix_1y_2 + ix_2y_1 - x_2y_2) \\ &= f(x_1y_1) + if(x_1y_2) + if(x_2y_1) - f(x_2y_2) \\ &= f(x_1)f(y_1) + if(x_1)f(y_2) + if(x_2)f(y_1) - f(x_2)f(y_2) \\ &= (f(x_1) + if(x_2))(f(y_1) + if(y_2)) \\ &= f(x_1 + ix_2))f(y_1 + iy_2) \\ &= f(x)f(y). \end{aligned}$$

□

Proposition 10.58 *Let $\mathfrak{A}$ be quasi-stellar and have positive Hermitian squares. Then every character of $\mathfrak{A}$ is an extreme point of $V_{\mathfrak{A}}$.*

Proof Given $u \in \Sigma_{\mathfrak{A}}$ and $\varepsilon > 0$, we first prove the following.

For each $x \in \mathfrak{A}$ there exists $\delta > 0$ such that $|f(x) - g(x)| < \varepsilon$ whenever $f, g \in V_{\mathfrak{A}}$ and $\left\| u - \frac{1}{2}(f+g) \right\|_w < \delta$ (*)

Consider the case $x \in \mathsf{Her}(\mathfrak{A})$. Choose $\delta > 0$ such that if $\alpha, \beta \in \mathfrak{A}_1^*$ and $\|\alpha - \beta\|_w < \delta$, then

$$\left|(\alpha - \beta)(x^2)\right| < \frac{\varepsilon^2}{8} \text{ and } |(\alpha - \beta)(x)| < \frac{\varepsilon^2}{16(1 + \|x\|)}.$$

If $f, g \in V_{\mathfrak{A}}$ and $\left\| u - \frac{1}{2}(f+g) \right\|_w < \delta$, then

$$\left| u(x^2) - \frac{1}{2}(f(x^2) + g(x^2)) \right| < \frac{\varepsilon^2}{8}$$

and

$$\begin{aligned} \left| u(x)^2 - \frac{1}{4}(f(x) + g(x))^2 \right| &= \left| u(x) + \frac{1}{2}(f(x) + g(x)) \right| \left| u(x) - \frac{1}{2}(f(x) + g(x)) \right| \\ &\leq 2\|x\| \left| u(x) - \frac{1}{2}(f(x) + g(x)) \right| < \frac{\varepsilon^2}{8}. \end{aligned}$$

Hence, by Lemma 10.53,

$$\begin{aligned} \frac{1}{2}(f(x)^2 + g(x)^2) &\leq \frac{1}{2}(f(x^2) + g(x^2)) \\ &< u(x^2) + \varepsilon^2/8 \\ &= u(x)^2 + \varepsilon^2/8 \\ &< \frac{1}{4}(f(x) + g(x))^2 + \frac{\varepsilon^2}{8} + \frac{\varepsilon^2}{8} \\ &= \frac{1}{4}f(x)^2 + \frac{1}{2}f(x)g(x) + \frac{1}{4}g(x)^2 + \frac{\varepsilon^2}{4} \end{aligned}$$

and therefore

$$\frac{1}{4}(f(x) - g(x))^2 = \frac{1}{4}f(x)^2 - \frac{1}{2}f(x)g(x) + \frac{1}{4}g(x)^2 < \frac{\varepsilon^2}{4}.$$

Since $f(x), g(x) \in \mathbb{R}$, it follows that $|f(x) - g(x)| < \varepsilon$. This disposes of (*) in the special case of Hermitian x. In the general case there exist $x_1, x_2 \in \mathsf{Her}(\mathfrak{A})$ with $x = x_1 + ix_2$. By the first part of this proof, for each $\varepsilon > 0$ there exists $\delta_1 > 0$ such that if $f, g \in V_{\mathfrak{A}}$ and $\left\| u - \frac{1}{2}(f+g) \right\|_w < \delta_1$, then $|f(x_k) - g(x_k)| < \varepsilon/2$ for $k = 1, 2$. For such f, g we have

$$|f(x) - g(x)| \leq |f(x_1) - g(x_1)| + |f(x_2) - g(x_2)| < \varepsilon.$$

This completes the proof of (*).

Finally, let $(a_n)_{n\geq 1}$ be the dense sequence in $\mathfrak{A}_1$ with respect to which $\| \ \|_w$ is defined. Given $\varepsilon > 0$, choose N such that $\sum_{n=N+1}^{\infty} 2^{-n} < \varepsilon/4$, and choose $\delta_2 > 0$ such that $|f(a_n) - g(a_n)| < \varepsilon/2N$ $(1 \leq n \leq N)$ whenever $f, g \in V_{\mathfrak{A}}$ and $\left\| u - \frac{1}{2}(f+g) \right\|_w < \delta_2$. For such f and g, since $|f(a_n) - g(a_n)| \leq 2$ for $n > N$, we have

$$\|f - g\|_w \leq \sum_{n=1}^{N} 2^{-n} |f(a_k) - g(a_k)| + 2 \sum_{n=N+1}^{\infty} 2^{-n} < N\frac{\varepsilon}{2N} + \frac{\varepsilon}{2} = \varepsilon.$$

Since $\varepsilon > 0$ is arbitrary, it follows that u is an extreme point of $V_{\mathfrak{A}}$. □

Theorem 10.59 *Let $\mathfrak{A}$ be an orderly, quasi-stellar Banach algebra with firm state space and positive Hermitian squares. Then the following are equivalent conditions on $f \in \mathfrak{A}^*$:*

(i) *f is a classical extreme point of $V_{\mathfrak{A}}$;*
(ii) *f is a character of $\mathfrak{A}$;*
(iii) *f is an extreme point of $V_{\mathfrak{A}}$.*

Proof Apply Corollary 10.57 and Proposition 10.58, and recall that extreme points are classical extreme points. □

An element x of a linear space X is a *convex combination* of elements of the subset K of X if there exist finitely many points $x_1, \ldots, x_n$ of K and non-negative numbers $\lambda_1, \ldots, \lambda_n$ with sum 1, such that $x = \lambda_1 x_1 + \cdots + \lambda_n x_n$. The set $\mathrm{co}(K)$ of convex combinations of elements of K is called the *convex hull* of K.

The fundamental result about extreme points is *the Krein–Milman theorem:*

> *If K is a compact convex subset of a separable normed space over $\mathbb{R}$, then $\mathrm{co}(K)$ is dense in K* [2, Chapter 7, Theorem (7.5)].

Proposition 10.60 *Let $\mathfrak{A}$ be orderly and quasi-stellar, with firm state space and positive Hermitian squares. Then $\Sigma_{\mathfrak{A}}$ is inhabited, and $\mathrm{co}(\Sigma_{\mathfrak{A}})$ is weak* dense in $V_{\mathfrak{A}}$.*

Proof By the Krein–Milman theorem, the convex combinations of the extreme points of $V_{\mathfrak{A}}$ are weak* dense in $V_{\mathfrak{A}}$. Since $V_{\mathfrak{A}}$, being firm, is compact and therefore inhabited, it follows that it has extreme points. By Theorem 10.59, those extreme points are precisely the elements of $\Sigma_{\mathfrak{A}}$. □

Lemma 10.61 *If $\mathrm{co}(\Sigma_{\mathfrak{A}})$ is weak* dense in the state space of $\mathfrak{A}$, then for each $x \in \mathfrak{A}$, each $f \in V_{\mathfrak{A}}$, and each $\varepsilon > 0$ there exist characters $u_1, \ldots, u_n$ of $\mathfrak{A}$, and non-negative numbers $\lambda_1, \ldots, \lambda_n$ with sum 1, such that*

$$\left| f(x) - \sum_{k=1}^{n} \lambda_k u_k(x) \right| < \varepsilon.$$

Proof Choose $\delta > 0$ such that $|f(x) - g(x)| < \varepsilon$ whenever $f, g \in \mathfrak{A}_1^*$ and $\|f - g\|_w < \delta$. There exist finitely many elements $u_1, \ldots, u_n$ of $\Sigma_{\mathfrak{A}}$, and non-negative numbers $\lambda_1, \ldots, \lambda_n$ with sum 1, such that $\|f - \sum_{k=1}^n \lambda_k u_k\|_w < \delta$ and therefore $|f(x) - \sum_{k=1}^n \lambda_k u_k(x)| < \varepsilon$. □

Proposition 10.62 *Let $\mathfrak{A}$ have firm state space in which $\mathsf{co}(\Sigma_{\mathfrak{A}})$ is weak* dense, and let $x \in \mathfrak{A}$. Then $x \in \mathsf{Her}(\mathfrak{A})$ (respectively, $x \in \mathsf{Pos}(\mathfrak{A})$) if and only if $u(x) \in \mathbb{R}$ (respectively, $u(x) \geq 0$) for each character u of $\mathfrak{A}$.*

Proof We deal only with the Hermitian case. Since $\Sigma_{\mathfrak{A}} \subset V_{\mathfrak{A}}$, 'only if' is a consequence of Proposition 10.37. Conversely, suppose that $u(x) \in \mathbb{R}$ for each $u \in \Sigma_{\mathfrak{A}}$. Let $f \in V_{\mathfrak{A}}$ and $\varepsilon > 0$. Using Lemma 10.61, construct characters $u_1, \ldots, u_n$ of $\mathfrak{A}$, and non-negative numbers $\lambda_1, \ldots, \lambda_n$ with sum 1, such that $|f(x) - \sum_{k=1}^n \lambda_k u_k(x)| < \varepsilon$. Then $u_k(x) \in \mathbb{R}$ for each k, so $\mathrm{Im} \sum_{k=1}^n \lambda_k u_k(x) = 0$ and therefore

$$|\mathrm{Im}\, f(x)| = \left| \mathrm{Im} \left(f(x) - \sum\nolimits_{k=1}^{n} \lambda_k u_k(x) \right) \right| \leq \left| f(x) - \sum\nolimits_{k=1}^{n} \lambda_k u_k(x) \right| < \varepsilon.$$

Since $\varepsilon > 0$ is arbitrary, it follows that for each $f \in V_{\mathfrak{A}}$, $\mathrm{Im}\, f(x) = 0$ and therefore $f(x) \in \mathbb{R}$; whence $x \in \mathsf{Her}(\mathfrak{A})$, by Proposition 10.37. □

Proposition 10.63 *Let $\mathfrak{A}$ have firm state space in which $\mathsf{co}(\Sigma_{\mathfrak{A}})$ is weak* dense. Then*

(i) *$\mathfrak{A}$ is orderly and has positive Hermitian squares.*
(ii) *The product of two Hermitian elements of $\mathfrak{A}$ is Hermitian.*

Proof For each $u \in \Sigma_{\mathfrak{A}}$,

- if $x, y \in \mathsf{Pos}(\mathfrak{A})$, then $u(x), u(y) \geq 0$, so $u(xy) = u(x)u(y) \geq 0$;
- if $x \in \mathsf{Her}(\mathfrak{A})$, then $u(x) \in \mathbb{R}$, so $u(x^2) = u(x)^2 \geq 0$;
- if $x, y \in \mathsf{Her}(\mathfrak{A})$, then $u(x), u(y) \in \mathbb{R}$, so $u(xy) = u(x)u(y) \in \mathbb{R}$.

The desired conclusions follow from this and Proposition 10.62. □

Proposition 10.64 *If $\mathfrak{A}$ has firm state space in which $\mathsf{co}(\Sigma_{\mathfrak{A}})$ is weak* dense, then $\mathfrak{A}$ is commutative.*

Proof Given $x, y \in \mathfrak{A}$, for each $f \in V_{\mathfrak{A}}$ and each $\varepsilon > 0$ construct elements $u_1, \ldots, u_n$ of $\Sigma_{\mathfrak{A}}$, and non-negative numbers $\lambda_1, \ldots, \lambda_n$, such that

$$\left| f(xy - yx) - \sum_{k=1}^{n} \lambda_k u_k(xy - yx) \right| < \varepsilon.$$

Since $u_k(xy - yx) = 0$ for each k, we have $|f(xy - yx)| < \varepsilon$. It follows that $f(xy - yx) = 0$ for all $f \in V_{\mathfrak{A}}$, and hence, by Proposition 10.37, that $0 \leq xy - yx \leq 0$. Therefore $xy - yx = 0$, by Corollary 10.47. □

Theorem 10.65 *The following are equivalent conditions on a quasi-stellar Banach algebra $\mathfrak{A}$ with firm state space.*

(i) *$\mathfrak{A}$ is orderly and has positive Hermitian squares.*
(ii) *$\mathsf{co}(\Sigma_{\mathfrak{A}})$ is weak* dense in $V_{\mathfrak{A}}$.*

If either of these conditions holds, then $\mathfrak{A}$ is commutative and $\mathsf{Her}(\mathfrak{A})$ is closed under multiplication.

Proof This follows from Propositions 10.60, 10.63, and 10.64. □

Let the Banach algebra $\mathfrak{A}$ be quasi-stellar and have firm state space. It follows from Proposition 10.50 that there exist unique functions $h_1, h_2 : \mathfrak{A} \to \mathsf{Her}(\mathfrak{A})$ such that $x = h_1(x) + ih_2(x)$ for each $x \in \mathfrak{A}$; moreover, those functions are linear. The mapping $x \rightsquigarrow x^* \equiv h_1(x) - ih_2(x)$ then has the properties $(x^*)^* = x$ and $(\alpha x + \beta y)^* = \overline{\alpha} x^* + \overline{\beta} y^*$, where $x, y \in \mathfrak{A}$ and $\alpha, \beta \in \mathbb{C}$. If $\mathfrak{A}$ is also commutative then $x \rightsquigarrow x^*$ satisfies the final condition for being an involution: $(xy)^* = y^* x^*$.

In view of Proposition 10.64, if $\mathfrak{A}$ is orderly and quasi-stellar, with firm state space and positive Hermitian squares, then $x \rightsquigarrow x^*$ is an involution. *Is it also the case that $\|x^* x\| = \|x\|^2$ for each $x \in \mathfrak{A}$? In other words, does this involution turn $\mathfrak{A}$ into a C*-algebra?* This question remains open.

References

[1] Bishop, E. 1967. *Foundations of Constructive Analysis*. New York: McGraw-Hill.

[2] Bishop, E., and Bridges, D. S. 1985. *Constructive Analysis*. Grundlehren der Math. Wiss., no. 279. Heidelberg: Springer-Verlag.

[3] Bonsall, F. F., and Duncan, J. 1973. *Complete Normed Algebras*. Ergebnisse der Mathematik und ihrer Grenzgebiete, no. 10. Berlin: Springer-Verlag.

[4] Bridges, D. S., and Havea, R. S. 2000. A constructive analysis of a proof that the numerical range is convex. *London Math. Soc. J. Math. Comput.*, **3**, 191–206.

[5] Bridges, D. S., and Havea, R. S. 2005. Approximating the numerical range in a Banach algebra. Pages 293–303 of: Crosilla, L., and Schuster, P. (eds.), *From Sets and Types to Topology and Analysis*. Oxford Logic Guides, no. 48. Oxford: Clarendon Press.

[6] Bridges, D. S., and Havea, R. S. 2012. Square roots and powers in constructive Banach algebra theory. Pages 68–77 of: Cooper, S. B., Dawar A., and Löwe, B. (eds), *How the World Computes – Turing Centenary Conference on Computability in Europe, CiE 2012*. Lecture Notes in Computer Science 7318. Berlin: Springer-Verlag.

[7] Bridges, D. S., and Vîţă, L. S. 2006. *Techniques of Constructive Analysis*. Universitext. New York: Springer.

[8] Bridges, D. S., Havea R. S., and Schuster, P. M. 2006. Finitely generated Banach algebras and local Nullstellensätze. *Publ. Math. Debrecen*, **69**(1–2), 171–184.

[9] Bridges, D. S., Richman F., and Wang, Y. 1996. Sets, complements and boundaries. *Proc. Koninklijke Nederlandse Akad. Wetenschappen (Indag. Math.)*, **7**(4), 425–445.

[10] Gustafson, K. E., and Rao, D. K. M. 1997. *Numerical Range*. New York: Springer.

[11] Havea, R. S. 2001. *Constructive Spectral and Numerical Range Theory*. Ph.D. thesis, University of Canterbury, Christchurch, New Zealand.

[12] Holmes, R. B. 1975. *Geometric Functional Analysis and its Applications*. Graduate Texts in Mathematics, vol. 24. New York: Springer.

[13] Jacobson, N. 1974. *Basic Algebra I*. San Francisco: W. H. Freeman.

[14] Kadison, R. V., and Ringrose, J. R. 1983. *Fundamentals of the Theory of von Neumann Algebras*, Vol. 1. New York: Academic Press.

[15] Takamura, H. 2005. An introduction to the theory of C^*-algebras in constructive mathematics. Pages 280–292 of: Crosilla, L., and Schuster, P. (eds.), *From Sets and Types to Topology and Analysis*. Oxford Logic Guides, no. 48. Oxford: Clarendon Press.

11

Constructive Convex Optimisation

Josef Berger and Gregor Svindland

11.1 Introduction

This chapter is a survey of our research on a constructive approach to convex optimisation. The results we present are taken from [3, 4, 5, 7, 8, 9]. We refer to [6] for an earlier detailed survey of [3, 4, 5] in which we also present essential parts of the underlying theory in Bishop's constructive mathematics (BISH). In this contribution, however, we assume that the reader is familiar with basic terminology and results from constructive analysis such those as presented in [12]. We will only briefly introduce some notation, conventions, and notions related to convexity in Section 11.2. Many of the results we discuss will not be proved here; we only refer to the respective papers. Nevertheless, where proofs are not too tedious, we will present them, in particular to illustrate applications of our main results. Section 11.3 considers results on existence of infima and minima for convex functions whereas Section 11.4 provides the corresponding background in the framework of Brouwer's Fan Theorem. Section 11.5 discusses some recent results on Lemmas of Alternatives.

11.2 Some Definitions and Notation

Throughout this article $\|\cdot\|$ will denote the Euclidean norm on $\mathbb{R}^n$. For $x \in \mathbb{R}^n$ we denote by x_i, $i = 1, \ldots, n$, the ith coordinate of x, that is $x = (x_1, \ldots, x_n)$. Moreover, for x, y in $\mathbb{R}^n$ we write

$$x \cdot y = \sum_{i=1}^{n} x_i y_i,$$

where $x, y \in \mathbb{R}^n$ for the Euclidean scalar product. If $A = (a_{ij}) \in \mathbb{R}^{m \times n}$ is a real matrix and $x \in \mathbb{R}^n$ and $y \in \mathbb{R}^m$, then $A \cdot x$ is the vector in $\mathbb{R}^m$ with ith coordinate

$$(A \cdot x)_i = \sum_{j=1}^{n} a_{ij} x_j,$$

whereas $y \cdot A$ is the vector in $\mathbb{R}^n$ with jth coordinate

$$(y \cdot A)_j = \sum_{i=1}^{m} a_{ij} y_i.$$

Whenever $C \subseteq \mathbb{R}^n$ is located,

$$d(x, C) := \inf\{\|x - y\| \mid y \in C\}$$

denotes the distance from $x \in \mathbb{R}^n$ to C. In this chapter, located sets are always inhabited. Also totally bounded sets, and thus compact sets, are always assumed to be inhabited. A set $C \subseteq \mathbb{R}^n$ is *convex* if it is inhabited and if

$$\forall x, y \in C \, \forall \lambda \in [0,1] \, (\lambda x + (1-\lambda) y \in C).$$

This in fact implies that C is closed under finite convex combinations. Let

$$\mathcal{X}_m := \left\{ \lambda \in \mathbb{R}^m \mid \lambda_i \geq 0 \, (i = 1, \ldots, m), \sum_{i=1}^{m} \lambda_i = 1 \right\}.$$

For m points $x^1, \ldots, x^m \in \mathbb{R}^n$ we define the *convex hull*

$$\mathrm{co}(x^1, \ldots, x^m) := \left\{ \sum_{i=1}^{m} \lambda_i x^i \mid \lambda \in \mathcal{X}_m \right\},$$

the *convex cone*

$$\mathrm{cone}(x^1, \ldots, x^m) := \left\{ \sum_{i=1}^{m} \lambda_i x^i \mid \lambda \in \mathbb{R}^m, \lambda_i \geq 0 (i = 1, \ldots, m) \right\},$$

and the *linear space*

$$\mathrm{span}(x^1, \ldots, x^m) := \left\{ \sum_{i=1}^{m} \lambda_i x^i \mid \lambda \in \mathbb{R}^m \right\}$$

generated by $x^1, \ldots, x^m$. Let $C \subseteq \mathbb{R}^n$ be convex. A function $f : C \to \mathbb{R}$ is called *convex* if

$$\forall x, y \in C \, \forall \lambda \in [0,1] \; f(\lambda x + (1-\lambda) y) \leq \lambda f(x) + (1-\lambda) f(y)$$

and *quasi-convex* if

$$\forall x, y \in C \, \forall \lambda \in [0,1] \; f(\lambda x + (1-\lambda) y) \leq \max\{f(x), f(y)\}.$$

Clearly, convex functions are quasi-convex.

11.3 Convexity and Existence of Infima and Minima

In this section we give an overview of results on infima and minima of quasi-convex functions. We recall that a uniformly continuous function $f : C \to \mathbb{R}$ on a compact set C always admits an infimum; see Corollary 2.2.7 in [12]. However, if $f : C \to \mathbb{R}^+ := (0, \infty)$, the statement that $\inf f > 0$ is equivalent to Brouwer's Fan Theorem; see [1, 13]. The same holds for the statement that if f admits at most one minimum it has a minimum point; see [1, 2]. Nevertheless, it turns out that if we add convexity to the picture, suddenly those statements are constructively verifiable. The underlying reason is that, in fact, Brouwer's Fan Theorem is constructively verifiable for so-called co-convex bars: bars in $\{0,1\}^*$, the set of all finite binary sequences, possessing a convexity property which we will discuss in Section 11.4.

Theorem 11.1 (See Theorem 1 in [4] and Proposition 1 in [3]) *If $C \subseteq \mathbb{R}^n$ is compact and convex and $f : C \to \mathbb{R}^+$ is quasi-convex and uniformly continuous, then* $\inf f > 0$.

As a first consequence we obtain the following version of Theorem 11.1 for convex hulls which, in contrast to classical mathematics, cannot be verified to be closed, and thus compact in general. Only special cases such as $\mathcal{X}_m$ are indeed compact.

Corollary 11.2 (See Corollary 1 in [3]) *Let $x^1, \ldots, x^m \in \mathbb{R}^n$ and suppose that $f : \operatorname{co}(x^1, \ldots, x^m) \to \mathbb{R}^+$ is quasi-convex and uniformly continuous. Then* $\inf f > 0$.

Proof Consider the function

$$\kappa : \mathcal{X}_m \to \operatorname{co}(x^1, \ldots, x^m), \ \ \lambda \mapsto \sum_{i=1}^{m} \lambda_i x^i.$$

The composition $f \circ \kappa$ satisfies the requirements of Theorem 11.1. □

Theorem 11.1 also implies the following separating hyperplane result for disjoint convex sets which does not require locatedness of the algebraic difference $Y - C$ such as in Theorem 5.2.9 in [12].

Theorem 11.3 (See Theorem 2 in [4]) *Let C and Y be subsets of $\mathbb{R}^n$ and suppose that*

(i) *C is convex and compact,*
(ii) *Y is convex, complete, and located,*
(iii) *$\|c - y\| > 0$ for all $c \in C$ and $y \in Y$.*

Then there exist $p \in \mathbb{R}^n$ *and reals* α, β *such that*

$$p \cdot c < \alpha < \beta < p \cdot y$$

for all $c \in C$ *and* $y \in Y$. *In particular, the sets* C *and* Y *are strictly separated by the hyperplane*

$$H = \{x \in \mathbb{R}^n \mid p \cdot x = \gamma\},$$

with $\gamma = \frac{1}{2}(\alpha + \beta)$.

Next we consider existence of minima. A function $f : C \to \mathbb{R}$, where $C \subseteq \mathbb{R}^n$ is inhabited, is said to have *at most one minimum point* if $\inf f$ exists and

$$\forall x, y \in C \; (\|x - y\| > 0 \Rightarrow f(x) > \inf f \vee f(y) > \inf f).$$

Theorem 11.4 (See Theorem 1 in [7]) *Let* C *be a convex and compact subset of* $\mathbb{R}^n$. *Then every quasi-convex, uniformly continuous function* $f : C \to \mathbb{R}$ *with at most one minimum point has a minimum point, that is*

$$\exists x \in C \; f(x) = \inf f.$$

As consequence we obtain supporting hyperplanes for compact, strictly convex sets; see Proposition 11.6 below. To this end, note that an inhabited subset C of $\mathbb{R}^n$ is *strictly convex* if

$$\lambda x + (1 - \lambda)y \in C^{\circ}$$

for all $x, y \in C$ with $\|x - y\| > 0$ and all $\lambda \in (0, 1)$. Here the set C°, the *interior* of C, is defined as usual:

$$x \in C^{\circ} \Leftrightarrow \exists \varepsilon > 0 \, \forall y \in \mathbb{R}^n \, (\|y - x\| < \varepsilon \Rightarrow y \in C)$$

Lemma 11.5 *Fix a subset* C *of* $\mathbb{R}^n$.

(i) *If* C *is convex and open, then it is strictly convex.*
(ii) *If* C *is strictly convex and closed, then it is convex.*

Proof We only prove (ii). Let $x, y \in C$ and $\lambda \in (0, 1)$. Define an increasing sequence $(a_n)_{n \in \mathbb{N}} \in \{0, 1\}^{\mathbb{N}}$ such that $a_n = 0$ implies $\|x - y\| < \frac{1}{n}$ whereas $a_n = 1$ implies $\|x - y\| > 0$. Next define a sequence

$$x_n = \begin{cases} x & \text{if } a_n = 0 \\ \lambda x + (1 - \lambda)y & \text{if } a_n = 1 \end{cases}, \quad n \in \mathbb{N}.$$

Note that $(x_n)_{n \in \mathbb{N}} \subseteq C$ is a Cauchy sequence. As C is closed, its limit $\lambda x + (1 - \lambda)y$ lies in C. For general $\lambda \in [0, 1]$ choose a sequence $(\lambda_n)_{n \in \mathbb{N}} \subseteq (0, 1)$ such that $\lambda_n \to \lambda$ and note that by closedness of C we have

$$\lambda x + (1 - \lambda)y = \lim_{n \to \infty} \lambda_n x + (1 - \lambda_n)y \in C. \qquad \square$$

Proposition 11.6 (See Proposition 1 in [7]) *Let $C \subseteq \mathbb{R}^n$ be a compact and strictly convex set. Let $g : \mathbb{R}^n \to \mathbb{R}$ be a linear function such that $g(v) > 0$ for some $v \in \mathbb{R}^n$. Then the restriction of g to C has a minimum point w.*

Proof Let f denote the restriction of g to C. Note that linear functions are quasi-convex. We will prove that f has at most one minimum point. To this end, consider $x, y \in C$ with $\|x - y\| > 0$. Set $z = (x + y)/2$. Since C is strictly convex, there exists $\delta > 0$ such that $z - \delta v \in C$. We obtain

$$f(z - \delta v) < f(z) \leq \max\{f(x), f(y)\}.$$

Thus f has at most one minimum point. By Theorem 11.4, f has a minimum point. □

In the situation of Proposition 11.6, the set

$$\mathcal{H} := \{x \in \mathbb{R}^n \mid g(x) = g(w)\}$$

is called a *supporting hyperplane* of C. Indeed, C lies on one side of $\mathcal{H}$, since $\forall x \in C \; g(x) \geq g(w)$, and $\mathcal{H}$ touches C in the point w.

Another consequence of Theorem 11.4 is that a strictly quasi-convex function defined on a convex compact set possesses a (unique) minimum point. To this end, note that a function $f : C \to \mathbb{R}$ defined on a convex set $C \subseteq \mathbb{R}^n$ is called *strictly quasi-convex* if

$$f(\lambda x + (1 - \lambda)y) < \max\{f(x), f(y)\}$$

for all $\lambda \in (0, 1)$ and $x, y \in C$ such that $\|x - y\| > 0$. Since strictly quasi-convex functions have at most one minimum point, the following result follows from Theorem 11.4.

Proposition 11.7 (See Proposition 2 in [7]) *Let $C \subseteq \mathbb{R}^n$ be convex and compact. Then every strictly quasi-convex, uniformly continuous function $f : C \to \mathbb{R}$ has a minimum point.*

Another application of Theorems 11.1 and 11.4 in game theory will be given at the end of Section 11.5.

11.4 Convexity and Brouwer's Fan Theorem

In this section we give a résumé on the deeper reason why statements equivalent to Brouwer's Fan Theorem become constructively verifiable once we add some convexity assumption. We will see in Theorem 11.9 that in fact the Fan Theorem is constructively verifiable for so-called co-convex bars. Before we can state this result we need to introduce some further notions and notation related to the Fan Theorem.

We write $\{0,1\}^*$ for the set of all finite binary sequences u, v, w. Let $\emptyset$ be the empty sequence and let $\{0,1\}^{\mathbb{N}}$ be the set of all infinite binary sequences α, β, γ. For every u let $|u|$ be the *length* of u, that is $|\emptyset| = 0$, and for $u = (u_0, \ldots, u_{n-1})$ we have $|u| = n$. For $u = (u_0, \ldots, u_{n-1})$ and $v = (v_0, \ldots, v_{m-1})$ the *concatenation* $u * v$ of u and v is defined by

$$u * v = (u_0, \ldots, u_{n-1}, v_0, \ldots, v_{m-1}).$$

A subset B of $\{0,1\}^*$ is *closed under extension* if $u * v \in B$ for all $u \in B$ and for all v. The *restriction* $\overline{\alpha}n$ of α to n bits is given by $\overline{\alpha}0 = \emptyset$ and

$$\overline{\alpha}n = (\alpha_0, \ldots, \alpha_{n-1}) \quad \text{whenever } n \geq 1.$$

For u with $n \leq |u|$, the restriction $\overline{u}n$ is defined analogously. A sequence α *hits* B if there exists n such that $\overline{\alpha}n \in B$. B is a *bar* if every α hits B. B is a *uniform bar* if there exists N such that for every α there exists $n \leq N$ such that $\overline{\alpha}n \in B$. Often one requires B to be *detachable*, that is, for every u the statement $u \in B$ is decidable. *Brouwer's Fan Theorem* for detachable bars is as follows.

FAN Every detachable bar is a uniform bar.

FAN is neither provable nor falsifiable in BISH; see Section 3 of Chapter 5 in [11]. In their seminal paper [13], Julian and Richman established a correspondence between FAN and functions on $[0,1]$ as follows.

Proposition 11.8 *For every detachable subset B of $\{0,1\}^*$ there exists a uniformly continuous function $f : [0,1] \to [0,\infty)$ such that*

(1) *B is a bar $\Leftrightarrow$ f is positive valued,*
(2) *B is a uniform bar $\Leftrightarrow$ $\inf f > 0$.*

Conversely, for every uniformly continuous function $f : [0,1] \to [0,\infty)$ there exists a detachable subset B of $\{0,1\}^*$ such that (1) and (2) hold.

Consequently, FAN is equivalent to the statement that every uniformly continuous, positive-valued function defined on the unit interval has positive infimum. In fact, in the latter statement the unit interval may be replaced by compact subsets of $\mathbb{R}^n$; see [1]. Now, in view of Theorem 11.1, the question arises whether there is a constructively valid convex version of Brouwer's Fan Theorem. To this end, we define

$$u < v :\Leftrightarrow |u| = |v| \wedge \exists k < |u| \, (\overline{u}k = \overline{v}k \wedge u_k = 0 \wedge v_k = 1)$$

and

$$u \leq v :\Leftrightarrow u = v \vee u < v.$$

A subset B of $\{0,1\}^*$ is *co-convex* if, for every α which hits B, there exists n such that either

$$\{v \mid v \leq \overline{\alpha}n\} \subseteq B \quad \text{or} \quad \{v \mid \overline{\alpha}n \leq v\} \subseteq B\,.$$

Note that for detachable B co-convexity follows from the convexity of the complement of B, where $C \subseteq \{0,1\}^*$ is *convex* if for all u, v, w we have

$$u \leq v \leq w \wedge u, w \in C \Rightarrow v \in C.$$

The following is the already-advertised fan theorem for co-convex bars:

Theorem 11.9 (See Theorem 2.1 in [5]) *Every co-convex bar is a uniform bar.*

Proof Fix a co-convex bar B. We can and will assume that B is closed under extension; see [5] for the details. Define

$$C = \{u \mid \exists n\, \forall w \in \{0,1\}^n\, (u * w \in B)\}\,.$$

Note that C consists of the set of nodes beyond which B is uniform, $B \subseteq C$, and that C is closed under extension as well. Moreover, B is a uniform bar if and only if there exists n such that $\{0,1\}^n \subseteq C$.

First, we show that

$$\forall u\, \exists i \in \{0,1\}\, (u * i \in C)\,. \tag{11.1}$$

Fix u. For

$$\beta = u * 1 * 0 * 0 * 0 * \ldots$$

there exists an l such that either

$$\left\{v \mid v \leq \overline{\beta}l\right\} \subseteq B,$$

or

$$\left\{v \mid \overline{\beta}l \leq v\right\} \subseteq B.$$

Since B is closed under extension, we can assume that $l > |u| + 1$. Let $m = l - |u| - 1$. If $\left\{v \mid v \leq \overline{\beta}l\right\} \subseteq B$, we can conclude that

$$u * 0 * w \in B$$

for every w of length m, which implies that $u * 0 \in C$. If $\left\{v \mid \overline{\beta}l \leq v\right\} \subseteq B$, we obtain

$$u * 1 * w \in B$$

for every w of length m, which implies that $u * 1 \in C$. This concludes the proof of (11.1).

By countable choice, there exists a function $F : \{0,1\}^* \to \{0,1\}$ such that

$$\forall u \, (u * F(u) \in C) \, .$$

Define α by

$$\alpha_n = 1 - F(\overline{\alpha}n).$$

Next, we show by induction on n that

$$\forall n \geq 1 \, \forall u \in \{0,1\}^n \, (u \neq \overline{\alpha}n \Rightarrow u \in C) \, . \tag{11.2}$$

The case $n = 1$ is easily verified. Now fix some $n \geq 1$ such that (11.2) holds. Moreover, fix $w \in \{0,1\}^{n+1}$ such that $w \neq \overline{\alpha}(n+1)$.

Case 1 $\overline{w}n \neq \overline{\alpha}n$. Then $\overline{w}n \in C$ and therefore $w \in C$.

Case 2 $w = \overline{\alpha}n * (1 - \alpha_n) = \overline{\alpha}n * F(\overline{\alpha}n)$. This implies that $w \in C$. So we have established (11.2).

There exists n such that $\overline{\alpha}n \in B$. Applying (11.2) to this n, we can conclude that every u of length n is an element of C, thus B is a uniform bar. □

Remark 11.10 Note that we do not need to require that the co-convex bar in Theorem 11.9 is detachable.

In order to bring convexity into the context of Proposition 11.8, we introduce the notion of weakly convex functions. Let S be an inhabited subset of $\mathbb{R}$. A function $f : S \to \mathbb{R}$ is *weakly convex* if for all $t \in S$ with $f(t) > 0$ there exists $\varepsilon > 0$ such that either

$$\forall s \in S \, (s \leq t \Rightarrow f(s) \geq \varepsilon) \quad \text{or} \quad \forall s \in S \, (t \leq s \Rightarrow f(s) \geq \varepsilon) \, .$$

Note that weak convexity is a generalisation of convexity in that uniformly continuous (quasi-) convex functions $f : [0,1] \to \mathbb{R}$ are weakly convex; see Lemma 3.3 in [5]. For convex functions we can even drop uniform continuity.

Proposition 11.11 (See Proposition 3 in [8]) *Every convex function $f : [0,1] \to \mathbb{R}$ is weakly convex.*

The following generalisation of Proposition 11.8 links Theorem 11.1 with Theorem 11.9.

Theorem 11.12 (See Theorem 3.4 in [5]) *For every detachable subset B of $\{0,1\}^*$ which is closed under extension there exists a uniformly continuous function $f : [0,1] \to \mathbb{R}$ such that*

(1) *B is a bar* $\Leftrightarrow$ *f is positive-valued,*
(2) *B is a uniform bar* $\Leftrightarrow$ $\inf f > 0$,
(3) *B is co-convex* $\Leftrightarrow$ *f is weakly convex.*

Conversely, for every uniformly continuous function $f : [0,1] \to \mathbb{R}$ there exists a detachable subset B of $\{0,1\}^*$ which is closed under extension such that (1), (2), and (3) hold.

Hence, by Theorems 11.9 and 11.12, uniformly continuous and weakly convex functions $f : [0,1] \to \mathbb{R}^+$ have positive infimum. This is a generalisation of Theorem 11.1 in the one-dimensional case. Indeed, an inspection of the proof of Theorem 11.1 given in [4] shows that one can replace the required quasi-convexity by the even weaker notion of weak convexity in that proof to directly obtain Theorem 11.1 for weakly convex functions, which then in conjunction with Theorem 11.12 again implies Theorem 11.9. Without studying the fan theorem for bars in $\{0,1\}^*$ we would, however, never have spotted weak convexity as the essential property behind positive infima; see also the discussion at the end of [5].

11.5 Lemmas of the Alternative and Consequences

Lemmas of the alternative, such as Farkas' lemma, play an important role in convex optimisation. Constructively valid versions of those results are therefore of great interest. In this section we will present two types of such constructive versions. The first type of results replaces the alternatives by equivalences and make some stronger assumptions on the appearing objects. These are useful in applications such as solvability criteria for systems of linear equations; see Propositions 11.15, 11.16, and Corollary 11.17. The second type of constructively valid versions concludes the classical formulation as alternatives from the detachability of a suitable set from $\{1,\ldots,k\}$ for some $k \in \mathbb{N}$. We will call these results conditionally constructive. More precisely, a formula ν is *conditionally constructive* if there exist a $k \in \mathbb{N}$ and a subset M of $\{1,\ldots,k\}$ such that the detachability of M from $\{1,\ldots,k\}$ implies ν. The rule of intuitionistic propositional logic

$$((\varphi \vee \neg\varphi) \Rightarrow \neg\psi) \Rightarrow \neg\psi$$

implies that conditionally constructive formulas ν may be used to prove negated statements:

$$(\nu \Rightarrow \neg\psi) \Rightarrow \neg\psi \tag{11.3}$$

(see [9]). This observation is very useful because lemmas of the alternative often come into play when we wish to derive falsum. Indeed, motivated by this, we will prove a constructive version of optimality criteria in linear programming

(see Proposition 11.19), and we will also provide a simple proof of a constructive version of the von Neumann minimax theorem which first appeared in [10]. As regards the lemmas of the alternative, as in [9], our main reference point will be Farkas' lemma. Throughout this section we will need the following notation. For $x, y \in \mathbb{R}^n$ we
write

$$x \leq y \;:\Leftrightarrow\; \forall i \in \{1, \ldots, n\}\,(x_i \leq y_i)\,, \quad y \geq x \;:\Leftrightarrow\; x \leq y,$$
$$x < y \;:\Leftrightarrow\; \forall i \in \{1, \ldots, n\}\,(x_i < y_i)\,, \quad y > x \;:\Leftrightarrow\; x < y,$$

and

$$x \lneq y \;:\Leftrightarrow\; x \leq y \wedge \exists i \in \{1, \ldots, n\}\,(x_i < y_i)\,, \quad x \gneq y \;:\Leftrightarrow\; y \lneq x.$$

Farkas' lemma in its classical formulation states the following. For any real matrix $A \in \mathbb{R}^{m \times n}$ and $b \in \mathbb{R}^m$ we have

FAR(A, b) Exactly one of the following statements is true:

(i) $\exists \xi \in \mathbb{R}^m\,(\xi \cdot A \geq 0 \;\wedge\; \xi \cdot b < 0)$,
(ii) $\exists q = (q_1, \ldots q_n) \in \mathbb{R}^n\,(q_i \geq 0\,(i = 1, \ldots, n) \;\wedge\; A \cdot q = b)$.

Farkas' lemma is not constructively verifiable; indeed, we have the following proposition.

Proposition 11.13 (See Proposition 2 in [9]) *The following are equivalent:*

(i) FAR : $\forall A \in \mathbb{R}^{m \times n}\; \forall b \in \mathbb{R}^m$ FAR(A, b),
(ii) LPO.

The following two propositions are constructive versions of Farkas' lemma of the first kind discussed previously. We write $\mathrm{cone}(A)$ for the convex cone generated by the columns of A, that is $\mathrm{cone}(A) = \mathrm{cone}(a^1, \ldots, a^n)$, where $a^i \in \mathbb{R}^m$ denotes the ith column of A. Similarly we will write $\mathrm{span}(A)$ for the linear space generated by the columns of A.

Proposition 11.14 (See Proposition 3 in [9]) *Fix a matrix $A \in \mathbb{R}^{m \times n}$ and $b \in \mathbb{R}^m$. If $\mathrm{cone}(A)$ is located, the following are equivalent:*

(i) $\exists \xi \in \mathbb{R}^m\,(\xi \cdot A \geq 0 \;\wedge\; \xi \cdot b < 0)$,
(ii) $d(b, \mathrm{cone}(A)) > 0$.

Proof (i) $\Rightarrow$ (ii) As $\mathbb{R}^m \ni x \mapsto \xi \cdot x$ is continuous and $\xi \cdot b < 0$, there exists $\delta > 0$ such that

$$\forall x \in \mathbb{R}^m\,(\|b - x\| < \delta \;\Rightarrow\; \xi \cdot x < 0)\,.$$

Fix $x \in \operatorname{cone}(A)$. If $\|b - x\| < \delta$, we can conclude that $\xi \cdot x < 0$, a contradiction. Thus,

$$\forall x \in \operatorname{cone}(A) \, (\|b - x\| \geq \delta) \, .$$

This implies (ii).

(ii) $\Rightarrow$ (i) Set $d := d(b, \operatorname{cone}(A))$. By Lemma 6 in [3], there exists $\xi \in \mathbb{R}^n$ such that

$$\forall x \in \operatorname{cone}(A) \left(\xi \cdot (x - b) \geq d^2 \right) .$$

Thus,

$$\forall x \in \operatorname{cone}(A) \left(\xi \cdot x \geq d^2 + \xi \cdot b \right) .$$

Since $0 \in \operatorname{cone}(A)$, we conclude that $\xi \cdot b < 0$. Finally, $\operatorname{cone}(A)$ being a cone implies

$$\forall x \in \operatorname{cone}(A) \, (\xi \cdot x \geq 0) \, . \qquad \square$$

Proposition 11.15 (See Proposition 4 in [9]) *Fix $A \in \mathbb{R}^{m \times n}$ and $b \in \mathbb{R}^m$. If* $\operatorname{cone}(A)$ *is located and closed, then the following are equivalent:*

(i) $\forall \xi \in \mathbb{R}^m \, (\xi \cdot A \geq 0 \; \Rightarrow \; \xi \cdot b \geq 0)$,
(ii) $\exists q \in \mathbb{R}^n \; (q_i \geq 0 (i = 1, \ldots, n) \; \wedge \; A \cdot q = b)$.

Proof Since $\operatorname{cone}(A)$ is located and closed, (ii) is equivalent to $d(b, \operatorname{cone}(A)) = 0$; that is $\neg(d(b, \operatorname{cone}(A)) > 0)$. Thus the assertion follows from Proposition 11.14. $\square$

For instance, Proposition 11.15 implies the following constructive version of the so-called Fredholm alternative.

Proposition 11.16 (See Proposition 8 in [9]) *Let $A \in \mathbb{R}^{m \times n}$ and $b \in \mathbb{R}^m$. Suppose that* $\operatorname{span}(A)$ *is located and closed. The following are equivalent:*

(i) $\forall \xi \in \mathbb{R}^m \, (\xi \cdot A = 0 \Rightarrow \xi \cdot b = 0)$,
(ii) $\exists x \in \mathbb{R}^n \, (A \cdot x = b)$.

Proof Consider the matrix $B := (A \;\; -A)$. Then $\operatorname{cone}(B) = \operatorname{span}(A)$ is closed and located. Hence, by Proposition 11.15 the following are equivalent:

(1) $\forall \xi \in \mathbb{R}^m \, (\xi \cdot B \geq 0 \Rightarrow \xi \cdot b \geq 0)$,
(2) $\exists q \in X_{2n} \, (B \cdot q) = b$.

Now (i) is equivalent to (1) and (ii) is equivalent to (2). $\square$

As a consequence, we obtain a constructive version of the Fredholm alternative for solvability of systems of linear equations.

Corollary 11.17 (See Corollary 2 in [9]) *Let $A \in \mathbb{R}^{m\times n}$ and $b \in \mathbb{R}^m$. Suppose* $\mathrm{span}(A)$ *is located and closed. If the homogeneous equation $\xi \cdot A = 0$ has a unique solution, then there exists a solution to the system of linear equations $A \cdot x = b$.*

Proof The unique solution to $\xi \cdot A = 0$ is of course $\xi = 0$, so (i) of Proposition 11.16 is satisfied, which implies (ii). □

Recall that a formula φ is conditionally constructive if there exists a $k \in \mathbb{N}$ and a subset M of $\{1, \ldots, k\}$ such that the detachability of M from $\{1, \ldots, k\}$ implies φ.

Theorem 11.18 (See Propositions 5, 7, 11, and 13 in [9]) *Fix $A \in \mathbb{R}^{m\times n}$ and $b \in \mathbb{R}^m$. The following are conditionally constructive.*

(i) $\mathrm{FAR(A, b)}$.
(ii) *The Fredholm alternative for A and b – exactly one of the following statements is true:*
  (a) $\exists \xi \in \mathbb{R}^m \,(\xi \cdot A = 0 \wedge |\xi \cdot b| > 0)$,
  (b) $\exists x \in \mathbb{R}^n \,(A \cdot x = b)$.
(iii) *Stiemke's lemma for A and b – exactly one of the following alternatives is true:*
  (a) $\exists \xi \in \mathbb{R}^m \,(\xi \cdot A \gneq 0)$,
  (b) $\exists p \in \mathcal{X}_n \,(p_i > 0 \,(i = 1, \ldots, n) \;\wedge\; (A \cdot p = 0))$.
(iv) *Exactly one of the following statements is true:*
  (a) $\exists p \in \mathcal{X}_m \,(p \cdot A \geq 0)$,
  (b) $\exists q \in \mathcal{X}_n \,(A \cdot q < 0)$.

Theorem 11.18 allows us to derive a number of constructive versions of prominent classical results from convex programming; see [9]. We review a few of those in the following. The first result is on optimality criteria for linear programming. To this end, consider the following linear optimisation problems. Let $A \in \mathbb{R}^{m\times n}$, $b \in \mathbb{R}^m$, and $c \in \mathbb{R}^n$. The primary problem is

$$(P) \quad \text{minimise } c \cdot x \quad \text{subject to } x \in \mathcal{P} := \{y \in X_n \mid A \cdot y = b\},$$

whereas the dual problem is

$$(D) \quad \text{maximise } b \cdot u \quad \text{subject to } u \in \mathcal{D} := \{v \in \mathbb{R}^m \mid v \cdot A \leq c\}.$$

Proposition 11.19 (See Proposition 10 in [9]) *Consider the $(m + 1) \times n$-matrix*

$$A' = \begin{pmatrix} A \\ c \end{pmatrix}$$

and suppose that cone(A') *is closed and located. If there is a solution* u *to* (D), *then there exists a solution* x *to (P) and* $c \cdot x = b \cdot u$.

The well-known von Neumann minimax theorem states that for any matrix $A \in \mathbb{R}^{m\times n}$

$$\max_{p\in\mathcal{X}_m} \min_{q\in\mathcal{X}_n} p \cdot A \cdot q = \min_{q\in\mathcal{X}_n} \max_{p\in\mathcal{X}_m} p \cdot A \cdot q.$$

A thorough discussion of this result in BISH is given in [10]. In that article also the following constructive version of von Neumann's minimax theorem was introduced; see Theorem 2.3 in [10]. Here, to illustrate the applicability of conditionally constructive statements, we provide a short proof of this result based on Theorem 11.18 and (11.3).

Proposition 11.20 (See Proposition 14 in [9]) *Let* $A \in \mathbb{R}^{m\times n}$. *Then*

$$\sup_{p\in\mathcal{X}_m} \inf_{q\in\mathcal{X}_n} p \cdot A \cdot q = \inf_{q\in\mathcal{X}_n} \sup_{p\in\mathcal{X}_m} p \cdot A \cdot q.$$

Proof Note that

$$\inf_{q\in\mathcal{X}_n} p \cdot A \cdot q = \min\{(p \cdot A)_i \mid i = 1, \dots, n\},$$

$p \in \mathcal{X}_m$, and

$$\sup_{p\in\mathcal{X}_m} p \cdot A \cdot q = \max\{(A \cdot q)_j \mid j = 1, \dots, m\},$$

$q \in \mathcal{X}_n$, and that the functions

$$\mathcal{X}_m \ni p \mapsto \inf_{q\in\mathcal{X}_n} p \cdot A \cdot q \quad \text{and} \quad \mathcal{X}_n \ni q \mapsto \sup_{p\in\mathcal{X}_m} p \cdot A \cdot q$$

are uniformly continuous, whence

$$\sup_{p\in\mathcal{X}_m} \inf_{q\in\mathcal{X}_n} p \cdot A \cdot q \quad \text{and} \quad \inf_{q\in\mathcal{X}_n} \sup_{p\in\mathcal{X}_m} p \cdot A \cdot q$$

exist; see Corollary 2.2.7 in [12]. Clearly,

$$\sup_{p\in\mathcal{X}_m} \inf_{q\in\mathcal{X}_n} p \cdot A \cdot q \le \inf_{q\in\mathcal{X}_n} \sup_{p\in\mathcal{X}_m} p \cdot A \cdot q,$$

so it remains to show that

$$\neg\Big(\sup_{p\in\mathcal{X}_m} \inf_{q\in\mathcal{X}_n} p \cdot A \cdot q < \inf_{q\in\mathcal{X}_n} \sup_{p\in\mathcal{X}_m} p \cdot A \cdot q\Big).$$

Suppose

$$\sup_{p\in\mathcal{X}_m} \inf_{q\in\mathcal{X}_n} p \cdot A \cdot q < \inf_{q\in\mathcal{X}_n} \sup_{p\in\mathcal{X}_m} p \cdot A \cdot q.$$

Without loss of generality, by suitable translation, we may assume that there exists $\iota > 0$ such that

$$\sup_{p\in\mathcal{X}_m}\inf_{q\in\mathcal{X}_n} p\cdot A\cdot q \leq -\iota \quad \text{and} \quad \iota \leq \inf_{q\in\mathcal{X}_n}\sup_{p\in\mathcal{X}_m} p\cdot A\cdot q. \tag{11.4}$$

As we aim at proving falsum, by (11.3) and Theorem 11.18 (see [9] for the details) it suffices to consider the cases

$$\exists p\in\mathcal{X}_m\,(p\cdot A\geq 0) \quad \text{or} \quad \exists q\in\mathcal{X}_n\,(A\cdot q<0)\,.$$

In the first case

$$\sup_{p\in\mathcal{X}_m}\inf_{q\in\mathcal{X}_n} p\cdot A\cdot q\geq 0>-\iota,$$

a contradiction, and in the second case

$$\inf_{q\in\mathcal{X}_n}\sup_{p\in\mathcal{X}_m} p\cdot A\cdot q\leq 0<\iota,$$

also a contradiction. □

Now on the basis of Proposition 11.20 and as a further consequence of Theorem 11.4 we obtain the following existence result for solutions to two-person zero-sum games; see, for instance, [14] for a classical discussion of such games.

Proposition 11.21 (See Proposition 15 in [9]) *Let $A\in\mathbb{R}^{m\times n}$. Suppose that*

$$f_A:\ \mathcal{X}_n\ni q\mapsto \sup_{p\in\mathcal{X}_m} p\cdot A\cdot q$$

admits at most one minimum point, and that

$$g_A:\ \mathcal{X}_m\ni p\mapsto \inf_{q\in\mathcal{X}_n} p\cdot A\cdot q$$

admits at most one maximum point, that is $-g_A$ admits at most one minimum point. Then there exists $(\hat{p},\hat{q})\in\mathcal{X}_m\times\mathcal{X}_n$ such that

$$\hat{p}\cdot A\cdot\hat{q}=\sup_{p\in\mathcal{X}_m}\inf_{q\in\mathcal{X}_n} p\cdot A\cdot q=\inf_{q\in\mathcal{X}_n}\sup_{p\in\mathcal{X}_m} p\cdot A\cdot q.$$

Proof Note that $\mathcal{X}_n$ and $\mathcal{X}_m$ are compact and that f_A is convex whereas g_A is concave, that is $-g_A$ is convex. Hence, according to Theorem 11.4, there exists a minimiser $\hat{q}\in\mathcal{X}_n$ of f_A and a minimiser $\hat{p}\in\mathcal{X}_m$ of $-g_A$, namely $\hat{p}$ is a maximiser of g_A. We have

$$\sup_{p\in\mathcal{X}_m}\inf_{q\in\mathcal{X}_n} p\cdot A\cdot q=\inf_{q\in\mathcal{X}_n}\hat{p}\cdot A\cdot q\leq\hat{p}\cdot A\cdot\hat{q}\leq\sup_{p\in\mathcal{X}_m} p\cdot A\cdot\hat{q}=\inf_{q\in\mathcal{X}_n}\sup_{p\in\mathcal{X}_m} p\cdot A\cdot q.$$

Now apply Proposition 11.20. □

Saddle points $(\hat{p}, \hat{q})$ as in Proposition 11.21 are called solutions to the two-person zero-sum game given by A. The following Corollary 11.22 generalises Theorem 3.2 of [10] and verifies the conjecture as regards existence of solutions to two-person zero-sum games made at the end of [10]. We will apply Theorem 11.1 in its proof.

Corollary 11.22 (See Corollary 3 in [9]) *Let $A \in \mathbb{R}^{m\times n}$, and suppose that the associated two-person zero-sum game has at most one solution in the sense of [10], that is, denoting*

$$\alpha := \sup_{p\in\mathcal{X}_m} \inf_{q\in\mathcal{X}_n} p \cdot A \cdot q = \inf_{q\in\mathcal{X}_n} \sup_{p\in\mathcal{X}_m} p \cdot A \cdot q,$$

we have, for any pairs $(p,q),(p',q') \in \mathcal{X}_m \times \mathcal{X}_n$ with $\|p-p'\| + \|q-q'\| > 0$, either $|p \cdot A \cdot q - \alpha| > 0$ or $|p' \cdot A \cdot q' - \alpha| > 0$. Then the game has a unique solution; that is, there exists a unique $(\hat{p}, \hat{q}) \in \mathcal{X}_m \times \mathcal{X}_n$ such that

$$\hat{p} \cdot A \cdot \hat{q} = \alpha.$$

Proof For uniqueness, assume that $(p,q),(p',q') \in \mathcal{X}_m \times \mathcal{X}_n$ are two solutions to the game. Then, as the game has at most one solution, $\|p-p'\| + \|q-q'\| > 0$ is absurd, which implies $(p,q) = (p',q')$.

As regards existence of solutions, we show that the function f_A defined in Proposition 11.21 admits at most one minimum. Note that $\inf_{q\in\mathcal{X}_n} f_A(q) = \alpha$ and

$$\forall\delta > 0\, \forall q \in \mathcal{X}_n\, \exists p \in \mathcal{X}_m\, (|p \cdot A \cdot q - f_A(q)| < \delta)\,. \tag{11.5}$$

Fix $q, q' \in \mathcal{X}_n$ and suppose that $\|q - q'\| > 0$. The function

$$\begin{array}{rcccl} h & : & \mathcal{X}_m \times \mathcal{X}_m & \to & \mathbb{R} \\ & & (p,p') & \mapsto & |p \cdot A \cdot q - \alpha| + |p' \cdot A \cdot q' - \alpha| \end{array}$$

is uniformly continuous, convex, and positive valued. The latter follows from the assumption that the game has at most one solution. Thus, according to Theorem 11.1 there exists $\varepsilon > 0$ such that

$$\inf_{(p,p')\in\mathcal{X}_m\times\mathcal{X}_m} h(p,p') > \varepsilon. \tag{11.6}$$

We have either $f_A(q) < \alpha + \varepsilon/4$ or $f_A(q) > \alpha$ and either $f_A(q') < \alpha + \varepsilon/4$ or $f_A(q') > \alpha$. Assume that

$$f_A(q) < \alpha + \frac{\varepsilon}{4} \quad \text{and} \quad f_A(q') < \alpha + \frac{\varepsilon}{4}.$$

Then there are $p, p' \in \mathcal{X}_m$ such that

$$|p \cdot A \cdot q - \alpha| < \frac{\varepsilon}{2} \quad \text{and} \quad |p' \cdot A \cdot q' - \alpha| < \frac{\varepsilon}{2}.$$

This is a contradiction to (11.6). Thus, either

$$f_A(q) > \alpha \quad \text{or} \quad f_A(q') > \alpha.$$

Similarly, one verifies that g_A defined in Proposition 11.21 admits at most one maximum. Hence, the assertion follows from Proposition 11.21. □

References

[1] Berger, J., and Ishihara, I. 2005. Brouwer's fan theorem and unique existence in constructive analysis. *Math. Logic Q.*, **51**(4), 360–364.
[2] Berger, J., Bridges, D., and Schuster, P. 2006. The fan theorem and unique existence of maxima. *J. Symbol. Logic*, **71**(2), 713–720.
[3] Berger, J., and Svindland, G. 2016. A separating hyperplane theorem, the fundamental theorem of asset pricing, and Markov's principle. *Ann. Pure Appl. Logic*, **167**, 1161–1170.
[4] Berger, J. and Svindland, G. 2016. Convexity and constructive infima. *Arch. Math. Logic*, **55**, 873–881.
[5] Berger, J. and Svindland, G. 2018. Brouwer's Fan Theorem and convexity. *J. Symbol. Logic*, **83**(4), 1363–1375.
[6] Berger, J. and Svindland, G. 2018. Constructive convex programming. *Proof and Computation: Digitalization in Mathematics, Computer Science, and Philosophy*. Singapore: World Scientific.
[7] Berger, J. and Svindland, G. 2019. Convexity and unique minimum points. *Arch. Math. Logic*, **58**(1–2), 27–34.
[8] Berger, J. and Svindland, G. 2019. Constructive proofs of negated statements. In: Centrone, S., Negri, S., Sarikaya, D., and Schuster, P. M. (eds.), *Mathesis Universalis, Computability and Proof.* Cham: Springer.
[9] Berger, J. and Svindland, G. 2021. On Farkas' Lemma and related propositions in BISH. Preprint available at https://arxiv.org/pdf/2101.03424.pdf.
[10] Bridges, D. 2004. First steps in constructive game theory. *Math. Logic Q.*, **50**, 501–506.
[11] Bridges, D. and Richman, F. 1987. *Varieties of Constructive Mathematics.* London Mathematical Society Lecture Notes 97. Cambridge: Cambridge University Press.
[12] Bridges, D. and Vîţă, L. 2006. *Techniques of Constructive Analysis.* Universitext. New York: Springer-Verlag.
[13] Julian, W. H. and Richman, F. 1984. A uniformly continuous function on $[0, 1]$ that is everywhere different from its infimum. *Pacific J. Math.*, **111**(2), 333–340.
[14] Gale, D. 1989. *The Theory of Linear Economic Models.* Chicago: University of Chicago Press.

12

Constructive Mathematical Economics

Matthew Hendtlass and Douglas Bridges

12.1 Introduction

Although mathematical economics may seem a somewhat esoteric subject to the mainstream mathematician, as we aim to show in this relatively brief survey chapter it gives rise to interesting, technically nontrivial analysis within the framework of Bishop's constructive mathematics (BISH) [5]. Taken with Chapter 11 by Berger and Svindland in this volume, our chapter should give the reader a sound introduction to constructive aspects of non-physical applied mathematics.

From a non-philosophical viewpoint, BISH can be regarded as mathematics with intuitionistic logic and some appropriate set- or type-theoretic foundation that includes the axiom of dependent choice[1] [1, 2, 23]; see also Chapter 2 in this volume. We shall not work formally, preferring to adopt the normal style of the analyst such as Bishop himself.

We assume some basic knowledge of the constructive theory of the real line $\mathbf{R}$ and metric spaces, as can be found in [5, 6, 14].

12.2 Preference and Utility

We begin with a question studied by several pioneers of mathematical economics: When is a preference relation represented by a utility function?

In traditional microeconomic theory, it is customary to assume (i) that the individual consumer imposes on his consumption set X a relation $\succ$ ranking his preferences; and (ii) that the preference relation $\succ$ can be represented by a real-valued utility function u, so that $x \succ y$ (x is preferred to y) if and only if $u(x) > u(y)$. Usually X will have some topological structure that allows further assumptions of continuity and differentiability of u; typically, X will be a convex subset of the Euclidean space $\mathbf{R}^N$.

1 Some mathematicians – notably Richman [29] – prefer not to include the axiom of countable choice, let alone dependent choice, in their practice of BISH.

Over the past century, assumptions (i) and (ii) have given rise to a sizeable literature, in which mathematicians and mathematical economists have endeavoured to find necessary and sufficient conditions on a topological space X, and a preference relation $\succ$ on X, such that $\succ$ can be represented by a continuous utility function [15, 16, 17, 28, 30, 31]. In pure-mathematical terms, this problem reduces to that of finding conditions under which there is a continuous order isomorphism from $(X, \succ)$ to $(\mathbf{R}, >)$, where X is a topological space, $\succ$ is a strict weak order on X, and $>$ is the usual 'greater than' relation on $\mathbf{R}$. As an indication of the economic origin of the mathematical problems with which we shall be concerned here, we prefer the terms 'preference relation' and 'utility function', instead of 'strict weak order' and 'order isomorphism'.

From our constructive viewpoint, it seems extremely unlikely that a continuous utility function can be constructed, as in [4, pp. 82–87], solely from the transitivity, connectedness, and continuity of the preference relation. Such a construction would mean that, starting with a numerically vague notion of preference and using the utility function, we could obtain a numerically precise notion of degree of preference: to be exact, we could say that x is preferred to y by an amount greater than ε if $u(x) > u(y) + \varepsilon$. Considerations like these, together with knowledge of the appropriate constructive properties of real numbers, led to a definition of 'x is preferred to y by an amount greater than ε' and a corresponding construction, à la Arrow–Hahn, of a utility function [7]. This was later superseded by a construction of utility from preference data without any prior notion of preference by a given amount [8, 10].

A close scrutiny of the classical Arrow–Hahn proof of existence of utility functions reveals several places where non-constructive arguments are employed. For example, its demonstration that any inhabited compact subset of X contains a least preferred point. We cannot expect to prove that constructively, as such a proof would readily lead to one that every uniformly continuous, real-valued mapping on $[0, 1]$ attains its supremum, a classical theorem that implies the lesser limited principle of omniscience (LLPO).

We shall describe the later constructive version of the Arrow–Hahn approach to the existence of utility functions. But in order to circumvent the non-constructive arguments we have already discussed, we need definitions of certain notions that are constructively stronger than, but classically equivalent to, their classical counterparts.

Let X be an inhabited set, let $\succ$ be a binary relation on X, and let $\succcurlyeq$ be the binary relation defined by

$$x \succcurlyeq y \text{ if and only if } \forall_{z \in X}(y \succ z \Rightarrow x \succ z).$$

We say that $\succ$ is

- *asymmetric* if $x \succ y \Rightarrow \neg(y \succ x)$;
- *cotransitive* if $x \succ y \Rightarrow \forall_{z \in X}(x \succ z \vee z \succ y)$;
- *non-trivial* if there exist x, y in X such that $x \succ y$;
- a *preference relation* or *strict weak order* if it is asymmetric, cotransitive, and non-trivial.

We define the *upper contour set*, and the *strict upper contour set*, for $\succ$ at a in X by

$$[a, \rightarrow) \equiv \{x \in X : x \succcurlyeq a\},$$
$$(a, \rightarrow) \equiv \{x \in X : x \succ a\}$$

respectively, and the corresponding *lower and strict lower contour sets* by

$$(\leftarrow, a] \equiv \{x \in X : a \succcurlyeq x\},$$
$$(\leftarrow, a) \equiv \{x \in X : a \succ x\}.$$

If $\succ$ is a preference relation, then we call $\succcurlyeq$ the corresponding *preference–indifference relation*. In that case, we also have these properties: $(x \succ x)$ is contradictory; both $\succ$ and $\succcurlyeq$ are transitive; if $x \succcurlyeq y \succ z$ or $x \succ y \succcurlyeq z$, then $x \succ z$; $x \succcurlyeq y$ if and only if $\neg(y \succ x)$. Note that we cannot expect to prove that $\neg(y \succcurlyeq x)$ implies $x \succ y$: consider the preference relation $>$ on $\mathbf{R}$.

By a *utility function* for, or representing, a preference relation $\succ$ on X we mean a mapping $u : X \rightarrow \mathbf{R}$ such that $x \succ y$ if and only if $u(x) > u(y)$. Such a mapping is also called an *order isomorphism* from $(X, \succ)$ to $\mathbf{R}$.

We now confine our attention to a preference relation $\succ$ on an inhabited metric space (X, ρ). For each $S \subset X$ we define

$$x \succ S \text{ if and only if } \forall_{s \in X}(x \succ s)$$

and

$$\rho(x, S) < \varepsilon \text{ if and only if } \exists_{s \in S}(\rho(x, s) < \varepsilon).$$

(Note that the statement '$\rho(x, S) < \varepsilon$' does not imply that S is located.) We denote the open and closed balls in X with centre x and radius r by $B(x, r)$ and $\overline{B}(x, r)$, respectively, and we denote the diameter (when it exists) of a subset S of X by $\operatorname{diam} S$.

We say that $\succ$ is *continuous* if, for each $x \in X$, both $(a, \rightarrow)$ and $(\leftarrow, a)$ are open sets in X, in which case $[a, \rightarrow)$ and $(\leftarrow, a]$ are closed in X. We shall need the stronger condition of *uniform continuity of* $\succ$ *on compact sets*.

UC *For all* $a, b \in X$ *with* $a \succ b$*, and each compact* $K \subset X$*, there exists* $r > 0$ *such that for all* x, y *in* K *with* $\rho(x, y) < r$*, either* $a \succ x$ *or* $y \succ b$*.*

This property, a strengthening of cotransitivity, is classically equivalent to the continuity of $\succ$. If X is locally compact (that is, every bounded subset of X is contained in a compact set), then a classical argument using sequential compactness shows that continuity and **UC** are equivalent conditions on $\succ$.[2] Moreover, a simple adaptation of the proof of Proposition 1 in [8] shows that if X is locally compact and $\succ$ is represented by a utility function that is uniformly continuous on compact sets, then $\succ$ has the property **UC**.

We say that $\succ$ is

- *locally non-satiated at* $a \in X$ if for each $\varepsilon > 0$ there exists $x \in X$ such that $x \succ a$ and $\rho(a, x) < \varepsilon$;
- *locally non-satiated* if it is locally nonsatiated at each point of X; and
- *locally non-satiated at the subset* S of X if for each $\varepsilon > 0$ there exists $x \in X$ with $\rho(x, S) < \varepsilon$ and $x \succ S$.

On a locally compact metric space, local non-satiation at each compact subset is classically equivalent to, but constructively stronger than, local non-satiation.[3]

We begin the theory proper with an illustration of how local non-satiation can help prove things for which it is a classically redundant hypothesis.

Proposition 12.1 *Let* $\succ$ *be a preference relation on a metric space* X*, let* $a \in X$*, and let* b *be a local non-satiation point of* X *that is bounded away from* $[a, \rightarrow)$*. Then* $a \succ b$*.*

Proof Pick $r > 0$ such that $\rho(x, b) \geq r$ whenever $x \succcurlyeq a$, and then $y \succ b$ with $\rho(y, b) < r$. Then for each $x \succ a$, $\rho(x, y) \geq \rho(x, b) - \rho(y, b) > 0$; whence $\neg(y \succ a)$, so $a \succcurlyeq y \succ b$ and therefore $a \succ b$. □

In order to prove the locatedness of strict upper contour sets, we next have a general result.

Lemma 12.2 *Let* Y *be a subset of a metric space* (X, ρ)*, and let* $a \in X$*. Suppose that for each* $\varepsilon > 0$ *there exist points* $y_1, \ldots, y_N$ *of* Y *such that for each* $y \in Y$*, there exists* i *with* $\rho(a, y_i) > \rho(a, y) - \varepsilon$*. Then* $\rho(a, Y)$ *exists.*

Proof Consider any real numbers α, β with $\alpha > \beta$, and set $\varepsilon = \frac{1}{2}(\alpha - \beta)$. Choose $y_1, \ldots, y_N$ as above. Either $\rho(a, y_i) < \alpha$ for some i, or else $\rho(a, y_i) > \beta + \varepsilon$ for all i. In the latter case, we have $\rho(a, y) > \beta$ for all $y \in Y$, by our choice

[2] This equivalence does not hold constructively: see [8, Example 1].

[3] There is a recursive example of a locally non-satiated preference relation on the locally compact, convex subset $X = [0, 1] \times [0, \infty)$ of $\mathbf{R}^2$ that is not locally non-satiated at some compact subsets of X: see [8, Example 2].

of the points y_n. Thus $\rho(a,Y) = \inf\{\rho(a,y) : y \in Y\}$ exists, by the constructive least-upper-bound principle [14, Theorem 2.1.18]. □

Proposition 12.3 *Let X be a locally compact metric space, and let $\succ$ be a preference relation on X that is locally non-satiated at each compact set. Then the strict upper contour sets of $\succ$ are located in X.*

Proof Fixing $a \in X$, let $x \in X$ and $\varepsilon > 0$, and let $R > \rho(a,x)$ be such that $K \equiv \overline{B}(a,R)$ is compact. Cover K by finitely many compact subsets $B_1, \ldots, B_N$, each of diameter less than $\varepsilon/2$ [6, Chapter 4, equation (4.8)]. For each i, pick $y_i \succ B_i$ such that $\rho(y_i, B_i) < \varepsilon/4$, and then $z_i \succ y_i$ such that $\rho(z_i, y_i) < \varepsilon/4$; then $\rho(z_i, B_i) < \varepsilon/2$. Using the cotransitivity of $\succ$, we can write $\{1, \ldots, N\} = P \cup Q$, where $i \in P$ entails $z_i \succ x$, and $i \in Q$ entails $x \succ y_i$. Given ξ in $[x, \rightarrow) \cap K$, choose i such that $\xi \in B_i$. If $i \in Q$, then $x \succ y_i \succ \xi_i$, so $x \succ \xi$, a contradiction. Hence $i \in P$. Moreover,

$$\rho(\xi_i, z_i) \leq \rho(z_i, B_i) + \mathsf{diam}(B_i) < \frac{\varepsilon}{2} + \frac{\varepsilon}{2} = \varepsilon.$$

Now consider any $\xi \in [a, \rightarrow)$. Either $\rho(a,\xi) > \rho(a,x)$ or else $\rho(a,\xi) < R$. In the latter case, $\xi \in K$, so, according to the preceding paragraph, there exists $i \in P$ with $\rho(\xi, z_i) < \varepsilon$ and therefore $\rho(a,\xi) > \rho(a, z_i) - \varepsilon$. Since $\varepsilon > 0$ is arbitrary, it follows from Lemma 12.2 that $\rho(a, Y)$ exists. □

We now introduce the condition of *uniform local non-satiation at each compact set*.

> **ULN** *For each compact $K \subset X$ and each $\varepsilon > 0$, there exists $\delta > 0$ such that for all $a, b, x \in K$ with $\rho(a,b) < \delta$ and $x \succcurlyeq a$, there exists $y \succ b$ such that $\rho(x,y) < \varepsilon$.*

Proposition 12.4 *The following are equivalent conditions on a preference relation $\succ$ on a locally compact metric space X:*

(a) *$\succ$ is uniformly continuous on, and locally non-satiated at, each compact subset of X;*

(b) *$\succ$ is pointwise continuous on X, and satisfies* **ULN.**

Proof See [10, Proposition 2]. □

Lemma 12.5 *Let X be a locally compact, convex subset of $\mathbf{R}^N$, and let $\succ$ be a preference relation on X that is uniformly continuous on compact sets. Let $C \subset X$ be compact, and let a, b be points of X with $a \succ b$. Then there exists $\alpha > 0$ such that if $x \succcurlyeq a$, $\xi \in C$, and $b \succcurlyeq \xi$, then $\|x - \xi\| > \|y - \xi\| + \alpha$ for some $y \succ b$.*

Proof Choose $s > \operatorname{diam} C$ and $R > 2s$ such that $\|x\| \leq R$ whenever $\rho(x, C) \leq 3s$. By the uniform continuity of $\succ$ on compact sets, there exists $r \in (0, 4s)$ such that if $x, y \in \overline{B}(0, 2R)$ and $\|x - y\| < r$, then either $a \succ x$ or $y \succ b$. Let $t = r/4R$ and $\alpha = tr/2$; note that $0 < t < 1$ and $0 < \alpha < s$. Consider any $x \in [a, \rightarrow)$ and any $\xi \in C$ with $b \succcurlyeq \xi$. Either $\|x\| > R$ or $\|x\| < 2R$. In the former case,

$$\begin{aligned}\|x - \xi\| &\geq \|x - a\| - \|a - \xi\| \geq \rho(x, C) - \operatorname{diam} C \\ &\geq 3s - s = 2s > s + \operatorname{diam} C > \alpha + \|a - \xi\| .\end{aligned}$$

In the case $\|x\| < 2R$, setting $y = (1 - t)x + t\xi$, we have $y \in X$ (by convexity) and

$$\|x - y\| = t\,\|x - \xi\| \leq t\,(\|x\| + \|\xi\|) \leq 3Rt < r.$$

Since $x \succcurlyeq a$, our choice of r ensures that $y \succ b$. If $\|x - \xi\| < r$, then either $a \succ x$ or $\xi \succ b$, which is absurd. Hence $\|x - \xi\| \geq r$ and therefore

$$\|x - \xi\| = \|y - \xi\| + t\,\|x - \xi\| > \|y - \xi\| + tr > \|y - \xi\| + \alpha. \qquad \square$$

This brings us to the constructive version of the *Arrow–Hahn utility representation theorem.*

Theorem 12.6 *Let X be a locally compact, convex subset of $\mathbf{R}^N$, and let $\succ$ be a preference relation on X that is uniformly continuous on, and locally non-satiated at, each compact subset of X. Then:*

(a) *$[x, \rightarrow)$ is locally compact for each $x \in X$;*
(b) *if $x \in X$ and $K \subset X$ is compact, then*

$$u(K, x) \equiv \sup\{\rho(\xi, [x, \rightarrow)) : \xi \in K\}$$

exists;
(c) *if $X = \bigcup_{n \geq 1} K_n$, where each K_n is compact, then*

$$u(x) \equiv \sum_{n=1}^{\infty} \frac{2^{-n} u(K_n, x)}{1 + u(K_n, x)}$$

defines a continuous utility function $u : X \rightarrow [0, 1]$ representing $\succ$.

Proof The proof, which we shall not give in detail, consists of six steps.

Step 1 If $K \subset X$ is compact, then the map $(\xi, x) \rightsquigarrow \rho(\xi, [x, \rightarrow))$ is uniformly continuous on $K \times K$. (This mapping is well defined, in view of Proposition 12.3.)

Step 2 For each compact $K \subset X$ and each $x \in X$, $u(K, x)$ exists.

Step 3 If $K \subset X$ is compact, then the mapping $u(K, .)$ of X into the non-negative real line is uniformly continuous on compact sets.

Step 4 If $K \subset X$ is compact, $z \in K$, and $a \succ b \succcurlyeq z$, then $u(K,a) > u(K,b)$. To show this, apply Lemma 12.5 to a compact superset C of $K \cup \{a,b\}$, to find $\alpha > 0$ such that if for $\xi \in K$, $b \succcurlyeq \xi$, and $x \succcurlyeq a$, then there exists $y \succ b$ with[4]

$$\|x - \xi\| > \|y - \xi\| + \alpha \geq \rho(\xi, [b, \rightarrow)) + \alpha.$$

For all $\xi \in K$ we then have

$$u(K, a) \geq \rho(\xi, [a, \rightarrow)) \geq \rho(\xi, [b, \rightarrow)) + \alpha.$$

Hence

$$\begin{aligned} u(K, a) &\geq \sup\{\rho(\xi, [b, \rightarrow)) : \xi \in K\} + \alpha \\ &= u(K, b) + \alpha \end{aligned}$$

and therefore $u(K,a) > u(K,b)$.

Step 5 If $K \subset X$ is compact and $u(K,a) > u(K,b)$, then $a \succ b$.

Step 6 With $(K_n)_{n \geq 1}$ and u as in hypothesis (c), it follows from the uniform convergence of the series defining $u(x)$ that u is uniformly continuous on each compact subset of X. Consider x, y in X with $x \succ y$. If either $u(K_n, y) > 0$ or $y \in K_n$, then $y \succcurlyeq \xi$ for some $\xi \in K_n$, so $u(K_n, x) > u(K_n, y)$, by Step 4. Thus for each n, since $u(K_n, x) \geq 0$, we have $u(K_n, x) \geq u(K_n, y)$. But $y \in K_n$ for some n, so $u(x) > u(y)$. Conversely, if $u(x) > u(y)$, then $u(K_n, x) > u(K_n, y)$ for some n; so $x \succ y$, by Step 5. This completes the proof. □

Let us agree that a preference relation $\succ$ on a locally compact metric space X is *admissible* if it is uniformly continuous on, and locally non-satiated at, each compact subset of X. We have the following.

Proposition 12.7 *If $\succ$ is an admissible preference relation on a locally compact space X, then the corresponding preference–indifference relation is a locally compact subspace of $X \times X$.*

Proof See [8, Proposition 5]. □

In the constructive context, we should expect continuity-in-parameters for our preference relations. To fulfil that expectation, we need to define appropriate topologies.

[4] Recall here Proposition 12.3.

Given a locally compact space (X,ρ), we define a metric ϱ on the set $A(X)$ of admissible preference relations on X as follows. Referring to [6, pp. 112–115], let (Z,d) be a one-point compactification of $X \times X$, with point at infinity ω and inclusion map i. Then for each locally compact subset S of $X \times X$, the closure $I(S)$ of $i(S) \cup \{\omega\}$ in Z is compact. In view of Proposition 12.7, we can define

$$\varrho(\succ,\succ') \equiv d(I(\succcurlyeq), I(\succcurlyeq'))$$

whenever $\succ,\succ'$ belong to $A(X)$, $\succcurlyeq$ and $\succcurlyeq'$ are the corresponding preference–indifference relations, and d is the usual Hausdorff metric on the set of compact subsets of Z [14, p. 46]. Then ϱ is a metric on $A(X)$, and the corresponding topology is independent of the one-point compactification of $X \times X$ that we picked to be Z. Classically, ϱ induces the topology of closed convergence on $A(X)$.

Since X is locally compact, it can be expressed as a union of compact subsets as in hypothesis (c) of Theorem 12.6. We define a corresponding metric on the set $\mathcal{C}(X,[0,1])$ of continuous mappings from X into $[0,1]$ by

$$\rho(f,g) \equiv \sum_{n=1}^{\infty} 2^{-n} \sup\{|f(x)-g(x)| : x \in K_n\}.$$

The topology induced on $\mathcal{C}(X,[0,1])$ by this metric is independent of the choice of the compact sets K_n whose union is X.

Theorem 12.8 *Let X be a locally compact, convex subset of $\mathbf{R}^N$, and let $(K_n)_{n\geq1}$ be a sequence of compact sets whose union is X. For each admissible preference relation $\succ$ on X, and each compact $K \subset X$, define*

$$u(\succ,K,x) \equiv \sup\{\rho(\xi,[x,\leftarrow)_{\succ}) : \xi \in X\},$$

where $[x,\leftarrow)_{\succ}$ denotes the upper contour set of X with respect to $\succ$. Define also $u : A(X) \times X \to [0,1]$ by

$$u(\succ,x) \equiv \sum_{n=1}^{\infty} 2^{-n}\frac{u(\succ,K_n,x)}{1+u(\succ,K_n,x)}.$$

Then for each $\succ$ in $A(X)$, $u(\succ,\cdot)$ is a utility function representing $\succ$, and is uniformly continuous on compact subsets of X. Moreover, the following two mappings are pointwise continuous:

(a) *the mapping $\succ \rightsquigarrow u(\succ,\cdot)$ of $A(X)$ into $\mathcal{C}(X,[0,1])$;*
(b) *the mapping u of $A(X) \times X$ into $[0,1]$.*

Proof See [8, Theorem 2]. □

Next we define an *admissible set of preference relations* on X to be a subset Σ of $A(X)$ with the following property: for each compact $K \subset X$ there exists a sequence $(\delta_n)_{n\geq1}$ of positive numbers such that for each positive integer n, if

$\succ \in \Sigma$, a, b, x belong to K, $\rho(a,b) < \delta_n$, and $x \succcurlyeq a$, then there exists $y \succ b$ with $\rho(x,y) < 2^{-n}$.

Under the hypotheses of Theorem 12.8, a classical sequential compactness argument shows that if X is a locally compact, convex subset of $\mathbf{R}^N$, then every compact subset of $A(X)$ is an admissible set. This last statement can also be proved intuitionistically, but its status within BISH remains unclear.

Careful adaptation of the proof of Theorem 12.8 brings uniform continuity-in-parameters into the picture.

Theorem 12.9 *Under the hypotheses of* Theorem 12.8, *let Σ be an admissible set of preference relations on X, and S a compact subset of X. Then the following mappings are uniformly continuous:*

(a) *the mapping $\succ \rightsquigarrow u(\succ, \cdot)$ of Σ into $\mathcal{C}(X, [0,1])$;*
(b) *the mapping u of $\Sigma \times S$ into $[0,1]$.*

Proof See [8, Theorem 3]. □

More information and detail of the constructive Arrow–Hahn approach to the existence and continuity of utility functions can be found in [8, 10]. A recent book chapter [12] shows how the Arrow–Hahn approach can be generalised in the context of a locally convex metric space. This generalisation allows the possibility of preferred points, and introduces an apparently new notion – the uniform ball–centre distance property on compact sets – that circumvents the lack of convexity in the abstract metric setting.

Recalling Debreu's classical theorem

> If X is a topological space that is either separable and connected or else has a countable base of open sets, any continuous preference relation is represented by a continuous utility function

it is fair to ask why we chose the Arrow–Hahn route to the construction of utility functions. There are two reasons for this choice. First, the original proof of Arrow and Hahn in [4] had the appearance of algorithmic substance (even though the proof of our final, constructive product, Theorem 12.6, is rather loosely connected to it). But a second, more important reason is provided as a corollary of the following result from [13].

Proposition 12.10 *Let S be a set of natural numbers. Then there exists a preference relation $\succ$ on $[0,1]$ that is uniformly continuous on each compact subset of $[0,1]$ and that is represented by a utility function if and only if S is countable.*

In the recursive interpretation of BISH, there exists a subset S of $\mathbf{N}$ that is not recursively enumerable. Applying Proposition 12.10 to this set, we obtain a recursive counterexample to Debreu's theorem.

12.3 Demand Functions

We now turn to the construction of a consumer's (continuous) demand function that under a given price system assigns the consumer their most preferred bundle of goods that they can afford. We take, as our starting point, the standard configuration in microeconomics consisting of a consumer whose consumption set X is a compact, convex subset of $\mathbf{R}^n$ endowed with a strictly ordered preference relation $\succ$. For a given price vector $p \in \mathbf{R}^n$ and a given initial endowment $w \in \mathbf{R}$, the consumer's *budget set*

$$\beta(p, w) = \{x \in X : p \cdot x \leqslant w\},$$

where the dot denotes inner product, is the collection of all consumption bundles available to the consumer. The collection of $\succ$-maximal elements of $\beta(p, w)$ form the consumers *demand set* for price p and initial endowment w.

Lemma 12.11 *If $p > 0$ and $\beta(p, w)$ is inhabited, then $\beta(p, w)$ is compact and convex.*

Proof Convexity is clear. See [9] for a proof that $\beta(p, w)$ is compact. □

As detailed in [9], it is easy to show classically that if $\succ$ is strictly convex and $\beta(p, w)$ is non-empty, then the demand set is a singleton; that is, there exists a unique ($\succ$*-maximal*) point $\xi_{p,w} \in \beta(p, w)$ with $\xi_{p,w} \succcurlyeq \beta(p, w)$. Let T be the set of pairs consisting of a price vector p and an initial endowment w for which $\beta(p, w)$ is inhabited. If the preference relation $\succ$ is continuous, then a sequential compactness argument gives the sequential, and hence pointwise, continuity of the *demand function* F on T that sends (p, w) to the maximal element $\xi_{p,w}$ of $\beta(p, w)$ (see, for example, [32, Chapter 2, Section D]).

Our first question is: under what conditions can we compute the demand function F?

A preference relation $\succ$ on X is said to be *strictly convex* if X is convex and $tx + (1 - t)x' \succ x$ or $tx + (1 - t)x' \succ x'$ whenever $x, x' \in X$, $x \neq x'$, and $t \in (0, 1)$. The next theorem shows, together with Lemma 12.11, that the demand function $F : \mathbf{R}^n \times \mathbf{R} \to X$ is a well-defined function when $\succ$ is strictly convex.

Theorem 12.12 *Let $\succ$ be a continuous, strictly convex preference relation on an inhabited, compact subset X of $\mathbf{R}^n$. Then there exists a unique $\xi \in X$ such that $\xi \succcurlyeq x$ for all $x \in X$.*

We sketch the proof; details can be found in [20].

Proof The proof proceeds by induction on the dimension. The case $n = 1$ is proved as follows.

(i) For $X = [0, 1]$, by applying the strict convexity of $\succ$ to $1/4 \in (0, 3/4)$, $1/2 \in (1/4, 3/4)$, and $3/4 \in (1/2, 1)$ we can show that either $1/2 \succcurlyeq [0, 1/4)$ or $1/2 \succcurlyeq (3/4, 1]$.

(ii) Using (i) repeatedly, we can inductively construct a sequence of nested intervals $[\underline{\xi}_n, \overline{\xi}_n]$ with diameter converging to 0 such that for each $x \in [0, 1] \setminus [\underline{\xi}_n, \overline{\xi}_n]$, there exists $y \in [\underline{\xi}_n, \overline{\xi}_n]$ such that $y \succcurlyeq x$. The unique point of intersection of these intervals is then $\succ$-maximal in $[0, 1]$.

(iii) We then use the λ-technique [14, Chapter3] to extend this to $X = [a, b]$ with $a \leqslant b$.

Now suppose we have proved the result for dimension $n - 1$ and consider a strictly convex preference relation $\succ$ on a compact, convex subset X of $\mathbf{R}^n$. Let $\pi_1 : (x_1, \ldots, x_n) \rightsquigarrow x_1$ be the projection function onto the first dimension. We define a preference relation $\succ'$ on $\pi_1(X) = [a, b]$ by

$$s \succ_i' t \quad \Leftrightarrow \quad \exists_{x \in X} \forall_{y \in X} \left(\pi_1(x) = s \text{ and if } \pi_1(y) = t, \text{ then } x \succ y \right).$$

It is straightforward to show that $\succ'$ is strictly convex and continuous. We apply the $n = 1$ case to construct a maximal element ξ_1 of $(\pi_1(X), \succ')$, and then the induction hypothesis to construct a maximal element ξ_2 of

$$S = \{x \in X : \pi_1(x) = \xi_1\}$$

with respect to $\succ$. Then $\xi = \xi_1 \times \xi_2$ is a $\succ$-maximal element of X. □

Corollary 12.13 *Under the conditions of* Theorem 12.12, *if $x \in X$ and $x \neq \xi$, then $\xi \succ x$.*

Proof Let $y = (x + \xi)/2$. Then either $y \succ x$ or $y \succ \xi$. Since $\xi \succcurlyeq y$ the former must obtain, so $\xi \succcurlyeq y \succ x$. □

We now turn to the problem of establishing when the demand function is continuous. If we were working in Brouwer's intuitionistic mathematics, then we would not need to make any further assumptions on $\succ$.

Theorem 12.14 *Suppose Brouwer's full fan theorem holds. If $\succ$ is continuous and strictly convex, then F is Bishop continuous.*

Proof See Theorem 9 of [20]. □

Returning to BISH, we will make use of two notions from [9].

- A metric space X is *uniformly rotund* if for each $\varepsilon > 0$ there exists $\delta > 0$ such that for all $x, x' \in X$, if $\|x - x'\| \geqslant \varepsilon$, then

$$\left\{ \frac{1}{2} \left(x + x' \right) + z : z \in B(0, \delta) \right\} \subset X.$$

- A preference relation $\succ$ on X is *uniformly rotund* if X is uniformly rotund and for each $\varepsilon > 0$ there exists $\delta > 0$ such that if $\|x - x'\| \geqslant \varepsilon$ for $x, x' \in X$, then for each $z \in B(0, \delta)$ either $\frac{1}{2}(x + x') + z \succ x$ or $\frac{1}{2}(x + x') + z \succ x'$.

A uniformly rotund preference relation is strictly convex. Proposition 14 of [20] gives a partial converse.

Proposition 12.15 *Assume Brouwer's fan theorem. If $\succ$ is continuous and strictly convex, then $\succ$ is uniformly rotund.*

We study the continuity of F by looking at the map Γ, on the set T of all inhabited, compact, convex subsets of X (in particular all inhabited demand sets of a continuous, strictly convex preference relation), taking X to the unique $\succ$-maximal element of X. We give T the Hausdorff metric: for located subsets A, B of a metric space Y

$$d(A, B) = \max\{\sup\{\rho(a, B) : a \in A\}, \sup\{\rho(b, A) : b \in B\}\}.$$

Theorem 12.12 shows that Γ is well defined and the next lemma shows how studying Γ allows us to show the continuity of F.

Lemma 12.16 *If Γ is pointwise continuous, then F is pointwise continuous. If Γ is uniformly continuous, then for each $p \in \mathbf{R}^n$, $w \rightsquigarrow F(p, w)$ is uniformly continuous, and for each $w \in \mathbf{R}$, $p \rightsquigarrow F(p, w)$ is Bishop continuous.*

Theorem 12.17 *If $\succ$ is a uniformly rotund preference relation, then Γ is uniformly continuous.*

Proof Let S, S' be compact, convex subsets of X and let ξ, ξ' be their $\succ$-maximal points. Fix $\varepsilon > 0$ and let $\delta' > 0$ be such that if $\|x - x'\| \geqslant \varepsilon$ $(x, x' \in X)$, then for each $z \in B(0, \delta')$ either $\frac{1}{2}(x + x') + z \succ x$ or $\frac{1}{2}(x + x') + z \succ x'$, and set $\delta = \min\{\varepsilon, \delta'\}/2$.

Let S, S' be such that $d(S, S') < \delta$ and suppose that $\|\xi - \xi'\| > \varepsilon$. Since S, S' are convex

$$S \cap B((\xi + \xi')/2, \delta) \text{ and } S' \cap B((\xi + \xi')/2, \delta)$$

are both inhabited; let z be an element of the former set and let z' be an element of the latter. By the maximality of $\xi \in S$ and our choice of δ, $z \succ \xi'$; similarly, $z' \succ \xi$. Therefore

$$\xi \not\succ z \succ \xi' \not\succ z' \succ \xi,$$

which is absurd. Hence $\|\xi - \xi'\| \leqslant \varepsilon$. □

Theorem 12.18 *Let $\succ$ be a uniformly rotund preference relation on a compact subset X of $\mathbf{R}^n$ and let S be a subset of $\mathbf{R}^n \times \mathbf{R}$ such that $\beta(p, w)$ is inhabited for* *each $(p, w) \in S$. Then for each $p \in \mathbf{R}^n$, the function $w \rightsquigarrow F(p, w)$ is uniformly continuous, and for each $w \in \mathbf{R}$, the function $p \rightsquigarrow F(p, w)$ is Bishop continuous. In particular, F is Bishop continuous.*

Proof The result follows directly from Lemma 12.16 and Theorem 12.17. □

12.4 Economic Equilibrium

In this section we consider the construction of an economic equilibrium – a state in which an economy's supply and demand are balanced – as introduced independently by McKenzie [24] and Arrow and Debreu [3] in 1954. In Section 12.4.1 we discuss the fixed-point theorems of Brouwer and Kakutani in BISH and in Section 12.4.2 we give a constructive proof of an approximate version of one of the classical equilibrium results of mathematical economics, using our constructive version of Kakutani's fixed-point theorem.

12.4.1 Fixed-Point Theorems

We begin our discussion on the construction of an economic equilibrium with the construction of fixed points, since an economic equilibrium is at heart a fixed point, as demonstrated by the Uzawa equivalence theorem [33] which links Brouwer's fixed-point theorem and Walras's theorem on the existence of an economic equilibrium. See [18] for a constructive treatment of Walras's existence theorem and Uzawa's equivalence theorem.

Let X be a metric space and let f be a function from X to X. We say that

- $x \in X$ is a *fixed point* of f if $f(x) = x$, and
- f has *approximate fixed points* if for each $\varepsilon > 0$, there exists $x \in X$ with $\|f(x) - x\| < \varepsilon$.

If X is compact and f has approximate fixed points, then a sequential compactness argument can be used to show classically that f has a fixed point. Adding a uniqueness condition on any fixed point of f can reduce or eliminate the nonconstructive principle required for this argument.

One of the fundamental results in fixed-point theory, and indeed topology, is

Brouwer's fixed-point theorem Every continuous function f from a convex, compact set X to itself has a fixed point.

The most elementary proofs of Brouwer's fixed-point theorem use a combinatorial result (such as Sperner's lemma) to show the existence of approximate fixed points of f and then apply sequential compactness to assert the existence of a fixed point. While the latter step is not constructively valid, these proofs do establish constructively an approximate Brouwer fixed-point theorem. However, since there is an explicit recursive example of a pointwise continuous mapping f of the unit square in $\mathbf{R}^2$ that moves every point of the square [27], Brouwer's fixed-point theorem itself is not provable in BISH.

The existence of approximate fixed points can be established constructively using a form of continuity that is weaker than uniform continuity: we say that a function $f : X \to X$ is *uniformly sequentially continuous* if for all sequences $(x_n)_{n\geqslant 1}, (y_n)_{n\geqslant 1}$ in X, if $\rho(x_n, y_n) \to 0$ as $n \to \infty$, then $\rho(f(x_n), f(y_n)) \to 0$ as $n \to \infty$.

Theorem 12.19 *Let S be a totally bounded subset of Euclidean space with convex closure. Then every uniformly sequentially continuous function $f : S \to S$ has approximate fixed points.*

Proof See [19, Theorem 5]. □

The main tool in our constructive proof of the existence of an (approximate) economic equilibrium, and in our results on game theory in Section 12.5, is an approximate version of Kakutani's fixed-point theorem, which extends Brouwer's fixed-point theorem to set-valued functions. Let U be a function from a metric space X into the class $\mathcal{P}^*(X)$ of inhabited subsets of X; U is said to be a *set-valued mapping on* X. We say that U is *convex* (*compact*, *closed*, etc.) if $U(x)$ is convex (compact, closed, etc.) for each $x \in X$. A mapping $U : X \to \mathcal{P}^*(X)$ is said to be *sequentially upper hemi-continuous* if for each pair of sequences $(x_n)_{n\geqslant 1}, (y_n)_{n\geqslant 1}$ in X converging to points x, y in X respectively, if $y_n \in U(x_n)$ for each n, then $\rho(y, U(x)) = 0$; in particular, if U is closed, then $y \in U(x)$. If U is closed, then U is sequentially upper hemi-continuous if and only if the *graph*

$$G(U) = \bigcup_{x\in X} \{x\} \times U(x)$$

of U is closed. A point $x \in S$ such that $x \in U(x)$ is called a *fixed point of* U. Kakutani's fixed-point theorem is the following.

> **Kakutani's fixed-point theorem** Let S be a compact, convex subset of $\mathbf{R}^n$ and let $U : S \to \mathcal{P}^*(S)$ be a closed, convex, sequentially upper hemi-continuous mapping. Then U has a fixed point.

Given a function $f : X \to X$, define $U_f : X \to \mathcal{P}^*(X)$ by $U_f(x) = \{f(x)\}$. If f is sequentially continuous, then U_f is sequentially upper hemi-continuous, and a fixed point of U_f is a fixed point of f; thus Kakutani's fixed-point theorem is indeed a generalisation of Brouwer's fixed-point theorem. Since some form of uniform continuity is required for all known constructive proofs of Brouwer's fixed-point theorem, we need a more uniform version of sequential upper hemi-continuity on U, one that at least implies that f is (sequentially) uniformly continuous for $U = U_f$.

A natural path to take is to define a pointwise version of upper hemi-continuity, then take the uniform version of this property. This, however, seems to inevitably lead to a notion equivalent to uniform continuity for functions $U : X \to \mathcal{P}^*(X)$, where $\mathcal{P}^*(X)$ is equipped with the Hausdorff metric, which is too strong to give an interesting generalisation of Brouwer's fixed-point theorem.

We circumvent the difficulty of finding a suitable strengthening of sequential upper hemi-continuity – one that is both strong enough to prove a constructive version of Kakutani's fixed-point theorem and weak enough for that result to be of interest – by focussing on the graph of U. We say that a mapping $U : X \to \mathcal{P}^*(X)$ is *locally approximable* if, for each $x \in X$ and each $\varepsilon > 0$, there exists $\delta > 0$ such that if $y, y' \in B(x, \delta)$, $u \in U(y)$, $u' \in U(y')$, and $t \in [0, 1]$, then

$$\rho\left((z_t, u_t), G(U)\right) < \varepsilon,$$

where $z_t = ty + (1 - t)y'$ and $u_t = tu + (1 - t)u'$. Note that we do not require $G(U)$ to be located here: recall that '$\rho(x, S) < \varepsilon$' is shorthand for 'there exists $s \in S$ such that $\rho(x, s) < \varepsilon$'.

Proposition 12.20 *Every convex, pointwise, upper hemi-continuous set-valued mapping on a linear metric space is locally approximable.*

Proof See [21, Proposition 3]. □

We say that $U : X \to \mathcal{P}^*(X)$ is *approximable* if it satisfies the uniform version of local approximability: for each $\varepsilon > 0$, there exists $\delta > 0$ such that if $x, x' \in X$, $\|x - x'\| < \delta$, $u \in U(x)$, $u' \in U(x')$, and $t \in [0, 1]$, then

$$\rho\left((z_t, u_t), G(U)\right) < \varepsilon,$$

where $z_t = tx + (1 - t)x'$ and $u_t = tu + (1 - t)u'$. If f is continuous then U_f is locally approximable, and if f is uniformly continuous then U_f is approximable.

A mapping U is approximable if for each $\varepsilon > 0$ there exists $\delta > 0$ such that the convex hull of any two points in the graph of U that are separated by less than δ never strays more than ε from the graph of U; our next lemma shows that if U is approximable, then we can generalise this from any two points of $G(U)$ to any finite subset of $G(U)$. This will allow us to give a constructive version of Kakutani's fixed-point theorem that is classically equivalent to the classical version. We write

$$\Delta_n = \left\{ \mathrm{t} \in [0, 1]^n : \sum_{i=1}^{n} t_i = 1 \right\}.$$

Lemma 12.21 *Let $U : X \to \mathcal{P}^*(X)$ be an approximable function. Then for each positive integer n and each $\varepsilon > 0$ there exists $\delta > 0$ such that for all $x_1, \ldots, x_n, u_1, \ldots, u_n \in X$ and all $\mathrm{t} \in \Delta_n$, if $u_i \in U(x_i)$ for each i and*

$$\max\{\|x_i - x_j\| : 1 \leqslant i, j \leqslant n\} < \delta,$$

then

$$\rho((z_{\mathrm{t}}, u_{\mathrm{t}}), G(U)) < \varepsilon,$$

where $z_{\mathrm{t}} = \sum_{i=1}^{n} t_i x_i$ *and* $u_{\mathrm{t}} = \sum_{i=1}^{n} t_i u_i$.

Proof We proceed by induction; the case $n = 1$ is trivial. Suppose that we have shown the result for $n = k-1$. Let $\mathrm{t} \in \triangle_k$ and $u_1, \ldots, u_k$ be as in the statement of the lemma, and let $\delta > 0$ be such that for all $x_1, \ldots, x_{k-1} \in X^n$ and all $\mathrm{t} \in \triangle_{k-1}$, if $u_i \in U(x_i)$ for each i and

$$\max\{\|x_i - x_j\| : 1 \leqslant i, j \leqslant k-1\} < \delta,$$

then $\rho((z_{\mathrm{t}}, u_{\mathrm{t}}), G(U)) < \varepsilon/2$. Let t' be the $(k-1)$-dimensional vector with ith component $t_i / \sum_{j=1}^{k-1} t_j$. Then

$$\rho\left((z_{\mathrm{t}'}, u_{\mathrm{t}'}), G(U)\right) < \varepsilon/2.$$

Picking $(x, u) \in G(U)$ with $\rho\left((z_{\mathrm{t}'}, u_{\mathrm{t}'}), (x, u)\right) < \varepsilon/2$ and $t \in [0,1]$ such that $\rho((z_{\mathrm{t}}, u_{\mathrm{t}}), (tx + (1-t)x_n, tu + (1-t)u_n)) < \varepsilon/2$, we have that

$$\begin{aligned}\rho\left((z_{\mathrm{t}}, u_{\mathrm{t}}), G(U)\right) &< \rho((z_{\mathrm{t}}, u_{\mathrm{t}}), (tx + (1-t)x_n, tu + (1-t)u_n)) + \varepsilon/2 \\ &< \varepsilon/2 + \varepsilon/2 = \varepsilon.\end{aligned}$$

This completes the induction. □

Theorem 12.22 *Let S be a totally bounded subset of $\mathbf{R}^n$ with convex closure and let U be an approximable set-valued mapping on S. Then for each $\varepsilon > 0$ there exists $x \in S$ such that $\rho(x, U(x)) < \varepsilon$.*

Proof Fix $\varepsilon > 0$ and let $\delta > 0$ be such that for all $x_1, \ldots, x_k \in X^n$ and all $\mathrm{t} \in [0,1]^n$, if $u_i \in U(x_i)$ for each i, $\sum_{i=1}^{n} t_i = 1$, and $\max\{\|x_i - x_j\| : 1 \leqslant i, j \leqslant k\} < \delta$, then

$$\rho((z_{\mathrm{t}}, u_{\mathrm{t}}), G(U)) < \varepsilon/3.$$

Let $S' = \{x_1, \ldots, x_l\}$ be a discrete δ-approximation to S. For each $x_i \in S'$, pick $u_i \in U(x_i)$; let g be the uniformly continuous affine function on S that takes the value u_i at x_i for each $1 \leqslant i \leqslant l$. By the approximate Brouwer fixed-point theorem, there exists $y \in S$ such that $\rho(y, g(y)) < \varepsilon/3$. By our choice of δ, there exists $(x, u) \in G(U)$ such that $\rho((y, g(y)), (x, u)) < \varepsilon/3$. Therefore

$$\begin{aligned}\rho(x, u) &\leqslant \rho(x, y) + \rho(y, g(y)) + \rho(g(y), u) \\ &< \varepsilon/3 + \varepsilon/3 + \varepsilon/3 = \varepsilon,\end{aligned}$$

so $\rho(x, U(x)) < \varepsilon$. □

As with the approximate version of Brouwer's fixed-point theorem, a sequential compactness argument can be used to recover an exact fixed point.

Theorem 12.22 gives a very simple and intuitive constructive version of Kakutani's fixed-point theorem. An examination of the proof shows that we, in fact, only require our set-valued mapping U to satisfy the following condition, weaker than approximability and often much easier to verify. A set-valued mapping U on a metric space X is said to be *weakly approximable* if for each $\varepsilon > 0$, there exist

- a positive real number $\delta < \epsilon$,
- a $\delta/2$-approximation S of X, and
- a function V from S into $\mathcal{P}^*(X)$ with $G(V) \subset G(U)$,

such that if $x, x' \in S$, $\|x - x'\| < \delta$, $u \in V(x)$, $u' \in V(x')$, and $t \in [0, 1]$, then

$$\rho\left((z_t, u_t), G(U)\right) < \varepsilon.$$

If V can be chosen independent of ε, in which case S is a dense subset of X, then U is said to be *weakly approximable with respect to* V.

The proofs of Lemma 12.21 and Theorem 12.22 readily extend to give the following result.

Theorem 12.23 *Let S be a totally bounded subset of $\mathbf{R}^n$ with convex closure and let U be a weakly approximable set-valued mapping on S. Then for each $\varepsilon > 0$ there exists $x \in S$ such that $\rho(x, U(x)) < \varepsilon$.*

12.4.2 Existence of an Economic Equilibrium

The results from this section are drawn from [22]. We work in the economic model used by McKenzie [25, 26]; we have N commodities, n producers, and m consumers. To each producer we associate a production set $Y_i \subset \mathbf{R}^N$; and to each consumer we associate a consumption set $X_i \subset \mathbf{R}^N$ endowed with a preference relation $\succ_i$. Further we assume that each consumer has no initial endowment, and we write $x \sim_i x'$ for $x \not\succ_i x' \wedge x' \not\succ_i x$. When the price $\mathrm{p} \in \mathbf{R}^N$ prevails, we denote the budget set of consumer i by $\beta_i(\mathrm{p})$ and the demand set of consumer i by $D_i(\mathrm{p})$.

A *competitive equilibrium* of an economy consists of a price vector $\mathrm{p} \in \mathbf{R}^N$, points $\xi_1, \ldots, \xi_i \in \mathbf{R}^N$, and a vector η in the *aggregate production set* $Y = Y_1 + \cdots + Y_n$, satisfying

E1 $\xi_i \in D_i(\mathrm{p})$ for each $1 \leqslant i \leqslant m$.
E2 $\mathrm{p} \cdot \mathrm{y} \leqslant \mathrm{p} \cdot \eta = 0$ for all $\mathrm{y} \in Y$.
E3 $\sum_{i=1}^{m} \xi_i = \eta$.

An economy is said to have *approximate competitive equilibria* if for all $\varepsilon > 0$ there exist a price vector $\mathrm{p} \in \mathbf{R}^N$, points $\xi_1, \ldots, \xi_i \in \mathbf{R}^N$, and a vector η satisfying **E1**, **E3**, and

AE $\mathrm{p} \cdot \eta > -\varepsilon$ and $p \cdot y \leqslant 0$ for each $y \in Y$.

Alternatively, an economy has approximate equilibria if

$$\sup\{\mathrm{p}\cdot(\xi_1,\ldots,\xi_m):\xi_i\in D_i(\mathrm{p})\text{ for each }1\leqslant i\leqslant m\}=0.$$

In an approximate equilibrium each consumer maximises his utility while, in contrast, each firm only approximately maximises its profits. Why not demand that profit is maximised, and allow consumers' utility to deviate from the optimal? Our task is to construct a price vector p satisfying our equilibrium condition, for once this is done the ξ_i are given by Theorem 12.18. Thus it is **E2** which requires the construction of a fixed point, and hence is not possible constructively. So we are forced, at an approximate equilibrium, to allow firms to make losses; however, these losses can be made arbitrarily small, and a loss of one millionth of a cent, for instance, is no loss at all.

Let F_i denote the demand function on $(X_i, \succ_i)$. A subset S of a normed space is said to be a *convex cone* if $\lambda y \in S$ and $y + y' \in S$ whenever $y, y' \in S$ and $\lambda \geqslant 0$. We use S° to denote the interior of a subset S of a metric space.

We can now state our constructive version of McKenzie's theorem on the existence of a competitive equilibrium.

Theorem 12.24 *Suppose that*

(i) *each X_i is compact and convex;*
(ii) *each $\succ_i$ is continuous and uniformly rotund;*
(iii) *$(X_i \cap Y)^\circ$ is inhabited for each i;*
(iv) *Y is a closed, located convex cone;*
(v) *$Y \cap \{(x_1, \ldots, x_N) : x_i \geqslant 0$ for each $i\} = \{0\}$; and*
(vi) *for each $\mathrm{p} \in \mathbf{R}^N$ and each i, if $\sum_{i=1}^m F_i(\mathrm{p}) \in Y$, then there exists $\mathrm{x}_i \in X_i$ such that $\mathrm{x}_i \succ_i F_i(\mathrm{p})$.*

Then there exist approximate competitive equilibria.

Our result has two differences to the classical result proved by McKenzie [26]. First, we assume that the preference relations $\succ_i$ $(i = 1, \ldots, m)$ are uniformly rotund instead of merely strictly convex. This allows us to construct Bishop continuous demand functions for each consumer. Secondly, we assert only the existence of an approximate competitive equilibrium.

Our proof follows the standard classical proof via Kakutani's fixed-point theorem (see [26]) as closely as possible; typical of constructive mathematics, it has a distinctly geometric character.

The *polar* of a subset S of $\mathbf{R}^N$ is the set

$$S^{\mathrm{pol}} = \left\{\mathrm{p} \in \mathbf{R}^N : \mathrm{p}\cdot\mathrm{x} \leqslant 0 \text{ for all } \mathrm{x} \in S\right\}.$$

In general the statement

the polar of the polar of a set is equal to its convex conic closure

is equivalent to the law of excluded middle, but it does hold constructively for closed, located convex cones.

Proposition 12.25 *Let S be a closed, located convex cone in $\mathbf{R}^N$. Then the polar of the polar of S equals S.*

Proof Let S be a closed, located convex cone in $\mathbf{R}^N$ and let $x \in \mathbf{R}^N$ be such that $d = \rho(x, S) > 0$. By the Riesz representation theorem and the separation theorem [14, Theorems 4.3.6 and 5.2.9] there exists $a \in \mathbf{R}^N$ such that $a \cdot x > a \cdot s + d/2$ for all $s \in S$. Since S is a convex cone, for each $r \geqslant 0$ we have $a \cdot x > a \cdot rs + d/2$ and thus

$$a \cdot s < \frac{a \cdot x - d/2}{r}$$

for all $s \in S$. Letting $r \to \infty$ we see that $a \cdot s \leqslant 0$ for all $s \in S$; that is, $a \in S^{\mathrm{pol}}$. Since $0 \in S$, $a \cdot x > d/2 > 0$ and thus $x \notin \left(S^{\mathrm{pol}}\right)^{\mathrm{pol}}$. Hence if $x \in \left(S^{\mathrm{pol}}\right)^{\mathrm{pol}}$, then $\rho(x, S) = 0$ and, since S is closed, $x \in S$. The converse is straightforward. □

For each i we fix $\overline{\xi}_i \in (X_i \cap Y)^\circ$ and let $\overline{\xi} = \sum_{i=1}^m \overline{\xi}_i$; without loss of generality, each term of $\overline{\xi}$ is non-zero. The proof of Theorem 12.24 proceeds by an application of our approximate version of Kakutani's fixed-point theorem (Theorem 12.23) to the set

$$P = \left\{\mathrm{p} \in Y^{\mathrm{pol}} : \mathrm{p} \cdot \overline{\xi} = -1\right\}$$

of normalised price vectors. First, however, we require a number of lemmas. For the remainder of this subsection we assume that the hypotheses of Theorem 12.24 hold.

Lemma 12.26 *If $\mathrm{y} \in Y^\circ$, then $\mathrm{p} \cdot \mathrm{y} < 0$ for all non-zero $\mathrm{p} \in Y^{\mathrm{pol}}$. Moreover,* $\sup\{\mathrm{p} \cdot \mathrm{y} : \mathrm{p} \in Y^{\mathrm{pol}}, \|p\| = 1\} < 0$.

Proof Let p be a non-zero element of Y^{pol}; pick $1 \leqslant i \leqslant N$ such that $p_i \neq 0$, and fix $r > 0$ such that $\overline{B}(\mathrm{y}, r) \subset Y$. Then

$$\mathrm{y}' \equiv \mathrm{y} + (\mathrm{sign}(p_i)r)\mathrm{e}_i \in Y,$$

where $\mathbf{e}_i$ is the ith basis vector. Hence

$$\mathrm{p} \cdot \mathrm{y} < \mathrm{p} \cdot \mathrm{y} + |rp_i| = \mathrm{p} \cdot \mathrm{y}' \leqslant 0.$$

If $\|p\| = 1$, then we may suppose that $|p_i| > 1/2\sqrt{N}$; thus $\mathrm{p} \cdot \mathrm{y} < -|rp_i| < -r/2\sqrt{N}$. □

Let S be a subset of a metric space X. The *complement* of S is

$$\sim\!S = \{x \in X : x \neq s \text{ for all } s \in S\}.$$

If S is located, then the *apartness complement* of S is the set

$$-S = \{x \in X : \rho(x, S) > 0\}.$$

Lemma 12.27 *For each i the demand function F_i for X_i maps into $\sim\!(Y^\circ)$.*

Proof Suppose that $F(\mathrm{p}) \in Y^\circ$. Then, by Lemma 12.26, $\mathrm{p} \cdot F(\mathrm{p}) < 0$, which contradicts Theorem 12.18. □

Let C be a located, convex subset of a Banach space X. Then for each $\xi \in C^\circ$ and each $z \in -C$ there exists a unique point $h(\xi, z)$ in the intersection of the *interval*

$$[\xi, z] = \{t\xi + (1-t)z : t \in [0, 1]\}$$

and the boundary ∂C of C; moreover, the mapping $(\xi, z) \rightsquigarrow h(\xi, z)$ – the *boundary crossing map* of C – is pointwise continuous on $C^\circ \times -C$ [14, Proposition 5.1.5]. The next lemma shows that, for a fixed $\xi \in C^\circ$, this mapping is uniformly continuous.

Lemma 12.28 *Let X be a bounded convex subset of $\mathbf{R}^N$ and let $\xi \in X^\circ$. Then the function $h : \mathbf{R}^N \to \overline{X}$ which fixes each point of $\overline{X}$ and sends $y \in \sim\!X$ to the unique intersection point of $[\xi, y]$ and ∂X is uniformly continuous.*

Proof Without loss of generality we suppose that $\xi = 0$. Let $N > 0$ be such that $X \subset B(0, N)$ and let $r > 0$ be such that $B(0, r) \subset X$. Since the function mapping a point $x \neq 0$ to the unique intersection point of

$$\mathbf{R}x = \{rx : r \in \mathbf{R}\}$$

and $\partial B(0, N)$ is uniformly continuous on $-B(0, r/2)$, it suffices to show that h is uniformly continuous on $\partial B(0, N)$.

Given $\delta > 0$, set

$$\begin{aligned}
\theta &= \cos^{-1}(1 - (\delta^2/2N^2)) \\
\beta &= \cos^{-1}(\delta/2N), \text{ and} \\
\alpha &= \sin^{-1}(r/N).
\end{aligned}$$

Define

$$\varphi(\delta) = \frac{\delta|\sin(\beta)|}{|\sin(\alpha + \theta)|}.$$

The function φ is constructed as a worst-case scenario given that X contains $B(0, r)$ and is strictly contained in $B(0, N)$; see Fig. 12.1.

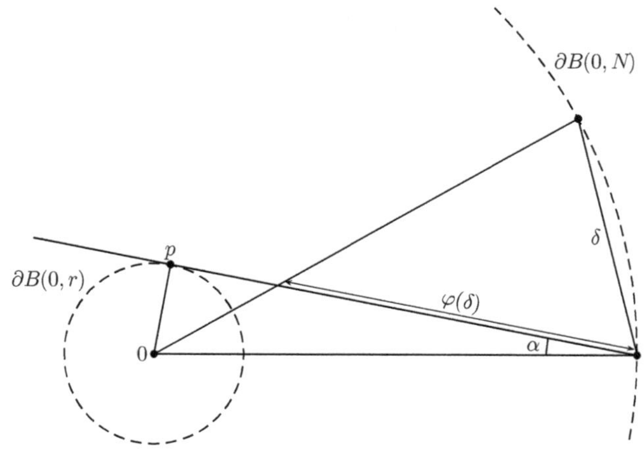

Figure 12.1

Fix $a, b \in \partial B(0, N)$ with $0 < \|a - b\| < \delta$, and let $x \in [0, a] \cap X$ and $y \in [0, b] \cap X$ such that $\|x - y\| > \varphi(\delta)$; without loss of generality, $\|x\| < \|y\|$. It suffices to show that it cannot occur that both $x, y \in \partial X$, for then the assumption that $\|h(x) - h(y)\| > \varphi(\delta)$ leads to a contradiction. By the construction of φ, the unique line passing through x and y must intersect $B(0, r)$. It follows that

$$x \in (\mathrm{conhull}\,(B(0, r) \cup \{y\}))^\circ \subset X^\circ,$$

where conhull(S) is the convex hull of S. Hence if $\|a - b\| < \delta$, then $\|h(a) - h(b)\| \leqslant \varphi(\delta)$.

It remains to show that for each $\varepsilon > 0$ we can find a $\delta > 0$ such that $\varphi(\delta) < \varepsilon$. From elementary calculations we have that

$$\begin{aligned}\varphi(\delta) &= \frac{\delta\sqrt{4N^2 - \delta^2}}{2r(1 - (\delta^2/2N^2)) + 2\delta\sqrt{(1 - (r^2/N^2))(1 - (\delta^2/2N^2))}} \\ &\leq \frac{\delta\sqrt{4N^2 - \delta^2}}{2r(1 - (r^2/2N^2)) + 2\delta(1 - (r^2/N^2))} \longrightarrow 0\end{aligned}$$

as $\delta \to 0$. □

Lemma 12.29 *Let X, Y be convex subsets of a normed space such that X, Y are both totally bounded, and $(X \cap Y)^\circ$ is inhabited. Then $X \cap Y$ is totally bounded.*

Proof Let $\xi \in (X \cap Y)^\circ$ and let $R > 0$; without loss of generality, $\xi \in B(0, R)$. Let $Y' = Y \cap B(0, R)$ and let h be the uniformly continuous function which fixes X and maps each point y in $-X$ to the unique point in $[\xi, z] \cap \partial X$. Fix $\varepsilon > 0$ and let $\delta \in (0, \varepsilon/4)$ be such that if $\|y - y'\| < \delta$, then $\|h(y) - h(y')\| < \varepsilon/4$. Let

$\{y_1, \ldots, y_k\}$ be a $\delta/2$-approximation of Y and partition $\{1, \ldots, k\}$ into disjoint sets P, Q such that

$$i \in P \Rightarrow \rho(y_i, X) < \delta;$$
$$i \in Q \Rightarrow \rho(y_i, X) > \delta/2.$$

If $i \in P$, then there exists $x \in X$ such that $\rho(x, y_i) < \delta$. Then

$$\|y_i - h(y_i)\| \leqslant \|y_i - x_i\| + \|x_i - h(y_i)\| < \varepsilon/4 + \varepsilon/4 = \varepsilon/2$$

and, since Y is convex, $h(y_i) \in X \cap Y$. The set

$$S = \{h(y_i) : i \in P\}$$

is an ε-approximation of $X \cap Y \cap B(0, R) = X \cap Y'$: fix $z \in X \cap Y$ and pick $1 \leqslant i \leqslant k$ such that

$$\|z - y_i\| < \delta/2.$$

Then $i \in P$, so $h(y_i) \in S$ and

$$\|z - h(y_i)\| \leqslant \|z - y_i\| + \|y_i - h(y_i)\| < \delta/2 + \varepsilon/2 < \varepsilon. \qquad \square$$

Lemma 12.30 *P is compact and convex.*

Proof It is straightforward to show that P is closed and convex; it just remains to show that P is totally bounded. By the bilinearity of the mapping $(\mathrm{p}, \mathrm{x}) \rightsquigarrow \mathrm{p} \cdot \mathrm{x}$, both Y^{pol} and

$$\{\mathrm{p} \in \mathbf{R}^N : \mathrm{p} \cdot \overline{\xi} = -1\}$$

are locally totally bounded. Since P is the intersection of these two sets, P is locally totally bounded, by Lemma 12.29. It remains to show that P is bounded: by Lemma 12.26,

$$M = \sup\{\mathrm{p} \cdot \mathrm{y} : \mathrm{p} \in Y^{\mathrm{pol}}, \|p\| = 1\} < 0.$$

Suppose that there exists $\mathrm{p} \in P$ such that $\|\mathrm{p}\| > -1/M$. Then $\mathrm{p}/\|\mathrm{p}\| \in Y^{\mathrm{pol}}$ and $(\mathrm{p}/\|\mathrm{p}\|) \cdot \overline{\xi} = -1/\|\mathrm{p}\| > M$ – a contradiction. $\square$

Lemma 12.31 *For each* $\mathrm{y} \in Y$ *and each* $r > 0$ *there exists* $\mathrm{p} \in P$ *such that* $\mathrm{p} \cdot \mathrm{y} > -r$.

Proof Fix $\mathrm{y} \in \partial Y$. Suppose that

$$\sup\{\mathrm{p} \cdot \mathrm{y} : \mathrm{p} \in P\} < 0;$$

this supremum exists since P is totally bounded and $(\mathrm{p}, \mathrm{x}) \rightsquigarrow \mathrm{p} \cdot \mathrm{x}$ is uniformly continuous. Then there exists $\mathrm{z} \in -Y$ such that $\mathrm{p} \cdot \mathrm{z} < 0$ for all $\mathrm{p} \in P$. But

$$\mathrm{z} \in P^{\mathrm{pol}} = (Y^{\mathrm{pol}})^{\mathrm{pol}} = Y.$$

This contradiction ensures that $\sup\{\mathrm{p}\cdot\mathrm{y} : \mathrm{p}\in P\} = 0$, from which the result follows. □

Lemma 12.32 *The composition of a weakly approximable mapping with a uniformly continuous function is weakly approximable.*

Proof The proof is straightforward. □

Lemma 12.33 *For a fixed $r > 0$ and for each $\mathrm{z} \in \partial Y$, define*

$$g_r(\mathrm{z}) = \{\mathrm{p} \in P : \mathrm{p}\cdot\mathrm{z} > -r\}.$$

Then $g_r(\mathrm{z})$ is inhabited and located for each $r > 0$, and g_r is weakly approximable.

Proof That $g_r(\mathrm{z})$ is inhabited for each z follows from Lemma 12.31. Fix $\varepsilon > 0$ and let $\delta > 0$ be such that for all $\mathrm{z},\mathrm{z}' \in \mathbf{R}^N$, if $\|\mathrm{z}-\mathrm{z}'\| < \delta$, then $\|\mathrm{p}\cdot\mathrm{z}-\mathrm{p}\cdot\mathrm{z}'\| < r/2$ for all $\mathrm{p} \in P$ – such a δ exists since the mapping $(\mathrm{p},\mathrm{x}) \rightsquigarrow \mathrm{p}\cdot\mathrm{x}$ is uniformly continuous and P is totally bounded. Let $\mathrm{z},\mathrm{z}' \in \mathbf{R}^N$ be such that $\|\mathrm{z}-\mathrm{z}'\| < \delta$ and let $\mathrm{p} \in g_{r/2}(\mathrm{z})$ and $\mathrm{p}' \in g_{r/2}(\mathrm{z}')$. For each $t \in [0,1]$, let $\mathrm{p}_t = t\mathrm{p} + (1-t)\mathrm{p}'$ and $\mathrm{z}_t = t\mathrm{z} + (1-t)\mathrm{z}'$. Then for all $t \in [0,1]$ we have

$$\begin{aligned}\mathrm{p}_t\cdot\mathrm{z}_t &= (t\mathrm{p} + (1-t)\mathrm{p}')\cdot(t\mathrm{z} + (1-t)\mathrm{z}')\\ &= t^2\mathrm{p}\cdot\mathrm{z} + t(1-t)(\mathrm{p}\cdot\mathrm{z}' + \mathrm{p}'\cdot\mathrm{z}) + (1-t)^2\mathrm{p}'\cdot\mathrm{z}'\\ &> -t^2 r/2 - 2t(1-t)r - (1-t)^2 r/2 \;=\; -r.\end{aligned}$$

Hence g_r is weakly approximable with respect to $g_{r/2}$. That $g_r(\mathrm{z})$ is located for each $\mathrm{z} \in \partial Y$ follows from [6, Theorem 4.9, p. 98] and the uniform continuity of the mapping $\mathrm{p} \rightsquigarrow \mathrm{p}\cdot\mathrm{y}$ on P. □

Finally we have the following.

Proof of Theorem 12.24 Let F_i be the demand function for the ith consumer and let

$$F = \sum_{i=1}^{m} F_i.$$

Fix $\varepsilon > 0$ and let $\delta > 0$ be such that for all $\mathrm{p} \in P$, if $\|\mathrm{x}-\mathrm{x}'\| < \delta$, then $\|\mathrm{p}\cdot\mathrm{x}-\mathrm{p}\cdot\mathrm{x}'\| < \varepsilon/2$. Set

$$m = \min\left\{\frac{\varepsilon}{2}, \frac{\delta}{\sup\{\|\overline{\xi}-\eta\| : \eta \in F(P)\}}\right\}.$$

For each $r > 0$, define a set-valued mapping Φ_r on P by

$$\Phi_r = g_r \circ h \circ F,$$

where h, g_r are as in Lemma 12.28 (for $X = \sum_{i=1}^{m} X_i$) and Lemma 12.33 respectively; Φ_r is well defined by Lemma 12.27. By Lemmas 12.28, 12.30, 12.32, 12.33, and Theorem 12.18, Φ_r is weakly approximable for each $r > 0$ and P is compact and convex. Using Theorem 12.23, construct $\mathrm{p} \in P$ such that

$$\mathrm{p} \in g_m \circ h \circ F(\mathrm{p}).$$

Set $\xi_i = F_i(\mathrm{p})$ for each i, and set $\eta = F(\mathrm{p})$. Then, by definition, the ξ_i satisfy condition **E1**, and η satisfies **E3**. Pick $t \in [0, 1)$ such that

$$\zeta \equiv h(F(\mathrm{p})) = t\overline{\xi} + (1 - t)\eta.$$

Since $\mathrm{p} \in g_m(\zeta)$ and $\eta \in Y$,

$$-m < \mathrm{p} \cdot \zeta = t\mathrm{p} \cdot \overline{\xi} + (1 - t)\mathrm{p} \cdot \eta \leqslant -t,$$

so $t < m$; whence $\|\zeta - \eta\| < \delta$. By our choice of δ, it follows that $\|\mathrm{p} \cdot \eta - \mathrm{p} \cdot \zeta\| < \varepsilon/2$. Thus $\mathrm{p} \cdot \eta > -\varepsilon$, so **AE** is satisfied. □

Classically we can recover the existence of an exact competitive equilibrium in the conclusion of Theorem 12.24. First, repeatedly apply Theorem 12.24 to construct sequences $(\mathrm{p}_n)_{n\geqslant 1}, (\xi_{1,n})_{n\geqslant 1}, \ldots, (\xi_{m,n})_{n\geqslant 1}, (\eta_n)_{n\geqslant 1}$ in $\mathbf{R}^N$ such that $\mathrm{p}_n, \xi_{1,n}, \ldots, \xi_{m,n}, \eta_n$ satisfy **E1, E3** and $\mathrm{p}_n \cdot \eta_n > -1/n$ for each n. With $m + 2$ applications of sequential compactness we can construct an increasing sequence $(k_n)_{n\geqslant 1}$ and points $\mathrm{p}, \xi_1, \ldots, \xi_m, \eta \in \mathbf{R}^N$ such that $\mathrm{p}_n \to \mathrm{p}, \xi_{i,n} \to \xi_i$ $(1 \leqslant i \leqslant m)$, and $\eta_n \to \eta$ as $n \to \infty$. The continuity of the demand functions, the inner product, and summation ensure that $\mathrm{p}, \xi_1, \cdots, \xi_m, \eta \in \mathbf{R}^N$ is a competitive equilibrium.

By Proposition 12.15, if we assume Brouwer's full fan theorem, then we can replace (ii) in our approximate version of McKenzie's existence theorem (Theorem 12.24) with the classical assumption that each $\succ_i$ is continuous and strictly convex. In this case we have the same hypothesis as the classical result, having only weakened the conclusion from the existence of a competitive equilibrium to the existence of approximate competitive equilibria.

12.5 Game Theory

In this final section we prove approximate versions of two of the foundational results of game theory: von Neumann's minimax theorem and the existence of Nash equilibria for a finite game with mixed strategies.

12.5.1 The Minimax Theorem

Kakutani's first application of his fixed-point theorem was to give a simple proof of a generalisation of von Neumann's minimax theorem, which guarantees the existence of saddle points for particular functions and is a fundamental result in game theory. Since the minimax theorem implies LLPO (see [11]), we construct only approximate saddle points.

Theorem 12.34 *Let $f : [0,1]^n \times [0,1]^m \to \mathbf{R}$ be a continuous function such that for each $x_0 \in [0,1]^n$, each y_0 in $[0,1]^m$, and each real number r the sets*

$$\{y \in [0,1]^m : f(x_0, y) \leqslant r\} \text{and}$$
$$\{x \in [0,1]^n : f(x, y_0) \geqslant r\}$$

are convex. Then

$$\sup_{x\in[0,1]^n} \inf_{y\in[0,1]^m} f(x,y) = \inf_{y\in[0,1]^m} \sup_{x\in[0,1]^n} f(x,y).$$

Throughout the remainder of this subsection we fix a uniformly continuous function $f : [0,1]^n \times [0,1]^m \to \mathbf{R}$ satisfying the conditions of Theorem 12.34, and for each $\varepsilon > 0$ we set

$$V_\varepsilon = \left\{ (x_0, y_0) \in [0,1]^n \times [0,1]^m : f(x_0, y_0) \leqslant \inf_{y\in[0,1]^m} f(x_0, y) + \varepsilon \right\},$$

$$W_\varepsilon = \left\{ (x_0, y_0) \in [0,1]^n \times [0,1]^m : f(x_0, y_0) \geqslant \sup_{x\in[0,1]^n} f(x, y_0) - \varepsilon \right\}.$$

In order to prove the minimax theorem, we extend, in the obvious way, the definitions of approximable, weakly approximable with respect to, and weakly approximable to functions which take points from a metric space X to subsets of a second metric space Y; we call such a function a *set-valued mapping from X into Y*. We associate V_ε and W_ε with the set-valued mappings given by

$$V_\varepsilon(x) = \{y \in [0,1]^m : (x,y) \in V\} \text{ and } W_\varepsilon(y) = \{x \in [0,1]^n : (y,x) \in W\};$$

note that $V_\varepsilon, W_\varepsilon$ are convex valued. For set-valued mappings U_i $(i = 1, 2)$ from X_i into Y_i $(i = 1, 2)$, the *product* of U_1 and U_2, written $U_1 \times U_2$, is the set-valued mapping from $X_1 \times X_2$ to $Y_1 \times Y_2$ given by

$$U_1 \times U_2(x_1, x_2) = U_1(x_1) \times U_2(x_2).$$

We omit the straightforward proof of the next lemma.

Lemma 12.35 *Let U_i be a set-valued mapping from X_i into Y_i $(i = 1, 2)$. If U_1, U_2 are (weakly) approximable, then $U_1 \times U_2$ is (weakly) approximable, and if U_1, U_2 are weakly approximable with respect to V_1, V_2 respectively, then $U_1 \times U_2$ is weakly approximable with respect to $V_1 \times V_2$.*

Lemma 12.36 *For each $\varepsilon > 0$, V_ε is weakly approximable with respect to $V_{\varepsilon/2}$ and W_ε is weakly approximable with respect to $W_{\varepsilon/2}$.*

Proof We only give the proof for V_ε; the proof for W_ε is entirely analogous. Since f is uniformly continuous, there exists $\delta > 0$ such that $(V_{\varepsilon/2})_\delta$ is contained in V_ε. Let x, x' be points of $[0,1]^n$ such that $\|x - x'\| < \delta$ and fix y, y' such that $(x,y), (x',y') \in V_{\varepsilon/2}$. Then $(\overline{x}, y'), (\overline{x}, y) \in V_\varepsilon$, where $\overline{x} = (x + x')/2$. Since $V_\varepsilon(\overline{x})$ is convex valued, $u_t = ty + (1-t)y' \in V_\varepsilon$ for each $t \in [0,1]$; whence

$$\begin{aligned}\rho((\mathrm{z}_t, \mathbf{u}_t), G(u)) &\leqslant \rho((\mathrm{z}_t, \mathbf{u}_t), \{\overline{x}\} \times V_\varepsilon(\overline{x})) \\ &= \rho(tx + (1-t)x', \overline{x}) \; < \; \delta.\end{aligned}$$

Since δ can be chosen to be arbitrarily small, this completes the proof. □

Here is the proof of Theorem 12.34.

Proof of Theorem 12.34 Let $f : [0,1]^n \times [0,1]^m \to \mathbf{R}$ be as in the statement of the theorem. It is easy to see that

$$\sup_{x \in [0,1]^n} \inf_{y \in [0,1]^m} f(x,y) \leqslant \inf_{y \in [0,1]^m} \sup_{x \in [0,1]^n} f(x,y).$$

Fix $\varepsilon > 0$ and let $\delta > 0$ be such that $\|f(x,y) - f(x',y')\| < \varepsilon/4$ whenever $\|(x,y) - (x',y')\| < \delta$. By Lemmas 12.35 and 12.36 the set-valued mapping U on $[0,1]^{n+m}$ given by

$$U = W_{\varepsilon/2} \times V_{\varepsilon/2}$$

is approximable with respect to $W_{\varepsilon/4} \times V_{\varepsilon/4}$. By Theorem 12.23, there exists $(x_0, y_0) \in [0,1]^{n+m}$ such that $\rho((x_0,y_0), U(x_0,y_0)) < \delta$. It follows from the definition of U and our choice of δ, that

$$\begin{aligned} f(x_0,y_0) &< \inf_{y \in [0,1]^m} f(x_0, y) + \varepsilon, \text{ and} \\ f(x_0,y_0) &> \sup_{x \in [0,1]^n} f(x, y_0) - \varepsilon. \end{aligned}$$

Hence

$$\begin{aligned}\inf_{y \in [0,1]^m} \sup_{x \in [0,1]^n} f(x,y) &\leqslant \sup_{x \in [0,1]^n} f(x_0, y_0) \\ &< f(x_0,y_0) + \varepsilon \\ &< \inf_{y \in [0,1]^m} f(x_0, y) + 2\varepsilon \\ &\leqslant \sup_{x \in [0,1]^n} \inf_{y \in [0,1]^n} f(x,y) + 2\varepsilon.\end{aligned}$$

Since $\varepsilon > 0$ is arbitrary, this completes the proof. □

12.5.2 Nash Equilibria

We consider a *finite game* with players $1, \ldots, n$ each with a finite set S_i of *pure strategies* and a *payoff* function $u_i : S_1 \times \cdots \times S_n \to \mathbf{R}$. A *mixed strategy* for player i is a distribution over $S_i = \{s_1^i, \ldots, s_{k_i}^i\}$; we represent a given mixed strategy by

$$\mathrm{p}^i = (p_1^i, \ldots, p_{k_i}^i) \in \triangle_{k_i},$$

where p_j^i is the probability of selecting strategy s_j^i. We can then extend the payoff functions to the set

$$\Sigma = \triangle_{k_1} \times \cdots \times \triangle_{k_n}$$

of mixed-strategy profiles by setting

$$u_i(\mathrm{p}^1, \ldots, \mathrm{p}^n) = \sum_{i_1, \ldots, i_n} \left(\prod_{j=1}^n p_{i_j}^j \right) u_i(s_{i_1}^1, \ldots, s_{i_n}^n),$$

where the summation is over all combinations of $1 \leqslant i_j \leqslant k_j$ for $j = 1, \ldots, n$. To simplify later arguments we give Σ the norm

$$\|\sigma - \sigma'\| = \max\{\|\sigma[i] - \sigma'[i]\| : 1 \leqslant i \leqslant n\},$$

where $\sigma[i]$ denotes the ith element of σ.

A *Nash equilibrium* is $\sigma \in \Sigma$ such that for each i and all $w \in \triangle_{k_i}$, $u_i(\sigma) \geqslant u_i(\sigma[w, i])$, where $\sigma[w, i] \in \Sigma$ is given by replacing the ith element of σ with w. Nash proved that any finite game with mixed strategies has a Nash equilibrium; however, constructively Nash's existence theorem implies LLPO.

To show this, let $a \in \mathbf{R}$ and consider the mixed-strategy one-player game with $S_1 = \{s_1, s_2\}$, $u_1(s_1) = 0$, and $u_1(s_2) = a$. A Nash equilibrium of this game corresponds to $t \in [0, 1]$ such that $ta = \max\{0, a\}$. Either $t > 0$, in which case $a \geqslant 0$, or $t < 1$, in which case $a \leqslant 0$. Hence the existence of Nash equilibria implies $\forall_{a \in \mathbf{R}}(a \leqslant 0 \vee a \geqslant 0)$, which in turn implies LLPO.

An *ε-approximate Nash equilibrium* is $\sigma \in \Sigma$ such that $u_i(\sigma) \geqslant u_i(\sigma[w, i]) - \varepsilon$ for each i and all $w \in \triangle_{k_i}$. A finite game is said to have *approximate Nash equilibria* if it has ε-approximate Nash equilibria for all $\varepsilon > 0$. We give a constructive proof of the existence of approximate Nash equilibria.

Lemma 12.37 *Each payoff function u_i is Lipschitz.*

Proof For any $\mathrm{p}, \mathbf{q} \in \Sigma$,

$$\left|u_i(\mathrm{p}^1, \ldots, \mathrm{p}^n) - u_i(\mathbf{q}^1, \ldots, \mathbf{q}^n)\right| = \sum_{i_1, \ldots, i_n} \left(\prod_{j=1}^n p_{i_j}^j - \prod_{j=1}^n q_{i_j}^j \right) u_i(s_{i_1}^1, \ldots, s_{i_n}^n)$$

$$\leqslant M \sum_{i_1,\dots,i_n} \left(\prod_{j=1}^{n} p^j_{i_j} - \prod_{j=1}^{n} q^j_{i_j} \right),$$

where $M = \max\{u_i(s_1, \dots, s_n) : s_j \in S_j\}$. It follows that u_i is Lipschitz since each $p_j, q_j \in [0, 1]$ $(1 \leqslant j \leqslant n)$. □

For each player i we define a family of functions $r^i_\alpha : \Sigma \to \mathcal{P}^*(\triangle_{k_i})$, for $\alpha \geqslant 0$, by

$$r^i_\alpha(\sigma) = \{v \in \triangle_{k_i} : \forall_{w \in \triangle_{k_i}} (u_i(\sigma[v, i]) \geqslant u_i(\sigma[w, i]) - \alpha)\}.$$

Lemma 12.38 *For each player i, each mixed-strategy profile $\sigma \in \Sigma$, and each $\alpha > 0$, $r^i_\alpha(\sigma)$ is inhabited and convex.*

Proof Each $r^i_\alpha(\sigma)$ is inhabited since u_i is uniformly continuous and is convex since

$$u_i(\sigma[tv + (1 - t)v', i]) = tu_i(\sigma[v, i]) + (1 - t)u_i(\sigma[v', i])$$

for all $t \in [0, 1]$. □

We define a set-valued mapping $r_\alpha : \Sigma \to \mathcal{P}^*(\Sigma)$ by

$$r_\alpha : \Sigma \to \mathcal{P}^*(\Sigma) = r^1_\alpha \times \cdots \times r^n_\alpha.$$

To construct approximate equilibria, we will apply the approximate Kakutani fixed-point theorem to r_α.

Lemma 12.39 *For each $\alpha > 0$, r_α is approximable.*

Proof Fix α and $\varepsilon \in (0, 1)$; pick $\xi \in (0, \alpha\varepsilon)$ and let $\delta > 0$ be such that if $||\sigma - \sigma'|| < \delta$, then $|u_i(\sigma) - u_i(\sigma')| < \xi$. It suffices to show that r^i_α is approximable for each i. To this end, let u, u', σ, σ' be such that $u \in r^i_\alpha(\sigma), u' \in r^i_\alpha(\sigma')$, and $||\sigma - \sigma'|| < \delta$; let $t \in [0, 1]$ and set $u_t = tu + (1 - t)u'$ and $\sigma_t = t\sigma - (1 - t)\sigma'$. Then for any w we have

$$\begin{aligned} u_i(\sigma_t[u_t, i]) &= u_i(\sigma_t[tu - (1 - t)u', i]) \\ &= tu_i(\sigma_t[u, i]) + (1 - t)u_i(\sigma_t[u', i]) \\ &> tu_i(\sigma[u, i]) + (1 - t)u_i(\sigma'[u', i]) - \xi \\ &\geqslant tu_i(\sigma[w, i]) + (1 - t)u_i(\sigma'[w, i]) - \alpha - \xi \\ &= u_i(\sigma_t[w, i]) - (\alpha + \xi). \end{aligned}$$

Now let $v \in r^i_{\alpha-\xi/\varepsilon}(u_t)$ and set $v' = \varepsilon v + (1-\varepsilon)u_t$. For all w we have

$$\begin{aligned} u_i(\sigma_t[v', i]) &= \varepsilon u_i(\sigma_t[v, i]) + (1-\varepsilon)u_i(\sigma_t[u_t, i]) \\ &> \varepsilon(u_i(\sigma_t[w, i]) - (\alpha - \xi/\varepsilon)) + (1-\varepsilon)(u_i(\sigma_t[w, i]) - (\alpha + \xi)) \\ &> u_i(\sigma_t[w, i]) - \alpha; \end{aligned}$$

whence $v' \in r^i_\alpha(\sigma_t)$. Since

$$||v' - u_t|| \;=\; \varepsilon||v - u_t|| \;\leqslant\; \varepsilon,$$

we have that $\rho((u_t, \sigma_t), G(r^i_\alpha)) \leqslant \varepsilon$. This completes the proof. □

Theorem 12.40 *Every finite game with mixed strategies has approximate Nash equilibria.*

Proof Choose $\delta > 0$ such that for all $\sigma, \sigma' \in \Sigma$, if $\|\sigma - \sigma'\| < \delta$, then $\|u_i(\sigma) - u_i(\sigma')\| < \varepsilon/2$ for each i. Applying the approximate Kakutani fixed-point theorem (Theorem 12.22) to $r_{\varepsilon/2}$ we construct $\sigma \in \Sigma$ such that $\rho(\sigma, G(r_{\varepsilon/2})) < \delta$. It follows from our choice of δ, that $\sigma \in r_\varepsilon(\sigma)$ and hence that σ is an ε-approximate Nash equilibrium. □

References

[1] Aczel, P., and Rathjen, M. J. 2010. *Notes on Constructive Set Theory.* `http://www1.maths.leeds.ac.uk/~rathjen/book.pdf`.

[2] Alps, R. A., and Bridges, D. S. *Morse Set Theory as a Foundation for Constructive Mathematics.* (Monograph in preparation.)

[3] Arrow, K. J., and Debreu, G. 1954. The existence of an equilibrium for a competitive economy. *Econometrica*, **22**(3), 265–290. `https://doi.org/10.2307/1907353`.

[4] Arrow, K. J., and Hahn, F. H. 1971. *General Competitive Analysis.* Edinburgh: Oliver and Boyd.

[5] Bishop, E. 1967. *Foundations of Constructive Analysis.* New York: McGraw-Hill.

[6] Bishop, E., and Bridges, D. S. 1985. *Constructive Analysis.* Grundlehren der Math. Wiss. 279. Heidelberg: Springer-Verlag.

[7] Bridges, D. S. 1982. Preference and utility – a constructive development. *J. Math. Econ.*, **9**, 165–185.

[8] Bridges, D. S. 1989. The constructive theory of preference relations on a locally compact space. *Proc. Koninklijke Nederlandse Akad. Wetenschappen (Indag. Math.)*, **92**(2), 141–165.

[9] Bridges, D. S. 1992. The construction of a continuous demand function for uniformly rotund preferences. *J. Math. Econ.*, **21**, 217–227.

[10] Bridges, D. S. 1994. The constructive theory of preference relations on a locally compact space – II. *Math. Soc. Sci.*, **27**, 1–9.

[11] Bridges, D. S. 2004. First steps in constructive game theory. *Math. Logic Q.*, **50(4–5)**, 501–506.

[12] Bridges, D. S. 2020. The Arrow–Hahn construction in a locally compact metric space. Pages 281–299 of: Bosi, B., Campión, M., Candeal, J. C., and Indurain, E. (eds.), *Mathematical Topics on Representations of Ordered Structures and Utility Theory*. Studies in Systems, Decision, and Control 263. Cham: Springer Nature Switzerland AG.

[13] Bridges, D. S., and Richman, F. 1991. A recursive counterexample to Debreu's theorem on the existence of a utility function. *Math. Soc. Sci.*, **21**, 179–182.

[14] Bridges, D. S., and Vîta, L. S. 2006. *Techniques of Constructive Analysis*. New York: Springer-Verlag.

[15] Debreu, G. 1964. Continuity properties of Paretian utility. *Int. Econ. Review*, **5**, 285–293.

[16] Eilenberg, S. 1941. Ordered topological spaces. *Amer. J. Math.*, **63**, 39–45.

[17] Fleischer, I. 1961. Numerical representation of utility. *SIAM Journal*, **9**, 48–50.

[18] Hendtlass, M. 2011. The computational content of Walras' existence theorem. *Appl. Math. Comput.*, **217**(13), 6185–6191.

[19] Hendtlass, M. 2012. Fixed point theorems in constructive mathematics. *J. Log. Anal.*, **4**, 20 pages, paper 10. `https://doi.org/10.4115/jla.2012.4.10`.

[20] Hendtlass, M. 2016a. Constructing the demand function of a strictly convex preference relation. `https://doi.org/10.48550/arXiv.1611.02542`.

[21] Hendtlass, M. 2016b. Kakutani's fixed point theorem in constructive mathematics. `https://doi.org/10.48550/arXiv.1611.02531`.

[22] Hendtlass, M., and Miheisi, N. 2016. On the construction of general equilibria in a competitive economy. `https://doi.org/10.48550/arXiv.1611.02534`.

[23] Martin-Löf, P. 1984. *Intuitionistic Type Theory*. Notes by Giovanni Sambin of a series of lectures given in Padua, June 1980. Napoli: Bibliopolis.

[24] McKenzie, L. W. 1954. On equilibrium in Graham's model of world trade and other competitive systems. *Econometrica*, **22**(2), 147–161.

[25] McKenzie, L. W. 1959. On the existence of general equilibrium for a competitive market. *Econometrica*, **27**(1), 54–71.

[26] McKenzie, L. W. 1961. On the existence of general equilibrium for a competitive market: some corrections. *Econometrica*, **29**(1), 247–248.

[27] Orevkov, V. P. 1963. A constructive map of the square into itself which moves every constructive point. *Dokl. Aka. Nauk. SSSR*, **152**, 55–58.

[28] Rader, J. T. 1963. The existence of a utility function to represent preferences. *Rev. Econ. Stud.*, **30**, 229–232.

[29] Richman, F. 2001. Constructive mathematics without choice. Pages 199–205 of: Schuster, P., Berger, U., and Osswald, H. (eds.), *Reuniting the Antipodes – Constructive and Nonstandard Views of the Continuum*. Synthese Library 306. Alphen aan den Rijn, Netherlands: Kluwer.

[30] Richter, M. K. 1980. Continuous and semi-continuous utility. *Int. Econ. Rev.*, **21**, 293–299.

[31] Sondermann, D. 1980. Utility representations for partial orders. *J. Econ. Theory*, **23**, 183–188.

[32] Takayama, A. 1974. *Mathematical Economics*. Hinsdale, IL: The Dryden Press.

[33] Uzawa, H. 1962. Walras' existence theorem and Brouwer's fixed point theorem. *Econ. Stud. Q.*, **8**, 59–62.

13

A Leisurely Random Walk Down the Lane of a Constructive Theory of Stochastic Processes

Yuen-Kwok Chan

13.1 Stochastic Process, in a Nutshell

Contrary to the common practice of authors of mathematical papers, in this chapter I will sometimes use the first-person singular, to make it perfectly clear that when I make a silly remark or a plea of ignorance, I speak only for myself.

The present chapter is an overview of some recent research on constructive probability theory, as an outgrowth of the constructive mathematics[1] in the seminal text [4] and its subsequent expansion and revision [5].

Specifically, it is a brief and leisurely overview of the book [8]. That book is an attempt to build a comprehensive tool set for probability theory, at once constructive and in terms familiar to traditional probabilists, to provide a foundation sufficient for further constructive developments in any active research topic in probability theory, stochastic processes, and their applications.

Other works in constructive probability along the lines of Bishop are in [4], [17], and [19].

Predating these, the works by A. N. Kolmogorov, the grandfather of a rigorous theory of probability and stochastic processes, are all constructive in the sense of [4] and [5].[2] Other masters of the field, K. L. Chung and P. Levy, to name just two, are constructivists in practice if not in name.

The book by Chan [8] is also an attempt to revive the tradition of Kolmogorov and these other masters.

In addition, constructive theorems in probability permeate the more recent traditional literature, although in an unorganized manner and intertwined with nonconstructive results.

[1] The term "constructive mathematics" will be explained presently. In the meantime, the reader can simply take it to mean mathematics of the mental creation of objects and operations, as opposed to classical mathematics which is of the mental discovery of objects and relations.

[2] In addition, Kolmogorov made a formal system wherein the symbols and rules of inference are to be interpreted with constructive meanings, as opposed to Hilbert's meaning-free formalism. See [23].

The purpose of the present overview is for the reader to leave with some understanding and warm feeling for mathematics, constructive mathematics, and probability theory.

13.1.1 Basic Tools from Calculus and Linear Algebra

Probability is a vast subject which draws upon most, if not all, tools that mathematics can provide. The text [5] supplies most of the basic tools needed in a constructive theory of probability, including real analysis, metric spaces, and measure spaces. Other basic tools are, by and large, available from advanced-level undergraduate calculus and linear algebra, with minimal cleanup when needed.

One surprise is that rigorous classical proofs, let alone constructive proofs, of the Change-of-Integration-Variables Theorem for higher dimensions, are hard to find. After an admittedly hasty search, I found an acceptable proof only in [1]. Other text books at the undergraduate level or first-year graduate level are hand-waving, or rely, explicitly or implicitly, on theorems from axiomatic Euclidean geometry learned in high school, or treat only one or two dimensions. The proof in [1], which relies on a theory of Jordan measure, is complicated, and not obviously constructive (though it is). So I wrote a proof, using Lebesgue integration in the place of Jordan measure, with [1] as a template. Alas, what I wrote is just as complicated, if not more so. Eventually I put in the Appendix of [8] only what I actually use elsewhere in that book, in order to have an immediate and readable reference.

Subsequently, an analyst suggested the method of differential forms. The desired Change-of-Integration-Variables Theorem can indeed be found in the popular "baby Rudin" text book [20], with rigorous and elegant proofs using differential forms. Perhaps the theory of differential forms merits a going-over by some constructive analyst.

But I digress.

13.1.2 Basic Tools from Probability Theory

The basic tools from probability theory, the concepts and theorems regarding Chebychev's inequality, the Borel–Cantelli lemmas, conditional expectations, independence, characteristic functions, and normal distributions for one or higher dimension, have been well developed in classical texts such as [9], [12], and [13]. These tools have been gathered in Chapter 5 of [8], with a minimal amount of modifications just to remove any doubts about subsequent applications being on a firm constructive footing. The adaptation of materials in Chung's text book is especially easy, because his works are mostly constructive. As a probabilist once complained (or complimented), "Chung beats the epsilon-deltas to death."[3]

[3] The classical assumption of general existence of conditional expectations did require some additional thoughts and/or some workarounds.

Next, I will mention two superficial dissimilarities between the constructive measure theory in [4] and its traditional counterpart.

Daniell Integration

In typical traditional text books, one postulates a family of measurable subsets of the underlying set Ω of samples, defines a measure function P on this family, and builds up the families of measurable functions and integrable functions. Then a family L of integrable functions, and an integration function E thereon, are made from the measurable subsets and the measure function P.

Even though such a formulation is possible in a constructive theory, as shown in [6], the earlier book [4] adopted the much more convenient Daniell-integral approach. In this approach, one postulates a family L of integrable functions alongside its associated integration function E on this family, then builds up the families of measurable functions, and then defines a measure function P.

In the case where the measure of the underlying set Ω is one, the function P is called a *probability* measure function, and the function E is called an *expectation* function.

The end products, the integration function or measure function, from the two different approaches, are equivalent. The dissimilarity of the two approaches is superficial and moot.

Probabilists think of P and E as two sides of the same coin.[4] Incidentally, probabilists often refer to one and/or the other as a *distribution* when the underlying set Ω is endowed also with a metric. It is when Ω has a metric that the convenience of the Daniell-integral approach is most evident.

Sometimes I write the triple (Ω, L, E) in the place of Ω, to emphasize the roles of the family L of integrable functions and of the integration function E.

Functions Defined Only Almost Everywhere

A question arises. Mathematicians observe that functions defined constructively *everywhere* on the unit interval $[0, 1]$ seem always to be continuous. The space of continuous functions is too narrow to hold many interesting integrable functions. Surely we want a simple one-step function on the unit interval $[0, 1]$ to be Lebesgue integrable?

The easy way out, provided by Bishop and Bridges [4] and [5], is to admit integrable functions which are not necessarily defined everywhere on the underlying set Ω, but are defined on domain-subsets whose complements have measure zero. In short, it is to admit functions which are defined *almost everywhere* (a.e.). Luckily, probabilists are used to ignoring anything in a set with measure zero; a function

[4] No pun intended.

defined a.e. is as good as one defined on the entire underlying set Ω. As a practical matter, the dissimilarity is, again, moot to probabilists.

For the constructivist, this approach does add the small task of checking, whenever we create an integrable function from prior ones, that the condition of being defined a.e. is duly inherited. That small task is easily dispatched for all the standard methods of such creation.

13.1.3 Applications of Stochastic Process

In probability theory, a *complete sample* in a random experiment is modeled as a point in the underlying set Ω of all imaginable outcomes. The latter is endowed with a probability measure function P. This probability measure function P assigns a measure, called probability, to each measurable subset of Ω. It is required to assign probability one to the entire underlying set Ω.

Tossing a Fair Coin Once

The simplest random experiment is tossing a fair coin once. Then we can use the metric space consisting of the set $\{H, T\}$, or equivalently, $S \equiv \{0, 1\}$,[5] as the underlying set Ω, equipped with the probability measure P which assigns probability $P(\{x\}) = 2^{-1}$ to each $x \in S$. [6]

Mathematical Finance

A somewhat more interesting game is to toss a fair coin n times. At each toss, the partial observation of a sample is a point in S. A complete sample is a point in the set

$$\Omega \equiv S^n \equiv \underbrace{\{0,1\} \times \cdots \times \{0,1\}}_{n}$$

$$\equiv \{(\omega_1, \omega_2, \ldots, \omega_n) : \omega_i \in S \quad \text{for each } i = 1, \ldots, n\}$$

which consists of 2^n elements. Each sample $\omega \in \Omega$ is assigned the probability of $P(\{\omega\}) \equiv 2^{-n}$. Formally, we define the set $Q \equiv \{1, 2, \ldots, n\}$, and define a function

$$X : Q \times \Omega \to S$$

by $X(t, \omega) \equiv \omega_t$ for each $t \in Q$ and $\omega \equiv (\omega_1, \omega_2, \ldots, \omega_n) \in \Omega$. Then, for each $t \in Q$, the measurable function $X_t \equiv X(t, \cdot)$ on Ω is called a random variable (r.v.) with values in the state space S.[7] The set Q is called a parameter set, and Ω

[5] In this chapter, I will write the expression $x \equiv y$ to mean "x is defined as y," "x, which is defined as y," "x, which has been defined earlier as y," or any other grammatical variation depending on the context.
[6] Here, $x \in S$ is short for "x is an element of the set S" or "x in S."
[7] Thus $X_t(\omega) \equiv X(\cdot, \omega)$.

is called a sample space. There we have a simple example of a stochastic process. This model is considered a good one if each relevant subset of $\{X_t : t \in Q\}$ has the same statistical characteristics of its counterpart of real-life observables.

The reader may be amused to hear that, as late as 20 years ago, a model based on a linear transformation of this simple stochastic process of n coin-tosses, with coefficients calibrated to each day's Treasury bond market, was found to be adequate for the calculation of the fair trade prices of many interest-rate derivatives.[8] The payoff of a derivatives contract at maturity time $t = n$ would be a predetermined function f of realized values of a targeted set of market observables. With that in mind, the model computes, at the trading time $t = 1$, the expected value of this function f of one or more of the r.v.s X_t, and uses the expected value as a fair trade price.[9] In this conceptually simple model, the expected value, or the integral, is simply the average value.

Since then, this model has undergone continual refinements, by various modelers, for calibration to an expanding set of liquidly traded benchmark instruments, for accuracy and for speed of computation, sometimes with encouragements from irate traders.

Clinical Trials

Applications of stochastic processes to many other fields predate the above simple financial models by decades, although the reader of the present chapter may be amused to note that L. Bachelier (*c.* 1900) developed a theory of financial option pricing with a first mathematical model for the all-important Brownian motion process.

Two examples of these other applications are for decisions with not only financial, but literally life-and-death consequences. One is in the clinical trials of drugs.

The process leading from scientific discovery, to a successful FDA-approved medicine, to widespread use of a drug, is fraught with uncertainty, and is horrendously expensive in terms of time, financial investment, human resources, low probability of success, and missed opportunities.

The long chain of steps in the development of a drug involves a statistical design and a calculation of probabilities.

The public, acting through the FDA, does not want to approve a drug whose apparent efficacy in the trials may be merely due to blind luck. Hence a drug is not approved if, under the hypothesis that its population-wide statistics were the same as no treatment or existing treatment, the probability of the drug's good performance in the trials due to chance exceeds, say, 5 percent.[10] With that constraint, and

8 Examples of applications to mathematical finance are just that, for illustration and not as advice for investments.

9 Each party in the trade would demand or yield a bit commensurate with his or her labor and exposure to risk.

10 Fritz Scholz points out that there is a raging debate in statistics circles and beyond over the cutoff threshold of 5 percent.

after already a large investment and more to come, the sponsoring pharmaceutical company wants a design which minimizes the probability for the trials to show no improvement, even under the hypothesis that the drug would perform acceptably well on the wider population. With these two probabilities in mind, the statistician needs to design a procedure, modeled by a stochastic process, and then put on his or her probabilist's hat, and calculate the two probabilities under the respective hypotheses.[11] The stochastic process modeling the progression of the trials is a significant, albeit small, link in the scheme of things.

Airplane in Turbulence

Another problem which needs a stochastic process model is an airplane flying through turbulence. Interacting with turbulence are the airplane's aerodynamics and control dynamics. A headwind suddenly changing to a tailwind can be disastrous during takeoff or landing, because then the lift can suddenly disappear. The target probability bound for such disaster, of, say, no more than one in a million, cannot be meaningfully inferred from a limited number of actual flight tests.

The next best thing is a theory[12] that yields statistical characteristics of turbulence that match reasonably well with observations. A stochastic process can then be built by combining the statistical turbulence model with an analytic aerodynamics and control-dynamics model. Then a probability of disaster can be estimated from the stochastic process model.

13.1.4 Ingredients for a Stochastic Process

The aforementioned simple finance model contains all the essential features of a stochastic process: (i) an underlying set Ω, containing all the imaginable outcomes of the experiment, equipped with a probability measure function P, or, equivalently, an expectation/integration function E, (ii) a parameter set Q, usually, but not always, a time interval or a countable subset of a time interval, (iii) a state space S, which is usually a metric space, and is, in most cases, a locally compact metric space, (iv) an r.v. X_t on the probability space Ω, with values in S, representing a glimpse of the sample at time t, for each $t \in Q$, and (v) real-valued functions f of one or more of the r.v.s X_t, whose related expectations and probabilities can be computed. These real-valued functions f, through their dependence on the r.v.s X_t, are r.r.v.s.[13]

Oftentimes, it is neither necessary nor possible to obtain a complete sample ω; it is sufficient to compute the expectations of the r.r.v.s. For example, a financial contract might specify that the payout at settlement is the sample average of the last three days' bond prices. Then the model is to supply the expectation of $(X_{n-2} + X_{n-1} +$

[11] These two probabilities are, respectively, called the significance and the power of the test.
[12] Pioneered by Kolmogorov.
[13] The term real-valued r.v. is abbreviated as r.r.v.

$X_n)/3$. As far as the trader is concerned, the sample space Ω exists only in the imagination of the mathematicians, aka nerds. I was told that similarly constructive views are held by most physicists, that there is no reality other than the observables, and by some artists, that there is no message beyond the medium.[14]

We will come back to probability and stochastic processes after a discussion of general constructive mathematics.

13.2 Constructive Mathematics, in a Nutshell

13.2.1 Finite Search and Infinite Search

Chan [8] explains constructive mathematics by focusing on two commonly used theorems. The first, the principle of finite search, states that, given a finite sequence each of whose members is either a 0 or a 1, either all members of the sequence are equal to 0, or there exists a member which is equal to 1. We use this theorem because a finite search would determine the result.

The second theorem, which Chan [8] calls the principle of infinite search, states that, given an infinite sequence each of whose members is either a 0 or a 1, either all members of the sequence are equal to 0, or there exists a member which is equal to 1. The name "infinite search" is perhaps unfair, but it brings into sharp focus that the computational meaning of this theorem is not clear. The second theorem is akin to an infinite loop in computer programming, with no assurance of exit after some finite number of steps.

The following sections will elaborate.[15]

13.2.2 Constructive Math

Most mathematicians prefer a constructive proof, that is a proof using only the principle of finite search, to one using the principle of infinite search, but use the latter as a powerful tool to prove theorems, with the belief that a constructive proof will surface in whatever special cases needed.

Contrary to this belief, many classical theorems proved directly or indirectly via the principle of infinite search are actually equivalent to the latter, and, as such, do not have a constructive proof. A blatant example is the principle of infinite search itself; others are not as obvious. Oftentimes, not even the computational meaning of the theorems in question is clear.

[14] These views are, incidentally, akin to the constructive mathematician's view that a real number is no more and no less than a sequence of rational numbers that approximates it to arbitrary precision.

[15] For more comprehensive reviews of constructive math, see [18] and [22]. See also the many articles, some lesisurely, in Fred Richman's home page `http://math.fau.edu/richman/`

We[16] believe that, for constructive formulations and proofs, the most direct way is also the easiest way. We use only finite searches, and we quantify each mathematical object and each theorem, starting with natural numbers.

Each natural number is accompanied by an operation which can reduce it to its decimal representation[17] *such as* 1, 2, 3, *etc.*[18]

For example, we say that 2 + 11 is a natural number; we can use the operation of addition, as learned in elementary school, to calculate its decimal representation, namely 13.

Natural numbers are also sometimes called finite numbers, or simply numbers.

Definition 13.1 (Sequences) Now we define a finite *sequence* of n mathematical objects to mean an operation that can produce an object for each natural number k which is less than or equal to n. Similarly, an infinite sequence of objects means an operation which can produce an object for each natural number k.[19] The operation should be a clearly spelled-out finite procedure which can be carried out[20] by a mechanical, electronic, or human computer.

We bootstrap to other objects and operations. Starting out, the only mathematical objects we know are the natural numbers, and the only operation we know is addition thereof. Then, successively, we construct objects like integers, rational numbers, real numbers, functions, metric spaces, measure spaces, and so on, by a bootstrapping process on the previous ones. Concomitantly, we construct the operations of subtraction, multiplication, division, square roots, limits in metric spaces, probability measures, and so on. This constructive program is carried out for analysis in [4] and [5], covering topics ranging from calculus to commutative Banach algebras.

An apt analogy is the development of an expanding library of computer software. Each object is defined numerically, tracing back to the natural numbers. To prove that an object exists is to provide a finite procedure to calculate it.

The references cited in the introduction of the present article show that this systematic approach is not only possible, but fruitful. Other publications include [15].

16 By "we," I am referring to constructivists along the lines of [4], including myself.

17 Some prefer the unary representation, as in Peano's axioms.

18 Here I am describing what we mean by natural numbers and what we do with them. I am not defining them. Hence there is no circular definition.

19 Note that this definition of infinite sequence says nothing about the existence of "a number called infinity."

20 Carried out in principle, if not in practice, due to limitation of time and resources. I will touch upon this limitation later.

13.2.3 Classical Math

We do not claim that theorems whose proofs require the principle of infinite search are untrue or incorrect. They are correct and consistent derivations from commonly accepted axioms. Indeed, we often study such classical theorems alongside their constructive counterparts. The term "nonconstructive mathematics" is not meant to be derogatory. We use, in its place, the more positive term "classical mathematics."[21]

13.2.4 Constructivists Use Also Classical Math

Reports that constructive mathematicians reject classical math are mistaken.[22]

Given a classical theorem, we first ask what its computational meaning is. Sometimes the theorem is partly constructive, and partly not. We extract the constructive part for our library for future reference. Then we consider the possible computational interpretations, often more than one, for the other part. Finally, perhaps with additional assumptions, or with weaker but still useful conclusions, we give a constructive proof.

Sometimes all we need is an additional condition (plus some work to devise a computational procedure) which was previously obscure but, once called out, is obviously satisfied according to classical thinking. Then we have a happy occasion when we have a constructive theorem, and the classical mathematicians think that the constructive theorem is new and cost free in terms of additional assumptions.

Constructivists even find a double-negative result useful. It can inspire a constructive proof even though we cannot use it as one. At the same time, a classical proof that a certain object does not exist serves as a loud warning that we should not waste time in looking for a construction of the object.

To continue a previous analogy, the developer of a computational software library does not reject, but consults and adds to, mathematics.[23]

No, we don't reject classical math, but seek to enrich it.

13.2.5 Constructivists Use Logic Only in the Common Sense

Moreover, it is simply not true that constructive mathematicians use a different system of logic. We do use symbols and formulas for abbreviation and succinct communication, but only after explaining what these symbols and formulas mean.

[21] Bishop used also the term "idealistic mathematics" for "classical mathematics."

[22] Or taken out of context.

[23] As we proceed, we may also solve some problems new even in the classical sense.

The only logic we use is everyday logic; no formal system is needed as a *firm foundation.* [24]

13.2.6 Constructive Math is Easy to Understand by Classical Mathematicians

Since a constructively valid argument is also correct from the classical view point, a reader of the classical persuasion should have no difficulties understanding constructive proofs. Proofs using only finite searches are surely agreeable to any reader who is accustomed to infinite searches.

13.2.7 Constructivists' Use of Some Symbols and Terminology

Classical mathematicians sometimes are bothered by the usage of certain symbols and terms by constructivists. This occurs when a classical term or statement admits more than one computational interpretation, and the constructivist adopts those with a stronger computational meaning.

The Notion "Complement"

Consider the simple notion of the complement of a subset B of the unit interval $[0, 1]$. Classically, this is usually defined as the "subset B' consisting of those members a of $[0, 1]$ such that the assumption that a is a member of B leads to a contradiction." A second definition can be "the subset B_c consisting of those members a of $[0, 1]$ such that $|a - b| > 0$ for each member b of B." A third definition can be "a measurable subset B^c whose measurable indicator function 1_{B^c} is such that $1_{B^c} + 1_B = 1$ on a subset of probability 1." The first two are the same to a classical mathematician, because he or she, equipped with the principle of infinite search, can prove that they are equal.

The three notions can be defined in the more general setting of a metric space endowed with a probability measure. Chan [8] calls the three complements described above, respectively, the set-theoretic complement, the metric complement, and the measure-theoretic complement. To the constructivist, all three are acceptable – as long as it is clear which is being used whenever the term appears. But they mean different things to a constructivist.

In the book [8], both the second and third are used, sometimes to describe two objects in one theorem. That book being on probability, the term "measure-theoretic complement" is used more often. Hence it is abbreviated to "complement," unless otherwise specified. A constructivist writing a book on analysis might, instead,

[24] I consider myself a mathematician who is ignorant of the formalization of mathematics, classical or constructive. I do use symbols and statements as tools for communication with others. In that sense, I use a formal system, much as I drive a car to go places while ignorant of, nay intimidated by, the proper arrangement of things under the hood.

abbreviate the term "metric complement" to "complement," unless otherwise specified. If the book covers only natural numbers, or only rational numbers, the three notions are equivalent and the author might, like his or her classical colleagues, use the term "complement" to mean any one of the three.

The Symbol " $\neq$ "

A similar situation arises for the statement "$a \neq b$," where a, b are real numbers. A classical mathematician can take it to mean that "the assumption $a = b$ leads to a contradiction." To a constructive mathematician, there is a second interpretation of this statement, namely, "we can compute a positive integer n such that $|a - b| > 2^{-n}$." The two definitions are classically equivalent, but the second is constructively stronger; it imposes a more stringent requirement that we can compute an *apartness* of a and b. Because we use it most often, the second interpretation is the default meaning of the symbol $\neq$ in the context of a metric space. We set aside, not reject, the first definition in our discussions.

Another possible arrangement could be to use the symbol $\neq$ when the first meaning is intended, and use full verbal description in lieu of any symbols when the second meaning is intended. But that would be too cumbersome because the second is so prevalent in constructive mathematics.

A similar situation is for the relations "$<$" and "$>$"; a computation of some apartness is required.

Note that the interpretation with the stronger meaning is equivalent to the alternative interpretations, as far as a classical mathematician is concerned. Hence the constructive usage should be at least as satisfactory as the classical usage, even to the classical mathematician.

Again, note also that the resolved ambiguity does not arise, in the first place, for integers or rational numbers. It is moot; for integers, the two interpretations are used interchangeably in constructive math.

The Connective Word "or"

Now the plot thickens – but is getting boring. Classically the statement "a exists or b exists" can come in many guises. Oftentimes it is taken to mean that "if a does not exist, then b does." A second possible meaning is that "we have an operation which, when carried out, would produce the object a, or produce the object b, depending on the input objects." Note the computational tone and substance in the second interpretation. It is naturally preferred by the constructivist. Many classical mathematicians, who are used to the first interpretation, would be dismayed to hear that we actually mean the second.

But, again, they don't need to hear it, because any constructive proof with the second interpretation of the word "or" is correct also when that word is taken to mean the first interpretation. This section and the next are for the benefit of the reader who wants to know *how* we do constructive math, as opposed to those readers who want only to *understand* constructive math after the latter is produced.

A Combo

To illustrate, consider the unit interval $[0, 1]$. Let B be its subset consisting of members x such that $x \leq \frac{1}{2}$. Consider the statement "there exists some x in B or there exists some y in the complement of B."

A constructivist would first ask the speaker of that statement to clarify "complement": (i) set-theoretic complement, (ii) metric complement, (iii) measure-theoretic complement, or (iv) something else, to be explained. Suppose said speaker replies "metric complement." Then he or she would be asked to clarify the intended meaning of "x exists": (i′) the assumption that x does not exist would lead to a contradiction, (ii′) there is a finite procedure to compute x to any given desired accuracy, or (iii′) something else, to be explained. Suppose the speaker selects (ii′). Then he or she would be prompted to make a similar selection also for "y exists." Then, finally the speaker would be asked about the intended meaning of the connective word "or" in the original statement.

Only then can the discussion be meaningfully continued. Each selection is respected as a legitimate choice.

A classical mathematician would regard all the selections at each step as equivalent; the declaration of selections is redundant and annoying. A constructivist author would therefore make the default selection of meanings for the reader, and make a loud and prominent declaration of this default selection at the beginning of each book or article, followed by subsequent re-declaration if and when a different meaning is intended in a later passage. This way, the long and distracting sequences of clarifications, page after page, are obviated.[25]

13.2.8 Recognizing Nonconstructive theorems

A quick way to test if a given theorem is constructive is to see if it implies the principle of infinite search. If it does, then the theorem, as given, does not have a constructive proof – and then there is the interesting task of examining it for constructivization of a part or the whole, or the task of finding a constructive substitute of the theorem which will serve all future purposes in its stead.

For example, consider the theorem which says that "for each real number a in the unit interval $[0, 1]$, either $a > 0$ or $a = 0$." A constructive proof of this theorem will not be possible because, as will be shown next, that would mean we can do an infinite search by means of a finite search, after all.

To elaborate, let us call this Theorem A. Suppose Theorem A is constructive. Let $k_1, k_2, \ldots$ be an arbitrary infinite sequence each of whose members is either a 0 or

[25] And we save ink, chalk, and annoyance.

a 1. Define the real number a to be the limiting value of the series $(k_1 2^{-1})$, $(k_1 2^{-1} + k_2 2^{-2}), \ldots$. Then the limiting value can be computed to any specified precision, and is in the unit interval $[0, 1]$. In other words, a is a well-defined real number in the unit interval $[0, 1]$.[26] Now, therefore, according to Theorem A, either (i) $a > 0$, or (ii) $a = 0$.

In the first case, we can find some positive integer n such that $a > 2^{-n}$. A finite search of the first n numbers in the given sequence $k_1, k_2, \ldots$ will therefore produce a member which is equal to 1.

In the second case, where $a = 0$, then it is impossible for any member of the given infinite sequence to be 1. Hence all numbers in the given sequence must be 0. Thus, by means of Theorem A, we would have reduced the principle of *infinite* search to the principle of *finite* search.

We conclude that Theorem *A* itself is not constructive.

The reader who is dismayed by the last sentence can be reassured that Theorem *A* has the constructive substitute Theorem *B* which says that "for each real number a in the unit interval $[0, 1]$, and for each positive real number ε, either $a > 0$ or $a < \varepsilon$." Theorem B simply recognizes the reality that any computation of the real number a should be allowed some arbitrarily prespecified positive inprecision ε.[27] Theorem B is adequate in constructive analysis.

To go one step further, any theorem that implies Theorem A cannot itself be constructive. This observation provides a second useful quick test.

13.2.9 Limitation of Constructive Math – Inattention to Computational Practicality

We prefer a finite construction to a nonconstructive existence proof. We would be happier to see a systematic and general development of mathematics which is not only constructive, but also computationally efficient. That admirable goal will, however, be left to abler hands.

Bishop pointed out, in that regard, that some impractical computations become practical if one compromises on accuracy. His example is the folly of counting the national debt in pennies. That exercise becomes practical when done in billions of dollars.[28]

I would also argue that ad hoc improvements for speed of computation are easier with constructive theorems than without.[29]

[26] As remarked in footnote [14], to the constructivist a real number is no more and no less than a sequence of rational numbers that approximates it to arbitrary precision. Such a sequence is called a Cauchy sequence.

[27] Engineers would call ε the "tolerance."

[28] The late Senator Everett Dirksen reportedly said, "A billion here and a billion there, and pretty soon you are talking real money." Fred Richman corrected my previous erroneous attribution of this quote to the late Senator Sam Ervin.

[29] An anonymous referee of the present chapter informs me that "the problem of computational efficiency of constructive mathematics is addressed mainly by Martin-Löf type-theorists, like Palmgren and Coquand, who

13.3 Stochastic Processes, in a Bigger Nutshell

Next is a brief view of the parts, then the wholes, of stochastic processes from the constructive perspective.

13.3.1 Parts of a Stochastic Process

State Space

We take, for the state space, a locally compact metric space (S, d), defined as a complete metric space in which every bounded subset of S is contained in some compact subset. The constructive study of locally compact metric spaces is well developed in [5]. Chan [8] adds some details and emphasis to the approximations of the locally compact metric space (S, d), and to the associated partition of unity. Very roughly speaking, a partition of unity is a sequence of finite subsets of the linear space of continuous functions of compact support which spans, in the limit, said linear space. It provides a convenient vehicle for the subsequent metrization of the space of distributions on (S, d).[30]

Sample Space

The sample space is a probability integration space Ω. In other words, it is a set Ω equipped with a probability measure function P for measurable subsets or, equivalently, a probability expectation function E defined on the linear space of integrable functions L .

Parameter Set

The parameter set Q is usually, but not always, a time interval or a countable subset of a time interval, with the Euclidean metric and natural ordering, so we can talk about continuous sample functions. Sometimes it is endowed with a measure itself so we can talk about measurable sample functions.

The Observable Random Variables and the Finite Joint Distributions

A stochastic process is a set $\{X_t : t \in Q\}$ of r.v.s with values in the state space (S, d). Of main interest are the expectations of real-valued functions f which depend on a finite or an infinite number of these r.v.s X_t.

We are given these expectations for the finite cases. We are to piece them together, to construct a stochastic process such that

$$Ef(X_{t(1)}, \ldots, X_{t(m)}) = F_{t(1),\ldots,t(m)} f \tag{13.1}$$

developed many parts of Bishop-style constructive analysis and algebra within formal Martin-Löf Type Theory, and also implemented their work, mainly in Agda or Coq . . . ." Incidentally, P. Martin-Löf is a thesis student of Kolmogorov.

30 Skorokhod first used partitions of unity to study probability measures and their convergence.

for each continuous function f of compact support on S^m, for each finite sequence $t_1, \ldots, t_m$ of points in Q , for each integer $m \geq 1$.[31]

The right-hand side of this equality are given expected values, perhaps previously inferred from actual observations in a large number of past experiments. Each given function $F_{t(1),\ldots,t(m)}$ has all the properties of a distribution on S^m, and is called a finite joint distribution, or f.j.d. for short.

The condition (13.1) can be extended to integrable functions. Thus

$$Ef(X_{t(1)}, \ldots, X_{t(m)}) = F_{t(1),\ldots,t(m)}f \tag{13.2}$$

for each function f of on S^m such that the function $f(X_{t(1)}, \ldots, X_{t(m)})$ is integrable, for each finite sequence $t_1, \ldots, t_m$ of points in Q, for each integer $m \geq 1$.

Note that a condition on the f.j.d.s on the right-hand side of Equation (13.2) is a condition on the left-hand side.

The Process

The set $\{X_t : t \in Q\}$ of r.v.s is equivalent to a function

$$X : Q \times (\Omega, L, E) \to (S, d)$$

defined by $X(t, \omega) \equiv X_t(\omega)$ for each point $(t, \omega) \in Q \times \Omega$ such that ω is in the domain of the r.v. X_t. Thus X is a function on $Q \times \Omega$ whose domain is a subset of $Q \times \Omega$, even as the function $X(t, \cdot) \equiv X_t$ is defined a.e. for each $t \in Q$.

We will call this function X a *random field*. In the background are the other ingredients Q, Ω, L, E, S, the integrable functions f in L, and the family of f.j.d.s.

In the usual cases where the parameter set Q is a subset of the real line R, the random field is called a *stochastic process*, or simply a *process*.

In the present chapter, all random fields will sometimes loosely be called stochastic processes.

One question we ask is, what kind of a family of f.j.d.s is acceptable as input data for the construction of a process with this or that property?

A Movie Analogy

To draw an analogy, the stochastic process is a movie, and each f.j.d. is an envelope containing a number of frames. Each envelope has some random characteristics. In other words, seeing several frames may give the viewer some idea about previous or subsequent frames, but, except for a very dull movie, does not tell the viewer about the unseen frames with any certainty.

Given a multitude of such envelopes, the movie producer's immediate task is to make a movie that tells the story. The important part of the production is the movie and the story.

[31] To lessen the burden on subscripts, I write t_j and $t(j)$ interchangeably.

In mathematical terms, the probabilist is given a family of f.j.d.s, with the task of putting them together to construct a stochastic process. The most important part of the constructed stochastic process is the function

$$X : Q \times (\Omega, L, E) \to (S, d).$$

Like some movie producers, a mathematician is not satisfied by producing the product only once; he or she would make a theorem which automatically produces the stochastic process from each admissible input family of f.j.d.s.

13.3.2 Construction of Stochastic Process from f.j.d.s

After a process X is constructed from admissible f.j.d.s, we can then compute the probabilities and expectations, if they exist, of a given function of all (not just a finite number) of the r.v.s in the set $\{X_t : t \in Q\}$. An example of such a function is $f \equiv \max_{t\in[0,1]} X_t$, provided that the constructed process $X : [0,1] \times \Omega \to R$ has the property that $X(\cdot, \omega)$ is uniformly continuious on $[0,1]$, for each $\omega \in \Omega$,

The computed probabilities help to answer the kind of questions in the example applications in Section 13.1 of the present chapter.

But what kind of input families of f.j.d.s are admissible?

Consistency of a Family of f.j.d.s

To continue the movie analogy, the multitude of envelopes containing the frames should be self-consistent. Imagine an envelope contains frames at time 3, 4, 7, and a second envelope contains frames at time $1, 3, 4, 7, 12$. Then frames 3, 4, 7 in the second envelope should look exactly like frames 3, 4, 7, respectively, in the first envelope.

An analogous mathematical condition is required on any input family of f.j.d.s. For a precise mathematical definition of this consistency condition, see any text books on stochastic processes, or see [8].

Continuity of a Family of f.j.d.s

A second condition can be continuity in probability. Assume that the parameter space Q is $[0,1]$. Roughly speaking in the movie analogy, this condition rules out jumps from one frame to the next at a predetermined time, for example, the hundredth frame, while still allowing jumps at unpredictable times due to chance developments in the story. For example, from one frame to the next, a scary character could suddenly appear and cause the hero's hair to shoot up, but the probability of such a scary scene at any predetermined time is arbitrarily small.

Mathematically, such a condition on the family of f.j.d.s is called *continuity in probability*, and is often imposed. It still allows the sample functions $X(\cdot, \omega)$ to be, individually, discontinuous functions, but the probability of $d(X_t, X_s) \geq \varepsilon$ is arbitrarily small if t and s are sufficiently close to each other, for any $\varepsilon > 0$.

Again, for a precise mathematical definition of continuity in probability, see any text book on stochastic processes, or see [8].

Continuity in probability is used in [8] as a default condition on the admissible families of f.j.d.s, because it is convenient and it allows all cases of interest.

13.4 Constructive Theory of Stochastic Processes, in an Even Bigger Nutshell

13.4.1 Existence, Construction, and Continuity of Construction

Classical Proofs of Existence of Stochastic Processes

An encyclopedic classical text book [10] proves the existence of processes corresponding to a given family of f.j.d.s which is consistent, provided that the parameter set contains a countable dense subset. The book [10] then studies the sample properties of important classes of processes with additional "nice" class-specific properties, including processes with countable parameter sets, almost surely continuous processes, processes with independent increments, Markov processes, martingales, and processes whose sample paths are right continuous with left limits.

Construction of Stochastic Processes

A slightly different tack is taken in [8]. First, just as in the classical development, the case is treated where the parameter set Q_0 is countable. In this case, there is a constructive theorem, the Daniell–Kolmogorov Theorem, which constructs a process $Z : Q_0 \times \Omega \to S$, from a consistent family of f.j.d.s.

Then, for the case of a more general parameter set Q, the following are assumed:

(i) the parameter set Q is a metric space containing a countable dense subset Q_0,

(ii) we are given a class of families of f.j.d.s which are consistent and continuous in probability, and

(iii) additional conditions specific to said class of families of f.j.d.'s

Then, for each family of f.j.d.s in a given class,

(iv) the Daniell–Kolmogorov Theorem is used to construct a process $Z : Q_0 \times \Omega \to S$,

(v) from the class-specific condition on the f.j.d.s, corresponding "nice" properties are inferred of the sample function $Z(\cdot, \omega)$ of the process Z, for each ω in Ω,

(vi) these nice properties of Z enable the extension of the sample functions $Z(\cdot, \omega)$ to a function $X(\cdot, \omega)$ on a domain subset of the full parameter set Q, for a.e. ω in Ω,

(vii) it is proved that the function X is a process with the family of f.j.d.s, and

(viii) it is proved that the process X inherits from its progenitor Z the "nice" class-specific properties.

Classic Publications on Stochastic Processes

Chan [8] uses [10] and other classical texts, including [2], [3], [7], [11] and [16] as sources and templates, and extracts, with acknowledgment, results which are constructive. Results and methods of Kolmogorov, Levy, and Chung, are used wherever possible.[32]

Continuity of Construction

A byproduct of the constructive approach in [8] is that the set of input families of f.j.d.s and the set of output processes can routinely be metrized, and that each construction is continuous relative to the metrics, in epsilon-delta terms. In short, each construction theorem is accompanied by a corresponding metrical continuity theorem.

Continuous dependence of stochastic processes on the input family of f.j.d.s is not always discussed in classical text books. When it is, it is usually in terms of sequential weak convergence of distributions, which is a weaker result than metrical continuity.

13.4.2 Sample Properties

After the construction of the stochastic process, the probabilist can now look at it as an organic whole, as opposed to the family of f.j.d.s which is a collection of finite sets of partial glimpses. We can start to talk about the "nice" sample properties which may involve an infinite set of the r.v.s X_t. We can discuss the movie, with its twists and turns and surprises, as opposed to a discussion of the finite sets of frames.

A simple example of sample properties is when the parameter set is $Q = \{1, 2, \ldots\}$. We can then ask: Can we find a condition on the f.j.d.s which is sufficient for the sequence X_n of r.v.s to converge *almost surely* (a.s.), in the sense that X_n converges as $n \to \infty$, for each sample ω in a set of probability one?

For a second example, assume $Q = [0, 1]$. We can then ask: Can we find a condition on the f.j.d.s which is sufficient for the construction of a process such that, for each sample ω in a set with probability one, the sample function $X(\cdot, \omega)$ is a continuous function on Q? A process with this property is said to be *a.s. continuous*.

[32] I, too, want to stand on the shoulders of giants.

Emphasis on Almost Uniform Sample Properties

Chan [8] asks these same questions with a computational emphasis. To elaborate only on the second example, define almost uniform (a.u.) sample continuity as follows.

Definition 13.2 (An a.u. continuous process) Suppose, for each $\varepsilon > 0$, there exist $\delta(\varepsilon) > 0$ and a *measurable* set D with $P(D^c) < \varepsilon$ such that

$$d(X(t,\omega), X(s,\omega)) \leq \varepsilon,$$

for each $s, t \in [0,1]$ with $|t - s| < \delta(\varepsilon)$, and each $\omega \in D$. Then the process X is said to be *a.u. continuous*, with the operation δ as a *modulus of a.u. continuity*.

In classical probability theory, this definition is equivalent to the a.s. continuity mentioned in Section 13.4.1. In a constructive theory, it is the default definition because it quantifies the continuity property of the process by a modulus of continuity. A modulus of a property/condition is simply a numerical description of that property/condition.[33]

Where classical probabilists talk about processes with this or that property a.s., constructivists talk about processes with this or that property a.u., with this or that modulus of the property.[34]

13.4.3 Examples of Processes, from the Constructive Perspective

Let (S, d) be a locally compact metric space, not necessarily a linear space or ordered. It serves as the state space for all the processes studied in [8].

Along the lines of Subsection 13.1.4 and using a locally compact metric space (S, d) as a state space, Chan [8] constructs and studies (i) processes with countable parameter sets, (ii) measurable random fields with a compact metric parameter space Q equipped with a measure, (iii) martingales, (iv) a.u. continuous processes on $[0,1]$ and on $[0,\infty)$, (v) a.u. càdlàg (continue à droite, limite à gauche) process on $[0,1]$ and on $[0,\infty)$, (vi) a.u. càdlàg strong Markov processes on $[0,\infty)$, (vii) a.u. càdlàg Feller processes on $[0,\infty)$, and (viii) Brownian motions on $[0,\infty)$ with values in R^m.

We conclude this chapter with the following brief discussion of the classes (i),(iii), (iv), and (v).

33 When informed that Juliet had love in her heart, Romeo, were he a constructivist, would ask "What is the modulus of that love?," meaning "How much?" or "How strong?".

34 Doob said that one can tell a measure theorist from a probabilist by noticing that the former says "almost everywhere" and the latter "almost surely." One might add that a constructive probabilist says "almost uniformly" and supplies a modulus.

Process with Countable Parameter Set

Thanks to the Daniell–Kolmogorov Theorem[35], a process $Z : Q_0 \times \Omega_0 \to S$, with an arbitrary given countable parameter set Q_0, can be constructed from each consistent family of f.j.d.s. This is the usual starting point to prove existence or construction of other processes.

Martingale

Roughly speaking, a martingale is a real-valued stochastic process X to model the accumulated fortune of a player in a fair game of chance, one bet after another. For simplicity of the present discussion, we will assume that $Q = \{1, 2, \ldots\}$. The player makes one bet at each time step $t \in Q$, resulting in accumulated fortune X_t at time $t \in Q$.

Doob first formulated the theory of martingales in terms of probability theory and, at once, developed it into a powerful tool with applications in probability theory and beyond.

Definition 13.3 (Definition of a martingale; a fair game) Mathematically, a process X is a *martingale* if, for each $t \in Q$, the the newly revised expectation of fortune X_{t+1} after the next bet is exactly the same as the current observed fortune X_t. The newly revised expectation is based on all information up to and including the current time and the current fortune X_t. In that sense, the process models a fair game.

One can ask what additional conditions on the f.j.d.s would guarantee that X_n converges a.u. as $n \to \infty$. In that regard, Doob proved the following theorem.

Theorem 13.4 (Classical martingale convergence theorem) *Suppose $E|X_n|$ converges to some real number K, as $n \to \infty$. Then X_n converges to some r.r.v. X a.s. as $n \to \infty$.*

Doob then shows how to deduce the classical Strong Law of Large Numbers (SLLN) from this martingale convergence theorem.

However, Chan [8] shows that the above theorem implies the principle of infinite search, and is therefore not constructive. Chan [8] proves a constructive substitute. To elaborate, define a special convex function $\overline{\lambda} : R \to R$ by

$$\overline{\lambda}(x) \equiv 2x + (e^{-|x|} - 1 + |x|)$$

for each $x \in R$. Chan [8] proves the following theorem.

Theorem 13.5 (Maximal inequality) *Let $\varepsilon > 0$ be arbitrary. Let $t_0 \le t_1 \le \cdots \le t_n$ be an arbitrary finite sequence in the parameter set Q. Suppose that*

[35] Or a version of it that I call the Daniell–Kolmogorov–Skorokhod Theorem in [8].

$$E\overline{\lambda}(X_{t(n)}) - E\overline{\lambda}(X_{t(0)}) < \frac{1}{6}\varepsilon^3 \exp(-3(E|X_{t(0)}| \vee E|X_{t(n)}|)\varepsilon^{-1}).$$

Then

$$P(\bigvee_{k=0}^{n} |X_{t(k)} - X_{t(0)}| > \varepsilon) < \varepsilon.$$

Note that the last probability bound ε is independent of the number n of terms in the sequence $t_0 \le t_1 \le \cdots \le t_n$. This maximal inequality is a modification of a version in [4], and was previously unknown. Doob's book [10] contains a maximal inequality for the L_p-integable case which, however, requires $p > 1$.[36]

From the maximal inequality involving the special function $\overline{\lambda}$, Chan [8] deduces, à la Doob, the following martingale convergence theorem.

Theorem 13.6 (Constructive martingale convergence theorem) *Suppose $E\overline{\lambda}(X_n)$ converges to some real number K, as $n \to \infty$. Then X_n converges to some r.r.v. X a.u. as $n \to \infty$, with some modulus of a.u. convergence δ.*

Classically, $E|X_n|$ converges if and ony if $E\overline{\lambda}(X_n)$ converges. Hence, classically, the constructive convergence theorem introduces no extra conditions. At the same time, the modulus of a.u. convergence δ, given in [8], is new. A classical mathematician may regard the modulus of a.u. convergence δ as cost-free.

Following Doob, Chan [8] deduces the SLLN from the martingale convergence theorem, the latter being constructive this time. Thus a constructive proof of the SLLN is obtained.[37]

A.u. Continuous Processes on Parameter Set $[0, 1]$

The a.u. continuity for processes with parameter set $Q = [0, 1]$ has been defined earlier. What is a condition on the f.j.d.s that would lead to an a.u. continuous process $X : [0, 1] \times \Omega \to (S, d)$?

Chan [8] specifies a necessary and sufficient condition: There is an operation δ which serves as a modulus of a.u. continuity for each subprocess $Z : \{t_1, \ldots, t_n\} \times \Omega \to (S, d)$ obtained by sampling only on a finite subset $\{t_1, \ldots, t_n\}$ of dyadic rationals in $[0, 1]$. A dyadic rational in [0; 1] means a real number of the form $i2^{-k}$ where i and k are nonnegative integers with $i \le 2^k$.

Note two important points. First, the modulus of a.u. continuity δ in the condition is independent of the finite subset $\{t_1, \ldots, t_n\}$ of sampling times. Second, being a condition on finite subsets $\{Z_{t(1)}, \ldots, Z_{t(n)}\}$ of the set $\{X_t : t \in [0, 1]\}$, the condition is indeed a condition on f.j.d.s.

This rather obvious condition is presented here, partly for comparison to the next example.

[36] The book [10] contains also a maximal inequality for the $L \log L$-integrable case.

[37] Separately, [9] contains a constructive proof of the SLLN, not using martingales.

The a.u. càdlàg Process on Parameter Set $[0, 1]$

The set of continuous functions on $[0, 1]$ is too narrow to hold all interesting functions, as pointed out previously.

Definition 13.7 (The càdlàg function) For that reason, Doob and others introduced the set of functions on $[0; 1]$ that are right continuous, with a left limit at each point $t \in [0, 1]$.[38] Chan [8] recasts this definition in stronger numerical terms, and, following common practice of probabilists, calls these functions by the French acronym càdlàg. The definition used in [8] is essentially due to Skorokhod [21], duly constructivized. Among other things, the implicit requirement that functions on $[0; 1]$ are defined everywhere is dispensed of. In analogy to continuity, the càdlàg property of a function is quantified by a modulus of càdlàg.

Skorokhod also introduced a metric d_D on the set of càdlàg functions. That, too, is constructivized in [8].

A prototypical example of a càdlàg function is the step function $f_{a,x,y}$ whose domain is $[0, a) \cup [a, 1]$, and whose values on $[0, a)$ and $[a, 1]$ are x and y in S, respectively. Here the real number $a \in [0, 1]$, and the elements $x, y \in (S, d)$ are arbitrary. The Skrokhod metric assigns a short distance $d_D(f_{a,x,y}, f_{a',x',y'})$ between two such simple càdlàg functions, $f_{a,x,y}$ and $f_{a',x',y'}$, if

$$\max\{|a - a'|, d(x, x'), d(y, y')\}$$

is small. Thus the Skorokhod metric recognizes the reality that measurements in time, just like measurements of distance in the space (S, d), can be imprecise, and that some allowance for the imprecision has to be made. The last displayed expression involving the maximum function would be sufficient for the prototypical step functions. The full formulation of the Skorokhod metric d_D needed in general is not presented here.

Definition 13.8 (An a.u. càdlàg process) In analogy to a.u. continuous processes, an a.u. càdlàg process can be defined. Suppose, for each $\varepsilon > 0$, there exist $\delta(\varepsilon) > 0$ and a *measurable* set D with $P(D^c) < \varepsilon$ such that

$$d_D(X(\cdot, \omega), X(\cdot, \omega)) \leq \varepsilon,$$

for each $\omega \in D$. Then the process X is said to be *a.u. càdlàg*, with the operation δ as a *modulus of a.u. càdlàg*.

In analogy to a.u. continuous processes, Chan [8] specifies a necesary and sufficient condition for a family of f.j.d.s to yield an a.u. càdlàg process. It also yields some simple sufficient conditions that are subsequently applied to construct a.u. càdlàg Markov processes from semigroups.

[38] See [2] and [3].

In this and the last subsection, the parameter interval is assumed to be $[0, 1]$. The results can easily be repeated for the definition and construction of a sequence of a.u. càdlàg processes on the intervals $[0, 1]$, $[1, 2]$, $[2, 3]$, ... which are then stitched together, to make an a.u. càdlàg process $X : [0; \infty) \times \Omega \to (S, d)$ with parameter set $[0, \infty)$.

13.5 Concluding Remarks

Constructive math is easy to understand. It does not reject classical math, but seeks to add computational content where missing. It uses only common-sense logic, and requires only a natural language for communication. A standard tool set has been developed constructively which is sufficient as a basis to do any research in probability theory in the constructive framework of [5], and in the constructive tradition of Kolmogorov and other masters of the field.

References

[1] Apostol, T. 1957. *Mathematical Analysis.* Boston, MA: Addison-Wesley.
[2] Billingsley, P. 1968. *Convergence of Probability Measures.* John Wiley & Sons.
[3] Billingsley, P. 1999. *Convergence of Probability Measures*, 2nd ed. John Wiley & Sons.
[4] Bishop, E., 1967. *Foundations of Constructive Analysis.* New York, San Francisco, St. Louis, Torronto, London, Sydney: McGraw–Hill.
[5] Bishop, E., and Bridges, D. 1985. *Constructive Analysis.* Berlin, Heidelberg, New York, Tokyo: Springer.
[6] Bishop, E., and Cheng, H. 1972. *Constructive Measure Theory*, AMS Memoirs no. **116**.
[7] Blumenthal, R. M., and Getoor, R. K. 1968. *Markov Processes and Potential Theory.* New York, London: Academic Press.
[8] Chan, Y. K. 2021. *Foundations of Constructive Probability Theory.* Encyclopedia of Mathematics and Its Applications, vol. 177. Cambridge: Cambridge University Press.
[9] Chung, K. L. 1968. *A Course in Probability Theory.* New York, Chicago, San Francisco, Atlanta: Harcourt, Brace & World.
[10] Doob, J. L. 1953. *Stochastic Processes.* John Wiley & Sons.
[11] Durrett, R. 1984. *Brownian Motion and Martingales in Analysis.* Wadsworth.
[12] Feller, W. 1971a. *An Introduction to Probability and its Applications*, vol. 1, 3rd ed. Wiley & Sons.
[13] Feller, W. 1971b. *An Introduction to Probability and its Applications*, vol. 2, 2nd ed. John Wiley & Sons.

[14] Kolmogorov, A .N. 1956. *Foundations of the Theory of Probability*, 2nd English ed. Mineola, NY: Dover Publications.

[15] Mines, R., Richman, F., and Ruitenburg, W. 1988. *A Course in Constructive Algebra.* New York: Springer.

[16] Neveu, J. 1965. *Mathematical Foundations of the Calculus of Probability.* (translated by A. Feinstein). San Francisco, London, Amsterdam: Holden–Day.

[17] Nuber, J. A. 1972. A constructive ergodic theorem. *Trans. Amer. Math. Soc.*, **164**, 115–137.

[18] Richman, F. 1982. Meaning and information in constructive mathematics. *Amer. Math. Monthly*, **89**, 385–388.

[19] Richman F. and Winkowska-Nowak, K. 2009. Transient limits. *Appl. Anal. Discr. Math.*, **3**(1), 52–63.

[20] Rudin, W. 2013. *Principles of Mathematical Analysis*, 3rd ed. McGraw–Hill.

[21] Skorokhod, A. V. 1956. *Limit theorems for stochastic processes. Theory of Probability and Its Applications* I, no. 3, pp. 261–289. Philadelphia, PA: SIAM.

[22] Stolzenberg, G. 1970. Review of 'Foundations of Constructive Analysis'. *Bull. Amer. Math. Soc.*, **76**, 301–323.

[23] Troelstra, A. S. 1991. *History of Constructivism in the 20th Century*, Institute for Language, Logic, and Information, Prepublication Series, ML-91-05.

PART IV

TOPOLOGY

14

Bases of Pseudocompact Bishop Spaces

Iosif Petrakis

14.1 The Problem of Constructivising General Topology

In his seminal book *Foundations of Constructive Analysis* [3] the great analyst Errett Bishop (1928–1983) redeveloped constructively a large part of mathematical analysis. He used intuitionistic logic, instead of classical logic, a new set theory, very different from the classical Zermelo–Fraenkel set theory, and an innovative approach to the definition of mathematical concepts, in order to be consistent with classical mathematics. Although Brouwer was the first who developed mathematics within intuitionistic logic, who employed a new constructive set theory, the theory of spreads and species, and who found for many classical mathematical concepts the (classically) equivalent formulation that suits best to constructive study, it was Bishop who managed to elaborate a system of informal constructive mathematics, which is known as BISH (see [11] and [12]), that did not contradict the informal system of classical mathematics CLASS. Despite the fundamental differences between BISH and CLASS, in [3] Bishop presented mathematics in a way remarkably friendly to classical mathematicians, his main target group.

In [3], Bishop developed the constructive theory of metric spaces, of measure and integration of normed and Banach spaces, of Hilbert spaces, of measure and integration, of locally compact Abelian groups, and of commutative Banach algebras. Although he did not elaborate a constructive counterpart to general topology, in [3] he proposed two, quite different, constructive alternatives to the notion of topological spaces, the notion of *neighbourhood space* and the notion of *function space*. The latter we call a *Bishop space*, and it is the subject matter of this chapter. Bishop did not elaborate the theory of either neighbourhood spaces or function spaces.

In [7, p. 28], Bishop described two obstacles in the constructivisation of general topology:

> The constructivization of general topology is impeded by two obstacles. First, the classical notion of a topological space is not constructively viable. Second, even for metric spaces the classical notion of a continuous function is not constructively viable; the reason is that there is no constructive proof that a (pointwise) continuous function from a compact metric space to $\mathbb{R}$ is uniformly continuous. Since uniform continuity for functions on a compact space is the useful concept, pointwise continuity (no longer useful for proving uniform continuity) is left with no useful function to perform. Since uniform continuity cannot be formulated in the context of a general topological space, the latter concept also is left with no useful function to perform.

On the First Obstacle The classical notion of topological space is not constructively viable. First of all, its definition is impredicative, in the sense that it requires quantification over the powerset of a set, an object that from a predicative point of view is, in general, a proper class. A topology T on a set X is defined classically as a subset of its powerset $\mathcal{P}(X)$ that contains X and $\emptyset$, and it is closed under finite intersections and arbitrary unions, that is, for every $S \subseteq T$ the set-theoretic union $\bigcup S$ is in T. Such vicious circles in the definition of mathematical concepts had already been criticised by Poincaré [55] and Russell [56], and the corresponding definitions are called *impredicative* (see, e.g., [21] for an analysis of *predicativity*).

A bottom-up approach to the notion of topology on X is based on a set-indexed family of *basic open* sets $(\nu(i))_{i \in I}$ that satisfies the following covering (NS_1) and neighbourhood (NS_2) conditions:

(NS_1) $\bigcup_{i \in I} \nu(i) = X$.

(NS_2) $\forall_{x \in X} \forall_{i,j \in I} \big[x \in \nu(i) \cap \nu(j) \Rightarrow \exists_{k \in I} \big(x \in \nu(k) \ \& \ \nu(k) \subseteq \nu(i) \cap \nu(j)\big)\big]$.

A subset G of X is called *ν-open*, if

$$\forall_{x \in G} \exists_{i \in I} \big(x \in \nu(i) \ \& \ \nu(i) \subseteq G\big).$$

One can then define the topology T_ν induced by the family of basic open sets $(\nu(i))_{i \in I}$ by

$$T_\nu := \big\{G \in \mathcal{P}(X) \mid G \text{ is } \nu\text{-open}\big\},$$

and show that T_ν satisfies the defining conditions of a topology on X. Let us denote by (X, I, ν) the corresponding neighbourhood structure. The definition of T_ν is based on the use of the axiom scheme of separation and the classical treatment of $\mathcal{P}(X)$ as a set. Constructively, the use of $\mathcal{P}(X)$ as a set is a source of impredicativity that a constructive reconstruction of a mathematical theory should avoid (see also [32, p. 354]). This is even more evident in Bishop's set theory ([3, chapter 5] and [8]), where a subset of a set X is defined as a couple (A, i_A), where A is an arbitrary set and $i_A : A \to X$ is an injection. Hence, if we follow the bottom-up approach to the concept of topology within BISH, we have that $G \in T_\nu$ if G is such a pair that satisfies the ν-openness condition, and quantifying over $\mathcal{P}(X)$ amounts to quantifying over the universe of sets.

If predicativism is part of constructivism, and in our opinion it should be, a constructive reconstruction of general topology should avoid the use of $\mathcal{P}(X)$ as a set.[1] The result of this predicative stance, though, makes the enterprise of constructivisation of general topology even more difficult. Classically, if T is a topology on X and S is a topology on Y, the set

$$\{U \times V \mid U \in T \,\&\, V \in S\}$$

is a base for the product topology $T \times S$ on $X \times Y$. If (X, I, ν) and (Y, J, μ) are structures of covering and neighbourhood basic families on X and Y, respectively, then, since T_ν and T_μ cannot be defined predicatively, we cannot reproduce the classical definition of their product $T_\nu \times T_\mu$. Since a structure (X, I, ν) is a formal topology in the sense of Sambin (see [57], [58], and [59]), it is not a surprise that a similar problem occurs in the definition of the product of formal topologies, where the definition of the product is predicatively possible only when the formal topologies are inductively generated[2] (see [16]).

Formal topology is an abstract, formal approach to point-free general topology that has its origin in the early work [31] of Martin-Löf, and can be seen as a constructive and predicative representation of the theory of locales (see [2] and [35]). It is a characteristic example of a reaction to the first obstacle that is based on replacing the main object of study of topology, a concrete topological space, by some abstract, formal, and predicative codification of it.

In the framework of BISH, one could replace T_ν by smaller, but predicative approximations of it. For example, one could define inductively the least set of subsets $T_\nu(\mathbb{N})$ of X that includes the basic open sets $\nu(i)$, for every $i \in I$, and is closed under finite unions and countable unions. If the index set I is an appropriate set of functions from X to $\mathbb{R}$, and the neighbourhood family of basic open sets on I is defined by

$$\nu(f) := \{x \in X \mid f(x) > 0\}$$

for every $f \in I$, the topology $T_\nu(\mathbb{N})$ is a subset of the inductively defined algebra of Borel sets over $(\nu(f))_{f \in I}$ (see also [41]). One could possibly replace the closure

[1] Bishop did not follow consistently this attitude in his work. In [3] he generally avoids the use of $\mathcal{P}(X)$ as a set. In his measure theory with Cheng [6], an enriched version of which is found in [8], the use of $\mathcal{P}(X)$ as a set is evident from the very starting definition of an integration space. A predicative approach to the Bishop–Cheng measure theory, along the general, predicative lines of the Bishop measure theory in [3], is attempted in [47] and [54]. In Myhill's formalisation of BISH in [32] the powerset axiom is avoided, and the exponentiation axiom is used instead, according to which the totality of functions $\mathbb{F}(A, B)$ from the set A to the set B is a set. Constructively, the set $\mathbb{F}(X, \{0, 1\})$ can be identified with the so-called 'detachable', or decidable subsets of X, and not with $\mathcal{P}(X)$.

[2] The product of function spaces is also defined inductively by Bishop in [3, p. 73], and it seems that it is not possible to define it non-inductively.

under countable unions by the closure under A-unions, where A is any given set, and get in this way an A-approximation $T_\nu(A)$ of the impredicatively defined set T_ν. In this case though, the corresponding inductive rule has A-many premisses, while the inductive definitions found in [3] have rules only with countably many premisses. Again, the inductive generation of $T_\nu(\mathbb{N})$, or $T_\nu(A)$, allows the inductive definition of the corresponding product. The fruitfulness of such a reaction to the first obstacle within BISH needs to be examined further.

The constructive non-viability of the notion of topological space is also corroborated by the fact that many classical topological phenomena, such as the duality between open and closed sets, are compatible only with classical logic. In a straightforward, constructive translation of general topology we cannot accept that the set-theoretic complement of a closed set is open. For example, $\{0\}$ is a closed subset of $\mathbb{R}$, with respect to the topology on $\mathbb{R}$ induced by its standard metric, while its complement cannot be accepted constructively as open, since that would lead to the implication

$$\neg(x = 0) \Rightarrow (x > 0 \vee x < 0),$$

which is (constructively) equivalent to the constructively unacceptable principle of Markov (see [11, p. 15]). The standard use of negative definitions in classical topology does not permit a smooth translation of classical topology to a constructive framework. Maybe this is why the intuitionistic development of general topology (see [25], [26], and [61]), which is mainly based on the use of intuitionistic logic in the study of the standard topological notions, is not under active current development.

On the Second Obstacle The classical uniform continuity theorem (UCT), according to which a pointwise continuous function $f : [a, b] \to \mathbb{R}$ is uniformly continuous, is not provable in BISH, since UCT is equivalent (over BISH) to the fan theorem (FAN), which is classically equivalent to König's lemma and whose intuitionistic proof was not accepted by Bishop. In order to be compatible with CLASS, Bishop defined a function $f : [a, b] \to \mathbb{R}$ to be continuous if it is uniformly continuous, and he defined a function $\phi : \mathbb{R} \to \mathbb{R}$ to be continuous, if it uniformly continuous on every interval $[-n, n]$, where $n \geq 1$. This is why, in his presentation of the second obstacle, Bishop considers uniform continuity for functions on a compact space as the right notion of continuity. Since the formulation of uniform continuity requires the language of metric spaces, and cannot be formulated, at least in a direct way, in the language of general topological spaces, Bishop seems to insinuate in [7] that the constructivisation of general topology is somehow impossible.

Later results though, in formal topology and, as we explain in this chapter, in the theory of Bishop spaces, imply that there are constructive alternatives to the notion of topological space and to the notion of a continuous function that overcome Bishop's second obstacle. Namely, in many interesting cases the continuous functions between these topological spaces are *reduced* to uniform continuous functions, although uniform continuity is not directly formulated in their language. For example, in [34] Palmgren shows that the continuity of functions from the formal reals to the formal reals agrees with Bishop's notion of a continuous function $\phi : \mathbb{R} \to \mathbb{R}$. Moreover, using a construction of Vickers, Palmgren defined in [35] a full and faithful functor from the category of locally compact metric spaces (in the sense of Bishop) into the category of formal topologies. In the theory of Bishop spaces and Bishop morphisms, various reducibility results of the notion of Bishop morphisms to that of uniform continuity are included in [10], [36], and [39]. These reducibility results are based on the presence of uniform continuity in the definition of a Bishop continuous function $\phi : \mathbb{R} \to \mathbb{R}$, a notion that plays a crucial role in the definition of a Bishop space, and consequently in the definition of a Bishop morphism. It is remarkable that both these solutions to the problem of constructivising topology are closely connected to Bishop's own proposed alternatives to the notion of a topological space.

Bishop's contributions to constructive general topology can be summarised as follows.

(i) In [3, Chapter 3], Bishop defined the notion of *neighbourhood space* $\mathcal{N} := (X, I, \nu)$, where X, I are sets, and $(\nu_i)_{i \in I}$ is a family of subsets of X indexed by I that satisfies the above conditions (NS_1) and (NS_2). A ν-open set is defined as above, and a ν-closed set F is not defined negatively, as the complement of a ν-open set, but positively by the condition

$$\forall_{x \in X}\big(\forall_{i \in I}(x \in \nu(i) \Rightarrow \nu(i) \between F) \Rightarrow x \in F\big),$$

where, if A, B are subsets of X, then $A \between B :\Leftrightarrow \exists_{y \in X}(y \in A \cap B)$. The interior and the closure of A are defined in the expected way, while if (Y, J, μ) is another neighbourhood space, a function $h : X \to Y$ is *neighbourhood continuous*, if $h^{-1}(\mu(j))$ is ν-open, for every $j \in J$. The concept of neighbourhood space was proposed as a set-theoretic alternative to the notion of topological space, and it is a formal topology in the sense of Sambin.

(ii) In [3, Chapter 3], Bishop defined the notion of *function space* $\mathcal{F} := (X, F)$, where X is a set and F is a subset of $\mathbb{F}(X)$, the real-valued functions on X, that satisfies some natural closure conditions. Bishop called F a *topology* (of functions) on X. The set $\mathrm{Bic}(\mathbb{R})$ of Bishop-continuous functions $\phi : \mathbb{R} \to \mathbb{R}$ is the canonical topology of functions on $\mathbb{R}$. Bishop also defined inductively[3]

[3] This inductive definition, together with the inductive definition of the notion of the least algebra of Borel sets generated by a family of complemented subsets of X, relative to a given set of real-valued functions on X, are

the *least topology* of functions on X that includes a given subset F_0 of $\mathbb{F}(X)$. The concept of function space was proposed as a function-theoretic alternative to the notion of topological space, and it is the notion studied here.

(iii) In his unpublished manuscript [4] and in Exercises 17–20 on page 110 of [3] Bishop gave the basic definitions and examples of a category with objects certain uniform spaces given by pseudometrics.

In [8, p. 77], Bishop and Bridges expressed in a clear way the superiority of the function-theoretic notion of function space to the set-theoretic notion of neighbourhood space.

> Proximity is introduced into X classically not by giving a family of functions, but by giving a family of subsets, either open sets or neighborhoods. Classically, this is equivalent to giving a family of functions from X to $\{0, 1\}$. Constructively, there is a vast difference: since functions are sharply defined, whereas most sets are fuzzy around the edges, only the all-too-rare detachable subsets of X correspond to functions from X to $\{0, 1\}$. The fuzziness of sets is another reason to focus attention on function spaces instead of neighborhood spaces.

Since the notion of neighborhood-continuous function is the standard generalisation of pointwise continuity, it is not expected to reflect uniform continuity. For example, as Palmgren remarks in [34], the composition of two neighbourhood-continuous functions $f : \mathbb{R} \to X$ and $g : X \to \mathbb{R}$ need not be in $\mathrm{Bic}(\mathbb{R})$. As Bridges and Palmgren remark in [14], 'little appears to have been done' in the theory of neighbourhood spaces. Ishihara has worked in [27] (and with co-authors in [30]) on their connections to the apartness spaces of Bridges and Vîţă (see [13]), and in [28] on their connections to Bishop's function spaces, while in [29] Ishihara and Palmgren studied the notion of quotient topology in neighbourhood spaces.

Regarding his unpublished attempt [4], in [7, p. 28], Bishop writes the following:

> In [3] I was able to get along by working mostly with metric spaces and using various ad hoc definitions of continuity: one for compact (metric) spaces, another for locally compact spaces, and another for the duals of Banach spaces. The unpublished[4] manuscript [5] was an attempt to develop constructive general topology systematically. The basic idea is that a topological space should consist of a set X, endowed with both a family of metrics and a family of boundedness notions, where a boundedness notion on X is a family S of

the main inductive definitions found in [3, Chapter 3]. The notion of the least algebra of Borel sets is avoided in [6] and [8], and the notion of the least topology is not developed either in [3] or in [8].

[4] As Douglas Bridges suggested to me, the two manuscripts [4] and [5] are probably the same. In [4] one can find all these definitions mentioned in [7], but neither he nor I have ever seen [5].

subsets of X (called bounded subsets), whose union is X, closed under finite unions and the formation of subsets:

Bishop was not satisfied with his reconstruction of topology in [4], and maybe this is why he never published this work. He found his theory too involved and not too broad to include properly the notion of a ball space (see [7, p. 29]). Thus, in discussing in [7, p. 29], some of the the tasks that constructive mathematics faces, he refers to constructive topology, right after mentioning the primary importance of the constructive reconstruction of algebra, as follows:

> Less critical, but also of interest, is the problem of a convincing constructive foundation for general topology, to replace the ad hoc definitions in current use. It would also be good to see a constructivization of algebraic topology actually carried through, although I suspect this would not pose the critical difficulties that seem to be arising in algebra.

14.2 Overview of Recent Work on Bishop Spaces

Bishop and Bridges kept the section of function spaces in [8] almost unchanged. The only new reference to function spaces between Bishop's book of 1967 and Bridges's paper[10] that we know of is a comment of Myhill in [32, p. 377], regarding the inductive definition of the least function space and the place of inductive definitions in the formalisation of Bishop-style constructive mathematics.

Bridges talked on Bishop's function spaces at the first workshop on formal topology in 1997, and revived the subject of function spaces in [10]. There he defined the morphisms between function spaces, the fundamental point–point apartness and the set–set apartness relation induced by a topology of functions, and he provided some important reducibility results.

Motivated by Bridges's paper, Ishihara showed in [28] the existence of an adjunction between the category of neighbourhood spaces and the category of Φ-closed pre-function spaces, where a pre-function space is an extension of the notion of a function space.

In [36]–[52] we try to develop the theory of function spaces, or Bishop spaces, as we call them. Next we give an overview of our elaboration of the theory of Bishop spaces.

(1) In [36] a first presentation of the theory of Bishop spaces as a whole is given. There the approach to continuity as a primitive notion is motivated, several examples and constructions on Bishop spaces are included, and many constructive counterparts to fundamental theorems of classical topology are proven, like the Urysohn lemma for the zero sets of a Bishop space, the Urysohn extension theorem for Bishop spaces (see also [38] and [45]), and the Tychonoff embedding

theorem (see also [37]). Moreover, the notion of 2-compactness, as a function-theoretic notion of compactness, is studied (see also [39]). The basics of a homotopy theory of Bishop spaces is also included in [36].

(2) In [42] the basic constructive theory of uniformities given by pseudometrics and its relation to the constructive theory of Bishop topologies is presented. After motivating the constructive study of uniformities of pseudometrics, a Stone–Čech theorem is shown for them. The $\mathfrak{f}$-uniform spaces are introduced and a Tychonoff embedding theorem is also shown for them. The uniformity of pseudometrics generated by some Bishop topology, and the pseudocompact Bishop topology generated by some uniformity of pseudometrics, are studied.

(3) In [47] and in [48] the theory of set-indexed families of Bishop sets (see also [53]) is applied to the theory of spectra of Bishop spaces and their limits, a constructive counterpart to the classical theory of spectra of topological spaces and their limits (see, e.g., [19, Appendix Two]).

(4) The measure theory of Bishop spaces is developed first in [41], where the Borel sets and the Baire sets as inductively defined sets of F-complemented subsets of X are studied, in [46], where the functions of Baire class one over a Bishop topology F are shown to form a new Bishop topology that includes F, and in [52], where Bishop's integration theory of locally compact metric spaces is generalised to the integration theory of Bishop spaces.

(5) Bishop topological groups are introduced in [51], where fundamental facts on closed subsets in Bishop topological groups are shown without the use of choice and with a clear algorithmic content.

(6) In [50] a Chu representation of the category of Bishop spaces $\mathbf{Bis}$ in $\mathbf{Chu}(\mathbf{Set}, \mathbb{R})$ is included. In contrast to the standard Chu representation of the category of topological spaces $\mathbf{Top}$ in $\mathbf{Chu}(\mathbf{Set}, 2)$, classical logic is avoided in the Chu representation of $\mathbf{Bis}$.

(7) In [23] connections between the theory of Bishop spaces and the theory of C-spaces of Escardó and Xu, developed by [20] and [62], are studied.

14.3 Structure of the Technical Part of this Chapter

We structure the rest of this chapter as follows.

(a) In Section 14.4 we include the basic definitions of a Bishop space, a Bishop morphism, a subbase of a Bishop topology, and of a pseudocompact Bishop space. We give several examples of Bishop spaces and some fundamental constructions of new Bishop spaces from given ones. All these notions are also found, for example, in [36]. What is added here in the definition of Bishop space, the notion of the least Bishop space generated by a given set of real-valued functions, and in the formulation of the corresponding induction principle, is the extensionality of a Bishop topology F as a subset of all real-valued functions $\mathbb{F}(X)$ on a set X. This property of F is crucial to many results, and it is not explicitly mentioned by Bishop.

(b) In Section 14.5 we define the notion of a base of a Bishop topology and prove some of its fundamental properties. The results of this section are new. Although Lemma 14.12 appears as the well-definability lemma in [36, p. 56], its proof here avoids choice, as a modulus of surjectivity is added in its hypotheses.

(c) In Section 14.6 we prove the first base theorem for pseudocompact Bishop spaces. This result is also found in [36, Section 3.5].

(d) In Section 14.7 we prove the second base theorem, a theorem of Stone–Weierstrass-type for pseudocompact Bishop spaces. This result is also found in [36, Section 4.2].

(e) In Section 14.8 we apply the second base theorem in order to generalise results of Bishop around the Stone–Weierstrass theorem for compact metric spaces to pseudocompact Bishop spaces. Corollary 14.21, Proposition 14.23, and Proposition 14.30 are also presented in [36]. The proof of Corollary 14.24 is included here but not in [36]. All remaining results are new. Of special importance is the new proof of the density of the Lipschitz real-valued functions in the uniformly continuous ones (Corollary 14.33) as a corollary of the second base theorem. The results of this section indicate that the theory of pseudocompact Bishop spaces is a natural constructive counterpart to the classical theory of the ring $C^*(X)$ of bounded, continuous, real-valued functions on a topological space X.

We work within Bishop's informal system BISH*, which is Bishop's informal system of mathematics BISH extended with inductive definitions with rules of countably many premisses. An elaborate study of the theory of sets underlying BISH, which updates the earlier approach in [43], is given in [47]. A formal system for BISH* is Myhill's formal system CST* with dependent choice, where CST* is Myhill's extension of his formal system of constructive theory CST with inductive definitions (see [32]). Another formal system for BISH*, which was motivated by Myhill's system CST*, is Aczel's constructive Zermelo–Fraenkel set theory CZF extended with a weak version of Aczel's regular extension axiom (REA), to accommodate the inductive definitions of BISH* (see [1]).

14.4 Basic Notions in the Theory of Bishop Spaces

A Bishop topology F on a set X is a set of real-valued functions on X that behaves as an abstract version of the ring of real-valued continuous functions $C(X)$ on a topological space X. Hence F is a constructive, function-theoretic alternative to the classical notion of a topology of open sets, and a Bishop morphism between Bishop spaces is the corresponding function-theoretic notion of continuous function between such spaces. As in Spanier's theory of quasitopological spaces in [60], or in the much earlier theory of limit spaces of Fréchet in [22], the continuity of a certain family of functions is a primitive notion. Although many concepts, questions, and results from the classical theory of $C(X)$ (see, e.g., [24]) can be translated into the theory of Bishop spaces, the use of intuitionistic logic does not permit a direct translation of the former theory to the latter. In this section we give the basic definitions and facts on Bishop spaces and the morphisms between them. For all concepts and results from constructive real analysis and Bishop's set theory that we use here without further explanation we refer to [8]. For all proofs that are not included in this section we refer to [36].

Definition 14.1 If X is a set and $\mathbb{R}$ is the set of real numbers, we denote by $\mathbb{F}(X)$ the set of functions from X to $\mathbb{R}$, by $\mathbb{F}^*(X)$ the bounded elements of $\mathbb{F}(X)$, and by $\mathrm{Const}(X)$ the subset of $\mathbb{F}(X)$ of all constant functions on X. If $a \in \mathbb{R}$, we denote by $\overline{a}^X$ the constant function on X with value a. We denote by $\mathbb{N}^+$ the set of non-zero natural numbers. A function $\phi : \mathbb{R} \to \mathbb{R}$ is called *Bishop continuous*, or simply continuous, if for every $n \in \mathbb{N}^+$ there is a function $\omega_{\phi,n} : \mathbb{R}^+ \to \mathbb{R}^+$, $\epsilon \mapsto \omega_{\phi,n}(\epsilon)$, which is called a *modulus of continuity* of ϕ on $[-n, n]$, such that the following condition is satisfied[5]

$$\forall_{x,y\in[-n,n]}\big(|x - y| < \omega_{\phi,n}(\epsilon) \Rightarrow |\phi(x) - \phi(y)| \leq \epsilon\big),$$

for every $\epsilon > 0$ and every $n \in \mathbb{N}^+$. We denote by $\mathrm{Bic}(\mathbb{R})$ the set of continuous functions from $\mathbb{R}$ to $\mathbb{R}$, which is equipped with the pointwise equality inherited from $\mathbb{F}(\mathbb{R})$.

Note that we could have defined the modulus of continuity $\omega_{\phi,n}$ as a function from $\mathbb{N}^+$ to $\mathbb{N}^+$, 'identifying' ϵ with $\frac{1}{2^n}$. In the constructive literature, a continuous $\phi\colon \mathbb{R} \to \mathbb{R}$ is uniformly continuous on every bounded subset of $\mathbb{R}$. This is an impredicative formulation of uniform continuity, as it requires quantification over the proper class 'powerset of $\mathbb{R}$'. The definition of uniform continuity in Definition 14.1 is predicative, as it requires quantification over the sets $\mathbb{N}^+$, $\mathbb{F}(\mathbb{R}^+, \mathbb{R}^+)$ and $[-n, n]$.

Definition 14.2 If X is a set, $f, g \in \mathbb{F}(X)$, $\epsilon > 0$, and $\Phi \subseteq \mathbb{F}(X)$, let

$$U(X; g, f, \epsilon) :\Leftrightarrow \forall_{x\in X}\big(|g(x) - f(x)| \leq \epsilon\big),$$

$$U(X; \Phi, f) :\Leftrightarrow \forall_{\epsilon>0}\exists_{g\in\Phi}\big(U(g, f, \epsilon)\big).$$

[5] The inequality is not strict in the conclusion, as in order to show constructively in $\mathbb{R}$ the inequality $x \leq y$, it suffices to show $\neg(x > y)$. The negation of $x \geq y$ though, does not, in general, imply constructively the strict inequality $x < y$.

If the set X is clear from the context, we write simply $U(f, g, \epsilon)$ and $U(\Phi, f)$, respectively. We denote by Φ^* the bounded elements of Φ, and its *uniform closure* $\overline{\Phi}$ is defined by

$$\overline{\Phi} := \{f \in \mathbb{F}(X) \mid U(\Phi, f)\}.$$

A Bishop topology on X is a certain subset of $\mathbb{F}(X)$. As the Bishop topologies considered here are all extensional[6] subsets of $\mathbb{F}(X)$, we do not mention the embedding $i_F^{\mathbb{F}(X)} : F \hookrightarrow \mathbb{F}(X)$, which is given in all cases by the identity map rule. The uniform closure $\overline{\Phi}$ of Φ is an extensional subset of $\mathbb{F}(X)$.

Definition 14.3 A *Bishop space* is a pair $\mathcal{F} := (X, F)$, where F is an extensional subset of $\mathbb{F}(X)$, which is called a *Bishop topology*, or a *topology* of functions on X, that satisfies the following conditions:

(BS_1) If $a \in \mathbb{R}$, then $\overline{a}^X \in F$.

(BS_2) If $f, g \in F$, then $f + g \in F$.

(BS_3) If $f \in F$ and $\phi \in \mathrm{Bic}(\mathbb{R})$, then $\phi \circ f \in F$

$$\begin{array}{ccc} X & \xrightarrow{\;f\;} & \mathbb{R} \\ & {\scriptstyle F \ni \phi \circ f} \searrow & \downarrow {\scriptstyle \phi \in \mathrm{Bic}(\mathbb{R})} \\ & & \mathbb{R}. \end{array}$$

(BS_4) $\overline{F} = F$.

If $\mathcal{F} := (X, F)$ is a Bishop space, then $\mathcal{F}^* := (X, F^*)$ is the Bishop space of bounded elements of F. The space Const(X) of constant functions is the *trivial* topology on X, while $\mathbb{F}(X)$ is the *discrete* topology on X. Clearly, if F is a topology on X, then $\mathrm{Const}(X) \subseteq F \subseteq \mathbb{F}(X)$, and the set of its bounded elements F^* is also a topology on X. It is straightforward to see that the pair $\mathcal{R} := (\mathbb{R}, \mathrm{Bic}(\mathbb{R}))$ is a Bishop space, which we call the *Bishop space of reals*. If X is a metric space, the set $C_p(X)$ of all weakly continuous functions of type $X \to \mathbb{R}$, as it is defined in [8, p. 76], is the set of pointwise continuous ones. It is easy to see that the pair $\mathcal{W}(X) = (X, C_p(X))$ is Bishop space. Bishop calls $C_p(X)$ the weak topology on X, but here we avoid this term, since in [36] we use this term for the Bishop topology that corresponds to the weak topology of open sets, and we call $C_p(X)$ the *pointwise* topology on X. If X is a compact metric space, the set $C_u(X)$ of all uniformly continuous functions of type $X \to \mathbb{R}$ is a topology, called by Bishop

[6] If X is a set and P is an extensional property on X, namely $P(x)$ & $x =_X y \Rightarrow P(y)$, for every $x, y \in X$, the extensional subset X_P of X is defined by separation, $X_P = \{x \in X \mid P(x)\}$, its equality is inherited from the equality of X and the embedding of X_P into X is defined by the identity rule (see [47, Definition 2.2.3]).

the *uniform* topology on X. We call $U(X) = (X, C_u(X))$ the uniform space. If X is a locally compact metric space, the set $\mathrm{Bic}(X)$ of Bishop continuous functions from X to $\mathbb{R}$—that is, functions uniformly continuous on every[7] bounded subset of X—is a Bishop topology on X.

A Bishop topology F is a ring and a lattice; since $|\mathrm{id}_{\mathbb{R}}| \in \mathrm{Bic}(\mathbb{R})$, where $\mathrm{id}_{\mathbb{R}}$ is the identity function on $\mathbb{R}$, by BS_3 we get that if $f \in F$ then $|f| \in F$. By BS_2 and BS_3, and using the following equalities

$$f \cdot g = \frac{(f+g)^2 - f^2 - g^2}{2} \in F,$$

$$f \vee g = \max\{f, g\} = \frac{f + g + |f - g|}{2} \in F,$$

$$f \wedge g = \min\{f, g\} = \frac{f + g - |f - g|}{2} \in F,$$

we get similarly that if $f, g \in F$, then $f \cdot g, f \vee g, f \wedge g \in F$. Turning the definitional clauses of a Bishop topology into inductive rules, Bishop defined in [3, p. 72] the least topology including a given subbase F_0. This inductive definition, which is also found in [8, p. 78], is crucial to the definition of new Bishop topologies from given ones.

Definition 14.4 The *Bishop closure* of F_0, or the *least topology* $\bigvee F_0$ generated by some $F_0 \subseteq \mathbb{F}(X)$, is defined by the following inductive rules:

$$\frac{f_0 \in F_0}{f_0 \in \bigvee F_0}, \quad \frac{a \in \mathbb{R}}{\overline{a}^X \in \bigvee F_0}, \quad \frac{f, g \in \bigvee F_0}{f + g \in \bigvee F_0}, \quad \frac{f \in \bigvee F_0,\ g =_{\mathbb{F}(X)} f}{g \in \bigvee F_0},$$

$$\frac{f \in \bigvee F_0,\ \phi \in \mathrm{Bic}(\mathbb{R})}{\phi \circ f \in \bigvee F_0}, \quad \frac{\left(g \in \bigvee F_0,\ U(g, f, \epsilon)\right)_{\epsilon > 0}}{f \in \bigvee F_0}.$$

We call $\bigvee F_0$ the *Bishop closure* of F_0, and F_0 a *subbase* of $\bigvee F_0$.

[7] As in the definition of $\mathrm{Bic}(\mathbb{R})$, at first sight it seems that this definition also requires quantification over the power set of X, that is,

$$\mathrm{Bic}(X)(f) \Leftrightarrow \forall_{B \in \mathcal{P}(X)}(\mathrm{bounded}(B) \Rightarrow f_{|B} \text{ is uniformly continuous}).$$

Here, a bounded subset B of an inhabited metric space X is a triplet (B, x_0, M), where $x_0 \in X, B \subseteq X$, and $M > 0$ is a bound for $B \cup \{x_0\}$. To avoid such a quantification, if x_0 inhabits X, then for every bounded subset (B, x_0', M) of X we have that there is some $n \in \mathbb{N}$ such that $n > 0$ and $B \subseteq [d_{x_0} \leq \overline{n}] = \{x \in X \mid d(x_0, x) \leq n\}$. If $x \in B$, then $d(x, x_0) \leq d(x, x_0') + d(x_0', x_0) \leq M + d(x_0', x_0)$, and therefore $x \in [d_{x_0} \leq \overline{n}]$, for some $n > M + d(x_0', x_0)$. Hence,

$$\mathrm{Bic}(X)(f) \Leftrightarrow \forall_{n \in \mathbb{N}}(f_{|[d_{x_0} \leq \overline{n}]} \text{ is uniformly continuous}),$$

since $[d_{x_0} \leq \overline{n}] = \{x \in d(x_0, x) \leq n\}$ is trivially a bounded subset of X.

If F_0 is inhabited, (BS_1) is provable by (BS_3). The last rule above can be replaced by the rule

$$\frac{g_1 \in \bigvee F_0 \,\&\, U\big(g_1, f, \frac{1}{2}\big),\ \ g_2 \in \bigvee F_0 \,\&\, U\big(g_2, f, \frac{1}{2^2}\big),\ \ldots}{f \in \bigvee F_0},$$

which is a rule with countably many premisses. The corresponding induction principle $\mathrm{Ind}_{\bigvee F_0}$ is given by the following formula:

$$\Big[\forall_{f_0 \in F_0}\big(P(f_0)\big) \,\&\, \forall_{a \in \mathbb{R}}\big(P(\overline{a}^X)\big) \,\&\, \forall_{f,g \in \bigvee F_0}\big(P(f) \,\&\, P(g) \Rightarrow P(f+g)\big)$$

$$\&\, \forall_{f \in \bigvee F_0}\forall_{g \in \mathbb{F}(X)}\big(g =_{\mathbb{F}(X)} f \Rightarrow P(g)\big)$$

$$\&\, \forall_{f \in \bigvee F_0}\forall_{\phi \in \mathrm{Bic}(\mathbb{R})}\big(P(f) \Rightarrow P(\phi \circ f)\big)$$

$$\&\, \forall_{f \in \bigvee F_0}\big(\forall_{\epsilon > 0}\exists_{g \in \bigvee F_0}\big(P(g) \,\&\, U(g, f, \epsilon)\big) \Rightarrow P(f)\big)\Big]$$

$$\Rightarrow \forall_{f \in \bigvee F_0}\big(P(f)\big),$$

where P is a *bounded formula*, that is, a formula that involves quantifications only over sets. Next we define the notion of a Bishop morphism between Bishop spaces, which is the notion of morphism in the category of Bishop spaces $\mathbf{Bis}$.

Definition 14.5 If $\mathcal{F} := (X, F)$ and $\mathcal{G} = (Y, G)$ are Bishop spaces, a function $h : X \to Y$ is called a *Bishop morphism* if $\forall_{g \in G}(g \circ h \in F)$

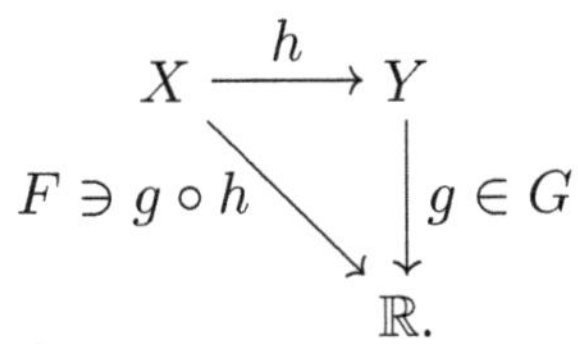

We denote by $\mathrm{Mor}(\mathcal{F}, \mathcal{G})$ the set of Bishop morphisms from $\mathcal{F}$ to $\mathcal{G}$. As F is an extensional subset of $\mathbb{F}(X)$, $\mathrm{Mor}(\mathcal{F}, \mathcal{G})$ is an extensional subset of $\mathbb{F}(X, Y)$. A Bishop morphism $h \in \mathrm{Mor}(\mathcal{F}, \mathcal{G})$ is a *Bishop isomorphism* if it is an isomorphism in the category $\mathbf{Bis}$; while h is called *open* if

$$\forall_{f \in F}\exists_{g \in G}(f = g \circ h).$$

We write $\mathcal{F} \cong \mathcal{G}$ to denote that $\mathcal{F}$ and $\mathcal{G}$ are Bishop isomorphic. If $h \in \mathrm{Mor}(\mathcal{F}, \mathcal{G})$, the *induced mapping* $h^* \colon G \to F$ from h is defined by $h^*(g) := g \circ h$, for every $g \in G$. A property P on $\mathbb{F}(X)$ is $\bigvee$-lifted from a subbase F_0 of F to F if $\forall_{f_0 \in F_0}\big(P(f_0)\big) \Rightarrow \forall_{f \in \bigvee F_0}\big(P(f)\big)$.

It follows inductively that the property 'f is bounded' is $\bigvee$-lifted. If $\mathcal{F} := (X, F)$ is a Bishop space, then $F = \mathrm{Mor}(\mathcal{F}, \mathcal{R})$, and if $\mathcal{G} := (Y, \bigvee G_0)$, then $h : X \to Y \in \mathrm{Mor}(\mathcal{F}, \mathcal{G})$ if and only if $\forall_{g_0 \in G_0}(g_0 \circ h \in F)$

$$\begin{array}{ccc} X & \xrightarrow{h} & Y \\ & {\scriptstyle F \ni g_0 \circ h} \searrow & \downarrow {\scriptstyle g_0 \in G_0} \\ & & \mathbb{R}. \end{array}$$

We call this fundamental fact the $\bigvee$-*lifting of morphisms*. If $h \in \mathrm{Mor}(\mathcal{F}, \mathcal{G})$ is a bijection, then h is a Bishop isomorphism if and only h is open.

Definition 14.6 Let $\mathcal{F} := (X, F), \mathcal{G} := (Y, G)$ be Bishop spaces, and $(A, i_A) \subseteq X$ inhabited.[8] The *product* Bishop space $\mathcal{F} \times \mathcal{G} := (X \times Y, F \times G)$, the *relative* Bishop space $\mathbb{F}_{|A} := (A, F_{|A})$ on A, and the *pointwise exponential* $\mathcal{F} \to \mathcal{G} := (\mathrm{Mor}(\mathcal{F}, \mathcal{G}), F \to G)$ are defined, respectively, by

$$F \times G := \bigvee [\{f \circ \mathtt{pr}_X, \mid f \in F\} \cup \{g \circ \mathtt{pr}_Y \mid g \in G\}] =: \bigvee_{f \in F}^{g \in G} f \circ \mathtt{pr}_X, g \circ \mathtt{pr}_Y,$$

$$F_{|A} := \bigvee \{f_{|A} \mid f \in F\} =: \bigvee_{f \in F} f_{|A}$$

$$A \overset{i_A}{\hookrightarrow} X \xrightarrow{f} \mathbb{R}, \qquad f_{|A} : A \to \mathbb{R}$$

$$F \to G := \bigvee \{\phi_{x,g} \mid x \in X, g \in G\} =: \bigvee_{x \in X}^{g \in G} \phi_{x,g},$$

$$\phi_{x,g} : \mathrm{Mor}(\mathcal{F}, \mathcal{G}) \to \mathbb{R}, \quad \phi_{x,g}(h) := g(h(x)); \quad x \in X,\ g \in G.$$

One can show inductively the following $\bigvee$-liftings

$$\bigvee F_0 \times \bigvee G_0 := \bigvee [\{f_0 \circ \mathtt{pr}_X, \mid f_0 \in F_0\} \cup \{g_0 \circ \mathtt{pr}_Y \mid g_0 \in G_0\}]$$
$$=: \bigvee_{f_0 \in F_0}^{g_0 \in G_0} f_0 \circ \mathtt{pr}_X, g_0 \circ \mathtt{pr}_Y,$$

[8] For simplicity, in the rest we avoid mentioning the embedding of a subset of a set X into X.

$$\Big(\bigvee F_0\Big)_{|A} = \bigvee\{f_{0|A} \mid f_0 \in F_0\} =: \bigvee_{f_0 \in F_0} f_{0|A},$$

$$F \to \bigvee G_0 = \bigvee \big\{\phi_{x,g_0} \mid x \in X, g_0 \in G_0\big\} =: \bigvee_{x \in X}^{g_0 \in G_0} \phi_{x,g_0}.$$

The relative topology F_A is the least topology on A that makes i_A a Bishop morphism, and $F \times G$ is the least topology on $X \times Y$ that makes the projections $\texttt{pr}_X$ and $\texttt{pr}_Y$ Bishop morphisms. The countable product of Bishop spaces is defined similarly. The topology $F \to G$ behaves like the the classical topology of pointwise convergence on $C(X, Y)$, the set of continuous functions from a topological space X to a topological space Y. The weak and the quotient Bishop topologies are also defined in [36]. Next follows the Bishop space analogue to a pseudocompact topological space.

Definition 14.7 A Bishop space (X, F), or its topology F, is called *pseudocompact* if $F = F^*$, that is, if every element of F is bounded.

Proposition 14.8 *Let $\mathcal{F} = (X, F), \mathcal{G} = (Y, G)$ be Bishop spaces, $A \subseteq X$ inhabited, $F_0 \subseteq F$ and $h \in \mathrm{Mor}(\mathcal{F}, \mathcal{G})$ such that h is onto Y.*

(i) *If F is pseudocompact, then G is pseudocompact.*

(ii) *If $F_0 \subseteq F^*$, then $\bigvee F_0$ is a pseudocompact topology.*

(iii) *The topologies F and G are pseudocompact if and only if $F \times G$ is pseudocompact.*

(iv) *If F is pseudocompact, then $F_{|A}$ is pseudocompact.*

Proof (i) If $g \in G$, then $g \circ h \in F$, and since h is onto Y, we have that a bound of $\{|(g \circ h)(x)| \mid x \in X\}$ is a bound of $\{|g(y)| \mid y \in Y\}$.
(ii) It follows immediately from the $\bigvee$-lifting of boundedness.
(iii) If F, G are pseudocompact topologies, then all the elements of the subbase $(F \circ \pi_1) \cup (G \circ \pi_2)$ of $F \times G$ are bounded, and we use (ii). Conversely, if $F \times G$ is pseudocompact, then the elements of F and G are bounded, since a bound of some $f \circ \pi_1$ is a bound of f for every $f \in F$, and a bound of some $g \circ \pi_2$ is a bound of g for every $g \in G$.
(iv) If $f \in F$, then a bound of f is also a bound of $f_{|A}$, and we use (ii). □

Proposition 14.8(ii) reveals the difference between pseudocompact Bishop spaces and pseudocompact topological spaces, as the product of two pseudocompact topological spaces is not in general pseudocompact. If the two topological spaces are also completely regular and T_1, then their product is pseudocompact. Note that the induced topology of open sets on X by some Bishop topology of functions on X has as base the sets of the form

$$U(f) = \{x \in X \mid f(x) > 0\}; \quad f \in F$$

that is, (X, F, U), where $f \mapsto U(f)$, is the *neighbourhood space induced by* (X, F). One can show classically that the induced-by-F topology of open sets is completely regular (see [36, Proposition 3.7.6]). The countable product of pseudocompact Bishop spaces is similarly shown to be pseudocompact (this is necessary to Corollary 14.24).

14.5 Bases of Bishop Spaces

The notions of a subbase, defined Section 14.4, and of a base of a Bishop topology, studied in this section, are the Bishop space-theoretic analogues to the classical notions of a subbase and a base of a uniform structure given by pseudometrics (see [24, p. 217]). As in classical topology the set-theoretical notions of a base and a subbase of open (closed) sets are frequently more useful than the topology of open sets itself, and the function-theoretic notions of a base and a subbase of a Bishop topology are often easier to work with rather than the Bishop topology itself.

Definition 14.9 If X is an inhabited set, a property P on $\mathbb{F}(X)$ is *uniformly lifted* from $\Phi \subseteq \mathbb{F}(X)$ to its uniform closure $\overline{\Phi}$ if $\forall_{\phi \in \Phi}(P(\phi)) \Rightarrow \forall_{f \in \overline{\Phi}}(P(f))$.

Because of condition BS_4, if P is an $\bigvee$-lifted property from Φ to $\bigvee \Phi$, then P is lifted from Φ to $\overline{\Phi}$. The converse is not true. For example, let $P(\phi) :\Leftrightarrow |\phi(x) - \phi(y)| \geq c$, where $x, y \in X$ and $c > 0$. Clearly, P is not $\bigvee$-lifted from Φ to $\bigvee \Phi$, but it is lifted from Φ to $\overline{\Phi}$; if $f \in \overline{\Phi}$, we consider some $\phi \in \Phi$ such that $U(\phi, f, \frac{\epsilon}{2})$. Since $c \leq |\phi(x) - \phi(y)| \leq |\phi(x) - f(x)| + |f(x) - f(y)| + |f(y) - \phi(y)|$, we get that $c - \epsilon \leq |f(x) - f(y)|$, and since $\epsilon > 0$ is arbitrary, we conclude that $c \leq |f(x) - f(y)|$.

The following definition of a base for a Bishop topology is in complete analogy to the definition of a base for a uniform structure on a set given by a family of pseudometrics (see [24, p. 217]).

Definition 14.10 A subset Φ of a Bishop topology F on X is a *base* for F if $\overline{\Phi} = F$.

The inhabitedness of F implies the inhabitedness of a base Φ for F; if $f \in F$ and $\epsilon > 0$, there exists by definition some $g \in \Phi$ such that $U(g, f, \epsilon)$.

Proposition 14.11 *Suppose that $\mathcal{F} = (X, F), \mathcal{G} = (Y, G)$ are Bishop spaces, $h \in \mathrm{Mor}(\mathcal{F}, \mathcal{G})$ such that h is onto Y, $\Phi \subseteq F$ and $\Theta \subseteq G$. Let $h^* : G \to F$ be the induced mapping of h, and let $h^*(\Theta) = \{h^*(\theta) = \theta \circ h \mid \theta \in \Theta\}$. Then*

(i) $h^*(\overline{\Theta}) \subseteq \overline{h^*(\Theta)}$.

(ii) *If h is open and Θ is a base for G, then $h^*(\Theta)$ is a base for F.*

(iii) $\overline{(h^*)^{-1}(\Phi)} \subseteq (h^*)^{-1}(\overline{\Phi})$.

(iv) *If h is open and Φ is a base of F, then $(h^*)^{-1}(\Phi)$ is a base for G.*

Proof (i) Let $g \in \overline{\Theta}$ i.e., $\forall_{\epsilon>0}\exists_{\theta\in\Theta}(U(\theta, g, \epsilon))$. Since h is onto Y, we have that $U(\theta, g, \epsilon) \Rightarrow U(\theta \circ h, g \circ h, \epsilon)$, for every $\epsilon > 0$, therefore $h^*(g) \in \overline{h^*(\Theta)}$.
(ii) If h is open, then h^* is onto F, and by (i) we have that $F = h^*(G) = h^*(\overline{\Theta}) \subseteq \overline{h^*(\Theta)} \subseteq F$, therefore $\overline{h^*(\Theta)} = F$.
(iii) Let

$$g \in \overline{(h^*)^{-1}(\Phi)} \Leftrightarrow \forall_{\epsilon>0}\exists_{g'\in h^{*-1}(\Phi)}(U(g', g, \epsilon)),$$

hence $\forall_{\epsilon}\exists_{g'\circ h\in\Phi}\,(U(g' \circ h,\ g \circ h, \epsilon))$. Consequently, $g \circ h \in \overline{\Phi}$, that is, $g \in (h^*)^{-1}(\overline{\Phi})$.
(iv) Let $g \in G$. Since $g \circ h \in F$ we have that $\forall_{\epsilon>0}\exists_{\phi\in\Phi}(U(\phi, g \circ h, \epsilon))$. Since h is open we have that every ϕ in the previous formula can be written as $g_\phi \circ h$, where $g_\phi \in G$. Since

$$\forall_{\epsilon>0}\exists_{g_\phi\in h^{*-1}(\Phi)}(U(g_\phi \circ h, g \circ h, \epsilon)) \Leftrightarrow \forall_{\epsilon>0}\exists_{g_\phi\in h^{*-1}(\Phi)}(U(g_\phi, g, \epsilon)),$$

we get $g \in \overline{(h^*)^{-1}(\Phi)}$. □

Next we show the uniform lifting of openness that is used here only in the proof of Proposition 14.26, but it has many other applications in the theory of Bishop spaces. The $\bigvee$-lifting of openness can also be shown similarly. First we prove a necessary lemma.[9]

Lemma 14.12 *Let X, Y be inhabited sets, $\Theta \subseteq \mathbb{F}(Y)$ and $f : X \to \mathbb{R}$. Let $h : X \to Y$ be onto Y, and let $s\colon Y \to X$ be a modulus of surjectivity for h, that is, $h \circ s = \mathrm{id}_Y$. Let $f^{\#}\colon Y \to \mathbb{R}$ be defined by $f^{\#} = f \circ s$. If $f \in \overline{\Theta \circ h}$, where $\Theta \circ h = \{\theta \circ h \mid \theta \in \Theta\}$, then the following hold.*

(i) $\forall_{x_1,x_2\in X}(h(x_1) = h(x_2) \Rightarrow f(x_1) = f(x_2))$.

(ii) *The following right triangle also commutes, that is, $f^{\#} \circ h = f$*

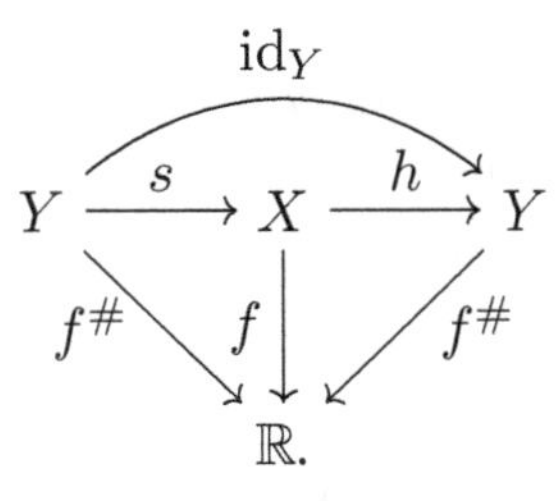

[9] A different formulation of this lemma is found in [36] and [37]. There we defined the function $f^{\#}$ using tacitly some choice; that is, we showed that the mapping $f^{\#} : Y \to \mathbb{R}$, where $f^{\#}(y) = f^{\#}(h(x)) := f(x)$, for every $y \in Y$, is well defined. Here we use the notion of a modulus of surjectivity, in order to avoid choice.

Proof (i) We fix $x_1, x_2 \in X$ such that $h(x_1) = h(x_2) = y_0$, and some $\epsilon > 0$. By our hypothesis on f there exists some $\theta : Y \to \mathbb{R}$ in Θ such that $\forall_{x \in X}(|(\theta \circ h)(x) - f(x)| \leq \frac{\epsilon}{2})$. Hence, $|\theta(h(x_1)) - f(x_1)| = |\theta(y_0) - f(x_1)| \leq \frac{\epsilon}{2}$ and $|\theta(h(x_2)) - f(x_2)| = |\theta(y_0) - f(x_2)| \leq \frac{\epsilon}{2}$. Consequently, $|f(x_1) - f(x_2)| \leq |f(x_1) - \theta(y_0)| + |\theta(y_0) - f(x_2)| \leq \frac{\epsilon}{2} + \frac{\epsilon}{2} = \epsilon$. Since ϵ is arbitrary, we get that $|f(x_1) - f(x_2)| \leq 0$, therefore $f(x_1) = f(x_2)$.
(ii) If $x' = s(h(x))$, then $h(x') = (h \circ s)(h(x)) = \mathrm{id}_Y(h(x)) = h(x)$. By case (i) we have that

$$f(x) = f(x') = f(s(h(x))) = [(f \circ s) \circ h](x) = (f^{\#} \circ h)(x),$$

and since $x \in X$ is arbitrary, we get $f^{\#} \circ h = f$. □

Proposition 14.13 (Uniform lifting of openness) *Suppose that X, Y are inhabited sets, $\Phi \subseteq \mathbb{F}(X)$, $\Theta \subseteq \mathbb{F}(Y)$, and $h : X \to Y$ is onto Y with $s \colon Y \to X$ a modulus of surjectivity for h.*
(i) If $\forall_{f \in \Phi} \exists_{\theta \in \Theta}(f = \theta \circ h)$, then $\forall_{f \in \overline{\Phi}} \exists_{\theta \in \overline{\Theta}}(f = \theta \circ h)$ and for every $f \in \overline{\Phi}$ we have that $f^{\#} \in \overline{\Theta}$.
(ii) If $(X, F), (Y, G)$ are Bishop spaces, $h \in \mathrm{Mor}(\mathcal{F}, \mathcal{G})$, and Φ is a base of F such that $\forall_{f \in \Phi} \exists_{g \in G}(f = g \circ h)$, then h is open, $s \in \mathrm{Mor}(\mathcal{G}, \mathcal{F})$, and $(h^*)^{-1}(\Phi)$ is a base for G.

Proof (i) If $f \in \overline{\Phi}$ and $\epsilon > 0$, there is $\phi \in \Phi$ such that $U(\phi, f, \epsilon)$. Hence there is $\theta \in \Theta$ such that $U(\theta \circ h, f, \epsilon)$. As this is the case for every $\epsilon > 0$, we get $f \in \overline{\Theta \circ h}$. By Lemma 14.12(ii) we get $f = f^{\#} \circ h$ and since h is onto Y the relation $U(X; \theta \circ h, f^{\#} \circ h, \epsilon)$ implies $U(Y; \theta, f^{\#}, \epsilon)$. As this is the case for every $\epsilon > 0$, we get $f^{\#} \in \overline{\Theta}$. Clearly, $f^{\#}$ is an element of $\overline{\Theta}$ for which we have that $f = f^{\#} \circ h$.
(ii) If we apply (i) on $\Theta = G$, we get the openness of h. By definition $s \in \mathrm{Mor}(\mathcal{G}, \mathcal{F}) \Leftrightarrow \forall_{f \in F}(f \circ s \in G)$. If $f \in F$, then $f \circ s = f^{\#} \in \overline{\Theta} = \overline{G} = G$. That $(h^*)^{-1}(\Phi) = \{g \in G \mid g \circ h \in \Phi\}$ is a base for G follows from Proposition 14.11(iv). □

Definition 14.14 If $\Phi \subseteq \mathbb{F}(X)$ and $a \bowtie_{\mathbb{R}} b :\Leftrightarrow |a - b| > 0$ is the point-point apartness relation on $\mathbb{R}$, if $x, y \in X$ and $A, B \subseteq X$, we define the following point–point and set–set apartness relations on X:[10]

$$x \bowtie_{\Phi} y :\Leftrightarrow \exists_{f \in \Phi}(f(x) \bowtie_{\mathbb{R}} f(y)),$$

$$A \bowtie_{\Phi} B :\Leftrightarrow \exists_{f \in \Phi} \forall_{a \in A} \forall_{b \in B}(f(a) = 0 \wedge f(b) = 1),$$

[10] See Chapter 18 below.

$$A \bowtie_\Phi B :\Leftrightarrow \exists_{f \in \Phi} \exists_{c>0} \forall_{a \in A} \forall_{b \in B} (|f(a) - f(b)| \geq c).$$

We write $x \bowtie_\theta y$, $A \bowtie_\theta B$ and $A \bowtie_\theta B$, if $\theta \in \Phi$ realises $x \bowtie_\Phi y$, $A \bowtie_\Phi B$ and $A \bowtie_\Phi B$, respectively. We say that Φ *separates the points of* X, if

$$\forall_{x,y \in X} \left(\forall_{f \in \Phi} (f(x) = f(y)) \Rightarrow x = y \right).$$

Recall that $a \bowtie_{\mathbb{R}} b$ is *tight*, that is, $\neg(a \bowtie_{\mathbb{R}} b) \Rightarrow a = b$, for every $a, b \in \mathbb{R}$. It is easy to see that $a \bowtie_{\mathbb{R}} b \Leftrightarrow a \bowtie_{\mathrm{Bic}(\mathbb{R})} b$, for every $a, b \in \mathbb{R}$, and $\bowtie_\Phi \subseteq \bowtie_\Phi$. Classically, the separation of points is expressed by contraposition of the above implication. Next we see how the apartness (separation) of points or subsets by some Bishop topology F implies their apartness (separation) by some base for F.

Proposition 14.15 *Let Φ a base for F, $x, y \in X$, and $A, B \subseteq X$.*

(i) $x \bowtie_F y \Rightarrow x \bowtie_\Phi y$.

(ii) *If F separates the points of X, then Φ separates the points of X.*

(iii) $A \bowtie_F B \Rightarrow A \bowtie_\Phi B$.

(iv) $\forall_{c>0} \exists_{\theta \in \Phi} \forall_{x \in X} (\theta(x) \geq c)$.

Proof (i) If $x \bowtie_f y$ and $U(\theta, f, \frac{\epsilon_0}{4})$, where $f \in F, \theta \in \Phi$ and $\epsilon_0 = |f(x) - f(y)|$, then

$$\epsilon_0 \leq |f(x) - \theta(x)| + |\theta(x) - \theta(y)| + |\theta(y) - f(y)| \Leftrightarrow$$
$$\epsilon_0 \leq \frac{\epsilon_0}{4} + |\theta(x) - \theta(y)| + \frac{\epsilon_0}{4} \Leftrightarrow$$
$$0 < \frac{\epsilon_0}{2} \leq |\theta(x) - \theta(y)|.$$

(ii) If $x, y \in X$ such that $\forall_{\theta \in \Phi}(\theta(x) = \theta(y))$, it suffices to show that $\forall_{f \in F}(f(x) = f(y))$. We fix some $f \in F$ and by the tightness of $a \bowtie_{\mathbb{R}} b$ it suffices to show that $\neg(f(x) \bowtie_{\mathbb{R}} f(y))$. If $f(x) \bowtie_{\mathbb{R}} f(y)$, then we set $\epsilon_0 = |f(x) - f(y)| > 0$. If $\theta \in \Phi_0$ such that $U(\theta, f, \frac{\epsilon_0}{4})$, we work as in case (i), where the term $|\theta(x) - \theta(y)|$ by our hypothesis vanishes, and we reach the required absurdity.
(iii) Let $f \in F$ such that $A \bowtie_f B$ and $\theta \in \Phi$ such that $U(\theta, f, \frac{c}{2})$, where $2c = |f(A) - f(B)|$. Working as in case (i) we get that $|\theta(a) - \theta(b)| \geq c$, for every $a \in A$ and $b \in B$.
(iv) If $c > 0$ and $\theta \in \Phi_0$ such that $U(\theta, \overline{2c}, c)$, then for every $x \in X$ we have that $|\theta(x) - 2c| \leq c \Leftrightarrow -c \leq \theta(x) - 2c \leq c \Leftrightarrow c \leq \theta(x) \leq 3c$. □

14.6 The First Base Theorem

Classically, a base of a uniform structure given by pseudometrics is generated by a subbase by taking all suprema of finite subsets of the subbase (see [24, p. 217]). The generation of a base of a Bishop topology from a subbase of it is described in the following theorem, under the hypothesis that all elements of the subbase are bounded, real-valued functions.

Definition 14.16 A *pseudo-Bishop space* is a pair $\mathcal{F}_0 = (X, F)$ satisfying conditions (BS_1)–(BS_3) in the definition of a Bishop space. If $F_0 \subseteq \mathbb{F}(X)$, the *least pseudo-Bishop space* $\bigvee_0 F_0$ generated by F_0 is defined by the following inductive rules:

$$\frac{f_0 \in F_0}{f_0 \in \bigvee_0 F_0}, \quad \frac{a \in \mathbb{R}}{\overline{a}^X \in \bigvee_0 F_0},$$

$$\frac{f, g \in \bigvee_0 F_0}{f + g \in \bigvee_0 F_0}, \quad \frac{f \in \bigvee_0 F_0, \ g =_{\mathbb{F}(X)} f}{g \in \bigvee F_0}, \quad \frac{f \in \bigvee_0 F_0, \ \phi \in \mathrm{Bic}(\mathbb{R})}{\phi \circ f \in \bigvee_0 F_0},$$

which induce the following induction principle $\mathrm{Ind}_{\mathcal{F}_0}$ on $\bigvee_0 F_0$

$$\Big[\forall_{f_0 \in F_0}(P(f_0)) \ \& \ \forall_{a \in \mathbb{R}}(P(\overline{a})) \ \& \ \forall_{f,g \in \bigvee_0 F_0}(P(f) \ \& \ P(g) \Rightarrow P(f+g))$$

$$\& \ \forall_{f \in \bigvee_0 F_0} \forall_{\phi \in \mathrm{Bic}(\mathbb{R})}(P(f) \Rightarrow P(\phi \circ f))\Big] \Rightarrow \forall_{f \in \bigvee_0 F_0}(P(f)),$$

where P is a bounded formula. Next we describe the base of a topology F with respect to a given subbase of bounded elements of F.

Theorem 14.17 (First base theorem) *If F_0 is a subset of $\mathbb{F}(X)$ such that every element of F_0 is bounded, then $\bigvee_0 F_0$ is a base of $\bigvee F_0$.*

Proof It is clear that $\overline{\bigvee_0 F_0} \subseteq \bigvee F_0$. Next we show inductively that $\bigvee F_0 \subseteq \overline{\bigvee_0 F_0}$. Of course, $F_0 \subseteq \bigvee_0 F_0$ and $\mathrm{Const}(X) \subseteq \bigvee_0 F_0$. If $f_1, f_2 \in \bigvee F_0$ such that $f_1, f_2 \in \overline{\bigvee_0 F_0}$, that is, for every $\epsilon > 0$ there exist $g_1, g_2 \in \bigvee_0 F_0$ such that $U(g_1, f_1, \frac{\epsilon}{2})$ and $U(g_2, f_2, \frac{\epsilon}{2})$, then $U(g_1 + g_2, f_1 + f_2, \epsilon)$. Since $g_1 + g_2 \in \bigvee_0 F_0$ and $\epsilon > 0$ is arbitrary, we get that $f_1 + f_2 \in \overline{\bigvee_0 F_0}$. If $f \in \bigvee F_0$ such that $f \in \overline{\bigvee_0 F_0}$, and if $g =_{\mathbb{F}(X)} f$, then $g \in \overline{\bigvee_0 F_0}$, as the uniform closure $\overline{\bigvee_0 F_0}$ of $\bigvee_0 F_0$ is always an extensional subset of $\mathbb{F}(X)$. Suppose next that $f' = \phi \circ f$, where $\phi \in \mathrm{Bic}(\mathbb{R})$ and $f \in \overline{\bigvee_0 F_0}$. Since by the $\bigvee$-lifting of boundedness every element of $\bigvee F_0$ is bounded, let $M > 0$ such that $|f| \leq \overline{M}$. Without loss of generality we assume that $M > 1$. There is no loss of generality, if for every bounded subset B of $\mathbb{R}$ and for every $\epsilon > 0$ we assume that $\omega_{\phi,B}(\epsilon) < 1$, as we may use the modulus $\omega^*_{\phi,B} = \omega_{\phi,B} \wedge \overline{\frac{1}{2}}$. Clearly, $2M > M + 1 > M + \omega_{\phi,B}(\epsilon)$. Let the bounded subset $[-2M, 2M]$ of $\mathbb{R}$, and suppose that $g \in \bigvee_0 F_0$ such that $U(g, f, \omega_{\phi,[-2M,2M]}(\epsilon))$ i.e., $\forall_{x \in X}(|g(x) - f(x)| \leq \omega_{\phi,[-2M,2M]}(\epsilon))$. Since

$$|g(x)| \leq |g(x) - f(x)| + |f(x)| \leq \omega_{\phi,[-2M,2M]}(\epsilon) + M < 1 + M < 2M,$$

for every $x \in X$, and since $|f| \leq \overline{M}$, we conclude that $g(x), f(x) \in [-2M, 2M]$, for every $x \in X$. Therefore, the hypothesis $U(g, f, \omega_{\phi,[-2M,2M]}(\epsilon))$ implies that $U(\phi \circ g, \phi \circ f, \epsilon)$. Since $\phi \circ g \in \bigvee_0 F_0$ and $\epsilon > 0$ is arbitrary, we get that $\phi \circ f \in \overline{\bigvee_0 F_0}$. Finally, we suppose that

$$\forall_{\epsilon>0}\exists_{g\in\bigvee F_0}\left(U(g, f, \epsilon) \;\&\; g \in \overline{\bigvee_0 F_0}\right).$$

Let $\epsilon > 0$, $g \in \bigvee F_0$ such that $U\big(g, f, \frac{\epsilon}{2}\big)$ and $h \in \bigvee_0 F_0$ such that $U\big(h, g, \frac{\epsilon}{2}\big)$. Since

$$U\big(h, g, \frac{\epsilon}{2}\big) \;\&\; U\big(g, f, \frac{\epsilon}{2}\big) \Rightarrow U(h, f, \epsilon),$$

$h \in \bigvee_0 F_0$, and $\epsilon > 0$ is arbitrary, we conclude that $f \in \overline{\bigvee_0 F_0}$. □

Within a classical theory of ordinals, if we define $\Phi : \mathrm{On} \to V$ by

$$\Phi_0 = F_0, \quad \Phi_{\alpha+1} = \overline{\bigvee_0 \Phi_\alpha}, \quad \Phi_\lambda = \bigcup_{\alpha<\lambda} \Phi_\alpha,$$

where λ is a limit ordinal, then, it can be shown (see [36, Proposition 3.5.6]) that if $F_0 \subseteq \mathbb{F}(X)$, and not necessarily $F_0 \subseteq \mathbb{F}^*(X)$, then $\bigvee F_0 = \Phi_{\omega_1}$, where ω_1 is the first uncountable limit ordinal. By the first base theorem, if $\bigvee F_0$ is a pseudocompact Bishop space, then $\bigvee F_0 = \overline{\bigvee_0 F_0} = \Phi_1$.

14.7 The Second Base Theorem

According to the classical Stone–Weierstrass theorem, if $(X, \mathcal{T})$ is a compact Hausdorff topological space and $A \subseteq C(X)$ such that (i) A separates the points of X, that is, $\forall_{x,y\in X}(x \neq y \to \exists_{f\in A}(f(x) \neq f(y)))$, (ii) $\mathrm{Const}(X) \subseteq A$, and (iii) A is a subalgebra of $C(X)$, then the uniform closure of A is $C(X)$. According to a slightly stronger formulation (see [33]), a subalgebra $A \subseteq C(X)$ is uniformly dense in $C(X)$ if and only if (i) A separates the points of X, and (ii) A contains a function f *bounded away from zero*, that is,

$$\inf\{|f(x)| \mid x \in X\} > 0.$$

Consequently, if X, Y are compact Hausdorff topological spaces, $h \in C(X \times Y)$ and $\epsilon > 0$, there are functions $f_1, \ldots, f_n \in C(X)$ and $g_1, \ldots, g_n \in C(Y)$ such that

$$U\left(\sum_{i=1}^{n}(f_i \circ \pi_1)(g_i \circ \pi_2), h, \epsilon\right),$$

that is, the function $h_\epsilon(x, y) = \sum_{i=1}^{n} f_i(x)g_i(y)$, for every $(x, y) \in X \times Y$, is uniformly ϵ-close to h. More generally, see [9, p. 314], if $(X_i, \mathcal{T}_i)_{i\in I}$ is a family of compact Hausdorff topological spaces and $\mathcal{T}$ is the product topology on $X = \prod_{i \in I} X_i$, then for every $f \in C(X)$ and $\epsilon > 0$, there exists $\phi \in \Sigma(X)$ such that $U(\phi, f, \epsilon)$, where

$$\Sigma_0(X) := \left\{ \prod_{j \in J} (f_j \circ \pi_j) \mid J \subseteq^{\text{fin}} I, f_j \in C(X_j), j \in J \right\},$$

$$\Sigma(X) := \left\{ \sum_{k=1}^{n} \phi_k \mid n \in \mathbb{N}, \phi_k \in \Sigma_0(X), 1 \leq k \leq n \right\}.$$

In [8, p. 108], the previous fact is shown within BISH for a sequence of compact spaces, that is, if $(X_n, d_n)_{n \in \mathbb{N}}$ is a sequence of compact metric spaces, where d_n is bounded by 1, for every $n \in \mathbb{N}$, then

$$C_u\left(\prod_{i \in \mathbb{N}} X_i \right) = \overline{\bigoplus_{i \in \mathbb{N}} C_u(X_i)},$$

where $\prod_{i \in \mathbb{N}} X_i$ is endowed with the product metric σ_∞, where

$$\sigma_\infty \left((x_i)_{i \in \mathbb{N}}, (y_i)_{i \in \mathbb{N}} \right) := \sum_{i \in \mathbb{N}} \frac{d_i(x_i, y_i)}{2^i}.$$

As a special case, if (X, d) and (Y, ρ) are compact metric spaces, then

$$C_u(X \times Y) = \overline{C_u(X) \oplus C_u(Y)},$$

where $X \times Y$ is endowed with the product metric σ_2, where

$$\sigma_2 \left((x_i)_{i=1}^2, (y_i)_{i=1}^2 \right) := \sum_{i=1}^{2} d_i(x_i, y_i).$$

While classically these are corollaries of the Stone–Weierstrass theorem, constructively these are lemmas for the proof of Bishop's version of the Stone–Weierstrass theorem for compact metric spaces.[11] We generalise these facts to pseudocompact Bishop spaces (Corollaries 14.24 and 14.21 of Theorem 14.19, respectively). First we show some expected uniform liftings. The next proposition incorporates the algebraic part of Lemma 5.11 in [8, p. 105], which refers to compact metric spaces, into the more general setting of pseudocompact Bishop spaces.

Lemma 14.18 *If $\Phi \subseteq \mathbb{F}^*(X)$ such that Φ is closed under addition, multiplication by reals and multiplication, then $\overline{\Phi} \subseteq \mathbb{F}^*(X)$ is closed under addition, multiplication by reals and multiplication. Moreover, if Φ is closed under $|.|$, then $\overline{\Phi}$ is closed under $|.|$.*

[11] An alternative approach to the constructive Stone–Weierstrass theorem for compact metric spaces is given in [40].

Proof Suppose that $f_1, f_2 \in \overline{\Phi}$ and $\theta_1, \theta_2 \in \Phi$. If $\epsilon > 0$, then

$$U\left(\theta_1, f_1, \frac{\epsilon}{2}\right) \;\&\; U\left(\theta_2, f_2, \frac{\epsilon}{2}\right) \Rightarrow U(\theta_1 + \theta_2, f_1 + f_2, \epsilon),$$

and since $\epsilon > 0$ is arbitrary and $\theta_1 + \theta_2 \in \Phi$ by hypothesis, we conclude that $f_1 + f_2 \in \overline{\Phi}$. For the closure of $\overline{\Phi}$ under multiplication by reals we fix some $\lambda \in \mathbb{R}$ and $f \in \overline{\Phi}$. Since there is some $\theta \in \Phi$ such that $U\big(\theta, f, \frac{\epsilon}{|\lambda|+\sigma}\big)$, where $\epsilon, \sigma > 0$, we have that

$$|\lambda\theta(x) - \lambda f(x)| = |\lambda||\theta(x) - f(x)| \leq |\lambda|\frac{\epsilon}{|\lambda| + \sigma} < 1\epsilon = \epsilon,$$

that is, $U(\lambda\theta, \lambda f, \epsilon)$. Since by our hypothesis $\lambda\theta \in \Phi$, we get that $\lambda f \in \overline{\Phi}$. For the closure of $\overline{\Phi}$ under multiplication and because of the elementary equality $f \cdot g = \frac{1}{2}((f+g)^2 - f^2 - g^2)$ it suffices to show that $f \in \overline{\Phi} \Rightarrow f^2 \in \overline{\Phi}$. For that we need to show first that $\overline{\Phi} \subseteq \mathbb{F}^*(X)$. If $f \in \overline{\Phi}$, $\epsilon > 0$ and $\theta \in \Phi$ such that $U(\theta, f, \epsilon)$, then $\forall_{x \in X}(|f(x)| \leq |\theta(x) - f(x)| + |\theta(x)|)$, hence if $M_\theta > 0$ is a bound of θ, then $\epsilon + M_\theta$ is a bound of f. Let $f \in \overline{\Phi}$ and M_f a bound of f such that $M_f > 1$. Next we fix some $\epsilon \leq 1$ and we show that $U(\theta', f^2, \epsilon)$, for some $\theta' \in \Phi$. Note that there is no loss of generality in our choice of ϵ, since $U(\theta', f^2, \epsilon \wedge 1) \Rightarrow U(\theta', f^2, \epsilon)$, for arbitrary $\epsilon > 0$, where $\epsilon \wedge 1 = \min\{\epsilon, 1\}$. If $\theta \in \Phi$ such that $U\big(\theta, f, \frac{\epsilon}{3M_f}\big)$, then for every $x \in X$ we get

$$|\theta(x)| \leq |\theta(x) - f(x)| + |f(x)| \leq \frac{\epsilon}{3M_f} + M_f < \epsilon + M_f \leq 1 + M_f < 2M_f,$$

$$\begin{aligned}
|\theta^2(x) - f^2(x)| &= |\theta(x) - f(x)||\theta(x) + f(x)| \\
&\leq |\theta(x) - f(x)|(|\theta(x)| + |f(x)|) \\
&\leq |\theta(x) - f(x)|(2M_f + M_f) \\
&= |\theta(x) - f(x)|3M_f \\
&\leq \frac{\epsilon}{3M_f}3M_f \\
&= \epsilon.
\end{aligned}$$

Hence $U(\theta^2, f^2, \epsilon)$, and since by our hypothesis $\theta^2 \in \Phi$, we get that $f^2 \in \overline{\Phi}$. Moreover, if for every $\theta \in \Phi$ we have that $|\theta| \in \Phi$, then if $f \in \overline{\Phi}$ and $U(\theta, f, \epsilon)$, for some $\epsilon > 0$, then $U(|\theta|, |f|, \epsilon)$, since $||\theta(x)| - |f(x)|| \leq |\theta(x) - f(x)| \leq \epsilon$ for every $x \in X$. Since ϵ is arbitrary, we get $|f| \in \overline{\Phi}$. □

The next theorem is a theorem of Stone–Weierstrass-type for pseudocompact Bishop spaces.[12]

[12] Theorem 14.19 corresponds to Nel's classical theorem on the approximation of bounded continuous functions (see [33]). According to it, if X is a topological space and A is a subalgebra of $C^*(X)$, where (i) A separates

Theorem 14.19 (Second base theorem) *If $\mathcal{F} = (X, \bigvee F_0)$ is a Bishop space such that every element of F_0 is bounded, and if $\Phi \subseteq \bigvee F_0$ is such that*

(i) $F_0 \subseteq \Phi$,

(ii) $\mathrm{Const}(X) \subseteq \Phi$,

(iii) Φ *is closed under addition and multiplication,*

then Φ is a base for $\bigvee F_0$. We call a subset Φ of $\bigvee F_0$ satisfying conditions (i)–(iii) *an* SW*-base of the pseudocompact topology $\bigvee F_0$.*

Proof By the first base theorem we know that $\bigvee_0 F_0$ is a base of F. We show that $\overline{\Phi} = \overline{\bigvee_0 F_0} = F$. Since $\Phi \subseteq F$, we get that $\overline{\Phi} \subseteq F$. Thus we need to show that $\overline{\bigvee_0 F_0} \subseteq \overline{\Phi}$. As the hypothesis $\bigvee_0 F_0 \subseteq \overline{\Phi}$ implies that $\overline{\bigvee_0 F_0} \subseteq \overline{\overline{\Phi}} = \overline{\Phi}$, it suffices to show $\bigvee_0 F_0 \subseteq \overline{\Phi}$. For this inclusion we use the induction principle associated to the definition of $\bigvee_0 F_0$. By hypothesis (i) and (ii) we have that $F_0 \subseteq \Phi \subseteq \overline{\Phi}$ and $\mathrm{Const}(X) \subseteq \Phi \subseteq \overline{\Phi}$, respectively. If $f, g \in \bigvee_0 F_0$ such that $f, g \in \overline{\Phi}$, then by Lemma 14.18 we get $f + g \in \overline{\Phi}$. If $f \in \bigvee_0 F_0$ such that $f \in \overline{\Phi}$, and $g =_{\mathbb{F}(X)} f$, then $g \in \overline{\Phi}$, as $\overline{\Phi}$ is an extensional subset of $\mathbb{F}(X)$. If $\phi \in \mathrm{Bic}(\mathbb{R})$ and $f \in \bigvee_0 F_0$ such that $f \in \overline{\Phi}$, we show that $\phi \circ f \in \overline{\Phi}$. Let $M_f > 0$ be a bound of f. Since $|f| \leq M_f$ we get that $\phi \circ f = \phi_{|[-M_f, M_f]} \circ f$, where by the definition of $\mathrm{Bic}(\mathbb{R})$ the function $\phi_{|[-M_f, M_f]}$ is uniformly continuous on $[-M_f, M_f]$. If $\epsilon > 0$, by the Weierstrass approximation theorem (see [8, p. 109]) there is a real polynomial p such that $U(p, \phi_{|[-M_f, M_f]}, \epsilon)$. Hence

$$U(p \circ f, \phi_{|[-M_f, M_f]} \circ f, \epsilon) \Leftrightarrow U(p \circ f, \phi \circ f, \epsilon).$$

By hypothesis (iii) we have that $p \circ f \in \overline{\Phi}$; since Φ is closed under multiplication and includes $\mathrm{Const}(X)$, Φ is closed under multiplication by reals, therefore by Lemma 14.18 $\overline{\Phi}$ is closed under addition, multiplication by reals and multiplication. Since $\epsilon > 0$ is arbitrary, we conclude that $\phi \circ f \in \overline{\Phi}$. □

14.8 Applications of the Second Base Theorem

We present some fundamental corollaries of the second base theorem. As in the case of the classical Stone–Weierstrass theorem, with it we can determine canonical bases of the product of pseudocompact Bishop spaces.

Definition 14.20 If (X, F) and (Y, G) are Bishop spaces, we define the subset $F \oplus G$ of $F \times G$ by

disjoint zero sets, and (ii) A contains a function f which is bounded away from zero, then A is uniformly dense in $C^*(X)$.

$$F \oplus G := \left\{ \sum_{i=1}^{n} (f_i \circ \pi_1)(g_i \circ \pi_2) \mid n \in \mathbb{N}, f_i \in F, g_i \in G, 1 \leq i \leq n \right\}.$$

If $\mathcal{F}_n = (X_n, F_n)$ is a sequence of Bishop spaces, we define the subset $\bigoplus_{n\in\mathbb{N}} F_n$ of $\prod_{n\in\mathbb{N}} F_n$ by

$$\Sigma_0 := \left\{ \prod_{i=1}^{n} (f_k \circ \pi_k) \mid n \in \mathbb{N}, f_k \in F_k, 1 \leq k \leq n \right\},$$

$$\bigoplus_{n\in\mathbb{N}} F_n := \left\{ \sum_{j=1}^{m} \phi_j \mid m \in \mathbb{N}, \phi_j \in \Sigma_0, 1 \leq j \leq m \right\}.$$

Corollary 14.21 *If (X, F) and (Y, G) are pseudocompact Bishop spaces, then $F \oplus G$ is a base for $F \times G$.*

Proof By Theorem 14.19 it suffices to show that $F \oplus G$ includes $F \circ \pi_1$, $G \circ \pi_2$, $\mathrm{Const}(X \times Y)$, and it is closed under addition and multiplication. If $f \in F$ and $g \in G$, then $f \circ \pi_X = (f \circ \pi_X)(\overline{1} \circ \pi_Y) \in F \oplus G$ and $g \circ \pi_Y = (\overline{1} \circ \pi_X)(g \circ \pi_Y) \in F \oplus G$. If $\overline{a} \in \mathrm{Const}(X \times Y)$, then $\overline{a} = \overline{a} \circ \pi_X \in F \oplus G$, where we use the same notation for $\overline{a} \in \mathrm{Const}(X) \subseteq F$. The closure of $F \oplus G$ under addition is shown by the equality

$$\begin{aligned}\sum_{i=1}^{n} (f_i \circ \pi_X)(g_i \circ \pi_Y) &+ \sum_{j=1}^{m} \left(f_j{}' \circ \pi_X\right)\left(g_j{}' \circ \pi_Y\right) \\ &= \sum_{k=1}^{n+m} \left(f_k{}'' \circ \pi_X\right)\left(g_k{}'' \circ \pi_Y\right) \in F \oplus G,\end{aligned}$$

where $f''{}_k = f_k$ if $1 \leq k \leq n$, $f''{}_k = f_j{}'$ if $n+1 \leq k = n+j \leq n+m$, and $g''{}_k$ is defined similarly. If $f_i, f_j{}' \in F$ and if $(x, y) \in X \times Y$, we have that

$$\left[(f_i \circ \pi_X)\left(f_j{}' \circ \pi_Y\right)\right](x, y) = f_i(x) f_j{}'(x) = \left(f_i f_j{}' \circ \pi_X\right)(x, y).$$

Similarly we show that if $g_i, g_j{}' \in G$, then $(g_i \circ \pi_Y)(g_j{}' \circ \pi_Y) = g_i g_j{}' \circ \pi_Y$. The closure of $F \oplus G$ under multiplication follows by the equalities

$$\begin{aligned}\theta \cdot \theta' &= \sum_{i=1}^{n} (f_i \circ \pi_X)(g_i \circ \pi_Y) \sum_{j=1}^{m} \left(f_j{}' \circ \pi_X\right)\left(g_j{}' \circ \pi_Y\right) \\ &= \sum_{j=1}^{m} \left[\sum_{i=1}^{n} (f_i \circ \pi_X)(g_i \circ \pi_Y)\right] \left(f_j{}' \circ \pi_X\right)\left(g_j{}' \circ \pi_Y\right)\end{aligned}$$

$$= \sum_{j=1}^{m}\sum_{i=1}^{n}(f_i \circ \pi_X)\,(f_j{}' \circ \pi_X)\,(g_i \circ \pi_Y)\,(g_j{}' \circ \pi_Y)$$

$$= \sum_{j=1}^{m}\sum_{i=1}^{n}\,(f_i f_j{}' \circ \pi_X)\,(g_i g_j{}' \circ \pi_Y) \in F \oplus G,$$

since F and G are rings, $\sum_{i=1}^{n}(f_i f_j{}' \circ \pi_X)\,(g_i g_j{}' \circ \pi_Y) \in F \oplus G$ by the definition of $F \oplus G$, and $F \oplus G$ is closed under addition. □

Proposition 14.22 *If (X, F) and (Y, G) are pseudocompact Bishop spaces, then $F \oplus G = \mathcal{A}(\Phi_0)$, where $\Phi_0 = (F \circ \pi_1) \cup (G \circ \pi_2)$ and $\mathcal{A}(\Phi_0)$ is the least set of real-valued functions on X including Φ_0 that is closed under addition, multiplication by reals and multiplication.*[13]

Proof First we show that $\mathcal{A}(\Phi_0) \subseteq F \oplus G$. Since $\Phi_0 \subseteq F \oplus G$, we have that $\mathcal{A}(\Phi_0) \subseteq \mathcal{A}(F \oplus G) = F \oplus G$, since, as we showed in the proof of Corollary 14.21, $F \oplus G$ is an algebra. For the converse inclusion we have that if $f_i \in F$ and $g_i \in G$ for every $i \in \{1, \ldots, n\}$, then $f_i \circ \pi_1 \in F \circ \pi_1$, $g_i \circ \pi_2 \in G \circ \pi_2$, and $(f_i \circ \pi_1)(g_i \circ \pi_2) \in \mathcal{A}(\Phi_0)$ for every $i \in \{1, \ldots, n\}$. Since $\mathcal{A}(\Phi_0)$ is closed under addition, we get that the arbitrary element $\sum_{i=1}^{n}(f_i \circ \pi_1)(g_i \circ \pi_2)$ of $F \oplus G$ is in $\mathcal{A}(\Phi_0)$. □

As $\mathrm{Const}(X \times Y) = \mathrm{Const}(X) \circ \pi_1$, the set $\Phi_0 = (F \circ \pi_1) \cup (G \circ \pi_2)$ includes $\mathrm{Const}(X \times Y)$. If Φ is a base of $F \times G$, then $\Phi_Y \subseteq F$ and $\Phi_X \subseteq Y$, where

$$\Phi_Y := \{\theta_y \mid \theta \in \Phi, y \in Y\}, \text{ and } \theta_y(x) = \theta(x, y), \text{ for every } x \in X,$$

$$\Phi_X := \{\theta_x \mid \theta \in \Phi, x \in X\}, \text{ and } \theta_x(y) = \theta(x, y), \text{ for every } y \in Y.$$

Proposition 14.23 *If (X, F) and (Y, G) are Bishop spaces and Φ is a base for $F \times G$, then Φ_Y is a base for F and Φ_X is a base for G.*

Proof We only show that Φ_Y is a base of F, as the proof for Φ_X is similar. If $f \in F$ and $\epsilon > 0$, we find $\theta_y \in \Phi_Y$ such that $U(\theta_y, f, \epsilon)$. Since $f \circ \pi_1 \in F \times G$, there exists $\theta \in \Phi$ such that $U(\theta, f \circ \pi_1, \epsilon) \leftrightarrow \forall_{(x,y) \in X \times Y}(|\theta(x, y) - f(x)| \leq \epsilon)$. If y_0 inhabits Y, we consider the function θ_{y_0}, for which we get that $\forall_{x \in X}(|\theta_{y_0}(x) - f(x)| \leq \epsilon)$, that is, $U(\theta_{y_0}, f, \epsilon)$. □

According to Definition 14.20, if $\theta_1, \theta_2 \in \bigoplus_{n \in \mathbb{N}} F_n$, then

$$\theta_1 = \sum_{j=1}^{m}\prod_{k=1}^{n_j}(f_{k,j} \circ \pi_k), \quad \theta_2 = \sum_{w=1}^{s}\prod_{u=1}^{t_w}(f_{u,w} \circ \pi_u),$$

[13] This set is defined non-inductively in [8, p. 105].

where $f_{k,j} \in F_k$ for every $k \in \{1, \ldots, n_j\}$ and $f_{u,w} \in F_u$ for every $u \in \{1, \ldots, t_w\}$. Their sum $\theta_1 + \theta_2$ is in $\bigoplus_{n\in\mathbb{N}} F_n$, since

$$\theta_1 + \theta_2 = \sum_{b=1}^{m+s} \prod_{c=1}^{a_b} (f'_{c,b} \circ \pi_c),$$

where if $1 \leq b \leq m$, then $a_b = n_b$ and $f'_{c,b} = f_{c,b}$ for every $1 \leq c \leq n_b$, and if $m+1 \leq b = m+w \leq m+s$, then $a_b = t_{b-w}$ and $f'_{c,b} = f_{c,b-m}$ for every $1 \leq c \leq t_w$. As $\bigoplus_{n\in\mathbb{N}} F_n$ is closed under addition, we get that their product $\theta_1\theta_2$ is in $\bigoplus_{n\in\mathbb{N}} F_n$, since

$$\begin{aligned}
\theta_1\theta_2 &= \sum_{w=1}^{s} \left[\sum_{j=1}^{m} \prod_{k=1}^{n_j} (f_{k,j} \circ \pi_k) \right] \prod_{u=1}^{t_w} (f_{u,w} \circ \pi_u) \\
&= \sum_{w=1}^{s} \sum_{j=1}^{m} \prod_{k=1}^{n_j} (f_{k,j} \circ \pi_k) \prod_{u=1}^{t_w} (f_{u,w} \circ \pi_u) \\
&= \sum_{w=1}^{s} \sum_{j=1}^{m} \prod_{k=1}^{m_{j,w}} (f_{k,j} \circ \pi_k)(f_{k,w} \circ \pi_k) \prod_{\sigma=1}^{M_{j,w}} (g_{\sigma,u} \circ \pi_\sigma) \\
&= \sum_{w=1}^{s} \sum_{j=1}^{m} \prod_{k=1}^{m_{j,w}} (f_{k,j} f_{k,w} \circ \pi_k) \prod_{\sigma=1}^{M_{j,w}} (g_{\sigma,u} \circ \pi_\sigma),
\end{aligned}$$

where $m_{j,w} := \min\{n_j, t_w\}$ and $M_{j,w} := \max\{n_j, t_w\}$, and if $t_w \geq n_j$, then $u = w$ and $g_{\sigma,u} = f_{\sigma,u}$ for every $n_j + 1 \leq \sigma \leq t_w$, while if $t_w \leq n_j$, then $u = j$ and $g_{\sigma,u} = f_{\sigma,u}$ for every $t_w + 1 \leq \sigma \leq n_j$. Since $\bigoplus_{n\in\mathbb{N}} F_n$ includes $\mathrm{Const}(X)$ and $F_n \circ \pi_n$ for every $n \in \mathbb{N}$, by the second base theorem we get the following corollary.

Corollary 14.24 *If $\mathcal{F}_n = (X_n, F_n)$ is a sequence of pseudocompact Bishop spaces and $\mathcal{F} = (X, F)$, where $X = \prod_{n\in\mathbb{N}} X_n$ and $F = \prod_{n\in\mathbb{N}} F_n$, then $\bigoplus_{n\in\mathbb{N}} F_n$ is a base for F.*

Similarly to Proposition 14.22, we get the following.

Proposition 14.25 *If $\mathcal{F}_n = (X_n, F_n)$ is a sequence of pseudocompact Bishop spaces, then $\bigoplus_{n\in\mathbb{N}} F_n = \mathcal{A}(\Phi_0)$, where $\Phi_0 = \bigcup_{n\in\mathbb{N}}(F_n \circ \pi_n)$.*

Next we get a base and a subbase of the codomain of an open morphism, given a base of the domain Bishop space, while the topology of the codomain Bishop space is not given beforehand as a Bishop closure of a subbase.

Proposition 14.26 *Suppose that $\mathcal{F} = (X, \bigvee F_0)$ is a pseudocompact Bishop space, $\mathcal{G} = (Y, G)$ is a Bishop space, $h : X \to Y \in \mathrm{Mor}(\mathcal{F}, \mathcal{G})$ is onto Y with $s \colon Y \to X$ a modulus of surjectivity for h. If $\Phi \subseteq F$ and satisfies*

(i) $F_0 \subseteq \Phi$,

(ii) $\mathrm{Const}(X) \subseteq \Phi$,

(iii) Φ *is closed under addition and multiplication,*

(iv) $\forall_{f \in \Phi} \exists_{g \in G} (f = g \circ h)$,

then $\Theta = (h^*)^{-1}(\Phi)$ is is a base for G and $G_0 = (h^*)^{-1}(F_0)$ is a subbase for G, where $h^* \colon G \to F$ is the induced mapping from h (see Definition 14.5).

Proof By hypothesis (i) we get that $G_0 \subseteq \Theta$, and by (ii) $\mathrm{Const}(Y) \subseteq \Theta$. If $g_1, g_2 \in \Theta$, then by hypothesis (iii) we have that $(g_1 + g_2) \circ h = (g_1 \circ h) + (g_2 \circ h) \in \Phi$, and $(g_1 \cdot g_2) \circ h = (g_1 \circ h) \cdot (g_2 \circ h) \in \Phi$, that is, Θ is closed under addition and multiplication. Since $h \in \mathrm{Mor}(\mathcal{F}, \mathcal{G})$ is a surjection, the Bishop space $\mathcal{G}$ is also pseudocompact, and by Theorem 14.19 we have that Θ is a base of $\bigvee G_0$. By the uniform lifting of openness (Proposition 14.13) we have that h is open, since by Theorem 14.19 we get that Φ is a base of $\bigvee F_0$, while hypothesis (iv) implies that $\forall_{f \in \overline{\Phi} = F} \exists_{g \in \overline{G} = G} (f = g \circ h)$. By Proposition 14.11(iv) we have that Θ is a base for G too. Since Θ is a base for $\bigvee G_0$ and G, we conclude that $G = \bigvee G_0$. □

Next we translate the non-algebraic part of Lemma 5.11 in [8, p. 105], and Lemma 5.12 in [8, p. 106], into the general setting of pseudocompact Bishop spaces. The lattice part of Proposition 14.27 translates the classical fact that if A is a subring of $C^*(X)$ including the constant functions, then its uniform closure is a sublattice of $C^*(X)$ (see [24, p. 241]). We avoid the use of supremum and infimum in our formulation, as in [8] these exist by the compactness property of the related metric spaces.

Proposition 14.27 *If F is a pseudocompact topology on X and $\Phi \subseteq F$ such that*

(i) $\mathrm{Const}(X) \subseteq \Phi$, *and*

(ii) Φ *is closed under addition and multiplication,*

then the following hold:

(a) $\overline{\Phi}$ *is closed under* $|.|$, $\vee$ *and* $\wedge$;

(b) *if $f \in \overline{\Phi}$ such that $\forall_{x \in X}(|f(x)| \geq c)$, for some $c > 0$, then $\frac{1}{f} \in \overline{\Phi}$.*

Proof (a) Suppose that $f \in \overline{\Phi}$. If $M > 0$ is a bound of f, we find $\lambda > 0$ such that λf has 1 as a bound; just take $\lambda = \frac{1}{M}$. By Lemma 14.18 we have that $\lambda f = \overline{\lambda f} \in \overline{\Phi}$. By Lemma 5.9 in [8, pp. 104–105], for every $\epsilon > 0$ there exists a strict polynomial $p : \mathbb{R} \to \mathbb{R}$ such that

$$U([-1,1]; p, |.|, \epsilon),$$

which implies that

$$U([-1,1]; p \circ \lambda f, |\lambda f|, \epsilon).$$

By Lemma 14.18 we have that $p \circ \lambda f \in \overline{\Phi}$, and since $\epsilon > 0$ is arbitrary, we get that $|\lambda f| = \lambda |f| \in \overline{\Phi}$; therefore $|f| \in \overline{\Phi}$. Consequently, $f \vee g = \frac{1}{2}(f+g+|f-g|)$, $f \wedge g = \frac{1}{2}(f+g-|f-g|) \in \overline{\Phi}$.
(b) Let $M > 0$ be a bound of f, and without loss of generality we take $M > 1$. If $\lambda(x) = \frac{f(x)^2}{M^2}$, for every $x \in X$, we get that $\frac{c^2}{M^2} \leq \lambda(x) \leq 1$ and $|1 - \lambda(x)| = 1 - \lambda(x) = 1 - \frac{f(x)^2}{M^2} < 1$, for every $x \in X$. Therefore, the function

$$h(x) := \sum_{n=0}^{\infty} \frac{f(x)}{M^2}(1 - \lambda(x))^n$$

is well defined, since it is given through geometric series. Moreover,

$$h(x) = \frac{f(x)}{M^2} \sum_{n=0}^{\infty} (1 - \lambda(x))^n = \frac{f(x)}{M^2} \frac{1}{1 - (1 - \lambda(x))} = \frac{f(x)}{M^2} \frac{1}{\lambda(x)} = \frac{1}{f(x)}.$$

By Lemma 14.18 we have that all partial sums of h are in $\overline{\Phi}$, therefore by their uniform convergence to h we get $h \in \overline{\Phi}$. □

Definition 14.28 If (X, d) is a metric space, let $U_0(X) := \{d_{x_0} \mid x_0 \in X\}$, where we define $d_{x_0}(x) := d(x, x_0)$ for every $x \in X$.

In [8, p. 108], the following is a corollary of Bishop's version of the Stone–Weierstrass theorem.

Corollary 14.29 *If (X, d) is a compact metric space with positive diameter, then $\mathcal{A}(U_0(X))$ is dense in $C_u(X)$.*

Proposition 14.30 *If (X, d) is a compact metric space with positive diameter, then $\bigvee U_0(X) = C_u(X)$.*

Proof Since $U_0(X) \subseteq C_u(X)$, by the $\bigvee$-lifting of uniform continuity for the bounded elements of $U_0(X)$ (see [36, Proposition 3.4.9]) we get $\bigvee U_0(X) \subseteq C_u(X)$. For the converse inclusion, and using Corollary 14.29, it suffices to show that $\overline{\mathcal{A}(U_0(X))} \subseteq \bigvee U_0(X)$. Of course, $U_0(X) \subseteq \bigvee U_0(X)$ and, as a Bishop topology is always closed under addition, multiplication by reals and multiplication, we get $\mathcal{A}(U_0(X)) \subseteq \bigvee U_0(X)$. By condition (BS_4) we get $\overline{\mathcal{A}(U_0(X))} \subseteq \overline{\bigvee U_0(X)} = \bigvee U_0(X)$. □

Consequently, if X is a compact metric space with positive diameter, then

$$\Phi = \mathcal{A}\big(U_0(X) \cup \{\overline{1}\}\big)$$

is an SW-base of the uniform topology $C_u(X)$.

Definition 14.31 Let (X, d) and (Y, ρ) be metric spaces. The set of *Lipschitz functions* from X to Y is defined by

$$\mathrm{Lip}(X, Y) := \bigcup_{\sigma \geq 0} \mathrm{Lip}(X, Y, \sigma),$$

$$\mathrm{Lip}(X, Y, \sigma) := \{f \in \mathbb{F}(X, Y) \mid \forall_{x,y \in X}(\rho(f(x), f(y)) \leq \sigma d(x, y))\}.$$

If $Y = \mathbb{R}$, we write $\mathrm{Lip}(X)$ and $\mathrm{Lip}(X, \sigma)$, respectively.

Clearly, $\mathrm{Lip}(X, Y) \subseteq C_u(X, Y)$. If $f \in \mathrm{Lip}(X, Y)$, then f sends a bounded subset of X to a bounded subset of Y, which is not generally the case for some $g \in C_u(X, Y)$; the identity map $\mathrm{id} : \mathbb{N} \to \mathbb{R}$, where $\mathbb{N}$ is equipped with the discrete metric, is in $\mathrm{C_u}(\mathbb{N}) \setminus \mathrm{Lip}(\mathbb{N})$ and $\mathrm{id}(\mathbb{N}) = \mathbb{N}$ is unbounded in $\mathbb{R}$.

Proposition 14.32 *If X is a metric space, the set* $\mathrm{Lip}(X)$ *includes the sets* $U_0(X)$, $\mathrm{Const}(X)$, *and it is closed under addition and multiplication by reals. Moreover, if every element of* $C_u(X)$ *is a bounded function, then* $\mathrm{Lip}(X)$ *is closed under multiplication.*

Proof All parts of the proof are straightforward. If M_f is a bound for f, then if $f \in \mathrm{Lip}(X, \sigma)$, then $f^2 \in \mathrm{Lip}(X, 2M_f\sigma)$. Since the closure of $\mathrm{Lip}(X)$ under addition is trivial to show, its closure under multiplication follows. □

The next result is an immediate corollary of the second base theorem.

Corollary 14.33 *If X is a compact metric space with a positive diameter, then* $\mathrm{Lip}(X)$ *is an* SW*-base of* $C_u(X)$.

Proof The uniform topology $C_u(X)$ on the compact space X is pseudocompact (see Corollary 4.2 in [8, p. 95]). By Proposition 14.30 its subbase $U_0(X)$ is included in $\mathrm{Lip}(X)$, hence by Proposition 14.32 we conclude that $\mathrm{Lip}(X)$ is an SW-base of $C_u(X)$. □

What is remarkable with this corollary is that it follows from the second base theorem for pseudocompact Bishop spaces without involving any direct proof of the density of $\mathrm{Lip}(X)$ in $C_u(X)$. Such a direct proof for totally bounded metric spaces is given in [40], while another formulation of the latter density, shown in [46], is analogous to the formulation of the McShane–Whitney extension theorem for Lipschitz functions. One can generalise Corollary 14.33 to totally bounded metric spaces.

14.9 Concluding Remarks

In the technical part of this chapter we presented some fundamental results on bases of pseudocompact Bishop spaces, which indicate that the theory of pseudocompact Bishop spaces is a natural, constructive counterpart to the the classical theory of the ring $C^*(X)$ of bounded, continuous, real-valued functions on a topological space X, as this is developed, for example, in the classic book [24].

While in the classical theory one refers to compact topological spaces and uses the Stone–Weierstrass theorem, within the theory of Bishop spaces we refer to pseudocompact Bishop spaces and we use only the Weierstrass approximation theorem.[14] The classical version of the Stone–Weierstrass theorem for compact topological spaces requires that Φ separate the points of X, a condition that is not used here. On the other hand, here we used the hypothesis that Φ includes a subbase of the corresponding Bishop topology. Clearly, a Stone–Weierstrass theorem for compact Bishop spaces will depend on the notion of compactness that is used for them. Such a notion of compactness should include that of pseudocompactness, as it is the case for the notion of a 2-*compact* Bishop space, developed in [36] and in [39].

A more interesting approach to compactness is developed in [49], where the notion of a *Comfort-compact* Bishop topology is introduced. There, a Bishop topology F is called Comfort-compact if it is pseudocompact and every non-zero *character* π of F, that is, every non-zero ring-homomorphism $\pi\colon F \to \mathbb{R}$, is *fixed*, namely there is some $x \in X$ such that

$$\pi = \pi_x, \quad \pi_x(f) = f(x)$$

for every $f \in F$. This function-theoretic notion of compactness seems more suitable to the theory of Bishop spaces than 2-compactness.

In [15], Comfort used the notion of maximal ideal in the formulation of his notion of topological compactness. Comfort's motivation was his need to avoid the use of the axiom choice in the construction of the Stone–Čech compactification and the proof of the Tychonoff theorem. Comfort defined a topological space to be compact if it is a completely regular Hausdorff space for which each *maximal ideal* $\mathcal{M}$ in $C^*(X)$ is *fixed*, that is, there is some $x \in X$ such that

$$\mathcal{M} = \mathcal{M}_x, \quad \mathcal{M}_x := \{f \in C^*(X) \mid f(x) = 0\}.$$

Using classical logic, but avoiding the axiom of choice, Comfort showed many expected properties for his notion of compactness. For example, based on a theorem

[14] The Weierstrass approximation theorem is shown in [8] as a corollary of the Stone–Weierstrass theorem, but it can be proven constructively without using the whole machinery of the proof of the Stone–Weierstrass theorem.

of Stone–Čech-type he proved the corresponding Tychonoff theorem. In our elaboration of the theory of characters of pseudocompact Bishop topologies in [49], the use of the second base theorem is crucial, as one can extend uniquely a ring homomorphism $\Phi \to \mathbb{R}$, where Φ is a Stone–Weierstrass base of F, to a character of F. Our work [49] relies heavily on the base theorems for pseudocompact Bishop spaces presented here, and extends their study.

The description of bases of general, not necessarily pseudocompact, Bishop topologies, and the proof of results, similar to the two base theorems for pseudocompact Bishop topologies, for them, is to be a non-trivial open problem. Although the uniform continuity of a function $\phi \in \mathrm{Bic}(\mathbb{R})$ on every bounded subset of $\mathbb{R}$ is perfectly matched to the boundedness of an element of a pseudocompact Bishop topology, its relation to some unbounded element of a Bishop topology is more difficult to grasp. We hope that we, or somebody else, will tackle this problem in the theory of Bishop spaces in the near future.

The results presented here, like all fundamental theorems of the theory of Bishop spaces found in the given literature, are clearly related to the corresponding theorems of classical topology. By that we mean that their reading and understanding from a classical mathematician is almost straightforward, given the basic principles of Bishop-style constructive mathematics BISH. In this sense, the theory of Bishop spaces satisfies Bishop's need to develop constructive mathematics in a way friendly to a classical mathematician. Such a friendly coexistence between a constructive and a classical theory is not always possible within other, set-based approaches to constructive topology.

Moreover, the results presented here show how successfully non-trivial theorems of classical topology can be approached constructively, when suitable constructive concepts are chosen. Bishop's 'Copernican revolution' to suggest the notion of continuity of real-valued functions as the primary object of study, rather than the structure of open sets, reflects his deep insight that functions are suited better to constructive study rather than sets. In classical topology it has become clear after the pioneering work of Stone, Čech, Kolmogoroff, and Gelfand that the ring of continuous functions is an important tool in the study of the topological space itself. A similar attitude was taken in the influential classical integration theory of Daniell (see [18]). We find it most appropriate to close this chapter with Bishop's words in [3, p. 154], explaining his choice also to use functions as a starting point to his integration theory.

> The original approach was to let a measure be a function defined for sets. More recently it has been popular to let it be a function defined for functions. This approach fits our philosophy, that functions (at least continuous ones) should be preferred to sets as the primary objects of investigation whenever there is a choice.

Acknowledgements

Our research was partially supported by LMUexcellent, funded by the Federal Ministry of Education and Research (BMBF) and the Free State of Bavaria under the Excellence Strategy of the Federal Government and the Länder.

References

[1] Aczel, P., and Rathjen, M. 2010. *Constructive Set Theory.* Book draft Available at http://www1.maths.leeds.ac.uk/~rathjen/book.pdf.

[2] Aczel, P., and Fox, C. 2005. Separation Properties in Constructive Topology. Pages 176–192 of [17].

[3] Bishop, E. 1967. *Foundations of Constructive Analysis.* New York: McGraw-Hill.

[4] Bishop, E. 1971. The neat category of stratified spaces. University of California, San Diego. Unpublished manuscript.

[5] Bishop, E. 1971. Notes on constructive topology. University of California, San Diego. Unpublished manuscript.

[6] Bishop, E., and Cheng, H. 1972. *Constructive Measure Theory.* Memoirs of the American Mathematical Society, no. 116.

[7] Bishop, E. 1973. Schizophrenia in contemporary mathematics. *American Mathematical Society Colloquium Lectures.* Missoula: University of Montana.

[8] Bishop, E., and Bridges, D. S. 1985. *Constructive Analysis*, Grundlehren der math. Wissenschaften 279. Heidelberg, Berlin, New York: Springer-Verlag.

[9] Bourbaki, N. 1966. *General Topology.* Part 2, Chapters 5–10, Vol. 4. Boston, MA: Addison-Wesley.

[10] Bridges, D. S. 2012. Reflections on function spaces. *Ann. Pure Appl. Logic* **163**, 101–110.

[11] Bridges, D. S., and Richman, F. 1987. *Varieties of Constructive Mathematics.* Cambridge: Cambridge University Press.

[12] Bridges, D. S., and Vîţă, L. S. 2006. *Techniques of Constructive Analysis.* Universitext. New York: Springer.

[13] Bridges, D. S., and Vîţă, L. S. 2011. *Apartness and Uniformity: A Constructive Development.* CiE series "Theory and Applications of Computability". Berlin, Heidelberg: Springer-Verlag.

[14] Bridges, D. S., Palmgren, E., and Ishihara H. 2022. Constructive mathematics; approaches to constructive topology. In Zalta, E. N. (ed.), *Stanford Encyclopedia of Philosophy.*

[15] Comfort, W. W. 1968. A theorem of Stone–Čech type, and a theorem of Tychonoff type, without the axiom of choice; and their realcompact analogues. *Fund. Math.*, **63**(1), 97–110.

[16] Coquand, T., Sambin, G., Smith, J., and Valentini, S. 2003. Inductively generated formal topologies. *Ann. Pure Appl. Logic*, **124**, 71–106.

[17] Crosilla, L., and Schuster, P. (eds.) 2005. *From Sets and Types to Topology and Analysis, Towards Practicable Foundations for Constructive Mathematics*. Oxford: Clarendon Press.

[18] Daniell, P. J. 1918. A general form of integral. *Ann. Math., Second Series*, **19**(4), 279–294.

[19] Dugundji, J. 1966. *Topology*. Boston, MA: Allyn and Bacon.

[20] Escardó, M. H., and Xu, C. 2016. A constructive manifestation of the Kleene–Kreisel continuous functionals. *Ann. Pure Appl. Logic*, **167**(9), 770–793.

[21] Feferman, S. 2005. Predicativity. Pages 590–624 of: Shapiro, S. (ed.), *Oxford Handbook of Philosophy of Mathematics and Logic*. Oxford: Oxford University Press.

[22] Fréchet, M. 1906. Sur quelques points du calcul functionnel. *Rend. Circ. Mat. di Palermo*, **22**.

[23] Geins, J. 2018. Bridges between the Theory of Bishop Spaces and the Theory of C-Spaces. Master's Thesis, Mathematics Institute LMU.

[24] Gillman, L., and Jerison, M. 1960. *Rings of Continuous Functions*. New York: Van Nostrand.

[25] Grayson, R. J. 1981. Concepts of general topology in constructive mathematics and in sheaves. *Ann. Math. Logic*, **20**, 1–41.

[26] Grayson, R. J. 1982. Concepts of general topology in constructive mathematics and in sheaves, II. *Ann. Math. Logic*, **23**, 55–98.

[27] Ishihara, H. 2012. Two subcategories of apartness spaces. *Ann. Pure Appl. Logic*, **163**, 132–139.

[28] Ishihara, H. 2013. Relating Bishop's function spaces to neighborhood spaces. *Ann. Pure Appl. Logic*, **164**, 482–490.

[29] Ishihara, H., and Palmgren, E. 2006. Quotient topologies in constructive set theory and type theory. *Ann. Pure Appl. Logic*, **141**, 257–265.

[30] Ishihara, H., Mines, R., Schuster, P., and Vîţă, L. S. 2006. Quasi-apartness and neighborhood spaces. *Ann. Pure Appl. Logic*, **141**, 296–306.

[31] Martin-Löf, P. 1970. *Notes on Constructive Mathematics*. Stockholm: Almqvist and Wiksell.

[32] Myhill, J. 1975. Constructive set theory. *J. Symbol. Logic*, **40**, 347–382.

[33] Nel, L. D. 1968. Theorems of Stone–Weierstrass type for non-compact spaces. *Math. Zeitschr.*, **104**, 226–230.

[34] Palmgren, E. 2005. Continuity on the real line and in formal spaces. Pages 165–175 of [17].

[35] Palmgren, E. 2007. A constructive and functorial embedding of locally compact metric spaces into locales. *Topol. Appl.*, **154**, 1854–1880.

[36] Petrakis, I. 2015a. *Constructive topology of Bishop spaces. Ph.D. thesis*, Ludwig Maximilians University of Munich.

[37] Petrakis, I. 2015b. Completely regular Bishop spaces. Pages 302–312 of: *Evolving Computability*, CiE 2015, LNCS 9136. Berlin: Springer.

[38] Petrakis, I. 2016a. The Urysohn extension theorem for Bishop spaces. Pages 299–316 of: *Symposium on Logical Foundations of Computer Science 2016*, LNCS 9537. Berlin: Springer.

[39] Petrakis, I. 2016b. A constructive function-theoretic approach to topological compactness. Pages 605–614 of: *Proceedings of the 31st Annual ACM–IEEE Symposium on Logic in Computer Science* (LICS 2016).

[40] Petrakis, I. 2016c. A direct constructive proof of a Stone–Weierstrass theorem for metric spaces. Pages 364–374 of: *Pursuit of the Universal*, CiE 2016, LNCS 9709. Berlin: Springer.

[41] Petrakis, I. 2019a. Borel and Baire sets in Bishop spaces. Pages 240–252 of: *Computing with Foresight and Industry*, CiE 2019, LNCS 11558. Berlin: Springer.

[42] Petrakis, I. 2019b. Constructive uniformities of pseudometrics and Bishop topologies. *J. Logic Anal.*, **11**, (FT2), 1–44.

[43] Petrakis, I. 2019c. Dependent sums and dependent products in Bishop's set theory. Article no. 3 of *TYPES 2018*, LIPIcs, vol. 130.

[44] Petrakis, I. 2020a. McShane–Whitney extensions in constructive analysis. *Logical Meth. Comput. Sci.*, **16**(2020), 18:1–18:23.

[45] Petrakis, I. 2020b. Embeddings of Bishop spaces. *Journal of Logic Computation,* **30**(1), 2020, 349–379, https://doi.org/10.1093/logcom/exaa015.

[46] Petrakis, I. 2020c. Functions of Baire class one over a Bishop topology. Pages 215–227 of: *Beyond the Horizon of Computability*, CiE 2020, LNCS 12098. Berlin: Springer.

[47] Petrakis, I. 2020d. Families of sets in Bishop set theory. Habilitation Thesis, LMU Munich.

[48] Petrakis, I. 2021a. Direct spectra of Bishop spaces and their limits. *Logical Meth. Comput. Sci.*, **17**(2), 4:1–4:50.

[49] Petrakis, I. 2021b. Constructive Comfort-compactness. Preprint.

[50] Petrakis, I. 2021c. Chu representations of categories related to constructive mathematics. ArXiv:2106.01878v1.

[51] Petrakis, I. 2021d. Closed subsets in Bishop topological groups. ArXiv:2103.04718v1.

[52] Petrakis, I. 2021e. Integration theory of Bishop spaces. Preprint.

[53] Petrakis, I. 2022f. Proof-relevance in Bishop-style constructive mathematics. *Math. Struct. Comp. Sci.* (in the press).

[54] Petrakis, I., and Zeuner, M. 2021. Pre-measure spaces and pre-integration spaces in predicative Bishop–Cheng measure theory. ArXiv:2207.08684.

[55] Poincaré, H. 1906. Les mathématiques et la logique. *Rev. métaphysique morale*, **14**, 294–317.

[56] Russell, B. 1906. On some difficulties in the theory of transfinite numbers and order types. *Proc. Lond. Math Soc.*, **4**, 29–53.

[57] Sambin, G. 1987. Intuitionistic formal spaces – a first communication. Pages 187–204 of: Skordev, D. (ed.), *Mathematical Logic and its Applications*. New York: Plenum Press.

[58] Sambin, G. 2003. Some points in formal topology. *Theoret. Comput. Sci.*, **305**, 347–408.

[59] Sambin, G. 2021. *The Basic Picture: Structures for Constructive Topology*. Oxford: Oxford University Press (in the press).

[60] Spanier, E. 1963. Quasi-topologies. *Duke Math. J.*, **30**(1), 1–14.

[61] Troelstra, A. S. 1966. Institutionistic general topology. Ph.D. thesis, University of Amsterdam.

[62] Xu, C. 2015. A continuous computational interpretation of type theories. Ph.D. thesis, University of Birmingham.

15

Bishop Metric Spaces in Formal Topology

Tatsuji Kawai

15.1 Introduction

In Bishop constructive mathematics, pointwise continuous functions between compact metric spaces need not be uniformly continuous: if every pointwise continuous function between compact metric spaces were uniformly continuous, Brouwer's fan theorem would follow, to which recursive counterexample is known [36, Chapter 4, 7.6]. This seems to have led Bishop to adopt uniform continuity on compact subsets as a fundamental notion of continuity in his development of constructive analysis [2].

However, the gap between the notion of continuity in point-set topology and that of Bishop's theory of metric spaces has been a major obstacle to finding the right notion of general topology which naturally extends the theory of metric spaces. In particular, in the usual point-set topology, a metric space is endowed with a pointwise topology generated by open balls, and the topological notion of continuity between them is that of pointwise continuity. Moreover, the pointwise topology associated with a compact metric space, a complete and totally bounded space, need not be topologically compact constructively because compactness of that topology would imply Brouwer's fan theorem. In particular, the unit interval $[0, 1]$ need not be topologically compact constructively.

Independently of Bishop's development of constructive mathematics, Sambin [32] initiated *formal topology*, a pointfree topology [18] in constructive and predicative mathematics. Formal topology allows some important results in classical topology to be constructivised, which would not be possible in the point-set setting. Examples include the Heine–Borel theorem [6], Tychonoff's theorem [9, 27, 39], and Stone–Čech compactification [15]. From the viewpoint of Bishop constructive mathematics, however, the following questions should be raised: how are the notions and results in formal topology related to those of Bishop constructive mathematics [2, 3]? Are the results obtained in formal topology compatible with Bishop's own development of mathematics, or are they just as incompatible as classical point-set topology?

In this chapter, we review two strands of work which relate the notions of uniform continuity and compactness in Bishop's theory of metric spaces to the corresponding notions in formal topology.

One of the strands originates with the work by Palmgren [28]. He observed that morphisms between formal topology of real numbers correspond to functions between real numbers which are not only pointwise continuous, as they are usually treated in (classical) pointfree topology [18], but also locally uniformly continuous, that is, continuous in the sense of Bishop. In his subsequent work, Palmgren [29] extended this correspondence to Bishop locally compact metric spaces using Vickers's localic completions of metric spaces [38].

The second strand originates with the work by Spitters [34]. He showed that the metric notion of locatedness and the pointfree notion of overtness are intimately related through the localic completion. The notion of located subset of metric spaces [2, Chapter 4] is fundamental in constructive mathematics and plays an important role in its development.[1] On the other hand, the notion of overtness originates in topos theory [19], and it has been part of formal topology from its inception [32]. In [34], Spitters introduced a pointfree notion of located subset for compact regular formal topologies and showed that located subsets correspond to overt closed subtopologies. Moreover, he showed that on localic completions of compact metric spaces as mentioned previously, the pointfree notion of located subset exactly corresponds to the metric notion of located closed subset.

The purpose of this chapter is to summarise the above two strands of work and show how the two strands lead to purely pointfree characterisations of Bishop compact metric spaces and locally compact metric spaces. Our exposition is based on the work that has been carried out by Kawai [21, 22, 23, 24]. In this chapter, most of the proofs are omitted; the interested reader is referred to the above-mentioned papers for further details.

This chapter is organised as follows: Section 15.2 gives background on formal topology that will be needed in the later sections. Section 15.3 summarises Palmgren's work on the embedding of Bishop's theory of locally compact metric spaces into formal topology. Section 15.4 elaborates on the relation between locatedness and overtness due to Spitters. Building on these sections, Section 15.5 gives a pointfree characterisation of Bishop compact metric spaces, and Section 15.6 extends this characterisation to locally compact metric spaces. Section 15.7 discusses the relation between Bishop metric spaces and formal topology beyond the setting of

[1] To give one example which shows the importance and constructive peculiarity of locatedness, a closed subset of a compact metric space is compact if and only if it is located.

locally compact metric spaces. Section 15.8 closes the chapter with some remarks on related works.

We work informally in Bishop-style constructive mathematics. However, the work described here should be formalisable in Aczel's constructive set theory CZF [1] extended with the regular extension axiom (REA) and dependent choice. REA allows us to carry out generalised inductive definitions used in the theory of inductively generated formal topologies. Constructive choice principles such as dependent choice are not needed for the most part of formal topology. However, for the purpose of relating formal topology to some point-set theories such as Bishop's metric spaces, some choice principles seem to be indispensable.

Notation 15.1 We write $\mathbb{N}, \mathbb{Q}, \mathbb{R}$ for the sets of natural numbers, rational numbers, and real numbers, respectively. The unit interval of $\mathbb{R}$ is denoted $[0, 1]$. $\mathrm{Pow}(S)$ denotes the collection of subsets of a set S. $\mathrm{Fin}(S)$ denotes the set of finitely enumerable subsets of S, where a set X is finitely enumerable if there exists a surjection $f\colon \{0, \ldots, n-1\} \to X$ for some $n \in \mathbb{N}$.

15.2 Formal Topology

Pointfree topology takes the notion of open subsets as primitive. This allows us to develop large parts of topology under intuitionistic logic [18]. A formal topology is a variant of pointfree topology which takes the notion of bases (basic opens) of topology as primitive. It is particularly suited to predicative mathematics such as Bishop constructive mathematics, where the lattice of open subsets need not form a set in general.

In this section, we recall some basic notions of formal topology that will be used in the later sections. For further details of formal topology, the reader is referred to Coquand et al. [12], Fox [17], and Sambin [32, 33].

15.2.1 Formal Topologies

Definition 15.2 A *formal topology* is a triple $\mathcal{S} = (S, \leq, \lhd)$ where $(S, \leq)$ is a preordered set and $\lhd$ is a relation between S and $\mathrm{Pow}(S)$ such that

$$\mathcal{A}U \overset{\mathrm{def}}{=} \{a \in S \mid a \lhd U\}$$

is a set for each $U \subseteq S$ and that

(i) $a \in U \implies a \lhd U$,
(ii) $a \lhd U \;\&\; U \lhd V \implies a \lhd V$,
(iii) $a \lhd U \;\&\; a \lhd V \implies a \lhd U \downarrow V$,
(iv) $a \leq b \implies a \lhd \{b\}$

for each $a, b \in S$ and $U, V \subseteq S$, where

$$U \lhd V \overset{\text{def}}{\iff} (\forall a \in U)\, a \lhd V,$$
$$U \downarrow V \overset{\text{def}}{=} \{c \in S \mid (\exists a \in U)\,(\exists b \in V)\, c \leq a \,\&\, c \leq b\}.$$

Later, we also write $a \downarrow U$ for $\{a\} \downarrow U$ and $a \downarrow b$ for $\{a\} \downarrow \{b\}$. The set S is called the *base* of $\mathcal{S}$, and the relation $\lhd$ is called a *cover* on $(S, \leq)$, or the cover of $\mathcal{S}$. For each $U, V \subseteq S$, we write $U =_{\mathcal{S}} V$ for $\mathcal{A}U = \mathcal{A}V$.

Note that the original definition of formal topology by Sambin [32] contains a positivity predicate Pos, which is here called an overt formal topology; see Definition 15.27.

Notation 15.3 The letters $\mathcal{S}, \mathcal{S}', \ldots$ denote formal topologies. If $\mathcal{S}$ is a formal topology, then the symbols $S, \leq$, and $\lhd$ will be used to denote the base, the preorder, and the cover of $\mathcal{S}$, respectively. Subscripts or superscripts are sometimes attached to these symbols for clarity. For example, the base, the preorder, and the cover of a formal topology $\mathcal{S}'$ are denoted by $S', \leq'$, and $\lhd'$, respectively.

In formal topology, the notion of continuity is described by certain relations between bases.

Definition 15.4 Let $\mathcal{S}$ and $\mathcal{S}'$ be formal topologies. A *formal topology map* from $\mathcal{S}$ to $\mathcal{S}'$ is a relation $r \subseteq S \times S'$ such that

(FTM1) $S \lhd r^{-}S'$,
(FTM2) $r^{-}\{a\} \downarrow r^{-}\{b\} \lhd r^{-}(a \downarrow' b)$,
(FTM3) $a \lhd' U \implies r^{-}\{a\} \lhd r^{-}U$

for each $a, b \in S'$ and $U \subseteq S'$, where

$$r^{-}U = \{a \in S \mid (\exists b \in U)\, a \; r \; b\}.$$

Two formal topology maps $r, s \colon \mathcal{S} \to \mathcal{S}'$ are defined to be *equal* if $r^{-}\{a\} =_{\mathcal{S}} s^{-}\{a\}$ for each $a \in S'$.

The formal topologies and formal topology maps between them form a category **FTop**. The composition of two formal topology maps is the composition of the underlying relations of these maps. The identity morphism on a formal topology is the identity relation on its base.

The formal topology $\mathbf{1} \overset{\text{def}}{=} (\{*\}, \in, =)$ is a terminal object in **FTop**. Then, a global point $r : \mathbf{1} \to \mathcal{S}$ corresponds to the following notion.

Definition 15.5 Let $\mathcal{S}$ be a formal topology. A *formal point* of $\mathcal{S}$ is a subset $\alpha \subseteq S$ such that

(P1) α is inhabited,

(P2) $a, b \in \alpha \implies \alpha \between (a \downarrow b)$,

(P3) $a \in \alpha \,\&\, a \lhd U \implies \alpha \between U$

for each $a, b \in S$ and $U \subseteq S$, where $\alpha \between U$ means that the intersection $\alpha \cap U$ is inhabited. We write $\mathrm{Pt}(\mathcal{S})$ for the collection of formal points of $\mathcal{S}$.[2]

The correspondence between formal points of $\mathcal{S}$ and global points of $\mathcal{S}$ is as follows: each formal point $\alpha \in \mathrm{Pt}(\mathcal{S})$ determines a global point $r_\alpha \colon \mathbf{1} \to \mathcal{S}$ by $* \; r_\alpha \; a \overset{\text{def}}{\iff} a \in \alpha$. Conversely, each global point $r \colon \mathbf{1} \to \mathcal{S}$ determines a formal point $\alpha_r \in \mathrm{Pt}(\mathcal{S})$ by $\alpha_r \overset{\text{def}}{=} \{a \in S \mid * \; r \; a\}$.

A formal topology map $r \colon \mathcal{S} \to \mathcal{S}'$ induces a function $\mathrm{Pt}(r) \colon \mathrm{Pt}(\mathcal{S}) \to \mathrm{Pt}(\mathcal{S}')$ defined by

$$\mathrm{Pt}(r)(\alpha) \overset{\text{def}}{=} \{b \in S' \mid (\exists a \in \alpha)\, a \; r \; b\}.$$

The function $\mathrm{Pt}(r)$ is pointwise continuous with respect to a topology on $\mathrm{Pt}(\mathcal{S})$ generated by basic opens of the form $a_* \overset{\text{def}}{=} \{\alpha \in \mathrm{Pt}(\mathcal{S}) \mid a \in \alpha\}$ ($a \in S$). In some special cases, however, $\mathrm{Pt}(r)$ enjoys a stronger property; see Theorem 15.22.

15.2.2 Inductively Generated Formal Topologies

The notion of inductively generated topology allows us to reason about formal topologies using selected sets of axioms. Most of the formal topologies which arise in practice are inductively generated, including those treated in this chapter.[3] Moreover, some constructions on formal topologies can be carried out only for the class of inductively generated formal topologies. A typical example is the construction of products (see Section 15.3.3), which plays a key role in the pointfree proof of Tychonoff's theorem [9, 27, 39].

Definition 15.6 An *axiom-set* on a set S is a pair (I, C), where $(I(a))_{a \in S}$ is a family of sets indexed by S, and C is a family $(C(a,i))_{a \in S, i \in I(a)}$ of subsets of S indexed by $\sum_{a \in S} I(a)$.

The following is a fundamental theorem of inductively generated formal topologies.

[2] In a predicative constructive mathematics, $\mathrm{Pt}(\mathcal{S})$ need not be a set; see, for example, Curi [15, Corollary 7.1].

[3] See Coquand et al. [12, Section 4.7] for some important examples of formal topologies which are not inductively generated.

Theorem 15.7 (Coquand et al. [12]) *Let $(S, \leq)$ be a preordered set and (I, C) be an axiom-set on S. Then, there exists a cover $\lhd_{I,C}$ on $(S, \leq)$ inductively generated by the following rules:*

$$\frac{a \in U}{a \lhd_{I,C} U}\ (\textit{reflexivity}) \qquad \frac{a \leq b \quad b \lhd_{I,C} U}{a \lhd_{I,C} U}\ (\leq\textit{-left})$$

$$\frac{a \leq b \quad i \in I(b) \quad a \downarrow C(b,i) \lhd_{I,C} U}{a \lhd_{I,C} U}\ (\leq\textit{-infinity}).$$

The relation $\lhd_{I,C}$ is the least cover on $(S, \leq)$ which satisfies ($\leq$-left) and $a \lhd_{I,C} C(a,i)$ for each $a \in S$ and $i \in I(a)$.

The cover $\lhd_{I,C}$ defined in the above theorem is called the cover *generated by* (I, C). A formal topology $\mathcal{S} = (S, \leq, \lhd)$ is said to be *inductively generated* if there exists an axiom-set (I, C) on S which generates the cover $\lhd$ on $(S, \leq)$ as in Theorem 15.7.

Remark 15.8 Let $r\colon \mathcal{S} \to \mathcal{S}'$ be a formal topology map, where $\mathcal{S}'$ is inductively generated by an axiom-set (I, C) on S'. In the presence of (FTM2), the condition (FTM3) becomes equivalent to the following two conditions:

(FTM3a) $a \leq' b \implies r^{-}\{a\} \lhd r^{-}\{b\}$,
(FTM3b) $r^{-}\{a\} \lhd r^{-}C(a,i)$

for each $a, b \in S'$ and $i \in I(a)$. In particular, a formal point of an inductively generated formal topology such as $\mathcal{S}'$ can be characterised using an axiom-set (I, C). Specifically, the condition (P3) can be replaced with the following two conditions:

(P3a) $a \leq' b \;\&\; a \in \alpha \implies b \in \alpha$,
(P3b) $a \in \alpha \implies \alpha \between C(a,i)$

for each $a, b \in S'$ and $i \in I(a)$.

Example 15.9 The topology on the real numbers in formal topology is a typical example of an inductively generated formal topology. Let $(S_{\mathcal{R}}, \leq_{\mathcal{R}})$ be the preorder on the set $S_{\mathcal{R}} \overset{\text{def}}{=} \{(p,q) \in \mathbb{Q} \times \mathbb{Q} \mid p < q\}$ defined by

$$(r,s) \leq_{\mathcal{R}} (p,q) \overset{\text{def}}{\iff} p \leq r \;\&\; s \leq q.$$

We also define a strict variant $<_{\mathcal{R}}$ by

$$(r,s) <_{\mathcal{R}} (p,q) \overset{\text{def}}{\iff} p < r \;\&\; s < q$$

for all $(p,q),(r,s) \in S_{\mathcal{R}}$. The *formal topology $\mathcal{R}$ of real numbers (formal reals* for short) is a formal topology $(S_{\mathcal{R}}, \leq_{\mathcal{R}}, \lhd_{\mathcal{R}})$ where the cover $\lhd_{\mathcal{R}}$ is inductively generated by the following axioms:

(R1) $(p,q) \lhd_{\mathcal{R}} \{(r,s) \in S_{\mathcal{R}} \mid (r,s) <_{\mathcal{R}} (p,q)\}$,
(R2) $(p,q) \lhd_{\mathcal{R}} \{(p,s),(r,q)\}$ for each $(r,s) <_{\mathcal{R}} (p,q)$.

It is well known that the formal points of $\mathcal{R}$ are homeomorphic to the Dedekind reals defined as the pairs of left and right cuts; see Negri and Soravia [26, Proposition 9.2] or Fourman and Grayson [16, Example 1.2 (4)].

15.3 Functorial Embedding of Locally Compact Metric Spaces

In this section, we recall the construction of an embedding of the category of locally compact metric spaces into that of formal topologies by Palmgren [29]. The embedding is based on a certain representation of a complete metric space in formal topology, called a localic completion, which was introduced by Vickers [38]. The idea behind the construction of localic completions is to represent the space of Cauchy sequences on a metric space in a pointfree way.

The above-mentioned embedding preserves compactness and local compactness of metric spaces. Moreover, for locally compact metric spaces, the morphisms between localic completions bijectively correspond to continuous maps in Bishop's theory of locally compact metric spaces. This allows us to transfer results between Bishop's theory of metric spaces and formal topology.

15.3.1 Localic Completions of a Metric Space

Definition 15.10 Let $X = (X, \rho)$ be a metric space with a metric ρ on X, and let $\mathbb{Q}^{>0}$ be the set of positive rationals. Define a set

$$M_X \stackrel{\text{def}}{=} X \times \mathbb{Q}^{>0}$$

of *formal balls* of X. A formal ball is denoted by $\mathsf{b}(x,\varepsilon)$ ($x \in X$, $\varepsilon \in \mathbb{Q}^{>0}$). The set of formal balls are ordered by

$$\mathsf{b}(x,\varepsilon) \leq_X \mathsf{b}(y,\delta) \stackrel{\text{def}}{\iff} \rho(x,y) + \varepsilon \leq \delta. \tag{15.1}$$

Similarly, we define a relation $<_X$ on M_X, replacing $\leq$ with $<$ in (15.1).

The *localic completion* of X is a formal topology $\mathcal{M}(X) \stackrel{\text{def}}{=} (M_X, \leq_X, \lhd_X)$, where $\lhd_X$ is inductively generated by the following axioms:

(M1) $a \lhd_X \{b \in M_X \mid b <_X a\}$,
(M2) $a \lhd_X \mathcal{C}_\varepsilon$ for each $\varepsilon \in \mathbb{Q}^{>0}$ where $\mathcal{C}_\varepsilon \stackrel{\text{def}}{=} \{\mathsf{b}(x,\varepsilon) \in M_X \mid x \in X\}$.

The collection $\mathrm{Pt}(\mathcal{M}(X))$ of formal points admits a metric $\widetilde{\rho}\colon \mathrm{Pt}(\mathcal{M}(X)) \times \mathrm{Pt}(\mathcal{M}(X)) \to \mathbb{R}^{\geq 0}$, which can be defined using Dedekind upper cuts:

$$\widetilde{\rho}(\alpha, \beta) \stackrel{\text{def}}{=} \{q \in \mathbb{Q}^{>0} \mid (\exists\, \mathsf{b}(x, \varepsilon) \in \alpha)\, (\exists\, \mathsf{b}(y, \delta) \in \beta)\ \rho(x, y) + \varepsilon + \delta < q\}.$$

A localic completion can be seen as a pointfree completion of a metric space in the following sense.

Proposition 15.11 (Palmgren [29, Theorem 2.4]) *The space* $(\mathrm{Pt}(\mathcal{M}(X)), \widetilde{\rho})$ *is metrically isomorphic to the completion of* (X, ρ).

Specifically, the function $j_X \colon X \to \mathrm{Pt}(\mathcal{M}(X))$ defined by

$$j_X(x) \stackrel{\text{def}}{=} \{\mathsf{b}(y, \varepsilon) \in M_X \mid \rho(x, y) < \varepsilon\} \tag{15.2}$$

is a metric completion of X. Hence, j_X is a metric isomorphism if and only if X is complete. Since j_X is a metric completion, the collection $\mathrm{Pt}(\mathcal{M}(X))$ forms a set, which bijectively corresponds to the set of Cauchy sequences on X with a suitable equivalence relation (see [29, Theorem 2.4]).

Example 15.12 (Palmgren [29, Example 2.2]) The localic completion of $\mathbb{Q}$ is the formal reals $\mathcal{R}$ (see Example 15.9). To see this, let d be the metric on $\mathbb{Q}$ inherited from $\mathbb{R}$. Then, the set $M_{\mathbb{Q}}$ of formal balls with the relations $\leq_{\mathbb{Q}}$ and $<_{\mathbb{Q}}$ is isomorphic (in the obvious sense) to the underlying structure $(S_{\mathcal{R}}, \leq_{\mathcal{R}}, <_{\mathcal{R}})$ of the formal reals $\mathcal{R}$. By definition, the cover of $\mathcal{M}(\mathbb{Q})$ is generated by the following axioms:

(Q1) $(p, q) \lhd_{\mathbb{Q}} \{(r, s) \in S_{\mathcal{R}} \mid (r, s) <_{\mathcal{R}} (p, q)\}$,
(Q2) $(p, q) \lhd_{\mathbb{Q}} \mathcal{C}_{\varepsilon}$ for each $\varepsilon \in \mathbb{Q}^{>0}$.

Since the axioms (R1) and (Q1) are the same, it suffices to show that the axioms (Q2) and (R2) are derivable in $\mathcal{R}$ and $\mathcal{M}(\mathbb{Q})$, respectively. First, assume (R2), and let $(p, q) \in S_{\mathcal{R}}$ and $\varepsilon \in \mathbb{Q}^{>0}$. Then, by (R2), we have

$$(p, q) \lhd_{\mathcal{R}} \{(p, (p + 2q)/3), ((2p + q)/3, q)\}.$$

By repeating this process sufficiently many times for each element on the right-hand side, we obtain a subset $U \in \mathrm{Fin}(S_{\mathcal{R}})$ such that $(p, q) \lhd_{\mathcal{R}} U$ and $s - r < \varepsilon$ for each $(r, s) \in U$. Thus $(p, q) \lhd_{\mathcal{R}} \mathcal{C}_{\varepsilon}$. Conversely, assume (Q2), and let $(p, q), (r, s) \in S_{\mathcal{R}}$ be such that $(r, s) <_{\mathcal{R}} (p, q)$. Choose $\varepsilon \in \mathbb{Q}^{>0}$ such that $\varepsilon < s - r$. Then $(p, q) \lhd_{\mathbb{Q}} \mathcal{C}_{\varepsilon} \downarrow (p, q) \leq_{\mathcal{R}} \{(p, s), (r, q)\}$, and so $(p, q) \lhd_{\mathbb{Q}} \{(p, s), (r, q)\}$.

The following plays an important role in obtaining pointfree characterisations of compact metric spaces and locally compact metric spaces (see Lemma 15.51).

Theorem 15.13 (Palmgren [29, Theorem 2.7]) *Let X be a metric space and Y be a dense subset of X. Then $\mathcal{M}(X) \cong \mathcal{M}(Y)$.*

For example, the localic completions of $\mathbb{Q}$ and $\mathbb{R}$ are isomorphic; thus $\mathcal{M}(\mathbb{R}) \cong \mathcal{M}(\mathbb{Q}) \cong \mathcal{R}$ (see Example 15.12). Moreover, the localic completion of a separable metric space is isomorphic to a formal topology $\mathcal{S}$ with a countable base.

We close this subsection by introducing some notations which will be used in later sections. First, we associate each formal ball $\mathsf{b}(x, \varepsilon)$ with the corresponding open ball $\mathsf{b}(x, \varepsilon)_*$ defined by

$$\mathsf{b}(x, \varepsilon)_* \stackrel{\text{def}}{=} B(x, \varepsilon) = \{y \in X \mid \rho(x, y) < \varepsilon\}.$$

The notation $(-)_*$ is extended to subsets of M_X by $U_* \stackrel{\text{def}}{=} \bigcup_{a \in U} a_*$. Conversely, we associate each point $x \in X$ with the set $\Diamond x$ of its open neighbourhoods:

$$\Diamond x \stackrel{\text{def}}{=} \{a \in M_X \mid x \in a_*\}.$$

Again, the notation $\Diamond(-)$ is extended to subsets of X by $\Diamond Y \stackrel{\text{def}}{=} \bigcup_{y \in Y} \Diamond y$.

15.3.2 The Functorial Embedding

The main result of Palmgren [29, Sections 4 and 5] says that the category of Bishop locally compact metric spaces embeds into that of locally compact regular formal topologies.

We begin by recalling some notions concerning metric spaces [2, Chapter 4].

Definition 15.14 A metric space (X, ρ) is *totally bounded* if for each $\varepsilon \in \mathbb{Q}^{>0}$, there exists a finitely enumerable subset $Y = \{x_0, \ldots, x_{n-1}\}$ such that $X = \bigcup_{i<n} B(x_i, \varepsilon)$. In this case, Y is called an *ε-net*. A metric space is *compact* if it is complete and totally bounded. A metric space X is *locally compact* if each open ball $B(x, \varepsilon)$ is contained in a compact subset of X. In particular, every compact metric space is locally compact, and every locally compact metric space is complete.

A function $f \colon X \to Y$ between locally compact metric spaces is *continuous* if it is uniformly continuous on each open ball of X. We write **LComp** for the category of locally compact metric spaces and continuous functions.

Remark 15.15 Bishop [2] defined a metric space to be totally bounded if for each $\varepsilon \in \mathbb{Q}^{>0}$ there exists an inhabited ε-net. With this definition, the empty set is not totally bounded. However, our deviation from his definition is not essential because we can decide whether a totally bounded space in our sense is inhabited or not. On the other hand, Bishop requires a locally compact metric space to be inhabited. This seems to be essential for such a space to become separable. In the following, an inhabited locally compact metric space will be called a *Bishop locally compact metric space*.

Next, we recall the notion of compactness in pointfree topology.

Definition 15.16 A formal topology $\mathcal{S}$ is *compact* if

$$S \lhd U \implies (\exists A \in \mathrm{Fin}(U))\, S \lhd A. \tag{15.3}$$

The condition (15.3) is a pointfree way of expressing the usual covering compactness in point-set topology: a space is compact if and only if every open cover has a finite subcover. However, there is a crucial difference between the pointfree notion of compactness and the corresponding point-set notion. Specifically, if one regards Bishop constructive mathematics as the core of the varieties of constructive mathematics, the point-set notion cannot be accepted as it would imply Brouwer's fan theorem (see Section 15.1). In contrast, the pointfree notion is compatible with that of Bishop's metric spaces if one regards the localic completion as a primary representation of topology on a metric space.

Theorem 15.17 (Palmgren [29, Theorem 4.20]) *If X is a complete metric space, then $\mathcal{M}(X)$ is compact if and only if X is totally bounded.*

In particular, the localic completion of a compact metric space is compact in the sense of Definition 15.16. The above theorem has been extended to locally compact metric spaces (see Theorem 15.20). To this end, we recall the pointfree notion of local compactness.

Definition 15.18 A formal topology $\mathcal{S}$ is *locally compact* if there exists a function $\mathsf{wb}\colon S \to \mathrm{Pow}(S)$ such that

(i) $a \lhd \mathsf{wb}(a)$,
(ii) $(\forall b \in \mathsf{wb}(a))\, [a \lhd U \implies (\exists A \in \mathrm{Fin}(U))\, b \lhd A]$

for each $a \in S$ and $U \subseteq S$.

Roughly speaking, $b \in \mathsf{wb}(a)$ if b is compact relative to the covers of a.

Example 15.19 Compact regular formal topologies are important examples of locally compact formal topologies. Here, a formal topology $\mathcal{S}$ is *regular* if

$$a \lhd \{b \in S \mid b \lll a\}$$

where $b \lll a \overset{\text{def}}{\iff} S \lhd b^* \cup \{a\}$ for $b^* \overset{\text{def}}{=} \{c \in S \mid c \downarrow b \lhd \emptyset\}$. For example, the localic completion of a metric space is regular since we have $a <_X b \implies a \lll b$. Now, if $\mathcal{S}$ is a compact regular formal topology, then a function $\mathsf{wb}\colon S \to \mathrm{Pow}(S)$ defined by $\mathsf{wb}(a) \overset{\text{def}}{=} \{b \in S \mid b \lll a\}$ serves as a witness of its local compactness.

Theorem 15.20 (Palmgren [29, Theorem 4.19]) *If X is a locally compact metric space, then $\mathcal{M}(X)$ is locally compact. Specifically, we can define* $\mathsf{wb}\colon M_X \to \mathrm{Pow}(M_X)$ *by* $\mathsf{wb}(a) \overset{\text{def}}{=} \{b \in M_X \mid b <_X a\}$.

For a locally compact metric space X, the cover of $\mathcal{M}(X)$ admits an explicit description in terms of a certain structure on X.

Proposition 15.21 (Palmgren [29, Theorem 4.17]; Kawai [21, Proposition 4.9]) *If X is a locally compact metric space, then*

$$a \lhd_X U \iff (\forall b <_X a)\,(\exists V \in \mathrm{Fin}(M_X))\, b \sqsubseteq V <_X U,$$

where

$$V <_X U \overset{\text{def}}{\iff} (\forall a \in V)\,(\exists b \in U)\, a <_X b,$$
$$b \sqsubseteq V \overset{\text{def}}{\iff} (\exists \varepsilon \in \mathbb{Q}^{>0})\, b \downarrow \mathcal{C}_\varepsilon \leq_X V.$$

In the definition of $b \sqsubseteq V$, the relation $V \leq_X U$ is defined similarly as $V <_X U$.

The above proposition shows that for locally compact metric spaces, the construction of their localic completions does not require generalised inductive definitions; hence it can be carried out in CZF without using REA. In particular, the localic completion of $\mathbb{R}$—namely, the formal reals $\mathcal{R}$—can be constructed in CZF (see [6, Section 3]).

Another important observation by Palmgren is that formal topology maps correspond to Bishop continuous functions between locally compact metric spaces. This can be expressed using the language of categories.

Theorem 15.22 (Palmgren [29, Theorem 5.8]) *The localic completion extends to a full and faithful functor* $\mathcal{M}\colon$ **LComp** $\to$ **FTop**.

Here, by an abuse of notation, we have used the same symbol for localic completions and the functor $\mathcal{M}\colon$ **LComp** $\to$ **FTop** induced by them. The functor $\mathcal{M}$ sends a continuous function $f\colon X \to Y$ in **LComp** to a formal topology map $\mathcal{M}(f)\colon \mathcal{M}(X) \to \mathcal{M}(Y)$ defined by

$$a\, \mathcal{M}(f)\, b \overset{\text{def}}{\iff} (\exists c <_Y b)\, f[a_*] \subseteq c_*,$$

where $f[a_*]$ is the forward image of the open ball a_*. Then, for any formal topology map $r\colon \mathcal{M}(X) \to \mathcal{M}(Y)$ where X and Y are locally compact, there exists a unique continuous function $f\colon X \to Y$ such that $\mathcal{M}(f) = r$ and which makes the following diagram commute:

$$\begin{array}{ccc} \mathrm{Pt}(\mathcal{M}(X)) & \xrightarrow{\mathrm{Pt}(r)} & \mathrm{Pt}(\mathcal{M}(Y)) \\ {\scriptstyle j_X}\uparrow\cong & & {\scriptstyle j_Y}\uparrow\cong \\ X & \xrightarrow{\quad f \quad} & Y. \end{array} \tag{15.4}$$

Here, $j_X\colon X \to \mathrm{Pt}(\mathcal{M}(X))$ is the metric isomorphism defined in (15.2).

Since $\mathcal{M}$ is full and faithful, $\mathcal{M}$ gives rise to an equivalence between **LComp** and its image. In Section 15.6, we give a pointfree characterisation of the image which does not make any reference to metric spaces.

15.3.3 Preservation of Products

In this subsection, we see that the embedding $\mathcal{M}\colon \mathbf{LComp} \to \mathbf{FTop}$ preserves countable products of inhabited compact metric spaces. This will play an important role in our characterisation of compact metric spaces in formal topology.

First, we recall the construction of a product of inductively generated formal topologies [39].[4] Let $(\mathcal{S}_i)_{i\in I}$ be a set-indexed family of inductively generated formal topologies, where each $\mathcal{S}_i = (S_i, \leq_i, \lhd_i)$ is generated by an axiom-set (I_i, C_i). Define a preorder $(S_\Pi, \leq_\Pi)$ on the set $S_\Pi = \mathrm{Fin}\left(\sum_{i\in I} S_i\right)$ by

$$A \leq_\Pi B \overset{\text{def}}{\iff} (\forall (i,b) \in B)\, (\exists (j,a) \in A)\, i = j \,\&\, a \leq_i b.$$

The axiom-set on S_Π is defined by

(S1) $S_\Pi \lhd_\Pi \{\{(i,a)\} \in S_\Pi \mid a \in S_i\}$ for each $i \in I$,
(S2) $\{(i,a),(i,b)\} \lhd_\Pi \{\{(i,c)\} \in S_\Pi \mid c \leq_i a \,\&\, c \leq_i b\}$,
(S3) $\{(i,a)\} \lhd_\Pi \{\{(i,b)\} \in S_\Pi \mid b \in C_i(a,k)\}$ for each $k \in I_i(a)$.

Let $\prod_{i\in I} \mathcal{S}_i = (S_\Pi, \leq_\Pi, \lhd_\Pi)$ be the formal topology inductively generated by the above axiom-set. For each $i \in I$, the projection $p_i\colon \prod_{i\in I} \mathcal{S}_i \to \mathcal{S}_i$ is defined by

$$A \, p_i \, a \overset{\text{def}}{\iff} A = \{(i,a)\}.$$

Then, the family of projections $\left(p_i\colon \prod_{i\in I} \mathcal{S}_i \to \mathcal{S}_i\right)_{i\in I}$ is a product of $(\mathcal{S}_i)_{i\in I}$. Thus, each family $(r_i\colon \mathcal{S} \to \mathcal{S}_i)_{i\in I}$ of formal topology maps determines a unique formal topology map $r\colon \mathcal{S} \to \prod_{i\in I} \mathcal{S}_i$ such that $r_i = p_i \circ r$ for each $i \in I$, which is defined by

$$a \, r \, A \overset{\text{def}}{\iff} (\forall (i,b) \in A)\, a \lhd r_i^- \{b\}.$$

[4] Constructions of products are only known for inductively generated formal topologies.

Next, a product of a sequence $(X_n, \rho_n)_{n\in\mathbb{N}}$ of metric spaces consists of a set-theoretic product $\prod_{n\in\mathbb{N}} X_n$ and a metric ρ on $\prod_{n\in\mathbb{N}} X_n$ defined by

$$\rho\left((x_n)_{n\in\mathbb{N}}, (y_n)_{n\in\mathbb{N}}\right) \stackrel{\text{def}}{=} \sum_{n=0}^{\infty} \frac{\rho_n(x_n, y_n)}{2^{n+1}}.$$

Without loss of generality, we may assume that each metric ρ_n is bounded by 1. Then, the family $\pi_n \colon \prod_{n\in\mathbb{N}} X_n \to X_n$ of projection functions forms a product of the metric spaces. A countable product of inhabited compact metric spaces is compact (see [2, Chapter 4, Proposition 6]).

Now, let $(X_n, \rho_n)_{n\in\mathbb{N}}$ be a sequence of inhabited compact metric spaces, and let $\prod_{n\in\mathbb{N}} \mathcal{M}(X_n)$ be the product of their localic completions $(\mathcal{M}(X_n))_{n\in\mathbb{N}}$. Then, the family of projections $\pi_n \colon \prod_{n\in\mathbb{N}} X_n \to X_n$ gives rise to a family $(\mathcal{M}(\pi_n) \colon \mathcal{M}(\prod_{n\in\mathbb{N}} X_n) \to \mathcal{M}(X_n))_{n\in\mathbb{N}}$ of formal topology maps, which induces a (unique) formal topology map $r \colon \mathcal{M}(\prod_{n\in\mathbb{N}} X_n) \to \prod_{n\in\mathbb{N}} \mathcal{M}(X_n)$.

Proposition 15.23 *The formal topology map $r \colon \mathcal{M}(\prod_{n\in\mathbb{N}} X_n) \to \prod_{n\in\mathbb{N}} \mathcal{M}(X_n)$ is an isomorphism of formal topologies. Hence, the functor $\mathcal{M}$:* **LComp** $\to$ **FTop** *preserves countable products of inhabited compact metric spaces.*

The proof of Proposition 15.23 in Kawai [22, Section 5] goes by showing that the morphism r is an embedding of a formal topology and that the image of the embedding coincides with $\prod_{n\in\mathbb{N}} \mathcal{M}(X_n)$.

15.4 Located Subsets in Formal Topology

In Bishop constructive mathematics, the notion of located subset plays an important role. The pointfree counterpart of this notion was introduced by Spitters [34]. He showed that his pointfree notion corresponds to the metric notion in the case of localic completions of metric spaces. In the same work, he related the pointfree notion of located subset to that of overt and closed subtopology. The notions of overt subtopology and closed subtopology were well known in pointfree topology; however, the importance of overt and closed subtopologies has not been recognised very well. In this section, we establish some properties of overt closed subtopologies and relate that notion to Bishop's notion of located subsets of metric spaces.

15.4.1 Overt Closed Subtopologies

A *subtopology* of a formal topology $\mathcal{S}$ is a formal topology $\mathcal{S}' = (S, \leq, \lhd')$, where $\lhd'$ is a cover on S and $(S, \leq)$ is the underlying preorder of $\mathcal{S}$ such that $\lhd \subseteq \lhd'$. The relation of being a subtopology of another formal topology is a partial order.

Thus, the class of subtopologies of a given formal topology forms a poset with this relation.[5]

Important examples of subtopologies are open and closed subtopologies.

Definition 15.24 Let $\mathcal{S}$ be a formal topology and let $V \subseteq S$ be a subset.

(i) The *open subtopology* of $\mathcal{S}$ determined by V is a subtopology $\mathcal{S}_V$ of $\mathcal{S}$ with a cover $\lhd_V$ defined by $a \lhd_V U \overset{\text{def}}{\iff} a \downarrow V \lhd U$.
(ii) The *closed subtopology* of $\mathcal{S}$ determined by V is a subtopology $\mathcal{S}^{\mathcal{S}-V}$ of $\mathcal{S}$ with a cover $\lhd^{\mathcal{S}-V}$ defined by $a \lhd^{\mathcal{S}-V} U \overset{\text{def}}{\iff} a \lhd V \cup U$.

The open subtopology $\mathcal{S}_V$ characterises the cover restricted to V, while the closed subtopology $\mathcal{S}^{\mathcal{S}-V}$ is the pointfree complement of $\mathcal{S}_V$ (see [40, Section 3]). Thus, the notion of closed subtopology captures that of the complement of an open subset.

Example 15.25 Let $\mathcal{R}$ be the formal reals (see Example 15.9). The *formal unit interval* $\mathcal{I}[0,1]$ is a closed subtopology of $\mathcal{R}$ determined by a subset $(-\infty, 0) \cup (1, \infty)$, where $(-\infty, 0) \overset{\text{def}}{=} \{(p, q) \in S_{\mathcal{R}} \mid q = 0\}$ and $(1, \infty) \overset{\text{def}}{=} \{(p, q) \in S_{\mathcal{R}} \mid p = 1\}$.

In Bishop constructive mathematics, a closed subset of a metric space is usually defined as the collection of its limit points (see (15.6)), not as the complement of an open subset. The pointfree analogue of this notion of closed subset is the following (See [8, 33]).

Definition 15.26 A subset $V \subseteq S$ of a formal topology $\mathcal{S}$ is said to be *formal closed* if

$$a \in V \mathbin{\&} a \lhd U \implies V \between U.$$

We write $\mathrm{Red}(\mathcal{S})$ for the collection of formal closed subsets of $\mathcal{S}$.[6]

Verification of formal closedness requires reference to the entire relation $\lhd$. This can be avoided in inductively generated formal topologies. Specifically, if $\mathcal{S}$ is a formal topology inductively generated by an axiom-set (I, C), then a subset $V \subseteq S$ is formal closed if and only if

(i) $a \leq b \mathbin{\&} a \in V \implies b \in V$,
(ii) $a \leq b \mathbin{\&} a \in V \implies V \between (a \downarrow C(b, i))$

for each $a, b \in S$ and $i \in I(b)$.

Formal closed subsets are closely related to overt formal topologies.

[5] The class of subtopologies of a given formal topology determines another pointfree space (see [18, Chapter II, 2.5]).

[6] In the literature, formal closed subsets are often called *splitting subsets* [7, 8, 33], where the term *formal closed subset* is reserved for a subset fixed by a (binary) positivity relation. Our use of terminology follows that of Kawai [22, 23].

Definition 15.27 A *positivity predicate* (*positivity* for short) on a formal topology $\mathcal{S}$ is a formal closed subset $\mathrm{Pos} \subseteq S$ such that

$$a \lhd \{a\} \cap \mathrm{Pos}\,. \tag{Pos}$$

A formal topology is *overt* if it is equipped with a positivity predicate.[7] A formal topology is *inhabited* if it is overt and its positivity is inhabited.

For example, the localic completion $\mathcal{M}(X)$ of a metric space is always overt and its positivity is the whole base; thus $\mathcal{M}(X)$ is inhabited if and only if X is inhabited. In more general settings, a positivity predicate allows us to express in a pointfree way that certain formal opens (i.e., elements of the base) are non-trivial, or inhabited. Thus, overt formal topologies are more expressive than general formal topologies.

In an inductively generated formal topology, formal closed subsets bijectively correspond to overt weakly closed subtopologies.

Definition 15.28 A subtopology of a formal topology $\mathcal{S}$ is *overt weakly closed* if it is overt (say with a positivity Pos) and is the largest overt subtopology of $\mathcal{S}$ among the subtopologies of $\mathcal{S}$ with the same positivity Pos.

The above mentioned bijection between formal closed subsets and overt weakly closed subtopologies is as follows. Let $\mathcal{S}$ be an inductively generated formal topology. If $V \subseteq S$ is a formal closed subset of $\mathcal{S}$, then the formal topology generated by the axiom-set of $\mathcal{S}$ together with an extra axiom (Pos) is an overt weakly closed subtopology of $\mathcal{S}$ with a positivity V. Conversely, the positivity of an overt weakly closed subtopology of $\mathcal{S}$ is a formal closed subset of $\mathcal{S}$.

Classically, every formal topology is overt; indeed, we can simply define $\mathrm{Pos} \stackrel{\text{def}}{=} \{a \mid \neg\,(a \lhd \emptyset)\}$. This is not the case constructively; for example, there is a closed subtopology which is not overt [20, Example 2.3.10]. Thus, the notion of formal closed subset (or overt weakly closed subtopology) and that of closed subtopology are different constructively. However, we have the following correspondence.

Proposition 15.29 *Every overt closed subtopology is an overt weakly closed subtopology. Thus, the class of overt closed subtopologies coincides with that of overt weakly closed subtopologies which are also closed.*

Since an overt weakly closed subtopology is completely determined by its positivity, a question arises as to which subclass of formal closed subsets corresponds to that of overt closed subtopologies.

[7] In locale theory [18], an overt formal topology is sometimes called *open* locale [19]. The term *overt* is attributed to Taylor [35].

Lemma 15.30 *Let $\mathcal{S}$ be a formal topology. There exists an order preserving bijection between overt closed subtopologies of $\mathcal{S}$ and formal closed subsets V of $\mathcal{S}$ satisfying*

$$a \lhd \neg V \cup (V \cap \{a\}) \tag{15.5}$$

for each $a \in S$, where $\neg V \overset{\text{def}}{=} \{b \in S \mid b \notin V\}$.

Example 15.31 The formal unit interval $\mathcal{I}[0, 1]$ (see Example 15.25) can be defined as an overt weakly closed subtopology of $\mathcal{R}$. Specifically, $\mathcal{I}[0, 1]$ is determined by a formal closed subset Pos of $\mathcal{R}$ defined as

$$\mathrm{Pos} \overset{\text{def}}{=} \{(p, q) \in S_{\mathcal{R}} \mid 0 < q \,\&\, p < 1\}.$$

15.4.2 Located Subtopologies

For locally compact formal topologies, overt closed subtopologies can be captured by the pointfree notion of located subsets.

Definition 15.32 (Compare Spitters [34, Definition 63]) Let $\mathcal{S}$ be a locally compact formal topology, and let $\mathsf{wb}\colon S \to \mathrm{Pow}(S)$ be the function which witnesses the local compactness of $\mathcal{S}$. Then, a subset $V \subseteq S$ is said to be *located* if V is a formal closed subset of $\mathcal{S}$ such that

$$a \in \mathsf{wb}(b) \implies a \notin V \vee b \in V$$

for each $a, b \in S$. A subtopology $\mathcal{S}'$ of $\mathcal{S}$ is said to be *located* if $\mathcal{S}'$ is a closed subtopology of the form $\mathcal{S}^{\mathcal{S}-\neg V}$ for some located subset V of $\mathcal{S}$.

Lemma 15.33 (Kawai [22, Proposition 4.11]) *Let $V \subseteq S$ be a formal closed subset of a locally compact formal topology $\mathcal{S}$. Then, V is located if and only if it satisfies* (15.5).

Theorem 15.34 (Kawai [22, Theorem 4.13]) *For each locally compact formal topology $\mathcal{S}$, there exists an order-preserving bijection $V \mapsto \mathcal{S}^{\mathcal{S}-\neg V}$ between located subsets of $\mathcal{S}$ and overt closed subtopologies of $\mathcal{S}$.*

In a compact regular formal topology, the notion of closed subtopology and that of compact subtopology coincide [18, Chapter III, Proposition 1.2]. Thus, Theorem 15.34, when specialised to compact regular formal topologies, yields the following.

Corollary 15.35 (Spitters [34, Theorem 74]) *For any compact regular formal topology $\mathcal{S}$, there exists an order-preserving bijection between located subsets of $\mathcal{S}$ and compact overt subtopologies of $\mathcal{S}$.*

In the following, we relate the pointfree notion of locatedness to the corresponding notion in Bishop's theory of metric spaces. To this end, we first relate the metric notion of closed subset to that of formal closed subset.

As in Bishop [2, Chapter 4], a subset Y of a metric space X is *closed* if

$$\left(\forall \varepsilon \in \mathbb{Q}^{>0}\right) B(x, \varepsilon) \between Y \implies x \in Y \tag{15.6}$$

for each $x \in X$. We write $\mathrm{Cl}(X)$ for the collection of closed subsets of X.

The following lemma plays a crucial role when relating some metric notions and pointfree notions. The proof requires dependent choice. Since this is one of the few places where we need a choice principle, we provide its proof.

Lemma 15.36 (Coquand et al. [13, Lemma 3.2]) *Let (X, ρ) be a metric space and V be a formal closed subset of its localic completion $\mathcal{M}(X)$. For each $a \in V$, there exists a formal point $\alpha \in \mathrm{Pt}(\mathcal{M}(X)_V)$ such that $a \in \alpha$, where $\mathcal{M}(X)_V$ is the overt weakly closed subtopology of $\mathcal{M}(X)$ determined by V.*

Proof Define a relation $R \subseteq V \times V$ by

$$\mathsf{b}(x, \varepsilon) \; R \; \mathsf{b}(y, \delta) \overset{\text{def}}{\iff} \mathsf{b}(y, \delta) <_X \mathsf{b}(x, \varepsilon) \;\&\; \delta \leq \varepsilon/2.$$

Let $a_0 \in V$. Since V is formal closed and $a \lhd_X \{b \in M_X \mid b <_X a\} \downarrow \mathcal{C}_{\varepsilon/2}$ for each $a \in M_X$ and $\varepsilon \in \mathbb{Q}^{>0}$, dependent choice yields a function $f \colon \mathbb{N} \to V$ such that $f(0) = a_0$, and $f(n) \; R \; f(n+1)$ for each $n \in \mathbb{N}$. Now, define

$$\alpha \overset{\text{def}}{=} \{a \in M_X \mid (\exists n \in \mathbb{N}) \, f(n) \leq_X a\}.$$

Then $a_0 \in \alpha$, and since V is upward closed, we have $\alpha \subseteq V$. Moreover, α is a formal point of $\mathcal{M}(X)$ by the definition of R. Hence $\alpha \in \mathrm{Pt}(\mathcal{M}(X)_V)$ (see Remark 15.8). □

For complete metric spaces, we have a bijective correspondence between the metric notion of closed subset and that of formal closed subset.

Proposition 15.37 *For any complete metric space X, there exists a bijection $\varphi \colon \mathrm{Cl}(X) \to \mathrm{Red}(\mathcal{M}(X))$ between closed subsets of X and formal closed subsets of $\mathcal{M}(X)$ defined by*

$$\varphi(Y) \overset{\text{def}}{=} \Diamond Y.$$

The inverse $\psi \colon \mathrm{Red}(\mathcal{M}(X)) \to \mathrm{Cl}(X)$ of φ is defined by

$$\psi(V) \overset{\text{def}}{=} \{x \in X \mid \Diamond x \subseteq V\}.$$

Note that showing that φ is surjective requires Lemma 15.36 and hence dependent choice (see [22, Theorem 4.3]).

Proposition 15.37 can be restricted to a bijection between located closed subsets of a metric space and those of its localic completion. We recall the metric notion of locatedness [2, Chapter 4]. The following definition is adopted from Spitters [34, Definition 4] with a slight generalisation.

Definition 15.38 A subset Y of a metric space (X, ρ) is *located* if the distance $\rho(x, Y) \stackrel{\text{def}}{=} \inf \{\rho(x, y) \mid y \in Y\}$ from each $x \in X$ exists as an extended Dedekind real number.

In other words, $Y \subseteq X$ is located if and only if for each $x \in X$, the set

$$U_x \stackrel{\text{def}}{=} \{q \in \mathbb{Q}^{>0} \mid (\exists y \in Y)\, \rho(x, y) < q\}$$

is an extended Dedekind upper cut, that is,

$$\left(\forall p, q \in \mathbb{Q}^{>0}\right) p < q \rightarrow p \notin U_x \vee q \in U_x.$$

Remark 15.39 In the usual definition of located subset [2, Chapter 4], the distance $\rho(x, Y)$ is required to be a Dedekind real, that is, the subset U_x is also required to be inhabited. With this definition, the empty set is excluded from the located subsets. We drop this condition in order to obtain a smoother correspondence between the metric notion and the pointfree notion. This approach also goes well with our decision to include the empty set among the totally bounded subsets (see Remark 15.15).

The pointfree notion of locatedness (Definition 15.32) was first introduced for localic completions of metric spaces.

Definition 15.40 (Spitters [34, Definition 44]) Let $\mathcal{M}(X)$ be the localic completion of a metric space X. A subset $V \subseteq M_X$ is said to be *located* if it is formal closed and satisfies

$$a <_X b \rightarrow a \notin V \vee b \in V.$$

A subtopology $\mathcal{S}'$ of $\mathcal{M}(X)$ is *located* if $\mathcal{S}'$ is of the form $\mathcal{M}(X)^{\mathcal{M}(X) - \neg V}$ for some located subset V of $\mathcal{M}(X)$.

Remark 15.41 It is easy to see that every located subset in the sense of Definition 15.40 satisfies (15.5). Thus a located subtopology $\mathcal{M}(X)^{\mathcal{M}(X) - \neg V}$ is closed and overt, and it is an overt weakly closed subtopology of $\mathcal{M}(X)$ determined by V. Moreover, for a locally compact metric space, the notion of locatedness in Definition 15.40 is equivalent to the one given in Definition 15.32. This follows from Theorem 15.20.

For each metric space X, we write $\mathrm{LCl}(X)$ for the collection of its located closed subsets; we write $\mathrm{LRed}(\mathcal{M}(X))$ for the collection of located subsets of its localic completion.

Theorem 15.42 (Kawai [22, Theorem 4.9]) *For a complete metric space X, the bijection between $\mathrm{Cl}(X)$ and $\mathrm{Red}(\mathcal{M}(X))$ in Theorem 15.37 restricts to a bijection between $\mathrm{LCl}(X)$ and $\mathrm{LRed}(\mathcal{M}(X))$.*

In particular, for a locally compact metric space, the metric notion of located closed subset exactly corresponds to the pointfree notion of overt closed subtopology (see Theorem 15.34). In this case, the bijection of Theorem 15.42 can be further restricted to compact subsets.[8]

Theorem 15.43 (Kawai [22, Theorem 4.16]) *For a locally compact metric space X, the bijection between $\mathrm{Cl}(X)$ and $\mathrm{Red}(\mathcal{M}(X))$ in Theorem 15.37 restricts to a bijection between compact subsets of X and compact overt subtopologies of $\mathcal{M}(X)$.*

Let X be a locally compact metric space, $Y \subseteq X$ be a compact subset, and V be a located subset of $\mathcal{M}(X)$. An inspection of the proof of Theorem 15.43 shows $\mathcal{M}(X)^{\mathcal{M}(X)-\neg\varphi(Y)} \cong \mathcal{M}(Y)$ and $\psi(V) \cong \mathrm{Pt}(\mathcal{M}(X)^{\mathcal{M}(X)-\neg V})$.

Thus, we have the following characterisation of localic completions of compact metric spaces.

Proposition 15.44 *The following are equivalent for a formal topology $\mathcal{S}$:*

(i) *$\mathcal{S}$ is isomorphic to a localic completion of some compact metric space.*
(ii) *$\mathcal{S}$ is isomorphic to a compact overt subtopology of a localic completion of some locally compact metric space.*

Example 15.45 The formal unit interval $\mathcal{I}[0,1]$ (see Example 15.25) is isomorphic to the localic completion of $[0,1]$. Indeed, $[0,1]$ is a compact subset of $\mathbb{R}$, where the latter is locally compact. Hence, $\mathcal{M}([0,1])$ is isomorphic to a compact overt subtopology $\mathcal{S}$ of the formal real $\mathcal{R}$, which is determined by a located subset Pos defined by

$$\mathrm{Pos} \stackrel{\mathrm{def}}{=} \varphi([0,1]) = \{(p,q) \in S_{\mathcal{R}} \mid 0 < q \;\&\; p < 1\}.$$

In Example 15.31, we have seen that Pos determines $\mathcal{I}[0,1]$. Hence $\mathcal{M}([0,1]) \cong \mathcal{I}[0,1]$. Now, since the localic completion preserves countable products, we have

$$\mathcal{M}\Big(\prod_{n\in\mathbb{N}}[0,1]\Big) \cong \prod_{n\in\mathbb{N}}\mathcal{M}([0,1]) \cong \prod_{n\in\mathbb{N}}\mathcal{I}[0,1],$$

that is, the localic completion sends the metric cube to the formal cube.

[8] To see that Theorem 15.43 is a restriction of Theorem 15.42, recall that a compact subset of a metric space is closed and located [2, Chapter 4, Proposition 9], and that a compact subtopology of a regular formal topology is closed [18, Chapter III, Proposition 1.2].

15.5 Pointfree Characterisation of Compact Metric Spaces

Proposition 15.44, together with the well-known theory of compact completely regular formal topologies [18, Chapter IV], allows us characterise the notion of compact metric space in a pointfree way.

First, we recall some technical notions concerning completely regular formal topologies (see [14]). In the following, we write $\mathbb{I}$ for the rational interval:

$$\mathbb{I} \stackrel{\text{def}}{=} \{q \in \mathbb{Q} \mid 0 \leq q \leq 1\}.$$

Definition 15.46 Let $\mathcal{S}$ be a formal topology. Given $U, V \subseteq S$, a *scale* from U to V is a family $(U_q)_{q \in \mathbb{I}}$ of subsets of S such that

(i) $U \lhd U_0$ and $U_1 \lhd V$,
(ii) $p < q \implies U_p \lll U_q$ for each $p, q \in \mathbb{I}$,

where $U \lll V \stackrel{\text{def}}{\iff} S \lhd U^* \cup V$ for $U^* \stackrel{\text{def}}{=} \{a \in S \mid a \downarrow U \lhd \emptyset\}$ (cf. Example 15.19). Then, U is said to be *really covered* by V, denoted by $U \twoheadleftarrow\!\!\!\lll V$, if there is a scale from U to V.

The following is a fundamental property of $\twoheadleftarrow\!\!\!\lll$. In classical point-set topology, this would imply that for any disjoint closed subsets U, V of a normal space X, there is a continuous function $f : X \to [0,1]$ such that $f(U) = 0$ and $f(V) = 1$.

Proposition 15.47 (Johnstone [18, Chapter IV, Proposition 1.4]) *Let $\mathcal{S}$ be a formal topology. For each $U, V \subseteq S$, the following are equivalent:*

(i) $U \twoheadleftarrow\!\!\!\lll V$.
(ii) *There exists a formal topology map $r : \mathcal{S} \to \mathcal{I}[0,1]$ such that*

$$r^-(0,\infty) \downarrow U \lhd \emptyset, \qquad r^-(-\infty,1) \lhd V,$$

where

$$(0,\infty) \stackrel{\text{def}}{=} \{(p,q) \in S_{\mathcal{R}} \mid p = 0\},$$
$$(-\infty,1) \stackrel{\text{def}}{=} \{(p,q) \in S_{\mathcal{R}} \mid q = 1\}.$$

For subsets $U, V \subseteq S$, the collection of scales from U to V does not form a set in general. Hence, we introduce the following predicative notion.

Definition 15.48 Let $U, V \subseteq S$ be subsets of a formal topology $\mathcal{S}$. A scale $(U_q)_{q \in \mathbb{I}}$ from U to V is said to be *finitary* if U_q is finitely enumerable for each $q \in \mathbb{I}$. The set of finitary scales from U to V is denoted by $\mathrm{Sc}_{\mathrm{Fin}}(U,V)$. For each $a, b \in S$, we write $a \twoheadleftarrow\!\!\!\lll_{\mathrm{Fin}} b$ if there exists a finitary scale from $\{a\}$ to $\{b\}$.

The pointfree version of Urysohn's lemma [18, Chapter IV, Proposition 1.6] implies that in a compact regular formal topology, $a \lll b$ implies $a \twoheadleftarrow\!\!\!\lll_{\mathrm{Fin}} b$. This motivates the following definition, which can be seen as a notion of separable compact regular formal topologies.

Definition 15.49 A formal topology $\mathcal{S}$ is *compact enumerably completely regular* if it is compact and is equipped with functions $\mathsf{wc}\colon S \to \mathrm{Pow}(S)$ and $\mathsf{sc} \in \prod_{(a,b)\in\overline{\mathsf{wc}}} \mathrm{Sc}_{\mathrm{Fin}}(\{a\},\{b\})$ such that $\overline{\mathsf{wc}} \overset{\mathrm{def}}{=} \{(a,b) \mid a \in \mathsf{wc}(b)\}$ is a countable set and $a \lhd \mathsf{wc}(a)$ for each $a \in S$. The function sc is called a *choice of scale* for wc.[9]

Remark 15.50 The notion of enumerably completely regular formal topologies [14] can be obtained by omitting the compactness and the finitary condition on scales from Definition 15.49. See Curi [14] and Kawai [23, Section 5] for further properties of enumerably completely regular formal topologies.

We are now ready to show that overt compact enumerably completely regular formal topologies are the pointfree counterparts of Bishop compact metric spaces.

Lemma 15.51 *The localic completion $\mathcal{M}(X)$ of a compact metric space X is isomorphic to an overt compact enumerably completely regular formal topology.*

Proof Let (X,ρ) be a compact metric space and $Y \subseteq X$ be a countable dense subset of X. By Theorem 15.13, $\mathcal{M}(X)$ and $\mathcal{M}(Y)$ are isomorphic; thus, we may assume that X is countable. Since $M_X = \bigcup_{\varepsilon\in\mathbb{Q}^{>0}} \mathcal{C}_\varepsilon$, the set M_X is countable. By (M1), we have

$$\mathsf{b}(x,\varepsilon) \lhd_X \left\{\mathsf{b}(x,\delta) \in M_X \mid \delta \in \mathbb{Q}^{>0} \;\&\; \delta < \varepsilon\right\}$$

for each $\mathsf{b}(x,\varepsilon) \in M_X$. Since $a <_X b$ implies $a \lll b$ for each $a, b \in M_X$, we define a function $\mathsf{wc}\colon M_X \to \mathrm{Pow}(M_X)$ by

$$\mathsf{wc}(\mathsf{b}(x,\varepsilon)) \overset{\mathrm{def}}{=} \left\{\mathsf{b}(x,\delta) \in M_X \mid \delta \in \mathbb{Q}^{>0} \;\&\; \delta < \varepsilon\right\}.$$

It is straightforward to show that the relation $\overline{\mathsf{wc}}$ is countable and that wc admits a choice of scale. Since X is compact, $\mathcal{M}(X)$ is compact by Theorem 15.17. Hence $\mathcal{M}(X)$ is an overt compact enumerably completely regular formal topology. □

Theorem 15.52 (Kawai [22, Theorem 6.10]) *The following are equivalent for a formal topology $\mathcal{S}$:*

(1) *$\mathcal{S}$ is isomorphic to an overt compact enumerably completely regular formal topology.*

[9] Assuming countable choice, one may omit sc by requiring $\mathsf{wc}\colon S \to \mathrm{Pow}(S)$ to be a countable set (viewed as a relation $\overline{\mathsf{wc}}$) that satisfies $a \in \mathsf{wb}(b) \to a \twoheadleftarrow\!\!\!\lll_{\mathrm{Fin}} b$. However, the choice function sc is required in the proof ($1 \to 2$) of Theorem 15.52, and hence it is included as a part of the definition. A similar remark applies to the notion of enumerably locally compact formal topology in Definition 15.60.

(2) *$\mathcal{S}$ is isomorphic to a compact overt subtopology of* $\prod_{n\in\mathbb{N}} \mathcal{I}[0,1]$.
(3) *$\mathcal{S}$ is isomorphic to a localic completion of some compact metric space.*

Proof (1 → 2) This is a direct corollary of the localic Tychonoff's embedding theorem [18, Chapter IV, Theorem 1.7], which implies that any enumerably completely regular formal topology can be embedded into $\prod_{n\in\mathbb{N}} \mathcal{I}[0,1]$. Specifically, countability of the relation $\overline{\text{wc}}$ (Definition 15.49) together with Proposition 15.47 allows us to embed any enumerably completely regular formal topology into $\prod_{n\in\mathbb{N}} \mathcal{I}[0,1]$. Note also that the image of an overt formal topology is overt (see e.g., [17, Section 3.5.1]); moreover, compactness is preserved by a formal topology map.

(2 → 3) This follows from $\prod_{n\in\mathbb{N}} \mathcal{I}[0,1] \cong \mathcal{M}(\prod_{n\in\mathbb{N}}[0,1])$ and Proposition 15.44.
(3 → 1) By Lemma 15.51. □

Theorem 15.52 can be seen as a pointfree version of Urysohn's metrisation theorem for overt compact enumerably completely regular formal topologies. The classical Urysohn's metrisation theorem says that a second countable Hausdorff regular space is metrisable if and only if it can be embedded into the cube. Classically, a direct pointfree translation of Urysohn's characterisation would imply that a compact regular formal topology is metrisable if and only if it can be embedded into $\prod_{n\in\mathbb{N}} \mathcal{I}[0,1]$ as a closed subtopology. Constructively, we further require these closed subtopologies to be overt, or equivalently located.

To close this section, we summarise how the pointfree characterisation of compact metric spaces has been obtained. First, the class of compact overt subtopologies of the localic completions of locally compact metric spaces is shown to represent the images of compact metric spaces under the functor $\mathcal{M}$: **LComp** → **FTop** (Proposition 15.44). Since every compact metric space can be embedded into $\prod_{n\in\mathbb{N}}[0,1]$ and the localic completion preserves countable products of inhabited compact metric spaces (Proposition 15.23), it is in fact enough to characterise compact overt subtopologies of the formal cube $\prod_{n\in\mathbb{N}} \mathcal{I}[0,1]$. This has been done in Definition 15.49 and Theorem 15.52.

15.6 Pointfree Characterisation of Locally Compact Metric Spaces

We extend the pointfree characterisation of compact metric spaces obtained in the previous section to Bishop locally compact metric spaces. The approach taken in this section is similar to that of the previous section; here, instead of characterising the compact overt subtopologies of localic completions of compact metric spaces, we characterise open complements of located subtopologies, which would include all the images of Bishop locally compact metric spaces under the functor $\mathcal{M}$: **LComp** → **FTop**. Our insight comes from Bishop's one-point compactification of locally compact metric spaces [2, Chapter 4, Proposition 9], by which each locally compact metric space is embedded into some compact subset X of $\prod_{n \in \mathbb{N}}[0,1]$ as a metric complement (of some located subset Y of X).

15.6.1 Open Complements of Located Subtopologies

As a first step, we give a sufficient condition in which a formal topology becomes isomorphic to a localic completion of a Bishop locally compact metric space. To this end, we exploit the theory of open complements of locally compact metric spaces due to Palmgren [30].

Definition 15.53 The category **OLCM** of open complements of locally compact metric spaces consists of the following data: an object is a pair (X, U), where X is a locally compact metric space and U is an open subset of X; a morphism $f\colon (X,U) \to (Y,V)$ of **OLCM** is a function $f\colon U \to V$ such that for each inhabited compact subset $K \Subset U$, where

$$K \Subset U \overset{\text{def}}{\iff} (\exists r \in \mathbb{Q}^{>0})\, \{x \in X \mid \rho_X(x,K) \leq r\} \subseteq U,$$

we have that f is uniformly continuous on K and $f[K] \Subset V$.

Note that an inhabited totally bounded subset of a metric space is located and that the image of a totally bounded subset under a uniformly continuous function is totally bounded. Hence, the condition $f[K] \Subset V$ is well defined, that is, the distance $\rho_Y(y, f[K])$ exists for each $y \in Y$.

Palmgren [30, Section 4] showed that **OLCM** can be embedded into **FTop** via a full and faithful functor $\mathcal{OM}\colon \mathbf{OLCM} \to \mathbf{FTop}$. The functor $\mathcal{OM}$ assigns to each object (X,U) of **OLCM** the open subtopology $\mathcal{M}(X)_{H(U)}$ of $\mathcal{M}(X)$ determined by the subset

$$H(U) \overset{\text{def}}{=} \{\mathsf{b}(x,\varepsilon) \in M_X \mid B(x,\varepsilon) \subseteq U\}.$$

The category of Bishop locally compact metric spaces is embedded into **OLCM** via the inclusion $X \mapsto (X,X)$. Note that $\mathcal{OM}((X,X)) = \mathcal{M}(X)$ for each locally compact metric space X.

Let A be a located subset of a metric space (X,ρ). The *metric complement* [2, Chapter 4] of A is the open subset $X - A$ of X defined by

$$X - A \overset{\text{def}}{=} \{x \in X \mid \rho(x,A) > 0\}.$$

The corresponding pointfree notion is the following.

Definition 15.54 Let V be a located subset of a locally compact formal topology $\mathcal{S}$. The *open complement* of the located subtopology $\mathcal{S}^{S - \neg V}$ is the open subtopology $\mathcal{S}_{\neg V}$ determined by $\neg V$.

The embedding $\mathcal{OM}$: **OLCM** $\to$ **FTop** preserves metric complements and open complements of located subtopologies. The following proposition is a slight generalisation of Kawai [23, Proposition 4.8] with essentially the same proof.

Proposition 15.55 *Let X be a locally compact metric space, and let $\varphi\colon \mathrm{LCl}(X) \to \mathrm{LRed}(\mathcal{M}(X))$ be the bijection in Theorem 15.42. Then, for each $A \in \mathrm{LCl}(X)$ and $V \in \mathrm{LRed}(\mathcal{M}(X))$, we have*

$$H(X - A) = \neg\,\varphi(A), \qquad\qquad (\neg\,V)_* = X - \psi(V).$$

The assignments $U \mapsto H(U)$ and $W \mapsto W_*$ restrict to a bijection between the metric complements of located closed subsets of X and the open complements of located subtopologies of $\mathcal{M}(X)$.

Proposition 15.55 and Theorem 15.43 yield the following special case.

Corollary 15.56 *Let X be a compact metric space. For each located subset V of $\mathcal{M}(X)$, there exists a unique compact subset $A \subseteq X$ such that $\mathcal{OM}\,((X, X - A)) = \mathcal{M}(X)_{\neg V}$.*

A subset of a metric space which is not locally compact may become locally compact under a new metric. A prime example is the open interval $(0, 1)$: since $(0, 1)$ is not complete, it is not a Bishop locally compact subset of $[0, 1]$; however, it becomes locally compact under the metric $\rho(x, y) := |x - y| + |1/x - 1/y|$. This construction of a new metric can be generalised to any metric complement of compact metric spaces.[10]

Proposition 15.57 *Let (X, ρ) be a compact metric space and A be an inhabited compact subset of X. Then, the metric complement $Y = X - A$ equipped with a metric*

$$\rho'(x, y) \overset{\text{def}}{=} \rho(x, y) + \left| \frac{1}{\rho(x, A)} - \frac{1}{\rho(y, A)} \right| \tag{15.7}$$

is locally compact, and the pair (Y, Y) is isomorphic to $(X, X - A)$ in **OLCM.**

Proposition 15.57 implies that a formal topology of the form $\mathcal{M}(X)_{\neg V}$ where X is compact and V is located is in the image of the functor $\mathcal{M}$: **LComp** $\to$ **FTop**.

Proposition 15.58 (Kawai [23, Proposition 4.13]) *Let X be a compact metric space and V be a located subset of $\mathcal{M}(X)$. Then, there exists a locally compact metric space Y such that $\mathcal{M}(Y) \cong \mathcal{M}(X)_{\neg V}$.*

Proof By Corollary 15.56, there exists a unique compact subset A of X such that

$$\mathcal{OM}\,((X, X - A)) = \mathcal{M}(X)_{\neg V}.$$

[10] The construction of ρ' in (15.7) is suggested by the notion of compactifier [2, page 103].

If A is empty, we have $(Y, Y) \cong (X, X - A)$ for $Y = X$. If A is inhabited, then Proposition 15.58 implies there exists a locally compact metric space Y such that $(Y, Y) \cong (X, X - A)$ in **OLCM**. In either case, we have

$$\mathcal{M}(Y) = \mathcal{OM}((Y, Y)) \cong \mathcal{OM}((X, X - A)) = \mathcal{M}(X)_{\neg V}. \qquad \square$$

In the setting of Proposition 15.58, it can be shown that the open complement $\mathcal{M}(X)_{\neg V}$ is inhabited if and only if $(\neg V)_*$ is inhabited. This yields the following.

Corollary 15.59 *Let X be a compact metric space and V be a located subset of $\mathcal{M}(X)$ such that the open complement $\mathcal{M}(X)_{\neg V}$ is inhabited. Then, there exists a Bishop (i.e., inhabited) locally compact metric space Y such that $\mathcal{M}(Y) \cong \mathcal{M}(X)_{\neg V}$.*

In the next subsection, we will see that an inhabited formal topologies of the form $\mathcal{M}(X)_{\neg V}$, where X is compact and V is located, characterise the image of Bishop locally compact metric spaces under the functor $\mathcal{M}\colon$ **LComp** $\to$ **FTop**.

15.6.2 The Pointfree Characterisation

We begin with a description of a candidate for a pointfree notion of Bishop locally compact metric space. First, we define the notion of separable locally compact spaces in formal topology.

Definition 15.60 Let $U, V \subseteq S$ be subsets of a formal topology $\mathcal{S}$. A *wb-scale* from U to V is a family $(U_q)_{q \in \mathbb{I}}$ of subsets of S such that

(i) $U \lhd U_0$ and $U_1 \lhd V$,
(ii) $p < q \implies U_p \ll U_q$ for each $p, q \in \mathbb{I}$,

where

$$U \ll V \overset{\text{def}}{\iff} (\forall W \in \mathrm{Pow}(S))\,[V \lhd W \implies (\exists A \in \mathrm{Fin}(W))\, U \lhd A]\,.^{11}$$

We write $\mathrm{Sc}_{\ll}(U, V)$ for the collection of wb-scales from U to V.

A formal topology $\mathcal{S}$ is *enumerably locally compact* if it is equipped with functions $\mathsf{wb}\colon S \to \mathrm{Pow}(S)$ and $\mathsf{sc} \in \prod_{(a,b) \in \overline{\mathsf{wb}}} \mathrm{Sc}_{\ll}(\{a\}, \{b\})$ such that $\overline{\mathsf{wb}} \overset{\text{def}}{=} \{(a, b) \in S \times S \mid a \in \mathsf{wb}(b)\}$ is a countable set and $a \lhd \mathsf{wb}(a)$ for each $a \in S$. The function sc is called a *choice of wb-scale* for wb.

We now focus on overt enumerably locally compact regular formal topologies. These topologies admit a one-point compactification, which is analogous to Bishop's metric version.

[11] Note that the definition of $U \ll V$ is impredicative.

Definition 15.61 Let $\mathcal{S}$ be an overt enumerably locally compact regular formal topology. A *one-point compactification* of $\mathcal{S}$ is a triple $(\mathcal{T}, \omega, r)$ where $\mathcal{T}$ is an overt compact enumerably completely regular formal topology, ω is a formal point of $\mathcal{T}$, and $r : \mathcal{S} \to \mathcal{T}$ is an embedding such that the image of r is isomorphic to the open complement $\mathcal{T}_{\neg\omega}$ of the located subtopology determined by ω.

Note that each formal point of a locally compact regular formal topology is located and thus determines a located subtopology.

Theorem 15.62 (Kawai [23, Theorem 6.5]) *Every overt enumerably locally compact regular formal topology has a one-point compactification.*

First, it should be noted that the proof requires countable choice. The proof, which we now briefly explain, is similar to that of the metric one-point compactification by Bishop and Bridges [3, Chapter 4, Theorem 6.8]. Given an overt enumerably locally compact regular formal topology $\mathcal{S}$, we construct an embedding $r : \mathcal{S} \to \prod_{n \in \mathbb{N}} \mathcal{I}[0, 1]$ using Tychonoff's embedding theorem (cf. Theorem 15.52). This is possible because every enumerably locally compact regular formal topology is enumerably completely regular (see [23, Lemma 6.3]). Then, we adjoin a point ω of $\prod_{n \in \mathbb{N}} \mathcal{I}[0, 1]$, which is a pointfree analogue of the constant sequence of 1, to the image of r. The resulting formal topology $\mathcal{T}$ together with r and ω satisfies the requirement of a one-point compactification of $\mathcal{S}$.

The construction of a one-point compactification (Theorem 15.62) is a crucial step in obtaining a pointfree characterisation of Bishop locally compact metric spaces. First, by Theorem 15.52, Proposition 15.58, and Theorem 15.62, we obtain the following.

Proposition 15.63 *For any overt enumerably locally compact regular formal topology $\mathcal{S}$, there exists a locally compact metric space X such that $\mathcal{M}(X) \cong \mathcal{S}$.*

It is easy to see that the image of each inhabited formal topology under a formal topology map is inhabited (see [17, Section 3.5.1]). Hence, Corollary 15.59 yields the following.

Corollary 15.64 *For any inhabited enumerably locally compact regular formal topology $\mathcal{S}$, there exists a Bishop locally compact metric space X such that $\mathcal{M}(X) \cong \mathcal{S}$.*

As in the proof of Lemma 15.51, one can show that the localic completion of a Bishop locally compact metric space is isomorphic to an inhabited enumerably locally compact regular formal topology. Hence, we obtain the following pointfree characterisation of Bishop locally compact metric spaces.

Theorem 15.65 (Kawai [23, Theorem 7.5]) *A formal topology $\mathcal{S}$ is isomorphic to an inhabited enumerably locally compact regular formal topology if and only if $\mathcal{S}$ is isomorphic to a localic completion of some Bishop locally compact metric space.*

15.7 Beyond Locally Compact Metric Spaces

For the class of metric spaces more general than locally compact spaces, it is not obvious how one can relate Bishop's theory of metric spaces to formal topology. This is because there seems to be no fully satisfactory notion of 'continuous function' between general metric spaces which would naturally extend Bishop's notion of continuity on locally compact spaces. See, for example, Bridges [5, Section 3] for further discussion on this issue.

Nevertheless, Bishop proposed one notion of continuity for general metric spaces which is sometimes adopted in practice (see Definition 15.66). The aim of this section is to compare this notion with that of formal topology in the setting of complete metric spaces.

15.7.1 Strong Continuity vs FT-Continuity

Based on Bishop's unpublished manuscript, Bridges [4] proposed the following notion of continuous function.[12]

Definition 15.66 A subset L of a metric space X is a *compact image* if there exist a compact metric space K and a uniformly continuous function $f\colon K \to X$ with $L = f[K]$. A function $f\colon X \to Y$ between metric spaces (X, ρ_X) and (Y, ρ_Y) is *strongly continuous* (or uniformly continuous near each compact image) if for each compact image $L \subseteq X$ and $\varepsilon > 0$, there is $\delta > 0$ such that

$$x \in L \mathbin{\&} \rho_X(x,u) < \delta \implies \rho_Y(f(x), f(u)) < \varepsilon$$

for each $x, u \in X$.

For locally compact metric spaces, the notion of strong continuity between them agrees with Bishop's notion of continuous map (Definition 15.14).

For complete metric spaces, localic completions of metric spaces provide another candidate for a good notion of continuity between them. From the pointfree perspective, this notion is satisfactory as it would allow us to extend the functorial embedding of locally compact metric spaces (cf. Section 15.3) to complete metric spaces.

[12] The term *strongly continuous function* adopted here is taken from Troelstra and van Dalen [37, Chapter 7]. The notion of strongly continuous function is attributed to Bishop by Bridges [5, Section 3], where he refers to these functions as *B-continuous* functions. In Bridges [4], these functions are simply called *continuous* maps.

Definition 15.67 (Palmgren [31, Section 2]) A function $f\colon X \to Y$ between complete metric spaces is *FT-continuous* if there exists a formal topology map $r\colon \mathcal{M}(X) \to \mathcal{M}(Y)$ between the localic completions of X and Y which makes the diagram (15.4) commute.

For locally compact metric spaces, FT-continuity and strong continuity agree: this follows from the fullness and faithfulness of the functor $\mathcal{M}\colon$ **LComp** $\to$ **FTop**. The question is whether this correspondence carries over to non-locally compact metric spaces. Palmgren [31, Theorem 2.3] showed that every FT-continuous function between complete metric spaces is strongly continuous. This implies that FT-continuity is stronger than strong continuity; in fact, the former is strictly stronger than the latter (see [24, Section 3]). Thus, from the pointfree perspective, strong continuity is not strong enough beyond the context of locally compact metric spaces. However, even for the restricted class of complete metric spaces, finding a natural notion of continuity between them that would agree with FT-continuity seems quite a challenge. Nevertheless, Section 15.7.2 deals with one positive example.

15.7.2 FT-Continuous Functions on the Baire Space

For the Baire space $\mathbb{N}^{\mathbb{N}}$, FT-continuity corresponds to the following intuitionistic notion of continuity.

Definition 15.68 (Kreisel and Troelstra [25, Section 3]; Troelstra and van Dalen [36, Chapter 4, Section 8.4]) Let $\mathbb{N}^*$ be the set of finite sequences of $\mathbb{N}$, and let $\mathbb{F}(\mathbb{N}^*, \mathbb{N})$ be the set of functions from $\mathbb{N}^*$ to $\mathbb{N}$. The class $K \subseteq \mathbb{F}(\mathbb{N}^*, \mathbb{N})$ of *Brouwer operations* is inductively defined by the following clauses:

$$\frac{n \in \mathbb{N}}{\lambda a.n + 1 \in K} \qquad\qquad \frac{\gamma(\langle\,\rangle) = 0 \quad (\forall n \in \mathbb{N})\, \gamma_n \in K}{\gamma \in K}.$$

Here $\gamma_n \stackrel{\text{def}}{=} \lambda a.\gamma(\langle n \rangle * a)$ for each $n \in \mathbb{N}$, where $\langle n \rangle * a$ denotes the sequence a preposed by n. We say that a function $f\colon \mathbb{N}^{\mathbb{N}} \to \mathbb{N}$ is *induced by a Brouwer operation* $\gamma \in K$ if

$$\gamma(\overline{\alpha}n) > 0 \implies \gamma(\overline{\alpha}n) = f(\alpha) + 1$$

for each $\alpha \in \mathbb{N}^{\mathbb{N}}$ and $n \in \mathbb{N}$, where $\overline{\alpha}n$ is the initial segment of α of length n. Similarly, we say that a function $f\colon \mathbb{N}^{\mathbb{N}} \to \mathbb{N}^{\mathbb{N}}$ is induced by a Brouwer operation if the composition $\pi_n \circ f$ with each projection $\pi_n\colon \mathbb{N}^{\mathbb{N}} \to \mathbb{N}$ is induced by a Brouwer operation.

The following establishes a precise correspondence between FT-continuous functions and functions induced by Brouwer operations.

Theorem 15.69 (Kawai [24, Theorem 4.11]) *A function from Baire space to itself is FT-continuous if and only if it is induced by a Brouwer operation.*

15.8 Related Works

Vickers [41] generalised the localic completion of metric spaces to various hyperspaces, that is, spaces of certain classes of subsets of a given metric space.[13] Among the hyperspaces that Vickers treated is the space of compact subsets (of a given space X) with the Hausdorff metric (see [2, page 94]), which corresponds to the Vietoris powerlocale of the localic completion of X in pointfree topology [18, Chapter III, Section 4]. The localic completion can also be extended to uniform spaces [21]. Here, the notion of uniform space in question is that where the uniform structure is derived from a family of pseudometrics [2, Chapter 4, Problem 17]. By the localic completion, the category of locally compact uniform spaces can be full and faithfully embedded into that of locally compact regular formal topologies. However, it seems impossible to obtain a pointfree characterisation of locally compact uniform spaces along the lines of this chapter. This is because the proof for metric spaces relies crucially on the separability of Bishop locally compact metric spaces.

In this chapter, we have focussed only on the connection between Bishop's theory of metric spaces and formal topology. As for the connection between Bishop's constructive analysis in general and formal topology, we should mention important work by Coquand and Spitters [11]: using a pointfree representation of spectrum of Riesz spaces [10], they obtained the Stone–Yoshida representation theorem of Riesz spaces. From this, they derived the Gelfand representation theorem in Bishop constructive analysis.

References

[1] Aczel, P., and Rathjen, M. 2000/2001. *Notes on Constructive Set Theory*. Technical Report 40, Institut Mittag-Leffler.

[2] Bishop, E. 1967. *Foundations of Constructive Analysis*. New York: McGraw-Hill.

[3] Bishop, E., and Bridges, D. 1985. *Constructive Analysis*. Berlin: Springer.

[4] Bridges, D. 1979. *Constructive Functional Analysis*. Research Notes in Mathematics, **28**. London: Pitman.

[5] Bridges, D. 2012. Reflections on function spaces. *Ann. Pure Appl. Logic*, **163**(2), 101–110.

[13] Vickers uses the notion of a generalised metric, which takes its values in the upper reals and where the only assumptions are the zero self-distance law and the triangle inequality.

[6] Cederquist, J., and Negri, S. 1996. A constructive proof of the Heine–Borel covering theorem for formal reals. Pages 62–75 of: Berardi, S., and Coppo, M. (eds.), *Types for Proofs and Programs*. Lecture Notes in Computer Science, **1158**. Berlin, Heidelberg: Springer.

[7] Ciraulo, F., and Sambin, G. 2018. Embedding locales and formal topologies into positive topologies. *Arch. Math. Logic*, **57**, 755–768.

[8] Ciraulo, F., and Vickers, S. 2016. Positivity relations on a locale. *Ann. Pure Appl. Logic*, **167**(9), 806–819.

[9] Coquand, T. 1992. An intuitionistic proof of Tychonoff's theorem. *J. Symbol. Logic*, **57**(1), 28–32.

[10] Coquand, T. 2005. About Stone's notion of spectrum. *J. Pure. Appl. Algebra*, **197**(1–3), 141–158.

[11] Coquand, T., and Spitters, B. 2005. Formal topology and constructive mathematics: the Gelfand and Stone–Yosida representation theorems. *J. UCS*, **11**(12), 1932–1944.

[12] Coquand, T., Sambin, G., Smith, J., and Valentini, S. 2003. Inductively generated formal topologies. *Ann. Pure Appl. Logic*, **124**(1–3), 71–106.

[13] Coquand, T., Palmgren, E., and Spitters, B. 2011. Metric complements of overt closed sets. *Math. Log. Q.*, **57**(4), 373–378.

[14] Curi, G. 2003. Constructive metrisability in point-free topology. *Theoret. Comput. Sci.*, **305**(1–3), 85–109.

[15] Curi, G. 2007. Exact approximations to Stone–Čech compactification. *Ann. Pure Appl. Logic*, **146**(2–3), 103–123.

[16] Fourman, M. P., and Grayson, R. J. 1982. Formal spaces. Pages 107–122 of: Troelstra, A. S., and van Dalen, D. (eds.), *The L.E.J. Brouwer Centenary Symposium*. Amsterdam: North-Holland.

[17] Fox, C. 2005. Point-set and point-free topology in constructive set theory. Ph.D. thesis, University of Manchester.

[18] Johnstone, P. T. 1982. *Stone Spaces*. Cambridge: Cambridge University Press.

[19] Joyal, A., and Tierney, M. 1984. An extension of the Galois theory of Grothendieck. *Mem. Amer. Math. Soc.*, **51**(309), vii+71.

[20] Kawai, T. 2015. Bishop compactness in formal topology. Ph.D. thesis, Japan Advanced Institute of Science and Technology.

[21] Kawai, T. 2017a. Localic completion of uniform spaces. *Log. Meth. Comput. Sci.*, **13**(3:22), 1–39.

[22] Kawai, T. 2017b. Point-free characterisation of Bishop compact metric spaces. *J. Log. Anal.*, **9**(5), 1–30.

[23] Kawai, T. 2017c. A point-free characterisation of Bishop locally compact metric spaces. *J. Log. Anal.*, **9**(c2), 1–41.

[24] Kawai, T. 2018. Formally continuous functions on Baire space. *Math. Log. Q.*, **64**(3), 192–200.

[25] Kreisel, G., and Troelstra, A. S. 1970. Formal systems for some branches of intuitionistic analysis. *Ann. Math. Logic*, **1**(3), 229–387.

[26] Negri, S., and Soravia, D. 1999. The continuum as a formal space. *Arch. Math. Logic*, **38**(7), 423–447.

[27] Negri, S., and Valentini, S. 1997. Tychonoff's theorem in the framework of formal topologies. *J. Symbol. Logic*, **62**(4), 1315–1332.

[28] Palmgren, E. 2005. Continuity on the real line and in formal spaces. Pages 165–175 of: Crosilla, L., and Schuster, P. (eds.), *From Sets and Types to Topology and Analysis: Towards Practicable Foundations for Constructive Mathematics*. Oxford Logic Guides, **48**. Oxford: Oxford University Press.

[29] Palmgren, E. 2007. A constructive and functorial embedding of locally compact metric spaces into locales. *Topol. Appl.*, **154**, 1854–1880.

[30] Palmgren, E. 2010. Open sublocales of localic completions. *J. Log. Anal.*, **2**, 1–22.

[31] Palmgren, E. 2014. Formal continuity implies uniform continuity near compact images on metric spaces. *Math. Log. Q.*, **60**(1–2), 66–69.

[32] Sambin, G. 1987. Intuitionistic formal spaces – a first communication. Pages 187–204 of: Skordev, D. G. (ed.), *Mathematical Logic and its Applications*, **305**. New York: Plenum Press.

[33] Sambin, G. 2003. Some points in formal topology. *Theoret. Comput. Sci.*, **305**(1–3), 347–408.

[34] Spitters, B. 2010. Locatedness and overt sublocales. *Ann. Pure Appl. Logic*, **162**(1), 36–54.

[35] Taylor, P. 2010. A lambda calculus for real analysis. *J. Log. Anal.*, **2**(5), 1–115.

[36] Troelstra, A. S., and van Dalen, D. 1988a. *Constructivism in Mathematics: An Introduction. Volume I*. Studies in Logic and the Foundations of Mathematics, **121**. Amsterdam: North-Holland.

[37] Troelstra, A. S., and van Dalen, D. 1988b. *Constructivism in Mathematics: An Introduction. Volume II*. Studies in Logic and the Foundations of Mathematics, **123**. Amsterdam: North-Holland.

[38] Vickers, S. 2005a. Localic completion of generalized metric spaces I. *Theory Appl. Categ.*, **14**(15), 328–356.

[39] Vickers, S. 2005b. Some constructive roads to Tychonoff. Pages 223–238 of: Crosilla, L., and Schuster, P. (eds.), *From Sets and Types to Topology and Analysis: Towards Practicable Foundations for Constructive Mathematics*. Oxford Logic Guides, **48**. Oxford: Oxford University Press.

[40] Vickers, S. 2007. Sublocales in formal topology. *J. Symbol. Logic*, **72**(2), 463–482.

[41] Vickers, S. 2009. Localic completion of generalized metric spaces II: Powerlocales. *J. Log. Anal.*, **1**(11), 1–48.

16

Subspaces in Pointfree Topology: Towards a New Approach to Measure Theory

Francesco Ciraulo

16.1 Introduction

A colleague of mine once told me that after attending some lectures about logic and foundations he felt he no longer knew what a real number was! The same feeling occurred to me when I wanted to replace my pointwise view of the real line with its pointfree version. As undergraduate students, we all realized that the open sets, not the points, are those who play the main role in topology.[1] Therefore, the idea of doing topology by taking the structure of open sets as a primitive (which is what is called *pointfree topology*) should not sound so weird, but quite natural instead. Such an idea, which was already present in some work by McKinsey and Tarski in the 1940s, was taken up seriously in the 1950s by Ehresmann and Bénabou, who started regarding *frames* (see below) as generalized topological spaces.

To the algebraist's eyes, the open sets of a topological space X form a **frame** ΩX, that is, a complete lattice in which binary meets distribute over arbitrary (set-indexed) joins.[2] And a continuous map $f : X \to Y$ induces a frame homomorphism $f^{-1} : \Omega Y \to \Omega X$, that is, a map which preserves joins and finite meets.[3] From a categorical point of view, we thus have a contravariant functor Ω from the category of topological spaces to the category of frames. This gives, of course, a covariant functor from topological spaces to the opposite of the category of frames; the latter

[1] Different opinions about that are legitimate, of course. For instance, one could argue that the class of continuous functions is what really counts, or even a subclass of it, for example, uniformly continuous real-valued functions (on a metric space). The latter, by the way, is the sense in which the word topology is understood in other chapters of this handbook.

[2] In the case of the open sets, meets and joins are given by the set-theoretic operations of intersection and union, of course. It is well-known that a frame is the same thing as a complete Heyting algebra, although this fact is 'foundations-dependent': in a predicative framework, a frame with only set-indexed joins need not have an implication.

[3] Actually, every function $f : X \to Y$ induces a function $f^{-1} : \mathrm{Pow}(Y) \to \mathrm{Pow}(X)$ between the corresponding powersets, and such a map preserves intersections and joins. However, f is continuous precisely when f^{-1} restricts to a map from ΩY to ΩX, that is to say, the image of ΩY under f^{-1} is contained in ΩX.

is usually referred to as the category of **locales**. A locale (or frame) of the form ΩX is called **spatial**.

The topological and categorical properties of frames were studied extensively by Dowker and Papert Strauss in the 1960s and 1970s, but it was Isbell [8] who claimed that the category of locales is more convenient than the category of topological spaces in many respects. This is nicely expressed in [10], but it is only after seeing some technical results, for instance after reading a good part of [9], that one becomes convinced that Isbell's claim could be true.

The aim of this chapter is to present a couple of examples in which the pointfree approach to topology looks convenient, either because it requires fewer foundational assumptions or because it produces stronger results (or both). Our treatment will be limited to the real line; for greater generality we refer the reader to [9] (where the interested reader can also find a detailed history of the topic and an extended bibliography) or to [15] for a quite different treatment not relying upon category-theoretic notions.

In Section 16.2 we shall encounter the notion of a *sublocale* of the real line, which extends the familiar notion of a subspace. So there are more sublocales than subspaces, and such a richness can be fruitfully used in practice. In particular, we shall discuss an application to measure theory (Section 16.3) that was proposed in [19]: the failure of the additivity property for the exterior measure on subspaces disappears as soon as sublocales are considered in the computation of intersections (see, in particular, Subsection 16.3.3); this is possible because disjoint subspaces could happen to share a common part, a 'pointless' sublocale.

Throughout the chapter we shall try to work constructively, essentially in the sense of Bishop-style mathematics (as in most of this book). In particular, we shall make free use of countable choice. Some work has still to be done in order to reach a satisfactory constructive treatment of some of the matter presented in Section 16.3. Even more, if we want to move to a pure pointfree perspective (in which a real number, an integrable set, a measure, and so on, have to be understood as 'points' in a corresponding pointfree space) and to a predicative framework. A first necessary step is described in Section 16.4, while some discussion about what is still to be done can be found in our concluding remarks in Section 16.5.

16.2 Pointfree Parts of the Real Line

Let $\mathbb{R}$ be the real line, that is, the set of real numbers with their natural topology, as we all[4] know them; and let $\Omega\mathbb{R}$ be the frame of open subsets of $\mathbb{R}$. We adopt the standard 'Bishop's style' definition of an open set as the union of a family of

[4] Except, perhaps, that colleague of mine I mentioned in the introduction!

neighbourhoods, where a neighbourhood is an open finite interval with rational endpoints.[5]

16.2.1 Sublocales of the Real Line

A subspace of the real line is a subset $D \subseteq \mathbb{R}$ equipped with the induced topology $\Omega D = \{U \cap D \mid U \in \Omega\mathbb{R}\}$; in other words, a basic neighbourhood in D has the form $(a, b) \cap D$ with $a, b \in \mathbb{R}$ ($a, b \in \mathbb{Q}$ is enough, of course). And ΩD is a frame too, although it is not a subframe of $\Omega\mathbb{R}$, in general.[6] Actually, ΩD is a quotient of the frame $\Omega\mathbb{R}$. In other words the map $U \mapsto U \cap D$ from $\Omega\mathbb{R}$ to ΩD is onto, and preserves, arbitrary unions and finite intersections. A category theorist would say that $\Omega\mathbb{R} \to \Omega D$ is an epimorphism, actually a regular epimorphism, in the category of frames. This corresponds to the fact that the inclusion $D \hookrightarrow \mathbb{R}$ is a regular monomorphism in the category of topological spaces (not just an injective, continuous map[7]). This is the content of the following proposition.

Proposition 16.1 *Let X be a T_0 space and let $e : X \to \mathbb{R}$ be continuous. Then the following are equivalent:*

(i) *X is homeomorphic, via e, to the subspace $e(X)$ of $\mathbb{R}$;*
(ii) *the induced frame homomorphism $e^{-1} : \Omega\mathbb{R} \to \Omega X$ is surjective; that is, $\Omega X = \{e^{-1}(U) \mid U \in \Omega\mathbb{R}\}$;*
(iii) *e is an equalizer in the category of topological spaces.*

Proof It is clear that (i) implies (ii).

To show the converse, it is sufficient to check that e is injective if (ii) holds. Indeed, if $e(x_1)$ and $e(x_2)$ coincide, then they have the same open neighbourhoods; so x_1 and x_2 have the same open neighbourhood as well, by (ii), and hence they coincide (because X is T_0).

In order to show that (i) implies (iii) we consider two particular continuous functions $f, g : \mathbb{R} \to \text{Pow}(\{0\})$, where $\text{Pow}(\{0\})$, the powerset of a singleton, is equipped with the trivial topology, f is the constant function with value $\{0\}$, and $g(r) = \{z \in \{0\} \mid r \in e(X)\}$. So $f \circ e = g \circ e$. If $t : Y \to \mathbb{R}$ is another continuous map with $f \circ t = g \circ t$, then $t(y) \in e(X)$ for all $y \in Y$ and the map $h : Y \to X$ defined as $h(y) = e^{-1}(t(y))$ is the unique continuous map h such that $t = e \circ h$.

[5] So $\mathbb{R}$ is a *neighbourhood space* or, modulo some size issue, a *concrete space* in the sense of [18]. Note that in some constructive approaches a more restrictive definition is adopted; for instance, Martin-Löf [14] defines an open to be, essentially, the union of a *recursively enumerable* family of neighbourhoods.

[6] In fact, ΩD is a subframe of $\Omega\mathbb{R}$ if, and only if, D is open.

[7] For instance, the identity function from $\mathbb{R}$ (with the euclidean topology) to the reals equipped with the trivial topology is injective and continuous, but it does not give a subspace inclusion.

On the other hand, it is easy to check that the equalizer of two continuous functions $f, g : \mathbb{R} \to Z$ is (up to homeomorphism) the subspace of $\mathbb{R}$ induced on the subset $\{x \in \mathbb{R} \mid f(x) = g(x)\}$. □

Summing up, every subspace $D \subseteq \mathbb{R}$ corresponds to a particular quotient ΩD of the frame $\Omega\mathbb{R}$. Isbell's basic insight was to consider *every* quotient frame of $\Omega\mathbb{R}$ as a generalized subspace of $\mathbb{R}$; this is what is called a **sublocale** of $\mathbb{R}$. Usually we identify a subspace D of $\mathbb{R}$ with the corresponding (spatial) sublocale ΩD of $\Omega\mathbb{R}$.[8] Accordingly, by an **open sublocale** we mean a sublocale of the form ΩU with U an open subset.

As one can guess from the couple of examples we shall meet in this chapter, the notion of a sublocale turns out to be quite well behaved. And this is somehow not a surprise, given that sublocales correspond to regular subobjects in the category of locales. From now on, let $\text{Sub}(\mathbb{R})$ be the collection of sublocales of $\mathbb{R}$ (not to be confused with $\text{Pow}(\mathbb{R})$, the collection of subsets/subspaces of $\mathbb{R}$). It is well-known that $\text{Sub}(\mathbb{R})$ is a complete lattice, actually, a *coframe* (binary joins distribute over arbitrary meets; details can be found in [9]); the ordering is the standard one between (regular) subobjects in a category and so is an extension of the usual ordering between subspaces, that is, inclusion.

Sublocales (of $\Omega\mathbb{R}$) can be explicitly represented in a number of possible ways, each having some peculiar advantage: in one representation, for instance, it is easy to compute meets of sublocales, in another representation it is easy to compute joins; moreover, one of the representations works well in a more general context (such as the σ-locales of Section 16.2.3 below).

Nuclei Perhaps the most standard way to describe a sublocale is by means of a **nucleus**. A nucleus on $\Omega\mathbb{R}$ is a mapping $j : \Omega\mathbb{R} \to \Omega\mathbb{R}$ with the following properties:

(i) $U \subseteq jU$
(ii) $jjU = jU$
(iii) $j(U \cap V) = jU \cap jV$

for all $U, V \in \Omega\mathbb{R}$. The correspondence between nuclei and quotient frames is as follows.

Given a nucleus j, the set

$$Fix(j) \;=\; \{U \in \Omega\mathbb{R} \mid jU = U\} \;=\; \{jU \mid U \in \Omega\mathbb{R}\} \;=\; j(\Omega\mathbb{R})$$

is a quotient of $\Omega\mathbb{R}$ via the mapping $U \mapsto jU$. This could seem a little bit misleading, since $Fix(j)$ is a subset of $\Omega\mathbb{R}$: an element of $Fix(j)$ should be understood as a

[8] From a constructive point of view, there is some subtlety here which we shall discuss at the beginning of Section 16.2.2.

canonical representative of an equivalence class of open sets. Order and finite meets in $\mathrm{Fix}(j)$ are inherited from $\Omega\mathbb{R}$; the join of a family $\{jU_i\}_{i\in I}$, instead, is given by $j(\bigcup_{i\in I} U_i)$.

On the other hand, if $f : \Omega\mathbb{R} \to L$ is an epimorphisms of frames, then one can construct a nucleus on $\Omega\mathbb{R}$ as the composition $g \circ f$, where $g : L \to \Omega\mathbb{R}$ is the right adjoint of f.[9]

The nucleus j_D corresponding to a subspace $D \subseteq \mathbb{R}$ turns out to be $j_D(U) = \mathsf{int}(D \to U)$, where int is the topological interior operator, and $D \to U$ is the set-theoretic implication $\{x \in \mathbb{R} \mid \{x\} \cap D \subseteq U\}$ (classically, this is equivalent to $-D \cup U$, where $-$ is the set-theoretic complement).

Note that the meet $U \wedge V$ of two open sublocales U and V is induced by their intersection $U \cap V$. Indeed, if W is such that $\mathsf{int}(U \to W) = W = \mathsf{int}(V \to W)$, then $W = \mathsf{int}(U \to \mathsf{int}(V \to W)) = \mathsf{int}((U \cap V) \to W)$; on the other hand, an open set of the form $\mathsf{int}((U\cap V) \to W)$ can be rewritten as $\mathsf{int}(U \to \mathsf{int}(V \to W))$ and also as $\mathsf{int}(V \to \mathsf{int}(U \to W))$, and hence it is both of the form $\mathsf{int}(U \to \_)$ and of the form $\mathsf{int}(V \to \_)$.[10]

In general, the meet $D \wedge D'$ of two subspaces is strictly larger than (the sublocale induced by) their intersection $D \cap D'$, as we will see.

Congruences A natural way to define a quotient frame is by means of a **congruence**, that is, an equivalence relation which respects finite meets and arbitrary joins.[11] The nuclei-congruences dictionary is as follows: given a congruence $\simeq$, the corresponding nucleus is $jU = \bigcup\{V \in \Omega\mathbb{R} \mid U \simeq V\}$, the largest element in the equivalence class of U; on the other hand, given a nucleus j, the corresponding congruence is $U \simeq V$ iff $jU = jV$. In particular, the congruence corresponding to a subspace D is simply $U \simeq_D V$ iff $U \cap D = V \cap D$.

Congruences make the computation of joins of sublocales easy. Indeed, the set-theoretic intersection of a family of congruences is again a congruence, which corresponds to taking the join in $\mathrm{Sub}(\mathbb{R})$. In particular, the lattice of frame congruences on $\Omega\mathbb{R}$ is the opposite of the lattice $\mathrm{Sub}(\mathbb{R})$ of sublocales.

Note that the join, in the sense of sublocales, of a family $\{D_i\}_i$ of subspaces is the sublocale induced by their union, that is, it can be identified with the subspace $\bigcup_i D_i$. Indeed, $U \cap D_i = V \cap D_i$ holds for all i precisely when $U \cap \bigcup_i D_i = V \cap \bigcup_i D_i$.[12]

9 This means that $f(U) \leq x$ if and only if $U \subseteq g(x)$, for $x \in L$ and $U \in \Omega\mathbb{R}$. Such a g exists because f preserves joins, and it is $g(x) = \bigcup\{U \in \Omega\mathbb{R} \mid f(U) \leq x\}$.

10 This is clearer, I think, when interpreted in terms of the Heyting algebra structure ($\mathsf{int}(\_ \to \_)$ is the implication in $\Omega\mathbb{R}$) or, equivalently, in terms of intuitionistic logic: if γ is equivalent both to $\varphi \to \gamma$ and to $\psi \to \gamma$, then it is also equivalent to $\varphi \wedge \psi \to \gamma$; on the other hand, $\varphi \wedge \psi \to \gamma$ can be rewritten both as $\varphi \to (\psi \to \gamma)$ and as $\psi \to (\varphi \to \gamma)$.

11 This is, of course, an instance of a general construction in (universal) algebra.

12 One direction follows from the distributivity of intersection over union. As for the opposite direction, it is enough to intersect both members of $U \cap \bigcup_i D_i = V \cap \bigcup_i D_i$ with D_i.

As we will see in Subsection 16.2.3, this way of presenting sublocales extends with minor changes to the more general context of σ-locales (in which only countable joins are considered). For this reason, we shall usually work with congruences and with the strictly related notion of a congruence preorder.[13]

Congruence Preorders and Covers A congruence is completely determined by the preorder it induces. More explicitly, given a congruence $\simeq$, let us put $U \preceq V$ when $U \simeq (U \cap V)$. This gives a binary relation on $\Omega\mathbb{R}$ with the following properties

(i) $U \preceq U$;
(ii) if $U' \subseteq U \preceq V \subseteq V'$, then $U' \preceq V'$;
(iii) if $U \preceq V \preceq W$, then $U \preceq W$;
(iv) if $U \preceq V_1$ and $U \preceq V_2$, then $U \preceq (V_1 \cap V_2)$;
(v) if $U_i \preceq V$ for all $i \in I$, then $(\bigcup_{i \in I} U_i) \preceq V$.

Note that conditions (i) and (ii) together can be replaced (in the presence of (iii)) with the following single condition

(i–ii) if $U \subseteq V$, then $U \preceq V$

and, also, condition (iv) can be replaced with the following

(vi) if $U \preceq V$, then $(U \cap W) \preceq (V \cap W)$.

These are called **congruence preorders** in [20] and are essentially the same thing as **covers** on $\Omega\mathbb{R}$ in the sense of [17] (see also Section 16.4 below): for this reason we shall use the term 'cover' as a synonym for 'congruence preorder'. Clearly, the congruence from which a cover originated can be recovered by putting $U \simeq V$ iff $U \preceq V$ and $V \preceq U$.

Note that the cover corresponding to a subspace D (that is, the cover corresponding to $\simeq_D$) turns out to be just $U \preceq_D V$ iff $U \cap D \subseteq V$.

Sub-'algebras' The collections of open sets of the form $Fix(j)$, with j a nucleus, are precisely the subsets $S \subseteq \Omega\mathbb{R}$ which satisfy the following two conditions:

(i) if $\{U_i\}_i$ is a family of elements of S, then also $\mathsf{int}(\bigcap_i U_i) \in S$;
(ii) if $V \in S$, then $\mathsf{int}(U \to V) \in S$ for every $U \in \Omega\mathbb{R}$.

13 Such a preference for congruences is well expressed by a passage of [11]: '*Quotients of κ-frames cannot generally be described by means of nuclei. This chapter shows that nonetheless a sizeable chunk of frame theory goes through for κ-frames with no hitches. I think that this is a bit of a surprise, given the amount of fanfare which has frequently accompanied discussions of the Heyting algebra structure and the nucleus construction*'.

Note that $\mathsf{int}(\bigcap_i U_i)$ is the infinite meet in $\Omega\mathbb{R}$, seen as a complete lattice, and $\mathsf{int}(U \to V)$ is the implication in $\Omega\mathbb{R}$, seen as a Heyting algebra. On the other hand, given such an S, the corresponding nucleus j can be recovered as $jU = \mathsf{int}(\bigcap\{V \in S \mid U \subseteq V\})$.

This approach is the one adopted in [15]. One of its advantages is that it makes the computation of meets easy: given a family $\{S_i\}_i$ of sublocales in this sense, their meet is just the sublocale represented by $\bigcap_i S_i$. In particular, $S_1 \leq S_2$ in $\mathrm{Sub}(\mathbb{R})$ if and only if $S_1 \subseteq S_2$.

Example: Regular Opens A particularly important example of a sublocale (which is not a subspace) is given by Reg, the collection of all regular open subsets of $\mathbb{R}$ (an open set is regular if it equals the interior of its own closure). The corresponding nucleus is the function that maps each open set U to the interior of its closure $\mathsf{int}\,\mathsf{cl}\,U$.

As a sublocale, Reg is contained in every dense subspace D of $\mathbb{R}$.[14] To see this, it is sufficient to check that $j_D(U) \subseteq \mathsf{cl}\,U$ for all $U \in \Omega\mathbb{R}$. So take $x \in \mathsf{int}(D \to U)$ and let V be any open neighbourhood of x; we must show that $V \cap U$ is inhabited. Since D is dense, it overlaps with every inhabited open set; therefore $V \cap \mathsf{int}(D \to U) \cap D$ is inhabited and we are done, because $V \cap \mathsf{int}(D \to U) \cap D \subseteq V \cap (D \to U) \cap D \subseteq V \cap U$.

From this we learn two things, at least: that there are more sublocales than subspaces (there is no subspace like Reg); and that the meet of two subspaces in $\mathrm{Sub}(\mathbb{R})$ can be strictly larger than their intersection (for instance, Reg is contained in the meet of $\mathbb{Q}$ and $\mathbb{R} \setminus \mathbb{Q}$ as sublocales).

16.2.2 Sublocales and Fitness

Classically, the map which assigns to every subspace D of $\mathbb{R}$ the corresponding sublocale is injective, so that $\mathrm{Pow}(\mathbb{R})$ embeds into $\mathrm{Sub}(\mathbb{R})$. Here is a sketchy proof. For every $x \in \mathbb{R}$, consider the open $U = -\{x\}$; if $\simeq_D = \simeq_{D'}$, then we have $x \notin D$ iff $D \subseteq U$ iff $U \cap D = \mathbb{R} \cap D$ iff $U \cap D' = \mathbb{R} \cap D'$ iff $D' \subseteq U$ iff $x \notin D'$. Although there seems to be no constructive proof of this fact available, we can consistently keep the idea that there are more sublocales than subspaces. This explains, at least intuitively, why the theory of sublocales will look smoother than that of subsbaces in some respects.

In virtually all concrete cases, distinct subspaces give rise to distinct sublocales. This happens, for instance, with the open subspaces (the proof is trivial). Hence,

[14] Actually, Reg is contained in every dense sublocale of $\mathbb{R}$, where a sublocale is dense if the corresponding nucleus satisfies $j\emptyset = \emptyset$. Such a definition of density is justified by the fact that if $D \subseteq \mathbb{R}$ is a dense subset, then $j_D(\emptyset) = \mathsf{int}(D \to \emptyset) = \mathsf{int}(-D) \subseteq -\mathsf{cl}\,D = -\mathbb{R} = \emptyset$; and the converse holds, classically.

without danger of confusion, we can identify an open set with the corresponding open sublocale.

Classically, every subset $D \subseteq \mathbb{R}$ is an intersection of open sets, namely the intersection of all its open neighbourhoods. For if $x \in U$ for all open $U \supseteq D$, but $x \notin D$, then $U = -\{x\}$ would be an open neighbourhood of D such that $x \notin U$; a contradiction. Classically, this property characterize T_1 topological spaces; see [2]. Constructively, $\mathbb{R}$ does not satisfy it, although $\mathbb{R}$ is T_1 (in many reasonable senses). In general, we only have $D \subseteq \bigcap\{U \in \Omega\mathbb{R} \mid D \subseteq U\} =: \widetilde{D}$, although D and $\widetilde{D}$ are topologically indistinguishable (they have the same open neighbourhoods). When passing from $\mathrm{Pow}(\mathbb{R})$ to $\mathrm{Sub}(\mathbb{R})$ we are essentially identifying each set D with the set $\widetilde{D}$. Note that $x \in \widetilde{D}$ iff for every positive n there is some $d \in D$ with $|x - d| < \frac{1}{n}$; and if D is *located*, then $x \in \widetilde{D}$ means just $d(x, D) = 0$.

The fact that D and $\widetilde{D}$ are topologically indistinguishable suggests that every D, seen as a sublocale, is likely to be the meet of the open sublocales above it. In fact, this is provable for all sublocales, not only for those induced by a subspace, as we now see: this is an example of a pointfree theorem whose pointwise counterpart fails constructively.

Proposition 16.2 *Every sublocale of* $\mathbb{R}$ *is the meet in* $\mathrm{Sub}(\mathbb{R})$ *of the open sublocales above it.*[15]

Proof Let X be a sublocale of $\mathbb{R}$. For U open, $X \leq U$ in $\mathrm{Sub}(\mathbb{R})$ means that $V \cap U = W \cap U$ implies $V \simeq_X W$ for all $V, W \in \Omega\mathbb{R}$. This is equivalent to saying just that $U \simeq_X \mathbb{R}$ (or even $\mathbb{R} \preceq_X U$): one direction holds because $U \cap U = \mathbb{R} \cap U$; the opposite direction holds because $V = V \cap \mathbb{R} \simeq_X V \cap U = W \cap U \simeq_X W \cap \mathbb{R} = W$ provided that $U \simeq_X \mathbb{R}$ and $V \cap U = W \cap U$.

We claim that $\bigwedge\{U \text{ open } \mid X \leq U\} \leq X$, the other direction being trivial. So let Y be any other sublocale such that $Y \leq U$ for all open $U \geq X$. This means that $\mathbb{R} \simeq_X U$ implies $\mathbb{R} \simeq_Y U$ for every $U \in \Omega\mathbb{R}$. We must show that $Y \leq X$. So let $V, W \in \Omega\mathbb{R}$ be such that $V \simeq_X W$; we must check that $V \simeq_Y W$, that is, $V \preceq_Y W$ and $W \preceq_Y V$. Let us show that $V \preceq_Y W$. Since $V = \bigcup\{(a, b) \mid (a - 1/n, b + 1/n) \subseteq V \text{ for some positive } n\}$ it is sufficient to show that $(a, b) \preceq_Y W$ for all such intervals. Now $V \cup (-\infty, a) \cup (b, +\infty) = \mathbb{R}$, $V \cup (-\infty, a) \cup (b, +\infty) \simeq_X \mathbb{R}$ and hence $W \cup (-\infty, a) \cup (b, +\infty) \simeq_X \mathbb{R}$; thus $W \cup (-\infty, a) \cup (b, +\infty) \simeq_Y \mathbb{R}$. By intersecting both sides with (a, b) we get $W \cap (a, b) \simeq_Y (a, b)$, that is, $(a, b) \preceq_Y W$ as claimed. □

In other words, $\Omega\mathbb{R}$ is **fitted**, in the terminology of [19], or **fit**, in the terminology of [2] and [8]. The previous proposition suggests that it could make sense to define an outer measure on sublocales; this will be done in Section 16.3.

[15] This is a special case of a more general result; namely, that every regular locale is fit; see [8].

16.2.3 σ-Sublocales

In Section 16.3 we shall need still more subobjects. Since we are going to deal with (countably additive) measures, it is quite natural to regard $\Omega\mathbb{R}$ as a **σ-frame** (or $\aleph_1$-frame in the terminology of [11]), that is, a lattice with finite meets and countable joins where binary meets distribute over countable joins.[16] Accordingly, we shall consider **σ-sublocales** of $\mathbb{R}$; they are obtained simply by adjusting the notions of congruence and congruence preorder. In the latter case, for instance, one gets the following 'σ-version' of a cover.

A **σ-cover** is a binary relation $\preceq$ on $\Omega\mathbb{R}$ such that:

(i) if $U \subseteq V$, then $U \preceq V$;
(ii) if $U \preceq V \preceq W$, then $U \preceq W$;
(iii) if $U \preceq V$, then $(U \cap W) \preceq (V \cap W)$;
(iv) if $U_i \preceq V$ for all $i \in I$ with I countable, then $(\bigcup_{i\in I} U_i) \preceq V$;

the only difference with a cover lies on the countability requirement in the last clause. It is sometimes convenient to split the fourth condition into two, namely:

(iv) (a) if $U_1 \preceq V$ and $U_2 \preceq V$, then $(U_1 \cup U_2) \preceq V$;
(b) if $\{U_i\}_{i\in I}$ is a countable, non-decreasing chain and $U_i \preceq V$ for all $i \in I$, then $(\bigcup_{i\in I} U_i) \preceq V$.

Perhaps surprisingly, the lattice of all σ-sublocales of $\mathbb{R}$ is still complete (see [11] or [19]); that is, it has all joins and meets, not only countable ones; and, as in the case of sublocales, it is a co-frame.

16.3 A Measure on σ-Sublocales

Among probabilists, mention of sample points in an argument has always been bad form. A fully probabilistic argument must be pointless.

This is how Rota expresses himself in a passage from [16]. In this section we are going to present some essential aspects of a quite recent approach to (measure theory and) probability as proposed by Simpson in [19], which is certainly deeply pointfree in nature.

Picado's review on *MathSciNet* describes Simpson's work as '*an exciting application of locale theory to the problem of measuring subsets of the Euclidean space* $\mathbb{R}^n$'. The idea is, in some sense, surprisingly simple: instead of restricting ourselves to measurable sets, we enlarge the class of subobjects (by taking the σ-sublocales of

[16] Classically, of course, this makes no difference: every union of open sets in $\mathbb{R}$ can be reduced to a countable union, simply because there is a countable base for the topology... and a subset of a countable set is countable, a fact which fails constructively.

$\mathbb{R}$) and we realize that the outer measure is in fact a measure on them (in particular, every subspace becomes measurable), provided that the set-theoretic operations are replaced by the lattice operations between sublocales. In the end, what makes it all possible is the fact that disjoint subsets need not be disjoint as sublocales.

In what follows, we limit our treatment to probability measures. Thus, from now on, we restrict our attention to the open unit interval $(0, 1)$. What we have said so far about $\mathbb{R}$ also applies to $(0, 1)$ with no difficulty, of course.

16.3.1 The Outer Measure on σ-Sublocales

Let $\mathcal{M}$ be the collection of all integrable open sets[17] in $(0, 1)$. For $U \in \mathcal{M}$, let $\mu(U) \in [0, 1]$ be the measure of U.

Classically, $\mathcal{M} = \Omega(0, 1)$, the collection of all open subsets of $(0, 1)$, of course. Constructively, an open set need not be integrable (see [14]): there is a recursively enumerable set of pairwise disjoint open intervals (with rational endpoints) whose measure, if it existed, would be the limit of a Specker sequence.[18]

What follows depends on the following assumptions about $\mathcal{M}$ and μ:

(i) $\emptyset \in \mathcal{M}$ and $\mu(\emptyset) = 0$;
(ii) $(0, 1) \in \mathcal{M}$ and $\mu((0, 1)) = 1$;
(iii) if $U \in \mathcal{M}$ and $V \in \mathcal{M}$, then $U \cap V \in \mathcal{M}$ and $U \cup V \in \mathcal{M}$ as well, and $\mu(U) + \mu(V) = \mu(U \cap V) + \mu(U \cup V)$;
(iv) if $\{U_n\}_{n \geq 0}$ is a non-decreasing chain in $\mathcal{M}$, then $\bigcup_n U_n \in \mathcal{M}$ if and only if $\sup_n \mu(U_n)$ exists, in which case $\mu(\bigcup_n U_n) = \sup_n \mu(U_n)$.[19]

In particular, μ is monotone (because the maximum of two real numbers always exists) and so μ satisfies the properties of a σ-*continuous valuation* on the (integrable) open sets, in the sense of [19].[20]

For future reference we note the following fact. Given $U, V, W \in \mathcal{M}$, we have

$$\mu(U \cap W) - \mu(U \cap V \cap W) \leq \mu(U) - \mu(U \cap V)\,. \tag{16.1}$$

[17] In [1], measure theory is carried on for *complemented sets*. A complemented set is a pair (U, V) of subsets of $\mathbb{R}$ such that every $u \in U$ is different (that is, has positive distance) from any $v \in V$. So 'U is integrable' means that U is the left component of an integrable complemented set (which makes sense since the measure of a complemented set does not depend on the second component).

[18] There is a way out of this situation, namely, to consider the measure of an open set to be a *lower* real. A lower real is to a real as an inhabited, downwards closed, bounded, open subset of $\mathbb{Q}$ is to a Dedekind cut (classically they are essentially the same thing). The supremum of a bounded sequence of reals is a lower real, in general. The importance of lower reals for a pointfree approach to measure theory (especially for those foundational frameworks in which countable choice may fail) has been emphasized in [3], [5], and [21], for instance.

[19] Recall from [1] that the supremum b of a set A of real numbers is an upper bound of A such that for each $\epsilon > 0$ there exists $a \in A$ with $a > b - \epsilon$. Since $\{\mu(U_n)\}_n$ is a non-decreasing sequence bounded by 1, $\sup_n \mu(U_n)$ always exists classically, and it always exists as a lower real, constructively.

[20] The usual definition of a continuous valuation (without σ-) requires extending (iv) to directed families.

Indeed, $\mu(U\cap W)+\mu(U\cap V) = \mu((U\cap W)\cup(U\cap V))+\mu((U\cap W)\cap(U\cap V)) \leq \mu(U) + \mu(U \cap V \cap W)$. Also note that (16.1) is equivalent to the following:

$$\text{if } U' \subseteq U, \text{ then } \mu(U \cap W) - \mu(U' \cap W) \leq \mu(U) - \mu(U') \tag{16.2}$$

or, equivalently, $\mu(U') - \mu(U' \cap W) \leq \mu(U) - \mu(U \cap W)$.

Definition 16.3 For a σ-sublocale X of $(0, 1)$, let

$$\mathcal{N}(X) \;=\; \{U \in \mathcal{M} \mid X \leq U\}$$

be the collection of all integrable open neighbourhoods of X.[21]

If $\inf_{U\in\mathcal{N}(X)} \mu(U)$ exists, we put

$$\mu^*(X) = \inf_{U\in\mathcal{N}(X)} \mu(U)$$

and we call it the **outer measure** of X.[22]

If, in addition, $X = \bigwedge \mathcal{N}(X)$ in the lattice of sublocales, then we say that X is **fine**; and X is **very fine** if, whenever $X \leq V$ with V open, there exists $U \in \mathcal{N}(X)$ such that $U \subseteq V$; that is, if every open neighbourhood of X can be refined to get an integrable open neighbourhood.[23]

Every $U \in \mathcal{M}$ is fine and $\mu^*(U) = \mu(U)$.

If X is very fine, then $\bigwedge\mathcal{N}(X) \leq \bigwedge\{V \textit{ open} \mid X \leq V\} = X$ (because Proposition 16.2 works for σ-sublocales too, the crucial step being the possibility of writing any open set as a countable union of open intervals), and so X is fine.

The following proposition extends the familiar properties of the Lebesgue outer measure on subsets to the more general framework of σ-sublocales.

Proposition 16.4 *The following hold for the σ-sublocales of* $(0, 1)$.

(i) *If $\mu^*(X)$ and $\mu^*(Y)$ exist, and $X \leq Y$, then $\mu^*(X) \leq \mu^*(Y)$.*
(ii) *If $\mu^*(X)$, $\mu^*(Y)$, $\mu^*(X \vee Y)$, and $\mu^*(X \wedge Y)$ all exist, then*

$$\mu^*(X \vee Y) \leq \mu^*(X) + \mu^*(Y) - \mu^*(X \wedge Y)\,.$$

(iii) *Let $\{X_n\}_{n\geq 0}$ be a non-decreasing chain of σ-sublocales such that all $\mu^*(X_n)$ exist. If $\sup_{n\geq 0} \mu^*(X_n)$ exists and $\bigvee_n X_n$ is very fine, then*

$$\mu^*(\bigvee_n X_n) \;=\; \sup_{n\geq 0} \mu^*(X_n)\,.$$

[21] Recall (from the proof of Proposition 16.2) that $X \leq U$ means $U \simeq_X (0, 1)$ or, equivalently, $(0, 1) \preceq_X U$.

[22] According to the usual constructive definition of an infimum, this means that for each $\epsilon > 0$ there exists $U \in \mathcal{N}(X)$ with $\mu(U) < \mu^*(X) + \epsilon$.

[23] Classically, every σ-sublocale is (very) fine, of course

Proof Item (i) holds because $\mathcal{N}(Y) \subseteq \mathcal{N}(X)$ whenever $X \leq Y$.

To show (ii), let $U \in \mathcal{N}(X)$ and $V \in \mathcal{N}(Y)$. Thus $X \vee Y \leq U \vee V = U \cup V$ and $X \wedge Y \leq U \wedge V = U \cap V$. Therefore $\mu^*(X \vee Y) \leq \mu^*(U \cup V) = \mu(U \cup V)$ and $\mu^*(X \wedge Y) \leq \mu^*(U \cap V) = \mu(U \cap V)$. Hence $\mu^*(X \vee Y) + \mu^*(X \wedge Y) \leq \mu(U \cup V) + \mu(U \cap V) = \mu(U) + \mu(V)$. As U and V were arbitrary, $\mu^*(X \vee Y) + \mu^*(X \wedge Y) \leq \inf_{U \in \mathcal{N}(X), V \in \mathcal{N}(Y)}(\mu(U) + \mu(V)) = \inf_{U \in \mathcal{N}(X)} \mu(U) + \inf_{V \in \mathcal{N}(Y)} \mu(V) = \mu^*(X) + \mu^*(Y)$.

As for (iii), note that $\sup_n \mu^*(X_n) \leq \mu^*(\bigvee_n X_n)$ holds by (i). To show the converse inequality, fix any $\epsilon > 0$. By countable choice, for each $n \geq 0$ we can choose $U_n \in \mathcal{N}(X_n)$ with $\mu(U_n) < \mu^*(X_n) + \epsilon/2^{n+1}$ because $\mu^*(X_n) = \inf_{U \in \mathcal{N}(X_n)} \mu(U)$ exists by assumption. Since $\bigvee_n X_n \leq \bigcup_n U_n$ and $\bigvee_n X_n$ is very fine, there must exists $W \in \mathcal{N}(\bigvee_n X_n)$ with $\bigvee_n X_n \leq W \subseteq \bigcup_n U_n$. Let us consider the non-decreasing chain $\{W_n\}_{n \geq 0}$ where $W_n = W \cap V_n$ with $V_n = \bigcup_{i=0}^n U_i \in \mathcal{M}$. Clearly $W_n \in \mathcal{M}$ and $W = \bigcup_n W_n$; so $\sup_n \mu(W_n)$ exists and equals $\mu(W)$ by our fourth assumption on $\mathcal{M}$. We now claim that $\mu(W) \leq \sup_n \mu^*(X_n) + \epsilon$. Let us check, by induction, that $\mu(V_n) < \mu^*(X_n) + (1 - \frac{1}{2^{n+1}})\epsilon$. Indeed, $\mu(V_0) = \mu(U_0) < \mu^*(X_0) + \epsilon/2$ and $\mu(V_{n+1}) = \mu(V_n \cup U_{n+1}) = \mu(V_n) + \mu(U_{n+1}) - \mu(V_n \cap U_{n+1}) < \mu^*(X_n) + (1 - \frac{1}{2^{n+1}})\epsilon + \mu^*(X_{n+1}) + \frac{\epsilon}{2^{n+2}} - \mu^*(X_n) = \mu^*(X_{n+1}) + (1 - \frac{1}{2^{n+2}})\epsilon$, where the penultimate step is justified by the fact that $V_n \cap U_{n+1} \geq U_n \wedge X_{n+1} \geq X_n$. Therefore $\mu(W_n) \leq \mu(V_n) < \mu^*(X_n) + \epsilon$ and hence $\mu(W) = \sup_n \mu(W_n) \leq \sup_n \mu^*(X_n) + \epsilon$, which proves our claim. Summing up, for every $\epsilon > 0$, there exists $W \in \mathcal{N}(\bigvee_n X_n)$ such that $\mu(W) \leq \sup_n \mu^*(X_n) + \epsilon$. So $\mu^*(\bigvee_n X_n) \leq \sup_n \mu^*(X_n)$. □

As a corollary of the previous proposition (and under suitable, similar hypotheses), we get the following extension of the familiar σ-subadditivity property:

$$\mu^*\left(\bigvee_n X_n\right) \leq \sum_n \mu^*(X_n).$$

16.3.2 The Outer Measure is Additive!

The surprising thing about the outer measure on the σ-sublocales is that it is more than just σ-subadditive: it is in fact σ-additive. In order to show this, what we still lack is the equality $\mu^*(X) + \mu^*(Y) = \mu^*(X \vee Y) + \mu^*(X \wedge Y)$.[24] Equivalently, thanks to item (ii) of Proposition 16.4, all we need to show is the inequality $\mu^*(X) + \mu^*(Y) - \mu^*(X \vee Y) \leq \mu^*(X \wedge Y)$. The following lemma goes in the right direction.

[24] Once we have that, a routine proof will give us the σ-additivity law $\mu^*(\bigvee_n X_n) = \sum_n \mu^*(X_n)$ for pairwise disjoint σ-sublocales, provided that all the needed assumptions hold, as in Proposition 16.4.

Lemma 16.5 *If $\mu^*(X)$, $\mu^*(Y)$ and $\mu^*(X \vee Y)$ exist, then*

$$\mu^*(X) + \mu^*(Y) - \mu^*(X \vee Y) \leq \mu(U \cap V)$$

for all $U \in \mathcal{N}(X)$ and $V \in \mathcal{N}(Y)$.

Proof Given $\epsilon > 0$ there is $W \in \mathcal{N}(X \vee Y)$ with $\mu(W) < \mu^*(X \vee Y) + \epsilon$. So $\mu^*(X) + \mu^*(Y) \leq \mu(W \cap U) + \mu(W \cap V) = \mu((W \cap U) \cup (W \cap V)) + \mu(W \cap U \cap V) \leq \mu(W) + \mu(U \cap V) < \mu^*(X \vee Y) + \mu(U \cap V) + \epsilon$. □

Note that this does not show that $\mu^*(X) + \mu^*(Y) - \mu^*(X \vee Y) \leq \mu^*(X \wedge Y)$ because an elements of $\mathcal{N}(X \wedge Y)$ need not have the form $U \cap V$ with $U \in \mathcal{N}(X)$ and $V \in \mathcal{N}(Y)$, although every such $U \cap V$ is in $\mathcal{N}(X \wedge Y)$.

Let X and Y be two σ-sublocales. The idea in [19] is to use the opens of the form $U \cap V$, with $U \in \mathcal{N}(X)$ and $V \in \mathcal{N}(Y)$, to construct a suitable σ-sublocale XY which turns out to have the same measure of $X \wedge Y$. The corresponding σ-cover $\preceq_{XY}$ on $\Omega(0,1)$ is defined as follows

$$W_1 \preceq_{XY} W_2 \quad \stackrel{\text{def}}{\iff} \quad \begin{array}{l} \text{for all } n > 0 \\ \text{there exist } U \in \mathcal{N}(X), V \in \mathcal{N}(Y) \text{ such that} \\ \mu(W_1 \cap U \cap V) - \mu(W_1 \cap U \cap V \cap W_2) < \frac{1}{n} \end{array} \tag{16.3}$$

(classically, $W_1 \preceq_{XY} W_2$ holds when $\mu((W_1 \setminus W_2) \cap (U \cap V))$ can be made arbitrary small). Unfortunately, this definition does not make sense constructively because W_1 and W_2 need not be measurable. What we can do is read $\preceq_{XY}$ as a binary relation on $\mathcal{M}$ and then show (Lemma 16.6) a list of properties which makes $\preceq_{XY}$ a σ-cover on $\mathcal{M} = \Omega(0,1)$ in the classical case.

Lemma 16.6 *Let X and Y be two σ-sublocales of $(0,1)$. Then $\preceq_{XY}$, seen as a binary relation on $\mathcal{M}$, satisfies the following:*

(i) *if $W_1 \subseteq W_2$, then $W_1 \preceq_{XY} W_2$;*
(ii) *if $W_1 \preceq_{XY} W_2 \preceq_{XY} W_3$, then $W_1 \preceq W_3$;*
(iii) *if $W_1 \preceq_{XY} W_2$, then $(W_1 \cap W_3) \preceq_{XY} (W_2 \cap W_3)$;*
(iv) *if $W_1 \preceq_{XY} W_3$ and $W_2 \preceq_{XY} W_3$, then $(W_1 \cup W_2) \preceq_{XY} W_3$*

for every $W_1, W_2, W_3 \in \mathcal{M}$.

If, in addition, the infimum of all $\mu(U \cap V)$ for $U \in \mathcal{N}(X)$ and $V \in \mathcal{N}(Y)$ exists, then $\preceq_{XY}$ satisfies also the following:

(v) *if $W_n \preceq_{XY} W$ for all n, then also $(\bigcup_n W_n) \preceq_{XY} W$*

for every $W \in \mathcal{M}$ and every non-decreasing chain $\{W_n\}_{n \geq 0}$ in $\mathcal{M}$ such that $\bigcup_n W_n \in \mathcal{M}$.

Proof (i) As $\mu(W_1 \cap (0,1) \cap (0,1)) - \mu(W_1 \cap (0,1) \cap (0,1) \cap W_2) = 0$ by monotonicity of μ, we get $W_1 \preceq_{XY} W_2$.

(ii) Given $n > 0$, there are $U', U'' \in \mathcal{N}(X)$ and $V', V'' \in \mathcal{N}(Y)$ such that
$\mu(W_1 \cap U' \cap V') - \mu(W_1 \cap U' \cap V' \cap W_2) < \frac{1}{2n}$ and
$\mu(W_2 \cap U'' \cap V'') - \mu(W_2 \cap U'' \cap V'' \cap W_3) < \frac{1}{2n}$.
By applying fact (16.2), the previous two inequalities hold a fortiori if we substitute $U = U' \cap U''$ in place of U', U'', and $V = V' \cap V''$ in place of V', V''. Thus $\mu(W_1 \cap U \cap V) - \mu(W_1 \cap U \cap V \cap W_3) = \mu(W_1 \cap U \cap V) - \mu(W_1 \cap U \cap V \cap W_2) + \mu(W_1 \cap U \cap V \cap W_2) + - \mu(W_1 \cap U \cap V \cap W_2 \cap W_3) + \mu(W_1 \cap U \cap V \cap W_2 \cap W_3) - \mu(W_1 \cap U \cap V \cap W_3) < \frac{1}{2n} + \mu(U \cap V \cap W_2) - \mu(U \cap V \cap W_2 \cap W_3) + 0 < \frac{1}{n}$, where the second to last step is justified by fact (16.1). So $W_1 \preceq_{XY} W_3$.

(iii) We have $\mu(W_1 \cap W_3 \cap U \cap V) - \mu(W_1 \cap W_3 \cap U \cap V \cap W_2 \cap W_3) \leq \mu(W_1 \cap U \cap V) - \mu(W_1 \cap U \cap V \cap W_2)$ by fact (16.1). Therefore $W_1 \preceq_{XY} W_2$ implies $W_1 \cap W_3 \preceq_{XY} W_2 \cap W_3$.

(iv) Given $n > 0$, there are $U_1, U_2 \in \mathcal{N}(X)$ and $V_1, V_2 \in \mathcal{N}(Y)$ such that
$\mu(W_1 \cap U_1 \cap V_1) - \mu(W_1 \cap U_1 \cap V_1 \cap W_3) < \frac{1}{2n}$ and
$\mu(W_2 \cap U_2 \cap V_2) - \mu(W_2 \cap U_2 \cap V_2 \cap W_3) < \frac{1}{2n}$.
Both inequalities hold at the same time for $U = U_1 \cap U_2$ in place of U_1, U_2, and $V = V_1 \cap V_2$ in place of V_1, V_2.
Thus $\mu((W_1 \cup W_2) \cap U \cap V) - \mu((W_1 \cup W_2) \cap U \cap V \cap W_3) = \mu((W_1 \cap U \cap V) \cup (W_2 \cap U \cap V)) - \mu((W_1 \cap U \cap V \cap W_3) \cup (W_2 \cap U \cap V \cap W_3)) = \mu(W_1 \cap U \cap V) + \mu(W_2 \cap U \cap V) - \mu(W_1 \cap W_2 \cap U \cap V) + - \mu(W_1 \cap U \cap V \cap W_3) - \mu(W_2 \cap U \cap V \cap W_3) + \mu(W_1 \cap W_2 \cap U \cap V \cap W_3) < \frac{1}{2n} + \frac{1}{2n} + 0 = \frac{1}{n}$. So $W_1 \cup W_2 \preceq_{XY} W_3$.

(v) Let l be the infimum of all $\mu(U \cap V)$ for $U \in \mathcal{N}(X)$ and $V \in \mathcal{N}(Y)$. We claim that $\mu(W_n \cap U \cap V) - \mu(W_n \cap U \cap V \cap W) \leq \mu(U \cap V) - l$ for all $U \in \mathcal{N}(X)$, $V \in \mathcal{N}(Y)$ and $n > 0$.
Indeed, for every $m > 0$ we choose $U_n \in \mathcal{N}(X)$ and $V_n \in \mathcal{N}(Y)$ with
$\mu(W_n \cap U_n \cap V_n) - \mu(W_n \cap U_n \cap V_n \cap W) < \frac{1}{m}$.
Thus $\mu(W_n \cap U \cap U_n \cap V \cap V_n) - \mu(W_n \cap U \cap U_n \cap V \cap V_n \cap W) < \frac{1}{m}$ by fact (16.1). Again by fact (16.1), we get
$\mu(W_n \cap U \cap V) - \mu(W_n \cap U \cap U_n \cap V \cap V_n) \leq \mu(U \cap V) - \mu(U \cap U_n \cap V \cap V_n)$
and hence $\mu(W_n \cap U \cap V) - \mu(W_n \cap U \cap V \cap W) = \mu(W_n \cap U \cap V) - \mu(W_n \cap U \cap U_n \cap V \cap V_n) + \mu(W_n \cap U \cap U_n \cap V \cap V_n) + - \mu(W_n \cap U \cap U_n \cap V \cap V_n \cap W) + \mu(W_n \cap U \cap U_n \cap V \cap V_n \cap W) + - \mu(W_n \cap U \cap V \cap W) < \mu(U \cap V) - \mu(U \cap U_n \cap V \cap V_n) + \frac{1}{m} + 0 \leq \leq \frac{1}{m} + \mu(U \cap V) - l$ because $U \cap U_n \in \mathcal{N}(X)$ and $V \cap V_n \in \mathcal{N}(Y)$.
Since m was arbitrary, this proves our claim. A fortiori, $\mu(W_n \cap U \cap V) \leq \sup_n \mu(W_n \cap U \cap V \cap W) + \mu(U \cap V) - l$ and hence, as this holds for every

n, we get $\sup_n \mu(W_n \cap U \cap V) \leq \sup_n \mu(W_n \cap U \cap V \cap W) + \mu(U \cap V) - l$, that is, $\mu((\bigcup_n W_n) \cap U \cap V) - \mu((\bigcup_n W_n) \cap U \cap V \cap W) \leq \mu(U \cap V) - l$. Since $\mu(U \cap V) - l$ can be made arbitrary small, we conclude that $\bigcup_n W_n \preceq_{XY} W$. □

Lemma 16.7 *Let X and Y be two σ-sublocales of $(0,1)$.*
If $\mu^*(X)$, $\mu^*(Y)$ and $\mu^*(X \vee Y)$ exist, then

$$\mu^*(X) + \mu^*(Y) - \mu^*(X \vee Y) \leq \mu(W)$$

for all $W \in \mathcal{M}$ such that $(0,1) \preceq_{XY} W$.

Proof By (16.3), for every $n > 0$ there are $U \in \mathcal{N}(X)$ and $V \in \mathcal{N}(Y)$ with $\mu((0,1) \cap U \cap V) - \mu((0,1) \cap U \cap V \cap W) < \frac{1}{n}$. By Lemma 16.5, we have $\mu^*(X) + \mu^*(Y) - \mu^*(X \vee Y) \leq \mu(U \cap V) < \mu(U \cap V \cap W) + \frac{1}{n} \leq \mu(W) + \frac{1}{n}$. □

Note that $(0,1) \preceq_{XY} W$ means that $W \in \mathcal{N}(XY)$, provided that the σ-sublocale XY exists (a fact which is not automatically true since, as we said, $\preceq_{XY}$ need not be a σ-cover on $(0,1)$, constructively). In view of this, the meaning of the previous lemma is that

$$\mu^*(X) + \mu^*(Y) - \mu^*(X \vee Y) \leq \mu^*(XY)$$

when all objects involved exist.

To conclude the proof of the σ-additivity property, we now have to assume the existence of the σ-sublocale XY. Therefore, the significance of the following proposition is better seen in the classical case.

Proposition 16.8 *Let X and Y be two σ-sublocales of $(0,1)$ such that there exists a σ-sublocale XY of $(0,1)$ whose underlying σ-cover coincides with the relation $\preceq_{XY}$ defined in* (16.3).
If X and Y are fine, and $\mu^*(X \vee Y)$ and $\mu^*(X \wedge Y)$ exist, then

$$\mu^*(X) + \mu^*(Y) - \mu^*(X \vee Y) \leq \mu^*(X \wedge Y).$$

Proof The statement follows from the previous discussion and the fact that $XY \leq X \wedge Y$ provided that X and Y are fine. Let us check that $XY \leq X$. Since $X = \bigwedge \mathcal{N}(X)$, we only have to check that $XY \leq W$, that is, $(0,1) \preceq_{XY} W$, for all $W \in \mathcal{N}(X)$. This is trivial, because if we choose $U = W$ and $V = (0,1)$, then $\mu((0,1) \cap U \cap V) - \mu((0,1) \cap U \cap V \cap W) = 0$. Analogously, one checks that $XY \leq Y$. □

Thus $\mu^*(X) + \mu^*(Y) - \mu^*(X \vee Y) = \mu^*(X \wedge Y)$, provided that XY exists, in which case XY has necessarily the same measure of $X \wedge Y$.

16.3.3 Subadditivity and 'Hidden Mass'

In view of the previous results, the fact that the (Lebesgue) outer measure on subsets fails to be additive, in general, can be seen in a new light. (In the following discussion we are going to use classical reasoning.)

Let V be a non-measurable set of reals. Thanks to the Carathéodory's characterization of measurable sets, there must exist $D \subseteq \mathbb{R}$ such that $\mu^*(D) > \mu^*(D \cap V) + \mu^*(D \setminus V)$. In other words, by choosing $X = D \cap V$ and $Y = D \setminus V$, we get two disjoint subsets X and Y of $\mathbb{R}$ such that $\mu^*(X \cup Y) - \mu^*(X) - \mu^*(Y) > 0$. Recall that the join, as sublocales, of two subspaces is just (the sublocale induced by) their union. Therefore we also have $\mu^*(X \cup Y) - \mu^*(X) - \mu^*(Y) = \mu^*(X \wedge Y)$ where $X \wedge Y$ is the meet of X and Y as sublocales. Therefore, $X \wedge Y$ has positive measure, although $X \cap Y = \emptyset$.

In cases like this, $\mu^*(X \cup Y) - \mu^*(X) - \mu^*(Y)$ can be understood as the measure of a 'dark' pointfree part which, despite being hidden from a pointwise perspective, has a concrete topological existence and a perfectly determined mass.

16.4 The Pointfree Approach to the Real Line

In the previous sections, the frame $\Omega\mathbb{R}$ was considered as given. If one wants to extend the treatment above to a 'choicefree' framework, such as the internal language of a generic topos or Maietti–Sambin minimalist type theory, as presented in [12] and [13], then a different definition of $\Omega\mathbb{R}$ is required, that cannot rely on the existence of *enough* real numbers. Indeed, Cauchy sequences of rational numbers are not enough to determine $\Omega\mathbb{R}$ in the absence of countable choice, as shown in [7].

The aim of the present section is to recall how to describe the topology of $\mathbb{R}$, that is, the frame $\Omega\mathbb{R}$, without mentioning points. Such a goal could look self-contradictory (how can one define sets of reals without previously defining the reals?) but it is not.

The open intervals with rational endpoints form a convenient base for the topology of $\mathbb{R}$, that is, each element of $\Omega\mathbb{R}$ is the union of all the open intervals with rational endpoints which are contained in it. Now, we can identify each open interval in the base with the pair of its rational endpoints. Thus we start from the set $S = \{(a, b) \in \mathbb{Q} \times \mathbb{Q} \mid a < b\}$ as our abstract base (extended intervals could be used as well for convenience). Every element of $\Omega\mathbb{R}$ is then a union of elements of S, although each open set can be written as a union of basic elements in several different ways. This means that there exists a surjective mapping from $\mathrm{Pow}(S)$, the powerset of S, onto $\Omega\mathbb{R}$; in other words, $\Omega\mathbb{R}$ can be realized as a quotient of $\mathrm{Pow}(S)$. The idea is clear: two subsets U and V of S must be equivalent precisely if they give rise to the same open set; that is, precisely when $\bigcup_{(a,b)\in U}(a,b) = \bigcup_{(a,b)\in V}(a,b)$.[25] How can we define such an equivalence without referring to its intended (spatial) meaning?

Instead of trying to characterize the equality $\bigcup U = \bigcup V$ directly, we could start with statements of the form $\bigcup U \subseteq \bigcup V$; we write $U \lhd V$ in this case, which we read as 'U is covered by V'. Clearly we can further reduce ourselves to the case $(a,b) \subseteq \bigcup V$; for the sake of simplicity, we write $(a,b) \lhd V$ in lieu of $\{(a,b)\} \lhd V$, and so $U \lhd V$ means that $(a,b) \lhd V$ for all $(a,b) \in U$. What are the abstract properties of such a **cover** relation? The solution is due to André Joyal (unpublished). Essentially, $\lhd$ is the smallest subset of $S \times \mathrm{Pow}(S)$ that satisfies the following conditions:

(i) if $(a,b) \in V$, then $(a,b) \lhd V$;
(ii) if $(a,d) \lhd V$ and $(c,b) \lhd V$ with $a \leq c < d \leq b$, then $(a,b) \lhd V$;
(iii) if $(c,d) \lhd V$ for all $a < c < d < b$, then $(a,b) \lhd V$.

It is not hard to check that $\lhd$ satisfies the following further properties (see [4]):

(i) if $(a,b) \lhd U$ and $U \lhd V$, then $(a,b) \lhd V$;
(ii) if $(a,b) \lhd U$ and $(a,b) \lhd V$, then $(a,b) \lhd U \wedge V$,
where $U \wedge V$ is the set of all pairs of the form $(\max\{u,v\}, \min\{u',v'\})$ with $(u,u') \in U$, $(v,v') \in V$ provided that $\max\{u,v\} < \min\{u',v'\}$.

In particular, the relation $\lhd \subseteq \mathrm{Pow}(S) \times \mathrm{Pow}(S)$ is a pre-order (reflexive and transitive), and hence the relation $U =_{\lhd} V$ defined as $U \lhd V$ and $V \lhd U$ is an equivalence on $\mathrm{Pow}(S)$.

The quotient $\mathrm{Pow}(S)/=_{\lhd}$ is a frame which is isomorphic to $\Omega\mathbb{R}$ classically. Constructively, that isomorphism is equivalent to the 'open cover' compactness of $[0,1]$ which is known to fail in some sheaf model for analysis; see [6] and [7] for further on this topic.

16.5 Concluding Remarks

Since its origins, the pointfree approach to topology has shown itself fruitful, especially in those intuitionistic foundations which lack the axiom of (countable) choice. For instance, a direct description of the frame $\Omega\mathbb{R}$ is better behaved than the concrete collection of open sets of Cauchy reals.

Even classical mathematics can benefit from a pointfree approach: an example is the case of measure theory as described above. It turns out that the usual (Lebesgue)

[25] Here the symbol (a,b) sometimes stands for a pair, sometimes for an open interval; but there is no danger of confusion, of course.

outer measure becomes in fact additive when it is extended to all pointfree subobjects (that is, sublocales) of $\mathbb{R}$.

A fully satisfactory 'constructivization' of the results we presented is still to come. One main problem is to cope with the classical and somehow impredicative definition of the σ-sublocale XY as in display (16.3). Another major issue is to understand what notion of a real number is the most appropriate for developing measure theory for sublocales (evidence in favour of lower reals has been provided in [3], [5], and [21], for instance). Finally, it would be nice to build on [3] in order to understand the complemented sets (which are at the base of the measure theory in [1]) from a pointfree perspective.

References

[1] Bishop, E., and Bridges, D. 1985. *Constructive Analysis*. Volume 279 of: *Grundlehren der Mathematischen Wissenschaften [Fundamental Principles of Mathematical Sciences]*. Berlin: Springer-Verlag.

[2] Clementino, M. M., Picado, J.M. and Pultr, A. 2018. The other closure and complete sublocales. *Appl. Categ. Struct.*, **26**(5), 891–906.

[3] Coquand, T., and Palmgren, E. 2002. Metric Boolean algebras and constructive measure theory. *Arch. Math. Logic*, **41**(7), 687–704.

[4] Coquand, T., Sambin, G., Smith, J., and Valentini, S. 2003. Inductively generated formal topologies. *Ann. Pure Appl. Logic*, **124**(1–3), 71–106.

[5] Coquand, T., and Spitters, B. 2009 Integrals and valuations. *J. Log. Anal.*, **1**. **Paper 3**, 22.

[6] Fourman, M. P., and Grayson, R. J. 1982. Formal spaces. Pages 107–122 of: Troelstra, A. S., and van Dalen, D. (eds.), *The L. E. J. Brouwer Centenary Symposium (Noordwijkerhout, 1981)*. Studies in Logic and the Foundations of Mathematics, vol. 110. Amsterdam: North-Holland.

[7] Fourman, M. P., and Hyland, J. M. E. 1979. Sheaf models for analysis. Pages 280–301 of: Fourman, M. P., Mulvey, C. J., and Scott, D. S. (eds.), *Applications of Sheaves (Proc. Res. Sympos. Appl. Sheaf Theory to Logic, Algebra and Anal., University of Durham, Durham, 1977)*. Lecture Notes in Mathematics, vol. 753. Berlin: Springer.

[8] Isbell, J. R. 1972. Atomless parts of spaces. *Math. Scand.*, **31**, 5–32.

[9] Johnstone, P. T. 1982. *Stone Spaces*. Volume 3 of: *Cambridge Studies in Advanced Mathematics*. Cambridge: Cambridge University Press.

[10] Johnstone, P. T. 1983. The point of pointless topology. *Bull. Amer. Math. Soc. (N.S.)*, **8**(1), 41–53.

[11] Madden, J. J. 1981. κ-frames. *J. Pure Appl. Algebra*, **70**, 107–127.

[12] Maietti, M. E. 2009. A minimalist two-level foundation for constructive mathematics. *Ann. Pure Appl. Logic*, **160**(3), 319–354.

[13] Maietti, M. E., and Sambin, G. 2005. Toward a minimalist foundation for constructive mathematics. Pages 91–114 of: Crosilla, L., and Schuster, P. (eds.), *From Sets and Types to Topology and Analysis.* Oxford Logic Guides, vol. 48. Oxford: Oxford University Press.

[14] Martin-Löf, P. 1970. *Notes on Constructive Mathematics.* Stockholm: Almqvist & Wiksell.

[15] Picado, J., and Pultr, A. 2012. *Frames and Locales. Topology without Points.* Frontiers in Mathematics. Basel: Birkhäuser/Springer Basel AG.

[16] Rota, G.-C. 2001. Twelve problems in probability no one likes to bring up. Pages 57–93 of: Crapo, H., and Senato, D. (eds.), *Algebraic Combinatorics and Computer Science. A Tribute to Gian-Carlo Rota.* Milan: Springer Italia.

[17] Sambin, G. 1987. Intuitionistic formal spaces – a first communication. Pages 187–204: Skordev, D. G. (ed.), of *Mathematical Logic and its Applications, Proceedings of the Advanced International Summer School and Conference (Druzhba, 1986).* New York: Plenum.

[18] Sambin, G. 200. Some points in formal topology. *Theor. Comput. Sci.*, **305**(1–3), 347–408.

[19] Simpson, A. 2012. Measure, randomness and sublocales. *Ann. Pure Appl. Logic*, **163**(11), 1642–1659.

[20] Vickers, S. 2007. Sublocales in formal topology. *J. Symbol. Logic*, **72**(2), 463–482.

[21] Vickers, S. 2008. A localic theory of lower and upper integrals. *Math. Logic Q.*, **54**(1), 109–123.

17

Synthetic Topology

Davorin Lešnik

17.1 Introduction

Synthetic topology is a way to study topology by making it an intrinsic property of sets, rather than an additional structure. That is, simply constructing a set already yields a topology on it, and in nice models of synthetic topology this ***intrinsic topology*** matches the most standard one (discrete for natural numbers, euclidean for reals, etc.).

The intrinsic topology is induced by the background logic which makes the subject of synthetic topology firmly rooted in constructivism. One can of course apply this theory in a classical setting (see [8, Section 5.1]), but if we can positively decide for any two points whether they are equal or distinct, then every point is isolated, and so classically every set has discrete topology, making this an uninteresting example.

On the other hand, constructively we generally cannot decide equality of points, which yields a rich topological structure. The more difficult it is to decide equality on a set, the further away the intrinsic topology of the set is from being discrete. For example, if we rescind the Limited Principle of Omniscience (LPO), then we cannot in general separate a binary sequence from others, and under reasonable assumptions the topology of $\mathbf{2}^{\mathbb{N}}$ is the usual topology of the Cantor space.

17.1.1 Setting

The definition of synthetic topology is usually given in terms of toposes. A topos [7] is a category with enough structure to interpret higher-order intuitionistic logic (with bounded quantifiers), as well as the usual set-theoretic constructions, such as products, coproducts, solution sets of equations, quotients, exponentials and powersets. This allows us to reason *internally* in a topos, by deriving results in a suitable intuitionistic set theory [3]. The alternative is to reason *externally*, that is, interpreting a statement of a topos to a statement on the metalevel, which is generally much more complex than the original statement.

The most fruitful approach (at least in the case of synthetic topology) seems to be to reason internally when proving general results (ones that are independent from the choice of a topos), and when proving results for a specific topos, use external reasoning to establish basic principles, from which one afterwards derives more complicated theorems internally. This is the approach that we take in the chapter on synthetic topology. We hope that it will be clear from context when we use internal reasoning and when we use external reasoning, and we ask the reader to forgive minor abuses, which we use in the interest of a clearer presentation.[1] When reasoning internally, we refer to objects of the topos as 'sets' and to morphisms as 'maps'. For a map f, we use the notation f_* for its direct image, and f^* for its preimage.

17.1.2 Intrinsic Topology

Every topos has a ***subobject classifier*** Ω which internally behaves as the set of truth values. The set Ω is a complete lattice, the partial order being given by the implication ($p \leq q$ whenever $p \Rightarrow q$). The top element is the ***truth***, denoted by $\top$, and the bottom is ***falsehood***, denoted by $\bot$. These are the nullary meet and join, respectively. The binary meet is the conjunction, the binary join is the disjunction. The equivalence of truth values is just the equality.

The set of ***decidable*** truth values is $\mathbf{2} := \{\top, \bot\}$. We always have $\mathbf{2} \subseteq \Omega$; the converse $\Omega \subseteq \mathbf{2}$ states that every truth value is decidable, that is, it is equivalent to the Law of Excluded Middle (LEM).

The set of ***stable*** truth values is $\Omega_{\neg\neg} := \{p \in \Omega \mid \neg\neg p \Rightarrow p\}$. More generally, for any $\Gamma \subseteq \Omega$ we denote $\Gamma_{\neg\neg} := \Gamma \cap \Omega_{\neg\neg}$. We also denote $\neg\Gamma := \{\neg p \mid p \in \Gamma\}$.

The defining property of Ω is that every subset has a (unique) characteristic map. That is, given any set X, for every subset $S \subseteq X$ there exists a unique map $\chi_S \colon X \to \Omega$ such that for all $x \in X$, $\chi_S(x)$ holds precisely when $x \in S$. This means that χ_S is defined as $\chi_S(x) := x \in S$.

We may recover the subset S from its characteristic map by $S = {\chi_S}^*(\top)$. Thus we have an isomorphism between the powerset $\mathcal{P}(X)$ and the set Ω^X of characteristic functions on X (elements of Ω^X are also called ***predicates*** on X). In particular, for a singleton $\mathbf{1} = \{*\}$ we have $\mathcal{P}(\mathbf{1}) \cong \Omega^{\mathbf{1}} \cong \Omega$.

[1] To give an example, in Proposition 17.12 we verify that a set X is compact if and only if X-indexed intersections of opens of Y are open for *every* set Y. Technically the latter is not a well-formed statement in a topos (the quantifier must be bounded to a specific object, it cannot quantify over all objects), but it should be clear how to state this precisely. Compactness of X is implied by the intersection property on a specific set, namely a singleton. Conversely, we have one statement for each object Y of a topos, namely that compactness of X implies the intersection property on Y.

Synthetic topology is given by a choice of a subobject $\Sigma \subseteq \Omega$. *All* subsets of a set X are classified by Ω; *some* will be classified by Σ. The latter are declared ***open***, and their collection forms the topology $\mathcal{O}(X)$. It follows that $\mathcal{O}(X)$ and Σ^X are isomorphic (elements of Σ^X are called ***open predicates*** on X). In particular $\mathcal{O}(1) \cong \Sigma^1 \cong \Sigma$, so choosing a topology of a singleton induces the topology on every set.

The source of the idea that topology on X can be represented by Σ^X comes from the following observation from classical topology. Let $\mathbb{S}$ be the ***Sierpiński space***, namely the two-point space, in which one point is open and the other is not. Then the sets $\mathcal{O}(X)$ and $C(X, \mathbb{S})$ (the set of continuous maps from X to $\mathbb{S}$) are isomorphic. Indeed, one can view $\mathbb{S}$ as the set of classical truth values with the topology, in which truth is open, and $C(X, \mathbb{S})$ is then the set of characteristic maps of open subsets of X.

Within the category of topological spaces *all* maps are continuous. When the exponential $\mathbb{S}^X$ exists, it is the set $C(X, \mathbb{S})$, together with the suitable[2] topology. In this sense we may identify $\mathcal{O}(X)$ with $\mathbb{S}^X$. Going back to synthetic topology, if we have an object Σ which has similar properties to the Sierpiński space, then we can expect that the exponentials Σ^X will indeed behave like some sort of topologies.

We now give the formal definitions.

Definition 17.1 (Synthetic Topology) A ***model of synthetic topology*** is a topos, together with a choice of a subobject $\Sigma \subseteq \Omega$ which is closed under finite meets of the subobject classifier Ω. The object Σ is called the ***Sierpiński object*** (or when reasoning in the internal set theory, the ***Sierpiński set***).

A Sierpiński object induces a notion of openness.

Definition 17.2 (Openness) A truth value $p \in \Omega$ is ***open*** when $p \in \Sigma$. A subset $U \subseteq X$ is ***open*** in X when it is classified by open truth values, that is, it satisfies any of the following equivalent statements:

- the characteristic map $\chi_U \colon X \to \Omega$ restricts to a map $X \to \Sigma$,
- the image of the characteristic map χ_U is a subset of Σ,
- for every $x \in X$ the truth value of the statement $x \in U$ is open.

The collection of the open subsets of a set X is the ***intrinsic topology*** of X, denoted by $\mathcal{O}(X)$. When we want to emphasize that $U \subseteq X$ is open with regard to the intrinsic (as opposed to some other) topology, we say that U is ***intrinsically open*** in X.

We also need the notion of closed sets. Classically, closed sets are just the complements of open sets, or equivalently, are the sets, of which complements are open.

[2] Specifically, the Isbell topology [4].

Constructively these two definitions do not coincide, and more problematically, neither proves suitable for practice.

Complementation is induced by negation, and it is a standard constructive practice to treat negation as a special case of implication: $\neg p$ is equivalent to $p \Rightarrow \bot$. It turns out that for closed sets to works well synthetically, we need them to be compatible with general implication, not just negation.[3]

Definition 17.3 (Closedness) The set of ***closed truth values*** is defined as

$$\mathrm{Z} := \{p \in \Omega \mid p \Rightarrow u \text{ is an open truth value for every } u \in \Sigma\}$$

(Z is the capital Greek letter zeta). A subset $F \subseteq X$ is ***closed*** in X when it is classified by closed truth values, that is, it satisfies any of the following equivalent statements:

- the characteristic map $\chi_F \colon X \to \Omega$ restricts to a map $X \to \mathrm{Z}$,
- the image of the characteristic map χ_F is a subset of Z,
- for every $x \in X$ the truth value of the statement $x \in F$ is closed,
- for every $x \in X$ and $U \in \mathcal{O}(X)$ the truth value of the statement $x \in F \implies x \in U$ is open,
- for every $U \in \mathcal{O}(X)$ the set $\{x \in X \mid x \in F \implies x \in U\}$ is open in X.

The collection of closed subsets of a set X is denoted by $\mathcal{Z}(X)$. When we want to emphasize that $F \subseteq X$ is closed with regard to the intrinsic (rather than some other) topology, we also say that F is ***intrinsically closed*** in X.

Just as Σ is analogous to the Sierpiński space $\mathbb{S}$ in classical topology, Z is analogous to the 'inverted Sierpiński space', in which falsehood is open and truth is not, as continuous maps from a space X to it are characteristic maps of closed subsets of X.

Because sets, defined by an implication, repeatedly crop up in the context of closed sets, it is useful to have a notation for them. Take any $A, B \subseteq X$. Following the example of the well-established notation $A \cap B = \{x \in X \mid x \in A \wedge x \in B\}$ and $A \cup B = \{x \in X \mid x \in A \vee x \in B\}$, where the pointy tips are smoothed when going from logical connectives to set operators, it makes sense to define

$$A \Rrightarrow_X B := \{x \in X \mid x \in A \implies x \in B\}.$$

Technically every powerset has its own operator of intersection and union, and formally one should write $\cap_X$, $\cup_X$, but in the case of the intersection and union there is no danger of misinterpretation. In the case of the implication however, writing the index X in $\Rrightarrow_X$ is important, and we shall do so consistently.

[3] The insight that one should define closedness via implication is originally due to Martín Escardó.

It follows easily from the laws of intuitionistic logic that Z is closed under finite meets and finite joins (i.e., it is a bounded sublattice of Ω).

Let us consider a few simple examples of Σ.

The largest possible Sierpiński object is $\Sigma = \Omega$. In that case also $\mathrm{Z} = \Omega$, that is, every subset is open and closed, so every set has discrete topology.

The smallest possible Sierpiński object is $\Sigma = \{\top\}$, in which case the only open subset of any set X is X itself. We cannot separate points with open sets in any way, so in this sense sets have indiscrete (trivial) topology. Note, however, that $\mathrm{Z} = \Omega$, so every subset is closed.

Classically the indiscrete topology is the one where the only open (and closed) subsets of X are $\emptyset$ and X. Synthetically we cannot have such a situation with all sets: if $\mathcal{O}(\mathbf{1}) = \{\emptyset, \mathbf{1}\}$, then $\Sigma = \{\top, \bot\} = \mathbf{2}$, in which case all decidable subsets are open. In particular $\mathcal{O}(\mathbf{2}) = \{\emptyset, \{\top\}, \{\bot\}, \mathbf{2}\} \neq \{\emptyset, \mathbf{2}\}$.

If we do take $\Sigma = \mathbf{2}$, then the open truth values are those for which the LEM holds: $\Sigma = \{p \in \Omega \mid p \vee \neg p\}$. In this case the closed truth values are those, for which the Weak Law of Excluded Middle (WLEM) holds: $\mathrm{Z} = \{p \in \Omega \mid \neg\neg p \vee \neg p\}$. If we take $\Sigma = \{p \in \Omega \mid \neg\neg p \vee \neg p\}$, then also $\mathrm{Z} = \{p \in \Omega \mid \neg\neg p \vee \neg p\}$.

A much more useful example of Σ is the set of ***semidecidable truth values***

$$\Sigma_1^0 := \left\{ p \in \Omega \;\middle|\; \exists \alpha \in \mathbf{2}^{\mathbb{N}} .\, p \Leftrightarrow \exists n \in \mathbb{N}.\, \alpha_n \right\},$$

also called the ***Rosolini dominance***. Determining what Z is in this case is more involved, though if we postulate Markov's Principle, the answer is given by Proposition 17.49. This choice of Σ is useful in the realizability examples of synthetic topology – see Subsection 17.1.3.

Since openness and closedness are given via truth values, synthetic topological statements often reduce to simple logical facts. Here is an example.

Proposition 17.4 *Every map is continuous with respect to the intrinsic topology, in the sense that preimages of open subsets are open, and preimages of closed subsets are closed.*

Proof Because $x \in f^*(S) \iff f(x) \in S$. □

As this is the first such example, let us elaborate on this proof. Let $f \colon X \to Y$ be a map and $S \subseteq Y$. If S is open in Y, by definition the truth value of the statement $y \in S$ is open for every $y \in Y$. In particular, $f(x) \in S$ is open for every $x \in X$. The truth values of $f(x) \in S$ and $x \in f^*(S)$ are the same, so the latter is open also. Since $x \in X$ was arbitrary, $f^*(S)$ is open in X. Likewise if S is closed.

This proposition tells us that $\mathcal{O}$ and $\mathcal{Z}$ are (contravariant) functors: $\mathcal{O}(f)$ and $\mathcal{Z}(f)$ are restrictions of the preimage map f^* to open and closed subsets, respectively.

In a topos we have the usual set-theoretic constructions available. If we have some sets and have a precise description of their intrinsic topologies, what can we say about sets constructed from them?

For colimits, the answer is easy. Exponentiation (with a fixed basis) is adjoint to the same exponentiation in the opposite category. Right adjoints preserve limits and left adjoints preserve colimits. We have $\mathcal{O}(X) \cong \Sigma^X$. It follows that the topology of a colimit is the corresponding limit of the topologies: $\mathcal{O}\big(\varinjlim X_i\big) \cong \varprojlim \mathcal{O}(X_i)$, where the isomorphism is given by the restriction of opens from $\varinjlim X_i$ to all X_i. The same is true for closed sets, as $\mathcal{Z}(X) \cong \mathrm{Z}^X$.

For example, we have $\mathcal{O}(X + Y) \cong \mathcal{O}(X) \times \mathcal{O}(Y)$. The isomorphism expresses the fact that an open subset of the sum (coproduct) is uniquely determined by its restrictions to individual summands. This is true for general coproducts: $\mathcal{O}\Big(\sum_{i \in I} X_i\Big) \cong \prod_{i \in I} \mathcal{O}(X_i)$.

Applying the preservation property of adjoints to coequalizers tells us that quotients have the expected quotient topology. Actually, in a topos every surjection is isomorphic to a quotient map, so any surjection induces the quotient topology on its codomain.

Proposition 17.5 *Let $f : X \to Y$ be a surjection. Then*

$$\mathcal{O}(Y) = \big\{V \subseteq Y \;\big|\; f^*(V) \in \mathcal{O}(X)\big\}.$$

While the topology of colimits is clear, we have no such luck in the case of limits (or other constructions, such as exponentials). For example, a product $X \times Y$ may or may not have the usual product topology (see [8, Section 2.3, Footnote 18]). That said, the topology of the product must be at least as strong as the product topology, since projections are continuous.

Likewise, equalizers need not have the expected topology. In a topos, any subset is an equalizer of its characteristic map and the constant $\top$. From classical topology, we would expect that the open subsets of $S \subseteq X$ are open subsets of X, restricted to S.

Definition 17.6 A subset $S \subseteq X$ has the ***inherited topology*** from X when

$$\mathcal{O}(S) = \big\{U \cap S \;\big|\; U \in \mathcal{O}(X)\big\}.$$

In general the topology of a subset is as least as strong as the inherited topology since the inclusion map is continuous. It can be strictly stronger, however. For example, in nice synthetic topological models the reals $\mathbb{R}$ have the euclidean topology (see Proposition 17.74), but the rationals $\mathbb{Q}$ do not ($\mathbb{Q}$ has decidable equality, hence discrete topology).

Decidable subsets and retracts do have inherited topology, though.

Proposition 17.7 *Every decidable subset S has the inherited topology (both in terms of open and in terms of closed subsets). More precisely:*

- *if $W \subseteq S \subseteq X$, S is decidable in X and W is open (respectively closed) in S, then $W = S \cap V$ for $V = W \cup S^{\complement}$, where V is open (respectively closed) in X ($S^{\complement} = X \setminus S$ denotes the complement of S),*
- *if $\bot \in \Sigma$, then we may also take $V = W$ (i.e., open/closed subsets of a decidable subset are open/closed in the whole set).*

Proof Take any $x \in X$. If $x \notin S$, then the truth value of $x \in V$ is $\top$ in the first case and $\bot$ in the second; either way, it is in Σ and Z. If $x \in S$, then the statement $x \in V$ is open (resp. closed) because it is equivalent to $x \in W$ and W is open (respectively closed) in S. □

Proposition 17.8 *Suppose $S \subseteq X$ and there exists a retraction $r \colon X \to S$ (that is, r is the left inverse to the inclusion $S \hookrightarrow X$). Then S has the inherited topology from X.*

Proof Take any $U \in \mathcal{O}(S)$. Then $r^*(U) \in \mathcal{O}(X)$ (every map is intrinsically continuous) and $U = S \cap r^*(U)$. A similar argument also proves the inheritance of closed subsets. □

17.1.3 Models

In this subsection we recall some typical models of synthetic topology. Of course, by definition, any topos with the choice of any bounded meet-subsemilattice of the subobject classifier is a model. We are interested in the ones with practical applications and which have a natural choice of $\Sigma \subseteq \Omega$. These are primarily sheaf and realizability toposes.

A Grothendieck topos [10] is the topos of sheaves over a site (a category with a coverage). When the site is the topology of a particular topological space (or more generally a locale), we get a so-called ***petit topos*** or ***little topos***. If a site consists of several topological spaces (or locales), we get a ***gros topos*** or ***big topos***. In these cases, sheaves are a certain type of contravariant functors from a collection of topological spaces (or locales) to the category of sets, and there is a natural choice of the Sierpiński object: the topology functor $\mathcal{O}$ itself, that is, the one which assigns to a topological space its topology (or to a locale its corresponding frame), and to a continuous map its preimage map (or to a locale morphism its corresponding frame morphism).

Given a realizability (or computability) theory (described, for example, via assemblies), one can construct a topos around it, called a ***realizability topos*** (for that theory) [14]. In this case the notion of semidecidability behaves similarly to openness in topology, so it makes sense to take $\Sigma = \Sigma^0_1$.

Following [8, Chapter 5], we mention two instances for each of the two aforementioned types of toposes, one which behaves poorly from the perspective of synthetic topology, and one which behaves nicely (meaning, the intrinsic topologies and the topological properties of object are what we would expect).

Generally, the more spaces there are in the site, the nicer the topology behaves in the sheaf topos over it. Taking a trivial example, such as the sheaf topos over the topology of a singleton, gives us a topos, equivalent to the category of classical sets – in other words, a classical mathematical universe [8, Section 5.1]. In this case every subset is open, so every set is discrete. On the other hand, the site of separable metric spaces is large enough that the sheaf topos over it has nice topological properties [8, Section 5.4].

The first construction of a realizability topos was the effective topos by Martin Hyland [6] which captures the notion of type 1 computability (i.e., the one, based on ordinary, or type 1, Turing machines), also called number realizability. From the topological point of view this model is poorly behaved (though a rich source of counterexamples); for example, no metric space with an accumulation point can be compact, nor does the usual metric topology match the intrinsic one [8, Section 5.2]. In contrast, the realizability topos for type 2 computability (based on type 2 Turing machines, or Weihrauch computability, or function realizability) has very nice topological properties [8, Section 5.3].

17.2 Topological Properties

Because topology is given via truth values, we can expect topological properties to be as well. We consider some examples in this section.

17.2.1 Hausdorffness and Discreteness

Mathematical notions generally have many classically but not constructively equivalent definitions, and constructivists have learned from experience that the 'default' classical definition is often not the correct constructive one. This is also the case with Hausdorffness in synthetic setting. The usual definition that distinct points have disjoint neighbourhoods turns out to not be particularly useful. However, classically a topological space is Hausdorff if and only if its diagonal is closed. Taking this as a synthetic topological definition works well.

Definition 17.9 A set X is ***Hausdorff*** when it satisfies any of the following equivalent statements:

- the diagonal $\Delta_X := \{(x, x) \mid x \in X\}$ is intrinsically closed in $X \times X$,
- equality is a closed relation, that is, for all $x, y \in X$ the truth value of $x = y$ is closed,
- singleton subsets of X are closed, that is, for all $x \in X$ the set $\{x\}$ is closed in X.[4]

It follows from the definition of Z that decidable truth values are closed. Hence, any set with decidable equality, such as $\emptyset$, $\mathbf{1}$, $\mathbf{2}$ or $\mathbb{N}$, is Hausdorff.

By switching closedness and openness, we obtain a dual property.

Definition 17.10 A set X is ***discrete*** when it satisfies any of the following equivalent statements:

- the diagonal Δ_X is intrinsically open in $X \times X$,
- equality is an open relation, that is, for all $x, y \in X$ the truth value of $x = y$ is open,
- singleton subsets of X are open, that is, for all $x \in X$ the set $\{x\}$ is open in X.

We say that $x \in X$ is an ***isolated point*** when $\{x\}$ is open in X. Thus a set is discrete if and only if every point is isolated.

The empty set and singletons are always discrete. Usually we have $\bot \in \Sigma$, in which case sets with decidable equality are discrete. In fact, under certain assumptions (see Proposition 17.49), a set has decidable equality precisely when it is both discrete and Hausdorff.

Contrary to the classical case, discreteness does not generally imply Hausdorffness. For any $u \in \Sigma$ the set $\{\top, u\}$ is discrete. If it was also Hausdorff for all u, we would get $\Sigma \subseteq \mathrm{Z}$. This can happen for sufficiently trivial choices of Σ, but it becomes impossible if we postulate Phoa's principle and that complements of open subsets are closed (see Subsections 17.3.3 and 17.3.5) – two reasonable assumptions.

17.2.2 Compactness

The usual classical definition also does not work in the case of compactness ('every open cover has a finite subcover'). Recall [4] that if X is a topological space, then the natural choice of a topology on $\mathcal{O}(X)$ is the Scott topology, and X is compact if and only if the singleton $\{X\}$ is Scott open in $\mathcal{O}(X)$. It is this version of compactness that proves suitable synthetically.

Definition 17.11 A set X is ***compact*** when it satisfies any of the following equivalent statements:

[4] This is different from classical topology, where singletons being closed is equivalent to the T_1 property.

- the set $\{X\}$ is open in $\mathcal{O}(X)$,
- for every $U \in \mathcal{O}(X)$ the truth value of $X \subseteq U$ is open,
- for every $U \in \mathcal{O}(X)$ the truth value of $\forall x \in X.\, x \in U$ is open,
- for every $\phi \in \Sigma^X$ the truth value of $\forall x \in X.\, \phi(x)$ is open.

That is, a set is compact when 'universal quantification over it preserves openness', that is, when the universal quantifier $\forall_X : \Omega^X \to \Omega$ restricts to $\Sigma^X \to \Sigma$.

$$\begin{array}{ccc} \Omega^X & \xrightarrow{\forall_X} & \Omega \\ \uparrow & & \uparrow \\ \Sigma^X & \dashrightarrow & \Sigma \end{array}$$

Yet another way to phrase this is that the intersection of any family of open subsets, indexed by a compact set, is open.

Proposition 17.12 *The following statements are equivalent for any set X.*

(i) *X is compact.*
(ii) *For every set Y and every map $X \to \mathcal{O}(Y)$, $i \mapsto U_i$, the intersection $\bigcap_{i \in X} U_i$ is open in Y.*
(iii) *For every map $X \to \mathcal{O}(\mathbf{1})$, $i \mapsto U_i$, the intersection $\bigcap_{i \in X} U_i$ is open in $\mathbf{1}$.*

Proof The implication $(1 \Rightarrow 2)$ follows from the observation

$$y \in \bigcap_{i \in X} U_i \iff \forall i \in X.\, y \in U_i$$

for any $y \in Y$. The implication $(2 \Rightarrow 3)$ is obvious. To get the implication $(3 \Rightarrow 1)$, note that item 3 is just the definition of compactness, restated via the isomorphism $\mathcal{O}(\mathbf{1}) \cong \Sigma$. □

Since Σ is closed under finite meets by definition, every topology $\mathcal{O}(X)$ is closed under finite intersections. Hence, finite sets are compact.

In sufficiently nice synthetic topological models the expected sets are compact ($\mathbb{R}_{[0,1]}$, $\mathbf{2}^{\mathbb{N}}$,..., but not $\mathbb{N}$, $\mathbb{R}$,...), although in general this need not be the case, as we have a lot of freedom when choosing Σ. For example, if $\Sigma = \Omega$, then *every* set is compact. See Subsection 17.3.7 for further discussion.

In classical topology, compactness is usually treated as an absolute property – as opposed to, say, openness, which is relative: whether a set is open depends in which superset we consider it. However, compactness can also be seen as a relative property: call a subset $S \subseteq X$ ***subcompact*** in the topological space X when every family of open subsets in X which covers S has a finite subcover of S. Clearly a space is compact if and only if it is subcompact in itself.

Subcompactness is rarely emphasized in classical topology: insofar as this property is mentioned, it is immediately shown that it is equivalent to compactness (i.e., a space is compact if and only if it is subcompact in every larger space). However, the crucial part of the proof is that the subset is assumed to have the inherited topology. Without this assumption we clearly do not have the equivalence. For example, if we equip $\mathbb{R}$ with the euclidean topology and $\mathbb{R}_{[0,1]}$ with the discrete topology, then the inclusion $\mathbb{R}_{[0,1]} \hookrightarrow \mathbb{R}$ represents a subobject in the category of topological spaces and $\mathbb{R}_{[0,1]}$ is subcompact in $\mathbb{R}$, but not compact by itself.

Synthetically we have no guarantee in general that a subset has the inherited topology. Hence we need subcompactness as a separate notion.

Definition 17.13 A subset $S \subseteq X$ is ***subcompact*** in X when it satisfies any of the following equivalent statements:

- the set $\uparrow S := \{U \in \mathcal{O}(X) \mid S \subseteq U\}$ is open in $\mathcal{O}(X)$,
- for every $U \in \mathcal{O}(X)$ the truth value of $S \subseteq U$ is open,
- for every $U \in \mathcal{O}(X)$ the truth value of $\forall x \in S.\, x \in U$ is open,
- for every $\phi \in \Sigma^X$ the truth value of $\forall x \in S.\, \phi(x)$ is open.

The collection of all subcompact subsets of X is denoted by $\mathtt{Cmp}(X)$.

There is a reason to generalize this definition further. Sometimes we need to index something by a compact set (we already had an example in Proposition 17.12). However, we might not actually require the domain of the map to be compact, only its image (i.e., what it actually indexes). A minor modification in the proof of Proposition 17.12 shows that as long as the image of the map $X \to \mathcal{O}(Y)$, $i \mapsto U_i$ is compact (or even just subcompact), the intersection $\bigcap_{i \in X} U_i$ is open.

One can replace an indexing with the inclusion of the image, but the other way around is problematic constructively: in general, we cannot recover indices from the indexed objects at once without the axiom of choice. We might need to work with a particular indexing, rather than with a subset of elements, and we need a notion of compactness for maps, rather than just subsets.

Definition 17.14 A map $f \colon X \to Y$ is ***compact*** when for every $\phi \in \Sigma^Y$ the truth value of

$$\forall x \in X.\, \phi\big(f(x)\big)$$

is open. If a compact map f is viewed as an indexing, we call it a ***compact indexing***, and we say that the elements in its image are ***compactly indexed*** by f.

Proposition 17.15 *The various notions of compactness are connected in the following way.*

(i) *A set is compact if and only if it is subcompact in itself.*

(ii) *If $S \subseteq X$ is compact, it is subcompact in X.*

(iii) *If $S \subseteq X$ is subcompact in X and has the inherited topology, it is compact.*

(iv) *A map $f\colon X \to Y$ is compact if and only if its image is subcompact in Y.*

(v) *A set X is compact if any only if Id_X is a compact map.*

(vi) *A subset $S \subseteq X$ is subcompact if any only if the inclusion $S \hookrightarrow X$ is a compact map.*

Proof Easy. □

Compact sets (and maps) have the expected properties.

Proposition 17.16 *A surjective image of a compact set is compact. More generally, the image of a subcompact subset is subcompact in the codomain. More generally, precomposing any map with a compact map yields a compact map.*[5] *In particular,* $\mathtt{Cmp}$ *is a functor:*

$$\mathtt{Cmp}\big(f\colon X \to Y\big) \quad = \quad \mathtt{Cmp}(X) \xrightarrow{K \mapsto f_*(K)} \mathtt{Cmp}(Y).$$

Proof Let $f\colon X \to Y$ and $g\colon Y \to Z$ be maps and $\phi \in \Sigma^Z$. The statement follows since

$$\forall x \in X.\, \phi\big((g \circ f)(x)\big) \iff \forall x \in X.\, (\phi \circ g)\big(f(x)\big).$$ □

Proposition 17.17 *A closed subset of a compact set is subcompact. More generally, the intersection of a subcompact and a closed subset is subcompact. More generally, if $f\colon X \to Y$ is compact and $F \in \mathcal{Z}(Y)$, then the diagonal map in the following pullback diagram is compact.*

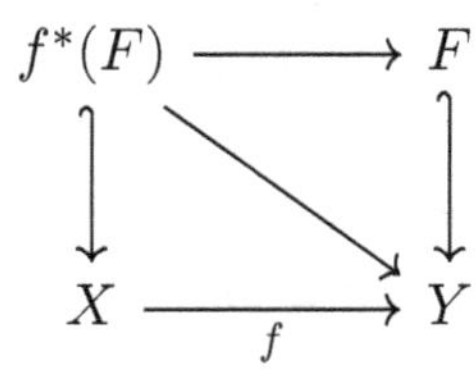

Proof Let $d\colon f^*(F) \to Y$ denote the diagonal map. Take any $\phi \in \Sigma^Y$. Then

$$\forall x \in f^*(F)\,.\, \phi\big(d(x)\big) \iff \forall x \in X.\, \Big(f(x) \in F \implies \phi\big(f(x)\big)\Big).$$ □

Proposition 17.18 *A subcompact subset of a Hausdorff set is closed.*

Proof Let X be Hausdorff, $K \subseteq X$ subcompact, $U \in \mathcal{O}(X)$ and $x \in X$. Then

$$\big(x \in K \implies x \in U\big) \iff \forall y \in K.\, \big(x = y \implies y \in U\big).$$ □

[5] That is, compact maps form a cosieve.

Proposition 17.19 *A finite product of compact sets is compact. More generally, a finite product of subcompact subsets is subcompact. More generally, a finite product of compact maps is a compact map.*

Proof It suffices to check this for nullary and binary products. The nullary product is clear: $\mathbf{1}$ (and $\mathrm{Id}_{\mathbf{1}}$) is compact.

Let $f\colon A \to B$, $g\colon C \to D$ be maps and $\phi \in \Sigma^{C \times D}$. Then

$$\forall x \in A \times B.\, \phi\big((f \times g)(x)\big) \iff \forall a \in A.\, \forall b \in B.\, \phi\big(f(a), g(b)\big). \qquad \square$$

Unlike the classical topology, compactness need not be preserved by *arbitrary* products. This is not surprising – even classically, Tychonoff's theorem requires the full axiom of choice. But the reasons are deeper than that.

In classical topology products of topological spaces are always set-indexed. If one were to represent such indexings with a map inside the category of topological spaces, one would only get products, indexed by discrete spaces. For different kinds of products, Tychonoff's theorem does not hold. For example, if we equip $\mathbb{I} := \mathbb{R}_{[0,1]}$ with the euclidean topology, the exponential $\mathbb{I}^{\mathbb{I}}$ exists (it is the set of continuous maps $\mathbb{I} \to \mathbb{I}$ with the topology of uniform convergence), and we can view it as the $\mathbb{I}$-indexed product of intervals $\mathbb{I}$. However, it is not compact (use the Arzelà–Ascoli theorem, or just note that the sequence of maps $(x \mapsto x^n)_{n\in\mathbb{N}}$ does not have an accumulation point). Similarly, in synthetic topological models products of compact sets (or maps), indexed by a non-discrete set, are generally not compact.

But even when the indexing set is discrete we, in general, have no guarantee that the product is compact. For example, the Cantor set $\mathbf{2}^{\mathbb{N}}$ is not compact in the effective topos [8].

Preservation of compactness by binary products is a special case of the more general fact that compact sums of compact sets are compact.

Proposition 17.20 *Let I be a compact set and $(X_i)_{i\in I}$ a family of compact sets. Then the sum $\coprod_{i\in I} X_i$ is compact.*

Proof Denote $X = \coprod_{i\in I} X_i$. Then for any $\phi \in \Sigma^X$

$$\forall x \in X.\, \phi(x) \iff \forall i \in I.\, \forall x \in X_i.\, \phi(x). \qquad \square$$

17.2.3 Overtness

As discussed, compactness means that universal quantification preserves openness or, in other words, that compactly indexed intersections of opens are open. This suggests the existence of a dual property – one where existential quantification preserves openness, and which informs which unions of opens are open. This property is called ***overtness***.

In classical topology all unions of opens are open. In this sense every space is overt, making this a vacant property. When defining Σ, we insisted it be closed under finite meets, which means that finite intersections of opens are open. If one were to follow the classical example, one would require that Σ is also closed under arbitrary joins.

However, synthetically we do not want everything to be overt. If Σ was closed under arbitrary joins (and since $\top \in \Sigma$), then $\Sigma = \Omega$, because $p = \bigvee \{\top \mid p\}$ for every $p \in \Omega$. This is an uninteresting example of synthetic topology where everything is discrete.

Despite topologies not being closed under arbitrary unions synthetically, we generally still recover the familiar topological behaviour by having enough overt sets. It is common that every set that can be equipped with a complete separable metric, or is the image of such, is overt[6]; in particular, every countable set is overt.

We now essentially repeat the definitions and results from compactness, just make them dual. The proofs are very similar to those in Subsection 17.2.2, so we omit them.

Definition 17.21 A set X is ***overt*** when it satisfies any of the following equivalent statements:

- the set $\{U \in \mathcal{O}(X) \mid U \text{ inhabited}\}$ is open in $\mathcal{O}(X)$,
- for every $U \in \mathcal{O}(X)$ the truth value of 'U is inhabited' is open,
- for every $U \in \mathcal{O}(X)$ the truth value of $\exists x \in X \,.\, x \in U$ is open,
- for every $\phi \in \Sigma^X$ the truth value of $\exists x \in X \,.\, \phi(x)$ is open.

That is, a set is overt when 'existential quantification over it preserves openness', that is, when the existential quantifier $\exists_X : \Omega^X \to \Omega$ restricts to $\Sigma^X \to \Sigma$.

$$\begin{array}{ccc} \Omega^X & \xrightarrow{\exists_X} & \Omega \\ \uparrow & & \uparrow \\ \Sigma^X & \dashrightarrow & \Sigma \end{array}$$

Singletons are overt; whether other sets are overt, depends on the model of synthetic topology. We discuss this more in detail in Subsection 17.3.2.

Proposition 17.22 *The following statements are equivalent for any set X.*

(i) *X is overt.*
(ii) *For every set Y and every map $X \to \mathcal{O}(Y)$, $i \mapsto U_i$, the union $\bigcup_{i \in X} U_i$ is open in Y.*

[6] An example when that happens is given by Proposition 17.62. For another example, combine Theorem 17.99 or 17.100 with Proposition 17.88.

(iii) *For every map* $X \to \mathcal{O}(\mathbf{1})$, $i \mapsto U_i$, *the union* $\bigcup_{i \in X} U_i$ *is open in* $\mathbf{1}$.

Let $A \between B$ denote that sets A and B have an inhabited intersection.[7]

Definition 17.23 A subset $S \subseteq X$ is ***subovert*** in X when it satisfies any of the following equivalent statements:

- the set $\{U \in \mathcal{O}(X) \mid S \between U\}$ is open in $\mathcal{O}(X)$,
- for every $U \in \mathcal{O}(X)$ the truth value of $S \between U$ is open,
- for every $U \in \mathcal{O}(X)$ the truth value of $\exists x \in S.\, x \in U$ is open,
- for every $\phi \in \Sigma^X$ the truth value of $\exists x \in S.\, \phi(x)$ is open.

The collection of all subovert subsets of X is denoted by $\mathtt{Ov}(X)$.

Definition 17.24 A map $f\colon X \to Y$ is ***overt*** when for every $\phi \in \Sigma^Y$ the truth value of

$$\exists x \in X.\, \phi\big(f(x)\big)$$

is open. If an overt map f is viewed as an indexing, we call it an ***overt indexing***, and we say that the elements in its image are ***overtly indexed*** by f.

Proposition 17.25 *The various notions of overtness are connected in the following way.*

(i) *A set is overt if and only if it is subovert in itself.*
(ii) *If* $S \subseteq X$ *is overt, it is subovert in* X.
(iii) *If* $S \subseteq X$ *is subovert in* X *and has the inherited topology, it is overt.*
(iv) *A map* $f\colon X \to Y$ *is overt if and only if its image is subovert in* Y.
(v) *A set* X *is overt if any only if* Id_X *is an overt map.*
(vi) *A subset* $S \subseteq X$ *is subovert if any only if the inclusion* $S \hookrightarrow X$ *is an overt map.*

Proposition 17.26 *A surjective image of an overt set is overt. More generally, the image of a subovert subset is subovert in the codomain. More generally, precomposing any map with an overt map yields an overt map.*[8] *In particular,* $\mathtt{Ov}$ *is a functor:*

$$\mathtt{Ov}\big(f\colon X \to Y\big) \quad = \quad \mathtt{Ov}(X) \xrightarrow{O \mapsto f_*(O)} \mathtt{Ov}(Y).$$

Proposition 17.27 *An open subset of an overt set is subovert. More generally, the intersection of a subovert and an open subset is subovert. More generally, if* $f\colon X \to Y$ *is overt and* $U \in \mathcal{O}(Y)$, *then the diagonal map in the following pullback diagram is overt.*

[7] The 'overlap symbol' $\between$ was popularized by Giovanni Sambin.
[8] That is, overt maps form a cosieve.

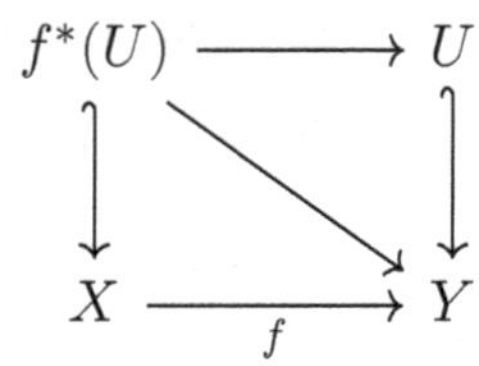

Proposition 17.28 *A subovert subset of a discrete set is open.*

We emphasize that synthetically it is not the case that every subset of a discrete set is open (this fails in the case of a singleton: if every subset of $\mathbf{1}$ is open, then $\Sigma = \Omega$). What we have is that if X is overt discrete, then subovert and open subsets match (just as if X is compact Hausdorff, then subcompact and closed subsets match).

Proposition 17.29 *A finite product of overt sets is overt. More generally, a finite product of subovert subsets is subovert. More generally, a finite product of overt maps is an overt map.*

Proposition 17.30 *Let I be an overt set and $(X_i)_{i \in I}$ a family of overt sets. Then the sum $\coprod_{i \in I} X_i$ is overt.*

17.2.4 Topological Bases

As we know from topological practice, actual topologies might be very complicated objects, but they might have a simple description via a basis. Classically, a basis of a given topological space is any collection of open sets, called basic, such that every open set can be expressed as a union of basic ones.

To get a suitable synthetic notion of a basis, we need to make two adjustments. The first is the one we already discussed in the case of compactness. When indexing a family, we cannot freely pass between indices and indexed members without the axiom of choice, and bases are typically given by a non-injective indexing. For example, in metric spaces we want the basis of open balls, indexed by centres and radii, but balls with different centres and/or radii can be the same as sets. We would be rather limited, using the basis of balls as sets, without being able to uniformly access their indices (centres and radii). As such, the indexing should be entailed in the definition of a basis – that is, a basis should be a map, rather than just a subset of the topology (compare with how we wanted compactness and overtness to be properties of maps, not just of subsets).

Definition 17.31 A ***weak basis*** of a set X is a map $I \to \mathcal{O}(X)$, $i \mapsto B_i$, which satisfies any of the following equivalent properties:

- for every $U \in \mathcal{O}(X)$ and every $x \in U$ there exists $i \in I$ such that $x \in B_i \subseteq U$,
- for every $U \in \mathcal{O}(X)$ there exists $J \subseteq I$ such that $U = \bigcup_{j \in J} B_j$,
- for every $U \in \mathcal{O}(X)$ we have $U = \bigcup\{B_i \mid i \in I, B_i \subseteq U\}$.

We call this a *weak* basis because there is another issue to be resolved: a basis has to determine the topology. Classically this is the case because opens are closed under arbitrary unions, so the opens are precisely the unions of basics. Every topology should be a basis for itself, but arbitrary unions of members in $\mathcal{O}(\mathbf{1})$ amount to the whole $\mathcal{P}(\mathbf{1})$ which matches the topology only in the uninteresting case $\Sigma = \Omega$, when everything is discrete.

Synthetically, open sets are closed under overtly indexed unions. Therefore the definition of a basis should say that every open subset is an overtly indexed union of basic subsets. This is opposed to the weak basis, where an open was a union of precisely the basic subsets it contains, we might not have a canonical choice of an overtly indexed union for an open subset. Hence we present two versions of a basis.

Definition 17.32 A map $I \to \mathcal{O}(X), i \mapsto B_i$ is

- a ***basis*** for the set X when for every $U \in \mathcal{O}(X)$ there exists $J \in \mathtt{Ov}(I)$ such that $U = \bigcup_{j \in J} B_j$,
- a ***canonical basis*** for X when there exists a map $o\colon \mathcal{O}(X) \to \mathtt{Ov}(I)$ such that $U = \bigcup_{j \in o(U)} B_j$ for all $U \in \mathcal{O}(X)$.

Note that if $J \in \mathtt{Ov}(I)$, then the inclusion $J \hookrightarrow I$ is an overt map, and since precomposing with an overt map yields an overt map, the union $\bigcup_{j \in J} B_j$ is overtly indexed. Thus a (canonical) basis fully determines the topology: a subset is open if and only if it can be written as an overtly indexed union of basic subsets.

In a more diagrammatic form: a map $B\colon I \to \mathcal{O}(X)$ is a basis for X when the slanted arrow in the commutative diagram

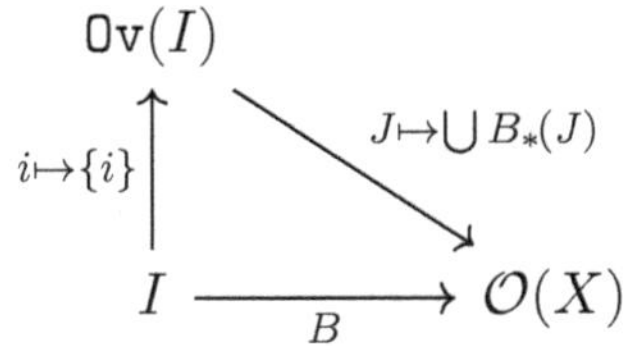

is a surjection, and B is a canonical basis when the slanted map is a split surjection (to get a weak basis, replace $\mathtt{Ov}$ with $\mathcal{P}$).

The main application of bases in this chapter will be in Subsection 17.3.8, where we discuss the match between the metric and the intrinsic topology. In that case the basis is given by open balls: $X \times \mathbb{R} \to \mathcal{O}(X)$, $(x, r) \mapsto B(x, r)$.

17.3 Principles

The definition of synthetic topology requires only that $\Sigma \subseteq \Omega$ is closed under finite meets. This is enough for the general definitions and properties to make sense, but is broad enough to allow almost arbitrarily ugly topologies which do not yield useful results. To get a more compelling theory, we need to postulate stronger assumptions.

However, the interesting models of synthetic topology are diverse enough that the additional axioms jump-starting the theory are not always the same, which is why we do not add them to the definition of synthetic topology. Instead, as is common in constructive mathematics, we present them as ***principles*** that might hold in certain situations.

In this section we exhibit the most common synthetic topological principles, as well as some of their immediate consequences. For reasons of space, we omit most of the proofs. Many of the omitted proofs can be found in [8], and the rest are left as an exercise to the reader.

17.3.1 Inheritance Principles

Recall that a subset $S \subseteq X$ need not have the inherited topology from X. It is useful to have some assurance that certain subsets do. Propositions 17.7 and 17.8 already told us that decidable subsets and retracts inherit topology; here is another proposition on this topic.

Proposition 17.33 *Let $S \subseteq X$.*

(i) *If X is discrete and S subovert, then S has the inherited topology if and only if it is overt.*
(ii) *If X is Hausdorff and S subcompact, then S has the inherited topology if and only if it is compact.*

Aside from these propositions, we cannot say much about the topological inheritance in general. Consequently we formulate the inheritance of topology for certain kinds of subsets as principles.

Requiring the inheritance for open and/or closed sets is a relatively mild assumption since if we can extend their open subsets to opens of the whole set, we can do so in a canonical way.

Lemma 17.34 *Let X be a set and $S \subseteq X$. Let $U \in \mathcal{O}(S)$ be 'inherited from X', that is, there exists $V \in \mathcal{O}(X)$ such that $U = S \cap V$. If S is open or closed in X, then we can choose V canonically. Specifically:*

(i) *if S is open in X, then U is open in X, and $U = S \cap U$,*
(ii) *if S is closed in X, then $S \Rrightarrow_X U$ is open in X, and $U = S \cap (S \Rrightarrow_X U)$.*

Proposition 17.35 *The following statements are equivalent.*

(i) *The* **dominance axiom** *holds for* Σ*:*

$$\forall u \in \Sigma.\, \forall p \in \Omega.\, \Big(\big(u \Rightarrow (p \in \Sigma)\big) \implies (u \wedge p) \in \Sigma\Big).$$

(ii) *Openness is a transitive property: for all* $U \subseteq V \subseteq X$*, if* U *is open in* V *and* V *is open in* X*, then* U *is open in* X*.*
(iii) *Openness is a transitive property on a singleton: for all* $U \subseteq V \subseteq \mathbf{1}$*, if* U *is open in* V *and* V *is open in* $\mathbf{1}$*, then* U *is open in* $\mathbf{1}$*.*
(iv) *All open subsets have the inherited topology.*
(v) *Every open subset of* $\mathbf{1}$ *has the inherited topology from* $\mathbf{1}$*.*

Definition 17.36 The Sierpiński object Σ is a ***dominance***[9] when the equivalent statements in Proposition 17.35 are satisfied.

The dual of Proposition 17.35 leads to the principle that all closed sets have the inherited topology; in particular, the dual of the dominance axiom (the ***codominance axiom***) states

$$\forall f \in \mathrm{Z}.\, \forall p \in \Omega.\, \Big(\big(f \Rightarrow (p \in \Sigma)\big) \implies (f \Rightarrow p) \in \Sigma\Big).$$

That said, general closed sets having the inherited topology is neither commonly needed nor widely available. It is more common for *stable* closed sets (those that are equal to their double complement) to have the inherited topology (later we see that stable closed truth values might have a simple description – see Propositions 17.42 and 17.49).

Proposition 17.37 *The following statements are equivalent.*

(i) $\forall f \in \mathrm{Z}_{\neg\neg}.\, \forall p \in \Omega.\, \big(\big(f \Rightarrow (p \in \Sigma)\big) \implies (f \Rightarrow p) \in \Sigma.\big)$
(ii) *For all* $U \subseteq F \subseteq X$*, if* F *is stable closed in* X *and* U *is open in* F*, then* $F \Rrightarrow_X U$ *is open in* X*.*
(iii) *For all* $U \subseteq F \subseteq \mathbf{1}$*, if* F *is stable closed in* $\mathbf{1}$ *and* U *is open in* F*, then* $F \Rrightarrow_{\mathbf{1}} U$ *is open in* $\mathbf{1}$*.*
(iv) *All stable closed subsets have the inherited topology.*
(v) *Every stable closed subset of* $\mathbf{1}$ *has the inherited topology from* $\mathbf{1}$*.*

If the conditions of Proposition 17.49 below are satisfied ($\emptyset$ *and* $\mathbf{2}$ *are overt, and* $\Sigma \subseteq \Omega_{\neg\neg}$*), then these statements are further equivalent to the following ones.*

(vi) *Stable closedness is a transitive property: for all* $F \subseteq G \subseteq X$*, if* F *is stable closed in* G *and* G *is stable closed in* X*, then* F *is stable closed in* X*.*

[9] The notion of a dominance originally comes from [12].

(vii) *Stable closedness is a transitive property on a singleton: for all $F \subseteq G \subseteq \mathbf{1}$, if F is stable closed in G and G is stable closed in $\mathbf{1}$, then F is stable closed in $\mathbf{1}$.*

Definition 17.38 We say that the ***paradominance axiom*** holds, or that $\mathrm{Z}_{\neg\neg}$ is a ***paradominance***, when the equivalent statements of Proposition 17.37 hold.

17.3.2 Overtness Principles

As discussed, overtness tells us which joins Σ is closed under (and thus which unions topologies are closed under). Singletons are trivially overt, but that is as much as we can say for general Σ. If we want the intrinsic topologies to look more like classical topologies, we need to postulate more overt sets.

For ease of reading and reference to overtness principles, let us use the notation $X : \mathtt{Overt}$ for the statement 'X is overt'. When we postulate overtness of several sets, we just list them: $X, Y, Z, \ldots : \mathtt{Overt}$.

Generally we postulate overtness of a few very basic sets, then derive overtness of more complicated sets. We examine a few instances here.

- The principle $\emptyset : \mathtt{Overt}$ is equivalent to $\bot \in \Sigma$, meaning that decidable subsets are open.
- The principle $\mathbf{2} : \mathtt{Overt}$ means that Σ is closed under binary disjunctions, so topologies are closed under binary unions. This implies closure under all inhabited finite joins. Hence $\emptyset, \mathbf{2} : \mathtt{Overt}$ means that Σ and topologies are closed under all finite joins. This makes Σ a bounded sublattice of Ω (it is closed under finite meets by definition).
- The principle[10] $\mathbb{N} : \mathtt{Overt}$ means that inhabited countable[11] sets are overt, so Σ and topologies are closed under inhabited countable joins; in particular, this implies $\mathbf{2} : \mathtt{Overt}$. The principle $\emptyset, \mathbb{N} : \mathtt{Overt}$ means that *all* countable sets are overt,[12] that is, Σ and topologies are closed under arbitrary countable joins.

One way to verify that a set is overt is to find a dense subovert subset.

Definition 17.39 A subset $S \subseteq X$ is ***dense*** in X when it intersects every inhabited open subset of X. A set X is ***separable*** when it has a countable dense subset. (We

[10] There is a notion of a ***natural numbers object*** (NNO) in any category with finite products. A general topos need not possess a NNO, but when it does, it behaves internally as the set of natural numbers $\mathbb{N}$, subject to the usual Peano axioms and having the usual algebraic operations, including addition and multiplication. When we state $\mathbb{N} : \mathtt{Overt}$, we are implicitly postulating that the NNO $\mathbb{N}$ *exists*, then saying that it is overt.

[11] A set X is ***countable*** when there exists a map $\mathbb{N} \to X + \mathbf{1}$ such that its image contains X. The reason for the summand $\mathbf{1}$ is that inhabitedness should not be a prerequisite of countability – when enumerating elements of X, choosing $* \in \mathbf{1}$ allows us to 'skip' enumerating an element. If X is inhabited, it is countable if and only if there exists a surjection $\mathbb{N} \to X$: instead of choosing $*$ in the enumeration, choose a specific element of X.

[12] By Proposition 17.26, since a countable set is an image of a decidable subset of $\mathbb{N}$, which is overt by Propositions 17.27, 17.7, and 17.25.

use the terms ***intrinsically dense*** and ***intrinsically separable*** when we want to emphasize that we refer to the intrinsic topology.)

Lemma 17.40 *If a set has a subovert dense subset, it is overt.*

Corollary 17.41 *If* $\emptyset, \mathbb{N} : \mathtt{Overt}$*, then every separable set is overt.*

All specific models of synthetic topology that we mentioned in Subsection 17.1.3 satisfy the principle $\emptyset, \mathbb{N} : \mathtt{Overt}$. In this sense overtness can be seen as a generalization of separability, and one can often generalize results (that do not rely on countable choice) from separable to overt sets.

17.3.3 Closedness Principle

If follows from the definition that decidable truth values (and therefore decidable subsets) are closed, but aside from that, closed truth values might be scarce. In particular, negations of open truth values (or complements of open subsets) need not be closed: there is no reason, in general, why $\neg u \Rightarrow v$ would be open for all $u, v \in \Sigma$.

Nevertheless, it is the case that complements of open subsets are closed in all examples of synthetic topology, considered in Subsection 17.1.3. It is useful to have this as a principle; we can summarize it as $\neg\Sigma \subseteq \mathrm{Z}$.

We start by applying this principle to obtain a simple description of stable closed truth values and subsets.

Proposition 17.42 *Assume* $\emptyset : \mathtt{Overt}$ *and* $\neg\Sigma \subseteq \mathrm{Z}$*. Then* $\neg\Sigma = \mathrm{Z}_{\neg\neg}$*. Hence the negation map restricts to a split surjection* $\neg \colon \Sigma \to \mathrm{Z}_{\neg\neg}$*, and further restricts to an isomorphism* $\Sigma_{\neg\neg} \cong \mathrm{Z}_{\neg\neg}$*. Consequently, stable closed subsets are precisely the complements of open subsets, and complementation yields a bijective correspondence between stable open and stable closed subsets.*

The principle $\neg\Sigma \subseteq \mathrm{Z}$ can be seen as saying that 'we have enough closed subsets'. In particular it ensures that common classical Hausdorff topological spaces, such as $\mathbb{R}$, have synthetic counterparts which have Hausdorff intrinsic topology. In fact, one of the crucial applications of the principle $\neg\Sigma \subseteq \mathrm{Z}$ is that it (together with $\emptyset, \mathbb{N} : \mathtt{Overt}$) allows us properly to define real numbers in the synthetic setting.

Let us postulate $\neg\Sigma \subseteq \mathrm{Z}$ and $\emptyset, \mathbb{N} : \mathtt{Overt}$ for the remainder of the subsection. If we have $\mathbb{N}$, we can construct $\mathbb{Z}$ and $\mathbb{Q}$ in the usual way. Assuming countable choice, the reals $\mathbb{R}$ can be constructed as equivalence classes of Cauchy sequences of rationals, but this does not work without choice – the Cauchy reals need not even be Cauchy complete [9].

The Dedekind reals work well even in the choiceless environment.[13] Recall that $L \subseteq \mathbb{Q}$ is a ***left cut*** when it is inhabited and lower rounded, that is, $\forall x \in L.\, \exists q \in L.\, x < q$. Analogously, $R \subseteq \mathbb{Q}$ is a ***right cut*** when it is inhabited and upper rounded, that is, $\forall x \in R.\, \exists q \in R.\, q < x$. A ***Dedekind cut*** is a pair (L, R) where L is a left cut, R is a right cut and additionally $L \cap R = \emptyset$ and the pair (L, R) is ***located***, that is, $\forall q, r \in \mathbb{Q}.\, q < r \implies q \in L \lor r \in R$.

However, even if we assume the reals to be given by Dedekind cuts, synthetically there is another hurdle. If we want to do analysis, we need the so-called open (respectively closed) balls in metric spaces to genuinly be (intrinsically) open (respectively closed). This is equivalent to the order relations $<$ and $\leq$ on $\mathbb{R}$ being open and closed, respectively. Under our assumptions this is further equivalent to openness of both components of all Dedekind cuts.

This can happen; in fact, it happens in all models of synthetic topology we consider in Subsection 17.1.3. If countable choice holds, there is no problem at all, as follows from the next proposition.

Proposition 17.43 *For any Cauchy real, the cuts of its corresponding Dedekind real are open.*

Because so often all Dedekind cuts are open, the need for openness of cuts is frequently invisible even for constructivists. Still, in general, we need this openness, so we *define* the reals $\mathbb{R}$ to be only the (ones, represented by) open Dedekind cuts.[14] These reals still contain $\mathbb{Q}$ (the cuts representing rationals are decidable, therefore open), are Cauchy complete even in the absence of choice, are closed under the usual algebraic operations, and they have the required properties of the order relations.

Proposition 17.44 *The relation $<$ on $\mathbb{R}$ is open and the relation $\leq$ on $\mathbb{R}$ is (stable) closed. It follows that $\mathbb{R}$ is Hausdorff. Also, the relation $\leq$ induces a lattice structure on $\mathbb{R}$.*

Once we have real numbers, we can define metric spaces in the usual way. Given a metric space X with a metric $d\colon X \times X \to \mathbb{R}$, we denote the open and closed ball with the centre $a \in X$ and radius $r \in \mathbb{R}$ by

$$B_X(a, r) := \{x \in X \mid d(a, x) < r\}\,,$$
$$\overline{B}_X(a, r) := \{x \in X \mid d(a, x) \leq r\}\,.$$

When the metric space X is understood, we drop the subscripts, writing only $B(a, r)$ and $\overline{B}(a, r)$.

[13] As long as we can define them – in a sufficiently predicative environment, we might not be able to. However, in a topos there is no problem.

[14] This insight and the definition are originally due to Paul Taylor, albeit in the context of Abstract Stone duality [2].

Corollary 17.45 *Open (respectively closed) balls in metric spaces are open (respectively closed) in the intrinsic topology. Moreover, every set that can be equipped with a metric is Hausdorff.*

A subset $S \subseteq X$ is ***metrically dense*** in a metric space X when $\forall a \in X.\, \forall r \in \mathbb{R}_{>0}.\, S \between B(a,r)$, and X is ***metrically separable*** when it has a countable metrically dense subset. We emphasize the designation 'metrically' because metric and intrinsic denseness are not equivalent.[15] Still, 'metrically separable metric space' is a bit unwieldy, so whenever we use the term 'separable' specifically together with 'metric space', it is meant in the metric, not intrinsic, sense.

It is common in constructive study of metric spaces that we need to restrict to separable ones for many applications. However, synthetically we have a generalization: often it suffices to just have a *subovert* dense subset, not necessarily a *countable* one.

Definition 17.46 A metric space is ***quasi-overt*** when it possesses a metrically dense subovert subset.

Compare this definition with Lemma 17.40: if a set has an *intrinsically* dense subovert subset, it is outright overt. In that case, it is quasi-overt for every metric we equip it with.

Quasi-overtness enables various metric constructions, especially in an environment without countable choice. Completion is a case in point.

The usual contruction of a metric completion via equivalence classes of Cauchy sequences does not work without countable choice, so it is useful to have a definition of completion, independent from a specific construction. If X and Y are metric spaces, equipped with metrics d_X, d_Y respectively, then a map $f\colon X \to Y$ is an ***isometry*** (or an ***isometric embedding***) when $d_Y\big(f(x), f(y)\big) = d_X(x,y)$ for all $x, y \in X$, and it is a ***dense isometry*** when additionally its image is metrically dense in Y.

Informally speaking, the completion of a metric space X is a dense isometry with domain X and the largest possible codomain. More precisely, a dense isometry $c\colon X \to \widehat{X}$ is a ***completion*** of X when for every dense isometry $i\colon X \to Y$ there exists a unique dense isometry $i'\colon Y \to \widehat{X}$, such that $c = i' \circ i$. In other words, if $\mathcal{M}$ is the category of metric spaces and dense isometries, then the completion of X is the terminal object in the coslice category $X \backslash \mathcal{M}$. As usual, we say that a metric space is ***complete*** when its completion exists and is an isomorphism.

Proposition 17.47 *Every quasi-overt metric space has a completion.*

[15] For example, diadic rationals are metrically dense in rationals (in the usual euclidean metric), but not intrinsically dense since $\mathbb{Q}$ has discrete topology.

Proof We need to use a version of metric completion which works without countable choice. We adopt the construction via locations from [11].

Let X be a quasiovert metric space with the metric $d\colon X \times X \to \mathbb{R}$ and a metrically dense subovert subset S. A ***location*** on X is any map $\ell\colon X \to \mathbb{R}$ satisfying

$$\forall x, y \in X.\, \big|\ell(x) - d(x,y)\big| \leq \ell(y) \qquad \text{and} \qquad \forall \epsilon \in \mathbb{R}_{>0}.\, \exists x \in X.\, \ell(x) < \epsilon.$$

For every $a \in X$ the map $x \mapsto d(a,x)$ is a location; the idea is that locations are precisely distances from points in the completion.

Let $\mathscr{L}$ be the set of locations on X. For any $\ell', \ell'' \in \mathscr{L}$ define

$$D(\ell', \ell'') := \Big(\{q \in \mathbb{Q} \mid \exists x \in S.\, q < |\ell'(x) - \ell''(x)|\},$$
$$\{q \in \mathbb{Q} \mid \exists x \in S.\, q > \ell'(x) + \ell''(x)\}\Big).$$

Subovertness of S assures us that this is a pair of open subsets of $\mathbb{Q}$. The rest of the proof is standard [8, Section 3.3][11, Theorem 3], showing that $D(\ell', \ell'')$ is a Dedekind cut, so we have a map $D\colon \mathscr{L} \times \mathscr{L} \to \mathbb{R}$, which is a metric on $\mathscr{L}$, and the map $X \to \mathscr{L}$, $a \mapsto (x \mapsto d(a,x))$, is the terminal dense isometry. □

If we are in a setting where all Dedekind cuts are open (such as if we assume countable choice), then quasi-overtness is not necessary for the proof and we get that all metric spaces have a completion. Classically, quasi-overtness is a vacant property (just like overtness), and it is often the case that a classical theorem stating something for *all* metric spaces has a synthetic counterpart for *quasi-overt* metric spaces.

Here is another application of quasi-overtness. Constructively, duals of normed vector spaces need not be normed, but they are still locally convex. If X is a normed vector space, then every $x \in X$ induces the seminorm $p_x(f) := |f(x)|$, and the family of these seminorms determines the locally convex structure on the dual of X.

Proposition 17.48 *The dual of any quasi-overt normed vector space is Hausdorff.*

We have seen that the principle $\neg\Sigma \subseteq \mathrm{Z}$ (in the presence of $\emptyset, \mathbb{N} : \mathtt{Overt}$) provides the foundation for basic analysis. To bring the discussion about this principle full circle, we claim that it is also in a sense necessary for this basic analysis. Specifically, we claim the following: assume that we have some way to reasonably define real numbers, metric spaces, normed vector spaces and duals of quasi-overt normed vector spaces, and all of these are Hausdorff. Then at least if $\emptyset, \mathbb{N} : \mathtt{Overt}$ and Σ is a dominance, this implies the principle $\neg\Sigma \subseteq \mathrm{Z}$.

Take any $u \in \Sigma$. Define $X_u := \{x \in \mathbb{R} \mid x = 0 \vee u\}$ and equip it with the normed vector structure, inherited from $\mathbb{R}$. The set $\{q \in \mathbb{Q} \mid q = 0 \vee u\}$ is open in the overt $\mathbb{Q}$, so subovert in $\mathbb{Q}$. Owing to the dominance axiom it has the inherited topology, so it is overt. Therefore it is subovert also in X_u. Clearly it is also metrically dense there, so X_u is quasi-overt, and has a Hausdorff dual by assumption, that is, the equality between functionals is closed. The inclusion $X_u \hookrightarrow \mathbb{R}$ is a member of the dual, and its equality to the zero functional is equivalent to $\neg u$ which must therefore be closed as well.

17.3.4 Stability Principle

Markov's principle (MP), a familiar constructive principle, states

$$\forall \alpha \in \{0,1\}^{\mathbb{N}}.\,(\neg\forall n \in \mathbb{N}.\,\alpha_n = 1) \implies \exists n \in \mathbb{N}.\,\alpha_n = 0.$$

In can be equivalently restated as $\Sigma_1^0 \subseteq \Omega_{\neg\neg}$. Recall that the Rosolini dominance Σ_1^0 (the set of semidecidable truth values) is a reasonable choice for a Sierpiński object. Thus the principle $\Sigma \subseteq \Omega_{\neg\neg}$ can be seen as a generalized Markov's principle.

Stable truth values are not closed under even binary joins, much less more general ones. However, open truth values are closed under arbitrary overtly indexed joins. The main usefulness of the principle $\Sigma \subseteq \Omega_{\neg\neg}$ is that we can prove certain joins via contradiction.

Of course, this is only useful if Σ has enough interesting joins, that is, if enough sets (or maps) are overt. Finite joins (meaning that finite sets are overt) are already sufficient that we can tell a lot about topology.

Proposition 17.49 *Assume $\emptyset, \mathbf{2} : \mathtt{Overt}$ and $\Sigma \subseteq \Omega_{\neg\neg}$. Then the following statements hold.*

(i) *Σ and Z determine each other via negation:*

$$\mathrm{Z} = \{p \in \Omega \mid \neg p \in \Sigma\} \qquad \text{and} \qquad \Sigma = \neg\mathrm{Z}.$$

Moreover, $\neg\neg\mathrm{Z} = \neg\Sigma = \mathrm{Z}_{\neg\neg}$. That is, closed sets are precisely those whose complements are open, open sets are precisely complements of closed sets, and complements of open sets are precisely stable closed sets. In particular, the principle $\neg\Sigma \subseteq \mathrm{Z}$ holds and the negations

$$\Sigma \underset{\neg}{\overset{\neg}{\rightleftarrows}} \mathrm{Z}_{\neg\neg}$$

are (antitone) isomorphisms.

(ii) *$\Sigma \cap \mathrm{Z} = \mathbf{2} = \Sigma \cap \mathrm{Z}_{\neg\neg}$. This means that decidable sets are precisely those which are both open and closed (or open and stable closed).[16] In particular, a set has decidable equality if and only if it is discrete and Hausdorff.*

[16] When interpreted in number realizability, this captures the following fact: recursive subsets are precisely those which are recursively enumerable and corecursively enumerable.

This proposition is broadly useful; in fact, all models of synthetic topology in Subsection 17.1.3 satisfy its conditions.

Below are a few more applications of stability of opens.

Lemma 17.50 *Assume $\Sigma \subseteq \Omega_{\neg\neg}$ and let X be a set.*

(i) *If $S \subseteq X$ has the inherited topology and $S^{\mathtt{CC}} = X$, then the inclusion $S \hookrightarrow X$ induces an isomorphism $\mathcal{O}(S) \cong \mathcal{O}(X)$.*
(ii) *Define the equivalence relation on X by $x \sim y := \neg\neg(x = y)$. Then the quotient map $X \to X/_\sim$ induces an isomorphism $\mathcal{O}(X) \cong \mathcal{O}(X/_\sim)$.*

Lemma 17.51 *Assume $\Sigma \subseteq \Omega_{\neg\neg}$. Let $S \subseteq X$ satisfy $S^{\mathtt{CC}} = X$.*

(i) *If S is subcompact in X, then X is compact.*
(ii) *If S is subovert in X, then X is overt.*

Recall that a binary relation $\#$ on a set X is an ***apartness*** [13] when the following holds for all $x, y, z \in X$: $\neg(x \# x)$ (ireflexivity), $x \# y \implies y \# x$ (symmetry) and $x \# y \implies x \# z \vee z \# y$ (cotransitivity). The complement of an apartness is always an equivalence relation. When this complement is the equality, the apartness is called ***tight***.

Proposition 17.52 *Assume $\emptyset, \mathbf{2} : \mathtt{Overt}$ and $\Sigma \subseteq \Omega_{\neg\neg}$. Let X be Hausdorff. Then the inequality relation $\neq$ (the negation of equality) is an open apartness relation. Moreover, it is the largest apartness relation (the closest to being tight) and it is tight if and only if X has stable equality, in which case it is the unique open tight apartness on X.*

17.3.5 Phoa's Principle

Recall that the classical analogue of Σ is the Sierpiński space, $\mathbb{S}$. Informally speaking, Phoa's principle states that Σ behaves similarly to $\mathbb{S}$, and therefore a synthetic topological model has similar properties to classical topology.

Since $\mathbb{S}$ serves as the codomain of the characteristic maps of open subsets, it has some logical structure. The part that is continuous (that is, inside the topological category), is that $\mathbb{S}$ is a bounded lattice (it has finite meets and joins). Furthermore, the natural choice of the topology on $\mathbb{S}$ is the Scott topology; in particular, open subsets are upper. Since $\mathcal{O}(\mathbb{S}) \cong C(\mathbb{S}, \mathbb{S})$, this can be restated that every continuous map $\mathbb{S} \to \mathbb{S}$ is monotone.

Classically $\mathbb{S}$ has exactly two points, but synthetically we do not literally have $\Sigma = \mathbf{2}$. Still, we somehow need to capture the fact that $\mathbb{S}$ is a two-element topological space, if we want a property that says that Σ looks like $\mathbb{S}$. The point is, *as far as topology is concerned*, Σ should look like a two-element set. Since $\mathcal{O}(\Sigma) \cong \Sigma^\Sigma$, we are saying that maps $\Sigma \to \Sigma$ should be fully determined by their values on $\mathbf{2}$, namely, $\forall f, g \in \Sigma^{\Sigma}.\ f|_{\mathbf{2}} = g|_{\mathbf{2}} \implies f = g$.

Proposition 17.53 *Suppose $\emptyset, \mathbf{2} : \mathtt{Overt}$ (i.e., Σ is closed under finite joins). The following statements are equivalent.*

(i) *Every map $\Sigma \to \Sigma$ is monotone and fully determined by its values on* $\mathbf{2}$.
(ii) *For every $f : \Sigma \to \Sigma$ and $x \in \Sigma$ we have $f(x) = \big(f(\top) \wedge x\big) \vee f(\bot)$.*
(iii) *For every $f : \Sigma \to \Sigma$ and $x \in \Sigma$ we have $f(x) = f(\top) \wedge \big(x \vee f(\bot)\big)$.*

When $f(\top) \geq f(\bot)$, we can directly see that $\big(f(\top) \wedge x\big) \vee f(\bot) = f(\top) \wedge \big(x \vee f(\bot)\big)$ – if Σ is a sublattice of Ω, it is a modular lattice since Ω is (even a distributive lattice). Because of this and because this kind of expression appears often, we drop the bracketing and simply write $f(\top) \wedge x \vee f(\bot)$.

Definition 17.54 ***Phoa's principle*** states $\emptyset, \mathbf{2} : \mathtt{Overt}$ and the equivalent statements of Proposition 17.53 hold, that is, $f(x) = f(\top) \wedge x \vee f(\bot)$ for all $f : \Sigma \to \Sigma$ and $x \in \Sigma$.

Phoa's principle has connections with other principles. We examine some now.

Lemma 17.55 *If Phoa's principle holds, then the principles $\neg\Sigma \subseteq \mathrm{Z}$ (complements of open subsets are closed) and $\Sigma \subseteq \Omega_{\neg\neg}$ (open subsets are stable) are equivalent.*

Lemma 17.56 *Under the assumptions from Proposition 17.49 (i.e., $\emptyset, \mathbf{2} : \mathtt{Overt}$ and $\Sigma \subseteq \Omega_{\neg\neg}$), the following statements are equivalent.*

(i) *Phoa's principle holds.*
(ii) *Every map $\Sigma \to \Sigma$ is monotone.*
(iii) *We have $f(\top) \geq f(\bot)$ for every $f : \Sigma \to \Sigma$.*
(iv) *$\top$ is the only isolated point of Σ.*[17]

Phoa's principle implies monotonicity of maps between arbitrary topologies.

Lemma 17.57 *If Phoa's principle holds, then every map $\Sigma^X \to \Sigma^Y$ (equivalently, $\mathcal{O}(X) \to \mathcal{O}(Y)$) is monotone.*

Corollary 17.58 *If Phoa's principle holds, then every topology is overt and compact.*

[17] $\top$ is always an isolated point in Σ – it is classified by the identity on Σ.

17.3.6 Finite Approximation Principles

This subsection deals with principles involving the set of natural numbers. For the results to be of use, we need countable sets to be overt. Consequently we postulate $\emptyset, \mathbb{N} : \mathtt{Overt}$ throughout the subsection.

We will make use of the following sets:

$$\begin{aligned}
\mathbb{N}^\bullet &:= \big\{ s \in \{0,1\}^{\mathbb{N}} \;\big|\; \forall n \in \mathbb{N}.\, s_n \geq s_{n+1} \big\}, \\
\overline{\underrightarrow{\mathbb{N}}} &:= \big\{ L \in \mathcal{O}(\mathbb{N}) \;\big|\; \forall m, n \in \mathbb{N}.\, n \in L \wedge m \leq n \implies m \in L \big\}, \\
\underrightarrow{\mathbb{N}} &:= \big\{ L \in \overline{\underrightarrow{\mathbb{N}}} \;\big|\; L \text{ inhabited} \big\}.
\end{aligned}$$

The set $\mathbb{N}^\bullet$ plays the same role synthetically as the one-point compactification of $\mathbb{N}$ does classically. We treat $\mathbb{N}$ as a subset of $\mathbb{N}^\bullet$ (write a natural number in its unary notation); its complement is the singleton containing the constantly 1 sequence which we denote by ∞.

The set $\mathbb{N}^\bullet$ is isomorphic to the set of inhabited decidable lower subsets of $\mathbb{N}$; in this sense we view it as a subset of $\underrightarrow{\mathbb{N}}$ and $\overline{\underrightarrow{\mathbb{N}}}$.

We can extend the order relations $<$ and $\leq$ from $\mathbb{N}$ to $\mathbb{N}^\bullet$ by

$$s < t \;:=\; \exists n \in \mathbb{N}.\, s_n = 0 \wedge t_n = 1 \qquad \text{and} \qquad s \leq t \;:=\; \neg(t < s).$$

The relation $<$ is open and the partial order $\leq$ (for which $\mathbb{N}^\bullet$ is a lattice) is closed, which in particular implies that $\mathbb{N}^\bullet$ is Hausdorff. We do not have a meaningful open relation $<$ on $\underrightarrow{\mathbb{N}}$ or $\overline{\underrightarrow{\mathbb{N}}}$ (although they are still lattices for the partial order $\leq$, defined as the subset inclusion), nor are they Hausdorff: Phoa's principle implies that all open subsets of $\underrightarrow{\mathbb{N}}$ and $\overline{\underrightarrow{\mathbb{N}}}$ are upper.

Addition and multiplication can be extended from $\mathbb{N}$ to $\mathbb{N}^\bullet$, $\underrightarrow{\mathbb{N}}$ and $\overline{\underrightarrow{\mathbb{N}}}$ in the obvious way, making $\mathbb{N}^\bullet$ and $\underrightarrow{\mathbb{N}}$ into semirings (the algebraic structure of $\overline{\underrightarrow{\mathbb{N}}}$ is not nice; $\emptyset$ is an absorbing element for both addition and multiplication).

The sets $\underrightarrow{\mathbb{N}}$ and $\overline{\underrightarrow{\mathbb{N}}}$ allow us to measure some quantities which are classically, but not constructively, natural numbers. For example, matrix ranks are elements of $\underrightarrow{\mathbb{N}}$, and polynomial degrees are elements of $\overline{\underrightarrow{\mathbb{N}}}$ (for which the formula $\deg(p \cdot q) = \deg(p) + \deg(q)$ holds in $\overline{\underrightarrow{\mathbb{N}}}$).

We now use these sets to define certain synthetic topological principles. Some principles are simply a way to positively state that we do not have the trivial situation $\Sigma = \Omega$ where everything is discrete. Phoa's principle was an example, assuring us that $\top$ is the only isolated point of Σ (Lemma 17.56). We now consider the principle which essentially states that given a convergent sequence, its limit is an accumulation point with regard to the intrinsic topology.

Definition 17.59 The ***principle WSO*** ('weakly sequentially open') [1] states that for every $U \in \mathcal{O}(\mathbb{N}^\bullet)$, if $\infty \in U$, then there exists $n \in \mathbb{N}$ which is in U. In short:

$$\forall U \in \mathcal{O}(\mathbb{N}^\bullet).\, \infty \in U \implies U \between \mathbb{N}.$$

Proposition 17.60 *WSO holds if and only if* $\mathbb{N}$ *is dense in* $\mathbb{N}^\bullet$.

The set $\mathbb{N}^\bullet$ allows a simple characterization of some constructive principles. As mentioned, $\mathbb{N}^{\mathtt{C}} = \{\infty\}$. Markov's principle (MP) is equivalent to $\{\infty\}^{\mathtt{C}} = \mathbb{N}$, and LPO is equivalent to $\mathbb{N}$ being a decidable subset of $\mathbb{N}^\bullet$ (so $\mathbb{N}^\bullet \cong \mathbb{N}+\mathbf{1}$). Clearly LPO contradicts WSO; in fact, WSO can be seen as a positive way to state that LPO fails.

Proposition 17.61 *WSO implies ¬LPO. If* $\Sigma \subseteq \Omega_{\neg\neg}$, *the converse also holds.*

Here are some quick applications of WSO.

Proposition 17.62 *Assume countable choice and WSO (as well as* $\emptyset, \mathbb{N} : \mathtt{Overt}$ *and* $\neg\Sigma \subseteq \mathrm{Z}$, *so that we can discuss metric spaces, as per Subsection 17.3.3). Then every complete separable metric space is overt.*

Lemma 17.63 *Suppose* $\Sigma \subseteq \Omega_{\neg\neg}$. *Then WSO implies Phoa.*

The principle WSO can be weakened by replacing $\mathbb{N}^\bullet$ in the definition by the larger sets $\underrightarrow{\mathbb{N}}$ or $\overline{\underrightarrow{\mathbb{N}}}$ (as sets, they are isomorphic, so it does not matter which we take). The weaker principle is still sufficient for most applications.

Definition 17.64 The ***vergence principle*** states that for every $U \in \mathcal{O}(\underrightarrow{\mathbb{N}})$ (or equivalently $U \in \mathcal{O}(\overline{\underrightarrow{\mathbb{N}}})$), if $\infty \in U$, then there exists $L \in U$, which is ***bounded*** (there is $n \in \mathbb{N}$ with $L \subseteq \mathbb{N}_{\leq n}$).

In applications we use vergence for $\underrightarrow{\mathbb{N}}$ or for $\overline{\underrightarrow{\mathbb{N}}}$, whichever is more convenient.

Proposition 17.65 *WSO implies vergence. If* $\neg\Sigma \subseteq \mathrm{Z}$ *and* $\mathbb{N}^\bullet$ *is compact, the converse also holds.*

The true significance of the vergence and WSO principles is that they state that infinite elements can be approximated by finite ones in some sense, from which the domain theorists can recognize a connection with relative compactness [5]. In full generality of synthetic topology compactness is not saying much – for example, if $\Sigma = \Omega$, then *all* sets are compact. Vergence essentially says that compact sets in many ways behave synthetically like we are used to from classical topology. Here are some examples.

Proposition 17.66 *If the vergence principle holds, then any countable open cover of a compact set has a finite subcover.*

Recall the classically, a real-valued continuous function, defined on a compact domain, is bounded, and attains a maximum and a minimum if the domain is inhabited. Here is the synthetic version of this result.

Proposition 17.67 *Assume the vergence principle and let X be any set.*

(i) *Any compact map $X \to \mathbb{R}$ is bounded.*
(ii) *Assume $\Sigma \subseteq \Omega_{\neg\neg}$ and that X is inhabited. Then the image of any compact overt map $X \to \mathbb{R}$ has a supremum and an infimum in $\mathbb{R}$.*

Corollary 17.68 *If the vergence principle holds, then any (sub)compact subset of $\mathbb{R}^n$ is bounded.*

There are several other instances of usage of the vergence principle in Subsections 17.3.7 and 17.3.8, whereas in the remainder of this subsection we examine the connection to another principle which also posits a form of finite approximation – Scott's principle, which essentially states that $\mathcal{O}(\mathbb{N})$ has the Scott topology.

Let $\mathcal{F}(X)$ denote the set of finite[18] subsets of X. Since $\mathbb{N}$ is discrete, $\mathcal{F}(\mathbb{N}) \subseteq \mathcal{O}(\mathbb{N})$. Also, finite sets are compact, so for every $F \in \mathcal{F}(\mathbb{N})$ the set $\uparrow F := \{U \in \mathcal{O}(\mathbb{N}) \mid F \subseteq U\}$ is open in $\mathcal{O}(\mathbb{N})$, that is, it is an element of $\mathcal{O}\mathcal{O}(\mathbb{N})$.

Definition 17.69 ***Scott's principle*** states that for every $\mathcal{U} \in \mathcal{O}\mathcal{O}(\mathbb{N})$ and every $U \in \mathcal{U}$ there exists $F \in \mathcal{F}(\mathbb{N})$ such that $U \in \uparrow F \subseteq \mathcal{U}$.

In other words, Scott's principle states that the sets $\uparrow F$ form a weak basis for the topology on $\mathcal{O}(\mathbb{N})$, but we can equivalently require a stronger form of a basis.

Proposition 17.70 *If Scott's principle holds, then the map $\mathcal{F}(\mathbb{N}) \to \mathcal{O}\mathcal{O}(\mathbb{N})$, $F \mapsto \uparrow F$, is a canonical basis.*

We see that Scott's principle gives us a simple description of the second topology of $\mathbb{N}$. Let us now connect it to the previous principles.

Lemma 17.71 *The vergence principle is equivalent to the following weaker version of Scott's principle: for every $\mathcal{U} \in \mathcal{O}\mathcal{O}(\mathbb{N})$ and $U \in \mathcal{U}$ there exist $n \in \mathbb{N}$ and $V \in \mathcal{O}(\mathbb{N})$ such that $V \subseteq \mathbb{N}_{\leq n}$, $V \subseteq U$ and $V \in \mathcal{U}$.*

Proposition 17.72 *Scott's principle holds if and only if vergence and Phoa's principles hold.*

Corollary 17.73 *WSO implies Scott's principle, assuming at least one of the following:*

(i) $\Sigma \subseteq \Omega_{\neg\neg}$,
(ii) *Phoa,*
(iii) *countable choice and $\Sigma = \Sigma^0_1$.*

[18] A set X is ***finite*** when there exists $n \in \mathbb{N}$ and a surjection $\mathbb{N}_{<n} \to X$.

17.3.7 Compactness Principles

We have already discussed in Subsection 17.3.2 that generality of Σ can leave us with few overt sets, so we generally need to postulate overtness of some common sets. The situation is similar when it comes to compactness. We know that finite sets are compact (essentially by definition), but beyond that we cannot say much in general. For example, in the effective topos, if a metric space has an accumulation point (in the metric sense), it cannot be compact [8, Theorem 5.7].

Still, compactness is a very useful property, and it is convenient to postulate compactness of certain common sets. Following the example from overtness, let $X : \mathtt{Compact}$ stand for the assumption 'X is compact'. In this subsection we will consider compactness of sets which would classically be considered some of the prototypical compact topological spaces: the Cantor set $\mathbf{2}^{\mathbb{N}}$, the closed unit interval $\mathbb{I} := \mathbb{R}_{[0,1]}$, the Hilbert cube $\mathbb{I}^{\mathbb{N}}$, and complete totally bounded metric spaces in general. In order to discuss real numbers, their intervals, and metric spaces, we postulate $\emptyset, \mathbb{N} : \mathtt{Overt}$ and $\neg\Sigma \subseteq \mathrm{Z}$ for the remainder of the subsection (as per the discussion in Subsection 17.3.3).

Since we typically view $\mathbb{I}$ as a subset of $\mathbb{R}$, the first question is whether we actually want $\mathbb{I}$ to be compact, or only subcompact in $\mathbb{R}$. But there is no difference: $\mathbb{I}$ is a retract of $\mathbb{R}$ (via the map $x \mapsto \inf\{\sup\{x, 0\}, 1\}$), and retracts inherit the topology by Proposition 17.8.

Next, if $\mathbb{I}$ is compact, then so is every interval $\mathbb{R}_{[a,b]}$, where $a, b \in \mathbb{R}$, $a \leq b$. This is because $\mathbb{R}_{[a,b]}$ is a surjective image of $\mathbb{I}$ (for example, via the map $x \mapsto (1-x)a + xb$).

Proposition 17.74 *Assume $\mathbb{I} : \mathtt{Compact}$ and the vergence principle.*

(i) *The sets $\mathbb{R}$ and $\mathbb{I}$ are overt and have the euclidean topology.*
(ii) *For any $n \in \mathbb{N}$ a subset of $\mathbb{R}^n$ is subcompact if and only if it is closed and bounded.*

Proposition 17.75

(i) $\mathbb{I}^{\mathbb{N}} : \mathtt{Compact}$ *implies* $\mathbb{I} : \mathtt{Compact}$.
(ii) *Assume that $\mathrm{Z}_{\neg\neg}$ is a paradominance. Then $\mathbb{I} : \mathtt{Compact}$ implies $\mathbf{2}^{\mathbb{N}}$: $\mathtt{Compact}$.*

Classically a metric space is compact if and only if it is complete and totally bounded. We now examine a synthetic version of this theorem.

Synthetically it is not the case that a compact metric space is automatically complete. This is obvious if we allow ludicrous choices for Σ, such as $\Sigma = \Omega$, where all sets are compact. However, even assuming the vergence principle which makes compact sets behave nicely, the statement can be false. For example, if $\mathbb{I}$ is compact, then so is $\mathbb{R}_{[-1,0]} \cup \mathbb{R}_{[0,1]}$, being a union of two compact sets. This union is complete if and only if it is equal to $\mathbb{R}_{[-1,1]}$ which implies LLPO (Lesser Limited

Principle of Omniscience). However, we can prove the next best thing: a compact metric space has empty complement in its completion.

Lemma 17.76 *Assume the vergence principle. Let X and Y be metric spaces, equipped with metrics d_X and d_Y, respectively. Let $f\colon X \to Y$ be a compact map which has a metrically dense image in Y.*

(i) *The complement of the image of f in Y is empty.*
(ii) *If additionally $\Sigma \subseteq \Omega_{\neg\neg}$, then Y is compact.*

Corollary 17.77 *Assume vergence. Then any compact metric space has empty complement in its completion. More generally, any metric space, for which the completion is a compact map, has empty complement in its completion.*

Classically a metric space is totally bounded when for any $\epsilon \in \mathbb{R}_{>0}$ it can be covered by finitely many open balls with radius ϵ. Constructively this definition still works if we assume countable choice, but not in general – without choice we might not even be able to show that a totally bounded metric space is separable. To avoid this, we need to add a 'modulus of total boundedness' to the choiceless definition. The following definition of total boundedness is equivalent to the usual one in the presence of countable choice.

Definition 17.78 (Choiceless Synthetic Definition of Total Boundedness) A ***totally bounded metric space*** is a metric space X, together with maps $s\colon \mathbb{N} \to X + \mathbf{1}$ and $a\colon \mathbb{N} \to \underleftarrow{\mathbb{N}}$ such that

$$\forall n \in \mathbb{N}.\, \forall k \in a_n.\, X \subseteq \bigcup \left\{B\left(s_i, 2^{-n}\right) \mid i \in \mathbb{N}_{\leq k} \cap s^*(X)\right\}$$

(the map a is called the ***modulus of total boundedness***).

Remark 17.79 In the definition of total boundedness we restricted ourselves to radii of the form 2^{-n} which made it so that the domain of a was $\mathbb{N}$. However, the codomain $\underleftarrow{\mathbb{N}}$ allows us to extend a to all positive reals. Explicitly, if $s\colon \mathbb{N} \to X + \mathbf{1}$ and $a\colon \mathbb{N} \to \underleftarrow{\mathbb{N}}$ witness total boundedness of X, define $\tilde{a}\colon \mathbb{R}_{>0} \to \underleftarrow{\mathbb{N}}$ by

$$\tilde{a}(\epsilon) := \left\{k \in \mathbb{N} \mid \exists n \in \mathbb{N}.\, 2^{-n} < \epsilon \wedge k \in a_n\right\}.$$

Then

$$\forall \epsilon \in \mathbb{R}_{>0}.\, \forall k \in \tilde{a}(\epsilon).\, X \subseteq \bigcup \left\{B(s_i, \epsilon) \mid i \in \mathbb{N}_{\leq k} \cap s^*(X)\right\}.$$

In other words, we can equivalently define totally bounded metric spaces with this expanded modulus.

We recall some of the typical totally bounded metric spaces.

- The real numbers $\mathbb{R}$, equipped with the euclidean metric $d_E(x, y) = |x - y|$, are a complete separable metric space, though not a totally bounded one. However, the interval $\mathbb{I} = \mathbb{R}_{[0,1]}$ with the restricted euclidean metric is a complete totally bounded metric space.
- There are many reasonable metrics which make the Hilbert cube $\mathcal{H} := \mathbb{I}^{\mathbb{N}}$ into a complete totally bounded metric space, for example $d_{\mathcal{H}}(\alpha, \beta) := \sup\{2^{-n} \cdot |\alpha_n - \beta_n| \mid n \in \mathbb{N}\}$.
- The Baire space is the set $\mathbb{N}^{\mathbb{N}}$, equipped with the ***comparison metric*** $d_C \colon \mathbb{N}^{\mathbb{N}} \times \mathbb{N}^{\mathbb{N}} \to \mathbb{R}$,

$$\begin{aligned} d_C(\alpha, \beta) &:= \lim_{n \to \infty} 2^{-\inf(\{k \in \mathbb{N}_{\leq n} \mid \alpha_k \neq \beta_k\} \cup \{n\})} \\ &= \sup\big(\{2^{-n} \mid n \in \mathbb{N}, \alpha_n \neq \beta_n\} \cup \{0\}\big). \end{aligned}$$

  The Baire space is a complete separable metric space, but not totally bounded. However, we can restrict the comparison metric to $\mathbf{2}^{\mathbb{N}}$ and further to $\mathbb{N}^{\bullet}$. In both cases we get a complete totally bounded metric space.

Whenever we refer to these sets as metric spaces later in the chapter, we mean that they are equipped with the metrics, given here.

Proposition 17.80 *If X is a totally bounded metric space, then so is its completion. More precisely, if $c \colon X \to \widehat{X}$ is the completion of X and $s \colon \mathbb{N} \to X + \mathbf{1}$ and $a \colon \mathbb{N} \to \overleftarrow{\mathbb{N}}$ witness total boundedness of X, then $(c + \mathit{Id}_{\mathbf{1}}) \circ s \colon \mathbb{N} \to \widehat{X} + 1$ and $a \circ (n \mapsto n + 1) \colon \mathbb{N} \to \overleftarrow{\mathbb{N}}$ witness total boundedness of $\widehat{X}$.*

Lemma 17.81 *Assume vergence. Then any compact separable metric space is totally bounded. More generally, any separable metric space, the completion of which is a compact map, is totally bounded.*

Proposition 17.82 *For any totally bounded metric space its inhabitedness is decidable.*

Corollary 17.83 *A metric space X is totally bounded if and only if it is empty or there exist $s \colon \mathbb{N} \to X$ and $a \colon \mathbb{N} \to \overleftarrow{\mathbb{N}}$ such that*

$$\forall n \in \mathbb{N}.\, \forall k \in a_n.\, X \subseteq \bigcup\{B(s_i, 2^{-n}) \mid i \in \mathbb{N}_{\leq k}\}.$$

Proposition 17.84 *Let X be an inhabited totally bounded metric space with a metric d. Then its* **diameter** $\operatorname{diam}(X) := \sup d_*(X \times X)$ *is a real number.*

Theorem 17.85 *Assume vergence and at least one of the following:*

(i) $\mathbb{I}^{\mathbb{N}} : \mathtt{Compact}$ *and $\mathrm{Z}_{\neg\neg}$ is a paradominance, or*
(ii) $\mathbf{2}^{\mathbb{N}} : \mathtt{Compact}$ *and countable choice holds, or*
(iii) $\mathbf{2}^{\mathbb{N}} : \mathtt{Compact}$, *$\Sigma \subseteq \Omega_{\neg\neg}$ and $\mathrm{Z}_{\neg\neg}$ is a paradominance.*

Then a complete separable metric space is compact if and only if it is totally bounded.

The following version of this result is closer to the classical formulation that a metric space is compact if and only if it is complete totally bounded.

Corollary 17.86 *Assume vergence,* $\mathbf{2}^{\mathbb{N}} : \texttt{Compact}$*,* $\Sigma \subseteq \Omega_{\neg\neg}$ *and at least one of the following: countable choice or the paradominance axiom. Then the following statements are equivalent for every separable metric space* X*:*

(i) X *is subcompact in its completion* $\widehat{X}$ *(more precisely, the completion* $X \to \widehat{X}$ *is a compact map),*
(ii) X *is totally bounded and has empty complement in its completion.*

17.3.8 Metrization Principles

In this subsection we discuss when the topology induced by a metric is the same as the intrinsic topology. Because we use metric spaces, we postulate $\emptyset, \mathbb{N} : \texttt{Overt}$ and $\neg\Sigma \subseteq \mathrm{Z}$ throughout the subsection, as per discussion in Subsection 17.3.3.

Definition 17.87 (Metrization) Let X be a metric space and $B \colon X \times \mathbb{R} \to \mathcal{O}(X)$ the map which takes any pair $(x, r) \in X \times \mathbb{R}$ to the open ball $B(x, r)$. We say that X is (***weakly***, ***canonically***) ***metrized*** when B is a (weak, canonical) basis.

Let us unpack this definition. Consider the following commutative diagram.

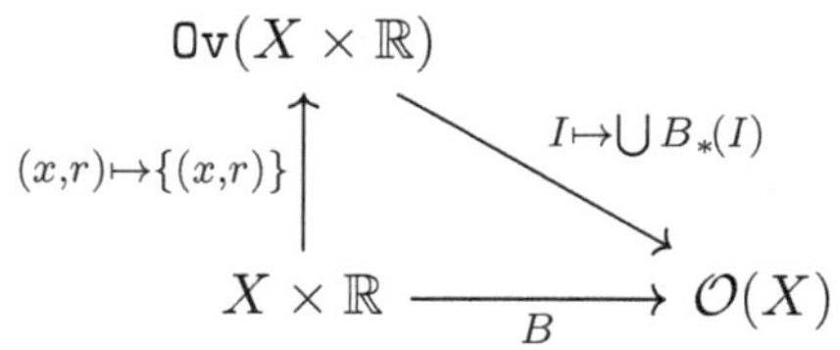

We call elements of the image of the slanted map ***metrically open*** subsets of X. Let $\mathcal{M}(X)$ denote the set of all metrically open subsets of X; call it the ***metric topology*** of X. We always have $\mathcal{M}(X) \subseteq \mathcal{O}(X)$; a metric space X is metrized when $\mathcal{M}(X) = \mathcal{O}(X)$, that is, when the slanted map is a surjection. Similarly, X is canonically metrized when the slanted map is a split surjection (to get a diagrammatic version for weak metrization, replace $\texttt{Ov}$ with $\mathcal{P}$).

Proposition 17.88 *Every metrized metric space is overt (and in particular quasi-overt).*

Proposition 17.89 *A retract of a (weakly, canonically) metrized metric space is (weakly, canonically) metrized.*

Proposition 17.90 *Any overt set with decidable equality is canonically metrized by the discrete metric.*

Hence the sets $\mathbb{N}$ and $\mathbb{N}_{<n}$ (for all $n \in \mathbb{N}$) are metrized by the discrete metric, and obviously also by metrics equivalent to it, such as the euclidean metric. This also makes it clear that it cannot happen that all metric spaces are metrized – there are sets that can be equipped with non-equivalent metrics. For instance, $\mathbb{Q}$ is isomorphic to $\mathbb{N}$ as a set, so it is also metrized by the discrete metric, but is not metrized by the euclidean metric. On the other hand, $\mathbb{R}$ may be metrized by its euclidean metric (see Proposition 17.74), but it might also not be, as in the effective topos (see [8, Theorem 5.7]).

It is useful to have some sufficient conditions that certain metric spaces are metrized, as that gives us a simple description of their topology. The following theorem is a significant generalization of both Propositions 17.90 and 17.74.

Theorem 17.91 *Assume vergence. Let X be a quasi-overt metric space, for which there exists an overt map $I \to X \times \mathbb{R}$, $i \mapsto (x_i, r_i)$ which covers X in the sense that $X \subseteq \bigcup_{i \in I} B(x_i, r_i)$, and additionally $\overline{B}(x_i, r_i)$ is subcompact in X for every $i \in I$.*[19] *Then X is canonically metrized.*

Corollary 17.92 *Any compact quasi-overt metric space is canonically metrized.*

Consequently, if $\mathbf{2}^{\mathbb{N}}$ or $\mathbb{I}^{\mathbb{N}}$ is compact and some other assumptions hold (recall Theorem 17.85), then all complete totally bounded metric spaces are metrized. However, what if we are in a setting where the Cantor set and the Hilbert cube are not compact – or if they are, but we are looking for a way to prove that? In that case it is good to have a more direct way to verify compactness and metrization.

Definition 17.93 A metric space X is ***Lebesgue*** when, for every overt map $I \to X \times \mathbb{R}$, $i \mapsto (x_i, r_i)$, if $X \subseteq \bigcup_{i \in I} B(x_i, r_i)$, then there exists $\epsilon \in \mathbb{R}_{>0}$ such that $X \subseteq \bigcup_{i \in I} B(x_i, r_i - \epsilon)$ (equivalently, there exists $n \in \mathbb{N}$ such that $X \subseteq \bigcup_{i \in I} B(x_i, r_i - 2^{-n})$).

In short, a metric space is Lebesgue when, in every overt cover by open balls, we can shrink the radii by some amount and it remains a cover. The reason for the name is that this is essentially saying that overt covers by balls have a Lebesgue number. Also note that to verify that a metric space is Lebesgue, we do not need to check the condition for all overt maps into $X \times \mathbb{R}$ (which do not form a set), it suffices to check the inclusions of subovert subsets of $X \times \mathbb{R}$.

Theorem 17.94 *Assume vergence. Then a separable metric space X is compact if and only if it is totally bounded, Lebesgue, and metrized.*

[19] In classical topology the existence of such a map is equivalent to X being locally compact. The theorem therefore essentially says that quasi-overt locally compact metric spaces are metrized.

Corollary 17.95 *The following statements are equivalent.*

(i) $\mathbb{N}^\bullet$ *is compact and WSO holds.*
(ii) $\mathbb{N}^\bullet$ *is compact and the vergence principles holds.*
(iii) $\mathbb{N}^\bullet$ *is canonically metrized.*
(iv) $\mathbb{N}^\bullet$ *is metrized.*

If Σ *is a dominance, these statements are further equivalent to the following ones.*

(v) $\mathbb{N}^\bullet$ *is weakly metrized.*
(vi) *For every* $U \in \mathcal{O}(\mathbb{N}^\bullet)$*, if* $\infty \in U$*, then there exists* $n \in \mathbb{N}$ *such that* $\mathbb{N}^\bullet_{\geq n} \subseteq U$*.*

Corollary 17.96 *The following statements are equivalent.*

(i) $\mathbf{2}^{\mathbb{N}}$ *is compact and the vergence principle holds.*
(ii) $\mathbf{2}^{\mathbb{N}}$ *is metrized and every subovert bar is uniform.*

If countable choice holds and $\Sigma = \Sigma^0_1$*, then 'every subovert bar is uniform' can be replaced by the Brouwer's Fan principle 'every decidable bar is uniform'.*[20]

Proposition 17.89 tells us that metrization transfers onto retracts. We can generalize this in two ways – transfer of metrization via quotients and via embeddings.

The idea is the following. Metrization is the match between the intrinsic and the metric topology. Surjections and quotients preserve the intrinsic topology (Proposition 17.5); those that also preserve the metric topology will allow the transfer of metrization from a metric space to its quotient. On the other hand, subsets of a metric space inherit the metric; if they also inherit the intrinsic topology, we can have the transfer of metrization from a metric space to its subspace. We now make this precise.

Definition 17.97 Let X and Y be metric spaces.

- A map $f\colon X \to Y$ is ***metrically continuous*** when for every metrically open $V \subseteq Y$ its preimage $f^*(V)$ is metrically open in X, that is, when f induces a well-defined map $\mathcal{M}(f)\colon \mathcal{M}(Y) \to \mathcal{M}(X)$, $\mathcal{M}(f)(V) := f^*(V)$.
- A map $f\colon X \to Y$ is ***metric semiquotient*** when for every $V \subseteq Y$ the implication $f^*(V) \in \mathcal{M}(X) \implies V \in \mathcal{M}(Y)$ holds.
- A map $f\colon X \to Y$ is ***metric quotient*** when it is surjective, metrically continuous and metric semiquotient.

Lemma 17.98 (Transfer of metrization) *Let* $f\colon X \to Y$ *be a map between metric spaces* X *and* Y*.*

(i) *If* f *is a metric quotient map (or at least metric semiquotient), then metrization of* X *implies metrization of* Y*.*

[20] For Brouwer's Fan principle and bars in general, see [13].

(ii) *If f is an isometry, X is quasi-overt and $f_*(X)$ inherits the intrinsic topology from Y, then metrization of Y implies metrization of X.*

If follows from this lemma that a principle positing metrization of a particular metric space can have far-reaching consequences, with many other metric spaces being metrized as well. In particular, we are interested when the classes of complete separable metric spaces and complete totally bounded metric spaces are metrized.

Theorem 17.99 *Assume countable choice.*

(i) *If the Baire space $\mathbb{N}^{\mathbb{N}}$ is metrized, then all complete separable metric spaces are metrized.*
(ii) *If the Cantor space $\mathbf{2}^{\mathbb{N}}$ is metrized, then all complete totally bounded metric spaces are metrized.*

Theorem 17.100 *Assume the paradominance axiom.*

(i) *If the Urysohn universal metric space $\mathbb{U}$ is metrized, then all complete separable metric spaces are metrized.*
(ii) *If the Hilbert cube $\mathbb{I}^{\mathbb{N}}$ is metrized, then all complete totally bounded metric spaces are metrized.*

References

[1] Bauer, A., and Lešnik, D. 2012. Metric spaces in synthetic topology. *Ann. Pure Appl. Logic*, **163**(2), 87–100.
[2] Bauer, A., and Taylor, P. 2009. The Dedekind reals in abstract Stone duality. *Mathematical Structures in Computer Science (MSCS)*, **19**(4), 757.
[3] Caramello, O. 2017. Topos-theoretic background. In: *Theories, Sites, Toposes*. Oxford: Oxford University Press.
[4] Escardó, M., and Heckmann, R. 2001. Topologies on spaces of continuous functions. *Topology Proc.*, **26**(2), 545–564.
[5] Gierz, G., Hofmann, K. H., Keimel, K., Lawson, et al. 2003. *Continuous Lattices and Domains*, vol. 93. Cambridge: Cambridge University Press.
[6] Hyland, J., and Martin E. 1982. The effective topos. Pages 165–216 of: *Studies in Logic and the Foundations of Mathematics*, vol. 110. Amsterdam: Elsevier.
[7] Johnstone, P. T. 2002. *Sketches of an Elephant: A Topos Theory Compendium – Volume 1*. Oxford: Clarendon Press.
[8] Lešnik, D. 2010. Synthetic topology and constructive metric spaces. Ph.D. thesis, University of Ljubljana.
[9] Lubarsky, R. S. 2007. On the Cauchy completeness of the constructive Cauchy reals. *Math. Logic. Q.*, **53**(4–5), 396–414.

[10] Mac Lane, S., and Moerdijk, I. 1992. *Sheaves in Geometry and Logic: A First Introduction to Topos Theory*. Berlin: Springer-Verlag.

[11] Richman, F. 2000. The fundamental theorem of algebra: a constructive development without choice. *Pacific J. Math.*, **196**(1), 213–230.

[12] Rosolini, G. 1986. Continuity and effectiveness in topoi. Ph.D. thesis, University of Oxford.

[13] Troelstra, A. S., and van Dalen, D. 1988. *Constructivism in Mathematics: An Introduction, Vol. I*. Volume 121 of: Studies in Logic and the Foundations of Mathematics. Amsterdam: Elsevier.

[14] Van Oosten, J. 2008. *Realizability: An Introduction to its Categorical Side*. Amsterdam: Elsevier.

18

Apartness on Lattices and Between Sets

Douglas Bridges

18.1 Introduction

Although Errett Bishop, in his ground-breaking work [1], expressed doubts about the potential for development of a constructive counterpart of classical topology, subsequent research has produced several approaches revealing substantial constructive content in that branch of mathematics. In Chapter 20 of this handbook, Maietti and Sambin describes their and others' research on point-free formal topology, and in Chapter 14 Petrakis presents his work on function spaces.[1] The approach to topology that we outline in this chapter is based on axioms for a relation of apartness between sets. The underlying idea is to abstract the essential qualities of apartness from the intuition that two subsets A, B of a metric space (X, ρ) should be considered apart if there exists $r > 0$ such that $\rho(x, y) \geq r$ for all $x \in A$ and $y \in B$. These essentials are captured in the following definitions.

Let X be an inhabited set (that is, it contains at least one element), with a binary relation $\neq_X$ of inequality satisfying the conditions:[2]

$$x \neq_X y \Rightarrow \mathsf{not}(x = y),$$
$$x \neq_X y \Rightarrow y \neq_X x.$$

We then define the **(set-theoretic) complement** of a subset S of X to be

$$\sim S \equiv \{x \in X : \forall_{s \in S}(x \neq_X s)\}, \tag{18.1}$$

Let $\bowtie$ be a binary relation on the power set[3] $P(X)$ of X, and for each $S \subset X$ define

$$-S \equiv \{x \in X : \{x\} \bowtie S\}.$$

[1] For more information on various approaches to constructive topology, see, for example, [13, 14, 19, 23, 24, 27].

[2] In order to avoid confusion with the lattice-theoretic symbols '$\wedge$' and '$\vee$', we shall use 'and', 'or', and 'not' to denote logical conjunction, disjunction, and negation, respectively.

[3] At the risk of upsetting some of our constructivist colleagues, we use the term *power set*, rather than *power class*. As presented, our theory is impredicative (but cf. [15]). For ways of rendering it predicative, the reader should consult [9, Section 2.6].

We call $\bowtie$ a **pre-apartness** on X if it satisfies the following axioms:

B1 $X \bowtie \varnothing$
B2 $-A \subset \sim A$
B3 $(A_1 \cup A_2) \bowtie (B_1 \cup B_2) \Leftrightarrow \forall_{i,j \in \{1,2\}} A_i \bowtie B_j$
B4 $-A \subset \sim B \Rightarrow -A \subset -B.$

We then call $(X, \bowtie)$, or, loosely, X itself), or, loosely, X itself, **pre-apartness space**, and $-S$ the **apartness complement** of the subset S of X. If, in addition, $\bowtie$ satisfies

B5 $x \in -A \Rightarrow \exists_{S \subset X}(x \in -S \text{ and } X = -A \cup S),$

we call it an **apartness** and X an **apartness space.**

In [9], we studied such relations in considerable detail, including their connection to the theory of uniform spaces. In the first half of this chapter we focus mostly on an abstraction from set–set apartness to apartness on a certain type of lattice. In Sections 18.2 and 18.3 we describe the fundamentals of lattices, frames, and apartness on frames. In Sections 18.4 and 18.5 we introduce topological structures related to apartness on frames, and some resulting notions of continuity. In Sections 18.6 and 18.7 we examine set–set apartness as a special case of apartness on frames, dealing there with uniform spaces and outlining some of the technical problems in relating the apartness-space notion of continuity with uniform continuity; and in Section 18.8 we discuss notions of compactness and precompactness in set–set apartness theory.

18.2 Lattices

Assuming some familiarity with the basics of lattice theory, as found in [18, Chapter 8], we consider a bounded, distributive lattice $\mathfrak{L}$ with top 1, bottom 0, join $\vee$, meet $\wedge$, and the corresponding partial order $\leqslant$ given by $x \leqslant y$ if and only if $x \wedge y = x$ or, equivalently, $x \vee y = y$.

By a **complementation function** on $\mathfrak{L}$ we mean a mapping $x \rightsquigarrow \sim x$ of $\mathfrak{L}$ into itself with the following properties:

C1 $x \leqslant \sim\sim x$
C2 $x \leqslant \sim y \Rightarrow x \wedge y = 0$
C3 $\sim(x \vee y) = \sim x \wedge \sim y.$

We say that $\sim x$ is the **complement** of the element $x \in \mathfrak{L}$ and that, taken with the complementation function, $\mathfrak{L}$ is a **complemented lattice.** If also

$$\forall_{x,y \in \mathfrak{L}}(x \wedge y = 0 \Rightarrow x \leqslant y),$$

then $\mathfrak{L}$ is said to be **strongly complemented.**

For elements x, y of a complemented lattice $\mathfrak{L}$ we have

$$x \leqslant y \Rightarrow \sim y \leqslant \sim x.$$

For if $x \leqslant y$, then $\sim y = \sim(x \vee y) = \sim x \wedge \sim y$, by **C3**. Moreover, by **C1** and **C2**, $x \wedge \sim x = 0$, so $\sim 1 = 1 \wedge \sim 1 = 0$; and, by **C1**, $1 \leqslant \sim\sim 1 \leqslant 1$, so $1 = \sim\sim 1$. Since $x \wedge y \leqslant x$, we have $\sim x \leqslant \sim(x \wedge y)$; likewise, $\sim y \leqslant \sim(x \wedge y)$, so $\sim x \vee \sim y \leqslant \sim(x \wedge y)$. Thus

$$1 = \sim 0 \vee 1 = \sim 0 \vee \sim\sim 1 \leqslant \sim(0 \wedge \sim 1) = \sim 0 \leqslant 1$$

and therefore $\sim 0 = 1$. Hence

$$\sim\sim(x \vee \sim x) = \sim(\sim x \wedge \sim\sim x) = \sim 0 = 1.$$

Lastly, $x \leqslant \sim x \vee y$, then $x \leqslant y$: for then, by distributivity,

$$x = x \wedge (\sim x \vee y) = (x \wedge \sim x) \vee (x \wedge y) = 0 \vee (x \wedge y) = x \wedge y.$$

The reader is invited to prove that

$$\forall_{x \in \mathfrak{L}}(x = \sim\sim x) \Leftrightarrow \forall_{x \in \mathfrak{L}}(x \vee \sim x = 1).$$

Our motivating examples of a bounded, complemented, distributed lattice are constructed as follows. Let X be an inhabited set with inequality relation $\neq_X$, and let $\mathfrak{L}$ be a set of subsets of X that contains $\varnothing$ and X and is closed under binary unions and intersections. Then $\mathfrak{L}$ becomes a bounded, distributive lattice with $\varnothing$, X, union, intersection, and $\subset$ playing, respectively, the roles of $0, 1, \vee, \wedge$ and the corresponding partial order. We have two simple cases of this construction:

- the **power set model**, in which $\mathfrak{L}$ is the entire power set $P(X)$ of X;
- the **metspace model**, in which X is equipped with a metric ρ, the inequality on X is defined by

  $$x \neq_X y \text{ if and only if } \rho(x, y) > 0,$$

  and $\mathfrak{L}$ is $P(X)$.

A related case is the **topspace model**, where X is equipped with a topology, $\mathfrak{L}$ is the set of all open subsets of X, and the meet of two elements A, B of $\mathfrak{L}$ is not simply their intersection but its *interior*. In future, we shall refer to these three models together as the **standard model trilogy**. The **standard complementation function**

- in the power set and metspace models maps $S \in P(X)$ to its set-theoretic complement;
- in the topspace model maps an open subset of X to the *interior* of its set-theoretic complement.

It is easily verified that in each case the complementation function satisfies **C1–C3**.

Taking the metspace model in which $X = \mathbf{R}$, with the standard inequality relation defined by

$$x \neq_{\mathbf{R}} y \text{ if and only if } |x - y| > 0,$$

we readily see that we cannot prove constructively either side of (18.1); nor can we prove that $\sim(x \wedge y) = \sim x \vee \sim y$.

Next, we define the **join and meet of an inhabited family** $(x_i)_{i \in I}$ of elements of a bounded lattice $\mathfrak{L}$ in the standard way:

$$x = \bigvee_{i \in I} x_i \text{ if and only if } \forall_{i \in I} (x_i \leqslant x) \text{ and } \forall_y (\forall_{i \in I} (x_i \leqslant y) \Rightarrow x \leqslant y),$$
$$x = \bigwedge_{i \in I} x_i \text{ if and only if } \forall_{i \in I} (x \leqslant x_i) \text{ and } \forall_y (\forall_{i \in I} (y \leqslant x_i) \Rightarrow y \leqslant x).$$

These elements need not exist when I is an infinite index set. Note that

$$\forall_{i \in I} (\bigwedge_{i \in I} x_i \leqslant x_i \leqslant \bigvee_{i \in I} x_i)$$

whenever the appropriate elements exist.

Let I be an inhabited subset of $\mathfrak{L}$, and for each $i \in I$ let $x_i = i$. We define $\bigvee I$ (respectively, $\bigwedge I$) to be $\bigvee_{i \in I} x_i$ (respectively, $\bigwedge_{i \in I} x_i$), if it exists. For example, if $y \in \mathfrak{L}$, then

$$I = \{y \wedge z : z \in \mathfrak{L} \text{ and } y \wedge z = 0\} = \{i \in \mathfrak{L} : \exists z \in \mathfrak{L}(i = y \wedge z \text{ and } y \wedge z = 0)\},$$

is inhabited and $\bigvee I = \bigvee_{i \in I} x_i$, where for each $i \in I$, there exists $z \in \mathfrak{L}$ such that $x_i = i = y \wedge z$ and $y \wedge z = 0$; thus each $x_i = 0$, and therefore $\bigvee I = 0$.

We say that a lattice $\mathfrak{L}$ is **complete** if the join exists for any inhabited family $(x_i)_{i \in I}$ of elements of $\mathfrak{L}$, in which case $\bigwedge_{i \in I} x_i$ exists and equals

$$\bigvee \{x \in \mathfrak{L} : \forall_{i \in I} (x \leqslant x_i)\}$$

and for each $x \in \mathfrak{L}$,

$$x = \bigvee \{z \in \mathfrak{L} : z \leqslant x\}.$$

The models in the standard trilogy are complete.

If $\mathfrak{L}$ is complete, and

$$x \wedge \bigvee_{i \in I} x_i = \bigvee_{i \in I} (x \wedge x_i)$$

for all $x \in \mathfrak{L}$ and all inhabited families $(x_i)_{i \in I}$ of elements of $\mathfrak{L}$, we say that $\wedge$ is **infinitely distributive** over $\bigvee$; in which case, $\mathfrak{L}$ is distributive.

Proposition 18.1 *Let $(x_i)_{i \in I}$ and $(y_j)_{j \in J}$ be inhabited families of elements of a complete lattice $\mathfrak{L}$. Then*

$$\bigvee_{i \in I} x_i \vee \bigvee_{j \in J} y_j = \bigvee_{i \in I, j \in J} (x_i \vee y_j) \text{ and } \bigvee_{i \in I, j \in J} (x_i \wedge y_j) \leqslant \bigvee_{i \in I} x_i \wedge \bigvee_{j \in J} y_j.$$

Dually,

$$\bigwedge_{i\in I} x_i \wedge \bigwedge_{j\in J} y_j = \bigwedge_{i\in I, j\in J}(x_i \wedge y_j) \text{ and } \bigwedge_{i\in I} x_i \vee \bigwedge_{j\in I} y_j \leqslant \bigwedge_{i\in I, j\in J}(x_i \vee y_j).$$

Proof We prove only the first of the dual set of formulae. Fixing j_0 in J, we see

$$\bigwedge_{i\in I, j\in J}(x_i \wedge y_j) \leqslant x_k \wedge y_{j_0} \leqslant x_k$$

for all $k \in I$ and therefore, by the definition of '$\bigwedge$', that $\bigwedge_{i\in I, j\in J}(x_i \wedge y_j) \leqslant \bigwedge_{i\in I} x_i$. Likewise, $\bigwedge_{i\in I, j\in J}(x_i \wedge y_j) \leqslant \bigwedge_{j\in J} y_j$, so

$$\bigwedge_{i\in I, j\in J}(x_i \wedge y_j) \leqslant \bigwedge_{i\in I} x_i \wedge \bigwedge_{j\in J} y_j.$$

On the other hand, for all $i' \in I$ and $j' \in J$ we have $\bigwedge_{i\in I} x_i \leqslant x_{i'}$ and $\bigwedge_{j\in J} y_j \leqslant y_{j'}$; whence $\bigwedge_{i\in I} x_i \wedge \bigwedge_{j\in J} y_j \leqslant x_{i'} \wedge y_{j'}$. It follows from the definition of '$\bigwedge$' that

$$\bigwedge_{i\in I} x_i \wedge \bigwedge_{j\in J} y_j \leqslant \bigwedge_{i\in I, j\in J}(x_i \wedge y_j).$$

Hence

$$\bigwedge_{i\in I} x_i \wedge \bigwedge_{j\in J} y_j = \bigwedge_{i\in I, j\in J}(x_i \wedge y_j).$$ □

Proposition 18.2 *Let $(x_i)_{i\in I}$ be a family of elements of a complete, bounded, complemented lattice $\mathfrak{L}$. Then $\sim\bigvee_{i\in I} x_i \leqslant \bigwedge_{i\in I} \sim x_i$ and $\bigvee_{i\in I} \sim x_i \leqslant \sim\bigwedge_{i\in I} x_i$.*

Proof For each $k \in I$, since $x_k \leqslant \bigvee_{i\in I} x_i$, we have $\sim\bigvee_{i\in I} x_i \leqslant \sim x_k$. It follows from the definition of '$\bigwedge$' that $\sim\bigvee_{i\in I} x_i \leqslant \bigwedge_{i\in I} \sim x_i$. The proof of the second part of the proposition is similar. □

Each element x of a complete, bounded lattice $\mathfrak{L}$ has a **logical complement**, or **pseudocomplement,**

$$\neg x \equiv \bigvee \{z \in \mathfrak{L} : z \wedge x = 0\},$$

which is the unique element u of $\mathfrak{L}$ such that

$$u \wedge x = 0 \text{ and } \forall_{y,z\in\mathfrak{L}}(z \wedge x = 0 \Rightarrow z \leqslant y) \Rightarrow u \leqslant y.$$

Then $\neg 0 = 1$, $\neg 1 = 0$, and $\neg x \leqslant \sim x$. Also, if $x \leqslant y$ and $z \wedge y = 0$, then $z \wedge x = 0$, from which it follows that $\neg y \leqslant \neg x$.

In the standard model trilogy, the logical complement of a subset S of the ambient set X is

$$\neg S = \{x \in X : \forall_{s\in S} \operatorname{not}(x = s)\}.$$

Classically, the logical complement of x is the unique element x' of $\mathfrak{L}$ such that $x \wedge x' = 0$ and $x \vee x' = 1$. In the metspace model with $X = \mathbf{R}$ we cannot prove that $x \vee \neg x = 1$, or that $\neg x = \sim x$, or that $x \vee y = \neg(\neg x \wedge \neg y)$.

By a **frame** we mean a complete, bounded, complemented lattice $\mathfrak{L}$ such that $\wedge$ is infinitely distributive over $\bigvee$. The models in the standard trilogy are frames.

Proposition 18.3 *Let $\mathfrak{L}$ be a frame. Then $x \rightsquigarrow \neg x$ is a strong complementation function on $\mathfrak{L}$.*

Proof Let $x, y \in \mathfrak{L}$. Using infinite distributivity, we obtain

$$\begin{aligned} x \leqslant \neg y \Rightarrow x &\leqslant \bigvee \{z \in \mathfrak{L} : z \wedge y = 0\} \\ &\Rightarrow x \wedge y \leqslant y \wedge \bigvee \{z \in \mathfrak{L} : z \wedge y = 0\} \\ &\Rightarrow x \wedge y \leqslant \bigvee \{y \wedge z : z \in \mathfrak{L} \text{ and } z \wedge y = 0\} \\ &\Rightarrow x \wedge y \leqslant 0 \\ &\Rightarrow x \wedge y = 0. \end{aligned}$$

Thus **C2** holds for '$\neg$'. On the other hand, it follows immediately from the definition of '$\neg$' that if $x \wedge y = 0$, then $x \leqslant \neg y$. As

$$\begin{aligned} x \wedge \neg x &= x \wedge \bigvee \{z \in \mathfrak{L} : z \wedge x = 0\} \\ &= \bigvee \{x \wedge z : z \in \mathfrak{L} \text{ and } z \wedge x = 0\} = 0, \end{aligned}$$

it follows that $x \leqslant \neg\neg x$, so **C1** holds. For **C3** we first note that $x \leqslant x \vee y$ and therefore $\neg(x \vee y) \leqslant \neg x$; likewise, $\neg(x \vee y) \leqslant \neg y$, so $\neg(x \vee y) \leqslant \neg x \wedge \neg y$. On the other hand, for all $z \leqslant \neg x \wedge \neg y$, we have $z \wedge x = 0$ and $z \wedge y = 0$, so

$$z \wedge (x \vee y) = (z \wedge x) \vee (z \wedge y) = 0$$

and therefore $z \leqslant \neg(x \vee y)$. It follows from this that

$$\begin{aligned} \neg x \wedge \neg y &= \bigvee \{z \in \mathfrak{L} : z \leqslant \neg x \wedge \neg y\} \\ &\leqslant \bigvee \{z \in \mathfrak{L} : z \leqslant \neg(x \vee y)\} = \neg(x \vee y). \end{aligned}$$

Thus $\neg x \wedge \neg y = \neg(x \vee y)$. □

Proposition 18.4 *Let $\mathfrak{L}$ be a strongly complemented frame. Then $\sim\bigvee_{i \in I} x_i = \bigwedge_{i \in I} \sim x_i$ for all inhabited families $(x_i)_{i \in I}$ in $\mathfrak{L}$.*

Proof For each $k \in I$,

$$x_k \wedge \bigwedge_{i \in I} \sim x_i \leqslant x_k \wedge \sim x_k = 0$$

and therefore $x_k \wedge \bigwedge_{i \in I} \sim x_i = 0$. Applying infinite distributivity, we obtain

$$0 = \bigvee_{k \in I} (x_k \wedge \bigwedge_{i \in I} \sim x_i) = \bigvee_{k \in I} x_k \wedge \bigwedge_{i \in I} \sim x_i$$

Since $\mathfrak{L}$ is strongly complemented, and in view of Proposition 18.2,

$$\bigwedge_{i \in I} \sim x_i \leqslant \sim\bigvee_{i \in I} x_i \leqslant \bigwedge_{i \in I} \sim x_i,$$

from which the result follows. □

Corollary 18.5 *If $(x_i)_{i \in I}$ is an inhabited family of elements of a frame $\mathfrak{L}$, then $\neg \bigvee_{i \in I} x_i = \bigwedge_{i \in I} \neg x_i$.*

Proof Apply Propositions 18.3 and 18.4. □

A bounded lattice $\mathfrak{L}$ may be equipped with a unary **habitation relation**, denoted by hab, whose axiomatic requirements mirror those of inhabitedness for sets:

H1 $\mathsf{hab}(x) \Rightarrow \mathsf{not}(x = 0)$.
H2 $\mathsf{hab}(x \wedge y) \Rightarrow \mathsf{hab}(x)$ and $\mathsf{hab}(y)$.
H3 The **join-existential property:** For any family $(x_i)_{i \in I}$ of elements of $\mathfrak{L}$, if $\bigvee_{i \in I} x_i$ exists and $\mathsf{hab}(\bigvee_{i \in I} x_i)$, then there exists $i \in I$ such that $\mathsf{hab}(x_i)$.
H4 $\mathsf{hab}(1)$.

In that case, we say that $\mathfrak{L}$ is **habitive** and that an element x of $\mathfrak{L}$ satisfying $\mathsf{hab}(x)$ is **inhabited.**[4] By **H2**,

$$x \leqslant y \Rightarrow x = x \wedge y \Rightarrow (\mathsf{hab}(x) \Rightarrow \mathsf{hab}(x \wedge y) \Rightarrow \mathsf{hab}(y)). \tag{18.2}$$

We define the relation $\neq_{\mathfrak{L}}$ on $\mathfrak{L}$ by

$$x \neq_{\mathfrak{L}} y \text{ if and only if } \mathsf{hab}(x \wedge {\sim} y) \text{ or } \mathsf{hab}({\sim} x \wedge y).$$

Since

$$x = y \text{ and } x \neq_{\mathfrak{L}} y \Rightarrow \mathsf{hab}(x \wedge {\sim} x) \Rightarrow \mathsf{hab}(0) \Rightarrow \mathsf{not}(0 = 0),$$

we see from **H1** that if $x \neq_{\mathfrak{L}} y$, then $\mathsf{not}(x = y)$. Thus $\neq_{\mathfrak{L}}$ is an inequality, the **standard lattice inequality**, on $\mathfrak{L}$. Also,

$$\begin{aligned} x \neq_{\mathfrak{L}} 0 &\Leftrightarrow \mathsf{hab}(x \wedge {\sim} 0) \text{ or } \mathsf{hab}({\sim} x \wedge 0) \\ &\Leftrightarrow \mathsf{hab}(x \wedge 1) \text{ or } \mathsf{hab}(0) \\ &\Leftrightarrow \mathsf{hab}(x), \end{aligned}$$

so, by **H4**, $1 \neq_{\mathfrak{L}} 0$.

In the standard model trilogy we have a natural habitation relation defined by $\mathsf{hab}(U)$ if and only if $U \subset X$ is inhabited in the usual set-theoretic sense.

The reader may wonder why, in view of Proposition 18.3 and when dealing with a frame $\mathfrak{L}$, we bother with a complementation function $\sim$ other than $\neg$. The reason is that in a set X the given inequality relation $\neq_X$ will not usually be the same as, in fact is usually stronger than, the **denial inequality** $\neq^D$ defined by[5]

$$x \neq^D y \text{ if and only if } \mathsf{not}(x = y).$$

4 The habitation relation corresponds to the notion of *openness* in a locale see [19, 27] and to that of *positivity* in formal topology [23, 24].

5 It is well known that the denial inequality and the metric inquality on a metric space are the same if and only if Markov's principle is accepted.

In consequence, although $\forall_{x\in\mathfrak{L}}(\neg x \leqslant \sim x)$ always holds, in general $\forall_{x\in\mathfrak{L}}(\neg x = \sim x)$ does not. These matters are significant for the theory of pre-apartness.

From now on we write $\neq$ rather than $\neq_{\mathfrak{L}}$ when it is clear which lattice $\mathfrak{L}$ we are dealing with.

18.3 Apartness on Frames

Acknowledging the importance and success of locale theory and formal topology as constructive foundations for topology, we now introduce something not found in standard presentations such as [19, 23, 24, 27].[6]

Let $\mathfrak{L}$ be a *habitive* frame and $\bowtie$ a binary relation $\mathfrak{L}$, and define an associated unary function $-$ on $\mathfrak{L}$ by

$$-x \equiv \bigvee \{y \in \mathfrak{L} : y \bowtie x\} .$$

For $\bowtie$ to be a **frame pre-apartness** we require that the following axioms hold:

AL1 $1 \bowtie 0$.
AL2 $-x \leqslant \sim x$.
AL3 $(x_1 \vee x_2) \bowtie (y_1 \vee y_2) \Leftrightarrow \forall_{i,j\in\{1,2\}} (x_i \bowtie y_j)$.

If $x \bowtie y$, we say that x and y are **apart**, and we call $-x$ the **apartness complement** of x. A habitive frame $\mathfrak{L}$ that carries a frame pre-apartness is called a **pre-apartness frame**.

Proposition 18.6 *The following hold in a pre–apartness frame* $\mathfrak{L}$.

(i) $-0 = 1$ *and* $-1 = 0$.
(ii) *If* $x_1 \leqslant y_1, x_2 \leqslant y_2$, *and* $y_1 \bowtie y_2$, *then* $x_1 \bowtie x_2$.
(iii) $-(x \vee y) = -x \wedge -y$.
(iv) $x \leqslant y \Rightarrow -y \leqslant -x$.

Proof For the proofs of (i)–(iii) see [8, Proposition 2]. For (iv), using (iii), we have

$$x \leqslant y \Rightarrow y = x \vee y \Rightarrow -x \wedge -y = -(x \vee y) = -y \Rightarrow -y \leqslant -x. \qquad \square$$

Everything we have done so far with lattices is point-free. We now introduce our counterpart to the notion of a point: an **atom** of a habitive lattice $\mathfrak{L}$ is an element x such that

$$x \neq 0 \text{ and } \forall_y (0 \neq y \leqslant x \Rightarrow y = x) .$$

We denote by $\mathsf{at}(\mathfrak{L})$ the set of all atoms of $\mathfrak{L}$, and we call $\mathfrak{L}$ **atomic** if

$$x \neq 0 \Rightarrow x = \bigvee\{y \in \mathsf{at}(\mathfrak{L}) : y \leqslant x\}.$$

6 Note also [9, Postlude, pages 183–185] and the paper [22], in each of which the authors demonstrate connections between formal topology and the theory of apartness spaces.

It should be clear that atoms correspond to the singleton subsets in the power set and metspace models. For future reference we now state some elementary properties of atoms.

Lemma 18.7 *If $\mathfrak{L}$ is a habitive, distributive lattice, $x \in \mathsf{at}(\mathfrak{L})$, and $x \leqslant u \vee v$, then $x \leqslant u$ or $x \leqslant v$.*

Lemma 18.8 *Let $(u_i)_{i\in I}$ be a family of elements of a pre-apartness frame $\mathfrak{L}$, and let x be an atom of $\mathfrak{L}$ such that $x \leqslant \bigvee_{i\in I} u_i$. Then there exists i such that $x \leqslant u_i$.*

Proof We have

$$0 \neq x = x \wedge \bigvee_{i\in I} u_i = \bigvee_{i\in I} (x \wedge u_i),$$

so, by (18.2), $\mathsf{hab}(\bigvee_{i\in I} (x \wedge u_i))$. Hence, by **H3**, there exists i such that $x \wedge u_i \neq 0$. Since $x \wedge u_i \leqslant x$ and x is an atom, $x \wedge u_i = x$ and therefore $x \leqslant u_i$. □

Proposition 18.9 *If x, y are elements of a pre-apartness frame $\mathfrak{L}$ such that $x \in \mathsf{at}(\mathfrak{L})$ and $x \leqslant -y$, then $x \bowtie y$.*

Proof Since

$$x \leqslant -y = \bigvee \{z \in \mathfrak{L} : z \bowtie y\},$$

it follows from Lemma 18.8 that there exists $z \in \mathfrak{L}$ such that $x \leqslant z$ and $z \bowtie y$; whence $x \bowtie y$, by Proposition 18.6(ii). □

For all subsets S, T of a metric space (X, ρ) define

$$S \bowtie_\rho T \text{ if and only if } \exists_{r>0} \forall_{x\in S} \forall_{y\in T} (\rho(x, y) \geq r)$$

and

$$-_\rho S \equiv \{x \in X : \exists_{r>0} \forall_{y\in S} (\rho(x, y) \geq r)\}.$$

Then $\bowtie_\rho$ is a pre-apartness on X, and $-_\rho S$ is the corresponding apartness complement of S. This **metric pre-apartness** has the following two additional properties, which we state for a general pre-apartness frame $\mathfrak{L}$:

AL4 Local decomposability: $-x = \bigvee \{-y : y \in \mathfrak{L} \text{ and } -x \vee y = 1\}$.
AL5 The **Lodato property:**[7] $-x \leqslant {\sim}y \Rightarrow -x \leqslant -y$.

If a frame pre-apartness satisfies **AL4**, then both it and the frame itself are said to be **locally decomposable**.

Proposition 18.10 *Every locally decomposable pre-apartness frame has the Lodato property.*

[7] For the connection between the constructive and classical Lodato properties see [9, page 21].

Proof Let $\mathfrak{L}$ be an a-frame, and let x, y be elements of $\mathfrak{L}$ such that $-x \leqslant {\sim}y$. For each $z \in \mathfrak{L}$ with $-x \vee z = 1$ we have

$$y = y \wedge (-x \vee z) = (y \wedge -x) \vee (y \wedge z) = 0 \vee (y \wedge z) = y \wedge z,$$

that is, $y \leqslant z$; whence, by Proposition 18.6(iv), $-z \leqslant -y$. It follows that

$$-x = \bigvee \{-z : z \in \mathfrak{L} \text{ and } -x \vee z = 1\} \leqslant -y.$$

Thus the Lodato condition holds in $\mathfrak{L}$. □

If the pre-apartness on a frame is locally decomposable (and therefore Lodato), then it is called an **apartness** and the frame is called an **a-frame.**

Proposition 18.11 *Let $\mathfrak{L}$ be an a-frame, x an atom of $\mathfrak{L}$, and u an element of $\mathfrak{L}$ such that $x \leqslant -u$. Then there exists $v \in \mathfrak{L}$ such that $x \leqslant -v$ and $-u \vee v = 1$.*

Proof Since $\mathfrak{L}$ is locally decomposable, $x \leqslant -u = \bigvee\{-v : -u \vee v = 1\}$. It remains to apply Lemma 18.8. □

Proposition 18.12 *Let $\mathfrak{L}$ be an a-frame, x an atom of $\mathfrak{L}$, and u an element of $\mathfrak{L}$ such that $x \leqslant -u$. Then for each atom y of $\mathfrak{L}$, either $y \leqslant -u$ or $x \neq_{\mathfrak{L}} y$.*

Proof By Proposition 18.11, there exists $v \in \mathfrak{L}$ such that $x \leqslant -v$ and $-u \vee v = 1$. Lemma 18.7 shows us that for each $y \in \mathsf{at}(\mathfrak{L})$, either $y \in -u$ or else $y \leqslant v$. In the latter event, since (by **AL2**) $x \leqslant -v \leqslant {\sim}v$, we have $y \leqslant v \leqslant {\sim}{\sim}v \leqslant {\sim}x$; whence, y being an atom, $0 \neq y = y \wedge {\sim}x$ and therefore $x \neq_{\mathfrak{L}} y$. □

18.4 Frame Topologies

A family τ of elements of a frame $\mathfrak{L}$ is called a **frame topology** on $\mathfrak{L}$ if it has the following three properties.

TL1 $0 \in \tau$ and $1 \in \tau$.
TL2 If $(u_i)_{i \in I}$ is a family of elements of τ, then $\bigvee_{i \in I} u_i \in \tau$.
TL3 If $u, v \in \tau$, then $u \wedge v \in \tau$.

In that case, $(\mathfrak{L}, \tau)$ is called a **topological frame**.

If X is an inhabited topological space, then its topology $\mathcal{T}$ is a frame topology on the power set frame $P(X)$. Regarded as a topspace model of a frame, $\mathcal{T}$ is also a frame topology on itself.

Now consider a pre-apartness frame $(\mathfrak{L}, \bowtie)$. The **nearly open** elements of $\mathfrak{L}$ are those of the form $\bigvee_{i \in I} - u_i$ for some index set I. The join of any family of nearly open elements is nearly open; both 0 and 1 are nearly open; and if $a = \bigvee_{i \in I} - u_i$

and $b = \bigvee_{j\in J} - v_j$ are nearly open, then, since by the generalised distributive law and Proposition 18.6(iii),

$$\bigvee_{i\in I} - u_i \wedge \bigvee_{j\in J} - v_j = \bigvee_{i\in I, j\in J} (-u_i \wedge -v_j) = \bigvee_{i\in I, j\in J} - (u_i \vee v_j),$$

we see that $a \wedge b$, and hence the meet of each finite family of nearly open sets, is nearly open. Thus the nearly open elements of $\mathfrak{L}$ constitute a frame topology, denoted by $\tau_{\bowtie}$ and called the **apartness frame topology generated by** $\bowtie$.

We now show how to move in the reverse direction, from frame topology to pre-apartness.

Proposition 18.13 *Let $(\mathfrak{L}, \tau)$ be a topological frame. Define a binary relation $\bowtie_\tau$ and a unary relation $-_\tau x$ on $\mathfrak{L}$ by*

$$x \bowtie_\tau y \text{ if and only if } \exists_{u\in\tau}(x \leqslant u \leqslant \sim y)$$

and

$$-_\tau x \equiv \bigvee \{z \in \mathfrak{L} : z \bowtie_\tau x\}.$$

Then $\bowtie_\tau$ is a Lodato pre-apartness on $\mathfrak{L}$ for which $x \rightsquigarrow -_\tau x$ is the corresponding apartness complement function.

Proof First observe that, since $1 \leqslant 1 = \sim 0$ and $0, 1 \in \tau$, we have $1 \bowtie_\tau 0$; that is, **AL1**. Next, if $z \bowtie_\tau x$, then there exists $u \in \tau$ such that $z \leqslant u \leqslant \sim x$, so $z \leqslant \sim x$; whence $-x \leqslant \sim x$ and **AL2** holds. To deal with **AL3,** suppose first that $(x_1 \vee x_2) \bowtie_\tau (y_1 \vee y_2)$. Then there exists $u \in \tau$ such that

$$x_1 \vee x_2 \leqslant u \leqslant \sim (y_1 \vee y_2) = \sim y_1 \wedge \sim y_2.$$

Hence $x_i \leqslant u \leqslant \sim y_j$ and therefore $x_i \bowtie_\tau y_j$. Now suppose, conversely, that $x_i \bowtie_\tau y_j$ for $i, j = 1, 2$. For such i, j there exists $u_{ij} \in \tau$ such that $x_i \leqslant u_{ij} \leqslant \sim y_j$. Then

$$x_i \leqslant u_{i1} \wedge u_{i2} \leqslant \sim y_1 \wedge \sim y_2 = \sim(y_1 \vee y_2),$$

so

$$x_1 \vee x_2 \leqslant (u_{11} \wedge u_{12}) \vee (u_{21} \wedge u_{22}) \leqslant \sim(y_1 \vee y_2).$$

Since $(u_{11} \wedge u_{12}) \vee (u_{21} \wedge u_{22}) \in \tau$, we now see that $(x_1 \vee x_2) \bowtie_\tau (y_1 \vee y_2)$.

It remains to verify the Lodato property. Let $-_\tau x \leqslant \sim y$ and $z \bowtie_\tau x$. There exists $u \in \tau$ such that $z \leqslant u \leqslant \sim x$. Since $u \leqslant u \leqslant \sim x$, we have $u \bowtie_\tau x$ and therefore $u \leqslant -_\tau x \leqslant \sim y$. Hence $z \leqslant u \leqslant \sim y$, so $z \bowtie_\tau y$ and therefore $z \leqslant -_\tau y$. It follows that $-_\tau x = \bigvee \{z \in \mathfrak{L} : z \bowtie_\tau x\} \leqslant -_\tau y$. □

We call $\bowtie_\tau$ the **topological pre-apartness induced** on the topological frame $(\mathfrak{L}, \tau)$, and when we regard a topological frame as a pre-apartness frame, we have in mind that pre-apartness. We define also the τ-**interior** of an element x of $\mathfrak{L}$ to be

$$\mathsf{int}_\tau(x) \equiv \bigvee \{z \in \tau : z \leqslant x\}.$$

Then $\mathsf{int}_\tau(x) \in \tau$, by **TL2**, and $\mathsf{int}_\tau(x) \leqslant x$. If $x \in \tau$, then $x \leqslant \mathsf{int}_\tau(x)$ and therefore $x = \mathsf{int}_\tau(x)$.

Proposition 18.14 *Let $(\mathfrak{L}, \tau)$ be a topological frame, and let $\bowtie_\tau$ and $-_\tau$ be the corresponding topological pre-apartness and apartness complement. Then $-_\tau x = \mathsf{int}_\tau(\sim x)$ for each $x \in \mathfrak{L}$.*

Proof For each $z \bowtie_\tau x$ there exists $u \in \tau$ such that $z \leqslant u \leqslant \sim x$ and therefore $z \leqslant u \leqslant \mathsf{int}_\tau(\sim x)$. Hence

$$-_\tau x = \bigvee \{z \in \mathfrak{L} : z \bowtie_\tau x\} \leqslant \mathsf{int}_\tau(\sim x).$$

Conversely, for each $u \in \tau$ such that $u \leqslant \sim x$, we have $u \bowtie_\tau x$; so

$$\begin{aligned}\mathsf{int}_\tau(\sim x) &= \bigvee \{u \in \tau : u \leqslant \sim x\} \\ &\leqslant \bigvee \{u \in \tau : u \bowtie_\tau x\} \\ &\leqslant \bigvee \{u \in \mathfrak{L} : u \bowtie_\tau x\} = -_\tau x.\end{aligned}$$

□

Lemma 18.15 *For each x in an a-frame $\mathfrak{L}$, $-x = \bigvee\{y \in \mathfrak{L} : -x \vee \sim y = 1\}$.*

Proof Let $x \in \mathfrak{L}$ and

$$z = \bigvee\{y \in \mathfrak{L} : -x \vee \sim y = 1\}.$$

For each $y \in \mathfrak{L}$ with $-x \vee \sim y = 1$ we have $y = (y \wedge -x) \vee (y \wedge \sim y) = y \wedge -x$, so $y \leqslant -x$. Hence $z \leqslant -x$. For the reverse inequality, consider any $y \in \mathfrak{L}$ such that $-x \vee y = 1$. We have $-y \leqslant \sim y$, so $\sim\sim y \leqslant \sim - y$. Hence

$$1 = -x \vee y \leqslant -x \vee \sim\sim y \leqslant -x \vee \sim - y \leqslant 1$$

and therefore $-x \vee \sim - y = 1$. Thus $-y \leqslant z$. It follows from this and local decomposability (we are dealing with an a-frame) that

$$\begin{aligned}-x &= \bigvee\{-y : y \in \mathfrak{L} \text{ and } -x \vee y = 1\} \\ &\leqslant \bigvee\{-y : y \in \mathfrak{L} \text{ and } -x \vee \sim - y = 1\} \\ &\leqslant \bigvee\{t \in \mathfrak{L} : -x \vee \sim t = 1\} = z.\end{aligned}$$

□

Proposition 18.16 *Let $(\mathfrak{L}, \bowtie)$ be a strongly complemented a-frame, and $z \in \tau_\bowtie$. Then for each z in $\tau_\bowtie$, $\neg z = \sim z$.*

Proof It is enough to prove that $\neg z \leqslant {\sim}z$. Pick a family $(a_i)_{i\in I}$ of elements of $\mathfrak{L}$ such that $z = \bigvee_{i\in I} - a_i$, and consider any $y \in \mathfrak{L}$ with $y \wedge z = 0$. For any $i \in I$ and any t with $-a_i \vee {\sim}t = 1$ we have $y \wedge -a_i = 0$ and therefore

$$y = y \wedge (-a_i \vee {\sim}t) = (y \wedge -a_i) \vee (y \wedge {\sim}t) = y \wedge {\sim}t \leqslant {\sim}t.$$

It follows from this, Lemma 18.15, and Proposition 18.2 that

$$\begin{aligned} y &\leqslant \bigwedge\{{\sim}t : t \in \mathfrak{L} \text{ and } -a_i \vee {\sim}t = 1\} \\ &= {\sim}\bigvee\{t \in \mathfrak{L} : -a_i \vee {\sim}t = 1\} \\ &= {\sim} - a_i. \end{aligned}$$

Hence

$$y \leqslant \bigwedge\nolimits_{i\in I} {\sim} - a_i = {\sim}\bigvee\nolimits_{i\in I} - a_i = {\sim}z,$$

and therefore $\neg z = \bigvee \{y : y \wedge z = 0\} \leqslant {\sim}z$. □

We have seen that a frame pre-apartness $\bowtie$ on a habitive frame $\mathfrak{L}$ gives rise to the frame topology $\tau_{\bowtie}$. In turn, $\tau_{\bowtie}$ gives rise to a new frame pre-apartness $\bowtie_{\tau_{\bowtie}}$ on $\mathfrak{L}$ with the Lodato property. Dually, a frame topology τ on a frame $\mathfrak{L}$ gives rise to a frame pre-apartness $\bowtie_\tau$ on $\mathfrak{L}$, which, in turn, provides a new frame topology $\tau_{\bowtie_\tau}$ on $\mathfrak{L}$. What connections can we establish in each of these two situations?

Proposition 18.17 *Let $(\mathfrak{L}, \bowtie)$ be a pre-apartness frame, let $\tau_{\bowtie}$ be the corresponding frame topology, and let $\bowtie_{\tau_{\bowtie}}$ be the pre-apartness induced on $\mathfrak{L}$ by $\tau_{\bowtie}$. For all $x, y \in \mathfrak{L}$, if $x \bowtie y$, then $x \bowtie_{\tau_{\bowtie}} y$. Conversely, if $\bowtie$ has the Lodato property and $x \bowtie_{\tau_{\bowtie}} y$, then $x \leqslant -y$; if also $x \in \mathsf{at}(\mathfrak{L})$, then $x \bowtie y$.*

Proof If $x \bowtie y$ in $\mathfrak{L}$, then $x \leqslant -y \leqslant {\sim}y$, so $x \bowtie_{\tau_{\bowtie}} y$. Conversely, if $x \bowtie_{\tau_{\bowtie}} y$, then there exists a family $(a_i)_{i\in I}$ of elements of $\mathfrak{L}$ such that $x \leqslant \bigvee_{i\in I} - a_i \leqslant {\sim}y$. If $\mathfrak{L}$ has the Lodato property, then for each $i \in I$, since $-a_i \leqslant {\sim}y$, we have $-a_i \leqslant -y$; whence $x \leqslant -y$. Reference to Proposition 18.9 completes the proof. □

Proposition 18.18 *Let $(\mathfrak{L}, \tau)$ be a topological frame, $\bowtie_\tau$ the topological pre-apartness induced by τ, and $\tau_{\bowtie_\tau}$ the apartness frame topology generated by $\bowtie_\tau$. Then $\tau_{\bowtie_\tau} \subset \tau$.*

Proof Let $-_\tau$ be the apartness complement function corresponding to $\bowtie_\tau$. By Proposition 18.14, for each u in $\mathfrak{L}$, $-_\tau u = \mathsf{int}_\tau({\sim}u)$ and so belongs to τ. It follows that every nearly open element of $(\mathfrak{L}, \bowtie_\tau)$, that is, every element of $\tau_{\bowtie_\tau}$, is a join of elements of τ, and is therefore in τ. □

In the preceding proposition, if $u \in \tau$, then classically,

$$u = \mathsf{int}_\tau({\sim}u) = {\sim}{\sim}u = \mathsf{int}_\tau({\sim}{\sim}u) = -_\tau{\sim}u$$

(the last step using Proposition 18.14) and so Proposition 18.18 holds with '$\subset$' replaced by '$=$'. It is shown in [9, pages 32–33] that if this replacement holds constructively, then we can derive the law of excluded middle.[8] If $\tau_{\bowtie_\tau} = \tau$, then we call $(\mathfrak{L}, \tau)$ **topologically consistent.** We have the following interconnections, established in [8, Propositions 12–14].

Proposition 18.19 *Let $(\mathfrak{L}, \tau)$ be a topological frame such that*

$$\forall_{u\in\tau}(u = \bigvee\{v \in \tau : u \vee {\sim}v = 1\}).$$

Then $\mathfrak{L}$ is topologically consistent, and $(\mathfrak{L}, \bowtie_\tau)$ is an a-frame.

Proposition 18.20 *If $(\mathfrak{L}, \bowtie)$ is an a-frame, then $(\mathfrak{L}, \tau_{\bowtie})$ is topologically consistent.*

18.5 Join Homomorphisms and Continuity

Let $\mathfrak{L}, \mathfrak{M}$ be (perforce habitive) frames with respective pre-apartness relations $\bowtie_{\mathfrak{L}}, \bowtie_{\mathfrak{M}}$ and apartness frame topologies generated by $\bowtie_{\mathfrak{L}}, \bowtie_{\mathfrak{M}}$. We will use the symbols $\neq$, $-$, and $\sim$ to denote, respectively, the inequality, apartness complement, and complementation function in both $\mathfrak{L}$ and $\mathfrak{M}$.

A mapping $f : \mathfrak{L} \to \mathfrak{M}$ is a **join homomorphism**[9] if

J1 $f(\bigvee_{i\in I} x_i) = \bigvee_{i\in I} f(x_i)$ for each family $(x_i)_{i\in I}$ of elements of $\mathfrak{L}$, and
J2 for each $x \in \mathfrak{L}$, $\mathsf{hab}(x)$ if and only if $\mathsf{hab}(f(x))$.

Such a map f is order preserving:

$$x \leqslant y \Rightarrow x = x \vee y \Rightarrow f(x) = f(x \vee y) = f(x) \vee f(y) \Rightarrow f(x) \leqslant f(y).$$

Hence

$$f(\bigwedge_{i\in I} x_i) \leqslant \bigwedge_{i\in I} f(x_i)$$

for each family $(x_i)_{i\in I}$ in $\mathfrak{L}$. Moreover, defining the mapping $f^{-\infty} : \mathfrak{M} \to \mathfrak{L}$ by[10]

$$f^{-\infty}(v) \equiv \bigvee\{x \in \mathfrak{L} : f(x) \leqslant v\} \quad (v \in \mathfrak{M}),$$

we see that $f(f^{-\infty}(v)) \leqslant v$ for each $v \in \mathfrak{M}$, and that if $g : \mathfrak{M} \to \mathfrak{N}$ is also a join homomorphism, then $(g \circ f)^{-\infty} = f^{-\infty} \circ g^{-\infty}$. In the standard model trilogy, for

[8] The paper [17], where a generalisation of point–set apartness spaces is considered, has a particularly illuminating discussion of the constructive plurality of topologies compatible with a given point–set pre-apartness.

[9] Comparing our notion of *join homomorphism* with the standard notion of *frame homomorphism* [19, page 39], we see that the former is, in one respect, more general, as it does not require the mapping to preserve finite meets; but in another, it is less general, as it requires that non-zero elements be mapped to non-zero elements.

[10] We have introduced the notation $f^{-\infty}(v)$ in order to avoid confusion with $f^{-1}(v)$, which equals $\{x \in \mathfrak{L} : f(x) = v\}$.

each subset S of the ambient set X, $f^{-\infty}(S)$ is just $f^{-1}(S)$, the inverse image of S under f.

We now introduce three notions of continuity for a join homomorphism f between a-frames $\mathfrak{L}$ and $\mathfrak{M}$. We say that f is

- **continuous** if $\forall_{x\in\mathfrak{L}}(f^{-\infty}(-f(x)) \leqslant -x)$;
- **topologically continuous** if $\forall_{v\in\tau_{\mathfrak{M}}}(f^{-\infty}(v) \in \tau_{\mathfrak{L}})$;
- **strongly continuous** if $\forall_{x,y\in\mathfrak{L}}(f(x) \bowtie_{\mathfrak{M}} f(y) \Rightarrow x \bowtie_{\mathfrak{L}} y)$.

We say that the inequality on a pre-apartness frame $\mathfrak{L}$ is **zero-tight** if

$$\forall_{x\in\mathfrak{L}}(\mathsf{not}(x \neq 0) \Rightarrow x = 0).$$

Proposition 18.21 *Let $\mathfrak{L}$ be a strongly complemented, Lodato pre-apartness frame with zero-tight inequality. Then every topologically continuous join homomorphism from $\mathfrak{L}$ to a pre-apartness frame is continuous.*

Proof Let f be a topologically continuous join homomorphism of $\mathfrak{L}$ into a pre-apartness frame $\mathfrak{M}$. Given $a \in \mathfrak{L}$, let $b = f^{-\infty}(-f(a))$. Since $f(b) \leqslant -f(a)$,

$$0 \leqslant f(a \wedge b) \leqslant f(a) \wedge f(b) = 0,$$

so $f(a \wedge b) = 0$ and therefore, in view of **J2**, $\mathsf{not}(\mathsf{hab}(a \wedge b))$. Thus $\mathsf{not}(a \wedge b \neq 0)$; whence, by zero tightness, $a \wedge b = 0$ and therefore, since $\mathfrak{L}$ is strongly complemented, $b \leqslant {\sim}a$. On the other hand, by topological continuity, $b \in \tau_{\mathfrak{L}}$, so there exists an inhabited family $(u_i)_{i\in I}$ of elements of $\mathfrak{L}$ such that $\bigvee_{i\in I} - u_i = b \leqslant {\sim}a$. For each $i \in I$, $-u_i \leqslant {\sim}a$ and therefore, by the Lodato property, $-u_i \leqslant -a$. Hence $b = \bigvee_{i\in I} - u_i \leqslant -a$. □

A mapping $f : \mathfrak{L} \to \mathfrak{M}$ between lattices is **atom-preserving** if for each atom a of $\mathfrak{L}$, $f(a)$ is an atom in $\mathfrak{M}$. For the remainder of this section we concentrate on continuity properties of atom-preserving frame homomorphisms on an atomic pre-apartness frame.

Proposition 18.22 *Let $\mathfrak{L}$ be an atomic pre-apartness frame, and let f be an atom-preserving frame homomorphism of $\mathfrak{L}$ into a pre-apartness frame $\mathfrak{M}$. Then f is continuous if and only if:*

(*) *For all $x \in \mathsf{at}(\mathfrak{L})$ and all $a \in \mathfrak{L}$, $f(x) \leqslant -f(a) \Rightarrow x \leqslant -a$.*

Proof Since 'if' holds trivially, we need only prove 'only if'. Suppose that (*) holds and consider $x, a \in \mathfrak{L}$ with $f(x) \leqslant -f(a)$. Choose a family $(x_i)_{i\in I}$ of atoms of $\mathfrak{L}$ such that $x = \bigvee_{i\in I} x_i$. Since f is atom- and order-preserving, for each $i \in I$

we have $f(x_i) \in \mathsf{at}(\mathfrak{M})$ and $f(x_i) \leqslant f(x) \leqslant -f(a)$, so $x_i \leqslant -a$, by (*). Hence $x \leqslant -a$. Thus

$$f^{-\infty}(-f(a)) = \bigvee\{x \in \mathfrak{L} : f(x) \leqslant -f(a)\} \leqslant -a. \qquad \square$$

Corollary 18.23 *Let $\mathfrak{L}$ be an atomic pre-apartness frame $\mathfrak{L}$, and let f be a strongly continuous, atom-preserving join homomorphism of $\mathfrak{L}$ into a pre-apartness frame $\mathfrak{M}$. Then f is continuous.*

Proof For each $a \in \mathfrak{L}$ and each $x \in \mathsf{at}(\mathfrak{L})$ with $f(x) \leqslant -f(a)$, since $f(x) \in \mathsf{at}(\mathfrak{M})$, by Proposition 18.9 we have $f(x) \bowtie_{\mathfrak{M}} f(a)$, so $x \bowtie_{\mathfrak{L}} a$ and therefore $x \leqslant -a$. Hence, by Proposition 18.22, f is continuous. □

Proposition 18.24 *Let $\mathfrak{L}$ be an atomic pre-apartness frame $\mathfrak{L}$, and let f be a continuous, atom-preserving join homomorphism of $\mathfrak{L}$ into an a-frame $\mathfrak{M}$. Then f is topologically continuous.*

Proof Let $v = \bigvee_{i\in I} -s_i$ be an element of $\tau_{\mathfrak{M}}$. If $x \leqslant f^{-\infty}(v)$, then $f(x) \in \mathsf{at}(\mathfrak{M})$ and $f(x) \leqslant v$. Lemma 18.8 shows that there exists $i \in I$ such that $f(x) \leqslant -s_i$. By Proposition 18.11, there exists $t \in \mathfrak{M}$ such that $f(x) \leqslant -t$ and $-s_i \vee t = 1$. Setting $y = f^{-\infty}(t)$, we have $f(y) \leqslant t$, so $f(x) \leqslant -t \leqslant -f(y)$ and therefore, by the continuity of f, $x \leqslant f^{-\infty}(-f(y)) \leqslant -y$. Next, consider any $z \in \mathsf{at}(\mathfrak{L})$ with $z \leqslant -y$. Since $\mathsf{not}(z \leqslant y)$ and $y = f^{-\infty}(t)$, we have $\mathsf{not}(f(z) \leqslant t)$. But $f(z) \in \mathsf{at}(\mathfrak{M})$ and $f(z) \leqslant 1 = -s_i \vee t$, so $f(z) \leqslant -s_i$, by Lemma 18.7, and therefore $z \leqslant f^{-\infty}(-s_i)$. Hence

$$-y = \bigvee\{z \in \mathfrak{L} : z \text{ is an atom and } z \leqslant -y\} \leqslant f^{-\infty}(-s_i) \leqslant f^{-\infty}(v).$$

We have now proved that for each $x \in \mathsf{at}(\mathfrak{L})$ with $x \leqslant f^{-\infty}(v)$ there exists $y \in \mathfrak{L}$ such that $x \leqslant -y \leqslant f^{-\infty}(v)$. It follows that

$$\begin{aligned} f^{-\infty}(v) &= \bigvee\left\{x : x \in \mathsf{at}(\mathfrak{L}) \text{ and } x \leqslant f^{-\infty}(v)\right\} \\ &\leqslant \bigvee\left\{-y : y \in \mathfrak{L} \text{ and } -y \leqslant f^{-\infty}(v)\right\} \\ &\leqslant \bigvee\left\{z : z \in \mathfrak{L} \text{ and } z \leqslant f^{-\infty}(v)\right\} \\ &\leqslant f^{-\infty}(v). \end{aligned}$$

Hence

$$f^{-\infty}(v) = \bigvee\left\{-y : y \in \mathfrak{L} \text{ and } -y \leqslant f^{-\infty}(v)\right\},$$

which belongs to $\tau_{\mathfrak{L}}$. □

Let $\mathfrak{L}$ be an *atomic* pre-apartness frame. We say that $a \in \mathfrak{L}$ is **near** an element b of $\mathfrak{L}$, and write $\mathsf{near}(a, b)$, if

$$a \in \mathsf{at}(\mathfrak{L}) \text{ and } \forall_{u \in \mathfrak{L}}(a \leqslant -u \Rightarrow \mathsf{hab}(b \wedge -u)).$$

In that case, since $a \leqslant 1 = -0$, we have $\mathsf{hab}(b)$. We say that a join homomorphism $f : \mathfrak{L} \to \mathfrak{M}$ between *atomic* pre-apartness frames is **nearly continuous** if

$$\forall_{a,b \in \mathsf{at}(\mathfrak{L})}(\mathsf{near}(a, b) \Rightarrow \mathsf{near}(f(a), f(b))).$$

In a power set pre-apartness model, a point x of the ambient set X is near the subset B of X if and only if each neighbourhood of x in the apartness topology intersects B; that is, if x is in the closure of B in the apartness topology.

A pre-apartness frame $\mathfrak{L}$ is said to be T_1 if

$$\forall_{x,y \in \mathsf{at}(\mathfrak{L})}(x \neq y \Rightarrow x \bowtie_{\mathfrak{L}} y).$$

Lemma 18.25 *Let $\mathfrak{L}$ be a T_1 pre-apartness frame, let $a \in \mathfrak{L}$, and let $x, y \in \mathsf{at}(\mathfrak{L})$ be such that $\mathsf{near}(x, a)$ and $x \neq y$. Then $\mathsf{hab}(a \wedge -y)$.*

Proof Since x, y are atoms, $x \neq y$, and $\mathfrak{L}$ is T_1, we have $x \bowtie_{\mathfrak{L}} y$, so $0 \neq x \leqslant -y$. Since $\mathsf{near}(x, a)$, it follows that $\mathsf{hab}(a \wedge -y)$. □

Proposition 18.26 *Let $\mathfrak{L}$ be an atomic, T_1 pre-apartness frame, let $\mathfrak{M}$ be a pre-apartness frame, and let $f : \mathfrak{L} \to \mathfrak{M}$ be an atom-preserving, topologically continuous join homomorphism. Then f is nearly continuous.*

Proof Let $a \in \mathsf{at}(\mathfrak{L})$ and $b \in \mathfrak{L}$ be such that $\mathsf{near}(a, b)$. If $v \in \mathfrak{M}$ and $f(a) \leqslant -v$, then $a \leqslant f^{-\infty}(-v)$ and, by topological continuity, $f^{-\infty}(-v) \in \tau_{\mathfrak{L}}$. Thus there exists a family $(u_i)_{i \in I}$ in $\mathfrak{L}$ such that $a \leqslant \bigvee_{i \in I} - u_i = f^{-\infty}(-v)$. Since a is an atom, there exists $j \in I$ such that $0 \leqslant a \leqslant -u_j$; whence $\mathsf{hab}(b \wedge -u_j)$ by nearness, and therefore, by **J2**, $\mathsf{hab}(f(b \wedge -u_j))$. Moreover

$$\begin{aligned} f(b \wedge -u_j) &\leqslant f(b) \wedge f(-u_j) \\ &\leqslant f(b) \wedge \textstyle\bigvee_{i \in I} f(-u_i) \\ &= f(b) \wedge f(\textstyle\bigvee_{i \in I} - u_i) \\ &= f(b) \wedge f(f^{-\infty}(-v)) \leqslant f(b) \wedge -v \end{aligned}$$

and therefore $\mathsf{hab}(f(b) \wedge -v)$. It now follows that $\mathsf{near}(f(a), f(b))$. □

A join homomorphism $f : \mathfrak{L} \to \mathfrak{M}$ between pre-apartness frames is said to be **atom-strongly extensional** if $x \neq y$ whenever x, y are *atoms* of $\mathfrak{L}$ and $f(x) \neq f(y)$.

Proposition 18.27 *Let $\mathfrak{L}$ be an atomic pre-apartness frame, and f a nearly continuous, atom-preserving join homomorphism of $\mathfrak{L}$ into a T_1 pre-apartness frame $\mathfrak{M}$. Then f is atom-strongly extensional.*

Proof Given $x, y \in \mathsf{at}(\mathfrak{L})$ with $f(x) \neq f(y)$, let

$$a = \bigvee\{z \in \mathsf{at}(\mathfrak{L}) : z = x \text{ or } (z = y \text{ and } x \neq y)\}$$

and pick a family $(a_i)_{i \in I}$ of atoms of $\mathfrak{L}$ such that $a = \bigvee_{i \in I} a_i$. Note that, by Lemma 18.8,

$$z \in \mathsf{at}(\mathfrak{L}) \Rightarrow (z \leqslant a \Leftrightarrow z = x \text{ or } (z = y \text{ and } x \neq y)). \tag{18.3}$$

Clearly, $0 \neq x \leqslant a$. For each $u \in \mathfrak{L}$ with $y \leqslant -u$ we see from Proposition 18.12 that either $x \neq y$ or $x \leqslant -u$. In the first case, by (18.3), $y \in a$, so $0 \neq y \leqslant a \wedge -u$; in the second case, $0 \neq x \leqslant a \wedge -u$. Hence

$$\forall_{u \in \mathfrak{L}}(y \in -u \Rightarrow \mathsf{hab}(a \wedge -u)),$$

that is, $\mathsf{near}(y, a)$. By near continuity, $\mathsf{near}(f(y), f(a))$. Since f is atom-preserving, both $f(x)$ and $f(y)$ are atoms of $\mathfrak{M}$. It follows from Lemma 18.25 that $\mathsf{hab}(f(a) \wedge -f(x))$. But

$$\begin{aligned} f(a) \wedge -f(x) &= f(\textstyle\bigvee_{i \in I} a_i) \wedge -f(x) \\ &= \textstyle\bigvee_{i \in I} f(a_i) \wedge -f(x) \\ &= \textstyle\bigvee_{i \in I}(f(a_i) \wedge -f(x)) \\ &\leqslant \textstyle\bigvee_{i \in I}(f(a_i) \wedge {\sim} f(x)) \end{aligned}$$

so

$$\mathsf{hab}(\textstyle\bigvee_{i \in I}(f(a_i) \wedge {\sim} f(x)))$$

and therefore, by **H3**, there exists $j \in I$ such that $\mathsf{hab}(f(a_j) \wedge {\sim} f(x))$. Hence $f(a_j) \neq f(x)$ and therefore $\mathsf{not}(a_j = x)$. But $a_j \in \mathsf{at}(\mathfrak{L})$ and $a_j \leqslant a$, so, by (18.3),

$$a_j = x \text{ or } (a_j = y \text{ and } x \neq y),$$

from which we deduce that $x \neq y$. □

From the preceding two propositions we obtain the following.

Corollary 18.28 *Let $\mathfrak{L}, \mathfrak{M}$ be T_1 pre-apartness frames with $\mathfrak{L}$ atomic, and $f : \mathfrak{L} \to \mathfrak{M}$ be a topologically continuous, atom-preserving join homomorphism. Then f is atom-strongly extensional.*

18.6 Set–Set Pre-apartness

Moving from the abstraction of apartness on lattices, we now return to the original, concrete setting of the theory of apartness.

Let X be an inhabited set equipped with an inequality relation $\neq$ between its elements and with a pre-apartness $\bowtie$ satisfying the axioms **B1–B4** from Section 18.1. Then the power set $P(X)$ is a habitive atomic frame with top X, bottom $\varnothing$, join $\cup$, and meet $\cap$, such that $\mathsf{hab}(S)$ if and only if S is inhabited in the usual set-theoretic sense, the atoms are the singleton subsets of X, and the complement and the logical complement of $S \subset X$ are given respectively by

$$\begin{aligned} {\sim}S &= \{x \in X : \forall_{y \in S}(x \neq y)\}, \\ \neg S &= \bigcup \{T \subset X : T \cap S = \varnothing\}. \end{aligned}$$

Note that, in this case, $\neg S = \{x \in X : x \notin S\}$.

What in Section 18.1 we called a pre-apartness on X is actually a frame pre-apartness turning $P(X)$ into a pre-apartness frame which, in view of **B4**, has the Lodato property. When this frame pre-apartness has some property that we have defined earlier, we refer to the original pre-apartness on X as having that property. For example, we say that the pre-apartness $\bowtie$ on X is locally decomposable if it is so as a frame pre-apartness on $P(X)$; this happens precisely when $\bowtie$ satisfies axiom **B5** in Section 18.1. Hence $\bowtie$ is a set–set apartness on X if and only if it is a frame apartness on the a-frame $P(X)$. For another example, we refer to the frame topology $\tau_{\bowtie}$ generated on $P(X)$ by the frame pre-apartness $\bowtie$ as the **apartness topology**, also denoted by $\tau_{\bowtie}$, generated on X by $\bowtie$; the open sets in the latter topology are precisely the nearly open sets in the former frame topology.

If f is a mapping between set–set pre-apartness spaces $(X, \bowtie)$ and $(X', \bowtie')$, then, for each $A \in P(X)$ defining

$$\phi_f(A) \equiv f(A) = \{f(x) : x \in A\},$$

we obtain a join homomorphism between the pre-apartness frames $P(X)$ and $P(X')$. The mapping $f \rightsquigarrow \phi_f$ is a one–one correspondence: given a join homomorphism $F : P(X) \to P(Y)$, we have $F = \phi_f$ where $f(x) = F(\{x\})$ for each $x \in X$. In future, we shall identify f and ϕ_f, denoting both by f.

Applying the theory of the preceding sections to the pre-apartness frame $P(X)$, we obtain theorems in the original theory of set–set apartness. But rather than push farther into the axiomatic theory of set–set apartness spaces, for which we refer the reader to [9, Chapter 3], we now discuss[11] the important connection between set–set (pre-)apartness and (quasi-)uniform structures.[12] To that end, let X be an

[11] For the remainder of this chapter we omit proofs of most results, for which we refer the reader to external sources.

[12] For the classical theory of uniform spaces, see [2, Chapter II], [20, Chapter 6], or [25, Part II].

inhabited set, and let U,V be subsets of the Cartesian product $X \times X$. We define certain associated subsets as follows:

$$U \circ V \equiv \{(x, y) : \exists_{z \in X} ((x, z) \in U \text{ and } (z, y) \in V)\},$$
$$U^1 \equiv U, \quad U^{n+1} \equiv U \circ U^n \quad (n = 1, 2, \ldots),$$
$$U^{-1} \equiv \{(x, y) : (y, x) \in U\},$$

and, for $x \in X$,

$$U[x] \equiv \{y \in X : (x, y) \in U\}.$$

We say that U is **symmetric** if $U = U^{-1}$.

A family $\mathcal{U}$ of subsets of X is a **quasi-uniform structure**, or a **quasi-uniformity**, on X if the following conditions hold.

U1 (i) Every finite intersection of sets in $\mathcal{U}$ belongs to $\mathcal{U}$.
(ii) Every subset of $X \times X$ that is a superset of a member of $\mathcal{U}$ belongs to $\mathcal{U}$.
U2 For all x,$y \in X$, $x = y$ if and only $(x, y) \in U$ for each $U \in \mathcal{U}$.
U3 For each $U \in \mathcal{U}$ there exists $V \in \mathcal{U}$ such that $V^2 \subset U$.
U4 For each $U \in \mathcal{U}$ there exists $V \in \mathcal{U}$ such that $X \times X = U \cup \neg V$.[13]

The elements of $\mathcal{U}$ are called the **entourages** of (the quasi-uniform structure on) X, and the pair $(X,\mathcal{U})$, or, loosely, X itself, is called a **quasi-uniform space**. We call $\mathcal{U}$ a **uniform structure**, or **uniformity**, and X a **uniform space**, if, in addition to **U1–U4**, for each $U \in \mathcal{U}$ we have $U^{-1} \in \mathcal{U}$; in that case, $U \cap U^{-1}$ is a symmetric entourage.

A **quasi-metric space** is a pair (X, ρ) comprising an inhabited set X and a **quasi-metric** $\rho : X \times X \to \mathbf{R}$ such that for all x, y, z in X,

$$\rho(x, y) = 0 \Leftrightarrow x = y,$$
$$\rho(x, y) > 0 \Rightarrow \rho(y, x) > 0,$$
$$\rho(x, z) \leq \rho(x, y) + \rho(y, z).$$

Then $\rho(x, y) \geq 0$ for all $x, y \in \mathbf{R}$. A quasi-metric space is a quasi-uniform space in which the quasi-uniformity consists of all supersets, in $X \times X$, of sets of the form

$$U_\varepsilon \equiv \{(x, y) : \rho(x, y) < \varepsilon\}$$

with $\varepsilon > 0$. It is straightforward to verify axioms **U1–U3**. To verify **U4**, we use the fact that for each $\varepsilon > 0$ and all x,$y \in X$, either $\rho(x$,$y) > \varepsilon/2$ or $\rho(x$,$y) < \varepsilon$. If ρ is symmetric and therefore a metric on X, then the corresponding quasi-uniformity is a uniformity.

[13] Classically, property **U4** always holds with $V = U$. It is important to postulate it in the constructive theory, since it is the only uniform-space axiom that provides us with alternatives that enable proof- and definition-by-cases.

For the rest of this section, unless we state otherwise, $(X,\mathcal{U})$ is a quasi-uniform space. We define the **(uniform) inequality** on $(X,\mathcal{U})$ by

$$x \neq y \text{ if and only if } \exists_{U\in\mathcal{U}}\left((x,y)\notin U\right) \text{ or } \exists_{U\in\mathcal{U}}\left((y,x)\notin U\right).$$

In view of **U2**, it is easily seen that $\neq$ is indeed an inequality relation on X. If $\mathcal{U}$ is a uniformity, then $x \neq y$ if and only if there exists $U \in \mathcal{U}$ with $(x,y) \notin U$.

The quasi-uniform structure $\mathcal{U}$ induces a quasi-uniform structure $\mathcal{U}_Y$ on an inhabited subset Y of X: the entourages of $\mathcal{U}_Y$ are the sets $U \cap (Y \times Y)$ with $U \in \mathcal{U}$. Taken with the inequality and quasi-uniform structure induced by those on X, the set Y is called a **quasi-uniform subspace** of X. The quasi-uniform structure $\mathcal{U}_Y$ is called the **subspace quasi-uniform structure** on Y.

In the context of a quasi-uniform space, the following lemma is a valuable partial substitute for the classical decidability of equality.

Lemma 18.29 *If $(X,\mathcal{U})$ is quasi-uniform space and $U \in \mathcal{U}$, then for all $x,y \in X$ either $x \neq y$ or $(x,y) \in U$.*

Proof By **U4**, there exists $V \in \mathcal{U}$ such that $X \times X = U \cup \neg V$. If $(x,y) \in \neg V$, then $(x,y) \notin V$, so $x \neq y$, by definition of the uniform inequality on X. □

The quasi-uniform structure $\mathcal{U}$ on X gives rise to

- the **quasi-uniform topology** $\tau_{\mathcal{U}}$, in which the sets $U[x]$, with $U \in \mathcal{U}$, form a base of neighbourhoods of the point x; and
- the **quasi-uniform pre-apartness**, defined by

$$S \bowtie_{\mathcal{U}} T \Leftrightarrow \exists_{U\in\mathcal{U}}(S \times T \subset {\sim}U). \tag{18.4}$$

If $\mathcal{U}$ is a uniform structure, then we drop 'quasi-' from the names of $\tau_{\mathcal{U}}$ and $\bowtie_{\mathcal{U}}$.

Proposition 18.30 *Let S,T be subsets of a quasi-uniform space $(X,\mathcal{U})$. Then $S \bowtie_{\mathcal{U}} T$ if and only if there exists $V \in \mathcal{U}$ such that $S \times T \subset \neg V$* [9, Proposition 3.2.11].

Next we introduce the **Efremovič property** for a frame pre-apartness $\bowtie$ on a frame $\mathfrak{L}$:

EF For all $a,b \in \mathfrak{L}$, if $a \bowtie b$, then there exists $e \in \mathfrak{L}$ such that $a \bowtie \neg e$ and $e \bowtie b$.

Proposition 18.31 *If $(X,\mathcal{U})$ is a quasi-uniform space, then $\bowtie_{\mathcal{U}}$, defined at* (18.4) *is a locally decomposable set–set apartness on X with the Efremovič property* [9, Proposition 3.2.24].

In the classical theory of proximity spaces, for each set–set apartness space $(X,\bowtie)$ with the Efremovič property there is at least one quasi-uniform structure $\mathcal{U}$

on X that induces $\bowtie$, in the sense that $\bowtie = \bowtie_{\mathcal{U}}$ [21, page 71]. In the constructive theory, however, this is definitely not the case.

Proposition 18.32 *Let X be an inhabited set with the denial inequality. Then*

$$\forall_{A,B \in P(X)} (A \bowtie B \Leftrightarrow (A = \varnothing \text{ or } B = \varnothing))$$

defines a symmetric set–set apartness on X with the Efremovič property. If there exists a uniform structure $\mathcal{U}$ on X such that $\bowtie = \bowtie_{\mathcal{U}}$, then the weak law of excluded middle, not $P \vee$ not not P, *holds* [9, Corollary 3.2.30].

The proof of this proposition depends on a classically vacuous result whose proof we include for the reader's entertainment.

Proposition 18.33 *Suppose there exists a set–set pre-apartness space X with the following two properties.*

(i) $A \bowtie B \Rightarrow \forall_{x \in X} (x \notin A \text{ or } x \notin B)$.
(ii) Any two disjoint subsets of a singleton subset of X are apart; that is, if $x \in X, A \subset \{x\}, B \subset \{x\}$, and $A \cap B = \varnothing$, then $A \bowtie B$.

Then not $P \vee$ not not P *holds.*

Proof Fixing $\xi \in X$, consider any statement P and let

$$A = \{x \in X : x = \xi \text{ and } P\},$$
$$B = \{x \in X : x = \xi \text{ and not } P\}.$$

Then, by hypothesis (ii), $A \bowtie B$. By hypothesis (i), either $\xi \notin A$, in which case not P holds, or else $\xi \notin B$ and therefore not not P holds. □

If $(X, \mathcal{U})$ is a quasi-uniform space, then the eccentric hypothesis (ii) of this last proposition holds for the apartness $\bowtie_{\mathcal{U}}$: for if $x \in X$ and A, B are disjoint subsets of $\{x\}$ and $U \in \mathcal{U}$, then we have $A \times B = \varnothing \subset \neg U$.

18.7 Strong and Uniform Continuity

We now concentrate on the connection between strong continuity and uniform continuity for a mapping f between quasi-uniform apartness spaces $(X, \mathcal{U})$ and $(Y, \mathcal{V})$. Bearing in mind our identification of mappings from X to Y with join homomorphisms between $(P(X), \bowtie_{\mathcal{U}})$ and $P(Y, \bowtie_{\mathcal{V}})$, we see that

$$f^{-\infty}(A) = \bigcup \{A \subset X : f(S) \subset A\} = f^{-1}(A) = \{x \in X : f(x) \in A\}.$$

Since $(X, \bowtie_{\mathcal{U}})$ is atomic, the atoms being the points of x, f is atom-preserving. Proposition 18.22 shows that f is continuous if and only if

$$\forall_{x \in X} \forall_{A \subset X} (f(x) \in -f(A) \Rightarrow x \in -A).$$

Also, recalling our identification of pre-apartness properties on X and on $P(X)$, we see that f is strongly continuous if and only if

$$\forall_{A,B \subset X} (f(A) \bowtie_{\mathcal{V}} f(B) \Rightarrow A \bowtie_{\mathcal{U}} B).$$

We say that f is **uniformly continuous** if for each $V \in \mathcal{V}$ there exists $U \in \mathcal{U}$ such that $(f(x), f(y)) \in V$ for all (x, y) in U; in that case, f is strongly continuous [9, Proposition 3.3.2]. When is a strongly continuous f uniformly continuous? We aim to show that the answer is: (almost) when X or Y is totally bounded.

Proposition 18.34 *If U is an entourage of a quasi-uniform space X, then for each integer $n \geq 2$ there exists an entourage V such that $V^n \subset U$ and $X \times X = U \cup {\sim} V$. If X is a uniform space, then this conclusion holds for some symmetric entourage V* [9, Proposition 3.2.6 and Corollary 3.2.7].

For each positive integer n we define an **n-chain of entourages** of a quasi-uniform space $(X, \mathcal{U})$ to be an n-tuple $(U_1, \ldots, U_n)$ of entourages such that $U_k^2 \subset U_{k-1}$ and $X \times X = U_{k-1} \cup \neg U_k$ for each $k \geq 2$. The existence of such objects follows from Proposition 18.34.

Given U in $\mathcal{U}$, by a **U-approximation** to a subset S of the quasi-uniform space X we mean an inhabited subset A of S such that

$$S = \bigcup_{x \in A} U[x].$$

We say that S is **totally bounded** if, for each $U \in \mathcal{U}$, there exists a finitely enumerable U-approximation to S.

Proposition 18.35 *Let X and Y be uniform apartness spaces, with Y totally bounded, and let f be a strongly continuous mapping of X* onto *Y. Then f is uniformly continuous.*

Proof Given an entourage V of Y, construct a 4-chain (V_1, V_2, V_3, V_4) of entourages of Y such that $V_2^3 \subset V_1 = V$, V_4 is symmetric, and $V_4^3 \subset V_3$ Choose $x_1, \ldots, x_m$ in X such that $Y = Y_1 \cup \cdots \cup Y_m$, where $Y_i \equiv V_4 [f(x_i)]$; then set $X_i \equiv f^{-1}(Y_i)$. For $1 \leqslant i, j \leqslant m$ construct c_{ij} such that

$$c_{ij} = 0 \Rightarrow (f(x_i), f(x_j)) \in V_2,$$
$$c_{ij} = 1 \Rightarrow (f(x_i), f(x_j)) \in \neg V_3.$$

We prove the following:

(i) If $c_{ij} = 0$, then $Y_i \times Y_j \subset V$.
(ii) If $c_{ij} = 1$, then $Y_i \bowtie Y_j$.

Let $c_{ij} = 0$ and consider (x, x') with $(f(x), f(x')) \in Y_i \times Y_j$. We have

$$(f(x), f(x_i)) \in V_4^{-1} = V_4$$

and $(f(x_j), f(x')) \in V_4$, so since $V_4 \subset V_2$,

$$\left(f(x), f(x')\right) \in V_4 \circ V_2 \circ V_4 \subset V_2^3 \subset V.$$

This proves (i). It follows that if $c_{ij} = 0$ for all i and j, then $(f(x), f(x')) \in V$ for all $x, x' \in X$. Thus we may assume that there exist i, j with $c_{ij} = 1$. For such i and j, consider an element (y, y') of $Y_i \times Y_j$, and suppose that $(y, y') \in V_4$. Then $(f(x_i), y) \in V_4$ and $(y', f(x_j)) \in V_4^{-1} = V_4$, so $(f(x_i), f(x_j)) \in V_4^3 \subset V_3$, a contradiction. Hence $(y, y') \in \neg V_4$ It follows that $Y_i \times Y_j \subset \neg V_4$ and therefore that $Y_i \bowtie Y_j$. This proves (ii). Moreover, we have $X_i \bowtie X_j$, by the strong continuity of f; so, by Proposition 18.30, there exists an entourage U_{ij} of X with $X_i \times X_j \subset \neg U_{ij}$. Let

$$U \equiv \bigcap \{U_{ij} : c_{ij} = 1\},$$

which is an entourage of X. Consider points x, x' of X with $(x, x') \in U$. Choose i, j such that $f(x) \in Y_i$ and $f(x') \in Y_j$. If $c_{ij} = 1$, then

$$\left(x, x'\right) \in X_i \times X_j \subset \neg U_{ij} \subset \neg U,$$

a contradiction. Hence $c_{ij} = 0$ and therefore, by (i), $(f(x), f(x')) \in V$. Since $V \in \mathcal{U}_Y$ is arbitrary, this completes the proof that f is uniformly continuous. □

A subset $\mathcal{B}$ of a quasi-uniformity $\mathcal{U}$ is called a **base of entourages** if, for each $U \in \mathcal{U}$, there exists $B \in \mathcal{B}$ with $B \subset U$. Every metric space has such a base, comprising the sets U_ε with $\varepsilon > 0$.[14]

Theorem 18.36 *Let X be a totally bounded uniform space with a countable base of entourages. Then every strongly continuous mapping from X into a uniform space is uniformly continuous* [9, Theorem 3.3.18].

Recall the following definitions relating to a subset S of the set $\mathbf{N}^+$ of positive integers: we say that

- $n \in S$ **eventually**, or **for all sufficiently large** n, if there exists $N \in \mathbf{N}^+$ such that $n \in S$ whenever $n \geq N$;
- $n \in S$ **infinitely often** if for each $n \in \mathbf{N}^+$ there exists $m > n$ such that $m \in S$.

[14] Classically, a uniform space is pseudometrisable if and only if it has a countable base of entourages. It seems extremely unlikely that 'if' holds constructively.

In the latter case, using dependent choice, we can construct a strictly increasing sequence $(n_k)_{k\geq 1}$ of positive integers such that $n_k \in S$ for each k.

Two sequences $(x_n)_{n\geq 1}$,$(x'_n)_{n\geq 1}$ in a uniform space $(X, \mathcal{U})$ are said to be **eventually close** if for each U in $\mathcal{U}$ we have $(x_n, x'_n) \in U$ for all sufficiently large n. A mapping f of X into a uniform space Y is **uniformly sequentially continuous** if the sequences $(f(x_n))_{n\geq 1}$, $(f(x'_n))_{n\geq 1}$ are eventually close in Y whenever $(x_n)_{n\geq 1}$,$(x'_n)_{n\geq 1}$ are eventually close in X.

Theorem 18.37 *A strongly continuous mapping $f : X \to Y$ between uniform spaces is uniformly sequentially continuous* [9, Theorem 3.3.11].

The converse of this theorem holds classically. But it cannot be proved constructively: for example, even the statement

> *Every uniformly sequentially continuous mapping of a complete, separable metric space into a metric space is uniformly continuous*

is equivalent to Ishihara's principle BD-**N** [11, Theorem 11].

The proof of Theorem 18.37 depends on a purpose-built technique (reminiscent of Ishihara's tricks [10, 16]), which we now outline.

Lemma 18.38 *Let S be an inhabited set, and let H be a set of sequences in S such that if $s = (s_n)_{n\geq 1} \in H$, then each subsequence of s belongs to H. Let T be a subset of S with the following property: for all $P,Q \subset \mathbf{N}^+$, if*

$$s \in H,\ \mathbf{N}^+ = P \cup Q,\ \textit{and}\ s_n \in T\ \textit{for each}\ n \in Q, \tag{18.5}$$

then either $n \in P$ for all n or else there exists $n \in Q$. If (18.5) *obtains, then either $n \in P$ eventually or else $n \in Q$ infinitely often* [9, Lemma 3.3.12].

Our next lemma may seem bizarre, since it shows that under certain hypotheses we can prove the non-constructive limited principle of omniscience, LPO: for each binary sequence $(a_n)_{n\geq 1}$, either $a_n = 0$ for all n or else there exists n such that $a_n = 1$. However, the lemma enables us to use LPO to rule out the unwanted second alternative in the conclusion of Lemma 18.38. This idea is used on a number of occasions in constructive proofs; see, for example, [7, pages 195–196].

Lemma 18.39 *Let S, H, and T be as in* Lemma 18.38. *Let $s = (s_n)_{n\geq 1} \in H$, let $\mathbf{N}^+ = P \cup Q$, and suppose that $s_n \in T$ for each $n \in Q$. If $n \in Q$ infinitely often, then* LPO *holds* [9, Lemma 3.3.13].

In order to apply the foregoing, we need to set up suitable S, H, T; show that if $s \in H$, $\mathbf{N}^+ = P \cup Q$, and $s_n \in T$ for each $n \in Q$, then either $n \in P$ for all n or else there exists $n \in Q$; and use LPO to show that it is impossible for Q to be infinite. It will then follow that $n \in P$ eventually. The suitable objects are introduced in the following lemma.

Lemma 18.40 *Let X, Y be uniform spaces, $f : X \to Y$ a strongly continuous function, and V an entourage of Y. Let S be the space $X \times X$,* $\mathsf{sequ}(S)$ *the set of sequences in S,*

$$H = \{s \in \mathsf{sequ}(S) : \forall_{U \in \mathcal{U}_X} \exists_N \forall_{n \geq N} (s_n \in U)\},$$

and

$$T = \{(x, x') \in S : (f(x), f(x')) \in \neg V\}.$$

If $s \in H$, $\mathbf{N}^+ = P \cup Q$, and $s_n \in T$ for each $n \in Q$, then either $n \in P$ for all n or else there exists $n \in Q$ [9, Lemma 3.3.15].

Lemmas 18.38 and 18.39 are the key not only to the proof of Theorem 18.37, but also to that of a theorem relating an apartness-theoretic notion of *proximal convergence* to one of *uniform sequential convergence* for sequences of mappings from an inhabited set into a uniform space [9, Section 3.4].

18.8 Compactness

Of several approaches to compactness in the context of set–set apartness, that introduced by Diener ([12]; see also [9, Section 3.10]) and based on open covers, and the sequential compactness one in [4] are perhaps the most satisfactory.[15] We sketch these two approaches, beginning with Diener's.

Let X be a set–set pre-apartness space with $\bowtie$ symmetric. An ordered pair (S, T) of subsets of X is a **neat cover** of X if there exist subsets S', T' of X such that $S' \bowtie T'$, $X = S \cup S'$, and $X = T \cup T'$. For example, if $\bowtie$ is the metric apartness on X, $\xi \in X$, $0 \leq \alpha < \beta$, and

$$S = \{x \in X : \rho(\xi, x) > \alpha\},$$
$$T = \{x \in X : \rho(\xi, x) < \beta\},$$

then (S, T) is a neat cover of X, since the requirements of the definition are satisfied by taking

$$S' = \{x \in X : \rho(\xi, x) < \alpha + \tfrac{1}{3}(\beta - \alpha)\},$$
$$T' = \{x \in X : \rho(\xi, x) > \beta - \tfrac{1}{3}(\beta - \alpha)\}.$$

A subset A of the pre-apartness space X is

- **neatly located** (in X) if for each neat cover (S, T) of X, either $A \subset S$ or else $A \cap T$ is inhabited;

[15] For a different type of compactness see [26].

- **neatly compact** if it is neatly located and satisfies **Diener's condition**: if LPO holds, then there does not exist a sequence $(V_n)_{n\geq 1}$ of $\tau_{\bowtie}$-open elements of X such that $V_1 \subset V_2 \subset \cdots$, $A \subset \bigcup_{n\geq 1} V_n$, and $A \cap \neg V_n$ is inhabited for each n.

Diener's rather strange-looking condition is a strong form of the assertion that, under LPO, no countable open cover of A contains a finite subcover.

If (X, ρ) is a metric apartness space, then every neatly located subset S of X is located [9, Proposition 3.10.3]. If X is a uniform apartness space, then every totally bounded subset of X is neatly located [9, Proposition 3.10.4], and every separable, neatly compact uniform subspace of X is totally bounded [9, Proposition 3.10.12].

Diener also introduced notions of *neat Cauchyness, neat convergence*, and *neat completeness* for nets in a set–set pre-apartness space; we omit the details. His fundamental theorem is as follows.

Theorem 18.41 *A uniform apartness space with a countable base of entourages is totally bounded and complete*[16] *if and only if it is separable, neatly compact, and neatly complete* [9, Theorem 3.10.18].

In particular, a separable metric apartness space is totally bounded and complete (that is, compact according to Bishop's definition) if and only if it is neatly complete and neatly compact. Moreover, Diener's notion of neat compactness has desirable properties under strongly continuous mappings; for example, if f is a strongly continuous mapping of a neatly compact pre-apartness space X into a uniform space Y, then $f(X)$ is neatly compact; if also $Y = \mathbf{R}$, then $\sup_{x\in X} f(x)$ exists [9, Corollaries 3.10.20 and 3.10.21].

We now turn to the alternative theory of compactness found in [4]. Let X be a set–set pre-apartness space with $\bowtie$ symmetric, and let $x \in X$. We say that a sequence $(x_n)_{n\geq 1}$ is

- **eventually bounded away from the point** $x \in X$ if there exists N such that $x \in -\{x_n : n \geq N\}$;
- **converges to** x if for each $U \in \tau_{\bowtie}$ with $x \in U$, there exists N such that $x_n \in U$ for all $n \geq N$.

We say that X is

- **sequentially almost compact** if LPO implies that every sequence in X contains a subsequence that converges in X;
- **weak-sequentially almost compact** if LPO implies that there is no sequence in X that is eventually bounded away from each point of X.

[16] See [25, pages 133–136] for a discussion of completeness for uniform spaces.

Sequential almost compactness implies Diener's condition, which implies weak-sequential almost compactness[17] [4, Proposition 5].

In [4, page 522] we introduce a special set $\mathcal{B}_w$ of subsets of $X \times X$. (Classically, if X satisfies the Efremovič condition, then $\mathcal{B}_w$ is a base for the unique totally bounded uniform structure that is compatible with $\bowtie$.) We then define notions of *$\mathcal{B}_w$-Cauchyness for nets* and *$\mathcal{B}_w$-completeness*, and say that X is

- **sequentially compact** if it is $\mathcal{B}_w$-complete and sequentially almost compact;
- **weak-sequentially compact** if it is $\mathcal{B}_w$-complete and weak-sequentially almost compact.

Theorem 18.42 *The following are equivalent conditions on a uniform apartness space X with a countable base of entourages.*

(i) *X is separable, neatly located, and weak-sequentially compact.*
(ii) *X is complete and totally bounded.*
(iii) *X is separable, neatly located, and sequentially compact* [4, Theorem 13].

Sequential compactness and its weak counterpart are preserved under strongly continuous maps.

Proposition 18.43 *Let f be a strongly continuous mapping of a separable, neatly located, weak-sequentially almost compact apartness space into a uniform space. Then $f(X)$ is totally bounded* [4, Corollary 16].

Taken with Proposition 18.35, this gives us a weak version of the uniform continuity theorem.

Proposition 18.44 *Let X be a separable, neatly located, weak-sequentially almost compact uniform space. Then every strongly continuous mapping of X into a uniform space is uniformly continuous.*

Next we ask: is there a viable constructive counterpart of total boundedness for apartness spaces that are not uniform spaces?

We say that an apartness space $(X, \bowtie)$ is **precompact** if there exists a strongly continuous mapping f from a dense subset of Cantor space $2^{\mathbf{N}}$ into X such that the range of f is $\tau_{\bowtie}$-dense in X. The following results, among several others, can be found in [6].

Proposition 18.45 *A precompact uniform space is totally bounded.*

Proposition 18.46 *If f is a strongly continuous mapping of a precompact apartness space into $\mathbf{R}$, then $f(X)$ is totally bounded, and* $\sup f, \inf f$ *exist.*

[17] The proof of this in [4, Proposition 5] is correct, even though the definition of *eventually bounded away from the point x* is given there only for the case where X is a metric space.

Proposition 18.47 *A metric space is precompact if and only if it is totally bounded.*

Proposition 18.48 *The union of two precompact apartness spaces is precompact.*

The definition and properties of the product of two apartness spaces are given in [9, Section 3.7]. With those properties at hand we can also prove the following.

Proposition 18.49 *The product of two apartness spaces X and Y is precompact if and only if both X and Y are precompact* [6, Proposition 3.3].

18.9 Concluding Remarks

In this chapter there is not enough space to elaborate on further developments on the general theory of pre-apartness frames, such as: the axiomatic theory of products [3]; the construction of such products [5]; proximity (nearness) of sets [9, Section 3.9]; and possible approaches to compactness in frames. Even in the context of set–set apartness and uniform spaces there remain many topics about which we have said nothing. However, we hope to have convinced the reader that the theory of (pre-)apartness, between sets or on lattices, sheds a new light on constructive topology and is a worthy alternative to other approaches to that rich, important branch of mathematics.

Acknowledgement

This chapter is an amplified, corrected version of research carried out with Luminiţa Vîţă over several years. I am most grateful to her for her collaboration, and for her kindly scrutinising the draft of this present work.

References

[1] Bishop, E. 1997. *Foundations of Constructive Analysis.* New York: McGraw-Hill.

[2] Bourbaki, N. 1971. *General Topology (Part 1).* Reading, MA: Addison-Wesley. Re-published by Springer Nature Switzerland AG, 1989.

[3] Bridges, D. S. 2008. Product a-frames and proximity. *Math. Logic Q.*, **54**(1), 12–25.

[4] Bridges, D. S. 2012a. Compactness notions for an apartness space. *Arch. Math. Logic*, **51**, 517–534.

[5] Bridges, D. S. 2012b. How to construct the product of a-frames. *Math. Logic Q.*, **58**(4–5), 281–293.

[6] Bridges, D. S. 2012c. Precompact apartness spaces. *Logic. Meth. Comp. Sci.*, **8**(2:15), 1–10.

[7] Bridges, D. S., and Vîţă, L. S. 2006. *Techniques of Constructive Analysis.* New York: Springer.

[8] Bridges, D. S., and Vîţă, L. S. 2009. A constructive theory of apartness on lattices. *Scientiae Math. Jap.*, **69**(2), 187–206.

[9] Bridges, D. S., and Vîţă, L. S. 2011. *Apartness and Uniformity – A Constructive Development*. Berlin: Springer-Verlag.

[10] Bridges, D. S., van Dalen, D., and Ishihara, H. 2003. Ishihara's proof technique in constructive analysis. *Proc. Koninklijke Nederlandse Akad. Wetenschappen (Indag. Math.) N.S.*, **14**(2), 163–168.

[11] Bridges, D. S., Ishihara, H., Schuster, P., and Vîţă, L. S. 2005. Strong continuity implies uniform sequential continuity. *Arch. Math. Logic*, **44**, 887–895.

[12] Diener, H. 2008. Compactness under constructive scrutiny. Ph.D. thesis, University of Canterbury, Christchurch, New Zealand.

[13] Grayson, R. J. 1981. Concepts of general topology in constructive mathematics and in sheaves I. *Ann. Math. Logic*, **20**, 1–41.

[14] Grayson, R. J. 1982. Concepts of general topology in constructive mathematics and in sheaves II. *Ann. Math. Logic*, **23**, 55–98.

[15] Hedin, A. 2010. *A note on set-presentable apartness spaces*. Technical report University of Uppsala, Sweden.

[16] Ishihara, H. 1991. Continuity and nondiscontinuity in constructive mathematics. *J. Symbol. Logic*, **56**(4), 1349–1354.

[17] Ishihara, H., Mines, R., Schuster, P. M., and Vîţă, L. S. 2006. Quasi-apartness and neighbourhood spaces. *Ann. Pure Appl. Logic*, **141**, 296–306.

[18] Jacobson, N. 2009. *Basic Algebra I*, 2nd ed. Mineola, New York: Dover Publications.

[19] Johnstone, P. T. 1983. The point of pointless topology. *Bull. Amer. Math. Soc.*, **8**(1), 41–53.

[20] Kelley, J. L. 1955. *General Topology*. Princeton, NJ: van Nostrand. Re-published as Graduate Text in Mathematics 27, Springer Verlag, Heidelberg, Germany, 1975.

[21] Naimpally, S. A., and Warrack, B. D. 1970. *Proximity Spaces*. Cambridge Tracts in Mathematics and Mathematical Physics 59. Cambridge: Cambridge University Press.

[22] Palmgren, E., and Schuster, P. M. 2006. Apartness and formal topology. *New Zealand J. Math.*, **35**, 77–84.

[23] Sambin, G. 2003. Some points in formal topology. *Theor. Comp. Sci.*, **305**, 347–408.

[24] Sambin, G. 2023. *The Basic Picture: Structures for Constructive Topology*. Oxford Logic Guides. Oxford: Clarendon Press (In the press.).

[25] Schubert, H. 1968. *Topology*. London: Macdonald Technical and Scientific.

[26] Steinke, T. A. 2011. Constructive notions of compactness in apartness spaces. M.Phil. thesis, University of Canterbury.

[27] Vickers, S. J. 1988. *Topology via Logic*. Cambridge Tracts in Theoretical Computer Science 5. Cambridge: Cambridge University Press.

PART V

LOGIC AND FOUNDATIONS

19

Countable Choice

Fred Richman

19.1 Axioms of Choice

Let $\mathbb{N}$ be the set of natural numbers, X a set, and $S \subseteq \mathbb{N} \times X$. Consider the two statements:

(i) For each $n \in \mathbb{N}$ there is $x \in X$ such that $(n, x) \in S$.
(ii) There is a function $a : \mathbb{N} \to X$ such that $(n, a_n) \in S$ for all $n \in \mathbb{N}$.

The **axiom of countable choice** (CC) says that (i) implies (ii). Note that (i) says that S describes an arbitrary sequence of nonempty subsets of X, namely $S_n = \{x \in X : (n, x) \in S\}$, while (ii) constructs a sequence of elements $a_n \in S_n$.

A set P is called **projective** if whenever f maps a set A onto a set B, then each map from P to B factors through f. Let P be a set such that if P is nonempty (has an element), then P is projective. Then P is projective. The **axiom of choice** (AC) says that every set is projective.

Theorem 19.1 *Countable choice implies* $\mathbb{N}$ *is projective.*

Proof Let π map A onto B, and $f : \mathbb{N} \to B$. We want a map $g : \mathbb{N} \to A$ such that $\pi g = f$. Define $A_n = \pi^{-1} f(n)$ and note that it is a nonempty subset of A because π is onto. So there exist $a_1, a_2, a_3, \ldots$ in A with $a_n \in A_n$. Define $g(n) = a_n$. □

A subset D of a set S is said to be **detachable** (from S) if for each $s \in S$, either $s \in D$ or not $s \in D$. Equivalently, there exists a function f from S to $\{0, 1\}$ such that $D = \{s \in S : f(s) = 1\}$. The subset $C = \{s \in S : f(s) = 0\}$ is then also detachable, and S is the categorical coproduct (direct sum) of D and C. We say that C is the **complement** of D. Note that a subset is detachable if and only if it is a summand.

We say a set is **countable** if it is the range of a function f from a detachable subset D of $\mathbb{N}$. A nonempty countable set S is an image of $\mathbb{N}$ (extend f to $\mathbb{N}$ by defining it to be constant on the complement of D). A set is **discrete** if, given elements x and y in it, either $x = y$ or not $x = y$.

The usual reason constructivists give for rejecting AC is the **Goodman–Myhill theorem**, which says that AC implies the law of excluded middle. It's hard not to be disappointed by the Goodman–Myhill proof [4]. They consider the set $S = \{0, 1\}$ with the usual equality, and the set $T = \{0, 1\}$ with equality defined by $0 = 1$ if P, where P is an arbitrary proposition. There is a natural function f from S onto T. The full axiom of choice says that there is a function $g : T \to S$ such that fg is the identity on T. If $g(0) = 1$ or if $g(1) = 0$, then P holds. Otherwise, $\neg P$ holds.

The sets S and T in the Goodman–Myhill proof are classically finite, yet the whole point of the classical axiom of choice is that it deals with an infinite number of choices. Curiously, if we had constructed such a Brouwerian counterexample to some other mathematical statement, we would immediately have tried to reformulate that statement in a classically equivalent form so that the Brouwerian counterexample didn't apply.

Note that the set T in the proof of the Goodman–Myhill theorem is countable (indeed it is an image of $\mathbb{N}$ itself) but not discrete. To circumvent the Goodman–Myhill theorem, we could have the constructive version of AC be that *discrete* sets are projective, although this does not look like a particularly fruitful direction in which to go.

Constructivists typically reject the full axiom of choice but embrace the countable one. I'm more inclined to go along with Lebesgue who, in a letter to Borel [7], wrote, "I agree completely with Hadamard when he states that to speak of an infinity of choices without giving a rule presents a difficulty that is just as great whether or not the infinity is denumerable."

The following theorem gives two conditions equivalent to the statement that $\mathbb{N}$ is projective.

Theorem 19.2 *The following conditions are equivalent:*

(i) *The set $\mathbb{N}$ is projective,*

(ii) *detachable subsets of $\mathbb{N}$ are projective,*

(iii) *discrete countable sets are projective.*

Proof Obviously (iii) implies (ii) implies (i). To show that (i) implies (iii), suppose f is a function from a detachable subset D of $\mathbb{N}$ onto a discrete set S. We may assume that S is nonempty, so we can extend f to $\mathbb{N}$. To see that S is projective, we show that it is a summand of $\mathbb{N}$. Given an element s in S, the preimage of s in $\mathbb{N}$ is nonempty and detachable, hence has a least element $g(s)$. Note that fg is the identity on S, so S is a summand of the projective set $\mathbb{N}$, whence S is projective. □

19.2 Living without Countable Choice

I have long been interested in doing constructive mathematics without countable choice. I've always disliked proofs that use countable choice although it took me some time to realize why I disliked them. I also have an antipathy to relying on sequences. If you do mathematics without countable choice, you find that there are many situations where you can't use sequences because you don't have the main tool for constructing them. But you can argue against using sequences themselves, even classically.

For example, one semester, I taught out of a book in elementary real analysis for undergraduates. The author talked about limits of sequences and proved some theorems about them. Then he addressed what he called "functional limits": the limit of $f(x)$ as x goes to 0. In order to prove the same theorems for functional limits, he reduced them to sequential limits, presumably for efficiency. Of course that can't be done constructively, nor can it be done classically in a larger context. But surely you are not supposed to think of a functional limit in terms of sequences, whatever the context. The good idea, that $f(x)$ is close to L if x is small, should not be reduced to the bad idea, that if a sequence s_n converges to 0, then the sequence $f(s_n)$ converges to L.

The rejection of countable choice seems natural from a computational point of view: if you write a program to compute something, you don't normally allow your computer any discretion in choosing among alternatives.

In the absence of the countable axiom of choice, you find that you have little reason to use sequences at all. For example, if you define a real number to be a Cauchy sequence of rational numbers, then you will be unable to show that every Cauchy sequence of real numbers converges. So you will be inclined to abandon that definition in favor of one that is not based on sequences, and you will be led to a nonsequential definition of completeness also.

Being forced to downplay sequences is actually an argument in favor of rejecting countable choice, perhaps the most important one. That is, I have come around to thinking that sequences are undesirable, therefore I don't want to use countable choice, rather than that I don't want to use countable choice, therefore I must abandon sequences. This also addresses the idea that when we ostensibly appeal to the axiom of countable choice, we are really not invoking that axiom but are instead relying on a particular view of how mathematics is done. That's the best interpretation I can give to Bishop's notorious statement [1, page 9] that "A choice function exists in constructive mathematics, because a *choice is implied by the very meaning of existence*."

Brouwer was also bothered by countable choice. His solution was to adopt a more general notion of a sequence that allowed choices. In effect, he simply defined this more general notion of a sequence by stipulating that countable choice holds. Those sequences that did not allow choices, what we might call *deterministic sequences*, were said to be "lawlike." I find that move a little mysterious, but I appreciate what the motivation might be. Most constructivists reject the idea that every sequence of integers should be assumed Turing computable, so why insist that such a sequence be based on a rule? Brouwer's choice sequences are abstractions from the notion of a rule-based sequence. The internals of a sequence are not something we should be concerned about. That's why Brouwer's notion of an infinitely proceeding sequence is so attractive. As long as we know that for each n we get an integer m, the mechanism for getting that integer m is a matter of indifference to us.

Bishop's sequences are definitely lawlike: a sequence is a function on the positive integers, and a function is a well-defined finite procedure that assigns to each element of its domain an element of its codomain. So how can Bishop make moves that appear to be applications of the axiom of countable choice while using only lawlike sequences? I have some ideas, as do other people, but the question is unimportant. We don't need to know what Bishop's rationale was; we can explain his moves as appeals to countable choice even if he did not believe he was doing that. There is no reason to doubt that he believed that the axiom of countable choice is valid. One of the virtues of the axiomatic method is that you can describe the behavior of mathematicians without inquiring into their thought processes.

The benefits of abandoning countable choice are like the benefits of abandoning the law of excluded middle. The prevalence of separability hypotheses and sequential arguments in constructive mathematics is due in large part to the adoption of countable choice. Without this axiom, there is very little you can do with sequences that you cannot do more generally. Rejecting countable choice forces you to formulate things better, and it makes separability hypotheses pointless for lack of consequences.

19.3 The Fundamental Theorem of Algebra

The **fundamental theorem of algebra** (which is not a theorem in algebra) says that you can construct a complex root of a monic polynomial f of degree greater than 0 with complex coefficients. Without countable choice we can construct only isolated roots, or isolated multiple roots; indeed, we can't construct a root of the polynomial $X^2 - a$, for an arbitrary complex number a (the problem occurs when a is small). Rather than constructing an approximation to each root of f, we construct an arbitrarily close approximation to the (finite) multiset of all roots of f [9].

An n-**multiset** of complex numbers is a sequence $s_1, \ldots, s_n$ of complex numbers. We equip the set of all n-multisets of complex numbers with the pseudometric

$$d(s,t) = \inf_{\pi} \sup_{i=1}^{n} (|s_i - t_{\pi i}|),$$

where π ranges over the permutations of $1, \ldots, n$. We then approximate f by a polynomial g whose coefficients are Gaussian numbers, and compute the set of roots, including multiplicities, of g in the discrete field of algebraic numbers. This multiset then approximates the multiset of roots of f in a precise sense.

For example, if we want to approximate the multiset of roots of $X^2 - a$ within $\varepsilon > 0$, we use the multiset $\{0, 0\}$ if $|a| < \varepsilon^2$ (so if $x^2 = a$, then $|x - 0| < \varepsilon$), while if $|a| > 0$, then we can factor $X^2 - a = (X - s)(X - t)$ and use (an approximation to) the multiset $\{s, t\}$.

Such an approach is simpler, and more closely aligned with applied practice, than the traditional constructive theory. It focuses on finding an approximate linear factorization of the polynomial, rather than on finding separate approximations to each of its roots – that is, the difference is between an approximation to the set of roots and a set of approximations to each of the roots.

19.4 Completions

Without countable choice, the completion of a metric space via Cauchy sequences need not be complete. Given a Cauchy sequence of Cauchy sequences, you need to be able to extract a sequence that converges to an appropriate limit, and that normally requires choice. The most practical way to get around this is to let an element of the **completion** be a regular sequence of nonempty sets as suggested in [12]. A sequence of nonempty subsets S_n of a metric space is said to be **regular** if $d(S_m, S_n) \leq 1/m + 1/n$ for all m and n. (To say, for two subsets S and T, that $d(S,T) \leq q$, means simply that $d(s,t) \leq q$ for all $s \in S$ and $t \in T$.) Note that a regular sequence of *elements* of the metric space is obtained by choosing an element out of each nonempty subset S_n.

The first completion one runs across is the construction of the real numbers from the rational numbers. A minimalist, axiomatic approach to the real numbers, in the absence of choice, is given in [6]. The starting point is a general theory of (not necessarily discrete) linearly ordered sets. To convey the flavor of this paper, I'll mention one (early) result [6, Theorem 13]: *An ordered set X is upper complete if and only if it is lower complete.* Here X is **upper complete** if every upper located subset with an upper bound has a supremum in X. A subset S is **upper located** if for any $a < b$ in X either $a < s$ for some $s \in S$, or $S \leq u < b$ for some $u \in X$.

19.5 The Ascending Tree Condition

Consider the classical theorem that you can diagonalize any matrix M over a principal ideal domain; there exist invertible matrices A and B such that all the entries off the diagonal of AMB are 0, and each diagonal entry divides the next. That is, every principal ideal domain is an **elementary divisor ring**. The usual classical definition of a principal ideal domain is useless from a constructive point of view because even for the two-element field, you can't prove that every ideal is principal. The algorithmic part of the definition is that *finitely generated* ideals are principal. The ring of integers satisfies this condition, as does the polynomial ring in one variable over a field, because they both have a division algorithm – they are Euclidean rings. An integral domain that satisfies this condition is called a **Bezout domain**.

The theorem usually states that each diagonal entry in the resulting diagonal matrix divides the next, to make the ideals generated by diagonal entries unique: the two-by-two diagonal matrices $2, 3$ and $6, 0$ are equivalent yet do not even have the same rank. This last step, which we leave to the reader, can be done constructively over an arbitrary Bezout domain.

It's enough to diagonalize a two-by-two matrix. Using the Bezout property, you can multiply on the left by an invertible matrix to get a zero in the lower left corner and the greatest common divisor (gcd) of the first column in the upper left. You can then do the same on the right to get a zero in the upper right corner and the gcd of the (modified) first row in the upper left. In this way you generate a sequence of elements in the upper left corner, $a_0, a_1, \ldots$ such that $Ra_0 \subseteq Ra_1 \subseteq \cdots$. To finish the proof, you need the ascending chain condition on principal ideals to conclude that $Ra_j = Ra_{j+1}$ for some j; so matrix $j+1$ is diagonal because a_j divided everything in the first row and column of matrix j.

From a choiceless point of view there's a problem constructing the sequence $a_0, a_1, \ldots$. There are choices made in the generation of the sequence of upper-left-corner elements. That's because in a Bezout domain, the s and t such that $sa + tb = d$ are not singled out – there may be many. Indeed, the d itself is not unique, so there is a choice involved in the construction of the upper-left-corner element to which we want to apply the divisor chain condition.

One way to get around this is the *ascending tree condition* [10]. By a **tree** we mean a partially ordered set T with a smallest element such that:

(i) For each t in T, the set $\{s \in T : s \leq t\}$ is a finite chain.
(ii) For each s in T there is t in T such that $s \leq t$ and $s \neq t$. We write $s < t$.

Condition (ii) is unambiguous because (i) tells us what $s \neq t$ means if $s \leq t$. Note that the natural numbers $\mathbb{N}$ form a tree. A (deterministic) sequence is a family of elements indexed by $\mathbb{N}$. A Brouwerian choice sequence can be thought of as a family of elements indexed by a tree.

Let P be a partially ordered set, for example, the finitely generated ideals in some ring. A family of elements I_t of P, indexed by T, is **increasing** if $s < t$ implies $I_s \leq I_t$. An increasing family I_t **halts** if there are $s < t$ in T such that $I_s = I_t$.

A partially ordered set P satisfies the **ascending chain condition** (ACC) if every increasing family of elements of P indexed by $\mathbb{N}$ halts. It satisfies the **ascending tree condition** (ATC) if every increasing family of elements indexed by a tree halts. In the presence of the axiom of dependent choices, the two notions are equivalent [10, Theorem 1].

The ATC on finitely generated ideals holds without choice for Euclidean domains, and even for Dedekind–Hasse domains. If you define a **Noetherian ring** to be one that satisfies the ATC, rather than the ACC, on finitely generated ideals, then you can prove the Hilbert basis theorem, without choice, in the form: If R is coherent and Noetherian, then so is $R[X]$ [10, Theorem 9]. Moreover, an arbitrary quotient of a Noetherian module is Noetherian, as is the extension of a Noetherian module by a Noetherian module, [10, Theorems 3 and 4].

It is thus natural to define a **PID** (a principal ideal domain) to be a Noetherian Bezout domain. We can then show, without choice, that a PID is an elementary divisor ring. I once thought that this analysis of the proof that a PID is an elementary divisor ring established the superiority of ATC over ACC in a choiceless environment. In fact, it turned out that I was just looking at the wrong proof of that theorem. Helmer, in [5], also worries about using ACC to prove that a ring is an elementary divisor ring. He defines the notion of an **adequate** Bezout domain that does not refer to a chain condition. Here is a slightly stronger and simpler definition: given a and c, we can write $a = rs$ such that $\gcd(r, c) = 1$ and s divides c^n for some n.

Theorem 19.3 *A Bezout domain with ACC is adequate [10, Theorem 10]. An adequate Bezout domain is an elementary divisor ring [10, Theorem 11].*

19.6 Bishop's Principle and the λ-Technique

Lemma 7 on page 177 of [1], known as **Bishop's lemma** [3], or **Bishop's principle** [2], says that if S is a complete located subset of a metric space X, and $x \in X$, then there exists $s \in S$ such that if $x \neq s$, then x is bounded away from S. In its proof, Bishop introduces a binary sequence λ with the property that

$$\lambda_n = 0 \Rightarrow r < \frac{1}{n}$$
$$\lambda_n = 1 \Rightarrow r > \frac{1}{2n}$$

where r is a given real number (the distance from x to S). He also requires that the sequence λ be increasing. I have taken the liberty of interchanging the roles of 0 and 1, and replacing the original "decreasing" by "increasing," because we normally look for a 1 in an increasing binary sequence rather than look for a 0 in a decreasing binary sequence. The construction of such a binary sequence λ is called **the λ-technique** in [3, page 64].

Ignoring the increasing requirement for the moment, what principle would allow the construction of the sequence λ? We are given two sequences of propositions, P_n and Q_n, such that $P_n \vee Q_n$ holds for all n, and we want to construct a binary sequence λ such that

$$\lambda_n = 0 \Rightarrow P_n$$
$$\lambda_n = 1 \Rightarrow Q_n.$$

To be given two such sequences of proposition is equivalent to being given a sequence S_n of nonempty subsets of $\{0, 1\}$, where $S_n = \{0 : P_n\} \cup \{1 : Q_n\}$. The condition on λ says that $\lambda_n \in S_n$. So this is countable choice for a sequence of nonempty subsets of $\{0, 1\}$, that is, we restrict X to be $\{0, 1\}$ in the definition of countable choice at the start of the paper.

The additional condition that λ be increasing, which suggests that dependent choice is being invoked, is harmless because, in this case, Q_n implies Q_{n+1} for each n. Thus, after the initial construction, we may modify λ so that if $\lambda_n = 0$, then $\lambda_{n+1} = 0$ and still maintain the displayed condition.

If we choose Q_n to be $r > 1/(n+1)$ instead of $r > 1/(2n)$, then either S_n or $S_{n'}$ is a singleton whenever $n \neq n'$. Indeed, that choice is made in the proof of Bishop's lemma in [3]. So there is at most one real choice to be made in selecting an element of S_n for each n. This kind of weak countable choice (WCC) principle was formulated in [2] as follows.

WCC If S_n is a sequence of nonempty subsets of a set X, such that either S_n or $S_{n'}$ is a singleton whenever $n \neq n'$, then there is a sequence s_n such that $s_n \in S_n$ for all n.

Note that X is not required to be $\{0, 1\}$ here. In [2], it is shown that WCC suffices for Bishop's lemma and for the fundamental theorem of algebra (in its traditional form).

19.6.1 When is $\mathbb{R}a$ Closed in $\mathbb{R}$?

If $a = 0$, then $\mathbb{R}a = \{0\}$, while if $a \neq 0$, then $\mathbb{R}a = \mathbb{R}$. Both $\{0\}$ and $\mathbb{R}$ are closed in $\mathbb{R}$, so if $a = 0$ or $a \neq 0$, then $\mathbb{R}a$ is closed in $\mathbb{R}$. What about the converse? That's proved in [8], but countable choice is used. Actually, the appeal to countable choice for the converse can be eliminated. That's a little embarrassing for me, but I'm glad that I finally realized it. We state the theorem more generally.

Theorem 19.4 *Let $a \leq b$ be real numbers, and $\{a, b\} \subseteq \overline{K} \subseteq [a, b]$. Let V be the vector subspace $\langle K \rangle$ of $\mathbb{R}$ generated by K. If V is closed in $\mathbb{R}$, then V is finite dimensional.*

Taking $a = b$ we get the converse mentioned in the first paragraph. We can also take K to be any totally bounded subset of $\mathbb{R}$, letting $a = \inf K$ and $b = \sup K$. In particular, K can be a finitely enumerable subset of $\mathbb{R}$. Note that if V is finite dimensional, then $V = \{0\}$ or $V = \mathbb{R}$.

First we establish a lemma characterizing the elements of $\overline{V}$.

Lemma 19.5 $x \in \overline{V}$ *if and only if* $x \neq 0 \Rightarrow a \neq 0$ *or* $b \neq 0$.

Proof If $x \in \overline{V}$ and $x \neq 0$, then there is $y \in K$ such that $y \neq 0$, so $a \neq 0$ or $b \neq 0$. Conversely, let $\varepsilon > 0$. Either $|x - 0| < \varepsilon$ or $x \neq 0$. In the latter case $a \neq 0$ or $b \neq 0$ so $V = \mathbb{R}$ whence $x \in \mathbb{R}$. In either case $d(x, V) < \varepsilon$. □

A second lemma gives a criterion for a real number to be either 0 or different from 0.

Lemma 19.6 *If $s \geq 0$, and $\sqrt{s} \leq ms$ for some integer m, then $s = 0$ or $s \neq 0$.*

Proof If $s \neq 0$, then $1 \leq m^2 s$, so $s \geq 1/m^2$. Thus, if $s < 1/m^2$, then $s = 0$. But $s < 1/m^2$ or $s > 0$. □

Now for the proof of Theorem 19.4.

Proof Let $s = |a| + |b|$. Then $\sqrt{s} \in \overline{V} = V$, by Lemma 19.5, so $\sqrt{s} = \sum_i r_i t_i$ where $a \leq t_i \leq b$. Choose $m \in \mathbb{Z}^+$ so that $\sum |r_i| \leq m$. Then $\sqrt{s} \leq m \sup |t_i| \leq ms$ because $|t_i| \leq s$ for all i. Thus $s = 0$ or $s \neq 0$ by Lemma 19.6. In the first case $V = \{0\}$; in the second $V = \mathbb{R}$. □

An extension of Theorem 19.4 to a more general setting is proved, using choice, in [11]. The more general theorem in [11] differs from Theorem 19.4 in that it is not classically trivial. Moreover, its proof appeals to the Baire category theorem and the constructive Hahn–Banach theorem. It looks like a formidable challenge to reformulate it appropriately and prove it without countable choice. As a start, one might try to generalize Theorem 19.4 to $\mathbb{R}^2$ with K a finitely enumerable set.

References

[1] Bishop, E. 1967. *Foundations of Constructive Analysis*. New York: McGraw-Hill.

[2] Bridges, D. S., Richman, F., and Schuster, P. 2000. A weak countable choice principle. *Proc. Amer. Math. Soc.*, **128**, 2749–2752.

[3] Bridges, D. S., and Vîţă, L. 2006. *Techniques of Constructive Analysis*. Berlin: Springer.

[4] Goodman, N., and Myhill, J. 1978. Choice implies excluded middle. *Zeit. Math. Log.*, **23**, 461.

[5] Helmer, O. 1943. The elementary divisor theorem for certain rings without chain condition. *Bull. Amer. Math. Soc.*, **49**, 225–236.

[6] Joseph, J. S. 2018. A constructive theory of ordered sets and their completions. Ph.D. dissertation, Florida Atlantic University.

[7] Moore, G. H. 1983 *Zermelo's Axiom of Choice, its Origins, Development and Influence*. Berlin: Springer.

[8] Richman, F. 1982. Meaning and information in constructive mathematics. *Amer. Math. Mon.*, **89**, 385–388.

[9] Richman, F. 2000. The fundamental theorem of algebra: a constructive development without choice. *Pacific J. Math.*, **196**, 213–230.

[10] Richman, F. 2003. The ascending tree condition: constructive algebra without countable choice. *Comm. Algebra*, **31**, 1992–2002.

[11] Richman, F., Bridges, D., Calder, A., Julian, W., and Mines, R. 1981. Compactly generated Banach spaces. *Archiv Math.*, **36**, 239–243.

[12] Stolzenberg, G. 1988. *Sets as Limits*. Preprint

20

The Minimalist Foundation and Bishop's Constructive Mathematics

Maria Emilia Maietti and Giovanni Sambin

20.1 Introduction

A central aspect of Bishop's constructive mathematics in [10, 12] emphasized in [15] is that of being a generalization of classical mathematics. Indeed, contrary to other constructive approaches, such as Brouwer's intuitionistic mathematics or Markov's recursive mathematics, in his mathematical development Bishop did not use any principle incompatible with classical mathematics as that formalizable in Zermelo–Fraenkel set theory. In this way Bishop produced an analysis of mathematical concepts that is finer than in other approaches.

Bishop himself in [10, 11] and in unpublished notes sketched a foundation for his mathematics. Many proposals of a formal system apt to founding his constructive mathematics followed afterwards in the style of axiomatic set theory in [1, 2, 3, 23, 58] and in that of type theory by Martin-Löf in [54, 60].

Most notably, the so-called notion of 'setoid' over Martin-Löf's type theory appears to be close to the idea of 'set' sketched in [10] as well as the notion of 'type-theoretic function', which appears to be an adequate representation of Bishop's notion of 'operation' because it explicitly shows its computational contents or 'numerical meaning'.

Then the model of setoids formalized over Martin-Löf's type theory appears to be a suitable framework to formalize Bishop's constructive mathematics. A whole study of its categorical structure as a quotient completion had been started and it is still ongoing (see, for example, [45, 46, 49, 60, 63]), and many different kinds of setoid models have been considered (see, for example, [6, 27]).

The main drawback of the formalization of mathematics in the setoid models is that it is very far from the language used in the informal mathematical practice of constructive proofs, including that in Bishop's literature and even more so that of classical mathematics. This is because the formalization in this model, and more generally in Martin-Löf's type theory, requires us to handle lots of computational details useful for the extraction of programs from proofs but apparently useless to develop the constructive proofs themselves.

To overcome this problem, in [75] it was proposed that Martin-Löf's type theory should be extended with some abstract concepts, like that of 'proof-irrelevant proposition' and that of 'subset', as soon as they satisfy the *forget–restore principle* introduced by the second author of the present chapter. This principle states that one can abstract away from irrelevant computational information when this information can be restored in the process of extracting a program from a constructive proof.

Pushing forward the idea of the forget–restore principle, in [47] we introduced the notion of *two-level foundation* for constructive mathematics. Such a foundation should consist of:

- one theory acting as the *extensional level* written in a language close to the usual mathematical practice of proofs;
- another theory acting as the *intensional level* written in a type-theoretic language suitable for extraction of programs from proofs;
- an interpretation of the extensional level in (a model of) the intensional level showing that the extensional level has been obtained from the intensional one following the forget–restore principle.

The introduction of a two-level foundation was also motivated by the need to build a new foundation for constructive mathematics. Indeed, since 2005 with [47], we embarked on the project of building a *Minimalist Foundation* where the mathematics developed in it turns out to be compatible with the different approaches to constructivism, and also with classical mathematics. To formalize Bishop's mathematics we intended to build an intuitionistic and predicative foundation finer than the formal systems available in the literature, and characterized by the lack of whatsoever choice principle, including the so-called *axiom of unique choice*.

In [39] a full formal system, called the Minimalist Foundation, here named **MF** for short, was proposed.

In parallel, a new approach to constructivism, called 'dynamic', was also put forward in [67, 69, 70, 71, 72, 73]. This was inspired by the constructive approach originating explicitly with Brouwer at the beginning of the twentieth century and revived in the 1960s and 1970s (see, for example, [10, 53], among others). The first chapter of [74] will contain a detailed introduction to dynamic constructivism.

In the following we are going to describe what aspects of our minimalist approach and **MF** have in common with Bishop's one, called `BISH`, and what differ.

The main common aspects include the following:

- the compatibility with classical mathematics via a language close to that of usual mathematical practice;

- the need to compile this language in a strictly algorithmic language to extract the computational contents of constructive proofs.

Both aspects are fulfilled in **MF** by crucially employing its two-level structure.

Indeed, compatibility with the standard Zermelo–Fraenkel foundation for classical mathematics is fulfilled at the extensional level of **MF**, while the extraction of programs from proofs is at its intensional level. In particular, the intensional level of **MF** and of its extensions with point-free topological inductive and coinductive definitions can be interpreted in realizability semantics extending the *Kleene realizability of intuitionistic arithmetic* as shown by Ishihara, Maietti, Maschio, and Streicher in [31] for **MF** and by Maietti, Maschio and Rathjen in [51] and [52] for extensions of MF. This fact has two main consequences which emphasize the constructivity of the whole **MF** and for its extensions with point-free topological inductive and coinductive definitions.

The first is that the intensional level of **MF** is consistent with *full axiom of choice* and *formal Church thesis* as advocated in [47]. This characteristic is generally not satisfied by the other constructive intensional foundations in the literature such as the extension of Martin-Löf type theory called Homotopy Type Theory in [78] (because it satisfies the function extensionality principle).

The second consequence is that the extensional level of **MF** turns out to be consistent with the formal Church thesis (see [44]) via its interpretation at the intensional level.

Furthermore, the intensional level of **MF** could serve as a base for a minimalist proof-assistant whose formalized proofs can, a priori, be reused in proof-assistants based on the many extensions. This would be a practical application of the fact that **MF** can well serve as a basic theory to compare the different approaches to mathematics and their proofs.

We can underline some major peculiarities of **MF** not present in Bishop's conception of mathematics `BISH`.

One main difference is about the concept of function. As in `BISH`, in **MF** we have both the notion of *operation* with the meaning of representing a computable function, and that of functional relation. However, contrary to `BISH` and other type-theoretic foundations for `BISH`, in **MF** these two notions are kept well distinct. In fact, while operations between two sets do form a set, functions do not generally do. This distinction is guaranteed by the lack of the general validity of choice principles in both levels of **MF** (see [41]). Indeed it is enough to add a rule of unique choice to both levels of **MF** to guarantee the validity of the axiom of unique choice which makes the two notions coincide.

There is a major consequence of the absence of choice principles from **MF** combined with its predicative nature (even à la Feferman see [31, 43, 44]) when adopting **MF** to develop topology. It is that the constructive pointfree approach to topology introduced by Martin-Löf and the second author in the 1980s in [66] under the name of *formal topology* constitutes not only a valid alternative to pointwise approaches for constructive analysis by Brouwer (see [77]) and Bishop (see [10, 12]), but it appears to be compulsory. The main reason, as sketched in [40, 42], is that in **MF** real numbers, either as Dedekind cuts or as Cauchy sequences, cannot be proven to be sets. Also, choice sequences of Baire and Cantor spaces do not form a set. All this is a consequence of the fact that in **MF** functional relations between two sets do not generally form a set. Instead, a priori, using Martin-Löf's type theory in [60] as a foundation, both pointwise approaches and pointfree ones could seem legitimate. In fact, one can define a pointwise topology on Dedekind real numbers, because these are in bijective correspondence with Cauchy sequences, and the latter can be represented in Martin-Löf's type theory as a setoid. Also, in the predicative foundation of Aczel's constructive set theory **CZF** in [1, 2, 3] both Dedekind reals and Cauchy reals form a set and a pointwise approach is possible.

In this chapter we recall the basic definitions of formal topology necessary to introduce the previously mentioned example of real numbers and Baire and Cantor spaces by underlying how they are formalized in **MF**.

The way constructive topology is formalized in **MF** agrees well with our minimalist attitude, especially if we want to work in a constructive foundation compatible with classical predicativity where we can distinguish the real (effective) structure of a topology from a corresponding ideal (infinitary) structure of formal points.

A major benefit from developing pointfree topology in **MF** in the form of formal topology is that we gain in clarity and in an analysis of topological concepts finer than in other foundations.

Finally, we conclude by describing an extension of **MF**, actually of its extensional level, which appears closer to `BISH`.

This extension of **MF** is characterized by the validity of choice principles including the axiom of unique choice and the axiom of countable choices. It should also be interpretable in Martin-Löf's type theory to form a two-level foundation by extending the interpretation in [39]. But a proof of this is left to future work.

20.2 Why Adopt a Minimalist Foundation?

A plurality of philosophical reasons for a constructive approach to mathematics has been proposed, both before and after Brouwer and Bishop.

Presently, various logical systems to formalize constructive mathematics are available in the literature. They range from axiomatic set theories, as Aczel's **CZF** in [1, 2, 3] or Friedman's **IZF** in [8], to the internal theory of categorical universes as topoi or pretopoi in [33, 35, 37], to type theories such as Martin-Löf's type theory in [60] or Coquand's Calculus of Inductive Constructions in [19, 21]. No existing constructive foundation has yet superseded the others as the standard one, as Zermelo–Fraenkel axiomatic set theory did for classical mathematics.

Various machine-aided proof development systems are also available to implement mathematics (see, for example, [80]). Many of those for constructive mathematics are based on type systems which are also paradigms of (functional) programming languages with the possibility of extracting the computational contents of constructive mathematical proofs. Some of these, for example Coq in [18] or Matita in [5], are based on impredicative typed systems, while some others, for example Agda in [13] and Nuprl in [4], are based on predicative ones.

Beginning with [47], we embarked on the project of developing a foundation with minimal assumptions. The main reason for this choice is to support our general attitude to preserve all effective notions and conceptual distinctions as much as possible, with no a priori exception. The result is a foundation which is minimalist also in the sense that it becomes a common core among the most relevant constructive foundations. Thus we expect that such a minimalist foundation should be useful not only to constructive mathematicians but also to logicians, for example as a base system to do constructive reverse mathematics, and also to computer scientists, as a base for a *minimalist* proof-assistant suitable for formalizing reusable proofs and for program extraction from proofs.

20.2.1 Founding Constructive Mathematics on a Two-Level Theory

In our opinion, a constructive foundation should make evident those key aspects which differentiate constructive mathematics from classical mathematics. For example, a typical characteristic of constructive proofs, contrary to classical ones, is the possibility of extracting programs computing witnesses of true existential statements occurring in them.

Even better, any proof in a constructive system should be seen as a program. Hence, a foundation for constructive mathematics should be at the same time a theory of sets, in which to formalize mathematical theorems, and a programming language, in which to extract the computational contents of mathematical proofs.

In [47] we argued that such a constructive foundation (validating Heyting arithmetics at least) should be a two-level theory consisting of the following.

- A level, called *extensional*, which should be an extensional set theory (with undecidable equality of sets and elements) formulated in a language close to that used in the common practice of developing mathematics.
- Another level, called *intensional*, which should be an intensional theory (with decidable equality of sets and elements) enjoying extraction of programs from proofs; according to [47] this level should possibly be a proofs-as-programs theory, that is, a theory consistent with the axiom of choice

$$\text{(AC)} \qquad \forall x \in A\, \exists y \in B\; R(x,y) \quad \longrightarrow \quad \exists f \in A \to B\; \forall x \in A\; R(x, f(x))$$

for A, B sets and $R(x,y)$ a logical relation, and the formal Church thesis for functions between natural numbers denoted with the symbol $Fun(Nat, Nat)$

$$\text{(CT)} \qquad \forall f \in Fun(Nat, Nat) \qquad \exists e \in Nat$$
$$(\,\forall x \in Nat\; \exists y \in Nat\; T(e,x,y)\;\&\; U(y) = f(x)),$$

where Nat is the set of natural numbers and $T(e,x,y)$ is the Kleene predicate expressing that y is the computation executed by the program numbered e on the input x and $U(y)$ is the output of the computation y.

- Then, in order to guarantee the extraction of programs even from proofs written at the extensional level, we required that the extensional level should be obtained as an abstraction of the intensional level according to the forget–restore principle proposed by the second author of the present chapter in [75].

The link between the two levels was then made more technical in [39], by requiring that the extensional level should be interpreted in the intensional one by means of a quotient completion of the latter, that is, the extensional level should be seen as (a fragment of) the internal language of a quotient completion built on the intensional one.

This kind of link captures what happens in the practice of computer-aided formalization of mathematics in an intensional type theory, which makes use of the so-called model of 'setoids' built on it (see [6, 27]). Actually another motivation behind the notion of two-level foundation in [39, 47] is the desire to make explicit the extensional theory validated in the quotient model chosen to formalize mathematical proofs in intensional-type theory.

Our two-level structure where the intensional level is consistent with axiom of choice and formal Church thesis fully agrees with Bishop's need to exhibit the computational contents of constructive proofs, in particular of existential statements whose witness can be chosen computationally (see [10, Chapter 1] and [11]).

20.3 The Minimalist Foundation

In [39] we presented a two-level formal system which satisfies the requirements in [47] of a two-level foundation for constructive mathematics. We call this system *the two-level minimalist foundation*, or **MF** for short. We are aware, however, that a specific formal system, which is static by definition, cannot fully capture the dynamics of the minimalist approach to constructivism, started in [47, 69, 70, 71].

The two levels of **MF** are both given by a type theory à la Martin-Löf: the intensional level, called **mTT**, is an intensional type theory including aspects of Martin-Löf theory in [60] (and extending the set-theoretic version in [47] with collections), and its extensional level, called **emTT**, is an extensional type theory including aspects of extensional Martin-Löf's theory in [55]. Then a quotient model of setoids à la Bishop in [6, 10, 27, 61] is used in [39] to interpret the extensional level in the intensional one. A categorical study of this quotient model has been carried out in [45, 46, 49] and is related to the construction of Hyland's effective topos in [28, 29].

In the following, we explain the main characteristics of the extensional level **emTT** and of **mTT** viewed more as a many-sorted logic than as a type theory. This is because both levels of **MF** are given by a type theory that includes a primitive notion of proposition, which allows us to control the validity of choice principles.

Need for Two Types of Entities: Sets and Collections A minimalist foundation for constructive mathematics should certainly be based on intuitionistic predicate logic and include at least the axioms of Heyting arithmetic. Hence we could expect to build it starting from a many-sorted logic, such as Heyting arithmetic of finite types in [77], where sorts, which we call *types*, include the basic sets we need to represent our mathematical entities.

However, in order to develop topology in an intuitionistic and predicative way, we need a foundation with two kinds of types: sets and collections. The main reason is that the *power of a non-empty set*, namely the discrete topology over a non-empty set, fails to be a set in a predicative foundation, and it is only a *collection*.

Need for Two Types of Propositions In parallel with the presence of sets and collections, to keep the system predicative we also need to distinguish two types of propositions: those closed under quantifications on sets, called here *small propositions* as in [39] (and *proper propositions* in [74]), from those closed under any kind of quantification, called here simply *propositions* as in [39] (and *improper propositions* in [74]). Both kinds of propositions include propositional equalities which are small propositions only if they refer to elements of a set.

Need for Two Types of Functions It is well known that by *adding the principle of excluded middle* to some constructive foundations, such fas Aczel's **CZF** or Martin-Löf's type theory, one can derive that *power-collections become sets* and thus get an impredicative theory. In both such theories this is due to the fact that the collection of functions from a set A to the boolean set $\{0, 1\}$, called *exponentiation of the boolean set over A*, forms a set, too. Therefore, if we wish to have compatibility with classical theories where the power of a non-empty set is not a set as in Feferman's predicative theories in [23], we need to avoid exponentiation of functions.

A drastic solution is to drop all axioms yielding any form of exponentiation. What we propose is to allow exponentiation only of a certain kind, as happens in

[23]. To this purpose, we introduce a primitive notion of *operation*, represented by certain functional terms

$$f(x) \in B\ [x \in A]$$

in a set B with a free variable in the set A. These operations can be defined as *type-theoretic functions* of a type theory, like in Martin-Löf's type theories in [55, 60]. Clearly any operation $f(x) \in B\ [x \in A]$ must give rise to a functional relation $f(x) =_B y\ [x \in A, y \in B]$, namely what is usually called function. What we do not wish to guarantee is the converse. Our idea is then that *only exponentiation of operations from a set A to a set B forms a set.*

20.3.1 The Main Types of the Extensional Level of the Minimalist Foundation

The formal system **emTT** of the extensional level of the Minimalist Foundation in [39] is written in the style of Martin-Löf's type theory in [60] by means of the following four kinds of judgements:

$$A\ type\ [\Gamma] \quad A = B\ type\ [\Gamma] \quad a \in A\ [\Gamma] \quad a = b \in A\ [\Gamma];$$

that is, the type judgement (expressing that something is a specific type), the type equality judgement (expressing that two types are equal), the term judgement (expressing that something is a term of a certain type), and the term equality judgement (expressing the *definitional equality* between terms of the same type), respectively, all under a context Γ.

The word *type* is used as a meta-variable to indicate four kinds of entities: collections, sets, propositions, and small propositions, namely

$$type \in \{coll, set, prop, prop_s\}.$$

Therefore, in **emTT** types are actually formed by using the following judgements:

$$A\ set\ [\Gamma] \qquad B\ coll\ [\Gamma] \qquad \phi\ prop\ [\Gamma] \qquad \psi\ prop_s\ [\Gamma]$$

saying that A is a set, that B is a collection, that ϕ is a proposition, and that ψ is a small proposition.

Here, contrary to [39] where we use only capital latin letters as meta-variables for all types, we use greek letters ψ, ϕ as meta-variables for propositions and capital italic latin letters A, B as meta-variables for sets or collections, and small italic latin letters a, b, c as meta-variables for terms, that is, elements of the various types.

Observe that for a set A, when we say that

$$a \in A\ [\Gamma]$$

is derivable in **emTT**, we actually mean that the term a *is an element of the set* A *under the context* Γ and hence the symbol $\in$ stands for a *set membership*. As usual in type theory, *equality of sets* is given *primitively* and is not defined by equating sets with the same elements. This is indeed a main difference between a set theory defined as a typed system in the style of Martin-Löf's type theory in [60] and an axiomatic set theory à la Zermelo–Fraenkel.

We now proceed by briefly describing the various kinds of types in **emTT**, starting from small propositions and propositions, then sets, and finally collections.

Small propositions in **emTT** include all the logical constructors of intuitionistic predicate logic with equality and quantifications restricted to sets:

$$\begin{aligned} \phi\ prop_s \ \equiv\ & \bot \mid \phi \wedge \psi \mid \phi \vee \psi \mid \phi \to \psi \mid \\ & \forall x \in A\, \phi(x) \mid \exists x \in A\, \phi(x) \mid x =_A y \end{aligned}$$

provided that A *is a set*. Here we use the more familiar $x =_A y$ for the *extensional* equality type $\mathsf{Eq}(A, a, b)$ of Martin-Löf type theory in [55].

Then, *propositions* of **emTT** include all the logical constructors of intuitionistic predicate logic with equality and quantifications on all kinds of types, namely sets and collections. Of course, small propositions are also propositions:

$$\begin{aligned} \phi\ prop \ \equiv\ & \phi\ prop_s \mid \phi \wedge \psi \mid \phi \vee \psi \mid \phi \to \psi \mid \\ & \forall x \in B\, \phi(x) \mid \exists x \in B\, \phi(x) \mid x =_B y. \end{aligned}$$

In order to close sets under comprehension, for example to include the set of positive natural numbers $\{x \in \mathrm{N} \mid x \geq 1\}$, and to define operations on such sets, we need to think of propositions as types of their proofs: small propositions are seen as sets of their proofs while generic propositions are seen as collections of their proofs. That is, we add to **emTT** the following rules:

$$(\textbf{prop}_s\textbf{-into-set})\ \frac{\phi\ prop_s}{\phi\ set} \qquad\qquad (\textbf{prop-into-coll})\ \frac{\phi\ prop}{\phi\ coll}.$$

The difference between the notion of set and collection will be explained later in this section.

A key feature of the extensional typed system **emTT** is proof-irrelevance of propositions. This means that in **emTT** a proof of a proposition, if it exists, is *unique* and equal to a canonical proof term called true thanks to the following rules:

$$(\textbf{prop-mono})\ \frac{\phi\ prop\ [\Gamma] \qquad p \in \phi\ [\Gamma] \quad q \in \phi\ [\Gamma]}{p = q \in \phi\ \ [\Gamma]}$$

$$(\textbf{prop-true})\ \frac{\phi\ prop \qquad p \in \phi}{\mathsf{true} \in \phi}.$$

Proof-irrelevance of propositions justifies the introduction of a judgement asserting that *a proposition ϕ is true under a context Γ assuming propositions $\psi_1, \ldots, \psi_m$ are true* as in [55, 56]. This judgement can be directly interpreted in **emTT** as follows:

$$\phi\ true\ [\Gamma; \psi_1\ true, \ldots, \psi_m\ true] \ \equiv\ \mathsf{true} \in \phi\ [\Gamma,\ y_1 \in \psi_1, \ldots, y_m \in \psi_m].$$

In **emTT** *sets* are characterized as inductively generated types and they include the following:

$$A\ set \ \equiv\ \phi\ prop_s \mid N_0 \mid N_1 \mid List(A) \mid \Sigma_{x\in A}B(x) \mid A+B \mid \Pi_{x\in A}\, B(x) \mid A/\rho,$$

where the notation N_0 stands for the empty set, N_1 stands for the singleton set, $List(A)$ stands for the set of lists on the set A, $\Sigma_{x\in A}B(x)$ stands for the indexed sum of the family of sets $B(x)\ set\ [x \in A]$ indexed on the set A, $A + B$ stands for the disjoint sum of the set A with the set B, $\Pi_{x\in A}B(x)$ for the product type of the family of sets $B(x)\ set\ [x \in A]$ indexed on the set A, and A/ρ stands for the quotient set provided that ρ is a small equivalence relation $\rho\ prop_s\ [x \in A, y \in A]$. Moreover, we call N the set of natural numbers represented by $List(N_1)$.

The notion of set in **emTT** agrees with that in [10] and in [53]. According to them, sets must have an effective nature which is mostly forgotten in any axiomatic approach where a universe of sets closed under certain properties is implicitely assumed as the underlying range of the set variables. In fact, each set A must be specified by providing a finite number of rules to construct all its elements (see the rules of **emTT** forming elements of sets in [39]). It is understood that the rules defining a set are inductive, that is, their application can be iterated any finite number of times. The infinite is only potential, and in a certain sense it is always reduced to a finite description, at a higher order: not a finite number of elements, but a finite number of rules to generate (the infinite number of) them. In particular, the elements of the product type $\Pi_{x\in A}B(x)$ are only terms

$$b(x) \in B(x)\ [x \in A].$$

In the case the family $B(x)\ set\ [x \in A]$ is just a constant set B indexed on the set A, we indicate the product type simply as

$$A \to B \equiv \Pi_{x\in A}B$$

and its elements are just *operations*

$$b(x) \in B\ [x \in A].$$

Hence, in **emTT** *operations between two sets form a set*, but generic functions between them do not.

Finally, *collections* in **emTT** include the following types:

$$B\ coll\ \equiv\quad A\ set\ \mid\ \phi\ prop\ \mid\ \mathcal{P}(1)\ \mid A \to \mathcal{P}(1)\ \mid \Sigma_{x\in B}\, C(x),$$

where $\mathcal{P}(1)$ and $A \to \mathcal{P}(1)$ stand for the power-collections of the singleton and of a set A respectively, and $\Sigma_{x\in B}\, C(x)$ stands for the indexed sum of the family of collections $C(x)$ *col* $[x \in B]$ indexed on the collection B. Actually, for a set A, we will use the common abbreviation of power-collection

$$\mathcal{P}(A)\ \equiv\ A \to \mathcal{P}(1).^{1}$$

Elements of the power-collections rely on the notion of subset, which in **emTT** is inspired by that in [75] put on top of Martin-Löf's type theory. A *subset of a set* A is defined as the equivalence class of a *small* predicates $\phi(x)$ depending on one argument in A with respect to the equivalence relation of equiprovability. This is the minimum we must require in order to close subsets under comprehension. Indeed, for any small predicate $\phi(x)\ prop_s\ [x \in A]$ on a set A we can define its subset comprehension as

$$\{\, x \in A \mid \phi(x)\,\} \in \mathcal{P}(A).$$

Moreover, two equiprovable small predicates give rise to the same subset, that is, in **emTT** we can derive

$$\frac{\phi_1(x)\ \leftrightarrow\ \phi_2(x)\ true\ [x \in A]}{\{\, x \in A \mid \phi_1(x)\,\} =_{\mathcal{P}(A)} \{\, x \in A \mid \phi_2(x)\,\}\ true}.$$

In the following we indicate subsets of a set A with capital letters $U, V, W \ldots$.

Associated with the notion of subset we have also a *subset membership* indicated with the symbol ϵ, which we distinguish from the primitive set membership $\in$ used to say that an element belongs to a certain set. Given a subset $U \subseteq A$ of a set A, that is, $U \in \mathcal{P}(A)$, for any $a \in A$ we define a new small proposition

$$a \,\epsilon\, U\ prop_s.$$

We can prove in **emTT** that

$$U = \{\, x \in A \mid x \,\epsilon\, U\,\} \in \mathcal{P}(A)$$

and also that, for any small predicate $\phi(x) \in \mathsf{prop_s}\ [x \in A]$ on the set A and for any element $a \in A$,

$$a \,\epsilon\, \{\, x \in A \mid \phi(x)\,\}\ \leftrightarrow\ \phi(a)\ \ true.$$

[1] The notation $A \to \mathcal{P}(1)$ for the power-collection $\mathcal{P}(A)$ is used to remember that its elements are operations from a set A to the power-collection on the singleton.

The subset equality is equivalent to usual extensional equality with respect to membership ϵ, namely we can derive in **emTT** that

$$\forall x \in A\,(x \,\epsilon\, U \leftrightarrow x \,\epsilon\, W) \;\leftrightarrow\; U =_{\mathcal{P}(A)} W \;\; true$$

and, of course, that

$$\{\, x \in A \mid \phi(x) \,\} =_{\mathcal{P}(A)} \{\, x \in A \mid \psi(x) \,\} \quad \leftrightarrow \quad \forall_{x \in A}\,(\phi(x) \leftrightarrow \psi(x)) \;\; true.$$

In particular, $\mathcal{P}(1)$ denotes the power-collection of the singleton N_1 and its elements are equivalence classes of small propositions closed under the equivalence relation of equiprovability.

The fact that subset equality corresponds to usual extensional equality of sets suggests that we can view the subset theory in **emTT** as a *local set theory* where subsets of a set A can be considered *local sets* in [9] in the style of Zermelo–Fraenkel set theory. Then, membership and extensional equality via elements becomes a local property restricted to a given set A. To this purpose, observe that among subsets of A, there is A itself thought of as the subset

$$\{\, x \in A \mid \mathtt{tt} \,\},$$

where tt is any tautology. Moreover, we can define *quantifiers relativized to a subset*: this means that, if $U \subseteq A$ and φ is a small predicate (or propositional operation) with an argument in A, we write $\exists x \,\epsilon\, U\; \varphi$ as an abbreviation for the formula $\exists x \in A\; (x \,\epsilon\, U \;\; \& \;\; \varphi)$, and $\forall x \,\epsilon\, U\; \varphi$ as an abbreviation for the formula $\forall x \in A\; (x \,\epsilon\, U \rightarrow \varphi)$. A consequence of these definitions is that all laws of many-sorted intuitionistic logic regarding quantifiers extend to quantifiers relativized to a subset.

Note that the membership relation ϵ between terms and subsets is crucial in **emTT** to obtain an embedding of subsets into sets, which associates the set

$$\Sigma_{x \in A}\; x \,\epsilon\, U \;\; set$$

to a subset $U \subseteq A$. In this way an operation from $U \subseteq A$ to a set B can be represented as an operation in $\Sigma_{x \in A}\; x \,\epsilon\, U \rightarrow B$.

The **emTT** distinction between set and collection is analogous to the distinction between set and class in axiomatic set theory. But while in axiomatic set theory the distinction is mainly due to problems with consistency (or size), here it is motivated by quality of information and preservation of predicativity. Indeed, sets are kept distinct from collections to be able to maintain a distinction between computable, effective domains (represented by sets) and non-computable ones (represented by collections). This distinction is also extended to propositions in **emTT** by selecting small propositions as those propositions closed only under quantifications over sets and only under propositional equality only on sets. Then, to avoid an impredicative power-collection of a set, a subset must be defined as an equivalence class of small predicates and not of generic ones.

An important conceptual reason why even the power-collection $\mathcal{P}(1)$ of the singleton is only a collection and not a set is that in **emTT** we intend the notion of small proposition to be *open*. The same is done for that of proposition, of set, and of collection. Indeed, although we have fixed the system **emTT**, new sets or collections can be introduced at any time. This implies in particular that the collection of small propositions (quotiented under equiprovability) is not a set. Indeed, each time we fix our propositions or sets by fixing a formal system, both notions become inductively generated. However, we cannot support an induction principle inside the formal system, given that the number of inductive hypotheses should change any time we introduce a new set or proposition. This is different from the induction principle on the set of natural numbers, which has only two hypothesis: what we do on the number zero, and with any successor number.

20.3.2 The Main Types of the Intensional Level of the Minimalist Foundation

Here we briefly describe the main types of the formal system **mTT** of the intensional level of the Minimalist Foundation in [39] by simply pointing out the differences with those of **emTT**.

In essence **mTT** is a dependent type theory which provides a *predicative version* of Coquand's Calculus of Constructions in [19]. It is written in the style of *intensional* Martin-Löf's type theory in [60] by means of the following four kinds of judgements:

$$A\ type\ [\Gamma] \quad A = B\ type\ [\Gamma] \quad a \in A\ [\Gamma] \quad a = b \in A\ [\Gamma].$$

Like **emTT**, **mTT** includes *small propositions* and *propositions* which are closed under the same type constructors as those in **emTT** except that the propositional equality type is written $\mathsf{Id}(A, a, b)$ and has proper rules specifying its elements. There are also the rules stating that small propositions are propositions, that small propositions are sets and that propositions are collections. A main difference with respect to **emTT** is that in **mTT** the rules (**prop-mono**) and (**prop-true**) are omitted. As a consequence, all propositions in **mTT** are seen as types of their proofs which are *not in general unique* as usual in intensional type theory. Moreover, as in the intensional version of Martin-Löf's type theory, in **mTT** the definitional equality of terms of the same type given by the judgement

$$a = b \in A\ [\Gamma],$$

which should be computable, is *no longer equivalent* to the propositional equality type

$$\mathsf{Id}(A, a, b)\ prop\ [\Gamma],$$

which is not necessarily computable and not necessarily equipped with only one proof.

Sets in **mTT** are closed under the same constructors as those in **emTT** with the exception of the quotient set constructor A/ρ. As in **emTT**, in **mTT** there is also the rule stating that sets are collections.

Finally, *collections* in **mTT** include the same constructors as those of **emTT** except that the power-collection of the singleton $\mathcal{P}(1)$ is replaced by the universe of small propositions $\mathsf{prop_s}$ and the power-collection constructor $A \to \mathcal{P}(1)$ on a set A is replaced by the collection $A \to \mathsf{prop_s}$ of predicates or propositional operations depending on the set A.

The dependent type theory **mTT** was designed in order to serve as a base for a proof-assistant.

20.3.3 On the Extraction of Programs from Proofs in MF

Here we describe how Bishop's desire to compile a foundation for constructive mathematics in a programming language is fulfilled for **MF**.

First of all, **MF** was structured as a two-level theory to interpret constructive proofs done at its extensional level **emTT** to proofs done at its intensional level **mTT** from which to extract the computational contents in the form of programs. However, the extraction of the computational contents of proofs in **mTT** cannot be performed in **mTT** itself as shown in [41] but in a stronger theory or in the realizability semantics in [31]. One could then think of enlarging the intensional level to become the stronger theory needed, but this would not satisfy the forget–restore principle according to which the entities at the extensional level should be obtained by abstraction from the intensional ones, or more concretely as quotients of intensional entities.

A priori, the intensional level **mTT** itself could serve as a programming language to compile proofs done at the extensional level. Indeed, **mTT** is a dependent type theory where we can construct a correct and terminating program as a typed term meeting a certain specification defined as its type. But to extract programs from constructive proofs it is desirable that from a proof of an existential statement under hypothesis

$$p(x) \in \exists y \in B\ R(x, y)\ [x \in A]$$

for generic types A and B, one may extract a functional program $f \in A \to B$ whose graph is contained in the graph of $R(x, y)$, namely for which we can prove that there exists a proof-term $q(x)$ such that we can derive

$$q(x) \in R(x, f(x)) \; [x \in A]$$

This property is called *choice rule*.

In all the versions of Martin-Löf dependent type theory in [55, 60] the choice rule is valid thanks to the identification of the **MLTT**-existential quantifier with the *the strong indexed sum of a set family*, which characterizes the so-called *propositions-as-sets isomorphism*. Then the *axiom of choice*

$$\text{(AC)} \qquad \forall x \in A \, \exists y \in B \; R(x, y) \quad \longrightarrow \quad \exists f \in A \to B \, \forall x \in A \; R(x, f(x))$$

is valid for generic types A and B.

However, in **mTT** the existential quantifier is not identified with the strong indexed sum type whilst it is still a type of its proofs. The result is that the choice rule in Definition 20.1 is not valid.

Definition 20.1 The dependent type theory **mTT** satisfies the *choice rule* if for every small proposition $R(x, y) \; prop_s \; [x \in A, y \in B]$ derivable in **mTT**, for any derivable judgement in **mTT** of the form

$$p(x) \in \exists_{y \in B} \; R(x, y) \; [x \in A],$$

there exists in **mTT** a typed term

$$f(x) \in B[x \in A]$$

for which we can find a proof-term $q(x)$ and derive in **mTT**

$$q(x) \in R(x, f(x)) \; [x \in A].$$

Proposition 20.2 *In* **mTT** *the* choice rule *is not valid.*

Proof See [41]. □

Hence, when proving a statement of the form

$$\forall_{x \in A} \exists y \in B \; R(x, y)$$

in the dependent typed theory **mTT**, we cannot always extract a functional term $f \in A \to B$ computing the witness of the existential quantification depending on a $x \in A$ within the theory itself but we need to find it in a more expressive proofs-as-programs theory.

For **mTT** we can use Martin-Löf's type theory, for short **MLTT**, in [60] as the more expressive theory where to perform the mentioned witness extraction. Indeed, we can interpret **mTT** inside **MLTT** as shown in [39] by preserving the meaning of its entities.

This extraction is done by first embedding the proof-term

$$p \in \forall_{x \in A} \exists y \in B\ R(x, y)$$

derived in **mTT** and then using **MLTT**-projections to extract f.

The other possibility is to perform this witness extraction in the realizability model of **mTT** in [31]. This realizability model guarantees that the intensional level **mTT** of **MF** is a 'proofs-as-programs theory' in the sense of [47], namely that **mTT** is consistent with the axiom of choice (AC) and the formal Church thesis (CT) by identifying $Fun(Nat, Nat)$ in **mTT** with the type of *functional relations between natural numbers.*

$$\Sigma_{R \in \mathcal{P}(\mathrm{N},\mathrm{N})}\ \forall_{x \in A} \exists ! y \in B\ \langle x, y \rangle \varepsilon R,$$

where $\exists ! y \in B\ \langle x, y \rangle \varepsilon R\ \equiv\ \forall_{y_1 \in B} \forall_{y_2 \in B}\ R(x, y_1)\ \&\ R(x, y_1)\ \rightarrow\ \mathsf{Id}(B, y_1, y_2)$

Actually in [31] **mTT** is shown to be consistent with (AC) and the *formal Church thesis for operations between natural numbers*

$$(\mathrm{CT}_{\mathrm{tt}})^2 \qquad \forall f \in \mathrm{N} \rightarrow \mathrm{N}$$
$$\exists e \in \mathrm{N} \qquad \forall x \in \mathrm{N}\, \exists y \in \mathrm{N}\ (T(e, x, y) \wedge U(y) =_N f(x))$$

stating that all operations between natural numbers are recursive.

The consistency of **mTT** with (AC) and ($\mathrm{CT}_{\mathrm{tt}}$) implies the consistency of **mTT** with (AC) and (CT) since we can easily show the following.

Lemma 20.3 *In* **mTT** *extended with (AC) the formal Church thesis for functional relations CT is equivalent to the formal Church thesis for operations* CT_{tt}.

Therefore we conclude as follows.

Proposition 20.4 *The intensional level* **mTT** *of* **MF** *in [39] is consistent with (CT) and the axiom of choice in the form*

$$(\mathrm{AC}) \qquad \forall x \in A\, \exists y \in B\ R(x, y)\ \longrightarrow\ \exists f \in A \rightarrow B\, \forall x \in A\ R(x, f(x))$$

for A, B *collections and* $R(x, y)$ *any proposition in* **mTT**.

Proof It follows from Lemma 20.3 and [31]. □

As a consequence of the interpretation of the extensional level of **MF** into its intensional level in [39], we can also deduce that the extensional level of **MF** is consistent with the formal Church thesis.

Proposition 20.5 *The extensional level* **emTT** *of* **MF** *in [39] is consistent with* (CT).

Proof A proof of this proposition can be obtained in various ways. For example, it also follows from the realizability interpretations of **mTT** in [43, 44], or from the fact that **emTT** can be interpreted by preserving the meaning of its entities both in the internal theory of a topos and then in Hyland's Effective Topos in [28] or in Aczel's set theory in [1]. It would be also possibile to interpret **emTT** in the predicative version of Hyland's Effective Topos in [48]. □

Observe that the consistency requirement of a proofs-as-programs theory with (AC) and (CT) just guarantees that from proofs of existential statements on natural numbers under hypothesis

$$\exists y \in \mathrm{N}\, R(x,y)\ true\ [x \in Nat]$$

we can extract of a computable choice function

$$f \in \mathrm{N} \to \mathrm{N}$$

producing a witness under hypothesis such that we can find a proof of

$$R(x, f(x))\ true\ [x \in \mathrm{N}].$$

It is then clear that our proofs-as-programs requirement does not fully capture the idea of a foundational theory that is at the same time a programming language satisfying the choice rule, such as **MLTT** in [60].

It is still an open problem whether **MLTT** enjoys our proofs-as-programs requirement or, equivalently, whether **MLTT** is consistent with the formal Church thesis (CT).

Our purpose with the proofs-as-programs requirement in [47] was to single out a property characterizing theories which are interpretable in extensions of Kleene realizability semantics for Heyting Arithmetics with finite types (see [77]).

20.3.4 Benefits of Distinguishing Operations from Functions

Inspired by Brouwer's difference between lawlike and choice sequences in [77], in **MF** contrary to `BISH` we can define *choice sequences* from the set of natural numbers N to a set B as *functions* (in the sense of functional relations, that is, total and single-valued relations), and *lawlike sequences* as *operations*.

Definition 20.6 (Choice and lawlike sequences) Given a set A, a *choice sequence* from the set N of natural numbers to A is a function defined by a small functional relation $\alpha(x,y)\ prop_s\ [x \in \mathrm{N}, y \in A]$ in **emTT**.

A *lawlike sequence* from the set N of natural numbers to A is an operation

$$f \in \mathrm{N} \to A$$

in **emTT**, or equivalently, thanks to the rules in [39] defining elements in $\mathrm{N} \to A$, an **emTT**-term $f(x) \in A\ [x \in \mathrm{N}]$.

It is possible to keep a distinction between choice sequences and lawlike sequences because in **emTT** the *axiom of unique choice*

$$(\mathrm{AC!}_{N,N}) \qquad \forall x \in \mathrm{N}\, \exists! y \in \mathrm{N}\, R(x,y) \;\longrightarrow\; \exists f \in \mathrm{N} \to \mathrm{N}\, \forall x \in \mathrm{N}\, R(x, f(x)),$$

which turns a function between natural numbers into an operation, is not valid, as shown in [40, 41]. Our distinction allows us to clarify and compare results about choice sequences in the literature, since choice sequences are sometimes identified with our functions, for example in [65], and sometimes with our operations, for example in [77].

Another consequence of the distinction between operations and functions is that we can refine the notion of decidable subset of the set of natural numbers N. In constructive mathematics it is common to say that a subset $U \subseteq \mathrm{N}$ is decidable if $\forall x\ (\ x \,\epsilon\, U \vee x \,\not\epsilon\, U\ )$ holds. In our theory we can distinguish three notions.

Definition 20.7 A subset U of the set N is said to be:

- *complemented*, if $\forall x\ (\ x \,\epsilon\, U \vee (x \,\not\epsilon\, U)\ )$ holds, and in this case U is classified by a function from N to the boolean set Bool

$$\chi_U(x,y) \;\equiv\; (x \,\epsilon\, U \,\&\, y =_{\mathsf{Bool}} 1) \;\vee\; (x \,\not\epsilon\, U \,\&\, y =_{\mathsf{Bool}} 0);$$

- *detachable*, if the subset U is classified by an operation, namely we can derive

$$\exists_{f \in \mathrm{N} \to \mathsf{Bool}}\ \forall x \in \mathrm{N}\, ((x \,\epsilon\, U \,\&\, f(x) =_{\mathsf{Bool}} 1) \;\vee\; (x \,\not\epsilon\, U \,\&\, f(x) =_{\mathsf{Bool}} 0));$$

- *decidable*, if U is classified by a computable operation, namely we can derive

$$\begin{aligned} \exists_{f \in \mathrm{N} \to \mathsf{Bool}}\ & (\forall x \in \mathrm{N}\, ((x \,\epsilon\, U \,\&\, f(x) =_{\mathsf{Bool}} 1) \;\vee\; (x \,\not\epsilon\, U \,\&\, f(x) =_{\mathsf{Bool}} 0)) \\ & \&\quad \exists e \in \mathrm{N} \quad \forall x \in \mathrm{N}\, \exists y \in \mathrm{N}\, (T(e,x,y) \wedge U(y) =\ _{\mathrm{N}} f(x))), \end{aligned}$$

  where $T(e,x,y)$ is the Kleene predicate expressing that y is the computation executed by the program numbered e on the input x and $U(y)$ is output of the computation y.

Observe that, classically, all subsets are complemented. Of course, in the presence of the axiom of unique choice, functions and operations coincide and hence complemented and detachable subsets coincide, too, as for example in Martin-Löf's type theory.

All the three kinds of subsets coincide in the Kleene realizability interpretation of Heyting arithmetic. This interpretation is in some sense the intended interpretation of the arithmetic fragment of a constructive foundation. Hence, the identification of the name 'decidable' with our notion of complemented subsets (that we do not follow here, however), has its own (plausible) justification.

In [40, 42], we observed that if we extend **emTT** with the principle of excluded middle then we can prove the existence of a power-set of detachable subsets, which do not necessarily coincide with all subsets, that is, with complemented ones. This option of restricting exponentiation as a set to lawlike sequences opens the way to build a theory compatible with classical predicativity as those in [23].

20.4 Why Adopt the Pointfree Approach to Develop Topology in MF?

Bishop, like Brouwer, developed constructive analysis by adopting a pointwise approach which presented some difficulties solved by them in different ways (see [62, 77]). When developing topology in **MF** we need to adopt the pointfree approach. The most important reason is that, when working in **MF**, the pointwise approach is not suitable because relevant examples of classical topologies (real numbers both as Dedekind cuts or Cauchy sequences, Baire space, Cantor space, etc.) do not give rise to a pointwise topology since their points do not form a set.

A solution is to work with the pointfree topology associated to each of these spaces. The constructive approach to pointfree topology given by formal topology has provided evidence that most important results of constructive analysis (see, for example, [53, 62]) can be reached in a compatible way with classical mathematics as in Bishop's constructive approach, but without assuming further principles, such as the Fan Theorem adopted by Brouwer in his pointwise approach and in [14, 15, 30, 77].

Before entering into details, we briefly review a constructive notion of topological space and then the main concepts of formal topology.

20.4.1 A Predicative Constructive Notion of Topological Space

Considering that in a predicative foundation the discrete topology on a given non-empty set is not a set but a collection, we need to review the concept of topological space by distinguishing what belongs to the realm of sets from what belongs to the realm of collections.

At first, one could think of simply keeping the traditional definition of topological space $(X, \mathcal{O}X)$ by just declaring the topology $\mathcal{O}X$ to be only a subcollection of the power of X which is a suplattice, that is, a complete join-semilattice, with finite distributive meets. This approach is compulsory in order to include the discrete topology among topologies. Even more, as shown in [22], there is no non-trivial suplattice, and hence no non-trivial topology, which is a set.

One should then define suplattices as collections closed under sups of *set-indexed* families. However, as in [7] and [66], suplattices are easier to handle by restricting to the notion of *set-based* suplattice, namely a semilattice that is generated by taking sups from a set(-indexed family) of elements, called *generators*.

Topologically this means that we need to assume that the collection of opens of a space has a base that is a set. To make this assumption rigorous, we require that for a given set of points X we have a set S together with a family of subsets $\mathsf{ext}\,(a) \subseteq X\ [a \in S]$ acting as a base for the topology on X. Elements a of S act as names of basic opens of X; they are called *formal basic neighbourhoods* or simply *observables*.

Then, following [68], we define a subset of X to be open if it is equal to $\mathsf{ext}\,U \equiv \cup_{a\epsilon U}\,\mathsf{ext}\,a$ for some subset $U \subseteq S$. It is immediate to see that open subsets are closed under unions of set-indexed families.

Then we need to require closure of open subsets under intersection. To this purpose, it is convenient to start from *basic neighbourhoods*, that is, subsets of X of the form $\mathsf{ext}\,a$ for some $a \in S$. For all $a, b \in S$, the intersection $\mathsf{ext}\,a \cap \mathsf{ext}\,b$ is open, that is, it is equal to $\mathsf{ext}\,W$ for some $W \subseteq S$, if and only if

B0 $\quad \mathsf{ext}\,a \cap \mathsf{ext}\,b = \mathsf{ext}\,(a \downarrow b)$ for all $a, b \in S$,

where

$$a \downarrow b \equiv \{c \in S : \mathsf{ext}\,c \subseteq \mathsf{ext}\,a \cap \mathsf{ext}\,b\}.$$

In fact, $\mathsf{ext}\,(a \downarrow b)$ is by its definition the greatest open subset contained in $\mathsf{ext}\,a \cap \mathsf{ext}\,b$. Then, from B0, by two applications of distributivity in $\mathcal{P}X$, we can easily obtain

B1 $\quad \mathsf{ext}\,U \cap \mathsf{ext}\,V = \mathsf{ext}\,(U \downarrow V)$ for all $U, V \subseteq S$,

where

$$U \downarrow V \equiv \cup_{a\epsilon U} \cup_{b\epsilon V}\, a \downarrow b.$$

Finally, to obtain that the whole space is open we need to add the requirement

B2 $\quad X = \mathsf{ext}\,S.$

It is clear that for any family of subsets of X indexed by the set S, that is, for any $\mathsf{ext}\,a \subseteq X$ for $a \in S$, satisfying B1 and B2, the collection of subsets $\mathsf{ext}\,U \subseteq X$ for $U \in \mathcal{P}S$ is closed under set-indexed unions and finite intersections.

Therefore we can give the following constructive version of topological spaces (see [68]).

Definition 20.8 A *concrete space* is a structure $\mathcal{X} = (X, \mathsf{ext}, S)$ where X, S are sets and $\mathsf{ext}(a) \subseteq X\ [a \in S]$ is a set-indexed family of subsets satisfying:

B1 $\mathsf{ext}\, U \cap \mathsf{ext}\, V = \mathsf{ext}(U \downarrow V)$ for all $U, V \subseteq S$,

B2 $X = \mathsf{ext}\, S$.

In an impredicative foundation with power-sets, this is just a reformulation of the common notion of topological space.

The notion of concrete space is present in [10], under the name of neighbourhood space. The discrete topology on a set X is obviously an example of concrete space with X itself as base and $\mathsf{ext}(x) \equiv \{x\}$ for $x \in X$.

A useful example of concrete space is given by the set $\mathbb{Q}$ of rational numbers with the topology produced by the base of open intervals.

In more detail, the base is the set $\mathbb{Q} \times \mathbb{Q}$ of pairs $\langle p, q \rangle$ of rational numbers, and the basic neighbourhood with index $\langle p, q \rangle$ is the subset

$$\mathsf{ext}(\langle p, q \rangle) \equiv \{r \in \mathbb{Q} \mid p < r < q\}$$

for all $p, q \in \mathbb{Q}$.

In other constructive and predicative foundations, such as Aczel's **CZF** and Martin-Löf's type theory in [60], another example of concrete space is that of real numbers. It is not so in our **MF**, as we shall see later. Even when the topology of real numbers provides an example of concrete space, it is well known from Brouwer that a constructive pointwise development of analysis fails to get important properties (see [77]), such as compactness of the closed interval $[0, 1]$, unless further principles, for example the Fan Theorem, are assumed or some basic topological notions are changed as in Bishop's approach (see [10, 12, 15]). An alternative approach to constructive topology, and analysis, is offered by formal topology.

20.4.2 The Predicative Constructive Pointfree Approach of Formal Topology

The approach of formal topology to pointfree topology was introduced by Per Martin-Löf and the second author in the 1980s; the first published account is [66]. The intended foundation was then Martin-Löf's type theory MLTT in [60]. However, as underlined in the introduction of [66], to the second author it was already clear that it was necessary to work with an explicit notion of subset, and with a primitive notion of proposition using the judgement that a proposition is true without any reference to proof-terms in [56]. Such a conception of subsets and propositions was later specified in [75] as a tool to be added on top of type theory. As noticed in [36, 37], working with existential quantifiers with no proof-terms means that the axiom of choice no longer holds. This is different from **MLTT** where existential quantifications are identified with indexed sums, according to the proposition-as-set isomorphism, thus making the axiom of choice (AC) derivable. Moreover, this explains why in formal topology, as developed by the second author, every use of the axiom of choice was explicit. Given that the notion of subset in [75] and a primitive notion of proof-irrelevant propositions have been incorporated in our Minimalist Foundation, all the main definitions and results on formal topology (by the second author) can be carried in it. Actually the combination of the tool of *extensional* subsets with the *intensional* **MLTT** partly anticipated the notion of two-level theory in [47], because subsets are not formally included in **MLTT**.

The main idea of formal topology is to replace the notion of concrete space with an abstract axiomatization of the structure of open subsets, and then to recover its points in a *formal way* as suitable subsets of opens. The precise definition is reached by describing the structure of the set S of basic neighbourhoods in a concrete space $(X, \mathsf{ext}\,, S)$ with no mention of the set X of points.

While in the concept of concrete space $(X, \mathsf{ext}\,, S)$ points in X are given at the same time as the formal basic neighbourhoods in S and both form a set, in formal topology only the structure of opens is described, starting from the set S of formal basic neighbourhoods and from a new primitive relation $a \lhd U$, called *formal cover*, between formal basic neighbourhoods $a \in S$ and subsets $U \subseteq S$. A formal cover relation is the abstract counterpart of

$$\mathsf{ext}\, a \subseteq \mathsf{ext}\, U,$$

which expresses in a concrete space that the open $\mathsf{ext}\, U$ is a covering of the basic neighbourhood $\mathsf{ext}\, a$. Then, the notion of *formal topology* extends that of formal cover with the addition of a primitive predicate $Pos(a)$ for $a \in S$, which is the abstract counterpart of the assertion that the basic neighbourhood $\mathsf{ext}\, a$ is inhabited. Details of the definitions are now presented.

Definition 20.9 (Formal cover) A *formal cover* $\mathcal{A} = (S, \lhd)$ is given by a set S and a relation $\lhd \subseteq S \times \mathcal{P}(S)$ between elements and subsets of S that satisfies the following rules for every $a \in S$ and $U, V \subseteq S$:

$$\frac{a \,\epsilon\, U}{a \lhd U}\ \mathit{reflexivity} \qquad\qquad \frac{a \lhd U \quad U \lhd V}{a \lhd V}\ \mathit{transitivity}$$

$$\frac{a \lhd U \quad a \lhd V}{a \lhd U{\downarrow_{\mathcal{A}}}V}\ \mathit{convergence},$$

where $U \lhd V \overset{def}{\Longleftrightarrow} (\forall b \,\epsilon\, U)\ (b \lhd V)$ and $U \downarrow_{\mathcal{A}} V = \{a \in S : (\exists u \,\epsilon\, U)(a \lhd u)\ \&\ (\exists v \,\epsilon\, V)(a \lhd v)\}$.

This definition provides a predicative counterpart of the impredicate notion of pointfree topology called *locale* in [32, 35]. In fact, to any formal cover $\mathcal{A} = (S, \lhd)$

we can associate an operator $\mathcal{A}$ on $\mathcal{P}(S)$, that is, an operation $\mathcal{A} : \mathcal{P}(S) \to \mathcal{P}(S)$ (that by abuse of notation we call the formal cover itself!), by putting

$$\mathcal{A}U \stackrel{def}{=} \{a \in S \mid a \lhd U\} \tag{20.1}$$

for any $U \subseteq S$. Then, reflexivity and transitivity of the cover means that the operator $\mathcal{A}$ is a *saturation* (or closure operator) and convergence means that $\mathcal{A}$ satisfies

$$\mathcal{A}(U \downarrow_{\mathcal{A}} V) \;=\; \mathcal{A}U \cap \mathcal{A}V.$$

The collection $Sat(\mathcal{A})$ of all fixed points of the operator $\mathcal{A}$ (i.e., all subsets U of S satisfying $\mathcal{A}(U) = U$) with the order given by inclusion form a locale. See [17] for an account and discussion on the several variants of the definitions of formal cover.

Then a formal topology is defined as follows.

Definition 20.10 A *formal topology* $\mathcal{S} = (S, \lhd, \mathsf{Pos})$ is a formal cover $(S, \lhd)$ equipped with a *positivity* predicate, that is a predicate $\mathsf{Pos}(a)$ for $a \in S$ which satisfies the conditions

$$(\text{monotonicity})\ \frac{\mathsf{Pos}(a) \qquad a \lhd U}{(\exists u \,\epsilon\, U)\ \mathsf{Pos}(u)} \qquad (\text{positivity})\ \frac{\mathsf{Pos}(a) \to a \lhd U}{a \lhd U}.$$

Formal topologies provide a predicative counterpart of the impredicative notion of *open locale* in [34].

Formal covers, as well as formal topologies, can be inductively generated from a set-indexed family of axioms of the form $a \lhd U$.

Definition 20.11 Given a set S, an *axiom-set* is a pair I, C, given by a family of sets $I(a)$ for each $a \in S$ and a family of subsets $C(a, i) \subseteq S$ for $a \in S$ and $i \in I(a)$ (with the intended meaning that $a \lhd C(a, i)$ holds).

The definition of inductively generated formal cover was introduced in [20] and for our purposes we just recall the following.

Definition 20.12 Given a preordered set $(S, \leq)$ and an axiom-set I, C, the *inductively generated formal cover (formal topology)* $(S, \lhd_{I,C})$ is a formal cover (formal topology) satisfying:

(i) $a \lhd_{I,C} C(a, i)$ for every $a \in S$ and $i \in I(a)$;
(ii) if $\lhd'$ is another formal cover (formal topology) such that $a \lhd' C(a, i)$ for all $a \in S$ and $i \in I(A)$, then $a \lhd_{I,C} U \;\to\; a \lhd' U$ holds for all $a \in S$ and $U \subseteq S$.

Observe that in generating the formal topology the preorder on the set of basic neighbourhoods $(S, \leq)$ is essential to produce a distributive lattice of open subsets (see [17] for a detailed explanation).

In the Minimalist Foundation we can define some inductively generated formal topologies but not all, as shown in [50]. For example, in **MF** we can represent the pointfree topology of real numbers or that of the Cantor space by reproducing the argument used in [79] to define inductively generated formal covers, and formal topologies, in an extension of Martin-Löf's type theory with ordinals.

Therefore we assume the existence of an inductively generated cover when needed both at the extensional and at the intensional levels of **MF**. A proper two-level extension of the Minimalist Foundation with generic inductively generated formal topologies satisfying the requirements in [48] was built in [50] and called $\mathbf{MF}_{ind}$. In particular, in [50] the intensional level is shown to be consistent with the formal Church thesis and the axiom of choice by extending the Kleene realizability interpretation of intuitionistic arithmetic as was done in [31] but in a constructive metatheory.

We now recall the notion of a formal point. Given any formal topology $\mathcal{S}$, a formal point over $\mathcal{S}$ is a subset α of the set S such that it makes sense to think of $a \epsilon \alpha$ as meaning that the observable a is an approximation of α. To obtain a precise definition, one considers the case in which $\mathcal{S}$ is the topology of a concrete space X and takes some pointfree properties of the subset $\{ a \in A \mid x \epsilon \mathsf{ext}(a) \}$, which is the trace on S of a concrete point $x \in X$, as the conditions to define a subset $\alpha \subseteq S$ to be a formal point.

Definition 20.13 (Formal point) Let $\mathcal{A} \equiv (A, \lhd)$ be a formal cover. An inhabited subset α of A is a *formal point* if, for any $a, b \in A$ and any $U \subseteq A$, it satisfies the following conditions:

$$(\alpha \text{ is filtering})\ \frac{a \epsilon \alpha \quad b \epsilon \alpha}{(\exists c \epsilon \{a\} \downarrow_{\mathcal{A}} \{b\})\, c \epsilon \alpha} \qquad (\alpha \text{ splits the cover})\ \frac{a \epsilon \alpha \quad a \lhd U}{(\exists u \epsilon U)\, u \epsilon \alpha}.$$

Then, one can take the collection of $Pt(\mathcal{A}) \equiv \{\alpha \in \mathcal{P}(S) \mid \alpha \text{ formal point}\}$ and make it a formal space as follows.

Definition 20.14 (Formal topology and formal space) For any formal cover $\mathcal{A} \equiv (S, \lhd)$, the collection $Pt(\mathcal{A})$ of formal points of $\mathcal{A}$ with the topology generated by the basic neighbourhoods of the form $\mathsf{Ext}(a) \equiv \{ \alpha \in Pt(\mathcal{A}) \mid a \epsilon \alpha \}$ for $a \in S$ defines the *formal space* of points of $\mathcal{A}$ (that by abuse of notation we still call $Pt(\mathcal{A})$).

In an impredicative foundation, where power-collections are sets, it is clear that $Pt(\mathcal{A})$ defines a concrete space for any formal cover $\mathcal{A}$. Hence, as is well known, impredicatively one can prove the existence of an adjunction between formal covers and concrete spaces (see [32, 35]). This impredicative adjunction associates to a formal cover its formal space, and conversely to a concrete space (X, ext, S) the formal cover $(S, \lhd_X)$ defined by

$$a \lhd_X U \equiv \mathsf{ext}\, a \subseteq \mathsf{ext}\, U.$$

But, not all formal covers arise from concrete spaces in this way. Moreover, the formal cover induced by a formal space is not necessarily equal to the starting formal cover $\mathcal{A}$, that is, not all formal covers are *spatial*. And even more, the formal space of a formal cover arising from a concrete space is not necessarily equivalent to the starting concrete space, that is, not all concrete spaces are *sober*.

In a constructive and predicative foundation like our minimalist one, such an adjunction is no longer available, because the collection $Pt(\mathcal{A})$ is *not necessarily* a set.

Here we will see at least three relevant examples of proper formal spaces, that is, formal spaces whose formal points cannot form a set in our **MF**: the formal space of real numbers, Cantor, and Baire spaces. In all these examples, we will see how our foundation allows us to distinguish points which are given effectively, namely concrete points identified with lawlike sequences, from points which are only ideally so, that is, formal points, which are identified with choice sequences. It is a predicative foundation which allows one, and, in the same time compels one, to take care of this distinction between an effective or real structure, such as that of open basic neghbourhoods, from an ideal or non-effective structure such as that of formal points.

So in a constructive approach to topology such as our minimalist one, formal topologies and formal points are not just an option to describe something which is there in any case. They are introduced also as the only way to treat those spaces which otherwise would be constructively unreachable.

20.4.3 Examples of Pointfree Topologies whose Formal Points do not Form a Set

The first example of topology whose formal points do not form a set in the Minimalist Foundation is the formal topology of real numbers, such as Dedekind cuts.

Definition 20.15 (Formal topology of real numbers) The *formal topology of real numbers* $\mathcal{R} \equiv (\mathbb{Q} \times \mathbb{Q}, \lhd_{\mathcal{R}}, \mathsf{Pos}_{\mathcal{R}})$ is an inductively generated formal topology defined as follows. The base is $\mathbb{Q} \times \mathbb{Q}$ and the basic neighbourhoods are pairs of rational numbers, $\langle p, q\rangle$ with $p, q \in \mathbb{Q}$. A preorder on $\mathbb{Q} \times \mathbb{Q}$ is defined as follows

$$\langle p, q\rangle \leq \langle p', q'\rangle \equiv p' \leq p \leq q \leq q'$$

for p, q, p', q' in $\mathbb{Q}$. The cover is defined inductively by the following rules (which are a formulation in our context of Joyal axioms, cf. [32], pp. 123–124):

$$\frac{q \le p}{\langle p,q\rangle \lhd_{\mathcal{R}} U} \qquad \frac{\langle p,q\rangle \in U}{\langle p,q\rangle \lhd_{\mathcal{R}} U} \qquad \frac{p' \le p < q \le q' \qquad \langle p',q'\rangle \lhd_{\mathcal{R}} U}{\langle p,q\rangle \lhd_{\mathcal{R}} U}$$

$$\frac{p \le r < s \le q \qquad \langle p,s\rangle \lhd_{\mathcal{R}} U \quad \langle r,q\rangle \lhd_{\mathcal{R}} U}{\langle p,q\rangle \lhd_{\mathcal{R}} U} \qquad \mathsf{wc}\ \frac{wc(\langle p,q\rangle) \lhd_{\mathcal{R}} U}{\langle p,q\rangle \lhd_{\mathcal{R}} U}$$

where in the last axiom we have used the abbreviation

$$wc(\langle p,q\rangle) \equiv \{\, \langle p',q'\rangle \in \mathbb{Q} \times \mathbb{Q} \mid p < p' < q' < q\}$$

(wc stands for 'well-covered').

The positivity predicate is $\mathsf{Pos}_{\mathcal{R}}(\,\langle p,q\rangle\,) \equiv p < q$, expressing that the pair of rationals represents a non-empty interval.

As shown in [59], formal points of the formal topology $\mathcal{R}$ are in bijection with the collection of Dedekind cuts on the rationals. The proof carries over to our foundation after observing that $\mathcal{R}$ can be defined in **emTT** as described in [50].

Definition 20.16 A *Dedekind cut* on the rationals is a pair (L,U) with inhabited $L,U \subseteq \mathbb{Q}$ satisfying the following properties:

(disjointness)	$\forall q \in \mathbb{Q}\ \neg(q \,\epsilon\, U \;\&\; q \,\epsilon\, L)$
(L-openess)	$\forall p \,\epsilon\, L\ \exists q \,\epsilon\, L\ p < q$
(U-openess)	$\forall q \,\epsilon\, U\ \exists p \,\epsilon\, U\ p < q$
(L-monotonicity)	$\forall q \,\epsilon\, L\ \forall p \in \mathbb{Q}\ (p < q \rightarrow p \,\epsilon\, L)$
(U-monotonicity)	$\forall p \,\epsilon\, U\ \forall q \in \mathbb{Q}\ (p < q \rightarrow q \,\epsilon\, U)$
(locatedness)	$\forall q \in \mathbb{Q}\, \forall p \in \mathbb{Q}\ (p < q \rightarrow p \,\epsilon\, L \vee q \,\epsilon\, U)$.

Proposition 20.17 *In* **emTT** *the formal points of the inductively generated formal topology $\mathcal{R}$ are in bijection with the collection of Dedekind cuts on the rationals.*

Proof Given a formal point $\alpha \in Pt(\mathcal{R})$ we can build the following Dedekind cut:

$$L_\alpha \equiv \{p \in \mathbb{Q} \mid \exists_{q\in\mathbb{Q}} \langle p,q\rangle \,\epsilon\, \alpha\} \qquad U_\alpha \equiv \{q \in \mathbb{Q} \mid \exists_{p\in\mathbb{Q}} \langle p,q\rangle \,\epsilon\, \alpha\}.$$[3]

Conversely, given a Dedekind cut (L,U) we can define the following formal point

$$\alpha_{(L,U)} \equiv \{\langle p,q\rangle \in \mathbb{Q}\times\mathbb{Q} \mid p \,\epsilon\, L \;\&\; q \,\epsilon\, U\}.$$

In [59] it is proved that formal points of $\mathcal{R}$, or Dedekind cuts, are also in bijective correspondence with Cauchy sequences à la Bishop in [10]. This correspondence does not work in **emTT**: only Cauchy sequences à la Bishop can be shown to be formal points of $\mathcal{R}$. To make this point clear, we recall the notion of Cauchy sequence à la Bishop. In the following with N^+ we mean the set of positive natural numbers. □

[3] Note that the base of our topology $\mathcal{R}$ does not contain $+\infty, -\infty$ like that in [59].

Definition 20.18 (Cauchy sequence à la Bishop) A function $R(n, x)$ $props_s$ $[n \in \mathrm{N}^+, x \in \mathbb{Q}]$, indicated with the usual notation $(x_n)_{n \in N^+}$, is a *Cauchy sequence* in **emTT** if we can prove for any $n, m \in \mathrm{N}^+$

$$\mid x_n - x_m \mid \leq 1/n + 1/m.^4$$

As in [59], we can also prove in **emTT** that any Cauchy sequence $(x_n)_{x \in \mathrm{N}}$ determines a formal point α of the formal topology $\mathcal{R}$ if we define it by:

$$\alpha \equiv \{\langle p, q\rangle \in \mathbb{Q} \times \mathbb{Q} \mid \exists n \in \mathrm{N}^+ \ p < x_n - 2/n < x_n + 2/n < q\}.$$

Conversely, given a formal point α, we can prove in **emTT** that within α there exists a countable number of strictly decreasing intervals as follows:

$$\begin{aligned} &\forall n \in \mathrm{N}^+ \ \exists \langle x_n, y_n\rangle \in \mathbb{Q} \times \mathbb{Q} \ \ (((\langle x_n, y_n\rangle \ \epsilon \ \alpha \ \& \ \mid x_n - y_n \mid < (2/3)^n) \\ &\qquad \& \ \exists \langle x_{n+1}, y_{n+1}\rangle \in \mathbb{Q} \times \mathbb{Q} \ (x_n \leq x_{n+1} < y_{n+1} \leq y_n \ \& \\ &\qquad\qquad ((\langle x_{n+1}, y_{n+1}\rangle \ \epsilon \ \alpha \ \& \ \mid x_{n+1} - y_{n+1} \mid < (2/3)^{n+1}))). \end{aligned}$$

This is proved by induction; in fact, for $n \in \mathrm{N}$ and $\langle x_n, y_n\rangle \ \epsilon \ \alpha$ we can find a covering

$$\langle x_n, y_n\rangle \lhd_{\mathcal{R}} \{ l_{n_1}, l_{n_2} \}$$

such that $l_{n_1} \equiv \langle x_n, z_{n_1}\rangle$ and $l_{n_2} \equiv \langle z_{n_2}, y_n\rangle$ with $z_{n_1} \equiv x_n + (2 \cdot (y_n - x_n))/3$ and $z_{n_2} \equiv x_n + (y_n - x_n)/3$. Since the formal point α splits the cover, we can prove

$$\forall n \in N \ \exists i \ (i \varepsilon \{1, 2\} \ \& \ l_{ni} \ \epsilon \ \alpha).$$

However, such l_{ni} is *not necessarily unique*, because the formal point can be contained in both!

Classically, one can define a function $\mathcal{L}(n)$ for $n \in N$ by cases by putting

$$\mathcal{L}(n) \equiv \begin{cases} l_{n1} & \text{if } l_{n_1} \ \epsilon \ \alpha \\ l_{n2} & \text{if } l_{n_2} \ \epsilon \ \alpha \ \& \ \neg l_{n_1} \ \epsilon \ \alpha. \end{cases}$$

Constructively this does not work because α is not complemented. But, if we work in a foundation such as Martin-Löf's type theory MLTT, actually in the setoid model over it, by using the axiom of dependent choice on $\mathbb{Q} \times \mathbb{Q}$, we can even extract an operation $l(n) \in \mathbb{Q} \times \mathbb{Q}$ $[n \in \mathrm{N}^+]$ such that for any $n \in \mathrm{N}^+$

$$l(n) \ \epsilon \ \alpha$$

and, after naming $l(n) =_{Q \times Q} \langle x_n, y_n\rangle$ the values of the operation on each natural

[4] This is formally written as $\forall p \in \mathbb{Q} \, \forall q \in \mathbb{Q} \ (R(n, p) \ \& \ R(m, q) \rightarrow \mid q - p \mid \leq 1/n + 1/m)$, where the definition of module is the usual one.

number, the conditions $x_n - y_n \leq (2/3)^n$ and $x_n \leq x_{n+1} < y_{n+1} \leq y_n$ hold for each natural number n. Then, a Cauchy sequence can be defined by taking the first components $(x_n)_{n \in N}$ or the second components $(y_n)_{n \in N}$. Hence, any Dedekind cut or formal point of $\mathcal{R}$ corresponds to a lawlike Cauchy sequence à la Bishop in **MLTT**.

Since in our foundation no axiom of choice is available, this proof cannot be carried out. At a closer look, it does not appear constructively justified to be able to extract a choice of the interval where the formal point is, with no extra information.

What actually happens in **MLTT** is that the splitting of points is *already given* with an operation choosing an interval where the point is, and hence from this choice a definition by cases can be given similarly to that done classically. This example explains why in Martin-Löf's type theory real numbers as formal points are only the lawlike ones, namely those for which we can extract a lawlike Cauchy sequence.

This is not true in our foundation. In fact, the property that a formal point splits the cover is expressed through an existential quantifier $\exists_{x \in A}\, \phi(x)\ [w \in \Gamma]$ under a context Γ, which does not necessarily provide an operation $\mathsf{wit}(d) \in A\ [w \in \Gamma]$ (depending on the context Γ) returning a witness for the existential statement when this holds. Such an operation is available only in a Kleene realizability interpretation of our foundation. As expected, in **emTT** real numbers as formal points of the formal topology $\mathcal{R}$ *cannot coincide* with lawlike Cauchy sequences (see [40, 42]). Even more, *real numbers such as formal points of* $\mathcal{R}$*, and hence such as Dedekind cuts, do not form a set*. Analogously, *real numbers as Cauchy sequences à la Bishop do not form a set*. These results are obtained through a realizability interpretation of **emTT** which interprets **emTT**-sets as countable subsets of natural numbers and **emTT**-collections as entities which are not necessarily countable.

Now we describe two other examples of formal topologies whose formal points do not form a set in **emTT**. These are Cantor and Baire formal topologies, which are defined as instances of the more general notion of formal topology on the tree over a set. In order to define such formal topologies, we need to represent the tree over a set A, which we identify with the nodes labelled by lists of elements in a set A, using the abbreviation $A^* \equiv List(A)$. We write $[l, x]$ for the list obtained by appending $x \in A$ to the list $l \in A^*$ and $[l, t]$ for the list obtained by appending the list $t \in A^*$ to the list $l \in A^*$.

Definition 20.19 The *tree formal topology over a set* A is the formal topology $A^N \equiv (A^*, \lhd_{A^N}, \mathsf{Pos}_{A^N})$ where $\lhd_{A^N}$ is inductively generated by the following rules

$$\text{rfl}\ \frac{l \epsilon V}{l \lhd_{A^N} V} \qquad \leq \frac{s \sqsubseteq l \qquad l \lhd_{A^N} V}{s \lhd_{A^N} V} \qquad \text{tr}\ \frac{\forall x \in A\ \ [l, x] \lhd_{A^N} V}{l \lhd_{A^N} V},$$

where $s \sqsubseteq l \equiv \exists_{t \in A^*}\, s =_{A^*} [l, t]$, i.e. l is an initial segment of s.

The positivity predicate is true on any element, that is, $\mathsf{Pos}_{A^N}(l) \equiv \mathtt{tt}$ for any $l \in A^*$.

Among tree formal topologies, we distinguish Cantor and Baire formal topologies as follows

Definition 20.20 (Cantor and Baire formal topologies) The tree formal topology when $A \equiv \{0, 1\}$, namely

$$\{0,1\}^N \equiv (\{0,1\}^*, \lhd_{\{0,1\}^N}, \mathsf{Pos}_{\{0,1\}^N})$$

is called *Cantor formal topology*.

The tree formal topology when $A \equiv \mathrm{N}$, namely

$$\mathrm{N}^N \equiv (\mathrm{N}^*, \lhd_{N^N}, \mathsf{Pos}_{N^N})$$

is called *Baire formal topology*.

Note that the Cantor formal topology is definable in **MF** but the Baire formal topology is not as specified in [50].

Formal points of such topologies coincide with choice sequences of Definition 20.6.

Proposition 20.21 *Formal points $Pt(A^N)$ of the tree formal topology over a set A are in bijective correspondence with choice sequences on the tree A^*.*

Proof Given a formal point α, we define a function $R_\alpha(n, x)\ prop_s\ [n \in \mathrm{N}, x \in A]$ as follows:

$$R_\alpha(n, a) \equiv \exists\, l \,\epsilon\, \alpha\ \ l_{n+1} =_A a,$$

where l_n is the nth component of l. □

Conversely, given a function $R(n, x)\ prop_s\ [n \in \mathrm{N}, x \in A]$ the subset

$$\alpha_R \equiv \{\, l \in A^* \mid \forall n \in \mathrm{N}\, (n < \mathsf{lh}(l) \to R(n, l_{n+1}))\},$$

where $\mathsf{lh}(l)$ is the length of l, turns out to be a formal point.

An alternative proof follows after noting, as observed in [76], that the tree formal topology over a set A is isomorphic to the exponential formal topology of the discrete formal topology of N over the discrete formal topology on the set A (see [38] for a predicative treatment of exponentiation). Therefore its formal points are in bijection with functions, because every function between discrete topologies is continuous. This explains why we denote the tree formal topology with the symbol A^N.

The realizability interpretation in [40, 42], showing that real numbers (both as

Dedekind cuts or Cauchy sequences) do not form a set, also shows that *choices sequences as formal points of Cantor or Baire formal topology do not form a set either* and hence they only form a proper collection. Therefore predicatively we can only work with the pointfree topologies of usual Cantor and Baire spaces.

The equivalence of each of our pointfree topologies with the corresponding pointwise topology of their formal spaces as in Definition 20.14, namely spatiality, is not generally valid in **emTT**. Indeed, spatiality of our tree formal topologies amounts to the well-known principle of Bar Induction, as first observed in [24].

Definition 20.22 (Bar Induction in topological form) In **emTT** extended with the inductive definitions necessary to define the formal topologies $\lhd_{A^N}$ for any set A, the principle of Bar Induction is the following statement: for any given set A

(BI(A))

$$\forall l \in A^* \, \forall V \in \mathcal{P}(A^*) \;\; (\forall \alpha \,\epsilon\, Pt(\lhd_{A^N}) \;\; (\, l \,\epsilon\, \alpha \;\rightarrow\; \alpha \between V \,) \rightarrow l \lhd_{A^N} V)$$

where

$$V \between W \;\equiv\; \exists a \in A \, (a \,\epsilon\, V \;\wedge\; a \,\epsilon\, W)$$

expresses that two subsets V, W of a set A overlap (see [68]).

The above formulation of BI(A) means that the topology put on the formal points of the tree A^* that are its choice sequences, coincides with the pointfree one. Hence, Bar Induction implies that we can reason topologically on choice sequences by induction on finite sequences, given that the pointfree topologies are inductively generated (see [69, 74]).

The usual Fan Theorem in [77] is then an instance of Bar Induction (see [24, 25]).

Definition 20.23 (Fan Theorem) We call *Fan Theorem* the formulation BI($\{\, 0, 1 \,\}$) of BI(A) on Cantor formal topology, namely when $A \equiv \{\, 0, 1 \,\}$.

Spatiality of Cantor formal topology allows us to derive compactness of Cantor space in [24].

In [40, 42] it is shown that **emTT** is compatible with the described principle of Bar Induction BI(A) for any set A, and the identification of lawlike sequences with recursive ones. Indeed, there exists a realizability interpretation showing that real numbers and choice sequences do not form a set validating *the formal Church thesis for operations between natural numbers* ($\mathrm{CT_{tt}}$).

Hence, the realizability interpretation in [40, 42] shows that **emTT** is compatible with constructive foundations where Bar Induction or the Fan Theorem are used as in Brouwer's constructive pointwise approach of topology, by keeping a computable interpretation of operations between natural numbers with the validity of $\mathrm{CT_{tt}}$.

Actually a motivation to develop our **MF** was exactly to study a development of topology in the presence of these extra axioms.

Observe that compatibility with Bar Induction and $\mathrm{CT}_{\mathrm{tt}}$ is not possible for Martin-Löf's type theory, because the axiom of choice, and hence also the axiom of unique choice, is valid in there. To see this, first observe that in our **MF** we can prove the well-known result by Kleene [77] that formal Church thesis for choice sequences is contradictory with the Fan Theorem, and hence also with Bar Induction. Observe then that this result can be formulated by saying that *there is no model of* **emTT** $+\lhd_{A^N}$ + FT + $\mathrm{CT}_{\mathrm{tt}}$ + $\mathrm{AC}_{\mathrm{N,N}}$. Therefore a theory validating the axiom of unique choice cannot keep together a computational interpretation of operations and Bar Induction. Hence, the consistency of **emTT** with Bar Induction and $\mathrm{CT}_{\mathrm{tt}}$ explains why the axiom of unique choice, and a fortiori the axiom of choice, is not valid in **emTT**.

20.5 Extending MF with choice principles

The aim of this section is to show an extension of **MF** with choice principles in the spirit of BISH.

Ideally one would choose – as the most adequate basic foundation for Bishop's mathematics – a theory which fully axiomatizes the setoid model on Martin-Löf's type theory in [60]. But Martin-Löf's type theory is intended as a full-scale theory for formalizing constructive mathematics only if it is left opened to further extensions with all the needed inductive definitions. Currently only the first-order fragment of the setoid model has been axiomatized categorically in [64]. It would be desirable to have a type-theoretic presentation of the internal theory of the setoid model over **MLTT** as an extension of **emTT**, or better as an extension of the internal type theory of an arithmetic locally cartesian closed pretopos in [37].

Here we just consider an extension of **emTT** with the forms of axiom of choice which are acceptable constructively. Indeed, it is well known [26] that the full axiom of choice is not constructively acceptable in foundations with extensional principles.

The two-level structure of **MF** makes very evident the connection established in [57] between Zermelo's choice axiom and the type-theoretical axiom of choice in showing that the formula of the full axiom of choice

$$\text{(AC)} \qquad \forall x \in A\, \exists y \in B\, R(x,y) \quad \longrightarrow \quad \exists f \in A \to B\, \forall x \in A\, R(x, f(x))$$

for A, B sets and $R(x,y)$ a small proposition, written at the extensional level of **MF** gets interpreted at the intensional level of **MF** as *Martin-Löf's extensional axiom of choice* in [57] represented by the formula

$$(\mathrm{AC}_{ext})\quad \forall x \in A\ \exists y \in B\ R(x,y) \longrightarrow \exists f \in A \to B\ \ (\mathbf{Ext(f)}_{\simeq_\mathbf{A}}^{\simeq_\mathbf{B}}\ \ \&\ \ \forall x \in A\ R(x, f(x))),$$

where A , B are sets and

$$\mathbf{Ext}(f)_{\simeq_A}^{\simeq_B} \equiv \forall x_1 \in A\, \forall x_2 \in A\, (x \simeq_A y \to f(x_1) \simeq_B f(x_2)),$$

provided that $R(x,y)$ is a small proposition preserving given equivalence relations $\simeq_A$ and $\simeq_B$ in the sense that we can find in **mTT** a proof of

$$\forall x_1 \in A\, \forall x_2 \in A\, \forall y \in B\, (x_1 \simeq_A x_2 \to (R(x_1,y) \leftrightarrow R(x_2,y)))$$

and of

$$\forall x \in A\, \forall y_1 \in B\, \forall y_2 \in B\, (y_1 \simeq_B y_2 \to (R(x,y_1) \leftrightarrow R(x,y_2))).$$

As expected, the extensional axiom of choice (AC_{ext}) is not constructively acceptable since it implies the law of the excluded middle (see [16, 39, 57]).

Therefore, to develop constructive mathematics we cannot use the axiom of choice on all sets if we work in an extensional foundation such as **emTT**. On the contrary, as shown in Section 20.3.3, the intensional level **mTT** is consistent with the full axiom of choice

$$(\mathrm{AC})\qquad \forall x \in A\, \exists y \in B\, R(x,y) \longrightarrow \exists f \in A \to B\, \forall x \in A\, R(x, f(x)),$$

which is called *intensional* by Martin-Löf in [57].

The different status of (AC) in the extensional and intensional levels of **MF** reflects the different status of the axiom of choice in axiomatic set theory and in type theory. This fact was another motivation for building a two-level theory where to distinguish the various kinds of choice principles. Indeed, as described previously, the two forms of (AC) can only be visible in an intensional theory, like **mTT** or **MLTT**, but the fact that they express the same formula in different contexts is only visible in the two-level structure of **MF**.

20.5.1 The Extension emtt$_{ac}$ *of* MF

We extend the extensional level of **MF** to form a theory that we call emtt$_{ac}$ for short.

This theory emtt$_{ac}$ is simply obtained from **emTT** by adding the axiom of unique choice for all types and the axiom of choice for what we call *intensional sets*.

Intensional sets are sets of **emTT** whose interpretations in **mTT** given in [39] turn out to be 'faithful copies' of them in **mTT** (with their terms and their propositional equality), and they include the following:

$$A\ set_i \equiv \quad N_0 \mid N_1 \mid List(A) \mid \Sigma_{x \in A} B(x) \mid A + B,$$

provided that A and B are intensional sets, as well as $B(x)$ $[x \in A]$ is a family of intensional sets on an intensional set A.

Then, formally the theory emtt$_{ac}$ is obtained by extending **emTT** with:

- the *axiom of unique choice* written as follows

  (AC!$_R$)

$$\frac{R(a,b)\ prop\ [a \in A, b \in B]}{\forall x \in A\, \exists! y \in B\, R(x,y) \quad \longrightarrow \quad \exists f \in A \to B\, \forall x \in A\, R(x, f(x))\ true}$$

  for all A and B types in **emTT** where as usual

$$\exists! y \in A\, R(x,y) \equiv \exists y \in B\, R(x,y) \wedge \forall y_1, y_2 \in B\, (R(x,y_1) \wedge R(x,y_1) \\ \to y_1 =_B y_2);$$

- the axiom of choice on intensional sets

  (iAC$_R$)

$$\frac{A\ set_i \quad B\ set \qquad\qquad R(a,b)\ prop_s\ [a \in A, b \in B]}{\forall x \in A\, \exists y \in B\, R(x,y) \quad \longrightarrow \quad \exists f \in A \to B\, \forall x \in A\, R(x, f(x))\ true}.$$

By adding the axiom of unique choice we guarantee that the graph of each functional relation is also a graph of an operation and hence has a computational meaning in accordance with BISH.

Following the interpretation in [39] emtt$_{ac}$ can be interpreted in **mTT** extended with a proof term p_{ac}

(AC$_R$)

$$\frac{A\ set \quad B\ set \qquad R(a,b)\ prop_s\ [a \in A, b \in B]}{\mathsf{p}_{\mathsf{ac}} \in \forall x \in A\, \exists y \in B\, R(x,y) \quad \longrightarrow \quad \exists f \in A \to B\, \forall x \in A\, R(x, f(x))}.$$

In turn this extension of **mTT** naturally interprets in Martin-Löf's type theory **MLTT** where we can also extract programs from its proofs. Hence, **MLTT** could serve for the intensional level of a two-level foundation with emtt$_{ac}$ for the extensional level. But a direct extension of the interpretation in [39] for emtt$_{ac}$ into **MLTT** is left to future work.

20.6 Concluding Remarks

In our opinion the existence in **MF** of proper formal spaces, such as the space of real numbers both as Dedekind cuts or Cauchy sequences, constitutes an advantage which could be useful for performing reverse Bishop constructive mathematics. In fact, a positive and practical motivation of the minimalist approach is to provide a finer grid to look at reality, in particular topology, and thus preserve pieces of information, structures, conceptual distinctions (for example, all that is necessary to be able to instruct a computer) which would be lost, and actually are not even considered, in a classical or impredicative foundation. For instance, our Minimalist Foundation allows us to distinguish infinitary or ideal topological concepts not enjoying induction principles, like Brouwer's choice sequences, from inductive or real ones, for example, lawlike sequences. On the other hand, the minimalist attitude means that all results in the Minimalist Foundation about topology are also valid for the most relevant constructive and classical foundations.

Another important motivation supporting the pointfree approach to topology is given by a recent result showing that pointfree topology can be seen as a generalization of topology with points. To obtain this, one first has to introduce the notion of *positive topology*, that is an enrichment of formal topologies by the addition of a suitable primitive notion of closed subset. Then one can show that the category of concrete spaces can be embedded in the category of positive topologies. A full book [74] on positive topologies and their developments will soon be published.

For the future, we plan to investigate topologies of real numbers and of choice sequences using the more powerful tool provided by positive topologies.

Acknowledgements

We heartily thank Per Martin-Löf, Michael Rathjen, and Thomas Streicher for fruitful discussions on the topics of this chapter.

References

[1] Aczel, P. 1978. The type theoretic interpretation of constructive set theory. In: *Logic Colloquium '77 (Proceedings of the Conference, Wrocław, 1977)*. Studies in Logic and the Foundations of Mathematics, vol. 96. Amsterdam, New York: North-Holland.

[2] Aczel, P. 1982. The type-theoretic interpretation of constructive set theory: choice principles. In: Troelstra, A., and van Dalen, D (eds.), *The L.E.J. Brouwer Centenary Symposium (Noordwijkerhout, 1981)*. Studies in Logic and the Foundations of Mathematics, vol. 110. Amsterdam, New York: North-Holland.

[3] Aczel, P. 1986. The type-theoretic interpretation of constructive set theory: inductive definitions. In: *Logic, Methodology and Philosophy of Science, VII (Salzburg, 1983)*. Studies in Logic and the Foundations of Mathematics, vol. 114. Amsterdam, New York: North-Holland.

[4] Allen, S. F., Bickford, M., Constable, R. L., et al. 2006. Innovations in computational type theory using Nuprl. *J. Appl. Logic*, **4**(4), 428–469.

[5] Asperti, A., Ricciotti, W., Coen, C. Sacerdoti, and Tassi, E. 2011. The Matita interactive theorem prover. In: Bjørner, N., and Sofronie-Stokkermans, V. (eds.), *Proceedings of the 23rd International Conference on Automated Deduction (CADE-2011), Wroclaw, Poland.* Lecture Notes in Computer Science, vol. 6803. Cham: Springer.

[6] Barthes, G., Capretta, V., and Pons, O. 2003. Setoids in type theory. *J. Funct. Programming*, **13**(2), 261–293. Special issue on Logical Frameworks and Metalanguages.

[7] Battilotti, G., and Sambin, G. 2006. Pretopologies and a uniform presentation of sup-lattices, quantales and frames. Pages 30–61 of: Banaschewski, B., Coquand, T., and Sambin, G. (eds), *Special Issue: Papers presented at the 2nd Workshop on Formal Topology (2WFTop 2002).* Annals of Pure and Applied Logic, vol. **137**. Amsterdam: Elsevier.

[8] Beeson, M. 1985. *Foundations of Constructive Mathematics.* Berlin: Springer-Verlag.

[9] Bell, J. L. 1988. *Toposes and Local Set Theories: An Introduction.* Oxford: Clarendon Press.

[10] Bishop, E. 1967. *Foundations of Constructive Analysis.* New York: McGraw-Hill.

[11] Bishop, E. 1970. Mathematics as a numerical language. Pages 53–71 of: Kino, A., Myhill, J., and Vesley, R. E. (eds.), *Intuitionism and Proof Theory: Proceedings of the Summer Conference at Buffalo N.Y. 1968.* Studies in Logic and the Foundations of Mathematics, vol. 60. Amsterdam: Elsevier.

[12] Bishop, E., and Bridges, D. S. 1985. *Constructive Analysis.* Berlin: Springer.

[13] Bove, A., Dybjer, P., and Norell, U. 2009. A brief overview of Agda – a functional language with dependent types. Pages 73–78 of: Berghofer, S., Nipkow, T., Urban, C., and Wenzel, M. (eds.), *Theorem Proving in Higher Order Logics, 22nd International Conference, TPHOLs 2009.* LNCS, vol. 5674. Berlin: Springer.

[14] Bridges, D. 2008. A reverse look at Brouwer's fan theorem. Pages 316–325 of: *One Hundred Years of Intuitionism (1907-2007).* Basel: Birkäuser.

[15] Bridges, D., and Richman, F. 1987. *Varieties of Constructive Mathematics.* London Mathematical Society Lecture Note Series, vol. 97. Cambridge: Cambridge University Press.

[16] Carlström, J. 2004. $\mathrm{EM} + \mathrm{Ext}_{-} + \mathrm{AC}_{\mathrm{int}}$ is equivalent to $\mathrm{AC}_{\mathrm{ext}}$. *Math. Logic Q.*, **50**(3), 236–240.

[17] Ciraulo, F., Maietti, M. E., and Sambin, G. 2013. Convergence in formal topology: a unifying presentation. *Logic Anal.*, **5**(2), 1–45.

[18] Coq Development Team. 2010. *The Coq Proof Assistant Reference Manual: release 8.3.* Orsay: INRIA.

[19] Coquand, T. 1990. Metamathematical investigation of a calculus of constructions. Pages 91–122 of: Odifreddi, P. (ed.), *Logic in Computer Science.* New York: Academic Press.

[20] Coquand, T., Sambin, G., Smith, J., and Valentini, S. 2003. Inductively generated formal topologies. *Ann. Pure Appl. Logic*, **124**(1–3), 71–106.

[21] Coquand, T., and Paulin-Mohring, C. 1990. Inductively defined types. Pages 50–66 of: Martin-Löf, P., and Mints, G. (eds.), *Proceedings of the International Conference on Computer Logic (Colog '88)*. LNCS, vol. 417. Berlin: Springer.

[22] Curi, G. 2010. On some peculiar aspects of the constructive theory of point-free spaces. *Math. Logic. Q.*, **56**(4), 375–387.

[23] Feferman, S. 1979. Constructive theories of functions and classes. Pages 159–224 of: *Logic Colloquium '78 (Mons, 1978)*. Studies in Logic and the Foundations of Mathematics Amsterdam, New York: North-Holland.

[24] Fourman, M., and Grayson, R. J. 1982. Formal spaces. Pages 107–122 of: *The L.E.J. Brouwer Centenary Symposium* (Noordwijkerhout, 1981). Studies in Logic and the Foundations of Mathematics, vol. 110. Amsterdam, New York: North-Holland.

[25] Gambino, N., and Schuster, P. 2007. Spatiality for formal topologies. *Math. Struct. Comput. Sci.*, **17**(1), 65–80.

[26] Goodman, N., and Myhill, J. 1978. Choice implies excluded middle. *Z. Math. Logik Grundlag. Math.*, **24**, 461.

[27] Hofmann, M. 1997. *Extensional Constructs in Intensional Type Theory*. Distinguished Dissertations. Berlin: Springer.

[28] Hyland, J. M. E. 1982. The effective topos. Pages 165–216 of: *The L.E.J. Brouwer Centenary Symposium* (Noordwijkerhout, 1981). Studies in Logic and the Foundations of Mathematics, vol. 110. Amsterdam, New York: North-Holland.

[29] Hyland, J. M. E., Johnstone, P. T., and Pitts, A. M. 1980. Tripos theory. *Bull. Austral. Math. Soc.*, **88**, 205–232.

[30] Ishihara, H. 2005. Constructive reverse mathematics: compactness properties. Pages 245–267 of: Crosilla, L., and Schuster, P. (eds.), *From Sets and Types to Topology and Analysis: Practicable Foundations for Constructive Mathematics*. Oxford Logic Guides, no. 48. Oxford University Press.

[31] Ishihara, H., Maietti, M. E., Maschio, S., and Streicher, Th. 2018. Consistency of the intensional level of the Minimalist Foundation with Church's thesis and axiom of choice. *Arch. Math. Log.*, **57**(7–8), 595–668.

[32] Johnstone, P. T. 1982. *Stone Spaces*. Cambridge: Cambridge University Press.

[33] Joyal, A., and Moerdijk, I. 1995. *Algebraic Set Theory*. London Mathematical Society Lecture Note Series, vol. 220. Cambridge: Cambridge University Press.

[34] Joyal, A., and Tierney, M. (1984). An extension of the Galois theory of Grothendieck. *Memoirs of the American Mathematical Society*, **309**. Providence, RI: American Mathematical Society.

[35] Mac Lane, S., and Moerdijk, I. 1992. *Sheaves in Geometry and Logic. A First Introduction to Topos Theory*. Berlin: Springer Verlag.

[36] Maietti, M. E. 1998 (February). The Type Theory of Categorical Universes. Ph.D. thesis, University of Padova.

[37] Maietti, M. E. 2005a. Modular correspondence between dependent type theories and categories including pretopoi and topoi. *Math. Struct. Comp. Sci.*, **15**(6), 1089–1149.

[38] Maietti, M. E. 2005b. Predicative exponentiation of locally compact formal topologies over inductively generated ones. Pages 202–222 of: *From Sets and Types to Topology and Analysis: Practicable Foundation for Constructive Mathematics*. Oxford Logic Guides, vol. 48. Oxford: Oxford University Press.

[39] Maietti, M. E. 2009. A minimalist two-level foundation for constructive mathematics. *Ann. Pure Appl. Logic*, **160**(3), 319–354.

[40] Maietti, M. E. 2012. *Consistency of the Minimalist Foundation with Church Thesis and Bar Induction*. Preprint (available via `www.math.unipd.it/~maietti/pubb.html`).

[41] Maietti, M. E. 2017. On choice rules in dependent type theory. Pages 12–23 of: *Proc. Theory and Applications of Models of Computation – Proceedings 14th Annual Conference, TAMC 2017, Bern, Switzerland, April 20–22, 2017*.

[42] Maietti, M. E. 2018. The continuum from the predicative point of view of the Minimalist Foundation. Invited talk at "Das Kontinuum – 100 years later", University of Leeds, 11–14/9.

[43] Maietti, M. E., and Maschio, S. 2014. An extensional Kleene realizability semantics for the Minimalist Foundation. Pages 162–186 of: *Proc. 20th International Conference on Types for Proofs and Programs, TYPES 2014, May 12–15, 2014, Paris, France*.

[44] Maietti, M. E., and Maschio, S. 2016. A predicative variant of a realizability tripos for the Minimalist Foundation. *IfCoLog J. Logics Appl.*, **3**(4), 873–888.

[45] Maietti, M. E., and Rosolini, G. 2013a. Elementary quotient completion. *Theory Appl. Categ.*, **27**, 445–463.

[46] Maietti, M. E., and Rosolini, G. 2013b. Quotient completion for the foundation of constructive mathematics. *Log. Univers.*, **7**(3), 371–402.

[47] Maietti, M. E., and Rosolini, G. 2015. Unifying exact completions. *Appl. Categ. Structures*, **23**, 43–52.

[48] Maietti, M. E., and Sambin, G. 2005. Toward a minimalist foundation for constructive mathematics. Pages 91–114 of: Crosilla, L., and Schuster, P. (eds.), *From Sets and Types to Topology and Analysis: Practicable Foundations for Constructive Mathematics*. Oxford Logic Guides, no. 48. Oxford: Oxford University Press.

[49] Maietti, M. E., and Maschio, S. 2021. A predicative variant of Hyland's Effective Topos. *J. Symbol. Logic*, **86**(2), 433–447.

[50] Maietti, M. E., Maschio, S., and Rathjen, M. 2021. A realizability semantics for inductive formal topologies, Church's Thesis and Axiom of Choice. *Log. Meth. Comp. Sci.*, **17**(2), 21-1–21-21.

[51] Maietti, M. E., Maschio, S., and Rathjen, M. 2021. A realizability semantics for inductive formal topologies, Church's Thesis and Axiom of Choice. *Log. Methods Comput. Sci.*, **17**(2). https://dblp.org/rec/journals/lmcs/MaiettiMR21.bib

[52] Maietti, M. E., Maschio, S., and Rathjen, M. 2022. Inductive and Coinductive Topological Generation with Church's Thesis and Axiom of Choice. *Log. Methods Comput. Sci.*, **18**(4).

[53] Martin-Löf, P. 1970. *Notes on Constructive Mathematics*. Stockholm: Almqvist & Wiksell.

[54] Martin Löf, P. 1975. An intuitionistic theory of types: predicative part. Pages 73–118 of: *Logic Colloquium '73 (Bristol)*. Studies in Logic and the Foundations of Mathematics, vol. 80. Amsterdam: North-Holland.

[55] Martin-Löf, P. 1984. *Intuitionistic Type Theory*. Notes by G. Sambin of a Series of Lectures given in Padua, June 1980. Naples: Bibliopolis.

[56] Martin-Löf, P. 1985. On the meanings of the logical constants and the justifications of the logical laws. Pages 203–281 of: *Proceedings of the Conference on Mathematical Mogic (Siena, 1983/1984)*, vol. 2. Reprinted in: *Nordic J. Philosophical Logic*, **1**(1), no. 1, 11–60.

[57] Martin-Löf, P. 2006. 100 years of Zermelo's axiom of choice: What was the problem with it? *Comp. J.*, **49**(3), 10–37.

[58] Myhill, J. 1975. Constructive set theory. *J. Symbol. Logic*, **40**(3), 347–382.

[59] Negri, S., and Soravia, D. 1999. The continuum as a formal space. *Arch. Math. Logic*, **38**(7), 423–447.

[60] Nordström, B., Petersson, K., and Smith, J. 1990. *Programming in Martin Löf's Type Theory*. Oxford: Clarendon Press.

[61] Palmgren, E. 2005a. *Bishop's set theory*. Slides for lecture at the TYPES summer school.

[62] Palmgren, E. 2005b. Continuity on the real line and in formal spaces. Pages 165–175 of: Crosilla, L., and Schuster, P. (eds.), *From Sets and Types to Topology and Analysis: Practicable Foundations for Constructive Mathematics*. Oxford Logic Guides, no. 48. Oxford: Oxford University Press.

[63] Palmgren, E., and Wilander, O. 2014. Constructive categories and setoids of setoids in type theory. *Logical Meth. Comp. Sci.*, **10**(3), 1–14.

[64] Palmgren, Erik. 2012. Constructivist and structuralist foundations: Bishop's and Lawvere's theories of sets. *Ann. Pure Appl. Logic*, **163**(10), 1384–1399.

[65] Rathjen, M. 2005. Constructive set theory and Brouwerian principles. *J. UCS*, **11**(12), 2008–2033.

[66] Sambin, G. 1987. Intuitionistic formal spaces – a first communication. Pages 187–204 of: Skordev, D. (ed.), *Mathematical Logic and its Applications*. New York: Plenum.

[67] Sambin, G. 2002. Steps towards a dynamic constructivism. Pages 261–284 of: Gärdenfors, P., Wolenski, J., and Kijania-Placek, K. (eds.), *In the Scope of Logic, Methodology and Philosophy of Science. Volume One of the XI*

International Congress of Logic, Methodology and Philosophy of Science, Cracow, August 1999. Alphen aan den Rijn: Kluwer.

[68] Sambin, G. 2003. Some points in formal topology. *Theoretical Computer Science*, **305**, 347–408.

[69] Sambin, G. 2008. Two applications of dynamic constructivism: Brouwer's continuity principle and choice sequences in formal topology. Pages 301–315 of: van Atten, M., Boldini, P., Bourdeau, M., and Heinzmann, G. (eds.), *One Hundred Years of Intuitionism (1907–2007): The Cerisy Conference*. Basel: Birkhäuser.

[70] Sambin, G. 2011. A minimalist foundation at work. Pages 69–96 of: DeVidi, D., Hallett, M., and Clark, P. (eds.), *Logic, Mathematics, Philosophy, Vintage Enthusiasms. Essays in Honour of John L. Bell*. The Western Ontario Series in Philosophy of Science, vol. 75. Berlin: Springer.

[71] Sambin, G. 2012. Real and ideal in constructive mathematics. Pages 69–85 of: Dybjer, P., Lindström, S., Palmgren, E., and Sundholm, G. (eds.), *Epistemology versus Ontology, Essays on the Philosophy and Foundations of Mathematics in honour of Per Martin-Löf*. Logic, Epistemology and the Unity of Science, vol. 27. NewYork/Dordrecht: Springer.

[72] Sambin, G. 2017. C for Constructivism. Beyond clichés. *Lett. Mat. Int.*, **5**, 87–91. https://doi.org/10.1007/s40329-017-0169-1.

[73] Sambin, G. 2019. Dynamics in foundations: what does it mean in the practice of mathematics? Pages 455–494 of: Centrone, S., Kant, D., and Sarikaya, D. (eds.), *Reflections on the Foundations of Mathematics: Univalent Foundations, Set Theory and General Thoughts*. Synthese Library, vol. 407. Cham: Springer. `https://doi.org/10.1007/978-3-030-15655-8_21`.

[74] Sambin, G. 2022. *Positive Topology and the Basic Picture. New Mathematics Emerging from Dynamic Constructivism*. Oxford: Oxford University Press. (In the press.)

[75] Sambin, G., and Valentini, S. 1998. Building up a toolbox for Martin-Löf's type theory: subset theory. Pages 221–244 of: Sambin, G., and Smith, J. (eds.), *Twenty-Five Years of Constructive Type Theory*. Proceedings of a Congress held in Venice, October 1995. Oxford: Oxford University Press.

[76] Sigstam, I. 1995. Formal spaces and their effective presentations. *Arch. Math. Logic*, **34**, 211–246.

[77] Troelstra, A. S., and van Dalen, D. 1988. *Constructivism in Mathematics, an Introduction*, vol. I and II. *Studies in Logic and the Foundations of Mathematics*. Amsterdam: North-Holland.

[78] Univalent Foundations Program, 2013. *Homotopy Type Theory: Univalent Foundations of Mathematics*. Institute for Advanced Study: `https://homotopytypetheory.org/book`.

[79] Valentini, S. 2007. Constructive characterizations of bar subsets. *Ann. Pure Appl. Logic*, **145**(3), 368–378.

[80] Wiedijk, F. 2006. *The Seventeen Provers of the World*. LNCS, vol. 3600. Berlin: Springer.

21

Identity, Equality, and Extensionality in Explicit Mathematics

Gerhard Jäger

21.1 Introduction

In the mid 1970's, Solomon Feferman introduced a new formalism – he called it *explicit mathematics* – motivated by the aim to set up a proper formal framework for *Bishop-style constructive mathematics*, see [2]. In three milestone articles [4, 5, 6] he sketched the main ideas of explicit mathematics, compared his approach with other axiomatic frameworks for constructive mathematics, and started to look at explicit mathematics from the broader perspective of proof theory and generalized recursion theory.

This is one of the striking features of explicit mathematics: Although originally designed as a framework for constructive mathematics (and based on intuitionistic logic) it soon became evident that systems of explicit mathematics (now based on intuitionstic <u>or</u> classical logic) play an important role in many other parts of logic (such as, for example, subsystems of second-order arithmetic and set theory, reductive proof theory, and generalized/abstract recursion theory).

For me personally, explicit mathematics offers a convincing basis for large parts of mathematics. And for me it is not primarily a question of whether one uses intuitionistic or classical logic. Instead, I think that the distinction between concrete and abstract entities is of central importance.

The starting point for us (in this chapter) is an open-ended universe of objects. The exact nature of these objects is intentionally left open; you may think of them as (constructive) operations of any kind, bit strings, computer programs, or whatever. To give these objects a name, we call them *operations*. We only claim that they form a partial combinatory algebra, *PCA*.

Mathematical examples of such universes are Kleene's first and second models, the graph model as well as variants of term models. But one can also imagine situations where the application is understood as a more general kind of interaction between objects.

Our universe is open-ended in the sense that new constants can be added whenever needed or useful. The (partial) terms that can be built upon this partial combinatory algebra are our "first-class citizens." Because of the properties of combinatory algebras we have quite some expressive power and a basic computation theory over this base universe.

Classes, on the other hand, are abstractly given collections of operations. But we make sure – by means of a naming relation – that there is a deep connection between the concretely given elements of the *PCA* at the bottom and the class structure on top of it: every class has to have a name that belongs to the *PCA*. This naming relation is crucial in our operational approach.

One of its features is that some higher-type operations become possible. In a nutshell: Level 0 comprises the elements of the *PCA*; level $n+1$ consists of names of classes whose elements are objects of levels n; and it is obvious how to proceed into the transfinite.

However, from a conceptual point of view it is more important that this sort of naming allows us to look at classes – which are per se extensional objects – also from an *intensional perspective* and as such to reconcile set/class-theoretic *Platonism* with a sort of *conceptualism*. The "price" we have to pay is that we have to work with several equality relations. But the "gain" we make is a rich and ontologically interesting structure.

In this chapter we confine ourselves to a simple system of explicit mathematics with the class of *natural numbers* and *elementary comprehension* as its central class existence principle. Induction on the natural numbers will be restricted to classes so that we stay conservative over Peano arithmetic.

In Section 21.2 we introduce the basic axiomatic operational framework. Then in Section 21.3 we discuss some first ontological consequences. This is followed in Section 21.4 by some considerations that are related to the identities and equality relations of our formalism. In Section 21.5 we treat so-called *abstract data structures* on an abstract level before we turn to specific examples in Section 21.6: the number systems $\mathbb{N}$, $\mathbb{Z}$, and $\mathbb{Q}$. Section 21.7 is dedicated to some remarks about the representation of the structure of the real numbers $\mathbb{R}$ in our formalism.

21.2 The Basic Axiomatic Operational Framework

Before turning to the classes in explicit mathematics we have to look more closely at the general operational framework. We confine ourselves to a relatively weak core operational theory. The basic idea is simple: The universe of discourse is a partial combinatory algebra; its elements are operations and they share the following properties.

- Operations may be partial, they may freely be applied to each other, and self-application of operations is permitted.
- As a consequence, the general theory of operations is type-free. Later classes of operations will be added with the purpose of partly structuring the universe.
- Operations are intensional objects; extensionality of operations is assumed or claimed axiomatically only in very special situations.

The need to work with possibly undefined objects has also some impact on the logic that we will use in this chapter.

The system of explicit mathematics with which we will work in the following is formulated in the second-order language $\mathbb{L}$ for individuals and classes. It comprises individual variables $a, b, c, f, g, h, u, v, w, x, y, z, \dots$ as well as class variables $U, V, W, X, Y, Z, \dots$ (both possibly with subscripts). $\mathbb{L}$ also includes the individual constants k, s (combinators), $\mathsf{p}, \mathsf{p}_0, \mathsf{p}_1$ (pairing and projections), 0 (zero), s_N (successor), p_N (predecessor), d_N (definition by numerical cases), and additional individual constants, called generators, which will be used for the uniform naming of classes, namely nat (natural numbers), id (identity), co (complement), un (union), dom (domain), and inv (inverse image).

There is one binary function symbol $\cdot$ for (partial) application of individuals to individuals. Further, $\mathbb{L}$ has unary relation symbols $\downarrow$ (defined), N (natural numbers) as well as three binary relation symbols $\in$ (membership), $=$ (equality), and $\Re$ (naming, representation).

The *individual terms* $(r, s, t, r_0, s_0, t_0, \dots)$ of $\mathbb{L}$ are built up from individual variables and individual constants by means of our function symbol $\cdot$ for application. In the following, $\cdot(r, s)$ is usually written as $(r \cdot s)$, (rs) or – if no confusion arises – simply as rs. The convention of association to the left is also adopted so that $r_1 r_2 \dots r_n$ stands for $(\dots(r_1 r_2)\dots r_n)$, and we often also write $s(r_1, \dots, r_n)$ for $sr_1 \dots r_n$. Note that $\langle r, s\rangle$ stands for the pair $\mathsf{p}(r, s)$, and general n-tupling is defined by induction on $n \geq 1$ as follows:

$$\langle r_1\rangle := r_1 \quad \text{and} \quad \langle r_1, \dots, r_{n+1}\rangle := \langle\langle r_1, \dots, r_n\rangle, r_{n+1}\rangle.$$

If n is a natural number, we write $\overline{n}$ for the corresponding numeral, that is, for the closed term given recursively by $\overline{0} :\equiv 0$ and $\overline{n+1} :\equiv \mathsf{s}_\mathsf{N}\overline{n}$.

The atomic formulas of $\mathbb{L}$ are the formulas $(r\downarrow)$, $\mathsf{N}(r)$, $(r = s)$, $(r \in U)$, and $\Re(r, U)$. Since we work with a logic of partial terms, it is not guaranteed that all terms have values, and thus $(r\downarrow)$ is used to express that r *is defined* or r *has a value*. Moreover, $\mathsf{N}(r)$ and $(r \in U)$ say that r is a natural number and an element of class U, respectively. Finally, the formula $\Re(r, U)$ is used to express that the individual r *represents* the class U or is a *name* of U.

The *formulas* $(A, B, C, A_0, B_0, C_0, \ldots)$ of $\mathbb{L}$ are generated from these atomic formulas by closing them under the usual propositional connectives and quantification over individuals and classes. We will often omit parentheses if there is no danger of confusion.

An $\mathbb{L}$ formula A is called *stratified* iff the relation symbol $\Re$ does not occur in A; it is called *elementary* iff it is stratified and does not contain bound class variables.

The partial equality relation $\simeq$ is introduced, following Kleene, by

$$(r \simeq s) \;:=\; ((r{\downarrow} \vee s{\downarrow}) \;\rightarrow\; r = s),$$

and $(r \neq s)$ is written for $(r{\downarrow} \wedge s{\downarrow} \wedge \neg(r = s))$. Hence $(r \neq s)$ is not the logical negation of $(r = s)$.

Since we will be dealing with possibly undefined objects, it is convenient to work with Beeson's *logic of partial terms*, LPT, see [1], which can be based on classical or intuitionistic logic. It corresponds to the E^+-logic with equality and strictness of [15], where $E(r)$ is written instead of $r{\downarrow}$.

Space does not permit us to describe LPT in detail, but let us mention that, in LPT, the axioms for quantification over individuals have the form

$$A[r] \wedge r{\downarrow} \;\rightarrow\; \exists x A[x] \quad \text{and} \quad \forall x A[x] \wedge r{\downarrow} \;\rightarrow\; A[r].$$

The *definedness axioms* imply that all individual variables and constants are defined and that

$$A[r_1, \ldots, r_n] \;\rightarrow\; r_1{\downarrow} \wedge \ldots \wedge r_n{\downarrow}$$

for any atomic A. The *equality axioms* for individuals are as usual, formulated, however, only for variables.[1] Equality for classes is defined (see Section 21.3) and not an axiom.

In the following, we work with the classical version of LPT. However, the central arguments go through in intuitionistic LPT as well; if necessary, small adjustments must be made.

Before turning to our systems BON^+ and EC^+ we introduce some useful shorthand notation:

$$\begin{aligned}
r \in \mathsf{N} \;&:=\; \mathsf{N}(r),\\
(\exists x \in \mathsf{N})A[x] \;&:=\; \exists x(x \in \mathsf{N} \wedge A[x]),\\
(\forall x \in \mathsf{N})A[x] \;&:=\; \forall x(x \in \mathsf{N} \rightarrow A[x]),\\
r \in (\mathsf{N}^n \rightarrow \mathsf{N}) \;&:=\; (\forall x_1, \ldots, x_n \in \mathsf{N})(r(x_1, \ldots, x_n) \in \mathsf{N}).
\end{aligned}$$

Hence $r \in (\mathsf{N}^n \rightarrow \mathsf{N})$ says that the individual r represents an n-ary function from N to N.

[1] For example, $u = u$ is an axiom whereas $r = r$ is false if r is not defined.

All the "usual" systems of explicit mathematics comprise the axioms of a partial combinatory algebra as well as the standard axioms about pairing and projections. In addition, we have some canonical axioms for the natural numbers with successor and predecessor. The following theory BON^+ is the extension of the *basic theory of operations and numbers* BON introduced in [8] by the schema of induction on the natural numbers for all elementary formulas.

Partial Combinatory Algebra

- $\mathsf{k}(x, y) = x$,
- $\mathsf{s}(x, y)\downarrow \;\land\; \mathsf{s}(x, y, z) \simeq x(z, yz)$.

Pairing and Projections

- $\mathsf{p}_0\langle x, y\rangle = x \;\land\; \mathsf{p}_1\langle x, y\rangle = y$.

Natural Numbers

- $0 \in \mathsf{N} \;\land\; \mathsf{s_N} \in (\mathsf{N} \to \mathsf{N})$,
- $\mathsf{s_N}x \neq 0 \;\land\; \mathsf{p_N}0 = 0 \;\land\; (\forall x \in \mathsf{N})(\mathsf{p_N}(\mathsf{s_N}x) = x)$,
- $A[0] \;\land\; (\forall x \in \mathsf{N})(A[x] \to A[\mathsf{s_N}x]) \;\to\; (\forall x \in \mathsf{N})A[x]$

for all elementary formulas A. Of course, we could also allow induction for arbitrary $\mathbb{L}$ formulas. However, elementary induction is sufficient for our purposes in this article.

Definition by Cases on N

- $x \in \mathsf{N} \land y \in \mathsf{N} \land x = y \;\to\; \mathsf{d_N}(a, b, x, y) = a$,
- $x \in \mathsf{N} \land y \in \mathsf{N} \land x \neq y \;\to\; \mathsf{d_N}(a, b, x, y) = b$.

Since BON^+ comprises the axioms of a partial combinatory algebra, we clearly have λ-abstraction and the usual fixed point theorem. This was mentioned in [4] and proved in detail in, for example, [1] and [16]. Induction on the natural numbers is not needed for these two lemmas.

Lemma 21.1 (λ-abstraction) *For each variable x and term t we can construct a term $\lambda x\,.\,t$ whose free variables are those of t, excluding x, such that* BON^+ *proves*

$$\lambda x\,.\,t\downarrow \;\land\; (\lambda x\,.\,t)x \simeq t.$$

Lemma 21.2 (Fixed point) *There exists a closed term* fix *such that* BON^+ *proves*

$$\mathsf{fix}(f)\downarrow \;\land\; (g = \mathsf{fix}(f) \;\to\; \forall x(gx \simeq f(g, x))).$$

Following [13] we can easily show that BON^+ provides a good framework for dealing with the primitive recursive functions. We write $\mathbf{N}$ for the set of natural numbers. Given a (possibly partial) function $\mathcal{F}$ from $\mathbf{N}^k$ to $\mathbf{N}$ we say that a closed term t *numeralwise represents* $\mathcal{F}$ in BON^+ iff

$$\mathcal{F}(m_1,\ldots,m_k) \simeq n \quad \iff \quad \mathsf{BON}^+ \vdash t(\overline{m_1},\ldots,\overline{m_k}) \simeq \overline{n}$$

for all $m_1,\ldots,m_k,n \in \mathbf{N}$. However, this does not guarantee the expected behavior of t on nonstandard natural numbers. To impose such a condition we have to assume that it is described by formulas, for example, by equations. Let us consider an unary function $\mathcal{G}$ that is defined by primitive recursion from a natural number n_0 and a binary function $\mathcal{F}$ as

$$\mathcal{G}(0) = n_0 \quad \text{and} \quad \mathcal{G}(m+1) = \mathcal{F}(m, \mathcal{G}(m))$$

for all natural numbers m. Then, if the terms r and s represent the functions $\mathcal{F}$ and $\mathcal{G}$, respectively, we want the conditional equations

$$s0 \simeq \overline{n_0} \quad \text{and} \quad (\forall x \in \mathsf{N})(s(\mathsf{s_N} x) \simeq r(x, sx)).$$

If the defining formula of a function $\mathcal{F}$ is provable for a term t in BON^+, we say that t *definitionally represents* $\mathcal{F}$ in BON^+. The following is immediate from [16].

Theorem 21.3 *For any (definition of a) k-ary primitive recursive function $\mathcal{F}$, there exists a closed term $prim_{\mathcal{F}}$ that numeralwise and definitionally represents $\mathcal{F}$ in BON^+ and for which BON^+ proves $prim_{\mathcal{F}} \in (\mathsf{N}^k \to \mathsf{N})$.*

BON^+ is also a reasonable basis for general recursion theory. However, in [13] a point is made for introducing a new operation τ_{N}, called *truncation*, to obtain a natural formalization of partial recursive functions and semidecidability notions in our applicative framework. Adding the non-constructive minimum operator and the Suslin operator allows the move to higher recursion theory; see [8], [12], and [14].

21.3 Adding Elementary Classes

In this section we turn to classes. As mentioned in the introduction (Section 21.1), classes are collections of individuals and will be treated extensionally. Consequently, two classes are considered equal if the have the same elements. Therefore, we define

$$U \subseteq V \ := \ (\forall x \in U)(x \in V) \quad \text{and} \quad U = V \ := U \subseteq V \wedge V \subseteq U,$$

where $(\forall x \in U)A[x]$ is short for $\forall x(x \in U \to A[x])$.

In our explicit world every class must have a name or – if you prefer this point of view – every class can be addressed via an individual. This is achieved with the help of the relation $\Re$. We claim, in addition, that there are no homonyms and that $\Re$ respects the extensional equality of classes. This gives the next group of axioms.

Explicit Representation and Equality

- $\exists x \Re(x, U)$,
- $(\Re(r, U) \wedge \Re(r, V)) \to U = V$,
- $(U = V \wedge \Re(r, U)) \to \Re(r, V)$.

We say that an individual r is a *name* iff there exists a class X which is named by r; an individual r "belongs" to an individual s iff s is the name of a class that contains r. Also, r is equal in this new sense to s iff r and s name the same class. From now on we use the following notation:

$$\begin{aligned}
\Re(r) &:= \exists X \Re(r, X), \\
r \mathrel{\dot{\in}} s &:= \exists X(\Re(s, X) \wedge r \in X), \\
r \doteq s &:= \exists X(\Re(r, X) \wedge \Re(s, X)), \\
(\exists x \mathrel{\dot{\in}} r)A[x] &:= \exists x(x \mathrel{\dot{\in}} r \wedge A[x]), \\
(\forall x \mathrel{\dot{\in}} r)A[x] &:= \forall x(x \mathrel{\dot{\in}} r \to A[x]), \\
r \mathrel{\dot{\subseteq}} s &:= (\forall x \mathrel{\dot{\in}} r)(x \mathrel{\dot{\in}} s).
\end{aligned}$$

So, if $\Re(r)$ and $\Re(s)$, we clearly have

$$r \doteq s \leftrightarrow (r \mathrel{\dot{\subseteq}} s \wedge s \mathrel{\dot{\subseteq}} r).$$

If the vector $\vec{r}$ consists of the individual terms $r_1, \ldots, r_n$ and the vector $\vec{U}$ of the class variables $U_1, \ldots, U_n$, then

$$\Re(\vec{r}, \vec{U}) := \Re(r_1, U_1) \wedge \cdots \wedge \Re(r_n, U_n).$$

Instead of $\Re(r)$ we often write $r \in \Re$. However, this somewhat sloppy notation must not give the impression that $\Re$ is a class. Classes (more precisely: elementary definable classes) are generated by the following axioms.

Basic Class Existence Axioms In the following we provide a finite axiomatization of uniform elementary comprehension.

Natural Numbers

- $\mathsf{nat} \in \Re$,
- $\forall x(x \mathrel{\dot{\in}} \mathsf{nat} \leftrightarrow \mathsf{N}(x))$.

Identity

- $\mathsf{id} \in \Re$,
- $\forall x(x \mathrel{\dot\in} \mathsf{id} \leftrightarrow \exists y(x = \langle y, y\rangle))$.

Complements

- $a \in \Re \leftrightarrow \mathsf{co}(a) \in \Re$,
- $a \in \Re \rightarrow \forall x(x \mathrel{\dot\in} \mathsf{co}(a) \leftrightarrow \neg(x \mathrel{\dot\in} a))$.

Unions

- $a, b \in \Re \leftrightarrow \mathsf{un}(a, b) \in \Re$,
- $a, b \in \Re \rightarrow \forall x(x \mathrel{\dot\in} \mathsf{un}(a, b) \leftrightarrow x \mathrel{\dot\in} a \vee x \mathrel{\dot\in} b)$.

Domains

- $a \in \Re \leftrightarrow \mathsf{dom}(a) \in \Re$,
- $a \in \Re \rightarrow \forall x(x \mathrel{\dot\in} \mathsf{dom}(a) \leftrightarrow \exists y(\langle x, y\rangle \mathrel{\dot\in} a))$.

Inverse Images

- $a \in \Re \leftrightarrow \mathsf{inv}(a, f) \in \Re$,
- $a \in \Re \rightarrow \forall x(x \mathrel{\dot\in} \mathsf{inv}(a, f) \leftrightarrow fx \mathrel{\dot\in} a)$.

These are the generators and axioms if we work in classical logic. In the intuitionistic case more such generators and axioms are needed, for example, a generator and axioms for intersections.

Our theory EC^+ for elementary classes is the theory BON^+ extended by the axioms for explicit representation and equality plus the basic class existence axioms. The plus signifies that induction on N for elementary formulas is available. Accordingly, EC is EC^+ without the schema of induction on N for elementary formulas.

That the naming "elementary classes" is justified becomes evident in view of the following theorem. It is taken from [9] and states that (informally written) the class $\{x : A[x, \vec{v}, \vec{W}]\}$ can be uniformly named by a term $t_A(\vec{v}, \vec{w})$, provided that $\vec{w}$ names the classes $\vec{W}$.

Theorem 21.4 (Elementary comprehension) *For every elementary formula $A[u, \vec{v}, \vec{W}]$ with at most the indicated free variables there exists a closed term t_A such that one can prove in* EC*:*

(1) $\Re(\vec{w}, \vec{W}) \rightarrow t_A(\vec{v}, \vec{w}) \in \Re$,
(2) $\Re(\vec{w}, \vec{W}) \rightarrow \forall x(x \mathrel{\dot\in} t_A(\vec{v}, \vec{w}) \leftrightarrow A[x, \vec{v}, \vec{W}])$.

In view of this theorem, EC^+ is equivalent to EC plus the axiom of induction on N,

$$\forall X(0 \in X \wedge (\forall y \in \mathsf{N})(y \in X \rightarrow \mathsf{s_N} y \in X) \rightarrow (\forall y \in \mathsf{N})(y \in X)). \quad (\mathsf{C\text{-}I_N})$$

It also follows from [9] that EC^+ is a conservative extension of Peano arithmetic. For the rest of this chapter we will concentrate on EC and EC^+, discuss some basic ontological properties of this framework, and show that (and how) abstract data structures present themselves there.

21.4 About Some Ontological Aspects of EC and EC$^+$

There are two sorts of ontological properties of EC: (i) those referring to the first-order part of EC^+, and (ii) those that have to do with classes. We begin with the first group and consider three additional principles.

- *Operational extensionality* $(Op\text{-}Ext)$: $\forall x(fx \simeq gx) \rightarrow f = g$.
- *Totality* (T): $\forall x, y(xy\downarrow)$.
- *All operations are numbers* (N): $\forall x(x \in \mathsf{N})$.

Each of these principles makes strong ontological claims which will not be considered natural in most cases. Nevertheless, it is interesting to see how they relate to each other. The following assertions are spread over the literature and can be found, for example, in [11].

Theorem 21.5 *None of the principles* $(Op\text{-}Ext)$, (T), *or* (N) *is provable in* EC^+. *In addition, we have the following.*

(1) $\mathsf{EC}^+ + (Op\text{-}Ext) + (T)$ *is consistent.*
(2) $\mathsf{EC} + (Op\text{-}Ext) + (N)$ *is inconsistent.*
(3) $\mathsf{EC} + (T) + (N)$ *is inconsistent.*

There are further first-order principles that deserve attention[2] but we cannot discuss them here and refer the reader to the literature; see, for example, [11] and [13]. Instead, we turn to ontological properties that have to do with the class structure of EC.

Let us begin these considerations with the observation that EC plus full comprehension is inconsistent. It is proved by mimicking the usual Russell argument.

Theorem 21.6 EC *plus full comprehension is inconsistent.*

[2] Full definition by cases and truncation are such principles.

Proof Given full comprehension, there exists a class U such that

$$\forall x(x \in U \leftrightarrow \exists X(\Re(x,X) \wedge x \notin X)).$$

According to our axioms about explicit representation and equality, U has a name, say u. These axioms also tell us that all classes named u are extensionally equal, therefore

$$\exists X(\Re(u,X) \wedge u \notin X) \leftrightarrow u \notin U.$$

Together with the previous equivalence, this yields that $u \in U$ iff $u \notin U$. A contradiction. □

Of course, this result is not surprising. It can be shown, however, that EC plus comprehension for stratified formulas is consistent.

Now we turn to one of the most central properties of the naming relation. It says that the names of a class never form a class. The following is a generalization of Theorem 3 of [10].

Theorem 21.7 $\mathsf{EC} \vdash \forall X \neg \exists Y \forall z(z \in Y \leftrightarrow \Re(z,X))$.

Proof Working informally in EC, let U be an arbitrary class and u one of its names. We consider the elementary formula

$$A[x,y,V,W] := (x \notin V \wedge y \notin W) \vee (x \in V \wedge y \in W).$$

In view of Theorem 21.4 there exists a closed term t_A such that

$$\Re(v,V) \wedge \Re(w,W) \to t_A(y,v,w) \in \Re, \tag{21.1}$$

$$\Re(v,V) \wedge \Re(w,W) \to \forall x(x \mathrel{\dot{\in}} t_A(y,v,w) \leftrightarrow A[x,y,V,W]). \tag{21.2}$$

We set

$$s := \lambda w.t_A(w,u,w)$$

and conclude from (21.1) and (21.2) that, for any w and W,

$$\Re(w,W) \quad \to \quad sw \doteq \begin{cases} u & \text{if } w \in W, \\ \mathsf{co}(u) & \text{if } w \notin W. \end{cases} \tag{21.3}$$

Now assume that the names of U form a class Z_0, that is,

$$\forall x(x \in Z_0 \leftrightarrow \Re(x,U)),$$

and that Z_1 is introduced by elementary comprehension such that

$$\forall x(x \in Z_1 \leftrightarrow sx \notin Z_0).$$

Therefore,

$$\forall x(x \in Z_1 \leftrightarrow \neg\Re(sx, U)). \tag{21.4}$$

In addition, let z_1 be a name of Z_1. Then (21.3) yields

$$sz_1 \doteq \begin{cases} u & \text{if } z_1 \in Z_1, \\ \mathsf{co}(u) & \text{if } z_1 \notin Z_1. \end{cases} \tag{21.5}$$

From (21.4) and (21.5) we obtain that $z_1 \in Z_1$ iff $z_1 \notin Z_1$. This is a contradiction, implying that the names of U cannot form a class. □

Let U and V be classes with the names u and v, respectively. Then it is clear that $u = v$ implies $U = V$ and thus $u \doteq v$. However, in general it is possible that a class U has two different names, that is, we may have $u \neq v$ but $u \doteq v$. By adding class extensionality, we would rule this possibility out, but, as it turns out, class extensionality is not compatible with EC.

Theorem 21.8 EC *plus* class extensionality

$$(\forall x, y \in \Re)(x \doteq y \to x = y) \tag{Cl-Ext}$$

is inconsistent.

Proof Pick, for example, the class of natural numbers. From $(Cl\text{-}Ext)$ we could derive that all names of this class are identical to nat and thus form a class (by elementary comprehension), contradicting our previous theorem. □

Although the names of a class never form a class, it is consistent with EC^+ to claim that there exists the class of all names. This can be seen by extending the model construction for EC that is presented in detail in [6].

Theorem 21.9 *The assertion* $\exists X \forall x(x \in X \leftrightarrow x \in \Re)$ *is consistent with* EC^+, *but not provable in* EC^+.

With some additional effort even a strengthening of this result is possible: We can consistently assume in EC that all objects are names.

Power classes provide some problems in explicit mathematics. The naive approach is to claim that for every class X there exists a class Y such that Y contains exactly the names of <u>all</u> subclasses of X. We call it the *strong power class axiom*:

$$\forall X \exists Y \forall z(z \in Y \leftrightarrow \exists Z(\Re(z, Z) \wedge Z \subseteq X)). \tag{SP}$$

The *weak power class axiom* asks for less: It only claims that for each class X there exists a class Y such that each element of Y names a subclass of X and for any subclass of X at least one of its names belongs to Y,

$$\forall X \exists Y ((\forall z \in Y)(\exists Z \subseteq X)\Re(z, Z) \wedge (\forall Z \subseteq X)(\exists z \in Y)\Re(z, Z)). \quad \text{(WP)}$$

Both the strong and the weak power class axioms are problematic. By Theorem 21.7 we know that in EC the names of the empty class cannot form a class, and thus the strong power class of the empty class cannot exist.

Corollary 21.10 (SP) *is inconsistent with* EC.

The weak power class axiom is less problematic in the sense that it is consistent with EC. However, one may rightly argue whether it is in the spirit of explicit mathematics.

Unlike in ordinary mathematics, explicit mathematics does not permit the identification of power classes with the collection of its characteristic operations. Given a class U neither the strong nor the weak power class of U make much sense. On the other hand,

$$\{0, 1\}^U := \{f : (\forall x \in U)(fx = 0 \vee fx = 1)\}$$

is a class by elementary comprehension. In most cases it is the adequate "substitute" for the power class of U. This is also the approach taken in Section 21.7.

21.5 Abstract Data Structures

As we have seen, there are already two equality relations on our basic universe of individuals.

- $a = b$ means that a and b are identical in the sense of the underlying partial combinatory algebra. It is the strongest form of equality in explicit mathematics. Whenever $a = b$ then a will be equal to b also in the sense of any equality relation to be introduced later.
- $a \doteq b$ means that a and b name the same class. So $\doteq$ is inherited from the extensional equality of classes: If a is a name of class U and b is a name of class V then $a \doteq b$ iff U and V are extensionally equal as classes.

These two equality relations are "global" equality relations on the universe. However, in many situations "local" equalities are required. This is exactly what Bishop and Bridges do when they introduce their notion of set; see [3], pp. 15–16:

> *The totality of all mathematical objects constructed in accordance with certain requirements is called a* set. *The requirements of the construction, which vary with the set under consideration, determine the set. Thus the integers form a*

set, the rational numbers form a set, and ... the collection of all sequences of integers is a set.

Each set will be endowed with a binary relation of equality. *This relation is a matter of convention, except that it must be an* equivalence relation.

This approach can be implemented in EC in a very natural way. Given $a \in \Re$, we write $a \times a$ for a name of the class of all pairs $\langle x, y \rangle$ with $x, y \mathrel{\dot{\in}} a$. Such a name is provided by elementary comprehension.

Definition 21.11 A *set* is defined to be a pair $a = \langle a_1, a_2 \rangle$ such that $a_1, a_2 \in \Re$ and $a_2 \mathrel{\dot{\subseteq}} a_1 \times a_1$ is an equivalence relation on a_1. In this case we call a_1 the *universe* and a_2 the *equality* of a.

In the following we let lower-case boldface Latin letters range over sets. Then $|\mathbf{a}|$ stands for the universe of $\mathbf{a}$ and $=_{\mathbf{a}}$ for its equality; that is,

$$\mathbf{a} \;=\; \langle |\mathbf{a}|, =_{\mathbf{a}} \rangle.$$

Clearly, we write $x =_{\mathbf{a}} y$ instead of $\langle x, y \rangle \mathrel{\dot{\in}} =_{\mathbf{a}}$ for all $x, y \mathrel{\dot{\in}} |\mathbf{a}|$. Also, if $\vec{u}$ and $\vec{v}$ are the lists $u_0, \ldots, u_n$ and $v_1, \ldots, v_n$ of elements of $|\mathbf{a}|$, then $\vec{u} =_{\mathbf{a}} \vec{v}$ is short for

$$u_1 =_{\mathbf{a}} v_1 \;\wedge\; \cdots \;\wedge u_n =_{\mathbf{a}} v_n.$$

Keep in mind that sets are represented via their names. In general, we can have that $|\mathbf{a}| \doteq |\mathbf{b}|$ and, for any $x, y \mathrel{\dot{\in}} |\mathbf{a}|$,

$$x =_{\mathbf{a}} y \;\leftrightarrow\; x =_{\mathbf{b}} y,$$

although $\mathbf{a}$ and $\mathbf{b}$ are different as names.

Therefore, a higher equality on the level of sets can be introduced: For sets $\mathbf{a}$ and $\mathbf{b}$ we set

$$\mathbf{a} =_{set} \mathbf{b} \quad := \quad \begin{cases} |\mathbf{a}| \doteq |\mathbf{b}| \;\wedge \\ (\forall x, y \mathrel{\dot{\in}} |\mathbf{a}|)(x =_{\mathbf{a}} y \;\leftrightarrow\; x =_{\mathbf{b}} y). \end{cases}$$

However, for the following we do not need this equality.

There are two operations on sets that are particularly important: *Cartesian products* and *Cartesian powers*. In view of Theorem 21.4 it is easy to see that there is a closed term $\mathsf{carprod}$ such that $\mathsf{carprod}(\mathbf{a}, \mathbf{b})$ is, for all sets $\mathbf{a}$ and $\mathbf{b}$, the pair $\langle |\mathbf{a}| \times |\mathbf{b}|, =_{\mathbf{a} \times \mathbf{b}} \rangle$, where $=_{\mathbf{a} \times \mathbf{b}}$ stands for the binary relation on $|\mathbf{a}| \times |\mathbf{b}|$ given by, for all $x_1, x_2 \mathrel{\dot{\in}} |\mathbf{a}|$ and $y_1, y_2 \mathrel{\dot{\in}} |\mathbf{b}|$,

$$\langle x_1, y_1 \rangle =_{\mathbf{a} \times \mathbf{b}} \langle x_2, y_2 \rangle \;\leftrightarrow\; x_1 =_{\mathbf{a}} x_2 \wedge y_1 =_{\mathbf{b}} y_2.$$

In the following we simply write $\mathbf{a} \times \mathbf{b}$ for $\mathsf{carprod}(\mathbf{a}, \mathbf{b})$. Then $\mathbf{a}^2$ is defined to be the set $\mathbf{a} \times \mathbf{a}$, and $\mathbf{a}^{n+1} = \mathbf{a}^n \times \mathbf{a}$.

The introduction of the Cartesian power requires a bit more care. First we make use of Theorem 21.4 to find a closed term efun[3] such that $\mathsf{efun}(\mathbf{a},\mathbf{b})$ names the class

$$\{f : (\forall x \mathrel{\dot\in} |\mathbf{a}|)(f(x) \in |\mathbf{b}|) \wedge (\forall x, y \mathrel{\dot\in} |\mathbf{a}|)(x =_{\mathbf{a}} y \to f(x) =_{\mathbf{b}} f(y))\}$$

of all functions from $|\mathbf{a}|$ to $|\mathbf{b}|$ that respect the equalities of $\mathbf{a}$ and $\mathbf{b}$. The standard equivalence relation on $\mathsf{efun}(\mathbf{a},\mathbf{b})$ is then given by

$$f =_{\mathbf{a}\to\mathbf{b}} g \;\leftrightarrow\; (\forall x \mathrel{\dot\in} |\mathbf{a}|)(f(x) =_{\mathbf{b}} g(x))$$

and is thus elementary. We write $\mathbf{b}^{\mathbf{a}}$ for the pair $\langle \mathsf{efun}(\mathbf{a},\mathbf{b}), =_{\mathbf{a}\to\mathbf{b}}\rangle$ and observe that $\mathbf{b}^{\mathbf{a}}$ is a set.

In a next step we look at functions and relations acting on a given set. We call f an *n-extensional function* on set $\mathbf{a}$ if $f \in (|\mathbf{a}|^n \to |\mathbf{a}|)$ and for all strings $\vec{x}$ and $\vec{y}$ of length n the following condition is satisfied:

$$\vec{x} \mathrel{\dot\in} |\mathbf{a}| \wedge \vec{x} =_{\mathbf{a}} \vec{y} \;\to\; f(\vec{x}) =_{\mathbf{a}} f(\vec{y}).$$

Accordingly, r is called an *n-extensional relation* on set $\mathbf{a}$ if $r \mathrel{\dot\subseteq} |\mathbf{a}|^n$ and for all strings $\vec{x}$ and $\vec{y}$ of length n the following condition is satisfied:

$$\vec{x} =_{\mathbf{a}} \vec{y} \wedge \langle\vec{x}\rangle \mathrel{\dot\in} r \;\to\; \langle\vec{y}\rangle \mathrel{\dot\in} r.$$

Thus the n-extensional functions on a set $\mathbf{a}$ are exactly those n-ary total functions on $\mathbf{a}$ which respect its equality; analogously for the n-extensional relations on $\mathbf{a}$. Such functions and relations together with the underlying set form what we call an abstract data structure.

Definition 21.12 An *abstract data structure* consists of a set $\mathbf{a}$ and finitely many functions and relations of various arities that are extensional on $\mathbf{a}$ and thus is a tuple of the form $\langle \mathbf{a}, f_1, \ldots, f_m, r_1, \ldots, r_n\rangle$.

Rather than studying abstract data structures on a general level we now turn to a series of specific such structure.

21.6 The Number Systems $\mathbb{N}$, $\mathbb{Z}$, and $\mathbb{Q}$ as Abstract Data Structures

In the case of the natural numbers the situation is simple: The natural numbers themselves are directly given in explicit mathematics, and the corresponding equality relation is the identity relation of our language $\mathbb{L}$. Consequently, the *set of natural numbers* in our sense is given as the pair

$$\mathbf{nat} := \langle \mathsf{nat}, \mathsf{id}\rangle.$$

[3] It stands for “extensional functions.”

As far as functions and relations on $\mathbf{nat}$ are concerned, we can work with the usual primitive recursive machinery, which is available in EC^+ according to Theorem 21.3. Thus the algebraic properties of the natural numbers are summarized in the abstract data structure

$$\mathbb{N} := \langle \mathbf{nat}, 0, 1, +_\mathsf{N}, \cdot_\mathsf{N}, <_\mathsf{N} \rangle$$

that describes an ordered commutative semiring without zero divisors in which 0 is the least element and 1 its successor. It should be evident that $(\mathsf{C}\text{-}\mathsf{I}_\mathsf{N})$ is sufficient for proving the basic arithmetic properties of $\mathbb{N}$.

Any integer is coded as a pair $\langle x, y \rangle$ of natural numbers, intended to represent the difference $x - y$. Again we proceed by elementary comprehension and fix a constant term int such that

$$\mathsf{int} \doteq \{\langle x, y \rangle : \mathsf{N}(x) \wedge \mathsf{N}(y)\},$$

and a constant term $=_\mathsf{Z}$ for integer equality, satisfying

$$\langle x_1, y_1 \rangle =_\mathsf{Z} \langle x_2, y_2 \rangle \;\leftrightarrow\; x_1 +_\mathsf{N} y_2 = x_2 +_\mathsf{N} y_1$$

for all $x_1, x_2, y_1, y_2 \in \mathsf{N}$. It is obvious that

$$\mathbf{int} := \langle \mathsf{int}, =_\mathsf{Z} \rangle$$

is a set. The zero element 0_Z and unit element 1_Z for $\mathbf{int}$ are defined by $\langle 0, 0 \rangle$ and $\langle 1, 0 \rangle$, respectively. The addition $+_\mathsf{Z}$, subtraction $-_\mathsf{Z}$, and multiplication $\cdot_\mathsf{Z}$ are 2-extensional functions on $\mathbf{int}$ defined such that, for $x_1, x_2, y_1, y_2 \mathrel{\dot{\in}} \mathsf{nat}$,

$$\begin{aligned}
\langle x_1, y_1 \rangle +_\mathsf{Z} \langle x_2, y_2 \rangle &= \langle x_1 +_\mathsf{N} x_2, y_1 +_\mathsf{N} y_2 \rangle, \\
\langle x_1, y_1 \rangle -_\mathsf{Z} \langle x_2, y_2 \rangle &= \langle x_1 +_\mathsf{N} y_2, y_1 +_\mathsf{N} x_2 \rangle, \\
\langle x_1, y_1 \rangle \cdot_\mathsf{Z} \langle x_2, y_2 \rangle &= \langle (x_1 \cdot_\mathsf{N} x_2) +_\mathsf{N} (y_1 \cdot_\mathsf{N} y_2), (y_1 \cdot_\mathsf{N} x_2) +_\mathsf{N} (x_1 \cdot_\mathsf{N} y_2) \rangle.
\end{aligned}$$

Finally, there is an elementary definable and 2-extensional less-relation on $\mathbf{int}$ with

$$\langle x_1, y_1 \rangle <_\mathsf{Z} \langle x_2, y_2 \rangle \;\leftrightarrow\; x_1 +_\mathsf{N} y_2 <_\mathsf{N} x_2 +_\mathsf{N} y_1$$

again for all $x_1, x_2, y_1, y_2 \in \mathsf{N}$. Then it can be seen easily that the abstract data structure

$$\mathbb{Z} := \langle \mathbf{int}, 0_\mathsf{Z}, 1_\mathsf{Z}, +_\mathsf{Z}, \cdot_\mathsf{Z}, <_\mathsf{Z} \rangle$$

is an ordered integral domain. We have an injection of $\mathbb{N}$ into $\mathbb{Z}$ by mapping any $x \mathrel{\dot{\in}} \mathsf{nat}$ onto $\langle x, 0 \rangle$.

The rational numbers are introduced by following a similar path. Any rational number is coded as a pair $\langle x, y\rangle$ with $x, y \mathrel{\dot{\in}} \mathsf{int}$ and y different from 0_Z. As before, we work in EC, fix a constant rat such that

$$\mathsf{rat} \ \doteq\ \{\langle x, y\rangle : x, y \mathrel{\dot{\in}} \mathsf{int} \wedge y \neq_\mathsf{Z} 0_\mathsf{Z}\}$$

and introduce a further constant $=_\mathsf{Q}$ for the equality relation on int, satisfying

$$\langle x_1, y_1\rangle =_\mathsf{Q} \langle x_2, y_2\rangle \ \leftrightarrow\ x_1 \cdot_\mathsf{Z} y_2 =_\mathsf{Z} y_1 \cdot_\mathsf{Z} x_2$$

for all $x_1, x_2, y_1, y_2 \mathrel{\dot{\in}} \mathsf{int}$ where y_1 and y_2 are different from 0_Z. Observe that

$$\mathbf{rat} \ :=\ \langle \mathsf{rat}, =_\mathsf{Q}\rangle$$

is a set with the corresponding zero element $0_\mathsf{Q} := \langle 0_\mathsf{Z}, 1_\mathsf{Z}\rangle$ and unit element $1_\mathsf{Q} := \langle 1_\mathsf{Z}, 1_\mathsf{Z}\rangle$.

As before it is now straightforward to introduce 2-extensional functions for addition, subtraction, and multiplication, and a 2-extensional less-relation on $\mathbf{rat}$ that satisfy that

$$\langle x_1, y_1\rangle +_\mathsf{Q} \langle x_2, y_2\rangle \ =\ \langle (x_1 \cdot_\mathsf{Z} y_2) +_\mathsf{Z} (x_2 \cdot_\mathsf{Z} y_1), y_1 \cdot_Z y_2\rangle,$$

$$\langle x_1, y_1\rangle -_\mathsf{Q} \langle x_2, y_2\rangle \ =\ \langle (x_1 \cdot_\mathsf{Z} y_2) -_\mathsf{Z} (x_2 \cdot_\mathsf{Z} y_1), y_1 \cdot_Z y_2\rangle,$$

$$\langle x_1, y_1\rangle \cdot_\mathsf{Q} \langle x_2, y_2\rangle \ =\ \langle x_1 \cdot_\mathsf{Z} x_2, y_1 \cdot_Z y_2\rangle,$$

$$\langle x_1, y_1\rangle <_\mathsf{Q} \langle x_2, y_2\rangle \ \leftrightarrow\ \begin{cases} (x_1 \cdot_\mathsf{Z} y_2 <_\mathsf{Z} x_2 \cdot_\mathsf{Z} y_1 \wedge 0_\mathsf{Z} <_\mathsf{Z} y_1 \cdot_\mathsf{Z} y_2) \ \vee \\ (x_2 \cdot_\mathsf{Z} y_1 <_\mathsf{Z} x_1 \cdot_\mathsf{Z} y_2 \wedge y_1 \cdot_\mathsf{Z} y_2 <_\mathsf{Z} 0_\mathsf{Z}), \end{cases}$$

where $\langle x_1, y_1\rangle$ and $\langle x_2, y_2\rangle$ range over the elements of rat. Now it can be verified in EC^+ that

$$\mathbb{Q} \ :=\ \langle \mathbf{rat}, 0_\mathsf{Q}, 1_\mathsf{Q}, +_\mathsf{Q}, \cdot_\mathsf{Q}, <_\mathsf{Q}\rangle$$

is an ordered field. $\mathbb{Z}$ is embedded into $\mathbb{Q}$ by sending x to $\langle x, 1_\mathsf{Z}\rangle$. In particular, for every natural number x greater than 0 its quotient $1/x$ is easily represented in $\mathbb{Q}$ by setting

$$x^{-1} \ :=\ \langle 1_\mathsf{Z}, \langle x, 0\rangle\rangle.$$

If x is 0, then x^{-1} does of course not belong to rat.

Before turning to the representation of the real numbers in Section 21.7, we have to address the treatment of *absolute values* of integers and rationals. For the integers this is achieved by the 1-extensional function $|.|_\mathsf{Z}$ on $\mathbf{int}$ with

$$|\langle x,y\rangle|_{\mathsf{Z}} \;=\; \begin{cases} \langle x,y\rangle & \text{if } y <_{\mathsf{N}} x, \\ \langle y,x\rangle & \text{if } y \not<_{\mathsf{N}} x \end{cases}$$

for all $x, y \mathrel{\dot{\in}} \mathsf{nat}$. This function is easily defined by means of definition by integer cases and the closed term $\mathsf{cut_{off}}$ for the binary primitive-recursive cut-off subtraction; simply set

$$|u|_{\mathsf{Z}} \;:=\; \mathsf{d_N}(\langle \mathsf{p_1}u, \mathsf{p_0}u\rangle, u, \mathsf{cut_{off}}(\mathsf{p_0}u, \mathsf{p_1}u), 0).$$

The absolute value of a rational number $\langle x, y\rangle$ is given by

$$|\langle x,y\rangle|_{\mathsf{Q}} \;:=\; \langle |x|_{\mathsf{Z}}, |y|_{\mathsf{Z}}\rangle.$$

It should be clear the the functions $|.|_{\mathsf{Z}}$ and $|.|_{\mathsf{Q}}$ are extensional on **int** and **rat**, respectively.

In the following, when we work with functions extensional on **int** or **rat** and when there is no ambiguity as determined by the context, we drop the subscripts "Z" and "Q".

21.7 Representing the Real Numbers

So far the build up of the number systems $\mathbb{N}$, $\mathbb{Z}$, and $\mathbb{Q}$ has been completely canonical – perhaps with the small peculiarity that we make use of specific equality relations rather than moving to equivalence classes – and there has not been much choice how to proceed. When it comes to the real numbers the situation is different and one can take different paths, for example:

- Dedekind cuts,
- Cauchy sequences,
- nestings of intervals with rational bounds,
- completion of the topological group of rational numbers.

As it turns out, the approach to the reals in [3] can be reproduced very well in explicit mathematics. This has been worked out up to a certain extent in [7].[4] In order to illustrate the conceptual power of the explicit framework, we recall a few central definitions.

Let nat^+ be a name of the class of all positive natural numbers,

$$\mathsf{nat}^+ \;\doteq\; \{x : \mathsf{N}(x) \wedge x \neq 0\}.$$

A *sequence* of elements of a class named a is an operation $x \in (\mathsf{nat}^+ \to a)$. In this case we often write x_n for $x(n)$ and $(x_n : n \mathrel{\dot{\in}} \mathsf{nat}^+)$ for x.

[4] Unfortunately, this research note is not available via Feferman's homepage. If you are interested in it, please contact me.

Definition 21.13

(1) A sequence of rational numbers is called *regular* iff for all $m, n \mathrel{\dot\in} \mathsf{nat}^+$,

$$|x_m - x_n| \leq m^{-1} + n^{-1}.$$

(2) By a *real number* we mean a regular sequence of rational numbers,

$$\mathsf{real} \ \doteq \ \{x : x \mathrel{\dot\in} (\mathsf{nat}^+ \to \mathsf{rat}) \wedge x \text{ is regular}\}.$$

(3) For the equality relation on real we introduce a further constant $=_\mathsf{R}$ such that

$$x =_\mathsf{R} y \ \leftrightarrow \ (\forall n \in \mathsf{nat}^+)(|x_n - y_n| \leq 2 \cdot n^{-1})$$

for all $x, y \mathrel{\dot\in} \mathsf{real}$.

(4) Finally, we introduce the pair **real** and observe that it is a set,

$$\mathbf{real} \ := \ \langle \mathsf{real}, =_\mathsf{R} \rangle.$$

We end this section by introducing a series of functions that are extensional on the set **real**. These definitions are taken from [7] and are based on [3].

Associated with each real number $x = (x_n : n \mathrel{\dot\in} \mathsf{nat}^+)$ is a *canoniocal bound* $\mathbf{k}_x$ such that

$$|x_n| \ < \ \mathbf{k}_x$$

for all $n \mathrel{\dot\in} \mathsf{nat}^+$; $\mathbf{k}_x$ may simply be taken to be the least integer greater than $|x_1|+2$. Working informally, one may now proceed by defining, for all $x, y \mathrel{\dot\in} \mathsf{real}$:

$$\begin{aligned}
x + y \ &:= \ (x_{2n} + y_{2n} : n \mathrel{\dot\in} \mathsf{nat}^+), \\
x \cdot y \ &:= \ (x_{2kn} \cdot y_{2kn} : n \mathrel{\dot\in} \mathsf{nat}^+) \text{ for } k = \max(\mathbf{k}_x, \mathbf{k}_y), \\
\max(x, y) \ &:= \ (\max(x_n, y_n) : n \mathrel{\dot\in} \mathsf{nat}^+), \\
-x \ &:= \ (-x_n : n \mathrel{\dot\in} \mathsf{nat}^+), \\
|x| \ &:= \ \max(x, -x).
\end{aligned}$$

It has to be checked that the sequences $x + y$, $x \cdot y$, $\max(x, y)$, and $-x$ belong to real and that these functions are extensional on **real**. It is a routine matter to verify that this can be done in EC^+.

It should be clear how to embed the rational numbers into the reals: Simply send a $q \mathrel{\dot\in} \mathsf{rat}$ to the regular sequence $q^* = (q, q, \ldots)$. Then operations like addition and multiplication are easily seen to be preserved under this embedding.

One aspect that is distinctive to constructivity and goes beyond explicit representations is the definition of positivity of real numbers. The real number x is defined to be *positive* iff for some positive integer k, $k^{-1} < x_k$. On the other hand, a real number x is called *nonnegative* iff $-n^{-1} < x_n$ for all $n \dot{\in} \mathsf{nat}^+$.

Based on that, the ordering on the reals is given by

$$y < x \ := \ (x - y) \text{ is positive},$$

$$y \leq x \ := \ (x - y) \text{ is nonnegative}.$$

Calling, in addition, a real number x *nonzero* iff its absolute value is positive, completes the picture.

We are now in the realm of "ordinary" constructive analysis and set

$$\mathbb{R} \ := \ \langle \mathbf{real}, 0^*, 1^*, +, \cdot, \max, \leq, <, \ldots \rangle$$

for the reals and the usual operations on them. For more details about the ordering of real numbers, the question of nonzeroness, the least upper bound principle, and continuous functions, we refer to [7].

One should consider the conceptual considerations above and the technical work [7] as a proof of principal that a reasonable part of elementary analysis can be developed in a weak fragment of explicit mathematics such as EC^+ – a conservative extension of Peano arithmetic. There are also many interesting extensions of EC^+ by, for example, the join axiom, inductive generation, higher functionals and/or strong induction principles that provide suitable frameworks for stronger parts of mathematics such as complex analysis or measure theory.

References

[1] Beeson, M. J. 1985. *Foundations of Constructive Mathematics*. Ergebnisse der Mathematik und ihrer Grenzgebiete, vol. 3/6. Berlin: Springer.

[2] Bishop, E. 1967. *Foundations of Constructive Analysis*. McGraw-Hill Series in Higher Mathematics. New York: McGraw-Hill.

[3] Bishop, E., and Bridges, D. S. 1985. *Constructive Analysis*. Grundlehren der mathematischen Wissenschaften, vol. 279. Berlin: Springer.

[4] Feferman, S. 1975. A language and axioms for explicit mathematics. Pages 87–139 of: Crossley, J. N. (ed.), *Algebra and Logic*. Lecture Notes in Mathematics, vol. 450. Berlin: Springer.

[5] Feferman, S. 1978. Recursion theory and set theory: a marriage of convenience. Pages 55–98 of: Fenstad, J. E., Gandy, R. O., and Sacks, G. E. (eds.), *Generalized Recursion Theory II, Oslo 1977*. Studies in Logic and the Foundations of Mathematics, vol. 94. Amsterdam: Elsevier.

[6] Feferman, S. 1979. Constructive theories of functions and classes. Pages 159–224 of: Boffa, M., van Dalen, D., and McAloon, K. (eds.), *Logic Colloquium '78*. Studies in Logic and the Foundations of Mathematics, vol. 97. Amsterdam: Elsevier.

[7] Feferman, S. 2012. Constructive analysis in a *BON* system. *Technical note*, Stanford University.

[8] Feferman, S., and Jäger, G. 1993. Systems of explicit mathematics with non-constructive μ-operator. Part I. *Ann. Pure Appl. Logic*, **65**(3), 243–263.

[9] Feferman, S., and Jäger, G. 1996. Systems of explicit mathematics with non-constructive μ-operator and join. Part II. *Ann. Pure Appl. Logic*, **79**(1), 37–52.

[10] Jäger, G. 1997. Power types in explicit mathematics? *J. Symbol. Logic*, **62**(4), 1142–1146.

[11] Jäger, G. 2017. The operational penumbra: some ontological aspects. Pages 253–283 of: Jäger, G., and Sieg, W. (eds.), *Feferman on Foundations: Logic, Mathematics, Philosophy*. Outstanding Contributions to Logic, vol. 13. Berlin: Springer.

[12] Jäger, G., and Probst, D. 2011. The Suslin operator in applicative theories: Its proof-theoretic analysis via ordinal theories. *Ann. Pure Appl. Logic*, **162**(8), 647–660.

[13] Jäger, G., Rosebrock, T., and Sato, K. 2018. Truncation and semi-decidability notions in applicative theories. *J. Symbol. Logic*, **83**(3), 967–990.

[14] Jäger, G., and Strahm, T. 2002. The proof-theoretic strength of the Suslin operator in applicative theories. Pages 270–292 of: Sieg, W., Sommer, R., and Talcott, C. (eds.), *Reflections on the Foundations of Mathematics: Essays in Honor of Solomon Feferman*. Lecture Notes in Logic, vol. 15. Storrs, CT: Association for Symbolic Logic.

[15] Troelstra A. S., and van Dalen, D. 1988a *Constructivism in Mathematics, I*. Studies in Logic and the Foundations of Mathematics, vol. 121. Amsterdam: Elsevier.

[16] Troelstra A. S., and van Dalen, D. 1988b *Constructivism in Mathematics, II*. Studies in Logic and the Foundations of Mathematics, vol. 123. Amsterdam: Elsevier.

22

Inner and Outer Models for Constructive Set Theories

Robert S. Lubarsky

22.1 Introduction

A mathematician who is classically trained, as most are, could well wonder, when confronted with constructivism, what sense it could make. How could Excluded Middle possibly fail, as well as other classical validities? One can give philosophical motivations, say about increasing knowledge over time or the centrality of constructions, or proof-theoretic demonstrations of the underivability of these principles, perhaps by cut-elimination leading to normal form theorems coupled with the observation that some such principle has no normal form proof. Maybe someone could be convinced to accept some anticlassical principle, some principle contradicting classical logic and set theory, such as Church's Thesis or a Brouwerian continuity axiom. Ultimately, those approaches are likely to fail. Most mathematicians will remain unmoved by philosophical arguments, especially regarding their mathematics; proof-theoretic nonderivability is rather formal, and would leave many cold as mere symbol manipulation; anyone wondering how classical logic could fail is unlikely to embrace a principle violating it.

I got a sense of the coherence of anticlassical principles only via model theory. A model falsifying a classical principle was good; one satisfying an anti classical principle was even better; best of all was a theorem that a model satisfied a principle iff the model was of such-and-such a form. For instance, consider the ultimate principle at play here, Excluded Middle, in its propositional form: $p \vee \neg p$. The easiest way to have this fail is in a Kripke model. (For the definition and semantics of Kripke models, see Section 22.3, or consult the general literature, such as [33].) In the two-node Kripke model, with nodes $\bot$ and $\top$, with $\bot < \top$ and p true only at $\top$, $\bot$ will not satisfy $p \vee \neg p$. This seemed like an intuitive example of how classical logic could fail. Its limitation was that classical logic was not false, because the top node $\top$, with no extension, must satisfy classical logic. Considering that $\neg\neg(p \vee \neg p)$ is constructively provable, the best we can hope for by way of satisfying the negation of Excluded Middle would be to model $\neg\forall p(p \vee \neg p)$. That could be done by a kind of iteration of the first model: A Kripke model in which the nodes are indexed by the natural numbers, and at every node there is a proposition which becomes true only at the next node. Indeed, any Kripke model validating $\neg\forall p(p \vee \neg p)$ would have to embed the model described above: There can be no terminal nodes, since they would satisfy classical logic, and every node would have to have an extension at which some proposition p was not true, yet became true at some later extension.

The motivation of this chapter is to present the kinds of model constructions that are already known, and what they are typically useful for. While the content will always remain about models of constructive theories, in which classical logic fails, it turns out that many can be viewed as variants of well-known classical constructions. In particular, many of the constructions of current-day set theory are forcing extensions of V, and inner models of V, and elementary embeddings which can be viewed as going to simultaneously an extension and an inner model of V. Hence the focus here will be on models in which V embeds, and extensions and inner models thereof.

In practice, there are three known basic techniques for building constructive models: Kripke models, Heyting (or Heyting-valued) models, and realizability. In a realizability model, though, there is no embedding of V which preserves Δ_0 formulas. (There is an embedding and a translation of formulas from V, it's just that the translation does not preserve even Δ_0 formulas.) So here we restrict attention to the former two models and their variants. To be sure, one can easily enough build Kripke and Heyting models which do not embed V, depending upon just what one means by those names; the justification for this choice to study them is that there are standard constructions of Kripke and Heyting models which are usually what one wants, and they naturally embed V. We consider Kripke and Heyting-valued models as two separate techniques, even though the partial order which is an essential part of a Kripke model is a topological space (and hence a Heyting algebra), because in practice there are things you do with Kripke models that you don't do with Heyting models.

It should be observed that the choice of a set-theoretic, ZF-style framework is personal preference, and has little to do with the actual content of the results or constructions. Frequently enough, category or type theory is used to build models, and there could be other ways to do it that have yet to be discovered. While there are some differences among what can be expressed most naturally and easily, all these formalisms are powerful and flexible enough that they can simulate each other. One reason to want a set-theoretic development is that ZFC is nominally the gold standard for mathematics, and so something in that style might be more accessible to general mathematicians and in particular to logicians. To anyone well-versed in a different foundational paradigm, it should be easy enough to translate the material here.

The goals here include providing an up-to-date introduction to the subject, serving as a reference, introducing some uniform terminology and notation, bringing together in one place various constructions that are scattered in the literature, and formulating some questions for further development of the subject. One limitation of this work is that it takes insufficient account of some of the earlier literature. In addition to those listed in the references below, researchers such as Beeson, de Jongh, Fitting, H. Friedman, Goodman, Moerdijk, J. R. Moschovakis, Scowcroft, Smorynski, and S. Weinstein, among others, produced work directly relevant to the topic at hand, in some cases current work. A more thorough study would incorporate more of this earlier material.

22.2 Heyting Models, or Constructive Forcing

A Boolean algebra is a complemented distributive lattice (with top and bottom elements), and as such is exactly what one needs to model the truth values of classical logic: The meet and join of the lattice interpret "and" and "or" respectively, and the complement "not". A Heyting algebra performs the same function for constructive logic. In a Heyting algebra, instead of a complement there is a relative pseudo-complement: The relative pseudo-complement of a with respect to b is the largest element x such that $a \wedge x \leq b$. This can model implication, and so standard notation for this x is $a \rightarrow b$. A common example of a Heyting algebra which is typically not a Boolean algebra is the collection of open sets of a topological space. In this example, $\mathcal{O} \rightarrow \mathcal{U}$ is the interior of the union of $\mathcal{U}$ and the complement of $\mathcal{O}$. While this suffices to model propositional logic, in order to model predicate logic, in particular the quantifiers $\forall$ and $\exists$, one requires that the algebra (Boolean or Heyting) be complete, that is, closed under arbitrary meets and joins.

From the name Heyting or Heyting-valued model, one might think of any model in which the truth value of an assertion is a member of a given Heyting algebra. This indeed is the meaning used in [33], where they need such general models to show that Heyting models in their sense form a complete semantics for constructive logic. Historically and practically, though, this is not how Heyting models have been used. Instead, the models that have actually been used have assigned members of a Heyting algebra to basic properties (like set membership) of only standard objects, as opposed to nonstandard objects like infinitesimal reals or nonstandard integers, and not imposed any other restrictions on the objects to be considered, resulting in what is called below the *full model*.[1] Once this Heyting-algebra construction was

[1] The first topological interpretations of constructive systems [31, 32] were for propositional logic, so the question of the kind of objects allowed was not yet relevant. When extended to predicate logic [27], the objects needed

applied to sets, it was quickly realized that this is the constructive analogue of how Boolean algebras are used classically: Boolean-valued models, more commonly called forcing. To try to convey the difference between the two in practice, when forcing with a partial order if something is true densely then it's just outright true (i.e., the truth value is the top element), whereas a dense open set in a topology may still not be the whole space.

By way of an example, consider what must be the simplest forcing partial order of all, $2^{<\omega}$, the set of finite binary sequences. The empty condition forces $\exists n\ G(n) = 1$, where G stands for the generic, because every condition can be extended to a condition with a 1 in it. In fact, the set of such conditions is the entirety of $2^{<\omega}$, save for those nodes of the form 0^n, finite sequences of 0s. If we now view such finite binary sequences as names for open sets within Cantor space, then the open sets forcing a particular occurrence of 1 cover the entire space, except for the one point 0^ω. Because it's not the entire space which is so covered, the empty condition does not force $\exists n\ G(n) = 1$ under the Heyting semantics.

Another difference between forcing and Heyting models bears mention. When forcing with a cBa (complete Boolean algebra) $\mathcal{B}$ typically the class $V^{\mathcal{B}}$ of names, or terms, is defined first; then a $\mathcal{B}$-generic G induces an evaluation function from $V^{\mathcal{B}}$ to a (classical, two-valued) universe $V[G]$. In fact, the second step is, in principle, unnecessary. One could work entirely on the level of names. That's where all the action is. It's just that most people find it easier to think of the generic as a particular set, and the universe as two-valued (i.e. any assertion is either true or it's false). With Heyting models, the first step (the universe of names) is the same, and the second (choosing a generic) is not an option. Indeed, it's hard to imagine what a generic G for a cHa (complete Heyting algebra) $\mathcal{H}$ would be, and, furthermore, even if we had such a thing, if we had started with a classical universe V, as we typically will, how $V[G]$ would be anything other than another classical universe, which is not what we want. Instead, one defines the class $V^{\mathcal{H}}$ of names, and then stays on that level. With some practice though, this difference disappears. With forcing, you say you have a particular generic G, but you never pick one. Rather, the discussion is always what happens if this condition is in G and that one isn't. Which is really just working on the level of names. Within a Heyting model there is a name G for a generic, and the same kind of reasoning is used.

In the remainder of this section, we define full Heyting-valued models, and give examples and intuition for topological models (Heyting models specialized to the

for the examples were standard objects. The first application of topological models to higher-order systems, Scott's interpretation of analysis [29, 30], uses full models up to the type level being considered. Grayson's extension to full set theory [9, 11] is where we first find the full model, as presented here, even though not with that name.

opens of a topological space as the Heyting algebra) and nonspatial models (Heyting models that are not topological).

22.2.1 Full Heyting-Valued Models

As we have described, the difference between classical and constructive forcing is in the interpretation of truth, and pointedly not in the choice of objects. So the following definition of the members of the full model is identical to that of the terms of the standard forcing language.

Definition 22.1 (The full Heyting model) The *(full) Heyting model* over a complete Heyting algebra $\mathcal{H}$ consists of the class of *names* or *terms*, defined inductively by

$$V_\alpha^{\mathcal{H}} = \bigcup\{\mathcal{P}(\mathcal{H} \times V_\beta^{\mathcal{H}}) \mid \beta \in \alpha\},$$
$$V^{\mathcal{H}} = \bigcup_{\alpha \in \mathrm{ORD}} V_\alpha^{\mathcal{H}}.$$

Given $\sigma \in V_\alpha^{\mathcal{H}}$, the meaning of $\langle h, \tau\rangle \in \sigma$ is that the truth-value or degree of truth of $\tau \in \sigma$ is (at least) h. The idea behind calling it the full model is that you throw in absolutely everything you can. Sometimes one does consider inner models of $V^{\mathcal{H}}$, which we also take to be Heyting models, hence the qualifier of fullness.

The standard embedding $\check{\cdot}$ of the ground model V into $V^{\mathcal{H}}$ is defined inductively by

$$\check{a} = \{\langle \top, \check{b}\rangle \mid b \in a\}.$$

A particularly important object, the *generic*, not from the ground model (except in degenerate cases), is given by the name

$$G = \{\langle h, \check{h}\rangle \mid h \in \mathcal{H}\}.$$

The generic is characterized by the equation

$$[\![\check{h} \in G]\!] = h,$$

where $[\![\phi]\!]$, as defined below, is, intuitively, the truth value of ϕ.

The semantics of the sentences with parameters is given inductively on the sentences, with the base cases of $\in$ and $=$ given by induction on the ranks of the parameters. There are two equivalent ways of doing this: defining the relation $h \Vdash \phi$, $h \in \mathcal{H}$ and ϕ a sentence, or defining the function $[\![\phi]\!]$ taking values in $\mathcal{H}$. The connection between those two is that $h \Vdash \phi$ iff $h \leq [\![\phi]\!]$. By analogy with the way forcing is usually developed, we do the former.

Definition 22.2 (Heyting semantics)

- $h \Vdash \sigma \in \tau$ iff h is covered in the sense of the Heyting algebra by some $H \subseteq \mathcal{H}$ (that is, $h \leq \bigvee H$), and for all $h' \in H$ there is an $\langle \bar{h}, \bar{\sigma} \rangle \in \tau$ such that $h' \leq \bar{h}$ and $h' \Vdash \sigma = \bar{\sigma}$.
- $h \Vdash \sigma = \tau$ iff for all $\langle \bar{h}, \bar{\sigma} \rangle \in \sigma$, $h \wedge \bar{h} \Vdash \bar{\sigma} \in \tau$, and vice versa.
- $h \Vdash \phi \wedge \psi$ iff $h \Vdash \phi$ and $h \Vdash \psi$.
- $h \Vdash \phi \vee \psi$ iff h is covered by some H, and for each $h' \in H$ either $h' \Vdash \phi$ or $h' \Vdash \psi$.
- $h \Vdash \phi \to \psi$ iff for all $h' \leq h$ if $h' \Vdash \phi$ then $h' \Vdash \psi$.
- $h \Vdash \bot$ iff $h = \bot$ in $\mathcal{H}$. (Equivalently, $\bot$ can be taken to be $\check{0} = \check{1}$.)
- $h \Vdash \forall x\, \phi(x)$ iff for all σ $h \Vdash \phi(\sigma)$.
- $h \Vdash \exists x\, \phi(x)$ iff h is covered by some H and for all $h' \in H$ there is some σ such that $h' \Vdash \phi(\sigma)$.

As usual, $\neg\phi$ is taken to be $\phi \to \bot$. To say that a proposition ϕ is *true*, or *satisfied*, in such a model means $\top \Vdash \phi$, otherwise A is said not to be true or to fail. Being false is a stronger property: ϕ is said to be *false* if $\neg\phi$ is true, or equivalently the only value forcing ϕ is $\bot$.

Of fundamental importance is the following result, the history of which was discussed toward the beginning of this section.

Proposition 22.3 *The axioms and rules of inference of constructive logic and the equality axioms are all true under this interpretation.*

Despite its centrality, we think of it as a proposition and not a theorem, because it is so well understood at this point. For the same reason, we refrain from even mentioning the corresponding fact for the other semantics introduced later.

Theorem 22.4 *Under IZF, the full model satisfies IZF [9]. Under CZF, the full model satisfies CZF [8].*

The proofs are mostly like the corresponding proofs for forcing, and we will see several proofs later of IZF, albeit for different semantics, so we refer the reader to the literature this time.

Question 22.5 Which axiom systems are self-realizing? That is, which natural theories T prove that the full Heyting model satisfies T?[2]

22.2.2 Topological Models: Examples

In most cases, $\mathcal{H}$ is the cHa of the open sets of some topological space $\mathcal{T}$, and G can be considered to be a new element of $\mathcal{T}$. Of course, G is not in the ground

[2] For instance, typically one does not expect that substituting Collection by Replacement will keep a theory self-realizing. The problem is that even if a model satisfies $\forall x \in X\ \exists! y\ \phi(x, y)$, the choice of a term for y may not be unique, so the ability to use Replacement in the meta-theory might not suffice.

model, so it would not be possible for G to be a member of $\check{\mathcal{T}}$. But typically $\mathcal{T}$ has an independent description, and if $\mathcal{T}$ is interpreted in $V[G]$ via that description, then G will be a member of $\mathcal{T}$. I know of no way to make precise the notion of $G \in \mathcal{T}^{V[G]}$ and to prove when that holds, so for now it seems that the only way to convey to the reader what's going on is via examples.

Question 22.6 How can we formulate the claim that in the full model over the topological space $\mathcal{T}$ the generic G is a member of $\mathcal{T}$? Once formulated, when is that true?

Let's look first at what is no doubt the simplest example, in which $\mathcal{T}$ is taken to be Cantor space , mentioned above. Being quite literal, G is forced to be a set of finite binary sequences. It is easy to see that $\bigcup G$ is forced to be an infinite binary sequence, interdefinable with G (the latter as the set of proper initial segments of $\bigcup G$), so we can identify G and $\bigcup G$. Even though G is not even in the ground model, as an infinite binary sequence it can be taken to be a member of Cantor space as interpreted in the extension.

Another simple example is letting $\mathcal{T}$ be the reals (with the standard topology) [7].[3] A basic open set is an open interval $\mathcal{I}$, which as a forcing condition means "the generic is in me." Since $\mathbb{R}$ is covered by intervals of arbitrarily small length $\epsilon > 0$, each of which approximates G to within ϵ, the generic is a Dedekind cut. How is this useful? The two standard ways of constructing $\mathbb{R}$ from $\mathbb{Q}$ are Dedekind cuts and Cauchy sequences. It is not hard to see that every Cauchy sequence yields a Dedekind cut. The other direction is a nice exercise for bachelors students. Invariably Excluded Middle (EM) is used in that construction, in the form of a case split (because the alternative is to use Countable Choice, and most would automatically avoid that as just too dishonest). The suspicion arises quickly that without Excluded Middle and Countable Choice this converse would not hold. The likely place to look for a counter-example would be a generic Dedekind cut, because as a generic it will have only the properties it is forced to have. As it turns out, this is exactly the case, as the generic over $\mathbb{R}$ is not (the limit of) a Cauchy sequence of rationals; the generic is not represented by a Cauchy sequence of rationals.

After this last example, one might ask whether the next step might be to take a generic Cauchy sequence, and whether that gets us anything [19]. What one typically wants from a Cauchy sequence, in order for it to be useful, is a modulus of convergence. That is, for a Cauchy sequence (x_n), for every $\epsilon > 0$ there is a spot beyond which the members of the sequence stay within ϵ of each other. A modulus of convergence is a function that on input $\epsilon = 1/n$ yields such an index. Once one starts to think that the existence of such a modulus might not be provable, the most

[3] Not only does [7] contain this example, it is the first work in the style of this article, containing many examples of Heyting models and general theorems about them.

likely place to look for a counter-example is a generic Cauchy sequence. A basic open set is a finite sequence of rational numbers, which is an initial segment of the generic, along with an open interval, which constrains the generic in that all further entries in the generic have to come from this interval, as well as the limit of the generic. As expected, the generic has no modulus of convergence. (It bears mention that this is a building block in the construction of a model in which the Cauchy reals, meaning equivalence classes of Cauchy sequences of rationals, are not Cauchy complete, that is, not closed under taking Cauchy sequences. For more on this, see Section 22.6.2 on permutation models.)

Sometimes care must be exercised in understanding G as a member of $\mathcal{T}$, because $\mathcal{T}$ can be understood in classically equal yet constructively different ways. Take, for instance, forcing with $\mathbb{R}$, as above. That would have to yield a generic real. But what is a real? If the meta-theory is classical, then $\mathbb{R}$ could just as well be taken to be Cauchy sequences as well as Dedekind cuts, so one might be misled to think of the generic as a generic Cauchy sequence. This confusion is not hard to resolve in this case, because what determines the generic is not so much the nature of the points in the topological space being used, but rather the topology on them. For the case at hand, a basic open set $\mathcal{I}$ is naturally partial information about a located cut, with the rationals less than $\mathcal{I}$ being in the lower part and those greater than $\mathcal{I}$ in the upper. In other cases, the right way to view $\mathcal{T}$ is not so clear.

Perhaps the cleanest example of that is letting $\mathcal{T}$ consist of finite subsets of $\mathbb{R}^2$ with the Vietoris topology, which is the topology induced by the Hausdorff metric [20]. To be explicit, a basic open neighborhood of $F \in \mathcal{T}$ consists of an open set containing all of F, as well as finitely many open sets each one of which contains at least one point from F; there is no requirement that these latter open sets must cover F. One might first guess that the generic would be a generic finite subset of $\mathbb{R}^2$. A moment's thought though should make clear that it is not possible to have a new finite subset of a ground model set. Perhaps though the generic is a finite subset of $\mathbb{R}^2$ as interpreted in the generic extension, and since there are new reals then there are also new points in $\mathbb{R}^2$. This also looks problematic, because the next question you should ask is: If the generic is to be a finite subset of $\mathbb{R}^2$, wherever interpreted, what's its size? None stand out as being the natural answer. In fact, when one goes to look for members of G, there aren't any. But G also isn't just the empty set. A clue is afforded by the fact that the topology is induced by the Hausdorff distance function. Each member F of $\mathcal{T}$ determines a distance function $d(z, F)$ for $z = (x, y) \in \mathbb{R}^2$, namely the shortest distance from z to any point in F. Furthermore, F is definable from d. So $\mathcal{T}$ can be viewed as a space of distance functions. As it turns out, that is the right way to view G, as a generic distance function. From this point of view, it's not so surprising that G has no points. That is, using distances given by G, we could triangulate, and determine regions of the plane that look as though they contain members of G. These regions are determined only up to ϵ. So what we cannot do is ever specify how many points are in each such region. There could be any finite number. As such, we can never get our hands on any individual one.

Another example of the need to interpret $\mathcal{T}$ correctly is the forcing to falsify BD-N [22]. BD-N states that every pseudo-bounded sequence of natural numbers is bounded. The space to do this is the set of bounded sequences, appropriately topologized. What one gets though is not a generic bounded sequence, which is good, because we want something which is not bounded. Rather, the generic is merely pseudo-bounded. This can happen because classically bounded and pseudo-bounded are the same. So it might be unclear at the beginning what the better way is to think about the space, as the bounded or as the pseudo-bounded sequences. The former is easier and more familiar, but as it turns out not the more useful way.

22.2.3 Nontopological Models: Examples

While the opens of a topological space form a Heyting algebra, not every Heyting algebra can be viewed as the opens of a topological space. There are interesting examples of forcing with nontopological or nonspatial Heyting algebras.

One aspect of forcing with spaces is that you can specialize to a point. That is, when examining, say, a particular object in the generic extension, you can look at what is forced by any neighborhood of a fixed point. As all such neighborhoods overlap, whatever they force must cohere. For example, suppose the term r is forced to be a real number. For any $\epsilon > 0$, the space is covered by opens that determine r to within ϵ. So all of the opens of $\mathcal{T}$ containing a fixed point x are enough to determine r exactly, at x. This is essentially the (well-known) proof that a (Dedekind) real in the extension is given by an arbitrary continuous function from $\mathcal{T}$ to $\mathbb{R}$ in the ground model.

The example of specializing to a point which is of immediate application has to do with the Fan Theorem, or FAN, which for now we assume holds in the meta-theory. Let T be forced to be a binary tree with no infinite paths. Consider in the meta-theory the tree T' of those binary sequences not forced by any neighborhood of a fixed $x \in \mathcal{T}$ to be out of T. That is, σ is in T' when no neighborhood of x forces σ not to be in T. If T' had an infinite branch B through it, then by the hypothesis on T, some neighborhood of x would force some $\sigma \in B$ to be out of T. But then that violates σ being on T'! So T' has no infinite path. By the fan theorem in the meta-theory, T' is finite, which means x has a neighborhood forcing T to be finite. This proves the fan theorem in all topological models [7]. Hence, to falsify FT, if a forcing model is possible at all, one needs at least a nontopological Heyting algebra.

As it turns out, this is possible [7]. Let $K(\mathcal{T})$ be the cHa of coperfect open sets of $\mathcal{T}$. This can be viewed either as the subset of the family of open sets consisting of only those open sets with complement a perfect set (with a different join operation), or as the quotient of the opens which identifies $\mathcal{O}$ with $\mathcal{O}\backslash\{x\}$. This is why $K(\mathcal{T})$ is not the opens of a space of points: Individual points don't count, in that if a sentence holds everywhere except a point, then it holds everywhere. Letting I be the unit interval, $K(I \times I)$ falsifies the fan theorem. (The same paper also examines another nonspatial model, $K(I)$.)

Point-free spaces can be used not only to make the fan theorem false, they can be used to make it partially true. The historically first model falsifying the fan theorem is based on the Kleene tree, an infinite computable tree with no (infinite) computable branch. Classically, the Kleene tree shows that the computable sets form a model in which Weak König's Lemma is false. Constructively, one can work under recursive realizability, also called Kleene's first model K_1, which (using later developments here) is a model of full IZF set theory based on computability. The (internalization of the) Kleene tree is in this model an infinite tree, and the fact that it has no computable paths translates in this model to it having no paths, thereby providing a counter-example to FAN. In fact, this example does even more for us. The computability of the Kleene tree translates into its internalization being decidable (meaning membership in the tree T is decidable: "$\forall\sigma\ \sigma \in T \vee \sigma \notin T$" holds in the model). So not only is the Fan Theorem violated here, so is a weak fragment of it, the Decidable Fan Theorem, or D-FAN: Every decidable bar is uniform. (A bar is a set of nodes such that every infinite path contains at least one of those nodes; a bar is uniform if the bar contains every node from some fixed level of the tree.) This is interesting, because some fragments of the fan theorem, including D-FAN, turn out to be equivalent to other statements that are interesting and natural [14]. Another such fragment is c-FAN , the fan theorem for c-bars (where $X \subseteq 2^{<\omega}$ is a c-set if there is a decidable set Y such that $\sigma \in X$ iff every extension of σ from $2^{<\omega}$ is in Y) [2]. Easily, c-FAN implies D-FAN, because every decidable bar is a c-bar. Is the converse implication true? Classical logic is no help here (in our context with sufficient set existence axioms), because that makes all versions of FAN true. Kleene realizability doesn't answer the question, because it makes all versions of FAN false. But it does give us a start. Is there a way to start from K_1 realizability and make D-FAN true that is simultaneously gentle enough to keep c-FAN false?

D-FAN can be stated as "for all decidable $X \subseteq 2^{<\omega}$, if X is a bar then X is uniform." (Easily, if X is uniform then X is a bar.) Plausibly one could make D-FAN true by turning decidable bars uniform – by going to a model with nonstandard integers, say, a tree that was infinite could be made uniform by killing it at some nonstandard level; this is the approach taken in [25]. A different way to make D-

FAN true is to consider decidable trees, which by their decidability can be readily exported into the meta-theory; if they are not uniform internally then they are not uniform externally, which means that they are infinite, which means that they actually do have branches (using WKL in the meta-theory); such a branch could then be included in the model, so that the tree (actually, its complement) no longer represents a bar. The choice of branch to include must be made carefully, to preserve IZF. The easiest way to do that is generically.

One's first guess might well be to force with the tree itself. After all, forcing with a tree like the full binary tree $2^{<\omega}$ produces an infinite path through it. That will not work though in the setting at hand. The trees with which we must be concerned are those like the Kleene tree, in which the terminal nodes are dense (as a moment's thought shows that any tree in which the terminal nodes are not dense already contains a path). Any such partial order forces "$\neg\neg$ the generic path is finite," so this does not help. Furthermore, suppose $\mathcal{T}$ is a topological space forcing "there is an infinite path, say P, through a ground model binary tree T." Then by specializing to a point $x \in \mathcal{T}$ we can build a path through T in the ground model: $\mathcal{T}$ forces that for all n there is a neighborhood of x forcing a necessarily unique sequence of length n to be in P, and stringing these sequences together produces an infinite path. So no spatial forcing will help get paths through Kleene-like trees. The construction that works is to take the tree you want to shoot a path through and to turn it into a Heyting algebra by modding out by all the nodes beyond which the tree is finite [23]. Of course, the forcing just described removes merely one counter-example to D-FAN. To make D-FAN true, all counter-examples must be removed; this can be accomplished by iterating this construction. This is described in Section 22.4.1.

22.3 Kripke Models

By way of introduction, we first describe Kripke semantics for first-order constructive logic[4] in general, and afterwards specialize to the particular techniques pertaining to Kripke models for set theory.

Fix a first-order language. A Kripke model for that language consists of a partial order $\mathcal{P}$, the elements of which will typically be referred to as nodes; a (classical) first-order structure M^p (for that language) at each node $p \in \mathcal{P}$; and, whenever $p < q$, a transition function k^{pq} from the universe of M^p to that of M^q. (To simplify notation, k^{pq} will be written as just k.) Furthermore, atomic truth is monotone: If $M^p \models R(x_1, \ldots, x_n)$, where R is an n-ary relation symbol, then $M^q \models R(k(x_1), \ldots, k(x_n))$. Also, the transition functions must cohere: If

[4] In Section 22.1, a Kripke model for propositional logic was described. Such models are the straightforward restriction of those for predicate logic to a propositional language. There is no longer a need for a universe of objects at each node. Rather, we just need to know which atomic propositions are true at which nodes, and this decision must be monotone.

$p < q < r$, then $k^{qr} \circ k^{pq} = k^{pr}$. By abuse of notation, we can consider k as applying to a formula ϕ with parameters, by applying k to the parameters. Given this, the relation $p \models \phi$ is defined inductively on ϕ, simultaneously for all $p \in \mathcal{P}$, where ϕ is a formula with parameters from M^p.

Definition 22.7 (Kripke semantics)

- $p \models R(x_1, \ldots, x_n)$ iff $M^p \models R(x_1, \ldots, x_n)$.
- $p \models \phi \wedge \psi$ iff $p \models \phi$ and $p \models \psi$.
- $p \models \phi \vee \psi$ iff either $p \models \phi$ or $p \models \psi$.
- $p \models \phi \to \psi$ iff for all $q \geq p$ if $q \models \phi$ then $q \models \psi$.
- $p \models \bot$ never.
- $p \models \forall x\ \phi(x)$ iff for all $q \geq p$ and x in the universe of M^q, $q \models \phi(x)$.
- $p \models \exists x\ \phi(x)$ iff for some x in the universe of M^p, $p \models \phi(x)$.

Lemma 22.8 *Truth is monotone: if $p < q$ and $p \models \phi$, then $q \models k(\phi)$.*

Proof By induction on ϕ. □

Perhaps from this brief definition one can see how Kripke models can provide a certain flexibility that Heyting models do not, namely the chance to change ground models. That is, in Heyting models, while different statements can become true when going to different Heyting truth values, the objects are always built over some fixed ground model. This ground model might be V, or an inner model like L, or an outer model like a forcing extension $V[G]$, or a constructive model like a realizability model, but once selected it doesn't change among the various Heyting values. With a Kripke model, the universe in which an object can be said to live can switch from node to node.

Let's illustrate this with an example. Consider the sentence "every Turing machine either converges or diverges." Its failure is most easily modeled with realizability. Is there another way to accomplish that? Heyting models as above will not do: Because the integers in such models are standard, convergence and divergence there are the same as in the ambient universe. But one can construct a Kripke model counter-example. Let e be (the code for) a machine the convergence of which is not decided by ZFC. (Such a machine exists, assuming the consistency of ZFC, by diagonalization.) Easily, $\{e\}$ diverges in V, because convergence would be witnessed by a natural number and hence ZFC-provable. Let M be a model of ZFC in which $\{e\}$ converges and in which V embeds (albeit not elementarily). Such a model can be seen to exist by considering the theory "ZFC + $\{e\}$ converges" along with the diagram of V. (The diagram of V is in the language expanded with a constant symbol for each member of V, and consists of all true atomic and negative atomic assertions.) This theory is consistent by Compactness, and so has a model by Completeness, which we take to be definable over V. Consider the Kripke model with nodes $\bot$ and $\top$, with $\bot < \top$. The structure at $\top$ is M. The structure at $\bot$ is built in V, and will be described in detail shortly as the full model with V at $\perp$ and M at $\top$;[5] the intuition is that sets at $\perp$ look like they come from V and can grow arbitrarily as they move to $\top$. At $\perp$, $\{e\}$ does not converge, because the integers are standard, but at $\top$ it does, because there we're in M.

What makes this example go is that some nodes have only standard sets and others have nonstandard sets. This ability to change the ambient universe is the main way Kripke models are flexible and Heyting models not.

A secondary difference between them is that even when the underlying universe is taken to be the same at all nodes of a Kripke model, it is natural to define submodels of the full model by using the nodes of the Kripke partial order. Typically this happens in the context of some kind of settling, under which sets are not allowed to grow throughout the whole partial order, but must remain constant, or settle down, at some point, examples of which are given below. Furthermore, in the presence of certain kinds of settling, it is natural to consider partial existence: if $p < q$ in the partial order of a Kripke model, there is no expectation that the universe at q is the image of that at p under the transition function; in fact, if the universe at p has all settled down by q, then we will definitely need new sets at q if the model is to violate classical logic. In contrast, the use of settling with topological models is different. It is possible to have immediate settling with partial existence, just as for Kripke models, while using topological ideas (for examples, see Section 22.4.2); but the settling comes into play only because the topology is placed on a tree, where there is a notion of a child, which is an essentially Kripke-style idea. A different kind of topological settling is exposited in Section 22.4.2; but there the settling can happen without even changing to a different open set, and you can change open sets arbitrarily without settling, an essentially different kind of settling property. Of course, one could consider submodels of the full topological model in which settling or partial existence is substantive, which is why this distinction is being called secondary. The fact remains that such constructions in the context of topological models have not yet appeared in nature, and so this still seems to be a distinction worth making.

22.3.1 Full Kripke Models

Analogously to the Heyting models above, there is a notion of a full Kripke model. The presentation here is a refinement of that in [12].

Let $\mathcal{P}$ be a partial order of nodes. For simplicity, we assume $\mathcal{P}$ has a least element $\perp$, although this is really not necessary. (At any node $p \in \mathcal{P}$, all nodes less than or incomparable with p are irrelevant. So one can restrict attention to those nodes that

[5] This structure does not have an ordinal embedding, given in Definition 22.9, and so is not covered by the development there. Nonetheless, the definition given there of the full model can be applied in the current case. The only issue is whether this is a model of IZF. It is easy enough to check directly that it is.

extend p, among which p is the least element.) Let $p \mapsto V_p$ be an assignment to nodes of models of ZF. Since these models are typically class models, this assignment cannot be understood as a set of ordered pairs; rather, it is given definitionally. That is, whether x is in V_p, as well as the membership relation and equality, are definable uniformly in x and p.

The full model (over this assignment) will be defined inductively. In order for this induction to work, since we are not assuming that the V_p's are actually well-founded, we need additional structure.

Definition 22.9 (Admissible assignment) An *ordinal embedding* is an order-preserving function f from the ordinals of V_p to those of V_q whenever $p < q$. (So actually f is an indexed set of functions f_{pq}, but the choice of p and q should be clear from the context, so we write it polymorphically as simply f.) Moreover, f must cohere, in that $f_{qr} \circ f_{pq} = f_{pr}$. Finally, whenever $p < r$, there is a finite sequence $p = q_0 < q_1 < \cdots < q_n = r$ such that each $f_{q_i q_{i+1}}$ is either an isomorphism or an elementary embedding. We say that such an assignment is *admissible* if for all p the entire structure beyond p (namely the set $\mathcal{P}^{\geq p}$, the assignment $q \mapsto V_q$ ($q \geq p$), and the restriction of f to $\mathcal{P}^{\geq p}$ (i.e. the family $f_{qr}, r > q \geq p$)) is definable in V_p. Furthermore, there is a uniform bound on the quantifier complexity of these definitions. That is, there is an n such that, for all p, all of the components of the assignment for $q \geq p$ are Σ_n.

When working with one fixed admissible assignment, we can without loss of generality assume that $V_\perp = V$, since we could confine ourselves to working in $V_\perp$ anyway. In contrast, when quantifying over admissible assignments, these assignments might include ones in which $V_\perp$ is an inner model of V, so we cannot assume $V_\perp$ is some fixed model.

Example 22.10 Let $\mathcal{P}$ be finite, and if $p < q$ choose V_q to be a specific inner model of V_p. For instance, $\mathcal{P} = \{\perp, \top\}$, $V_\perp = V[G]$ for some forcing generic G, and $V_\top = V$. In this case, the ordinal embedding can be taken to be the identity function. Note that, at $\perp$, that is, in $V[G]$, the definition of $V_\top$ as V could well be different from its definition in $V_\top$. Or let $V_\perp = V$ and let $V_\top$ be some ultrapower of V by an ultrafilter in V. Here the ordinal embedding would be the elementary ultrapower embedding. (In practice, the ultrafilters will not be countably complete, so that the ultrapower has non standard integers.) Again, the definitions of $V_\top$ within $V_\perp$ and $V_\top$ are different. For an example with $\mathcal{P}$ infinite, let $\mathcal{P}$ be ω and each V_n be V, with f as the identity.

Example 22.11 For an example of how admissibility could fail, let $V_\perp = V$ and $V_\top = V[G]$, where $\perp < \top$ ($V[G]$ is not definable in V). Or let $\mathcal{P}$ be ω, V_0 be V, V_1 be the ultrapower of V by some noncountably complete ultrafilter $\mathcal{U}$, and more generally V_{n+1} be the ultrapower of V_n by the image of $\mathcal{U}$ in V_n. The problem here is that $\mathcal{P}$ is not a set in any V_n once $n > 0$.

Remark 22.12 Even though the first counter-example of V going to $V[G]$ is being excluded here, it does express a major intuition behind constructivism, namely that of knowing more as time goes on. There are still ways to build a model in that situation though, just of a different flavor from the full model as defined below, so they will not be considered here. Also, one might well ask why the various V_ps are to be models of ZF and not of IZF. The reason IZF models are disallowed here is that they would bring up all sorts of additional issues. For one, there are various kinds of models of constructive theories, whereas classical theories have only one notion of a model. Beyond that, all of the constructive models have additional structure – realizability has realizers, topological models have open sets, Kripke models have nodes – which would have to be accounted for in the construction to come, adding to the complication. Again, it is possible to deal with this situation. For an example that simultaneously deals with both of these issues – later nodes being IZF models properly extending earlier nodes – see [23].

Given an admissible assignment over the partial order $\mathcal{P}$, we define the full model M over it. Note that M depends on $\mathcal{P}$, the assignment V_p, and the embeddings f, mention of which is suppressed in the simple notation M. At node p, the universe M^p will be the union of the M^p_αs as α ranges over the ordinals of V_p. In addition, the transition function k^{pq} from M^p to M^q $(q > p)$ will be defined as the union of the partial transition functions k^{pq}_α defined along the way, from M^p_α to $M^q_{f(\alpha)}$. Since these partial transition functions cohere, we will drop the mention of α. Similarly, we do not mention p and q and refer just to k, allowing k to act polymorphically.

Definition 22.13 (The full Kripke model) M^p_α consists of the functions g with the following properties:

- $\mathsf{dom}(g) = \mathcal{P}^{\geq p}$,
- $g \restriction \mathcal{P}^{\geq q} \in V_q$,
- $g(q) \subseteq \bigcup_{\beta < f(\alpha)} M^q_\beta$, and
- if $h \in g(q)$ and $q < r$ then $k(h) \in g(r)$.

The transition function k works by restricting the domain.

$M^p = \bigcup_{\alpha \in V_p} M^p_\alpha$. To complete the description of M^p as a first-order structure for the language of set theory, we must interpret the relation symbols $\in$ and $=$. $M^p \models g \in h$ if $g \in h(p)$, and $M^p \models g = h$ if $g = h$.

Remark 22.14 There is a sense in which sets grow under the action of k, and another sense in which they shrink. For growing, with $p < q$, if $p \models x \in z$, then by monotonicity $q \models k(x) \in k(z)$, yet if $q \models k(y) \in k(z)$, it might not be the case that $p \models y \in z$. So a set could not lose members, but it might gain them. At the same time, the way we have set it up, a set x at node p contains the information of not only the members of x at p, but also the members of $k(x)$ at later nodes q. That's why the domain of x is $\mathcal{P}^{\geq p}$. When we move to a later nodes q, we no longer need to keep track of the members of x at nodes between p and q. That's why k works by restricting the domain. So actually the action of k is to lose information. Strictly speaking, $k(x)$ is a proper subset of x. In that sense, k makes sets shrink. Hence, if you consider x at p as its members at p, k makes x grow; whereas if you consider x at p as its members at p as well as what x will be at future nodes, k makes x shrink.

Definition 22.13 is an induction on ordinals in various models, which could be nonstandard, and so needs further explanation beyond what was offered in [12]. There is no problem working at node p inductively on the ordinals of V_p, even if those are ill-founded, because the induction is taking place within V_p. However, in the inductive definition just given, the third clause refers also to M^q_β, for q possibly strictly extending p and $\beta < f(\alpha)$ in V_q. If the ordinals are the same in V_p and V_q this won't be a problem, as the induction in V_p applies just as well to V_q. Consider though the case in which f_{pq} is an elementary embedding. Within V_p, an induction at stage α could well assume that M^p_β is well-defined for all $\beta < \alpha$, but it is at best unclear how it would be legitimate to assume well-definedness of M^q_β for all $\beta < f(\alpha)$ in V_q. It does not do to use elementarity on the assertion "for all $\beta < \alpha$, M^p_β is well-defined." Yes, superficially it looks like applying f yields "for all $\beta < f(\alpha)$, M^q_β is well-defined," which is what we want. But recall what the notation M suppresses: M^p_β is an abbreviation for "the full Kripke model, up to height β, based on $\mathcal{P}^{\geq p}$ and the assignment V_q to $q \in \mathcal{P}^{\geq p}$ and f." With that realization, applying f to "for all $\beta < \alpha$, M^p_β is well-defined" would yield the well-definedness up to $f(\alpha)$ of the full model based on f applied to the system $\mathcal{P}^{\geq p}$ and its assignments (and even that much only if we extended the notion of f from an elementary embedding on the ordinals to one on all of V_p, or at least on enough of V_p to include $\mathcal{P}^{\geq p}$). There is no reason to think that $f(\mathcal{P}^{\geq p})$ is $\mathcal{P}^{\geq q}$ and other such problems. Rather, one must quantify over all partial orders and embeddings too. This same problem, incidentally, also comes up in the proof of $\in$-induction.

Lemma 22.15 *For all partial orders $\mathcal{P}$ with least element $\bot$, admissible assignment of ZF models V_p $(p \in \mathcal{P})$ with $ORD^{V_\bot} = ORD^V$ and ordinal embeddings f, for all ordinals $\alpha \in V$, $M^\bot_\alpha$ is defined.*

Proof Actually, in order even to be able to state the lemma, we must bound the quantifier complexity Σ_n of the assignment; the proof is uniform in n. Assume inductively that for all $\beta < \alpha$, for any partial order $\mathcal{P}$ and admissible assignment

V_p and f, $M^{\perp}_{\beta}$ is defined. Fix one such partial order $\mathcal{Q}$ and system V_q $(q \in \mathcal{Q})$ with ordinal embedding g; once we show that $M^{\perp}_{\alpha}$ is defined for that system, we will have proven the lemma.

For $q \in \mathcal{Q}$, let $\perp = p_0 < p_1 < \cdots < p_n = q$ be as given by the admissible assignment. We show inductively on $i \leq n$ that within V_{p_i} the inductive hypothesis holds up to $g(\alpha)$; that is, within V_{p_i}, for all $\beta < g(\alpha), \mathcal{P}$, and admissible assignment V_p and f, $M^{\perp}_{\beta}$ is defined. Recall that g is actually a system g_{pq} of ordinal embeddings for all $p \leq q$ in the partial order; for $\alpha \in ORD^V = ORD^{V_\perp}$, $g(\alpha)$ is actually $g_{\perp p_i}(\alpha)$. For $i = 0$, $g(\alpha) = \alpha$, and this is just the inductive hypothesis. Given the inductive assertion for a value $i < n$, $g_{p_i p_{i+1}}$ is either an isomorphism or an elementary embedding. In the former case, if within $V_{p_{i+1}}$ there were some counter-example $\beta < g(\alpha), \mathcal{P}$, and f, that would also be a counter-example within V_{p_i}, because $g(\alpha)$ in V_{p_i} is literally the same ordinal as $g(\alpha)$ in $V_{p_{i+1}}$ and $V_{p_{i+1}}$ is definable within V_{p_i}. In the latter case, the elementarity of g transfers the truth of the statement about $g(\alpha)$ from V_{p_i} to $V_{p_{i+1}}$. This completes the induction on $i \leq n$.

What we have just shown is that for any $q \in \mathcal{Q}$, interpreted within V_q it is true that for all $\beta < g(\alpha), \mathcal{P}$, and f, $M^{\perp}_{\beta}$ is defined. In particular, we can take $\mathcal{P}$ to be $\mathcal{Q}^{\geq q}$ and f to be g. For those choices, the interpretation within V_q of $M^{\perp}_{\beta}$ is exactly the interpretation within V of M^{q}_{β} for $\mathcal{P} = \mathcal{Q}$ and $f = g$. This is all that is needed for $M^{\perp}_{\alpha}$ to be defined; namely, the union in the third clause of the definition of M^{p}_{α} makes sense. □

Theorem 22.16 *The full model satisfies IZF.*

Proof The proof is basically the same as in [12], even if the context there is more limited. Namely, in [12], the embeddings f_{pq} all had to be elementary embeddings from the entire model, as opposed to here, where they are allowed to be elementary embeddings of the ordinals only, or the identity function. That makes no difference in the proof, except in the case of $\in$-Induction.

Empty Set: Let g be the function (with appropriate domain) that always returns the empty set.

Infinity: Within any V_p, one can define inductively on $n \in \omega^{V_p}$ an element n_p of the full model. To start, 0_σ is the set satisfying the Empty Set Axiom from above. Given m_p for $m < n$, let $n_p(q)$ be $\{k_{pq}(m_p) \mid m < n\}$. Notice that $k_{pq}(m_p) = m_q$. Let ω_p be such that $\omega_p(q) = \{m_q \mid m \in \omega^{V_p}\}$. Then ω_p suffices.

Pairing: Given g and h at node p, the desired pair is the function which, at node q, yields the set $\{k(g), k(h)\}$.

Union: Given g at node p, let h be such that $h(q) = \bigcup\{\hat{g}(q) \mid \hat{g} \in g(q)\}$.

Extensionality: Because equality is taken to be actual equality, it is easy to see that equal sets have the same members. On the other hand, if two sets are forced at a node to have the same members, then they are literally equal as functions from that node on. By Extensionality in the ambient universe, they are then equal.

$\in$-Induction: Suppose that $p \models \forall x\ (\forall y \in x\ \phi(y) \rightarrow \phi(x))$; we need to show that $p \models \forall x\ \phi(x)$. If not, then for some $q \geq p$ and $g \in M^q$, $q \not\models \phi(g)$. This state of affairs is a true statement in V_q: V_q satisfies "there is a partial order $\mathcal{Q}$, with bottom element $\perp$ and admissible ordinal assignment f and full model K, satisfying $\forall x\ (\forall y \in x\ \phi(y) \rightarrow \phi(x))$, with a counter-example g in $K^{\perp}$." Because V_q is a model of ZFC, and $K^{\perp}$ is defined via an induction along the ordinals of V_q, there is an example $\mathcal{Q}$, f, K, and g, with g of least possible rank, say α, among all such models. In this model, letting x from above be g, $\perp \models \forall y \in g\ \phi(y) \rightarrow \phi(g)$. Since g is a counter-example, there is an $r \geq \perp$ and an $h \in K^r$ such that $r \models h \in g$ yet $r \not\models \phi(h)$. Let $\perp = q_0 < q_1 < \cdots < q_n = r$ be as given by the admissible assignment. We show inductively on $i \leq n$ that $f(\alpha)$ is the least rank of a counter-example within V_{q_i} to induction for ϕ. For $i = 0$, $f(\alpha) = \alpha$ was chosen as the least such rank. For the inductive step, given the inductive hypothesis for a value $i < n$, $f_{q_i q_{i+1}}$ is either an isomorphism or an elementary embedding. In the former case, if $f(\alpha) \in V_{q_{i+1}}$ were not the least rank of a counter-example, then restricting everything ($\mathcal{Q}$ and the assignment) to the nodes q_{i+1} and above, we would have a counter-example of rank smaller than $f(\alpha)$ within V_{q_i}, contradicting the inductive hypothesis. In the latter case, the elementarity of f transfers the minimality of $f(\alpha)$ from V_{q_i} to $V_{q_{i+1}}$. Finally, letting i be n, $f(\alpha)$ is the least ordinal rank in V_r of a counter-example to ϕ. But g has rank $f(\alpha)$, and so h as a member of g has a smaller rank. This is a contradiction.

Power Set: Let $g \in M^p$. At any node $q \geq p$, if $q \models h \subseteq g$, then h is (forced at q to be) a function with domain $\mathcal{P}^{\geq q}$ such that $h(r) \subseteq g(r)$ (of course, also respecting the other condition on being in M^r). In fact, any such h at $q \geq p$ can be taken to be an object at p, by letting $h(r)$ be the empty set whenever r does not extend q. Noticing there are only set many such hs, the set of all of them is the power set of g.

Separation: Given ϕ and g at p, let $h \in M^p$ be $h(q) = \{g' \in g(q) \mid q \models \phi(g')\}$.

Replacement\Collection\Reflection: While those three axiom schemas are equivalent classically, they are apparently of strictly increasing strength constructively. That is, Friedman–Scedrov showed that Replacement does not imply Collection, and while a proof is wanting, presumably Collection does not imply Reflection. (The converse implications are all soft proofs.) Typically IZF is taken to be the theory with Collection. In fact, in our setting it is easy to see that Reflection holds true. Working in $V = V_p$, by Reflection there, let $V_\alpha \prec_{\Sigma_n} V$. Cutting off the construction of the model at α, which is M^p_α at p and $M^q_{f(\alpha)}$ at any other q, yields a Σ_n substructure of the entire model, in the sense that any property of M^p which is Σ_n expressible in the ambient V_p reflects. □

In the following examples, M is an ultrapower of V via a noncountably complete ultrafilter. That means V embeds elementarily into M, and M has nonstandard integers.

Example 22.17 WLPO does not imply MP [12]: The Limited Principle of Omniscience (LPO) states that every (binary) sequence either is all 0s or has a 1. Markov's Principle states that if a sequence is not all 0s then it has a 1. Trivially, LPO implies MP. Weak LPO is a slight weakening of LPO, that every sequence either is all 0s or it's not. To see that WLPO does not imply MP, let $\mathcal{P}$ consist of $\bot < \top$, and assign V to $\bot$ and M to $\top$, with the ordinal embedding being the standard elementary embedding. The full model over that structure satisfies WLPO but not MP. (Consider a sequence which is 0 at all standard places and has a 1 at a nonstandard place.)

Example 22.18 WKL does not imply WLPO [12]: Weak König's Lemma states that every infinite (binary) tree has an infinite branch. It is not hard to see that WLPO implies WKL: Using WLPO, any tree restricted to the descendants of a node is either finite or infinite, which can be used straightforwardly to build a branch. To see that the reverses implication fails, let $\mathcal{P}$ consist of a bottom node $\bot$ and two incomparable successors $\top_0$ and $\top_1$. Assign V to $\bot$ and to $\top_0$ and M to $\top_1$. The ordinal embeddings are the identity in the former case and the elementary embedding in the latter. The full model satisfies WKL: If $\bot \Vdash T$ is an infinite tree, then at $\top_1$ T is interpreted as an infinite tree T_M in M; letting B be an infinite branch through T_M, the object in the Kripke model which looks at $\bot$ and $\top_0$ like B restricted to the standard natural numbers and at $\top_1$ like just B is the infinite branch desired. WLPO fails, by considering a sequence which is all 0s at standard places and has a 1 in a nonstandard place.

22.3.2 Settling

What characterizes the full model is that sets can keep growing throughout the partial order. Under settling, the sets have to stop growing at some point.

The kind of settling in any particular model is determined by two aspects: the result of the settling, that is, the kind of object the set has to settle down to, and the location of the settling within the partial order, that is, what node(s) the set has to settle down by.

The Result of the Settling

We consider two different kinds of examples of what sets must settle down to. The first is external settling, under which sets must settle down to something from the ambient, external universe, or more precisely the internalization of such, such as something from $\check{V}$. The second is internal settle, under which sets must settle down, but not (necessarily) to something which is (the internalization of) an external set. Another way to view this distinction is whether, once a set has settled down, its members have also settled down (yes in the former case, no or, more accurately, not necessarily in the latter).

External Settling: We describe external settling via two examples.

Example 22.19 Class-based settling: Let $\mathcal{P}$ be the class of ordinals. Of course, one cannot think of an object in the Kripke model as a function in the standard sense with domain $\mathcal{P}$, because such a creature would have to be a proper class. But one could consider a function as being given by a definition. For instance, the function which at node κ looks like κ is an ordinal which is not the image $\check{\alpha}$ of any ordinal α from V. In this sense, we could speak of the full model over ORD with V assigned to each node. The model of interest now though is not this full model, but rather the one consisting of actual set-sized functions, with domain some ordinal κ. Beyond κ the set represented by this function does not change; another way to look at it is that it is the image $\check{x}$ of a ground model set $x \in V$. The Kripke set has settled down by κ. The reason to consider this model is that it shows that CZF does not prove Power Set. In fact, full Separation holds in this model, so it shows that IZF − Power Set + Subset Collection does not prove Power Set [18].

Example 22.20 Class-based settling to an inner model: This example is a lot like the previous one, with $\mathcal{P}$ being ORD and sets settling down by some ordinal to something in V, only here the functions are from $V[G]$, where G is generic for Cohen forcing over V. Normally Cohen forcing is thought of as giving a subset of $\mathbb{N}$, but by identifying $\mathbb{N}$ with $\mathbb{N} \times \mathbb{N}$, G can be thought of as a relation on $\mathbb{N}$. This model shows that IZF − Power Set + Exponentiation does not prove Subset Collection [18].

It bears observation that this kind of settling produces models which violate Power Set, by design. Consider a set X as a possible power set of 1. If X has settled down to an external set by node p, then the only subsets of 1 that X could contain there are 0 and 1. In any nontrivial model, there will be a set which looks like 0 at p but then ends up being 1 at some extension, witnessing that X is not the power set of 1 at p. So to model Power Set, a different kind of settling is needed, namely internal settling.

Internal Settling: First let's see via an example why some kind of full model will not work. Either of the examples from the section on full Kripke models would do; for specificity, we will consider the first, the two-node model separating WLPO and MP. In this model, MP fails, but MP is not false. At $\top$, classical logic holds. This is called a *weak separation*. To get a *strong separation*, one in which MP is false, we would like to iterate the construction. Your first guess as to how to do this might well be to assign V to $\perp$, M to $\top$, and however you went from V to M (say via an ultrafilter $\mathcal{U}$), do that to M (say via $f(\mathcal{U})$) to get a model nonstandard relative to M, place that model at some successor of $\top$, and then iterate this procedure through ω. With this assignment, you would want to take the full model. Indeed, WLPO would be true there, just because the partial order is linear, and MP would fail, because you're always getting new nonstandard integers. The problem is in defining the model. Admissibility is lost. If the partial order is the standard ω, then it does not exist in any of the associated models after $\perp$'s V. One could try to piece together the nonstandard ωs that appear along the way, but this is starting to get complicated.

A simpler approach is just to use immediate settling to an internal set [12]: Instead of taking the full model over ω, allow only those sets that settle down at the node after they appear. At that next node, there are new sets; that is, sets that are not in the range of the transition function from previous nodes. Those new sets can then grow at the node after that, but then they would have to settle down. This solves the problem of how the partial order, in this case ω, can get away with not being in the base models: All the information needed to be built into a set is how it changes once.

Of course, one is then left with the question of how Power Set could hold. After all, how could X ever be the power set of, say, $1 = \{0\}$, if X must eventually settle down, yet new subsets of 1 keep on being introduced? The answer is that X settles down to an internal set. Take the example above, where $\mathcal{P}$ is ω. At any node $n \in \omega$, what is the power set of $1 = \{0\}$? Viewed externally, at n, 1 has three subsets, namely $0 = \emptyset$, 1 itself, and the set that looks like 0 at n and then grows to 1 at $n + 1$, which we call $1_\top$. So at n, this power set looks like $\{0, 1, 1_\top\}$. Under the transition function, 0 goes to 0, 1 to 1, and $1_\top$ also goes to 1. But then a new $1_\top$ appears, and the power set remains settled, still having three elements, one of which in some sense was already in the power set at n and in another sense is new. The set $\{0, 1, 1_\top\}$ is not the internalization of an external set, but it is still settled if it goes to itself at the next node, even though not all of its members are settled yet.

With regard to the technical details, just as with the full Kripke model above, the exposition in [12] needs some refinement to be correct. The problem, as before, is getting the induction right in a context where there may be nonstandard models. The additional challenge in thinking of this as a Kripke model is that the underlying partial order, say ω as in the example above, is typically not in any of the V_ps (except $V_\perp = V$) since they are ω-non standard. The solution makes essential use of the fact that the settling is immediate, so that the model and the semantics can be defined locally, with reference only to the immediate successor nodes.

By way of providing the details, suppose we have a set Q, and a definable assignment $q \mapsto V_q$ of models of ZF to members q of Q, along with elementary embeddings $f_q : V \to V_q$, which we write polymorphically as simply f. Then we define the Kripke partial order $\mathcal{P}$ as certain finite sequences p, the assignment $p \mapsto V_p$ of ZF models to members of $\mathcal{P}$, and elementary embeddings $f : V_{p_0} \to V_{p_1}$ whenever p_0 is a proper initial segment of p_1 (f being taken as the identity function whenever $p_0 = p_1$), inductively on the length of p.

Notation: For $p = \langle q_0, ..., q_n \rangle, p^- = \langle q_0, ..., q_{n-1} \rangle$. Also, $\frown$ is the concatenation of a string with an element, so that $p^\frown q = \langle q_0, ..., q_n, q \rangle$.

Definition 22.21 (The partial order $\mathcal{P}$, assignment V_p, and elementary embeddings f) $\langle\rangle \in \mathcal{P}$, and $V_{\langle\rangle} = V$.

$\langle q \rangle \in \mathcal{P}$ whenever $q \in Q$, in which case $V_{\langle q \rangle} = V_q$ and $f : V_{\langle\rangle} \to V_{\langle q \rangle} = f_q$ as given.

$p \in \mathcal{P}$ whenever $p^- \in \mathcal{P}$, and, for f the embedding from V to $V_{p^-}, q_n \in f(Q)$. The assignment $q \mapsto V_q$ in V on Q goes under f to an assignment in V_{p^-} on $f(Q)$, of which q_n is a member; the model attached thereby to q_n is what we take to be V_p. Similarly, the elementary embeddings f from V to the V_qs, $q \in Q$, go under f to elementary embeddings from V, as interpreted in V_{p^-}, to the V_qs, $q \in f(Q)$; in particular, we have an elementary embedding from V_{p^-} to V_p. The embedding from $V_{\bar{p}}$, for $\bar{p}$ any shorter initial segment of p, is the composition of the one from $V_{\bar{p}}$ to V_{p^-} (inductively) followed by the one from V_{p^-} to V_p.

Given this system on $\mathcal{P}$ of ZF models and elementary embeddings, we can now define the universe M^p of the Kripke model at nodes $p \in \mathcal{P}$, the interpretations of membership and equality, and the transition functions k, as positive inductive definitions.

Definition 22.22 (The immediate settling model) M^p consists of functions g such that:

- the domain of g, $\mathrm{dom}(g)$, consists of $\langle\rangle$ and all q such that $p^\frown q \in \mathcal{P}$ (i.e. $f(Q)$),
- $g \in V_p$ and $g(q) \in V_{p^\frown q}$,
- inductively, $g(\langle\rangle) \subseteq M^p$ and $g(q) \subseteq M^{p^\frown q}$,
- if $h \in g(\langle\rangle)$ then $k(h) \in g(q)$, where k is the inductively defined transition function from M^p to $M^{p^\frown q}$, and
- if $h \in g(q)$ then $k(h) \in f(g(q))$, where k is the transition function from $M^{p^\frown q}$ to $M^{p^\frown q^\frown r}$ and f is the elementary embedding from $V_{p^\frown q}$ to $V_{p^\frown q^\frown r}$.

$p \models g \in h$ if $g \in h(\langle\rangle)$, and $p \models g = h$ if $g = h$.

The transition function k from $M^{\bar{p}}$ to M^p, when $\bar{p}$ is a proper initial segment of p, is the composition of the transition functions among nodes that are immediate successors. For p of length at least 1, $k : M^{p^-} \to M^p$ is defined via:

- $k(g)(\langle\rangle) = g(q_n)$, and
- $k(g)(q) = f(g(q_n))$, where f is the elementary embedding from V_p to $V_{p^\frown q}$.

Lemma 22.23 *The immediate settling model is well-defined.*

Proof Given M^p, the interpretation of $\in$ and $=$ within M^p is unproblematic. Similarly, given any particular member g of M^p, $k(g)$ is well-defined: q_n is in the domain of g, and f comes from the assignment of models and elementary embeddings.

Regarding the definition of M^p, the first two clauses are unproblematic, referencing only fixed entities. The last two clauses are unproblematic, because they depend only on the application of k to members of $g(\langle\rangle)$ and $g(q)$, which are members of M^p and $M^{p^\frown q}$ respectively, to which it has already been shown k can be applied. The only issue is with the third clause. One can take M^p as $\bigcup_{\alpha \in ORD^{V_p}} M^p_\alpha$, and also use M^p_β for all $\beta < \alpha$ to define M^p_α by working inductively in V_p. The issue though is that the construction of M^p_α makes reference not just to M^p_β for smaller β, but also to $M^{p^\frown q}$. An inductive definition of the latter would need to reference $p^\frown q^\frown r$, and so on.

What makes this work is all of the uniformities within the set-up. It might even be the case that $\mathcal{P}$ is undefinable in any V_p save for $V_{\langle\rangle}$, because $\mathcal{P}$ is based on the true ω, and all the other models might be ω-nonstandard. Nonetheless, all of the information at any p is about immediate successors of p.

We can define M^p_α just as M^p was defined, only restricting clause three to $g(\langle\rangle)$ being a subset of $\bigcup_{\beta<\alpha} M^p_\beta$ and $g(q)$ being a subset of $\bigcup_{\beta<f(\alpha)} M^{p^\frown q}_\beta$. For any γ for which M^p_γ is defined, $M^{p^\frown q}_{f(\gamma)} = f(M^p_\gamma)$. To argue inductively that M^p_α is well-defined, we can assume that within V_p for each $\beta < \alpha$ M^p_β is well-defined. By applying f, we have that within $V_{p^\frown q}$ for each $\beta < f(\alpha)$ $M^{p^\frown q}_\beta$ is well-defined. This is all we need to have M^p_α be well-defined. □

Lemma 22.24 *k is a transition function; that is, $k(g) \in M^p$.*

Proof We check the defining clauses of being a member of M^p, in order.

The domain of $k(g)$ is exactly what it's supposed to be, because that's exactly what $k(g)$ was defined on.

For the second clause, we first need to show $k(g) \in V_p$. For starters, since $g \in M^{p^-}$, by the second clause applied to g, $g(q_n) \in V_p$, which at least gets us to $k(g)(\langle\rangle) \in V_p$. Because we started with an admissible assignment, the system beyond p is definable in V_p, so $f(g(q_n))$ is also in V_p, which gets us that $k(g)(q) \in V_p$. Since $k(g)$ is easily definable from those two sets, $k(g) \in V_p$. Continuing with the second clause, we need to show that $k(g)(q) \in V_{p^\frown q}$. This is immediate, since $k(g)(q)$ is in the range of $f : V_p \to V_{p^\frown q}$.

For the third clause, we first need to show $k(g)(\langle\rangle) \subseteq M^p$. For that, apply the second part of clause three to g. We also need to show $k(g)(q) \subseteq M^{p^\frown q}$. Consider that $g(q_n) \subseteq M^p$, and apply f.

Regarding clause four, let $h \in k(g)(\langle\rangle)$, that is, $h \in g(q_n)$. Applying the fifth clause to g, $k(h) \in f(g(q_n))$, that is, $k(h) \in k(g)(q)$, as was to be shown.

Finally, let $h \in k(g)(q)$, that is, $h \in f(g(q_n))$. We need to show $k(h) \in f(k(g)(q))$, i.e. $k(h) \in f(f(g(q_n)))$. Apply clause five to g and q_n. That yields that for all $h \in g(q_n)$, $k(h) \in f(g(q_n))$. Applying f to that statement yields exactly what we need, using that k is given via its definition (which is uniform, and written polymorphically, without indices, like f, so that $f(k) = k$). □

Lemma 22.25 *Suppose that $g(q) = f(g(\langle\rangle))$ for all q. (This is the case, for instance, when g is in the image of k.) Then $k(g) = f(g)$.*

Proof By way of notation, $g \in M^p$ for some p, and k and f are functions from node p to some successor node $p^\frown r$. By the definition of k, $k(g)(\langle\rangle)) = g(r)$, which by hypothesis is $f(g(\langle\rangle)) = f(g)(\langle\rangle)$. Furthermore, for all q, $k(g)(q) = f(g(r))$. Applying f to the hypothesis of the lemma yields, for all q, $f(g)(q) = f(f(g)(\langle\rangle)) = f(g(r))$. □

Theorem 22.26 *The immediate settling model satisfies IZF.*

Proof Empty Set: Let g be the function (with appropriate domain) that always returns the empty set.

Infinity: Within any V_p, one can define inductively on $n \in \omega^{V_p}$ an element $n_{p'} \in M^{p'}$, for p' any extension of p, which, in $V_{p'}$, will be the nth successor of 0 in $M^{p'}$. To start, $0_{p'}$ is the set satisfying the Empty Set Axiom from above. Given $m_{p'}$ for $m < n$, let $n_{p'}(\langle\rangle)$ be $\{m_{p'} \mid m < n\}$, and $n_{p'}(q)$ be $\{m_{p'^\frown q} \mid m < n\}$. Let ω_p be such that $\omega_p(\langle\rangle) = \{n_p \mid n \in \omega^{V_p}\}$, and $\omega_p(q) = f(\omega_p(\langle\rangle))$. Then ω_p suffices at the node p.

Pairing: Given g and h at node p, the desired pair is the function which, on input $\langle\rangle$, yields the set $\{g, h\}$, and, on input q, yields $\{k(g), k(h)\}$. To show that this set is in M^p, most of the properties are simple; for the last, use Lemma 22.25.

Union: Given g at node p, $\bigcup g(\langle\rangle)$ is $\bigcup\{h(\langle\rangle) \mid h \in g(\langle\rangle)\}$, and $\bigcup g(q)$ is $\bigcup\{h(\langle\rangle) \mid h \in g(q)\}$.

Extensionality: Because equality is taken to be actual equality, it is easy to see that equal sets have the same members. On the other hand, if two sets are forced at a node to have the same members, then they are literally equal as functions. By Extensionality in the ambient universe, they are then equal.

$\in$-Induction: Because of the uniformities and simplicities in the current set-up, this proof can proceed by consideration of only this model, as opposed to the proof for full Kripke models. So suppose $\in$-Induction fails: $p \models \forall x(\forall y \in x\ \phi(y) \to \phi(x))$, yet $p \not\models \forall x\ \phi(x)$. Therefore, for some $p' \geq p$ there is a $g \in M^{p'}$, $p' \not\models \phi(g)$.

First we're going to consider the simple case when there is such a p' which strictly extends p. The advantage here is that, at p', "ϕ" is really $k(\phi)$, meaning that the parameters are actually of the form $k(h)$, for h an original parameter of ϕ at p. That means, by Lemma 22.25, that with reference to any extension p'' of p', $k(\phi) = f(\phi)$. Working in $V_{p'}$, let g be a set of least rank α such that $p' \not\models \phi(g)$. Anything forced at p' to be a member of g has smaller rank, and so is forced at p' to satisfy ϕ. Also, anything forced at any $p'' > p'$ to be a member of $k(g)$ has rank less than $f(\alpha)$. By elementarity, in $V_{p''}$, if h has rank less than $f(\alpha)$ then $p'' \models f(\phi)(h)$. As remarked above, $f(\phi) = k(\phi)$, so if $p'' \models h \in k(g)$, then $p'' \models \phi(h)$. So $p' \models \forall y \in g\ \phi(y)$, hence by the inductive hypothesis $p' \models \phi(g)$, contrary to the choice of g.

So there is no such p'; hence, if $p' > p$, then $p' \models \forall h\ \phi(h)$. Let g be of minimal rank such that $p \not\models \phi(g)$. If $p \models h \in g$ then h has smaller rank than g and so $p \models \phi(g)$. If $p' > \sigma$ and $h \in M^p$ then as already observed $p' \models \phi(h)$. So $p \models \forall y \in g\ \phi(y)$, hence by the inductive hypothesis $p \models \phi(g)$, contrary to the choice of g.

Power Set: This is likely the most interesting axiom to check. After all, any alleged power set must settle down after one step through $\mathcal{P}$, yet a possible subset might show up later that then has its own step in which to settle down. Here we present a general proof; it might help the reader to have looked through the discussion of Power Set toward the beginning of this section on Internal Settling, where we worked through the most basic example of the Power Set Axiom and why it holds there.

Let $g \in M^p$. Let $\wp(g)(\langle\rangle)$ consist of all functions $h \in M^p$ such that $h(\langle\rangle) \subseteq g(\langle\rangle)$ and, for all q in the common domain of g and h, $h(q) \subseteq g(q)$. Also, $\wp(g)(q)$ consists of all $h \in M^{p^\frown q}$ such that $h(\langle\rangle) \subseteq g(q)$ and, for all relevant r, $h(r) \subseteq f(g(q))$. Note there are only set-many such functions. It is an unenlightening technical exercise to show that $\wp(g) \in M^p$.

We show that $\wp(g)$ is as desired. First consider what happens at node p. If $p \models h \subseteq g$, then in particular $h(\langle\rangle) \subseteq g(\langle\rangle)$ and, for all relevant q, $h(q) \subseteq g(q)$, so that $h \in \wp(g)(\langle\rangle)$ by construction, and $p \models h \in \wp(g)$. Conversely, suppose

$p \models h \in \wp(g)$, that is, $h \in \wp(g)(\langle\rangle)$. We need to show that $p \models h \subseteq g$, that is, recalling that k is the transition function to an unspecified extension of p (wherein k is the identity when viewing p as extending to itself), $k(h)(\langle\rangle) \subseteq k(g)(\langle\rangle)$. When k goes to either p or an immediate extension, this is exactly the definition of $wp(g)$. For k going to a strict extension p' of $p^\frown q$, apply f to the assertion "$k(h)(\langle\rangle) \subseteq k(g)(\langle\rangle)$", where the k there goes to $p^\frown q$. This yields $f(k(h))(\langle\rangle) \subseteq f(k(g))(\langle\rangle)$, where f goes to p'. Now use Lemma 22.25.

Now consider node $p^\frown q$. In one direction, suppose $p^\frown q \models h \subseteq k(g)$; in particular, $h(\langle\rangle) \subseteq k(g)(\langle\rangle) = g(q)$, and, for all relevant r, $h(r) \subseteq k(g)(r) = f(g(q))$. We need to show $p^\frown q \models h \in k(\wp(g))$. That is, we must show $h \in k(\wp(g))(\langle\rangle)$. But that latter set is $\wp(g)(q)$. The definition of h getting into $\wp(g)(q)$ is that $h(\langle\rangle) \subseteq g(q)$ and $h(r) \subseteq f(g(q))$, which is exactly what we have. In the other direction, suppose $p^\frown q \models h \in k(\wp(g))$, that is, $h \in k(\wp(g))(\langle\rangle) = \wp(g)(q)$. That means that $h(\langle\rangle) \subseteq g(q)$, and $h(r) \subseteq f(g(q))$. To see that $p^\frown q \models h \subseteq k(g)$, what we need to show is $k_1(h)(\langle\rangle) \subseteq k_1(k(g))(\langle\rangle)$, where k_1 is a transition function from node $p^\frown q$. When k_1 is the identity, this reduces to $h(\langle\rangle) \subseteq (k(g))(\langle\rangle) = g(q)$, which we have already. If k_1 goes to an immediate extension, given by say r, this reduces to $h(r) \subseteq k(g)(r) = f(g)(q)$, which again we already have. If k_1 goes to a strict extension of $p^\frown q^\frown r$, decompose k_1 into a composition $k_2 \circ k_r$, where k_r goes from $p^\frown q$ to $p^\frown q^\frown r$. Then what we need to show becomes $k_2(k_r(h))(\langle\rangle) \subseteq k_2(k_r(k(g)))(\langle\rangle)$. Using Lemma 22.25, this becomes $f(k_r(h))(\langle\rangle) \subseteq f(k_r(k(g)))(\langle\rangle)$. And that is true, by applying f to what we have already established about k_r in the previous case.

For strict extensions of $p^\frown q$, argue similarly.

Separation: Given ϕ and g at p, let $Sep_{\phi,g} \in M^p$ be $Sep_{\phi,g}(\langle\rangle) = \{h \in g(\langle\rangle) \mid p \models \phi(h)\}$ and $Sep_{\phi,g}(q) = \{h \in g(q) \mid p^\frown q \models \phi(h)\}$. The crux of the matter is to show that, if p' strictly extends $p^\frown q$, and $h \in k(g)(\langle\rangle)$, and $p' \models k(\phi)(h)$, then $p' \models h \in k(Sep_{\phi,g})$. Decompose k as $k_1 \circ k_q$, where k_q goes to $p^\frown q$, apply that decomposed form to g, ϕ, and $Sep_{\phi,g}$, and appeal to Lemma 22.25.

Replacement\Collection\Reflection: As in the proof of IZF for the full Kripke model, it is easy to see that Reflection holds true. Working in V_p, by Reflection there, let $V_\alpha \prec_{\Sigma_n} V_p$. Cut off the construction of the immediate settling model at α. That is, let $M(\langle\rangle)$ be $M^{\langle\rangle}_\alpha$ and $M(q)$ be $M^{p^\frown q}_{f(\alpha)} = f(M^{\langle\rangle}_\alpha)$. This yields a Σ_n substructure of the entire model, in the sense that any property of M^p which is Σ_n expressible in the ambient V_p reflects to M. □

The Location of the Settling: Eventual, Immediate, Uniform, and Sideways

Eventual Settling All of the sets must settle down eventually. Examples are the two examples given above for external settling. The partial order was ORD, and each set had to settle down by some ordinal, with no restriction on which ordinal.

Immediate Settling Each node except the bottom has a predecessor, and every set which is the image under the transition function of a set from the predecessor node can no longer change. An example is given under Internal Settling, previously.

Uniform Settling Immediate settling is just a special case of the more general uniform settling. Under uniform settling, one starts with a partial order $\mathcal{P}$ in which the terminal nodes are dense (every node has an extension which is terminal). Then one places a copy of $\mathcal{P}$ (actually, $f(\mathcal{P})$, the image of $\mathcal{P}$ under the elementary embedding) at each such terminal node, and iterates that procedure ω-many times. Immediate settling is what you get when $\mathcal{P}$ consists merely of $\perp$ followed by a set of children (a tree of height 1). In contrast, examples of settling which are nonuniform are the examples given for external settling. Every set there settles down for sure, but the settling is not uniform: the objects at node 0 settle down at all possible ordinals.

Since the examples given for nonuniform settling are the same as those for external settling, and the examples for uniform settling are those given for internal settling, one might thereby make the mistake of identifying external with nonuniform settling and internal with uniform. We see no reason for this identification to be valid. Uniformity seems to be orthogonal to internality, in that there could be internal nonuniform models and external uniform ones. Consider for instance the Kripke model based on the partial order ω, with V associated to each node. Take all those sets that settle down to internal sets anywhere along the way. This is an example of non uniform settling to internal sets. Also, one can instead take those sets that settle down the node after they are introduced to something in V, for uniform settling to external sets. It may not be clear what holds in these models or why someone would be interested in them; the point remains, they are perfectly legitimate models.

Question 22.27 Are there interesting, useful examples of internal, non uniform settling, or of external, uniform settling?

Sideways Settling There is a different kind of settling that can be useful ([12, Theorem 5.7]). Consider the partial order with bottom node $\perp$ and children n for $n \in \omega$. Associate V with $\perp$, and the same M with each n. Take the submodel of the full model of those gs that are eventually constant (i.e., for some n and all $i > n$, $g(i) = g(n)$). (The principles separated by this model are some refined principles from constructive logic, the description of which would take us too far afield from this methods-oriented exposition.)

22.4 Heyting–Kripke Models

There are constructions that use a mix of ideas from both Heyting and Kripke models. Here we content ourselves with just describing some examples.

Question 22.28 Is it necessary to have a mix of Heyting and Kripke models to prove the theorems in this section? Or are there simpler models that use only one of those methods?

Question 22.29 Is there a general definition of or framework for Heyting–Kripke models? Any such definition should have at least some of the examples below as particular instances.

22.4.1 Iterating Heyting Models

Heyting models are the constructive version of forcing. An important forcing technique is iteration. What is the constructive analogue of iteration?

Iteration is used when you have to do more than one forcing. If all the partial orders of concern are in the ground model, then a simpler form, product forcing, suffices. If instead some of the partial orders needed for the forcing are only in the generic model for earlier forcings, then an actual iteration is necessary. Doing either a product or an iteration finitely often is unproblematic, constructively as well as classically. Issues arise only with an infinite number of forcings, which are naturally arranged along the order-type of some ordinal. Mostly it's a question of what to do at limits: At what places along this ordinal should a condition be allowed to be nontrivial? All of them? Only finitely many? Or what? Classically, the decision often involves set-theoretic concepts, such as countability, inaccessibility, or stationarity. Needless to say, these solutions are problematic constructively, even starting from the idea of organizing the forcings along a linear ordinal, to say nothing of the other, more advanced concepts.

Fortunately, the use of limits can typically be finessed constructively. Iteration comes up when you want a model satisfying an assertion of the form "every structure with property A has property B." This would typically come up when B implies A, and A and B are close, so one suspects it is consistent that A and B be equivalent. Suppose you had a forcing for any structure not satisfying B that would make it not satisfy A. Then you could just do those forcings, one after the other, for each such structure, including those that come up along the way, until the process closes off. At that point, you'd be left with a model in which the only structures left satisfying A are those you couldn't force not to have A, namely those with B. If your context is constructive, though, you don't actually have to do the forcing to kill A. It's enough to threaten to do so. If you're in a Kripke model, and some later node does the forcing to kill A, then at the current node it is false that A holds, since A fails later; this suffices to have the assertion not apply to the structure at hand.

A concrete example might be useful. Recall that D-FAN is "for all decidable $X \subseteq 2^{<\omega}$, if X is a bar then X is uniform." In Section 22.2.3 we described how to force to get a decidable, nonuniform set of nodes to be not a bar. To get D-FAN false, this forcing would in some sense have to be applied to all decidable, nonuniform sets, which seems to call for some kind of iteration. In [23] this iteration was organized as a Kripke-like model. To the bottom node is associated the K_1-realizability model. At any node p, the children are indexed by those Heyting algebras forced at p to have been constructed as above from a decidable, nonuniform tree, along with a value from this Heyting algebra. Over this partial order with associated models, take the full model. As above, D-FAN holds, because if at a node p a decidable tree is not uniform, some later node forces a path through it, so at p the tree couldn't have represented a bar.

What keeps this from being just a Kripke model is the semantics. For instance, in a Kripke model, a disjunction is true at a node when one of the disjuncts is true at that node. In the current model, topological (or Heyting) concepts play a role, in that there is a notion of a set of nodes covering a node, just as how the join of a set of Heyting values can be greater than another Heyting value. So in this model, a node validates a disjunction if it has a cover each member of which validates one of the disjuncts. Similar considerations apply to membership and the existential quantifier. It should not be a surprise that some such consideration is necessary, because the nodes came from Heyting algebras, yet if there were no consideration of such covering the Heyting structure would be completely lost.

It should be mentioned even if only briefly that due respect must be shown to the computational (K_1) nature of the setting. This does influence the construction of the model: The covers used in the semantics are stricter than arbitrary covers for the Heyting algebras, to make them computational.

22.4.2 Topological Settling

Settling is an idea that more naturally fits with Kripke models. The basic intuition behind settling is that there is some future stage at which some qualitative change happens. This is consistent in style with a Kripke model on a discrete order, such as a tree, where there is a notion of a next step, or in the case of nonuniform settling a notion of a step; it is inconsistent in style with the sense behind a topological space, where usually an increase in information is thought of as an open set gradually or continuously getting smaller. Of course one can come up with counter-examples to formal versions of those assertions, and we will even see some shortly. We still think this is fair as a stylistic description, one that is usually true. Coupled with the fact that the examples to be given were developed using a mix of topological and Kripke intuitions, it seems right to include this topic in the category of Heyting–Kripke models.

Topological Settling to Internal Sets

Suppose you had a topological space in which some singleton sets $\{x\}$ were open. In fact, suppose that the set of such points was dense. Then any classically valid assertion you might want to falsify, even if not true in the induced topological model, would not be false: it would be not not true. This would be a prime candidate for iteration, which in this case would be placing another copy of the space at each of these singleton opens, or some variant thereof. This is starting to look like the uniform settling discussed briefly above. Where such a construction has actually come up already [12, models 11 and 18 and Theorem 5.7], the space was simple, and so ended up looking more like immediate settling. We describe first an oversimplified version of these models, one that did not even need to appear in [12], for expository purposes, and then describe the modifications needed for the other models.

Let $\mathcal{U}$ be a nonprincipal ultrafilter on ω. Let $\mathcal{T}$ be the space with points $\omega \cup \{*\}$, the discrete topology on ω, and open neighborhoods of $*$ sets of the form $\{*\} \cup X$, where $X \in \mathcal{U}$. Perhaps it is useful to try to view the full topological model over $\mathcal{T}$ as a Kripke-like model. Let $\mathcal{P}$ have bottom node $*$ and successor nodes $n, n \in \omega$. Then the universe at n is simply V, and at $*$, inductively, the universe consists of functions g with domain $\mathcal{T}$ such that if $h \in g(*)$ then h is also in the universe at $*$, and $\{n \mid h(n) \in g(n)\} \in \mathcal{U}$. So this is like a Kripke model, except, even if something is a member of g at the bottom node $*$, that fact can be forgotten at some later node; all that must happen is for that membership fact to hold at ultrafilter-many later nodes. (Of course, this member might itself be interpreted differently at various later nodes.)

It is a simple matter to iterate this topology: Place a copy of $\mathcal{T}$ at each open point n. This can be formalized as follows. Consider the tree $\omega^{<\omega}$, finite sequences of natural numbers. A basic open set $\mathcal{O}$ contains a unique shortest sequence σ, and for all $\tau \in \mathcal{O}$, $\{n \mid \tau^\frown n \in \mathcal{O}\} \in \mathcal{U}$. The idea behind the immediate settling model is that at every node you place a copy of the full model on $\mathcal{T}$. So at a node σ, you have sets that can change when going to a child $\sigma^\frown n$ (as long as they respect the topology when doing so), at which point they have all settled down; but then there are new sets that themselves can change at the next nodes.

We now give the formal details. Since every node looks exactly like every other node, we can dispense with the tree of sequences, and work just with the space $\mathcal{T}$. The sets g are defined inductively, as are the functions k_i from the model to itself, which play the role of transition functions in a Kripke model from $*$ to i. Then g is a set in this model if

- g is a function with domain $\mathcal{T}$,
- $g(*)$ is a collection of sets from this model, as is $g(i)$ (where $i \in \omega$),
- if $h \in g(i)$ then $k_j(h) \in g(i)$ $(i, j \in \omega)$, and
- if $h \in g(*)$ then $\{j \mid k_j(h) \in g(j)\} \in \mathcal{U}$.

Also, $k_i(g)$ is the constant function with value $g(i)$.

For the semantics, we give the interpretation of a formula as a subset of $\mathcal{T}$:

- $[\![g \in h]\!] = \{q \mid \exists f \in h(q) \perp \in [\![k_q(g) = f]\!]\}$,
- $[\![g = h]\!] = \{q \in \omega \mid \text{for all } f \in g(q) \perp \in [\![f \in k_q(h)]\!], \text{ and vice versa}\} \cup$ $\{\perp \mid \forall f \in g(\perp) \perp \in [\![f \in h]\!] \text{ and vice versa, and } [\![g = h]\!] \in \mathcal{U}\}$,
- $[\![\phi \wedge \psi]\!] = [\![\phi]\!] \cap [\![\psi]\!]$,
- $[\![\phi \vee \psi]\!] = [\![\phi]\!] \cup [\![\psi]\!]$,
- $[\![\phi \rightarrow \psi]\!] = [\![\psi]\!] \cup (\omega \backslash [\![\phi]\!]) \cup \{\perp \mid \perp \notin [\![\phi]\!] \wedge [\![\psi]\!] \cup (\omega \backslash [\![\phi]\!]) \in \mathcal{U}\}$,
- $[\![\exists x\ \phi(x)]\!] = \{q \mid \exists h\ q \in [\![\phi(h)]\!]\}$,
- $[\![\forall x\ \phi(x)]\!] = \{q \in \omega \mid \text{for all } h \perp \in [\![k_q(\phi)(h)]\!]\} \cup$ $\{\perp \mid \text{for all } h \perp \in [\![\phi(h)]\!], \text{ and } [\![\forall x\ \phi(x).]\!] \in \mathcal{U}\}$.

The actual models used in [12] differ from the above in that ω is partitioned into infinite subsets, and a copy of $\mathcal{U}$ is applied to each of those slices. For instance, ω could be partitioned into two subsets, the evens and the odds, and, in terms of an open set $\mathcal{O}$ on $\omega^{<\omega}$, if $\sigma \in \mathcal{O}$, then both $\{k \mid \sigma^\frown 2k \in \mathcal{O}\}$ and $\{k \mid \sigma^\frown 2k+1 \in \mathcal{O}\}$ must be in $\mathcal{U}$.

Topological Settling to External Sets

This topic was first developed to find a model of CZF_{Exp}, that is, CZF with the Subset Collection axiom replaced by Exponentiation, in which the Dedekind reals do not form a set [26]. Since that theory does suffice to show that the Cauchy reals form a set, this is a strong way of separating the Dedekind and Cauchy reals. Two aspects of such a construction quickly come to mind. One is that the Cauchy and Dedekind reals would have to be unequal; the most natural model for getting that is the topological model over the reals. The other is that Subset Collection must fail; the model for getting that which we saw above is settling to external sets. So the obvious candidate for the task at hand is the topological model over the reals with external settling, whatever that might mean. In fact, that's exactly what works.

Not surprisingly, the definition of this model, based on the reals with external settling, can be extended to external settling over any topological space [21]. The idea is that not only can a set be specified more by shrinking the open set you're taking as truth, as usual in topological models, you can also specialize to a point in the set. At that moment, the universe of sets you were looking at becomes the ground model. At the same time, a new universe of variable sets appears, based on the same space. We give the formal details.

Definition 22.30 (Terms) For a topological space T, a term is a set of the form $\{\langle J_i, \sigma_i\rangle \mid i \in I\} \cup \{\langle r_h, \sigma_h\rangle \mid h \in H\}$, where each σ is (inductively) a term, each J is an open set, each r is a member of T, and H and I are index sets.

Definition 22.31 (Specialization to a point) For σ a term and $r \in T$, σ^r is defined inductively on the terms as $\{\langle T, \sigma_i^r\rangle \mid \langle J_i, \sigma_i\rangle \in \sigma \wedge r \in J_i\} \cup \{\langle T, \sigma_h^r\rangle \mid \langle r, \sigma_h\rangle \in \sigma\}$.

Definition 22.32 (The forcing relation) $J \Vdash \sigma = \tau$ iff for all $\langle J_i, \sigma_i\rangle \in \sigma$ $J \cap J_i \Vdash \sigma_i \in \tau$ and vice versa, and for all $r \in J$ $\sigma^r = \tau^r$

$J \Vdash \sigma \in \tau$ iff for all $r \in J$ there is a $\langle J_i, \tau_i\rangle \in \tau$ and $J_r \subseteq J_i$ containing r such that $J_r \Vdash \sigma = \tau_i$

$J \Vdash \phi \wedge \psi$ iff $J \Vdash \phi$ and $J \Vdash \psi$

$J \Vdash \phi \vee \psi$ iff for all $r \in J$ there is a $J_r \subseteq J$ containing r such that $J_r \Vdash \phi$ or $J_r \Vdash \psi$

$J \Vdash \phi \rightarrow \psi$ iff for all $J' \subseteq J$ if $J' \Vdash \phi$ then $J' \Vdash \psi$, and, for all $r \in J$, there is a $J_r \subseteq J$ containing r such that, for all $K \subseteq J_r$, if $K \Vdash \phi^r$ then $K \Vdash \psi^r$

$J \Vdash \exists x\ \phi(x)$ iff for all $r \in J$ there is a $J_r \subseteq J$ containing r and a σ such that $J_r \Vdash \phi(\sigma)$

$J \Vdash \forall x\ \phi(x)$ iff for all σ $J \Vdash \phi(\sigma)$, and for all $r \in J$ there is a $J_r \subseteq J$ containing r such that for all σ $J_r \Vdash \phi^r(\sigma)$.

In Definition 22.32, when $T = \mathbb{R}$, because of the homogeneity of the space, the case of $\rightarrow$ can be simplified to "for all $J' \subseteq J$ if $J' \Vdash \phi$ then $J' \Vdash \psi$, and, for all $r \in J$, if $\mathbb{R} \Vdash \phi^r$ then $\mathbb{R} \Vdash \psi^r$."

Theorem 22.33 *The model given by the semantics above satisfies Infinity, Pairing, Union, Extensionality, Set Induction, Bounded (i.e. Δ_0) Separation, and Collection. It also satisfies Eventual Power Set: For every X there is a C such that everything in C is a subset of X, and every subset of X is not unequal to everything in C. If T is locally connected then Exponentiation holds. If T is locally homogeneous then Full Separation holds.*

The Location of Topological Settling

The model in the subsection "Topological Settling to Internal Sets" is an example of immediate settling, of the same kind as the immediate settling from Section 22.3.2: A set at * can transition only to some $i \in \omega$, at which point it is settled. The model in the subsection "Topological Settling to External Sets" is an example of eventual settling, of a different kind to the eventual settling from Section 22.3.2: A set can remain a variable set (as opposed to becoming a settled set) as the open set J shrinks; it settles only when a point $r \in T$ is specified.

There is a topological version of the sideways settling model from Section 22.3.2, iterated via immediate settling ([12, Theorem 5.7]). Let $\mathcal{U}$ be a nonprincipal ultrafilter on ω; also, identify ω with $\omega \times \omega$. Let T be $\omega \cup \{\bot\}$. Take the discrete topology on ω; for neighborhoods of $\bot$, let $A \cup \{\bot\}$ be open exactly when each A_i is in $\mathcal{U}$, where A_i is the ith slice of A, using here the identification of ω with $\omega \times \omega$. Take the submodel of the full model of those sets g that eventually settle down on slices: For some j and all $i \geq j$, the value of g is constant on the ith section of ω.

22.5 Classical Outer Models

So far, we have been using the idea of an outer model as something into which V embeds. At the same time, almost all of the constructions and examples we have seen can be done within V. For example, a full Heyting-valued model can be taken to be an extension of V by the inclusion of a generic; at the same time, just like with Boolean-valued models, the entire structure can be developed within V. It is at this point that the classical mathematician has the option to call Boolean-valued models forcing, pull in a generic G from outside of V, and work in the two-valued model $V[G]$. The constructivist does not have this option. Nevertheless, we can still avail ourselves of this two-valued outer model $V[G]$, to help us in building our constructive models.

We have already seen an example of this: the second example in Section 22.3.2, about Kripke settling to external sets, when we took eventual settling of a set in $V[G]$, G Cohen generic over V, to something in V.

In Section 22.6.1, we will see another example of the use of V-generics in combination with inner models.

For the rest of this section, we consider an application of generics over V to independence results around the fan theorem.

The fan theorem and its variants In Sections 22.2.3 and 22.4.1, there were sketches of two models in which the fan theorem failed in various ways. Let's put this into context. The fan theorem says that every bar is uniform. The X-fan theorem, also known as X-FAN, says that every X-bar is uniform, for any choice of a property X. We have already seen D-FAN, the fan theorem for decidable bars, and c-FAN, the fan theorem for c-bars. Extending the notion of a c-bar, a Π^0_1 bar is a bar which is definable as a Π^0_1 set, which can be understood as a set given by a Π^0_1 formula, or as an intersection of countably many decidable sets. Then Π^0_1-FAN states that every Π^0_1 bar is uniform. Over IZF, the following implications are trivial:

$$\text{Full FAN} \Rightarrow \Pi^0_1\text{−FAN} \Rightarrow c\text{−FAN} \Rightarrow D\text{−FAN} \Rightarrow \text{IZF}.$$

The natural question is whether any of those implications reverse. Some have long been known not to. For instance, the recursive realizability model shows that IZF does not prove D-FAN. As observed in [25], the construction from [7] shows that Π^0_1-FAN does not imply the full fan theorem. In [2] it is shown that D-FAN does not prove c-FAN over a weak base theory. In [25] a technique is developed which shows all of these nonreversals, including that c-FAN does not imply Π^0_1-Fan. It is this technique, different from the two already discussed, that interests us here.

The failure of the fan theorem To develop some intuition, ask yourself, how could FAN fail? It is hard to see how FAN could fail without there being a specific counter-example. There are other instance in which a $\forall x \exists y$ statement fails without a counter-example, because there is just no uniform way of going from an x to a y. In the case of the fan theorem though, the y would be a finite level of the binary tree witnessing the uniformity of the bar. If there were such a y, it should be easy to find, by just going through the full binary tree level by level until the uniform bound is found. Instead, one can more readily imagine a counter-example, a set of nodes which is not uniform, but it still represents a bar, because one cannot find a path missing that set. How could that be the case? You can start at the root $\langle\rangle$; if only one of $\langle 0 \rangle$ and $\langle 1 \rangle$ is in the tree (i.e., is not in the bar), then your choice of successor is clear; you keep going until both children are in the tree – then what? If you happen to choose a child beyond which the entire tree dies, your goose is cooked. For X to be a nonuniform bar, there would have to be no way of predicting, when confronting a tree split, which if either of the two subtrees will die. That's the cleverness of the Kleene tree: Membership on this infinite tree is computable, yet it is not computable when a subtree will entirely die – there is no computable look-ahead mechanism. Hence if your universe consists only of computable objects, this tree provides a counter-example, actually a decidable counter-example, to the fan theorem.

Forcing the Failure of D-FAN It would be nice to get an example like the Kleene tree which is more amenable to manipulation, so that we can not only falsify D-FAN but also separate the varieties of FAN. What is the essence of the Kleene tree? From the point of view of computability, subtrees die out in seemingly random, unpredictable ways. The ultimate in set-theoretic unpredictability is forcing. A Kleene-like tree can be built generically. Let a forcing condition be an assignment to finitely many binary sequences of one of three labels: IN, meaning in the putative bar (as well as all of its descendants), OUT, meaning not in the bar although there's nothing to stop the bar from being uniform beneath it, and ∞, meaning the tree beneath that node is infinite. This infinitude beneath an ∞-labeled node is enforced in that at no time may a condition assign IN or OUT to all descendants at some fixed level; if all extensions of an ∞ node of length n are labeled, then at least one of those labels must be ∞. The generic G will be a tree with labels OUT sprinkled seemingly randomly beneath labels of ∞. Of course, one can easily find a path through this tree: At a node labeled ∞, take a child labeled ∞, which is guaranteed always to exist. What if we erase the labels ∞, and replace them with OUT? Notationally, this would be substituting G with $\text{proj}(G)$, the projection of G onto IN–OUT-valued nodes. This is exactly the information contained in a bar – not which nodes are part of infinite vs. finite subtrees, but which nodes are in vs. out of the alleged bar. Not surprisingly, no infinite path through this tree will be computable from the tree.

This is, of course, just a start. One can erase the ∞ labels from the tree easily, but it is another matter to erase the ∞ information from the model. The difference between those OUT nodes rooting finite subtrees and those rooting infinite ones must be obscured. At the same time, for the tree to remain decidable, we no longer have any choice about this labeling; once an IN–OUT decision has been made, we need to stick with it. The only solution seems to be to use nonstandard integers. By taking the ultrapower of this model using a countably incomplete ultrafilter , we get an elementary extension $\mathcal{M}$ with nonstandard integers. Now we can take the image of G, call it $\text{i}(G)$, in this ultrapower $\mathcal{M}$. There is nothing stopping us from changing finitely many ∞s to OUTs in $\text{i}(G)$, even on standard nodes, and of course changing all of their standard descendants to OUTs also, as this does not affect $\text{proj}(G)$. What it does do is free us up to change all of the descendants on some nonstandard level to INs. If we can do this coherently, then any branch that went through a node formerly labeled ∞ will indeed hit the alleged bar, albeit at a nonstandard level, but no matter.

Arranging the generic models into a Kripke model The way to do this coherently is to arrange all of the various ∞–OUT substitutions as nodes in a Kripke model, in the iterative style described in 22.4.1. The bottom node $\perp$ of this model is assigned the model $V[\text{proj}(G)]$, to use the Kripke terminology from above. The immediate successors of $\perp$ are indexed by the IN–OUT labeled trees H in $\mathcal{M}$ which arise from $\text{i}(G)$ as described, of course as determined in $V[G]$, since a distinction has to be made between standard and nonstandard nodes. Some of these trees H will end up being finite in the sense of $\mathcal{M}$; those yield terminal nodes in the Kripke model. Others will not, and the construction of the Kripke model continues on from there, with children of H's node being determined just the way the children of $\perp$ were.

In the end, the tree proj(G) will have no paths, because whatever a potential path might look like at some node, there will be some future node in which that path is already on a dead end.

Falsifying the variants of FAN As described, this model falsifies D-FAN. To get any of the other separations is just a matter of hiding the tree proj(G) better. For instance, proj(G) could be hidden as a c-tree (the complement of a c-bar), thereby falsifying c-FAN, and if this is done slyly enough, D-FAN will not be disturbed.

It should be mentioned that the model given does not make ¬D-FAN true, but merely makes D-FAN not true. The reason is that the terminal nodes of the Kripke model are dense, so ¬¬D-FAN is true. To get instead ¬D-FAN to hold, one could imagine iterating this construction from all of the terminal nodes. This might involve using settling as described above, although in a more intricate form. The partial order is more complicated, and we would constantly need to pull in generics from outside.

Question 22.34 The construction just sketched uses a classical generic to build a Kripke model in which D-FAN is not true, although ¬¬D-FAN is true. How could this be iterated to get ¬D-FAN to hold?

22.6 Inner Models

Classical set theorists have identified and studied different kinds of inner models: L-like inner models for large cardinals, permutation submodels of forcing extensions, HOD (the hereditarily ordinal definable sets). Here we will examine two, L and permutation models. It would be interesting to see what happens and what could be done with HOD.

Question 22.35 Develop HOD constructively. What properties does it have?

22.6.1 L: Constructibility Meets Constructivism

L is often called the universe of constructible sets, but this kind of constructibility has little to do with constructivism. Still, those notions are compatible, and it is an obvious question what happens with L constructively.

The Development of L

As first noticed by William Powell (in unpublished notes), the definition of L can remain essentially unchanged, modulo the standard way of avoiding the successor-or-limit case split: $L_\alpha = \bigcup_{\beta\in\alpha} def(L_\beta)$, where $def(X)$ is the collection of definable subsets of X. (For a published version see [16], which also contains the rest of this subsection.) One would then like to prove some of the basic theorems about L. It is at this point that the problems start.

The first goals would be to show that L is a model of IZF and that $L^L = L$ (a kind of absoluteness of L, from the ordinals). It plays a role just what the meta-theory is taken to be. Through most of this chapter, the meta-theory is for simplicity taken to be ZFC. If we were to do that when studying L, then one is left with classical L. Instead, what one wants is to develop L constructively, within IZF or something similar. The axioms at issue are the ones around Collection. Classically, over the other ZF axioms, Replacement, Collection, and Reflection are all equivalent. The soft proofs that Reflection implies Collection, which implies Replacement, go through constructively. The reverse implications do not. So it can make a difference which version of IZF you're working in: IZF$_{Rep}$ (with Replacement), IZF (with Collection), and IZF$_{Ref}$ (with Reflection).

Reflection is very strong, and you can usually prove what you want to with Reflection easily. For instance, IZF$_{Ref}$ shows that L satisfies IZF$_{Ref}$ pretty easily. In contrast, in the cases at hand, Replacement seems just not to be enough. The issue is that if X is in L then there is some L_α that X is definable over; there just may not be a canonical such α (like the least). Let's examine the effect that has on a sample case, the proof of Replacement in L. Suppose that, in L, $\forall x \in X\ \exists! y\ \phi(x, y)$. One needs a bounding set, an L_α such that for all $x \in X$ there is some $\beta \in \alpha$ such that the witness y for x is definable over L_β. In order to use Replacement in the meta-theory to bound some such set of βs, one would have to pick out some unique such β, which is not clear is possible. It is easy to see though that Collection suffices: If $\forall x \in X\ \exists y\ \phi(x, y)$, then $\forall x \in X\ \exists \beta_x\ (\exists y \in def(L_{\beta_x})\ \phi(x, y))$; using Collection one can then get a bounding set for the β_xs; of course, this bounding set would have to be turned into an ordinal α, by removing all of the nonordinals and then taking the transitive closure; this α suffices to produce a bounding set L_α.

Checking the other IZF axioms in L is straightforward for most of them. The only challenge is Separation. The standard classical argument for Separation in L goes via Reflection, to which we do not have access. Instead, one has to argue inductively on formulas. The only difficult cases are the quantifiers. For $\exists$, consider $\phi(x) = \exists y\ \psi(x, y)$ and $X \in L$. The collection we want, $A = \{x \in X \mid \phi^L(x)\}$, is in any case a set in V, using Separation there. By the construction of A, $\forall x \in A\ \exists y \in L\ \psi^L(x, y)$. Using arguments similar to the proof of Collection in L, there is a bounding set $Y \in L$ for the ys necessary: $\forall x \in A\ \exists y \in Y\ \psi^L(x, y)$. By induction, we can use Separation to get $\{\langle x, y\rangle \in X \times Y \mid \psi^L(x, y)\}$ in L. The projection of that latter set onto its first components is definable, and hence in L.

The case of $\forall$ is even a bit trickier: $\phi(x) = \forall y\ \psi(x, y)$, $X \in L$. With $\exists$, the goal was clear: Find a set big enough to include enough witnesses. With a universal statement, there are no witnesses. Instead, one considers the different subsets of X obtained by restricting the range of y: For each $C \in L$, let $A_C \in V$ be $\{x \in X \mid \forall y \in C\ \psi^L(x,y)\}$. (We use the notation A_L for the desired set $\{x \in X \mid \phi^L(x)\}$.) Notice that if $D \supseteq C$ then $A_D \subseteq A_C$. Consider, in V, $B = \{A_C \mid C \in L\}$. (Even though C here ranges over a class, B is a set, as a subcollection of the power set of X.) For each $b \in B$ there is an ordinal β such that $b = A_C$ for some $C \in L_\beta$. By Collection, there in an ordinal α including such a β for each $b \in B$. Moreover, since L_α is transitive, not only is any such C in L_α, but also $C \subseteq L_\alpha$. That means $A_{L_\alpha} \subseteq b$ for each $b \in B$. We would like to show that $A_{L_\alpha} = A_L$. For one direction, since $L \supseteq L_\alpha$, $A_L \subseteq A_{L_\alpha}$. In the other direction, suppose $x \in A_{L_\alpha}$. Let $y \in L$. Then $A_{\{y\}} \in B$. Since $A_{L_\alpha} \subseteq A_{\{y\}}$, $x \in A_{\{y\}}$, which means $\psi^L(x,y)$, as desired. Inductively, the set $E = \{\langle x,y\rangle \in X \times L_\alpha \mid \psi^L(x,y)\}$ is in L. From E, A_L is easily definable as $\{x \in X \mid \forall y \in L_\alpha\ \langle x,y\rangle \in E\}$.

Turning to the other goal, that $L^L = L$, the classical argument is that, inductively, α is definable over L_α as the collection of ordinals, so that ORD $\subseteq L$. Then definability over L_α is absolute, so that $L^L_\alpha = L_\alpha$, and you're done. Constructively this falls apart immediately. It is not (always) the case that α is the set of ordinals in L_α. For example, consider something as simple as the two-node Kripke model, with nodes $\bot$ and $\top$. (We do not distinguish notationally in the following between sets in V and their canonical internalizations.) Let $1_\top$ be the set with no members at $\bot$ and 0 as a member at $\top$, so that $\top \Vdash 1_\top = 1$. Notice that $L_{1_\top} = \{0, 1_\top\}$. Let $\alpha = \omega \cup \{1_\top\}$. Then $L_\alpha = L_\omega \cup \{1_\top\}$. The sets definable over L_α include not only α, but also, for any natural numbers $k < n$, those sets x that look like k at $\bot$ (i.e., $\bot \Vdash y \in x$ iff $\bot \Vdash y < k$) and are equal to n at $\top$, which are all ordinals. By way of notation, $x^+ = x \cup \{x\}$. Then L_{α^+} includes all of those funny ordinals just described. So one can certainly define over L_{α^+} the set of ordinals, but that will be a strict superset of α^+. In this case, one can still get α^+ definably over L_{α^+}, but it should be clear that with a more elaborate example even that would not be possible. For instance, throw in some of those funny ordinals into α^+, the ones that look like k and then grow to n, for rather randomly chosen ks and ns, calling the result β. Definably over L_β are all of those funny ordinals, and there is no good way to pick out exactly which got put into β. So definably over L_β we can get a superset of β, but not β itself. So it is not clear that the ordinal β is in L.

That much being understood, there still is a very different construction to show that, under IZF, L in the sense of L is L. The reason is not that L contains all the ordinals, but rather that for every α there is an α^* in L such that $L_\alpha = L_{\alpha^*}$. This latter fact is shown inductively. One works in a set X large enough to include all the β^*s for $\beta \in \alpha$. Within L, one cannot use α as a parameter to pick out exactly the β^*s, because α may not be in L. Instead, one uses L_α as a parameter, which is in L. Take α^* to be the subset of X of all γs such that $def(L_\gamma) \subseteq L_\alpha$, which is in L. By the choice of what goes in to α^*, $L_{\alpha^*} \subseteq L_\alpha$; since each β^* is in α^*, $L_{\alpha^*} \supseteq L_\alpha$.

Transferring Independence Results from V to L

Although the very basics of L carry over from ZF to IZF, the next level of results, AC and GCH, apparently do not. The problem seems to be that it is at best unclear how to do condensation arguments constructively. So ultimately it could be that the study of L is not so interesting constructively. If we do believe that, it would be nice to have at least some theorem or proof giving concrete evidence of such. One possibility is that there are constructions showing that it is not hard to get an arbitrary set into L by coding it into an ordinal (unpublished). The upshot is that constructively L might be a lot like V, even be V itself, regardless of how complicated V is. If it is so easy to get anything into L, even more so to get L to be V by expanding L, arguably that implies that there is no use in studying L for its own sake. Here we will sketch more modest examples tending in the same direction, translating classical independence results over V into constructive independence results over L [17].

The theories we will be considering are those around admissibility, or KP. Some gentle extensions of KP have been considered over the years. For instance, Π_2 Reflection implies the main KP axiom, Σ_1 Bounding. Also, Resolvability (see below for its statement) and Σ_1 Dependent Choice (as theories extending KP) each implies Π_2 Reflection. Over L, all of these theories are equivalent, but not in general: any implication that does not follow from what was just said does not hold, as can be demonstrated by forcing the appropriate reals and sets of reals (for details see [17]).

Our interest is of course in the constructive version of KP, namely IKP. This has not been studied much – the literature might well be limited to [1], [13], and [17]. Perhaps this is because of CZF, which is a significant extension of IKP yet has the same proof-theoretic strength. Still, IKP is a perfectly coherent theory, and one can ask about independence results over it. Trivially independence results over KP are also independence results over IKP. Of interest here is to transfer the cited independence proofs over KP to independence proofs over IKP in L. The original proofs are based on generic reals and sets of such; the technique to effect the transfer is to code the generics as ordinals.

We describe the simplest example of such, that Resolvability does not imply Σ_1 DC. Resolvability is the axiom that the universe is the union of the range of a Δ_1 definable function on the ordinals. Although it is not usually described this way, the model of Resolvability + $\neg\Sigma_1$ DC in [17] is very well known, being the standard permutation model for the failure of AC. That is, take countably many mutually generic Cohen reals $G_i, i \in \omega$, and the set G – not the sequence! – of these reals. If the ground model is L, then the permutation model $L(G)$ is the extension of L by each of the generics as well as the set G. The resolution is $L_\alpha[G]$ as α runs through the ordinals, and the failure of AC is actually a failure of Σ_1 DC.

The task is to get a model of IKP in which G and its members are in some sense reflected in ordinals, which then end up being in L. The model will be a Kripke model with underlying partial order $2^{<\omega}$. Given $\sigma \in 2^{<\omega}$, let 1_σ be the set which is forced to be 1 (i.e., $\{\emptyset\}$, as usual) by any τ extending or incompatible with σ, whereas any initial segment of σ does not force anything into 1_σ. A real (i.e., infinite binary sequence) R induces a branch B through the Kripke partial order. A first approximation to B is $\{1_\sigma \mid \sigma \text{ is an initial segment of } R\}$. Indeed, at any node σ along R that's exactly what we want. But if τ is incompatible with σ then at node τ the set just listed becomes $\{1\}$ and all information about B has been lost. So instead, at node τ, B is taken to consist of those sets 1_ρ such that, for all j between the lengths of τ and ρ, $\rho(j) = R(j)$. That is, the tail end of R, the part beyond the length of τ, is the path through the partial order (or rather the p.o.'s reflection in the ordinals) taken by B at τ. Including also 0 into B makes B an ordinal. Let B_i be the branch so induced by G_i. Then B_i is definable over L_{B_i}. Furthermore, letting β be the transitive closure of $\{B_i \mid i \in \omega\} \cup \{1_\sigma \mid \sigma \in 2^{<\omega}\}$, we get that $\{B_i \mid i \in \omega\}$ is definable over L_β. The ultimate model is gotten by iterating definability, starting with L_β, ω_1^{CK}-many times, which is enough to get IKP. Resolvability is, in essence, given by the very definition of the model, as an iteration along some ordinals of definability. The failure of Choice in $L[G]$ translates to the failure of Σ_1 DC here. It is easy to see that everything happening here stays in L.

The other models in [17] are of Π_2 Reflection plus the failures of Σ_1 DC and Resolvability, and of IKP plus the failure of Π_2 reflection, all in L of course.

22.6.2 Permutation Models

The first permutation model within constructive mathematics was Krol's [15], also described in [10] and modified in [28]. Here we will describe the permutation model from [19], because it has the expository advantage of being simpler.

The motivating question behind [19], from [6], was whether the Cauchy reals are Cauchy complete. Classically the reals can be construed many different ways, for instance as Cauchy sequences, or equivalence classes of such, or Dedekind cuts, which are all equivalent. But whatever you take them to be, the move from the rationals to the reals is a closure operator, in that it is idempotent: If you take Dedekind cuts of rationals, then a Dedekind cut of those things can be converted to an equivalent Dedekind cut of rationals. This all breaks down constructively. We have already seen (in Section 22.2.2) that the Dedekind and Cauchy reals might differ, and that a Cauchy sequence might have no modulus of convergence. It is not hard to imagine other related scenarios, such as a Cauchy sequence of Cauchy sequences which have no moduli of convergence, which itself is not equivalent to any Cauchy sequence of rationals. (One could try to diagonalize through the given sequence of sequences, but you don't know how far out to go in each one.) It was the purpose of [19] to go through all such possibilities, culminating in what seemed like the most difficult, the motivating question: a Cauchy sequence R_i, with modulus of convergence, of equivalence classes of pairs, each consisting of a Cauchy sequence r_{ij} of rationals and a modulus for it, which is inequivalent to any Cauchy sequence of rationals.

The limiting factor here is choosing a representative from each equivalence class. If one had that, then one could simply pick R_i within $\epsilon/2$ of the limit, and then within R_i pick r_{ij} within $\epsilon/2$ of its limit, to get a rational within ϵ of the ultimate limit. So the issue really is picking a representative of each equivalence class. This has a similar feel to the standard model of $\neg$AC, in which there is a set of reals with no canonical choice of member. So one is naturally led to think of permutation models. At the same time, there has to be more than that, since classically one can choose a representative from each equivalence class of Cauchy sequences. (For instance, use a case split as to whether the limit is rational or irrational; in the former case, a constant sequence will do, whereas in the latter, there is a unique rational of the form $1/n$ closest to the intended limit.) The solution, or at least a solution, is to take a permutation submodel of a topological mode. Generically, take a Cauchy sequence G_i of Cauchy sequences g_{ij} (the latter consisting of rational numbers, and each having the same fixed modulus of convergence). Then take the submodel of those sets X with support a finite set I of the indices i, meaning that arbitrary changes of the g_{ij}s for $i \notin I$ which do not affect any limits do not change X. For instance, replacing a sequence $\langle g_{ij} \rangle_j$ with its equivalence class $[\langle g_{ij} \rangle_j]$ of Cauchy sequences with the same limit yields a set with support $\emptyset$. Similarly, replacing the sequence $\langle G_i \rangle_i$ with $\langle [G_i] \rangle_i$ produces a set with null support, which is exactly the example we want. No Cauchy sequence with finite support I can have a limit equal to the limit of $\langle [G_i] \rangle_i$, because the former depends only on $\{G_i \mid i \in I\}$ whereas the latter does not.

22.7 A Final Example

As a culmination of this material, we present a construction that uses many of the ideas and techniques developed here.

Although the purpose here is not that the result is of particular importance, but rather that the construction itself is hopefully appealing, still there can be no model without it being a model of something, and so we give the content background. Building on work of [3] and Fred Richman (personal communication), [5] defines the following.

Definition 22.36 (Almost Cauchy) A sequence (x_n) in a metric space is *almost Cauchy* if for all $\epsilon > 0$ and strictly increasing $g : \mathbb{N} \to \mathbb{N}$ there is an $N \in \mathbb{N}$ such that, for all $n \geq N$, the diameter of $\{x_{g(n)}, x_{g(n)+1}, \ldots, x_{g(n+1)}\}$ is less than ϵ (meaning that, whenever $g(n) \leq i, j \leq g(n+1), d(x_i, x_j) < \epsilon$).

Easily, every Cauchy sequence is almost Cauchy. Whether every almost Cauchy sequence is Cauchy is an interesting question. It is implied by BD-N [3], but does not itself imply BD-N nor follow from set theory (in particular, IZF) alone [24].[6] Various similar formulations of this property are also discussed in [5], along with their implications among each other, some being equivalent to almost Cauchyness, others being merely implied by almost Cauchyness, all of which are mutually equivalent under Countable Choice. The obvious question is whether the equivalence holds in the absence of CC. The version we will discuss is the apparently weakest form, which in a display of unimaginativeness will be called here $<$-almost Cauchyness. This version considers not the entire subsequence indexed from $g(n)$ to $g(n+1)$, but rather just the endpoints.

Definition 22.37 ($<$-almost Cauchy) A sequence (x_n) in a metric space is *$<$-almost Cauchy* if for all $\epsilon > 0$ and strictly increasing $g : \mathbb{N} \to \mathbb{N}$ there is an $N \in \mathbb{N}$ such that, for all $n \geq N, d(x_{g(n)}, x_{g(n+1)}) < \epsilon$.

Theorem 22.38 *$<$-almost Cauchyness does not imply almost Cauchyness.*

We will provide a weak counter-example; that is, a model with a $<$-almost Cauchy sequence, which is not almost Cauchy in a weak sense: the sequence will not be almost Cauchy at the bottom node $\perp$ in a Kripke, or Kripke-like, model, but it will still be not not almost Cauchy. To get a strong separation, the construction would have to be iterated. The sequence will be called (G_n^+), because it will be the non-negative part max(0, G_n) of a generic sequence (G_n). The generic (G_n) will be based on a ground model sequence (x_n).

Outline For (G_n^+) to be a counter-example to almost Cauchyness, there must be some positive ϵ witnessing that. All positive numbers are roughly the same, so the witnessing ϵ will be 1. Also, the function g witnessing that (G_n^+) is not almost Cauchy for $\epsilon = 1$ must grow fast, because there would have to be enough room in the interval from $g(n)$ to $g(n+1)$ for the G^+-sequence to change significantly, without there being an h picking out indices in that interval that would contradict $<$-almost Cauchyness. It will turn out that the exponential function $g(n) = 2^n$ suffices. Then, for that choice of ϵ and g, to show that the diameter property of almost Cauchyness does not eventually hold ("$\exists N \; \forall n \geq N$"), it would seem at first blush that we

[6] Reference [24] discusses a slight variant of almost Cauchyness, called there partially Cauchyness, which was shown in [5] to be equivalent with almost Cauchyness.

must have unboundedly many examples n of its failure ("$\forall N\ \exists n \geq N$"), which would be difficult. It's easier to get one counter-example k, as long as that k is nonstandard. That is, we will get a nonstandard k such that the diameter of the set $\{x_{g(k)}, \ldots, x_{g(k+1)}\}$ is 1, which property will be approximately inherited by (G_n), in that the diameter of $\{G_{g(k)}, \ldots, G_{g(k+1)}\}$ will be approximately 1. The ultimate model will be a Kripke model based on a tree of height 1, with only standard integers at $\perp$ and including nonstandard integers at the successor nodes. Then at $\perp$ there will be no N that works, because any N at $\perp$ must be standard, and k will provide a counter-example larger than N at at least one of the successor nodes.

Thus far the discussion has been about falsifying almost Cauchyness. We must also validate $<$-almost Cauchyness. At this point, it's helpful to over-simplify matters and temporarily set our sights lower. Whereas $<$-almost Cauchyness is about all $\epsilon > 0$ and all increasing functions g, we will be concerned with only ϵs and gs from the ground model V, as opposed to ones from the nonstandard extension M. This helps because the counter-example k we want is nonstandard and the ground model has less finesse when it comes to manipulating nonstandard elements. In the end, though, we still have to account for elements from M. The idea for this is that the sequence (x_n) will be $<$-almost Cauchy in the limited sense of ground model ϵs and gs, and the derivation of (G_n^+) and its ambient model will be exactly so that the ground model ϵs and gs are the only ones that will be relevant.

To summarize, our intermediate goal will be a sequence (x_n) in a nonstandard model M, upon which (G_n) and (G_n^+) will later be based, that satisfies the $<$-almost Cauchyness condition for ϵ and g from V, and seems to violate almost Cauchyness, in that for some i, j, and k non-standard with $2^k \leq i < j \leq 2^{k+1}$, $d(x_i, x_j) = 1$.

Lemma 22.39 *There is an elementary embedding $j : L \to M$ such that M has nonstandard natural numbers k, a, and b with $2^k \leq a, a < b, b \leq 2^{k+1}$, $b - a$ nonstandard, and, for every increasing $h \in L$ from $\mathbb{N}$ to $\mathbb{N}$, if $j(h)(n)$ or $j(h)(n+1)$ is strictly between a and b, then $j(h)(n+1) - j(h)(n)$ is standard. (Notation: the image $j(h)$ of h in M will typically be referred to as simply h.)*

Once we have such a model, we would take (x_n) as being 0 outside of the interval $[a, b]$, as increasing from 0 at x_a up to 1 at the midpoint $(a+b)/2$ of $[a, b]$ by equal-sized, infinitesimally small steps $2/(b-a)$, and then back down again to 0 at x_b by the same-sized steps. This gives the promised violation of almost Cauchyness, with i as a and j as $(a+b)/2$, with the added benefit that each step from x_n to x_{n+1} is only an infinitesimal change. It helps validate $<$-almost Cauchyness, because if neither $h(n)$ nor $h(n+1)$ is in (a, b) then $|x_{h(n)} - x_{h(n+1)}| = 0$, and if one is then $|x_{h(n)} - x_{h(n+1)}|$ is infinitesimal, hence less than any standard $\epsilon > 0$.

Proof Work in L.

We can package up all the things we want into a type Ty, which will be realized in M. The free variables will be k, a, and b. For starters, include in Ty the set of axioms $2^k \leq a, a < b, b \leq 2^{k+1}$, and the infinite collection $b-a > 0, b-a > 1, b-a > 2$, etc. This is easily seen to be consistent by compactness.

For increasing $h : \mathbb{N} \to \mathbb{N}$ from L, let ϕ_h be the formula stating that h misses the interval (a, b) entirely:

$$\phi_h := \exists n\ h(n) \leq a \wedge b \leq h(n+1).$$

There will indeed be such instances true in our model, for example the function 2^n. This will not always be possible though, for example for the identity function. For that case, with β (for "bound") any standard natural number, let

$$\psi_{h,\beta} := \forall n(\text{if } h(n) \text{ or } h(n+1) \text{is in the interval } (a,b) \text{ then } h(n+1)-h(n) < \beta).$$

For each $h \in L$ we would like to validate either ϕ_h or $\psi_{h,\beta}$ for some standard β. There is not always a natural choice between those options. For instance, consider the function that enumerates all the elements between 2^n and 2^{n+1} whenever n is even and omits that interval entirely when n is odd. Compare that with the same kind of function but interchanging the parity of n. One of those functions will fall on the ϕ side, the other on the ψ, and the choice of which is arbitrary.

Finally, include in Ty a maximal consistent set of formulas of the form ϕ_h and $\psi_{h,\beta}$ (β standard).

Notice that Ty has the size of the continuum $\mathfrak{c}$. There are guaranteed to be realizers a, b, and k for Ty in a model M whenever every consistent set of formulas of size at most $\mathfrak{c}$ is realized in M. This property is called the $\mathfrak{c}^+$-saturation of M. It is a result of introductory model theory that an ultrapower is κ^+-saturated if the ultrafilter used to develop it is κ-regular, and that ZFC proves the existence of κ-regular ultrafilters for all κ (see, e.g., [4, Section 4.3, especially 4.3.5 and 4.3.14]). Pick such an M, a, b, and k, with elementary embedding $j : L \to M$.

We would like that for each $h \in L$ either ϕ_h or some $\psi_{h,\beta}$ is in Ty, whereas all we know so far is that Ty is maximal consistent. Toward this end, consider some such h. We will show that either ϕ_h or $\psi_{h,\beta}$ is consistent with Ty, and so by maximality will then be in Ty.

Say that n is *relevant* if $h(n)$ or $h(n+1)$ is in the interval (a, b), the point being that only relevant ns affect the truth of either ϕ_h or $\psi_{h,\beta}$. If no ns are relevant then ϕ_h is true in M and hence consistent with Ty. Else consider the (necessarily nonempty) set of *gaps*, which are the numbers $h(n+1) - h(n)$ whenever n is relevant. If all gaps are standard, then, since the set of gaps is definable in M, they have a standard bound, say β. Immediately, $\psi_{h,\beta}$ is true in M and so consistent with Ty. Else there is a nonstandard gap. Now suppose there is a nonstandard gap with both endpoints (i.e., $h(n)$ and $h(n+1)$) in the interval $[a, b]$: $a \leq h(n)$ and

$h(n+1) \leq b$. In this case, a and b could be reinterpreted to be $h(n)$ and $h(n+1)$ respectively. That interpretation would still satisfy Ty, and make ϕ_h true, hence consistent with Ty. If instead there is no such nonstandard gap, then all of the nonstandard gaps have to include one of the endpoints a or b: either $h(n) < a$ (and $a < h(n+1) < b$, in order for n to be relevant), or $b < h(n+1)$ (and similarly $a < h(n) < b$). For each of those two possibilities there is at most one such n. To summarize the current hypotheses, there is at least one and at most two nonstandard gaps, each of which contains one of the endpoints a or b. We will show what to do when there are two. This will call for a two-step procedure. If instead there is only one, then only one of those steps need be done.

Toward this end, let $h(n) < a < h(n+1) < b$. If $h(n+1) - a$ is non standard, then reinterpret b as $h(n+1)$ (and leave a as itself). Under this interpretation Ty remains true and ϕ_h becomes true, showing that ϕ_h is consistent with Ty. Else $h(n+1) - a$ is standard. Keep that in mind. Now consider the other nonstandard gap, $a < h(m) < b < h(m+1)$. If $b - h(m)$ is nonstandard, reinterpret a as $h(m)$. As above, Ty remains true and ϕ_h becomes true, so is consistent with Ty. Else $b - h(m)$ is standard, as is $h(n+1) - a$. In this case, reinterpret a as $h(n+1)$ and b as $h(m)$. Since the original distance between a and b was nonstandard, and both were changed by only a standard amount, the distance between their reinterpretations is still nonstandard. Hence all of Ty remains true. Furthermore, what had been the only nonstandard gaps are no longer gaps, as m and n are no longer relevant. So all gaps (if any) are standard, and we can argue as above to get either ϕ_h or some $\psi_{h,\beta}$ consistent with Ty. □

Now that we have a and b as desired, consider (x_n) as defined above. If we were to try to use (x_n) as our counter-example to almost Cauchyness, $<$-almost Cauchyness would fall flat on its face, because a and b, hence their mid-point, are readily definable from it. So we must hide a and b, by fuzzing (x_n) up. That calls for a topological model. The idea is to replace each value x_n for n between a and b with a small interval. Even with the other improvements we will make later, if that's all we do, a and b will still be definable as, roughly, the first and last places where the sequence is nonzero. (It will be case that the "first" will be either of two consecutive integers, and that between those two a can be picked out, using parity; similarly for "last" and b.) So we consider such a space based on any pair i, j with $a \leq i < j \leq b$. (It is an unfortunate overuse of notation that sometimes j is a natural number index, as it is here, and sometimes an elementary embedding from V to M. We leave it to the context to disambiguate these two uses.)

The construction of the model $\mathfrak{A}$ Work for the moment in M. For any i and j with $a \leq i < j \leq b$, let $(x_n^{i,j})$ be the sequence which is 0 outside of the open interval

(i, j), starting at i grows by $2/(b-a)$ at each step until the midpoint of $[i, j]$, then shrinks by the same amount until it hits 0 again at j. (Please note, we take $x_i^{i,j}$ and $x_j^{i,j}$ to be 0. If $j-i$ is odd, then the midpoint of $[i, j]$ is only a half-integer, in which case we take both neighboring integers to be "the" midpoint, and the maximum value of $x_n^{i,j}$ occurs at both of them.) Let $T^{i,j}$ consist of all sequences (y_n) which are 0 outside of $[i, j]$ and for which $|x_n^{i,j} - y_n| < 2/(b-a)$ for n in $[i, j]$. We place a topology upon $T^{i,j}$. A basic open set is given by restricting each y_n to an open interval. Let $(G_n^{i,j})$ be the generic sequence. From $(G_n^{i,j})$, the integers i and j can be defined as the first and last places where $(G_n^{i,j})$ is not 0. This could be used to falsify $<$-almost Cauchyness. Hence we need to erase some information from $(G_n^{i,j})$.

Let $G_n^{i,j,+}$ be max$(0, G_n^{i,j})$. For any n for which $x_n^{i,j} \geq 2/(b-a)$, there is no difference between $G_n^{i,j}$ and $G_n^{i,j,+}$. The only differences are in the ith and jth positions. The importance of the sequence $(G_n^{i,j,+})$ relative to $(G_n^{i,j})$, as we will see, is that there is an open set forcing $G_i^{i,j,+}$ and $G_j^{i,j,+}$ to be 0, namely the open set determined by restricting $G_i^{i,j}$ and $G_j^{i,j}$ to the interval $(-2/(b-a), 0)$. Another way to look at it is that $(G_n^{i,j,+})$ is the generic for a topological model built using a different topological space. This is by analogy with classical forcing: if $H \in V[G]$, where G is generic for a complete Boolean algebra $\mathcal{B}$, then there is a subalgebra of $\mathcal{B}$ for which H is generic. The same is true for forcing with complete Heyting algebras, as shown in (hopefully) forthcoming work. In the case at hand, the subalgebra $S^{i,j}$ of (the opens of) $T^{i,j}$ consists of those open sets for which the ith and jth components (independently of each other) include either all of $(-2/(b-a), 0)$ or none of it. Alternatively, $(G_n^{i,j,+})$ can be described as being generic for the quotient space $U^{i,j}$ of $T^{i,j}$ in which, at the ith and jth components, the interval $(-2/(b-a), 0)$ is collapsed to a single point.

The model $\mathfrak{A}$ is a Kripke model, or perhaps a Kripke-inspired model, since some of the nodes will carry a topological instead of a classical semantics. The Kripke partial order of $\mathfrak{A}$ consists of a bottom node $\bot$ and successor nodes indexed by i and j whenever $a \leq i < j \leq b$. The model at the node (i, j) has several different yet ultimately equivalent descriptions, corresponding to different ways of viewing $G_n^{i,j,+}$. One could work within $M^{T^{i,j}}$, define $G_n^{i,j,+}$ as max$(0, G_n^{i,j})$, and take for the model $L[(G_n^{i,j,+})]$. (This is well-defined because, briefly, under IZF, L can be defined much as it is classically, and forms a model of IZF. For details, see [16]. The extension to $L[X]$ is straightforward.) Alternatively, the model can be described as the Heyting-valued model $M^{S^{i,j}}$, or the topological model $M^{U^{i,j}}$.

It remains to define $\mathfrak{A}$ at $\bot$ (and the transition functions). The example at $\bot$ of a $<$-almost Cauchy sequence which is not almost Cauchy is a sequence we will call (G_n^+). The intention is that (G_n^+) becomes at node (i, j) (under application of the transition function) $(G_n^{i,j,+})$. Since the various $(G_n^{i,j,+})$'s for the various choices of i and j differ at some nonstandard entries, $\perp$ should contain only standard natural numbers n; we could then define G_n when n is standard as 0, since each such entry $G_n^{i,j,+}$ is 0. An obvious guess would then be to take some kind of full model. That doesn't work though. Recall that $(G_n^{i,j,+})$ was designed to be $<$-almost Cauchy in the limited sense of ground model ϵs and gs. The full model, in contrast, has ϵs that look like 0 at $\perp$ and become infinitesimal later, and gs that look like they come from V at $\perp$ but then do whatever later. We need to tie the model at $\perp$ more closely to V.

Toward this end, the idea is to take $L[(G_n^+)]$ at $\perp$. The issue with that is the transition function. We already have the elementary embedding j from L to M, and we know that (G_n^+) is supposed to go to $(G_n^{i,j,+})$. What is not clear is that j can be extended to accommodate (G_n^+). The problem is that a set in L could have different definitions that are extensionally the same but intensionally different. The way that would express itself in this context is that these different definitions would give rise to different sets in the model we're building. So the transition function cannot operate directly on the sets in L; rather, we have to go one level deeper, and analyze the definitions or the names of the sets. This is done inductively on the ordinals in V. It can be viewed as being carried out in V; it bears mention that each ordinal in V has a canonical image in the full model (with $L[(G_n^+)]$ at $\perp$), and so this induction can be done within the full model, save for the final step of taking the union over the ordinals of V.

For the limit case, suppose λ is a limit ordinal, and for each $\alpha < \lambda$ we have a structure $\mathfrak{A}_\alpha$ (for the language with $\in$ and a symbol for (G_n^+)). (If this work is viewed as being done in V, then it should be stated that these $\mathfrak{A}_\alpha$s are Kripke structures on the partial order from above. If rather this definition is being done in the full model, then the fact of its yielding a Kripke structure on that p.o. is inherited from the full model.) Also suppose there are canonical embeddings from $\mathfrak{A}_\alpha$ to $\mathfrak{A}_\beta$ (for $\alpha < \beta$). Then $\mathfrak{A}_\lambda$ is their union. Given such a structure $\mathfrak{A}_\alpha$, $\mathfrak{A}_{\alpha+1}$ is Def($\mathfrak{A}_\alpha$), with embeddings $\sigma \mapsto \{x \mid x \in \sigma\}$. (For the details of how to define the structure Def(X) from X, including $\in$ and $=$, see [17].) The structure at $\perp$ is then $\bigcup_{\alpha \in ORD} \mathfrak{A}_\alpha$. (Notice that this is a proper subclass of $L[(G_n^+)]$ as evaluated in the full model, since $\mathfrak{A}$ is a union over the ordinals of V, whereas $L[(G_n^+)]$, as evaluated in the full model, is a union over the ordinals of the full model.)

As for the transition function, any set is given by its definition over some $\mathfrak{A}_\alpha$; that means that it's of the form $\{x \in \mathfrak{A}_\alpha \mid \mathfrak{A}_\alpha \models \phi(x,p)\}$, where p is some parameter. The idea is that such a set, which can be thought of informally as a definition over $L_\alpha[(G_n^+)]$, gets sent to the same definition over $L_{j(\alpha)}[(G_n^+)]$, where the parameter gets carried along, inductively. More formally, let $(i,j)_\beta$ be the model at node (i, j) restricted to L-rank β. Since the model at (i, j) had several equivalent descriptions, the model so restricted has several equivalent descriptions. It can be taken to be $(L_\beta[(G_n^{i,j,+})])^{M^{T^{i,j}}}$, or $M_\beta^{S^{i,j}}$, or $M_\beta^{U^{i,j}}$. (By the notation M_β, we mean the sets, in M, of L-rank less than β, not V-rank. In order for these three descriptions to be equivalent, we may have to require that β be sufficiently closed, such as set-primitively recursively closed.) We define the action of the transition function f to the node (i, j) to take the set $\{x \in \mathfrak{A}_\alpha \mid \mathfrak{A}_\alpha \models \phi(x, p)\}$ to $\{x \in (i, j)_{j(\alpha)} \mid (i, j)_{j(\alpha)} \models \phi(x, f(p))\}$. (Perhaps it should be mentioned that this latter set is to be interpreted in some topological model M^T, as opposed to being a classical set.)

Lemma 22.40 $\bot \models$ *IZF.*

Proof This proof will be only sketched.

At successor nodes are topological models, which have already been seen to satisfy IZF. So much of the checking can be confined to $\bot$.

The empty set is definable over $\mathfrak{A}_0$, and the natural numbers over $\mathfrak{A}_\omega$. The union and pair of set(s) are easily definable from the set(s). Extensionality uses extensionality in the ambient V, and $\in$-induction uses Foundation in V. The proof of Power Set is like the proof of Power Set in L. Reflection holds via reflection in V; that is, letting α be such that $V_\alpha \prec_{\Sigma_n} V$, anything Σ_n expressible about $\mathfrak{A}$ is true in $\mathfrak{A}$ iff it's true in $\mathfrak{A}_\alpha$. Separation follows easily, since, given any formula ϕ, one must only let n be such that "$\bot \models \phi$" is Σ_n definable, and α as above, and then work definably over $\mathfrak{A}_\alpha$. □

The next pair of lemmas finishes the proof.

Lemma 22.41 $\bot \not\models (G_n^+)$ *is almost Cauchy.*

Proof At $\bot$, let $\epsilon = 1$ and $g(n) = 2^n$. For any $N \in \mathbb{N}$ at $\bot$, at the node (a, b) k provides a counter-example. □

Lemma 22.42 $\bot \models (G_n^+)$ *is* $<$*-almost Cauchy.*

Proof This is easy to see at any successor node, by letting N be $b + 1$, as the sequence (G_n^+) is 0 from then on out. Hence we must check only what happens at $\bot$.

If $\bot \Vdash \epsilon > 0$ then we can assume that ϵ is standard. (That's because, for instance, there is a rational between 0 and ϵ, and the naturals at $\bot$ are standard. That means the rational is standard too, and having $<$-almost Cauchyness for this rational implies the property for ϵ.) Let $\bot \Vdash g : \mathbb{N} \to \mathbb{N}$, and take N to be 0. If n is standard, that provides a confirming instance of $<$-almost Cauchyness, as $d(G_{g(n)}, G_{g(n+1)}) = 0$. For n nonstandard at some successor node, the entire construction has been so that

n is also a confirming instance, as long as $f(g)$ (f the transition function) gets interpreted at that node as $j(h)$, the image in M of some function h from L. Showing that will be a two-step process: first that $f(g)$, which prima facie is from $L[(G_n^{i,j,+})]$, is actually in M, and second, using that $f(g) \in M$, that $f(g)$ is the image $j(h)$ of something from L^V. In fact, we will see $f(g)$ is $j(g)$.

The model at node (i,j), $L[(G_n^{i,j,+})]$, is a submodel of the topological model $M^{T^{i,j}}$, built using the topological space $T^{i,j}$ over the ground model M. Hence any assertion about $L[(G_n^{i,j,+})]$ is seen within M to be forced by some open set from $T^{i,j}$. If any nonempty open set forces a value k for $f(g)(n)$, then the entire space $T^{i,j}$ forces that same value. The reason for this is a standard argument, leveraging the discreteness of $\mathbb{N}$ against the connectedness of $\mathbb{R}$, as used for example to show that in the topological model over $\mathbb{R}$, all functions from $\mathbb{N}$ to $\mathbb{N}$ are ground model functions. To repeat the argument here, $T^{i,j}$ is a cube in $(j-i+1)$-dimensional Euclidean space, and so is connected. The sets $[\![f(g)(n)=k]\!]$, as k ranges over the natural numbers of M, form a disjoint open cover of $T^{i,j}$. By connectedness, one of those open sets is the whole space, and the others are empty. Therefore, at node (i,j), $f(g)$ is forced to be a function from M.

Next we would like to see that the interpretation of $f(g)$ as a function from M does not depend on the choice of node (i,j). Toward this end, we will start by showing that, for $j<b$, the interpretation of $f(g)$ is the same at node (i,j) as at node $(i,j+1)$. An immediate hurdle is that the spaces $T^{i,j}$ and $T^{i,j+1}$ cannot so readily be compared, because they are of different dimension, the latter having a $j+1$st component and the former not. The role that that fact plays is that $T^{i,j} \Vdash G_{j+1}=0$, whereas $T^{i,j+1} \nVdash G_{j+1}=0$. However, that discrepancy can be repaired, by letting $\mathcal{O}$ be $T^{i,j+1}$ with the last component restricted to the open interval $(-2/(b-a),0)$; then $\mathcal{O} \Vdash G_{j+1}=0$.

Regarding the other components of (G_n^+), for s between i and j inclusive, the sth components of $T^{i,j}$ and $T^{i,j+1}$ have a nonempty intersection. Therefore we can view $T^{i,j} \cap T^{i,j+1}$ as an open subset of $T^{i,j}$, and, alternatively, as an open subset of $T^{i,j+1}$, the last component of which is $(-2/(b-a),0)$. Keep in mind that we are actually not working in the models $M^{T^{i,j}}$, but rather the submodels sufficient for $(G_n^{i,j,+})$, namely $M^{S^{i,j}}$ or $M^{U^{i,j}}$. Working beneath the open set $T^{i,j} \cap T^{i,j+1}$, whether in $S^{i,j}$ or $S^{i,j+1}$ the Heyting algebras are isomorphic, and so the models are identical. Returning to $f(g)$, when g is given by its definition $\{x \in \mathfrak{A}_\alpha \mid \mathfrak{A}_\alpha \models \phi(x,p)\}$, $f(g)$ at either node is given by $\{x \in (i,*)_{j(\alpha)} \mid (i,*)_{j(\alpha)} \models \phi(x,f(p))\}$, where $*$ is interpreted as either j or $j+1$. Between the nodes (i,j) and $(i,j+1)$, $j(\alpha)$ is interpreted the same, $(G_n^{i,j,+})$ and $(G_n^{i,j+1,+})$ are interpreted the same, and, inductively, p is interpreted the same, so $f(g)$ is interpreted the same too. Hence $f(g)$ as a function in M is the same at (i,j) as at $(i,j+1)$. Similarly, when changing

i by 1. Iterating this procedure of changing i or j by a value of 1, $f(g)$ is seen to be the same function from M at all nodes.

Finally, consider some node $(i, i+1)$. The space $T^{i,i+1}$ consists of two intervals of radius $2/(b-a)$ around 0, at i and $i+1$. Restrict each interval to $(-2/(b-a), 0)$; call that open set $\mathcal{O}$. Then $\mathcal{O}$ is a minimal nonempty open set in the Heyting algebra of $S^{i,i+1}$, and $\mathcal{O}$ forces (G_n^+) to be the constant 0 sequence, and hence the model $L[(G_n^+)]$ at that node to be L in the sense of M, i.e., M. Furthermore, by taking (G_n^+) to be the constant 0 sequence, j can be extended to a function from the universe at $\perp$ to M; namely, $j(\{x \in \mathfrak{A}_\alpha \mid \mathfrak{A}_\alpha \models \phi(x,p)\}) = \{x \in (L_{j(\alpha)})^M \mid (L_{j(\alpha)})^M \models \phi(x, j(p))\}$. It is easy to show inductively that $f(p) = j(p)$. In particular, $f(g) = j(g)$, as was to be shown. □

References

[1] Avigad, J. 2000. Interpreting classical theories in constructive ones. *J. Symbol. Logic*, **65**(4), 1785–1812.

[2] Berger, J. 2006. The logical strength of the uniform continuity theorem. Pages 35–39 of: *Logical Approaches to Computational Barriers, Proceedings of CiE 2006*, Beckman, A., Berger, U., Löwe, B., and Tucker,j. v. (eds.). LNCS 3988. Berlin: Springer.

[3] Berger, J., Bridges, D., Palmgren, E. 2012. Double sequences, almost cauchyness and BD-N. *Logic J. IGPL*, **20**(1), 349–354.

[4] Chang, C. C., and Keisler, H. J. 2013. *Model Theory*. New York: Dover Publications.

[5] Diener, H., and Lubarsky. R. 2018. Notions of Cauchyness and metastability. Pages 140–153 Pages 140–153 of: Artemov, S. N., and Nerode, A. (eds.),Proceedings of LFCS 2018, Lecture Notes in Computer Science 10703 Berlin:Springer.

[6] Escardo, M. H. and Simpson, A. K. 2001. A universal characterization of the closed Euclidean interval (extended abstract). Sixteenth Annual IEEE Symposium on Logic in Computer Science, PP. 115–125. Washington, DC: IEEE Computer Society.

[7] Fourman, M. P. and Hyland, J. M. 1979. Sheaf models for analysis. Pages 280–301 of: Fourman, M. P., Mulvey, C. J., and Scott, D. S. (eds)., *Applications of Sheaves*, Lecture Notes in Mathematics vol. 753. Berlin, Heidelberg, New York: Springer Verlag.

[8] Gambino, N. 2006. Heyting-valued interpretations for Constructive Set Theory. *Ann. Pure Appl. Logic*, **137**, 164–188.

[9] Grayson, R. J. 1979. Heyting-valued models for intuitionistic set theory. Pages 402–414 of: Fourman, M. P., Mulvey, C. J., and Scott, D. S. (eds.), *Applications*

of Sheaves, Lecture Notes in Mathematics, vol. 753 Berlin, Heidelberg, New York: Springer-Verlag.

[10] Grayson, R. J. 1981. Concepts of general topology in constructive mathematics and in sheaves. *Ann. Math. Logic*, **20**(1), 1–41.

[11] Grayson, R. J. 1984. Heyting-valued semantics. Pages 181–208 of: Lolli, G., Longo, G., and Marcja, A. (eds.), *Logic Colloquium '82*. Studies in Logic and the Foundations of Mathematics, vol. 112. Amsterdam, New York, Oxford: North-Holland.

[12] Hendtlass, M., and Lubarsky, R. 2016. Separating fragments of WLEM, LPO, and MP. *J. Symbol. Logic*, **81**(4), 1315–1343. https://doi.org/10.1017/jsl.2016.38

[13] Iemhoff, R., and Passmann R. 2021. Logics of intuitionistic Kripke–Platek set theory. *Ann. Pure Appl. Logic*, **172**(10), [103014]. https://doi.org/10.1016/j.apal.2021.103014

[14] Julian, W., and Richman, F. 1984. A uniformly continuous function on $[0, 1]$ that is everywhere different from its infimum. *Pac. J. Math.*, **111**(2), 333–340.

[15] Krol, M. D. 1978. A topological model for intuitionistic analysis with Kripke's scheme. *Zeitschr. Math. Logik Grundl. Math.*, **24**(25–30), 427–436.

[16] Lubarsky, R. 1993. Intuitionistic L. Pages 555–571 of: Crossley, J. N. et al. (eds.), *Logical Methods in Computer Science: The Nerode Conference*. Basel: Birkhauser.

[17] Lubarsky, R. 2002. IKP and friends. *J. Symbol. Logic*, **67**, 1295–1322.

[18] Lubarsky, R. 2005. Independence results around constructive ZF. *Ann. Pure Appl. Logic*, **132**(2–3), 209–225.

[19] Lubarsky, R. 2007. On the Cauchy completeness of the constructive Cauchy reals. *Math. Logic. Q.*, **53**(4–5), 396–414.

[20] Lubarsky, R. 2010. Geometric spaces with no points (and addendum). *J. Logic Anal.*, **2**(6), 1–10. https://doi.org/10.4115/jla2010.2.6

[21] Lubarsky, R. 2009. Topological forcing semantics with settling. Pages 820–830 of: Artemov, S. N., and Nerode, A. (eds.), *Proceedings of LFCS '09*, Lecture Notes in Computer Science. Berlin: Springer. Also *Ann. Pure Appl. Logic*, **163**, 820–830 (2012). https://doi.org/10.1016/j.apal.2011.09.014

[22] Lubarsky, R. 2013. On the failure of BD-N. *J. Symbol. Logic*, **78**(1), 39–56.

[23] Lubarsky, R. 2018. Separating the fan theorem and its weakenings II. Pages 242–255 of: Artemov, S., and Nerode, A. (eds.), *Proceedings of LFCS 2018*, Lecture Notes in Computer Science 10703 Cham: Springer. Also *J. Symbol. Logic* (in the press).

[24] Lubarsky, R., and Diener, H. 2013. Principles weaker than BD-N. *J. Symbol. Logic*, **78**(3), 873–885.

[25] Lubarsky, R., and Diener, H. 2013. Separating the fan theorem and its weakenings. Pages 280–295 of: Artemov, S. N., and Nerode, A. (eds.), *Proceedings of LFCS '13*, Lecture Notes in Computer Science 7734. Cham: Springer. Also *J. Symbol. Logic*, 79(3), 792–813 (2014). https://doi.org/10.1017/jsl.2014.9

[26] Lubarsky, R., and Rathjen, M. 2007. On the constructive Dedekind reals. Pages 349–362 of: Artemov, S. N., and Nerode, A. (eds.), *Proceedings of LFCS '07*, Lecture Notes in Computer Science 4514. Berlin: Springer. Also *Logic Anal.*, **1**(2), 131–152 (2008).

[27] Mostowski, A. W. 1948. Proofs of non-deducibility in intuitionistic functional calculus. *JSL, 13*, **13**, 204–207.

[28] Scedrov, A. 1981. Consistency and independence results in intuitionistic set theory. Pages 54–86 of: Richman, F. (ed.), *Constructive Mathematics, Proceedings of the Las Cruces Meeting, 1980*, Lecture Notes in Mathematics, vol. 873. Berlin: Springer.

[29] Scott, D. S. 1968. Extending the topological interpretation to intuitionistic analysis I. *Compos. Math.*, **20**, 194–210.

[30] Scott, D. S. 1970. Extending the topological interpretation to intuitionistic analysis II. Pages 235–255 of: Myhill, J., Kino, A., and Vesley, R. E. (eds.), *Intuitionism and Proof Theory*, Amsterdam: North Holland.

[31] Stone, M. H. 1937. Topological representations of distributive lattices and Brouwerian logics. *Casopis Pro Pestvovani Mathematiky a Fysiki Cast Matematicka*, **67**, 1–25.

[32] Tarski, A. 1938. Der Aussagenkalkül and die Topologie. *Fundam. Math.*, **31**, 103–134.

[33] Troelstra, A. S. and van Dalen, D. 1988. *Constructivism in Mathematics – An Introduction, Vol. II*. Studies in Logic and the Foundations of Mathematics, vol. 123. Amsterdam, New York, Oxford: North-Holland.

23

An Introduction to Constructive Reverse Mathematics

Hajime Ishihara

23.1 Introduction

The Friedman–Simpson program [39], called *classical reverse mathematics* here, has adopted, as its base formal system, a subsystem $\mathbf{RCA}_0$ of classical second-order arithmetic with a restricted induction schema and a very weak set existence axiom. Its main question is 'What set existence axioms are needed to prove the theorems of ordinary mathematics?', and many theorems have been classified by set existence axioms of various strengths over $\mathbf{RCA}_0$.

Bishop-style constructive mathematics [7, 9, 10] is called *neutral constructivism*, since it is based on intuitionistic logic and is a core of the varieties of mathematics in the sense that it could be extended not only to Brouwer's intuitionism [43, Section 1.1.6] and constructive recursive mathematics [43, Section 1.1.5], but also to classical mathematics practised by most mathematicians today. This suggests to us a possibility of generalising classical reverse mathematics, by using intuitionistic logic, so as to classify theorems in Brouwer's intuitionism or in constructive recursive mathematics which are inconsistent with classical mathematics and therefore classical reverse mathematics has never dealt with them.

The purpose of *constructive reverse mathematics* is to classify various theorems in Brouwer's intuitionism, constructive recursive mathematics and classical mathematics by logical principles, function (or set) existence axioms and their combinations. However, since Bishop-style constructive mathematics is informal and assumes some set existence axioms, that is, the axioms of countable and dependent choice, we could not adopt it as a base for constructive reverse mathematics which is precisely comparable with classical reverse mathematics.

Constructive reverse mathematics, based on formal systems that allow us to compare precisely the results there with those in classical reverse mathematics, was initiated in [21], and has been developed over recent years [6, 25, 27]. The aim of this chapter is to present an introduction to constructive reverse mathematics in this formal style.

In Section 23.2, we briefly review our base formal system $\mathbf{M}_3^\omega$ for constructive reverse mathematics, introduced in [25, 27]. Then we deal with Brouwer's continuity principle in its simple form by investigating various notions of weak continuity and relationships among them, based on [25], in Section 23.3. In Section 23.4, we deal with compactness properties of the Cantor space, such as the Heine–Borel theorem, the Cantor intersection theorem, the Bolzano–Weierstraß theorem, and sequential compactness, based on [21]. In Section 23.5, we consider the monotone completeness theorem based on [27]. We conclude the chapter with a figure showing implications among principles and axioms, and briefly reviewing other approaches to constructive reverse mathematics, including reverse mathematics in constructive set theory and in Bishop-style constructive mathematics.

23.2 A Formal System

In this section, we briefly review our base formal system $\mathbf{M}_3^\omega$ for constructive reverse mathematics.

23.2.1 Description of $\mathbf{M}_3^\omega$

The set $\mathcal{T}$ of *finite-type* symbols is inductively generated by the clauses:

$$0 \in \mathcal{T}, \quad \sigma, \tau \in \mathcal{T} \Rightarrow (\sigma \to \tau) \in \mathcal{T}.$$

We write 1 for $(0 \to 0)$, and $n+1$ for $(n \to 0)$; parentheses will be omitted following the convention of association to the right.

The *language* $\mathcal{L}(\mathbf{M}_3^\omega)$ of $\mathbf{M}_3^\omega$ is based on the standard language of many-sorted first-order predicate logic containing $\wedge, \vee, \to, \bot, \forall$, and $\exists$ as primitive logical operators. The language contains countably infinite variables of type σ, a binary predicate $=_\sigma$ for equality at type σ, and an application operator $\mathrm{Ap}^{\sigma,\tau}$ for each $\sigma, \tau \in \mathcal{T}$. Furthermore, the language contains the following constants for each $\rho, \sigma, \tau \in \mathcal{T}$. Here we write $c \in \sigma$ to denote that c is a constant of type σ. The combinators

$$\begin{aligned} \boldsymbol{k}^{\sigma,\tau} &\in \sigma \to \tau \to \sigma, \\ \boldsymbol{s}^{\rho,\sigma,\tau} &\in (\rho \to \sigma \to \tau) \to (\rho \to \sigma) \to \rho \to \tau, \end{aligned}$$

and

$$\begin{array}{lll} 0 \in 0, & S \in 1, & \mathrm{prd} \in 1; \\ + \in 0 \to 1, & \dot{-} \in 0 \to 1, & \times \in 0 \to 1; \\ \exp \in 0 \to 1, & \boldsymbol{\mu} \in (0 \to 1) \to 1. & \end{array}$$

Terms of $\mathbf{M}_3^\omega$ are inductively defined by

(i) variables and constants of type σ are terms of type σ;
(ii) if t is a term of type $\sigma \to \tau$ and s is a term of type σ, then $\mathrm{Ap}^{\sigma,\tau}(t, s)$ is a term of type τ.

Formulae of $\mathbf{M}_3^\omega$ are inductively defined by

(i) $\perp$ is a formula;
(ii) atomic formulae $t =_\sigma s$ are formulae, where t and s are terms of type σ;
(iii) if φ and ψ are formulae, then $(\varphi \wedge \psi)$, $(\varphi \vee \psi)$ and $(\varphi \to \psi)$ are formulae;
(iv) if φ is a formula and x^σ is a variable of type σ, then $(\forall x^\sigma \varphi)$ and $(\exists x^\sigma \varphi)$ are formulae.

Arithmetical formulae are formulae containing only atomic formulae of the form $t =_0 s$ and quantifiers for variables of type 0; *quantifier-free formulae* are arithmetical formulae without quantifiers. A Σ_0^0 formula (and Π_0^0 formula) is a quantifier free formula. A Σ_{n+1}^0 *formula* is either a Π_n^0 formula or a formula of the form $\exists x^0 \varphi$, where φ is a Σ_{n+1}^0 formula; a Π_{n+1}^0 *formula* is either a Σ_n^0 formula or a formula of the form $\forall x^0 \varphi$, where φ is a Π_{n+1}^0 formula.

Notational conventions In the following, the lower-case letters i, j, k, l, m, n are metavariables for variables of type 0, the Greek lower-case letters $\alpha, \beta, \gamma, \delta, \kappa$ for variables of type 1, and the upper-case letters F, G, H, M for variables of type 2. We use r, s, t for terms, and often provide them with a subscript and a type superscript: s_0, t_1^σ, etc. The application $\mathrm{Ap}^{\sigma,\tau}(t, s)$ will be written as (ts), and parentheses will be omitted following the convention of association to the left; we sometimes write $t(s_1, \ldots, s_n)$ for $ts_1 \ldots s_n$. To indicate variables occurring free in a term, we use square brackets: $s[x]$, $t[F, \alpha]$, etc. Type superscripts of variables, constants, and terms will be mostly omitted.

For terms t and s of type 0, the inequalities $t \leq s$ and $t < s$ are defined by $t \leq s \equiv t \dot{-} s = 0$ and $t < s \equiv St \dot{-} s = 0$, respectively, and we write $\forall k \leq t\varphi$ and $\exists k \leq t\varphi$ for $\forall k(k \leq t \to \varphi)$ and $\exists k(k \leq t \wedge \varphi)$, respectively.

Axioms of $\mathbf{M}_3^\omega$. The logical basis of $\mathbf{M}_3^\omega$ is many-sorted intuitionistic predicate logic **IQC** with equality.

(i) For equality at all types we assume

$$x = x, \qquad x = y \to y = x, \qquad x = y \wedge y = z \to x = z,$$
$$y = z \to xy = xz, \qquad x = y \to xz = yz.$$

(ii) The defining equations for the combinators:

$$\boldsymbol{k}xy = x, \qquad \boldsymbol{s}xyz = xz(yz).$$

(iii) The defining equations for the constants:

$$
\begin{aligned}
\neg(0 = Sn), &\qquad Sm = Sn \to m = n;\\
\mathrm{prd}0 = 0, &\qquad \mathrm{prd}(Sn) = n;\\
m + 0 = m, &\qquad m + (Sn) = S(m+n);\\
m \dot{-} 0 = m, &\qquad m \dot{-} (Sn) = \mathrm{prd}(m \dot{-} n);\\
m \times 0 = 0, &\qquad m \times (Sn) = (m \times n) + m;\\
\exp(m, 0) = S0, &\qquad \exp(m, Sn) = \exp(m, n) \times m.
\end{aligned}
$$

$$
\begin{aligned}
\forall z^{0\to 1}\forall n[(\forall k \le n(znk \ne 0) \to \boldsymbol{\mu} zn = Sn) \wedge\\
(\exists k \le n(znk = 0) \to zn(\boldsymbol{\mu} zn) = 0 \wedge \forall k < \boldsymbol{\mu} zn(znk \ne 0))].
\end{aligned}
$$

(iv) The induction schema for quantifier-free formulae QF-IND, that is, the induction schema:

IND $\qquad \varphi(0) \wedge \forall n(\varphi(n) \to \varphi(Sn)) \to \forall n \varphi(n)$

for quantifier-free formulae φ.

As in [44, Section 9.1.8], we have the following in $\mathbf{M}_3^\omega$.

Lemma 23.1 *For each term $t^\tau[x^\sigma]$, there exists a term $\lambda x^\sigma.t[x]$ of type $\sigma \to \tau$ such that $(\lambda x.t[x])s =_\tau t[s]$ for all term s^σ.*

We write **0** and **1** for $\lambda n.0$ and $\lambda n.S0$, respectively. For terms t and s of type 1, we write $t \# s$ for $\exists n(tn \ne sn)$, and $t \sim_1 s$ for $\neg t \# s$, that is, $\forall n(tn = sn)$.

There is no difficulty at all to establish basic theorems of arithmetic (on natural numbers) in $\mathbf{M}_3^\omega$, as in [43, Section 3.2].

23.2.2 Elementary Functions

Within $\mathbf{M}_3^\omega$, we are able to define functions belonging to the class $\mathcal{M}^3$ (and hence $\mathcal{E}^3$) of elementary functions in [38, Section 5.1], such as the absolute difference $|m - n| = (m \dot{-} n) + (n \dot{-} m)$ and the bounded maximum operator

$$
\max_{k \le n}\{\alpha(m,k) = 0\} = \begin{cases} \text{the greatest } k \le n \text{ with } \alpha(m,k) = 0, \text{ if it exists,}\\ 0, \text{ otherwise.}\end{cases}
$$

Especially, we can define the surjective *pairing function* $\mathrm{p}^{0\to 1}$, and its *projections* p_0^1 and p_1^1 such that

$$
\mathrm{p}_0(\mathrm{p}mn) = m, \qquad \mathrm{p}_1(\mathrm{p}mn) = n, \qquad \mathrm{p}(\mathrm{p}_0 n)(\mathrm{p}_1 n) = n;
$$

see [43, 3.1]. Using the pairing function and the projections, we are able to code a finite sequence $k_0, \ldots, k_{n-1}$ by a number

$$\langle k_0, \ldots, k_{n-1} \rangle,$$

and to define the *decoding function* $(a)_i$ computing the ith element k_i of finite sequence coded by $a = \langle k_0, \ldots, k_{n-1} \rangle$, the *length function* $|\text{-}|$, and the *concatenation function* $*$; see [43, 3.1] for details. Moreover, we are able to construct a *bounding function* $\mathsf{b}^{0 \to 1}$ such that

$$\langle k_0, \ldots, k_{n-1} \rangle \leq \mathsf{b}(m, n),$$

whenever $k_i \leq m$ for all $i < n$. Using the bounding function, we are able to define the *finite sum* and the *finite product* operators:

$$\sum_{k<n} \alpha(m, k), \qquad \prod_{k<n} \alpha(m, k),$$

and to construct a code

$$\overline{\alpha}n = \langle \alpha 0, \ldots, \alpha(n-1) \rangle$$

of initial segment of α as a function in n.

23.2.3 Sets and Setoids

A *set* A of type σ is given by a formula $\varphi(x^\sigma)$, and we then write $t \in A$ for $\varphi(t)$. Let $\mathbb{N}$ and $\{0, 1\}$ denote the sets of type 0 given by formulae

$$n =_0 n, \qquad n =_0 0 \vee n =_0 1;$$

and let $\mathbb{N}^*$ and $\{0, 1\}^*$ denote the sets of type 0 given by formulae

$$a =_0 a, \qquad \forall k < |a| [(a)_k \in \{0, 1\}];$$

let $\mathbb{N}^{\mathbb{N}}$ and $\{0, 1\}^{\mathbb{N}}$ denote that sets of type 1 given by formulae

$$\alpha =_1 \alpha, \qquad \forall k [\alpha k \in \{0, 1\}].$$

In the following, a, b, c are supposed to range over the set N^* or the set $\{0, 1\}^*$. For a set A, we write $\forall x \in A \varphi$ and $\exists x \in A \varphi$ for $\forall x (x \in A \to \varphi)$ and $\exists x (x \in A \wedge \varphi)$. respectively.

A *subset* B of a set A of type σ is given by a formula $\psi(x^\sigma)$, and we then write $t \in B$ for $t \in A \wedge \psi(t)$; a *detachable* subset D of a set A of type σ is given by a formula of the form $s(x^\sigma) =_0 0$, where s is a term of type $\sigma \to 0$. Note that a subset B of a set A of type σ is a set of type σ given by the formula $\varphi(x) \wedge \psi(x)$, $\{0, 1\}$ is a detachable subset of $\mathbb{N}$, and $\{0, 1\}^*$ is a detachable subset of $\mathbb{N}^*$. A *function* from a set A of type σ into a set B of type τ is given by a term $t^{\sigma \to \tau}$ such that $\forall x \in A (tx \in B)$; a *sequence* in a set A is a function from $\mathbb{N}$ into A.

A *setoid* $(A, =_A)$ is given by a set A of type σ and a formula $=_A(x^\sigma, y^\sigma)$ such that

(i) $\forall x \in A =_A(x, x)$,
(ii) $\forall xy \in A[=_A(x, y) \to =_A(y, x)]$,
(iii) $\forall xyz \in A[=_A(x, y) \wedge =_A(y, z) \to =_A(x, z)]$;

we then write $t =_A s$ for $=_A(t, s)$. Note that a set A of type σ induces a trivial setoid $(A, =_\sigma)$. A *mapping* of a setoid $(A, =_A)$ into a setoid $(B, =_B)$ is a function f from the set A into the set B such that

$$\forall xy \in A(x =_A y \to f(x) =_B f(y)).$$

Using the pairing function, we are able to code integers and (dyadic) rationals into natural numbers; define the setoid of integers $(\mathbb{Z}, =_\mathbb{Z})$ and the setoid of (dyadic) rationals $(\mathbb{Q}, =_\mathbb{Q})$; develop the elementary theory of arithmetical operations as mappings on $(\mathbb{Z}, =_\mathbb{Z})$ and $(\mathbb{Q}, =_\mathbb{Q})$; and prove their basic properties such as commutativity, associativity, etc., in $\mathbf{M}_3^\omega$.

Notes

The base system $\mathbf{RCA}_0$ for classical reverse mathematics is a subsystem of second-order arithmetic with the induction schema IND restricted to Σ_1^0 formulae and the *recursive comprehension axiom*:

$$\Delta_1^0\text{-CA} \qquad \forall n(\varphi(n) \leftrightarrow \psi(n)) \to \exists X \forall n(n \in X \leftrightarrow \varphi(n)),$$

where φ is a Σ_1^0 formula and ψ is a Π_1^0 formula, and X does not occur free in φ; see [39, Section II.1]. Kohlenbach [32] introduced the system $\mathbf{RCA}_0^\omega$ which is a conservative finite-type extension of $\mathbf{RCA}_0^2$, an inessential extension of $\mathbf{RCA}_0$. Within $\mathbf{RCA}_0$ and $\mathbf{RCA}_0^\omega$, we are able to define functions belonging to the class $\mathcal{E}^\omega$ of primitive recursive functions; see [38, Section 4.1]. The subsystem $\mathbf{RCA}_0^*$ of $\mathbf{RCA}_0$, in which definable functions are elementary functions, is obtained from $\mathbf{RCA}_0$ by adding a function symbol for exponentiation and its axioms, and restricting the induction scheme IND to quantifier-free formulae; see [39, Section X.4]. The systems $\mathbf{RCA}_0$, $\mathbf{RCA}_0^\omega$, and $\mathbf{RCA}_0^*$ are all classical, that is, use classical logic.

Berger et al. [6] introduced the system $\mathbf{EL}_0$ for constructive reverse mathematics, which is a subsystem of (intuitionistic) elementary analysis $\mathbf{EL}$ [43, Section 3.6] obtained by restricting the induction schema IND to quantifier-free formulae. Note that $\mathbf{EL}$ and hence $\mathbf{EL}_0$ have, as an axiom, the *quantifier-free axiom of choice* QF-AC_{00}, and that we are able to define primitive recursive functions by the recursor in $\mathbf{EL}_0$.

The system $\mathbf{M}_3^\omega$ is a subsystem of *intuitionistic finite-type arithmetic* $\mathbf{HA}^\omega$ in [44, 9.1] and a finite-type extension of $\mathbf{EL}_{\mathrm{ELEM}}$ in [21]. It is a variant of the intuitionistic Grzegorczyk arithmetic $\mathrm{G}_3\mathrm{A}_i^\omega$ of level 3 in all finite types introduced in [33, Section 3.5]. Note that $\mathbf{M}_3^\omega$ has an equality predicate $=_\sigma$ and the equality axioms at all type σ, whereas $\mathrm{G}_3\mathrm{A}_i^\omega$ has only the equality predicate $=_0$ and the equality axioms at type 0, and the quantifier-free extensionality rule $\mathrm{QF\text{-}ER}$; see [33, Section 3.12].

The axiom for the bounded minimum operator $\boldsymbol{\mu}$ can be replaced by the universal closures of formulae:

$$k \le n \wedge znk = 0 \to zn(\boldsymbol{\mu}zn) = 0, \qquad k < \boldsymbol{\mu}zn \to znk \neq 0,$$
$$(zn(\boldsymbol{\mu}zn) = 0 \wedge \boldsymbol{\mu}zn \le n) \vee \boldsymbol{\mu}zn = Sn;$$

see [33, Section 3.5]. In the previous papers [25, 27], the axiom of extensional equality for the type 1

$$\mathrm{EXT}_1 \qquad \forall\alpha\beta[\alpha \sim_1 \beta \to \alpha =_1 \beta]$$

was assumed in $\mathbf{M}_3^\omega$, whereas we drop it here.

Since the system $\mathbf{M}_3^\omega$ is a subsystem of $\mathbf{RCA}_0^\omega$, the results in the following sections hold also in $\mathbf{RCA}_0^\omega$. A comparison among systems for (constructive) reverse mathematics is dealt with in Chapter 24 in this Handbook.

23.3 Continuity Properties

In this section, we deal with a simple form of *Brouwer's continuity principle*:

$$\mathrm{BCP} \qquad \forall\alpha\exists n\forall\beta[\forall k \le n(\alpha k = \beta k) \to F\alpha = F\beta]$$

for all $F \in \mathbb{N}^\mathbb{N} \to \mathbb{N}$, where $\mathbb{N}^\mathbb{N} \to \mathbb{N}$ denotes the set of type 2 given by a formula

$$\forall\alpha\beta[\alpha \sim_1 \beta \to F\alpha = F\beta].$$

We do so by investigating various notions of weak continuity and relationships among them.

23.3.1 Non-discontinuity and Strong Extensionality

We start with introducing the notions of a discontinuous function and a non-discontinuous function.

Definition 23.2 A sequence $\alpha_{(\text{-})}^{0\to 1}$ *converges to* $\alpha \in \mathbb{N}^\mathbb{N}$ *with a modulus* γ if

$$\forall k\forall n[\gamma k \le n \to \forall i \le k(\alpha_n i = \alpha i)],$$

and we then write $\alpha_n \underset{\gamma}{\longrightarrow} \alpha$.

Definition 23.3 A function $F \in \mathbb{N}^{\mathbb{N}} \to \mathbb{N}$ is

- *discontinuous* if there exist α, $\alpha_{(\text{-})}$ and γ such that

$$\alpha_n \underset{\gamma}{\longrightarrow} \alpha \text{ and } \forall n(F\alpha_n \neq F\alpha);$$

- *non-discontinuous at* $\alpha \in \mathbb{N}^{\mathbb{N}}$ if for each $\alpha_{(\text{-})}$ and γ,

$$\alpha_n \underset{\gamma}{\longrightarrow} \alpha \text{ implies } \neg\forall n(F\alpha_n \neq F\alpha);$$

- *non-discontinuous* if it is non-discontinuous at each $\alpha \in \mathbb{N}^{\mathbb{N}}$.

Then we deal with discontinuity, non-discontinuity, and the following version of comprehension:

$\forall$-CA_1 $\quad \exists F \forall \alpha[F(\alpha) \neq 0 \leftrightarrow \neg \alpha \# \mathbf{0}]$.

We may regard the following theorem, characterising the existence of a discontinuous function, as a constructive version of Grilliot's trick [15]; see also [13, Section 7] and [32, Section 3].

Theorem 23.4 *The following are equivalent over* $\mathbf{M}_3^{\omega}$.

(i) *There exists a discontinuous* $F \in \mathbb{N}^{\mathbb{N}} \to \mathbb{N}$.
(ii) $\forall$-CA_1.

Proof See the proof of [25, Theorem 5], and note that F in the proof is in $\mathbb{N}^{\mathbb{N}} \to \mathbb{N}$. □

The following corollary characterises a very weak continuity principle.

Corollary 23.5 *The following are equivalent over* $\mathbf{M}_3^{\omega}$.

(i) *Every* $F \in \mathbb{N}^{\mathbb{N}} \to \mathbb{N}$ *is non-discontinuous.*
(ii) $\neg\forall$-CA_1.

We introduce the notions of a strongly extensional function and a strongly non-discontinuous function.

Definition 23.6 A function $F \in \mathbb{N}^{\mathbb{N}} \to \mathbb{N}$ is

- *strongly extensional at* $\alpha \in \mathbb{N}^{\mathbb{N}}$ if

$$\forall \beta(F\alpha \neq F\beta \to \alpha \# \beta);$$

- *strongly extensional* if it is strongly extensional at each $\alpha \in \mathbb{N}^{\mathbb{N}}$.

Definition 23.7 A function $F \in \mathbb{N}^{\mathbb{N}} \to \mathbb{N}$ is

- *strongly non-discontinuous at* $\alpha \in \mathbb{N}^{\mathbb{N}}$ if for each $\alpha_{(\text{-})}$ and γ,

$$\alpha_n \underset{\gamma}{\longrightarrow} \alpha \text{ implies } \exists n(F\alpha_n = F\alpha);$$

- *strongly non-discontinuous* if it is strongly non-discontinuous at each $\alpha \in \mathbb{N}^{\mathbb{N}}$.

The following proposition shows that strong extensionality fills the gap between non-discontinuity and strong non-discontinuity.

Proposition 23.8 *Let $F \in \mathbb{N}^{\mathbb{N}} \to \mathbb{N}$, and let $\alpha \in \mathbb{N}^{\mathbb{N}}$. Then the following are equivalent over* $\mathbf{M}_3^{\omega}$.

(i) *F is strongly non-discontinuous at α,*
(ii) *F is non-discontinuous at α and strongly extensional at α.*

Proof See the proof of [25, Proposition 10], and note that it works with the assumption that $F \in \mathbb{N}^{\mathbb{N}} \to \mathbb{N}$. □

Then we deal with strong extensionality and the following version of weak Markov's principle.

WMP$^-$ $\forall \alpha \beta \forall G[\forall \gamma[(G\gamma = 0 \to \neg\neg\alpha \# \gamma) \wedge (G\gamma \neq 0 \to \neg\neg\gamma \# \beta)] \to \alpha \# \beta]$

Theorem 23.9 *The following are equivalent over* $\mathbf{M}_3^{\omega}$.

(i) *Every $F \in \mathbb{N}^{\mathbb{N}} \to \mathbb{N}$ is strongly extensional.*
(ii) *Every non-discontinuous $F \in \mathbb{N}^{\mathbb{N}} \to \mathbb{N}$ is strongly non-discontinuous.*
(iii) WMP$^-$.

Proof See the proof of [25, Theorem 11], and note that S in the proof is in $\mathbb{N}^{\mathbb{N}} \to \mathbb{N}$. □

We introduce the notion of a quasi-continuous function which is a strongly non-discontinuous function with modulus.

Definition 23.10 A function $F \in \mathbb{N}^{\mathbb{N}} \to \mathbb{N}$ is

- *quasi-continuous at* $\alpha \in \mathbb{N}^{\mathbb{N}}$ if there exists a modulus M such that for each $\alpha_{(\text{-})}$ and γ,

$$\alpha_n \underset{\gamma}{\longrightarrow} \alpha \text{ implies } F\alpha_{M(\alpha_{(\text{-})},\gamma)} = F\alpha;$$

- *quasi-continuous* if it is quasi-continuous at each $\alpha \in \mathbb{N}^{\mathbb{N}}$.

We consider the following schema of quantifier-free choice for types 1 and 0:

QF-AC$_{10}$ $\qquad \forall \alpha \exists n \varphi(\alpha, n) \to \exists F \forall \alpha \varphi(\alpha, F\alpha)$

for quantifier-free φ. The following proposition shows that QF-AC$_{10}$ suffices to construct a modulus for strongly non-discontinuous function.

Proposition 23.11 *Let $F \in \mathbb{N}^{\mathbb{N}} \to \mathbb{N}$, and let $\alpha \in \mathbb{N}^{\mathbb{N}}$. Then the following are equivalent over* $\mathrm{M}_3^\omega + \mathrm{QF\text{-}AC}_{10}$.

(i) *F is strongly non-discontinuous at α,*
(ii) *F is quasi-continuous at α.*

Proof See the proof of [25, Proposition 14]. □

23.3.2 Sequential Continuity and Continuity

In this subsection, we first introduce the notion of a sequentially continuous function.

Definition 23.12 A function $F \in \mathbb{N}^{\mathbb{N}} \to \mathbb{N}$ is

- *sequentially continuous at* $\alpha \in \mathbb{N}^{\mathbb{N}}$ if there exists a modulus M such that for each $\alpha_{(\text{-})}$ and γ,

$$\alpha_n \underset{\gamma}{\longrightarrow} \alpha \text{ implies } \forall m(M(\alpha_{(\text{-})}, \gamma) \leq m \to F\alpha_m = F\alpha);$$

- *sequentially continuous* if it is sequentially continuous at each $\alpha \in \mathbb{N}^{\mathbb{N}}$.

The following proposition shows that the notions of a quasi-continuous function and a sequentially continuous function are equivalent.

Proposition 23.13 *Let $F \in \mathbb{N}^{\mathbb{N}} \to \mathbb{N}$, and let $\alpha \in \mathbb{N}^{\mathbb{N}}$. Then the following are equivalent over* M_3^ω.

(i) *F is quasi-continuous at α,*
(ii) *F is sequentially continuous at α.*

Proof See the proof of [25, Proposition 17]. □

Finally, we introduce the usual notion of a continuous function.

Definition 23.14 A function $F \in \mathbb{N}^{\mathbb{N}} \to \mathbb{N}$ is

- *continuous at* $\alpha \in \mathbb{N}^{\mathbb{N}}$ if there exists n such that

$$\forall \beta[\forall k \leq n(\alpha k = \beta k) \to F\alpha = F\beta];$$

- *continuous* if it is continuous at each $\alpha \in \mathbb{N}^{\mathbb{N}}$.

Proposition 23.15 *If $F \in \mathbb{N}^{\mathbb{N}} \to \mathbb{N}$ is continuous at $\alpha \in \mathbb{N}^{\mathbb{N}}$, then F is sequentially continuous at α.*

Proof See the proof of [25, Proposition 20]. □

We deal with the gap between sequential continuity and continuity, and a boundedness principle.

Definition 23.16 A sequence κ is *pseudo-bounded* if there exists a modulus H such that

$$\forall\gamma\forall m(H\gamma \leq m \to \kappa(\gamma m) < m);$$

bounded if there exists n such that $\kappa m < n$ for all m.

We consider the following boundedness principle.

BD-N$^-$ Every pseudo-bounded κ is bounded,

or formally

$$\forall\kappa\forall H[\forall\gamma\forall m(H\gamma \leq m \to \kappa(\gamma m) < m) \to \exists n\forall m(\kappa m < n)].$$

The following theorem shows that BD-N$^-$ fills the gap between sequential continuity and continuity.

Theorem 23.17 *The following are equivalent over* M_3^ω.

(i) *Every sequentially continuous* $F \in \mathbb{N}^\mathbb{N} \to \mathbb{N}$ *is continuous.*
(ii) BD-N$^-$.

Proof See the proof of [25, Theorem 24]. □

Now we conclude the section with the following theorem characterising Brouwer's continuity principle.

Theorem 23.18 *The following are equivalent over* $\mathrm{M}_3^\omega + \mathrm{QF\text{-}AC}_{10}$.

(i) BCP.
(ii) $\neg\forall\text{-CA}_1 + \mathrm{WMP}^- + \text{BD-N}^-$.

Proof Use Corollary 23.5, Theorem 23.9, Proposition 23.11, and Theorem 23.17. □

Notes

Kohlenbach [32] considered a version of comprehension:

$\exists$-CA$_1$ $\exists F\forall\alpha[F(\alpha) \neq 0 \leftrightarrow \alpha \# \mathbf{0}]$

(there $\exists$-CA$_1$ is called $\exists^2$) and various discontinuities in the classical system $\mathbf{RCA}_0^\omega$. Note that $\exists$-CA$_1$ is equivalent to $\forall$-CA$_1$ together with Markov's principle MP (or Σ_1^0-DNE):

MP $\forall\alpha[\neg\neg\alpha \# \mathbf{0} \to \alpha \# \mathbf{0}]$

which is an instance of *double negation elimination*:

DNE $\neg\neg\varphi \to \varphi;$

see [14] and [43, 4.5] for Markov's principle.

Mandelkern first considered the weak Markov principle (WMP) under the names ASP and WLPE in [35] and [36], respectively. Originally, it was stated as follows:

Every pseudo-positive real number is positive,

where a real number a is pseudo-positive if $\forall x \in \mathbb{R}[\neg\neg(0 < x) \vee \neg\neg(x < a)]$, and positive if $0 < a$. Here we consider the form

WMP $\qquad \forall\alpha\beta[\forall\gamma(\neg\neg\alpha \# \gamma \vee \neg\neg\gamma \# \beta) \to \alpha \# \beta]$,

which is equivalent to its special case in [14, Proposition 6]:

$$\forall\alpha[\forall\gamma(\neg\neg\mathbf{0} \# \gamma \vee \neg\neg\gamma \# \alpha) \to \alpha \# \mathbf{0}].$$

Note that WMP easily follows from Markov's principle MP. In [20], it was shown that WMP is derivable from Church's thesis for disjunctions, and it is known that WMP is a principle which holds in intuitionistic, constructive recursive and classical mathematics. However, Kohlenbach [31] showed, using monotone realisability, that WMP is underivable in $\mathbf{E}\text{-}\mathbf{HA}^{\omega} + \mathrm{AC} + \mathrm{CA}^{\omega}_{\neg}$; see also [33, Section 7.1]. Note that WMP^{-} above is weaker than and, in the presence of arithmetical AC_{10}, equivalent to WMP.

We may think of a notion of sequential continuity without modulus: a function F is *sequentially continuous at* $\alpha \in \mathbb{N}^{\mathbb{N}}$ *without modulus* if for each $\alpha_{(\text{-})}$ and γ,

$$\alpha_n \underset{\gamma}{\longrightarrow} \alpha \text{ implies } \exists n \forall m(n \leq m \to F\alpha_m = F\alpha).$$

Then a proof similar to the proof of Proposition 23.13 shows that for each $F \in \mathbb{N}^{\mathbb{N}} \to \mathbb{N}$ and $\alpha \in \mathbb{N}^{\mathbb{N}}$, the following are equivalent over $\mathbf{M}^{\omega}_3$.

(i) F is strongly nondiscontinuous at α,
(ii) F is sequentially continuous at α without modulus.

Therefore the notions of sequential continuity with modulus and without modulus are equivalent over $\mathbf{M}^{\omega}_3 + \mathrm{QF}\text{-}\mathrm{AC}_{10}$, by Proposition 23.11.

The following boundedness principle was first recognised in [19]:

BD-N Every countable pseudo-bounded set of natural numbers is bounded,

where a set A of natural numbers is pseudo-bounded if $\lim_{n\to\infty} a_n/n = 0$ for all sequences $(a_n)_{n\in\mathbb{N}}$ in A, and bounded if there exist n such that $a < n$ for all $a \in A$. Then, in [28, Lemma 3], it was shown that a set A of natural numbers is pseudo-bounded if and only if for each sequence $(a_n)_{n\in\mathbb{N}}$ in A, $a_n < n$ for all sufficiently large n. Here we consider the following form of BD-N by identifying a countable set of natural numbers with the range of a sequence κ of natural numbers:

BD-N $\qquad \forall\kappa[\forall\gamma\exists n\forall m(n \leq m \to \kappa(\gamma m) < m) \to \exists n\forall m(\kappa m < n)]$.

We know that BD-N is equivalent to various theorems in analysis (see, for example, [24, 28]), and that BD-N is a principle, like WMP, which holds in intuitionistic, constructive recursive and classical mathematics. However, Lietz and Streicher [34] showed, using realisability, that BD-N is underivable in $\mathbf{E}\text{-}\mathbf{HA}^{\omega}+\mathrm{AC}$, by invoking the fact that $\mathbf{E}\text{-}\mathbf{HA}^{\omega}+\mathrm{AC}+\mathrm{BCP}$ is inconsistent [44, 9.6.10]. Note that BD-N$^-$ above is weaker than and, in the presence of arithmetical AC_{10}, equivalent to BD-N.

23.4 Compactness Properties

In this section, we deal with compactness properties of the Cantor space $\{0,1\}^{\mathbb{N}}$, such as the Heine–Borel theorem, the Cantor intersection theorem, and the Bolzano–Weierstraß theorem.

23.4.1 The Heine–Borel Theorem

Here, we consider the *Heine–Borel theorem* and *Brouwer's fan theorem.*

Definition 23.19 A detachable subset B of $\{0,1\}^*$ is a *detachable bar* if for each $\alpha \in \{0,1\}^{\mathbb{N}}$ there exists $n \in \mathbb{N}$ such that $\overline{\alpha}n \in B$. A detachable bar B is *uniform* if there exists $n \in \mathbb{N}$ such that $\exists k \leq n(\overline{\alpha}k \in B)$ for all $\alpha \in \{0,1\}^{\mathbb{N}}$.

Brouwer's fan theorem for detachable bars is stated as:

$\mathrm{FAN}_{\mathrm{D}}$ Every detachable bar is uniform,

or formally,

$$\forall\alpha \in \{0,1\}^{\mathbb{N}}\exists n(\overline{\alpha}n \in B) \to \exists n\forall\alpha \in \{0,1\}^{\mathbb{N}}\exists k \leq n(\overline{\alpha}k \in B)$$

for all detachable subset B of $\{0,1\}^*$.

Definition 23.20 A *countable family* (or *sequence*) $(E_n)_{n\in\mathbb{N}}$ of *enumerable* subsets of $\{0,1\}^*$ is given by $t \in 0 \to 0 \to 1$ such that

$$a \in E_n \leftrightarrow a \in \{0,1\}^* \wedge \exists m(t(m,n,a) = 0)$$

for each a and n.

Definition 23.21 A countable family $(U_n)_{n\in\mathbb{N}}$ of *enumerably open* subsets of $\{0,1\}^{\mathbb{N}}$ is given by a countable family $(E_n)_{n\in\mathbb{N}}$ of enumerable subsets of $\{0,1\}^*$ such that

$$\alpha \in U_n \leftrightarrow \exists a \in E_n(\alpha \in a)$$

for each $\alpha \in \{0,1\}^{\mathbb{N}}$ and $n \in \mathbb{N}$. Such a family is a *cover* if

$$\forall\alpha \in \{0,1\}^{\mathbb{N}}\exists n \in \mathbb{N}(\alpha \in U_n),$$

and has a *finite subcover* if there exists $n \in \mathbb{N}$ such that

$$\forall \alpha \in \{0,1\}^{\mathbb{N}} \exists k \leq n(\alpha \in U_k).$$

The following theorem shows that the Heine–Borel theorem for countable enumerably open covers is equivalent to $\mathrm{FAN_D}$.

Theorem 23.22 *The following are equivalent over* $\mathbf{M}_3^\omega$.

(i) *Every countable enumerably open cover of* $\{0,1\}^{\mathbb{N}}$ *has a finite subcover.*
(ii) $\mathrm{FAN_D}$.

Proof The proof is a simplified version of the one in [21, Theorem 16.17]. □

23.4.2 The Cantor Intersection Theorem

In this subsection, we deal with the Cantor intersection theorem and the weak König lemma.

Definition 23.23 A *tree* is a detachable subset T of $\{0,1\}^*$ such that $\langle\rangle \in T$ and it is closed under restriction – that is, for each $a, b \in \{0,1\}^*$, if $a \preceq b$ and $b \in T$, then $a \in T$. A tree T is *infinite* if for each $n \in \mathbb{N}$ there exists $a \in \{0,1\}^*$ such that $a \in T$ and $|a| = n$. A *branch* of a tree T is $\alpha \in \{0,1\}^{\mathbb{N}}$ such that $\overline{\alpha}n \in T$ for all $n \in \mathbb{N}$.

The weak König lemma is stated as:

WKL Every infinite tree has a branch,

or formally,

$$\forall n \exists a \in \{0,1\}^*(a \in T \wedge |a| = n) \rightarrow \exists \alpha \in \{0,1\}^{\mathbb{N}} \forall n(\overline{\alpha}n \in T)$$

for all trees T.

The weak König lemma is decomposed into two parts:

- a logical principle, the *lesser limited principle of omniscience* LLPO (or Σ_1^0-DML):

  LLPO $\quad \neg(\alpha \# \mathbf{0} \wedge \beta \# \mathbf{0}) \rightarrow \neg\alpha \# \mathbf{0} \vee \neg\beta \# \mathbf{0}$

  which is an instance of *De Morgan's law*

  DML $\quad \neg(\varphi \wedge \psi) \rightarrow \neg\varphi \vee \neg\psi;$

- the function existence axiom (Π_1^0, Π_1^0)-$\mathrm{AC}_0^\vee$, the *axiom of countable choice for disjunctions*

  $\mathrm{AC}_0^\vee$ $\quad \forall m(\varphi(m) \vee \psi(m)) \rightarrow \exists\gamma\forall m[(\gamma m = 0 \wedge \varphi(m)) \vee (\gamma m \neq 0 \wedge \psi(m))]$

  for Π_1^0 formula φ and Π_1^0 formula ψ; see [44, Exercise 9.5.4].

Lemma 23.24 *The following are equivalent over* $\mathbf{M}_3^\omega$.

(i) WKL.
(ii) LLPO + (Π_1^0, Π_1^0)-$\mathrm{AC}_0^\vee$.

Proof For the equivalence between LLPO + (Π_1^0, Π_1^0)-$\mathrm{AC}_0^\vee$ and the *separation principle* SEP:

$$\begin{aligned}&\forall m \neg(\exists n(\alpha(m,n) \neq 0) \wedge \exists n(\beta(m,n) \neq 0))\\ &\quad \rightarrow \exists \gamma \forall m[(\gamma m = 0 \wedge \forall n(\alpha(m,n) = 0)) \vee (\gamma m \neq 0 \wedge \forall n(\beta(m,n) = 0))],\end{aligned}$$

see [21, Proposition 16.18]. For the equivalence between SEP and WKL, see [21, Theorem 16.21]. See also [5] for the decomposition. □

Definition 23.25 A sequence $(C_n)_{n\in\mathbb{N}}$ of *enumerably closed* subsets of $\{0,1\}^{\mathbb{N}}$ is given by a sequence $(E_n)_{n\in\mathbb{N}}$ of enumerable subsets of $\{0,1\}^*$ such that

$$\alpha \in C_n \leftrightarrow \forall a \in E_n(\alpha \notin a)$$

for all $\alpha \in \{0,1\}^{\mathbb{N}}$ and $n \in \mathbb{N}$. It

- is a sequence of *inhabited* enumerably closed subsets of $\{0,1\}^{\mathbb{N}}$ if

$$\forall n \exists \alpha \in \{0,1\}^{\mathbb{N}}(\alpha \in C_n);$$

- is *decreasing* if

$$\forall n \forall k \leq n \forall \alpha \in \{0,1\}^{\mathbb{N}}(\alpha \in C_n \rightarrow \alpha \in C_k);$$

- has *inhabited intersection* if

$$\exists \alpha \in \{0,1\}^{\mathbb{N}} \forall n(\alpha \in C_n).$$

The following theorem shows that the Cantor intersection theorem for sequences of enumerably closed subsets is equivalent to WKL.

Theorem 23.26 *The following are equivalent over* $\mathbf{M}_3^\omega$.

(i) *Every decreasing sequence of inhabited enumerably closed subsets of* $\{0,1\}^{\mathbb{N}}$ *has inhabited intersection.*
(ii) WKL.
(iii) LLPO + (Π_1^0, Π_1^0)-$\mathrm{AC}_0^\vee$.

Proof By a simplified proof of [21, Theorem 16.21]. □

23.4.3 The Bolzano–Weierstraß Theorem

We consider here the Bolzano–Weierstraß theorem and a comprehension axiom Σ_1^0-CA_0, the *comprehension axiom for numbers*:

$\mathrm{CA}_0 \qquad \exists\beta\forall m[\beta m \neq 0 \leftrightarrow \varphi(m)]$

for Σ_1^0 formula φ.

The comprehension axiom Σ_1^0-CA_0 is decomposed into two parts:

- a logical principle, the *limited principle of omniscience*LPO (or Σ_1^0-LEM)

  LPO $\qquad \alpha \# \mathbf{0} \vee \neg\alpha \# \mathbf{0}$

  which is an instance of the *law of excluded middle*

  LEM $\qquad \varphi \vee \neg\varphi$;
- the function existence axiom (Σ_1^0, Π_1^0)-$\mathrm{AC}_0^\vee$, the axiom of countable choice for disjunctions $\mathrm{AC}_0^\vee$ for Σ_1^0 formula φ and Π_1^0 formula ψ.

Lemma 23.27 *The following are equivalent over* $\mathbf{M}_3^\omega$.

(i) Σ_1^0-CA_0.
(ii) $\mathrm{LPO} + (\Sigma_1^0, \Pi_1^0)$-$\mathrm{AC}_0^\vee$.

Proof Straightforward. □

Definition 23.28 Let A be a subset of $\{0,1\}^{\mathbb{N}}$. Then A is *infinite* if there exists a function $f : \mathbb{N} \to A$ such that for each m and n,

$$m \neq n \leftrightarrow f(m) \# f(n);$$

an element α of $\{0,1\}^{\mathbb{N}}$ is an *accumulation point* of A if for each n there exists $\beta \in A$ such that

$$\neg(\alpha \sim_1 \beta) \text{ and } \forall k \leq n(\alpha k = \beta k).$$

The following theorem shows that the Bolzano–Weierstraß theorem is equivalent to Σ_1^0-CA_0.

Theorem 23.29 *The following are equivalent over* $\mathbf{M}_3^\omega$.

(i) *Every infinite subset of* $\{0,1\}^{\mathbb{N}}$ *has an accumulation point.*
(ii) Σ_1^0-CA_0.
(iii) $\mathrm{LPO} + (\Sigma_1^0, \Pi_1^0)$-$\mathrm{AC}_0^\vee$.

Proof By a simplified proof of [21, Theorem 16.27]. □

23.4.4 Sequential Compactness

In this subsection, we deal with the sequential compactness theorem, the comprehension axiom for numbers Σ_1^0-CA_0 and the *quantifier-free axiom of choice from numbers to numbers*:

$$\text{QF-AC}_{00} \qquad \forall m \exists n \varphi(m,n) \rightarrow \exists \alpha \forall m \varphi(m, \alpha m)$$

for quantifier-free φ.

Definition 23.30 A sequence $\alpha_{(\text{-})}$ in $\{0,1\}^{\mathbb{N}}$ *has a convergent subsequence* if there exist $\alpha \in \{0,1\}^{\mathbb{N}}$, κ, and γ such that

$$\forall n(\kappa n < \kappa(n+1)) \text{ and } \alpha_{\kappa n} \underset{\gamma}{\longrightarrow} \alpha.$$

The following theorem shows that the sequential compactness theorem is equivalent to Σ_1^0-CA_0 + QF-AC_{00}.

Theorem 23.31 *The following are equivalent over* $\mathbf{M}_3^\omega$.

(i) *Every sequence in* $\{0,1\}^{\mathbb{N}}$ *has a convergent subsequence.*
(ii) Σ_1^0-CA + QF-AC_{00}.
(iii) LPO + (Σ_1^0, Π_1^0)-$\mathrm{AC}_0^\vee$ + QF-AC_{00}.

Proof The proof is a simplified version of the one in [21, Theorem 16.31]. □

Notes

In classical reverse mathematics, the Heine–Borel theorem and the Cantor intersection theorem are both equivalent to WKL over $\mathbf{RCA}_0$ ([39, Theorem IV.1.2]), and the Bolzano–Weierstraß theorem and the sequential compactness theorem are both equivalent to the *arithmetical comprehension axiom*:

$$\text{ACA} \qquad \exists X \forall n(n \in X \leftrightarrow \varphi(n)),$$

where φ is any arithmetical formula in which X does not occur freely; see [39, Theorem III.2.2]. Note that, since $\mathrm{FAN_D}$ is a classical contraposition of WKL, we can not distinguish $\mathrm{FAN_D}$ from WKL in $\mathbf{RCA}_0$, and we can derive QF-AC_{00} from Δ_1^0-CA in $\mathbf{RCA}_0$.

In [21], the conditions

$$(\mathrm{E}_0) \qquad \forall \alpha \exists \beta \forall m[(\beta m \neq 0 \rightarrow \exists n(\alpha(m,n) \neq 0)) \wedge (\beta m = 0 \rightarrow \forall n(\alpha(m,n) = 0))],$$

$$(\mu) \qquad \forall \alpha \exists \beta \forall m[\exists n(\alpha(m,n) \neq 0)) \rightarrow \alpha(m, \beta m) \neq 0)],$$

which are equivalent to Σ^0_1-CA_0 and Σ^0_1-CA_0 + QF-AC_{00}, respectively, are considered instead.

The implication of $\mathrm{FAN_D}$ in WKL was shown in [18], and then a direct (and formal) proof in **EL**, which is easily translated into a proof of $\mathbf{M}^\omega_3$, was given in [23]. Berger [3, 4] investigated in detail relationships among WKL, $\mathrm{FAN_D}$, and the *uniform continuity theorem*:

UCT Every continuous $F \in \{0,1\}^{\mathbb{N}} \to \mathbb{N}$ is uniformly continuous,

where $F \in \{0,1\}^{\mathbb{N}} \to \mathbb{N}$ is *uniformly continuous* if

$$\exists n \forall \alpha\beta \in \{0,1\}^{\mathbb{N}}[\forall k \leq n(\alpha k = \beta k) \to F\alpha = F\beta].$$

Berger et al. [6] dealt, in $\mathbf{EL}_0$, with the *binary expansion* of real numbers in the unit interval:

BE Every real number in $[0,1]$ has a binary expansion,

and the *intermediate value theorem*:

IVT If $f : [0,1] \to \mathbb{R}$ is a uniformly continuous function with $f(0) < 0 < f(1)$, then there exists $x \in [0,1]$ such that $f(x) = 0$;

see [42, Section 5.9] for BE, and [9, Section 3.2.4] and [43, Section 6.1.2] for IVT. We are able to show that BE and IVT follow from WKL and LLPO follows from BE or IVT, and hence BE and IVT are in between WKL and LLPO. However, in classical reverse mathematics, the base system $\mathbf{RCA}_0$ proves BE and IVT; see [39, Theorem II.6.6]. Note that in $\mathbf{RCA}_0$, we cannot compare theorems which are provable in $\mathbf{RCA}_0$ (such as BE and IVT). Berger et al. [6] introduced the notions of a tree *having at most (or exactly) two nodes at each level* and a *convex* tree, and showed that BE is equivalent to WKL for trees having at most two nodes at each level; IVT is equivalent to WKL for convex trees; and IVT implies BE.

23.5 The Monotone Completeness Theorem

In this section, we deal with the monotone completeness theorem based on [27].

23.5.1 The Bounded Comprehension Axiom

We consider here Σ^0_1-BCA, the *bounded comprehension axiom for numbers*:

BCA $\forall m \exists b[|b| = m \wedge \forall k < m((b)_k \neq 0 \leftrightarrow \varphi(k))]$

for Σ^0_1 formula φ.

The bounded comprehension axiom Σ^0_1-BCA implies LPO and the induction schema for Σ^0_1 formulae.

Lemma 23.32 *The following are derivable in* $\mathbf{M}_3^\omega + \Sigma_1^0\text{-BCA}$.

(i) LPO.
(ii) Σ_1^0-IND.

Proof See the proof of [27, Lemma 3]. □

The following lemma shows that Σ_1^0-IND implies Σ_1^0-BCA with classical logic, and that LPO implies Σ_1^0-BCA with the full induction schema.

Lemma 23.33 *The following are derivable in* $\mathbf{M}_3^\omega$.

(i) $\mathrm{LEM} + \Sigma_1^0\text{-IND} \Rightarrow \Sigma_1^0\text{-BCA}$.
(ii) $\mathrm{LPO} + \mathrm{IND} \Rightarrow \Sigma_1^0\text{-BCA}$.

Proof See the proofs of [27, Lemma 4 and Lemma 5]. □

The bounded comprehension axiom $\Sigma_1^0\text{-BCA}_0$ is decomposed into the logical principle LPO and a bounded function existence axiom $(\Sigma_1^0, \Pi_1^0)\text{-ABC}^\vee$, the *axiom of bounded choice for disjunctions* $\mathrm{ABC}^\vee$:

$$\begin{aligned}&\forall m(\varphi(m) \vee \psi(m))\\ &\qquad \to \forall m \exists b[|b| = m \wedge \forall k < m(((b)_k = 0 \wedge \varphi(k)) \vee ((b)_k \neq 0 \wedge \psi(k)))],\end{aligned}$$

where φ is a Σ_1^0 formula and ψ a Π_1^0 formula.

Lemma 23.34 *The following are equivalent over* $\mathbf{M}_3^\omega$.

(i) Σ_1^0-BCA.
(ii) $\mathrm{LPO} + (\Sigma_1^0, \Pi_1^0)\text{-ABC}_0^\vee$.

Proof Straightforward. □

23.5.2 Monotone Completeness

Definition 23.35 We adopt a definition of real number with a fixed modulus: a *real number* is a sequence $(p_n)_{n\in\mathbb{N}}$ of rationals, that is, of type 1, such that

$$\forall mn \left(|p_m - p_n| <_{\mathbb{Q}} 2^{-m} + 2^{-n}\right). \tag{23.1}$$

The relations $<_{\mathbb{R}}$, $\leq_{\mathbb{R}}$, and $=_{\mathbb{R}}$ between real numbers $x = (p_n)_{n\in\mathbb{N}}$ and $y = (q_n)_{n\in\mathbb{N}}$ are defined by

$$x <_{\mathbb{R}} y \Leftrightarrow \exists n \left(2^{-n+2} <_{\mathbb{Q}} q_n - p_n\right),$$

$x \leq_{\mathbb{R}} y \Leftrightarrow \neg(y <_{\mathbb{R}} x)$, and $x =_{\mathbb{R}} y \Leftrightarrow x \leq_{\mathbb{R}} y \wedge y \leq_{\mathbb{R}} x$, respectively. A *sequence* $(x_n)_{n\in\mathbb{N}}$ *of reals* is a double sequence $(p_{m,n})_{m,n\in\mathbb{N}}$ of rationals, that is, of type $0 \to 1$, such that $(p_{m,n})_{m\in\mathbb{N}}$ is a real number for all n.

Let $\mathbb{R}$ be a set of type 1 given by the formula (23.1). Then $(\mathbb{R}, =_{\mathbb{R}})$ is a setoid. There is no trouble to define the arithmetical operations on the reals as mappings on $(\mathbb{R}, =_{\mathbb{R}})$, and to show basic theorems on them in $\mathbf{M}_3^\omega$; see [43, 5.2 and 5.3] and Chapter 8 of this Handbook.

Definition 23.36 A sequence $(x_n)_{n\in\mathbb{N}}$ of real numbers is *bounded* if there exists K such that $|x_n| <_{\mathbb{R}} K$ for all n; *non-decreasing* if $n \le m \to x_n \le_{\mathbb{R}} x_m$ for all n and m; *non-increasing* if $n \le m \to x_n \ge_{\mathbb{R}} x_m$ for all n and m; and *monotone* if it is non-decreasing or non-increasing.

Definition 23.37 A sequence $(x_n)_{n\in\mathbb{N}}$ of real numbers is a *weak Cauchy sequence* if

$$\forall k \exists n \forall m m' \ge n(|x_m - x_{m'}| <_{\mathbb{R}} 2^{-k}).$$

The following theorem shows that a version of the monotone completeness theorem is equivalent to Σ_1^0-BCA.

Theorem 23.38 *The following are equivalent over* $\mathbf{M}_3^\omega$.

(i) *Every bounded monotone sequence of rational numbers is a weak Cauchy sequence.*
(ii) *Every bounded monotone sequence of real numbers is a weak Cauchy sequence.*
(iii) Σ_1^0-BCA.
(iv) LPO $+$ (Σ_1^0, Π_1^0)-$\mathrm{ABC}_0^\vee$.

Proof By the proof of [27, Theorem 2] and Lemma 23.34. □

Notes

In classical reverse mathematics, the monotone completeness theorem is equivalent to ACA over $\mathbf{RCA}_0$; see [39, Theorem III.2.2]. Theorem 23.38 together with Lemma 23.33(i) shows that, since LEM and Σ_1^0-IND are derivable in $\mathbf{RCA}_0$, so is Σ_1^0-BCA, and hence the essence of the monotone completeness theorem in classical reverse mathematics is the Cauchy completeness of the reals:

Every weak Cauchy sequence of real numbers converges,

which is also classified into ACA [39, Theorem III.2.2]. Here we reserve the term *Cauchy sequence* for a sequence $(x_n)_{n\in\mathbb{N}}$ of real numbers with *Cauchy modulus* γ such that

$$\forall k \forall m m' \ge \gamma k(|x_m - x_{m'}| <_{\mathbb{R}} 2^{-k}).$$

The Cauchy completeness of the reals for Cauchy sequences is provable without assuming countable choice [43, Theorem 5.4.2]. Note that the proof is easily translated into a proof in $\mathbf{RCA}_0$ and in $\mathbf{M}_3^\omega$.

The statement of Theorem 23.38(ii) was investigated under the name PCM_{ar} in [30, 40]. Kohlenbach [30, Corollary 5.3] showed that

$$\mathbf{EA}^2 \vdash \Sigma_1^0\text{-IND} \leftrightarrow \mathrm{PCM}_{ar},$$

where $\mathbf{EA}^2$ is a restriction of the second-order fragment of the classical finite-type arithmetic $\mathbf{PA}^\omega$ to elementary functionals; see also [33, Section 13.4 and Section 13.4]. On the other hand, Toftdal [40, Proposition 6.6 and Theorem 6.8] showed that

$$\mathbf{HA}^\omega \vdash \mathrm{LPO} \leftrightarrow \mathrm{PCM}_{ar};$$

see also [41]. Since LEM is derivable in $\mathbf{EA}^2$, the former result follows from Theorem 23.38 and Lemma 23.32 together with Lemma 23.33(i), and, since IND is derivable in $\mathbf{HA}^\omega$, the latter result follows from them together with Lemma 23.33(ii).

The bounded comprehension axiom BCA was introduced in [39, Section II.3.8] and was shown that Σ_1^0-BCA is equivalent to Σ_1^0-IND over $\mathbf{RCA}_0^*$ in [39, Section X.4.4]; see also [16, Section I.2.6] for BCA, where it is stated as $\forall k \exists b\,(b$ is k-piece of $\varphi)$. Nemoto [37] has investigated BCA, including Σ_1^0-BCA, and its equivalents in detail.

23.6 Concluding Remarks

We conclude the chapter with a figure showing implications among principles and axioms, and briefly reviewing other approaches to constructive reverse mathematics.

23.6.1 Implications Among Principles and Axioms

Figure 23.1 shows implications among principles and axioms we have dealt with. Note that Σ_2^0-DNE + QF-AC_{00} implies BD-N. In fact, for each κ with

$$\forall \gamma \exists n \forall m (n \leq m \rightarrow \kappa(\gamma m) < m),$$

if $\forall n \exists m(\kappa m \geq n)$, then there exists γ such that $\kappa(\gamma n) \geq n$ for all n, by QF-AC_{00}, and this entails a contradiction. Therefore $\neg \forall n \exists m(\kappa m \geq n)$, which is equivalent to $\neg\neg \exists n \forall m(\kappa m < n)$, by Σ_2^0-DNE (or Σ_1^0-DNE), and so $\exists n \forall m(\kappa m < n)$, by Σ_2^0-DNE.

23.6.2 Other Approaches

Reverse mathematics in constructive set theory has been opened up by presenting statements equivalent to some fragments of the principle of non-deterministic

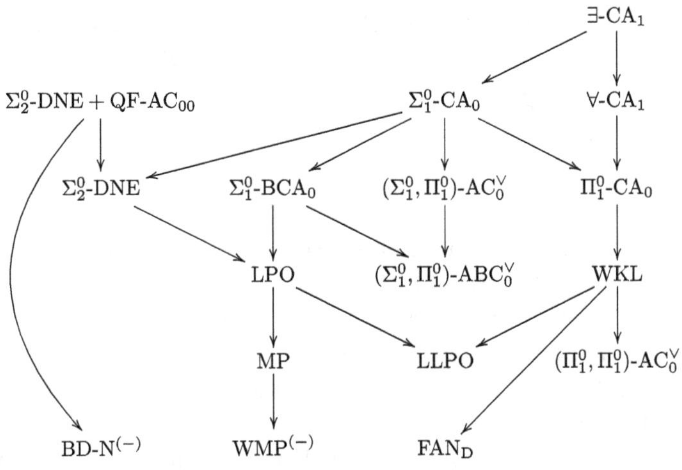

Figure 23.1 Implications among principles and axioms.

inductive definitions (NID) by van den Berg [45], working in a weak subsystem **ECST** of constructive Zermelo–Fraenkel set theory **CZF**; see [1, 2] for **CZF** and **ECST**. In [26], a principle, called nullary NID, was introduced, and it was shown that Fullness, an axiom of **CZF**, is equivalent to nullary NID over **ECST**. In [17], it was proved that several statements in constructive topology which were initially proved using NID are equivalent to the elementary and finitary NIDs; the finitary NID is equivalent to its binary fragment; and the elementary NID is equivalent to a variant of NID based on the notion of a biclosed subset.

The first result in constructive reverse mathematics working within Bishop-style (informal) constructive mathematics is [29, Theorem 2.4] by Julian and Richman on the equivalence between FAN$_D$ and the statement that *every uniformly continuous, positively valued function* $f : [0,1] \to \mathbb{R}$ *has a positive infimum.* Since then many people, including J. Berger, D. S. Bridges, H. Diener, M. Hendtlass, M. Mandelkern, F. Richman, and P. Schuster, have contributed to reverse mathematics in Bishop-style constructive mathematics; see [11, 12, 22] for overview. Since Bishop-style constructive mathematics is informal and assumes the axioms of countable and dependent choice – that is, set existence axioms in second-order arithmetic – it is difficult to precisely compare the results there with those in classical reverse mathematics which classifies theorems by set existence axioms.

Intuitionistic reverse mathematics by Veldman [46, 47] has adopted a formal system, Basic Intuitionistic Mathematics (BIM), based on intuitionistic logic, and has focussed on theorems in Brouwer's intuitionism.

Brattka and others have investigated and classified mathematical theorems into Weihrauch degrees, based on their main question 'which theorems can be computably transformed into others'; see [8] for an overview. However, they have used classical logic, and hence have not been able to deal with some logical principles such as Markov's principle.

References

[1] Aczel, P., and Rathjen, M. 2000/2001. *Notes on Constructive Set Theory*. Technical report 40. Institut Mittag-Leffler.

[2] Aczel, P., and Rathjen, M. August, 2010. *CST Book draft*. https://www1.maths.leeds.ac.uk/ rathjen/book.pdf

[3] Berger, J. 2008. The weak König lemma and uniform continuity. *J. Symbol. Logic*, **73**(3), 933–939.

[4] Berger, J. 2012. Aligning the weak König lemma, the uniform continuity theorem, and Brouwer's fan theorem. *Ann. Pure Appl. Logic*, **163**(8), 981–985.

[5] Berger, J., Ishihara, H, and Schuster, P. 2012. The weak Kőnig lemma, Brouwer's Fan theorem, de Morgan's law, and dependent choice. *Rep. Math. Logic*, **47**, 63–86.

[6] Berger, J., Ishihara, H., Kihara, T., and Nemoto, T. 2019. The binary expansion and the intermediate value theorem in constructive reverse mathematics. *Arch. Math. Logic*, **58**(1–2), 203–217.

[7] Bishop, E., and Bridges, D. 1985. *Constructive Analysis*. Grundlehren der Mathematischen Wissenschaften [Fundamental Principles of Mathematical Sciences], vol. 279. Berlin: Springer-Verlag.

[8] Brattka, V., Gherardi, G., and Pauly, A. 2021. Weihrauch complexity in computable analysis. In: Brattka, V., and Hertling, P. (eds.), *Handbook of Computability and Complexity*. Cham: Springer.

[9] Bridges, D., and Richman, F. 1987. *Varieties of Constructive Mathematics*. London Mathematical Society Lecture Note Series, vol. 97. Cambridge: Cambridge University Press.

[10] Bridges, D. S., and Vîţă, L. S. 2006. *Techniques of Constructive Analysis*. Universitext. New York: Springer.

[11] Diener, H. 2018. Constructive Reverse Mathematics. Habilitation Schrift, Universtät Siegen, `https://nbn-resolving.org/urn:nbn:de:hbz:467-13064`.

[12] Diener, H., and Ishihara, H. 2021. Bishop style constructive reverse mathematics. In: Brattka, V., and Hertling, P. (eds.), *Handbook of Computability and Complexity*. Cham: Springer.

[13] Escardó, M. H. 2013. Infinite sets that satisfy the principle of omniscience in any variety of constructive mathematics. *J. Symbol. Logic*, **78**(3), 764–784.

[14] Fujiwara, M., Ishihara, H., and Nemoto, T. 2015. Some principles weaker than Markov's principle. *Arch. Math. Logic*, **54**(7–8), 861–870.

[15] Grilliot, T. J. 1971. On effectively discontinuous type-2 objects. *J. Symbol. Logic*, **36**, 245–248.

[16] Hájek, P., and Pudlák, P. 1993. *Metamathematics of First-Order Arithmetic*. Perspectives in Mathematical Logic. Berlin: Springer-Verlag.

[17] Hirata, A., Ishihara, H., Kawai, T., and Nemoto, T. 2019. Equivalents of the finitary non-deterministic inductive definitions. *Ann. Pure Appl. Logic*, **170**(10), 1256–1272.

[18] Ishihara, H. 1990. An omniscience principle, the König lemma and the Hahn–Banach theorem. *Z. Math. Logik Grundlag. Math.*, **36**(3), 237–240.

[19] Ishihara, H. 1992. Continuity properties in constructive mathematics. *J. Symbol. Logic*, **57**(2), 557–565.

[20] Ishihara, H. 1993. Markov's principle, Church's thesis and Lindelöf's theorem. *Indag. Math. (N.S.)*, **4**(3), 321–325.

[21] Ishihara, H. 2005. Constructive reverse mathematics: compactness properties. Pages 245–267 of: Crosilla, L., and Schuster, P. (eds.), *From Sets and Types to Topology and Analysis*. Oxford Logic Guides, vol. 48. Oxford: Oxford University Press.

[22] Ishihara, H. 2006a. Reverse mathematics in Bishop's constructive mathematics. *Philosophia Scientiae, Cahier special*, **6**, 43–59.

[23] Ishihara, H. 2006b. Weak König's lemma implies Brouwer's fan theorem: a direct proof. *Notre Dame J. Formal Logic*, **47**(2), 249–252.

[24] Ishihara, H. 2012. The uniform boundedness theorem and a boundedness principle. *Ann. Pure Appl. Logic*, **163**(8), 1057–1061.

[25] Ishihara, H. 2018. On Brouwer's continuity principle. *Indag. Math. (N.S.)*, **29**(6), 1511–1524.

[26] Ishihara, H., and Nemoto, T. 2016. Non-deterministic inductive definitions and fullness. Pages 163–170 of: Probst, D., and Schuster, P. (eds.), *Concepts of Proof in Mathematics, Philosophy, and Computer Science*. Ontos Math. Log., vol. 6. Berlin: de Gruyter.

[27] Ishihara, H., and Nemoto, T. 2019. The monotone completeness theorem in constructive reverse mathematics. Pages 101–112 of: Centrone, S., Negri, S., Sarikaya, D., and Schuster, P. M. (eds.), *Mathesis Universalis, Computablility and Proof*. Synthese Library, vol. 412. Cham: Springer.

[28] Ishihara, H., and Yoshida, S. 2002. A constructive look at the completeness of the space $\mathcal{D}(\mathbf{R})$. *J. Symbol. Logic*, **67**(4), 1511–1519.

[29] Julian, W., and Richman, F. 1984. A uniformly continuous function on $[0, 1]$ that is everywhere different from its infimum. *Pac. J. Math.*, **111**(2), 333–340.

[30] Kohlenbach, U. 2000. Things that can and things that cannot be done in PRA. *Ann. Pure Appl. Logic*, **102**(3), 223–245.

[31] Kohlenbach, U. 2002. On weak Markov's principle. *MLQ Math. Log. Q.*, **48**(suppl. 1), 59–65. Dagstuhl Seminar on Computability and Complexity in Analysis, 2001.

[32] Kohlenbach, U. 2005. Higher order reverse mathematics. Pages 281–295 of: *Reverse Mathematics 2001*. Lecture Notes in Logic, vol. 21. La Jolla, CA: Association for Symbolic Logic.

[33] Kohlenbach, U. 2008. *Applied Proof Theory: Proof Interpretations and Their Use in Mathematics*. Springer Monographs in Mathematics. Berlin: Springer-Verlag.

[34] Lietz, P., and Streicher, T. 2012. Realizability models refuting Ishihara's boundedness principle. *Ann. Pure Appl. Logic*, **163**(12), 1803–1807.

[35] Mandelkern, M. 1983. Constructive continuity. *Mem. Amer. Math. Soc.*, **42**(277), v+117.

[36] Mandelkern, M. 1988. Constructively complete finite sets. *Z. Math. Logik Grundlag. Math.*, **34**(2), 97–103.

[37] Nemoto, T. 2020. Finite sets and infinite sets in weak intuitionistic arithmetic. *Arch. Math. Logic*, **59**(5–6), 607–657.

[38] Rose, H. E. 1984. *Subrecursion: Functions and Hierarchies*. Oxford Logic Guides, vol. 9. New York: The Clarendon Press, Oxford University Press.

[39] Simpson, S. G. 1999. *Subsystems of Second Order Arithmetic*. Perspectives in Mathematical Logic. Berlin: Springer-Verlag.

[40] Toftdal, Michael. 2004a. A calibration of ineffective theorems of analysis in a constructive context. M.Phil. thesis, Department of Computer Science, University of Aarhus.

[41] Toftdal, M. 2004b. A calibration of ineffective theorems of analysis in a hierarchy of semi-classical logical principles (extended abstract). Pages 1188–1200 of: *Automata, Languages and Programming*. Lecture Notes in Computational Science, vol. 3142. Berlin: Springer.

[42] Troelstra, A. S. 1977. Aspects of constructive mathematics. Pages 973–1052 of: *Handbook of Mathematical Logic*. Studies in Logic and the Foundations of Mathematics, vol. 90. Amsterdam: North-Holland.

[43] Troelstra, A. S., and van Dalen, D. 1988a. *Constructivism in Mathematics. Vol. I*. Studies in Logic and the Foundations of Mathematics, vol. 121. Amsterdam: North-Holland.

[44] Troelstra, A. S., and van Dalen, D. 1988b. *Constructivism in Mathematics. Vol. II*. Studies in Logic and the Foundations of Mathematics, vol. 123. Amsterdam: North-Holland.

[45] van den Berg, B. 2013. Non-deterministic inductive definitions. *Arch. Math. Logic*, **52**(1–2), 113–135.

[46] Veldman, W. 2009. Brouwer's approximate fixed-point theorem is equivalent to Brouwer's fan theorem. Pages 277–299 of: *Logicism, Intuitionism, and Formalism*. Synthese Library, vol. 341. Dordrecht: Springer.

[47] Veldman, W. 2014. Brouwer's fan theorem as an axiom and as a contrast to Kleene's alternative. *Arch. Math. Logic*, **53**(5–6), 621–693.

24

Systems for Constructive Reverse Mathematics

Takako Nemoto

24.1 Introduction

Constructive reverse mathematics is reverse mathematics over intuitionistic logic. Its formalization has been presented by several authors in different systems. In this chapter, we will see the variation of them and will clarify the relationship among them and systems of Friedman–Simpson reverse mathematics over classical logic.

Friedman–Simpson reverse mathematics, which is based on classical logic, started in the 1970s (cf. [7] and [8]). According to its main question "Which set existence axioms are needed to prove the theorems of ordinary, nonset-theoretic mathematics?," a classification of mathematical theorems is given. After a great deal of work, it turned out that most of theorems from undergraduate mathematics are provable in the standard base theory RCA_0, or equivalent to one of the characteristic axioms of "Big 5" systems. This is done in the systems of second-order arithmetic in the set-based language L_S, that is, a language for natural numbers and sets of them (cf. [21]).

In the 1970s, studies on formal systems of arithmetic over intuitionistic logic were also developed. Troelstra [23] gave a comprehensive view of studies on intuitionistic formal systems of arithmetic, including the first-order arithmetic $\mathbf{HA}$, the set-based second-order arithmetic $\mathbf{HAS}$, the function-based second-order arithmetic $\mathbf{EL}$,[1] the higher-order arithmetic $\mathbf{HA}^\omega$, and their variants, which inspired later researchers on formal constructive reverse mathematics.

Constructive reverse mathematics started in an informal way. The first known result was given in [10]. As presented in [5], it can classify not only mathematical theorems in classical mathematics but also Brouwerian intuitionistic mathematics and Russian recursive mathematics. Most of them used Bishop-style constructive mathematics as their base, where we can freely use the axiom of unique choice and induction schema.

[1] In [12], it is called "system of analysis."

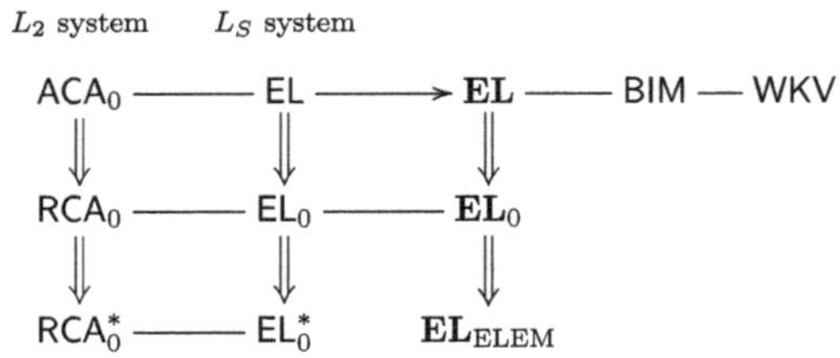

Figure 24.1 Relationship between systems.

In the 2000s, the study of formalized constructive reverse mathematics started. Loeb [13] used the system WKV of natural numbers and functions on them as the base theory. Since WKV has unrestricted axiom schema of unique choice for numbers and induction, it could be the simplest formalization of Bishop-style mathematics in this language. At almost the same time, some of the work in [25] was announced in the research community. Its base theory BIM has only a restricted version of the axiom of choice. We will see in Section 24.4, that BIM is closely related to $\mathbf{EL}$, which is a conservative extension of $\mathbf{HA}$ (cf. [12]). Since $\mathbf{HA}$ proves the same Π_2 sentence as $\mathbf{PA}$, and ACA_0 have the same consistency strength as $\mathbf{PA}$, $\mathbf{EL}$ and ACA_0 have the same consistency strength. Then it is natural to ask what intuitionistic system corresponds to the usual base theory RCA_0 and the weaker base theory RCA_0^* in Friedman–Simpson reverse mathematics. Ishihara [9] suggested a system $\mathbf{EL}_{\mathrm{ELEM}}$, which has all the elementary recursive functions and has only Δ_0^0 induction schema, which seems to correspond to RCA_0^*. In [16], the system EL_0^* was used as the base theory. Berger et al. [2] introduced the system $\mathbf{EL}_0$, which is obtained by restricting the induction schema of $\mathbf{EL}$ to Σ_1^0. The last one is apparently intended as the intuitionistic counterpart of RCA_0.

One of the advantages of using formal systems for reverse mathematics over intuitionistic logic is that it enables systematic comparison with Friedman–Simpson reverse mathematics. By a technical difficulty, formal constructive reverse mathematics tends to use systems in function-based languages, that is, languages for natural numbers and functions, which differs from the set-based language of Friedman–Simpson systems. In [17], systematic interpretations of them are given. We use them and will overview systems of constructive reverse mathematics known in the literature. Figure 24.1 illustrates the relationship between the systems we will see in this chapter. All the systems except for the left-most column are based on intuitionistic logic and systems in the same row have the same consistency strength. Systems connected with a simple line are mutually interpretable. The double implication $\mathsf{A} \Rightarrow \mathsf{B}$ means the system A proves the consistency of the system B. Using a formal system over intuitionistic logic gives quantification of classical principles used in proofs and refinement of the results over classical logic. In [2], the intermediate value theorem (IVT) is characterized by an axiom which is not provable in $\mathbf{EL}_0$. It is known that IVT is not provable in Bishop-style constructive mathematics (cf. [3]), but is provable in RCA_0 (cf. [21, Theorem II.6.6]).

This chapter is arranged as follows: Section 24.2 is preliminary. Subsection 24.2.1 explains the technique of interpretation among formal systems. Subsection 24.2.2 introduces intuitionistic systems of first-order arithmetic, which will be used as the common number sort part both in set-based systems and function-based systems. Subsection 24.2.3 introduces systems of Friedman–Simpson reverse mathematics to be compared with intuitionistic systems. Section 24.3 introduces a simple function-based language L_F. The relationship between set-based systems and interpretations between function-based systems will be also explained. In Section 24.4, 24.5, and 24.6, we will see systems of constructive reverse mathematics used in the literature and will see their relationship with Friedman–Simpson reverse mathematics. In Section 24.7, we give a proof that the interpretation in Section 24.3 has the appropriate property required in Sections 24.4–24.6.

24.2 Preliminary

24.2.1 Interpretation as a Tool to Compare Formal Systems

In this subsection, we introduce basic definitions and notions that we will need later.

Terms and formulas are defined as usual. We use the abbreviation $\exists ! x A(x)$ for $\exists x A(x) \wedge \forall x \forall y (A(x) \wedge A(y) \to x = y)$. For a term t and a formula A, $t[x_1/s_1, \ldots, x_n/s_n]$ and $A[x_1/s_1, \ldots, x_n/s_n]$ denote the result of simultaneous substitution of $s_1, \ldots, s_n$ for the variables $x_1, \ldots, x_n$ in t and A, respectively. Other basic definitions and notations can be found in [21] or [24].

In this chapter, a *system* $\mathbf{T}$ is a class of formulae based on a logic. For a system $\mathbf{T}$ and a formula A, we say $\mathbf{T}$ *proves* A, denoted by $\mathbf{T} \vdash A$, if A is derivable from the assumptions from $\mathbf{T}$ using rules and axioms in the logic which $\mathbf{T}$ is based on. Therefore if $\mathbf{T}$ is based on intuitionistic logic, then $\mathbf{T} \vdash A$ means that A is derivable from the axioms of $\mathbf{T}$ with intuitionistic logic, and if $\mathbf{T}$ is based on classical logic, then $\mathbf{T} \vdash A$ means that A is derivable from the axioms of $\mathbf{T}$ with classical logic. A schema $\mathcal{S}_0$ is equivalent to another schema $\mathcal{S}_1$ over a system $\mathbf{T}$ if $\mathbf{T} \cup \mathcal{S}_0 \vdash A$ for each instance A of $\mathcal{S}_1$ and vice versa. According to the terminology in [21], we call this $\mathbf{T}$ a base theory.

In the rest of this subsection, let $\mathbf{T}$ be a system of a language L and let $\mathbf{T}'$ be a system of a language L'.

In this chapter, an *interpretation* (or *translation*) from L' to L is a map -^I from the set of formulae in L' to the set of formulae in L. We say $\mathbf{T}'$ is *interpretable* into $\mathbf{T}$ by I if (1) $\mathbf{T} \vdash \bot^I \to \bot$, (2) $\mathbf{T} \vdash A^I$ for each axiom A of $\mathbf{T}'$, and (3) there is a function p from the set of derivations in $\mathbf{T}'$ with a single logical rule into the set of derivations in $\mathbf{T}$ such that if $\mathcal{D}(A, B, C)$ is a derivation from A and B to C by a single logical rule in $\mathbf{T}'$, then $p(\mathcal{D}(A, B, C))$ is a derivation in $\mathbf{T}$ from A^I and B^I to C^I. In particular, all interpretations in this chapter have polynomial-time computable functions with the property (3), which means interpretation can be executed in a weak meta theory.

Interpretations can be used for classification of systems with *consistency strength*. The consistency of the system $\mathbf{T}$ is stronger than or equal to the one of $\mathbf{T}'$ if the consistency of $\mathbf{T}$ implies the consistency of $\mathbf{T}'$ (cf. [19]). By (2) and (3) above, if $\mathbf{T}'$ is interpretable into $\mathbf{T}$, we can prove that $\mathbf{T}' \vdash A$ yields $\mathbf{T} \vdash A^I$ by induction on derivations in $\mathbf{T}'$. In particular, $\mathbf{T}' \vdash \bot$ yields $\mathbf{T} \vdash \bot$ by (1), and so the consistency $\mathbf{T}$ implies the consistency of $\mathbf{T}'$. Therefore if $\mathbf{T}$ and $\mathbf{T}'$ are mutually interpretable, then $\mathbf{T}$ and $\mathbf{T}'$ have the same consistency strength. Other benefits using interpretations for analyzing consistency strength is explained in [17, Section 1.2].

Let Γ be a class of formulae in L such that $\Gamma \subseteq L \cap L'$. Then $\mathbf{T}'$ is Γ*-preservingly interpretable in* $\mathbf{T}$ by I if the interpretation I satisfies $\mathbf{T} \vdash A^I \to A$ for any sentence A in Γ. If $L = L'$ and $\mathbf{T} \subseteq \mathbf{T}'$, then Γ-preserving interpretability of $\mathbf{T}'$ into $\mathbf{T}$ means that any theorem in Γ proved in the stronger system $\mathbf{T}'$ can be transferred as a theorem in $\mathbf{T}$. One of the most famous examples is the Π_2-preserving interpretability of $\mathbf{PA}$ into $\mathbf{HA}$. It could be used as a measure of how close the systems are in the sense of provability. In Sections 24.3–24.6, we will see that systems of Friedman–Simpson reverse mathematics are Π^0_2-preservingly mutually interpretable with some system of intuitionistic systems.

A special case of an interpretable system is *definitional extension*. Let L' be an extension of L with a family of new predicate symbols and function symbols. Then $\mathbf{T}'$ is a definitional extension of $\mathbf{T}$ if $\mathbf{T}'$ is interpretable into $\mathbf{T}$ by I with the following properties:

(i) $A^I \equiv A$ for A in L;

(ii) $\mathbf{T}' \vdash A \leftrightarrow A^I$;

(iii) $\mathbf{T}' \vdash A$ yields $\mathbf{T} \vdash A^I$;

(iv) -^I commutes with all logical operations in L.

In [24, Chapter 2, Section 7], it is shown that adding L-definable predicates and functions and extending $\mathbf{T}$ with defining axioms of them makes definitional extension of $\mathbf{T}$. This holds only in the case such that $\mathbf{T}$ has no restricted axiom schema like Σ^0_1 induction. In Subsection 24.3.2, we will see what kind of defining axioms make definitional extension.

A system $\mathbf{T}'$ is a *conservative extension* of $\mathbf{T}$ if $L \subseteq L'$, $\mathbf{T} \subseteq \mathbf{T}'$ and $\mathbf{T}' \vdash A$ yields $\mathbf{T} \vdash A$ for any sentence A in L. The above definitional extension is also a special case of conservative extension.

24.2.2 Systems of First-Order Arithmetic

In this subsection, we will present and overview of systems of first-order arithmetic over intuitionistic logic, which will be needed later sections.

Definition 24.1 (L_1 and $L_1(\exp)$) The one-sorted first-order language L_1 with equality consists of the following:

- the constant symbols 0 and 1;
- variables x, y, z, $\ldots$;
- binary function symbols $+$ and $\cdot$;
- a binary relation $<$.

The language $L_1(\exp)$ is the extension of L_1 with a binary function exp.

Terms and formulae are defined as usual. We use the abbreviations $(\forall x < t)A(x)$ for $\forall x(x < t \to A(x))$ and $(\exists x < t)A(x)$ for $\exists x(x < t \wedge A(x))$. The occurrences of number quantifiers of the form $\forall x < t$ and $\exists x < t$ are called *bounded* and the others are called *unbounded*.

Definition 24.2 (Classes of formuale in L_1) The class Δ_0 consists of all L formulae without unbounded quantifiers. Δ_0 is also called Σ_0 or Π_0. A formula is Σ_{n+1} if it is of the form $\exists x A(x)$ for some Π_n formula $A(x)$. A formula is Π_{n+1} if it is of the form $\forall x A(x)$ for some Σ_n formula $A(x)$.

Definition 24.3 (Γ induction) For a class Γ formulae, the schema Γ *induction* is all the axioms of the following form:

$$A(0) \wedge \forall x(A(x) \to A(x+1)) \to \forall x A(x),$$

where A is a formula in Γ.

Definition 24.4 (Γ bounding) For a class Γ formulae, the schema Γ *bounding* is all the axioms of the following form:

$$(\forall x < t)\exists u A(x,u) \to \exists v(\forall x < t)(\exists u < v)A(x,u),$$

where A is a formula in Γ.

If Γ is the class of all formulae in L, it is denoted by L. For example, L induction is the schema of all instances of the above form.

Definition 24.5 ($\mathbf{i\Delta_0}(\exp)$) The $L_1(\exp)$ system $\mathbf{i\Delta_0}(\exp)$ is based on intuitionistic logic with equality and consists of Δ_0 induction, and the following axioms.

(i) Basic arithmetic

$$\begin{array}{ll} x+0=x; & x+(y+1)=(x+y)+1; \\ x\cdot 0=0; & x\cdot(y+1)=(x\cdot y)+x; \\ \neg(x<x); & x<y\wedge y<z\to x<z; \\ x<x+1; & x<y+1\to x<y\vee x+1=y. \end{array}$$

(ii) Defining equations for the exponentiation function exp:

$$\exp(x,0)=1; \qquad \exp(x,y+1)=\exp(x,y)\cdot x.$$

Definition 24.6 ($\mathbf{b\Sigma}_1(\exp)$) The $L_1(\exp)$ system $\mathbf{b\Sigma}_1(\exp)$ is $\mathbf{i\Delta}_0(\exp)$ plus Σ_1 bounding.

Definition 24.7 ($\mathbf{i\Sigma}_1$) The L_1 system $\mathbf{i\Sigma}_1$ is based on intuitionistic logic with equality and consists of the axioms of basic arithmetic in Definition 24.5 (i) and Σ_1 induction.

Proposition 24.8 *The system* $\mathbf{i\Sigma}_1(\exp)$*, which is defined by* $\mathbf{i\Delta}_0(\exp)$ *plus* Σ_1 *induction schema, is a definitional extension of* $\mathbf{i\Sigma}_1$.

Proof See Remark 24.29 (ii). □

A function $f:\mathbb{N}^n\to\mathbb{N}$ is *bounded* Δ_0 definable in $\mathbf{T}$ when there is a Δ_0^0 formula $B(\vec{x},y)$ and a term $t(\vec{x})$ in the language of $\mathbf{T}$ such that (1) $\mathbf{T}\vdash\forall\vec{x}(\exists!y<t(\vec{x}))B(\vec{x},y)$ and (2) $f(\vec{x})=y$ if and only if $\mathbb{N}\models y<t(\vec{x})\wedge B(\vec{x},y)$ for any numerals $\vec{x}$ and y, where a numeral is a term of the form $\underbrace{1+\cdots+1}_{n\text{ times}}$ for some natural number n.

Remark 24.9 In $\mathbf{i\Delta}_0(\exp)$, pairs and sequences of natural numbers are coded by natural numbers. More precisely, we have a Δ_0 predicate $x\in\mathbb{N}^{<\mathbb{N}}$ in L_1, intended to express that x is a code for a finite sequence of natural numbers, and bounded Δ_0 definable functions pairing $x,y\mapsto(x,y)$ and projections $x\mapsto(x)_i$ for $i<2$, length $x\mapsto|x|$, ith element $[x,i]\mapsto x(i)$, i-length sequence $x\mapsto{}^ix$, concatenation $[x,y]\mapsto x*y$ and restriction $[x,i]\mapsto x\restriction i$, such that $\mathbf{i\Delta}_0(\exp)$ prove the following:

- $x\le(x,y)\wedge y\le(x,y)\wedge((x,y)=(x',y')\to x=x\wedge y=y')$;
- $((x,y))_0=x\wedge((x,y))_1=y\wedge\forall z(\exists!x<z)(\exists!y<z)((z)_0=x\wedge(z)_1=y)$;
- $x\in\mathbb{N}^{<\mathbb{N}}\wedge y\in\mathbb{N}^{<\mathbb{N}}\wedge|x|=|y|\wedge(\forall i<|x|)(x(i)=y(i))\to x=y$;
- $\exists x(x\in\mathbb{N}^{<\mathbb{N}}\wedge|x|=0)$;
- $x\in\mathbb{N}^{<\mathbb{N}}\wedge y\in\mathbb{N}^{<\mathbb{N}}\to$
  $x*y\in\mathbb{N}^{<\mathbb{N}}\wedge|x*y|=|x|+|y|\wedge$
  $\forall i((i<|x|\to(x*y)(i)=x(i))\wedge(|x|\le i<|x*y|\to(x*y)(i)=y(i-|x|)))$;

- $x \in \mathbb{N}^{<\mathbb{N}} \wedge i < |x| \to x(i) < x \wedge |x| < x$;
- $x \in \mathbb{N}^{<\mathbb{N}} \wedge i < |x| \to x\lceil i \in \mathbb{N}^{<\mathbb{N}} \wedge |x\lceil i| = i \wedge \forall j < i((x\lceil i)(j) = x(j))$;
- ${}^i x \in \mathbb{N}^{<\mathbb{N}} \wedge |{}^i x| = i \wedge (\forall j < i)(({}^i x)(j) = x)$.

Let $A(z)$ be a formula in $L_1(\exp)$. If a function $f(\vec{x})$ is bounded Δ_0 definable by a formula $\exists y < t(\vec{x})B(\vec{x}, y)$, then we write $A(f(\vec{x}))$ for the formula such that all of the atomic formulae $C(z)$ in $A(z)$ in which z occurs are replaced with $\exists z < t(\vec{x})(B(\vec{x}, z) \wedge C(z))$. Then $A(f(\vec{x}))$ has the same complexity as $A(z)$. Hence we can use such functions as if it were a part of $L_1(\exp)$ in $\mathbf{i\Delta}_0(\exp)$.

As mentioned in Proposition 24.8, $\mathbf{i\Sigma}_1(\exp)$ is a definitional extension of $\mathbf{i\Sigma}_1$ and all Σ_{n+1} formulae in $L_1(\exp)$ can be interpreted into some Σ_{n+1} formulae in L_1 by Remark 24.29 (ii). Therefore, again we can use these functions as a part of L_1 in $\mathbf{i\Sigma}_1$.

Definition 24.10 ($L_{\mathbf{HA}}$) The one-sorted first-order language $L_{\mathbf{HA}}$ is the language consisting of the following:

- a constant symbol 0;
- function symbols for all primitive recursive functions including the successor function S.

The term $S0$ is often written as 1.

Definition 24.11 (**HA**) The $L_{\mathbf{HA}}$ system **HA** is based on intuitionistic logic and consists of the equality axioms, $L_{\mathbf{HA}}$ induction, and the following axioms.

(i) Axiom of the successor: $\neg S0 = 0$.
(ii) Defining equations of all primitive recursive functions, such as

$$x + 0 = x; \quad x + (Sy) = S(x + y); \quad x \cdot 0 = 0; \quad x \cdot Sy = x \cdot y + x.$$

Remark 24.12 **HA** is a definitional extension of the system HA, which is defined by $\mathbf{i\Sigma}_1$ plus L_1 induction. See Remark 24.29 (ii).

24.2.3 Systems of Friedman–Simpson Reverse Mathematics

For systems and results of Friedman–Simpson reverse mathematics we have the encyclopedia [21]. Here we only introduce the language and systems in which we will see the relationship with the systems over intuitionistic logic in later sections.

Definition 24.13 (L_S and $L_S(\exp)$) The language L_S of Friedman–Simpson reverse mathematics is the two-sorted language, whose sorts are called *number* and *set*. It includes L_1 as the number sort, and has variables $X, Y, Z, \ldots$ for the set sort, and relation symbol $\in$ whose arity is one for number and one for set and written as $x \in X$. The language $L_S(\exp)$ is L_S plus binary numerical function symbol exp.

All the systems in this subsection are based on two sorted classical logic with equality for numbers. Note that it does not contain the equality for sets. The notation $X = Y$ is an abbreviation for $\forall x(x \in X \leftrightarrow x \in Y)$.

Definition 24.14 (Class of formulae in L_S) The class Δ^0_0 consists of formulae without set quantifiers and unbounded number quantifiers. Δ^0_0 is also called Σ^0_0 and Π^0_0. The class Σ^0_{n+1} consists of formulae of the form $\exists x A$ for some Π^0_n formula A. The class Π^0_{n+1} consists of formulae of the form $\forall x A$ for some Σ^0_n formula A. The class Δ^1_0 of *arithmetical formulae* consists of formulae without set quantifiers. Δ^1_0 is called also Σ^1_0 and Π^1_0. The class Σ^1_{n+1} consists of formulae of the form $\exists X A$ for some Π^1_n formula A. The class of Π^1_{n+1} consists of formulae of the form $\forall X A$ for some Σ^1_n formula A.

In systems ACA_0, RCA_0, and RCA^*_0 which we define below, each instance of their axiom schema may have set parameters unless otherwise stated.

Definition 24.15 (ACA_0) The L_S-system ACA_0 consists of the axioms of basic arithmetic in Definition 24.5 (i) and the schema of *arithmetical comprehension*, that is,

$$\exists X \forall x(x \in X \leftrightarrow A(x)),$$

for any arithmetical formula $A(x)$ without free occurrences of X, and the *induction axiom*

$$(0 \in X) \wedge \forall x(x \in X \to x + 1 \in X) \to \forall x(x \in X).$$

Definition 24.16 (RCA_0) The L_S-system RCA_0 consists of the axioms of basic arithmetic in Definition 24.5(i) and the schema of Σ^0_1 induction and Δ^0_1 *comprehension*

$$\forall x(A(x) \leftrightarrow B(x)) \to \exists X \forall x(x \in X \leftrightarrow A(x)),$$

for any Σ^0_1 formula $A(x)$ and any Π^0_1 formula $B(x)$ without free occurrences of X.

Definition 24.17 (RCA^*_0) The $L_S(\exp)$-system RCA^*_0 consists of the axioms of basic arithmetic and the defining equations for the exponentiation function in Definition 24.5 and the schema of Δ^0_0 induction and Δ^0_1 comprehension.

Remark 24.18 Most of the results of reverse mathematics show the equivalence over RCA_0 in [21]. By arithmetical comprehension, the existence of sets defined by any Σ^0_1 formula is proved in ACA_0. Therefore ACA_0 proves Σ^0_1 induction and so ACA_0 can be seen as the system RCA_0 plus arithmetical comprehension. In [21], it is presented that numbers of mathematical theorems, including the Bolzano–Weierstraß theorem and the existence of algebraic closure for countable fields, are equivalent to arithmetical comprehension over RCA_0. ACA_0 is sometimes used as the base theory for reverse mathematics. In [22], it is shown that Δ^0_2 determinacy is equivalent to Π^1_1 transfinite recursion over ACA_0. By the same way as proving that $\mathbf{i\Sigma}_1(\exp)$ is a definitional extension of $\mathbf{i\Sigma}_1$, we can prove that $\mathsf{RCA}_0^* + \Sigma^0_1$ induction is a definitional extension of RCA_0. By using RCA_0^* as the base theory, we can see the equivalence between mathematical statements with Σ^0_1 induction. Some of them can be found in [14], [20] or [21, X.4]. As for the consistency strength, ACA_0 proves the consistency of RCA_0 (cf. [21, Corollary VIII.1.7]). ACA_0 is a conservative extension of PA, which has the same axioms as HA but is based on classical logic (cf. [21, Section IX.1]). It is known that RCA_0 is a conservative extension of $\mathbf{I\Sigma}_1$, which is $\mathbf{i\Sigma}_1$ over classical logic (cf. [21, Section IX.1]), and RCA_0^* is a conservative extension of $\mathbf{B\Sigma}_1(\exp)$, which is $\mathbf{b\Sigma}_1(\exp)$ over classical logic. Since $\mathbf{i\Sigma}_1$ proves the consistency of $\mathbf{B\Sigma}_1(\exp)$ (cf. [17, Section 5.2]), RCA_0 proves the consistency RCA_0^*. In [20, Section 4], it is also shown that RCA_0^* and $\mathbf{I\Delta}_0(\exp)$, which is $\mathbf{i\Delta}_0(\exp)$ over classical logic, and hence $\mathbf{B\Sigma}_1(\exp)$ prove the same Π_2 sentences. See also [18, Theorem 11 and Corollary 12].

24.3 Function-Based Language and Systems

24.3.1 Systems of Natural Numbers and Functions on Them

Almost all known results in formal constructive reverse mathematics are given in systems of natural numbers and functions on them, but not sets. In this subsection, we will see one of the simplest such language L_F, which has the same number sort part as L_S. All systems in Sections 24.4–24.6 are variations of them.

The systems in this subsection are based on intuitionistic logic with equality for numbers. The fruit of using functions as primitive language is the decidability of atomic formulae. In general, we can prove $\alpha(x) = 0 \vee \neg\alpha(x) = 0$ for a function α on natural numbers by induction, but not $x \in X \vee \neg x \in X$ for a set X of natural numbers over intuitionistic logic in general.

Definition 24.19 (L_F) The language L_F has two sorts, called *number* and *function*. L_F contains L_1 as the number sort and additionally has the following:

- function variables $\alpha, \beta, \gamma, \ldots$;
- function symbol Ap with arity one of function and one of number, and with the value of number, which is called the *application operator*.

In what follows, terms of number sort are simply called *terms*, and terms of function sort are called *functor*, which are defined in each language. We use s and t for terms as in the previous section, and φ and ψ for functors. The terms in L_F are inductively defined as follows:

- the number constants 0 and 1 and number variables are terms;
- if s and t are terms, then $s + t$ and $s \cdot t$ are terms;
- if s is a term and α is a function variable, then $\mathrm{Ap}(\alpha, s)$ is a term.

The language L_F has no functors except for function variables.

For a term s and a functor φ, $\mathrm{Ap}(\varphi, s)$ is usually denoted by $\varphi(s)$ or even φs.

In L_F systems and other function-based systems in the following sections, the paring function is included in their language, or bounded Δ^0_0 definable. We will use abbreviations $\varphi(x, y)$ and $\varphi(x, y, z)$ for $\varphi((x, y))$ and $\varphi(((x, y), z))$, respectively.

Definition 24.20 (Class of formuale in L_F) The class Δ^0_0 consists of formulae without function quantifiers and unbounded number quantifiers. The class Δ^0_0 is also called Σ^0_0 and Π^0_0. The class Σ^0_{n+1} consists of formulae of the form $\exists x A$ for some Π^0_n formula A. The class Π^0_{n+1} consists of formulae of the form $\forall x A$ for some Σ^0_n formula A. The class Δ^1_0 of *arithmetical formulae* consists of formulae without set quantifiers. Δ^1_0 is also called Σ^1_0 and Π^1_0. The class Σ^1_{n+1} consists of formulae of the form $\exists \alpha A$ for some Π^1_n formula A. The class Π^1_{n+1} consists of formulae of the form $\forall \alpha A$ for some Σ^1_n formula A.

In the systems EL, EL_0, and EL^*_0 which we define below, each instance of axiom schema may contain free function parameters unless otherwise stated.

Definition 24.21 (EL) The L_F system EL is based on intuitionistic logic with the equality for numbers and consists of the axiom of basic arithmetic in Definition 24.5 (i), L_F induction, and the following schema of Δ^0_0 choice for numbers, denoted by Δ^0_0-AC^{00}:

$$\forall x \exists y A(x, y) \rightarrow \exists \alpha \forall x A(x, \alpha(x)),$$

where $A(x, y)$ is any Δ^0_0 formula without free occurrences of α.

Definition 24.22 (EL_0) The L_F-system EL_0 is obtained from EL by restricting the induction schema to Σ^0_1.

Definition 24.23 (L_{F^*}) The language L_{F^*} consists of L_F and

- the binary function symbol exp;
- the restriction operator Rest.

$\mathrm{Rest}(\alpha, x)$ is usually denoted $\alpha \lceil x$.

The terms in L_{F^*} is defined by the following clauses by adding those of L_1:

- if t and s are terms, then $\exp(t, s)$ is a term;
- if α is a function variable and t is a term, then $\alpha \lceil t$ is a term.

Functors in L_{F^*} are just function variables.

The system EL_0^* is used in [16] to show that the Baire Category Theorem is equivalent to Σ_1^0 induction.

Definition 24.24 (EL_0^*) The L_{F^*} system EL_0^* is based on intuitionistic logic with equality for numbers and consists of the axiom of basic arithmetic and the defining equation for exp in Definition 24.5, Δ_0^0 induction, Δ_0^0-AC^{00}, and the following *axiom for the restriction operator*:

$$\alpha\lceil 0 = \langle\rangle; \qquad\qquad \alpha\lceil(x+1) = (\alpha\lceil x) * \langle\alpha(x)\rangle,$$

where $\langle\rangle$ is the (code of the) empty sequence, $*$ is the concatenation operator, and $\langle t\rangle$ is the one-length sequence ${}^1t$ described in Remark 24.9.

$\Delta 0_0$-AC^{00} in EL, EL_0, and EL_0 can be enhanced as follows.

Lemma 24.25 *Each of* EL, EL_0, *and* EL_0^* *proves the schema of* Σ_1^0 *choice for numbers, denoted by* Σ_1^0-AC^{00}*:*

$$\forall x \exists y A(x, y) \to \exists \alpha \forall x A(x, \alpha(x)),$$

where A is any Σ_1^0 formula without free occurrences of α.

Proof Assume $\forall x \exists y A(x, y)$ for a Σ_1^0 formula $A(x, y) \equiv \exists z B(x, y, z)$, where $B(x, y, z)$ is a Δ_0^0 formula. By the assumption, we have $\forall x \exists y B(x, (y)_0, (y)_1)$, and so we have α such that $\forall x B(x, (\alpha(x))_0, (\alpha(x))_1)$ by Δ_0^0-AC^{00}. Then β defined by $\beta(x) = (\alpha(x))_0$ satisfies $\forall x A(x, \beta(x))$. □

24.3.2 Interpretation Between Function-Based Systems

In the following sections, we consider variations of function-based systems on intuitionistic logic. To analyze the relationship between them, we use interpretations such that the complexity of formulae in Σ_{n+1}^0 and Π_{n+1}^0 are not changed, which is crucial to interpret, for example, an instance of Σ_1^0 induction in the language L' of the original system $\mathbf{T}'$ into an instance of the same schema in the language L of the target system $\mathbf{T}$.

In this subsection, let L and L' be function-based languages, that is, two-sorted languages, where one sort is called number and the other is called function such that

- it has a numerical constant symbol 0;
- a function symbol Ap with arity one of function and one of number, and with the value of number is included.

The class of formulae are defined in the same way as the ones of L_F formulae.[2] The systems $\mathbf{T}$ and $\mathbf{T}'$ are based on intuitionistic logic with equality for number sort and have to satisfy the following properties:

- $\mathbf{i\Sigma}_1$ or $\mathbf{b\Sigma}_1$ is included in $\mathbf{T}$ and $\mathbf{T}'$, where $\mathbf{i\Sigma}_1$ is included in $\mathbf{T}$ means
  - there is a function -$^\triangle$ from the set of terms in L_1 to the one of L such that
    - $\triangle$ commutes with $+$ and $\times$;
    - $\mathbf{i\Sigma}_1 \vdash s = t$ if and only if $\mathbf{T} \vdash s^\triangle = t^\triangle$;
    - there is an atomic formula $A(x, y)$ in L such that $\mathbf{i\Sigma}_1 \vdash x < y$ if and only if $\mathbf{T} \vdash A(x^\triangle, y^\triangle)$ for any numeral x and y;
  - $\mathbf{T}$ proves Σ^0_1 induction in L;

  and $\mathbf{b\Sigma}_1$ is included in $\mathbf{T}$ means
  - there is a function $\triangle$ from the set of terms in L_1 to the one of L such that
    - $\triangle$ commutes with $+$, $\times$ and exp;
    - $\mathbf{b\Sigma}_1(\exp) \vdash s = t$ if and only if $\mathbf{T} \vdash s^\triangle = t^\triangle$;
    - there is an atomic formula $A(x, y)$ in L such that $\mathbf{b\Sigma}_1(\exp) \vdash x < y$ if and only if $\mathbf{T} \vdash A(x^\triangle, y^\triangle)$ for any numeral x and y;
  - $\mathbf{T}$ proves Σ^0_1 bounding in L;
- both $\mathbf{T}$ and $\mathbf{T}'$ proves Δ^0_0-LEM, that is, $A \vee \neg A$ for any Δ^0_0 formula A in L.

Proposition 24.26 *If there is a function $*$ from the sets of formulae L' and the one of L such that*

(i) $A^* \equiv A$ *for* A *in* $L \cap L'$;
(ii) $\mathbf{T}' \vdash A^* \leftrightarrow A$;
(iii) $*$ *commutes with all logical connectives and quantifiers;*
(iv) $\mathbf{T} \vdash A^*$ *for each axiom in* T',

then $\mathbf{T}'$ *is interpretable in* $\mathbf{T}$ *by* $*$.

Proof By conditions (iii) and (iv) we can show that $\mathbf{T}' \vdash A$ yields $\mathbf{T} \vdash A^*$ for any formula A in L' by induction on the derivation in $\mathbf{T}'$. □

[2] If any predicate symbol P of L does not take any function inputs, then we define the class Δ^0_0 in L as the class of formulae in L with no function quantifiers nor unbounded number quantifiers, and the classed Σ^i_n and Π^i_n are defined in the same way as before. Otherwise, we first define the class Δ^0_0 as the class of formulae with no function quantifiers, unbounded number quantifiers, nor predicate symbols with function inputs, and the complexity of atomic formulae $P(\vec{\alpha}, \vec{x})$ with function inputs $\vec{\alpha}$ is determined according to its defining formulae without function inputs. An example of such P is $\alpha =_1 \beta$ whose defining formula is $\forall x(\alpha(x) = \beta(x))$ in L_{BIM} in Subsection 24.4.3. We do not consider predicate symbols which do not have defining formula without function inputs in this chapter.

A function symbol in L (or L') is called a *term constructor* if its value is number sort and is called a *functor constructor* if its value is function sort. For a term constructor f, a functor constructor η and a predicate symbol P in L, *defining formulae* for f, η and P are formulae A_f, A_η, and A_P in L such that

$$\mathbf{T} \vdash \forall\vec{\alpha}\forall\vec{x}\forall y(f(\vec{\alpha},\vec{x}) = y \leftrightarrow \mathrm{A}_f(\vec{\alpha},\vec{x},y));$$
$$\mathbf{T} \vdash \forall\vec{\alpha}\forall\vec{x}\forall\beta\forall u(\eta(\vec{\alpha},\vec{x})(u) = \beta(u) \leftrightarrow \mathrm{A}_\eta(\vec{\alpha},\vec{x},\beta(u),u));$$
$$\mathbf{T} \vdash \forall\vec{\alpha}\forall\vec{x}(P(\vec{\alpha},\vec{x}) \leftrightarrow \mathrm{A}_P(\vec{\alpha},\vec{x})),$$

respectively. Note that $f(\vec{x}) = y$ itself is a defining formula for f.

In [24, Chapter 2, Section 7], the interpretation $*$ with Proposition 24.26 was given for general systems, such that $L \subseteq L'$ and $\mathbf{T}'$ is an extension of $\mathbf{T}$ with defining formulae for new function symbols. This $*$ does not always keep the complexity of formulae in general. If $\mathbf{T}$ has a restricted axiom schema such as Σ^0_1 induction and the axiom schema Δ^0_0 choice for numbers and if new symbols are defined by complex formulae in L, then an instance of axiom schema in L' may not be translated into an instance of axiom schema with the same complexity in L, and so the axiom schema in $\mathbf{T}$ may not be applied to the translated instance, which requires some other technique to show $\mathbf{T} \vdash A^*$ for each axiom A in $\mathbf{T}'$.

Lemma 24.27 *Assume that, for each term t and each functor φ in L', there are Σ^0_1 formulae A_t and A_φ in $L \cap L'$ such that*

$$\mathbf{T}' \vdash \forall\vec{\alpha}\forall\vec{x}\forall y(t = y \leftrightarrow \mathrm{A}_t(\vec{\alpha},\vec{x},y));$$
$$\mathbf{T} \vdash \forall\vec{\alpha}\forall\vec{x}\exists!y\forall u\mathrm{A}_t(\vec{\alpha},\vec{x},y);$$
$$\mathbf{T}' \vdash \forall\vec{\alpha}\forall\vec{x}\forall\beta\forall u(\varphi(u)=\beta(u) \leftrightarrow \mathrm{A}_\varphi(\vec{\alpha},\vec{x},\beta(u),u));$$
$$\mathbf{T} \vdash \forall\vec{\alpha}\forall\vec{x}\forall u\exists!y\mathrm{A}_\varphi(\vec{\alpha},\vec{x},y,u),$$
$$\mathbf{T} \vdash \forall\vec{\alpha}\forall\vec{x}\exists\beta\forall u\mathrm{A}_\varphi(\vec{\alpha},\vec{x},\beta(u),u),$$

and that each predicate symbol P in $L' \cup L$ has a Δ^0_0 defining formulae in $L' \cap L$ without function inputs. Then there are functions $\diamond$ and $*$ from the sets of formulae L' and the one of L such that

(1) $A^\diamond \equiv A^* \equiv A$ *for A in L;*
(2) $\mathbf{T}' \vdash A^\diamond \leftrightarrow A^* \leftrightarrow A$;
(3) $\mathbf{T} \vdash A^\diamond \leftrightarrow A^*$;
(4) $*$ *commutes with all logical connectives and quantifiers;*
(5) *if A is a Σ^i_{n+1} (Π^i_{n+1}) formula in L', then there is a Σ^i_{n+1} (Π^i_{n+1}) formula B in L such that $\mathbf{T} \vdash A^\diamond \leftrightarrow B$.*

Therefore, if the above $*$ additionally satisfies

(vi) $\mathbf{T} \vdash A^*$ *for each axiom A in $\mathbf{T}'$,*

or, equivalently,

$\mathbf{T} \vdash A^\diamond$ *for each axiom* A *in* $\mathbf{T}'$,

then $\mathbf{T}'$ *is interpretable in* $\mathbf{T}$ *by both* $*$ *and* $\diamond$.

We only describe how to define A^* and $A^\diamond$ here. In Section 24.7, we show these A^* and $A^\diamond$ have the required property.

For each term t and functor φ in L', let $\mathrm{A}_t(\vec{\alpha}, \vec{x}, y)$ and $\mathrm{A}_\varphi(\vec{\alpha}, \vec{x}, y, u)$ be Σ^0_1 formulae with the stated property whose all parameters are displayed, and, for each predicate symbol P in $L \cup L'$, let A_P be a Δ^0_0 defining formula in $L' \cap L$. Then, by the normal form theorem for Σ^0_1 formulae, which holds in $\mathbf{i\Sigma}_1$ and $\mathbf{b\Sigma}_1(\exp)$ (see Lemma 24.64), we can find Δ^0_0 formulae B_t and B_φ such that both $\mathbf{T}$ and $\mathbf{T}'$ prove

$$\forall\vec{\alpha}\forall\vec{x}\forall y(\mathrm{A}_t(\vec{\alpha}, \vec{x}, y) \leftrightarrow \exists w \mathrm{B}_t(\overrightarrow{\alpha{\restriction}w}, \vec{x}, y, w));$$
$$\forall\vec{\alpha}\forall\vec{x}\forall u\forall y(\mathrm{A}_\varphi(\vec{\alpha}, \vec{x}, y, u) \leftrightarrow \exists w \mathrm{B}_\varphi(\overrightarrow{\alpha{\restriction}w}, \vec{x}, y, u, w)).$$

Let $y_1, \ldots, y_n$ and w be fresh variables. Then, for each formula A, define $E'(A)$, $E(A)$, $U'(A)$ and $U(A)$

$$E'(A) \equiv \begin{cases} \mathrm{B}_{t_1}(\overrightarrow{\alpha{\restriction}w}, \vec{x}, y_1) \wedge \mathrm{B}_{t_2}(\overrightarrow{\alpha{\restriction}w}, \vec{x}, y_2) \wedge y_1 = y_2 & \\ \qquad\qquad \text{if } A \equiv t_1 = t_2; & \\ \mathrm{B}_{t_1}(\overrightarrow{\alpha{\restriction}w}, \vec{x}, y_1) \wedge \cdots \wedge \mathrm{B}_{t_n}(\overrightarrow{\alpha{\restriction}w}, \vec{x}, y_n) \wedge \mathrm{A}_P(y_1, \ldots, y_n) & \\ \qquad\qquad \text{if } A \equiv P(t_1, \ldots, t_n) \text{ for a predicate symbol } P \in L \cup L'; & \\ E'(B) \circ E'(C) & \text{if } A \equiv B \circ C \text{ for } \circ \in \{\wedge, \vee\}; \\ U'(B) \to E'(C) & \text{if } A \equiv B \to C; \\ QxE'(B) & \text{if } A \equiv QxB(x) \text{ for } Q \in \{\forall, \exists\}; \\ Q\alpha E'(B) & \text{if } A \equiv Q\alpha B(x) \text{ for } Q \in \{\forall, \exists\}, \end{cases}$$

$$E(A) \equiv \exists\vec{y}_A \exists\vec{w}_A E'(A),$$

$$U'(A) \equiv \begin{cases} \mathrm{B}_{t_1}(\overrightarrow{\alpha{\restriction}w}, \vec{x}, y_2) \wedge \mathrm{B}_{t_2}(\overrightarrow{\alpha{\restriction}w}, \vec{x}, y_n) \to y_1 = y_2 & \\ \qquad\qquad \text{if } A \equiv t_1 = t_2; & \\ \mathrm{B}_{t_1}(\overrightarrow{\alpha{\restriction}w}, \vec{x}, y_1) \wedge \cdots \wedge \mathrm{B}_{t_n}(\overrightarrow{\alpha{\restriction}w}, \vec{x}, y_n) \to \mathrm{A}_P(y_1, \ldots, y_n) & \\ \qquad\qquad \text{if } A \equiv P(t_1, \ldots, t_n) \text{ for a predicate symbol } P \in L \cup L'; & \\ U'(B) \circ U'(C) & \text{if } A \equiv B \circ C \text{ for } \circ \in \{\wedge, \vee\}; \\ E'(B) \to U'(C) & \text{if } A \equiv B \to C; \\ QxU'(B) & \text{if } A \equiv QxB(x) \text{ for } Q \in \{\forall, \exists\}; \\ Q\alpha E'(B) & \text{if } A \equiv Q\alpha B(x) \text{ for } Q \in \{\forall, \exists\}, \end{cases}$$

$$U(A) \equiv \forall\vec{y}_A \forall\vec{w}_A U'(A),$$

where $\vec{y}_A$ and $\vec{w}_A$ are the sets of variables y_is and w_is which are not bounded in $E'(A)$, respectively. For each L' formula A, define $A^\diamond$ and A^* by

$$A^\diamond \equiv \begin{cases} A & \text{if } A \text{ is a formula in } L; \\ E(A) & \text{if } A \text{ is } \Delta^0_0 \text{ formula;} \\ \exists x E(B) & \text{if } A \equiv \exists x B \text{ is a } \Sigma^0_1 \text{ formula and } B \text{ is } \Delta^0_0; \\ \forall x U(B) & \text{if } A \equiv \forall x B \text{ is a } \Pi^0_1 \text{ formula and } B \text{ is } \Delta^0_0; \\ B^\diamond \circ C^\diamond & \text{if } A \equiv B \circ C \text{ for } \circ \in \{\wedge, \vee, \rightarrow\} \\ & \text{and } B \text{ and } C \text{ are not } \Delta^0_0; \\ Q x B^\diamond & \text{if } A \equiv Q x B \text{ for } Q \in \{\exists, \forall\} \text{ and } B \text{ is not } \Delta^0_0; \\ Q \alpha B^\diamond & \text{if } A \equiv Q x B \text{ for } Q \in \{\exists, \forall\}. \end{cases}$$

$A^* \equiv A^\diamond$ for atomic formulae in L';

Note that $*$ commutes all logical connectives and quantifiers.

Lemma 24.28 shows that if each of the defining formulae has a certain complexity, we may construct an interpretation which keeps the complexity of formulae.

Lemma 24.28 *Assume that each term constructor f and each functor constructor η in $L' \cup L$ have Σ^0_1 defining formulae A_f and A_η in $L \cap L'$ with the following properties:*

$$\mathbf{T} \vdash \forall \vec{\alpha} \forall \vec{x} \exists ! y \mathrm{A}_f(\vec{\alpha}, \vec{x}, y)$$

$$\mathbf{T} \vdash \forall \vec{\alpha} \forall \vec{x} \forall u \exists ! y \mathrm{A}_\eta(\vec{x}, \vec{\alpha}, y, u), \qquad \mathbf{T} \vdash \forall \vec{\alpha} \forall \vec{x} \exists \beta \forall u \mathrm{A}_\eta(\vec{x}, \vec{\alpha}, \beta(u), u),$$

where all parameters in A_f and A_η are displayed. Then, for each term t and each functor φ in L', there are formulae A_t and A_φ in $L' \cap L$ such that

$$\mathbf{T}' \vdash \forall \vec{\alpha} \forall \vec{x} \forall y (t = y \leftrightarrow \mathrm{A}_t(\vec{\alpha}, \vec{x}, y));$$

$$\mathbf{T} \vdash \forall \vec{\alpha} \forall \vec{x} \exists ! y \mathrm{A}_t(\vec{\alpha}, \vec{x}, y);$$

$$\mathbf{T}' \vdash \forall \vec{\alpha} \forall \vec{x} \forall \beta \forall u (\varphi(u) = y \leftrightarrow \mathrm{A}_\varphi(\vec{\alpha}, \vec{x}, \beta(u), u));$$

$$\mathbf{T} \vdash \forall \vec{\alpha} \forall \vec{x} \forall u \exists ! y \mathrm{A}_\varphi(\vec{\alpha}, \vec{x}, y, u);$$

$$\mathbf{T} \vdash \forall \vec{\alpha} \forall \vec{x} \exists \beta \forall u \mathrm{A}_\varphi(\vec{\alpha}, \vec{x}, \beta(u), u).$$

We only give how to construct A_t and A_φ with the required property here. In Section 24.7, we show that A_t and A_φ have the appropriate property.

Again by Lemma 24.64, we can find Δ^0_0 formulae B_f and B_η whose parameters are displayed such that

$$\forall \vec{\alpha} \forall \vec{x} \forall y (\mathrm{A}_f(\vec{\alpha}, \vec{x}, y) \leftrightarrow \exists w \mathrm{B}_f(\overrightarrow{\alpha \restriction w}, \vec{x}, y, w));$$

$$\forall \vec{\alpha} \forall \vec{x} \forall u \forall y (\mathrm{A}_\eta(\vec{\alpha}, \vec{x}, y, u) \leftrightarrow \exists w \mathrm{B}_\eta(\overrightarrow{\alpha \restriction w}, \vec{x}, y, u, w));$$

for the defining formulae A_f for each function symbol f and A_η for each function constructors η in $L \cup L'$.

We assume that, for each formula in L', each of the occurrences of terms and functor constructors are indexed so that each of them are distinguished with the index. In particular, even distinct term constructors such as f and g in $L \cup L'$ have distinct indexes.

In what follows, the variables $y_0, y_1, \ldots, \beta_0, \beta_1, \ldots,$ and u_φ are supposed to be fresh. For each term t and each functor φ, set $\varepsilon(t)$, $\varepsilon(\varphi)$, A_t, and A_φ by induction on the construction of terms and functors as

$$\varepsilon(t) \equiv \begin{cases} t & \text{if } t \text{ is a term in } L; \\ y_i & \text{if } t \equiv f_i(\vec{\psi}, \vec{s}); \\ \varepsilon(\psi)(\varepsilon(s)) & \text{if } t \equiv \psi(s), \end{cases}$$

$$\varepsilon(\varphi) \equiv \begin{cases} \varphi & \text{if } \varphi \text{ is a functor in } L; \\ \beta_i & \text{if } \varphi \equiv \eta_i(\vec{\psi}, \vec{s}), \end{cases}$$

$$\mathrm{A}_t = \begin{cases} \top \quad \text{if } t \text{ is a term in } L; \\ \exists \vec{\beta}_t \exists \vec{y}_t \exists \vec{w}_t \exists \vec{v}_t \Big(\mathrm{B}_f(\overrightarrow{\varepsilon(\psi)\restriction w_i}, \overrightarrow{\varepsilon(s)}, y_i, w_i) \wedge \\ \qquad \bigwedge_{j<j'} \mathrm{A}_{s_j} \wedge \bigwedge_{l<l'} (\forall u_{\psi_l} < w_i)(\exists \vec{\mathbf{v}}_{\psi_l} < v_i) \mathrm{A}_{\psi_l} \Big) \\ \qquad \text{if } t = f_i(s_0, \ldots, s_{j'-1}, \psi_0, \ldots, \psi_{l'-1}); \\ \exists \vec{\beta}_t \exists \vec{y}_t \exists \vec{w}_t \exists \vec{v}_t \Big((\forall u_\psi < \varepsilon(s) + 1) \exists \vec{\mathbf{v}}_\psi < v_i \mathrm{A}_\psi \wedge \mathrm{A}_s \Big) \\ \qquad \text{if } t \equiv \psi(s), \end{cases}$$

$$\mathrm{A}_\varphi \equiv \begin{cases} \top \quad \text{if } \varphi \text{ is a functor in } L; \\ \exists \vec{\beta}_\varphi \exists \vec{y}_\varphi \exists \vec{w}_\varphi \exists \vec{v}_\varphi \Big(\mathrm{B}_\eta(\overrightarrow{\varepsilon(\psi)\restriction w_i}, \overrightarrow{\varepsilon(s)}, y_i, u_\varphi, w_i) \wedge \\ \qquad \bigwedge_{l<l'} (\forall u_{\psi_l} < w_i)(\exists \vec{\mathbf{v}}_{\psi_l} < v_i) \mathrm{A}_{\psi_l} \wedge \bigwedge_{j<j'} \mathrm{A}_{s_j} \Big) \\ \qquad \text{if } \varphi \equiv \eta_i(\psi_0, \ldots \psi_{l'-1}, s_0, \ldots, s_{j'-1}), \end{cases}$$

where

- for each term t, $\vec{\beta}_t$, $\vec{w}_t$, and $\vec{v}_t$ are the sets of variables β_is, w_is, and v_is which are not bounded in A_t, and $\vec{y}_t$ is the set of y_is except for $\varepsilon(t)$ which are not bounded in A_t, respectively;
- for each functor φ, $\vec{\beta}_\varphi$, $\vec{w}_\varphi$, and $\vec{v}_\varphi$ are the sets of variables y_is, w_is, and v_is which are not bounded in $\varepsilon(t)$, and $\vec{\beta}_\varphi$ is the set of β_is except for $\varepsilon(\varphi)$ which are not bounded in A_φ, respectively;
- for each functor φ, $\vec{\mathbf{v}}_\varphi$ is the set of variables y_is, w_is, and v_is which are not bounded in A_φ.

The intuitive idea is as follows. For a term t in L', if its outermost constructor is f_i, it is replaced with $\varepsilon(t) = y_i$, and A_t describes the condition which y_i should satisfy. Similarly, for a functor φ, if its outermost constructor is η_i, then it is replaced with β_i and A_φ gives the requirement for β_i. The variable w_i is a witness for Σ^0_1 defining formula $\exists w B_\square$ for f and η.

Remark 24.29 Lemma 24.27 and Lemma 24.28 can be modified as follows.

(i) **Interpretation keeping Δ^0_0 complexity** Assume that, for each term t and functor φ in L', there are Δ^0_0 formulae A_t and A_φ and terms s_t and s_φ in $L \cap L'$ such that

$$\begin{aligned}
&\mathbf{T}' \vdash \forall\vec{\alpha}\forall\vec{x}\forall y(t = y \leftrightarrow \mathrm{A}_t(\vec{\alpha}, \vec{x}, y));\\
&\mathbf{T} \vdash \forall\vec{\alpha}\forall\vec{x}(\exists! y < s_t)\mathrm{A}_t(\vec{\alpha}, \vec{x}, y);\\
&\mathbf{T}' \vdash \forall\vec{\alpha}\forall\vec{x}\forall\beta\forall u(\varphi(u) = \beta(u) \leftrightarrow \mathrm{A}_\varphi(\vec{x}, \vec{\alpha}, \beta(u), u));\\
&\mathbf{T} \vdash \forall\vec{\alpha}\forall\vec{x}\forall u(\exists! y < s_t)\mathrm{A}_\varphi(\vec{x}, \vec{\alpha}, y, u);\\
&\mathbf{T} \vdash \forall\vec{\alpha}\forall\vec{x}\exists\beta\forall u(\beta(u) < s_t \wedge \mathrm{A}_\varphi(\vec{x}, \vec{\alpha}, \beta(u), u));
\end{aligned}$$

and that each predicate symbol in L' has a Δ^0_0 defining formula in $L \cap L'$ without function inputs. Then there are functions $\diamond$ and $*$ from the sets of formulae L' to the one of L satisfying (i)–(v) of Lemma 24.27 by replacing $\mathrm{B}_\square$ and omitting $\vec{w}_A$ in the definition of $E'(A)$ and $U'(A)$ with $\mathrm{A}_\square$. Then we can show that $E(A)$ and $U(A)$ are equivalent some a Δ^0_0 formula in L.
Lemma 24.28 is modified as follows: If each term constructor f and each functor constructor η in $L' \cup L$ has bounded Δ^0_0 defining formulae A_f and A_φ, that is, there are Δ^0_0 formulae A_f and A_φ and terms s_f and s_η in $L \cap L'$ such that

$$\begin{aligned}
&\mathbf{T}' \vdash \forall\vec{\alpha}\forall\vec{x}\forall y(f(\vec{\alpha}, \vec{x}) = y \leftrightarrow \mathrm{A}_f(\vec{\alpha}, \vec{x}, y));\\
&\mathbf{T} \vdash \forall\vec{\alpha}\forall\vec{x}\exists! y < s_t \mathrm{A}_f(\vec{\alpha}, \vec{x}, y)\\
&\mathbf{T}' \vdash \forall\vec{\alpha}\forall\vec{x}\forall\beta\forall u(\eta(\vec{\alpha}, \vec{x})(u) = \beta(u) \leftrightarrow \mathrm{A}_\varphi(\vec{\alpha}, \vec{x}, \beta, u));\\
&\mathbf{T} \vdash \forall\vec{\alpha}\forall\vec{x}\forall u(\exists! y < s_\varphi)\mathrm{A}_\eta(\vec{x}, \vec{\alpha}, \beta(u), u),\\
&\mathbf{T} \vdash \forall\vec{\alpha}\forall\vec{x}\exists\beta\forall u(\beta(u) < s_\varphi \wedge \mathrm{A}_\eta(\vec{x}, \vec{\alpha}, \beta(u), u)),
\end{aligned}$$

where all parameters in A_f and A_η are displayed, then for each term t and each functor φ in L', there are Δ^0_0 formulae A_t and A_φ in $L' \cap L$ with the above property.

(ii) **Interpretations between first-order arithmetic** By omitting all function variables, we can construct an interpretation between systems of first order arithmetic. Proposition 24.8 is proved in this way.

(iii) **A simpler way of interpretation** If we do not require (v) in Lemma 24.27, we can relax the condition in Lemma 24.28 as any $L' \cap L$ definable term and functor constructor. For defining formulae $\mathrm{A}_f(\vec{\alpha}, \vec{x})$ and $\mathrm{A}_\eta(\vec{\alpha}, \vec{x})$, replace the definition A_t and A_φ with

$$\mathrm{A}_t = \begin{cases} \top \quad \text{if } t \text{ is a term in } L; \\ \exists \vec{\beta}_t \exists \vec{y}_t (\mathrm{A}_f(\overrightarrow{\varepsilon(\psi)}, \overrightarrow{\varepsilon(s)}, y_i) \wedge \bigwedge_{l<l'} \forall u_{\psi_l} \mathrm{A}_{\psi_l} \wedge \bigwedge_{j<j'} \mathrm{A}_{s_j}) \\ \qquad \text{if } t = f_i(s_0, \ldots, s_{j'-1}, \psi_0, \ldots, \psi_{l'-1}); \\ \exists \vec{\beta}_\psi \forall u_\psi \mathrm{A}_\psi \wedge \mathrm{A}_s \\ \qquad \text{if } t \equiv \psi(s), \end{cases}$$

$$\mathrm{A}_\varphi \equiv \begin{cases} \top \quad \text{if } \varphi \text{ is a functor in } L; \\ \exists \vec{\beta}_\varphi \exists \vec{y}_\varphi (\mathrm{A}_\eta(\overrightarrow{\varepsilon(\psi)}, \overrightarrow{\varepsilon(s)}, y_i, u_\varphi) \wedge \bigwedge_{l<l'} \forall u_{\psi_l} \mathrm{A}_{\psi_l} \wedge \bigwedge_{j<j'} \mathrm{A}_{s_j}) \\ \qquad \text{if } \varphi \equiv \eta_i(\psi_0, \ldots \psi_{l'-1} s_0, \ldots, s_{j'-1}), \end{cases}$$

and set $\vec{w}_t = \vec{w}_\varphi = \vec{w}_A = \vec{v}_t = \vec{v}_\varphi = \vec{v}_A = \emptyset$. Then $A^\diamond$ and A^* defined above satisfies the conditions (i)–(iv) in Lemma 24.27. The interpretation in [24, Chapter 2, Section 7] is similar to this interpretation.

24.3.3 Interpretation Between Set-Based Systems and Function-Based Systems

In this subsection, we will see how to construct interpretations between set-based systems defined in Subsection 24.2.3 and function-based systems over classical logic.

For a class Γ of formulae, Γ-LEM is the schema $A \vee \neg A$ for any Γ formula A which may have any function parameters. A system with L-LEM can be regarded as systems over classical logic.

Proposition 24.30 ACA_0 *and* $\mathsf{EL}_0 + L_F$-LEM $+ \Pi^0_1$-AC!^{00} *are mutually interpretable, where* Π^0_0-AC!^{00} *is*

$$\forall x \exists ! y A(x, y) \to \exists \alpha \forall x A(x, \alpha(x)),$$

for any Π^0_1 *formula* $A(x, y)$ *without free occurrences of* α. *Note that* $A(x, y)$ *may contain any function parameters other than* α.

Proposition 24.31 RCA_0 *and* $\mathsf{EL}_0 + L_F$-LEM *are mutually interpretable.*

Proposition 24.32 RCA^*_0 *and* $\mathsf{EL}^*_0 + L_{F^*}$-LEM *are mutually interpretable.*

Proposition 24.30, Proposition 24.31, and Proposition 24.32 are proved in the same way.

A typical way of interpretation from set-based language L_S to function-based language L_F is to interpret sets into their characteristic functions. For example, the interpretation $\mathfrak{ch}$ can be defined by

$$\begin{aligned}
&(\bot)^{\mathfrak{ch}}; \quad (t \in X)^{\mathfrak{ch}} \equiv \alpha_X(t) = 0;\\
&(sRt)^{\mathfrak{ch}} \equiv sRt \text{ for } R \in \{=, <\};\\
&(A \circ B)^{\mathfrak{ch}} \equiv A^{\mathfrak{ch}} \circ B^{\mathfrak{ch}} \text{ for } \circ \in \{\wedge, \vee, \to\};\\
&(QxA)^{\mathfrak{ch}} \equiv QxA^{\mathfrak{ch}} \text{ for } Q \in \{\exists, \forall\};\\
&(\forall XA)^{\mathfrak{ch}} \equiv \forall \alpha_X(\forall x(\alpha_X(x) < 2) \to A^{\mathfrak{ch}});\\
&(\exists XA)^{\mathfrak{ch}} \equiv \exists \alpha_X(\forall x(\alpha_X(x) < 2) \wedge A^{\mathfrak{ch}});
\end{aligned}$$

with a fixed injective assignment of a function variable α_X in L_F to each set variable X in L_S. Then, an L_S system $\mathbf{T}_S$ is interpreted into an L_F system $\mathbf{T}_F$ in the following procedure: By formalizing a set-based system $\mathbf{T}_S$ by sequent calculus and applying the standard cut elimination, we can reformulate each proof in $\mathbf{T}_S$ so that all the cut formulae are only instances of axioms of $\mathbf{T}_S$. Then we may assume that, for each theorem A in $\mathbf{T}_S$, there is a proof of A without any set parameters except for the ones in the consequence. By the induction on the derivation, we can prove that for $\mathbf{T}_S \vdash A$ implies $\mathbf{T}_F \vdash \forall \vec{\alpha}_X(\bigwedge \forall x(\alpha_X(x) < 2) \to A^{\mathfrak{ch}})$, where A has no set parameters other than $\vec{X}$. We can also see that $A^{\mathfrak{ch}}$ is Σ^i_n (Π^i_n) in L_F if A is Σ^i_n (Π^i_n) in L_S.

For the converse, functions over natural numbers are interpreted as their graphs. An intuitive idea is to send an atomic formula $\alpha(x) = y$ to $(x, y) \in X_\alpha$ for some set variable X_α in L_S. However, we need more tricks when function variables are nesting in a term like $\alpha(\beta(x)) = y$. For example, we can construct an interpretation from L_F (L_{F^*}) to L_S ($L_S(\exp)$) as follows: Fix an injective assignment of a set variable X_α in L_S to each function variable α in L_F. Define $\varepsilon(t)$ and A_t for each term t as

$$\varepsilon(t) \equiv \begin{cases} t & \text{if } t \text{ is a term in } L_S;\\ y & \text{otherwise;} \end{cases}$$

$$\mathrm{A}_t \equiv \begin{cases} \top & \text{if } t \text{ is a term in } L_S;\\ (\varepsilon(s), y) \in X_\alpha & \text{if } t = \alpha(s);\\ \varepsilon(t) \in \mathbb{N}^{<\mathbb{N}} \wedge |\varepsilon(t)| = \varepsilon(s) \wedge (\forall u < \varepsilon(s))(u, \varepsilon(t)(u)) \in X_\alpha & \\ & \text{if } t = \alpha \lceil s; \end{cases}$$

where y is a fresh variable. The intuitive idea for $\varepsilon(t)$ and A_t is the same as in the previous section. Then, for each Δ^0_0 formula A, define $E'(A)$, $U'(A)$, $E(A)$, and $U(A)$ as

$$E'(A) \equiv \begin{cases} \bigwedge_{j<2} \mathrm{A}_{s_j} \wedge \varepsilon(s_0) = \varepsilon(s_1) & \text{if } A \equiv s_0 = s_1; \\ \bigwedge_{j<2} \mathrm{A}_{s_j} \wedge \varepsilon(s_0) < \varepsilon(s_1) & \text{if } A \equiv s_0 < s_1; \\ E'(B) \circ E'(C) & \text{if } A \equiv B \circ C \text{ for } \circ \in \{\wedge, \vee\}; \\ U'(B) \to E'(C) & \text{if } A \equiv B \to C; \\ QxE'(B) & \text{if } A \equiv QxB(x) \text{ for } Q \in \{\forall, \exists\}; \end{cases}$$

$$U'(A) \equiv \begin{cases} \bigwedge_{j<2} \mathrm{A}_{s_j} \to \varepsilon(s_0) = \varepsilon(s_1) & \text{if } A \equiv s_0 = s_1; \\ \bigwedge_{j<2} \mathrm{A}_{s_j} \to \varepsilon(s_0) < \varepsilon(s_1) & \text{if } A \equiv s_0 < s_1; \\ U'(B) \circ U'(C) & \text{if } A \equiv B \circ C \text{ for } \circ \in \{\wedge, \vee\}; \\ E'(B) \to U'(C) & \text{if } A \equiv B \to C; \\ QxU'(B) & \text{if } A \equiv QxB(x) \text{ for } Q \in \{\forall, \exists\}; \end{cases}$$

$$E(A) \equiv \exists \vec{y}_A E'(A), \qquad U(A) \equiv \forall \vec{y}_A U'(A),$$

where $\vec{y}_A$ is the set of free number variables which is not in A but in $E'(A)$.

Then define $A^{\mathfrak{g}}$ as follows:

$$\begin{aligned}
&(\bot)^{\mathfrak{g}} \equiv \bot \\
&(sRt)^{\mathfrak{g}} \equiv E(sRt) \text{ for } R \in \{<, =\}; \\
&(B \circ C)^{\mathfrak{g}} \equiv B^{\mathfrak{g}} \circ C^{\mathfrak{g}} \text{ for } \circ \in \{\wedge, \vee, \to\}; \\
&(QxB)^{\mathfrak{g}} \equiv QxB^{\mathfrak{g}} \text{ for } Q \in \{\exists, \forall\}; \\
&(\forall \alpha B)^{\mathfrak{g}} \equiv \forall X_\alpha(\forall x \exists! y((x,y) \in X_\alpha) \to B^{\mathfrak{g}}); \\
&(\exists \alpha B)^{\mathfrak{g}} \equiv \exists X_\alpha(\forall x \exists! y((x,y) \in X_\alpha) \wedge B^{\mathfrak{ch}}).
\end{aligned}$$

We show that, for each Δ^0_0 formula $A(x)$ in L_F (L_{F^*}), $A^{\mathfrak{g}}(x)$ is equivalent to an atomic formula of the form $x \in X$ in $\mathbf{T}_S$ under appropriate assumptions. We can show that, for each Δ^0_0 formula $A(\vec{\alpha})$ in L_F whose function parameters are all displayed,

$$\mathbf{T}_S \vdash \forall \vec{X}_\alpha \left(\bigwedge \forall x \exists! y (x,y) \in X_\alpha \to (A^{\mathfrak{g}}(\vec{X}_\alpha) \leftrightarrow E(A)(\vec{X}_\alpha) \leftrightarrow U(A)(\vec{X}_\alpha)) \right).$$

By the definition, $E(A)$ is equivalent to a Σ^0_1 formula and $U(A)$ is equivalent to a Π^0_1 formula in L_S. Then Δ^0_1 comprehension in $\mathbf{T}_S$ yields

$$\forall \vec{X}_\alpha \left(\bigwedge_{\alpha \in \vec{\alpha}} \forall x \exists! y (x,y) \in X_\alpha \to \exists X (x \in X \leftrightarrow A^{\mathfrak{g}}(x)) \right).$$

As in the case of the interpretation from L_S to L_F, by formalizing the system $\mathbf{T}_F$ in L_F by sequent calculus, we can show that there is a proof of A in $\mathbf{T}_F$ without any free function variable except for the ones in A for each theorem A in $\mathbf{T}_F$.

Then, by induction on the derivation in $\mathbf{T}_F$, we can prove that $\mathbf{T}_F \vdash A(\vec{\alpha})$ yields $\mathbf{T}_S \vdash \forall \vec{X}_\alpha \left(\bigwedge_{\alpha \in \vec{\alpha}} \forall x \exists ! y(x,y) \in X_\alpha \to A^{\mathfrak{s}}(\vec{X}_\alpha)\right)$.

Remark 24.33 In the schema of Σ^0_1 *comprehension,*

$$\exists X \forall x (x \in X \leftrightarrow A(x)),$$

where $A(x)$ is a Σ^0_1 formula which may have any set parameter other than X. Therefore Σ^0_1 comprehension is equivalent to arithmetical comprehension over RCA_0 or RCA^*_0. For the proof of Proposition 24.30, we use this fact (cf. [17, Lemma 4.7 and Corollary 4.9]).

In [17], an interpretation method from L_F (L_{F^*})-systems over classical logic into systems over intuitionistic logic is developed. The following Proposition 24.34 is proved by Gödel–Gentzen's negative translation (cf. [24, Chapter 2, Section 3]), and Proposition 24.35 and Proposition 24.32 are proved by a combination of the negative translation and a version of Coquand–Hofmann forcing (cf. [1] and [4]).

Proposition 24.34 EL_0+L_F-LEM+Π^0_1-AC!^{00} *is* Δ^1_0 *preservingly interpretable in* EL_0+Σ^0_1-LEM+Π^0_1-AC!^{00}*. Therefore,* PA, ACA_0 *and* EL_0+Σ^0_1-LEM+Π^0_1-AC!^{00} *are mutually* L_1 *preservingly interpretable.*

Proof See [17, Corollary 4.9] □

Proposition 24.35 EL_0+L_F-LEM *is* Π^0_2 *preservingly interpretable in* EL_0*. Therefore,* $\mathbf{I\Sigma}_1$, RCA_0 *and* EL_0 *are mutually* Π_2 *preservingly interpretable.*

Proof It follows from [17, Corollary 4.15 and Lemma 4.16(ii)]. □

Proposition 24.36 EL^*_0 + L_{F^*}-LEM *is* Π^0_2 *preservingly interpretable in* EL^*_0*. Therefore* $\mathbf{B\Sigma}_1(\exp)$, RCA^*_0 *and* EL^*_0 *are mutually* Π_2 *preservingly interpretable.*

Proof It follows from [17, Corollary 4.15 and Lemma 4.16(ii)]. □

As for the relationship among L_F systems and L_1 systems over intuitionistic logic, we have the following. The proof of the following three propositions can be done in the same way as [23, Section 1.9.10], which shows that $\mathbf{EL}$ is a conservative extension of $\mathbf{HA}$.

Proposition 24.37 EL *is a conservative extension of* HA*. Therefore* EL *proves the same* Π_2 *sentences as* ACA_0.

Proposition 24.38 EL_0 *is a conservative extension of* $\mathbf{i\Sigma}_1$.

Proposition 24.39 EL^*_0 *is a conservative extension of* $\mathbf{b\Sigma}_1(\exp)$.

24.4 Base Theory with the Strength of ACA_0

In this section, we will see systems of constructive reverse mathematics whose consistency strength are the same as ACA_0, and relationship between them.

24.4.1 EL

The system **EL** was first presented in [12], and it plays an important role as an intuitionistic system of numbers and functions. In [6], it is used as the base theory to characterize the variation of the bar induction called the continuous bar induction.

Definition 24.40 ($L_{\mathbf{EL}}$) The language $L_{\mathbf{EL}}$ consists of the following:

- the number constant 0;
- number variables x, y, $z, \ldots$;
- function symbols for all primitive recursive functions, including the successor symbol S;
- function variables α, β, $\gamma, \ldots$;
- the application operator Ap;
- the abstract operator λ;
- the recursor $\boldsymbol{r}$.

Terms and functors in $L_{\mathbf{EL}}$ are defined as follows:

(i) the number constant 0 and each number variable are terms;
(ii) each unary function symbol and function variable are functors;
(iii) if $t_1, \ldots, t_n$ are terms and f is an n-ary function symbol, then $f(t_1, \ldots, t_n)$ is a term;
(iv) if t is a term and φ is a functor, then $\mathrm{Ap}(\varphi, t)$ is a term;
(v) if t is a term and x is a number variable, then $\lambda x.t$ is a functor;
(vi) if t and t' are terms and φ is a functor, then $\boldsymbol{r}(t, \varphi, t')$ is a term.

Definition 24.41 (**EL**) The system **EL** is based on intuitionistic logic with the equality for numbers and consists of the following axioms.

(i) Successor

$$\neg S0 = 0.$$

(ii) Defining equations for all primitive recursive functions, such as

$$x + 0 = x; \quad x + Sy = S(x + y); \quad x \cdot 0 = 0; \quad x \cdot Sy = x \cdot y + x.$$

(iii) λ-conversion

$$(\lambda x.t)s = t[x/s].$$

(iv) Axiom for recursor

$$\boldsymbol{r}(t, \varphi, 0) = t \qquad \boldsymbol{r}(t, \varphi, Ss) = \varphi(\boldsymbol{r}(t, \varphi, s), s).$$

(v) Quantifier-free axiom of choice for numbers

$$\forall x \exists y A(x, y) \to \exists \alpha \forall x A(x, \alpha(x)),$$

for any quantifier free formula $A(x, y)$ which may contain any function parameters but α.

(vi) $L_{\mathbf{EL}}$ induction

$$A(0) \wedge \forall x(A(x) \to A(Sx)) \to \forall x A(x)$$

for any quantifier free formula $A(x, y)$ which may contain function parameters.

24.4.2 WKV

The system WKV presented in [13] has the unrestricted axiom of unique choice for numbers, which is freely used in Bishop style constructive mathematics. In [13], it was shown that (a version of) Fan Theorem is equivalent to several theorems from analysis, including Heine–Borel Theorem and from logic, including Lindenbaum's Lemma.

Definition 24.42 (L_{WKV}) The language L_{WKV} consists of

- the number constant 0;
- number variables x, y, $z, \ldots$;
- the equality for numbers $=$;
- unary functions S (successor), K and L (projections);
- a binary function J (pairing);
- binary functions symbols $+$ and $\cdot$;
- function variables α, β, $\gamma, \ldots$;
- the application operator Ap;
- the abstraction operator λ.

Terms and functors are simultaneously defined as follows:

(i) the number constant 0 and each of the number variables are terms;
(ii) unary functions S, K, and L and function variables are functors;
(iii) if s and t are terms, then $s + t$, $s \cdot t$ and $J(s, t)$ are terms;
(iv) if t is a term and φ is a functor, then $\mathrm{Ap}(\varphi, t)$ is a term;
(v) if x is a number variable and t is a term, then $\lambda x.t$ is a functor.

Definition 24.43 (WKV) The system WKV is based on intuitionistic logic and consists of the following axioms.

(i) Equivalence axioms

$$x = y; \qquad x = y \to y = x; \qquad (x = y \wedge y = z) \to x = z.$$

(ii) Basic axioms of arithmetic

$$\begin{aligned} &\neg S(x) = 0; && S(x) = S(y) \to x = y; \\ &x + 0 = y; && x + S(y) = S(x + y); \\ &x \cdot 0 = 0; && x \cdot S(y) = x \cdot y + x. \end{aligned}$$

(iii) Paring axioms

$$K(J(x, y)) = x; \qquad L(J(x, y)) = y; \qquad J(K(x), L(x)) = x.$$

(iv) Function axiom

$$x = y \to \alpha(x) = \alpha(y).$$

(v) Conversion axiom schema

$$(\lambda x.s(x))(t) = s[x/t],$$

where $s(x)$ is a term and t is a term free for x in $s(x)$.

(vi) Axiom schema of induction

$$A(0) \wedge \forall x(A(x) \to A(Sx)) \to \forall x A(x),$$

for any formula $A(x)$ which may contain any function parameters.

(vii) Axiom schema of primitive recursion

$$\exists \beta(\beta(0) = t \wedge \forall y(\beta(Sy) = \varphi(J(y, \beta(y))))),$$

where t is a term and φ is a functor.

(viii) Axiom schema of unique choice for numbers

$$\forall x \exists! y A(x, y) \to \exists \alpha \forall x A(x, \alpha(x)),$$

for any formula $A(x, y)$ which may contain any function parameters but α.

Remark 24.44 Although the system WKV is based on intuitionistic logic without equality, it has an Equivalence axiom. Therefore we can regard it as a system based on intuitionistic logic with the equality for numbers.

24.4.3 BIM

The system BIM is introduced by [25]. In this paper, many equivalence between theorems from analysis and the Fan Theorem, and between *Kleene's alternative*, a version of contraposition of Fan Theorem were studied.

Definition 24.45 (L_{BIM}) The language L_{BIM} of BIM consists of

- the number constant 0;
- number variables x, y, $z, \ldots$;
- the binary paring function J;
- function variables α, β, $\gamma, \ldots$;
- the function constants $\underline{0}$, S (successor), K and L (projections);
- the equality for functions $=_1$;
- the application operator Ap.

Functors in L_{BIM} are function variables and function constants $\underline{0}$, S, K, and L.[3] Terms are inductively defined as follows.

(i) Number variables and the number constant 0 are terms.
(ii) If s and t are terms, then $S(t)$ and $J(s,t)$ are terms.
(iii) If s is a term and φ is a functor, then $\mathrm{Ap}(\varphi, s)$ is a term.

For a functor φ, we use abbreviations $\varphi(x,y)$ for $\varphi(J(x,y))$ and $\varphi(x,y,z)$ for $\varphi(J(J(x,y)),z)$, etc.

Definition 24.46 (BIM) The system BIM is based on intuitionistic logic with equality for each sorts and consists of the following axioms.

(i) Axiom of Extensionality

$$\forall\alpha\forall\beta(\alpha =_1 \beta \leftrightarrow \forall x(\alpha(x) = \beta(x))).$$

(ii) Definition of function constants

$$\begin{gathered}\forall x\neg(S(x) = 0) \wedge \forall x\forall y(S(x) = S(y) \to x = y)\wedge\\ \forall x(\underline{0}(x) = 0) \wedge \forall x\forall y(K(J(x,y)) = x \wedge L(J(x,y)) = y).\end{gathered}$$

(iii) Closure of the universe of functions under composition, primitive recursion and unbounded search

$$\begin{gathered}\forall\alpha\forall\beta\exists\gamma\forall x(\gamma(x) = \alpha(\beta(x)))\wedge\\ \forall\alpha\forall\beta\forall x\forall y\exists\gamma(\gamma(x,0) = \alpha(x) \wedge \gamma(x,S(y)) = \beta(x,y,\gamma(x,y)))\wedge\\ \forall\alpha(\forall x\exists y(\alpha(x,y) = 0) \to \exists\gamma\forall x(\alpha(x,\gamma(y)) = 0)).\end{gathered}$$

[3] In [25], they are called *function terms*.

(iv) Unrestricted Axiom Schema of Induction

$$A(0) \wedge \forall x(A(x) \to A(S(x))) \to \forall x A(x),$$

for any formula $A(x)$ which may contain any function parameters.

Remark 24.47 For each formula A of L_{BIM}, let A' be the formula obtained by replacing subformulae of the form $\varphi =_1 \psi$ with $\forall x(\varphi x = \psi x)$. By the Axiom of Extensionality, BIM proves $A \leftrightarrow A'$. Let BIM$'$ be the system whose language is $L_{\mathsf{BIM}} \setminus \{=_1\}$ and consists of all the axioms of BIM except the Axiom of Extensionality. Then it is easy to show that BIM is a definitional extension of BIM$'$.

24.4.4 Relationship Between Systems

In this subsection, we see the relationship between the systems mentioned in previous subsections.

Proposition 24.48 BIM *is interpretable in* **EL**.

Proof As in Remark 24.47, BIM is interpretable in BIM$'$. It is easy to see that each axiom of BIM$'$ is provable in **EL**, and so each proof in BIM$'$ is a proof in **EL**. □

Proposition 24.49 **EL** *is interpretable in* EL.

Proof First, note that each primitive recursive function symbol f has Σ_1 defining formula in L_1, that is, a L_1 formula A_f such that

$$\mathbb{N} \models \forall \vec{x} \forall y(f(x_0, \ldots, x_{j-1}) = y \leftrightarrow \mathrm{A}_f(x_0, \ldots, x_{j-1}, y)) \text{ and}$$
$$\mathsf{EL} \vdash \forall \vec{x} \forall y(f(x_0, \ldots, x_{j-1}) = y \leftrightarrow \mathrm{A}_f(x_0, \ldots, x_{j-1}, y)),$$

for any numerals $x_0, \ldots, x_{j-1}$ and y. The binary relation symbol $<$ in L_1 does not belong to $L_{\mathbf{EL}}$, but the characteristic function $\delta_<$ for the order relation $<$ such that, for any numeral x and y, $\mathbb{N} \models x < y$ if and only if $\mathbf{EL} \vdash \delta_<(x, y) = 0$. Hence, by regarding $s < t$ as the abbreviation of $\delta_<(s, t) = 0$, we have

$$\mathbf{EL} \vdash \forall \vec{x} \forall y(f(x_0, \ldots, x_{j-1}) = y \leftrightarrow \mathrm{A}_f(x_0, \ldots, x_{j-1}, y)).$$

EL also proves that

$$\forall \alpha \forall x \forall y \forall z(\boldsymbol{r}(x, \alpha, y) = z \leftrightarrow \mathrm{A}_{\boldsymbol{r}}(\alpha, x, y, z));$$

where

$$\mathrm{A}_{\boldsymbol{r}}(\alpha, x, y, z) \equiv \exists w(w \in \mathbb{N}^{<\mathbb{N}} \wedge |w| = Sy \wedge$$
$$w(0) = x \wedge (\forall j < y)(w(Sj) = \alpha(w(j), j)) \wedge w(y) = z).$$

It is easy to see that EL proves

$$\forall\vec{x}\exists!y\mathrm{A}_f(x_0,\dots,x_{j-1},y); \qquad \forall\alpha\forall x\forall y\exists!z\mathrm{A}_r(\alpha,x,y,z).$$

Furthermore, for each term t and s in $L_{\mathbf{EL}}$, if **EL** proves

$$\forall\vec{\alpha}\forall\vec{x}\forall y(t=y\leftrightarrow \mathrm{A}_t(\vec{\alpha},\vec{x},y)) \qquad \forall\vec{\alpha}\forall\vec{x}\forall y(s=y\leftrightarrow \mathrm{A}_s(\vec{\alpha},\vec{x},y))$$

for Σ^0_1 formulae A_t and A_s in L_F, where $\vec{x}$ are variables which may contain u, then **EL** proves

$$(\lambda u.t)s=w\leftrightarrow \exists z(\mathrm{A}_t(\vec{\alpha},\vec{x},w)[u/z]\wedge \mathrm{A}_s(\vec{\alpha},\vec{x},z))$$

for fresh variables w and z. Then, by induction on the construction of terms, we can show that for each term t and each functor φ, there exist Σ^0_1 formulae in L_F such that

$$\mathbf{EL}\vdash\forall\vec{\alpha}\forall\vec{x}\forall y(t=y\leftrightarrow \mathrm{A}_t(\vec{\alpha},\vec{x},y));$$
$$\mathsf{EL}\vdash\forall\vec{\alpha}\forall\vec{x}\exists!y\mathrm{A}_t(\vec{\alpha},\vec{x},y);$$
$$\mathbf{EL}\vdash\forall\vec{\alpha}\forall\vec{x}\forall\beta\forall u(\varphi(u)=\beta(u)\leftrightarrow \mathrm{A}_\varphi(\vec{\alpha},\vec{x},\beta,u));$$
$$\mathsf{EL}\vdash\forall\vec{\alpha}\forall\vec{x}\forall u\exists!y\mathrm{A}_\varphi(\vec{\alpha},\vec{x},y,u)\wedge\forall\vec{\alpha}\forall\vec{x}\exists\beta\forall u\mathrm{A}_\varphi(\vec{\alpha},\vec{x},\beta(u),u).$$

Then we can show that $\mathsf{EL}\vdash A^\diamond$ for any axiom of **EL** for the interpretation $\diamond$ constructed in Subsection 24.3.2 by using Lemma 24.25 and the fact that, for each Δ^0_0 formula A in $L_{\mathbf{EL}}$, $A^\diamond$ is equivalent to a Σ^0_1 formula in L_{EL}. Therefore **EL** is interpretable in EL by Lemma 24.27. □

Proposition 24.50 EL *is interpretable in* BIM.

Proof Let α_+, $\alpha_\times$ and $\alpha_<$ be fresh function variables. For each term t in L_F, set $t^\bullet$ by

$$t^\bullet\equiv t \text{ (if } t \text{ is term in } L_{\mathsf{BIM}}); \qquad (1)^\bullet\equiv S0;$$
$$(s+t)^\bullet\equiv\alpha_+(s^\bullet,t^\bullet); \qquad (s\cdot t)^\bullet\equiv\alpha_\times(s^\bullet,t^\bullet),$$

and define the interpretation $\bullet$ from the set for L_F formulae to the one of L_{BIM} formulae by

$$(s=t)^\bullet\equiv s^\bullet=t^\bullet; \qquad (s<t)^\bullet\equiv\alpha_<(s^\bullet,t^\bullet)=0;$$
$$(A\circ B)\equiv A^\bullet\circ B^\bullet \text{ for } \circ\in\{\wedge,\vee,\rightarrow\};$$
$$(QxA)^\bullet\equiv QxA^\bullet; \qquad (Q\alpha A)^\bullet\equiv Q\alpha A^\bullet.$$

Then we can show that $\mathsf{EL}\vdash A$ yields

$$\mathsf{BIM}\vdash\forall\alpha\forall\beta\forall\gamma(\mathrm{A}_+(\alpha)\wedge\mathrm{A}_\times(\alpha,\beta)\wedge\mathrm{A}_<(\gamma)\rightarrow A^\bullet),$$

where

$$\mathrm{A}_+(\alpha) \equiv \forall x \forall y(\alpha(x,0) = x \wedge \alpha(x,Sy) = S(\alpha(x,y)));$$
$$\mathrm{A}_\times(\alpha,\beta) \equiv \forall x \forall y(\mathrm{A}_+(\alpha) \wedge \beta(x,0) = 0 \wedge \beta(x,Sy) = \alpha(\beta(x,y),x));$$
$$\mathrm{A}_<(\alpha) \equiv \forall x \forall y(\alpha(x,Sx) = 0 \wedge (\alpha(x,Sy) = 0 \to \alpha(x,y) = 0 \vee x = y)).$$

□

The *realizability interpretation* based on continuous function application was introduced by [11], which gives a model of Brouwer's continuity axiom (cf. [24, Chapter 4, Section 6]). Nowadays, the most accessible literature is Troelstra [23, III, Section 3], which defines the following.

Definition 24.51 (Realizability interpretation) For a functor φ and a formula A in $L_{\mathbf{EL}}$, define $\varphi\mathbf{r}A$ as follows:

$$\begin{aligned}
\varphi\mathbf{r}A &\equiv A; \quad \text{for atomic } A;\\
\varphi\mathbf{r}A \wedge B &\equiv (\lambda x.K(\varphi(x))\mathbf{r}A) \wedge (\lambda x.L(\varphi(x))\mathbf{r}B);\\
\varphi\mathbf{r}A \to B &\equiv (\forall\psi(\psi\mathbf{r}A \to \varphi|\psi\downarrow \wedge \varphi|\psi\mathbf{r}B));\\
\varphi\mathbf{r}\forall xA &\equiv \forall x(\varphi|(\lambda u.x)\downarrow \wedge \varphi|(\lambda u.x)\mathbf{r}A);\\
\varphi\mathbf{r}\exists A &\equiv (\lambda x.L(\varphi(x)))\mathbf{r}A[x/K(\varphi(0))];\\
\varphi\mathbf{r}\forall\alpha A &\equiv \forall\alpha(\varphi|\alpha\downarrow \wedge \varphi|\alpha\mathbf{r}A);\\
\varphi\mathbf{r}\exists\alpha A &\equiv (\lambda x.K(\varphi(x)))\mathbf{r}A[\alpha/(\lambda x.L(\varphi(x)))],
\end{aligned}$$

where J, K and L primitive recursive pairing and projections satisfying axiom (iii) of WKV, where $\varphi|\psi$ is given by

$$(\varphi|\psi)(x) = \begin{cases} y & \text{if } \exists z(\forall v < z)(\varphi(\langle x\rangle * \psi\lceil v) = 0 \wedge \varphi(\langle x\rangle * \psi\lceil z) = y+1);\\ \text{undefined} & \text{otherwise,}\end{cases}$$

for functors φ and where ψ and $\varphi|\psi\downarrow$ is an abbreviation for $\forall x\exists y((\varphi|\psi)(x) = y)$.

An $L_{\mathbf{EL}}$ formula A is *almost negative* if it does not contain $\vee$ and $\exists$ appears only in front of the equations for terms.

Lemma 24.52 (i) *If A is a Π^0_2 formula in $L_{\mathbf{EL}}$, then there is an almost negative formula B such that*

$$\mathbf{EL} \vdash A \leftrightarrow B.$$

(ii) *If A is an almost negative formula, then*

$$\mathbf{EL} \vdash A \leftrightarrow \exists\alpha(\alpha\mathbf{r}A).$$

Proposition 24.53 WKV *is Π^0_2-preservingly interpretable in* **EL**.

Proof The *generalized continuous choice*, GC, is the following schema:

$$\forall\alpha(A(\alpha) \to \exists\beta B(\alpha,\beta)) \to \exists\gamma\forall\alpha(A(\alpha) \to \gamma|\alpha\downarrow \wedge B(\alpha,\gamma|\alpha)),$$

where A is almost negative, and where $A(\alpha)$ may contain any function parameters but β.

In [23, 3.3.11 Theorem], it is shown that

$$\mathbf{EL} + \mathrm{GC} \vdash A \leftrightarrow \mathbf{EL} \vdash \exists\alpha(\alpha\mathbf{r}A).$$

Since $L_{\mathsf{WKV}} \subseteq L_{\mathbf{EL}}$ and $\mathbf{EL}$ + GC proves all the axioms of WKV, we have the desired property by Lemma 24.52. □

Proposition 24.54 **EL** *is interpretable in* WKV.

Proof This can be proved in a similar way as Proposition 24.49. □

Corollary 24.55 WKV, **EL**, BIM *and* EL *are mutually interpretable.*

24.5 Base Theory with the Strength of RCA_0

In this section, we will see a subsystem of **EL** with the same consistency strength as the standard base theory RCA_0 and will see the relationship with a subsystem of EL.

24.5.1 $\mathbf{EL}_0$

The system $\mathbf{EL}_0$ is presented in [2]. In the paper, the equivalence between WKL_c, a kind of weak Kőnig's lemma, and the intermediate value theorem (IVT), etc., is presented.

Definition 24.56 ($\mathbf{EL}_0$) The system $\mathbf{EL}_0$ is obtained from **EL** by restricting the induction schema to Δ^0_0 formulae which may contain any free function variables.

Remark 24.57 $\mathbf{EL}_0$ proves Σ^0_1 induction. See [2, Proposition 1].

24.5.2 Relationship Between Systems

Proposition 24.58 $\mathbf{EL}_0$ *and* EL_0 *are mutually interpretable.*

Proof The interpretation from $\mathbf{EL}_0$ to EL_0 can be done in the same way as Proposition 24.49. Note that, for any Δ^0_0 formula A in $L_{\mathbf{EL}}$, $A^\diamond$ is equivalent to a Σ^0_1 formula in L_F. Therefore an instance of Δ^0_0 induction is interpreted as an instance of Σ^0_1 induction, which enables this interpretation. For the converse, for each L_F formula A, let A° be the formula whose subformula of the form $s < t$

with $\alpha_<(s,t) = 0$, where $\alpha_<$ is the characteristic function of the order relation $<$. Then we can prove that $\mathsf{EL}_0 \vdash A$ implies $\mathbf{EL}_0 \vdash A^\circ$. □

24.6 Base Theory with the Strength of RCA_0^*

In this section, we will see a base theory with the same consistency strength as RCA_0^* and the relationship between the systems.

24.6.1 $\mathbf{EL}_{\mathrm{ELEM}}$

The system $\mathbf{EL}_{\mathrm{ELEM}}$ is presented in [9], which does not have whole primitive recursive function symbols but only elementary recursive ones. As we will see in the next subsection, the consistency strength of $\mathbf{EL}_{\mathrm{ELEM}}$ is strictly weaker than EL_0 or WKV. The paper [9] shows that the equivalence between the Heine–Borel Theorem and a version of the Fan Theorem is provable even for this weaker base theory.

Definition 24.59 (The class ELEM of elementary functions) The class of Kalmár's *elementary functions*, ELEM, is generated by the following clauses.

(i) The following functions belong to ELEM: The constant zero function Z satisfying $Z(x) = 0$, the successor function S, the predecessor function P satisfying $P(0) = 0$ and $P(Sn) = n$, the addition $+$, the cut-off subtraction $\dot{-}$ satisfying $0 \dot{-} n = 0$ and $n \dot{-} Sm = P(n \dot{-} m)$, the multiplication $\cdot$, the projection function p_i^j satisfying $p_i^j(x_0, \ldots, x_j) = x_i$ for $i \leq j$.

(ii) ELEM is closed under composition: If f, g_i in ELEM with $f : \mathbb{N}^n \to \mathbb{N}$ and $g_i : \mathbb{N}^k \to \mathbb{N}$ $(1 \leq i \leq n)$, then there is an $h \in$ ELEM satisfying

$$h(\vec{x}) = f(g_1(\vec{x}), \ldots, g_n(\vec{x})).$$

(iii) ELEM is closed under bounded sum and bounded product: If $f \in$ ELEM with $f : \mathbb{N}^{n+1} \to \mathbb{N}$, then there are g, $h \in$ ELEM such that

$$g(0,\vec{x}) = 0; \qquad g(Sy,\vec{x}) = f(y,\vec{x}) + g(y,\vec{x});$$
$$h(0,\vec{x}) = 1; \qquad h(Sy,\vec{x}) = f(y,\vec{x}) \cdot h(y,\vec{x}).$$

We usually write $\sum_{z<y} f(z,\vec{x})$ and $\prod_{z<y} f(z,\vec{x})$ for g and h, respectively.

Definition 24.60 ($L_{\mathbf{EL}_{\mathrm{ELEM}}}$) The language $L_{\mathbf{EL}_{\mathrm{ELEM}}}$ is obtained from $\mathbf{EL}$ by restricting the primitive recursive function symbols to elementary function symbols and adding a function symbol $\boldsymbol{\mu}$ of arity one function and two numbers and of value number.

Terms and functors 40 in $L_{\mathbf{EL}_{\mathrm{ELEM}}}$ are defined by replacing (vi) in Definition 24.40 of terms and functors in L_{EL} with the following (vi)′:

(vi)′ If t and t' are terms and φ is a functor, then $\boldsymbol{\mu}(t, \varphi, t')$ is a term.

Definition 24.61 ($\mathbf{EL}_{\mathrm{ELEM}}$) The system $\mathbf{EL}_{\mathrm{ELEM}}$ is based on intuitionistic logic with the equality for numbers and consists of the following axioms.

(i) Successor

$$\neg S0 = 0.$$

(ii) Defining equations for all Kalmár's elementary functions.

(iii) λ-conversion

$$(\lambda x.t)s = t[x/s].$$

(iv) The axiom for the bounded minimum operator

$$\boldsymbol{\mu}(t, \varphi, 0) = \mathrm{sg}(\varphi(0, t)),$$
$$\boldsymbol{\mu}(t, \varphi, St') = \boldsymbol{\mu}(t, \varphi, t') + sg(\boldsymbol{\mu}(t, \varphi, t') \dot{-} t') \cdot \mathrm{sg}(\varphi(St', t)),$$

where $\mathrm{sg}(x)$ is an elementary function satisfying $\mathrm{sg}(x) = 0$ if $x = 0$ and $\mathrm{sg}(x) = 1$ otherwise. We usually write $\boldsymbol{\mu}(t, \varphi, t')$ as $\min\{x < t' : \varphi(x, t) = 0\}$.

(v) Δ^0_0 induction

$$A(0) \wedge \forall x(A(x) \to A(Sx)) \to \forall x A(x),$$

for any Δ^0_0 formula which may contain any function parameters.

24.6.2 Relationship between systems

Let EL_0^- be the system obtained from EL_0^* by replacing the Δ^0_0 axiom of choice with the following *axiom of Δ^0_0 bounded search*:

$$\exists\beta\forall x((\exists y < t(x))A(x, y) \to \beta(x) < t(x) \wedge A(x, \beta(x))),$$

where $A(x, y)$ is a Δ^0_0 formula which may contain any function parameters but β, and where $t(x)$ is any term in L_{F^*}.

Lemma 24.62 EL_0^- *is closed under compositions and bounded recursion of functions, that is,* EL_0^- *proves the following:*

$$\forall\alpha\forall\beta\exists\gamma\forall x(\gamma(x) = \alpha(\beta(x)));$$
$$\forall\alpha\forall\beta\forall\gamma\exists\delta\forall x\forall y(\delta(0, y) = \alpha(y) \wedge \delta(x + 1, y) = \min\{\beta(x, y), \gamma(\delta(x, y), x, y)\}).$$

Proof Since $\alpha(\beta(x)) + 1$ is a term in L_{F^*}, we have

$$\forall x(\exists y < \alpha(\beta(x)) + 1)(y = \beta(\alpha(x)));$$

by Δ_0^0 bounded search, we have γ with the desired property. The closure for bounded recursion is proved in a similar way by using additionally the axiom for restriction operator. □

Proposition 24.63 $\mathbf{EL}_{\mathrm{ELEM}}$ *and* EL_0^- *are mutually interpretable.*

Proof Since $\exp(x, y)$ is an elementary function, the interpretation $\bullet$ defined in Proposition 24.50 works also from EL_0^- to $\mathsf{EL}_{\mathrm{ELEM}}$ by setting $(\exp(s,t))^\bullet \equiv \exp(s^\bullet, t^\bullet)$.

For the converse direction, we can show that, for each elementary function symbol $f : \mathbb{N}^n \to \mathbb{N}$, there is a Δ_0^0 formula A_f and a term $t_f(\vec{x})$ in $L_1(\exp)$ such that

$$\begin{aligned}&\mathbb{N} \models \forall\vec{x}(f(x_0, \dots, x_{j-1}) = y \leftrightarrow \mathrm{A}_f(x_0, \dots, x_{j-1}, y));\\ &\mathrm{EL}_{\mathrm{ELEM}} \vdash \forall\vec{x}(f(x_0, \dots, x_{j-1}) = y \leftrightarrow \mathrm{A}_f(x_0, \dots, x_{j-1}, y));\\ &\mathsf{EL}_0^- \vdash \forall\vec{x}\exists! y < t_f \mathrm{A}_f(\vec{x}, y),\end{aligned}$$

by induction on the construction of the elementary functions. For atomic functions, that is, the constant zero function, etc., it is clear. If $f(x_0, \dots, x_{k-1}) = h(g_0(\vec{x}), \dots, g_{l-1}(\vec{x}))$, then $t_f(\vec{x}) \equiv t_h[x_0/t_{g_0}(\vec{x}), \dots, x_{l-1}/t_{g_{l-1}}(\vec{x})] + 1$ and

$$\mathrm{A}_f(\vec{x}, y) \equiv (\exists z_0 < t_{g_0}(\vec{x})) \dots (\exists z_{k-1} < t_{g_{k-1}}(\vec{x})) \left(\bigwedge_{j<k} \mathrm{A}_{g_j}(\vec{x}, z_j) \wedge \mathrm{A}_h(\vec{z}, y) \right)$$

satisfies the desired property.

Assume $f(z, \vec{x}) = \sum_{i \le z} g(i, \vec{x})$. Recall that there is a term ${}^i x$ for the code of the sequence v of length i and $v(j) = x$ for each x and that, for each sequence u and v, $|u| \le |v| \wedge (\forall j < |u|)(u(j) \le v(j))$ implies $u < v$. Then $t_f(z, \vec{x}) \equiv {}^{z+1}(t_g(z+1, \vec{x}))$ and

$$\begin{aligned}\mathrm{A}_f(z, \vec{x}, y) \equiv & \exists u < t_f(z, \vec{x})(u \in \mathbb{N}^{<\mathbb{N}} \wedge |u| = z + 1 \wedge \mathrm{A}_g(0, \vec{x}, u(0)) \wedge \\ & (\forall i{<}z)(\exists w{<}t_g(i, \vec{x}))(\mathrm{A}_g(i, \vec{x}, w) \wedge u(i{+}1){=}u(i){+}w) \wedge u(z){=}y)\end{aligned}$$

satisfy the desired property because we have the monotonicity of terms in $L_1(\exp)$, that is, $i \le z$ implies $t_g(i, \vec{x}) \le t_g(z, \vec{x})$.

If $f(z, x_0, \dots, x_{k-1}) = \prod_{i \le z} g(i, \vec{x})$, we can find $t_f(z, \vec{x},)$ and $A_f(z, \vec{x}, y)$ in a similar way.

For the term constructor $\boldsymbol{\mu}$, set $\mathrm{A}_{\boldsymbol{\mu}}(\alpha, x, y, z)$ as

$$\Big(z{\le}y \wedge \alpha(z) = 0 \wedge (\forall w{<}z)(\alpha(w) \ne 0) \Big) \vee \Big(z = y{+}1 \wedge (\forall w{\le}y)(\alpha(w) \ne 0) \Big).$$

Then we have

$$\begin{aligned}&\mathbf{EL}_{\mathrm{ELEM}} \vdash \forall\alpha\forall x\forall y\forall z(\mathrm{A}_{\boldsymbol{\mu}}(\alpha, x, y, z) \leftrightarrow \boldsymbol{\mu}(x, \alpha, y) = z);\\ &\mathsf{EL}_0^- \vdash \forall\alpha\forall x\forall y(\exists! z < y + 2)\mathrm{A}_{\boldsymbol{\mu}}(\alpha, x, y, z).\end{aligned}$$

As in the proof of Proposition 24.49, we can show that, for each term t and each functor φ in $L_{\mathbf{EL}_{\mathrm{ELEM}}}$, there are Δ^0_0 formulae A_t and A_φ and terms s_t and s_φ in L_{F^*} such that

$$\mathbf{EL}_{\mathrm{ELEM}} \vdash \forall\vec{\alpha}\forall\vec{x}\forall y(t = y \leftrightarrow \mathrm{A}_t(\vec{\alpha}, \vec{x}, y));$$
$$\mathsf{EL}_0^- \vdash \forall\vec{\alpha}\forall\vec{x}(\exists! y < s_t)\mathrm{A}_t(\vec{\alpha}, \vec{x}, y);$$
$$\mathbf{EL}_{\mathrm{ELEM}} \vdash \forall\vec{\alpha}\forall\vec{x}\forall\beta\forall u(\varphi(u) = y \leftrightarrow \mathrm{A}_\varphi(\vec{\alpha}, \vec{x}, \beta, u));$$
$$\mathsf{EL}_0^- \vdash \forall\vec{\alpha}\forall\vec{x}\forall u\exists! y(y < s_\varphi \wedge \mathrm{A}_\varphi(\vec{\alpha}, \vec{x}, y, u))$$
$$\mathsf{EL}_0^- \vdash \forall\vec{\alpha}\forall\vec{x}\exists\beta\forall u(\beta(u) < s_\varphi \wedge \mathrm{A}_\varphi(\vec{\alpha}, \vec{x}, \beta(u), u)).$$

By Remark 24.29(i), we can construct $*$ and $\diamond$ satisfying of Lemma 24.27 such that, for each Δ^0_0 formula A, $A^\diamond$ is equivalent to some Δ^0_0 formula in L_{F^*}. □

By the same method as Proposition 24.34, Proposition 24.35, and Proposition 24.36, we can show that $\mathsf{EL}_0^- + L_F$-LEM is Π^0_2-preservingly interpretable in EL_0^-. By the same method as Proposition 24.37, Proposition 24.38, and Proposition 24.39, we have also that EL_0^- is a conservative extension of $\mathbf{i\Delta}_0(\exp)$. Since $\mathsf{EL}_0^- + L_F$-LEM includes $\mathbf{I\Delta}_0(\exp)$ and is included by $\mathsf{EL}_0^* + L$-LEM, by Remark 24.18, it follows that RCA_0^*, $\mathbf{EL}_{\mathrm{ELEM}}$, EL_0^-, $\mathsf{EL}_0^- + L_F$-LEM, EL_0^*, $\mathsf{EL}_0^* + L_F$-LEM, $\mathbf{i\Delta}_0(\exp)$, $\mathbf{I\Delta}_0(\exp)$, $\mathbf{b\Sigma}_1(\exp)$, and $\mathbf{B\Sigma}_1(\exp)$ prove the same Π_2 sentences. In particular, they have the same consistency strength.

24.7 Appendix: Proof of Lemma 24.27 and Lemma 24.28

In this appendix, we show that the interpretations $\diamond$ and $*$ defined in Subsection 24.3.2 satisfy the conditions Lemma 24.27. Assume that L, L', $\mathbf{T}$, and $\mathbf{T}'$ satisfy the condition of Lemma 24.27.

We need several lemmata. The following is a Σ^0_1 *normal form theorem.*

Lemma 24.64 *For each Σ^0_1 formula $A(\vec{\alpha}, \vec{x})$ in L whose all parameters are displayed, there is a Δ^0_0 formula $B(\vec{u}, \vec{x})$ without function parameters such that*

$$\mathbf{T} \vdash \forall\vec{x}\forall\vec{\alpha}(A(\vec{\alpha}, \vec{x}) \leftrightarrow \exists w B(\overrightarrow{\alpha{\restriction}w}, \vec{x}, w));$$
$$\mathbf{T} \vdash \forall\vec{\alpha}\forall\vec{x}\forall w\forall v(B(\overrightarrow{\alpha{\restriction}w}, \vec{x}, w) \wedge v \geq w \rightarrow B(\overrightarrow{\alpha{\restriction}v}, \vec{x}, v).$$

Proof See [17, Lemma 2.13 and Theorem 2.14]. □

Lemma 24.65 *Both $\mathbf{T}$ and $\mathbf{T}'$ prove the following schema of Σ^0_1 bounding, that is,*

$$(\forall x < t)\exists u B(x, u) \rightarrow \exists v(\forall x < t)(\exists u < v)B(x, u)$$

for each Δ^0_0 formula B which may contain any function variables.

Proof Recall that $\mathbf{T}$ and $\mathbf{T}'$ include $\mathbf{b\Sigma}_1$ or $\mathbf{i\Sigma}_1$. It is clear that $\mathbf{b\Sigma}_1(\exp)$ proves Σ^0_1 bounding. Nemoto [15, Lemma 89.1] shows that $\mathbf{i\Sigma}_1$ yields Σ_1 bounding. □

Lemma 24.66 *For each Σ^0_1 formula $A(\vec{\alpha}, \vec{x})$, the formula $\exists\vec{\alpha}A(\vec{\alpha}, \vec{x})$ is equivalent to some Σ^0_1 formula.*

Proof By Lemma 24.64, for a Σ^0_1 formula $A(\vec{\alpha}, \vec{x})$, we can take a formula $\mathrm{B}(\vec{v}, x, w)$ such that $\mathbf{T}$ proves

(i) $\forall\vec{x}\forall\vec{\alpha}(A(\vec{\alpha}, \vec{x}) \leftrightarrow \exists w B(\overrightarrow{\alpha\restriction w}, \vec{x}, w))$, and
(ii) $\forall\vec{\alpha}\forall\vec{x}\forall w\forall v(B(\overrightarrow{\alpha\restriction w}, \vec{x}, w) \wedge v \geq w \rightarrow B(\overrightarrow{\alpha\restriction v}, \vec{x}, v)$.

Then $\exists\vec{\alpha}A(\vec{\alpha}, \vec{x}) \leftrightarrow \exists\vec{v}\exists w(\vec{v} \in \mathbb{N}^{<\mathbb{N}} \wedge B(\vec{v}, \vec{x}, w))$ holds. □

Lemma 24.67 *For each Π^0_1 formula $A(\vec{a}, \vec{x})$, the formula $\forall\vec{\alpha}A(\vec{\alpha}, \vec{x})$ is equivalent to some Π^0_1 formula.*

Proof By Lemma 24.64, for a Π^0_1 formula $A(\vec{\alpha}, \vec{x})$, we can take a formula $\mathrm{B}(\vec{v}, x, w)$ such that $\mathbf{T}$ proves

$$\forall\vec{x}\forall\vec{\alpha}(A(\vec{\alpha}, \vec{x}) \leftrightarrow \forall w B(\overrightarrow{\alpha\restriction w}, \vec{x}, w)).$$

Then $\forall\vec{\alpha}A(\vec{\alpha}, \vec{x}) \leftrightarrow \forall\vec{v}\forall w(\vec{v} \in \mathbb{N}^{<\mathbb{N}} \rightarrow B(\vec{v}, \vec{x}, w))$ holds. □

Lemma 24.68 *For each Δ^0_0 formula A in L', $E(A)$ is equivalent to some Σ^0_1 formula and $U(A)$ is equivalent to some Π^0_1 formula in* $\mathbf{T}$.

Proof This follows from Lemma 24.69, Lemma 24.66, and Lemma 24.67. □

Lemma 24.69 *For each Δ^0_0 formula A, both $U'(A)$ and $E'(A)$ are Δ^0_0 formulae in L and so $\mathbf{T} \vdash (U'(A) \vee \neg U'(A)) \wedge (E'(A) \vee \neg E'(A))$.*

Proof This is proved by induction on the construction of A. □

Lemma 24.70 *Assume that, for each term t and each functor φ in L', there are Σ^0_1 formulae A_t and A_φ in $L \cap L'$ such that*

$$\mathbf{T}' \vdash \forall\vec{\alpha}\forall\vec{x}\forall y(t = y \leftrightarrow \mathrm{A}_t(\vec{\alpha}, \vec{x}, y));$$
$$\mathbf{T} \vdash \forall\vec{\alpha}\forall\vec{x}\exists! y\forall u\mathrm{A}_t(\vec{\alpha}, \vec{x}, y);$$
$$\mathbf{T}' \vdash \forall\vec{\alpha}\forall\vec{x}\forall\beta\forall u(\varphi(u)=\beta(u) \leftrightarrow \mathrm{A}_\varphi(\vec{\alpha}, \vec{x}, \beta(u), u));$$
$$\mathbf{T} \vdash \forall\vec{\alpha}\forall\vec{x}\forall u\exists! < y\mathrm{A}_\varphi(\vec{\alpha}, \vec{x}, y, u);$$
$$\mathbf{T} \vdash \forall\vec{\alpha}\forall\vec{x}\exists\beta\forall u\mathrm{A}_\varphi(\vec{\alpha}, \vec{x}, \beta(u), u),$$

and that each predicate symbol P in $L' \cup L$ has a Δ^0_0 defining formula in $L' \cap L$ without function inputs. Then, for each Δ^0_0 formula A, $\mathbf{T}'$ proves the following:

(i) $E(A) \leftrightarrow A \leftrightarrow U(A)$;

(ii) $$\forall \vec{\alpha} \forall \vec{x} \forall \vec{\beta} \forall \vec{y} \left(\begin{array}{l} \bigwedge \forall u \mathrm{A}_{\psi_l}(\vec{\alpha}, \vec{x}, \beta_l(u), u) \wedge \bigwedge \mathrm{A}_{t_j}(\vec{\alpha}, \vec{x}, y_j) \to \\ \qquad \forall \vec{z}(E'(A(\vec{z}, \vec{t}, \vec{\psi})) \leftrightarrow A(\vec{z}, \vec{t}, \vec{\psi}) \leftrightarrow U'(A(\vec{z}, \vec{t}, \vec{\psi}))) \end{array} \right).$$

Proof (i) and (ii) are proved by simultaneous induction on the construction of A. We give only a sketch of a proof. If A is atomic, it is easy to show by noticing $t = s \leftrightarrow \exists x(t = x \wedge s = x)$ and $P(t_0, \ldots, t_{j-1}) \leftrightarrow \exists \vec{x}(A_P(x_0, \ldots, x_{j-1}) \wedge t_0 = x_0 \wedge \ldots \wedge t_{j-1} = x_{j-1})$ for a predicate symbol P in $L \cup L'$. If $A \equiv B \circ C$ for some $\circ \in \{\wedge, \vee, \to\}$, it is proved using Δ^0_0-LEM in $\mathbf{T}'$ and the induction hypothesis $E(B) \leftrightarrow B \leftrightarrow U(B)$, etc. If $A \equiv (\forall x < t)B$, (ii) follows from the induction hypothesis. The case of $U(A) \leftrightarrow A$ in (i) is easy to prove. For $E(A) \leftrightarrow A$, first assume $(\forall x < t)B$. Since, for each functor ψ_l and each term t_j whose free variables are contained in $\vec{x}$ and $\vec{\alpha}$, $\mathbf{T}'$ proves that $\exists y(t_j = y)$ and $\exists \beta \forall u(\psi_l(u) = \beta(u))$, take β_l and y_j such that $\forall u A_{\psi_l}(\vec{\alpha}, \vec{x}, \beta_l(u), u)$ and $A_{t_j}(\vec{\alpha}, \vec{x}, y_j)$ for each term t_j and functor ψ_l in $x < t \to B$. Then, by the induction hypothesis, we have $E'(x < t \to B) \leftrightarrow (x < t \to B)$ for all x and so we have $\forall x E'(x < t \to B)$, which implies

$$\exists \vec{\beta}_{x<t\to B} \exists \vec{y}_{x<t\to B} \forall x E'(x < t \to B) \equiv E((\forall x < t)B).$$

For the converse, assume $E((\forall x < t)B)$, which implies

$$\forall x \exists \vec{\beta}_{x<t\to B} \exists \vec{y}_{x<t\to B} E'(x < t \to B).$$

Hence we have the following by the induction hypothesis:

$$\begin{aligned} (\forall x < t)B \leftrightarrow & \forall x(x < t \to B) \\ \leftrightarrow & \forall x E(x < t \to B) \\ \equiv & \forall x \exists \vec{\beta}_{x<t\to B} \exists \vec{y}_{x<t\to B} E'(x < t \to B). \end{aligned}$$

Therefore we have the desired equivalence $(\forall x < t)B \leftrightarrow E((\forall x < t)B)$. If $A \equiv (\exists x < t)B$, (ii) follow from the induction hypothesis. The case of $E(A)$ in (i) is easy to prove. For $U(A)$, first we have the following:

$$\begin{aligned} (\exists x < t)B \leftrightarrow & \exists x(x < t \wedge B) \\ \leftrightarrow & \exists x U(x < t \wedge B) \\ \equiv & \exists x \forall \vec{\beta}_{x<t\wedge B} \forall \vec{y}_{x<t\wedge B} U'(x < t \wedge B), \end{aligned}$$

which implies $U((\exists x < t)B)$. For the converse, assume

$$U((\exists x < t)B) \equiv \forall \vec{\beta}_{x<t\wedge B} \forall \vec{y}_{x<t\wedge B} \exists x U'(x < t \wedge B).$$

As in the previous case, take $\vec{\beta}_{x<t\wedge B}$ and $\vec{y}_{x<t\wedge B}$ satisfying A_{ψ_l} and A_{s_j} for each functor ψ_l and each term s_j in $x < t \wedge B$, and then take x such that $U'(x < t \wedge B)$. By the induction hypothesis,

$$E'(x < t \wedge B) \leftrightarrow x < t \wedge B \leftrightarrow U'(x < t \wedge B),$$

and so we have $x < t \wedge B$, which implies $\exists x U(x < t \wedge B)$. □

Proof of Lemma 24.27. (i) is clear from the definition of $A^\diamond$ and A^*. (ii) is proved by induction on the definition of $A^\diamond$, using Lemma 24.70. (3) is proved by induction on the construction of formulae. (iv) is clear from the definition of $*$. (v) follows from Lemma 24.68. □

Although $A^\diamond$ is not always Δ^0_0 in L even if A is Δ^0_0 in L', we have the following.

Lemma 24.71 *If A is a Δ^0_0 formula in L', then $\mathbf{T}$ proves $A^\diamond \vee \neg(A^\diamond)$.*

Proof First, we show that if $\mathbf{T} \vdash \forall \vec{\alpha} \forall \vec{z} \forall x \exists ! y B(x, y, \vec{z}, \vec{\alpha})$, then

$$\mathbf{T} \vdash B(x, y, \vec{z}, \vec{\alpha}) \vee \neg B(x, y, \vec{z}, \vec{\alpha})$$

for any formula B. Take y such that $\mathbf{T} \vdash B(x, y, \vec{z}, \vec{\alpha})$. Then

$$\mathbf{T} \vdash \forall \vec{\alpha} \forall \vec{z} \forall y' (B(x, y', \vec{z}, \vec{\alpha}) \leftrightarrow y = y')$$

and so

$$\mathbf{T} \vdash \forall \vec{\alpha} \forall \vec{z} \forall y' [(B(x, y', \vec{z}, \vec{\alpha}) \vee \neg B(x, y', \vec{z}, \vec{\alpha})) \leftrightarrow (y = y' \vee \neg(y = y'))].$$

Since $\mathbf{T} \vdash y = y' \vee \neg(y = y')$, we have $\mathbf{T} \vdash B(x, y, \vec{z}, \vec{\alpha}) \vee \neg B(x, y, \vec{z}, \vec{\alpha})$.

The above implies that $\mathbf{T} \vdash C \vee \neg C$ for a formula C built up from formulae with the above property and Δ^0_0 formulae by means of $\wedge$, $\vee$, $\forall x < t$, and $\exists x < t$.

Then we can first prove that $\mathbf{T} \vdash (t = y)^\diamond \vee \neg(t = y)^\diamond$ for any term t in L'. By the induction on the construction of formulae, we can show that $\mathbf{T} \vdash A^\diamond \vee \neg(A^\diamond)$ for any Δ^0_0 formula A in L'. □

Next we show Lemma 24.28. For each term t, functor φ, and formula A in L', we define a function δ which gives the maximal nesting depth of function symbols and predicate symbols as follows:

$$\delta(t) = \begin{cases} 0 & \text{if } t \text{ is a term in } L; \\ \max\{\delta(\psi), \delta(\vec{s})\} + 1 & \text{if } t = f(\vec{\psi}, \vec{s}); \\ \max\{\delta(\psi), \delta(s)\} & \text{if } t = \psi(s); \end{cases}$$

$$\delta(\varphi) = \begin{cases} 0 & \text{if } \varphi \text{ is a functor in } L; \\ \max\{\delta(\vec{\psi}), \delta(\vec{s})\} + 1 & \text{if } \varphi \equiv \eta(\vec{\psi}, \vec{s}); \end{cases}$$

$$\delta(A) = \begin{cases} \max\{\delta(s_0), \delta(s_1)\} & \text{if } A \equiv s_0 = s_1; \\ \max\{\delta(s_0), \ldots, \delta(s_{j-1})\} + 1 & \\ & \text{if } A \equiv P(s_0, \ldots, s_{j-1}) \\ & \text{for some predicate symbol } P \text{ in } L \cup L'; \\ \max\{\delta(B), \delta(C)\} & \text{if } A \equiv B \circ C \text{ for some } \circ \in \{\wedge, \vee, \rightarrow\}; \\ \delta(B) & \text{if } A \equiv QxB(x) \text{ or } Q\alpha B \text{ for some } Q \in \{\exists, \forall\}, \end{cases}$$

where $\delta(\vec{s})$ is $\delta(s_0), \ldots, \delta(s_{k-1})$ and $\delta(\vec{\psi})$ is $\delta(\psi_0), \ldots, \delta(\psi_{l-1})$. Actually, $\varepsilon(t)$, $\varepsilon'(t)$, $\varepsilon(\psi)$, and $\varepsilon'(\psi)$, defined in Section 24.3.2, are given by induction according this δ.

Proof of Lemma 24.28 This is proved by simultaneous induction on $\delta(t)$ and $\delta(\varphi)$. If $\delta(t) = 0$ or $\delta(\varphi) = 0$, then t and β are in L and so it is trivial. □

Assume that $\delta(t) = n + 1$. For example, if $t = f_i(\vec{\psi}, \vec{s})$, then we can prove as follows: Note that each of term and functor constructors in t has a distinct index k, and so β_k, y_k, w_k, and v_k appear in only one of the $\varepsilon(s_j)$s and $\varepsilon(\psi_l)$s in $\varepsilon(t)$. Hence we have, for a formula $A(y_i, w_i)$ in which other y_k, w_k, v_k, and β_k do not appear,

$$\exists\vec{\beta}_t\exists\vec{y}_t\exists\vec{w}_t\exists\vec{v}_t\left(A(y_i, w_i) \wedge \bigwedge_{j<j'} \varepsilon(s_j) \wedge \bigwedge_{l<l'} \varepsilon(\psi_l)\right) \leftrightarrow$$

$$\exists y_i\exists w_i\left(A(y_i, w_i) \wedge \bigwedge_{j<j'} \exists\beta_{s_j}\exists\vec{y}_{s_j}\exists\vec{w}_{s_j}\exists\vec{v}_{s_j}\varepsilon(s_j) \wedge \exists v_i \bigwedge_{l<l'} \exists\vec{\beta}_{\psi_l}\exists\vec{y}_{\psi_l}\exists\vec{w}_{\psi_l}\exists\vec{v}_{\psi_l}\varepsilon(\psi_l)\right).$$

Then, with fresh variables $\vec{z}$ and $\vec{z'}$, we have the following:

$$\begin{aligned}
&E(t = x) \\
&\leftrightarrow \exists\vec{\beta}_t\exists\vec{y}_t\exists\vec{w}_t\exists\vec{v}_t\left(\begin{array}{l}\mathrm{B}_f(\overrightarrow{\varepsilon'(\psi)\lceil w_i}, \overrightarrow{\varepsilon'(s)}, y_i, w_i) \wedge \\ \quad \bigwedge_{j<j'} \varepsilon(s_j) \wedge \bigwedge_{l<l'} (\forall u_{\psi_l} {<} w_i)(\exists \mathbf{v}_{\psi_l} {<} v_i)\varepsilon(\psi_l) \wedge y_i{=}x\end{array}\right) \\
&\leftrightarrow \exists\vec{z'}\exists\vec{z}\exists\vec{\beta}_t\exists\vec{y}_t\exists\vec{w}_t\exists\vec{v}_t \\
&\quad \left(\begin{array}{l}\mathrm{B}_f(\vec{z'}, \vec{z}, y_i, w_i) \wedge \bigwedge_{j<j'} (\varepsilon(s_j) \wedge \varepsilon'(s_j) = z_j) \\ \quad \wedge \bigwedge_{l<l'} \left((\forall u_{\psi_l} < w_i)(\exists \mathbf{v}_{\psi_l} < v_i)\varepsilon(\psi_l) \wedge (\varepsilon'(\psi_l)\lceil w_i) = z'_j\right) \wedge y_i = x\end{array}\right) \\
&\leftrightarrow \exists\vec{z'}\exists\vec{z}\exists y_i\exists w_i \\
&\quad \left(\begin{array}{l}\mathrm{B}_f(\vec{z'}, \vec{z}, y_i, w_i) \wedge \bigwedge_{j<j'} \exists\vec{\beta}_{s_j}\exists\vec{y}_{s_j}\exists\vec{w}_{s_j}\exists\vec{v}_{s_j}(\varepsilon(s_j) \wedge \varepsilon'(s_j){=}z'_j) \\ \quad \wedge \exists v_i \bigwedge_{l<l'} \exists\vec{\beta}_{\psi_l} ((\forall u_{\psi_l} {<} w_i)(\exists \mathbf{v}_{\psi_l} {<} v_i)\varepsilon(\psi_l) \wedge (\varepsilon'(\psi_l)\lceil w_i){=}z_j) \wedge y_i{=}x\end{array}\right)
\end{aligned}$$

$$\leftrightarrow \exists \vec{z'} \exists \vec{z} \exists y_i \exists w_i \left(\mathrm{B}_f(\vec{z'}, \vec{z}, y_i, w_i) \wedge \bigwedge_{j<j'} s_j = z'_j \wedge \bigwedge_{l<l'} (\psi_l \upharpoonright w_i) = z_j \wedge y_i = x \right)$$

$$\leftrightarrow \exists y_i \exists w_i \left(\mathrm{B}_f(\overrightarrow{\psi \upharpoonright w_i}, \vec{s}, y_i, w_i) \wedge y_i = x \right) \leftrightarrow f(\vec{s}, \vec{\psi}) = x,$$

where $\varepsilon'(\psi_l \upharpoonright w_i) = z_j$ is an abbreviation of $(\forall w < w_i)(\varepsilon'(\psi_l)(w) = z_j(w))$. Other cases can be proved in a similar way.

References

[1] Avigad, J. 2000. Interpreting classical theories in constructive ones. *J. Symbol. Logic*, **65**, 1785–1812.

[2] Berger, J., Ishihara, H., Kihara, T., and Nemoto, T. 2019. The binary expansion and the intermediate value theorem in constructive reverse mathematics. *Arch. Math. Logic*, **58**, 203–217.

[3] Bishop, E. 1967. *Foundations of Constructive Analysis*. New York:McGraw-Hill.

[4] Coquand, T., and Hofmann, M. 1999. A new method for establishing conservativity of classical systems over their intuitionistic version. *Math. Struct. Comp. Sci.*, **9**, 323–333.

[5] Diener, H. 2018. Constructive Reverse Mathematics. Habilitation Schrift, Universtät Siegen.

[6] Fujiwara, M., and Kawai, T. 2020. A logical characterization of the continuous bar induction. Pages 15–33 of: Liu, F., Ono, H., and Yu, J. (eds.), *Knowledge, Proof and Dynamics: The Fourth Asian Workshop on Philosophical Logic*. Singapore: Springer.

[7] Friedman, H. 1975. Some systems of second order arithmetic and their use. Pages 235–242 of: *Proceedings of the International Congress of Mathematicians, Vancouver 1974, vol. 1*. Vancouver, BC: Canadian Mathematical Congress.

[8] Friedman, F. 1976. Systems of second order arithmetic with restricted induction I, II (abstracts). *J. Symbol. Logic*, **41**, 557–559.

[9] Ishihara, H. 2005. Constructive reverse mathematics: Compactness properties. Pages 245–267 of: Schuster, P., and Crosilla, L. (eds.), *From Sets and Types to Topology and Analysis: Towards Practicable Foundations for Constructive Mathematics*. Oxford: Oxford University Press.

[10] Julian, W., and Richman, F. 1984. A uniformly continuous function on $[0, 1]$ that is everywhere different from its infimum. *Pac. J. Math. Mathematics*, **111**, 333–340.

[11] Kleene, S. C., and Vesley R. E. 1965. *The Foundation of Intuitionistic Mathematics, Especially in Relation to Recursive Functions*. Amsterdam: North-Holland.

[12] Kreisel, G., and Troelstra, A. S. 1970. Formal systems for some branches of intuitionistic analysis. *Ann. Math. Logic*, **1**, 229–387.

[13] Loeb, I. 2005. Equivalents of the (Weak) Fan Theorem. *Ann. Pure Appl. Logic*, **132**, 51–66.

[14] Nemoto, T. 2009. Determinacy of Wadge classes and subsystems of second order arithmetic. *Math. Logic. Q.*, **55**, 154–176.

[15] Nemoto, T. 2020a. Finite sets and infinite sets in weak intuitionistic arithmetic. *Arch. Math. Logic*, **59**, 607–657.

[16] Nemoto, T. 2020b. A constructive proof of the dense existence of nowhere-differentiable functions in $C[0,1]$. *Computability*, **9**, 315–326.

[17] Nemoto, T., and Sato, K. 2022. A marriage of Brouwer's intuitionism and Hilbert's finitism I: Arithmetic. *J. Symbol. Logic*, **87**(2), 437–497.

[18] Paris, J. B. 1981. Some conservation results for fragments of arithmetic, Pages 251–262 of: Berline, C., McAloon, K., and Ressayre, J. P. (eds.), *Model Theory and Arithmetic*, Lecture Notes in Mathematics, vol. 890. Berlin, Heidelberg: Springer.

[19] Rathjen, M. 1999. The realms of ordinal analysis. Pages 219–279 of: Cooper, S., and Truss, J. (eds.),*Sets and Proofs*. Cambridge: Cambridge University Press.

[20] Simpson, S. G., and Smith, R. L. 1986. Factorization of polynomials and Σ^0_1 induction. *Ann. Pure Appl. Logic*, **31**, 289–306.

[21] Simpson, S. G. 2009. *Subsystems of Second Order Arithmetic*, 2nd edition. Cambridge: Cambridge University Press.

[22] Tanaka, K. 1990. Weak axioms of determinacy and subsystems of analysis I. *Z. Math. Logik Grundl. Math.*, **36**, 481–491.

[23] Troelstra, A. S. (ed.) 1973. *Metamathematical Investigation of Intuitionistic Arithmetic and Analysis*. Berlin, Heidelberg, New York: Springer.

[24] Troelstra, A. S., and van Dalen, D. 1988. *Constructivism in Mathematics: an Introduction, vol. 1*. Amsterdam: Elsevier Science.

[25] Veldman, W. 2014. Brouwer's fan theorem as an axiom and as a contrast to Kleene's alternative. *Arch. Math. Logic*, **53**, 621–693.

25

Brouwer's Fan Theorem

Josef Berger

25.1 Introduction

We analyse the relationship between Brouwer's fan theorem (FAN) for detachable bars, the weak König lemma (WKL), and the uniform continuity theorem (UC).

FAN Every detachable bar is a uniform bar.
WKL Every infinite tree has a path.
UC Every pointwise continuous function $F : \{0,1\}^{\mathbb{N}} \to \mathbb{N}$ is uniformly continuous.

Extending the well-known basic picture

$$\mathrm{WKL} \Rightarrow \mathrm{UC} \Rightarrow \mathrm{FAN},$$

we prove the implications of the following diagram:

$$\begin{array}{ccccccc}
\mathrm{WKL} & \Rightarrow & \mathrm{FAN} \wedge \mathrm{DEFU} & \Leftrightarrow & \text{c-FAN} & \Rightarrow & \mathrm{FAN} \\
 & & \Updownarrow & & \Updownarrow & & \\
 & & \mathrm{FAN} \wedge \mathrm{DECO} & \Leftrightarrow & \mathrm{UC} & &
\end{array}$$

The axiom c-FAN was introduced in [2] as a version of the fan theorem which is equivalent to UC. The axioms DEFU and DECO are auxiliary statements clarifying the gap between FAN and c-FAN, UC respectively.

Mosts results mentioned in this article have been published before, see

- references [9, 10] for results about WKL and FAN,
- references [4, 6] for results about unique existence, and
- references [1, 2, 8] for results about UC.

25.2 Notation

Let $\mathbb{N} = \{0, 1, 2, \ldots\}$ denote the set of natural numbers. We use the variables k, l, m, n, N, i, j for natural numbers. Let $\{0,1\}^*$ denote the set of all finite binary sequences, typically denoted by variables like u, v, w. The symbol ϕ denotes the empty sequence. Note that $\phi \in \{0,1\}^*$. Let $\{0,1\}^{\mathbb{N}}$ denote the set of all infinite binary sequences. We use the variables α, β, γ for elements of $\{0,1\}^{\mathbb{N}}$. The *length* of u is denoted by $|u|$. We define

$$\mathcal{N} = \{u \mid |u| > 0 \wedge \forall i < |u| \, (u_i = 0)\}$$

and

$$\mathcal{E} = \{u \mid |u| > 0 \wedge \forall i < |u| \, (u_i = 1)\} \, .$$

Let $\{0,1\}^n = \{u \mid |u| = n\}$ denote the set of all binary sequences of length n. The *concatenation* of u and v is denoted by $u * v$ and defined by

$$(u_0, \ldots, u_{n-1}) * (v_0, \ldots, v_{m-1}) = (u_0, \ldots, u_{n-1}, v_0, \ldots, v_{m-1}).$$

The *concatenation* of u and α is denoted by $u * \alpha$ and defined by

$$(u_0, \ldots, u_{n-1}) * \alpha = (u_0, \ldots, u_{n-1}, \alpha_0, \alpha_1, \alpha_2, \ldots).$$

If $n \leq |w|$, we denote the *restriction* of w to the first n elements by $\overline{w}n$. Note that $\overline{w}0 = \phi$. We denote the *restriction* of α to the first n elements by $\overline{\alpha}n$. We write $u < v$ for

$$|u| = |v| \wedge \exists k < |u| \, (\overline{u}k = \overline{v}k \wedge u_k = 0 \wedge v_k = 1) \, .$$

Fix a subset A of $\{0,1\}^*$. We define

- $A^\circ = \{u \mid \forall w \, (u * w \in A)\}$,
- $\overline{A} = \{u * w \mid u \in A \wedge w \in \{0,1\}^*\}$, and
- for any u, $A_u = \{w \mid u * w \in A\}$.

A *path* of A is a sequence α such that $\forall n \, (\overline{\alpha}n \in A)$. The set A has *at most one path* if

$$\forall \alpha, \beta \left(\exists n \, (\alpha_n \neq \beta_n) \Rightarrow \exists n \left(\overline{\alpha}n \notin A \vee \overline{\beta}n \notin A\right)\right) .$$

A *longest path* of A is a sequence α such that $\forall u \in A \, (\overline{\alpha}|u| \in A)$. Note that each path of A is a longest path of A but not necessarily vice versa.

The set A is said to be

- *detachable* (from $\{0,1\}^*$) if
$$\forall u \, (u \in A \vee u \notin A) \, ;$$
- a *c-set* if there exists a detachable set $D \subseteq \{0,1\}^*$ such that $A = D^\circ$;

- *closed under extension* if

$$\forall u \in A \, \forall w \, (u * w \in A)\,;$$

- *closed under restriction* if

$$\forall u, w \, (u * w \in A \Rightarrow u \in A)\,;$$

- a *tree* if it is detachable and closed under restriction;
- *infinite* if

$$\forall n \, \exists u \, (|u| = n \wedge u \in A)\,;$$

- a *bar* if

$$\forall \alpha \, \exists n \, (\overline{\alpha} n \in A)\,;$$

- a *uniform bar* if

$$\exists N \, \forall \alpha \, \exists n \leq N \, (\overline{\alpha} n \in A)\,;$$

- *convex* if

$$\forall u, w \in A \, \forall v \, (u < v < w \Rightarrow v \in A)\,;$$

- *co-convex* if $\{u \mid u \notin A\}$ is convex.

25.3 The Weak König Lemma

The *weak König lemma* is the following axiom:

WKL Every infinite tree has a path.

Lemma 25.1 *Let T be a tree. The following are equivalent.*

(a) *T is an infinite tree.*
(b) *Every longest path of T is a path of T.*

The notion of longest path is commonly used in graph theory. In the context of constructive mathematics, it was introduced in [10], along with the *longest path lemma*:

LPL Every tree has a longest path.

When working with trees and longest paths, it is convenient to have at hand an algorithm which transforms an arbitrary tree T into an infinite tree T', where T' is as closely related to T as possible. Given a tree T, we define subsets L_T and T' of $\{0,1\}^*$ by

$$u \in L_T \Leftrightarrow (u \in T \wedge \forall w \, ((|u| < |w| \vee u < w) \Rightarrow w \notin T) \vee (\phi \notin T \wedge u = \phi))$$

and

$$u \in T' \Leftrightarrow u \in T \cup \{\phi\} \vee \exists v \in L_T \, \exists w \in \mathcal{N} \, (u = v * w) \, .$$

Note that

- if $T = \emptyset$ then $L_T = \{\phi\}$ and $T' = \mathcal{N} \cup \{\phi\}$;
- if T is infinite, then $L_T = \emptyset$ and $T' = T$;
- if $\phi \in T$ and the set

$$\{|u| \mid u \in T\}$$

is bounded, then $L_T = \{u\}$, where u is an element of T of maximal length.

Proposition 25.2 *For every tree T there is an infinite tree T' such that*

(a) $T \subseteq T'$;
(b) *if T is infinite, then $T = T'$;*
(c) *if T is convex, then T' is convex;*
(d) *if T is finite, then T' has at most one path;*
(e) *every path of T' is a longest path of T;*
(f) *if T has at most one path, then T' has at most one path.*

The *lesser limited principle of omniscience* is the following.

LLPO For all α,

$$\forall n, m \, (\alpha_n = \alpha_m = 1 \Rightarrow n = m) \Rightarrow \forall n \, (\alpha_{2n} = 0) \vee \forall n \, (\alpha_{2n+1} = 0) \, .$$

Proposition 25.3 LLPO $\Rightarrow$ WKL.

Proof Let T be an infinite tree. Let $\zeta_T(u, m)$ be defined by

$$\zeta_T(u, m) \Leftrightarrow \exists w \in \{0, 1\}^m \, (u * w \in T) \, .$$

Moreover, set

$$Z_T = \{u \mid \forall m \, \zeta_T(u, m)\} \, .$$

Define β by

$$\beta_{2n} = 0 \Leftrightarrow \zeta_T((0), n)$$

and

$$\beta_{2n+1} = 0 \Leftrightarrow \zeta_T((1), n),$$

and define α by

$$\alpha_n = 1 \Leftrightarrow \beta_n = 1 \wedge \forall m < n \, (\beta_m = 0) \, .$$

By LLPO, either

$$\forall n\,(\alpha_{2n} = 0) \tag{25.1}$$

or else

$$\forall n\,(\alpha_{2n+1} = 0)\,. \tag{25.2}$$

If (25.1) holds, we can conclude that

$$\forall n\,(\beta_{2n} = 0)\,,$$

which implies that $(0) \in Z_T$. Analogously, we can show that (25.2) implies that $(1) \in Z_T$. Iterating this procedure yields a path γ of Z_T. Since $Z_T \subseteq T$, the sequence γ is also a path of T. □

Proposition 25.4 *The following are equivalent.*

(a) WKL.
(b) LPL.
(c) *Every convex tree has a longest path.*
(d) *Every infinite convex tree has a path.*
(e) LLPO.

Proof '(a) ⇒ (b)' This follows from Proposition 25.2.

'(c) ⇒ (d)' Let T be an infinite convex tree. By (c), the tree T has a longest path α. By Lemma 25.1, the sequence α is a path of T.

'(d) ⇒ (e)' Fix a sequence α with

$$\forall n, m\,(\alpha_n = \alpha_m = 1 \Rightarrow n = m)\,.$$

We define an infinite convex tree S by

$$S = \{\phi, (0), (1)\} \cup S_0 \cup S_1,$$

where

$$S_0 = \{(0) * w \mid w \in \mathcal{E} \wedge \forall i \leq |w|\,(\alpha_{2i} = 0)\}\,,$$

$$S_1 = \{(1) * w \mid w \in \mathcal{N} \wedge \forall i \leq |w|\,(\alpha_{2i+1} = 0)\}\,.$$

Let β be a path of S. If $\beta_1 = 0$, then $\forall i\,(\alpha_{2i} = 0)$. If $\beta_1 = 1$, then $\forall i\,(\alpha_{2i+1} = 0)$.

'(e) ⇒ (a)' This is Proposition 25.3. □

Proposition 25.3 and Proposition 25.4 (without the parts about convexity) were proved in [9] and [10].

25.4 The Fan Theorem

Brouwer's fan theorem is the following.

FAN Every detachable bar is a uniform bar.

Lemma 25.5 *If $B \subseteq \{0,1\}^*$ is closed under extension, the following are equivalent.*

(a) *B is a uniform bar.*
(b) $\exists n\,(\{0,1\}^n \subseteq B)$.

The next lemma enables us to switch freely between a set B itself and $\overline{B}$, when discussing bar-related properties. This technique was introduced in [10].

Lemma 25.6 *Let B be a subset $\{0,1\}^*$. Then*

(a) *$\overline{B}$ is closed under extension,*
(b) *$B \subseteq \overline{B}$,*
(c) *B is detachable $\Rightarrow$ $\overline{B}$ is detachable,*
(d) *B is a bar $\Rightarrow$ $\overline{B}$ is a bar,*
(e) *$\overline{B}$ is a uniform bar $\Rightarrow$ B is a uniform bar.*

The following weakening of the weak König lemma is useful for understanding the relationship between FAN and WKL.

WKL! Every infinite tree with at most one path has a path.

Proposition 25.7 *The following are equivalent.*

(a) WKL!
(b) *Each tree with at most one path has a longest path.*
(c) *Each detachable bar which is closed under extension is a uniform bar.*
(d) FAN.

Proof '(a) $\Rightarrow$ (b)' This follows from Proposition 25.2.

'(b) $\Rightarrow$ (c)' Let B be a detachable bar which is closed under extension. Note that the set

$$T = \{u \mid u \notin B\}$$

is a tree. Since B is a bar, the tree T has at most one path. Let α be a longest path of T. There exists n such that $\overline{\alpha}n \in B$. This implies that $T \cap \{0,1\}^n = \emptyset$ and therefore $\{0,1\}^n \subseteq B$. Thus, by Lemma 25.5, B is a uniform bar.

'(c) $\Rightarrow$ (d)' This follows from Lemma 25.6.

'(d) $\Rightarrow$ (a)' Fix an infinite tree T with at most one path. Define a set B by

$$u \in B \;:\Leftrightarrow\; (0) * u \notin T \vee \forall v \in \{0,1\}^{|u|}\,((1) * v \notin T)\,.$$

We show that the B is a bar. Fix a sequence α. We define $B_\alpha \subseteq \{0,1\}^*$ by

$$v \in B_\alpha :\Leftrightarrow ((0) * \overline{\alpha}|v| \notin T \vee (1) * v \notin T).$$

Note that B_α is closed under extension. Since T has at most one path, B_α is a bar and therefore a uniform bar. Thus, there exists m such that $\{0,1\}^m \subseteq B_\alpha$. Thus, $\overline{\alpha}m \in B$. This concludes the proof that B is a bar.

Since B is is a uniform bar and closed under extension, there exists an n such that $\{0,1\}^n \subseteq B$. We show that either

(i) $\forall u \in \{0,1\}^n\,((0) * u \notin T)$ or else
(ii) $\forall u \in \{0,1\}^n\,((1) * u \notin T)$.

If (i) fails to hold, there exists $u \in \{0,1\}^n$ such that $(0) * u \in T$. Since $u \in B$, this implies (ii).

Let Z_T be defined as in the proof of Proposition 25.3. If (i) holds, then $(1) \in Z_T$. If (ii) holds, then $(0) \in Z_T$. Iterating this construction yields a path γ of Z_T and therefore of T. □

The equivalence of FAN and WKL! was proved in [4].

Corollary 25.8 WKL $\Rightarrow$ FAN.

Corollary 25.8 was originally published in [10].

Lemma 25.9 *If $A \subseteq \{0,1\}^*$ is co-convex, then $\overline{A}$ is also co-convex.*

Proof Fix u, v, w such that $u < v < w$ and $u, w \notin \overline{A}$. We have to show that $v \notin \overline{A}$. To this end, fix $m \leq |v|$. Then $\overline{u}m \notin A$ and $\overline{w}m \notin A$. The co-convexity of A yields $\overline{v}m \notin A$. Since $m \leq |v|$ was chosen arbitrarily, this implies that $v \notin \overline{A}$. □

Proposition 25.10

(a) *Every infinite convex tree with at most one path has a path.*
(b) *Every convex tree with at most one path has a longest path.*
(c) *Every detachable co-convex bar is a uniform bar.*

Proof (a) Let T be an infinite convex tree with at most one path. Set

$$\alpha = (0,1,1,1,\ldots)$$

and

$$\beta = (1,0,0,0,\ldots).$$

Since T has at most one path, there exists n such that

$$\overline{\alpha}(n+1) \notin T \vee \overline{\beta}(n+1) \notin T.$$

Since T is convex, we can conclude that either

(i) $\forall u \in \{0,1\}^n\,((0) * u \notin T)$ or else
(ii) $\forall u \in \{0,1\}^n\,((1) * u \notin T)$.

In the first case, choose $\alpha_0 = 1$ and in the second case, choose $\alpha_0 = 0$. Iterating this construction leads to a path α of T.

(b) This follows from (a) and Proposition 25.2.

(c) Let B be a detachable, co-convex bar. By Lemma 25.9, we can assume that B is closed under extension. The set

$$T = \{u \mid u \notin B\}$$

is a convex tree. Since B is a bar, the tree T has at most one path. Let α be a longest path of T. There exists n such that $\overline{\alpha}n \in B$. Then $T \cap \{0,1\}^n = \emptyset$. Thus, $\{0,1\}^n \subseteq B$. This implies that B is a uniform bar. □

The notion of co-convex bars was introduced in [5]. In [12], the fan theorem for co-convex bars – every detachable co-convex bar is a uniform bar – is used for a case study in program extraction from proofs. For a treatment of convex trees in constructive reverse mathematics see [7].

25.5 The Uniform Continuity Theorem

Fix a function $F : \{0,1\}^{\mathbb{N}} \to \mathbb{N}$ and $u \in \{0,1\}^*$. We define

$$F(u) = F(u * \mathbf{0}),$$

where $\mathbf{0} = (0,0,0,\ldots)$. Moreover, we define a new function $F_u : \{0,1\}^{\mathbb{N}} \to \mathbb{N}$ by

$$F_u(\alpha) = F(u * \alpha).$$

A function $F : \{0,1\}^{\mathbb{N}} \to \mathbb{N}$ is

- *pointwise continuous* if

$$\forall \alpha\, \exists n\, \forall \beta\, \left(\overline{\alpha}n = \overline{\beta}n \;\Rightarrow\; F(\alpha) = F(\beta)\right);$$

- *uniformly continuous* if

$$\exists n\, \forall \alpha, \beta\, \left(\overline{\alpha}n = \overline{\beta}n \;\Rightarrow\; F(\alpha) = F(\beta)\right);$$

- *bounded* if

$$\exists N\, \forall \alpha\, (F(\alpha) \leq N);$$

- *constant* if

$$\exists n\, \forall \alpha\, (F(\alpha) = n).$$

Lemma 25.11 *If $F : \{0,1\}^{\mathbb{N}} \to \mathbb{N}$ is uniformly continuous,*

$$\forall\alpha\,(F(\alpha) = 0) \ \vee\ \neg\forall\alpha\,(F(\alpha) = 0)\,.$$

Lemma 25.12 *Fix $F : \{0,1\}^{\mathbb{N}} \to \mathbb{N}$.*

(a) *F is pointwise continuous* $\Leftrightarrow$ $\forall\alpha\,\exists n\,(F_{\overline{\alpha}n}$ *is constant).*
(b) *F is uniformly continuous* $\Leftrightarrow$ $\exists N\,\forall u \in \{0,1\}^N\,(F_u$ *is constant).*

Lemma 25.13 *Suppose that $F : \{0,1\}^{\mathbb{N}} \to \mathbb{N}$ is pointwise continuous. Then the set*

$$\{u \mid F(u) \leq |u|\}$$

is a bar.

Proof Fix α. There exists N such that

$$\forall\beta\,(\overline{\alpha}N = \overline{\beta}N \ \Rightarrow\ F(\alpha) = F(\beta))\,.$$

We may assume that $F(\alpha) \leq N$. Note that

$$F(\overline{\alpha}N) = F(\alpha) \leq N.$$

This implies that

$$\overline{\alpha}N \in \{u \mid F(u) \leq |u|\}\,.$$

□

Lemma 25.14 *Every uniformly continuous function $F : \{0,1\}^{\mathbb{N}} \to \mathbb{N}$ is bounded.*

We now consider the *uniform continuity theorem.*

UC Every pointwise continuous function $F : \{0,1\}^{\mathbb{N}} \to \mathbb{N}$ is uniformly continuous.

Proposition 25.15 UC $\Rightarrow$ FAN.

Proof Let B be a detachable bar. The function

$$F(\alpha) = \min\{n \mid \overline{\alpha}n \in B\} \tag{25.3}$$

is pointwise continuous. In view of UC, the function F is uniformly continuous and therefore, by Lemma 25.14, bounded. Thus, B is a uniform bar. □

Fix functions $F, M : \{0,1\}^{\mathbb{N}} \to \mathbb{N}$. The function M is a *modulus* for F if

(a) M is pointwise continuous;
(b) $\forall\alpha,\beta,\,(\overline{\alpha}M(\alpha) = \overline{\beta}M(\alpha) \ \Rightarrow\ F(\alpha) = F(\beta))$.

The following weakening of UC is useful for understanding the relationship between FAN and UC.

MUC Fix $F, M : \{0,1\}^{\mathbb{N}} \to \mathbb{N}$. Assume that M is a pointwise continuous modulus of F. Then F is uniformly continuous.

Proposition 25.16 MUC $\Leftrightarrow$ FAN.

Proof '$\Rightarrow$' Let B be a detachable bar. Then the function F, given by (25.3), is a pointwise continuous modulus of itself, which implies that F is uniformly continuous and therefore bounded. Thus, B is a uniform bar.

'$\Leftarrow$' Fix $F, M : \{0,1\}^{\mathbb{N}} \to \mathbb{N}$ and assume that M is a pointwise continuous modulus of F. By Lemma 25.13, the set

$$B = \{u \mid M(u) \leq |u|\}$$

is a bar and therefore, in view of FAN, a uniform bar. Thus, there exists N such that

$$\forall \alpha \, \exists n \leq N \, (M(\overline{\alpha}n) \leq n) \, .$$

This implies that

$$\forall \alpha \, (F_{\overline{\alpha}N} \text{ is constant}) \, .$$

By Lemma 25.12, we can conclude that F is uniformly continuous. □

The axiom MUC and Proposition 25.16 are taken from [1]. For more about the uniform continuity theorem and FAN see [8].

25.6 The Fan Theorem for c-sets

In Section 25.5, we matched uniform continuity with the fan theorem by replacing UC with its weaker version MUC. Another option is to strengthen FAN by relaxing the requirement that the bar be detachable. This leads us to c-FAN, which was introduced in [2].

c-FAN Every bar which is a c-set is a uniform bar.

Lemma 25.17 *For all subsets A of $\{0,1\}^*$ an all u we have*

(a) $u \in A^\circ \Leftrightarrow A_u = \{0,1\}^*$ *and*
(b) $(A_u)^\circ = (A^\circ)_u$.

Proof In order to prove (b), fix u, v and w and note that

$$\begin{aligned} w \in (A_u)^\circ \Leftrightarrow & \forall v \, (w * v \in A_u) \Leftrightarrow \\ & \forall v \, (u * (w * v) \in A) \Leftrightarrow \forall v \, ((u * w) * v \in A) \Leftrightarrow \\ & u * w \in A^\circ \Leftrightarrow w \in (A^\circ)_u . \end{aligned}$$

□

We introduce the auxiliary axioms DEFU (**de**cidability whether certain bars are **fu**ll) and DECO (**de**cidability whether certain functions are **co**nstant).

DEFU Let D be a detachable subset of $\{0,1\}^*$ such that D° is a bar. Then,

$$\exists u\,(u \notin D) \vee \neg\exists u\,(u \notin D)\,.$$

DECO Let $F : \{0,1\}^{\mathbb{N}} \to \mathbb{N}$ be pointwise continuous. Then,

$$\exists \alpha, \beta\,(F(\alpha) \neq F(\beta)) \vee \neg\exists \alpha, \beta\,(F(\alpha) \neq F(\beta))\,.$$

Lemma 25.18 DEFU $\Leftrightarrow$ DECO.

Proof '$\Rightarrow$' Fix a pointwise continuous function $F : \{0,1\}^{\mathbb{N}} \to \mathbb{N}$. Set

$$D = \{u \mid F(u) = F(u * (1))\}\,.$$

By Lemma 25.12, D° is a bar. Note that

$$\exists u\,(u \notin D) \Leftrightarrow \exists \alpha, \beta\,(F(\alpha) \neq F(\beta))\,.$$

'$\Leftarrow$' Fix $D \subseteq \{0,1\}^*$ and suppose that D° is a bar. Define

$$F(\alpha) = \min\{n \mid \forall m \geq n\,(\overline{\alpha}m \in D)\}$$

and note that

$$\exists u\,(u \notin D) \Leftrightarrow \exists \alpha\,(F(\alpha) > 0))\,. \tag{25.4}$$

In the presence of DECO, the right-hand side of (25.4) is decidable, which implies that the left-hand side of (25.4) is decidable. □

The diagram in Proposition 25.19 sheds some light on the relationship between bars and functions. Its core part is the equivalence between c-FAN and UC.

Proposition 25.19

$$\begin{array}{ccccccc} \text{WKL} & \Rightarrow & \text{FAN} \wedge \text{DEFU} & \Leftrightarrow & \text{c-FAN} & \Rightarrow & \text{FAN} \\ & & \Updownarrow & & \Updownarrow & & \\ & & \text{FAN} \wedge \text{DECO} & \Leftrightarrow & \text{UC} & & \end{array}$$

Proof 'WKL $\Rightarrow$ DEFU' Let D be a detachable subset D of $\{0,1\}^*$ such that D° is a bar. Define an infinite tree T by

$$u \in T \Leftrightarrow \forall w\,(|w| \leq |u| \wedge w \notin D \Rightarrow \exists k \leq |w|\,(\overline{u}k \notin D))\,.$$

For each $u \in T$ of length n we have

$$\{\overline{u}k \mid k \leq n\} \subseteq D \Leftrightarrow \{w \mid |w| \leq n\} \subseteq D.$$

Let α be a path of T. The following are equivalent.

(i) $\exists u\,(u \notin D)$.
(ii) $\exists n\,(\overline{\alpha}n \notin D)$.

There exists N such that $\overline{\alpha}N \in D^\circ$. Thus, (ii) is decidable, which yields the decidability of (i).

'FAN $\wedge$ DEFU $\Rightarrow$ c-FAN' Fix a detachable subset D of $\{0,1\}^*$ such that D° is a bar. We show that D° is detachable. To this end, fix some u. Note that $(D^\circ)_u$ is a bar and therefore $(D_u)^\circ$ is a bar. We can conclude that

$$D_u = \{0,1\}^* \vee \exists w\,(w \notin D_u)\,,$$

which implies that

$$u \in D^\circ \vee u \notin D^\circ.$$

'FAN $\wedge$ DECO $\Rightarrow$ UC' Let $F : \{0,1\}^{\mathbb{N}} \to \mathbb{N}$ be pointwise continuous. Set

$$B = \{u \mid F_u \text{ is constant}\}\,.$$

In view of DECO, the set B is detachable. Since F is pointwise continuous, B is a bar. Thus, B is a uniform bar, which implies the uniform continuity of F.

'UC $\Rightarrow$ FAN $\wedge$ DECO' The implication UC $\Rightarrow$ FAN holds by Proposition 25.15. The implication UC $\Rightarrow$ DECO follows from Lemma 25.12.

'c-FAN $\Rightarrow$ FAN' Let B be a detachable bar which is closed under extension. Then $B = B^\circ$, which implies that B is a c-set and therefore, in the presence of c-FAN, a uniform bar. □

Another equivalent of c-FAN is the so-called anti-Specker theorem; see [3]. For more on the fan theorem, see [11] and [13].

References

[1] Berger, J. (2005). The Fan theorem and uniform continuity. Pages 18–22 of: Cooper, S. B., Löwe, B., and Torenvliet, L. (eds.), *New Computational Paradigms. CiE 2005*. Lecture Notes in Computer Science, vol. 3526. Berlin, Heidelberg: Springer. https://doi.org/10.1007/11494645_3

[2] Berger, J. (2006). The logical strength of the uniform continuity theorem. Pages 35–39 of: Beckmann, A., Berger, U., Löwe, B., and Tucker, J. V. (eds.), *Logical Approaches to Computational Barriers. CiE 2006*. Lecture Notes in Computer Science, vol. 3988. Berlin, Heidelberg: Springer. https://doi.org/10.1007/11780342_4

[3] Berger, J., and Bridges, D. 2007. A fan-theoretic equivalent of the antithesis of Specker's theorem. *Indag. Mathematicae*, **18**, 195–202.

[4] Berger, J., and Ishihara, H. 2005. Brouwer's fan theorem and unique existence in constructive analysis. *Math. Logic Q.*, **51**(4), 360–364.
[5] Berger, J., and Svindland, G. 2018. Brouwer's fan theorem and convexity. *J. Symbol. Logic*, **83**(4), 1363–1375.
[6] Berger, J., Bridges, D., and Schuster, P. 2006. The fan theorem and unique existence of maxima. *J. Symbol. Logic*, **71**(2), 713–720.
[7] Berger, J., Ishihara, H., Kihara, T., and Nemoto, T. 2019. The binary expansion and the intermediate value theorem in constructive reverse mathematics. *Arch. Math. Logic*, **58**(1–2), 203–217.
[8] Bridges, D., and Richman, F. 1987. *Varieties of Constructive Mathematics*. Cambridge: Cambridge University Press.
[9] Ishihara, H. 1990. An omniscience principle, the König lemma and the Hahn–Banach theorem. *Z. Math. Logik Grundlagen Math.*, **36**(2), 237–240.
[10] Ishihara, H. 2006. Weak König's lemma implies Brouwer's fan theorem: A direct proof. *Notre Dame J. Formal Logic*, **47**(2), 249–252.
[11] Lubarsky, R. S., and Diener, H. 2014. Separating the fan theorem and its weakenings. *J. Symbol. Logic*, **79**(3), 792–813.
[12] Schwichtenberg, H. 2019. Program extraction from proofs: The fan theorem for uniformly coconvex bars. Pages 333–341 of: Centrone, S., Negri, S., Sarikaya, D., Schuster, P. M. (eds.), *Mathesis Universalis, Computability and Proof*. Synthese Library, vol 412. Cham: Springer.
[13] Veldman, W. 2014. Brouwer's fan theorem as an axiom and as a contrast to Kleene's alternative. *Arch. Math. Logic*, **53**(5–6), 621–693.

PART VI

ASPECTS OF COMPUTATION

26

Computational Aspects of Bishop's Constructive Mathematics

Helmut Schwichtenberg

Constructive mathematics, as its name says, puts an emphasis on the constructions involved in mathematical arguments. It therefore is tempting to have a closer look at these constructions, as advocated in Bishop [2]. The goal would be to extract them from mathematical proofs. Then one can view them as programs, obtained not by a programming effort but by searching for a perspicious mathematical argument, which of course has to use constructive logic. Clearly this requires formal proofs, but the benefit is that the program arising from a proof that can be machine checked for its correctness. In this sense we have a "certified" program.

One can even go a step further and ask for a formal proof that the extracted program correctly solves the original problem proved constructively. This can be expressed by the notion of realizability; see Troelstra [12] for a survey on this subject. We then need to view the extracted program as a term in a formal language (an extension of the system T of Gödel [6]), and give a formal proof that the term t realizes the formula A, written $t \mathbf{~r~} A$. Constructively to state A in a sense means the same as to say that A has a realizer. This statement $A \leftrightarrow \exists_x(x \mathbf{~r~} A)$ was called "to assert is to realize" in Feferman [3]. Here we call it the invariance axiom, since it expresses invariance of A under the realizability interpretation. Using the invariance axioms one can prove a soundness theorem, saying that for any proof M of a formula A one can find another proof that the term $\mathrm{et}(M)$ extracted from the proof M is a realizer of A, namely, $\mathrm{et}(M) \mathbf{~r~} A$. This step can be seen as a kind of reflection of what was done in the original (realizability-free) problem area. In this way we obtain a higher degree of reliability of the extracted term viewed as a program. We not only know that it came from a formal proof, but can even provide another formal soundness proof stating that the program satisfies its specification.

In the present chapter we describe the main steps to carry this program out. It involves the setup of an appropriate theoretical framework TCF (theory of computable functionals). To ensure that TCF is a meaningful theory it is designed to describe a particular model suitable to deal with computable higher type objects. For case studies we use the proof assistant Minlog[1] designed to support the generation of formal proofs in TCF.

26.1 Partial Continuous Functionals

Prior to the setup of a formal theory we define the model our theory is supposed to describe. It will be a model accomodating higher type objects, from a constructive point of view. The main idea is to view an object of an arbitrary type as given by its finite approximations. This approach has the advantage that the notion of computability of our functional objects is unproblematic: it means that the set of its finite approximations can be enumerated by an elementary function.

To allow for applications in exact real arithmetic with real numbers represented as streams of signed digits, we admit infinite data at base types. A benefit of this approach is that it brings down the type level of other concepts of constructive analysis, for instance continuity of real functions.

26.1.1 Information Systems

We aim at describing higher type functionals by their finite approximations. For this purpose we use Dana Scott's information systems. The basic idea is to provide an axiomatic setting to describe approximations of abstract objects (like functions or functionals) by concrete, finite ones. We take an arbitrary countable set A of "bits of data" or *tokens* as a basic notion to be explained axiomatically. In order to use such data to build approximations of abstract objects, we need a notion of *consistency*, which determines when the elements of a finite set of tokens are consistent with each other. We also need an *entailment* relation between consistent finite sets U of data and single tokens a, which intuitively expresses the fact that the information contained in U is sufficient to compute the bit of information a. The axioms below are a minor modification of Scott's [10], due to Larsen and Winskel [7].

Definition 26.1 An *information system* is a structure $(A, \mathrm{Con}, \vdash)$ where A is an at most countable nonempty set (the *tokens*), Con is a set of finite subsets of A (the *consistent* sets), and $\vdash$ is a subset of $\mathrm{Con} \times A$ (the *entailment* relation), which satisfy

$$\begin{aligned}
&U \subseteq V \in \mathrm{Con} \to U \in \mathrm{Con},\\
&\{a\} \in \mathrm{Con},\\
&U \vdash a \to U \cup \{a\} \in \mathrm{Con},\\
&a \in U \in \mathrm{Con} \to U \vdash a,\\
&U \in \mathrm{Con} \to \forall_{a \in V}(U \vdash a) \to V \vdash b \to U \vdash b.
\end{aligned}$$

[1] `http://minlog-system.de`

The elements of Con are called *formal neighborhoods*. We use U, V, W to denote *finite* sets, and write

$$\begin{array}{lll} U \vdash V & \text{for} & U \in \mathrm{Con} \wedge \forall_{a \in V}(U \vdash a), \\ a \uparrow b & \text{for} & \{a, b\} \in \mathrm{Con} \qquad (a, b \text{ are } \textit{consistent}), \\ U \uparrow V & \text{for} & \forall_{a \in U, b \in V}(a \uparrow b). \end{array}$$

Definition 26.2 (Objects) The *objects* (or *ideals*) of an information system $\boldsymbol{A} = (A, \mathrm{Con}, \vdash)$ are defined to be those subsets x of A which satisfy

$$\begin{array}{l} U \subseteq x \to U \in \mathrm{Con} \quad (x \text{ is } \textit{consistent}), \\ U \vdash a \to U \subseteq x \to a \in x \quad (x \text{ is } \textit{deductively closed}). \end{array}$$

For example the *deductive closure* $\overline{U} := \{ a \in A \mid U \vdash a \}$ of $U \in \mathrm{Con}$ is an object. The set of all objects of $\boldsymbol{A}$ is denoted by $|\boldsymbol{A}|$.

Example 26.3 Every countable set A can be turned into a "flat" information system by letting the set of tokens be A, $\mathrm{Con} := \{\emptyset\} \cup \{ \{a\} \mid a \in A \}$ and $U \vdash a$ mean $a \in U$. In this case the objects are just the elements of Con. For $A = \mathbb{N}$ we have the following picture of the Con-sets.

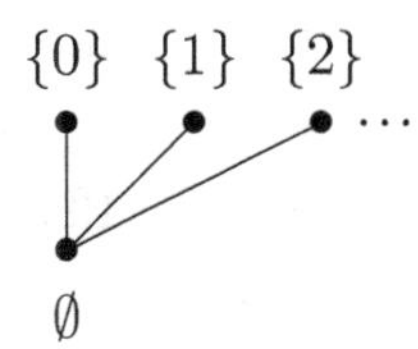

The main feature of information systems is that they admit the construction of function spaces.

Definition 26.4 Let $\boldsymbol{A} = (A, \mathrm{Con}_A, \vdash_A)$ and $\boldsymbol{B} = (B, \mathrm{Con}_B, \vdash_B)$ be information systems. Define $\boldsymbol{A} \to \boldsymbol{B} = (C, \mathrm{Con}, \vdash)$ by

$$\begin{array}{l} C := \mathrm{Con}_A \times B, \\ \{ (U_i, b_i) \mid i \in I \} \in \mathrm{Con} := \forall_{J \subseteq I}\Big(\bigcup_{j \in J} U_j \in \mathrm{Con}_A \to \{ b_j \mid j \in J \} \in \mathrm{Con}_B \Big). \end{array}$$

For the definition of the entailment relation $\vdash$ it is helpful to first define the notion of an *application* of $W := \{ (U_i, b_i) \mid i \in I \} \in \mathrm{Con}$ to $U \in \mathrm{Con}_A$:

$$\{ (U_i, b_i) \mid i \in I \} U := \{ b_i \mid U \vdash_A U_i \}.$$

From the definition of Con we know that this set is in Con_B. Now define $W \vdash (U, b)$ by $WU \vdash_B b$.

Remark 26.5 Clearly application is *monotone in the second argument*, in the sense that $U \vdash_A U'$ implies ($WU' \subseteq WU$, hence also) $WU \vdash_B WU'$. In fact, application is also *monotone in the first argument*, that is,

$$W \vdash W' \quad \text{implies} \quad WU \vdash_B W'U.$$

To see this let $W = \{\, (U_i, b_i) \mid i \in I \,\}$ and $W' = \{\, (U'_j, b'_j) \mid j \in J \,\}$. By definition $W'U = \{\, b'_j \mid U \vdash_A U'_j \,\}$. Now fix j such that $U \vdash_A U'_j$; we must show $WU \vdash_B b'_j$. By assumption $W \vdash (U'_j, b'_j)$, hence $WU'_j \vdash_B b'_j$. Because of $WU \supseteq WU'_j$ the claim follows.

Lemma 26.6 *If* $\boldsymbol{A}$ *and* $\boldsymbol{B}$ *are information systems, then so is* $\boldsymbol{A} \to \boldsymbol{B}$.

Lemma 26.7 *Let* $\boldsymbol{A}$ *and* $\boldsymbol{B}$ *be information systems and* $f\colon |\boldsymbol{A}| \to |\boldsymbol{B}|$ *monotone (i.e.,* $x \subseteq y \to f(x) \subseteq f(y)$*). Then the following are equivalent.*

(a) f *satisfies the "principle of finite support" PFS: If* $b \in f(x)$*, then* $b \in f(\overline{U})$ *for some* $U \subseteq x$.
(b) f *commutes with directed unions: for every directed* $D \subseteq |\boldsymbol{A}|$ *(i.e., for any* $x, y \in D$ *there is a* $z \in D$ *such that* $x, y \subseteq z$*)*

$$f\Big(\bigcup_{x \in D} x\Big) = \bigcup_{x \in D} f(x).$$

Note that in (b) the set $\{\, f(x) \mid x \in D \,\}$ is directed by monotonicity of f; hence its union is indeed an object in $|\boldsymbol{B}|$. Also from PFS and monotonicity of f it follows that if $V \subseteq f(x)$, then $V \subseteq f(\overline{U})$ for some $U \subseteq x$.

We call a function $f\colon |\boldsymbol{A}| \to |\boldsymbol{B}|$ continuous if it satifies the conditions in Lemma 26.7. Hence continuous maps $f\colon |\boldsymbol{A}| \to |\boldsymbol{B}|$ are those that can be completely described from the point of view of finite approximations of the abstract objects $x \in |\boldsymbol{A}|$ and $f(x) \in |\boldsymbol{B}|$: Whenever we are given a finite approximation V to the value $f(x)$, then there is a finite approximation U to the argument x such that already $f(\overline{U})$ contains the information in V; note that by monotonicity $f(\overline{U}) \subseteq f(x)$.

Clearly the identity and constant functions are continuous, and also the *composition* $g \circ f$ of continuous functions $f\colon |\boldsymbol{A}| \to |\boldsymbol{B}|$ and $g\colon |\boldsymbol{B}| \to |\boldsymbol{C}|$.

Theorem 26.8 *Let* $\boldsymbol{A} = (A, \mathrm{Con}_A, \vdash_A)$*,* $\boldsymbol{B} = (B, \mathrm{Con}_B, \vdash_B)$ *be information systems. Then the objects of* $\boldsymbol{A} \to \boldsymbol{B}$ *are in a natural bijective correspondence with the continuous functions from* $|\boldsymbol{A}|$ *to* $|\boldsymbol{B}|$*, as follows.*

(a) With any object x *of* $\boldsymbol{A} \to \boldsymbol{B}$ *we can associate a continuous function* $|x|\colon |\boldsymbol{A}| \to |\boldsymbol{B}|$ *by*

$$|x|(z) := \{\, b \in B \mid (U, b) \in x \text{ for some } U \subseteq z \,\}.$$

We call $|x|(z)$ *the* application *of* x *to* z.

(b) *Conversely, with any continuous function* $f\colon |\boldsymbol{A}| \to |\boldsymbol{B}|$ *we can associate an object* $\hat{f}$ *of* $\boldsymbol{A} \to \boldsymbol{B}$ *by*

$$\hat{f} := \{\, (U, b) \mid b \in f(\overline{U}) \,\}.$$

These assignments are inverse to each other, i.e., $f = |\hat{f}|$ *and* $x = \widehat{|x|}$.

26.1.2 Algebras and Types

We now consider concrete information systems, our basis for continuous functionals. Types will be built from base types by the formation of function types, $\tau \to \sigma$. As domains for the base types we choose nonflat free algebras, given by their constructors. The reason for taking nonflat base domains is that we want the constructors to be injective and with disjoint ranges. This generally is not the case for flat domains.

Definition 26.9 (Constructor types and algebra forms) *Constructor types* κ have the form

$$\vec{\alpha} \to (\xi)_{i<n} \to \xi$$

with all type variables α_i distinct from each other and from ξ. Iterated arrows are understood as associated to the right. An argument type of a constructor type is called a *parameter* argument type if it is different from ξ, and a *recursive* argument type otherwise. A constructor type κ is *nullary* if it has no recursive argument types. We call

$$\iota := \mu_\xi \vec{\kappa}$$

with $\vec{\kappa}$ not empty an *algebra form*. An algebra form is *explicit* if it does not have recursive argument types.

Examples 26.10 We list some parameter-free algebra forms, with standard names for the constructors added to each constructor type.

$$\begin{aligned}
\mathbb{U} &:= \mu_\xi(\mathrm{Dummy}\colon \xi) && \text{(unit)},\\
\mathbb{B} &:= \mu_\xi(\mathtt{tt}\colon \xi, \mathtt{ff}\colon \xi) && \text{(booleans)},\\
\mathbb{N} &:= \mu_\xi(0\colon \xi, S\colon \xi \to \xi) && \text{(natural numbers, unary)},\\
\mathbb{P} &:= \mu_\xi(1\colon \xi, S_0\colon \xi \to \xi, S_1\colon \xi \to \xi) && \text{(positive numbers, binary)},\\
\mathbb{Y} &:= \mu_\xi(-\colon \xi, \mathrm{Branch}\colon \xi \to \xi \to \xi) && \text{(binary trees)}.
\end{aligned}$$

Algebra forms with type parameters are

$$
\begin{array}{lll}
\mathbb{I}(\alpha) & := \mu_\xi(\mathrm{Id}\colon \alpha \to \xi) & \text{(identity)},\\
\mathbb{L}(\alpha) & := \mu_\xi(\mathrm{Nil}\colon \xi, \mathrm{Cons}\colon \alpha \to \xi \to \xi) & \text{(lists)},\\
\mathbb{S}(\alpha) & := \mu_\xi(\mathrm{SCons}\colon \alpha \to \xi \to \xi) & \text{(streams)},\\
\alpha \times \beta & := \mu_\xi(\mathrm{Pair}\colon \alpha \to \beta \to \xi) & \text{(product)},\\
\alpha + \beta & := \mu_\xi(\mathrm{InL}\colon \alpha \to \xi, \mathrm{InR}\colon \beta \to \xi) & \text{(sum)},\\
\mathrm{uysum}(\alpha) & := \mu_\xi(\mathrm{DummyL}\colon \xi, \mathrm{Inr}\colon \alpha \to \xi) & \text{(for } \mathbb{U} + \alpha),\\
\mathrm{ysumu}(\alpha) & := \mu_\xi(\mathrm{Inl}\colon \alpha \to \xi, \mathrm{DummyR}\colon \xi) & \text{(for } \alpha + \mathbb{U}).
\end{array}
$$

The default name for the ith constructor of an algebra form is $\mathcal{C}_i$.

Definition 26.11 (Type)

$$\rho, \sigma, \tau ::= \alpha \mid \iota(\vec{\rho}) \mid \tau \to \sigma,$$

where ι is an algebra form with $\vec{\alpha}$ its parameter type variables, and $\iota(\vec{\rho})$ is the result of substituting the (already generated) types $\vec{\rho}$ for $\vec{\alpha}$. Types of the form $\iota(\vec{\rho})$ are called *algebras*. An algebra is *closed* if it has no type variables. The *level* of a type is defined by

$$\mathrm{lv}(\alpha) := 0, \quad \mathrm{lv}(\iota(\vec{\rho})) := \max(\mathrm{lv}(\vec{\rho})), \quad \mathrm{lv}(\tau \to \sigma) := \max(\mathrm{lv}(\sigma), 1 + \mathrm{lv}(\tau)).$$

Base types are types of level 0, and a *higher* type has level at least 1.

26.1.3 The Model $(\mathfrak{C}_\tau)_\tau$

For every closed type τ we define an information system $\boldsymbol{C}_\tau = (C_\tau, \mathrm{Con}_\tau, \vdash_\tau)$. The definition is by induction on τ, and in case of an algebra $\iota(\vec{\rho})$ by a side inductive definition.

Definition 26.12 (Information system of type τ) *Case* $\iota(\tau)$. For simplicity assume that there is only one parameter type τ.

(a) *Tokens* $a \in C_{\iota(\tau)}$ are the type-correct constructor expressions $\mathcal{C}V a_1^* \dots a_n^*$ where a_i^* is an *extended token*, that is, a token or the special symbol $*$ which carries no information, and V is a consistent set of tokens in C_τ.

(b) A finite set U of tokens in $C_{\iota(\tau)}$ is *consistent* (i.e., $\in \mathrm{Con}_{\iota(\tau)}$) if all its elements start with the same constructor $\mathcal{C}$, say of arity $\tau \to \iota(\tau) \cdots \to \iota(\tau) \to \iota(\tau)$. Let $U = \{\mathcal{C}V_1 a_{11}^* \dots a_{1n}^*, \dots, \mathcal{C}V_m a_{m1}^* \dots a_{mn}^*\}$. Then we require that (i) $V_1 \cup \dots \cup V_m$ is consistent (i.e., $\in \mathrm{Con}_\tau$) and (ii) the sets U_i consisting of all (proper) tokens at the ith argument position of some token in U are consistent (i.e., $\in \mathrm{Con}_{\iota(\tau)}$).

(c) $\{\mathcal{C}V_1a^*_{11}\ldots a^*_{1n},\ldots,\mathcal{C}V_ma^*_{m1}\ldots a^*_{mn}\} \vdash_{\iota(\vec{\tau})} \mathcal{C}Va^*_1\ldots a^*_n$ if and only if (i) $V_1\cup\cdots\cup V_m \vdash_{\vec{\tau}} V$ and (ii) for each set U_i as in (b) we have $U_i \vdash_{\iota(\vec{\tau})} a^*_i$ (where $U_i \vdash *$ is taken to be true).

Case $\tau \to \sigma$ Tokens, consistency, and entailment for $\boldsymbol{C}_{\tau\to\sigma} := \boldsymbol{C}_\tau \to \boldsymbol{C}_\sigma$ are defined as done in Section 26.1.1 for arbitrary information systems.

Lemma 26.13 $\boldsymbol{C}_\tau := (C_\tau, \mathrm{Con}_\tau, \vdash_\tau)$ *is an information system.*

Definition 26.14 The objects $x \in |\boldsymbol{C}_\tau|$ are called *partial continuous functionals* of type τ. Since $\boldsymbol{C}_{\tau\to\sigma} = \boldsymbol{C}_\tau \to \boldsymbol{C}_\sigma$, the partial continuous functionals of type $\tau \to \sigma$ correspond to the continuous functions from $|\boldsymbol{C}_\tau|$ to $|\boldsymbol{C}_\sigma|$. A partial continuous functional $x \in |\boldsymbol{C}_\tau|$ is *computable* if it is recursively enumerable when viewed as a set of tokens. The *Scott–Ershov model* $\mathfrak{C}$ of partial continuous functionals is defined to be $(|\boldsymbol{C}_\tau|)_\tau$.

Definition 26.15 (Cototal and total objects of closed-base type) Let $\iota(\vec{\tau})$ be a closed-base type. Its tokens can be seen as constructor trees with some recursive argument positions occupied by $*$. An object x is *cototal* if for each of its tokens $P(*)$ with a distinguished occurrence of $*$ there is another token of the form $P(\mathcal{C}\vec{\emptyset}\vec{*})$ in x. Finite cototal objects are called *total*.

26.1.4 Cototality and Bisimilarity

For closed-ground types equality of cototal objects can be characterized by *bisimilarity*. As an example we consider the algebra $\mathbb{Y}$ of binary trees. We define bisimilarity $\approx_{\mathbb{Y}}$ as the largest relation on $C_{\mathbb{Y}}$ satisfying the *closure* axiom $\approx^-_{\mathbb{Y}}$:

$$\forall_{x,x'}(x \approx x' \to (x \equiv - \wedge x' \equiv -) \vee \exists_{x_1,x_2,x'_1,x'_2}(x_1 \approx x'_1 \wedge x_2 \approx x'_2 \wedge x \equiv \mathcal{C}x_1x_2 \wedge x' \equiv \mathcal{C}x'_1x'_2))$$

with $\mathcal{C}$ for the Branch constructor. Being the "largest" relation means that any other relation ("competitor") X satisfying the same closure property is below $\approx_{\mathbb{Y}}$, that is, we require the *greatest-fixed-point* property $\approx^+_{\mathbb{Y}}$:

$$\begin{aligned}\forall_{x,x'}(Xxx' \to {} & (x \equiv - \wedge x' \equiv -) \vee \\ & \exists_{x_1,x_2,x'_1,x'_2}((x_1 \approx x'_1 \vee Xx_1x'_1) \wedge (x_2 \approx x'_2 \vee Xx_2x'_2) \wedge \\ & x \equiv \mathcal{C}x_1x_2 \wedge x' \equiv \mathcal{C}x'_1x'_2))) \to \\ X \subseteq {\approx}. &\end{aligned}$$

Lemma 26.16 (Bisimilarity) $x \approx_{\mathbb{Y}} x'$ *implies* $x \equiv x'$, *for* $x, x' \in C_{\mathbb{Y}}$.

Proof Let a range over tokens for $\mathbb{Y}$, and define the *height* $|a^*|$ of an extended token a^* by $|*| := 0$, $|-| := 1$, $|\mathcal{C}a_1^*a_2^*| := 1+\max(|a_1^*|, |a_2^*|)$. By induction on the height $|a^*|$ of extended tokens a^* we prove that for all objects x, x' and extended tokens $a^* \in x$ we have $a^* \in x'$. It suffices to consider the case $\mathcal{C}a_1^*a_2^*$. From $x \approx_{\mathbb{Y}} x'$ we obtain by the closure axiom x_1, x_2, x_1', x_2' with

$$x_1 \approx x_1' \wedge x_2 \approx x_2' \wedge x \equiv \mathcal{C}x_1x_2 \wedge x' \equiv \mathcal{C}x_1'x_2'.$$

Then $a_i^* \in x_i$ (for $i = 1, 2$), and by IH $a_i^* \in x_i'$. Thus $\mathcal{C}a_1^*a_2^* \in x'$. □

From Lemma 26.16 we obtain the following characterization of $\approx_{\mathbb{Y}}$ on $\mathcal{C}_{\mathbb{Y}}$. We define ${}^{\mathrm{co}}T_{\mathbb{Y}}$ as the largest subset of $\mathcal{C}_{\mathbb{Y}}$ satisfying the *closure* axiom ${}^{\mathrm{co}}T_{\mathbb{Y}}^-$:

$$\forall_x(x \in {}^{\mathrm{co}}T \to x \equiv - \vee \exists_{x_1,x_2}(x_1 \in {}^{\mathrm{co}}T \wedge x_2 \in {}^{\mathrm{co}}T \wedge x \equiv \mathcal{C}x_1x_2)).$$

Again we require the *greatest-fixed-point* property ${}^{\mathrm{co}}T_{\mathbb{Y}}^+$:

$$\begin{aligned} &\forall_x(x \in X \to (x \equiv -) \vee \\ &\qquad \exists_{x_1,x_2}(x_1 \in {}^{\mathrm{co}}T \cup X \wedge x_2 \in {}^{\mathrm{co}}T \cup X \wedge x \equiv \mathcal{C}x_1x_2)) \to \\ &X \subseteq {}^{\mathrm{co}}T. \end{aligned}$$

For objects $x, x' \in \mathcal{C}_{\mathbb{Y}}$ we show the following.

Lemma 26.17 (Characterization of $\approx_{\mathbb{Y}}$)

$$x \approx_{\mathbb{Y}} x' \leftrightarrow x, x' \in {}^{\mathrm{co}}T_{\mathbb{Y}} \wedge x \equiv x', \textit{for } x, x' \in \mathcal{C}_{\mathbb{Y}}.$$

Proof "→". By Lemma 26.16 it remains to prove $x \approx_{\mathbb{Y}} x' \to x \in {}^{\mathrm{co}}T_{\mathbb{Y}}$. To this end we apply ${}^{\mathrm{co}}T_{\mathbb{Y}}^+$ with competitor $X := \{\, x \mid \exists_{x'}(x \approx_{\mathbb{Y}} x') \,\}$. It suffices to prove the premise. Fix x, x' with $x \approx_{\mathbb{Y}} x'$. The goal is

$$\begin{aligned} &(x \equiv -) \vee \\ &\exists_{x_1,x_2}((x_1 \in {}^{\mathrm{co}}T \vee \exists_{x_1'}(x_1 \approx x_1')) \wedge (x_2 \in {}^{\mathrm{co}}T \vee \exists_{x_2'}(x_2 \approx x_2')) \wedge x \equiv \mathcal{C}x_1x_2). \end{aligned}$$

By the closure property $\approx_{\mathbb{Y}}^-$ we have

$$(x \equiv - \wedge x' \equiv -) \vee \exists_{x_1,x_2,x_1',x_2'}(x_1 \approx x_1' \wedge x_2 \approx x_2' \wedge x \equiv \mathcal{C}x_1x_2 \wedge x' \equiv \mathcal{C}x_1'x_2')).$$

In the first case we have $x \equiv -$ and are done. In the second case we have x_1, x_2, x_1', x_2' with $x_1 \approx x_1'$, $x_2 \approx x_2'$ and $x \equiv \mathcal{C}x_1x_2$, and are done as well.

"←". We prove $x \in {}^{\mathrm{co}}T_{\mathbb{Y}} \to x \equiv x' \to x \approx_{\mathbb{Y}} x'$ by the greatest-fixed-point property $\approx_{\mathbb{Y}}^+$ with competitor $X := \{\, x, x' \mid x \in {}^{\mathrm{co}}T_{\mathbb{Y}} \wedge x \equiv x' \,\}$. It suffices to prove the premise. Fix x, x' with $x \in {}^{\mathrm{co}}T_{\mathbb{Y}} \wedge x \equiv x'$. The goal is

$$\begin{aligned} (x \equiv - \wedge x' \equiv -) \vee \exists_{x_1,x_2,x_1',x_2'}((x_1 \approx x_1' \vee (x_1 \in {}^{\mathrm{co}}T_{\mathbb{Y}} \wedge x_1 \equiv x_1')) \wedge \\ (x_2 \approx x_2' \vee (x_2 \in {}^{\mathrm{co}}T_{\mathbb{Y}} \wedge x_2 \equiv x_2')) \wedge \\ x \equiv \mathcal{C}x_1x_2 \wedge x' \equiv \mathcal{C}x_1'x_2')). \end{aligned}$$

By the closure property ${}^{\mathrm{co}}T^{-}_{\mathbb{Y}}$ applied to $x \in {}^{\mathrm{co}}T_{\mathbb{Y}}$ we have

$$(x \equiv -) \vee \exists_{x_1,x_2}(x_1 \in {}^{\mathrm{co}}T_{\mathbb{Y}} \wedge x_2 \in {}^{\mathrm{co}}T \wedge x \equiv Cx_1x_2).$$

In the first case we have $x \equiv -$ and are done, since $x \equiv x'$. In the second case we have $x_1, x_2 \in {}^{\mathrm{co}}T_{\mathbb{Y}}$ with $x \equiv Cx_1x_2$. Then we are done as well with $x'_1 := x_1$ and $x'_2 := x_2$, since again $x \equiv x'$. □

26.1.5 Constructors as Continuous Functions

Let ι be an algebra. Every constructor C generates the following object in the function space determined by the type of the constructor:

$$r_C := \{ (\vec{U}, C\vec{a^*}) \mid \vec{U} \vdash \vec{a^*} \}.$$

Here $(\vec{U}, a)$ abbreviates $(U_1, (U_2, \ldots (U_n, a) \ldots))$.

According to the general definition of a continuous function associated to an object in a function space, the continuous map $|r_C|$ satisfies

$$|r_C|(\vec{x}) = \{ C\vec{a^*} \mid \exists_{\vec{U} \subseteq \vec{x}}(\vec{U} \vdash \vec{a^*}) \}.$$

(For $\mathbb{N}$ we have $|r_S|(\{0\}) = \{S0, S*\}$ and $|r_S|(\{S0, S*\}) = \{SS0, SS*, S*\}$.) An immediate consequence is that the (continuous maps corresponding to) constructors are injective and their ranges are disjoint, which is what we wanted to achieve by associating nonflat rather than flat information systems with algebras.

Lemma 26.18 (Constructors are injective and have disjoint ranges) *Let ι be an algebra and C be a constructor of ι. Then*

$$|r_C|(\vec{x}) \subseteq |r_C|(\vec{y}) \leftrightarrow \vec{x} \subseteq \vec{y}.$$

If C_1, C_2 are distinct constructors of ι, then $|r_{C_1}|(\vec{x}) \neq |r_{C_2}|(\vec{y})$, since the two objects are nonempty and disjoint.

Proof Immediate from the definitions. □

Remark 26.19 Notice that neither property holds for flat information systems, since for them, by monotonicity, constructors need to be *strict* (i.e., if one argument is the empty object, then the value is as well). But then we have

$$|r_C|(\overline{\emptyset}, y) = \overline{\emptyset} = |r_C|(x, \overline{\emptyset}), \qquad |r_{C_1}|(\overline{\emptyset}) = \overline{\emptyset} = |r_{C_2}|(\overline{\emptyset})$$

where in the first case we have one binary and, in the second, two unary constructors.

26.2 A Term Language for Computable Functionals

We set up a system T^+ of typed terms, as an extension of Gödel's T [6]. Every closed term of type τ denotes an object of this type in the model $\mathfrak{C}$, that is, a partial continuous functional. This is in contrast to Martin-Löf style type theories like Coq's calculus of inductive constructions, where terms must be total. Dropping this restriction has the advantage that nonterminating operators like corecursion can directly be represented as constants. We will define constants by equations, in a pattern-matching style.

26.2.1 Constants, terms, and computation rules

For every algebra $\iota = \mu_\xi((\rho_{i\nu}(\xi))_{\nu<n_i} \to \xi)_{i<k}$ we have constants

$\mathcal{C}_{\iota,i}$	$(\rho_{i\nu}(\iota))_{\nu<n_i} \to \iota$	ith constructor
$\mathcal{R}_\iota^\alpha$	$\iota \to ((\rho_{i\nu}(\iota \times \alpha))_{\nu<n_i} \to \alpha)_{i<k} \to \alpha$	recursion
$\mathcal{D}_\iota$	${}^{\mathrm{co}}\iota \to \sum_{i<k} \prod_{\nu<n_i} \rho_{i\nu}({}^{\mathrm{co}}\iota)$	destructor
${}^{\mathrm{co}}\mathcal{R}_\iota^\alpha$	$\alpha \to (\alpha \to \sum_{i<k} \prod_{\nu<n_i} \rho_{i\nu}({}^{\mathrm{co}}\iota + \alpha)) \to {}^{\mathrm{co}}\iota$	corecursion.

It is convenient to write the type of the *recursion* operator $\mathcal{R}_{\mathbb{N}}^\tau$ in the form $\mathbb{N} \to \tau \to (\mathbb{N} \to \tau \to \tau) \to \tau$. The first argument is the recursion argument, the second one gives the base value, and the third gives the step function, mapping the recursion argument and the previous value to the next value. The *destructor* $\mathcal{D}_\iota$ disassembles a constructor-built object into its parts. The *corecursion* operator ${}^{\mathrm{co}}\mathcal{R}_\iota^\tau$ is used to construct a map from τ to ${}^{\mathrm{co}}\iota$.

From the constants above, typed variables, and possibly other typed constants D^τ we define *terms* by abstraction and application:

$$M, N ::= x^\tau \mid \mathcal{C}_{\iota,i} \mid \mathcal{R}_\iota \mid \mathcal{D}_\iota \mid {}^{\mathrm{co}}\mathcal{R}_\iota \mid D^\tau \mid (\lambda_{x^\tau} M^\sigma)^{\tau\to\sigma} \mid (M^{\tau\to\sigma} N^\tau)^\sigma.$$

For each term we want to define its denotation in the model $\mathfrak{C}$. To this end we use defining equations. Each constant C comes with a system of *computation rules* consisting of finitely many equations

$$C\vec{P}_i(\vec{y}_i) = M_i \qquad (i = 1, \dots, n \text{ where } n \geq 0) \tag{26.1}$$

with free variables of $\vec{P}_i(\vec{y}_i)$ and M_i among $\vec{y}_i$, where the arguments on the left-hand side must be "constructor patterns," that is, lists of applicative terms built from constructors and distinct variables. To ensure consistency of the defining equations, we require that, for $i \neq j$, $\vec{P}_i$ and $\vec{P}_j$ have disjoint free variables, and either $\vec{P}_i$ and $\vec{P}_j$ are nonunifiable (i.e., there is no substitution which identifies them), or else for the "most general unifier" ϑ of $\vec{P}_i$ and $\vec{P}_j$ we have $M_i\vartheta = M_j\vartheta$. Notice that the substitution ϑ assigns to the variables $\vec{y}_i$ in M_i constructor patterns $\vec{R}_k(\vec{z})$

($k = i, j$). A further requirement on a system of computation rules $C\vec{P}_i(\vec{y}_i) = M_i$ is that the lengths of all $\vec{P}_i(\vec{y}_i)$ are the same; this number is called the *arity* of C, denoted by $\mathrm{ar}(C)$. A substitution instance of a left-hand side of (26.1) is called a *C-redex.*

The computation rules for the constants $\mathcal{C}_{\iota,i}$, $\mathcal{R}_\iota$, $\mathcal{D}_\iota$, and ${}^{\mathrm{co}}\mathcal{R}_\iota$ are fixed, as follows. For the constructors no computation rules are necessary, since the model $\mathfrak{C}$ is built from them. For the recursion operator let

$$\alpha_0 \to \cdots \to \alpha_{m-1} \to (\xi)_{i<n} \to \xi$$

be the type of the ith constructor $\mathcal{C}_i$ of ι and consider a term $\mathcal{C}_i\vec{x}$ of type ι. We write $\vec{x}^P = x_0^P, \ldots, x_{m-1}^P$ for the *parameter arguments* $x_0^{\alpha_0}, \ldots, x_{m-1}^{\alpha_{m-1}}$ and $\vec{x}^R = x_0^R, \ldots, x_{n-1}^R$ for the *recursive arguments* $x_m^\iota, \ldots, x_{m+n-1}^\iota$. Writing $\mathcal{R}$ for $\mathcal{R}_\iota^\tau$ we take as its computation rules

$$\mathcal{R}(\mathcal{C}_i\vec{x})\vec{f} = f_i\vec{x}(\mathcal{R}x_0^R\vec{f}\,)\cdots(\mathcal{R}x_{n-1}^R\vec{f}\,).$$

In particular, $\mathcal{R}_{\mathbb{N}}^\tau$ is defined by the computation rules

$$\mathcal{R}_{\mathbb{N}}^\tau 0af = a, \qquad \mathcal{R}_{\mathbb{N}}^\tau(\mathcal{S}n)af = fn(\mathcal{R}_{\mathbb{N}}^\tau naf).$$

For example, $\mathcal{R}_{\mathbb{N}}^{\mathbb{N}}nm\lambda_{n,l}(\mathcal{S}l)$ defines addition $m + n$ by recursion on n. The computation rules for the *destructor* $\mathcal{D}_\iota$ are

$$\mathcal{D}_\iota(\mathcal{C}_i\vec{x}\,) = \langle\vec{x}\,\rangle.$$

To deal with *corecursion* we introduce some notation. For $f\colon \rho \to \tau$ and $g\colon \sigma \to \tau$ we denote $\lambda_x(\mathcal{R}_{\rho+\sigma}^\tau xfg)$ of type $\rho + \sigma \to \tau$ by $[f, g]$, and similary for ternary sumtypes, etc. The identity function id below is of type $\iota \to \iota$ with ι the respective algebra. The (single) computation rule for ${}^{\mathrm{co}}\mathcal{R}_\iota^\tau$ is

$${}^{\mathrm{co}}\mathcal{R}_\iota^\tau xf = [g_0, \ldots, g_{k-1}](fx),$$

where g_i of type $\prod_{\nu<n_i}\rho_{i\nu}(\iota + \tau) \to \iota$ is defined as

$$g_i := \lambda_{\vec{x}}(\mathcal{C}_i(N_\nu)_{\nu<n_i}) \qquad \text{with } x_\nu\colon \rho_{i\nu}(\iota + \tau),$$
$$N_\nu := \begin{cases} x_\nu & \text{if } \rho_{i\nu}(\xi) \text{ is a parameter arg. type,} \\ [\mathrm{id}^{\iota\to\iota}, \lambda_x({}^{\mathrm{co}}\mathcal{R}_\iota^\tau xf)]x_\nu^{\iota+\tau}) & \text{otherwise.} \end{cases}$$

Remark 26.20 It can be difficult to read the computation rules for corecursion operators. However, it helps if we know some properties of the "step" function f.

For instance we have

$$
{}^{\mathrm{co}}\mathcal{R}^{\tau}_{\mathbb{N}} x f = \begin{cases} 0 & \text{if } fx = \mathrm{DummyL}^{\mathbb{U}+(\mathbb{N}+\tau)} \\ \mathcal{S}n & \text{if } fx = \mathrm{Inr}(\mathrm{InL}^{\mathbb{N}\to\mathbb{N}+\tau} n) \\ \mathcal{S}({}^{\mathrm{co}}\mathcal{R}^{\tau}_{\mathbb{N}} x' f) & \text{if } fx = \mathrm{Inr}(\mathrm{InR}^{\tau\to\mathbb{N}+\tau} x') \end{cases}
$$

$$
{}^{\mathrm{co}}\mathcal{R}^{\tau}_{\mathbb{S}(\rho)} x f = \begin{cases} a :: u & \text{if } fx = \langle a, \mathrm{InL}^{\mathbb{S}(\rho)\to\mathbb{S}(\rho)+\tau} u\rangle \\ a :: {}^{\mathrm{co}}\mathcal{R}^{\tau}_{\mathbb{S}(\rho)} x' f & \text{if } fx = \langle a, \mathrm{InR}^{\tau\to\mathbb{S}(\rho)+\tau} x'\rangle. \end{cases}
$$

26.2.2 Denotational Semantics

We set up a connection between the term system T^+ and the model $\mathfrak{C}$. The main point is to clarify how the computation rules define an object z in a function space. The idea is to inductively define the set of tokens (U, a) that make up z. It is convenient to define the value $[\![\lambda_{\vec{x}} M]\!]$, where M is a term with free variables among $\vec{x}$. Since this value is a token set, we can define inductively the relation $(\vec{U}, a) \in [\![\lambda_{\vec{x}} M]\!]$.

For a constructor pattern $\vec{P}(\vec{x}\,)$ and a list $\vec{V}$ of the same length and types as $\vec{x}$ we define a list $\vec{P}(\vec{V})$ of formal neighborhoods of the same length and types as $\vec{P}(\vec{x}\,)$, by induction on $\vec{P}(\vec{x}\,)$. $x(V)$ is the singleton list V, and for $\langle\rangle$ take the empty list. $(\vec{P}, Q)(\vec{V}, \vec{w}\,)$ is covered by induction, and

$$
(\mathcal{C}\vec{P}\,)(\vec{V}) := \{\, \mathcal{C}\vec{a} \mid a_i \in P_i(\vec{V_i}) \text{ if } P_i(\vec{V_i}) \neq \emptyset, \text{ and } a_i = * \text{ otherwise} \,\}.
$$

We use the following notation. $(\vec{U}, a)$ means $(U_1, (U_2, \dots (U_n, a)) \dots)$, and $(\vec{U}, V) \subseteq [\![\lambda_{\vec{x}} M]\!]$ means $(\vec{U}, a) \in [\![\lambda_{\vec{x}} M]\!]$ for all (finitely many) $a \in V$.

Definition 26.21 (Inductive, of $(\vec{U}, a) \in [\![\lambda_{\vec{x}} M]\!]$)

$$
\frac{U_i \vdash a}{(\vec{U}, a) \in [\![\lambda_{\vec{x}} x_i]\!]}(V), \qquad \frac{(\vec{U}, V, a) \in [\![\lambda_{\vec{x}} M]\!] \qquad (\vec{U}, V) \subseteq [\![\lambda_{\vec{x}} N]\!]}{(\vec{U}, a) \in [\![\lambda_{\vec{x}} (MN)]\!]}(A).
$$

For every constructor $\mathcal{C}$ and defined constant D we have

$$
\frac{\vec{V} \vdash \vec{a}}{(\vec{U}, \vec{V}, \mathcal{C}\vec{a}\,) \in [\![\lambda_{\vec{x}} \mathcal{C}]\!]}(\mathcal{C}), \qquad \frac{(\vec{U}, \vec{V}, a) \in [\![\lambda_{\vec{x},\vec{y}} M]\!] \qquad \vec{W} \vdash \vec{P}(\vec{V})}{(\vec{U}, \vec{W}, a) \in [\![\lambda_{\vec{x}} D]\!]}(D)
$$

with one such rule (D) for every computation rule $D\vec{P}(\vec{y}\,) = M$.

This "denotational semantics" has good properties (see Schwichtenberg and Wainer [8, pp. 279–287]): $[\![\lambda_{\vec{x}} M]\!]$ is an object in the model, and Definition 26.21 of the denotation of a term is reasonable in the sense that it is not changed by an application of the standard (β- and η-) conversions or a computation rule.

26.3 A Theory of Computable Functionals

After becoming clear about the objects we intend to reason about, we now set up a theory to prove their properties. The main concepts are those of inductively and coinductively defined predicates. They can be declared to be either computationally relevant (c.r.) or else noncomputational (n.c.).

26.3.1 Formulas and their Computational Content

Assume an infinite supply of predicate variables, each of its own *arity* (a list of types). We distinguish two sorts of predicate variables, "computationally relevant" ones X^{c} and "noncomputational" ones X^{nc}, and use X for both.

Definition 26.22 (Clauses and predicate forms) *Clauses* K have the form

$$\forall_{\vec{x}}(\tilde{Y}^{\mathrm{c}} \to \tilde{Z}^{\mathrm{nc}} \to (\forall_{\vec{y}_i}(\tilde{W}_i^{\mathrm{nc}} \to \bar{X}_i))_{i<n} \to \bar{X})$$

with all predicate variables Y_i^{c}, Z_i^{nc}, W_i^{nc} occurring exactly once and distinct from each other and from X. By $\bar{X}$ we denote the result of applying the predicate variable X to a list of terms of fitting types, and by $\tilde{X}$ lists of those. Iterated implications are understood as associated to the right. A premise of a clause is called a *parameter* premise if X does not occur in it, and a *recursive* premise otherwise. A clause K is *nullary* if it has no recursive premises. We call $I^{\mathrm{c}} := \mu_{X^{\mathrm{c}}}\vec{K}$ and $I^{\mathrm{nc}} := \mu_{X^{\mathrm{nc}}}\vec{K}$ with $\vec{K}$ not empty *predicate forms* (and use I for both), and similarly with ${}^{\mathrm{co}}I$ for I and ν for μ.

Definition 26.23 (Algebra form of a predicate form) From every clause K we obtain a constructor type by (i) omitting quantifiers, (ii) dropping all n.c. predicates and from the c.r. predicates their arguments, and (iii) replacing the remaining predicate variables by type variables. That is, from the clause

$$\forall_{\vec{x}}(\tilde{Y}^{\mathrm{c}} \to \tilde{Z}^{\mathrm{nc}} \to (\forall_{\vec{y}_i}(\tilde{W}_i^{\mathrm{nc}} \to \bar{X}_i))_{i<n} \to \bar{X})$$

we obtain the constructor type $\vec{\alpha} \to (\xi)_{i<n} \to \xi$. With every predicate form $I^{\mathrm{c}} := (\mu/\nu)_{X^{\mathrm{c}}}\vec{K}$ we canonically associate the algebra form $\iota_{I^{\mathrm{c}}} := \mu_\xi\vec{\kappa}$.

Definition 26.24 (Predicates and formulas)

$$\begin{aligned} &P, Q ::= X \mid \{\vec{x} \mid A\} \mid I(\vec{\rho}, \vec{P}) \mid {}^{\mathrm{co}}I(\vec{\rho}, \vec{P}) && \text{(predicates)},\\ &A, B ::= P\vec{t} \mid A \to B \mid \forall_x A && \text{(formulas)} \end{aligned}$$

with $I/{}^{\mathrm{co}}I$ a predicate form. $(I/{}^{\mathrm{co}}I)(\vec{\rho}, \vec{P})$ is the result of substituting the types $\vec{\rho}$ and the (already generated) predicates $\vec{P}$ for its type and predicate variables. To take care of the difference between X^{c} and X^{nc} we define the final predicate of a predicate or formula by

$$\begin{aligned} \mathrm{fp}(X) &:= X, & \mathrm{fp}(P\vec{t}\,) &:= \mathrm{fp}(P), \\ \mathrm{fp}(\{\,\vec{x} \mid A\,\}) &:= \mathrm{fp}(A), & \mathrm{fp}(A \to B) &:= \mathrm{fp}(B), \\ \mathrm{fp}((I/{}^{\mathrm{co}}I)(\vec{\rho}, \vec{P}\,)) &:= I/{}^{\mathrm{co}}I, & \mathrm{fp}(\forall_x A) &:= \mathrm{fp}(A). \end{aligned}$$

We call a predicate or formula C *noncomputational* (n.c., or *Harrop*) if its final predicate $\mathrm{fp}(C)$ is of the form X^{nc} or I^{nc}, else *computationally relevant* (c.r.). All predicate substitutions involved in $(I/{}^{\mathrm{co}}I)(\vec{\rho}, \vec{P}\,)$ must substitute c.r. predicates for c.r. predicate variables and n.c. predicates for n.c. predicate variables. Such predicate substitutions are called *sharp*.

Predicates of the form $I(\vec{\rho}, \vec{P}\,)$ are called *inductive*, and predicates of the form ${}^{\mathrm{co}}I(\vec{\rho}, \vec{P}\,)$ *coinductive*.

The terms $\vec{t}$ are those introduced in Section 26.2.1, that is, typed terms built from typed variables and constants by abstraction and application, and (importantly) those with a common reduct are identified.

A predicate of the form $\{\,\vec{x} \mid C\,\}$ is called a *comprehension term*. We identify $\{\,\vec{x} \mid C(\vec{x}\,)\,\}\vec{t}$ with $C(\vec{t}\ )$. For a predicate C of arity $(\rho, \vec{\sigma}\,)$ we write Ct for $\{\,\vec{y} \mid Ct\vec{y}\,\}$.

Definition 26.25 (Type $\tau(C)$ and cotype $\varphi(C)$ of a c.r. predicate or formula C) Assume a global injective assignment of type variables ξ to c.r. predicate variables X:

$$\begin{aligned} \tau(X) &:= \xi, & \tau(P\vec{t}\,) &:= \tau(P), \\ \tau(\{\,\vec{x} \mid A\,\}) &:= \tau(A), & \tau(A \to B) &:= \begin{cases} \tau(A) \to \tau(B) & (A \text{ c.r.}) \\ \tau(B) & (A \text{ n.c.}) \end{cases} \\ \tau((I/{}^{\mathrm{co}}I)(\vec{r}, \vec{P}\,)) &:= \iota_I(\tau(\vec{P}^{\mathrm{c}})), & \tau(\forall_x A) &:= \tau(A) \end{aligned}$$

where $\vec{P}^{\mathrm{c}}$ are the c.r. predicates among $\vec{P}$ and ι_I is the algebra associated with the predicate $I/{}^{\mathrm{co}}I$. *Cotypes* are like types, but with algebra occurrences ι_I marked as ${}^{\mathrm{co}}\iota_I$ if they arise from a coinductive predicate: $\varphi({}^{\mathrm{co}}I(\vec{r}, \vec{P}\,)) := {}^{\mathrm{co}}\iota_I(\varphi(\vec{P}^{\mathrm{c}}))$.

Examples 26.26 (1) The *even numbers* are inductively defined by

$$\mathrm{Even} := \mu_{X^{\mathrm{c}}}(0 \in X^{\mathrm{c}}, \forall_n(n \in X^{\mathrm{c}} \to \mathcal{S}(\mathcal{S}n) \in X^{\mathrm{c}})).$$

The constructor types of $\tau(\mathrm{Even})$ are ξ and $\xi \to \xi$, hence $\tau(\mathrm{Even}) = \mathbb{N}$.

(2) *Leibniz equality* $\equiv$ is inductively defined by $\mathrm{EqD} := \mu_{X^{\mathrm{nc}}}(\forall_x X^{\mathrm{nc}}xx)$.

(3) The missing *logical connectives* $\vee, \wedge, \exists$ are nullary inductive predicates with parameters. For instance, disjunction is a special case of union

$$\mathrm{Cup}_{Y,Z} := \mu_{X^c}(\forall_{\vec{x}}(Y\vec{x} \to X^c\vec{x}),\ \forall_{\vec{x}}(Z\vec{x} \to X^c\vec{x})).$$

Since Y, Z can be chosen as either c.r. or n.c. we obtain the variants

$$\begin{aligned}
\mathrm{CupD}_{Y^c,Z^c} &:= \mu_{X^c}(\forall_{\vec{x}}(Y^c\vec{x} \to X^c\vec{x}),\ \forall_{\vec{x}}(Z^c\vec{x} \to X^c\vec{x})),\\
\mathrm{CupL}_{Y^c,Z^{nc}} &:= \mu_{X^c}(\forall_{\vec{x}}(Y^c\vec{x} \to X^c\vec{x}),\ \forall_{\vec{x}}(Z^{nc}\vec{x} \to X^c\vec{x})),\\
\mathrm{CupR}_{Y^{nc},Z^c} &:= \mu_{X^c}(\forall_{\vec{x}}(Y^{nc}\vec{x} \to X^c\vec{x}),\ \forall_{\vec{x}}(Z^c\vec{x} \to X^c\vec{x})),\\
\mathrm{CupU}_{Y^{nc},Z^{nc}} &:= \mu_{X^c}(\forall_{\vec{x}}(Y^{nc}\vec{x} \to X^c\vec{x}),\ \forall_{\vec{x}}(Z^{nc}\vec{x} \to X^c\vec{x})),\\
\mathrm{CupNc}_{Y,Z} &:= \mu_{X^{nc}}(\forall_{\vec{x}}(Y\vec{x} \to X^{nc}\vec{x}),\ \forall_{\vec{x}}(Z\vec{x} \to X^{nc}\vec{x})).
\end{aligned}$$

(D, L, R, U for "double," "left," "right," and "uniform"). Then by definition

$$\begin{aligned}
\tau(\mathrm{CupD}) &= \mu_\xi(\beta_0 \to \xi, \beta_1 \to \xi) = \beta_0 + \beta_1,\\
\tau(\mathrm{CupL}) &= \mu_\xi(\beta \to \xi, \xi) \qquad\quad = \beta + \mathbb{U},\\
\tau(\mathrm{CupR}) &= \mu_\xi(\xi, \beta \to \xi) \qquad\quad = \mathbb{U} + \beta,\\
\tau(\mathrm{CupU}) &= \mu_\xi(\xi, \xi) \qquad\qquad\ \ = \mathbb{B}.
\end{aligned}$$

In case of nullary predicates we write $A \vee^d B$ for $\mathrm{CupD}_{\{|A\},\{|B\}}$, and similarly for $\vee^l$, $\vee^r$, $\vee^u$, $\vee^{nc}$. Since the "decoration" is determined by the c.r./n.c. status of the two parameter predicates we can leave it out in $\vee^d$, $\vee^l$, $\vee^r$, $\vee^u$ and write $\vee$. However, in the final n.c.-variant we suppress even the information which clause has been used, and hence must keep the notation $\vee^{nc}$.

26.3.2 Axioms of TCF

The essential axioms of TCF are introduction and elimination axioms for (co)inductively defined predicates. To grasp the general form of these axioms it is convenient to write a clause

$$\forall_{\vec{x}}(\tilde{Y}^c \to \tilde{Z}^{nc} \to (\forall_{\vec{y}_i}(\tilde{W}_i^{nc} \to \bar{X}_i))_{i<n} \to \bar{X}) \quad \text{as} \quad \forall_{\vec{x}}((A_\nu(X))_{\nu<n} \to X\vec{t}\,).$$

Definition 26.27 For an inductive predicate $\mu_X(\forall_{\vec{x}_i}((A_{i\nu}(X))_{\nu<n_i} \to X\vec{t}_i))_{i<k} =:$ I we have k introduction axioms I_i^+ $(i < k)$ and one elimination axiom I^-:

$$I_i^+ : \forall_{\vec{x}_i}((A_{i\nu}(I))_{\nu<n_i} \to I\vec{t}_i), \tag{26.2}$$

$$I^- : (\forall_{\vec{x}_i}((A_{i\nu}(I \cap X))_{\nu<n_i} \to X\vec{t}_i))_{i<k} \to I \subseteq X, \tag{26.3}$$

where (26.3) expresses that every competitor X satisfying the same clauses contains I. We take all substitution instances of I_i^+, I^- (w.r.t. substitutions for type and predicate variables) as axioms.

In (26.3) a "strengthened" form of the "step formula" has been used, namely $\forall_{\vec{x}_i}(A_{i\nu}(I \cap X))_{\nu<n_i} \to X\vec{t}_i$ rather than $\forall_{\vec{x}_i}(A_{i\nu}(X))_{\nu<n_i} \to X\vec{t}_i$. In applications

this simplifies the proof of the "step," since we have an additional I-hypothesis available.

To understand the axioms for coinductive predicates note that the conjunction of the k clauses (26.2) of an inductive predicate I is equivalent to

$$\forall_{\vec{x}}\left(\bigvee_{i<k}\exists_{\vec{x}_i}\left(\bigwedge_{\nu<n_i}A_{i\nu}(I)\wedge\vec{x}\equiv\vec{t_i}\right)\to I\vec{x}\right).$$

Definition 26.28 For an inductive predicate $\mu_X(\forall_{\vec{x}_i}((A_{i\nu}(X))_{\nu<n_i}\to X\vec{t_i}))_{i<k} =: I$ we define its *closure axiom* ${}^{\mathrm{co}}I^-$ and its *greatest-fixed-point axiom* ${}^{\mathrm{co}}I^+$:

$${}^{\mathrm{co}}I^-:\forall_{\vec{x}}\left({}^{\mathrm{co}}I\vec{x}\to\bigvee_{i<k}\exists_{\vec{x}_i}\left(\bigwedge_{\nu<n_i}A_{i\nu}({}^{\mathrm{co}}I)\wedge\vec{x}\equiv\vec{t_i}\right)\right),\tag{26.4}$$

$${}^{\mathrm{co}}I^+:\forall_{\vec{x}}\left(X\vec{x}\to\bigvee_{i<k}\exists_{\vec{x}_i}\left(\bigwedge_{\nu<n_i}A_{i\nu}({}^{\mathrm{co}}I\cup X)\wedge\vec{x}\equiv\vec{t_i}\right)\right)\to X\subseteq{}^{\mathrm{co}}I,\tag{26.5}$$

where (26.5) expresses that every competitor X satisfying the closure axiom is contained in ${}^{\mathrm{co}}I$. We take all substitution instances of ${}^{\mathrm{co}}I^-$, ${}^{\mathrm{co}}I^+$ (w.r.t. type and predicate variables) as axioms.

Here again we have used a strengthened form of the step formula, with $A_{i\nu}({}^{\mathrm{co}}I\cup X)$ rather than $A_{i\nu}(X)$. In applications of (26.5) this simplifies the proof of the step, since its conclusion is weaker.

Example 26.29 The conjunction of the two clauses of Even is equivalent to

$$\forall_n(n\equiv 0\vee\exists_{n'}(n'\in\mathrm{Even}\wedge n\equiv\mathcal{S}(\mathcal{S}n'))\to n\in\mathrm{Even}).$$

Hence the closure and greatest-fixed-point axioms for its dual ${}^{\mathrm{co}}\mathrm{Even}$ are

$$\forall_n(n\in{}^{\mathrm{co}}\mathrm{Even}\to n\equiv 0\vee\exists_{n'}(n'\in{}^{\mathrm{co}}\mathrm{Even}\wedge n\equiv\mathcal{S}(\mathcal{S}n'))),$$
$$\forall_n(Xn\to n\equiv 0\vee\exists_{n'}(n'\in({}^{\mathrm{co}}\mathrm{Even}\cup X)\wedge n\equiv\mathcal{S}(\mathcal{S}n')))\to X\subseteq{}^{\mathrm{co}}\mathrm{Even}.$$

For n.c. inductive or coinductive predicates the axioms are formed as in the c.r. case, using $\vee^{\mathrm{nc}}$ for the closure axiom of ${}^{\mathrm{co}}I^{\mathrm{nc}}$. But there is an important restriction: For I^{nc} with more than one clause the elimination axiom $(I^{\mathrm{nc}})^-$ can only be used with a *noncomputational* competitor predicate. This is needed in the proof of the soundness theorem in Section 26.4.3. However, this restriction does not apply to I^{nc} defined by one clause only. Important examples of such *one-clause-nc* inductive predicates are Leibniz equality and the noncomputational variants of the existential quantifier and of conjunction.

Lemma 26.30 *$I\subseteq{}^{\mathrm{co}}I$, $I^{\mathrm{nc}}\subseteq{}^{\mathrm{co}}I^{\mathrm{nc}}$ and also $I\subseteq I^{\mathrm{nc}}$, ${}^{\mathrm{co}}I\subseteq{}^{\mathrm{co}}I^{\mathrm{nc}}$.*

From the definition of Leibniz equality we can deduce the property Leibniz used as a definition.

Lemma 26.31 (Compatibility of EqD) $\forall_{x,y}(x \equiv y \to A(x) \to A(y))$.

Proof By the elimination axiom with $X := \{\, x, y \mid A(x) \to A(y) \,\}$. □

Using compatibility of $\equiv$ one easily proves symmetry and transitivity. Define *falsity* by $\mathbf{F} := (\text{ff} \equiv \text{tt})$. Then we can prove the following "Ex-falso-quodlibet."

Theorem 26.32 *For every formula A we can derive $\mathbf{F} \to A$ from assumptions* $\mathrm{Ef}_Y : \forall_{\vec{x}}(\mathbf{F} \to Y\vec{x})$ *for predicate variables Y strictly positive in A, and* $\mathrm{Ef}_I : \forall_{\vec{x}}(\mathbf{F} \to I\vec{x})$ *for inductive predicates I without a nullary clause.*

Proof We first show $\mathrm{Ef}_{\mathrm{EqD}} : \mathbf{F} \to x^\rho \equiv y^\rho$. By the introduction axiom we have $\mathcal{R}^\rho_{\mathbb{B}}\text{ff}xy \equiv \mathcal{R}^\rho_{\mathbb{B}}\text{ff}xy$. Then from $\text{ff} \equiv \text{tt}$ we get $\mathcal{R}^\rho_{\mathbb{B}}\text{tt}xy \equiv \mathcal{R}^\rho_{\mathbb{B}}\text{ff}xy$ by compatibility. Now $\mathcal{R}^\rho_{\mathbb{B}}\text{tt}xy$ converts to x and $\mathcal{R}^\rho_{\mathbb{B}}\text{ff}xy$ converts to y. Hence $x^\rho \equiv y^\rho$, since we identify terms with a common reduct.

The claim can now be proved by induction on A. *Case $I\vec{s}$*. If I has no nullary clause take Ef_I. Otherwise let K_i be the nullary clause, with final conclusion $I\vec{t}$. By induction hypothesis from $\mathbf{F}$ we can derive all parameter premises. Hence $I\vec{t}$. From $\mathbf{F}$ we also obtain $s_i \equiv t_i$, by the remark above. Hence $I\vec{s}$ by compatibility. *Case ${}^{\mathrm{co}}I\vec{s}$*. Use Lemma 26.30. The cases $Y\vec{s}$, $A \to B$ and $\forall_x A$ are obvious. □

A crucial use of the equality predicate EqD is that it allows us to lift a boolean term $t^{\mathbb{B}}$ to a formula, using $\mathrm{atom}(t^{\mathbb{B}}) := (t^{\mathbb{B}} \equiv \text{tt})$. This opens up a convenient way to deal with equality on algebras. The computation rules ensure that, for instance, the boolean term $\mathcal{S}t =_{\mathbb{N}} \mathcal{S}s$, or more precisely $=_{\mathbb{N}}(\mathcal{S}t, \mathcal{S}s)$, is identified with $t =_{\mathbb{N}} s$. We can now turn this boolean term into the formula $(\mathcal{S}t =_{\mathbb{N}} \mathcal{S}s) \equiv \text{tt}$, which again is abbreviated by $\mathcal{S}t =_{\mathbb{N}} \mathcal{S}s$, but this time with the understanding that it is a formula. Then (importantly) the two formulas $\mathcal{S}t =_{\mathbb{N}} \mathcal{S}s$ and $t =_{\mathbb{N}} s$ are identified because the latter is a reduct of the first. Consequently there is no need to prove the implication $\mathcal{S}t =_{\mathbb{N}} \mathcal{S}s \to t =_{\mathbb{N}} s$ explicitly.

26.3.3 Equality and Extensionality

We first consider closed-base types and take the algebra $\mathbb{Y}$ of binary trees as an example. Totality $T_{\mathbb{Y}}$ is inductively defined by the axioms

$$(T_{\mathbb{Y}})^+_0 : - \in T_{\mathbb{Y}}, \qquad (T_{\mathbb{Y}})^+_1 : \forall_{t_1,t_2}(t_1, t_2 \in T_{\mathbb{Y}} \to \mathcal{C}t_1t_2 \in T_{\mathbb{Y}}),$$
$$T^-_{\mathbb{Y}} : - \in X \to \forall_{t_1,t_2}(t_1, t_2 \in T_{\mathbb{Y}} \cap X \to \mathcal{C}t_1t_2 \in X) \to T_{\mathbb{Y}} \subseteq X,$$

and cototality ${}^{\mathrm{co}}T_{\mathbb{Y}}$ coinductively by

$${}^{\mathrm{co}}T^-_{\mathbb{Y}} : \forall_t(t \in {}^{\mathrm{co}}T_{\mathbb{Y}} \to (t \equiv -) \vee \exists_{t_1,t_2}(t_1, t_2 \in {}^{\mathrm{co}}T_{\mathbb{Y}} \wedge t \equiv \mathcal{C}t_1t_2)),$$
$${}^{\mathrm{co}}T^+_{\mathbb{Y}} : \forall_t(t \in X \to (t \equiv -) \vee \exists_{t_1,t_2}(t_1, t_2 \in {}^{\mathrm{co}}T_{\mathbb{Y}} \cup X \wedge t \equiv \mathcal{C}t_1t_2)) \to X \subseteq {}^{\mathrm{co}}T_{\mathbb{Y}}.$$

As candidates for equality we define binary versions of $T_{\mathbb{Y}}$ and ${}^{\mathrm{co}}T_{\mathbb{Y}}$, called similarity $\sim_{\mathbb{Y}}$ and bisimilarity $\approx_{\mathbb{Y}}$, for instance by

$$- \sim_{\mathbb{Y}} -, \qquad \forall_{t_1,t_1'}(t_1 \sim_{\mathbb{Y}} t_1' \to \forall_{t_2,t_2'}(t_2 \sim_{\mathbb{Y}} t_2' \to \mathcal{C}t_1t_2 \sim_{\mathbb{Y}} \mathcal{C}t_1't_2')).$$

We aim at using $\sim_{\mathbb{Y}}$ and $\approx_{\mathbb{Y}}$ for a characterization of equality at $T_{\mathbb{Y}}$ and ${}^{\mathrm{co}}T_{\mathbb{Y}}$. This is useful because it gives us a tool (induction, coinduction) to prove equalities $t \equiv t'$, which otherwise would be difficult. We will need another axiom, the Bisimilarity Axiom, which is justified by the fact that it holds in our intended model (cf. Lemma 26.16).

Axiom 26.33 (Bisimilarity) $\forall_{t,t'}(t \approx_{\mathbb{Y}} t' \to t \equiv t')$.

Lemma 26.34 (Characterization of equality at $T_{\mathbb{Y}}$ and ${}^{\mathrm{co}}T_{\mathbb{Y}}$)

(a) $\forall_{t,t'}(t \sim_{\mathbb{Y}} t' \leftrightarrow t, t' \in T_{\mathbb{Y}} \land t \equiv t')$.
(b) $\forall_{t,t'}(t \approx_{\mathbb{Y}} t' \leftrightarrow t, t' \in {}^{\mathrm{co}}T_{\mathbb{Y}} \land t \equiv t')$.

Proof (b) The proof of Lemma 26.17 has been given in enough detail to make its formalization immediate. We need ${}^{\mathrm{co}}T_{\mathbb{Y}}^{\pm}$ and $\approx_{\mathbb{Y}}^{\pm}$.

(a) Similar to (b), using $T_{\mathbb{Y}}^{\pm}$, $\sim_{\mathbb{Y}}^{\pm}$ instead. For the proof of $t \sim_{\mathbb{Y}} t' \to t \equiv t'$ use (b) and $\sim_{\mathbb{Y}} \subseteq \approx_{\mathbb{Y}}$. □

Hence $\sim_{\mathbb{Y}}$ is a partial equivalence relation on $\mathfrak{C}_{\mathbb{Y}}$ with domain $T_{\mathbb{Y}}$, and similar for $\approx_{\mathbb{Y}}$ and ${}^{\mathrm{co}}T_{\mathbb{Y}}$.

At higher types we use *pointwise equality* of Gandy [4, 5] and Takeuti [11]. This notion is somewhat delicate in our setting, since we allow infinite-base type objects.

Definition 26.35 For every algebra form ι with type parameters $\vec{\alpha}$ we define two predicate forms $\sim_\iota, \approx_\iota$ (called *relative similarity* and *relative bisimilarity*) with type parameters $\vec{\alpha}$ and predicate parameters $\vec{Y}$ (where Y_i has arity (α_i, α_i)) as follows. Let $\vec{\alpha} \to (\xi)_{i<n} \to \xi$ be a constructor type. Take $(\mu/\nu)_Z(\vec{K})$, where the clause for the constructor type above is

$$Y_1u_1u_1' \to \cdots \to Y_nu_nu_n' \to Zv_1v_1' \to \cdots \to Zv_mv_m' \to Z(\mathcal{C}\vec{u}\vec{v}, \mathcal{C}\vec{u}'\vec{v})$$

with $\mathcal{C}$ the corresponding constructor of ι. (Absolute) similarity/bisimilarity predicates arise from the relative ones by substituting a similarity/bisimilarity predicate for Y.

Definition 26.36 (Cotype of a c.r. predicate or formula C) *Cotypes* $\varphi(C)$ are like types $\tau(C)$, but with algebra occurrences ι_I marked as ${}^{\mathrm{co}}\iota_I$ if they arise from a coinductive predicate: $\varphi({}^{\mathrm{co}}I(\vec{\tau}, \vec{P})) := {}^{\mathrm{co}}\iota_I(\varphi(\vec{P}^c))$.

Definition 26.37 (Pointwise equality $\doteq_\varphi$ w.r.t. a cotype φ)

$$\begin{aligned}
(x \doteq_\alpha y) &:= Yxy \quad \text{with } Y \text{ uniquely assigned to } \alpha,\\
(x \doteq_{\iota(\vec{\varphi})} y) &:= (x \sim y) \quad \text{with} \sim := \sim_\iota(\doteq_{\vec{\varphi}}),\\
(x \doteq_{^{\text{co}}\iota(\vec{\varphi})} y) &:= (x \approx y) \quad \text{with} \approx := \approx_\iota(\doteq_{\vec{\varphi}}),\\
(f \doteq_{\varphi\to\psi} g) &:= \forall_{x,y}(x \doteq_\varphi y \to fx \doteq_\psi gy).
\end{aligned}$$

Extensionality Ext_φ w.r.t. a cotype φ arises as a special case

$$(x \in \mathrm{Ext}_\varphi) := (x \doteq_\varphi x).$$

Of course extensionality is a desirable property, but in our model it does not hold generally. Here is an example of a functional F which is nonextensional w.r.t. $(\mathbb{N} \to \mathbb{N}) \to \mathbb{N}$. Define f, g of type $\mathbb{N} \to \mathbb{N}$ by the computation rules $fn = 0$ and $g0 = 0$, $g(\mathcal{S}n) = gn$. Then $f\bot_\mathbb{N} = 0$ because of the computation rules for f. For $g\bot_\mathbb{N}$ no computation rule fits, but by the inductive definition of $(\vec{U}, a) \in [\![\lambda_{\vec{x}} M]\!]$ (see Section 26.2) $[\![g\bot_\mathbb{N}]\!]$ is the empty object $[\![\bot_\mathbb{N}]\!]$. Hence $f \doteq g$, i.e., $\forall_{n,m}(n \doteq_\mathbb{N} m \to fn \doteq_\mathbb{N} gm)$, since $n \doteq_\mathbb{N} m$ implies $n \in T_\mathbb{N}$ and $n \equiv m$. Therefore F defined by $Fh = h\bot_\mathbb{N}$ maps the pointwise equal f, g to different values.

By Lemma 26.34 we know the equivalence of $\mathrm{Ext}_\mathbb{Y}$ and $T_\mathbb{Y}$ (and of $\mathrm{Ext}_{(^{\text{co}}\mathbb{Y})}$ and $^{\text{co}}T_\mathbb{Y}$); this also holds for arbitrary closed base cotypes. This equivalence can be extended to closed cotypes of level 1.

Lemma 26.38 *The predicates* Ext_φ *and* T_φ *are equivalent for closed cotypes of level* ≤ 1.

Proof For closed-base cotypes this has been proved in Lemma 26.34 (for the special case of the algebra $\mathbb{Y}$). In the case of level 1 we use induction on the height of the cotype. Let $\varphi \to \psi$ be a closed cotype of level 1. The following are equivalent.

$$\begin{aligned}
& f \in \mathrm{Ext}_{\varphi\to\psi}\\
& f \doteq_{\varphi\to\psi} f\\
& \forall_{x,y}(x \doteq_\varphi y \to fx \doteq_\psi fy)\\
& \forall_{x\in T_\varphi}(fx \doteq_\psi fx) \qquad \text{by Lemma 26.34, since } \mathrm{lv}(\varphi) = 0\\
& \forall_{x\in T_\varphi}(fx \in \mathrm{Ext}_\psi).
\end{aligned}$$

By induction hypothesis the final formula is equivalent to $f \in T_{\varphi\to\psi}$. □

Lemma 26.39 *For every closed cotype* φ *the relation* $\doteq_\varphi$ *is a partial equivalence relation with domain* Ext_φ.

Proof By induction on the height $|\varphi|$ of φ. *Case* $\iota(\vec{\varphi})/{}^{\mathrm{co}}\iota(\vec{\varphi})$. For $\sim_{\mathbb{Y}}$ and $\approx_{\mathbb{Y}}$ this was proved in Lemma 26.34. In the general case use the induction hypothesis and the inductive / coinductive definition of $\sim_\iota$ / $\approx_\iota$.

Case $\varphi \to \psi$. We first prove symmetry of $\doteq_{\varphi\to\psi}$. Let $f \doteq_{\varphi\to\psi} g$. The goal is $g \doteq_{\varphi\to\psi} f$. Assume $x \doteq_\varphi y$. The goal now is $gx \doteq_\psi fy$. From $x \doteq_\varphi y$ we obtain $y \doteq_\varphi x$ by symmetry of $\doteq_\varphi$, hence $fy \doteq_\psi gx$ from $f \doteq_{\varphi\to\psi} g$, hence $gx \doteq_\psi fy$ by symmetry of $\doteq_\psi$.

We finally prove transitivity of $\doteq_{\varphi\to\psi}$. Let $f \doteq_{\varphi\to\psi} g$ and $g \doteq_{\varphi\to\psi} h$. The goal is $f \doteq_{\varphi\to\psi} h$. Assume $x \doteq_\varphi y$. The goal now is $fx \doteq_\psi hy$. From $x \doteq_\varphi y$ we obtain $y \doteq_\varphi x$ by symmetry of $\doteq_\varphi$, hence $x \doteq_\varphi x$ by transitivity of $\doteq_\varphi$. Then $fx \doteq_\psi gx$ follows from $f \doteq_{\varphi\to\psi} g$. We also have $gx \doteq_\psi hy$ from $g \doteq_{\varphi\to\psi} h$. Using transitivity of $\doteq_\psi$ we obtain $fx \doteq_\psi hy$. □

26.4 Computational Content of Proofs

We define what it means for a term t to "realize" a c.r. formula A. From a proof M of A we extract a term $\mathrm{et}(M)$ and (formally) prove that it is a realizer of A. In this proof we need "invariance axioms" stating that every c.r. formula not involving realizability is invariant under realizability.

26.4.1 Realizability

Assume that we have a global assignment giving, for every c.r. predicate variable X of arity $\vec{\rho}$, an n.c. predicate variable $X^{\mathbf{r}}$ of arity $(\vec{\rho}, \xi)$ where ξ is the type variable associated with X. We will also introduce $I^{\mathbf{r}}/{}^{\mathrm{co}}I^{\mathbf{r}}$ for (co)inductive predicates $I/{}^{\mathrm{co}}I$. A formula or predicate C is called $\mathbf{r}$-free if it does not contain any of these $X^{\mathbf{r}}$, $I^{\mathbf{r}}$ or ${}^{\mathrm{co}}I^{\mathbf{r}}$. A derivation M is called $\mathbf{r}$-free if it contains $\mathbf{r}$-free formulas only.

Definition 26.40 ($C^{\mathbf{r}}$ for $\mathbf{r}$-free predicates and formulas C) For every $\mathbf{r}$-free predicate or formula C we define a predicate or formula $C^{\mathbf{r}}$. For n.c. C let $C^{\mathbf{r}} := C$. In the case where C is c.r., $C^{\mathbf{r}}$ is an n.c. predicate of arity $(\vec{\sigma}, \tau(C))$ with $\vec{\sigma}$ the arity of C. We often write $z \mathrel{\mathbf{r}} C$ for $C^{\mathbf{r}} z$ when C is a c.r. formula. For c.r. *predicates* X let $X^{\mathbf{r}}$ be the n.c. predicate variable provided, and

$$\{\vec{x} \mid A\}^{\mathbf{r}} := \{\vec{x}, z \mid z \mathrel{\mathbf{r}} A\}.$$

Now consider a c.r. (co)inductive predicate

$$I/{}^{\mathrm{co}}I := (\mu/\nu)_X((K_i(X))_{i<k}$$

with associated algebra form $\iota_I = \mu_\xi(\kappa_i(\xi))_{i<k}$ where $\kappa_i(\xi) := \tau(K_i(X))$. The ith

constructor of ι_I is $\mathcal{C}_i\colon \kappa_i(\iota_I)$. Let s be a variable of type $\tau(I)$ and ϑ the substitution $\xi \mapsto \tau(I)$, $X^{\mathbf{r}} \mapsto \{\vec{x}, s \mid Y\vec{x}s\}$. We define n.c. predicates $I^{\mathbf{r}}$ and ${}^{\mathrm{co}}I^{\mathbf{r}}$ by

$$I^{\mathbf{r}}/{}^{\mathrm{co}}I^{\mathbf{r}} := (\mu/\nu)_Y((\mathcal{C}_i\ \mathbf{r}\ K_i(X))\vartheta)_{i<k}.$$

The substitution ϑ is necessary since the arity of Y (and hence of $I^{\mathbf{r}}/{}^{\mathrm{co}}I^{\mathbf{r}}$) must be $(\vec{\rho}, \tau(I))$ and not $(\vec{\rho}, \xi)$. For c.r. *formulas* let

$$z\ \mathbf{r}\ P\vec{t} := P^{\mathbf{r}}\vec{t}z,$$

$$z\ \mathbf{r}\ (A \to B) := \begin{cases} \forall_w(w\ \mathbf{r}\ A \to zw\ \mathbf{r}\ B) & \text{if } A \text{ is c.r.} \\ A \to z\ \mathbf{r}\ B & \text{if } A \text{ is n.c.,} \end{cases}$$

$$z\ \mathbf{r}\ \forall_x A := \forall_x(z\ \mathbf{r}\ A).$$

As an example for the construction of $I^{\mathbf{r}}$ consider the predicate Even, defined by $\mu_X(K_0(X), K_1(X))$ with $K_0(X) := (0 \in X)$ and $K_1(X) := \forall_n(n \in X \to \mathcal{S}(\mathcal{S}n) \in X)$. The associated algebra form is $\mu_\xi(\kappa_0(\xi), \kappa_1(\xi))$ with $\kappa_0(\xi) := \xi$ and $\kappa_1(\xi) := \xi \to \xi$, that is, the algebra $\mathbb{N}$ with constructors $\mathcal{C}_0 := 0$ and $\mathcal{C}_1 := \mathcal{S}$. Let ϑ be the substitution $\xi \mapsto \mathbb{N}$, $X^{\mathbf{r}} \mapsto \{n, m \mid Ynm\}$. Since $\mathcal{S}\ \mathbf{r}\ K_1(X)$ is $\forall_{n,m}(X^{\mathbf{r}}nm \to X^{\mathbf{r}}(\mathcal{S}(\mathcal{S}n), \mathcal{S}m))$ we obtain

$$I^{\mathbf{r}} := \mu_Y(Y00, \forall_{n,m}(Ynm \to Y(\mathcal{S}(\mathcal{S}n), \mathcal{S}m))).$$

Lemma 26.41 *For closed-base types ι the following are equivalent.*

(a) $T_\iota^{\mathbf{r}}xy$,
(b) $x \sim_\iota^{\mathrm{nc}} y$,
(c) $x \in T_\iota^{\mathrm{nc}} \wedge x \equiv y$.

Proof (a) $\leftrightarrow$ (b) Both $T_\iota^{\mathbf{r}}xy$ and $x \sim_\iota^{\mathrm{nc}} y$ satisfy the same clauses. Use the respective elimination axiom in each of the two directions.

(b) $\leftrightarrow$ (c) Use Lemma 26.34. □

Lemma 26.42 *For closed-base types ι the following are equivalent.*

(a) ${}^{\mathrm{co}}T_\iota^{\mathbf{r}}xy$,
(b) $x \approx_\iota^{\mathrm{nc}} y$,
(c) $x \in {}^{\mathrm{co}}T_\iota^{\mathrm{nc}} \wedge x \equiv y$.

Proof As an example we give the proof for $\mathbb{N}$. Since we have n.c. goals only, decorations are omitted. For (a) $\to$ (b) apply $\approx_{\mathbb{N}}^-$ with ${}^{\mathrm{co}}T_{\mathbb{N}}^{\mathbf{r}}$ for X.

$$\approx_{\mathbb{N}}^-\colon \forall_{n,m}(Xnm \to (n \equiv 0 \wedge m \equiv 0) \vee \exists_{n',m'}((n' \approx_{\mathbb{N}} m' \vee Xn'm') \wedge n \equiv \mathcal{S}n' \wedge m \equiv \mathcal{S}m')) \to X \subseteq \approx_{\mathbb{N}}.$$

It suffices to prove the premise. Assume ${}^{\mathrm{co}}T^{\mathbf{r}}_{\mathbb{N}}nm$; the goal is

$$C := {}^{\mathrm{co}}T^{\mathbf{r}}_{\mathbb{N}}00 \vee \exists_{n',m'}((n' \approx_{\mathbb{N}} m' \vee {}^{\mathrm{co}}T^{\mathbf{r}}_{\mathbb{N}}n'm') \wedge n \equiv \mathcal{S}n' \wedge m \equiv \mathcal{S}m').$$

By the closure axiom $({}^{\mathrm{co}}T^{\mathbf{r}}_{\mathbb{N}})^-$ we have

$$(n \equiv 0 \wedge m \equiv 0) \vee \exists_{n',m'}({}^{\mathrm{co}}T^{\mathbf{r}}_{\mathbb{N}}n'm' \wedge n \equiv \mathcal{S}n' \wedge m \equiv \mathcal{S}m').$$

We argue by cases (i.e., use $\vee^-$).

Case 1*:* $n \equiv 0 \wedge m \equiv 0$ Go for the left-hand side of the disjunction C and show ${}^{\mathrm{co}}T^{\mathbf{r}}_{\mathbb{N}}00$. But this follows from the greatest-fixed-point axiom for ${}^{\mathrm{co}}T^{\mathbf{r}}_{\mathbb{N}}$ with competitor predicate $\{ n, m \mid n \equiv 0 \wedge m \equiv 0 \}$.

Case 2*:* $\exists_{n',m'}({}^{\mathrm{co}}T^{\mathbf{r}}_{\mathbb{N}}n'm' \wedge n \equiv \mathcal{S}n' \wedge m \equiv \mathcal{S}m')$ Go for the right-hand side of C.

(b) $\to$ (a) Recall ${}^{\mathrm{co}}T_{\mathbb{N}} := \nu_X(0 \in X,\ \forall_{n \in X}(\mathcal{S}n \in X))$, hence by definition

$${}^{\mathrm{co}}T^{\mathbf{r}}_{\mathbb{N}} := \nu_{X^{\mathbf{r}}}(X^{\mathbf{r}}00,\ \forall_{n,m}(X^{\mathbf{r}}nm \to X^{\mathbf{r}}(\mathcal{S}n)(\mathcal{S}m))).$$

To show $m \approx_{\mathbb{N}} n \to {}^{\mathrm{co}}T^{\mathbf{r}}_{\mathbb{N}}mn$, apply $({}^{\mathrm{co}}T^{\mathbf{r}}_{\mathbb{N}})^+$ with $\approx_{\mathbb{N}}$ for X; recall $({}^{\mathrm{co}}T^{\mathbf{r}}_{\mathbb{N}})^+$:

$$\forall_{n,m}(Xnm \to X00 \vee \exists_{n',m'}(n', m' \in ({}^{\mathrm{co}}T^{\mathbf{r}}_{\mathbb{N}} \cup X) \wedge n \equiv \mathcal{S}n' \wedge m \equiv \mathcal{S}m')) \to X \subseteq {}^{\mathrm{co}}T^{\mathbf{r}}_{\mathbb{N}}.$$

It suffices to prove the premise. Assume $n \approx_{\mathbb{N}} m$; the goal is

$$C := (0 \approx_{\mathbb{N}} 0) \vee \exists_{n',m'}((n', m' \in ({}^{\mathrm{co}}T^{\mathbf{r}}_{\mathbb{N}} \cup \approx_{\mathbb{N}}) \wedge n \equiv \mathcal{S}n' \wedge m \equiv \mathcal{S}m')).$$

By the closure axiom $(\approx_{\mathbb{N}})^-$ we have

$$n \approx_{\mathbb{N}} m \to (n \equiv 0 \wedge m \equiv 0) \vee \exists_{n',m'}(n' \approx_{\mathbb{N}} m' \wedge n \equiv \mathcal{S}n' \wedge m \equiv \mathcal{S}m').$$

We argue by cases (i.e., use $\vee^-$).

Case 1*:* $n \equiv 0 \wedge m \equiv 0$ Go for the left-hand side of the disjunction C and show $0 \approx_{\mathbb{N}} 0$. But this follows from $n \approx_{\mathbb{N}} m$.

Case 2*:* $\exists_{n',m'}(n' \approx_{\mathbb{N}} m' \wedge n \equiv \mathcal{S}n' \wedge m \equiv \mathcal{S}m')$. Go for the right-hand side of C.

(b) $\leftrightarrow$ (c) Use the Bisimilarity Axiom and Lemma 26.34. □

Lemma 26.43 (Realizers for $\exists$) *$z\ \mathbf{r}\ \exists_x A \leftrightarrow \exists_x(z\ \mathbf{r}\ A)$ for A c.r.*

Proof Recall $\mathrm{Ex}_Y := \mu_X(\forall_x(x \in Y \to X))$. Then

$$\mathrm{Ex}^{\mathbf{r}}_{Y^{\mathbf{r}}} := \mu_{X^{\mathbf{r}}}(\forall_{x,z}(Y^{\mathbf{r}}xz \to X^{\mathbf{r}}z)).$$

Now substituting $Y^{\mathbf{r}}$ by $\{ x, z \mid z\ \mathbf{r}\ A \}$ in the introduction axiom gives

$$\left(\mathrm{Ex}^{\mathbf{r}}_{\{x,z \mid z\mathbf{r}A\}}\right)^+_0 : \forall_{x,z}(z\ \mathbf{r}\ A \to z\ \mathbf{r}\ \exists_x A)$$

Conversely, the elimination axiom $(\mathrm{Ex}^{\mathbf{r}}_{Y^{\mathbf{r}}})^{-}$ is

$$\forall_z(z \in \mathrm{Ex}^{\mathbf{r}}_{Y^{\mathbf{r}}} \to \forall_{x,z}(Y^{\mathbf{r}}xz \to z \in X) \to z \in X),$$

which is equivalent to

$$\forall_z(z \in \mathrm{Ex}^{\mathbf{r}}_{Y^{\mathbf{r}}} \to \forall_z(\exists_x Y^{\mathbf{r}}xz \to z \in X) \to z \in X).$$

Substituting X by $\{\, z \mid \exists_x(Y^{\mathbf{r}}xz)\,\}$ makes the middle part provable. Thus with $\{\, x, z \mid z \;\mathbf{r}\; A \,\}$ for $Y^{\mathbf{r}}$ we obtain $\forall_z(z \;\mathbf{r}\; \exists_x A \to \exists_x(z \;\mathbf{r}\; A))$ from $\left(\mathrm{Ex}^{\mathbf{r}}_{\{x,z|z\mathbf{r}A\}}\right)^{-}$. □

Lemma 26.44 (Realizers for $\wedge$) *$z \;\mathbf{r}\; (A \wedge B)$ is equivalent to*

$$\begin{array}{ll} z \equiv \langle \mathrm{lft}(z), \mathrm{rht}(z)\rangle \wedge (\mathrm{lft}(z) \;\mathbf{r}\; A) \wedge (\mathrm{rht}(z) \;\mathbf{r}\; B) & \text{for } A \text{ c.r. and } B \text{ c.r.} \\ (z \;\mathbf{r}\; A) \wedge B & \text{for } A \text{ c.r. and } B \text{ n.c.} \\ A \wedge (z \;\mathbf{r}\; B) & \text{for } A \text{ n.c. and } B \text{ c.r.} \end{array}$$

Proof *Case* A, B c.r. Recall $\mathrm{AndD}_{X^{\mathrm{c}},Y^{\mathrm{c}}} := \mu_{Z^{\mathrm{c}}}(X^{\mathrm{c}} \to Y^{\mathrm{c}} \to Z^{\mathrm{c}})$. Then

$$\mathrm{AndD}^{\mathbf{r}}_{X^{\mathbf{r}},Y^{\mathbf{r}}} := \mu_{Z^{\mathbf{r}}}(\forall_x(x \in X^{\mathbf{r}} \to \forall_y(y \in Y^{\mathbf{r}} \to \langle x, y\rangle \in Z^{\mathbf{r}}))).$$

Substituting $X^{\mathbf{r}}$ by $\{\, x \mid x \;\mathbf{r}\; A \,\}$ and $Y^{\mathbf{r}}$ by $\{\, y \mid y \;\mathbf{r}\; B \,\}$ gives

$$\left(\mathrm{AndD}^{\mathbf{r}}_{\{x|x\mathbf{r}A\},\{y|y\mathbf{r}B\}}\right)^{+}_{0} : \forall_x((x \;\mathbf{r}\; A) \to \forall_y(y \;\mathbf{r}\; B \to \langle x, y\rangle \;\mathbf{r}\; (A \wedge B))).$$

This suffices for "$\leftarrow$". Conversely, the elimination axiom $\left(\mathrm{AndD}^{\mathbf{r}}_{X^{\mathbf{r}},Y^{\mathbf{r}}}\right)^{-}$ is

$$\forall_x(x \in X^{\mathbf{r}} \to \forall_y(y \in Y^{\mathbf{r}} \to \langle x, y\rangle \in Z)) \to \mathrm{AndD}^{\mathbf{r}}_{X^{\mathbf{r}},Y^{\mathbf{r}}} \subseteq Z.$$

Substitute Z by $\{\, z \mid z \equiv \langle \mathrm{lft}(z), \mathrm{rht}(z)\rangle \wedge (\mathrm{lft}(z) \;\mathbf{r}\; A) \wedge (\mathrm{rht}(z) \;\mathbf{r}\; B) \,\}$. With $\{\, x \mid x \;\mathbf{r}\; A \,\}$ for $X^{\mathbf{r}}$ and $\{\, y \mid y \;\mathbf{r}\; B \,\}$ for $Y^{\mathbf{r}}$ the premise is provable. Hence

$$\forall_z(z \;\mathbf{r}\; (A \wedge B) \to z \equiv \langle \mathrm{lft}(z), \mathrm{rht}(z)\rangle \wedge (\mathrm{lft}(z) \;\mathbf{r}\; A) \wedge (\mathrm{rht}(z) \;\mathbf{r}\; B)).$$

Case A c.r., B n.c. Recall $\mathrm{AndL}_{X^{\mathrm{c}},Y^{\mathrm{nc}}} := \mu_{Z^{\mathrm{c}}}(X^{\mathrm{c}} \to Y^{\mathrm{nc}} \to Z^{\mathrm{c}})$. Then

$$\mathrm{AndL}^{\mathbf{r}}_{X^{\mathbf{r}},Y^{\mathrm{nc}}} := \mu_{Z^{\mathbf{r}}}(\forall_z(z \;\mathbf{r}\; X \to Y^{\mathrm{nc}} \to z \in Z^{\mathbf{r}})).$$

Substituting $X^{\mathbf{r}}$ by $\{\, z \mid z \;\mathbf{r}\; A \,\}$ and Y^{nc} by B gives

$$\left(\mathrm{AndL}^{\mathbf{r}}_{\{z|z\mathbf{r}A\},B}\right)^{+}_{0} : \forall_z((z \;\mathbf{r}\; A) \to B \to z \;\mathbf{r}\; (A \wedge B)).$$

This suffices for "$\leftarrow$". Conversely, the elimination axiom $\left(\mathrm{AndL}^{\mathbf{r}}_{X,Y^{\mathrm{nc}}}\right)^{-}$ is

$$\forall_z(z \;\mathbf{r}\; X \to Y^{\mathrm{nc}} \to z \in Z) \to \mathrm{AndL}^{\mathbf{r}}_{X,Y^{\mathrm{nc}}} \subseteq Z.$$

Substitute Z by $\{ z \mid (z \ \mathbf{r} \ A) \wedge B \}$. Then with $\{ z \mid z \ \mathbf{r} \ A \}$ for X and B for Y^{nc} the premise is provable and we obtain

$$\forall_z (z \ \mathbf{r} \ (A \wedge B) \to (z \ \mathbf{r} \ A) \wedge B). \qquad \square$$

Recall that for the sum type $\rho + \sigma$ we had the constructors $(\mathrm{InL}_{\rho\sigma})^{\rho \to \rho+\sigma}$ and $(\mathrm{InR}_{\rho\sigma})^{\sigma \to \rho+\sigma}$. In the special situation that one of the two parameter types is the unit type $\mathbb{U}$ it is common to view the sum type $\mathbb{U} + \sigma$ as a unary algebra form, with constructors DummyL of type $\mathbb{U} + \sigma$ and Inr of type $\sigma \to \mathbb{U} + \sigma$. Similarly $\rho + \mathbb{U}$ is viewed as a unary algebra, with constructors Inl of type $\rho \to \rho + \mathbb{U}$ and DummyR of type $\rho + \mathbb{U}$.

Lemma 26.45 (Realizers for $\vee$) *$z \ \mathbf{r} \ (A \vee B)$ is equivalent to*

$$\begin{array}{ll}
\exists_x (x \ \mathbf{r} \ A \wedge z \equiv \mathrm{InL}(x)) \vee^{\mathrm{nc}} \exists_y (y \ \mathbf{r} \ B \wedge z \equiv \mathrm{InR}(y)) & \textit{for } A, B \textit{ c.r.} \\
\exists_x (x \ \mathbf{r} \ A \wedge z \equiv \mathrm{Inl}(x)) \vee^{\mathrm{nc}} (B \wedge z \equiv \mathrm{DummyR}) & \textit{for } A \textit{ c.r. and } B \textit{ n.c.} \\
(A \wedge z \equiv \mathrm{DummyL}) \vee^{\mathrm{nc}} \exists_y (y \ \mathbf{r} \ B \wedge z \equiv \mathrm{Inr}(y)) & \textit{for } A \textit{ n.c. and } B \textit{ c.r.} \\
(A \wedge z \equiv \mathtt{tt}) \vee^{\mathrm{nc}} (B \wedge z \equiv \mathtt{ff}) & \textit{for } A, B \textit{ n.c.}
\end{array}$$

Proof As an example consider the case A n.c., B c.r. Recall $\mathrm{OrR}_{X^{\mathrm{nc}}, Y^{\mathrm{c}}} := \mu_Z (X^{\mathrm{nc}} \to Z, Y^{\mathrm{c}} \to Z)$. Then

$$\mathrm{OrR}^{\mathbf{r}}_{X^{\mathrm{nc}}, Y^{\mathbf{r}}} := \mu_{Z^{\mathbf{r}}} (X^{\mathrm{nc}} \to \mathrm{DummyL} \in Z^{\mathbf{r}}, \forall_y (y \ \mathbf{r} \ Y \to \mathrm{Inr}(y) \in Z^{\mathbf{r}})).$$

Substituting X^{nc} by A and $Y^{\mathbf{r}}$ by $\{ y \mid y \ \mathbf{r} \ B \}$ gives

$$\left(\mathrm{OrR}^{\mathbf{r}}_{A, \{y \mid y \mathbf{r} B\}}\right)^{+}_{0} : A \to \mathrm{DummyL} \ \mathbf{r} \ (A \vee B),$$
$$\left(\mathrm{OrR}^{\mathbf{r}}_{A, \{y \mid y \mathbf{r} B\}}\right)^{+}_{1} : \forall_y (y \ \mathbf{r} \ B \to \mathrm{Inr}(y) \ \mathbf{r} \ (A \vee B)).$$

This suffices for "$\leftarrow$": if $A \wedge z \equiv \mathrm{DummyL}$, then from $\left(\mathrm{OrR}^{\mathbf{r}}_{A, \{y \mid y \mathbf{r} B\}}\right)^{+}_{0}$ we obtain $z \ \mathbf{r} \ (A \vee B)$, and if we have y with $y \ \mathbf{r} \ B$ and $z \equiv \mathrm{Inr}(y)$, then from $\left(\mathrm{OrR}^{\mathbf{r}}_{A, \{y \mid y \mathbf{r} B\}}\right)^{+}_{1}$ we again obtain $z \ \mathbf{r} \ (A \vee B)$.

Conversely, the elimination axiom $\left(\mathrm{OrR}^{\mathbf{r}}_{X^{\mathrm{nc}}, Y^{\mathbf{r}}}\right)^{-}$ is

$$(X^{\mathrm{nc}} \to \mathrm{DummyL} \in Z) \to \forall_y (y \ \mathbf{r} \ Y \to \mathrm{Inr}(y) \in Z) \to \mathrm{OrR}^{\mathbf{r}}_{X^{\mathrm{nc}}, Y^{\mathbf{r}}} \subseteq Z.$$

Substitute Z by $\{ z \mid (A \wedge z \equiv \mathrm{DummyL}) \vee^{\mathrm{nc}} \exists_y (y \ \mathbf{r} \ B \wedge z \equiv \mathrm{Inr}(y)) \}$. Then with A for X^{nc} and $\{ y \mid y \ \mathbf{r} \ B \}$ for $Y^{\mathbf{r}}$ the premises are provable. Hence

$$\forall_z (z \ \mathbf{r} \ (A \vee B) \to (A \wedge z \equiv \mathrm{DummyL}) \vee^{\mathrm{nc}} \exists_y (y \ \mathbf{r} \ B \wedge z \equiv \mathrm{Inr}(y))). \qquad \square$$

26.4.2 Extracted Terms

Let M be a proof in TCF of a c.r. formula A. Assume M is an $\mathbf{r}$-free proof, that is, M contains no realizability predicates $I^{\mathbf{r}}$ or ${}^{\mathrm{co}}I^{\mathbf{r}}$. We define its *extracted term* $\mathrm{et}(M)$, of type $\tau(A)$, with the aim of expressing M's computational content.

It will be a term built up from variables, constructors, recursion operators, destructors and corecursion operators by λ-abstraction and application.

Definition 26.46 (Extracted term) For an $\mathbf{r}$-free proof M of a c.r. formula A we define its extracted term $\mathrm{et}(M)$ by

$$\begin{aligned}
\mathrm{et}(u^A) &:= z_u^{\tau(A)} \quad (z_u^{\tau(A)} \text{ uniquely associated to } u^A),\\
\mathrm{et}((\lambda_{u^A} M^B)^{A\to B}) &:= \begin{cases} \lambda_{z_u}\mathrm{et}(M) & \text{if } A \text{ is c.r.}\\ \mathrm{et}(M) & \text{if } A \text{ is n.c.,}\end{cases}\\
\mathrm{et}((M^{A\to B}N^A)^B) &:= \begin{cases} \mathrm{et}(M)\mathrm{et}(N) & \text{if } A \text{ is c.r.}\\ \mathrm{et}(M) & \text{if } A \text{ is n.c.,}\end{cases}\\
\mathrm{et}((\lambda_x M^A)^{\forall_x A}) &:= \mathrm{et}(M),\\
\mathrm{et}((M^{\forall_x A(x)}t)^{A(t)}) &:= \mathrm{et}(M).
\end{aligned}$$

It remains to define extracted terms for the axioms. Consider a (c.r.) inductively defined predicate I. For its introduction and elimination axioms define $\mathrm{et}(I_i^+) := \mathcal{C}_i$ and $\mathrm{et}(I^-) := \mathcal{R}$, where both the constructor $\mathcal{C}_i$ and the recursion operator $\mathcal{R}$ refer to the algebra ι_I associated with I. For the closure and greatest-fixed-point axioms of ${}^{\mathrm{co}}I$ define $\mathrm{et}({}^{\mathrm{co}}I^-) := \mathcal{D}$ and $\mathrm{et}({}^{\mathrm{co}}I_i^+) := {}^{\mathrm{co}}\mathcal{R}$, where both the destructor $\mathcal{D}$ and the corecursion operator ${}^{\mathrm{co}}\mathcal{R}$ refer to the cotype ${}^{\mathrm{co}}\iota_I$ where ι_I is the algebra associated with I. For the elimination axiom $(I^{\mathrm{nc}})^-$ of a one-clause-nc inductive predicate with a c.r. competitor predicate the extracted term is the identity.

One can see easily that the identity realizes the elimination axiom $(I^{\mathrm{nc}})^-$ of a one-clause-nc inductive predicate with a c.r. competitor predicate. More work is needed to show that the extracted term of $I^\pm$, ${}^{\mathrm{co}}I^\pm$ realizes the respective axiom. We prove this for a special case only, the algebras of lists and streams of "signed digits." Such objects are of interest for the representation of (dyadic) rational numbers and of real numbers.

Let $\sim_{\mathbb{D}}$ be the similarity relation for the three-element algebra $\mathbb{D}$ of signed digits $1, 0, -1$ (written $\bar{1}$), defined by the three clauses $s \sim_{\mathbb{D}} s$ for s a signed digit. We will work with lists $\mathbb{L}(\mathbb{D})$ of signed digits and streams $\mathbb{S}(\mathbb{D})$ of signed digits, abbreviated $\mathbb{L}$ and $\mathbb{S}$. The similarity relation $\sim_{\mathbb{L}}$ has clauses

$$[] \sim_{\mathbb{L}} [], \qquad \forall_{s_1,s_2,\ell_1,\ell_2}(s_1 \sim_{\mathbb{D}} s_2 \to \ell_1 \sim_{\mathbb{L}} \ell_2 \to s_1 :: \ell_1 \sim_{\mathbb{L}} s_2 :: \ell_2)$$

and the elimination axiom $\sim_{\mathbb{L}}^{-}$:

$$X[][] \to \forall_{s_1,s_2,\ell_1,\ell_2}(s_1 \sim_{\mathbb{D}} s_2 \to \ell_1 \sim_{\mathbb{L}} \ell_2 \to X\ell_1\ell_2 \to X(s_1 :: \ell_1, s_2 :: \ell_2)) \to \\ \sim_{\mathbb{L}} \subseteq X.$$

For the first two claims we only consider the inductive predicate $\sim_{\mathbb{L}}$.

Lemma 26.47 *The constructors of* $\mathbb{L}$ *realize the clauses of* $\sim_{\mathbb{L}}$.

Proof We only consider the second constructor ::. We must show that :: realizes the following formula C equivalent to $(\sim_{\mathbb{L}})_1^{+}$:

$$\forall_{s_1,s_2}(s_1 \sim_{\mathbb{D}} s_2 \to \forall_{\ell_1,\ell_2}(\ell_1 \sim_{\mathbb{L}} \ell_2 \to s_1 :: \ell_1 \sim_{\mathbb{L}} s_2 :: \ell_2))$$

that is, $::\ \mathbf{r}\ C$. Pick s_1, s_2. The goal then is

$$::\ \mathbf{r}\ (s_1 \sim_{\mathbb{D}} s_2 \to \forall_{\ell_1,\ell_2}(\ell_1 \sim_{\mathbb{L}} \ell_2 \to s_1 :: \ell_1 \sim_{\mathbb{L}} s_2 :: \ell_2)).$$

Pick s with $\sim_{\mathbb{D}}^{\mathbf{r}}(s_1, s_2, s)$. The goal then is

$$::\ s\ \mathbf{r}\ \forall_{\ell_1,\ell_2}(\ell_1 \sim_{\mathbb{L}} \ell_2 \to s_1 :: \ell_1 \sim_{\mathbb{L}} s_2 :: \ell_2).$$

Pick ℓ_1, ℓ_2, ℓ with $\sim_{\mathbb{L}}^{\mathbf{r}}(\ell_1, \ell_2, \ell)$. The goal then is

$$(s :: \ell)\ \mathbf{r}\ (s_1 :: \ell_1 \sim_{\mathbb{L}} s_2 :: \ell_2), \quad \text{that is,} \quad \sim_{\mathbb{L}}^{\mathbf{r}}\ (s_1 :: \ell_1, s_2 :: \ell_2, s :: \ell)).$$

But this follows from what we have by the second clause of $\sim_{\mathbb{L}}^{\mathbf{r}}$:

$$\forall_{s_1,s_2,s,\ell_1,\ell_2,\ell}(\sim_{\mathbb{D}}^{\mathbf{r}}(s_1, s_2, s) \to \sim_{\mathbb{L}}^{\mathbf{r}}(\ell_1, \ell_2, \ell) \to \sim_{\mathbb{L}}^{\mathbf{r}}(s_1 :: \ell_1, s_2 :: \ell_2, s :: \ell)). \quad \square$$

Lemma 26.48 *The recursion operator* $\mathcal{R}_{\mathbb{L}}^{\alpha}$ *realizes* $\sim_{\mathbb{L}}^{-}$.

Proof We equivalently rewrite $\sim_{\mathbb{L}}^{-}$ as $C :=$

$$\forall_{\ell_1,\ell_2}(\ell_1 \sim_{\mathbb{L}} \ell_2 \to X[][] \to \forall_{s_1,s_2,\ell_1,\ell_2}(s_1 \sim_{\mathbb{D}} s_2 \to \ell_1 \sim_{\mathbb{L}} \ell_2 \to X\ell_1\ell_2 \to \\ X(s_1 :: \ell_1, s_2 :: \ell_2)) \to X\ell_1\ell_2)$$

to make its type the same as the one for $\mathcal{R}_{\mathbb{L}}^{\alpha}$:

$$\mathbb{L} \to \alpha \to (\mathbb{D} \to \mathbb{L} \to \alpha \to \alpha) \to \alpha.$$

We must show $\mathcal{R}_{\mathbb{L}}^{\alpha}\ \mathbf{r}\ C$. Pick ℓ_1, ℓ_2, ℓ, x with $\sim_{\mathbb{L}}^{\mathbf{r}}(\ell_1, \ell_2, \ell)$ and $X^{\mathbf{r}}[][]x$. The goal then is

$$\mathcal{R}_{\mathbb{L}}^{\alpha}\ell x\ \mathbf{r}\ (\forall_{s_1,s_2,\ell_1,\ell_2}(s_1 \sim_{\mathbb{D}} s_2 \to \ell_1 \sim_{\mathbb{L}} \ell_2 \to X\ell_1\ell_2 \to X(s_1 :: \ell_1, s_2 :: \ell_2)) \to \\ X\ell_1\ell_2).$$

Assume f **r** $\forall_{s_1,s_2,\ell_1,\ell_2}(s_1 \sim_{\mathbb{D}} s_2 \to \ell_1 \sim_{\mathbb{L}} \ell_2 \to X\ell_1\ell_2 \to X(s_1 :: \ell_1, s_2 :: \ell_2))$, which implies $\forall_{s_1,s_2,s,\ell_1,\ell_2,\ell,y}(\sim_{\mathbb{D}}(s_1,s_2,s) \to \sim_{\mathbb{L}}(\ell_1,\ell_2,\ell) \to X^{\mathbf{r}}\ell_1\ell_2 y \to X^{\mathbf{r}}(s_1 :: \ell_1, s_2 :: \ell_2, fs\ell y))$. Our goal is

$$X^{\mathbf{r}}(\ell_1,\ell_2,\mathcal{R}^{\alpha}_{\mathbb{L}}\ell x f) =: Q\ell_1\ell_2\ell.$$

To this end we use the elimination axiom for $\sim^{\mathbf{r}}_{\mathbb{L}}$:

$$\forall_{\ell_1,\ell_2,\ell}(\sim^{\mathbf{r}}_{\mathbb{L}}(\ell_1,\ell_2,\ell) \to Q[][][] \to \forall_{s_1,s_2,s,\ell_1,\ell_2,\ell}(\sim^{\mathbf{r}}_{\mathbb{D}}(s_1,s_2,s) \to \sim^{\mathbf{r}}_{\mathbb{L}}(\ell_1,\ell_2,\ell) \to Q\ell_1\ell_2\ell \to Q(s_1 :: \ell_1, s_2 :: \ell_2, s :: \ell)) \to Q\ell_1\ell_2\ell).$$

It suffices to prove the premises $Q[][][]$ and $\forall_{s_1,s_2,s,\ell_1,\ell_2,\ell}(\sim^{\mathbf{r}}_{\mathbb{D}}(s_1,s_2,s) \to \sim^{\mathbf{r}}_{\mathbb{L}}(\ell_1,\ell_2,\ell) \to Q\ell_1\ell_2\ell \to Q(s_1 :: \ell_1, s_2 :: \ell_2, s :: \ell))$. By a computation rule for $\mathcal{R}^{\alpha}_{\mathbb{L}}$ the former is $X^{\mathbf{r}}[][]x$, which we have. For the latter, assume s_1, s_2, s, ℓ_1, ℓ_2, ℓ, and its premises. We show $Q(s_1 :: \ell_1, s_2 :: \ell_2, s :: \ell)$, that is,

$$X^{\mathbf{r}}(s_1 :: \ell_1, s_2 :: \ell_2, \mathcal{R}^{\alpha}_{\mathbb{L}}(s :: \ell)xf).$$

By the computation rules for $\mathcal{R}^{\alpha}_{\mathbb{L}}$ this is the same as

$$X^{\mathbf{r}}(s_1 :: \ell_1, s_2 :: \ell_2, fs\ell(\mathcal{R}^{\alpha}_{\mathbb{L}}\ell x f)).$$

But with $y := \mathcal{R}^{\alpha}_{\mathbb{L}}\ell x f$ this follows from what we have. □

The bisimilarity relation $\approx_{\mathbb{S}}$ is defined by the closure axiom

$$\approx^{-}_{\mathbb{S}} : \forall_{u_1,u_2}(u_1 \approx_{\mathbb{S}} u_2 \to \exists_{s_1,s_2,u_1',u_2'}(s_1 \sim_{\mathbb{D}} s_2 \wedge u_1' \approx_{\mathbb{S}} u_2' \wedge u_1 \equiv s_1 :: u_1' \wedge u_2 \equiv s_2 :: u_2'))$$

and the greatest-fixed-point axiom $\approx^{+}_{\mathbb{S}}$:

$$\forall_{u_1,u_2}(Xu_1u_2 \to \exists_{s_1,s_2,u_1',u_2'}(s_1 \sim_{\mathbb{D}} s_2 \wedge (u_1' \approx_{\mathbb{S}} u_2' \vee Xu_1'u_2') \wedge u_1 \equiv s_1 :: u_1' \wedge u_2 \equiv s_2 :: u_2')) \to X \subseteq \approx_{\mathbb{S}}.$$

For the final two claims we only consider the coinductive predicate $\approx_{\mathbb{S}}$.

Lemma 26.49 *The destructor $\mathcal{D}_{\mathbb{S}}$ realizes the closure axiom $\approx^{-}_{\mathbb{S}}$.*

Proof Recall $\approx^{-}_{\mathbb{S}} : \forall_{u_1,u_2}(u_1 \approx_{\mathbb{S}} u_2 \to \exists_{s_1,s_2,u_1',u_2'}(s_1 \sim_{\mathbb{D}} s_2 \wedge u_1' \approx_{\mathbb{S}} u_2' \wedge u_1 \equiv s_1 :: u_1' \wedge u_2 \equiv s_2 :: u_2'))$ with cotype ${}^{\mathrm{co}}\mathbb{S} \to \mathbb{D} \times {}^{\mathrm{co}}\mathbb{S}$. The goal is $\mathcal{D}_{\mathbb{S}}$ **r** $\approx^{-}_{\mathbb{S}}$:

$$\forall_{u_1,u_2,u}(\approx^{\mathbf{r}}_{\mathbb{S}}(u_1,u_2,u) \to \mathcal{D}_{\mathbb{S}}u \; \mathbf{r} \; \exists_{s_1,s_2,u_1',u_2'}(s_1 \sim_{\mathbb{D}} s_2 \wedge u_1' \approx_{\mathbb{S}} u_2' \wedge u_1 \equiv s_1 :: u_1' \wedge u_2 \equiv s_2 :: u_2')).$$

Assume $\approx^{\mathbf{r}}_{\mathbb{S}}(u_1,u_2,u)$. We need to prove

$$\exists_{s_1,s_2,u_1',u_2'}(\mathcal{D}_{\mathbb{S}}u \; \mathbf{r} \; (s_1 \sim_{\mathbb{D}} s_2 \wedge u_1' \approx_{\mathbb{S}} u_2') \wedge u_1 \equiv s_1 :: u_1' \wedge u_2 \equiv s_2 :: u_2').$$

By $(\approx_{\mathbb{S}}^{\mathbf{r}})^-$ from $\approx_{\mathbb{S}}^{\mathbf{r}}(u_1, u_2, u)$ we obtain s_1, s_2, s, u_1', u_2', u' such that

$$\sim_{\mathbb{D}}^{\mathbf{r}}(s_1, s_2, s) \wedge \approx_{\mathbb{S}}^{\mathbf{r}}(u_1', u_2', u') \wedge u_1 \equiv s_1 :: u_1' \wedge u_2 \equiv s_2 :: u_2' \wedge u \equiv s :: u'.$$

Take s_1, s_2, u_1', u_2'. It remains to show $\mathcal{D}_{\mathbb{S}}u \ \mathbf{r} \ (s_1 \sim_{\mathbb{D}} s_2 \wedge u_1' \approx_{\mathbb{S}} u_2')$. By the computation rule of $\mathcal{D}_{\mathbb{S}}$ we know $\mathcal{D}_{\mathbb{S}}u \equiv \mathcal{D}_{\mathbb{S}}(s :: u') \equiv \langle s, u' \rangle$. Hence we must prove $\sim_{\mathbb{D}}^{\mathbf{r}}(s_1, s_2, s)$ and $\approx_{\mathbb{S}}^{\mathbf{r}}(u_1, u_2, u)$, which we both have. □

Lemma 26.50 *The corecursion operator* ${}^{\mathrm{co}}\mathcal{R}_{\mathbb{S}}^{\alpha}$ *realizes* $\approx_{\mathbb{S}}^{+}$.

Proof Equivalently rewrite $\approx_{\mathbb{S}}^{+}$ as $C := \forall_{u_1,u_2}(Xu_1u_2 \to \forall_{u_1,u_2}(Xu_1u_2 \to \exists_{s_1,s_2,u_1',u_2'}(s_1 \sim_{\mathbb{D}} s_2 \wedge (u_1' \approx_{\mathbb{S}} u_2' \vee Xu_1'u_2') \wedge u_1 \equiv s_1 :: u_1' \wedge u_2 \equiv s_2 :: u_2')) \to u_1 \approx_{\mathbb{S}} u_2)$ to make its cotype the same as the one for ${}^{\mathrm{co}}\mathcal{R}_{\mathbb{S}}^{\alpha}$:

$$\alpha \to (\alpha \to \mathbb{D} \times ({}^{\mathrm{co}}\mathbb{S} + \alpha)) \to {}^{\mathrm{co}}\mathbb{S}.$$

We show that ${}^{\mathrm{co}}\mathcal{R}_{\mathbb{S}}^{\alpha}$ realizes C, that is, ${}^{\mathrm{co}}\mathcal{R}_{\mathbb{S}}^{\alpha} \ \mathbf{r} \ C$. The goal then is

$$\begin{aligned}
&{}^{\mathrm{co}}\mathcal{R}_{\mathbb{S}}^{\alpha} \ \mathbf{r} \ (Xu_1u_2 \to \\
&\qquad \forall_{u_1,u_2}(Xu_1u_2 \to \exists_{s_1,s_2,u_1',u_2'}(s_1 \sim_{\mathbb{D}} s_2 \wedge (u_1' \approx_{\mathbb{S}} u_2' \vee Xu_1'u_2') \wedge \\
&\qquad\qquad\qquad\qquad u_1 \equiv s_1 :: u_1' \wedge u_2 \equiv s_2 :: u_2')) \to \\
&\qquad u_1 \approx_{\mathbb{S}} u_2).
\end{aligned}$$

Pick u with $X^{\mathbf{r}}u_1u_2u$ and f such that

$$\begin{aligned}
&\forall_{u_1,u_2,u}(X^{\mathbf{r}}u_1u_2u \to fu \ \mathbf{r} \ \exists_{s_1,s_2,u_1',u_2'}(s_1 \sim_{\mathbb{D}} s_2 \wedge (u_1' \approx_{\mathbb{S}} u_2' \vee Xu_1'u_2') \wedge \\
&\qquad\qquad\qquad\qquad u_1 \equiv s_1 :: u_1' \wedge u_2 \equiv s_2 :: u_2')).
\end{aligned}$$

Our goal is $\approx_{\mathbb{S}}^{\mathbf{r}}(u_1, u_2, {}^{\mathrm{co}}\mathcal{R}_{\mathbb{S}}^{\alpha}uf)$. To this end we use $(\approx_{\mathbb{S}}^{\mathbf{r}})^+$ in the form

$$\begin{aligned}
&\forall_{u_1,u_2,u}(Qu_1u_2u \to \\
&\qquad \forall_{u_1,u_2,u}(Qu_1u_2u \to \\
&\qquad\qquad \exists_{s_1,s_2,s,u_1',u_2',u'}(\sim_{\mathbb{D}}^{\mathbf{r}}(s_1, s_2, s) \wedge (\approx_{\mathbb{S}}^{\mathbf{r}}(u_1', u_2', u') \vee Qu_1u_2u') \wedge \\
&\qquad\qquad\qquad u_1 \equiv s_1{::}u_1' \wedge u_2 \equiv s_2{::}u_2') \wedge u \equiv s{::}u')) \to \\
&\qquad \approx_{\mathbb{S}}^{\mathbf{r}}(u_1, u_2, u))
\end{aligned}$$

with

$$\exists_{z'}(X^{\mathbf{r}}u_1u_2z' \wedge u \equiv {}^{\mathrm{co}}\mathcal{R}_{\mathbb{S}}^{\alpha}z'f) =: Qu_1u_2u.$$

It suffices to prove the closure property of Q. Let u_1, u_2, u, and also u' be given such that $X^{\mathbf{r}}u_1u_2u' \wedge u \equiv {}^{\mathrm{co}}\mathcal{R}_{\mathbb{S}}^{\alpha}u'f$. We need to show

$$\exists_{s_1,s_2,s,u_1',u_2',u'}(\sim^{\mathbf{r}}_{\mathbb{D}}(s_1,s_2,s)\wedge(\approx^{\mathbf{r}}_{\mathbb{S}}(u_1',u_2',u')\vee\exists_{u'}(X^{\mathbf{r}}u_1u_2u'\wedge u\equiv{}^{\mathrm{co}}\mathcal{R}^{\alpha}_{\mathbb{S}}u'f))\wedge u_1\equiv s_1::u_1'\wedge u_2\equiv s_2::u_2'\wedge u\equiv s::u'). \tag{26.6}$$

Since $\sim^{\mathbf{r}}_{\mathbb{D}}(s_1,s_2,s)$ is equivalent to $s_1\equiv s_2\equiv s$ and $X^{\mathbf{r}}u_1u_2u'$ we know

$$fu'\ \mathbf{r}\ \exists_{s_1,s_2,u_1',u_2'}(s_1\sim_{\mathbb{D}}s_2\wedge(u_1'\approx_{\mathbb{S}}u_2'\vee Xu_1'u_2')\wedge u_1\equiv s_1::u_1'\wedge u_2\equiv s_2::u_2').$$

Then $fu'\equiv\langle s,w\rangle$ with $\sim^{\mathbf{r}}_{\mathbb{D}}(s_1,s_2,s)$ and $w\ \mathbf{r}\ (u_1'\approx_{\mathbb{S}}u_2'\vee Xu_1'u_2')$, for some s_1,s_2,u_1',u_2' such that $u_1\equiv s_1::u_1'$ and $u_2\equiv s_2::u_2'$. Hence

$$\exists_{u'}(\approx^{\mathbf{r}}_{\mathbb{S}}(u_1',u_2',u')\wedge w\equiv\mathrm{InL}(u'))\vee\exists_{u''}(X^{\mathbf{r}}u_1'u_2'u''\wedge w\equiv\mathrm{InR}(u'')).$$

We distinguish cases on this disjunction. Recall

$${}^{\mathrm{co}}\mathcal{R}^{\alpha}_{\mathbb{S}}uf\equiv\begin{cases}s::u & \text{if } fu\equiv\langle s,\mathrm{InL}(u)\rangle,\\ s::{}^{\mathrm{co}}\mathcal{R}^{\alpha}_{\mathbb{S}}u'f & \text{if } fu\equiv\langle s,\mathrm{InR}(u')\rangle.\end{cases}$$

Case L. $\approx^{\mathbf{r}}_{\mathbb{S}}(u_1',u_2',u')\wedge w\equiv\mathrm{InL}(u')$ for some u'. Then (26.6) holds, since $u\equiv{}^{\mathrm{co}}\mathcal{R}^{\alpha}_{\mathbb{S}}u'f\equiv s::u'$.

Case R. $X^{\mathbf{r}}u_1'u_2'u''\wedge w\equiv\mathrm{InR}(u'')$ for some u''. Then again (26.6) holds with $u':={}^{\mathrm{co}}\mathcal{R}^{\alpha}_{\mathbb{S}}u''f$, since $u\equiv{}^{\mathrm{co}}\mathcal{R}^{\alpha}_{\mathbb{S}}u'f\equiv s::{}^{\mathrm{co}}\mathcal{R}^{\alpha}_{\mathbb{S}}u''f\equiv s::u'$. □

26.4.3 Soundness

Constructively to state A means the same as to say that A has a realizer. This statement $A\leftrightarrow\exists_x(x\ \mathbf{r}\ A)$ was called "to assert is to realize" in Feferman [3]. Here we call it invariance axiom, since it expresses invariance of A under the realizability interpretation. Using the invariance axioms we will prove the soundness theorem.

Axiom 26.51 (Invariance) For $\mathbf{r}$-free c.r. formulas A we require

$$\mathrm{InvAll}_A\colon\forall_z(z\ \mathbf{r}\ A\to A), \tag{26.7}$$

$$\mathrm{InvEx}_A\colon A\to\exists_z(z\ \mathbf{r}\ A). \tag{26.8}$$

Theorem 26.52 (Soundness) *Let M be an $\mathbf{r}$-free derivation of a formula A from assumptions $u_i\colon C_i$ ($i<n$). Then we can derive*

$$\begin{cases}\mathrm{et}(M)\ \mathbf{r}\ A & \textit{if A is c.r.}\\ A & \textit{if A is n.c.}\end{cases}$$

from assumptions

$$\begin{cases}z_{u_i}\ \mathbf{r}\ C_i & \textit{if } C_i \textit{ is c.r.}\\ C_i & \textit{if } C_i \textit{ is n.c.}\end{cases}$$

Proof By induction on M. The axiom cases have been done before, and from the remaining cases we only treat the ones using invariance axioms.

Case $(\lambda_{u^A} M^B)^{A \to B}$ with B n.c. and A c.r. We need a derivation of $A \to B$. By induction hypothesis we have a derivation of B from $z \mathbf{~r~} A$. Using the invariance axiom $A \to \exists_z (z \mathbf{~r~} A)$ we obtain the required derivation of B from A as follows.

$$\dfrac{\dfrac{A \to \exists_z(z \mathbf{~r~} A) \qquad A}{\exists_z(z \mathbf{~r~} A)} \qquad \begin{array}{c}[z \mathbf{~r~} A] \\ \mid \text{IH} \\ B\end{array}}{B}\, \exists^-$$

Case $(M^{A \to B} N^A)^B$ with B n.c. and A c.r. We need a derivation of B. By induction hypothesis we have derivations of $A \to B$ and of $\text{et}(N) \mathbf{~r~} A$. Using the invariance axiom $\forall_z (z \mathbf{~r~} A \to A)$ we obtain the required derivation from

$$\dfrac{\dfrac{\forall_z(z \mathbf{~r~} A \to A) \qquad \text{et}(N)}{\text{et}(N) \mathbf{~r~} A \to A} \qquad \begin{array}{c}\mid \text{IH} \\ \text{et}(N) \mathbf{~r~} A\end{array}}{A}$$

and the derivation of $A \to B$. □

26.4.4 Extensionality of Extracted Terms

Let I be an inductive predicate and ι_I its associated algebra. One can show that

- every constructor of ι_I is extensional w.r.t. its clause I_i^+,
- $\mathcal{R}_{\iota_I}^{\alpha}$ is extensional w.r.t. the least-fixed-point axiom I^-,
- the destructor of ι_I is extensional w.r.t. the closure axiom ${}^{\text{co}}I^-$, and
- ${}^{\text{co}}\mathcal{R}_{\iota_I}^{\alpha}$ is extensional w.r.t. the greatest-fixed-point axiom ${}^{\text{co}}I^+$.

We prove these claims for special cases only. For the first two claims we consider the inductive predicate $\sim_{\mathbb{L}}$.

Lemma 26.53 *The constructors of* $\mathbb{L}$ *are extensional w.r.t.* $\sim_{\mathbb{L}}$*'s clauses.*

Proof We only consider the second constructor $\mathcal{C}$. The goal is to show that $\mathcal{C}$ is extensional w.r.t. the cotype $\mathbb{D} \to \mathbb{L} \to \mathbb{L}$ of $\sim_{\mathbb{L}}$'s second clause, which by definition of $\doteq$ means

$$\forall_{s_1, s_2, \ell_1, \ell_2} (s_1 \sim_{\mathbb{D}} s_2 \to \ell_1 \sim_{\mathbb{L}} \ell_2 \to s_1 :: \ell_1 \sim_{\mathbb{L}} s_2 :: \ell_2).$$

But this is the second clause of $\sim_{\mathbb{L}}$. □

Lemma 26.54 $\mathcal{R}_{\mathbb{L}}^{\alpha}$ *is extensional w.r.t. the least-fixed-point axiom* $\sim_{\mathbb{L}}^-$.

Proof We equivalently rewrite $\sim_{\mathbb{L}}^{-}$ as $C :=$

$$\forall_{\ell_1,\ell_2}(\ell_1 \sim_{\mathbb{L}} \ell_2 \to X[][] \to \forall_{s_1,s_2,\ell_1,\ell_2}(s_1 \sim_{\mathbb{D}} s_2 \to \ell_1 \sim_{\mathbb{L}} \ell_2 \to X\ell_1\ell_2 \to X(s_1 :: \ell_1, s_2 :: \ell_2)) \to \ell_1\ell_2)$$

to make its cotype the same as the one for $\mathcal{R}_{\mathbb{L}}^{\alpha}$:

$$\mathbb{L} \to \alpha \to (\mathbb{D} \to \mathbb{L} \to \alpha \to \alpha) \to \alpha.$$

We must show $\mathcal{R}_{\mathbb{L}}^{\alpha} \doteq_C \mathcal{R}_{\mathbb{L}}^{\alpha}$ with $\alpha := \varphi(X)$. By definition of $\doteq_C$ this is equivalent to

$$\forall_{x_1,x_2,f_1,f_2,\ell_1,\ell_2}(x_1 \doteq_\alpha x_2 \to f_1 \doteq_{\mathbb{D}\to\mathbb{L}\to\alpha\to\alpha} f_2 \to \ell_1 \sim_{\mathbb{L}} \ell_2 \to \mathcal{R}_{\mathbb{L}}^{\alpha}\ell_1 x_1 f_1 \doteq_\alpha \mathcal{R}_{\mathbb{L}}^{\alpha}\ell_2 x_2 f_2).$$

Assume $x_1 \doteq_\alpha x_2$ and $f_1 \doteq_{\mathbb{D}\to\mathbb{L}\to\alpha\to\alpha} f_2$. Use the least-fixed-point axiom $\sim_{\mathbb{L}}^{-}$ (in its original form) with competitor predicate

$$X := \{\, \ell_1, \ell_2 \mid \mathcal{R}_{\mathbb{L}}^{\alpha}\ell_1 x_1 f_1 \doteq_\alpha \mathcal{R}_{\mathbb{L}}^{\alpha}\ell_2 x_2 f_2 \,\}.$$

Case []. By the computation rules for $\mathcal{R}_{\mathbb{L}}^{\alpha}$ the claim $X[][]$ follows from $x_1 \doteq_\alpha x_2$. *Case* ::. Assume $s_1 \sim_{\mathbb{D}} s_2$ and $f_1 \doteq_{\mathbb{D}\to\mathbb{L}\to\alpha\to\alpha} f_2$. Let $y_1 := \mathcal{R}_{\mathbb{L}}^{\alpha}\ell_1 x_1 f_1$ and $y_2 := \mathcal{R}_{\mathbb{L}}^{\alpha}\ell_2 x_2 f_2$. Then $y_1 \doteq_\alpha y_2$ by assumption. The goal $f_1 s_1 \ell_1 y_1 \doteq_\alpha f_2 \ell_2 x_2 y_2$ follows from $f_1 \doteq f_2$, $s_1 \sim s_2$, $\ell_1 \sim \ell_2$ and $y_1 \doteq_\alpha y_2$. □

For the final two claims we only consider the coinductive predicate $\approx_{\mathbb{S}}$.

Lemma 26.55 *The destructor $\mathcal{D}_{\mathbb{S}}$ is extensional w.r.t. $\approx_{\mathbb{S}}^{-}$.*

Proof The closure axiom $\approx_{\mathbb{S}}^{-}$ has cotype ${}^{\mathrm{co}}\mathbb{S} \to \mathbb{D} \times {}^{\mathrm{co}}\mathbb{S}$. The goal is $\mathcal{D}_{\mathbb{S}} \doteq_{({}^{\mathrm{co}}\mathbb{S}\to\mathbb{D}\times{}^{\mathrm{co}}\mathbb{S})} \mathcal{D}_{\mathbb{S}}$, which unfolds into

$$\forall_{u_1,u_2}(u_1 \approx_{\mathbb{S}} u_2 \to \mathcal{D}_{\mathbb{S}} u_1 \sim_{\mathbb{D}\times{}^{\mathrm{co}}\mathbb{S}} \mathcal{D}_{\mathbb{S}} u_2).$$

Assume $u_1 \approx_{\mathbb{S}} u_2$. By $\approx_{\mathbb{S}}^{-}$ we obtain s_1, s_2, u_1', u_2' with

$$s_1 \sim_{\mathbb{D}} s_2 \wedge u_1' \approx_{\mathbb{S}} u_2' \wedge u_1 \equiv s_1 :: u_1' \wedge u_2 \equiv s_2 :: u_2'.$$

By the computation rule for $\mathcal{D}_{\mathbb{S}}$ we have $\mathcal{D}_{\mathbb{S}} u_i \equiv \langle s_i, u_i' \rangle$. By the clause for $\sim_{\mathbb{D}\times{}^{\mathrm{co}}\mathbb{S}}$ this implies the claim $\mathcal{D}_{\mathbb{S}} u_1 \sim_{\mathbb{D}\times{}^{\mathrm{co}}\mathbb{S}} \mathcal{D}_{\mathbb{S}} u_2$. □

Lemma 26.56 *${}^{\mathrm{co}}\mathcal{R}_{\mathbb{S}}^{\alpha}$ is extensional w.r.t. the greatest-fixed-point axiom $\approx_{\mathbb{S}}^{+}$.*

Proof Equivalently rewrite $\approx_{\mathbb{S}}^{+}$ as $C := \forall_{u_1,u_2}(Xu_1u_2 \to \forall_{u_1,u_2}(Xu_1u_2 \to \exists_{s_1,s_2,u_1',u_2'}(s_1 \sim_{\mathbb{D}} s_2 \wedge (u_1' \approx_{\mathbb{S}} u_2' \vee Xu_1'u_2') \wedge u_1 \equiv s_1 :: u_1' \wedge u_2 \equiv s_2 :: u_2')) \to u_1 \approx_{\mathbb{S}} u_2)$ to make its cotype the same as the one for ${}^{\mathrm{co}}\mathcal{R}_{\mathbb{S}}^{\alpha}$:

$$\alpha \to (\alpha \to \mathbb{D} \times ({}^{\mathrm{co}}\mathbb{S} + \alpha)) \to {}^{\mathrm{co}}\mathbb{S}.$$

Call this cotype ψ. The goal is ${}^{\mathrm{co}}\mathcal{R}^{\alpha}_{\mathbb{S}} \doteq_{\psi} {}^{\mathrm{co}}\mathcal{R}^{\alpha}_{\mathbb{S}}$, which unfolds into

$$\forall_{x_1,x_2}(x_1 \doteq_{\alpha} x_2 \to \forall_{f_1,f_2}(f_1 \doteq_{\alpha\to\mathbb{D}\times({}^{\mathrm{co}}\mathbb{S}+\alpha)} f_2 \to {}^{\mathrm{co}}\mathcal{R}^{\alpha}_{\mathbb{S}} x_1 f_1 \approx_{\mathbb{S}} {}^{\mathrm{co}}\mathcal{R}^{\alpha}_{\mathbb{S}} x_2 f_2)).$$

Assume $x_1 \doteq_{\alpha} x_2$ and $f_1 \doteq_{\alpha\to\mathbb{D}\times({}^{\mathrm{co}}\mathbb{S}+\alpha)} f_2$. Let $u_1 := {}^{\mathrm{co}}\mathcal{R}^{\alpha}_{\mathbb{S}} x_1 f_1$ and $u_2 := {}^{\mathrm{co}}\mathcal{R}^{\alpha}_{\mathbb{S}} x_2 f_2$. To prove the goal $u_1 \approx_{\mathbb{S}} u_2$ we use coinduction, or more precisely $\approx^{+}_{\mathbb{S}}$ with competitor predicate

$$X := \{\, u_1, u_2 \mid \exists_{y_1,y_2}(u_1 \equiv {}^{\mathrm{co}}\mathcal{R} y_1 f_1 \wedge u_2 \equiv {}^{\mathrm{co}}\mathcal{R} y_2 f_2 \wedge y_1 \doteq_{\alpha} y_2) \,\}.$$

This means that we have to show

$$\begin{aligned}
&\exists_{s_1,s_2,u_1',u_2'}(\\
&s_1 \sim_{\mathbb{D}} s_2 \wedge (u_1' \approx_{\mathbb{S}} u_2' \vee \exists_{y_1,y_2}(u_1' \equiv {}^{\mathrm{co}}\mathcal{R} y_1 f_1 \wedge u_2' \equiv {}^{\mathrm{co}}\mathcal{R} y_2 f_2 \wedge y_1 \doteq_{\alpha} y_2)) \wedge \\
&u_1 \equiv s_1 :: u_1' \wedge u_2 \equiv s_2 :: u_2').
\end{aligned}$$

From $x_1 \doteq_{\alpha} x_2$ and $f_1 \doteq_{\alpha\to\mathbb{D}\times({}^{\mathrm{co}}\mathbb{S}+\alpha)} f_2$ we obtain $f_1 x_1 \sim_{\mathbb{D}\times({}^{\mathrm{co}}\mathbb{S}+\alpha)} f_2 x_2$. By definition of $\sim_{\times}$ this implies the existence of s_1, s_2, a_1, a_2 with

$$f_1 x_1 \equiv \langle s_1, a_1\rangle \wedge f_2 x_2 \equiv \langle s_2, a_2\rangle \wedge s_1 \sim_{\mathbb{D}} s_2 \wedge a_1 \sim_{({}^{\mathrm{co}}\mathbb{S}+\alpha)} a_2,$$

and by definition of $\sim_{+}$ from $a_1 \sim_{({}^{\mathrm{co}}\mathbb{S}+\alpha)} a_2$ we obtain the disjunction

$$\begin{aligned}
&(a_1 \equiv \mathrm{InL}(u_1') \wedge a_2 \equiv \mathrm{InL}(u_2') \wedge u_1' \approx_{\mathbb{S}} u_2') \vee \\
&(a_1 \equiv \mathrm{InR}(x_1') \wedge a_2 \equiv \mathrm{InR}(x_2') \wedge x_1' \doteq_{\alpha} x_2').
\end{aligned}$$

We argue by cases on this disjunction. Recall

$${}^{\mathrm{co}}\mathcal{R}^{\alpha}_{\mathbb{S}} x f \equiv \begin{cases} s :: u & \text{if } fx \equiv \langle s, \mathrm{InL}(u)\rangle, \\ s :: {}^{\mathrm{co}}\mathcal{R}^{\alpha}_{\mathbb{S}} x' f & \text{if } fx \equiv \langle s, \mathrm{InR}(x')\rangle. \end{cases}$$

Case L. Then we have s_1, s_2, u_1', u_2' with $s_1 \sim_{\mathbb{D}} s_2$ and $u_1' \approx_{\mathbb{S}} u_2'$ such that $f_i x_i \equiv \langle s_i, \mathrm{InL}(u_i')\rangle$. Hence $u_i := {}^{\mathrm{co}}\mathcal{R}^{\alpha}_{\mathbb{S}} x_i f_i \equiv s_i :: u_i'$, and the claim follows.

Case R. Then we have s_1, s_2, x_1', x_2' with $s_1 \sim_{\mathbb{D}} s_2$ and $x_1' \doteq_{\alpha} x_2'$ such that $f_i x_i \equiv \langle s_i, \mathrm{InR}(x_i')\rangle$. Hence $u_i := {}^{\mathrm{co}}\mathcal{R}^{\alpha}_{\mathbb{S}} x_i f_i \equiv s_i :: u_i'$ with $u_i' := {}^{\mathrm{co}}\mathcal{R}^{\alpha}_{\mathbb{S}} x_i' f_i$, and again the claim follows. □

We now prove compatibility of extracted terms with pointwise equality w.r.t. the cotype of the formula proved. For a convenient formulation we assume two more fixed assignments $u \mapsto z_u', z_u''$ of object variables to assumption variables.

Theorem 26.57 (Compatibility of extracted terms) *Let $M : A$ be a proof of a c.r. formula A, and $u_i : C_i$ $(i = 1, \dots, n)$ all free c.r. assumptions whose associated object variable z_{u_i} is free in $\mathrm{et}(M)$. Then we can find a proof of*

$$\mathrm{et}(M)(z'_{u_1},\ldots z'_{u_n}) \doteq_A \mathrm{et}(M)(z''_{u_1},\ldots z''_{u_n})$$

from assumptions $z'_{u_i} \doteq_{C_i} z''_{u_i}$ *for* $i = 1, \ldots, n$.

Proof By induction on M. *Case* $u\colon C$. Immediate. *Case* $c\colon A$ an axiom. This is clear in case the extracted term is the identity. For the axioms $I^\pm$ and ${}^{\mathrm{co}}I^\pm$ it was proved in Lemmas 26.55–26.58.

Case $(\lambda_{u^A} M^B)^{A\to B}$ with A c.r. For simplicity assume that u is the only assumption variable whose z_u is free in $\mathrm{et}(M)$. By IH we have a proof of $\mathrm{et}(M)(z'_u) \doteq_A \mathrm{et}(M)(z''_u)$ from $z'_u \doteq_A z''_u$. We want a proof of $\mathrm{et}(\lambda_u M) \doteq_{A\to B} \mathrm{et}(\lambda_u M)$, i.e., $\lambda_{z_u}\mathrm{et}(M)(z_u) \doteq_{A\to B} \lambda_{z_u}\mathrm{et}(M)(z_u)$, which is

$$\forall_{z'_u, z''_u}(z'_u \doteq_A z''_u \to \mathrm{et}(M)(z'_u) \doteq_B \mathrm{et}(M)(z''_u)).$$

Apply $\to^+$ and twice $\forall^+$ to the proof given by IH. In the case A n.c. the extracted term $\mathrm{et}(\lambda_u M)$ is $\mathrm{et}(M)$ and the claim is immediate.

Case $M^{A\to B}N^A$ with A c.r. For simplicity assume that there no assumption variables whose associated object variable is free in $\mathrm{et}(MN)$. By IH$_M$ we have a proof of $\mathrm{et}(M) \doteq_{A\to B} \mathrm{et}(M)$, that is,

$$\forall_{z'_u, z''_u}(z'_u \doteq_A z''_u \to \mathrm{et}(M)z'_u \doteq_B \mathrm{et}(M)z''_u).$$

By IH$_N$ we have a proof of $\mathrm{et}(N) \doteq_A \mathrm{et}(N)$. Applying an instance of the first proof to the second gives $\mathrm{et}(M)\mathrm{et}(N) \doteq_B \mathrm{et}(M)\mathrm{et}(N)$, as required. In the case A n.c. the extracted term $\mathrm{et}(MN)$ is $\mathrm{et}(M)$ and the claim is immediate.

Cases $\lambda_x M$, Mt. Obvious, since the extracted term does not change. □

Corollary 26.58 (Extensionality of extracted terms) *Let* $M\colon A$ *be a proof of a c.r. formula* A*, and* $u_i\colon C_i$ $(i = 1, \ldots, n)$ *all free c.r. assumptions whose associated object variable* z_{u_i} *is free in* $\mathrm{et}(M)$*. Then we can find a proof of* $\mathrm{et}(M) \doteq_A \mathrm{et}(M)$ *from assumptions* $z_{u_i} \doteq_{C_i} z_{u_i}$ *for* $i = 1, \ldots, n$.

Proof In the constructed proof substitute z'_{u_i}, z''_{u_i} by z_{u_i}. □

26.5 Applications

Space restrictions do not permit us to go into applications, which are mainly in constructive analysis.[2] We refer the reader to Berger et al. [1] and to Schwichtenberg and Wiesnet [9], for example.

References

[1] Berger, U., Miyamoto, K., Schwichtenberg, H., and Tsuiki, H. 2016. Logic for Gray-code computation. Pages 69–110 in: *Concepts of Proof in Mathematics, Philosophy, and Computer Science*, D. Probst and P. Schuster (eds). Berlin: De Gruyter.

[2] See `minlog/examples/analysis`

[2] Bishop, E. 1970. Mathematics as a numerical language. Pages 53–71 in: *Intuitionism and Proof Theory, Proceedings of the Summer Conference at Buffalo NY 1968*. J. Myhill, A. Kino, and R.E. Vesley (eds). Studies in Logic and the Foundations of Mathematics. Amsterdam: North-Holland.

[3] Feferman, S. 1979. Constructive theories of functions and classes. 1979 Pages 159–224 in: *Logic Colloquium 78*, K. McAloon, M. Boffa, and D. van Dalen (eds). Studies in Logic and the Foundations of Mathematics, vol. 97. Amsterdam: North-Holland.

[4] Gandy, R. 1953. On Axiomatic Systems in Mathematics and Theories in Physics. Ph.D. thesis, University of Cambridge.

[5] Gandy, R. 1956. On the axiom of extensionality – part I. *J. Symbol. Logic*, **21**(1), 36–48

[6] Gödel, K. 1958. Über eine bisher noch nicht benützte Erweiterung des finiten Standpunkts. *Dialectica*, **12**, 280–287.

[7] Larsen, K. G., and Winskel, G. 1991. Using information systems to solve recursive domain equations. *Information and Computation*, **91**, 232–258.

[8] Schwichtenberg, H., and Wainer, S. S. 2012. *Proofs and Computations*. Perspectives in Logic. Cambridge: Association for Symbolic Logic and Cambridge University Press.

[9] Schwichtenberg, H., and Wiesnet, F. 2021. Logic for exact real arithmetic. *Logic. Meth. Comp. Sci.*, **17**(2). `https://doi.org/10.23638/LMCS-17(2:7)2021`.

[10] Scott D. 1982. Domains for denotational semantics. Pages 577–613 in: *Automata, Languages and Programming*, E. Nielsen and E.M. Schmidt (eds). LNCS, vol. 140. Berlin, Heidelberg, NewYork : Springer Verlag.

[11] Takeuti, G. 1953. On a generalized logic calculus. *Japan. J. Math.*, **23**, 39–96.

[12] Troelstra, A. S. 1998. Realizability. Pages 408–473 in: *Handbook of Proof Theory*, S. Buss (ed). Amsterdam: Elsevier.

27

Application of Constructive Analysis in Exact Real Arithmetic

Kenji Miyamoto

27.1 Introduction

This chapter presents exact real arithmetic within Bishop's constructive mathematics (BISH). Exact real arithmetic is a field of studying algorithms involving computational real numbers and functions. Unlike floating-point arithmetic, the precision of real numbers in exact real arithmetic is not limited, and the same happens for functions. Here we present the following Rump's example which illustrates the fact that floating-point arithmeric with a high precision can give misleading results. Let $f(a,b) := 333.75b^6 + a^2(11a^2b^2 - b^6 - 121b^4 - 2) + 5.5b^8 + \frac{a}{2b}$ and evaluate $f(77\,617, 33\,096)$. Rump computed $1.172\,603\ldots$, $1.172\,603\,940\,0531\ldots$, and $1.172\,603\,940\,053\,178\ldots$, using single, double, and extended-precision arithmetic in IBM S/370, respectively, while the correct result is $-0.827\,396\,0599\ldots$. In order to correctly handle *Rump's example*, floating-point arithmetic with a dynamic range of at least 123 bits is required, hence even the 128-bit precision of IEEE-754, whose dynamic range is 112 bits, is still insufficient to correctly carry out the above computation [20]. Within exact real arithmetic, we always see a numerically correct result, although it is computationally more demanding than floating-point arithmetic. There are several different approaches to exact real arithmetic, and Bishop's constructive analysis is one of them. See Aberth, Edalat and Heckmann, and Weihrauch [1, 15, 35] for approaches other than Bishop's.

In constructive mathematics, the meanings of disjunction and existential quantifier are stronger than the classical ones as the BHK-interpretation states. It allows us to interpret a formula with a computational claim as a problem specification and, moreover, proving this formula is the same as finding an algorithmic solution to the problem. Therefore, we can learn a correct algorithmic solution to the problem by reading constructive proofs. We call such an algorithmic solution in a proof a computational content of the proof. Working within a suitable constructive system, we can extract from a proof a program code which is human-readable, executable on a computer, and also provably correct with respect to the proven formula. The procedure of generating such a code is called *program extraction. The Theory of Computable Functionals*, TCF in short, is a formal foundation for constructive mathematics [29]. The *Minlog* system is a proof-assistant which is an implementation of TCF [22]. Minlog has a feature of program extraction based on Kreisel's modified realizability and employs the term language T^+. This language is an extention of Gödel's T and its semantics is given via the Scott–Ershov domains, while the computability theory is left open in case of BISH. Although the presentation of this chapter is informal, all the mathematical results are done within TCF. The examples in Section 27.3 were formalized in Minlog to carry out program extraction.

In classical mathematics, there are many existence theorems of problems of exact real arithmetic. It often happens that the algorithmic solution to a problem is studied separately from the mathematical arguments, because classical proofs generally do not contain any computational ideas to develop an algorithm. Moreover, even a theorem statement which claims an existence of a noncomputable object may be proved in a classical setting. In a constructive setting on the other hand, any proof of an existence theorem contains a correct algorithm of finding the object of interest. We are also supposed to formulate the theorem statement in the way that it precisely specifies the computational problem, for example, explicitly stating the approximability. Constructive mathematics is one prospective approach to fill the gap between classical mathematics and empirical science, and this chapter is going to show how this works by the combination of exact real arithmetic and Bishop's constructive analysis.

The rest of this chapter is organized as follows. Section 27.2 is an introduction to the two subjects; the first one is Bishop's constructive analysis for exact real arithmetic, and the second is corecursion and coinduction, which are a preparation towards applications in Section 27.3. Section 27.3 describes three applications of constructive analysis in exact real arithmetic. The first one is the root-finding problem and inverse functions via the intermediate value theorem. Constructively studying the intermediate value theorem, one obtains an algorithm of finding an intermediate value, and it can be viewed as a solution of the root-finding problem. We further discuss an improvement of the computational efficiency which works for a special case, and use this variant of the intermediate value theorem to find an inverse function. The second one is ordinary differential equations. The well-known Cauchy–Euler approximation method comes as a computational content of constructive arguments on solutions of ordinary differential equations. The third one is a stream-based treatment of real numbers. The stream is an infinite data structure which is supported in modern programming languages. We show how this notion is used in exact real arithmetic through constructive arguments involving coinduction and corecursion, which offer an alternative perspective on infinite data processing. We study the average and multiplication functions in the stream-based real numbers. Section 27.4 concludes this chapter.

27.2 Preliminaries

Section 27.2.1 describes constructive elementary calculus in Bishop's style, covering fundamental notions such as Cauchy reals, uniformly continuous functions, differentiation, and integration, which will be used in Section 27.3 on applications of exact real arithmetic. The presentation follows the books by Bishop and by Bishop and Bridges, and the lecture note by Schwichtenberg [6, 7, 27]. In particular, we avoid using higher-type functions such as those of type-2, following the style of Schwichtenberg's lecture note. Section 27.2.2 describes corecursion and coinduction which are suitable devices to deal with infinite data of type-0, namely, of ground type. Corecursion is an important theoretical foundation for infinite data structures in modern programming languages as Python, Haskell, etc. Programming in exact real arithmetic is also benefited from it. In regard to logical foundation for program extraction, formal systems, and proof assistants to practice constructive mathematics, see [22, 29].

27.2.1 Exact Real Arithmetic

In classical mathematics, a rational sequence $(a_n)_n$ is a Cauchy real if it is converging, namely, $\forall_k \exists_l \forall_{n,m \geq l} |a_n - a_m| \leq 2^{-k}$. The idea is the following: For the required accuracy specified by k, choosing a sufficiently large l, we obtain a satisfactory approximation a_m for $m \geq l$. As a Cauchy real is a function object which we cannot directly observe, it is important that the approximation is always as accurate as required. From a computational viewpoint, a problem of the above classical Cauchy reals is that the convergence (Cauchyness in other words) of the rational sequence is not necessarily witnessed, namely, we do not necessarily know a number l such that $|a_n - a_m| \leq 2^{-k}$ if $n, m \geq l$. As a consequence, it is not always possible to approximate a classical Cauchy real up to a required error, hence a classical Cauchy real cannot be counted as an utterly numerical object. Within Bishop's constructive analysis, a Cauchy real is defined to be a rational sequence equipped with a Cauchy modulus which witnesses the convergence of the sequence. By means of the modulus, we recover the numerical meaning of real numbers.

Definition 27.1 (Cauchy real) A pair of a rational sequence $(a_n)_n$ and a weakly increasing function M, a *Cauchy modulus*, is a *Cauchy real* if $\forall_k \forall_{n,m \geq M(k)} (|a_n - a_m| \leq 2^{-k})$. A rational sequence is a *regular Cauchy real* if the above condition is satisfied for M being the identity function.

It often happens that we can constructivize a classical notion by supplying additional information. The notion of moduli is crucial in constructive analysis to witness not only the Cauchyness, but also other notions such as the uniform continuity, the uniform convergence, the differentiability, the increasement, and so on. We further see orders and arithmetical functions, then prove the approximate split property of Cauchy reals.

Definition 27.2 (Nonnegative and positive Cauchy reals) A Cauchy real $x = \langle (a_n)_n, M\rangle$ is nonnegative, written $x \in \mathbb{R}^{0+}$, if $-2^{-p} \leq a_{M(p)}$ for all $p \in \mathbb{Z}^+$, and is p-positive, written $x \in_p \mathbb{R}^+$ or $x \in \mathbb{R}^+$ in case p is omittable, if $2^{-p} \leq a_{M(p+1)}$.

Definition 27.3 (Equality on reals) Given Cauchy reals $\langle (a_n)_n, M\rangle$ and $\langle (b_n)_n, N\rangle$, the equality $x = y$ is defined to be $\forall_p(|a_{M(p+1)} - b_{N(p+1)}| \leq 2^{-p})$.

Lemma 27.4 (Bound of reals) *For any Cauchy real $x := \langle (a_n)_n, M\rangle$, there exists p_x such that $|a_n| \leq 2^{p_x}$ for any n.*

Proof Let n_0 be $M(1)$ and p_x satisfying that $\max\{|a_n| \mid n \leq n_0\} + \frac{1}{2} \leq 2^{p_x}$. Then $|a_n| \leq 2^{p_x}$ for all n. □

Definition 27.5 (Aritmetical functions) Let x and y be $\langle (a_n)_n, M\rangle$ and $\langle (b_n)_n, N\rangle$, then the following arithmetical functions are defined.

(i) $x + y$ is defined to be $\langle (a_n + b_n)_n, L\rangle$, where $L(p) := \max(M(p + 1), N(p + 1))$.
(ii) $-x$ is defined to be $\langle (-a_n)_n, M\rangle$.
(iii) $x \cdot y$ is defined to be $\langle (a_n \cdot b_n)_n, L\rangle$, where $L(p) := \max(M(p + 1 + p_y), N(p + 1 + p_x))$. Here, p_x and p_y are as given in Lemma 27.4.
(iv) $|x|$ is defined to be $\langle (|a_n|)_n, M\rangle$.
(v) Assuming $|x| \in_q \mathbb{R}^+$, $\frac{1}{x}$ is defined to be $\langle (c_n)_n, L\rangle$, where $c_n := \frac{1}{a_n}$ if $a_n \neq 0$ and otherwise $c_n := 0$, and $L(p) := \max(M(2(q + 1) + p), N(q + 1))$.

Definition 27.6 (Inequality) For Cauchy reals x, y, $x \leq y$ and $x <_k y$ are defined to be $y - x \in \mathbb{R}^{0+}$ and $y - x \in_k \mathbb{R}^+$, respectively. If there is no confusion, k is omitted.

Lemma 27.7 (LeIsNotGt) $x \leq y \leftrightarrow x \not> y$.

Proof Assume $x \leq y$ and $x > y$, from which we derive $x < x$, a contradiction. In the opposite direction, we prove $0 \not< z \to z \leq 0$ instead. Let z be given by $\langle (c_n)_n, M\rangle$ and assume $0 \not< z$. We now prove $\forall_p \exists_m \forall_{n \geq m}(c_n \leq 2^{-p})$. Assume p and let m be $M(p + 2)$, then for $n \geq M(p + 2)$, $c_n \leq c_{M(p+2)} + 2^{-(p+2)}$. The assumption $0 \not< z$ is same as $\neg\exists_q(2^{-q} \leq c_{M(q+1)})$ by the definition, which implies $c_{M(p+2)} \leq 2^{-(p+1)}$ for $q := p + 1$, hence $c_{M(p+2)} + 2^{-(p+2)} \leq 2^{-p}$. □

Unlike rational numbers, for a given $x, z \in \mathbb{R}$ we cannot generally determine either $x \le z$ or $z \le x$. Instead, we can compare a real number with an interval of a positive length.

Lemma 27.8 (ApproxSplit) *Assume x, y, z are Cauchy reals with $x < y$, then either $z \le y$ or $x \le z$.*

Proof Let x, y, and z be $\langle (a_n)_n, M\rangle$, $\langle (b_n)_n, N\rangle$, and $\langle (c_n)_n, L\rangle$, respectively. Assume $x <_k y$ for some k, namely, $2^{-k} \le b_n - a_n$ for $n := \max(M(k+2), N(k+2))$. Further let m be $\max(n, L(k+2))$, and then we make a case distinction. Case $c_m \le \frac{a_n+b_n}{2}$, $z \le y$ follows from $c_l \le b_l$ for $l \ge m$ as

$$c_l \le c_m + 2^{-(p+2)} \le \frac{a_n + b_n}{2} + \frac{b_n - a_n}{4} = b_n - \frac{b_n - a_n}{4} \le b_n - 2^{-(p+2)} \le b_l.$$

Case $c \not\le \frac{a_n+b_n}{2}$, $x \le z$ follows from $a_l \le c_l$ for $l \ge m$ as

$$a_l \le a_n + 2^{-(p+2)} \le a_n - \frac{b_n - c_n}{4} = \frac{a_n + b_n}{2} + \frac{b_n - a_n}{4} \le c_m - 2^{-(p+2)} \le c_l.$$

□

We discuss sequences of Cauchy reals and the completeness of reals.

Definition 27.9 (Cauchy sequence) A sequence $(x_n)_n$ of reals is a *Cauchy sequence* with modulus $M : \mathbb{Z}^+ \to \mathbb{N}$ whenever $|x_n - x_m| \le 2^{-p}$ for $n, m \ge M(p)$, and *converges* with modulus $M : \mathbb{Z}^+ \to \mathbb{N}$ to a real y, its *limit*, whenever $|x_n - y| \le 2^{-p}$ for $n \ge M(p)$.

Lemma 27.10 (RatCauchyConvMod) *Every modulated Cauchy sequence of rationals converges with the same modulus to the real number it represents.*

Proof Let $x := \langle (a_n)_n, N\rangle$. We prove $|a_n - x| \le 2^{-p}$ for all $n \ge M(p)$. Assume $n \ge M(p)$, then it suffices to show $|a_n - a_m| \le 2^{-p}$ for $m \ge M(p)$, which is trivial because M is a modulus. □

Theorem 27.11 (RealCompl) *For every Cauchy sequence of reals we can find a real to which it converges.*

Proof Let $(x_n)_n$ be a Cauchy sequence of reals with modulus M, where x_n can be denoted as $\langle (a_{nl})_l, N_n\rangle$. For each n and p, $|x_n - a_{nl}| \le 2^{-p}$ for all $l \ge N_n(p)$ by the previous lemma. Let $b_n := a_{nN_n(n)}$ for every n, so that $|x_n - b_n| \le 2^{-n}$ for all n. Then, $|b_m - b_n| \le |b_m - x_m| + |x_m - x_n| + |x_n - b_n| \le 2^{-m} + 2^{-(q+1)} + 2^{-n} \le 2^{-q}$ for all $m, n \ge \max(M(q+1), q+2)$. The real $y := (b_n)_n$ is a Cauchy sequence with modulus $L(q) := \max(M(q+1), q+2)$. Again by Lemma 27.10, $|x_n - y| \le |x_n - b_n| + |b_n - y| \le 2^{-n} + 2^{-(q+1)} \le 2^{-q}$ for all $n \ge L(q+1)$, namely, $(x_n)_n$ converges to y with the modulus $q \mapsto L(q+1)$. □

Lemma 27.12 (RealCauchyConvMod) *Every modulated Cauchy sequence of reals converges with the same modulus to its limit.*

Proof Let $(x_n)_n$ be a Cauchy sequence of reals with modulus M and y the limit of $(x_n)_n$. This means that $|x_n - x_m| \leq 2^{-p}$ for $n, m \geq M(p)$ and $|x_n - y| \leq 2^{-q}$ for $n \geq L(q)$. We prove $|x_n - y| \leq 2^{-p}$ for all $n \geq M(p)$. Assume $n \geq M(p)$ and let $q \in \mathbb{Z}^+$, then $|x_n - y| \leq |x_n - x_m| + |x_m - y|$ for $m \geq M(p), L(q)$, and furthermore $|x_n - x_m| + |x_m - y| \leq 2^{-p} + 2^{-q}$. Since q is arbitrary, we proved the claim. □

We introduce uniformly continuous functions on real numbers, which involves at most type-1 objects, namely, function objects whose arguments are of ground types. We also give the application and composition. It is possible to handle composition assuming Brouwer's fan theorem [11]. We require additional information to compute the lower and upper bounds, whose availability seems plausible in practice, so that we don't rely on the fan theorem.

Definition 27.13 (Uniformly continuous function) A uniformly continuous function $f : X \to \mathbb{R}$ is defined to be $\langle h, \alpha, \omega, N, M \rangle$:

(i) an approximating map $h^{(X \cap \mathbb{Q}) \to \mathbb{N} \to \mathbb{Q}}$,
(ii) a map $\alpha^{\mathbb{Q} \to \mathbb{Q} \to \mathbb{Z}^+ \to \mathbb{N}}$ such that each $(h(a, n)_n)_n$ for $c \leq a \leq d$ is a Cauchy sequence with $\lambda_p \alpha(c, d, p)$ as a Cauchy modulus,
(iii) a *modulus* $\omega^{\mathbb{Q} \to \mathbb{Q} \to \mathbb{Z}^+ \to \mathbb{Z}^+}$ *of uniform continuity* such that $\omega(c, d, p)$ satisfies for any $a, b \in [c, d]$, $|a - b| \leq 2^{-\omega(c,d,p)+1} \to |h(a, n) - h(b, n)| \leq 2^{-p}$ for $n \geq \alpha(c, d, p)$,
(iv) maps N, M of $\mathbb{Q} \to \mathbb{Q} \to \mathbb{Q}$ such that $N(c, d)$ and $M(c, d)$ are lower and upper bounds for $h(a, n)$ for $a \in [c, d]$,
(v) for $c' \leq c < d \leq d'$, α, ω, M, N satisfy $\alpha(c, d, p) \leq \alpha(c', d', p)$, $\omega(c, d, p) \leq \omega(c', d', p)$, $N(c', d') \leq N(c, d)$, and $M(c, d) \leq M(c', d')$.

The above definition is extended to functions on several variables. For $f : I_1 \times I_2 \to \mathbb{R}$ of two variables case, an approximating map h_f is of $(I_1 \cap \mathbb{Q}) \times (I_2 \cap \mathbb{Q}) \times \mathbb{N} \to \mathbb{Q}$, and the modulus of uniform continuity ω_f is to satisfy $|a - a'|, |b - b'| \leq 2^{-\omega_f(p)+1} \to |h_f(a, b, n) - h_f(a, b, n)| \leq 2^{-p}$ for $n \geq \alpha_f(p)$. In the rest of this chapter we can simply say "continuous function" and "modulus of continuity," omitting "uniformly" and "uniform."

Definition 27.14 (Application) Let a continuous function $f := \langle h, \alpha, \omega, \gamma, \delta \rangle$, a Cauchy real $x := \langle (a_n)_n, M \rangle$, and an interval $[c, d]$ be given, such that $[c, d] \cap \mathbb{Q} \subseteq X$ and $c \leq a_n \leq d$ for any n. The application $f(x)$ of f to x is defined to be the sequence $(h(a_n, n))_n$ with modulus $\max(\alpha(c, d, p + 2), M(\omega(c, d, p + 1) - 1))$.

Lemma 27.15 (ContAppCorr) *The application yields a Cauchy real.*

Proof For a given p, assume $n, m \geq \max(\alpha(c, d, p + 2), M(\omega(c, d, p + 1) - 1))$

and prove $|h(a_n, n)-h(a_m, m)| \leq 2^{-p}$. We can omit c and d. Assume $l \geq \alpha(p+2)$. Whatever the l is, $|h(a_n, n) - h(a_m, m)| \leq |h(a_n, n) - h(a_n, l)| + |h(a_n, l) - h(a_m, l)| + |h(a_m, l) - h(a_m, m)|$ holds. The first and the third terms are less than $2^{-(p+2)}$ because $n, m, l \geq \alpha(p+2)$. The second one is less than $2^{-(p+1)}$ because $l \geq \alpha(p+1)$ and $|a_n - a_m| \leq 2^{-\omega(p+1)+1}$ due to $n, m \geq M(\omega(p+1)-1)$. □

Definition 27.16 (Composition) Let $f : X \to \mathbb{R}$ and $g : Y \to \mathbb{R}$ be continuous functions, where $f = \langle h_f, \alpha_f, \omega_f, \gamma_f, \delta_f \rangle$ and $g = \langle h_g, \alpha_g, \omega_g, \gamma_g, \delta_g \rangle$. Assume that for $c \in \mathbb{Q}$ and $d \in \mathbb{Q}^+$ with $[c, d] \cap \mathbb{Q} \subseteq X$ we have $[\gamma_{cd}, \delta_{cd}] \subseteq Y$ where $\gamma_{cd} := \gamma_f(c, d)$ and $\delta_{cd} := \delta_f(c, d)$. The function $g \circ f : [c, d] \to \mathbb{R}$ is defined by

$$\begin{aligned}
&h_{g\circ f}(a, n) := h_g(h_f(a, n), n),\\
&\alpha_{g\circ f}(c, d, p) := \max(\alpha_g(\gamma_{cd}, \delta_{cd}, p+2), \alpha_f(c, d, \omega_g(\gamma_{cd}, \delta_{cd}, p+1) - 1)),\\
&\omega_{g\circ f}(c, d, p) := \omega_f(c, d, \omega_g(\gamma_{cd}, \delta_{cd}, p) - 1) + 1,\\
&\gamma_{g\circ f} := \gamma_g(\gamma_{cd}, \delta_{cd}), \qquad \delta_{g\circ f} := \delta_g(\gamma_{cd}, \delta_{cd}).
\end{aligned}$$

Lemma 27.17 (ContComposeCorr) *The composition is correctly defined.*

Proof For any $a \in [c, d] \cap \mathbb{Q}$, $\gamma_{cd} \leq h_f(a, n) \leq \delta_{cd}$, then Lemma 27.15 implies the Cauchyness of $\langle (h_{g\circ f}(a, n))_n, \alpha_{g\circ f} \rangle$. Next, we assume $c \leq a \leq d$, $c \leq b \leq d$, and $|a - b| \leq 2^{-\omega_{g\circ f}(c,d,p)}$, then $|h_f(a, n) - h_f(b, n)| \leq 2^{-\omega_g(\gamma_{cd},\delta_{cd},p)+1}$ for $n \geq \alpha_f(c, d, \omega_g(\gamma_{cd}, \delta_{cd}, p) - 1)$, and hence $|h_g(h_f(a, n), n) - h_g(h_f(b, n), n)| \leq 2^{-p}$ for $n \geq \alpha_g(\gamma_{cd}, \delta_{cd}, p)$. Finally, as $\gamma_{cd} \leq h_f(a, n) \leq \delta_{cd}$, we have $\gamma_g(\gamma_{cd}, \delta_{cd}) \leq h_g(h_f(a, n), n) \leq \delta_g(\gamma_{cd}, \delta_{cd})$. It is obvious that $(g \circ f)(x) = g(f(x))$. □

We define the uniform convergence of function sequences and show some properties.

Definition 27.18 (Uniform convergence) Let $f_n, f : [a, b] \to \mathbb{R}$ be continuous and their approximating maps and Cauchy moduli h_n, h and α_n, α, respectively. The sequence $(f_n)_n$ is uniformly convergent to f if there is a modulus δ of uniform convergence and $\forall_p \forall_{n\geq\delta(p)} \forall_{a\in I}(|h_n(a, \alpha_n(p+2)) - h(a, \alpha(p+2))| \leq 2^{-p})$.

Lemma 27.19 (UnifConvChar) *Let $f_n, f : I \to \mathbb{R}$ be continuous with approximating map h_n, h. The sequence $(f_n)_n$ is uniformly convergent to f if and only if $\forall_p \exists_{q_1} \forall_{n\geq q_1} \exists_{q_2} \forall_{k\geq q_2} \forall_{a\in I}(|h_n(a, k) - h(a, k)| \leq 2^{-p})$.*

Proof Assume the uniform convergence. Let p be given and feed $p + 1$ to the assumption, and let q_1 be $\delta(p+1)$ and for $n \geq q_1$, q_2 be $\max(\alpha_n(p+2), \alpha(p+2))$. Then, $|h_n(a, k) - h(a, k)| \leq |h_n(a, k) - h_n(a, \alpha_n(p+3))| + |h_n(a, \alpha_n(p+3)) - h(a, \alpha(p+3))| + |h(a, \alpha(p+3)) - h(a, k)| \leq 2^{-(p+3)} + 2^{-(p+1)} + 2^{-(p+3)} \leq 2^{-p}$. To prove the other direction, we assume p, $n \geq \delta(p)$, and $a \in I$. For $p + 1$, the premise gives q_1 and q_2 and satisfies $|h_n(a, k) - h(a, k)| \leq 2^{-(p+1)}$. Then,

$|h_n(a, \alpha_n(p+2)) - h(a, \alpha(p+2))| \leq |h_n(a, \alpha_n(p+2)) - h_n(a, k)| + |h_n(a, k) - h(a, k)| + |h(a, k) - h(a, \alpha(p+2))| \leq 2^{-(p+2)} + \frac{1}{p+1} + \frac{1}{p+2} \leq 2^{-p}$. □

Lemma 27.20 (UnifConvLim) *Let $f_n : I \to \mathbb{R}$ be continuous functions given by approximating functions h_n and moduli α of Cauchyness and ω of uniform continuity, where the moduli are independent of n. If there is a weakly increasing modulus δ of uniform convergence satisfying $|h_n(a, k) - h_m(a, k)| \leq 2^{-p}$ for $n, m \geq \delta(p)$, $k \geq \alpha(p)$, and all $a \in I$, then $(f_n)_n$ uniformly converges to $f : I \to \mathbb{R}$ given by $h_f(a, k) := h_k(a, k)$, $\alpha_f(p) := \max(\delta(p+1), \alpha(p+1))$, and $\omega_f := \omega$.*

Proof It is true that h_f, α_f, and ω_f can form a uniformly continuous function. To prove the uniform convergence, assume p and let $q := \max(\delta(p), \alpha(p))$. Then for $n, k \geq q$, $|h_n(a, k) - h_k(a, k)| \leq 2^{-p}$ holds, so we use Lemma 27.19. □

The last part of this section is on the derivation, the integration, and the fundamental theorem of calculus.

Definition 27.21 (Derivative) Let f, g of $I \to \mathbb{R}$ be uniformly continuous, then g is a *derivative* of f with a modulus of differentiability δ_f, if for $x, y \in I$ with $x < y$, $y \leq x + 2^{-\delta_f(p)} \to |f(y) - f(x) - g(x)(y - x)| \leq (y - x)2^{-p}$.

Definition 27.22 (Partition and mesh) A list $a_0, a_1, \ldots, a_n$ of rational numbers is a *partition* if $a_i \leq a_{i+1}$ for all $0 \leq i < n$. The *mesh* of the partition P is defined to be $\max\{a_{i+1} - a_i \mid 0 \leq i < n\}$. A partition $Q = b_0, b_1, \ldots, b_m$ is a *refinement* of P if $\forall_{i \geq n} \exists_{j \leq m} a'_j = a_i$. We say $P = a_0, a_1, \ldots, a_n$ is a partition of $[a, b]$ if $a = a_0$ and $b = a_n$.

Definition 27.23 (Riemannian sum) Assume $f : [a, b] \to \mathbb{R}$ is a uniformly continuous function given by h_f, α_f, and ω_f, and P is a partition $a_0, a_1, \ldots, a_n$ of $[a, b]$. We define the *Riemannian sum* $S(f, P)$ to be $\sum_{i=0}^{n-1} h_f(e_i, n) \cdot (a_{i+1} - a_i)$, assuming e_i is some rational in $[a_i, a_{i+1}]$. In particular for $a_i = a + \frac{i}{n}(b - a)$, an instance of Riemannian sum $S(f, a, b, n)$ is defined to be $\frac{b-a}{n} \sum_{i=0}^{n-1} h_f(a_i, n)$.

We define the definite integral as the limit of $S(f, a, b, n)$ for $n \to \infty$, and prove that it is a Cauchy real.

Definition 27.24 (Definite integral) Let $f : [a, b] \to \mathbb{R}$ be a continuous function with the moduli ω_f of uniform continuity and α_f of Cauchyness. We denote $\langle (S(f, a, b, n))_n, M \rangle$ by $\int_a^b f(x)dx$, where the modulus $M(p)$ is defined to be an integer at least $\max(2^{\omega_f(p+q+2)}(b-a), \alpha_f(p+q+2))$ and q is such that $b - a \leq 2^q$.

Theorem 27.25 *Assume $f : [a, b] \to \mathbb{R}$ is a continuous function, then $\int_a^b f(x)dx$ is a Cauchy real. Moreover, if $P = a_0, a_1, \ldots, a_n$ with $|a_{i+1} - a_i| \leq 2^{-\omega_f(p)}$, $a_i \leq a_{i+1}$, $a = a_0$, and $b = a_n$, then $|S(f, P) - \int_a^b f(x)dx| \leq (b - a)2^{-p}$*

Proof Given k and l, let partitions $P = a_0, a_1, \dots, a_n$ and $Q = b_0, b_1, \dots, b_m$ of $[a, b]$ be such that the mesh of $P \le 2^{-\omega_f(k+2)}$ and the one of $Q \le 2^{-\omega_f(l+2)}$. Let $R = c_0, c_1, \dots, c_r$ be a common refinement of P and Q. Then, we can derive $|S(f, P) - S(f, R)| \le 2^{-(k+1)}(b - a)$ (here we need $n \ge \alpha_f(k + 2)$). Similarly, $|S(f, R) - S(f, Q)| \le 2^{-(l+1)}(b - a)$. Hence $|S(f, P) - S(f, Q)| \le (2^{-(k+1)} + 2^{-(l+1)})(b-a)$. For a given p, assume $n, m \ge \max(2^{\omega_f(p+q+2)}(b-a), \alpha_f(p+q+2))$ where $b-a \le 2^q$. The mesh is bounded by $2^{-\omega_f(p+q+2)}$, hence $|S(f, a, b, n) - S(f, a, b, m)| \le 2^{-(p+q)}(b - a) \le 2^{-p}$ holds in particular. We further assume for a given l, $n \ge \max(2^{\omega_f(l+2)}(b - a), \alpha_f(l + 2))$, then $|S(f, P) - S(f, a, b, n)| \le (2^{-p} + 2^{-l})(b - a)$. Let $n \to \infty$ and $p \to \infty$ to obtain $|S(f, P) - \int_a^b f(t)dt| \le (b - a)2^{-l}$. □

The interaction between the integration and the limit is shown as follows.

Theorem 27.26 (IntLimit) *Let $f_n, f : [a, b] \to \mathbb{R}$ be continuous, and assume the moduli of Cauchyness and of uniform continuity for f_n are independent of n. If the sequence $(f_n)_n$ is uniformly convergent to f, $\lim_{n\to\infty} \int_a^b f_n(t)dt = \int_a^b f(t)dt$.*

Proof Let $a_{nk} := S(f_n, k)$ and $a_k := S(f, k)$, and further let $x_n := \int_a^b f_n(t)dt = (a_{nk})_k$ and $x := \int_a^b f(t)dt = (a_k)_k$. We prove that $|x_n - x| \le 2^{-p}$ if $n \ge M(p)$. For any k, $|x_n - x| \le |x_n - a_{nk}| + |a_{nk} - a_k| + |a_k - x|$ holds. According to the definition, the Cauchy modulus of $\int_a^b f_n(t)dt$ is $M_n(p) = \max(2^{\omega_{f_n}(p+q+1)}(b - a), \alpha_{f_n}(p + q + 2))$, where q satisfies $b - a \le 2^q$. Since α_{f_n} and ω_{f_n}a are both independent from n, we let $M(p) := M_n(p)$. For the first term, $|x_n - a_{nk}| \le 2^{-(p+2)}$ for $n \ge M(p + 2)$ by Lemma 27.10. For the second term, $|a_{nk} - a_k| \le 2^{-(p+1)}$, by definition $|a_{nk} - a_k| \le \frac{b-a}{k} \sum_{i=0}^{k-1} |h_n(a_i, k) - h(a_i, k)|$. Assume l with $b - a \le 2^l$, then by Lemma 27.19, there exists q_1, and assuming $n \ge q_1$ there exists q_2 (depending on n) such that for $k \ge q_2$, $|h_n(a_i, k) - h(a_i, k)| \le 2^{-(p+1+l)}$. Hence, $|a_{nk} - a_k| \le \frac{b-a}{k} \sum_{i=0}^{k-1} 2^{-(p+1+l)} \le (b - a)2^{-(p+1+l)} \le 2^{-(p+1)}$. For the third term, assume α' is a Cauchy modulus of $(a_k)_k$ and then by Lemma 27.10, $|a_k - x| \le 2^{-(p+2)}$ for $k \ge \alpha'(p + 2)$. We can safely assume $n \ge q_1$ and $k \ge \max(M(p + 2), \alpha'(p + 2), q_2)$, therefore $|x_n - x| \le 2^{-p}$. □

The undetermined integral follows the above-defined definite integral on rational ends. Assume $a_0 < c < b_0$ and a continuous function $f : [a_0, b_0] \to \mathbb{R}$. An undetermined integral of f is denoted by $\int_c^x f(t)dt =: F(x)$ and is defined to be $\langle h_F, \alpha_F, \omega_F, N_F, M_F\rangle$, where $h_F(a, n) := S(f, c, a, n)$, $\alpha_F(k) := \max(\alpha_f(0), 2^{\omega_f(k+1)})$, $\omega_F(k) := \max(p+k, \omega_f(k+1))$ for p with $h_f(b, n) \le 2^p$, $N_F := (b_0 - a_0)N_f$, and $M_F := (b_0 - a_0)M_f$. The uniform continuity is shown as follows. Assume $c < a < b$ without loss of generality, and let $a_i := c+\frac{i}{n}(a-c)$ and $b_i := c + \frac{i}{n}(b - c)$, where $[c, a]$ and $[c, b]$ are split into n pieces. Then, we can show that $|h_F(a,n) - h_F(b,n)| \le 2^{-p}$ for p with $h_f(b_i,n) \le 2^p$, provided $|a_i - b_i| \le 2^{-\omega_f(k+1)+1}$ and $|a-b| \le 2^{-p-k-1}$ as follows:

$$\begin{aligned}
|h_F(a,n) - h_F(b,n)| &= |S(f,c,a,n) - S(f,c,b,n)| \\
&= \frac{1}{n}|(a-c)\sum_{i=0}^{n-1} h_f(a_i,n) - (b-c)\sum_{i=0}^{n-1} h_f(b_i,n)| \\
&\le \frac{1}{n}\sum_{i=0}^{n-1}((a-c)|h_f(a_i,n) - h_f(b_i,n)| + |a-b|\cdot|h_f(b_i,n)|) \\
&\le \frac{1}{n}\sum_{i=0}^{n-1}((a-c)2^{k+1} + |a-b|2^p) \le 2^{-p}.
\end{aligned}$$

The fundamental theorem of calculus holds in our setting, too.

Lemma 27.27 *Let $f : [a,b] \to \mathbb{R}$ be continuous with modulus ω_f of continuity. For an arbitrarily given $c \in [a,b]$, there is a uniformly continuous function which coincides $\int_c^x f(t)dt$. This function, say $F(x)$, has f as a derivative, and $F - G$ is a constant if G is a function with $G' = f$,*

Proof The function f is a derivative of F, because $|F(y) - F(x) - f(x)(y-x)| = |\int_c^y f(t)dt - \int_c^x f(t)dt - f(x)(y-x)| \le \int_x^y |f(t) - f(x)|dt \le \int_x^y 2^{-p}dt \le (y-x)2^{-p}$ for $y \le x + 2^{\omega_f(k)}$. Assume G is a differentiable function on $[a,b]$ such that $G' = f$, then $(F-G)' = F' - G' = f - f = 0$, hence $F - G$ is a constant. □

Theorem 27.28 (Fundamental theorem of calculus) *Let $f : I \to \mathbb{R}$ be continuous and $F : I \to \mathbb{R}$ such that $F' = f$. Then for all $a, b \in I$, $\int_a^b f(t)dt = F(b) - F(a)$.*

Proof For $x \in I$ we define $F_0(x) = \int_a^x f(t)dt$. We have $F_0' = f$ by Lemma 27.27. Then, $F_0(a) = 0$ and $F_0(b) = \int_a^b f(t)dt$ hold, and hence for any $F : I \to \mathbb{R}$ with $F' = f$, the function $F - F_0$ is a constant. Therefore, $F(b) - F(a) = F_0(b) - F_0(a) = F_0(b) = \int_a^b f(t)dt$. □

27.2.2 Corecursion and Coinduction

This section describes (structural) corecursion and (structural) coinduction through examples. We start from the (structural) recursion operator and (structural) induction which we clearly know. The recursion operator is to iteratively inspect an object of ground type, while such an object is formed by constructors. As in Gödel's T, assume the constructors 0 and S for $\mathbf{N}$, the type of natural numbers, then the recursion operator $\mathcal{R}_{\mathbf{N}}^{\rho}$ from a natural number to ρ is defined as follows:

$$\mathcal{R}^{\rho}_{\mathbf{N}} : \mathbf{N} \to \rho \to (\mathbf{N} \to \rho \to \rho) \to \rho$$
$$\mathcal{R}^{\rho}_{\mathbf{N}}\, 0\, MN = M, \qquad \mathcal{R}^{\rho}_{\mathbf{N}}\, (S\, n)\, MN = Nn(\mathcal{R}^{\rho}_{\mathbf{N}}\, n\, MN).$$

We assume the constructors Nil and $::$ for $\mathbf{L}_\alpha$, the type of lists of objects of α, where the type α may be instantiated by any type. The recursion $\mathcal{R}^{\rho}_{\mathbf{L}_\alpha}$, which iteratively inspects an object of $\mathbf{L}_\alpha$ and outputs an object of an arbitrary type ρ, is given as follows:

$$\mathcal{R}^{\rho}_{\mathbf{L}_\alpha} : \mathbf{L}_\alpha \to \rho \to (\alpha \to \mathbf{L}_\alpha \to \rho \to \rho) \to \rho$$
$$\mathcal{R}^{\rho}_{\mathbf{L}_\alpha}\, \mathsf{Nil}\, M\, N = M, \qquad \mathcal{R}^{\rho}_{\mathbf{L}_\alpha}\, (x{::}xs)\, M\, N = N\, x\, (\mathcal{R}^{\rho}_{\mathbf{L}_\alpha}\, xs\, M\, N).$$

The primitive corecursion operator is to form a possibly infinite object of ground type. It takes two arguments, and determines how the outermost construction of the expression ${}^{\mathrm{co}}\mathcal{R}^{\rho}_{\mathbf{L}_\alpha} NM$ should look. The following is the case for $\mathbf{N}$:

$${}^{\mathrm{co}}\mathcal{R}^{\rho}_{\mathbf{N}} : \rho \to (\rho \to \mathbf{U} + (\mathbf{N} + \rho)) \to \mathbf{N}$$
$${}^{\mathrm{co}}\mathcal{R}^{\rho}_{\mathbf{N}} NM = [\mathbf{case}\ (MN)\ \mathbf{of}\ \mathsf{Inl}\, \cdot .0; \mathsf{Inr}\, x.S([\mathbf{case}\ x\ \mathbf{of}\ \mathsf{Inl}\, n.n; \mathsf{Inr}\, y.{}^{\mathrm{co}}\mathcal{R} y M])],$$

where $\mathbf{U}$ is an algebra which consists of the nullary constructor $\cdot$ and $\alpha + \beta$ is a sum type with the constructors $\mathsf{Inl} : \alpha \to \alpha + \beta$ and $\mathsf{Inr} : \beta \to \alpha + \beta$. The second argument M decides for the given N the outermost constructor is either 0 or S. If 0 is the case, the process of corecursion terminates. If S is the case, this process terminates with some n or it invokes another corecursion, feeding an updated y and the term M as the original one. The next one is the case for lists:

$${}^{\mathrm{co}}\mathcal{R}^{\rho}_{\mathbf{L}_\alpha} : \rho \to (\rho \to \mathbf{U} + (\alpha \times (\mathbf{L}_\alpha + \rho))) \to \mathbf{L}_\alpha$$
$${}^{\mathrm{co}}\mathcal{R}^{\rho}_{\mathbf{L}_\alpha} NM =$$
$$[\mathbf{case}\ (MN)\ \mathbf{of}\ \mathsf{Inl}\, \cdot .\mathsf{Nil}; \mathsf{Inr}\, x.\pi_0(x){::}[\mathbf{case}\ \pi_1(x)\ \mathbf{of}\ \mathsf{Inl}\, as.as; \mathsf{Inr}\, y.{}^{\mathrm{co}}\mathcal{R}^{\rho}_{\mathbf{L}_\alpha} y M]],$$

where $\alpha \times \beta$ is a product type and π_i for $i \in \{0, 1\}$ is the projection of type $\alpha_0 \times \alpha_1 \to \alpha_i$. The second argument M determines whether the whole term is of the form Nil or of the form $a :: as$. If Nil is the case, the process terminates. Otherwise, it fixes a for the head, and further determines either the tail as is given or invokes another corecursion with an updated input y. Apparently the above-mentioned processes are not necessarily terminating, hence corecursion can form an infinite object. We additionally consider the destructor on each ground type, which is a mean of inspection:

$$\mathcal{D}_{\mathbf{N}} : \mathbf{N} \to \mathbf{U} + \mathbf{N}, \qquad \mathcal{D}_{\mathbf{N}}\, 0 = \mathsf{Inl}\, \cdot, \qquad \mathcal{D}_{\mathbf{N}}(Sn) = \mathsf{Inr}\, n,$$
$$\mathcal{D}_{\mathbf{L}_\alpha} : \mathbf{L}_\alpha \to \mathbf{U} + \alpha \times \mathbf{L}_\alpha, \qquad \mathcal{D}_{\mathbf{L}_\alpha}\, \mathsf{Nil} = \mathsf{Inl}\, \cdot, \qquad \mathcal{D}_{\mathbf{L}_\alpha}(a :: as) = \mathsf{Inr}\langle a, as\rangle.$$

This makes our realizability logic look coherent, as the destructor and the corecursion operator realize the closure axiom and the greatest-fixed-point axiom, respectively, while the constructors and the recursion operator realize the introduction axioms and the least-fixed-point axiom, respectively. For example, ${}^{\mathrm{co}}\mathcal{R}^{\mathbf{N}}_{\mathbf{L_N}}\,0\,\lambda_n$ $(\mathsf{Inr}\langle n, \mathsf{Inr}(n+2)\rangle)$ defines the infinite ascending list of even numbers from 0. The same is given by $l(0)$, where $l^{\mathbf{N}\to\mathbf{L_N}}$ is equationally defined by $l(n) = n :: l(n+2)$. While the corecurion operator is necessary in order to carry out program extraction, the equational definition has an advantage in the readability.

The proof principle of induction can be set-theoretically illustrated through a monotone operator and its least-fixed-point. An operator $\Psi : 2^A \to 2^A$ is monotone if $X \subseteq Y$ implies $\Psi(X) \subseteq \Psi(Y)$ for any $X, Y \subseteq A$. As is known, a monotone operator Ψ has a least-fixed-point $\mu\Psi := \bigcap\{Y \mid Y \subseteq A \wedge \Psi(Y) \subseteq Y\}$. Then, $\Psi(\mu\Psi) \subseteq \mu\Psi$ holds and also $\Psi(\mu\Psi \cap X) \subseteq X$ (or a weakened one $\Psi(X) \subseteq X$) implies $\mu\Psi \subseteq X$ holds for any $X \subseteq A$, which are respectively called an introduction principle and an induction principle. In order to make our theory extendable, let us define an inductive predicate as a least-fixed-point of a formula $\Phi(X)$ where the occurrences of a predicate X are only strictly positive, and accept the corresponding principles $\forall_x(\Phi(X)x \to Xx)$ as the introduction axioms and $\forall_x(\mu\Phi x \to \forall_x(\Phi(X)x \to Xx) \to Xx)$ as the least-fixed-point axiom. The following is an inductive definition of a property "even" of natural numbers, taking $\Phi(X) := \{n \mid n = 0 \vee \exists_m(Xm \wedge n = m+2)\}$ and $\mathrm{Ev} := \nu\Phi$,

$$\mathrm{Ev}\,0, \qquad \forall_n(\mathrm{Ev}\,n \to \mathrm{Ev}(n+2)),$$
$$\forall_n(\mathrm{Ev}\,n \to Pn \to \forall_n(\mathrm{Ev}\,n \to Pn \to P(n+2)) \to Pn),$$

where P is an arbitrary predicate of the arity $(\mathbf{N})$ of Ev. The first two are the introduction axioms, and the third one is the least-fixed-point axiom or simply so-called induction. The least-fixed-point axiom is viewed as an inspector, such that it analyses how the introduction axioms have been used to derive the first premise $\mathrm{Ev}\,n$. Once introduction axioms are given, the formula of the least-fixed-point axiom is determined.

It is also known that the monotone operator Ψ has a greatest-fixed-point $\nu\Psi := \bigcup\{Y \mid Y \subseteq A \wedge Y \subseteq \Psi(Y)\}$, then $\nu\Psi \subseteq \Psi(\nu\Psi)$ and $X \subseteq \Psi(\nu\Psi \cup X)$ (or a weaker one $X \subseteq \Psi(X)$) implies $X \subseteq \nu\Psi$ for any $X \subseteq A$ hold, which are called the closure principle and the coinduction principle, respectively. We define a coinductive predicate as a greatest-fixed-point of a formula $\Phi(X)$, where a predicate X has only strictly positive occurrences in $\Phi(X)$, and accept the corresponding closure axiom and greatest-fixed-ponit axiom to extend our theory. Coinductively defining a predicate ${}^{\mathrm{co}}\mathrm{Ev} := \nu\Phi$ by means of the above Φ, we get the following axioms:

$$\forall_n({}^{\text{co}}\text{Ev}\, n \to n = 0 \vee \exists_m({}^{\text{co}}\text{Ev}\, m \wedge n = m + 2)),$$
$$\forall_n(P\, n \to \forall_n(P\, n \to n = 0 \vee \exists_m(({}^{\text{co}}\text{Ev}\, m \vee P\, m) \wedge n = m + 2)) \to {}^{\text{co}}\text{Ev}\, n).$$

The first one is *closure axiom* which is to inspect the derivation of ${}^{\text{co}}\text{Ev}\, n$ so that it tells us either n is zero or is of $m + 2$ for some m in ${}^{\text{co}}\text{Ev}$. The second one is the *greatest-fixed-point axiom* or simply *coinduction*. This derives a formula of the form ${}^{\text{co}}\text{Ev}\, n$ from an arbitrary $P\, n$, if it is guaranteed that $P\, n$ implies either $n = 0$ or n is of the form $m + 2$ for some m which is again in P, as the second premise of the axiom formula expresses. Once we give the closure axiom, the formula of the greatest-fixed-point axiom is determined. Coinductive definitions come from the way of observation, namely, a corresponding closure axiom, while inductive definitions come from the way of construction. This oppositeness is taken as a duality. The other example is possibly infinite lists of even numbers:

$$\forall_n({}^{\text{co}}\text{L}_{\text{Ev}}\, ns \to ns = \mathsf{Nil} \vee \exists_{m,ms}(\text{Ev}\, m \wedge {}^{\text{co}}\text{L}_{\text{Ev}} ms \wedge ns = m{::}ms)),$$
$$\forall_{ns}(P\, ns \to$$
$$\quad \forall_{ns}(P\, ns \to ns = \mathsf{Nil} \vee \exists_{m,ms}(\text{Ev}\, m \wedge (P\, ms \vee {}^{\text{co}}\text{L}_{\text{Ev}}\, ms) \wedge ns = m{::}ms)) \to$$
$$\quad {}^{\text{co}}\text{L}_{\text{Ev}}\, ns).$$

In informal arguments, we often use the weaker axioms for simplicity.

In the theory TCF, the introduction, least-fixed-point, closure, and greatest-fixed-point axioms are accommodated in the same framework of program extraction [29].

27.3 Applications

This section describes three applications of Bishop's constructive analysis in exact real arithmetic. The first one is a root-finding algorithm through the intermediate value theorem. Assuming an additional constraint on functions, we also derive an existence of an inverse function. The second one is an ordinary equation solver via the Cauchy–Euler approximation method. The solver comes from the existence proof of the exact solution, and on the other hand, the properties such as the uniqueness and the error estimation are shown. Those arguments are carried out in our constructive setting due to the fact that classical logic is a fragment of constructive logic. The third one is stream-based real numbers and its arithmetic. We make use of corecursion and coinduction to deal with algorithms on infinite data. These examples have been formalized in Minlog, and the results in Sections 27.3.1 and 27.3.3 are included in the latest Minlog package.

27.3.1 Root-Finding and Inverse Function

Intermediate value theorem is a basic result in (classical) analysis.

Theorem 27.29 (Intermediate value theorem, classical) *Let $I = [a, b]$ be a closed interval and $f : I \to \mathbb{R}$ a continuous function. For any u between $f(a)$ and $f(b)$, there is a $c \in I$ such that $f(c) = u$.*

As is known, the above theorem is shown to be constructively unacceptable due to the weak counter example argument [10, 31]. Constructive variants of the intermediate value theorem have been studied, that are classically equivalent to Theorem 27.29, but constructively not. In this section we study one of those variants which involves a clear arithmetic content in its proof. Then we consider a special case where we find a more efficient algorithm from the corresponding proof. As an application of such a variant of intermediate value theorem, we study the existence of inverse functions.

Theorem 27.30 (Approximate intermediate value theorem) *Let $f : I \to \mathbb{R}$ be a uniformly continuous function and a, b rational numbers in I such that $a < b$ and $f(a) \leq 0 \leq f(b)$, then for any p there exists c with $a < c < b$ such that $|f(c)| \leq 2^{-p}$.*

It is easily seen that the above statement expresses the approximate solution of the root-finding problem through the BHK-interpretation. In this section we start from constructively proving the above theorem, and then further discuss a way of finding a more efficient root-finding algorithm from proofs, posing a stronger assumption to the theorem statement.

Proof of Theorem 27.30 Let p be given and $\varepsilon := 2^{-p}$. We use Lemma 27.8 to determine whether $f(a) \leq -\frac{\varepsilon}{2}$ or $-\varepsilon \leq f(a)$ and also whether $\frac{\varepsilon}{2} \leq f(b)$ or $f(b) \leq \varepsilon$. If $-\varepsilon \leq f(a)$ or $f(b) \leq \varepsilon$, $|f(c)| < \varepsilon$ for $c = a$ or $c = b$, respectively. Otherwise, both $f(a) \leq -\frac{\varepsilon}{2}$ and $\frac{\varepsilon}{2} \leq f(b)$ hold. We use the modulus of uniform continuity of f, so that we find q such that if $|x - y| \leq 2^{-q}$ implies that $|f(x) - f(y)| \leq \varepsilon$ and form a partition of length m as $a = a_0 < a_1 < \cdots < a_m = b$ which divides $[a, b]$ with $|a_i - a_{i+1}| \leq 2^{-q}$ for $0 \leq i < m$. We compare each $f(a_i)$ with $-\frac{1}{\varepsilon}$ and $\frac{1}{\varepsilon}$ by Lemma 27.8 and find a minimal j such that $f(a_j) < -\frac{1}{\varepsilon}$ and $\frac{1}{\varepsilon} < f(a_j)$. We again use Lemma 27.8 to compare $f(a_j)$ with $-\varepsilon, -\frac{\varepsilon}{2}$ and $f(a_{j+1})$ with $\frac{\varepsilon}{2}, \varepsilon$. If $-\varepsilon \leq f(a_j)$, $|f(a_j)| \leq \varepsilon$. If $f(a_j) \leq \varepsilon$, $|f(a_{j+1})| \leq \varepsilon$. Otherwise, both $f(a_j) \leq -\frac{\varepsilon}{2}$ and $\frac{\varepsilon}{2} \leq f(a_{j+1})$ hold, hence $|f(a_{j+1}) - f(a_j)| > \varepsilon$, which contradicts the assumption $|a_j - a_{j+1}| \leq 2^{-q}$. □

Here we also prove the following strengthened form of the approximate intermediate value theorem. This finds the last approximate zero of the given function and we make use of it in Section 27.3.2 on ordinary differential equations.

Theorem 27.31 (LastApproxZero) *Let $f : I \to \mathbb{R}$ be continuous and $a < b$ rational numbers in I such that $f(a) \le 0 \le f(b)$. Then for every p we can find c with $a < c < b$ such that $f(c) \le 2^{-p}$ and $0 \le f(z)$ for all $z \in [c, b]$.*

The computational content of the above proof of Theorem 27.30 is not that efficient, since depending on the modulus of uniform continuity, it compares each $f(a_i)$ with $-\frac{\varepsilon}{2}$ and $\frac{\varepsilon}{2}$; much of those computations would not contribute to the final result. We assume the following special modulus, so that we find an improved computational content from a proof.

Definition 27.32 (Uniform modulus of increase) For a given uniformly continuous function f, q is its *uniform modulus of increase* if for all $c, d \in \mathbb{Q}$ and for all $p \in \mathbb{Z}^+$, $2^{-p} \le d - c \to 2^{-(p+q)} \le f(d) - f(c)$.

Lemma 27.33 (IVTAux) *Assume $f : I \to \mathbb{R}$ is a uniformly continuous function with a uniform modulus q of increase, a, b, c, d are rational numbers in I such that $a \le c < d \le b$, $2^{-p} < d - c$, and $f(c) \le 0 \le f(d)$. Then there exist c_1, d_1 with $d_1 - c_1 = \frac{2}{3}(d - c)$, such that $a \le c \le c_1 < d_1 \le d \le b$ and $f(c_1) \le 0 \le f(d_1)$.*

Proof Let $c_0 = \frac{2c+d}{3}$ and $d_0 = \frac{c+2d}{3}$ which split the interval $[c, d]$ into three. As $2^{-p} < d - c$ implies $2^{-(p+2)} < d_0 - c_0$, so $f(c_0) <_{p+2+q} f(d_0)$. Use Lemma 27.8 to compare 0 with $f(c_0)$ and $f(d_0)$. If $0 \le f(d_0)$, let $c_1 = c$ and $d_1 = d_0$. Otherwise, $f(c_0) \le 0$ holds and let $c_1 = c_0$ and $d_1 = d$. □

Theorem 27.34 (IVT) *Let $f : I \to \mathbb{R}$ be continuous with a uniform modulus of increase. Further assume a, b with $a < b$ are rational numbers in I such that $f(a) \le 0 \le f(b)$. Then there exists $x \in [a, b]$ such that $f(x) = 0$.*

Proof We construct two sequences $(c_n)_n$ and $(d_n)_n$ by iterating Lemma 27.33 such that:

$$\begin{aligned} &a = c_0 \le c_1 \le \cdots \le c_n < d_n \le \cdots \le d_1 \le d_0 = b, \\ &f(c_n) \le 0 \le f(d_n), \\ &d_n - c_n = \left(\frac{2}{3}\right)^n (b - a). \end{aligned}$$

Define Cauchy reals x and y by the sequences $(c_n)_n$ and $(d_n)_n$, supplying Cauchy moduli, respectively. As f is continuous and $x = y$, $f(x) = 0 = f(y)$. □

The following theorem is an application of Theorem 27.34.

Theorem 27.35 (Inv) *Assume $f : I \to \mathbb{R}$ is a uniformly continuous function with a uniform modulus q of increase and a, b are rational numbers in I such that $a \le b$ and $f(a) < f(b)$. Then there exists a uniformly continuous function g :*

$(f(a), f(b)) \to \mathbb{R}$ such that $f(g(y)) = y$ for all $y \in (f(a), f(b))$ and $g(f(x)) = x$ for all $x \in [a, b]$ such that $f(a) \le f(x) \le f(b)$.

Proof Let $q \in \mathbb{Z}^+$ be the assumed modulus of increase, such that for all $c, d \in [a, b]$ and $p \in \mathbb{Z}^+$, $2^{-p} \le d - c \to 2^{-(p+q)} \le f(d) - d(c)$. We construct a continuous $g : (f(a), f(b)) \to \mathbb{R}$. Assume $u, a', b' \in \mathbb{Q}$ with $u \in (a', b') \subseteq (f(a), f(b))$, then $f(a) - u \le a' - u \le 0$ and $0 \le b' - u \le f(b) - u$ hold. By Theorem 27.34, there exists an x such that $f(x) - u = 0$. Let $h_g(u, n)$, the approximate mapping of g, be the Cauchy sequence of x, say $(c_n)_n$. We define the Cauchy modulus α_g such that for $n \ge \alpha_g(p)$, $(\frac{2}{3})^n(b - a) \le 2^{-\omega_f(p+q+2)}$. Assume u, v, p are rationals such that $a' \le u < v \le b'$ and $p \in \mathbb{Z}^+$, and let $\omega_g(p)$ be $p + q + 2$. This ω_g is indeed a modulus of uniform continuity, because we can show that for any $n \ge \alpha_g(p)$,

$$|u - v| \le \frac{1}{2^{\omega_g(p)+1}} \to |h_g(u, n) - h_g(v, n)| \le \frac{1}{2^p}.$$

For $c_n^{(u)} := h_g(u, n)$ and $c_n^{(v)} := h_g(v, n)$, assume $|c_n^{(u)} - c_n^{(v)}| > 2^{-p}$, then we must show $|u - v| > 2^{-\omega_g(p)-1}$. As in the proof of Theorem 27.34, $d_n^{(u)}$ is also constructed and $d_n^{(u)} - c_n^{(u)} \le (\frac{2}{3})^n(b - a) \le 2^{-\omega_f(p+q+2)}$ for $n \ge \alpha_g(p)$. Because $f(c_n^{(u)}) - u \le 0 \le f(d_n^{(u)}) - u$ and also ω_f is the modulus of uniform continuity of f, $|f(c_n^{(u)}) - u| \le |(f(d_n^{(u)}) - u) - (f(c_n^{(u)}) - u)| = |f(d_n^{(u)}) - f(c_n^{(u)})| \le 2^{-(p+q+2)}$, and similarly $|f(c_n^{(v)}) - v| \le 2^{-(p+q+2)}$. The assumption $|c_n^{(u)} - c_n^{(v)}| > 2^{-p}$ implies $|f(c_n^{(u)}) - f(c_n^{(v)})| \ge 2^{-(p+q)}$, and hence $|u - v| \ge |f(c_n^{(u)}) - f(c_n^{(v)})| - |f(c_n^{(u)}) - u| - |f(c_n^{(v)}) - v| \ge 2^{-(p+q+2)} > 2^{-\omega_g(p)-1}$. Now we prove that g is the inverse, namely, $f(g(u)) = u$. It suffices to show $|f(g(u)) - u| \le |f(g(u)) - f(h_g(u, n))| + |f(h_g(u, n)) - u| \le 2^{-p}$ for sufficiently big n. The first inequality is trivial. By Lemma 27.12, $|g(u) - h_g(u, n)| \le 2^{-\omega_f(p+1)}$, assuming $n \ge \alpha_g(\omega_f(p+1))$, then $|f(g(u)) - f(h_g(u, n))| \le 2^{-(p+1)}$ holds because ω_f is the modulus of continuity of f. On the other hand, we prove that $|f(h_g(u, n)) - u| \le 2^{-(p+1)}$ for $M(\omega_f(p))$, where $M(p) := \log_{\frac{2}{3}} 2^{-p(b-a)}$. As for the Cauchy sequences $(c_n)_n$ and $(d_n)_n$ we have $f(c_n) \le 0 \le f(d_n)$ and $d_n - c_n = (\frac{2}{3})^n(b - a)$, for $n \ge M(\omega_f(p + 1))$, we have $d_n - c_n \le 2^{-\omega_f(p+1)}$ and hence $f(d_n) - f(c_n) \le 2^{-(p+1)}$. Therefore, $|f(h_g(u, n)) - u| = |f(c_n) - u| = u - f(c_n) \le f(d_n) - f(c_n) \le 2^{-(p+1)}$, and the claim follows. Since continuous functions are determined by their values on the rationals, $f(g(y)) = y$ for $y \in [a', b']$. For any $x \in [a, b]$ with $a' \le f(x) \le b'$, from $g(f(x)) < x$ we obtain the contradiction $f(x) = f(g(f(x))) < f(x)$ due to the property of increase due to the modulus q. The same for $g(f(x)) > x$, hence $g(f(x)) = x$. □

27.3.2 Ordinary Differential Equation Solver

We study ordinary differential equations, ODEs for short, of the form

$$y' = f(x, y) \tag{27.1}$$

and the Cauchy–Euler approximation method which solves ODEs of the above sort.

Theorem 27.36 (Cauchy–Euler approximation) *Let $f : D \to \mathbb{R}$ be uniformly continuous and $(a_0, b_0) \in D$ such that $(x, y) \in R$, where R is some rectangle in D given by $|x - a_0| \le a$ and $|y - b_0| \le b$. If $|f(x, y)| \le M$ for $(x, y) \in R$, then for any $k \in \mathbb{Z}$ there exists an approximate solution $\phi^{[a_0-h,a_0+h]\to\mathbb{R}}$ of (27.1) up to the error 2^{-k} such that $\phi(a_0) = b_0$, where $h := \min(a, b/M)$.*

Proof Assume k is arbitrarily given. From the modulus of uniform continuity of f, there exists l such that for any $(x, y), (\tilde{x}, \tilde{y}) \in D$ with $|x - \tilde{x}| \le 2^{-l}$ and $|y - \tilde{y}| \le 2^{-l}$, $|f(x, y) - f(\tilde{x}, \tilde{y})| \le 2^{-(k+1)}$. We construct an approximate solution on $[a_0, a_0 + h]$ in the following way. We find $a_0, a_1, \ldots, a_n$ to divide the interval $[a_0, a_0 + h]$ with $a_0 < a_1 < \cdots < a_{n-1} < a_n = a_0 + h$ and $a_i - a_{i-1} \le \min(2^{-l}, \frac{1}{M} \cdot 2^{-l})$ for $0 < i \le n$. For $i = 1, \ldots, n$, we recursively define $\phi(x) = b_{i-1} + (x - a_{i-1})s_{i-1}$ for $a_{i-1} \le a_i$, $b_i = \phi(a_i)$, and $-M \le s_{i-1} \le M$ such that $|s_{i-1} - f(a_{i-1}, b_{i-1})| \le 2^{-(k+1)}$. This ϕ is admissible, continuous, and there are piecewise derivatives $\phi'(x) = s_{i-1}$ for $a_{i-1} < x < a_i$. Assume $a_{i-1} < x < a_i$, then we have $|x - a_{i-1}| \le 2^{-l}$ and $|\phi(x) - b_{i-1}| \le |x - a_{i-1}| \cdot |s_{i-1}| \le \frac{1}{M} \cdot 2^{-l} \cdot M = 2^{-l}$, hence $|f(a_{i-1}, b_{i-1}) - f(x, \phi(x))| \le 2^{-(k+1)}$. Hence $|\phi'(x) - f(x, \phi(x))| \le 2^{-k}$, and therefore ϕ is an approximate solution up to 2^{-k}. □

In the rest of this subsection, we additionally assume the following *Lipschitz condition* and study the unique and exact solutions with an error analysis between an approximate solution and the exact one.

Definition 27.37 (Lipschitz condition) A function $f : D \to \mathbb{R}$, where $D \subseteq \mathbb{R}^2$, satisfies the *Lipschitz condition on the second argument* if there exists some $L > 0$, and for any $(x, y_1), (x, y_2) \in D$, $|f(x, y_1) - f(x, y_2)| \le L|y_1 - y_2|$ holds.

Lemma 27.38 (LinDiffIneq) *Let $\sigma : [a, b] \to \mathbb{R}$ be continuous with a piecewise continuous derivative σ' such that $\sigma'(x) \le L\sigma(x) + \varepsilon$ for all $x \in [a, b]$ where σ' is defined. Then $\sigma(x) \le e^{L(x-a)}\sigma(a) + \frac{\varepsilon}{L}(e^{L(x-a)} - 1)$ for all $x \in [a, b]$.*

Proof We have $\int_a^x e^{-Lt}(\sigma'(t) - L\sigma(t))dt \le \int_a^x \varepsilon e^{-Lt}dt$, since $\sigma'(x) - L\sigma(x) \le \varepsilon$ by the assumption. Notice that the left integrant has finite discontinuities but its indefinite integral is continuous. Easy calculation leads us to the conclusion. □

Lemma 27.39 (LipDiffApprox) *Let $f : D \to \mathbb{R}$ be a continuous function with a Lipschitz condition w.r.t. its second argument for the constant $L > 0$. Let $\phi, \psi : [a, b] \to \mathbb{R}$ be approximate solutions up to the errors $2^{-p}, 2^{-l}$ of (27.1), then $\psi'(x) - \phi'(x) \leq L|\psi(x) - \phi(x)| + 2^{-p} + 2^{-l}$.*

Proof For all points except finitely many we have $|\phi'(x) - f(x, \phi(x))| \leq 2^{-p}$ and $|\psi'(x) - f(x, \psi(x))| \leq 2^{-l}$, hence $\psi'(x) - \phi'(x) \leq |f(x, \psi(x)) - f(x, \phi(x))| + \varepsilon \leq L|\psi(x) - \phi(x)| + \varepsilon$ for $\varepsilon := 2^{-p} + 2^{-l}$. □

Theorem 27.40 (Fundamental inequality) *Let $f : D \to \mathbb{R}$ be a continuous function with a Lipschitz condition w.r.t. its second argument for the constant $L > 0$. Let $\phi, \psi : [a, b] \to \mathbb{R}$ be approximate solutions up to the errors $2^{-p}, 2^{-l}$ of (27.1), then $|\psi(x) - \phi(x)| \leq e^{L(x-a)}|\psi(x) - \phi(x)| + \frac{2^{-p}+2^{-l}}{L}(e^{L(x-a)} - 1)$.*

As Lemma 27.7 says, the theorem statement is a stabled formula, namely, the double negation of it is (constructively) logically equivalent to the original. The formulas of the forms $\neg(\neg A \wedge \neg B)$ and $\neg\forall_x \neg A$ are abbreviated as $A \tilde{\vee} B$ and $\tilde{\exists}_x A$, where $\tilde{\vee}$ and $\tilde{\exists}$ are called the *weak disjunction* and *weak existential quantifier*, respectively [29]. For example, for a negative formula C, $(A \tilde{\vee} B) \to (A \to C) \to (B \to C) \to C$ holds, and for any formula A, $\neg A \tilde{\vee} A$ holds. In the proof below, everything is carried out within our constructive setting, due to the fact that classical logic is a fragment of constructive logic [30, 31].

Proof Let $\sigma := \phi - \psi$ and $x \in [a, b]$, and then assume $0 \leq \sigma(x)$ without loss of generality. As $\neg(\sigma(y) < 0)$ is equivalent to $0 \leq \sigma(y)$, $\forall_{y \in (a,x]}(0 \leq \sigma(y)) \tilde{\vee} \tilde{\exists}_{y \in [a,x]}(\sigma(y) < 0)$ vacantly holds. From this weak disjunction formula, we prove $\sigma(y) \leq e^{L(y-a)}\sigma(a) + \frac{\varepsilon}{L}(e^{L(y-a)} - 1)$ which is also a negative formula. *Case 1.* Assume $\forall_{y \in (a,x]}(0 \leq \sigma(y))$. Use Lemma 27.39 to derive the goal $\sigma'(x) \leq L\sigma(x) + \varepsilon$, then by Lemma 27.38, the goal formula holds. *Case 2.* Assume $\tilde{\exists}_{y \in [a,x]}(\sigma(y) < 0)$. There exists $y \in [a, x]$ such that $\sigma(y) < 0$. By Lemma 27.31, for an arbitrary k we can find $c \in [a, b]$ such that $\sigma(c) \leq 2^{-k}$, and also $0 \leq \sigma(z)$ for all $z \in [c, b]$. We can choose k large enough to satisfy $\sigma(c) \leq \frac{1}{e^{L(x-a)}}2^{-k}$. By Lemma 27.39 we have $\sigma'(y) \leq L\sigma(y) + \varepsilon$ on $[c, x]$. Further using Lemma 27.38 for $[c, x]$, $\sigma(y) \leq e^{L(y-a)}\sigma(a) + \frac{\varepsilon}{L}(e^{L(y-a)} - 1)$ holds for all $y \in [c, x]$, hence also for x. We obtain $\sigma(x) \leq 2^{-k} + \frac{\varepsilon}{L}(e^{L(x-a)} - 1)$. Since k is arbitrary, the required inequality follows. □

Finally, we study the unique and exact solution.

Theorem 27.41 (Uniqueness) *Assume $f : D \to \mathbb{R}$ is a uniformly continuous function with the Lipschitz condition. Let $\phi, \psi : I \to \mathbb{R}$ be exact solutions of (27.1). If $\phi(a) = \psi(a)$ for some $a \in I$, $\phi(x) = \psi(x)$ for any $x \in I$.*

Proof We prove that $\phi(x) = \psi(x)$ for $a \leq x \leq a + \frac{1}{2L}, x \in I$. The same argument applies for $a - \frac{1}{2L} \leq x \leq a, x \in I$, hence the claim follows. From the two equations $\phi'(x) = f(x, \phi(x))$ and $\psi'(x) = f(x, \psi(x))$, $\phi'(x) - \psi'(x) = f(x, \phi(x)) - f(x, \psi(x))$. Integrating this formula from a to x, we obtain $\phi(x) - \psi(x) = \int_a^x (f(t, \phi(t)) - f(t, \psi(t)))dt$, since $\phi(a) = \psi(a)$. We also have $|\phi(x) - \psi(x)| \leq \int_a^x |f(t, \phi(t)) - f(t, \psi(t))|dt \leq L \cdot \int_a^x |\phi(t) - \psi(t)|dt$. Let M be the supremum of the range of $|\phi - \psi|$ on $[a, a + \frac{1}{2L}]$, then for $a \leq x \leq a + \frac{1}{2L}$, $|\phi(x) - \psi(x)| \leq L(x - a)M \leq \frac{1}{2}M$ holds, hence $M = 0$, therefore $\phi = \psi$ on $[a, a + \frac{1}{2L}]$. □

In classical mathematics, the existence of an exact solution is proven without assuming the Lipschitz condition. This result is due to Peano's existence theorem.

Theorem 27.42 (Peano's existence theorem, classical) *Assume $f : D \to \mathbb{R}$ is a function which is continuous on an open D and $(a_0, b_0) \in D$. Any initial value problem $\phi'(x) = f(x, \phi(x))$ with $\phi(x_0) = y_0$ has a solution $\psi : I \to \mathbb{R}$ for some neighborhood I of x_0, such that for any $x \in I$, $\psi'(x) = f(x, \psi(x))$.*

This theorem is shown to be constructively unacceptable through the weak counter example argument [1, 9]. Assuming the Lipschitz condition, we can constructively prove the existence of an exact solution.

Theorem 27.43 (Exact solutions) *Let $f : D \to \mathbb{R}$ be a uniformly continuous function satisfying the Lipschitz condition. Let $(a_0, b_0) \in D$ such that $(x, y) \in R$, where R is some rectangle in D given by $|x - a_0| \leq a$ and $|y - b_0| \leq b$. If $|f(x, y)| \leq M$ for $(x, y) \in R$, then for any $k \in \mathbb{Z}$ there exists an exact solution $\phi^{[a_0-h, a_0+h] \to \mathbb{R}}$ of (27.1) such that $\phi(a_0) = b_0$, where $h := \min(a, b/M)$.*

Proof By Theorem 27.36 there is a sequence of approximate solutions ϕ_n over $I := [a_0 - h, a_0 + h]$ up to the error 2^{-n}. By Lemma 27.20, the sequence $(\phi_n)_n$ uniformly converges over I to a uniformly continuous function ϕ. Hence the sequence $(f(x, \phi_n(x))$ uniformly converges to the function $f(x, \phi(x))$. By Theorem 27.26, $\lim_{n \to \infty} \int_a^b f(t, \phi_n(t))dt = \int_a^b f(t, \phi(t))dt$. We prove that this ϕ is an exact solution. For any $x \in I$, $|\phi'_n(x) - f(x, \phi_n(x))| \leq 2^{-n}$. Integrating this inequality from a_0 to x gives $| \int_{a_0}^x [\phi'_n(t) - f(t, \phi_n(t))]dt| \leq 2^{-n}(x - a_0) \leq 2^{-n}h$. Since ϕ_n is continuous, by Theorem 27.28, $|\phi_n(x) - \phi_n(a_0) - \int_{a_0}^x f(t, \phi(t))dt| \leq 2^{-n}h$. Taking the limit for $n \to \infty$, $\phi(x) - \phi(a_0) - \int_{a_0}^x f(t, \phi(t))dt = 0$ holds. Differentiation yields $\phi'(x) = f(x, \phi(x))$, and $\phi(a_0) = b_0$. □

We conclude this subsection with the following theorem, which tells us that an approximate solution of (27.1) up to the error 2^{-k} differs from the exact solution by a constant multiple of 2^{-k}.

Theorem 27.44 (Error estimation for approximate solutions) *Let $f : D \to \mathbb{R}$ be a uniformly continuous function satisfying the Lipschitz condition w.r.t. the second argument. Let $(a_0, b_0) \in D$ such that the rectangle R is given by $|x - a_0| \leq a$ and $|y - b_0| \leq b$ in D. Assume $|f(x, y)| \leq M$ for $(x, y) \in R$, and let $h := \min(a, b/M)$. Assume further $\phi : [a_0 - h, a_0 + h] \to \mathbb{R}$ is an exact solution of (27.1) such that $\phi(a_0) = b_0$. If ψ is an approximate solution of (27.1) up to the error 2^{-p} such that $\psi(a_0) = b_0$ and also $\phi \leq \psi$ or $\psi \leq \phi$, then there is a constant N independently from p such that $|\phi(x) - \psi(x)| \leq 2^{-p}N$ for $|x - a_0| \leq h$.*

Proof By Theorem 27.40, $|\phi(x) - \psi(x)| \leq 2^{-p}\frac{1}{L}(e^{Lh} - 1)$, hence N is defined to be $\frac{1}{L}(e^{Lh} - 1)$. □

27.3.3 Stream-Based Real Numbers

In this subsection we deal with the signed digit stream, which is a stream-based representation of real numbers in a closed interval $[-1, 1]$. We study the average and multiplication of such stream-based real numbers through corecursion and coinduction.

Definition 27.45 (SDS) The *signed digit* $\mathbf{SD}$ is defined to be the set $\{-1, 0, 1\}$ of integers. Among the list $\mathbf{L_{SD}}$ of signed digits, an infinite one is called *signed digit stream*, an SDS for short. We also define $\mathbf{SD}_2$ to be the set $\{i \in \mathbb{Z} \mid -6 \leq i \leq 6\}$.

For brevity, we let the variables d, e range over $\mathbf{SD}$, i, j over $\mathbf{SD}_2$, *ds*, *es* (individual symbols) over $\mathbf{L_{SD}}$, x, y, z over $\mathbb{R}$, and a, b over $\mathbb{Q}$. Informally speaking, an SDS $d_0 :: d_1 :: \ldots$ expresses a real number $\sum_{i=0} \frac{d_i}{2^{i+1}}$ in $[-1, 1]$. In this subsection we study the signed digit stream, SDS for short, which is a corecursive representation of real numbers in the unit interval $[-1, 1]$. We first see the equivalence of Cauchy reals in $[-1, 1]$ and SDSs, then secondly an average function of two SDSs.

Definition 27.46 (Intervals) Let $\mathbb{I}$ denote the rational interval $[-1, 1]$ and let $b \in \mathbb{I}$ denote $-1 \leq b \leq 1$ for $b \in \mathbb{Q}$. For given $x \in \mathbb{R}$, $b \in \mathbb{I}$, and $k \in \mathbb{Z}^+$, we define $x \in_k \mathbb{I}_b$ to be $a_{Mk} \in \mathbb{I}$ and $b - 2^{-k} \leq a_{Mk} \leq b + 2^{-k}$, where $x = \langle (a_n)_n, M \rangle$.

Definition 27.47 (Coinductive predicate ${}^{\mathrm{co}}I$) We define a coinductive predicate ${}^{\mathrm{co}}I$ of arity $(\mathbb{R})$ by the closure axiom ${}^{\mathrm{co}}Ix \to \exists_{y \in \mathbb{R}, d}({}^{\mathrm{co}}Iy \wedge x = \frac{y+d}{2} \wedge -1 \leq y \leq 1)$. The greatest-fixed-point axiom is given as $Px \to \forall_x(Px \to \exists_{y \in \mathbb{R}, d}(({}^{\mathrm{co}}Iy \vee Py) \wedge x = \frac{y+d}{2} \wedge -1 \leq y \leq 1)) \to {}^{\mathrm{co}}Ix$, where P is a predicate variable to be instantiated.

Because we are mainly interested in infinite lists, the case for the nil list is omitted, in contrast to the example in Section 27.2.2. We also assume simpler destructor and corecursion. For instance, $\mathcal{D}_{\mathbf{L_{SD}}}$ is of type $\mathbf{L_{SD}} \to \mathbf{SD} \times \mathbf{L_{SD}}$, and it computes the head and the tail of the given SDS.

The next proposition tells us that the approximability of a Cauchy real x, namely, $\forall_k \exists_b (x \in_k \mathbb{I}_b)$, is equivalent to ${}^{\mathrm{co}}Ix$. The computational content of the proof is a translator between a regular Cauchy real and an SDS.

Proposition 27.48 (Cauchy real and ${}^{\mathrm{co}}I$)

$$\forall_x (\mathrm{Real}\, x \wedge -1 \leq x \leq 1 \wedge \forall_p \exists_c (|x - c| \leq 2^{-p}) \leftrightarrow {}^{\mathrm{co}}Ix).$$

Proof Assume $\mathrm{Real}\, x \wedge -1 \leq x \leq 1 \wedge \forall_p \exists_c (|x - c| \leq 2^{-p}) \leq 2^{-p} =: Px$. Use the greatest-fixed-point axiom for Px, then it suffices to prove that Px implies $\exists_{y,d}(Py \wedge x = \frac{y+d}{2})$. By Lemma 27.8, $-1 \leq x \leq 0$, $-\frac{1}{2} \leq x \leq \frac{1}{2}$, or $0 \leq x \leq 1$, and corresponding to those three cases we let d be $-1, 0$, or 1. We prove $P(2x - d)$. Assume Px, then clearly $\mathrm{Real}(2x - d)$ and $-1 \leq 2x - d \leq 1$ hold. Assume p, then by the assumption there exists c such that $|x - c| \leq 2^{-(p+1)}$. Let $y := \langle (b_n)_n, N \rangle$ define $2x - d$, where $b_n = 2a_n - d$ and $Np = M(p + 1)$. It suffices to prove $|y - (2c - d)| \leq 2^{-p}$ as $|2x - d - (2c - d)| = 2 \cdot |x - c| \leq 2 \cdot 2^{-(p+1)} = 2^{-p}$.

For the opposite direction, prove $\forall_x ({}^{\mathrm{co}}Ix \to Px)$ by induction on p. Case $p = 1$. Assume ${}^{\mathrm{co}}Ix$, then $\mathrm{Real}\, x$ and $-1 \leq x \leq 1$ are obvious by the closure axiom. We prove $\exists_c |x - c| \leq \frac{1}{2}$. Let c be $a_{M(2)}$, then apparently $|x - a_{M2}| \leq \frac{1}{2}$. Case $k \mapsto k + 1$. Assuming ${}^{\mathrm{co}}I\ x$ and then by the closure axiom $x = \frac{y+d}{2}$ for some $y \in {}^{\mathrm{co}}I$ and d, hence by induction hypothesis $|y - c_0| \leq 2^{-p}$ for some c_0. Then, $|x - \frac{c_0+d}{2}| = |\frac{y+d}{2} - \frac{c_0+d}{2}| = \frac{1}{2} \cdot |y - c_0| \leq \frac{1}{2} \cdot 2^{-p} = 2^{-(p+1)}$. □

We study the average function on SDS. The average is closed in the unit interval, hence it is suitable for our stream-based investigation of real numbers. As a computational content of the proof, we witness a corecursive algorithm computing an average of two given SDSs. Given two SDSs $ds := d_0 :: d_1 :: \ldots$ and $es := e_0 :: e_1 :: \ldots$, we consider their average $d'_0 :: d'_1 :: \ldots$. The sum of d_0 and e_0, say the carry c_0, is in $[-2, 2]$. It is not always the case that we can determine the first digit d'_0 from c_0, but it becomes possible if we also read off d_1 and e_1. For example, assume $d_0 = 0$ and $e_0 = 1$. Just from those first digits, we can only say that d'_0 is either 0 or 1, because the average value may be anywhere in the interval $[-\frac{1}{4}, \frac{3}{4}]$ of the length 1. By reading the second digits d_1 and e_1, we find the interval of the length $\frac{1}{2}$ for the average, hence it is surely possible to determine the digit d'_0. Counting the significance of the digits d_0 and e_0 as twice of the one of d_1 and e_1, we make a case distinction on $i := 2c_0 + d_1 + e_1 \in [-6, 6]$ by the function K in Definition 27.49.

Definition 27.49 Define $J, K : \mathbb{Z} \to \mathbb{Z}$ to be

$$J(i) = i - 4K(i), \qquad K(i) = \begin{cases} -1 & \text{if } -6 \leq i \leq -2 \\ 0 & \text{if } -2 \leq i \leq 2 \\ 1 & \text{if } 2 \leq i \leq 6. \end{cases}$$

After determining d'_0, we compute the next carry c_1, which will be used together with d_2 and e_2 to compute d'_1. This carry is computed by the function J in Definition 27.49. The idea is that we find a c_1, such that the average of $d'_0 :: \hat{d}_1 :: d_2 :: d_3 :: \ldots$ and $d'_0 :: \hat{e}_1 :: e_2 :: e_3 :: \ldots$ is same as the average of ds and es for some $\hat{d}_1$ and $\hat{e}_1$ with $c_1 = \hat{d}_1 + \hat{e}_1$, so that we keep repeating the above procedure to compute the next digit, say d'_1, from c_1, d_2, and e_2.

The program of average comes from a proof of the following theorem.

Theorem 27.50 (Average) $\forall_{x,y}({}^{\mathrm{co}}Ix \to {}^{\mathrm{co}}Iy \to {}^{\mathrm{co}}I(\frac{x+y}{2}))$.

Lemma 27.51 (CoIAvToAvc) $\forall_{x,y \in {}^{\mathrm{co}}I} \exists_i \exists_{x',y' \in {}^{\mathrm{co}}I} (\frac{x+y}{2} = \frac{x'+y'+i}{4})$.

Proof Assume $x, y \in {}^{\mathrm{co}}I$, then by the closure axiom there exist $x', y' \in {}^{\mathrm{co}}I$ and $d, e \in \mathbf{SD}$ such that $x = \frac{x'+d}{2}$ and $y = \frac{y'+e}{2}$. Let i be $d + e$, that is in $\mathbf{SD}_2$, then for the above x' and y' the equality $\frac{x+y}{2} = \frac{x'+y'+i}{4}$ holds. □

Its computational content c_0 is expressed as $\mathsf{c}_0(d :: ds, e :: es) := \langle ds, es, d + e\rangle$.

Lemma 27.52 (CoIAvcSatCoICl) $\forall_{x,y \in {}^{\mathrm{co}}I} \forall_i \exists_{x',y' \in {}^{\mathrm{co}}I} \exists_{d,j} (x+y+i = \frac{x'+y'+j+4d}{2})$.

Proof Assume $x, y \in {}^{\mathrm{co}}I$ and $i \in \mathbf{SD}_2$, then by the closure axiom there exist $x', y' \in {}^{\mathrm{co}}I$ and $d_0, e_0 \in \mathbf{SD}$ such that $x = \frac{x'+d_0}{2}$ and $y = \frac{y'+e_0}{2}$. We let d and j be $K(d_0 + d_1 + 2i)$ and $J(d_0 + d_1 + 2i)$, respectively. The equality holds, since $d_0 + e_0 + 2i = j + 4d$. □

Its computational content c_1 is expressed as

$$\mathsf{c}_1(d :: ds, e :: es, i) := \langle ds, es, J(d + e + 2i), K(d + e + 2i)\rangle.$$

Lemma 27.53 (CoIAvcToCoI) $\forall_z \forall_{x,y \in {}^{\mathrm{co}}I, i} (z = \frac{x+y+i}{4} \to {}^{\mathrm{co}}Iz)$.

Proof Assume z and $\exists_{x,y \in {}^{\mathrm{co}}I, i}(z = \frac{x+y+i}{4})$, and use the greatest-fixed-point axiom for the competitor $Pz := \exists_{x,y \in {}^{\mathrm{co}}I, i}(z = \frac{x+y+i}{4})$. It suffices to prove $\forall_z(Pz \to \exists_{z' \in P, d' \in \mathbf{SD}}(z = \frac{z'+d'}{2}))$. Assume Pz, which means $z = \frac{x+y+i}{4}$ for some $x, y \in {}^{\mathrm{co}}I$ and $i \in \mathbf{SD}_2$. By Lemma 27.52, there exist $x', y' \in {}^{\mathrm{co}}I$, j, and d, such that $\frac{x+y+i}{4} = \frac{x'+y'+j+4d}{8} = \frac{1}{2} \cdot (\frac{x'+y'+j}{4} + d)$. Let $z' := \frac{x'+y'+j}{4} \in P$ and $d' := d$, then the equality holds. □

Its computational content c_2 is expressed via corecursive equation as

$$\mathsf{c}_2(d :: ds, e :: es, i) = K(d + e + 2i) :: \mathsf{c}_2(ds, es, J(d + e + 2i)),$$

where we implicitly used c_1 to compute $ds, es, K(d + e + 2i), J(d + e + 2i)$.

Proof of Theorem 27.50 We prove $\frac{x+y}{2} \in {}^{\mathrm{co}}I$ from $x, y \in {}^{\mathrm{co}}I$. Use Lemma 27.51, then $\frac{x+y}{2} = \frac{x'+y'+i}{4}$ for some $i \in \mathbf{SD}_2$ and $x', y' \in {}^{\mathrm{co}}I$, which is in ${}^{\mathrm{co}}I$ by Lemma 27.53. □

Let $i := d' + e' + 2(d + e)$, its computational content Av is expressed as

$$\mathsf{Av}(d :: d' :: ds, e :: e' :: es) := K(i) :: \mathsf{c}_2(ds, es, J(i)).$$

As the last application we study multiplication of real numbers.

Theorem 27.54 (Multiplication) $\forall_{x,y}({}^{\mathrm{co}}Ix \to {}^{\mathrm{co}}Iy \to {}^{\mathrm{co}}I(x \cdot y))$.

Lemma 27.55 (CoIMultToMultc) $\forall_{x,y \in {}^{\mathrm{co}}I} \exists_{i \in \mathbf{SD}_2, x',y',z' \in {}^{\mathrm{co}}I}(x \cdot y = \frac{x' \cdot y' + z' + i}{4})$.

Proof Assume $x, y \in {}^{\mathrm{co}}I$. By the Closure axiom, $x = \frac{x_0 + d_0}{2}$ and $y = \frac{y_0 + e_0}{2}$ for some $x_0, y_0 \in {}^{\mathrm{co}}I$ and d_0, e_0, then $x \cdot y = \frac{x_0 \cdot y_0 + x_0 \cdot e_0 + y_0 \cdot d_0 + d_0 \cdot e_0}{4}$, which is proven to be equal to $\frac{x' \cdot y' + z' + i}{2}$ for some i and $x', y', z' \in {}^{\mathrm{co}}I$. Obviously $x_0 \cdot e_0$ and $y_0 \cdot d_0$ are in ${}^{\mathrm{co}}I$, hence by Theorem 27.50, $\frac{x_0 \cdot e_0 + y_0 \cdot d_0}{2} \in {}^{\mathrm{co}}I$, therefore, by the Closure axiom, $\frac{x_0 \cdot e_0 + y_0 \cdot d_0}{2} = \frac{z_0 + d'}{2}$ for some d' and $z_0 \in {}^{\mathrm{co}}I$. We then let $x' := x_0$, $y' := y_0$, $z' := z_0$, and $i := d' + d_0 \cdot e_0$ to conclude the proof. □

Its computational content c_3 is expressed as

$$\mathsf{c}_3(d :: ds, e :: es) := \langle \mathsf{hd}(ds') + d \cdot e, ds, es, \mathsf{tl}(ds')\rangle\rangle,$$

where $ds' := \mathsf{Av}(d \cdot es, e \cdot ds)$ and hd and tl are the head and tail of a stream, easily defined by means of the destructor.

Lemma 27.56 (CoIMultcSatCoICl)

$$\forall_{x,y,z \in {}^{\mathrm{co}}I, i} \exists_{x_1, z_1 \in {}^{\mathrm{co}}I, d, j} \frac{x \cdot y + z + i}{4} = \frac{1}{2} \cdot (\frac{x_1 \cdot y + z_1 + j}{4} + d).$$

Proof Assume $x, y, z \in {}^{\mathrm{co}}I$. By the closure axiom, $x = \frac{x_0 + d_0}{2}$ and $z = \frac{z_0 + e_0}{2}$ for some $x_0, z_0 \in {}^{\mathrm{co}}I$ and d_0, e_0, and $\frac{x \cdot y + z + i}{4} = \frac{1}{2} \cdot \frac{x_0 \cdot y + i + e_0 + z_0 + d_0 \cdot y + i}{4}$. We prove that $\frac{z_0 + d_0 \cdot y + i}{4} = \frac{z_1 + e + 2e'}{4}$ for some $z_1 \in {}^{\mathrm{co}}I$. As $d_0 \cdot y \in {}^{\mathrm{co}}I$, $\frac{z_0 + d_0 \cdot y + i}{4} \in {}^{\mathrm{co}}I$ by Lemma 27.53, hence there exist some $z', z_1 \in {}^{\mathrm{co}}I$ and e', e such that $\frac{z_0 + d_0 \cdot y + i}{4} = \frac{z' + e'}{2} = \frac{\frac{z_1 + e}{2} + e'}{2} = \frac{z_1 + e + 2e'}{4}$ by the closure axiom. Using the proven equality, we can prove $\frac{1}{2} \cdot \frac{x_0 \cdot y + i + e_0 + z_0 + d_0 \cdot y + i}{4} = \frac{1}{2} \cdot \frac{x_0 \cdot y + z_1 + i + e_0 + e + 2e'}{4} = \frac{1}{2} \cdot (\frac{x_0 \cdot y + z_1 + j}{4} + d)$ for $j := J(i + e_0 + e + 2e')$ and $d := K(i + e_0 + e + 2e')$. □

Its computational content c_4 is expressed by

$$\mathsf{c}_4(d :: ds, ds', e :: es, i) := \langle ds, es', K(i + e_0 + e + 2e'), J(i + e_0 + e + 2e')\rangle,$$

where $e' :: e :: es' = \mathsf{c}_2(es_0, d_0 \cdot ds', i)$.

Lemma 27.57 (CoIMultcToCoI) $\forall_z(\forall_{x',y',z'\in{}^{\mathrm{co}}I,i\in\mathbf{SD}_2}(z = \frac{x'\cdot y'+z'+i}{4} \to {}^{\mathrm{co}}Iz))$.

Proof Assume z, and use the greatest-fixed-point axiom for the competitor $\exists_{x',y',z'\in{}^{\mathrm{co}}I,i\in\mathbf{SD}_2} z = \frac{x'\cdot y'+z'+i}{4} =: Pz$ to prove ${}^{\mathrm{co}}Iz$. It then suffices to prove $\forall_z(Pz \to \exists_{d_0\in\mathbf{SD},z_0\in P}(z = \frac{z_0+d_0}{2}))$. Assume Pz, namely, $z = \frac{x'\cdot y'+z'+i}{4}$ for some $x', y', z' \in {}^{\mathrm{co}}I$, then by Lemma 27.56, $\frac{x'\cdot y'+z'+i}{4} = \frac{1}{2}\cdot(\frac{x_1\cdot y+z_1+j}{4} + d)$ for some $x_1, z_1 \in {}^{\mathrm{co}}I$ and $j \in \mathbf{SD}_2$, where $\frac{x_1\cdot y+z_1+j}{4}$ is in P. □

Its computational content c_5 is corecursively expressed by

$$\mathsf{c}_5(ds, cs, es, i) = d :: \mathsf{c}_5(ds', cs, es', j),$$

where $\langle ds', es', d, j\rangle = \mathsf{c}_4(ds, cs, es, i)$.

From the above lemmas, the main theorem is proven.

Proof of Theorem 27.54 Assume $x, y \in {}^{\mathrm{co}}I$, and then by Lemma 27.55, there are $i \in \mathbf{SD}_2$ and $x', y', z' \in {}^{\mathrm{co}}I$ such that the equality $x\cdot y = \frac{x'\cdot y'+z'+i}{4}$ holds. We proceed by using Lemma 27.57, feeding $x \cdot y$ for z, then we conclude ${}^{\mathrm{co}}I(x\cdot y)$ which follows from the above equality. □

Its computational content Mul is read-off as follows:

$$\mathsf{Mul}(d :: ds, e :: es) := \mathsf{c}_5(ds, es, \mathsf{tl}(ds'), \mathsf{hd}(ds') + d\cdot e),$$

where $ds' := \mathsf{Av}(d\cdot es, e\cdot ds)$ and c_3 and c_4 are implicitly used.

27.3.4 Notes

Section 27.3.1 follows the treatment by Schwichtenberg [26, 27]. Its formalization within the Minlog system is available in the downloadable package [22]. Berger studies inverse functions and the modulus of increase [3]. Also in computable analysis, similar results are found (cf. Section 6.2 of [35]). The treatment of the Cauchy–Euler method in Section 27.3.2 is due to Bridges and Schwichtenberg and Weghorn. Bridges studies constructive Peano's existence theorem [9]. Weghorn studies the fundamental inequality and also practices program extraction from formalized proofs in the Minlog system, namely, within TCF, and Coq [27, 34]. With a similar motivation, Makarov and Spitters study the Picard algorithm within a higher-order type system and formalize their work in Coq [21]. In Section 27.3.3, the treatment of the average is due to Berger and Miyamoto and Schwichtenberg [4, 23], and of the multiplication is due to Überrück-Fries [33]. Wiedmer studies exact real arithmetic through streams [36, 37]. Ciaffaglione and di Gianantonio take a corecursive approach to stream-based exact real arithmetic and give formalization in Coq [12]. Berger et al. study (constructive) Gray code, which is an alternative stream-based real number representation [5]. Tsuiki studies (classical) Gray code in the context of real number computability [32]. Arithmetic for (constructive) Gray code is explored by Schwichtenberg and Wiesnet [28]. The average, multiplication, and division for both SDS and Gray code have been formalized and is available for program extraction in Minlog [22]. Ghani et al. study uniformly continuous functions from coalgebraic perspective, which yields a stream-based representation of uniformly continuous functions [16]. Their idea is applicable in the context of program extraction and has been implemented in the Minlog system [4, 24]. Concerning formalization in proof assistants, Bickford's work in NuPRL covers Chapter 2 of Bishop and Bridges' book [7, 25]. A formalization of constructive analysis is found in the CoRN project in Nijmegen [13, 14], which is used by the above-mentioned work by Makarov and Spitters. Anand and Knepper study certified programs for robots, where their reasoning about physics is due to CoRN's constructive analysis with exact reals [2].

27.4 Concluding Remarks

We have seen applications of Bishop's constructive analysis in exact real arithmetic. Working within a constructive setting, one ends up with a computationally precise formulation of a problem due to the BHK-interpretation and constructive proofs with algorithmic contents. It is practically beneficial for exact real arithmetic, because we can study both the method to solve the problem and the corresponding mathematics in an algorithmically integrated fashion. As a consequence, one has a clear mathematical understanding about the correctness of the algorithmic solution to the problem. By formalizing such constructive arguments in a suitable system, one can practice program extraction which generates from a proof a provably correct program to solve the problem specified by the corresponding statement. Such a human-readable and executable program is a helpful coproduct also in empirical science involving real number computation.

We conclude this chapter by mentioning future work in this field. Hendtlass studies various constructive fixed-point theorems such as Brouwer's, Schauder's, and Kakutani's theorems [17, 18]. These achievements lead us to further advancements of Section 27.3.1 and Section 27.3.2. The stream-based real number in Section 27.3.3 is not only useful in conjunction with modern programming languages, but also close to the real number representation in interval arithmetic. Interval arithmetic is a successful method in the field of validated numerical computation, and there are many applications of (classical) fixed-point theorems. Hisakado et al. point out the computational efficiency and applicability of Tsuiki's Gray-code for numerical computation in electronics via interval arithmetic [19].

References

[1] Aberth, O. 1970. Computable analysis and differential equations. Pages 47–52 of: *Intuitionism and Proof Theory*, Kino, A., Myhill, J., and Vesley, R. E. (eds.). *Proceedings of the Summer Conference at Buffalo, N.Y. 1968*. Studies in Logic and the Foundations of Mathematics. North-Holland.

[2] Anand, A., and Knepper, R. 2015. ROSCoq: Robots powered by constructive reals. Pages 34–50 of: *Interactive Theorem Proving*, Urban, C., and Zhang, X. (eds.). Springer.

[3] Berger, J. 2005. Exact calculation of inverse functions. *Math. Log. Q.*, **51**(2), 201–205.

[4] Berger, U. 2009. From coinductive proofs to exact real arithmetic. Pages 132–146 of: *Computer Science Logic*, Grädel, E., and Kahle, R. (eds.). Lecture Notes in Computer Science, vol. 5771. Berlin, Heidelberg, New York: Springer Verlag.

[5] Berger, U., Miyamoto, K., Schwichtenberg, H., and Tsuiki, H. 2016. Logic for Gray-code computation. Pages 69–110 of: *Concepts of Proof in Mathematics, Philosophy, and Computer Science*, Probst, D., and Schuster, P. (eds.). Berlin: De Gruyter.

[6] Bishop, E. 1967. *Foundations of Constructive Analysis*. McGraw-Hill.

[7] Bishop, E., and Bridges, D. 1985. *Constructive Analysis*. Grundlehren der mathematischen Wissenschaften, vol. 279. Berlin, Heidelberg, New York: Springer Verlag.

[8] Blazy, S., Paulin-Mohring, C., and Pichardie, D. (eds.) 2013. *Proc. Interactive Theorem Proving – 4th International Conference*, ITP 2013, Rennes, France, July 22-26, 2013. Proceedings. Lecture Notes in Computer Science, vol. 7998. Berlin, Heidelberg: Springer.

[9] Bridges, Douglas. 2012. Constructive solutions of ordinary differential equations. Pages 67–77 of: *Logic, Construction, Computation*, Berger, U., Diener, H., Schuster, P., and Seisenberger, M. (eds.). Heusenstamm: Ontos Verlag.

[10] Bridges, D., and Richman, F. 1987. *Varieties of Constructive Mathematics*. Cambridge: Cambridge University Press.

[11] Bridges, D. S., and Vîţă, L. S. 2007. *Techniques of Constructive Analysis*. Universitext. New York: Springer.

[12] Ciaffaglione, A., and Di Gianantonio, P. 1999. A co-inductive approach to real numbers. Pages 114–130 of: Coquand, T., Dybjer, P., Nordström, B., and Smith, J. (eds.), *Proceedings of the Workshop "Types 1999"*, Lecture Notes in Computer Science, vol. 1956. Berlin, Heidelberg, New York: Springer Verlag

[13] CoRN, Coq Repository at Nijmegen. `http://corn.cs.ru.nl`

[14] Cruz-Filipe, L. 2004 (April). *Constructive real analysis: a type-theoretical formalization and applications*. Ph.D. thesis, University of Nijmegen.

[15] Edalat, A., and Heckmann, R. 2000. Computing with real numbers. Pages 193–267 of: *APPSEM*. LNCS, vol. 2395. Berlin, Heidelberg: Springer.

[16] Ghani, N., Hancock, P., and Pattinson, D. 2006. Continuous functions on final coalgebras. In: *Proceedings CMCS 2006*, Power, J. (ed.). *Electr. Notes Theoret. Comp. Sci.* **164**(1) 141–155.

[17] Hendtlass, M. 2012. Fixed point theorems in constructive mathematics. *J. Logic & Anal.*, **4**. `https://doi.org/10.4115/jla.2012.4.10`.

[18] Hendtlass, M. 2016. Kakutani's fixed point theorem in constructive mathematics. ArXiv:1611.02531.

[19] Hisakado, T., Hamada, M., Yonemoto, A., and Okumura, K. 2004. Interval arithmetic using Gray code. Pages III–391–394 of: *Proc. 47th Midwest Symposium on Circuits and Systems*. Piscataway, NJ: IEEE.

[20] Loh, E., and Walster, G. W. 2002. Rump's example revisited. *Reliable Computing*, **8**(3), 245–248.

[21] Makarov, E., and Spitters, B. 2013. The Picard algorithm for ordinary differential equations in Coq. In: [8].

[22] Minlog. *The Minlog System*. `http://www.minlog-system.de/`

[23] Miyamoto, K., and Schwichtenberg, H. 2015. Program extraction in exact real arithmetic. *Math. Struct. Comp. Sci.*, **25**(8), 1692–1704.

[24] Miyamoto, K., Forsberg, F. N., and Schwichtenberg, H. 2013. Program extraction from nested definitions. In: [8].

[25] Nuprl. `http://www.nuprl.org/`

[26] Schwichtenberg, H. 2006. Inverting monotone continuous functions in constructive analysis. Pages 490–504 of: *Logical Approaches to Computational Barriers. (Proc. CiE 2006, Swansea)*, Beckmann, A., Berger, U., Löwe, B., and Tucker, J. V. (eds.). Lecture Notes in Computer Science, vol. 3988. Berlin, Heidelberg, New York: Springer Verlag.

[27] Schwichtenberg, H. 2016a. *Constructive analysis with witnesses*. Manuscript. `http://www.math.lmu.de/~schwicht/seminars/semws16/constr16.pdf`.

[28] Schwichtenberg, H. 2016b. *Logic for exact real arithmetic*. Manuscript. `http://www.mathematik.uni-muenchen.de/~schwicht/papers/bridges16/m4c.pdf`

[29] Schwichtenberg, H., and Wainer, S. S. 2012. *Proofs and Computations*. Perspectives in Logic. Cambridge: Association for Symbolic Logic and Cambridge University Press.

[30] Troelstra, A. S., and Schwichtenberg, H. 2000. *Basic Proof Theory*, 2nd ed. Cambridge: Cambridge University Press.

[31] Troelstra, A. S., and van Dalen, D. 1988. *Constructivism in mathematics. Vol. I*. Studies in Logic and the Foundations of Mathematics, vol. 121. North-Holland Publishing Co.

[32] Tsuiki, H. 2002. Real number computation through Gray code embedding. *Theor. Comput. Sci.*, **284**(2), 467–485.

[33] Überrück-Fries, T. 2016. Program Extraction for Exact Real Numbers: Stream Multiplication. Bachelor's thesis, Mathematisches Institut der Universität München.

[34] Weghorn, T. 2012. Verwendung klassischer Schlußweisen in der konstruktiven Analysis. M.Phil. Thesis, Mathematisches Institut der Universität München.

[35] Weihrauch, K. 2000. *Computable Analysis*. Berlin: Springer.

[36] Wiedmer, E. 1977. Exaktes Rechnen mit reellen Zahlen und anderen unendlichen Objekten. Ph.D. Thesis, ETH Zürich.

[37] Wiedmer, E. 1980. Computing with infinite objects. *Theor. Comp. Sci.*, **10**, 133–155.

28

Efficient Algorithms from Proofs in Constructive Analysis

Mark Bickford

28.1 Introduction

In Bishop's *Foundations of Constructive Analysis* (1967), and reprinted in Bishop and Bridges (1985), we find in Chapter 1 "A constructivist manifesto" the following:

> Any computation that can be performed by a finite intelligence – any computation that has a finite number of steps – is permissible. This does not mean that no value is to be placed on the efficiency of a computation. An applied mathematician will prize a computation for its efficiency above all else, whereas in formal mathematics much attention is paid to elegance and little to efficiency. Mathematics should and must concern itself with efficiency, perhaps to the detriment of elegance, but these matters will come to the fore only when realism has begun to prevail. Until then our first concern will be to put as much mathematics as possible on a realistic basis without close attention to questions of efficiency.

Bishop deferred the consideration of efficient algorithms until after analysis was at least done constructively. Today we have proof assistants like Nuprl, MINLOG, Coq, and Agda that can check formal proofs in constructive logic and produce from any proof an *extract*, its computational content. It would be the best of all possible worlds if, by simply carrying out in a proof assistant the elegant constructive proofs of Bishop, Bridges, and others, we would automatically extract efficient algorithms. The quotation from Bishop hints at the fact that we will not magically get efficient algorithms from every constructive proof, but, with some care, we often do get reasonably efficient algorithms from our constructive proofs. When more efficiency is desirable we can take our first construction to be the specification of a thing to be computed and then formalize the methods of numerical analysts and extract an alternate algorithm to compute that thing more efficiently.

We have formalized in Nuprl all of Chapter 2 of Bishop and Bridges, "Constructive analysis". In this chapter we will discuss methods we used in our formalization of basic real analysis to make its constructive content reasonably efficient. Similar methods have been used in MINLOG and Coq but we will not survey those works here.

In Nuprl, a version of Brouwer's continuity principle is true (it follows from the formal definition of Nuprl) and is also consistent with the rule of *bar induction*. Using these principles we can prove Ishihara's Weak Markov Principle for the real numbers, and prove Brouwer's theorem that all functions on a compact interval are uniformly continuous. These results allow us to state some of the theorems in Chapter 2 of Bishop and Bridges using the assumption that f is a "function" on interval $[a, b]$ (which means that it respects the equivalence relation on real numbers) rather than assuming that f is uniformly continuous on $[a, b]$. The constructive content of uniform continuity is a *modulus of uniform continuity*, so an algorithm (i.e., a theorem) that makes that assumption requires a modulus of continuity as an input. It turns out that the "function" assumption has no constructive content, so no additional input is needed for algorithms that make this assumption.

Does this mean that Brouwer's principles can make our algorithms simpler and more efficient "for free"? No – if a modulus of continuity is needed to prove anything with constructive content then a modulus would be constructed from the function f using the continuity principle and the fan theorem – and this construction will usually be horribly inefficient. So, the use of Brouwer's principles makes our theorems and proofs more elegant, but not more efficient. Because of this, real numbers such as $\sup\{f(x) \mid x \in [a, b]\}$, and $\int_a^b f(x)dx$ that are constructed in this manner will be taken to be the specifications of well-defined real numbers to be computed by other algorithms. The use of Brouwer's continuity principle and bar induction does not therefore affect the results we discuss in this article, and all of these results can be proved using only the methods used by Bishop.

Bishop's work was a major inspiration for the creation of the Nuprl proof assistant, and Nuprl, in turn, was an inspiration for proof assistants, like Coq and Agda, that are based on *constructive type theory*. Nuprl's concept of a *type* was initially inspired by Bishop's constructive concept of a set. Later the formal structure of Martin-Löf type theory was adopted by Nuprl, but additional computations and types that are not included in basic Martin-Löf type theory were added to Nuprl to facilitate Bishop-style mathematics. Using Coq to define the meta-theory, we have made a formal definition of Nuprl. This formal definition allows us to prove the soundness of Nuprl's rules (relative to the soundness of Coq) and it also allows us to add new computations and types when they are useful, and to verify the intuitionistic priniciples of continuity and bar induction.

28.2 Representation of Real Numbers

The thing that most affects the efficiency of algorithms extracted from proofs in constructive analysis is our choice of representation for the real numbers. We have experimented with several different representations and chosen one that we will define here.

Bishop defined a real number to be sequence $q_1, q_2, q_3 \dots$ of rational numbers such that

$$\forall n, m \colon \mathbb{N}^+ .\ |q_n - q_m| \leq n^{-1} + m^{-1} \qquad (\mathbb{N}^+ = \{1, 2, 3, \dots\}).$$

Such a real number is a Cauchy sequence of rational numbers with a fixed *modulus of convergence* given by $1/n$. Since we want efficient algorithms we might question the choice of $1/n$. Wouldn't a choice like $1/10^n$ lead to faster algorithms?

This was the idea behind the choice we made in our first formalization of Chapter 2 of Bishop and Bridges. We generalized Bishop's definition of a regular sequence to include a modulus that could be any sequence of positive rational numbers $\delta(n)$ with $\lim\limits_{n \to \infty} \delta(n) = 0$. This representation made the definitions of addition, multiplication, inverse, etc., more complicated, but, we thought, would allow some real numbers to be defined with a modulus like $\delta(n) = 1/10^n$ when convenient.

We found that this choice of representation was a mistake. Suppose that $x = \lambda n.q_n$ is a Bishop real number and $x' = \lambda n.r_n$ is a representation of an equivalent real number but with modulus $1/10^n$. To approximate this real number within $\epsilon > 0$ we can either find N with $1/N < \epsilon$ and use rational number $q_N = x(N)$ or we can find M with $1/10^M < \epsilon$ and use rational number $r_M = x'(M)$. Then M will be about $\log_{10}(N)$ so it will be smaller than N but why should that make the computation of r_M faster than the computation of q_N? If the sequences q and r were directly defined by primitive recursion, so that to compute q_N we first have to compute $q_1, q_2, \dots q_{N-1}$ and to compute r_M we first compute $r_1, \dots r_{M-1}$, then, perhaps, the computation of r_M might be faster than the computation of q_N, but we found that, in practice, real numbers are never actually defined by primitive recursion. Instead, if an iterative process is used to define the approximations to a real number, then the number of iterations needed to get an approximation within ϵ is computed first and then that number of iterations is carried out.

Bishop's definition with modulus $1/n$ is actually better than a definition with modulus $1/10^m$ because if we only need to compute a real number within $\epsilon = 1$ we can take $n = 1$ in Bishop's definition, whereas we would have to use $m = 1$ in the second definition and get an unnecessarily precise approximation that is within $\epsilon = 1/10$. By this line of argument, one might say that an even slower rate of convergence like $1/\log_{10}(n)$ could be better because sometimes we may only need to approximate a real number within $\epsilon = 10$. But this would add the overhead of finding an N for which $1/\log(N) < \epsilon$ and would complicate the definitions of the basic operations on real numbers.

Thus, for Cauchy sequences of rational numbers, we prefer Bishop's original definition and we used that for our second formalization of Chapter 2 of Bishop and Bridges. This meant that we could use the simpler definitions of addition, multiplication, inverse, etc., and use the proofs exactly as in the book. However, we found that this representation suffers from another problem and leads to algorithms that perform poorly.

To explain the problem, consider the computation of the first 50 decimal digits of the real number e^e. We need a rational approximation of e^e within $\epsilon = 10^{-50}$, and there should be one with a denominator of about 10^{50}, that is, with a denominator with about 50 digits. Using the estimates that come from the proof of convergence of the series $e^x = \sum_{n=0}^{\infty} x^n/n!$ we find that we need to approximate $\sum_{n=0}^{444} e^n/n!$ within $\epsilon/2$ so we will need each of the 445 terms in this sum within $\epsilon/890$. To approximate the last term $e^{444}/444!$ within $1/(890 * 10^{50})$ we will need to approximate e very accurately and this approximation $r = a/b$ will have a denominator b of at least 50 digits (and probably many more than 50 digits). Then using rational arithmetic, the rational number $(a/b)^{444}/444!$ will have a denominator with more than $444 * 444 * 50$ digits because b^{444} will have at least $444 * 50$ digits and 444! has more than 444 digits. Then since $(a/b) + (c/d)$ might have a denominator of $b * d$ (even if we reduce to lowest terms using the greatest common divisor (gcd)) when we add up all the 445 terms of the sum we could get a rational number with a denominator with more than $50 * (444)^3$ digits even though the approximation is only guaranteed accurate to 50 digits. Using rational arithmetic thus leads to intermediate terms that are grossly overprecise.

To remedy the problem of overprecise rational numbers we must somehow round off the intermediate results as we go to keep only the information needed for the final result to have the required accuracy. The c-CoRN library in Coq defines an abstract interface for the real numbers and implements it in various ways. One implementation uses approximate rational numbers with operations that round results to the desired accuracy.

Our first formalization effort taught us not to overgeneralize our representation, so rather than implement a new concept for approximate rational arithmetic we simply represent real numbers using sequences of *integers*. As we will see this leads naturally to the desired rounding operations.

28.3 Nuprl Representation of Real Numbers

We use Bishop's original definition of real numbers but normalize the nth rational q_n to have denominator $2n$ and keep only the numerators, so we have just a sequence of integers. Clearing the denominators in Bishop's regularity condition gives us the notion of a regular sequence of integers and we can use this to define the real numbers directly from the integers and we define all operations on reals using computations on integers. In our definitions each operation on reals rounds off the nth approximation so that it corresponds to a rational approximation with denominator $2n$. This rounding solves the problem of unnecessary accuracy and makes the algorithms more efficient.

28.3.1 Regular Sequences of Integers

Definition 28.1 A sequence $x_1, x_2, x_3, \ldots$ of integers is *regular* if

$$\forall n, m\!:\!\mathbb{N}^+.\ |m * x_n - n * x_m| \le 2(n+m).$$

The sequence is k-regular if

$$\forall n, m\!:\!\mathbb{N}^+.\ |m * x_n - n * x_m| \le 2k(n+m).$$

Definition 28.2 $\mathbb{R} = \{x : \mathbb{N}^+ \to \mathbb{Z} \mid x \text{is regular}\}$

To better understand what the regularity condition means we prove the following easy lemma.

Lemma 28.3 *Sequence x is k-regular if and only if for every $n, m \in \mathbb{N}^+$, the rational intervals $[\frac{x_n}{2kn} - \frac{1}{n}, \frac{x_n}{2kn} + \frac{1}{n}]$ and $[\frac{x_m}{2km} - \frac{1}{m}, \frac{x_m}{2km} + \frac{1}{m}]$ overlap (i.e. have a common member $q \in \mathbb{Q}$).*

Proof The intervals overlap if and only if $\frac{x_n}{2kn} + \frac{1}{n} \ge \frac{x_m}{2km} - \frac{1}{m}$ and $\frac{x_n}{2kn} - \frac{1}{n} \le \frac{x_m}{2km} + \frac{1}{m}$. Multiplying by $2knm$ to clear the denominators, these hold if and only if $m * x_n + 2km \ge n * x_m - 2kn$ and $m * x_n - 2km \le n * x_m + 2kn$, which is the same as $2kn + 2km \ge n * x_m - m * x_n$ and $m * x_n - n * x_m \le 2km + 2kn$. These hold if and only if $|m * x_n - n * x_m| \le 2k(n+m)$. □

Remark 28.4 The regularity condition does not say that the intervals $\frac{x_n}{2kn} \pm \frac{1}{n}$ are nested. It only says that they are pairwise overlapping. But if x is k-regular (even for just $n, m < B$) we can find a rational q that is a member of all the intervals $\frac{x_n}{2kn} \pm \frac{1}{n}$ for $n < B$, for we can take $q = \min\{\frac{x_n}{2kn} + \frac{1}{n} \mid n < B\}$.

28.3.2 Equivalence

Definition 28.5 Real numbers x and y are equivalent, $(x \equiv y)$, if and only if $\forall n{:}\mathbb{N}^{+}.\ |x_n - y_n| \leq 4$.

We show next that $(x \equiv y)$ is, in fact, an equivalence relation on $\mathbb{R}$. The proof is similar to Bishop's proof but introduces some concepts that we will need later. Following Bishop, we usually write the equivalence as $x = y$, and Nuprl displays the relation that way, but in this chapter we want to distinguish between equality of regular sequences and the equivalence relation on regular sequences. So we will use $x = y$ for the former and $x \equiv y$ for the latter.

Definition 28.6 The relation $\text{bdiff}(x, y) \equiv \exists B : \mathbb{N}.\ \forall n : \mathbb{N}^{+}.\ |x_n - y_n| \leq B$ says that the pointwise difference between x and y is bounded.

Lemma 28.7 *For all* $x, y \in \mathbb{R}$, $(x \equiv y) \iff \text{bdiff}(x, y)$

Proof If $(x \equiv y)$ then we may take $B = 4$. In the other direction, suppose $\forall n : \mathbb{N}^{+}.\ |x_n - y_n| \leq B$, then, for any $n, m \in \mathbb{N}^{+}$, if $5 \leq |x_n - y_n|$ then $5m \leq m|x_n - y_n| \leq |m{*}x_n - n{*}x_m| + |n{*}x_m - n{*}y_m| + |n{*}y_m - m{*}y_n|$, by the triangle inequality. By regularity and our assumption, the right side is $\leq 4(n + m) + nB$. Thus, $5m \leq 4(n+m) + nB$, so, $m \leq (4+B)n$ and we obtain a contradiction when $m = 1 + (4 + B)n$. Hence, $|x_n - y_n| \leq 4$. Note that this proof by contradiction is constructive because, since $|x_n - y_n| \in \mathbb{Z}$, we can prove $(5 \leq |x_n - y_n|) \vee (|x_n - y_n| \leq 4)$. □

Corollary 28.8 $\equiv$ *is an equivalence relation on* $\mathbb{R}$

Proof The bounded-difference relation, $\text{bdiff}(x, y)$ is clearly an equivalence relation on integer sequences. □

28.3.3 Acceleration

We use the symbol $\div$ for integer division; it satisfies

$$n = k(n \div k) + (n\text{rem } k) \quad (\text{with } |n\text{rem } k| < |k|).$$

Definition 28.9 The sequence $\text{accel}(k, x) = \lambda n.\ x_{2kn} \div 2k$ is called the k-acceleration of sequence x. It rounds off, using integer division, the result of computing x to a factor of $2k$ extra accuracy.

Lemma 28.10 *If sequence x is k-regular, then* $\text{accel}(k, x)$ *is regular, and* $\text{bdiff}(\text{accel}(k, x), x)$.

Proof Because x is k-regular, $2k|m * x_{2kn} - n * x_{2km}| \le 2k(2kn + 2km)$, so $|m * x_{2kn} - n * x_{2km}| \le (2kn + 2km)$. Thus,

$$\begin{aligned}
&2k|m * (x_{2kn} \div 2k) - n * (x_{2km} \div 2k)| \\
&\quad = |m * 2k * (x_{2kn} \div 2k) - n * 2k * (x_{2km} \div 2k)| \\
&\quad \le |m * x_{2kn} - n * x_{2km}| + m|x_{2kn} \operatorname{rem} 2k| + n|x_{2km} \operatorname{rem} 2k| \\
&\quad \le (2kn + 2km) + 2k(n + m) \\
&\quad = 4k(n + m)
\end{aligned}$$

and hence, $|m * \mathrm{accel}(k, x)_n - n * \mathrm{accel}(k, x)_m| \le 2(n + m)$, so $\mathrm{accel}(k, x)$ is regular.

To show that the difference between $\mathrm{accel}(k, x)$ and x is bounded,

$$\begin{aligned}
&2kn|\mathrm{accel}(k, x)_n - x_n| \\
&\quad = |n * 2k * x_{2kn} \div 2k - 2kn * x_n| \\
&\quad \le |n * x_{2kn} - 2kn * x_n| + n|x_{2kn} \operatorname{rem} 2k| \\
&\quad \le 2k(n + 2kn) + 2kn \\
&\quad = 2kn(2 + 2k)
\end{aligned}$$

hence, $|\mathrm{accel}(k, x)_n - x_n| \le (2 + 2k)$, so we can take $B = (2 + 2k)$ in the definition of $\mathrm{bdiff}(\mathrm{accel}(k, x), x)$. □

Corollary 28.11 *If $x \in \mathbb{R}$ and $k \in \mathbb{N}^+$ then* $\mathrm{accel}(k, x) \equiv x$.

Proof x is regular and therefore k-regular, so $y = \mathrm{accel}(k, x)$ is regular and $\mathrm{bdiff}(y, x)$. So, by Lemma 28.7, $y \equiv x$. □

28.3.4 Arithmetic and Ordering

Integers as Real Numbers For every integer k, sequence $k^* = \lambda n.\ 2nk$ is regular (note: $\frac{k^*_n}{2n} = k$). It is a real number, which we often write as k.

Addition If x and y are regular sequences then it is easy to see that the pointwise sum $\lambda n.\ x_n + y_n$ is a 2-regular sequence. So we define addition on real numbers to be

$$x + y = \mathrm{accel}(2, \lambda n.\ x_n + y_n).$$

If we use this definition to compute $(w + (x + (y + z)))_n$ we get

$$(w_{4n} + (x_{16n} + (y_{64n} + z_{64n}) \div 4) \div 4) \div 4.$$

The fact that the later terms in the sum are computed to higher accuracy than the earlier terms would not be desirable when summing many terms of a series like

$\Sigma_{n=0}^{444} e^n/n!$. So we define the sum of a list of real numbers not by iterating the binary operation but by

$$x_1 + \cdots + x_k = \text{accel}(k, \lambda n.\ (x_1)_n + \cdots + (x_k)_n).$$

With this operation $(w + x + y + z)_n$ is

$$(w_{8n} + x_{8n} + y_{8n} + z_{8n}) \div 8.$$

Every term in the sum is computed to the same accuracy and there is only one division.

Absolute Value, max, min, and Unary Minus These operations do not need acceleration. They define regular sequences when x and y are regular:

$$\begin{aligned} |x| &= \lambda n.\ |x_n| \\ \max(x, y) &= \lambda n.\ \max(x_n, y_n) \\ \min(x, y) &= \lambda n.\ \min(x_n, y_n) \\ -x &= \lambda n.\ (-x_n). \end{aligned}$$

Multiplication For multiplication, $\frac{x_n}{2n} * \frac{y_n}{2n} \approx \frac{z_n}{2n}$ suggests that $z = x * y$ should be defined by $z_n = (x_n * y_n) \div 2n$, but this sequence is only a $(k+1)$-regular sequence provided that for all n, $|x_n| \leq nk$ and $|y_n| \leq nk$. To compute such k we use the *canonical bound* operation:

$$\text{cBnd}(x) = |x_1| + 3$$

The reader can verify that for any regular sequence x for all n, $|x_n| \leq n * \text{cBnd}(x)$. So multiplication is

$$x * y = \text{accel}(\max(|x_1|, |y_1|) + 4, \lambda n.\ (x_n * y_n) \div 2n).$$

We see that multiplication can require much greater acceleration than addition and it is not obvious how to form the product of a list of real numbers in such a way that each number is computed to the same accuracy. However, there are two special cases of multiplication that can be simplified. First, for an integer k, the following real number is equivalent to $(k^*) * y$:

$$k(y) = \lambda n.\ \text{if } k < 0 \text{ then } -\left(y_{-(nk)}\right) \text{ else if } 0 < k \text{ then } y_{nk} \text{ else } 0.$$

Second, for a natural number k, and x a real number with $\text{cBnd}(x) = B$, the following real number x^k is equivalent to $x * \cdots * x$, (k-times):

$$x^k = \text{if } k = 0 \text{ then } 1^* \text{else accel}\left(K, \lambda n.\ (x_n)^k \div (2n)^{k-1}\right),$$

where $K = 1 + k * \left((B^{k-1} \div 2^{k-1}) + 1\right)$.

Division by a Positive Natural Number Division x/y is defined by $x * \mathrm{inv}(y)$ when real number y is separated from zero. So we have to define separation and inverse before we define division in the general case. But to divide by a positive natural number k we can use this simple definition

$$x/k = \lambda n.\, x_{2n} \div 2k.$$

For then we have $k(x/k) = \lambda n.\, x_{2nk} \div 2k = \mathrm{accel}(k, x)$. So, by Corollary 28.11, $k(x/k) \equiv x$.

Ordering The orderings, $x \leq y$, $x < y$, and $x\#y$ are defined as follows:

$$\begin{aligned} x \leq y &\Leftrightarrow \forall n{:}\mathbb{N}^+.\, x_n \leq y_n + 4 \\ x < y &\Leftrightarrow \exists n{:}\mathbb{N}^+.\, x_n + 4 < y_n \\ x\#y &\Leftrightarrow (x < y) \vee (y < x). \end{aligned}$$

When $(x\#y)$ we say that real numbers x and y are *separated*. All of these operations and relations must be shown to respect the equivalence relation. For example,

$$\begin{aligned} x_1 \equiv x_2 \wedge y_1 \equiv y_2 &\Rightarrow (x_1 + y_1) \equiv (x_2 + y_2) \\ x_1 \equiv x_2 \wedge y_1 \equiv y_2 &\Rightarrow (x_1 \leq y_1) \Leftrightarrow (x_2 \leq y_2). \end{aligned}$$

These proofs are straightforward using the definitions and Lemma 28.7.

We have now defined all the operations used in the following lemma, whose proof we leave to the reader.

Lemma 28.12 *If $x \in \mathbb{R}$ and $m \in \mathbb{N}^+$ then* $|x - ((x_m)^*/2m)| \leq 1^*/m$.

The following lemma is a companion to Lemma 28.12. We will write k^* as k from now on. We define $x \approx_k a/2m$ to be

$$|x - (a/2m)| \leq k/m.$$

Lemma 28.13 *For $x \in \mathbb{R}$, $y \in \mathbb{N}^+ \to \mathbb{Z}$, and $k \in \mathbb{N}^+$, if $\forall m{:}\mathbb{N}^+.\, x \approx_k y_m/2m$ then y is k-regular and* $\mathrm{accel}(k, y) \equiv x$.

Proof From $|x - (y_n/2n)| \leq k/n$ and $|x - (y_m/2m)| \leq k/m$ the triangle inequality (for $\mathbb{R}$) gives us $|(y_n/2n) - (y_m/2m)| \leq (k/n + k/m)$, which implies that y is k-regular. By Lemma 28.10, there is a bound B such that $\forall n.\, |(\mathrm{accel}(k, y))_n - y_n| \leq B$, so $\forall n.\, |((\mathrm{accel}(k, y))_n/2n) - (y_n/2n)| \leq (B/2n)$. Using the hypothesis and the triangle inequality, $\forall n.\, |((\mathrm{accel}(k, y))_n/2n) - x)| \leq ((B + 2k)/2n)$. So, by Lemma 28.12 and the triangle inequality,

$$\forall n.\, |(\mathrm{accel}(k, y) - x)| \leq ((B + 2k + 2)/2n).$$

Since this holds for all n, it implies $|\text{accel}(k, y) - x)| = 0$, and hence $\text{accel}(k, y) \equiv x$. □

Lemma 28.13 is useful when we have constructed a real number x but the construction does not result in an efficient program. If we can find another, more efficient program for a sequence y of approximations to x that satisfies the hypothesis of Lemma 28.13, then $\text{accel}(k, y)$ is equivalent to x.

It is instructive to see the proof of one more important lemma using this representation of the real numbers.

Lemma 28.14 *If $x < y$ then $\forall z:\mathbb{R}.\ (x < z) \vee (z < y)$.*

Proof Let n be such that $x_n + 4 < y_n$ and let $m = 12n + 1$. Since $x_n + 5 \le y_n$ we have $m*x_n + 5m \le m*y_n$. Then $n*x_m \le m*x_n + 2(n+m) \le m*y_n + 2n - 3m \le n*y_m + 2(n+m) + 2n - 3m = n*y_m + 4n - m = n*y_m - (8n+1)$. So, $n*(x_m + 8) < n*y_m$ and hence, $x_m + 8 < y_m$. So we have used regularity to widen the gap $x_n + 4 < y_n$ to $x_m + 8 < y_m$. Now we must have either $x_m + 4 < z_m$ or $z_m + 4 < y_m$ because otherwise $y_m \le z_m + 4 \le x_m + 8$. □

From Lemma 28.14 we get the fundamental *cotransitivity* property of separation:

Corollary 28.15 *(Co-transitivity of #) For all real numbers x and y*

$$x\#y \Rightarrow (\forall z:\mathbb{R}.\ (z\#x) \vee (z\#y)).$$

We will finish our discussion of our representation of the real numbers with the definitions of inverse and division and a look at the constructive content of completeness. But first we need a short interlude to introduce some type theory.

28.4 Some Type Theory

Suppose that for every $a \in A$, $B(a)$ is a type. Then we call B a *family of types*, indexed by type A. For any such family, we have the *dependent function type* $a : A \to B(a)$ and the *dependent pair type* $a : A \times B(a)$. A member f of $a : A \to B(a)$ is a function such that $f(a) \in B(a)$ whenever $a \in A$, and a member $\langle a, b\rangle$ of $a : A \times B(a)$ is a pair with first component $a \in A$ and second component $b \in B(a)$.

A constructive proof p of a statement $\forall a : A.\ P(a)$ constructs, for each $a \in A$, evidence that $P(a)$ is true. If we identify $P(a)$ with the type of evidence for $P(a)$, then the construction can be seen as a function p, defined for $a \in A$, such that $p(a) \in P(a)$. This is the same as saying $p \in a : A \to P(a)$. A constructive proof of $\exists a : A.\ P(a)$ must construct a witness $a \in A$ and evidence, $b \in P(a)$, that $P(a)$ is true. This is the same as saying that $\langle a, b\rangle \in a : A \times P(a)$.

We use the type Unit, which has only one member Ax, as the evidence for simple, checkable, true propositions about numbers, such as $3 < 7$, or $5 = 5$. The type Void, which has no members, is the evidence type for false propositions.

The logic of proof assistants like Nuprl and Coq is based on this "propositions as types" isomorphism.

$$\begin{aligned}
A \wedge B &\longleftrightarrow A \times B \\
A \Rightarrow B &\longleftrightarrow (A \to B) \\
A \vee B &\longleftrightarrow A + B \\
\forall a\!:\!A.\ B(a) &\longleftrightarrow a\!:\!A \to B(a) \\
\exists a\!:\!A.\ B(a) &\longleftrightarrow a\!:\!A \times B(a) \\
n = m &\longleftrightarrow \text{Unit, if } n = m, \text{ Void, otherwise} \\
n < m &\longleftrightarrow \text{Unit, if } n < m, \text{ Void, otherwise} \\
\text{False} &\longleftrightarrow \text{Void}
\end{aligned}$$

We often want to be even more precise when characterizing the evidence we have available and the evidence constructed by our proofs. Sometimes we know not only that for every $a \in A$ there is member of $B(a)$ but more: there is one thing, say t, that is evidence of $B(a)$ for all $a \in A$. In MINLOG this is expressed as $\forall^{nc} a\!:\!A.\ B(a)$. The superscript nc means *noncomputational* and in this case means that the function giving the evidence for $B(a)$ does not need to be provided with the input a.

In Nuprl, for any type family $B(a)$ indexed by type A, we also have the *intersection, union*, and *refinenment* types $\bigcap_{a:A} B(a)$, $\bigcup_{a:A} B(a)$, and $\{a\!:\!A \mid B(a)\}$. The intersection type $\bigcap_{a:A} B(a)$ corresponds to MINLOG's $\forall^{nc} a\!:\!A.\ B(a)$ since it contains those evidence terms that are in all the types $B(a)$ for $a \in A$. We call this the *uniform* $\forall$-quantifier.

The union and refinement types correspond to noncomputational versions of the existential quantifier in MINLOG. Members of the union type $\bigcup_{a:A} B(a)$ are all terms b that are the second component of a pair $\langle a, b\rangle$ in type $a : A \times B(a)$, and members of the refinement type $\{a : A \mid B(a)\}$ are the first components of such pairs. Using brackets to indicate the part of the evidence that is not provided, that is, is noncomputational, or *squashed*, we have the following propositional forms of these types and one additional definition:

$$\begin{aligned}
\forall [a\!:\!A].\ B(a) &\longleftrightarrow \bigcap_{a:A} B(a) \\
\exists [a\!:\!A].\ B(a) &\longleftrightarrow \bigcup_{a:A} B(a) \\
\exists a\!:\!A.\ [B(a)] &\longleftrightarrow \{a\!:\!A \mid B(a)\}
\end{aligned}$$

$$\downarrow P \longleftrightarrow \exists a\!:\!\text{Unit}.\ [P]$$

Evidence for $\downarrow P$ (called the *squash* of P) is just the term Ax $\in$ Unit but for which we know that there is evidence for P but we don't have that evidence. If there is a function that can construct the evidence for P under the assumption that such evidence exists, then we say that P is *squash stable*. Formally, P is squash stable if $\downarrow P \Rightarrow P$.

For the current discussion, when x and y are real numbers, an important example of a squash stable proposition is $x < y$. This is because $x < y$ is $\exists n\!:\!\mathbb{N}^{+}.\ x_n + 4 < y_n$, and if we know that it is true then we can find an integer k that is evidence for it using a search $k = \mu n.\ x_n + 4 < y_n$. This is not the same as Markov's principle which would imply that $\neg\neg(x < y) \Rightarrow x < y$. It only says that if we have proved that such an n exists but then thrown away the evidence from our proof, then we can reconstruct that evidence.

For $x < y$ we define a function rlessw(x, y), *witness real less*, that searches for a k such that $x_k + 4 < y_k$. If $y - x$ is very small then this search could take many steps. We write the function rlessw so that it tries $k = 1, 2, 4, 8, \ldots$ rather than trying $k = 1, 2, 3, 4, \ldots$. Even with this optimization, rlessw(x, y) can be inefficient, and because we need it in the implementation of division, division by very small real numbers can also be inefficient.

28.5 Extracts of Proofs by Induction

Using the uniform $\forall$, we can state the simple induction scheme for the natural numbers as the proposition:

$$\forall[P\!:\!\mathbb{N} \to \text{Type}].\ P(0) \Rightarrow (\forall n > 0.\ P(n-1) \Rightarrow P(n)) \Rightarrow (\forall n.\ P(n)).$$

Its witness is *primitive recursion* $\lambda b.\lambda f.\lambda n.$ primrec(n, b, f). Complete induction is

$$\forall[P\!:\!\mathbb{N} \to \text{Type}].\ (\forall i\!:\!\mathbb{N}\ (\forall j\!:\!\mathbb{N}_i.\ P(j)) \Rightarrow P(i)) \Rightarrow (\forall n\!:\!\mathbb{N}.\ P(n)).$$

Here $\mathbb{N}_i = \{j\!:\!\mathbb{N} \mid j < i\}$. The witness for complete induction is

$$\lambda F.\ \text{letrec}\ f(i) = F(i, f)\text{in}\ f,$$

which we can call *general recursion*. There is a uniform version of complete induction where all the $\forall$-quantifiers are uniform :

$$\forall[P\!:\!\mathbb{N} \to \text{Type}].\ (\forall[i\!:\!\mathbb{N}]\ (\forall[j\!:\!\mathbb{N}_i].\ P(j)) \Rightarrow P(i)) \Rightarrow (\forall[n\!:\!\mathbb{N}].\ P(n)).$$

Uniform complete induction has the beautiful witness $\lambda F.$ fix(F), where fix(F) $= F(\text{fix}(F))$.

When we prove a theorem or lemma by induction on natural numbers, we can choose which of these forms to use by invoking a *tactic* that instantiates the desired induction scheme. Then the extract of our proof will use one of the above induction forms. Choosing the right form of induction is part of the art of getting good extracts from our constructive proofs.

We illustrate the difference these kind of choices can make by proving the following lemma (which we will later use to prove that $\sin(x)$ converges very fast when $|x| \leq 1/2$).

Lemma 28.16 *For $a > 1$, let $f(k) = a^{2k+3} * (2k+3)!$. Then f is unbounded, that is:*

$$\forall N.\, \exists k.\, [N \leq f(k)].$$

Proof 1 By induction on N. If $N = 0$, take $k = 0$. Assuming that $N - 1 \leq f(k')$, if $N \leq f(k')$ then take $k = k'$, otherwise, clearly, $k = k' + 1$ works because $f(k'+1) = f(k') * (a^2 * (2k'+4) * (2k'+5))$, and the latter factor is at least 80.

Since this argument is so trivial, one would usually write *the proof is obvious*. But if we must use it to compute, then the proof matters; and this proof is terrible for computing. It has an extract of the form primrec$(N, 0, f)$, so to compute the k that works for $N = 10^{10}$ when $a = 2$, the algorithm will perform more than 10^{10} steps because it is using primitive recursion.

Proof 2 For given $a > 1$ and N, let $P(k) = (N \leq f(k))$. By complete induction on d we prove the claim

$$\forall d{:}\mathbb{N}.\, \forall k'{:}\mathbb{N}.\, \forall b{:}\{b{:}\mathbb{N} \mid N \leq (d+b) \wedge b = f(k')\}.\, \exists k.\, P(k).$$

We have $d, k', b \in \mathbb{N}$ and $N \leq d + b$ and $b = f(k')$. If $N \leq b$ then $P(k')$ and we are done. Otherwise, $d > 0$ so $d' = d - 1$ is less than d. We let $k'' = k' + 1$ and let $b' = a^2 * (2k'+4) * (2k'+5) * b$. Then $b' = f(k'')$ and we apply the induction hypothesis to d', k'', and b'. We only have to check that $N \leq d' + b'$, but since $b < b'$, $d + b \leq d' + b'$.

To prove the lemma from the claim: If $N \leq f(0)$ then take $k = 0$, otherwise let $d = N$, $k' = 0$, and $b = f(0)$ in the claim. This proof looks a bit strange because it introduces the extra variables d, k', and b, but the extract it produces is

$\lambda a \lambda N.$ let $b = 6a^3$in
  if $N \leq b$ then 0 else
  letrec $f(d, k', b) =$ let $k'' = k' + 1$in
    let $b' = a^2 * (2k'+4) * (2k'+5) * b$ in
    if $N \leq b'$ then k'' else $f(d-1, k'', b')$
  in $f(N, 0, b)$

which is far better than the extract of proof 1. For $a = 2$, $N = 10^{10}$ it will find $k = 4$ (because $10^{10} \leq 2^{11} * (11)!$) in very few steps.

Proof 3 We notice that the parameter d in the extract of Proof 2 gets decremented on each iteration of the recursion, but its value is never tested or used. If we use exactly the same proof as Proof 2, but use *uniform complete induction* to prove the claim

$$\forall[d{:}\mathbb{N}].\ \forall k'{:}\mathbb{N}.\ \forall b{:}\{b{:}\mathbb{N} \mid N \leq (d+b)\ \wedge\ b = f(k')\}.\ \exists k.\ P(k)$$

we get an extract with no parameter d, as follows.

$\lambda a \lambda N.$ let $b = 6a^3$in
 if $N \leq b$ then 0 else
 letrec $f(k', b) =$ let $k'' = k' + 1$in
 let $b' = a^2 * (2k' + 4) * (2k' + 5) * b$ in
 if $N \leq b'$ then k''else $f(k'', b')$
 in $f(0, b)$

The difference between the extract for Proof 3 and the extract for Proof 2 does not change the complexity but it does save a few steps and the resulting code is shorter and easier to read. In general, though, we don't need to read the code extracted from our proofs unless we want to understand how it could be improved. If it runs fast enough for our purposes, we never need to look at it.

Nuprl allows us to simply write the program we want and then prove that it realizes the theorem we intend it to realize. However, we find that that method can be tedious and error prone, so it pays to learn that one can get the desired extract by choosing which induction tactics to use and using the right versions of intermediate claims.

28.6 Inverse, Division, and Computation

28.6.1 Inverse and Division

If x is a member of the type $\{x : \mathbb{R} \mid x\#0\}$ then rlessw$(0, |x|)$ will find a k such that $4 < |x_k|$ and this implies (by regularity of x) that $0 < |x_n|$ for all $n \geq k$. Let $\overline{x} = \lambda n.$ if $n < k$ then 2 else x_n. Then bdiff$(x, \overline{x})$ so $\overline{x} \equiv x$ but $\overline{x}_n$ is never zero.

We want $\frac{x^{-1}(n)}{2n} \approx \frac{2n}{x(n)}$, so $x^{-1}(n)$ should be $4n^2 \div x(n)$, but we must not divide by zero, so we use $\overline{x}(n)$ instead. To get a regular sequence, we must also accelerate this sequence. It turns out that we can prove that accelerating by $4(4k^2 + 1)$ works.

Definition 28.17 The inverse of real $x \in \{x:\mathbb{R} \mid x\#0\}$ is
$x^{-1} = \text{accel}(4(4(\text{rlessw}(0,|x|))^2 + 1), \lambda n.\ 4n^2 \div x_n)$

We can prove that if $x \in \{x:\mathbb{R} \mid x\#0\}$ then $x^{-1} \in \mathbb{R}$ and $x * x^{-1} \equiv 1$. Division is then defined by

$$x/y = x * y^{-1}$$

28.6.2 Computation with Real Numbers

If x is any real number, by Lemma 28.12, for $m = (10^k) \div 2$, we have the approximation $|x - (x_m/10^k)| \leq 2/10^k$. So x_m gives us the first k decimal digits of x with only the last digit possibly off by one. So we define

$$\text{dd}(x,k) = x_{(10^k \div 2)}.$$

Nuprl allows us to compute such expressions in two ways. The first way, which is usually used in proofs, steps through the computation step by step. That is, it runs an interpreter for the Nuprl computation system. This can be very useful to see exactly how the computation unfolds and estimate its complexity, but it can be slow because it is only an interpreter and not a compiler.

The second way to compute relies on the fact that the Nuprl computation system has only about 20 primitive operations and these are all easily translated into LISP (which is the implementation language for the Nuprl kernel). The only issue is that LISP uses eager evaluation while Nuprl uses lazy evaluation by default. Thus it is possible that the LISP translation of a Nuprl expression could denote $\perp$ (namely, bottom) a diverging term even when the Nuprl term does not. However, if the LISP evaluation yields a canonical integer term, then it must denote the same value as the Nuprl expression. Thus for expressions like $dd(x,k)$ we can compute using LISP code that is compiled and runs much faster.[1]

To give some idea of how efficient such computations are we will write

$$\text{digits}(x) \mapsto N_1, N_2, N_3, N_4$$

if we can compute, using the compiled LISP for x, N_1 decimal digits of x in $1/25$ s, N_2 digits in $1/5$ s, N_3 digits in 1 s and N_4 digits in 5 s. Each bound is five times the previous bound; the sequence N_1, N_2, N_3, N_4 then gives a rough sense of the complexity.

[1] We have an example of a Nuprl expression that took 31 h to evaluate using the interpreter but less than 1 s using the compiled LISP code.

An Example from Siegfried Rump Define

$$f(a,b) = 21b^2 - 2a^2 + 55b^4 - 10a^2b^2 + (a/2b).$$

Then $f(a,b)$ is a real number as long as $b\#0$. We let $a = 77617$ and $b = 33096$ and compute 15 digits of $f(a,b)$ using $dd(f(a,b),15)$. The result -0.827396059946821 is computed fast using the interpreter and almost instantaneously using the compiler.

What is notable about this example is that making the same definition in OCaml (or MATLAB, or any other system) using the type float (which uses IEEE standard 754 double precision (64-bit) floating-point numbers) gives the answer 0.117260394 005317861 in which not even the first digit is correct!

Using the Nuprl code we have the following.

$$\text{digits}(f(77617, 33096)) \mapsto 6528, 14784, 32753, 73734$$

This shows that at least our basic arithmetic is fast since we can compute 6528 (correct!) digits of the example in 0.04 s, and each factor of five in time seems to give us more than double the number of digits.

To check that this speed does not overly depend on the simplicity of the inputs to the function, we will also check how many digits we can compute when the inputs are set to an irrational number[2], $\pi/4$:

$$\text{digits}(f(\pi/4, \pi/4)) \mapsto 255, 687, 1319, 2312.$$

As expected, we can compute fewer digits of this number within the given time bounds, but we are still getting roughly double the digits in each increment.

28.7 Completeness

Definition 28.18 A sequence X of reals is Cauchy if
$\forall k{:}\mathbb{N}^+.\ \exists N{:}\mathbb{N}.\ [\forall n, m{:}\mathbb{N}.\ (N \leq n \land N \leq m) \Rightarrow |X(n) - X(m)| \leq 1/k]$.
Sequence X converges to real x (written $\lim_{n\to\infty} X(n) = x$) if
$\forall k{:}\mathbb{N}^{\cdot}\ \exists N{:}\mathbb{N}.\ [\forall n{:}\mathbb{N}.\ (N \leq n) \Rightarrow |X(n) - x| \leq 1/k]$.
Sequence X converges ($X\downarrow$) if $\exists x{:}\mathbb{R}.\ \lim_{n\to\infty} X(n) = x$.

Note that we have used one of the squashed (noncomputational) forms of existential quantifier in the definition of Cauchy and the definition of converges. That is because we observe that since $(x \equiv y)$ is $\forall n{:}\mathbb{N}^+.\ |x_n - y_n| \leq 4$ and $|x_n - y_n| \leq 4$ is a primitive fact about integers, when $(x \equiv y)$ is true the evidence is $(\lambda n.\ \text{Ax})$. We say that such propositions have *no constructive content*. The proposition $x \leq y$ also has no constructive content. Squash-stable propositions like $x < y$ have constructive content, but it is trivial in the sense that it can be computed on demand from x

[2] We discuss the definition of π in Section 28.12.

and y. When stating that $\exists x:T.\ P(x)$ when $P(x)$ has no constructive content, or $P(x)$ is squash stable, then we use the form $\exists x:T.\ [P(x)]$ which has evidence in the type $\{x:T \mid P(x)\}$.

The evidence for $X$is Cauchy is therefore a function of type $\mathbb{N}^+ \to \mathbb{N}$ called the *modulus of Cauchyness*. Likewise the evidence for $\lim_{n\to\infty} X(n) = x$ is a function of type $\mathbb{N}^+ \to \mathbb{N}$ called the *modulus of convergence*, and the evidence for $X\downarrow$ is a pair $\langle x, c\rangle$ where x is a real number and c is a modulus of convergence for $\lim_{n\to\infty} X(n) = x$.

Theorem 28.19 (Completeness) $\forall X:\mathbb{N} \to \mathbb{R}.\ (X\downarrow) \Leftrightarrow (X\text{is Cauchy})$.

We won't give the proof of Theorem 28.19, which followed the proof by Bishop and Bridges. Instead we will examine the computational content extracted from the proof of this theorem. It is $\lambda X.\ \langle f, g\rangle$ where

$$f = \lambda p.\ \text{let } x, c = p \text{in } \lambda k.\ c(2k)$$

$$g = \lambda C.\ \langle \text{accel}(2, \lambda n.\ X(C(n))(n)), \lambda k.\ cy(4k) + 1\rangle\rangle.$$

Function f shows that given the modulus of convergence c, the modulus of Cauchyness is $\lambda k.\ c(2k)$ (independent of the x that X converges to). Function g shows that from X and its modulus of Cauchyness, C, we construct the real that X converges to by accelerating (by two) the "diagonal" sequence $\lambda n.\ X(C(n))(n)$, and the modulus of convergence is $\lambda k.\ C(4k) + 1$. (The factors of 2 and 4 that occur in the evidence, come from "$\epsilon/2$ arguments" used in the proof.)

Real numbers constructed using such a diagonal of a Cauchy sequence include the reals constructed to prove Cantor's theorem (that in any nontrivial interval there are uncountably many reals), and the constructive version of the intermediate value theorem (that constructs a real x where $|f(x) - c| < \epsilon$).

28.8 Constructing kth Roots

Bishop does not explicitly consider the function $\sqrt[k]{x}$ because it is $e^{\ln(x)/k}$, but, in general, the logarithm function is not very efficient and the exponential is also somewhat costly. The exact intermediate value theorem is true for polynomials (as proved in the exercises for Chapter 2 of Bishop and Bridges, and formalized in Nuprl) but using that theorem to construct kth roots is not very efficient, either. We can derive a much more efficient algorithm for $\sqrt[k]{x}$.

Definition 28.20 For $k \in \{2, 3, \dots\}$ and $z, r \in \mathbb{N}$ we say that r is the integer kth root of z (and write $r = \text{iroot}(k, z)$) if $r^k \le z < (r+1)^k$.

28.8.1 Fast Integer Roots

Every nonnegative integer has an integer kth root. The program generated from the following proof (by Christoph Kreitz) of this fact is very efficient.

Lemma 28.21 $\forall k\colon\{k\colon\mathbb{Z} \mid 2 \leq k\}\,.\,\forall z\colon\mathbb{N}.\ \exists r\colon\mathbb{N}.\ r^k \leq z < (r+1)^k.$

Proof We use complete induction on z, so we assume that every $z' < z$ has an integer kth root. If $z = 0$ then we take $r = 0$. Otherwise, let $z' = z \div 2^k$, then $z' < z$ so there is an s such that $s^k \leq z' < (s+1)^k$. Let $x = 2s$ and $y = 2s+1$. It is easy to check that if $y^k \leq z$ then y is the integer kth root of z and otherwise x is the integer kth root. □

The program generated from this proof is as follows.

$$
\begin{array}{rcl}
\lambda k.\ \text{let } b & := & 2^k \text{in} \\
\text{letrec } f(i) & = & \text{if } i = 0 \text{ then } 0 \\
 & & \text{else let } i' := i \div b \text{ in} \\
 & & \text{let } x := 2f(i') \text{ in} \\
 & & \text{let } y := x + 1 \text{in} \\
 & & \text{if } y^k \leq i \text{ then } y \text{ else } x \\
\text{in } \lambda z.\ f(z) & &
\end{array}
$$

This program runs in time proportional to $\frac{\log_2(z)}{k}$ so it is very fast. With $k = 2$ and $z = 2 * 10^{40}$ it computes to 141421356237309504880 in 1703 primitive computation steps.

28.8.2 A Simple kth-Root Program

If k is even, then we can only construct a kth root for real x when $0 \leq x$. So, we will first construct $z = \sqrt[k]{|x|}$ for any real x and any $k \in \{2, 3, \ldots\}$. What should the nth approximation $z(n)$ be? We want $\frac{(z(n))^k}{(2n)^k} \approx |x| \approx \frac{|x(n^k)|}{2n^k}$, so $(z(n))^k \approx 2^{k-1}|x(n^k)|$. This suggests that the following simple program might construct $\sqrt[k]{|x|}$

$$\text{simpleroot}(k, x) = \lambda n.\ \text{iroot}(k,\ 2^{k-1}|x(n^k)|) \stackrel{?}{=} \sqrt[k]{|x|}.$$

If this construction is correct, then simpleroot(k, x) is a regular sequence, whenever x is regular. If not, then maybe simpleroot(k, x) is a K-regular sequence for some $K = K(k, x)$ in which case the correct program is

$$\text{accel}(K, \text{simpleroot}(k, x)).$$

Initially, we did not know how to prove either of these conjectures, so we extracted a different, slower, program root(k, x) from a proof that constructed a kth root as the limit of a Cauchy sequence. Comparing the two programs root(k, x) and simpleroot(k, x) on many test cases we found that they never differed by more than one, so it was very likely that they define the same real number, and hence, simpleroot(k, x) is in fact regular. So we were encouraged to find the following proof that it is regular.

Before we present the proof, let's check how fast we can compute using this program.

$$\text{digits}\left(\sqrt{2}\right) \mapsto 1073, 1900, 3292, 5651$$
$$\text{digits}\left(\sqrt[12]{2}\right) \mapsto 255, 534, 930, 1558$$
$$\text{digits}\left(\sqrt{\pi/4}\right) \mapsto 409, 731, 1283, 2175$$
$$\text{digits}\left(\sqrt[12]{\pi/4}\right) \mapsto 69, 126, 221, 378$$

28.8.3 Proof of Simple Root Construction

We have to show that if x is a real number (i.e., a regular sequence) then $z =$ simpleroot(k, x) is a real number, and we have to show that $z^k = |x|$.

To show that z is regular we must show:

$$\left|m * \text{iroot}\left(k,\ 2^{k-1}\left|x\left(n^k\right)\right|\right) - n * \text{iroot}\left(k,\ 2^{k-1}\left|x\left(m^k\right)\right|\right)\right| \leq 2(n+m).$$

Let $s = 2^{k-1}$, $i = n^k$, $j = m^k$, $c = j * s * |x(i)|$, and $d = i * s * |x(j)|$. Let $a = m * \text{iroot}(k,\ s * |x(i)|)$ and $b = n * \text{iroot}(k,\ s * |x(j)|)$. Then we must prove $|a - b| \leq 2(n+m)$. Since

$$\text{iroot}(k,\ s * |x(i)|)^k \leq s * |x(i)| < (\text{iroot}(k,\ s * |x(i)| + 1)^k$$

we have, multiplying by $j = m^k$,

$$a^k \leq j * s * |x(i)| < (a+m)^k.$$

So, $a^k \leq c < (a+m)^k$, and similarly, $b^k \leq d < (b+n)^k$. By the regularity of $|x|$ we have $|c - d| \leq 2s(i+j)$. By symmetry, we may assume $a \leq b$. Then either $|a - b| \leq 2(n+m)$ (in which case we are done) or $(2n+m) + (a+m) < b$. In the latter case, $((2n+m) + (a+m))^k < b^k \leq d$, so

$$((2n+m) + (a+m))^k - c \leq d - c \leq |d - c| \leq 2^k(i+j).$$

Now, if $a = 0$, then also $c = 0$, and therefore

$$2^k(i+j) \leq 2^k(n+m)^k < b^k \leq d = |c - d| \leq 2^k(i+j),$$

which is impossible. Hence, $a \neq 0$, and this implies that $m \le a$, so $2m \le (a+m)$. The last term in the binomial expansion of $((2n+m)+(a+m))^k$ is $(a+m)^k$ which is greater than c, and all the other terms are $\ge$ the corresponding term in the binomial expansion of $((2n+m)+2m)^k$. Thus,

$$(2n+3m)^k - (2m)^k \le ((2n+m)+(a+m))^k - c \le 2^k(i+j) = (2n)^k + (2m)^k.$$

Then

$$(2n)^k + (3m)^k \le (2n+3m)^k \le (2n)^k + 2(2m)^k$$

so, $(3m)^k \le 2(2m)^k$, which implies $3^k \le 2^{k+1}$, and it is easy to prove that this is false when $k \ge 2$; so we have shown that $z = \text{simpleroot}(k, x)$ is a real number.

To show that $z^k = |x|$ we only have to show bdiff$(z^k, |x|)$. From the simplified definition of z^k we gave earlier, and Lemma 28.10, it is enough to find a bound B such that, for all $n \in \mathbb{N}$,

$$\left|(z(n))^k \div (2n)^{k-1} - |x(n)|\right| \le B.$$

Let $B_z = \text{cBnd}(z)$. Since z is regular we have, for all n, $z(n) \le 2nB_z$.

As above, set $s = 2^{k-1}$ and $i = n^k$; and, in addition, set $u = |x(i)|$, $v = |x(n)|$, and $R = z(n) = \text{iroot}(k,\ s * u)$.

Then, we have the following.

$$\begin{aligned} s * i * \left|R^k \div (2n)^{k-1} - v\right| &= \\ \left|n * \left(R^k - \left(R^k \text{ rem } (2n)^{k-1}\right)\right) - s * i * v\right| &\le \\ \left|nR^k - snu\right| + |snu - siv| + n\left|R^k \text{rem } (2n)^{k-1}\right| &\le \\ n\left|R^k - su\right| + 2s(i+n) + si &\le \\ n\left((R+1)^k - R^k\right) + 5si &\le \\ n * k * (R+1)^{k-1} + 5si &\le \\ n * k * (2R)^{k-1} + 5si &\le \\ si * k * (2B_z)^{k-1} + 5si & \end{aligned}$$

Thus, $\left|R^k \div (2n)^{k-1} - v\right| \le k * (2B_z)^{k-1} + 5$, so we can choose $B = k * (2B_z)^{k-1} + 5$.

That completes the proof that $\text{simpleroot}(k, x) = \sqrt[k]{|x|}$. We did not see this somewhat tricky proof at first, but our experimental results convinced us that the result was true, so we were able to work harder and find the proof. We extended this construction to get $\sqrt[k]{x}$ (i.e., to handle x, rather than $|x|$ when k is odd) but we won't give the details here.

28.9 Computing Power Series

Suppose $f(x) = \sum_{i=0}^{\infty} a_i x^i$ is a function defined as a limit of a power series. Let $S_k(x) = \sum_{i=0}^{k} a_i x^i$. We can use the modulus of convergence for the series, or some even better estimate, to find a function b such that for $N \in \mathbb{N}^+$

$$\left|f(x) - S_{b(N)}(x)\right| \leq \frac{1}{N}$$

If we can find $z \in \mathbb{N}^+ \to \mathbb{Z}$ such that $S_{b(N)}(x) \approx_1 z_N/2N$ then by Lemma 28.13, $f(x) \equiv \text{accel}(2, z)$, because $f(x) \approx_2 z_N/2N$. When the absolute value of x is small, we can compute the approximation more efficiently than by simply computing the Nth approximation $\left(S_{b(N)}(x)\right)_N$.

Lemma 28.22 *For $x, y \in \mathbb{R}$, $j, M \in \mathbb{M}^+$, and $u, v \in \mathbb{Z}$, if $|x| \leq 1/4$, $2|v| \leq M$, $x \approx_j u/2M$, and $y \approx_j v/2M$ then $x * y \approx_j (uv \div 2M)/2M$.*

Proof From $|x - (u/2M)| \leq j/2M$ and $|y - (v/2M)| \leq j/2M$ we derive $|xy - (uv/2M(2M))| \leq |x|j/2M + j|v|/2M^2$. We also have

$$|(uv/2M(2M)) - (uv \div 2M)/2M| \leq 1/2M$$

so $|xy - (uv \div 2M)/2M| \leq 1/2M + |x|j/2M + j|v|/2M^2 \leq 1/2M + j/8M + j/4M$
$\leq j/2M$. □

We use Lemma 28.22 to compute an approximation to $S_k(x)$ when $|x| \leq 1/4$ as follows. If $s_0 = a_k$ and $s_j = a_{k-j} + x * s_{j-1}$, then $s_k = S_k(x)$ so we want z such that $s_k \approx_1 z/2N$. To do that we let $M = 2(k+1)N$ and compute z_k such that $s_k \approx_{k+1} z_k/2M$, for then we can take $z = z_k \div 2(k+1)$. To get z_k we inductively construct z_j such that $s_j \approx_{j+1} z_j/2M$, so we start with $z_0 = (a_k)_M$. If we have z_{j-1} with $s_{j-1} \approx_j z_{j-1}/2M$ then we need z_j with $a_{k-j} + x * s_{j-1} \approx_{1+j} z_j/2M$. For that, it is enough to get p such that $x * s_{j-1} \approx_j p/2M$ for then we can take $z_j = (a_{k-j})_M + p$. According to Lemma 28.22 we can take $p = (u * z_{j-1} \div 2M)$ if we have $x \approx_j u/2M$ and $2|z_{j-1}| \leq M$. If we use $u = x_M$ then we have $x \approx_j u/2M$ for all $j > 0$ so we can use the same u in every step and compute it only once. Since $s_{j-1} \approx_j z_{j-1}/2M$, the condition $2|z_{j-1}| \leq M$ will hold when $|s_{j-1}| < 1$ and this will usually be the case when $|f(x)|$ is reasonably small, and especially for the earlier s_j which come from the later terms of the power series. When the condition is violated, then we have to compute p using the definition of multiplication, and that will require computing x to a higher accuracy than $u = x_M$.

28.10 $\sin(x)$, $\cos(x)$, and e^x

Bishop and Bridges construct the sine, cosine, and exponential functions using power series $\sum_{i=0}^{i=\infty} a_i * x^i$ that match the Taylor series they must have in order to satisfy the desired differential equations. The series for sine and cosine converge everywhere by the alternating series test, and the series for e^x converges everywhere by the ratio test. Having proved everything in Chapter 2 of Bishop and Bridges, we get programs for the functions sine(x), cosine(x), and exp(x).

$$\text{digits}(\text{sine}(2)) \mapsto 28, 63, 106, 163$$
$$\text{digits}(\text{cosine}(2)) \mapsto 29, 63, 106, 164$$
$$\text{digits}(\exp(2)) \mapsto 9, 18, 26, 37$$

This performance is reasonable but not good. Function exp(x) converges using the ratio test which does not give a modulus of convergence as good as for sine(x) and cosine(x) which converge by the alternating series test.

Using the method explained in Section 28.9, we made new versions sine$_s(x)$, cosine$_s(x)$, and exp$_s(x)$ of these functions for small x. For sine$_s(x)$ and cosine$_s(x)$ we only need $|x| \leq 1/2$, since they can be expressed using power series in x^2, while for exp(x) we need $|x| \leq 1/4$. We also need to find functions $b(N)$ to tell us how many terms of the power series we need in order to compute the function within $1/N$. For sine$_s(x)$, with $|x| \leq 1/2$, we can use $b(N)$ the number from Lemma 28.16 with $a = 2$ and for cosine$_s(x)$, with $|x| \leq 1/2$, a number computed similarly. For exp$_s(x)$, with $|x| \leq 1/4$, we use Taylor's theorem to prove that taking k terms of the series leaves a remainder R such that $R \leq 1/(4^k * 3 * k!)$ so $b(N)$ can be a k with $4^k * 3 * k! \geq N$ found in a manner similar to the proof of Lemma 28.16. Computing with these we get a dramatic improvement.

$$\text{digits}(\text{sine}_s(1/2)) \mapsto 911, 2591, 4657, 8227$$
$$\text{digits}(\text{cosine}_s(1/2)) \mapsto 950, 2608, 4658, 8193$$
$$\text{digits}(\exp_s(1/4)) \mapsto 759, 2080, 3439, 6399$$

For the general case of e^x we can decide, using Lemma 28.14, whether $x \in (-\infty, -1/5)$ or $x \in (-1/4, 1/4)$, or $x \in (1/5, \infty)$. In the middle case we use exp$_s(x)$; in the other two cases we use the canonical bound of $|4x|$ to find $N \in \mathbb{N}$ for which $|x/N| \leq 1/4$ and use $(\exp_s(x/N))^N$.

$$\text{digits}(e^2) \mapsto 1056, 2032, 3760, 6753$$
$$\text{digits}(e^{\pi/4}) \mapsto 337, 936, 1722, 3021$$
$$\text{digits}(e^{10}) \mapsto 504, 1680, 3330, 6041$$

To reduce sin(x) and cos(x) to cases where $|x| \leq 1/2$, we first use the periodicity and the identities sine$(x + \pi/2) =$ cosine(x) and cosine$(x + \pi/2) = -$sine(x) to reduce to computing a sine or cosine of an $x' \in (-2, 2)$, which is possible since $\pi/2 < 2$. Then we can decide of which interval $(-2, -99/1000)$, $(-1, -49/100)$, $(-1/2, 1/2)$,$(49/100, 1)$, or $(99/100, 2)$ is x' a member. For the middle interval, we use sine$_s(x)$ or cosine$_s(x)$. For the other intervals we use the double angle formulas for sine and cosine to express the function of x in terms of sine and cosine of $x/2$, or use them twice to get to sine and cosine of $x/4$.

$$\begin{aligned}
\text{digits}(\sin(2)) &\mapsto 255, 815, 2000, 3539 \\
\text{digits}(\sin(10^{10})) &\mapsto 387, 848, 1555, 2730 \\
\text{digits}(\cos(2)) &\mapsto 403, 1154, 2014, 3512 \\
\text{digits}(\cos(10^{10})) &\mapsto 235, 722, 1247, 2401
\end{aligned}$$

28.11 ln(x) and arcsin(x)

The functions ln(x) and arcsin(x) are defined as integrals and their basic properties are proved from these definitions using the fundamental theorem of calculus. If we try to compute these function using their definitions as integrals they are much too inefficient. Various Taylor series expansions for ln(x) are valid only in certain intervals, and they converge slowly.

We tried a version of the intermediate value theorem optimized for strictly increasing functions like e^x for which we can quickly compute a bound on the size of the derivative over a closed interval. Using this method, we can compute, for example, three decimal digits of ln(2) in 0.2 s, and four digits in 1.2 s, but five digits takes 3 s, and six digits takes 218 s! This method does not scale well because for high accuracy we have to evaluate the function e^x at many values, and our program for e^x, even though it has been optimized, is not fast enough for that. In order to get a reasonably fast program valid for ln(x) over its entire domain $(0, \infty)$ we used the following method.

For fixed $a \in (0, \infty)$, define the following function lgc(x) and find its first three derivatives:

$$\begin{aligned}
\text{lgc}(x) &= x + 2 * \frac{a - e^x}{a + e^x} \\
\text{lgc}'(x) &= \frac{(a - e^x)^2}{(a + e^x)^2} \\
\text{lgc}''(x) &= \frac{-4ae^x(a - e^x)}{(a + e^x)^3} \\
\text{lgc}'''(x) &= \frac{16a^2e^{2x} - 4a^3e^x - 4ae^{3x}}{(a + e^x)^4}.
\end{aligned}$$

At $c = \ln(a)$ we have $e^c = a$, so $\text{lgc}(c) = c$, $\text{lgc}'(c) = \text{lgc}''(c) = 0$, and $\text{lgc}'''(c) = 1/2$. Then the Taylor remainder theorem for the function $\text{lgc}(x)$ at c lets us prove

$$|x - \ln(a)| \leq 1 \Rightarrow |\text{lgc}(x) - \ln(a)| \leq (1/4)|x - \ln(a)|^3.$$

So, if we have an approximation x_0 with $|x_0 - \ln(a)| \leq 1/10$ then $x_1 = \text{lgc}(x_0)$ will satisfy $|x_1 - \ln(a)| \leq 1/4 * 10^3$. Even if x_0 is rational, x_1 will not be rational, so we will want to round off these approximations as we go along, but the extra factor of 4 gives us room to do that and easily define a sequence of approximations $\text{lgseq}(n)$ such that

$$|\text{lgseq}(n) - \ln(a)| \leq 10^{-(3^n)}.$$

Since $\text{lgc}(x)$ can also be written $x - 2 + \frac{4a}{a+e^x}$ we only need to compute e^x once on each iteration, using the fast version of e^x developed in Section 28.10, but we also have to do a division. This will give us our faster program for $\ln(a)$ provided that we can find the initial approximation x_0 that is within $1/10$ of $\ln(a)$. For that we use the slower version of the logarithm that uses the special case of the intermediate value theorem. Using the resulting alternate definition we obtain the following.

$$\text{digits}(\ln(2)) \mapsto 728, 728, 2186, 3664$$
$$\text{digits}(\ln(\pi/4)) \mapsto 242, 255, 728, 2186$$

With this algorithm the number of digits we can compute does not have a smooth dependence on time. If we can complete the next iteration, we triple the number of digits we have.[3] We can now use the optimized exponential and logarithm functions to compute $\sqrt{2} = e^{(\ln(2))/2}$ and compare with the results from Section 28.8 to see that the special square root is still faster.

$$\text{digits}(e^{(\ln(2))/2}) \mapsto 236, 722, 1919, 2179$$
$$\text{digits}(\sqrt{2}) \mapsto 1073, 1900, 3292, 5651$$

For $\arcsin(a)$ we use the same method of finding a contraction $\text{asc}(x)$ with iterates that converges cubically to $\arcsin(a)$.

$$\begin{aligned}
\text{asc}(x) &= x + a\cos(x) - \sqrt{1-a^2}\sin(x) \\
\text{asc}'(x) &= 1 - a\sin(x) - \sqrt{1-a^2}\cos(x) \\
\text{asc}''(x) &= -a\cos(x) + \sqrt{1-a^2}\sin(x) \\
\text{asc}'''(x) &= a\sin(x) + \sqrt{1-a^2}\cos(x)
\end{aligned}$$

At $c = \arcsin(a)$, $\sin(c) = a$, and $\cos(c) = \sqrt{1-a^2}$, so $\text{asc}(c) = c$, $\text{asc}'(c) = 0 =$

[3] Note that $728 = 3^6 - 1$ and $2186 = 3^7 - 1$. We get $6559 = 3^8 - 2$ decimal digits of $\ln(2)$ in 6.5 s.

$\mathrm{asc}''(c)$, and $\mathrm{asc}'''(c) = 1$. If $\mathrm{asc}^n(x)$ denotes iterating asc n times, then the Taylor remainder theorem lets us prove

$$|\mathrm{asc}^n(x) - \arcsin(a)| \leq |x - \arcsin(a)|^{3^n}.$$

We can prove that for $a \in (0, 3/4)$ we have $|a - \arcsin(a)| \leq 1/10$ so

$$|\mathrm{asc}^n(a) - \arcsin(a)| \leq 10^{-(3^n)}, \text{ for } a \in (0, 3/4).$$

For $a \in (0.73, 1)$ we can prove that $\sqrt{1-a^2} \in (0, 3/4)$ so $\mathrm{asc}^n(\sqrt{1-a^2})$ converges rapidly to $\arcsin(\sqrt{1-a^2}) = \pi/2 - \arcsin(a)$. Thus, for $a \in (0,1)$ we compute $\arcsin(a)$ within $\frac{1}{N}$ by deciding, using Lemma 28.14, whether $a < 3/4$ or $0.73 < a$ and use either $\mathrm{asc}^n(a)$ or $\pi/2 - \mathrm{asc}^n(\sqrt{1-a^2})$ for the n such that $10^{3^n} > N$. For $a \in (-1, 0)$ we use $\arcsin(-a) = -\arcsin(a)$ to reduce to the previous case, and for $a \leq \frac{1}{N+1}$, we can prove $\arcsin(a) \leq \frac{1}{N}$. Using this reasoning we can prove that there is a real number equivalent to $\arcsin(a)$ that computes faster.

Because $\frac{\sqrt{2}}{2} < \frac{3}{4}$ we do not need $\pi/2$ to compute $\arcsin\left(\frac{\sqrt{2}}{2}\right) = \pi/4$, so we could use $4 * \arcsin\left(\frac{\sqrt{2}}{2}\right)$ as means to compute π if it is fast enough.

$$\text{digits}\left(4 * \arcsin\left(\frac{\sqrt{2}}{2}\right)\right) \mapsto 28, 48, 77, 112$$

Since we get 48 digits of π in 0.2 s this method is reasonable, but there is a better method that computes π much faster, and we need a fast computation of π because our sine and cosine functions subtract multiples of $\pi/2$ from the input x in order to reduce it to $|x'| \leq 2$.

28.12 Computing π and arctan(x)

Bishop and Bridges define $\frac{\pi}{2}$ to be the limit of the sequence $x_1 = 1, x_{n+1} = x_n + \cos(x_n)$ and prove that the limit exists and is the first positive zero of the cosine function. If we compute with the original definition of $\frac{\pi}{2}$ that comes from the proof in Bishop and Bridges, we find the following.

$$\text{digits}\left(\frac{\pi}{2}\right) \mapsto 31, 47, 71, 101$$

So it computes at about the same speed as the formula using $\arcsin\left(\frac{\sqrt{2}}{2}\right)$. A faster method comes from Machin's formula

$$\frac{\pi}{4} = 4\arctan\left(\frac{1}{5}\right) - \arctan\left(\frac{1}{239}\right).$$

This formula is easy to prove once we establish the angle addition formula for the tangent function for $x, y \in (\frac{-\pi}{2}, \frac{\pi}{2})$, and $x + y \in (\frac{-\pi}{2}, \frac{\pi}{2})$

$$\tan(x+y) = \frac{\tan(x) + \tan(y)}{1 - \tan(x) * \tan(y)}.$$

For $\tan(a) = \frac{1}{5}$ we have $\tan(2a) = \frac{\frac{2}{5}}{1-(\frac{1}{5})^2} = \frac{5}{12}$ and $\tan(4a) = \frac{\frac{10}{12}}{1-(\frac{5}{12})^2} = \frac{120}{119}$. Then since $\tan(\frac{\pi}{4}) = 1$,

$$\tan\left(4a - \frac{\pi}{4}\right) = \frac{\tan(4a) - 1}{1 + \tan(4a)} = \frac{\frac{120}{119} - 1}{1 + \frac{120}{119}} = \frac{1}{239}.$$

Definition 28.23 (Definition of $\arctan(x)$) By following the methods used in Bishop and Bridges for $\arcsin(x)$ we define

$$\arctan(x) = \int_0^x \frac{dt}{1+t^2}$$

and we easily prove $\arctan(\tan(x)) = x = \tan(\arctan(x))$. Using the theorems about integration already proved in chapter two of Bishop and Bridges we prove for any $k \in \mathbb{N}$,

$$\int_0^x \left(\frac{1}{1+t^2} - \sum_{i=0}^{k} \left(-\left(t^2\right)\right)^i\right) dt = \int_0^x \frac{\left(-\left(t^2\right)\right)^{k+1})}{1+t^2}$$

So

$$|\arctan(x) - \sum_{i=0}^{k} \frac{(-1)^i x^{2i+1}}{2i+1}| \leq |\int_0^x (\frac{\left(-\left(t^2\right)\right)^{k+1})}{1+t^2}| \leq \frac{|x|^{2k+3}}{2k+3}$$

When $|x| \leq 1/2$ this gives a good polynomial approximation for $\arctan(x)$

$$P(k, x) = x * \sum_{i=0}^{k} \frac{(-1)^i x^{2i}}{2i+1}$$

which we can evaluate efficiently using the methods for power series from Section 28.9. Using this we made an efficient $\arctan_s(x)$ that is equivalent to $\arctan(x)$ for real numbers x with $|x| \leq 1/2$. Since that is the case for the $x = \frac{1}{5}$ and $x = \frac{1}{239}$ used in Machin's formula for $\frac{\pi}{4}$, we can use

$$\frac{\pi}{4} = 4\arctan_s\left(\frac{1}{5}\right) - \arctan - s\left(\frac{1}{239}\right) \text{ and then define } \pi = 4 * \frac{\pi}{4}.$$

We get the following speed.

$$\text{digits}(\pi) \mapsto 767, 1536, 2623, 4639$$

This is much faster than the first definition giving us 2623 digits in 1 s instead of only 71 digits.

To make an efficient version of $\arctan(x)$ for all x we make use of

$$\begin{aligned}
\arctan(x) &= -\arctan(-x)\\
\arctan(x) &= \frac{\pi}{4} + \arctan\left(\frac{x-1}{x+1}\right), \text{ for } -1 < x\\
\arctan(x) &= \frac{\pi}{2} - \arctan(1/x), \text{ for } 0 < x.
\end{aligned}$$

By Lemma 28.14, $x \in (-\infty, 0)$ or $x \in (-1/2, \infty)$, so using $-\arctan(-x)$ in the first case, we reduce to the case $x \in (-1/2, \infty)$. Then, similarly, x is in $(-1/2, 1/2)$, $(1/3, 3)$, or $(2, \infty)$. In the first case we use $\arctan_s(x)$. In the second case, $\frac{x-1}{x+1} \in \left(\frac{-1}{2}, \frac{1}{2}\right)$ so we use $\frac{\pi}{4} + \arctan_s\left(\frac{x-1}{x+1}s\right)$, and in the third case we use $\frac{\pi}{2} - \arctan_s(1/x)$.

$$\begin{aligned}
\text{digits}(\arctan(2)) &\mapsto 563, 987, 1806, 3167\\
\text{digits}\left(\arctan\left(\frac{\pi}{4}\right)\right) &\mapsto 479, 1056, 1854, 3247
\end{aligned}$$

28.13 Constructive Content of Brouwer's Principles

We don't have space to give all of the details, but we would like the reader to get some sense of why all functions $f \in [0, 1] \to \mathbb{R}$ are uniformly continuous and why a modulus of continuity for f can be computed from f but this computation is inefficient.

We call $p \in \mathbb{N}^+ \to \mathbb{Z}$ a *type 1 function*, and $F \in (\mathbb{N}^+ \to \mathbb{Z}) \to \mathbb{Z}$ a *type 2 functional*. Then $F(p)$ computes to an integer. Nuprl has an operation $(\nu e.\ G(e))$ that chooses a fresh name e that does not occur in G and then computes $G(e)$. Let

$$\overline{p}_{e,b}(n) = \text{if } n < b \text{ then } p(n) \text{ else exception(e)}$$

For a name e that is fresh, the functional F can not catch the named exception(e) so we can prove that $F(\overline{p}_{e,b})$ computes to $F(p)$ if the computation never applies function p to any $n \geq b$ and computes to exception(e) otherwise. We write a recursive program $G_{b,p}(e)$ that tries $F(\overline{p}_{e,1})$, $F(\overline{p}_{e,2})$, ..., $F(\overline{p}_{e,b})$ that returns inl$(F(p))$ if no exception is raised on some try, and increments the bound by one if the exception is raised, returning inr() if all the tries raise the exception. We call $M(F, p, b) = (\nu e.\ G_{b,p}(e))$ the *Kleene-M* function. It is the computational content of *Brouwer's continuity principle for numbers*. The function $M(F, p, b)$ tests whether b is greater than or equal to the modulus of continuity of the type 2 functional F at the point p, and we prove, in the meta-theory, that we may assume $\downarrow\exists b.\ M(F, p, b) = \text{inl}(F(p))$.[4]

Cantor space is the type $\mathcal{C} = (\mathbb{N}^+ \to \mathbb{N}_2)$, and each $p \in \mathcal{C}$ is a *branch* in the *tree* 2^ω and $s = (p \restriction b)$ is a *path* in $2^{<b}$. For a path s, the branch $s\overline{0}$ is the extension of s with all 0s. For $F \in \mathcal{C} \to \mathbb{N}$, let

$$v?(s) = M(F, s\overline{0}, \text{length}(s))$$

when the test $D(s) = \text{isl}(v?(s))$ is true then $F(q) = m = \text{outl}(v?(s))$ for all branches q that pass through the path s. By the above property of the Kleene-M function, D is a *decidable bar* on tree 2^ω. By *bar induction* we can then prove

$$\exists m.\ \forall p{:}\mathcal{C}.\ F(p) \le m.$$

This says that any integer-valued function on Cantor space is bounded. The computational content of this theorem combines the Kleene-M function and the *fan functional.*

There is a standard map c from $\mathcal{C}$ onto the closed interval $[0, 1]$. If f is a real-valued function with domain $[0, 1]$ then for each $k \in \mathbb{N}^+$, $F_k = \lambda p.\ (f(c(p))_k \in \mathcal{C} \to \mathbb{Z}$, so $M(F_k, p)$ computes the modulus of continuity for F_k at point p. Using the fan functional we can get a uniform bound b_k such that $M(F_k, p) < b_k$ for all $p \in \mathcal{C}$. Then the function $\lambda k.\ b_k$ is essentially a modulus of uniform continuity for the function f because for any $\epsilon > 0$ we choose k with $1/k < \epsilon$. If $x, y \in [0, 1]$ are within a certain distance $\delta = 1/g(b_k) > 0$ then there are paths $p, q \in \mathcal{C}$ such that $c(p) \equiv x$, $c(q) \equiv y$, and p and q agree on the first b_k inputs. Then $f(x) \equiv f(c(p))$ and $f(y) \equiv f(c(q))$ will be within ϵ of each other because $(f(c(p)))_k = (f(c(q)))_k$. This argument requires that f is a *function*, namely that $x \equiv x'$ implies $f(x) \equiv f(x')$.

This uniform continuity theorem implies, for example, that there is a number k such that

$$\forall x, y \in [0, 1].\ |x - y| < 1/k \Rightarrow |\sqrt{x} - \sqrt{y}| < 1/10.$$

But we could not compute k using the Nuprl code extracted from the proof of Brouwer's uniform continuity theorem, even for this simple example, because it is too inefficient.

28.14 Concluding Remarks

The formal definitions of Bishop's constructive analysis provide a solid foundation for computational mathematics of real numbers. Algorithms extracted form proofs

[4] The existence of the bound b has to be squashed, or, at least *quotiented*, because it is an *intensional* property of F. $M(F, p, b)$ and $M(G, p, b)$ may differ on extensionally equal functionals F and G.

can be made efficient with some care. Methods used by numerical analysts for floating-point numbers can be adapted to yield provably correct, efficient algorithms for operations on Bishop's real numbers. We have illustrated how this can be done for the functions defined in Chapter 2 of Bishop and Bridges, with the notable exception of general definite integrals.

We have given examples ($\ln(x)$ and $\arcsin(x)$) of functions defined as definite integrals, but those used special properties of the particular functions. General purpose methods for efficient evaluation of integrals is a major subject of numerical analysis. So called *quadrature* methods make use of families of orthogonal polynomials, such as the Legendre polynomials, and precompute tables of values for definite integrals of these polynomials. The quadrature methods then approximate a general function with one of these polynomials and the definite integral is then approximated efficiently using the table of precomputed values. We have begun the formalization of these methods in Nuprl, but space does not permit an exposition here. Furthermore, proving the exact error bounds needed to make these quadrature methods into provably correct algorithms in Bishop's analysis will be challenging.

29

On the Computational Content of Choice Principles

Ulrich Berger and Monika Seisenberger

29.1 Introduction

The axiom of choice is a controversial principle in the foundations of mathematics [1]. While having applications in many mathematical fields, it can neither be proven nor refuted in broadly accepted foundations such as Zermelo–Fraenkel set theory [21, 31], and it has some counter-intuitive consequences such as the Banach–Tarski paradox. In constructive dependent type theory [40] the axiom of choice has the form

$$(\forall x : A)(\exists y : P(x)\, Q(x,y)) \to (\exists f : (\forall x : A)P(x))(\forall x : A)Q(x, f(x)) \tag{29.1}$$

and is easily provable since the type-theoretic interpretation of the premise is almost identical to the conclusion (see [19], for a proof). In contrast, adding the axiom of choice to constructive *set* theory makes the law of excluded middle provable and hence the theory classical [26, 33].

The proof-theoretic strengths of various forms of the axiom of choice have been investigated in the context of fragments of first- and second-order arithmetic (see, e.g., [54]). Also here, the axiom of choice mocks us, for example, in the form of Weak König's Lemma (WKL) which states that every infinite binary tree has an infinite path: On the one hand, there is Kleene's computable infinite binary tree that has no computable infinite path, so WKL can't be valid, constructively. On the other hand, WKL is conservative over constructive arithmetic for Π^0_2 sentences (see, e.g., [55]), hence totality statements about computable functions that are proven with the help of WKL are constructively valid.

In Bishop constructive mathematics [17] one generally accepts the axioms of countable and dependent choice but tries to avoid choice principles altogether, if possible [20, 47]. This is natural, not only because of the puzzling phenomena highlighted above, but also, and more importantly, because in constructive mathematics one expects the existence of a choice function to be proven by a *construction* and not only asserted through reference to an axiom.

It is the aim of this chapter to give an overview of computational aspects of various choice principles, focusing on unexpected phenomena that appear within a general framework of program extraction from proofs. We aim to demonstrate that this framework provides a flexible means to formalize constructive mathematics and to exhibit, in a technically precise way, the constructivity or nonconstructivity of mathematical principles. It includes abstract and, to a limited extent, classical mathematical concepts, and its results are meaningful and correct from a constructive as well as classical point of view.

The chapter is organized as follows: Section 29.2 introduces the system known as IFP, explains program extraction in IFP, and sketches how Bishop constructive analysis can be formalized in this system. Section 29.3, discusses several realizable and unrealizable choice principles. Finally, Section 29.4 revisits the computational content of the classical axioms of countable choice and dependent choice, more precisely, the results of applying Gödel's negative translation followed by the Friedman–Dragalin A-translation to these axioms.

29.2 A Semi-constructive System with Computational Content

We briefly describe the logical system IFP (Intuitionistic Fixed Point Logic) we will be working with. It is similar to the system TCF (Theory of Computable Functionals) underlying the Minlog system [14, 50]. We highlight the most important aspects of these systems, concentrating on motivation and intuitive explanations; full details can be found in [10]. The relationship between the two systems will be explained in Section 29.2.4.

29.2.1 Expressivity

The IFP system is based on many-sorted first-order logic with equality extended by least and greatest fixed points of strictly positive predicate transformers. Least fixed points, also known as inductive definitions, can be used to specify common data structures. For example, given a constant 0 and a successor function $(+1)$, the natural numbers can be defined as the least set $\mathbf{N}$ that contains 0 and is closed under S:

$$\mathbf{N}(x) \stackrel{\mu}{=} x = 0 \lor \exists y\, (\mathbf{N}(y) \land x = y + 1). \tag{29.2}$$

Using greatest fixed points, also known as coinductive definitions, one can define infinite data. For example, giving a cons operation $*$, infinite streams of natural numbers are defined as the largest set $\mathbf{Str}$ whose elements are of the form $x * t$, where x is a natural number and t is a stream:

$$\mathbf{Str}(s) \stackrel{\nu}{=} \exists x, t\,(\mathbf{N}(x) \wedge \mathbf{Str}(t) \wedge s = x * t)\,.$$

Coinductive definitions can also be used to define infinitary properties in a quantifier-free way. For example in the real numbers (or any structure with an ordering $<$ (given as a predicate constant) and a predecessor function (-1)) the property ∞ of being infinite can be defined as

$$\infty(x) \stackrel{\nu}{=} 0 < x \wedge \infty(x-1)\,.$$

The Archimedean axiom

$$\forall x\, \exists y\, (\mathbf{N}(y) \wedge x < y) \tag{29.3}$$

can then be shown to be equivalent to $\neg\exists x\, \infty(x)$ using classical logic.

29.2.2 Semantics

The semantics of IFP has a classical part and a constructive part. The *classical* part is the usual classical semantics for first-order predicate logic, that is, sorts are interpreted as sets (called carrier sets), functions symbols as (set-theoretic) functions, and predicate constants as subsets of the corresponding cartesian products of carrier sets. No assumptions of computability or constructivity are made at this point. The *constructive* part consist of a realizability interpretation of formulas and (co)inductive predicates. The space of potential realizers is an effective Scott-domain D [30] which carries the structure of a PCA (partial combinatory algebra). In particular, it is closed under pairing, disjoint union, and the continuous function space construction, so that simple recursive types τ can be interpreted as subdomains $D(\tau)$ of D. To define the semantics of a formula A we first define a type $\tau(A)$ that represents the "propositional skeleton" of A. Then the semantics of A is defined as set $\mathbf{R}(A) \subseteq D(\tau(A))$ of realizers of A, essentially following Kleene [35] and Kreisel [38]. We will usually write "$a\,\mathbf{r}\,A$" ("a realizes A") for "$a \in \mathbf{R}(A)$." We don't give a complete definition of $\mathbf{R}(A)$ but only highlight the most important cases.

For atomic formulas $P(\vec{t})$ (P a predicate constant including $=$ and **False**, $\vec{t}$ a vector of first-order terms) realizability coincides with classical truth:

$$a\,\mathbf{r}\,P(\vec{t}) \stackrel{\text{Def}}{=} a = \mathbf{Nil} \wedge P(\vec{t}), \tag{29.4}$$

where $\mathbf{Nil}$ is an element of D indicating trivial computational content. Disjunction is interpreted in the usual constructive way:

$$c\,\mathbf{r}\,(A \vee B) \stackrel{\text{Def}}{=} \exists a\,(c = \mathbf{Left}(a) \wedge a\,\mathbf{r}\,A) \vee \exists b\,(c = \mathbf{Right}(b) \wedge b\,\mathbf{r}\,B)$$

(the appearance of "∃" and "∨" on the right-hand side can be avoided by appropriate computable discriminator and accessor functions). The interpretations of conjunction and implication are as expected but the definitions are simplified to avoid junk. For example, a realizer of a conjunction, $\mathbf{Pair}(a, b)$, is simplified to a if $b = \mathbf{Nil}$.

Inductive and coinductive definitions are realized by enriching the corresponding definitions with realizers. For example, the predicate $\mathbf{N}$ (29.2) characterizing natural numbers is realized as follows:

$$\begin{aligned} n\,\mathbf{r}\,\mathbf{N}(x) \overset{\mu}{=} \; & (n = \mathbf{Left}(\mathbf{Nil}) \wedge x = 0) \vee \\ & \exists m\,(n = \mathbf{Right}(m) \wedge \exists y\,(m\,\mathbf{r}\,\mathbf{N}(y) \wedge x = y + 1)). \end{aligned} \tag{29.5}$$

Therefore, $n\,\mathbf{r}\,\mathbf{N}(x)$ means that x is a natural number and n is the unary representation of x where zero and successor are modelled by $\mathbf{Left}(\mathbf{Nil})$ and $\mathbf{Right}(\_)$. Assuming suitable (purely universal) axioms for the basic arithmetic operations and relations one can easily show the following by induction on $\mathbf{N}$ and $\mathbf{R}(\mathbf{N})$.

Theorem 29.1 (1) $\mathbf{N}(x) \leftrightarrow \exists n\,(n\,\mathbf{r}\,\mathbf{N}(x))$;
(2) $n\,\mathbf{r}\,\mathbf{N}(x) \wedge m\,\mathbf{r}\,\mathbf{N}(y) \to (n = m \leftrightarrow x = y)$.

Quantifiers are interpreted uniformly, that is,

$$\begin{aligned} a\,\mathbf{r}\,\forall x\,A(x) &\overset{\text{Def}}{=} \forall x\,(a\,\mathbf{r}\,A(x)) \\ a\,\mathbf{r}\,\exists x\,A(x) &\overset{\text{Def}}{=} \exists x\,(a\,\mathbf{r}\,A(x))\,. \end{aligned}$$

Hence, the realizer of a universal formula does not depend on the quantified variable, and the realizer of an existential formula does not carry a witness. In our setting, this is inevitably so, since the variable x may range in an arbitrary set whose elements are not representable in the domain D. The usual realizability interpretation of **HA** (Heyting Arithmetic, that is, first-order arithmetic with intuitionistic logic [35, 57]) can be recovered by relativization. For example, the **HA**-formula

$$\forall n\,\exists m\,(n \leq m^2 < n + 1), \tag{29.6}$$

(where n, m range over natural numbers) is written in our system as

$$\forall x\,(\mathbf{N}(x) \to \exists y\,(\mathbf{N}(y) \wedge x \leq y^2 < x + 1))\,. \tag{29.7}$$

A realizer of (29.7) is a domain-theoretic function $f : \tau(\mathbf{N}) \to \tau(\mathbf{N})$ such that

$$\forall x\,\forall n \in \tau(\mathbf{N})\,(n\,\mathbf{r}\,\mathbf{N}(x) \to \exists y\,((f\,n)\,\mathbf{r}\,\mathbf{N}(y) \wedge x \leq y^2 < x + 1))\,.$$

Since, as we have seen, $n\,\mathbf{r}\,\mathbf{N}(x)$ means that n is the unary representation of the natural number x, and, by Theorem 29.1, natural numbers and their realizers are in a one-to-one correspondence, the above is equivalent to

$$\forall n \in \tau(\mathbf{N})\,(n \le (f\,n)^2 < n+1)\,,$$

which is the traditional realizability interpretation of (29.6). In general, however, the realizers of a property of an object are less tightly connected to the object. For example, consider the property of being a *Cauchy real*, that is, a real number that can be approximated by rational numbers,

$$\mathbf{C}(x) \stackrel{\mathrm{Def}}{=} \forall n\,(\mathbf{N}(n) \to \exists q\,(\mathbf{Q}(q) \wedge |x-q| \le 2^{-n})), \qquad (29.8)$$

where the predicate $\mathbf{Q}$ characterizes the rational numbers, for example,

$$\mathbf{Q}(x) \stackrel{\mathrm{Def}}{=} \exists y, z, u\,(\mathbf{N}(y) \wedge \mathbf{N}(z) \wedge \mathbf{N}(u) \wedge x\,(u+1) = y - z)\,. \qquad (29.9)$$

The formula $\mathbf{C}(x)$ is realized by any of the uncountably many rational Cauchy sequences converging quickly to x, and there might be no computable such. Based on the predicate $\mathbf{C}$ or other (often coinductive) characterizations of Cauchy reals, parts of constructive analysis [17] have been formalized [10, 41, 49, 50].

29.2.3 Extracting Programs

Program extraction is based on an intuitionistic proof calculus and a Soundness Theorem that asserts the constructive correctness of the calculus for the above semantics.

Theorem 29.2 (Soundness) *From an* IFP *proof of a formula A one can extract a closed program $M : \tau(A)$ such that $M\,\mathbf{r}\,A$ is provable.*

A detailed proof can be found in [10] and a corresponding proof for TCF is found in [50]. The extracted realizers are given as functional programs. It has turned out that programs extracted from proofs involving coinduction are most efficiently executed in a lazy programming language such as Haskell. Therefore, the Minlog system is equipped with a module that extracts to Haskell [42], instead of using Minlog's native programming language Scheme.

An important aspect of the proof calculus is the fact that it may involve any axioms for which effective realizers are available. In particular, disjunction-free formulas, which can only be realized by the constant **Nil** and whose realizability is equivalent to themselves, can be asserted as axioms provided they are true. We call such formulas *noncomputational (nc)*. Therefore, if one is willing to accept classical logic at the meta-level, one may assert arbitrary classically true nc formulas as axioms. For example, one can assert stability,

$$\forall \vec{x}\,(\neg\neg P(\vec{x}) \to P(\vec{x})),$$

for arbitrary predicate constants, including $=, <, \leq$, even if they concern objects, like real numbers, for which these relations are undecidable. One must, of course, *not* assert the law of exclude middle, $\forall \vec{x}\, (P(\vec{x}) \vee \neg P(\vec{x}))$ as an axiom. One may as well add constants for noncontinuous (and hence noncomputable) functions such as

$$\mathrm{sgn}(x) = \begin{cases} 1 & \text{if } x > 0 \\ 0 & \text{if } x = 0 \\ -1 & \text{if } x < 0 \end{cases}$$

together with characterizing axioms (that must be nc formulas). The class of admissible axioms can be extended to realizable *Harrop formulas*. A Harrop formula may contain disjunctions but not at a strictly positive position, that is, every occurrence must be part of the premise of an implication. For example, every negated formula is Harrop since we view negation $\neg A$ as implication to falsity, $A \to \bot$. Like nc formulas, Harrop formulas have trivial computational content, that is, **Nil** is the only possible realizer, however, they need not be equivalent to their realizability. For example, the negation of the Archimedean axiom (29.3) (which is Harrop but not nc since the definition of $\mathbf{N}$ contains a disjunction), fails in the real numbers. But its realizability interpretation (which states the unrealizability of (29.3)) is easily seen to be true.

Over the past 30 years, a substantial number of case studies on program extraction has been carried out, most of them using the Minlog system. Examples of areas covered are constructive analysis [7, 10, 41, 49], graph theory [11], lambda calculus [13], ordinals [4], satisfiability testing [15], and infinitary combinatorics, in particular well-quasi-orders [51].

29.2.4 Modeling $\mathbf{HA}^\omega$ and constructive analysis

The system TCF, which underlies Minlog, can be seen as an extension of $\mathbf{HA}^\omega$ (Heyting Arithmetic in finite types [57]) where variables carry a finite type over the base type $\mathbf{nat}$ of natural numbers and the realizability interpretation of quantifiers is *not* uniform (although TCF also has uniform quantifiers). This can be modeled in IFP by adding an application and a pairing operation and defining finite types as predicates, that is,

$$\begin{aligned} \mathbf{nat} &\stackrel{\text{Def}}{=} \mathbf{N} \\ (\rho \to \sigma)(f) &\stackrel{\text{Def}}{=} \forall x\, (\rho(x) \to \sigma(f(x))) \\ (\rho \times \sigma)(x) &\stackrel{\text{Def}}{=} \exists y, z\, (x = (y, z) \wedge \rho(y) \wedge \sigma(z)). \end{aligned}$$

Extensionality can be added in the usual way if desired.

To model Bishop constructive analysis in IFP one has two options. (a) The traditional way, as a subsystem of $\mathbf{HA}^\omega$ (plus additional axioms) where real numbers are modeled as special objects of type $\mathbf{nat} \to \mathbf{rat}$ (e.g., regular sequences) with $\mathbf{rat} \stackrel{\text{Def}}{=} \mathbf{nat} \times \mathbf{nat} \times \mathbf{nat}$ (according to (29.9)). (b) Using a sort for abstract real numbers and relativizing quantifiers to the predicate $\mathbf{C}$ as defined in (29.8). The basic properties of the arithmetic operations can be postulated as nc axioms (without relativization to $\mathbf{C}$). This means that Cauchy sequences feature only as realizers extracted from proofs of formulas of the form $\mathbf{C}(x)$. In (b), one must be a bit careful with primitive relations. For example, in Bishop constructive analysis the formula $0 < x$ implies the existence of a positive rational number below x. However, its relativization, $\mathbf{C}(x) \wedge 0 < x$, entails the existence of such a rational number only with the help of Markov's principle.[1] Therefore the correct translation of Bishop's $0 < x$ into IFP is $\mathbf{C}(x) \wedge \exists y\,(\mathbf{Q}(y) \wedge 0 < y < x)$.

29.3 Realizable and Unrealizable Choice Principles

In constructive mathematics the status of a choice principle depends heavily on its precise formulation. We consider various forms of the axiom of choice and assess them regarding their realizability. To be able to express choice principles we assume a sort for functions and an operation for function application.

29.3.1 Relativized Choice

In the language of predicate logic, the most general form of the axiom of choice appears to be *relativized choice*,

$$\forall x\,(A(x) \to \exists y\, B(x,y)) \to \exists f\, \forall x\,(A(x) \to B(x, f(x)))\,. \tag{29.10}$$

It includes the canonical translation of the type-theoretic axiom of choice (29.1) into predicate logic, which is

$$\begin{aligned}&\forall x\,(A(x) \to \exists y\,(P(x,y) \wedge Q(x,y))) \to \\ &\qquad \exists f\,(\forall x\,(A(x) \to P(x,f(x))) \wedge \forall x\,(A(x) \to Q(x,f(x))),\end{aligned} \tag{29.11}$$

where the formulas $A(x)$, $B(x,y)$, and $\forall x\,(A(x) \to P(x,f(x)))$ represent the typing statements $x : A$, $y : P(x)$, and $f : (\forall x : A)P(x)$. Clearly, (29.11) is equivalent to the instance of (29.10) where $B(x,y) \stackrel{\text{Def}}{=} P(x,y) \wedge Q(x,y)$.

[1] Markov's principle is not derivable in IFP; but it is realized by unbounded search and may therefore be considered as an axiom without spoiling program extraction.

29.3.2 Unrealizable Instances of Relativized Choice

We cannot expect relativized choice to be realizable in general. This becomes clear when expanding the statement "d realizes (29.10)," which yields, for the case that the predicates A and B are both non-Harrop,

$$\forall b\,(\forall x\,\forall a\,(a\,\mathbf{r}\,A(x) \to \exists y\,(b\,a)\,\mathbf{r}\,B(x,y)) \to \\ \exists f\,\forall x\,\forall a\,(a\,\mathbf{r}\,A(x) \to (d\,b\,a)\,\mathbf{r}\,B(x,f(x)))).$$

The problem is that the element y, the existence of which is guaranteed by the premise, may depend on a (and on x), whereas the witness $f(x)$ provided by the choice function f must be independent of a (see also [18, p. 100]). That this may indeed lead to unrealizability can be seen by relativizing the Archimedean axiom (29.3) to the Cauchy property (29.8)

$$\forall x\,(\mathbf{C}(x) \to \exists y\,(\mathbf{N}(y) \wedge x < y)). \tag{29.12}$$

This formula is provable (in IFP), hence effectively realizable. Now, if the corresponding instance of (29.10) were effectively realizable, then its conclusion,

$$\exists f\,\forall x \in \mathbf{C}\,(\mathbf{N}(f(x)) \wedge x < f(x))\,,$$

would be effectively realizable, too. Such a realizer would be a program d such that

$$\exists f\,\forall x\,\forall c\,(c\,\mathbf{r}\,\mathbf{C}(x) \to (d\,c)\,\mathbf{r}\,\mathbf{N}(f(x)) \wedge x < f(x))\,.$$

Since, by Theorem 29.1, natural numbers and their realizers are in a one-to-one correspondence, this would mean that d denotes a computable function that computes a natural upper bound to a real number from any Cauchy sequence converging to it in an extensional way, contradicting the well-known fact that all such extensional functions must be continuous and therefore constant.

29.3.3 Realizable Instances of Relativized Choice

In order to make relativized choice realizable, we impose certain conditions on the predicates A and B in (29.10). A binary relation R is called

- *functional* if $\forall x, y, z\,(R(x,y) \wedge R(x,z) \to y = z)$;
- *total* if $\forall x\,\exists y\,R(x,y)$;
- a *rectangle* if there are unary predicates P, Q such that $\forall x, y\,(R(x,y) \leftrightarrow P(x) \wedge Q(y))$, equivalently, $\forall x, x', y, y'\,(R(x,y') \wedge R(x',y) \to R(x,y))$;
- *functional on* A if $\forall x, y, z\,(A(x) \wedge R(x,y) \wedge R(x,z) \to y = z)$. This is equivalent to the relation $R'(x,y) := A(x) \wedge R(x,y)$ being functional.

Theorem 29.3 (Berger [8]) *Relativized choice (29.10) is realizable provided one of the following conditions (a)–(e) holds.*

(a) *A is a Harrop predicate.*
(b) *B is a Harrop predicate.*
(c) *$\lambda(x, a)\, a\, \mathbf{r}\, A(x)$ is functional.*
(d) *$\lambda(y, b)\, b\, \mathbf{r}\, B(x, y)$ is a rectangle for all x.*
(e) *The statement "B* is functional on A*" is realizable.*

In case (b) the realizer is **Nil***, in all other cases the realizer is the identity function. The proof uses relativized choice.*

Proof If A is Harrop, then the statement that the identity function realizes (29.10) is equivalent to the formula

$$\forall b\, (\forall x\, (\mathbf{Nil}\, \mathbf{r}\, A(x) \to \exists y\, b\, \mathbf{r}\, B(x, y)) \to \exists f\, \forall x\, (\mathbf{Nil}\, \mathbf{r}\, A(x) \to b\, \mathbf{r}\, B(x, f(x)))),$$

which is again an instance of relativized choice.

If B is Harrop, then (29.10) is Harrop and the statement that **Nil** realizes (29.10) is equivalent to the formula

$$\forall x\, (\exists a\, a\, \mathbf{r}\, A(x) \to \exists y\, \mathbf{Nil}\, \mathbf{r}\, B(x, y)) \to \exists f\, \forall x\, (\exists a\, a\, \mathbf{r}\, A(x) \to b\, \mathbf{r}\, B(x, f(x))),$$

which, yet again, is an instance of relativized choice.

For the remaining cases (c)–(e) we may assume that A and B are both non-Harrop. Assume d realizes $\forall x\, (A(x) \to \exists y\, B(x, y))$, that is,

$$\forall x\, \forall a\, (a\, \mathbf{r}\, A(x) \to \exists y\, (d\, a)\, \mathbf{r}\, B(x, y)). \tag{29.13}$$

We have to show that d realizes $\exists f\, \forall x\, (A(x) \to B(x, (f(x))))$, that is,

$$\exists f\, \forall x\, \forall a\, (a\, \mathbf{r}\, A(x) \to (d\, a)\, \mathbf{r}\, B(x, (f(x)))).$$

Set

$$E \stackrel{\text{Def}}{=} \lambda x\, \exists a\, a\, \mathbf{r}\, A(x)$$
$$P \stackrel{\text{Def}}{=} \lambda(x, y)\, \exists a\, (a\, \mathbf{r}\, A(x) \land (d\, a)\, \mathbf{r}\, B(x, y)).$$

By (29.13), $\forall x\, (E(x) \to \exists y\, P(x, y))$. Hence, by relativized choice, there exists f such that $\forall x\, (E(x) \to P(x, f(x)))$. We show that f is as required, provided one of the conditions (c), (e), (d) is satisfied. Assume

$$a\, \mathbf{r}\, A(x). \tag{29.14}$$

We have to show

$$(*) \qquad (d\, a)\, \mathbf{r}\, B(x, f(x)).$$

Since, by (29.14), $E(x)$ holds, we have $P(x, f(x))$, that is,

$$b\,\mathbf{r}\,A(x) \wedge (d\,b)\,\mathbf{r}\,B(x, f(x)) \tag{29.15}$$

for some b. Furthermore, from (29.13) and (29.14) it follows

$$(d\,a)\,\mathbf{r}\,B(x, y) \tag{29.16}$$

for some y. The remainder of the proof depends on the conditions (c)–(e).

(c) If $\lambda(a, x)\,a\,\mathbf{r}\,A(x)$ is functional, then $a = b$ and therefore $(*)$ holds, by (29.15).
(d) If $\lambda(y, a)\,a\,\mathbf{r}\,B(x, y)$ is a rectangle, say $\forall y\,\forall b\,(b\,\mathbf{r}\,B(x, y) \leftrightarrow Y(y) \wedge Z(b))$, then $Y(f(x))$ since $(d\,b)\,\mathbf{r}\,B(x, f(x))$, by (29.15). By (29.16), $Z(d\,a)$. Therefore, $(d\,a)\,\mathbf{r}\,B(x, f\,x)$, that is, (*) holds.
(e) Realizability of the statement "B is functional on A" can easily seen to be equivalent to the formula

$$\forall x\,\forall a\,(a\,\mathbf{r}\,A(x) \to \forall y, z\,\forall b, c\,(b\,\mathbf{r}\,B(x, y) \wedge c\,\mathbf{r}\,B(x, z) \to y = z)) \tag{29.17}$$

Combining (29.14), (29.15), and (29.16) with (29.17) we obtain $y = f(x)$. Hence (*), by (29.16). □

29.3.4 Unique Choice

A natural formulation of the axiom of unique choice is obtained by adding to the premise of relativized choice the condition that B is functional on A.

Unique choice

$$\begin{aligned}&\forall x, y, z\,(A(x) \wedge B(x, y) \wedge B(x, z) \to y = z) \wedge\\&\forall x\,(A(x) \to \exists y\,B(x, y))\\&\qquad\qquad\to \exists f\,\forall x\,(A(x) \to B(x, f(x))).\end{aligned} \tag{29.18}$$

To realize (29.18) means to find a realizer of (29.10) under the condition that the formula $\forall x, y, z\,(A(x) \wedge B(x, y) \wedge B(x, z) \to y = z)$ is realizable (by the trivial realizer **Nil**). But this condition is exactly condition (e) in Theorem 29.3. It follows that (29.18) is realized by the identity.

29.3.5 Countable and Dependent Choice

Countable choice is obtained from relativized choice by letting A be $\mathbf{N}$ in (29.10):

$$\forall x\,(\mathbf{N}(x) \to \exists y\,B(x, y)) \to \exists f\,\forall x\,(\mathbf{N}(x) \to B(x, f(x))). \tag{29.19}$$

Since, by Theorem 29.1, $\lambda(x, a)\,a\,\mathbf{r}\,\mathbf{N}(x)$ is functional, countable choice is realizable, by Theorem 29.3.

Dependent choice is the schema

$$A(y_0) \wedge \forall y\,(A(y) \to \exists z\,(A(z) \wedge B(y,z))) \to \\ \exists f\,(f(0) = y_0 \wedge \forall x\,(\mathbf{N}(x) \to A(f(x+1)) \wedge B(f(x), f(x+1))). \tag{29.20}$$

In the case that A is $\mathbf{N}$, dependent choice follows easily from countable choice and hence is realizable. However, its realizability is not restricted to this special case.

Theorem 29.4 *Dependent choice (29.20) is realizable for arbitrary formulas A and B.*

Proof We only look at the case that in (29.20) the predicates A and B are both non-Harrop (the other cases are easier). Hence, assume $a_0\,\mathbf{r}\,A(y_0)$ and $c\,\mathbf{r}\,\forall y\,(A(y) \to \exists z\,(A(z) \wedge B(y,z)))$. The latter means that there are d, e such that for all a, $c\,a = \mathbf{Pair}(d\,a, e\,a)$ and

$$a\,\mathbf{r}\,A(y) \to \exists z\,((d\,a)\,\mathbf{r}\,A(z) \wedge (e\,a)\,\mathbf{r}\,B(y,z)).$$

Define recursively

$$d'\,n = \mathbf{case}\,n\,\mathbf{of}\,\{0 \to a_0; m+1 \to d'\,m\}$$

and $c'\,n \stackrel{\text{Def}}{=} \mathbf{Pair}(d'\,n, e\,(d'\,n))$.

We show that c' realizes the conclusion of dependent choice. We have to find f such that $f(0) = y_0$ and

$$\forall x\,\forall n\,(n\,\mathbf{r}\,\mathbf{N}(x) \to (d'\,n)\,\mathbf{r}\,A(f(x+1)) \wedge (e\,(d'\,n))\,\mathbf{r}\,B(f(x), f(x+1))).$$

Set

$$A'(y) \stackrel{\text{Def}}{=} \exists n\,(d'\,n)\,\mathbf{r}\,A(y)$$
$$B'(y,z) \stackrel{\text{Def}}{=} \exists n\,(e\,(d'\,n))\,\mathbf{r}\,B(y,z).$$

Clearly, $A'(y_0)$ and $\forall y\,(A'(y) \to \exists z\,(A'(z) \wedge B'(y,z))$. By dependent choice, there exists f with $f(0) = y_0$ and

$$\forall x\,(\mathbf{N}(x) \to A'(f(x+1)) \wedge B'(f(x), f(x+1))).$$

Clearly, f is as required. □

In some applications, for example, Higman's Lemma (see Section 29.4.2), it is more convenient to work with a version of dependent choice that talks about finite sequences. We use the constant $[]$ (empty sequence) and operations $|s|$ (length), $x * s$ (add element x on the left), $s * x$ (add element x on the right), $\mathsf{last}(s)$ (last element), $\mathsf{lead}(s)$ (last element removed), s_i (ith element if $i < |s|$), and assume suitable nc axioms specifying them.

Dependent choice for sequences is the schema

$$E([]) \wedge \forall s\,(E(s) \to \exists y\, E(s * y)) \to \exists f\, \forall x\,(\mathbf{N}(x) \to E(\overline{f}(x))), \qquad (29.21)$$

where $\overline{f}(0) = []$ and $\overline{f}(x{+}1) = \overline{f}(x) * f(x)$, that is, $\overline{f}(x) = [] * f(0) * \cdots * f(x{-}1)$.

It is easy to see that the two forms of dependent choice are equivalent: (29.20) with E and $\lambda(s,t)\,.\,s = \mathsf{lead}(t) \wedge E(t)$ implies (29.21) for E, and (29.21) with $\lambda s\,.\,\forall i < |s|\,(A(s_i) \wedge (i > 0 \to B(s_{i-1}, s_i)))$ implies (29.20) for A and B.

One can now extract realizer of (29.21) from the proof that (29.20) implies (29.21). However, it is easier to provide a realizer directly, and, moreover, one that is simpler than the one that would be extracted.

Theorem 29.5 *Dependent choice for sequences, (29.21), is realized by plain iteration: Assume a_0 realizes $E([])$ and c realizes $\forall s\,(E(s) \to \exists y\, E(s * y))$, Then the program d realizes $\exists f\, \forall x\,(\mathbf{N}(x) \to E(\overline{f}(x)))$, where d is recursively defined by*

$$d\,n = \mathbf{case}\, n\, \mathbf{of}\, \{0 \to a_0; m+1 \to c\,(d\,m)\},$$

that is, $d\,n = c^n\, a_0$.

Proof Set

$$E'(s) \stackrel{\text{Def}}{=} d\,\mathbf{r}\,(\mathbf{N}(|s|) \to E(s)),$$

that is, $E'(s) \stackrel{\text{Def}}{=} \forall n\,(n\,\mathbf{r}\,\mathbf{N}(|s|) \to d(n)\,\mathbf{r}\,E(s))$. One easily verifies that $E'([])$ holds since a_0 realizes $E([])$, and $\forall s\,(E'(s) \to \exists y\, E'(s * y))$ holds since d realizes $\forall s\,(E(s) \to \exists y\, E(s * y))$. Therefore, by dependent choice for sequences, there exists f such that $\forall x\,(\mathbf{N}(x) \to E'(\overline{f}(x)))$. To see that d is as required, it suffices to show

$$\forall x\, \forall n\,(n\,\mathbf{r}\,\mathbf{N}(x) \to d(n)\,\mathbf{r}\,E(\overline{f}(x)))\,.$$

Hence, assume $n\,\mathbf{r}\,\mathbf{N}(x)$. Then $\mathbf{N}(x)$, by Theorem 29.1(1). It follows $E'(\overline{f}(x))$ and therefore $n\,\mathbf{r}\,\mathbf{N}(|\overline{f}(x)|) \to (d\,n)\,\mathbf{r}\,E(\overline{f}(x))$. Since $|\overline{f}(x)| = x$ and $n\,\mathbf{r}\,\mathbf{N}(x)$, the premise of this implication holds and it follows $(d\,n)\,\mathbf{r}\,E(\overline{f}(x))$. □

Remarks (1). In the proof of Theorem 29.5 we pedantically distinguished between an element x satisfying $\mathbf{N}(x)$ and a realizer n of this formula. Thanks to Theorem 29.1 we may be more relaxed in the following and identify the two.

(2) The results above do not show that choice principles can be eliminated, since in their proofs the very same principles are used. In fact, one cannot expect elimination of these principles since they involve choices of elements in abstract domains for which no computational methods are available.

29.4 Countable Choice and Classical Logic

It is well-known that **PA** (Peano Arithmetic, that is, **HA** with classical logic) is conservative over **HA** for Σ^0_1-formulas and hence for Π^0_2-formulas. This can be shown with the help of Gödel's functional or Dialectica interpretation [32], or by using Gödel's negative (or double-negation translation) followed by the so-called A-translation, due to Friedman [29] and independently to Dragalin [28]. This conservativity result implies that **PA** admits program extraction for Π^0_2-formulas.

What happens with this result in the presence of countable or dependent choice? This question was first answered by Spector [56] using the Dialectica interpretation. In the following, we revisit a solution for the case of dependent choice for sequences (29.21) based on the negative/A-translation [9, 53].

Note that **PA** extended by countable or dependent choice is as strong as full classical second-order arithmetic, since owing to classical logic and countable choice, every predicate defines a set in the form of its characteristic function. Therefore, we should not expect solutions as simple as in the constructive case where choice principles, if realizable at all, are realized by the identity or a simple iteration.

29.4.1 Computational Content of Classical Dependent Choice

In brief, the negative/A-translation works as follows: Given a proof of a Σ^0_1-formula A in **PA**, possibly using extra axioms, one replaces every atomic formula, disjunction and existentially quantified formula by its double negation (recall that $\neg B$ is $B \to \bot$), and then replaces $\bot$ by A. From the resulting proof one can easily build a new proof of A in **HA** that uses the negative/A-translated extra axioms. Therefore, the question whether a program realizing A can be extracted hinges on the question whether the translated axioms are realizable.

Since the negative/A-translation yields formulas B for which $A \to B$ is easily provable (even in minimal logic) the translation of dependent choice for sequences (29.21) is subsumed by the schema

$$(A \to \forall s\, E(s)) \land E([]) \land \forall s\, (E(s) \to \neg_A \neg_A \exists y\, E(s * y)) \to \\ \neg_A \neg_A \exists f\, \forall x\, (\mathbf{N}(x) \to E(\overline{f}(x))), \tag{29.22}$$

where $\neg_A B \stackrel{\text{Def}}{=} B \to A$ and A is the Σ^0_1-formula one wishes to prove. The question of whether the negative/A-translation of dependent choice for sequences is realizable now boils down to the question of whether (29.22) is effectively realizable.

Theorem 29.6 *Classical dependent choice for sequences (29.22) is effectively realizable.*

Proof Given

$$
\begin{aligned}
H &: \tau(A) \to \tau(E) \\
G_0 &: \tau(E) \\
G &: \tau(E) \to (\tau(E) \to \tau(A)) \to \tau(A) \\
Y &: (\tau(N) \to \tau(E)) \to \tau(A)
\end{aligned}
$$

such that

$$H \,\mathbf{r}\, (A \to \forall s\, E(s)) \tag{29.23}$$

$$G_0 \,\mathbf{r}\, E([]) \tag{29.24}$$

$$G \,\mathbf{r}\, \forall s\, (E(s) \to \neg_A \neg_A \exists y\, E(s * y)) \tag{29.25}$$

$$Y \,\mathbf{r}\, \neg_A \exists f\, \forall x\, (\mathbf{N}(x) \to E(\overline{f}(x))) \tag{29.26}$$

one has to compute a realizer of A. This realizer is obtained as $\Psi\,[]$ where $\Psi : \tau(E)^* \to \tau(A)$ ($\tau(E)^*$ = type of finite sequences of elements of $\tau(E)$) is defined recursively by

$$\Psi\, t = Y\, (\lambda k\,.\, \mathbf{if}\; k \leq |t|\; \mathbf{then}\; (G_0 * t)_k\; \mathbf{else}\; H\, (G\, (\mathsf{last}(G_0 * t))\, (\lambda z.\Psi(t * z)))). \tag{29.27}$$

The definition of Ψ is an instance of a form of bar recursion introduced in [9]. To prove the correctness of Ψ, it suffices to show that for all t (identifying i satisfying $\mathbf{N}(i)$ with the unique realizer of $\mathbf{N}(i)$)

$$(\exists f\, \forall i < |t|\; t_i\, \mathbf{r}\, E(\overline{f}(i+1))) \to (\Psi\, t)\, \mathbf{r}\, A\,. \tag{29.28}$$

This can be proven by a form of bar induction [9] that exploits the fact that the functional Y is continuous and, since A is a Σ^0_1-formula, the set of realizers of A is an open subset of the domain $\tau(A)$ of potential realizers. Continuity means that Y will inspect only finite initial segments of its arguments (which are infinite sequences of elements of type $\tau(E)$) to produce a finite amount of its output (which will suffice to realize A since $\mathbf{R}(A)$ is open). These finite segments form a wellfounded tree, and one sees that Ψ is defined by recursion on this wellfounded tree (since $\Psi\, t = Y\, (\lambda k\,.\, \mathbf{if}\; k \leq |t|\; \mathbf{then}\; (G_0 * t)_k\; \mathbf{else}\; \ldots)$ and Ψ is called recursively with end extensions of t). Bar induction permits induction on this tree. This means, that, when proving (29.28), one may assume, as induction hypothesis, that (29.28) holds if t is replaced by $t * z$ for any $z : \tau(E)$.

Now, to show (29.28), let $t : \tau(E)$ and assume we have f such that

$$\forall i < |t|\; t_i\, \mathbf{r}\, E(\overline{f}(i+1)). \tag{29.29}$$

We have to show $(\Psi\, t)\, \mathbf{r}\, A$. Since $\Psi\, t = Y\, \alpha$ where

$$\alpha = \lambda k\,.\, \mathbf{if}\; k \leq |t|\; \mathbf{then}\; (G_0 * t)_k\; \mathbf{else}\; H\, (G\, (\mathsf{last}(G_0 * t))\, (\lambda z.\Psi(t * z))),$$

it suffices, by the assumption (29.26) on Y, to show

$$\alpha\,\mathbf{r}\,\forall x\,(\mathbf{N}(x) \to E(\overline{f}(x))).$$

Therefore, we assume $\mathbf{N}(i)$ (more precisely $n\,\mathbf{r}\,\mathbf{N}(i)$ but we identify n and i) and have to show $(\alpha\,i)\,\mathbf{r}\,E(\overline{f}(i))$. If $i = 0$, then $\alpha\,i = G_0$ and we are done, by the assumption (29.24). If $1 \leq i \leq |t|$, then $(\alpha\,i)\,\mathbf{r}\,E(\overline{f}(i))$ holds by (29.29). If $i > |t|$, then, by the assumptions (29.23) and (29.25), it suffices to show that for $s \stackrel{\text{Def}}{=} \overline{f}(|t|)$,

(i) $\mathsf{last}(G_0 * t)$ realizes $E(s)$,
(ii) $\lambda z.\Psi(t * z)$ realizes $(\exists y E(s * y)) \to A$.

(i) follows from (29.24) and (29.29). To show (ii), we assume $z\,\mathbf{r}\,E(s * y)$ (for some y). Then $\forall i < |t * z|\; t_i\,\mathbf{r}\,E(\overline{g}(i+1))$ where g is like f but changed at $|t|$ to $g(|t|) = y$. By the bar induction hypothesis it follows $\Psi(t * z)\,\mathbf{r}\,A$. □

Remarks (1) The realizer (29.27) is slightly simpler than those in [9, 53], due to the uniform interpretation of quantifiers.

(2) In the proof of Theorem 29.6 one sees that the condition on A to be a Σ^0_1-formula can be weakened to the requirement that $\mathbf{R}(A)$ is an open set. This means, for example, that A is not required to be an existential quantification of a decidable formula and that atomic formulas do not need to be decidable. For example, any formula A of the form $\exists x\,(\mathbf{N}(x) \wedge B(x))$ for an arbitrary nc formula $B(x)$ meets the requirement. This is the case since then $\tau(A)$ is the domain of lazy natural numbers and $\mathbf{R}(A)$ is a set of natural numbers which is always open.

(3) As pointed out in [37], the negative/A-translation is not faithful since it increases the logical complexity of formulas, in particular the complexities of instances of induction are increased. This issue was resolved in [24] and adapted to the abstract setting (i.p. not requiring atomic formulas to be decidable) in [6]. On the other hand, proving Σ^0_1 conservativity of $\mathbf{PA}$ over $\mathbf{HA}$ via the Dialectica interpretation *is* faithful; however, it requires decidability of atomic formulas.

29.4.2 Case Studies

The negative/A-translation was applied in [53] to a classical proof of the infinite tape example, and in [52] to (a special case of) Higman's Lemma, based on Nash-Williams' minimal-bad-sequence argument [43]. Both case studies were carried out in Minlog, and a program was extracted with the help of a realizer of dependent choice similar to (29.27). The extracted program of the latter computes for every infinite sequence of binary strings two strings in the sequence such that the earlier one can be embedded in the later one. The program was not very efficient but still vastly better than the first program that was extracted from a similar translation of Nash-Williams' proof in [22], using second-order arithmetic.

Recently, our formalization was improved, mainly by making the formalization more concise and by not using Minlog's built-in term language, to run the extracted program, but automatically translating to the programming language Haskell. The extracted program is now one screen page long, and involves about 40 case distinctions. The improvement over Murthy's original version is due to several factors: (a) A refined A-translation [12] reduces the number of double-negations; (b) Minlog's optimized realizability interpretation makes sure that extracted programs are free of junk (see, e.g., Chapter 7 of [50]); and (c) lazy execution of programs in Haskell avoids unnecessarily repeated computations.

For completeness, we mention that Higman's Lemma also has a constructive proof where the minimal bad sequence argument is essentially turned into an inductive argument (see [23], [51]) and several other constructive proofs (see [51] for a detailed overview).

29.4.3 Related Work

There exist several other approaches to the extraction of computational content from classical countable choice. As mentioned before, the first, by [56], is not based on realizability, but on the Dialectica interpretation. To satisfy the functional interpretation of countable choice, Spector uses a form of bar recursion that does not require functionals to be continuous but only majorizable [16]. A special form of realizability and a different form of bar recursion are used to interpret countable and dependent choice in [2]. This was transferred in [5] to realizability as defined here using a special case of Raoult's open induction [45].[2] In [39] a combination of Gödel's negative translation and Kleene realizability (called "classical realizability") is used to interpret dependent choice.

29.5 Conclusion

We presented a formal system with a computationally meaningful semantics that integrates constructive, axiomatic and classical mathematics. We do not claim that the realizability semantics of this system fully captures the intended semantics of constructive mathematics nor that it is the only way of doing so,[3] but we are making the case that it faithfully represents its essential features and provides a simple and precise way of assessing the constructivity or nonconstructivity of mathematical principles that is intelligible from a constructive as well as classical point of view.

[2] It is an open problem whether Raoult's general open induction has a computational interpretation.

[3] Martin-Löf type theory is an alternative framework for formalizing Bishop constructive mathematics [19].

As prototypical examples of principles that lie at the borderline between constructive and classical mathematics we studied various forms of the axiom of choice and assessed their constructivity within our framework. We made no attempt to cover all constructive aspects of the axiom of choice on which there exists extensive research in, for example, set theory [46], constructive algebra [25, 44], proof interpretations [36], and constructive reverse mathematics [3, 27, 34, 48]. Instead, we focused on new phenomena that appear due to the uniform interpretation of quantifiers. For instance, relativized choice – though trivially provable in constructive type theory – is in general not realizable. We identified a number of syntactic and semantic conditions that guarantee realizability (Theorem 29.3) and include, for example, the realizability of countable choice. In addition, we assured ourselves that two common variants of dependent choice are realizable in our framework, the realizers being (as expected) simple iteration functionals (Theorem 29.4 and 29.5).

Finally, we revisited the well-known negative/A-translation of dependent choice using a bar recursive functional and found that in our more abstract framework the construction is considerably less complicated.

Comprehensive exposition of proof interpretations and their use for the extraction of computational content can be found in [37] and [50]. The latter also contains more examples of case studies in the Minlog system.

References

[1] Bell, J. L. 2008. The axiom of choice.Technical report *The Stanford Encyclopedia of Philosophy*.

[2] Berardi, S., Bezem, M., and Coquand, T. 1998. On the computational content of the axiom of choice. *J. Symbol. Logic*, **63**(2), 600–622.

[3] Berger, J., Ishihara, H., and Schuster, P. 2012. The weak König lemma, Brouwer's fan theorem, de Morgan's law, and dependent choice. *Rep. Math. Logic*, **47**, 47–63.

[4] Berger, U. 2001. Program extraction from Gentzen's proof of transfinite induction up to ϵ_0. Pages 68–77 of: Kahle, R., Schroeder-Heister, P., and Stärk, R. (eds.). *Proof Theory in Computer Science*. Lecture Notes in Computer Science, vol. 2138. Berlin, Heidelberg: Springer.

[5] Berger, U. 2004. A computational interpretation of open induction. Pages 326–334 of:Titsworth, F. (ed.). *Proceedings of the Nineteenth Annual IEEE Symposium on Logic in Computer Science*, Washington, DC: IEEE Computer Society.

[6] Berger, U. 2005. Uniform Heyting arithmetic. *Ann. Pure Appl. Logic*, **133**, 125–148.

[7] Berger, U. 2011. From coinductive proofs to exact real arithmetic: theory and applications. *Log. Meth. Comp. Sci.*, **7**(1), 1–24.

[8] Berger, U. 2019. On the constructive and computational content of abstract mathematics. Pages 55–73 of: Centrone, S., Negri, S., Sarikaya, D., and Schuster, P. (eds.),Mathesis Universalis, Computability and Proof. Synthese Library, vol. 412. Berlin, Heidelberg: Springer.

[9] Berger, U., and Oliva, P. 2001. Modified Bar recursion and classical dependent choice. Pages 89–107 of: Baaz, M., Friedman, S.-D., and Krajicek, J. (eds.), *Logic Colloquium 2001*. Lecture Notes in Logic. Wellesly: A K Peters Ltd.

[10] Berger, U., and Tsuiki, H. 2021. Intuitionistic fixed point logic. *Ann. Pure Appl. Logic*, **172**(3), 1–56.

[11] Berger, U., Schwichtenberg, H., and Seisenberger, M. 2001. The Warshall algorithm and Dickson's lemma: Two examples of realistic program extraction. *J. Autom. Reason.*, **26**(2), 205–221.

[12] Berger, U., Schwichtenberg, H., and Buchholz, W. 2002. Refined program extraction from classical proofs. *Ann. Pure Appl. Logic*, **114**, 3–25.

[13] Berger, U., Berghofer, S., Letouzey, P., and Schwichtenberg, H. 2006. Program extraction from normalization proofs. *Studia Logica*, **82**, 25–49.

[14] Berger, U., Miyamoto, K., Schwichtenberg, H., and Seisenberger, M. 2011. Minlog – a tool for program extraction for supporting algebra and coalgebra. Pages 393–399 of: *CALCO-Tools*. Lecture Notes in Computer Science, vol. 6859. Berlin, Hedelberg: Springer.

[15] Berger, U., Lawrence, A., Nordvall Forsberg, F., and Seisenberger, M. 2015. Extracting verified decision procedures: DPLL and resolution. *Log. Meth. Comp. Sci.*, **11**, 1–18.

[16] Bezem, M. 1985. Strongly majorizable functionals of finite type: a model for bar recursion containing discontinuous functionals. *J. Symbol. Logic*, **50**, 652–660.

[17] Bishop, E., and Bridges, D. 1985. *Constructive Analysis*. Grundlehren der Mathematischen Wissenschaften 279. Berlin, Heidelberg: Springer.

[18] Bridges, D. 1999. Constructive mathematics: a foundation for computable analysis. *Theor. Comp. Sci.*, **219**, 95–109.

[19] Bridges, D., and Reeves, S. 1999. Constructive mathematics in theory and programming practice. *Phil. Math.*, **7**(3), 65–104.

[20] Bridges, D., Richman, F., and Schuster, P. 1999. Linear independence without choice. *Ann. Pure Appl. Logic*, **101**(1), 95–102.

[21] Cohen, P. J. 1963. The independence of the continuum hypothesis. *Proc. U. S. Natl Acad. Sci.*, **50**, 1143–1148.

[22] Constable, R., and Murthy, C. 1991. Finding computational content in classical proofs. Pages 341–362 of: *Logical Frameworks*, Huet, G ., and Plotkin, G. (eds.). Cambridge: Cambridge University Press.

[23] Coquand, T., and Fridlender, D. 1993. A proof of Higman's lemma by structural induction. Unpublished manuscript.

[24] Coquand, T., and Hofmann, M. 1999. A new method for establishing conservativity of classical systems over their intuitionistic version. *Math. Struct. Comput. Sci.*, **9**, 323–333.

[25] Coquand, T., and Lombardi, H. 2006. A logical approach to abstract algebra. *Math. Struct. Comput. Sci.*, **16**, 885–900.

[26] Diaconescu, R. 1975. Axiom of choice and complementation. *Proc. Am. Math. Soc.*, **51**, 176–178.

[27] Diener, H. 2020. Constructive Reverse mathematics. PhD Thesis, Habilitationsschrift, Siegen. `https://arxiv.org/abs/1804.05495`

[28] Dragalin, A. 1980. New kinds of realizability and the Markov rule. *Soviet Mathematics, Doklady*, **251**(3), 534–537.

[29] Friedman, H. 1978. Classically and intuitionistically provably recursive functions. Pages 21–28 of: *Higher Set Theory*, Scott, D., and Müller, G. (eds.). Lecture Notes in Mathematics, vol. 669. Berlin, Heidelberg: Springer.

[30] Gierz, G., Hofmann, K. H., Keimel, K., et al. 2003. *Continuous Lattices and Domains*. Encyclopedia of Mathematics and its Applications, vol. 93. Cambridge: Cambridge University Press.

[31] Gödel, K. 1940. *The consistency of the axiom of choice and of the generalized continuum-hypothesis with the axioms of set theory. Ann. Math. Stud.*, **3**, 556–557.

[32] Gödel, K. 1958. Über eine bisher noch nicht benützte Erweiterung des finiten Standpunktes. *Dialectica*, **12**, 280–287.

[33] Goodman, N., and Myhill, J. 1978. Choice implies excluded middle. *Mathematical Logic Quarterly*, **24**, 461.

[34] Ishihara, H. 2006. Weak König's lemma implies Brouwer's fan theorem: a direct proof. *Notre Dame J. Form. Logic*, **47**(2), 249–252.

[35] Kleene, S. C. 1945. On the interpretation of intuitionistic number theory. *J. Symbol. Logic*, **10**, 109–124.

[36] Kohlenbach, U. 2002. On uniform weak König's lemma. *Ann. Pure Appl. Logic*, **114**, 103–116.

[37] Kohlenbach, U. 2008. *Proof Interpretations and their Use in Mathematics*. Springer Monographs in Mathematics. Berlin, Heidelberg: Springer.

[38] Kreisel, G. 1959. Interpretation of analysis by means of constructive functionals of finite types. *Construct. Math.*, pp. 101–128.

[39] Krivine, J.-L. 2003. Dependent choice, 'quote' and the clock. *Theor. Comp. Sci.*, **308**, 259–276.

[40] Martin-Löf, P. 1984. *Intuitionistic Type Theory*. Naples: Bibliopolis.

[41] Miyamoto, K. 2013. Program extraction from coinductive proofs and its application to extract real arithmetic. PhD Thesis, LMU München.

[42] Miyamoto, K., Forsberg, F., and Schwichtenberg, H. 2013. Program extraction from nested definitions. Pages 370–385 of: *Interactive Theorem Proving*, Blazy, S., Paulin-Mohring, C., and Pichardie, D. (eds.). Lecture Notes in Computer Science, vol. 7988. Berlin, Heidelberg: Springer.

[43] Nash-Williams, C. St. J. A. 1963. On well-quasi-ordering finite trees. *Proc. Cambridge Phil. Soc.*, **59**, 833–835.

[44] Powell, T., Schuster, P., and Wiesnet, F. 2021. A universal algorithm for Krull's theorem. *Inform. Comput.*, **287**(104761), 1–21.

[45] Raoult, J.-C. 1988. Proving open properties by induction. *Inform. Proc. Lett.*, **29**, 19–23.

[46] Rathjen, M. 2006. Choice principles in constructive and classical set theories. Pages 229–326 of: Chatzidakis, Z., Koepke, P., and Pohlers, W. (eds.). *Logic Colloquium '02*, Lecture Notes in Logic, vol. 27. Cambridge: Cambridge University Press.

[47] Richman, F. 2001. Constructive mathematics without choice. Pages 199–205 of: *Reuniting the Antipodes – Constructive and Nonstandard Views of the Continuum*, Schuster, P., Berger, U., and Osswald, H. (eds.). Synthese Library, vol. 306. Amsterdam: Kluwer.

[48] Schwichtenberg, H. 2005. A direct proof of the equivalence between Brouwer's fan theorem and König's lemma with a uniqueness hypothesis. *J. Universal Comp. Sci.*, **11**(12), 2086–2095.

[49] Schwichtenberg, H. 2008. Realizability interpretation of proofs in constructive analysis. *ACM Trans. on Comp. Sys.*, **43**(3–4), 583–602.

[50] Schwichtenberg, H., and Wainer, S. S. 2012. *Proofs and Computations*. Cambridge: Cambridge University Press.

[51] Schwichtenberg, H., Seisenberger, M., and Wiesnet, F. 2016. Higman's lemma and its computational content. In: *Advances in Proof Theory, Progress in Computer Science and Applied Logic*, vol. 28. Basel: Birkhäuser.

[52] Seisenberger, M. 2003. On the constructive content of proofs. PhD Thesis, LMU München.

[53] Seisenberger, M. 2008. Programs from proofs using classical dependent choice. *Ann. Pure Appl. Logic*, **153**(1–3), 97–110.

[54] Sieg, W. 1985. Fragments of arithmetic. *Ann. Pure Appl. Logic*, **28**, 33–71.

[55] Simpson, S. G. 1999. *Subsystems of Second Order Arithmetic*. Perspectives in Mathematical Logic. Berlin, Heidelberg: Springer.

[56] Spector, C. 1962. Provably recursive functionals of analysis: a consistency proof of analysis by an extension of principles in current intuitionistic mathematics. Pages 1–27 of: *Recursive Function Theory*, Dekker, F. D. E. (ed.). Proc. Symposia in Pure Mathematics, vol. 5. Providence, RI: AMS.

[57] Troelstra, A. S. 1973. *Metamathematical Investigation of Intuitionistic Arithmetic and Analysis*. Lecture Notes in Mathematics, vol. 344. Springer.

Index

www.ingramcontent.com/pod-product-compliance
Lightning Source LLC
LaVergne TN
LVHW081935200926
848484LV00009B/2440

* 9 7 8 1 3 1 6 5 1 0 8 6 5 *